YAMAHA

Personal Watercraft
2002-11 REPAIR MANUAL
All 4-STROKE MODELS

I0799319

SELOC®

Managing Partners Dean F. Morgantini
Barry L. Beck

Executive Editor Kevin M. G. Maher, A.S.E.

Production Managers Melinda Possinger
Ronald Webb

Manufactured in USA

104 Willowbrook Lane
West Chester, PA 19382
ISBN-13: 978-0-89330-082-1
ISBN-10: 0-89330-082-9
1234567890 0987654321

www.selocmarine.com
1-866-SELOC55

CONTENTS

CONTENTS

SAFETY NOTICE

Proper service and repair procedures are vital to the safe, reliable operation of all marine engines, as well as the personal safety of those performing repairs. This manual outlines procedures for servicing and repairing engines and drive systems using safe, effective methods. The procedures contain many NOTES, CAUTIONS and WARNINGS which should be followed, along with standard procedures, to minimize the possibility of personal injury or improper service which could damage the vehicle or compromise its safety.

It is important to note that repair procedures and techniques, tools and parts for servicing these engines, as well as the skill and experience of the individual performing the work, vary widely. It is not possible to anticipate all of the conceivable ways or conditions under which the engine may be serviced, or to provide cautions as to all possible hazards that may result. Standard and accepted safety precautions and equipment should be used during cutting, grinding, chiseling, prying, or any other process that can cause material removal or projectiles.

Some procedures require the use of tools specially designed for a specific task. Before substituting another tool or procedure, you must be completely satisfied that neither your personal safety, nor the performance of the vessel, will be endangered. All procedures covered in this manual requiring the use of special tools will be noted at the beginning of the procedure by means of an **OEM symbol** OEM

Additionally, any procedure requiring the use of an electronic tester or scan tool will be noted at the beginning of the procedure by means of a **DVOM symbol**

Although information in this manual is based on industry sources and is complete as possible at the time of publication, the possibility exists that some manufacturers made later changes which could not be included here. While striving for total accuracy, Seloc Publishing cannot assume responsibility for any errors, changes or omissions that may occur in the compilation of this data. We must therefore warn you to follow instructions carefully, using common sense. If you are uncertain of a procedure, seek help by inquiring with someone in your area who is familiar with these motors before proceeding.

PART NUMBERS

Part numbers listed in this reference are not recommendations by Seloc Publishing for any particular product brand name, simply iterations of the manufacturer's suggestions. They are also references that can be used with interchange manuals and aftermarket supplier catalogs to locate each brand supplier's discrete part number.

SPECIAL TOOLS

Special tools are recommended by the manufacturers to perform a specific job. Use has been kept to a minimum, but, where absolutely necessary, they are referred to in the text by the part number of the manufacturer if at all possible; and also noted at the beginning of each procedure with one of the following symbols: **OEM** or **DVOM.**

OEM

The **OEM** symbol usually denotes the need for a unique tool purposely designed to accomplish a specific task, it will also be used, less frequently, to notify the reader of the need for a tool that is not commonly found in the average tool box.

The **DVOM** symbol is used to denote the need for an electronic test tool like an ohmmeter, multi-meter or, on certain later engines, a scan tool.

These tools can be purchased, under the appropriate part number, from your local dealer or regional distributor, or an equivalent tool can be purchased locally from a tool supplier or parts outlet. Before substituting any tool for the one recommended, read the SAFETY NOTICE at the top of this page.

Providing the correct mix of service and repair procedures is an endless battle for any publisher of "How-To" information. Users range from first time do-it yourselfers to professionally trained marine technicians, and information important to one is frequently irrelevant to the other. The editors at Seloc Publishing strive to provide accurate and articulate information on all facets of marine engine repair, from the simplest procedure to the most complex. In doing this, we understand that certain procedures may be outside the capabilities of the average DIYer. Conversely we are aware that many procedures are unnecessary for a trained technician.

SKILL LEVELS

In order to provide all of our users, particularly the DIYers, with a feeling for the scope of a given procedure or task before tackling it we have included a rating system denoting the suggested skill level needed when performing a particular procedure. One of the following icons will be included at the beginning of most procedures:

EASY. These procedures are aimed primarily at the DIYer and can be classified, for the most part, as basic maintenance procedures; battery, fluids, filters, plugs, etc. Although certainly valuable to any experience level, they will generally be of little importance to a technician.

MODERATE. These procedures are suited for a DIYer with experience and a working knowledge of mechanical procedures. Even an advanced DIYer or professional technician will occasionally refer to these procedures. They will generally consist of component repair and service procedures, adjustments and minor rebuilds.

DIFFICULT. These procedures are aimed at the advanced DIYer and professional technician. They will deal with diagnostics, rebuilds and internal engine/drive components and will frequently require special tools.

SKILLED. These procedures are aimed at highly skilled technicians and should not be attempted without previous experience. They will usually consist of machine work, internal engine work and gear case rebuilds.

Please remember one thing when considering the above ratings—they are a guide for judging the complexity of a given procedure and are subjective in nature. Only you will know what your experience level is, and only you will know when a procedure may be outside the realm of your capability. First time DIYer, or life-long marine technician, we all approach repair and service differently so an easy procedure for one person may be a difficult procedure for another, regardless of experience level. All skill level ratings are meant to be used as a guide only! Use them to help make a judgement before undertaking a particular procedure, but by all means read through the procedure first and make your own decision—after all, our mission at Seloc is to make boat maintenance and repair easier for everyone whether you are changing the oil or rebuilding an engine. Enjoy boating!

ACKNOWLEDGMENTS

Seloc Publishing expresses appreciation to the following companies who supported the production of this book:

- Marine Mechanics Institute—Orlando, FL
- Belks Marine—Holmes, PA

Thanks to John Hartung and Judy Belk of Belk's Marine for for their assistance, guidance, patience and access to some of the motors photographed for this manual.

Seloc Publishing would like to express thanks to the fine companies who participate in the production of all our books:

- Hand tools supplied by Craftsman are used during all phases of our vehicle teardown and photography.
- Many of the fine specialty tools used in our procedures were provided courtesy of Lisle Corporation.
- Much of our shop's electronic testing equipment was supplied by Universal Enterprises Inc. (UEI).

1

GENERAL INFORMATION, SAFETY AND TOOLS

HOW TO USE THIS MANUAL

This manual is designed to be a handy reference guide to maintaining and repairing your Yamaha PWC. We strongly believe that regardless of how many or how few year's experience you may have, there is something new waiting here for you.

This manual covers the topics that a factory service manual (designed for factory trained mechanics) and a manufacturer owner's manual (designed more by lawyers than boat owners these days) covers. It will take you through the basics of maintaining and repairing your motor, step-by-step, to help you understand what the factory trained mechanics already know by heart. By using the information in this manual, any boat owner should be able to make better informed decisions about what they need to do to maintain and enjoy their motor.

Even if you never plan on touching a wrench (and if so, we hope that we can change your mind), this manual will still help you understand what a mechanic needs to do in order to maintain your engine.

Can You Do It?

If you are not the type who is prone to taking a wrench to something, NEVER FEAR. The procedures provided here cover topics at a level virtually anyone will be able to handle. And just the fact that you purchased this manual shows your interest in better understanding your motor.

You may even find that maintaining your motor yourself is preferable in most cases. From a monetary standpoint, it could also be beneficial. The money spent on hauling your boat to a marina and paying a tech to service the engine could buy you fuel for a whole weekend of boating. And, if you are really that unsure of your own mechanical abilities, at the very least you should fully understand what a marine mechanic does to your boat. You may decide that anything other than maintenance and adjustments should be performed by a mechanic (and that's your call), but if so you should know that every time you board your boat, you are placing faith in the mechanic's work and trusting him or her with your well-being, and maybe your life.

It should also be noted that in most areas a factory-trained mechanic will command a hefty hourly rate for off site service. If the tech comes to you this hourly rate is often charged from the time they leave their shop to the time that they return home. When service is performed at a boat yard, the clock usually starts when they go out to get the boat and bring it into the shop and doesn't end until it is tested and put back in the yard. The cost savings in doing the job yourself might be readily apparent at this point.

Of course, if even you're already a seasoned Do-It-Yourselfer or a Professional Technician, you'll find the procedures, specifications, special tips as well as the schematics and illustrations helpful when tackling a new job on a motor.

■ **To help you decide if a task is within your skill level, procedures will often be rated using a wrench symbol in the text. When present, the number of wrenches designates how difficult we feel the procedure to be on a 1-4 scale. For more details on the wrench icon rating system, please refer to the information under Skill Levels at the beginning of this manual.**

Where to Begin

Before spending any money on parts, and before removing any nuts or bolts, read through the entire procedure or topic. This will give you the overall view of what tools and supplies will be required to perform the procedure or what questions need to be answered before purchasing parts. So read ahead and plan ahead. Each operation should be approached logically and all procedures thoroughly understood before attempting any work.

Avoiding Trouble

Some procedures in this manual may require you to "label and disconnect . . . " a group of lines, hoses or wires. Don't be lulled into thinking you can remember where everything goes - you won't. If you reconnect or install a part incorrectly, the motor may operate poorly, if at all. If you hook up electrical wiring incorrectly, you may instantly learn a very expensive lesson.

A piece of masking tape, for example, placed on a hose and another on its fitting will allow you to assign your own label such as the letter "A", or a short name. As long as you remember your own code, you can reconnect the lines by matching letters or names. Do remember that tape will dissolve when saturated in some fluids (especially cleaning solvents). If a component is to be washed or cleaned, use another method of identification. A permanent felt-tipped marker can be very handy for marking metal parts; but remember that some solvents will remove permanent marker. A scribe can be used to carefully etch a small mark in some metal parts, but be sure NOT to do that on a gasket-making surface.

SAFETY is the most important thing to remember when performing maintenance or repairs. Be sure to read the information on safety in this manual.

Maintenance or Repair?

Proper maintenance is the key to long and trouble-free engine life, and the work can yield its own rewards. A properly maintained engine performs better than one that is neglected. As a conscientious boat owner, set aside a Saturday morning, at least once a month, to perform a thorough check of items that could cause problems. Keep your own personal log to jot down which services you performed, how much the parts cost you, the date, and the amount of hours on the engine at the time. Keep all receipts for parts purchased, so that they may be referred to in case of related problems or to determine operating expenses. As a do-it-yourselfer, these receipts are the only proof you have that the required maintenance was performed. In the event of a warranty problem (on new motors), these receipts can be invaluable.

It's necessary to mention the difference between maintenance and repair. Maintenance includes routine inspections, adjustments, and replacement of parts that show signs of normal wear. Maintenance compensates for wear or deterioration. Repair implies that something has broken or is not working. A need for repair is often caused by lack of maintenance.

For example: draining and refilling the gearcase oil is maintenance recommended by all manufacturers at specific intervals. Failure to do this can allow internal corrosion or damage and impair the operation of the motor, requiring expensive repairs. While no maintenance program can prevent items from breaking or wearing out, a general rule can be stated: MAINTENANCE IS CHEAPER THAN REPAIR.

Directions & Locations

◆ **See Figure 1**

Two basic rules should be mentioned here. First, whenever the Port side of the engine (or boat) is referred to, it is meant to specify the left side of the

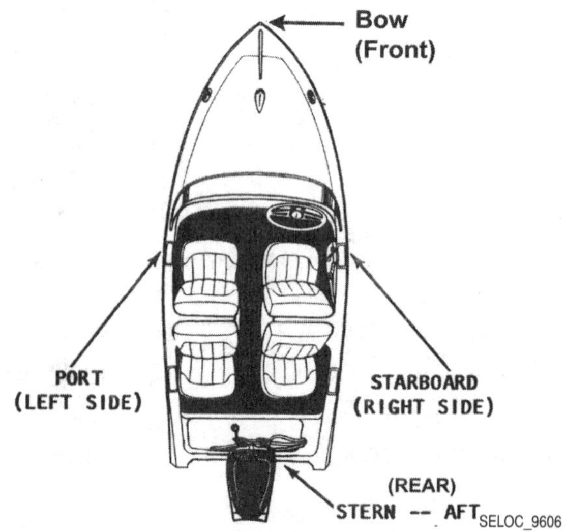

Fig. 1 Common terminology used for reference designation on boats of all size. These terms are used thru-out the text

engine when you are sitting at the helm. Conversely, the Starboard means your right side. The Bow is the front of the boat and the Stern or Aft is the rear.

Most screws and bolts are removed by turning counterclockwise, and tightened by turning clockwise. An easy way to remember this is: righty-tighty; lefty-loosey. Corny, but effective. And if you are really dense (and we have all been so at one time or another), buy a ratchet that is marked **ON** and **OFF** (like Snap-on® ratchets), or mark your own. This can be especially helpful when you are bent over backwards, upside down or otherwise turned around when working on a boat-mounted component.

Professional Help

Occasionally, there are some things when working on an motor that are beyond the capabilities or tools of the average Do-It-Yourselfer (DIYer). This shouldn't include most of the topics of this manual, but you will have to be the judge. Some engines require special tools or a selection of special parts, even for some basic maintenance tasks.

Talk to other boaters who use the same model of engine and speak with a trusted marina to find if there is a particular system or component on your engine that is difficult to maintain.

You will have to decide for yourself where basic maintenance ends and where professional service should begin. Take your time and do your research first (starting with the information contained within) and then make your own decision. If you really don't feel comfortable with attempting a procedure, DON'T DO IT. If you've gotten into something that may be over your head, don't panic. Tuck your tail between your legs and call a marine mechanic. Marinas and independent shops will be able to finish a job for you. Your ego may be damaged, but your boat will be properly restored to its full running order. So, as long as you approach jobs slowly and carefully, you really have nothing to lose and everything to gain by doing it yourself.

On the other hand, even the most complicated repair is within the ability of a person who takes their time and follows the steps of a procedure. A rock climber doesn't run up the side of a cliff, he/she takes it one step at a time and in the end, what looked difficult or impossible was conquerable. Worry about one step at a time.

Purchasing Parts

◆ **See Figures 2 and 3**

When purchasing parts there are two things to consider. The first is quality and the second is to be sure to get the correct part for your engine. To get quality parts, always deal directly with a reputable retailer. To get the proper parts always refer to the model number from the information tag on your engine prior to calling the parts counter. An incorrect part can adversely affect your engine performance and fuel economy, and will cost you more money and aggravation in the end.

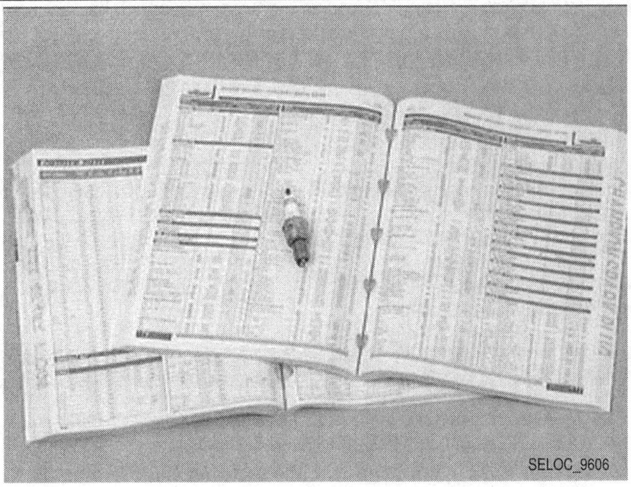

Fig. 3 Parts catalogs, giving application and part number information, are provided by manufacturers for most replacement parts

Fig. 2 By far the most important asset in purchasing parts is a knowledgeable and enthusiastic parts person

Just remember a tow back to shore will cost plenty. That charge is per hour from the time the towboat leaves their home port, to the time they return to their home port. Get the picture. . .$$$?

So whom should you call for parts? Well, there are many sources for the parts you will need. Where you shop for parts will be determined by what kind of parts you need, how much you want to pay, and the types of stores in your neighborhood.

Your marina can supply you with many of the common parts you require. Using a marina as your parts supplier may be handy because of location (just walk right down the dock) or because the marina specializes in your particular brand of engine. In addition, it is always a good idea to get to know the marina staff (especially the marine mechanic).

The marine parts jobber, who is usually listed in the yellow pages or whose name can be obtained from the marina, is another excellent source for parts. In addition to supplying local marinas, they also do a sizeable business in over-the-counter parts sales for the do-it-yourselfer.

Almost every boating community has one or more convenient marine chain stores. These stores often offer the best retail prices and the convenience of one-stop shopping for all your needs. Since they cater to the do-it-yourselfer, these stores are almost always open weeknights, Saturdays, and Sundays, when the jobbers are usually closed.

The lowest prices for parts are most often found in discount stores or the auto department of mass merchandisers. Parts sold here are name and private brand parts bought in huge quantities, so they can offer a competitive price. Private brand parts are made by major manufacturers and sold to large chains under a store label. And, of course, more and more large automotive parts retailers are stocking basic marine supplies.

Avoiding The Most Common Mistakes

There are 3 common mistakes in mechanical work:

1. Following the incorrect order of assembly, disassembly or adjustment. When taking something apart or putting it together, performing steps in the wrong order usually just costs you extra time; however, it CAN break something. Read the entire procedure before beginning disassembly. Perform everything in the order in which the instructions say you should, even if you can't immediately see a reason for it. When you're taking apart something that is very intricate, you might want to draw a picture of how it looks when assembled at one point in order to make sure you get everything back in its proper position. When making adjustments, perform them in the proper order; often, one adjustment affects another, and you cannot expect satisfactory results unless each adjustment is made only when it cannot be changed by subsequent adjustments.

■ **Digital cameras are handy. If you've got access to one, take pictures of intricate assemblies during the disassembly process and refer to them during assembly for tips on part orientation.**

2. Over-torquing (or under-torquing). While it is more common for over-torquing to cause damage, under-torquing may allow a fastener to vibrate loose causing serious damage. Especially when dealing with plastic and aluminum parts, pay attention to torque specifications and utilize a torque wrench in assembly. If a torque figure is not available, remember that if you are using the right tool to perform the job, you will probably not have to strain yourself to get a fastener tight enough. The pitch of most threads is so slight that the tension you put on the wrench will be multiplied many times in actual force on what you are tightening.

3. Cross-threading. This occurs when a part such as a bolt is screwed into a nut or casting at the wrong angle and forced. Cross-threading is more likely to occur if access is difficult. It helps to clean and lubricate fasteners, then to start threading with the part to be installed positioned straight inward. Always start a fastener, etc. with your fingers. If you encounter resistance, unscrew the part and start over again at a different angle until it can be inserted and turned several times without much effort. Keep in mind that some parts may have tapered threads, so that gentle turning will automatically bring the part you're threading to the proper angle, but only if you don't force it or resist a change in angle. Don't put a wrench on the part until it has been tightened a couple of turns by hand. If you suddenly encounter resistance, and the part has not seated fully, don't force it. Pull it back out to make sure it's clean and threading properly.

BOATING SAFETY

In 1971 Congress ordered the U.S. Coast Guard to improve recreational boating safety. In response, the Coast Guard drew up a set of regulations.

Aside from these federal regulations, there are state and local laws you must follow. These sometimes exceed the Coast Guard requirements. This section discusses only the federal laws. State and local laws are available from your local Coast Guard. As with other laws, "Ignorance of the boating laws is no excuse." The rules fall into two groups: regulations for your boat and required safety equipment on your boat.

Regulations For Your Boat

Most boats on waters within Federal jurisdiction must be registered or documented. These waters are those that provide a means of transportation between 2 or more states or to the sea. They also include the territorial waters of the United States.

DOCUMENTING OF VESSELS

A vessel of 5 or more net tons may be documented as a yacht. In this process, papers are issued by the U.S. Coast Guard as they are for large ships. Documentation is a form of national registration. The boat must be used solely for pleasure. Its owner must be a citizen of the U.S., a partnership of U.S. citizens, or a corporation controlled by U.S. citizens. The captain and other officers must also be U.S. citizens. The crew need not be.

If you document your yacht, you have the legal authority to fly the yacht ensign. You also may record bills of sale, mortgages, and other papers of title with federal authorities. Doing so gives legal notice that such instruments exist. Documentation also permits preferred status for mortgages. This gives you additional security, and it aids in financing and transfer of title. You must carry the original documentation papers aboard your vessel. Copies will not suffice.

REGISTRATION OF BOATS

If your boat is not documented, registration in the state of its principal use is probably required. If you use it mainly on an ocean, a gulf, or other similar water, register it in the state where you moor it.

If you use your boat solely for racing, it may be exempt from the requirement in your state. Some states may also exclude dinghies, while others require registration of documented vessels and non-power driven boats.

All states, except Alaska, register boats. In Alaska, the U.S. Coast Guard issues the registration numbers. If you move your vessel to a new state of principal use, a valid registration certificate is good for 60 days. You must have the registration certificate (certificate of number) aboard your vessel when it is in use. A copy will not suffice. You may be cited if you do not have the original on board.

NUMBERING OF VESSELS

A registration number is on your registration certificate. You must paint or permanently attach this number to both sides of the forward half of your boat. Do not display any other number there.

The registration number must be clearly visible. It must not be placed on the obscured underside of a flared bow. If you can't place the number on the bow, place it on the forward half of the hull. If that doesn't work, put it on the superstructure. Put the number for an inflatable boat on a bracket or fixture. Then, firmly attach it to the forward half of the boat. The letters and numbers must be plain block characters and must read from left to right. Use a space or a hyphen to separate the prefix and suffix letters from the numerals. The color of the characters must contrast with that of the background, and they must be at least three inches high.

In some states your registration is good for only one year. In others, it is good for as long as three years. Renew your registration before it expires. At that time you will receive a new decal or decals. Place them as required by state law. You should remove old decals before putting on the new ones. Some states require that you show only the current decal or decals. If your vessel is moored, it must have a current decal even if it is not in use.

If your vessel is lost, destroyed, abandoned, stolen, or transferred, you must inform the issuing authority. If you lose your certificate of number or your address changes, notify the issuing authority as soon as possible.

SALES & TRANSFERS

Your registration number is not transferable to another boat. The number stays with the boat unless its state of principal use is changed.

HULL IDENTIFICATION NUMBER

A Hull Identification Number (HIN) is like the Vehicle Identification Number (VIN) on your car. Boats built between November 1, 1972 and July 31, 1984 have old format HINs. Since August 1, 1984 a new format has been used.

Your craft's HIN must appear in two places. If it has a transom, the primary number is on its starboard side within 2 in. of its top. If it does not have a transom or if it was not practical to use the transom, the number is on the starboard side. In this case, it must be within 1 ft. of the stern and within 2 in. of the top of the hull side. On pontoon boats, it is on the aft crossbeam within one foot of the starboard hull attachment. Your boat also has a duplicate number in an unexposed location. This is on the boat's interior or under a fitting or item of hardware.

LENGTH OF BOATS

For some purposes, boats are classed by length. Required equipment, for example, differs with boat size. Manufacturers may measure a boat's length in several ways. Officially, though, your boat is measured along a straight line from its bow to its stern. This line is parallel to its keel.

The length does not include bowsprits, boomkins, or pulpits. Nor does it include rudders, brackets, outboard motors, outdrives, diving platforms, or other attachments.

CAPACITY INFORMATION

◆ **See Figure 4**

Manufacturers must put capacity plates on most recreational vessels less than 20 feet long. Sailboats, canoes, kayaks, and inflatable boats are usually exempt. PWCs must display the maximum permitted horsepower of their engines. The plates must also show the allowable maximum weights of the people on board. And they must show the allowable maximum combined weights of people, engine(s), and gear. Inboards and stern drives need not show the weight of their engines on their capacity plates. The capacity plate must appear where it is clearly visible to the operator when underway. This information serves to remind you of the capacity of your boat under normal circumstances. You should ask yourself, "Is my boat loaded above its recommended capacity" and, "Is my boat overloaded for the present sea and wind conditions?" If you are stopped by a legal authority, you may be cited if you are overloaded.

CERTIFICATE OF COMPLIANCE

◆ **See Figure 4**

Manufacturers are required to put compliance plates on motorboats greater than 20 feet in length. The plates must say, "This boat," or "This equipment complies with the U. S. Coast Guard Safety Standards in effect on the date of certification." Letters and numbers can be no less than 1/8 in. high. At the manufacturer's option, the capacity and compliance plates may be combined.

VENTILATION

A cup of gasoline spilled in the bilge has the potential explosive power of 15 sticks of dynamite. This statement, commonly quoted over 20 years ago, may be an exaggeration; however, it illustrates a fact. Gasoline fumes in the bilge of a boat are highly explosive and a serious danger. They are heavier than air and will stay in the bilge until they are vented out.

Because of this danger, Coast Guard regulations require ventilation on many powerboats. There are several ways to supply fresh air to engine and gasoline tank compartments and to remove dangerous vapors. Whatever the choice, it must meet Coast Guard standards.

■ **The following is not intended to be a complete discussion of the regulations. It is limited to the majority of recreational vessels. Contact your local Coast Guard office for further information.**

General Precautions

Ventilation systems will not remove raw gasoline that leaks from tanks or fuel lines. If you smell gasoline fumes, you need immediate repairs. The best device for sensing gasoline fumes is your nose. Use it! If you smell gasoline in a bilge, engine compartment, or elsewhere, don't start your engine. The smaller the compartment, the less gasoline it takes to make an explosive mixture.

Ventilation for Open Boats

In open boats, gasoline vapors are dispersed by the air that moves through them. So they are exempt from ventilation requirements.

To be "open," a boat must meet certain conditions. Engine and fuel tank compartments and long narrow compartments that join them must be open to the atmosphere." This means they must have at least 15 square inches of open area for each cubic foot of net compartment volume. The open area must be in direct contact with the atmosphere. There must also be no long, unventilated spaces open to engine and fuel tank compartments into which flames could extend.

Ventilation for All Other Boats

Powered and natural ventilation are required in an enclosed compartment with a permanently installed gasoline engine that has a cranking motor. A compartment is exempt if its engine is open to the atmosphere. Diesel powered boats are also exempt.

VENTILATION SYSTEMS

There are 2 types of ventilation systems. One is "natural ventilation." In it, air circulates through closed spaces due to the vessel's motion. The other type is "powered ventilation." In it, air is circulated by a motor-driven fan or fans.

Natural Ventilation System Requirements

A natural ventilation system has an air supply from outside the boat. The air supply may also be from a ventilated compartment or a compartment open to the atmosphere. Intake openings are required. In addition, intake ducts may be required to direct the air to appropriate compartments.

The system must also have an exhaust duct that starts in the lower third of the compartment. The exhaust opening must be into another ventilated compartment or into the atmosphere. Each supply opening and supply duct, if there is one, must be above the usual level of water in the bilge. Exhaust openings and ducts must also be above the bilge water. Openings and ducts must be at least three square inches in area or two inches in diameter. Openings should be placed so exhaust gasses do not enter the fresh air intake. Exhaust fumes must not enter cabins or other enclosed, non-ventilated spaces. The carbon monoxide gas in them is deadly.

Intake and exhaust openings must be covered by cowls or similar devices. These registers keep out rain water and water from breaking seas. Most often, intake registers face forward and exhaust openings aft. This aids the flow of air when the boat is moving or at anchor since most boats face into the wind when properly anchored.

Power Ventilation System Requirements

◆ **See Figure 5**

Powered ventilation systems must meet the standards of a natural system, but in addition, they must also have one or more exhaust blowers. The blower duct can serve as the exhaust duct for natural ventilation if fan blades do not obstruct the air flow when not powered. Openings in engine compartment, for carburetion are in addition to ventilation system requirements.

Fig. 4 A U.S. Coast Guard certification plate indicates the amount of occupants and gear appropriate for safe operation of the vessel

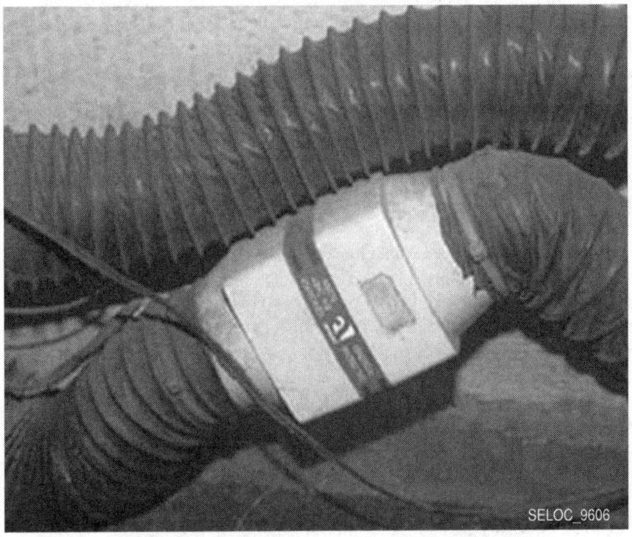

Fig. 5 Typical blower and duct system to vent fumes from the engine compartment

Required Safety Equipment

Coast Guard regulations require that your vessel have certain equipment aboard. These requirements are minimums. Exceed them whenever you can.

TYPES OF FIRES

There are four common classes of fires:

- Class A - fires are of ordinary combustible materials such as paper or wood.
- Class B - fires involve gasoline, oil and grease.
- Class C - fires are electrical.
- Class D - fires involve ferrous metals

One of the greatest risks to boaters is fire. This is why it is so important to carry the correct number and type of extinguishers onboard.

The best fire extinguisher for most vessels is a Class B extinguisher. Never use water on Class B or Class C fires, as water spreads these types of fires. Additionally, you should never use water on a Class C fire as it may cause you to be electrocuted.

FIRE EXTINGUISHERS

◆ **See Figure 6**

If your craft meets one or more of the following conditions, you must have at least 1 fire extinguisher aboard. The conditions are:

- Inboard or stern drive engines
- Closed compartments under seats where portable fuel tanks can be stored
- Double bottoms not sealed together or not completely filled with flotation materials
- Closed living spaces
- Closed stowage compartments in which combustible or flammable materials are stored
- Permanently installed fuel tanks
- Boat is 26 feet or more in length.

Contents of Extinguishers

Fire extinguishers use a variety of materials. Those used on boats usually contain dry chemicals, Halon, or Carbon Dioxide (CO2). Dry chemical extinguishers contain chemical powders such as Sodium Bicarbonate - baking soda.

Carbon dioxide is a colorless and odorless gas when released from an extinguisher. It is not poisonous but caution must be used in entering compartments filled with it. It will not support life and keeps oxygen from reaching your lungs. A fire-killing concentration of Carbon Dioxide can be lethal. If you are in a compartment with a high concentration of CO2, you will have no difficulty breathing. But the air does not contain enough oxygen to support life. Unconsciousness or death can result.

Halon Extinguishers

Some fire extinguishers and "built-in" or "fixed" automatic fire extinguishing systems contain a gas called Halon. Like carbon dioxide it is colorless and odorless and will not support life. Some Halons may be toxic if inhaled.

To be accepted by the Coast Guard, a fixed Halon system must have an indicator light at the vessel's helm. A green light shows the system is ready. Red means it is being discharged or has been discharged. Warning horns are available to let you know the system has been activated. If your fixed Halon system discharges, ventilate the space thoroughly before you enter it. There are no residues from Halon but it will not support life.

Although Halon has excellent fire fighting properties; it is thought to deplete the earth's ozone layer and has not been manufactured since January 1, 1994. Halon extinguishers can be refilled from existing stocks of the gas until they are used up, but high federal excise taxes are being charged for the service. If you discontinue using your Halon extinguisher, take it to a recovery station rather than releasing the gas into the atmosphere. Compounds such as FE 241, designed to replace Halon, are now available.

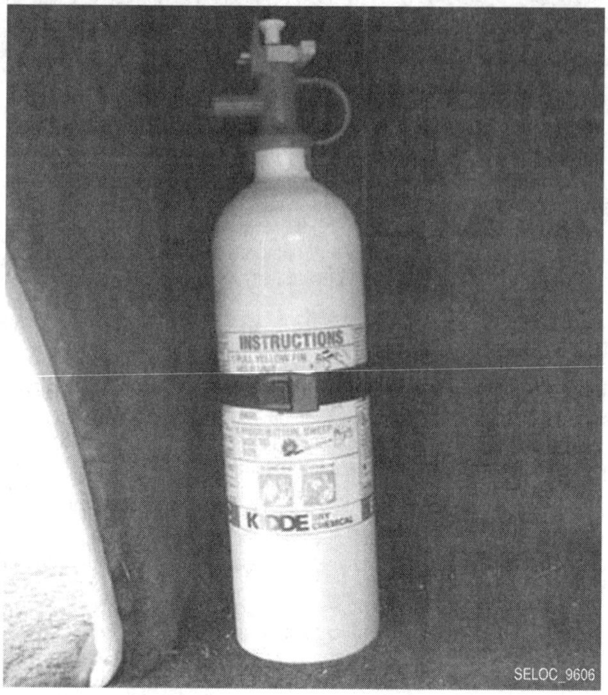

Fig. 6 An approved fire extinguisher should be mounted close to the operator for emergency use

Fire Extinguisher Approval

Fire extinguishers must be Coast Guard approved. Look for the approval number on the nameplate. Approved extinguishers have the following on their labels: "Marine Type USCG Approved, Size. . ., Type. . ., 162.208/," etc. In addition, to be acceptable by the Coast Guard, an extinguisher must be in serviceable condition and mounted in its bracket. An extinguisher not properly mounted in its bracket will not be considered serviceable during a Coast Guard inspection.

Care & Treatment

Make certain your extinguishers are in their stowage brackets and are not damaged. Replace cracked or broken hoses. Nozzles should be free of obstructions. Sometimes, wasps and other insects nest inside nozzles and make them inoperable. Check your extinguishers frequently. If they have pressure gauges, is the pressure within acceptable limits? Do the locking pins and sealing wires show they have not been used since recharging?

Don't try an extinguisher to test it. Its valves will not reseat properly and the remaining gas will leak out. When this happens, the extinguisher is useless.

Weigh and tag carbon dioxide and Halon extinguishers twice a year. If their weight loss exceeds 10 percent of the weight of the charge, recharge them. Check to see that they have not been used. They should have been inspected by a qualified person within the past six months, and they should have tags showing all inspection and service dates. The problem is that they can be partially discharged while appearing to be fully charged.

Some Halon extinguishers have pressure gauges the same as dry chemical extinguishers. Don't rely too heavily on the gauge. The extinguisher can be partially discharged and still show a good gauge reading. Weighing a Halon extinguisher is the only accurate way to assess its contents.

If your dry chemical extinguisher has a pressure indicator, check it frequently. Check the nozzle to see if there is powder in it. If there is, recharge it. Occasionally invert your dry chemical extinguisher and hit the base with the palm of your hand. The chemical in these extinguishers packs and cakes due to the boat's vibration and pounding. There is a difference of opinion about whether hitting the base helps, but it can't hurt. It is known that caking of the chemical powder is a major cause of failure of dry chemical extinguishers. Carry spares in excess of the minimum requirement. If you have guests aboard, make certain they know where the extinguishers are and how to use them.

Using a Fire Extinguisher

A fire extinguisher usually has a device to keep it from being discharged accidentally. This is a metal or plastic pin or loop. If you need to use your extinguisher, take it from its bracket. Remove the pin or the loop and point the nozzle at the base of the flames. Now, squeeze the handle, and discharge the extinguisher's contents while sweeping from side to side. Recharge a used extinguisher as soon as possible.

If you are using a Halon or carbon dioxide extinguisher, keep your hands away from the discharge. The rapidly expanding gas will freeze them. If your fire extinguisher has a horn, hold it by its handle.

Legal Requirements for Extinguishers

You must carry fire extinguishers as defined by Coast Guard regulations. They must be firmly mounted in their brackets and immediately accessible.

A motor vessel less than 26 feet long must have at least 1 approved hand-portable, Type B-1 extinguisher. If the craft has an approved fixed fire extinguishing system, you are not required to have the Type B-1 extinguisher. Also, if your vessel is less than 26 feet long, is propelled by an outboard motor, or motors, and does not have any of the first 6 conditions described at the beginning of this section, it is not required to have an extinguisher. Even so, it's a good idea to have one, especially if a nearby boat catches fire, or if a fire occurs at a fuel dock.

A motor vessel 26 feet to less than 40 feet long, must have at least 2 Type B-1 approved hand-portable extinguishers. It can, instead, have at least 1 Coast Guard approved Type B-2. If you have an approved fire extinguishing system, only one Type B-1 is required.

A motorboat 40 to 65 feet long must have at least 3 Type B-1 approved portable extinguishers. It may have, instead, at least 1 Type B-1 plus a Type B-2. If there is an approved fixed fire extinguishing system, 2 Type B-1 or 1 Type B-2 is required.

WARNING SYSTEM

Various devices are available to alert you to danger. These include fire, smoke, gasoline fumes, and carbon monoxide detectors. If your vessel has a galley, it should have a smoke detector. Where possible, use wired detectors. Household batteries often corrode rapidly on a boat.

There are many ways in which carbon monoxide (a by-product of the combustion that occurs in an engine) can enter your boat. You can't see, smell, or taste carbon monoxide gas, but it is lethal. As little as one part in 10,000 parts of air can bring on a headache. The symptoms of carbon monoxide poisoning - headaches, dizziness, and nausea - are like seasickness. By the time you realize what is happening to you, it may be too late to take action. If you have enclosed living spaces on your boat, protect yourself with a detector.

PERSONAL FLOTATION DEVICES

Personal Flotation Devices (PFDs) are commonly called life preservers or life jackets. You can get them in a variety of types and sizes. They vary with their intended uses. To be acceptable, PFDs must be Coast Guard approved.

Type I PFDs

A Type I life jacket is also called an offshore life jacket. Type I life jackets will turn most unconscious people from facedown to a vertical or slightly backward position. The adult size gives a minimum of 22 pounds of buoyancy. The child size has at least 11 pounds. Type I jackets provide more protection to their wearers than any other type of life jacket. Type I life jackets are bulkier and less comfortable than other types. Furthermore, there are only two sizes, one for children and one for adults.

Type I life jackets will keep their wearers afloat for extended periods in rough water. They are recommended for offshore cruising where a delayed rescue is probable.

Type II PFDs

◆ See Figure 7

A Type II life jacket is also called a near-shore buoyant vest. It is an approved, wearable device. Type II life jackets will turn some unconscious people from facedown to vertical or slightly backward positions. The adult size gives at least 15.5 pounds of buoyancy. The medium child size has a minimum of 11 pounds. And the small child and infant sizes give seven pounds. A Type II life jacket is more comfortable than a Type I but it does not have as much buoyancy. It is not recommended for long hours in rough water. Because of this, Type IIs are recommended for inshore and inland cruising on calm water. Use them only where there is a good chance of fast rescue.

Type III PFDs

Type III life jackets or marine buoyant devices are also known as flotation aids. Like Type IIs, they are designed for calm inland or close offshore water where there is a good chance of fast rescue. Their minimum buoyancy is 15.5 pounds. They will **not** turn their wearers face up.

Type III devices are usually worn where freedom of movement is necessary. Thus, they are used for water skiing, small boat sailing, and fishing among other activities. They are available as vests and flotation coats. Flotation coats are useful in cold weather. Type IIIs come in many sizes from small child through large adult.

Life jackets come in a variety of colors and patterns - red, blue, green, camouflage, and cartoon characters. From purely a safety standpoint, the best color is bright orange. It is easier to see in the water, especially if the water is rough.

Type IV PFDs

◆ See Figures 8 and 9

Type IV ring life buoys, buoyant cushions and horseshoe buoys are Coast Guard approved devices called throwables. They are made to be thrown to people in the water, and should not be worn. Type IV cushions are often used as seat cushions. But, keep in mind that cushions are hard to hold onto in the water, thus, they do not afford as much protection as wearable life jackets.

The straps on buoyant cushions are for you to hold onto either in the water or when throwing them, they are **NOT** for your arms. A cushion should never be worn on your back, as it will turn you face down in the water.

Type IV throwables are not designed as personal flotation devices for unconscious people, non-swimmers, or children. Use them only in emergencies. They should not be used for, long periods in rough water.

Ring life buoys come in 18, 20, 24, and 30 in. diameter sizes. They usually have grab lines, but you will need to attach about 60 feet of polypropylene line to the grab rope to aid in retrieving someone in the water. If you throw a ring, be careful not to hit the person. Ring buoys can knock people unconscious

Type V PFDs

Type V PFDs are of two kinds, special use devices and hybrids. Special use devices include boardsailing vests, deck suits, work vests, and others. They are approved only for the special uses or conditions indicated on their labels. Each is designed and intended for the particular application shown on its label. They do not meet legal requirements for general use aboard recreational boats.

Hybrid life jackets are inflatable devices with some built-in buoyancy provided by plastic foam or kapok. They can be inflated orally or by cylinders of compressed gas to give additional buoyancy. In some hybrids the gas is released manually. In others it is released automatically when the life jacket is immersed in water.

The inherent buoyancy of a hybrid may be insufficient to float a person unless it is inflated. The only way to find this out is for the user to try it in the water. Because of its limited buoyancy when deflated, a hybrid is recommended for use by a non-swimmer only if it is worn with enough inflation to float the wearer.

If they are to count against the legal requirement for the number of life jackets you must carry, hybrids manufactured before February 8, 1995 must be worn whenever a boat is underway and the wearer must not go below decks or in an enclosed space. To find out if your Type V hybrid must be worn to satisfy the legal requirement, read its label. If its use is restricted it will say, "REQUIRED TO BE WORN" in capital letters.

Hybrids cost more than other life jackets, but this factor must be weighed against the fact that they are more comfortable than Types I, II or III life jackets. Because of their greater comfort, their owners are more likely to wear them than are the owners of Type I, II or III life jackets.

Fig. 7 Type III PFDs are recommended for inshore/inland use on calm water (where there is a good chance of fast rescue)

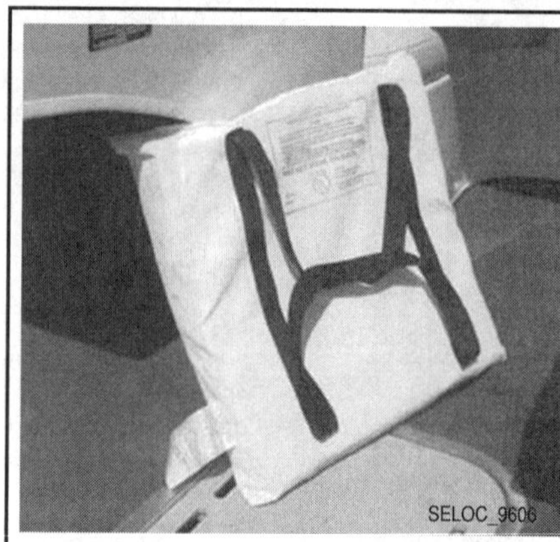

Fig. 8 Type IV buoyant cushions are thrown to people in the water. If you can squeeze air out of the cushion, it should be replaced

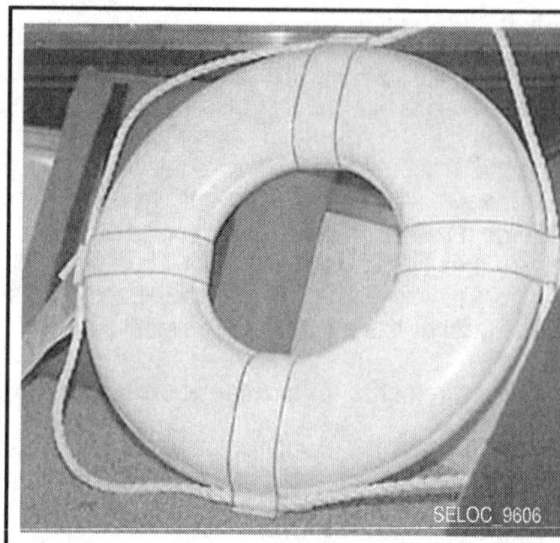

Fig. 9 Type IV throwables, such as this ring life buoy, are not designed for unconscious people, non-swimmers, or children

The Coast Guard has determined that improved, less costly hybrids can save lives since they will be bought and used more frequently. For these reasons, a new federal regulation was adopted effective February 8, 1995. The regulation increases both the deflated and inflated buoyancys of hybrids, makes them available in a greater variety of sizes and types, and reduces their costs by reducing production costs.

Even though it may not be required, the wearing of a hybrid or a life jacket is encouraged whenever a vessel is underway. Like life jackets, hybrids are now available in three types. To meet legal requirements, a Type I hybrid can be substituted for a Type I life jacket. Similarly Type II and III hybrids can be substituted for Type II and Type III life jackets. A Type I hybrid, when inflated, will turn most unconscious people from facedown to vertical or slightly backward positions just like a Type I life jacket. Type II and III hybrids function like Type II and III life jackets. If you purchase a new hybrid, it should have an owner's manual attached that describes its life jacket type and its deflated and inflated buoyancys. It warns you that it may have to be inflated to float you. The manual also tells you how to don the life jacket and how to inflate it. It also tells you how to change its inflation mechanism, recommended testing exercises, and inspection or maintenance procedures. The manual also tells you why you need a life jacket and why you should wear it. A new hybrid must be packaged with at least three gas cartridges. One of these may already be loaded into the inflation mechanism. Likewise, if it has an automatic inflation mechanism, it must be packaged with at least three of these water sensitive elements. One of these elements may be installed.

Legal Requirements

A Coast Guard approved life jacket must show the manufacturer's name and approval number. Most are marked as Type I, II, III, IV or V. All of the newer hybrids are marked for type.

You are required to carry at least one wearable life jacket or hybrid for each person on board your recreational vessel. If your vessel is 16 feet or more in length and is not a canoe or a kayak, you must also have at least one Type IV on board. These requirements apply to all recreational vessels that are propelled or controlled by machinery, sails, oars, paddles, poles, or another vessel. Sailboards are not required to carry life jackets.

You can substitute an older Type V hybrid for any required Type I, II or III life jacket provided:

1. Its approval label shows it is approved for the activity the vessel is engaged in
2. Its approved as a substitute for a life jacket of the type required on the vessel
3. Its used as required on the labels

and

4. It's used in accordance with any requirements in its owner's manual (if the approval label makes reference to such a manual.)

A water skier being towed is considered to be on board the vessel when judging compliance with legal requirements.

You are required to keep your Type I, II or III life jackets or equivalent hybrids readily accessible, which means you must be able to reach out and get them when needed. All life jackets must be in good, serviceable condition.

General Considerations

The proper use of a life jacket requires the wearer to know how it will perform. You can gain this knowledge only through experience. Each person on your boat should be assigned a life jacket. Next, it should be fitted to the person who will wear it. Only then can you be sure that it will be ready for use in an emergency. This advice is good even if the water is calm, and you intend to boat near shore.

Boats can sink fast. There may be no time to look around for a life jacket. Fitting one on you in the water is almost impossible. Most drownings occur in inland waters within a few feet of safety. Most victims had life jackets, but they weren't wearing them.

Keeping life jackets in the plastic covers they came wrapped in, and in a cabin, assure that they will stay clean and unfaded. But this is no way to keep them when you are on the water. When you need a life jacket it must be readily accessible and adjusted to fit you. You can't spend time hunting for it or learning how to fit it.

There is no substitute for the experience of entering the water while wearing a life jacket. Children, especially, need practice. If possible, give your guests this experience. Tell them they should keep their arms to their sides when jumping in to keep the life jacket from riding up. Let them jump in and see how the life jacket responds. Is it adjusted so it does not ride up? Is it the proper size? Are all straps snug? Are children's life jackets the right sizes for them? Are they adjusted properly? If a child's life jacket fits correctly, you can lift the child by the jacket's shoulder straps and the child's chin and ears will not slip through. Non-swimmers, children, handicapped persons, elderly persons and even pets should always wear life jackets when they are aboard. Many states require that everyone aboard wear them in hazardous waters.

Inspect your lifesaving equipment from time to time. Leave any questionable or unsatisfactory equipment on shore. An emergency is no time for you to conduct an inspection.

Indelibly mark your life jackets with your vessel's name, number, and calling port. This can be important in a search and rescue effort. It could help concentrate effort where it will do the most good.

Care of Life Jackets

Given reasonable care, life jackets last many years. Thoroughly dry them before putting them away. Stow them in dry, well-ventilated places. Avoid the bottoms of lockers and deck storage boxes where moisture may collect. Air and dry them frequently.

Life jackets should not be tossed about or used as fenders or cushions. Many contain kapok or fibrous glass material enclosed in plastic bags. The bags can rupture and are then unserviceable. Squeeze your life jacket gently. Does air leak out? If so, water can leak in and it will no longer be safe to use. Cut it up so no one will use it, and throw it away. The covers of some life jackets are made of nylon or polyester. These materials are plastics. Like many plastics, they break down after extended exposure to the ultraviolet light in sunlight. This process may be more rapid when the materials are dyed with bright dyes such as "neon" shades.

Ripped and badly faded fabrics are clues that the covering of your life jacket is deteriorating. A simple test is to pinch the fabric between your thumbs and forefingers. Now try to tear the fabric. If it can be torn, it should definitely be destroyed and discarded. Compare the colors in protected places to those exposed to the sun. If the colors have faded, the materials have been weakened. A life jacket covered in fabric should ordinarily last several boating seasons with normal use. A life jacket used every day in direct sunlight should probably be replaced more often.

SOUND PRODUCING DEVICES

All vessels are required to carry some means of making an efficient sound signal. Devices for making the whistle or horn noises required by the Navigation Rules must be capable of a four-second blast. The blast should be audible for at least one-half mile. Athletic whistles are not acceptable on boats 12 meters or longer. Use caution with athletic whistles. When wet, some of them come apart and lose their "pea." When this happens, they are useless.

If your vessel is 12 meters long and less than 20 meters, you must have a power whistle (or power horn) and a bell on board. The bell must be in operating condition and have a minimum diameter of at least 200mm (7.9 in.) at its mouth.

VISUAL DISTRESS SIGNALS

◆ **See Figure 10**

Visual Distress Signals (VDS) attract attention to your vessel if you need help. They also help to guide searchers in search and rescue situations. Be sure you have the right types, and learn how to use them properly.

It is illegal to fire flares improperly. In addition, they cost the Coast Guard and its Auxiliary many wasted hours in fruitless searches. If you signal a distress with flares and then someone helps you, please let the Coast Guard or the appropriate Search And Rescue (SAR) Agency know so the distress report will be canceled.

Recreational vessels less than 16 feet long must carry visual distress signals on coastal waters at night. Coastal waters are:

- The ocean (territorial sea)
- he Great Lakes
- Bays or sounds that empty into oceans
- Rivers over 2 miles across at their mouths upstream to where they narrow to 2 miles.

Recreational craft 16 feet or longer must carry VDS at all times on coastal waters. The same requirement applies to craft carrying 6 or fewer passengers for hire. Open sailboats less than 26 ft. long without engines are exempt in the daytime as are manually propelled boats. Also exempt are boats in organized races, regattas, parades, etc. Boats owned in the United States and operating on the high seas must be equipped with VDS.

A wide variety of signaling devices meet Coast Guard regulations. For pyrotechnic devices, a minimum of three must be carried. Any combination can be carried as long as it adds up to at least 3 signals for day use and at least 3 signals for night use. Three day/night signals meet both requirements. If possible, carry more than the legal requirement.

■ **The American flag flying upside down is a commonly recognized distress signal. It is not recognized in the Coast Guard regulations, though. In an emergency, your efforts would probably be better used in more effective signaling methods.**

Types of VDS

VDS are divided into 2 groups; daytime and nighttime use. Each of these groups is subdivided into pyrotechnic and non-pyrotechnic devices.

Daytime Non-Pyrotechnic Signals

A bright orange flag with a black square over a black circle is the simplest VDS. It is usable, of course, only in daylight. It has the advantage of being a continuous signal. A mirror can be used to good advantage on sunny days. It can attract the attention of other boaters and of aircraft from great distances. Mirrors are available with holes in their centers to aid in "aiming." In the absence of a mirror, any shiny object can be used. When another boat is in sight, an effective VDS is to extend your arms from your sides and move them up and down. Do it slowly. If you do it too fast the other people may think you are just being friendly. This simple gesture is seldom misunderstood, and requires no equipment.

Daytime Pyrotechnic Devices

Orange smoke is a useful daytime signal. Hand-held or floating smoke flares are very effective in attracting attention from aircraft. Smoke flares don't last long, and are not very effective in high wind or poor visibility. As with other pyrotechnic devices, use them only when you know there is a possibility that someone will see the display.

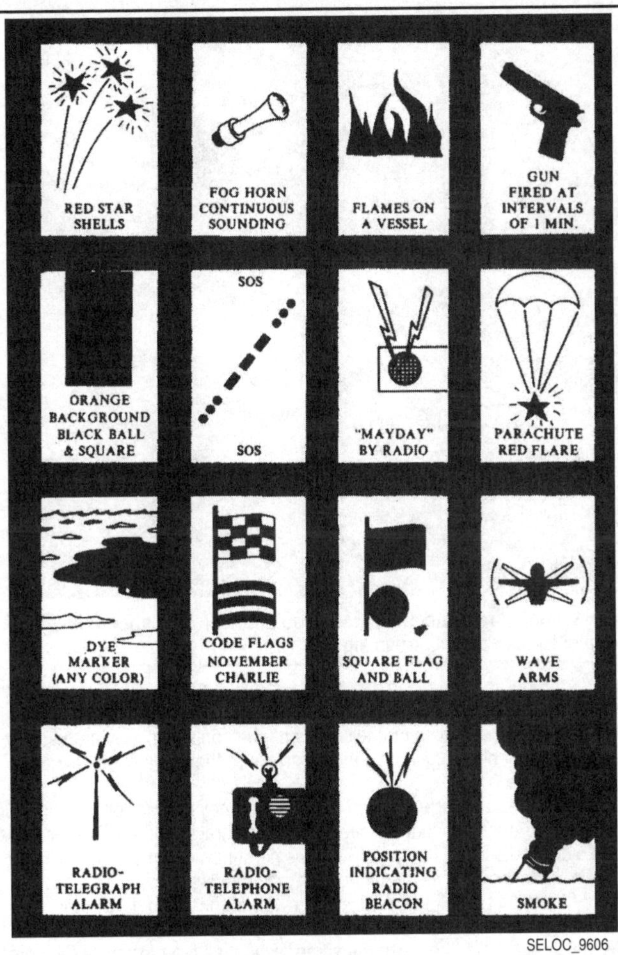

Fig. 10 Internationally accepted distress signals

To be usable, smoke flares must be kept dry. Keep them in airtight containers and store them in dry places. If the "striker" is damp, dry it out before trying to ignite the device. Some pyrotechnic devices require a forceful "strike" to ignite them.

All hand-held pyrotechnic devices may produce hot ashes or slag when burning. Hold them over the side of your boat in such a way that they do not burn your hand or drip into your boat.

Nighttime Non-Pyrotechnic Signals

An electric distress light is available. This light automatically flashes the international morse code SOS distress signal (••• — •••). Flashed four to six times a minute, it is an unmistakable distress signal. It must show that it is approved by the Coast Guard. Be sure the batteries are fresh. Dated batteries give assurance that they are current.

Under the Inland Navigation Rules, a high intensity white light flashing 50-70 times per minute is a distress signal. Therefore, use strobe lights on inland waters only for distress signals.

Nighttime Pyrotechnic Devices

◆ **See Figure 11**

Aerial and hand-held flares can be used at night or in the daytime. Obviously, they are more effective at night.

Currently, the serviceable life of a pyrotechnic device is rated at 42 months from its date of manufacture. Pyrotechnic devices are expensive. Look at their dates before you buy them. Buy them with as much time remaining as possible.

Like smoke flares, aerial and hand-held flares may fail to work if they have been damaged or abused. They will not function if they are or have

Fig. 11 Moisture-protected flares should be carried onboard any vessel for use as a distress signal

been wet. Store them in dry, airtight containers in dry places. But store them where they are readily accessible.

Aerial VDSs, depending on their type and the conditions they are used in, may not go very high. Again, use them only when there is a good chance they will be seen.

A serious disadvantage of aerial flares is that they burn for only a short time; most burn for less than 10 seconds. Most parachute flares burn for less than 45 seconds. If you use a VDS in an emergency, do so carefully. Hold hand-held flares over the side of the craft when in use. Never use a road hazard flare on a boat; it can easily start a fire. Marine type flares are specifically designed to lessen risk, but they still must be used carefully.

Aerial flares should be given the same respect as firearms since they are firearms! Never point them at another person. Don't allow children to play with them or around them. When you fire one, face away from the wind. Aim it downwind and upward at an angle of about 60 degrees to the horizon. If there is a strong wind, aim it somewhat more vertically. Never fire it straight up. Before you discharge a flare pistol, check for overhead obstructions that might be damaged by the flare. An obstruction might deflect the flare to where it will cause injury or damage.

Disposal of VDS

Keep outdated flares when you get new ones. They do not meet legal requirements, but you might need them sometime, and they may work. It is illegal to fire a VDS on federal navigable waters unless an emergency exists. Many states have similar laws.

Emergency Position Indicating Radio Beacon (EPIRB)

There is no requirement for recreational boats to have EPIRBs. Some commercial and fishing vessels, though, must have them if they operate beyond the 3 mile limit. Vessels carrying 6 or fewer passengers for hire must have EPIRBs under some circumstances when operating beyond the 3 mile limit. If you boat in a remote area or offshore, you should have an EPIRB. An EPIRB is a small (about 6-20 in. high), battery-powered, radio transmitting buoy-like device. It is a radio transmitter and requires a license or an endorsement on your radio station license by the Federal Communications Commission (FCC). EPIRBs are either automatically activated by being immersed in water or manually by a switch.

Courtesy Marine Examinations

One of the roles of the Coast Guard Auxiliary is to promote recreational boating safety. This is why they conduct thousands of Courtesy Marine Examinations each year. The auxiliarists who do these examinations are well-trained and knowledgeable in the field.

These examinations are free and done only at the consent of boat owners. To pass the examination, a vessel must satisfy federal equipment requirements and certain additional requirements of the coast guard auxiliary. If your vessel does not pass the Courtesy Marine Examination, no report of the failure is made. Instead, you will be told what you need to correct the deficiencies. The examiner will return at your convenience to redo the examination.

If your vessel qualifies, you will be awarded a safety decal. The decal does not carry any special privileges, it simply attests to your interest in safe boating.

BOATING EQUIPMENT (NOT REQUIRED BUT RECOMMENDED)

Although not required by law, there are other pieces of equipment that are good to have onboard.

Oar/Paddle (Second Means of Propulsion)

All vessels less than 16 ft. long should carry a 2nd means of propulsion. A paddle or oar can come in handy at times. For most small boats, a spare trolling or outboard motor is an excellent idea. If you carry a spare motor, it should have its own fuel tank and starting power. If you use an electric trolling motor, it should have its own battery.

Bailing Devices

All vessels should carry at least one effective manual bailing device in addition to any installed electric bilge pump. This can be a bucket, can, scoop, hand-operated pump, etc. If your battery "goes dead" it will not operate your electric pump.

First Aid Kit

◆ See Figure 12

All vessels should carry a first aid kit. It should contain adhesive bandages, gauze, adhesive tape, antiseptic, aspirin, etc. Check your first aid kit from time to time. Replace anything that is outdated. It is to your advantage to know how to use your first aid kit. Another good idea would be to take a Red Cross first aid course.

Anchors

◆ See Figure 13

All boats should have anchors, even PWCs. Choose one of suitable size for your vessel. Better still, have 2 anchors of different sizes. Use the smaller one in calm water or when anchoring for a short time to fish or eat. Use the larger one when the water is rougher or for overnight anchoring.

Carry enough anchor line, of suitable size, for your vessel and the waters in which you will operate. If your engine fails you, the first thing you usually should do is lower your anchor. This is good advice in shallow water where you may be driven aground by the wind or water. It is also good advice in windy weather or rough water, as the anchor, when properly affixed, will usually hold your bow into the waves.

VHF-FM Radio

Your best means of summoning help in an emergency or in case of a breakdown is a VHF-FM radio. You can use it to get advice or assistance from the Coast Guard. In the event of a serious illness or injury aboard your vessel, the Coast Guard can have emergency medical equipment meet you ashore.

■ **Although the VHF radio is the best way to get help, in this day and age, cell phones are a good back-up source, especially for boaters on inland waters. You probably already know where you get a signal when boating, keep the phone charged, handy and off (so it doesn't bother you when boating right?). Keep phone numbers for a local dockmaster, coast guard, tow service or maritime police unit handy on board or stored in your phone directory.**

Fig. 12 Always carry an adequately stocked first aid kit on board for the safety of the crew and guests

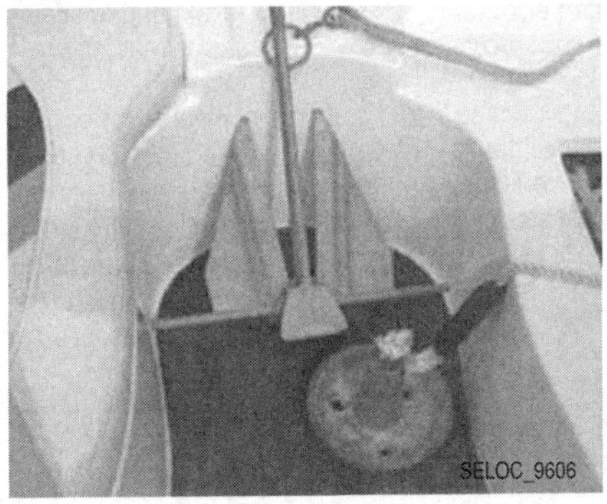

Fig. 13 Choose an anchor of sufficient weight to secure the boat without dragging

Fig. 14 Don't hesitate to spend a few extra dollars for a reliable compass

Compass

SELECTION

◆ **See Figure 14**

The safety of the boat and her crew may depend on her compass. In many areas, weather conditions can change so rapidly that, within minutes, a skipper may find himself socked in by a fog bank, rain squall or just poor visibility. Under these conditions, he may have no other means of keeping to his desired course except with the compass. When crossing an open body of water, his compass may be the only means of making an accurate landfall.

During thick weather when you can neither see nor hear the expected aids to navigation, attempting to run out the time on a given course can disrupt the pleasure of the cruise. The skipper gains little comfort in a chain of soundings that does not match those given on the chart for the expected area. Any stranding, even for a short time, can be an unnerving experience.

A pilot will not knowingly accept a cheap parachute. By the same token, a good boater should not accept a bargain in lifejackets, fire extinguishers, or compass. Take the time and spend the few extra dollars to purchase a compass to fit your expected needs. Regardless of what the salesman may tell you, postpone buying until you have had the chance to check more than one make and model.

Lift each compass, tilt and turn it, simulating expected motions of the boat. The compass card should have a smooth and stable reaction.

The card of a good quality compass will come to rest without oscillations about the lubber's line. Reasonable movement in your hand, comparable to the rolling and pitching of the boat, should not materially affect the reading.

INSTALLATION

◆ **See Figure 15**

Proper installation of the compass does not happen by accident. Make a critical check of the proposed location to be sure compass placement will permit the helmsman to use it with comfort and accuracy. First, the compass should be placed directly in front of the helmsman, and in such a position that it can be viewed without body stress as he sits or stands in a posture of relaxed alertness. The compass should be in the helmsman's zone of comfort. If the compass is too far away, he may have to bend forward to watch it; too close and he must rear backward for relief.

Second, give some thought to comfort in heavy weather and poor visibility conditions during the day and night. In some cases, the compass position may be partially determined by the location of the wheel, shift lever and throttle handle.

Third, inspect the compass site to be sure the instrument will be at least two feet from any engine indicators, bilge vapor detectors, magnetic instruments, or any steel or iron objects. If the compass cannot be placed at least two feet (six feet would be better but on a small craft, let's get real two feet is usually pushing it) from one of these influences, then either the compass or the other object must be moved, if first order accuracy is to be expected.

Once the compass location appears to be satisfactory, give the compass a test before installation. Hidden influences may be concealed under the cabin top, forward of the cabin aft bulkhead, within the cockpit ceiling, or in a wood-covered stanchion.

Move the compass around in the area of the proposed location. Keep an eye on the card. A magnetic influence is the only thing that will make the card turn. You can quickly find any such influence with the compass. If the influence cannot be moved away or replaced by one of non-magnetic material, test to determine whether it is merely magnetic, a small piece of iron or steel, or some magnetized steel. Bring the north pole of the compass near the object, then shift and bring the south pole near it. Both the north and south poles will be attracted if the compass is demagnetized. If the object attracts one pole and repels the other, then the compass is magnetized. If your compass needs to be demagnetized, take it to a shop equipped to do the job PROPERLY.

After you have moved the compass around in the proposed mounting area, hold it down or tape it in position. Test everything you feel might affect the compass and cause a deviation from a true reading. Rotate the wheel from hard over-to-hard over. Switch on and off all the lights, radios, radio direction finder, radio telephone, depth finder and, if installed, the shipboard intercom. Sound the electric whistle, turn on the windshield wipers, start the engine (with water circulating through the engine), work the throttle, and move the gear shift lever. If the boat has an auxiliary generator, start it.

If the card moves during any one of these tests, the compass should be relocated. Naturally, if something like the windshield wipers causes a slight deviation, it may be necessary for you to make a different deviation table to use only when certain pieces of equipment are operating. Bear in mind, following a course that is off only a degree or two for several hours can make considerable difference at the end, putting you on a reef, rock or shoal.

Check to be sure the intended compass site is solid. Vibration will increase pivot wear.

Fig. 15 The compass is a delicate instrument which should be mounted securely in a position where it can be easily observed by the helmsman

Now, you are ready to mount the compass. To prevent an error on all courses, the line through the lubber line and the compass card pivot must be exactly parallel to the keel of the boat. You can establish the fore-and-aft line of the boat with a stout cord or string. Use care to transfer this line to the compass site. If necessary, shim the base of the compass until the stile-type lubber line (the one affixed to the case and not gimbaled) is vertical when the boat is on an even keel. Drill the holes and mount the compass.

COMPASS PRECAUTIONS

◆ **See Figures 16, 17 and 18**

Many times an owner will install an expensive stereo system in the cabin of his boat. It is not uncommon for the speakers to be mounted on the aft bulkhead up against the overhead (ceiling). In almost every case, this position places one of the speakers in very close proximity to the compass, mounted above the ceiling.

You probably already know that a magnet is used in the operation of the speaker. Therefore, it is very likely that the speaker, mounted almost under the compass in the cabin will have a very pronounced effect on the compass accuracy.

Consider the following test and the accompanying photographs as proof:

First, the compass was read as 190° while the boat was secure in her slip.

Next, a full can of soda in an aluminum can was placed on one side and the compass read as 204°, a good 14° off.

Next, the full can was moved to the opposite side of the compass and again a reading was observed, this time as 189°, 11° off from the original reading.

Finally, the contents of the can were consumed, the can placed on both sides of the compass with NO effect on the compass reading.

Two very important conclusions can be drawn from these tests.

- Something must have been in the contents of the can to affect the compass so drastically.
- Keep even innocent things clear of the compass to avoid any possible error in the boat's heading.

■ **Remember, a boat moving through the water at 10 knots on a compass error of just 5 degrees will be almost 1.5 miles off course in only ONE hour. At night, or in thick weather, this could very possibly put the boat on a reef, rock or shoal with disastrous results.**

Tools & Spare Parts

◆ **See Figures 19 and 20**

Carry a few tools and some spare parts, and learn how to make minor repairs. Many search and rescue cases are caused by minor breakdowns that boat operators could have repaired. Carry spare parts such as propellers, fuses or basic ignition components (like spark plugs, wires or even ignition coils) and the tools necessary to install them.

Fig. 16 This compass is giving an accurate reading, right?

Fig. 17 ... well think again, as seemingly innocent objects may cause serious problems ...

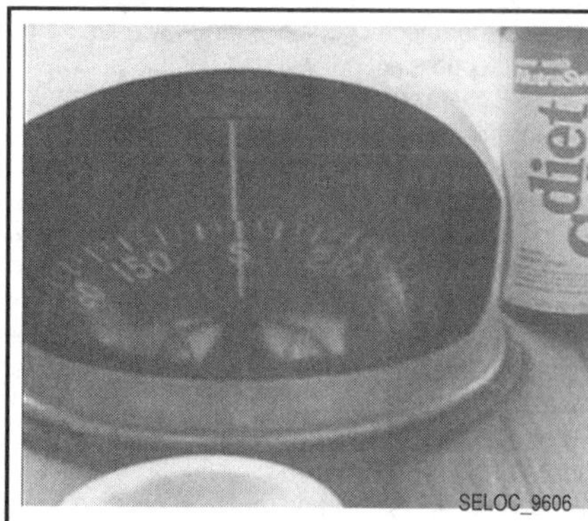

Fig. 18 ... a compass reading off by just a few degrees could lead to disaster

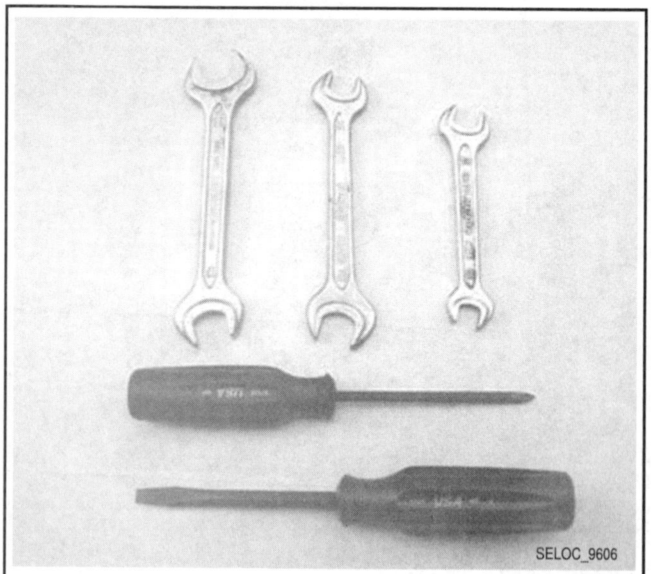

Fig. 19 A few wrenches, a screwdriver and maybe a pair of pliers can be very helpful to make emergency repairs

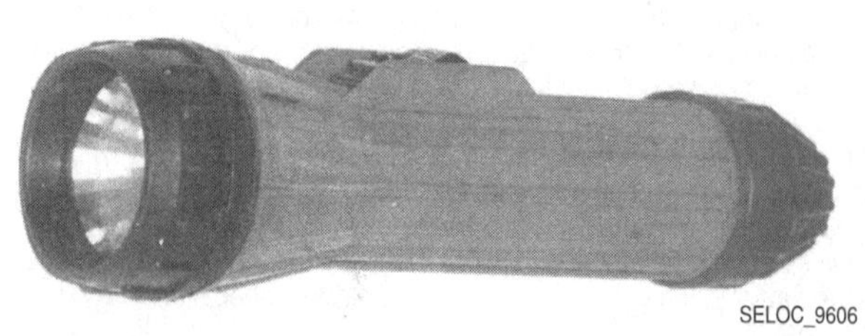

Fig. 20 A flashlight with a fresh set of batteries is handy when repairs are needed at night. It can also double as a signaling device

SAFETY IN SERVICE

It is virtually impossible to anticipate all of the hazards involved with maintenance and service, but care and common sense will prevent most accidents.

The rules of safety for mechanics range from "don't smoke around gasoline," to "use the proper tool(s) for the job." The trick to avoiding injuries is to develop safe work habits and to take every possible precaution. Whenever you are working on your boat, pay attention to what you are doing. The more you pay attention to details and what is going on around you, the less likely you will be to hurt yourself or damage your boat.

Do's

- Do keep a fire extinguisher and first aid kit handy.
- Do wear safety glasses or goggles when cutting, drilling, grinding or prying, even if you have 20-20 vision. If you wear glasses for the sake of vision, wear safety goggles over your regular glasses.
- Do shield your eyes whenever you work around the battery. Batteries contain sulfuric acid. In case of contact with the eyes or skin, flush the area with water or a mixture of water and baking soda; then seek immediate medical attention.
- Do use adequate ventilation when working with any chemicals or hazardous materials.
- Do disconnect the negative battery cable when working on the electrical system. The secondary ignition system contains EXTREMELY HIGH VOLTAGE. In some cases it can even exceed 50,000 volts. Furthermore, an accidental attempt to start the engine could cause the propeller or other components to rotate suddenly causing a potentially dangerous situation.
- Do follow manufacturer's directions whenever working with potentially hazardous materials. Most chemicals and fluids are poisonous if taken internally.
- Do properly maintain your tools. Loose hammerheads, mushroomed punches and chisels, frayed or poorly grounded electrical cords, excessively worn screwdrivers, spread wrenches (open end), cracked sockets, or slipping ratchets can cause accidents.
- Likewise, keep your tools clean; a greasy wrench can slip off a bolt head, ruining the bolt and often harming your knuckles in the process.
- Do use the proper size and type of tool for the job at hand. Do select a wrench or socket that fits the nut or bolt. The wrench or socket should sit straight, not cocked.
- Do, when possible, pull on a wrench handle rather than push on it, and adjust your stance to prevent a fall.
- Do be sure that adjustable wrenches are tightly closed on the nut or bolt and pulled so that the force is on the side of the fixed jaw. Better yet, avoid the use of an adjustable if you have a fixed wrench that will fit.
- Do strike squarely with a hammer; avoid glancing blows.
- Do use common sense whenever you work on your boat or motor. If a situation arises that doesn't seem right, sit back and have a second look. It may save an embarrassing moment or potential damage to your beloved boat.

Don'ts

- Don't run the engine in an enclosed area or anywhere else without proper ventilation - EVER! Carbon monoxide is poisonous; it takes a long time to leave the human body and you can build up a deadly supply of it in your system by simply breathing in a little every day. You may not realize you are slowly poisoning yourself.
- Don't work around moving parts while wearing loose clothing. Short sleeves are much safer than long, loose sleeves. Hard-toed shoes with neoprene soles protect your toes and give a better grip on slippery surfaces. Jewelry, watches, large belt buckles, or body adornment of any kind is not safe working around any craft or vehicle. Long hair should be tied back under a hat.
- Don't use pockets for toolboxes. A fall or bump can drive a screwdriver deep into your body. Even a rag hanging from your back pocket can wrap around a spinning shaft.
- Don't smoke when working around gasoline, cleaning solvent or other flammable material.
- Don't smoke when working around the battery. When the battery is being charged, it gives off explosive hydrogen gas. Actually, you shouldn't smoke anyway, it's bad for you. Instead, save the cigarette money and put it into your boat!
- Don't use gasoline to wash your hands; there are excellent soaps available. Gasoline contains dangerous additives that can enter the body through a cut or through your pores. Gasoline also removes all the natural oils from the skin so that bone dry hands will suck up oil and grease.
- Don't use screwdrivers for anything other than driving screws! A screwdriver used as a prying tool can snap when you least expect it, causing injuries. At the very least, you'll ruin a good screwdriver.

TROUBLESHOOTING

Troubleshooting can be defined as a methodical process during which one discovers what is causing a problem with engine operation. Although it is often a feared process to the uninitiated, there is no reason to believe that you cannot figure out what is wrong with a motor, as long as you follow a few basic rules.

To begin with, troubleshooting must be systematic. Haphazardly testing one component, then another, *might* uncover the problem, but it will more likely waste a lot of time. True troubleshooting starts by defining the problem and performing systematic tests to eliminate the largest and most likely causes first.

Start all troubleshooting by eliminating the most basic possible causes. Begin with a visual inspection of the boat and motor. If the engine won't crank, make sure that the kill switch or safety lanyard is in the proper position. Make sure there is fuel in the tank and the fuel system is primed before condemning the carburetor or fuel injection system. On electric start motors, make sure there are no blown fuses, the battery is fully charged, and the cable connections (at both ends) are clean and tight before suspecting a bad starter, solenoid or switch.

The majority of problems that occur suddenly can be fixed by simply identifying the one small item that brought them on. A loose wire, a clogged passage or a broken component can cause a lot of trouble and are often the cause of a sudden performance problem.

The next most basic step in troubleshooting is to test systems before components. For example, if the engine doesn't crank on an electric start motor, determine if the battery is in good condition (fully charged and properly connected) before testing the starting system. If the engine cranks, but doesn't start, you know already know the starting system and battery (if it cranks fast enough) are in good condition, now it is time to look at the ignition or fuel systems. Once you've isolated the problem to a particular system, follow the troubleshooting/testing procedures in the section for that system to test either subsystems (if applicable, for example: the starter circuit) or components (starter solenoid).

Basic Operating Principles

◆ **See Figures 21 and 22**

Before attempting to troubleshoot a problem with your motor, it is important that you understand how it operates. Once normal engine or system operation is understood, it will be easier to determine what might be causing the trouble or irregular operation in the first place. System descriptions are found throughout this manual, but the basic mechanical operating principles for both 2-stroke engines (like the motors covered here) and 4-stroke engines (like some outboards and like your car) are given here. A basic understanding of both types of engines is useful not only in understanding and troubleshooting your motor, but also for dealing with other motors in your life.

All motors covered by this manual (and probably *most* of the motors you own) operate according to the Otto cycle principle of engine operation. This means that all motors follow the stages of intake, compression, power and exhaust. But, the difference between a 2- and 4-stroke motor is in how many times the piston moves up and down within the cylinder to accomplish this. On 2-stroke motors (as the name suggests) the four cycles take place in 2 movements (one up and one down) of the piston. Again, as the name suggests, the cycles take place in 4 movements of the piston for 4-stroke motors.

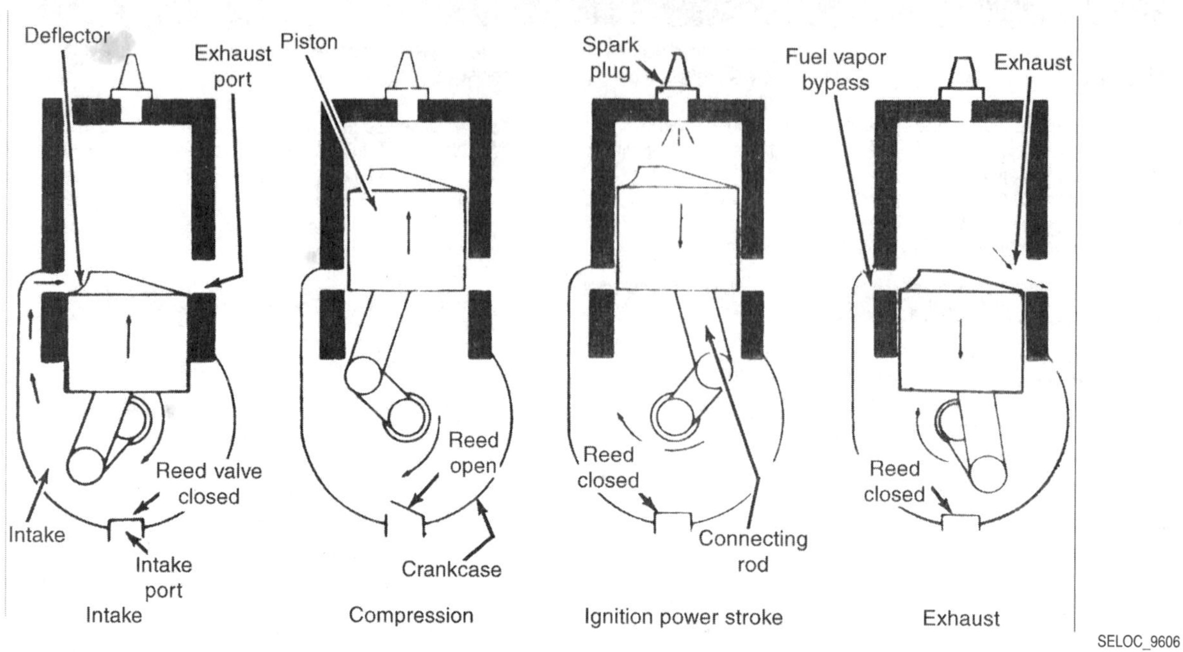

Fig. 21 The complete piston cycle of a 2-stroke motor (intake, compression, power and exhaust)

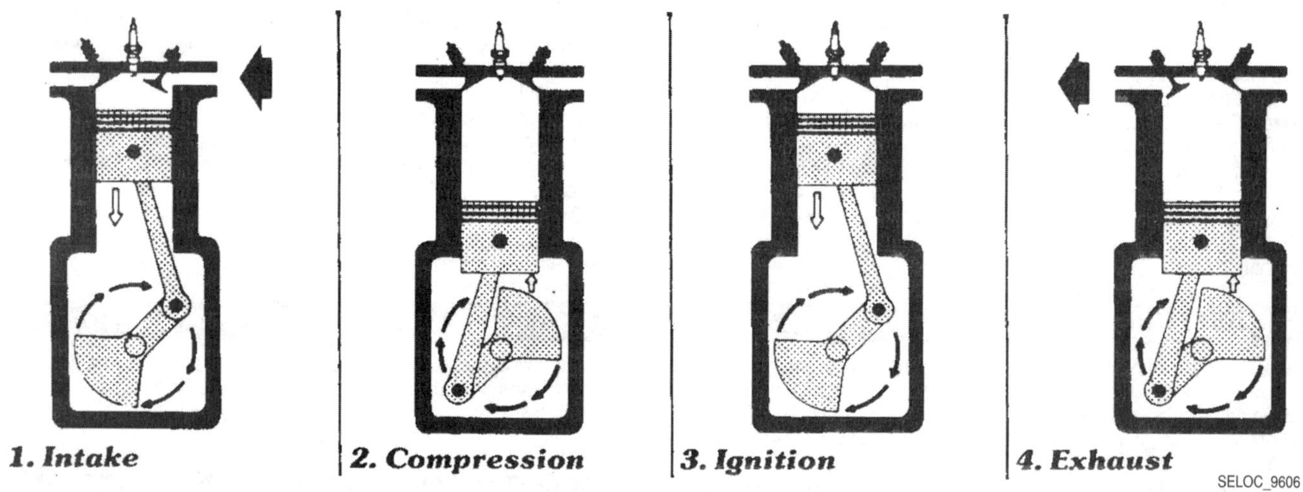

Fig. 22 The complete piston cycle of a 4-stroke motor (intake, compression, power and exhaust)

2-STROKE MOTORS

The 2-stroke engine differs in several ways from a conventional four-stroke (automobile or marine) engine.

1. The intake/exhaust method by which the fuel-air mixture is delivered to the combustion chamber.
2. The complete lubrication system.
3. The frequency of the power stroke.

Let's discuss these differences briefly (and compare 2-stroke engine operation with 4-stroke engine operation.)

Intake/Exhaust

◆ See Figures 23 thru 26

Two-stroke engines utilize an arrangement of port openings to admit fuel to the combustion chamber and to purge the exhaust gases after burning has been completed. The ports are located in a precise pattern in order for them to be open and closed off at an exact moment by the piston as it moves up and down in the cylinder. The exhaust port is located slightly higher than the fuel intake port. This arrangement opens the exhaust port first as the piston starts downward and therefore, the exhaust phase begins a fraction of a second before the intake phase.

Actually, the intake and exhaust ports are spaced so closely together that both open almost simultaneously. For this reason, some 2-stroke engines utilize deflector-type pistons. This design of the piston top serves two purposes very effectively.

First, it creates turbulence when the incoming charge of fuel enters the combustion chamber. This turbulence results in a more complete burning of the fuel than if the piston top were flat. The second effect of the deflector-type piston crown is to force the exhaust gases from the cylinder more rapidly.

Loop charged motors, or as they are commonly called "loopers", differ in how the air/fuel charge is introduced to the combustion chamber. Instead of the charge flowing across the top of the piston from one side of the cylinder to the other (CV) the use a looping action on top of the piston as the charge is forced through irregular shaped openings cut in the piston's skirt. In an LV motor, the charge is forced out from the crankcase by the downward motion of the piston, through the irregular shaped openings and transferred upward

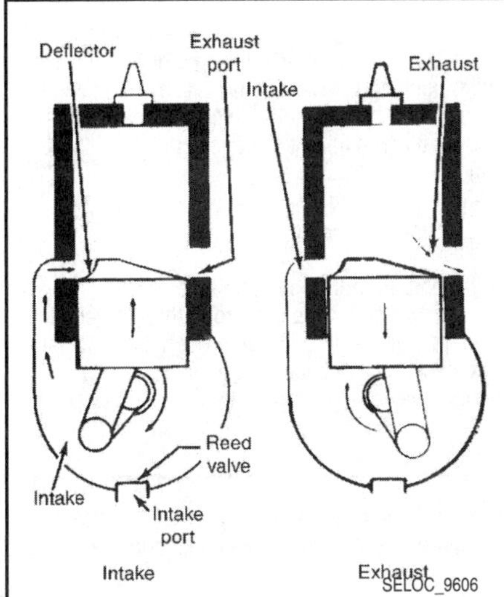

Fig. 23 The intake and exhaust cycles of a two-stroke engine—Cross flow (CV) design shown

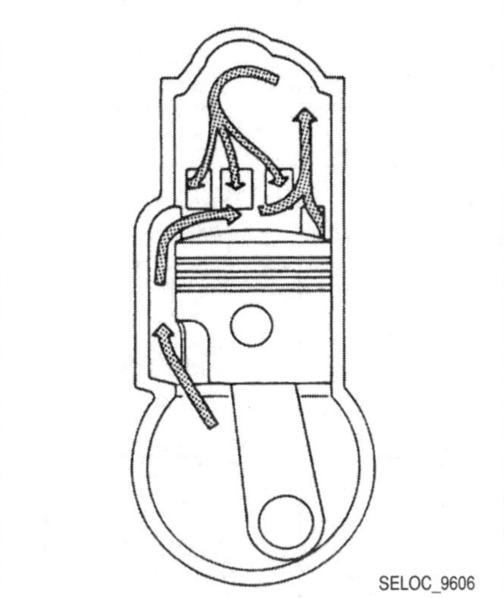

Fig. 24 Cross-sectional view of a typical loop-charged cylinder, showing charge flow while piston is moving downward

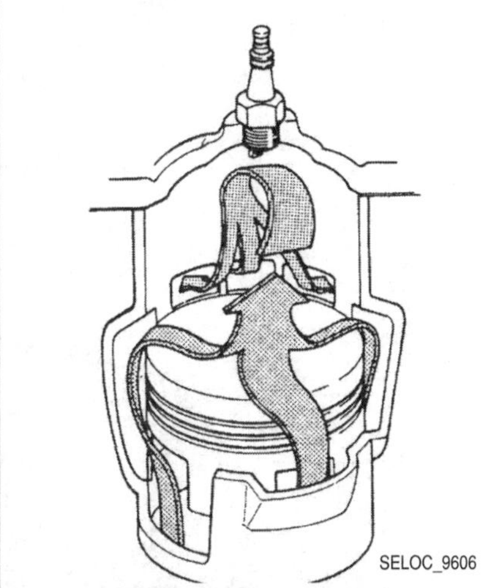

Fig. 25 Cutaway view of a typical loop-charged cylinder, depicting exhaust leaving the cylinder as the charge enters through 3 ports in the piston

by long, deep grooves in the cylinder wall. The charge completes its looping action by entering the combustion chamber, just above the piston, where the upward motion of the piston traps it in the chamber and compresses it for optimum ignition power.

Unlike the knife-edged deflector top pistons used in CV motors, the piston domes on Loop motors are relatively flat.

These systems of intake and exhaust are in marked contrast to individual intake and exhaust valve arrangement employed on four-stroke engines (and the mechanical methods of opening and closing these valves).

■ **It should be noted here that there are some 2-stroke engines that utilize a mechanical valve train, though it is very different from the valve train employed by most 4-stroke motors. Rotary 2-stroke engines use a circular valve or rotating disc that contains a port opening around part of one edge of the disc. As the engine (and disc) turns, the opening aligns with the intake port at and for a predetermined amount of time, closing off the port again as the opening passes by and the solid portion of the disc covers the port.**

Lubrication

A 2-stroke engine is lubricated by mixing oil with the fuel. Therefore, various parts are lubricated as the fuel mixture passes through the crankcase and the cylinder. In contrast, four-stroke engines have a crankcase containing oil. This oil is pumped through a circulating system and returned to the crankcase to begin the routing again.

Power Stroke

The combustion cycle of a 2-stroke engine has four distinct phases.

1. Intake
2. Compression
3. Power
4. Exhaust

The 4 phases of the cycle are accomplished with each up and down stroke of the piston, and the power stroke occurs with each complete revolution of the crankshaft. Compare this system with a four-stroke engine. A separate stroke of the piston is required to accomplish each phase of the cycle and the power stroke occurs only every other revolution of the crankshaft. Stated another way, 2 revolutions of the 4-stroke engine crankshaft are required to complete 1 full cycle, the 4 phases.

Physical Laws

◆ **See Figure 27**

The 2-stroke engine is able to function because of two very simple physical laws.

One: Gases will flow from an area of high pressure to an area of lower pressure. A tire blowout is an example of this principle. The high-pressure air escapes rapidly if the tube is punctured.

Two: If a gas is compressed into a smaller area, the pressure increases, and if a gas expands into a larger area, the pressure is decreased.

If these 2 laws are kept in mind, the operation of the 2-stroke engine will be easier understood.

Actual Operation

◆ **See Figure 21**

■ **The engine described here is of a carbureted type. EFI or DFI motors operate similarly for intake of the air charge and for exhaust of the unburned gasses. Obviously though, the very nature of fuel injection changes the actual delivery of the fuel/oil charge.**

Beginning with the piston approaching top dead center on the compression stroke: the intake and exhaust ports are physically closed (blocked) by the piston. During this stroke, the reed valve is open (because as the piston moves upward, the crankcase volume increases, which reduces the crankcase pressure to less than the outside atmosphere (creates a vacuum under the piston). The spark plug fires; the compressed fuel-air mixture is ignited; and the power stroke begins.

As the piston moves downward on the power stroke, the combustion chamber is filled with burning gases. As the exhaust port is uncovered, the gases, which are under great pressure, escape rapidly through the exhaust ports. The piston continues its downward movement. Pressure within the crankcase (again, under the piston) increases, closing the reed valves

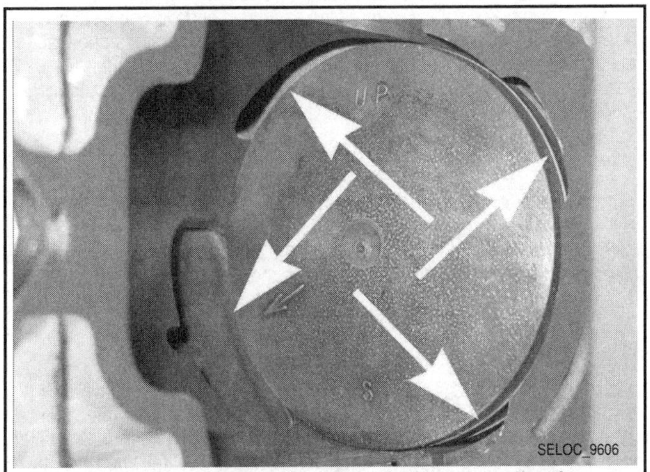

Fig. 26 The combustion chamber of a typical looper, notice the piston is far enough down the cylinder bore to reveal intake and exhaust ports

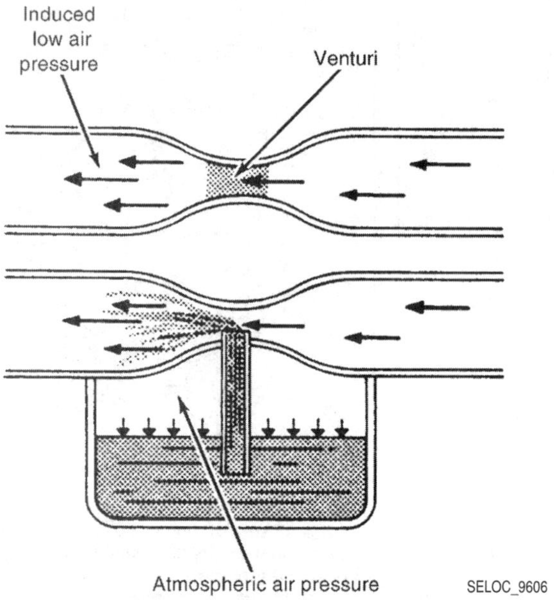

Fig. 27 Air flow principal for a modern carburetor

against their seats. The crankcase then becomes a sealed chamber so the air-fuel mixture becomes compressed (pressurized) and ready for delivery to the combustion chamber. As the piston continues to move downward, the intake port is uncovered. The fresh fuel mixture rushes through the intake port into the combustion chamber striking the top of the piston where it is deflected along the cylinder wall. The reed valve remains closed until the piston moves upward again.

When the piston begins to move upward on the compression stroke, the reed valve opens because the crankcase volume has been increased, reducing crankcase pressure to less than the outside atmosphere. The intake and exhaust ports are closed and the fresh fuel charge is compressed inside the combustion chamber.

Pressure in the crankcase (beneath the piston) decreases as the piston moves upward and a fresh charge of air flows through the carburetor picking up fuel. As the piston approaches top dead center, the spark plug ignites the air-fuel mixture, the power stroke begins and one complete Otto cycle has been completed.

4-STROKE MOTORS

◆ **See Figure 22**

The 4-stroke motor may be easier to understand for some people either because of its prevalence in automobile and street motorcycle motors today or perhaps because each of the four strokes corresponds to one distinct phase of the Otto cycle. Essentially, a 4-stroke engine completes one Otto cycle of intake, compression, ignition/power and exhaust using 2 full revolutions of the crankshaft and 4 distinct movements of the piston (down, up, down and up).

Intake

The intake stroke begins with the piston near the top of its travel. As crankshaft rotation begins to pull the piston downward, the exhaust valve closes and the intake opens. As volume of the combustion chamber increases, a vacuum is created that draws in the air/fuel mixture from the intake manifold.

Compression

Once the piston reaches the bottom of its travel, crankshaft rotation will begin to force it upward. At this point the intake valve closes. As the piston rises in the bore, the volume of the sealed combustion chamber (both intake and exhaust valves are closed) decreases and the air/fuel mixture is compressed. This raises the temperature and pressure of the mixture and increases the amount of force generated by the expanding gases during the Ignition/Power stroke.

Ignition/Power

As the piston approaches top dead center (the highest point of travel in the bore), the spark plug will fire, igniting the air/fuel mixture. The resulting combustion of the air/fuel mixture forces the piston downward, rotating the crankshaft (causing other pistons to move in other phases/strokes of the Otto cycle on multi-cylinder motors).

Exhaust

As the piston approaches the bottom of the Ignition/Power stroke, the exhaust valve opens. When the piston begins its upward path of travel once again, any remaining unburned gasses are forced out through the exhaust valve. This completes one Otto cycle, which begins again as the piston passes top dead center, the intake valve opens and the Intake stroke starts.

COMBUSTION

Whether we are talking about a 2- or 4-stroke engine, all Otto cycle, internal combustion engines require three basic conditions to operate properly,

1. Compression
2. Ignition (Spark)
3. Fuel

A lack of any one of these conditions will prevent the engine from operating. A problem with any one of these will manifest itself in hard-starting or poor performance.

Compression

An engine that has insufficient compression will not draw an adequate supply of air/fuel mixture into the combustion chamber and, subsequently, will not make sufficient power on the power stroke. A lack of compression in just one cylinder of a multi-cylinder motor will cause the motor to stumble or run irregularly.

But, keep in mind that a sudden change in compression is unlikely in 2-stroke motors (unless something major breaks inside the crankcase, but that would usually be accompanied by other symptoms such as a loud noise when it occurred or noises during operation). On 4-stroke motors, a sudden change in compression is also unlikely, but could occur if the timing belt or chain was to suddenly break. Remember that the timing belt/chain is used to synchronize the valve train with the crankshaft. If the valve train suddenly ceases to turn, some intake and some exhaust valves will remain open, relieving compression in that cylinder.

Ignition (Spark)

Traditionally, the ignition system is the weakest link in the chain of conditions necessary for engine operation. Spark plugs may become worn or fouled, wires will deteriorate allowing arcing or misfiring, and poor connections can place an undue load on coils leading to weak spark or even a failed coil. The most common question asked by a technician under a no-start condition is: "do I have spark and fuel" (as they've already determined that they have compression).

A quick visual inspection of the spark plug(s) will answer the question as to whether or not the plug(s) is/are worn or fouled. While the engine is shut **OFF** a physical check of the connections could show a loose primary or secondary ignition circuit wire. An obviously physically damaged wire may also be an indication of system problems and certainly encourages one to inspect the related system more closely.

If nothing is turned up by the visual inspection, perform the Spark Test provided in the Ignition System section to determine if the problem is a lack of or a weak spark. If the problem is not compression or spark, it's time to look at the fuel system.

Fuel

If compression and spark is present (and within spec), but the engine won't start or won't run properly, the only remaining condition to fulfill is fuel. As usual, start with the basics. Is the fuel tank full? Is the fuel stale? If the engine has not been run in some time (a matter of months, not weeks) there is a good chance that the fuel is stale and should be properly disposed of and replaced.

■ Depending on how stale or contaminated (with moisture) the fuel is, it may be burned in an automobile or in yard equipment, though it would be wise to mix it well with a much larger supply of fresh gasoline to prevent moving your driveability problems to that motor. But it is better to get the lawn tractor stuck on stale gasoline than it would be to have your boat motor quit in the middle of the bay or lake.

For hard starting motors, is the choke or primer system operating properly. Remember that the choke/prime should only be used for **cold** starts. A true cold start is really only the first start of the day, but it may be applicable to subsequent starts on cooler days, if the engine sat for more than a few hours and completely cooled off since the last use. Applying the primer to the motor for a hot start may flood the engine, preventing it from starting properly. One method to clear a flood is to crank the motor while the engine is at wide-open throttle (allowing the maximum amount of air into the motor to compensate for the excess fuel). But, keep in mind that the throttle should be returned to idle immediately upon engine start-up to prevent damage from over-revving.

Fuel delivery and pressure should be checked before delving into the carburetor(s) or fuel injection system. Make sure there are no clogs in the fuel line or vacuum leaks that would starve the motor of fuel.

Make sure that all other possible problems have been eliminated before touching the carburetor. It is rare that a carburetor will suddenly require an adjustment in order for the motor to run properly. It is much more likely that an improperly stored motor (one stored with untreated fuel in the carburetor) would suffer from one or more clogged carburetor passages sometime after shortly returning to service. Fuel will evaporate over time, leaving behind gummy deposits. If untreated fuel is left in the carburetor for some time (again typically months more than weeks), the varnish left behind by evaporating fuel will likely clog the small passages of the carburetor and cause problems with engine performance. If you suspect this, remove and disassemble the carburetor following procedures under Fuel System.

SHOP EQUIPMENT

Safety Tools

WORK GLOVES

◆ **See Figure 28**

Unless you think scars on your hands are cool, enjoy pain and like wearing bandages, get a good pair of work gloves. Canvas or leather gloves are the best. And yes, we realize that there are some jobs involving small parts that can't be done while wearing work gloves. These jobs are not the ones usually associated with hand injuries.

A good pair of rubber gloves (such as those usually associated with dish washing) or vinyl gloves is also a great idea. There are some liquids such as solvents and penetrants that don't belong on your skin. Avoid burns and rashes. Wear these gloves.

And lastly, an option. If you're tired of being greasy and dirty all the time, go to the drug store and buy a box of disposable latex gloves like medical professionals wear. You can handle greasy parts, perform small tasks, wash parts, etc. all without getting dirty! These gloves take a surprising amount of abuse without tearing and aren't expensive. Note however, that some people are allergic to the latex or the powder used inside some gloves, so pay attention to what you buy.

EYE & EAR PROTECTION

◆ **See Figures 29 and 30**

Don't begin any job without a good pair of work goggles or impact resistant glasses! When doing any kind of work, it's all too easy to avoid eye injury through this simple precaution. And don't just buy eye protection and leave it on the shelf. Wear it all the time! Things have a habit of breaking, chipping, splashing, spraying, splintering and flying around. And, for some reason, your eye is always in the way!

If you wear vision-correcting glasses as a matter of routine, get a pair made with polycarbonate lenses. These lenses are impact resistant and are available at any optometrist.

Often overlooked is hearing protection. Engines and power tools are noisy! Loud noises damage your ears. It's as simple as that! The simplest and cheapest form of ear protection is a pair of noise-reducing ear plugs. Cheap insurance for your ears! And, they may even come with their own, cute little carrying case.

More substantial, more protection and more money is a good pair of noise reducing earmuffs. They protect from all but the loudest sounds. Hopefully those are sounds that you'll never encounter since they're usually associated with disasters.

WORK CLOTHES

Everyone has "work clothes." Usually these consist of old jeans and a shirt that has seen better days. That's fine. In addition, a denim work apron is a nice accessory. It's rugged, can hold some spare bolts, and you don't feel bad wiping your hands or tools on it. That's what it's for.

When working in cold weather, a one-piece, thermal work outfit is invaluable. Most are rated to below freezing temperatures and are ruggedly constructed. Just look at what local marine mechanics are wearing and that should give you a clue as to what type of clothing is good.

Chemicals

There is a whole range of chemicals that you'll find handy for maintenance and repair work. The most common types are: lubricants, penetrants and sealers. Keep these handy. There are also many chemicals that are used for detailing or cleaning.

When a particular chemical is not being used, keep it capped, upright and in a safe place. These substances may be flammable, may be irritants or might even be caustic and should always be stored properly, used properly and handled with care. Always read and follow all label directions and be sure to wear hand and eye protection!

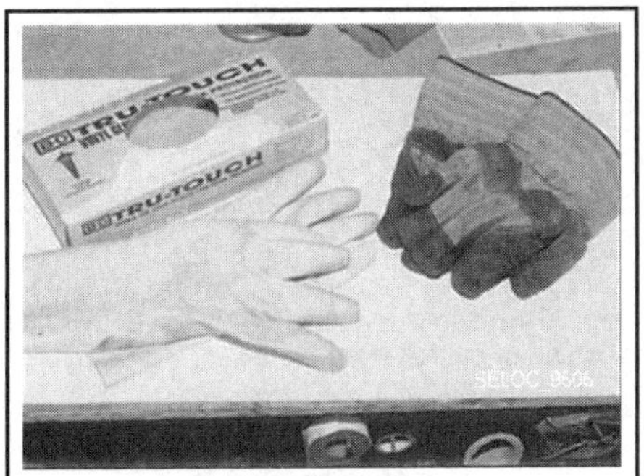

Fig. 28 Three different types of work gloves. The box contains latex gloves

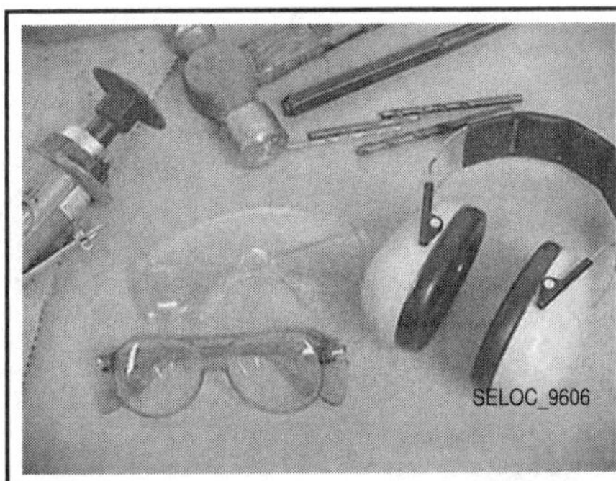

Fig. 29 Don't begin major repairs without a pair of goggles for your eyes and earmuffs to protect your hearing

Fig. 30 Things have a habit of, splashing, spraying, splintering and flying around during repairs

LUBRICANTS & PENETRANTS

◆ See Figure 31

Anti-seize is used to coat certain fasteners prior to installation. This can be especially helpful when two dissimilar metals are in contact (to help prevent corrosion that might lock the fastener in place). This is a good practice on a lot of different fasteners, BUT, NOT on any fastener that might vibrate loose causing a problem. If anti-seize is used on a fastener, it should be checked periodically for proper tightness.

Lithium grease, chassis lube, silicone grease or a synthetic brake caliper grease can all be used pretty much interchangeably. All can be used for coating rust-prone fasteners and for facilitating the assembly of parts that are a tight fit. Silicone and synthetic greases are the most versatile.

■ Silicone dielectric grease is a non-conductor that is often used to coat the terminals of wiring connectors before fastening them. It may sound odd to coat metal portions of a terminal with something that won't conduct electricity, but here is it how it works. When the connector is fastened the metal-to-metal contact between the terminals will displace the grease (allowing the circuit to be completed). The grease that is displaced will then coat the non-contacted surface and the cavity around the terminals, SEALING them from atmospheric moisture that could cause corrosion.

Silicone spray is a good lubricant for hard-to-reach places and parts that shouldn't be gooped up with grease.

Penetrating oil may turn out to be one of your best friends when taking something apart that has corroded fasteners. Not only can they make a job easier, they can really help to avoid broken and stripped fasteners. The most familiar penetrating oils are Liquid Wrench® and WD-40®. A newer penetrant, PB Blaster® works very well (and has become a mainstay in our shops). These products have hundreds of uses. For your purposes, they are vital!

Before disassembling any part, check the fasteners. If any appear rusted, soak them thoroughly with the penetrant and let them stand while you do something else (for particularly rusted or frozen parts you may need to soak them a few days in advance). This simple act can save you hours of tedious work trying to extract a broken bolt or stud.

SEALANTS

◆ See Figures 32 and 33

Sealants are an indispensable part for certain tasks, especially if you are trying to avoid leaks. The purpose of sealants is to establish a leak-proof bond between or around assembled parts. Most sealers are used in conjunction with gaskets, but some are used instead of conventional gasket material.

The most common sealers are the non-hardening types such as Permatex®) No.2 or its equivalents. These sealers are applied to the mating surfaces of each part to be joined, then a gasket is put in place and the parts are assembled.

■ A sometimes overlooked use for sealants like RTV is on the threads of vibration prone fasteners.

One very helpful type of non-hardening sealer is the "high tack" type. This type is a very sticky material that holds the gasket in place while the parts are being assembled. This stuff is really a good idea when you don't have enough hands or fingers to keep everything where it should be.

The stand-alone sealers are the Room Temperature Vulcanizing (RTV) silicone gasket makers. On some engines, this material is used instead of a gasket. In those instances, a gasket may not be available or, because of the shape of the mating surfaces, a gasket shouldn't be used. This stuff, when used in conjunction with a conventional gasket, produces the surest bonds.

RTV does have its limitations though. When using this material, you will have a time limit. It starts to set-up within 15 minutes or so, so you have to assemble the parts without delay. In addition, when squeezing the material out of the tube, don't drop any glops into the engine. The stuff will form and set and travel around a cooling passage, possibly blocking it. Also, most types are not fuel-proof. Check the tube for all cautions.

CLEANERS

◆ See Figures 34 and 35

There are 2 basic types of cleaners on the market today: parts cleaners and hand cleaners. The parts cleaners are for the parts; the hand cleaners are for you. They are *not* interchangeable.

There are many good, non-flammable, biodegradable parts cleaners on the market. These cleaning agents are safe for you, the parts and the environment. Therefore, there is no reason to use flammable, caustic or toxic substances to clean your parts or tools.

As far as hand cleaners go; the waterless types are the best. They have always been efficient at cleaning, but they used to all leave a pretty smelly odor. Recently though, most of them have eliminated the odor and added stuff that actually smells good. Make sure that you pick one that contains lanolin or some other moisture-replenishing additive. Cleaners not only remove grease and oil but also skin oil.

■ Most women already know to use a hand lotion when you're all cleaned up. It's okay. Real men DO use hand lotion too! Believe it or not, using hand lotion *before* your hands are dirty will actually make them easier to clean when you're finished with a dirty job. Lotion seals your hands, and keeps dirt and grease from sticking to your skin.

SELOC_9606

Fig. 31 Keep a supply of anti-seize, penetrating oil, lithium grease, electronic cleaner and silicone spray

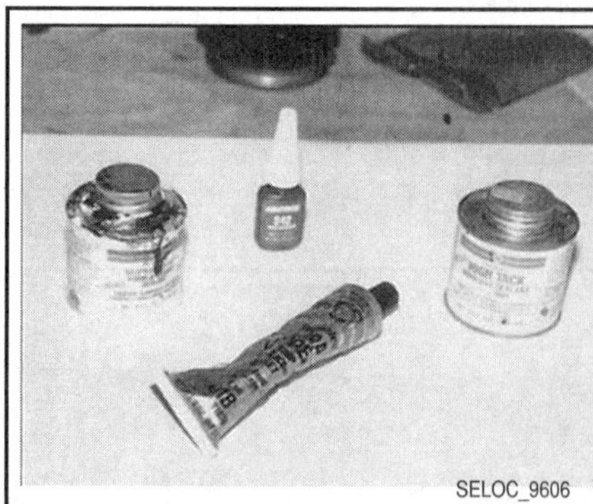

SELOC_9606

Fig. 32 Sealants are essential for preventing leaks

SELOC_9606

Fig. 33 On some engines, RTV is used instead of gasket material to seal components

Fig. 34 Citrus hand cleaners not only work well, but they smell pretty good too. Choose one with pumice for added cleaning power

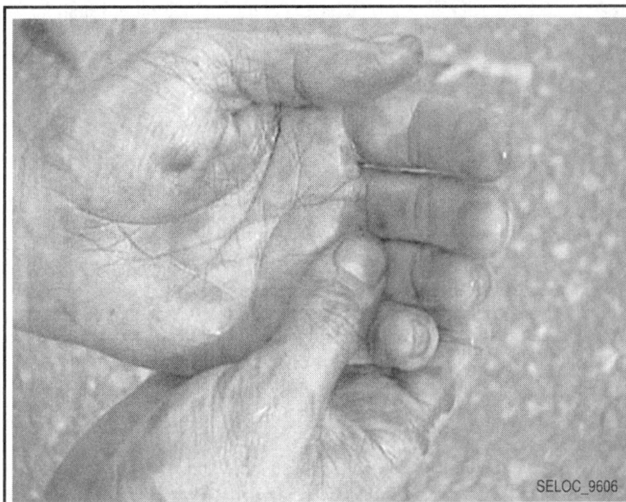

Fig. 35 The use of hand lotion seals your hands and keeps dirt and grease from sticking to your skin

TOOLS

◆ **See Figure 36**

Tools; this subject could fill a completely separate manual. The first thing you will need to ask yourself, is just how involved do you plan to get. If you are serious about maintenance and repair you will want to gather a quality set of tools to make the job easier, and more enjoyable. BESIDES, TOOLS ARE FUN!!!

Almost every do-it-yourselfer loves to accumulate tools. Though most find a way to perform jobs with only a few common tools, they tend to buy more over time, as money allows. So gathering the tools necessary for maintenance or repair does not have to be an expensive, overnight proposition.

When buying tools, the saying "You get what you pay for . . ." is absolutely true! Don't go cheap! Any hand tool that you buy should be drop forged and/or chrome vanadium. These two qualities tell you that the tool is strong enough for the job. With any tool, go with a name that you've heard of before, or, that is recommended buy your local professional retailer. Let's go over a list of tools that you'll need.

Most of the world uses the metric system. However, some American-built engines and aftermarket accessories use standard fasteners. So, accumulate your tools accordingly. Any good DIYer should have a decent set of both U.S. and metric measure tools.

■ **Don't be confused by terminology. Most advertising refers to "SAE and metric", or "standard and metric." Both are misnomers. The Society of Automotive Engineers (SAE) did not invent the English system of measurement; the English did. The SAE likes metrics just fine. Both English (U.S.) and metric measurements are SAE approved. Also, the current "standard" measurement IS metric. So, if it's not metric, it's U.S. measurement.**

Hand Tools

SOCKET SETS

◆ **See Figures 37 thru 43**

Socket sets are the most basic hand tools necessary for repair and maintenance work. For our purposes, socket sets come in three drive sizes: 1/4 in., 3/8 in. and 1/2 in. Drive size refers to the size of the drive lug on the ratchet, breaker bar or speed handle.

A 3/8 inch set is probably the most versatile set in any mechanic's toolbox. It allows you to get into tight places that the larger drive ratchets can't and gives you a range of larger sockets that are still strong enough for heavy-duty work. The socket set that you'll need should range in sizes from

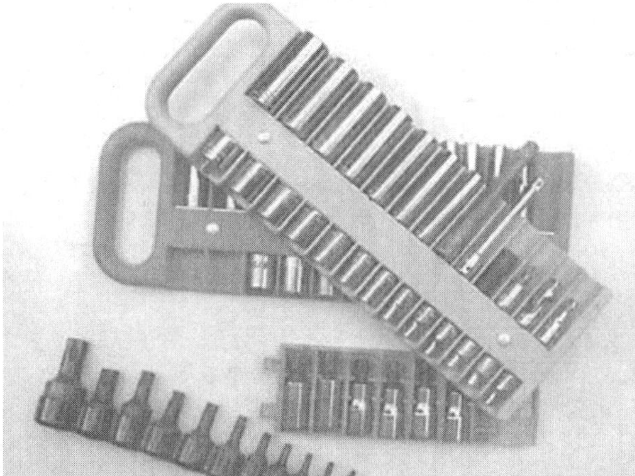

Fig. 36 Socket holders, especially the magnetic type, are handy items to keep tools in order

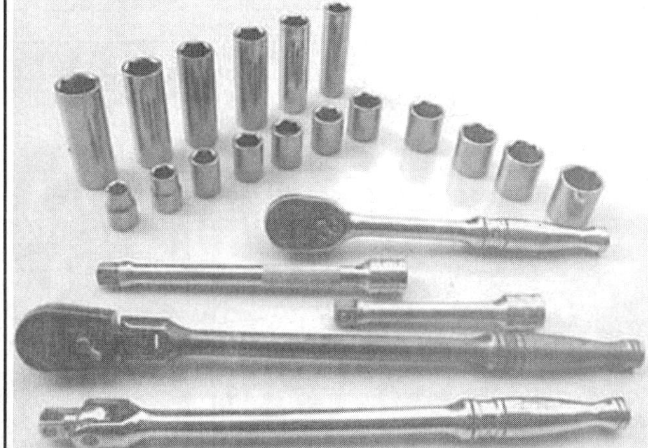

Fig. 37 A 3/8 in. socket set is probably the most versatile tool in any mechanic's tool box

1/4 in. through 1 in. for standard fasteners, and a 6mm through 19mm for metric fasteners.

You'll need a good 1/2 in. set since this size drive lug assures that you won't break a ratchet or socket on large or heavy fasteners. Also, torque wrenches with a torque scale high enough for larger fasteners are usually 1/2 in. drive.

Plus, 1/4 in. drive sets can be very handy in tight places. Though they usually duplicate functions of the 3/8 in. set, 1/4 in. drive sets are easier to use for smaller bolts and nuts.

As for the sockets themselves, they come in shallow (standard) and deep lengths as well as 6 or 12 point. The 6 and 12 points designation refers to how many sides are in the socket itself. Each has advantages. The 6 point socket is stronger and less prone to slipping which would strip a bolt head or nut. 12 point sockets are more common, usually less expensive and can operate better in tight places where the ratchet handle can't swing far.

Standard length sockets are good for just about all jobs, however, some stud-head bolts, hard-to-reach bolts, nuts on long studs, etc., require the deep sockets.

Most marine manufacturers use recessed hex-head fasteners to retain many of the engine parts. These fasteners require a socket with a hex shaped driver or a large sturdy hex key. To help prevent torn knuckles, we would recommend that you stick to the sockets on any tight fastener and leave the hex keys for lighter applications. Hex driver sockets are available individually or in sets just like conventional sockets.

More and more, manufacturers are using Torx(r) head fasteners, which were once known as tamper resistant fasteners (because many people did not have tools with the necessary odd driver shape). Since Torx(r) fasteners have become commonplace in many DIYer tool boxes, manufacturers designed newer tamper resistant fasteners that are essentially Torx(r) head bolts that contain a small protrusion in the center (requiring the driver to contain a small hole to slide over the protrusion. Tamper resistant fasteners are often used where the manufacturer would prefer only knowledgeable mechanics or advanced Do-It-Yourselfers (DIYers) work.

TORQUE WRENCHES

◆ **See Figure 44**

In most applications, a torque wrench can be used to ensure proper installation of a fastener. Torque wrenches come in various designs and most stores will carry a variety to suit your needs. A torque wrench should be used any time you have a specific torque value for a fastener. Keep in mind that because there is no worldwide standardization of fasteners, so charts or figure found in each repair section refer to the manufacturer's fasteners. Any general guideline charts that you might come across based on fastener size (they are sometimes included in a repair manual or with torque wrench packaging) should be used with caution. Just keep in mind that if you are using the right tool for the job, you should not have to strain to tighten a fastener.

Beam Type

◆ **See Figures 45 and 46**

The beam type torque wrench is one of the most popular styles in use. If used properly, it can be the most accurate also. It consists of a pointer attached to the head that runs the length of the flexible beam (shaft) to a scale located near the handle. As the wrench is pulled, the beam bends and the pointer indicates the torque using the scale.

Click (Breakaway) Type

◆ **See Figures 47 and 48**

Another popular torque wrench design is the click type. The clicking mechanism makes achieving the proper torque easy and most use a ratcheting head for ease of bolt installation. To use the click type wrench you pre-adjust it to a torque setting. Once the torque is reached, the wrench has a reflex signaling feature that causes a momentary breakaway of the torque

Fig. 38 A swivel (U-joint) adapter (left), a wobble-head adapter (center) and a 1/2 in.-to-3/8 in. adapter (right)

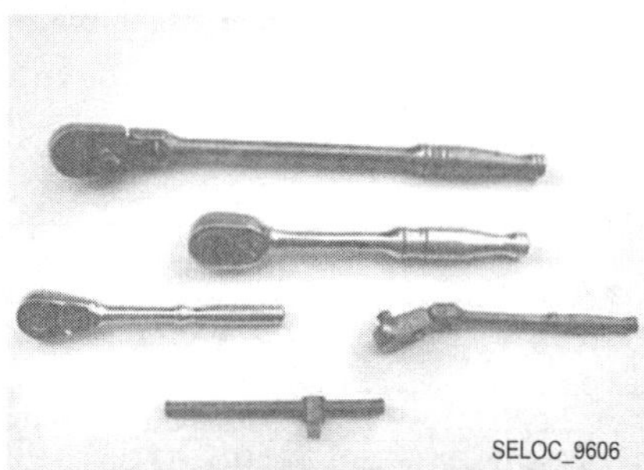

Fig. 39 Ratchets come in all sizes and configurations from rigid to swivel-headed

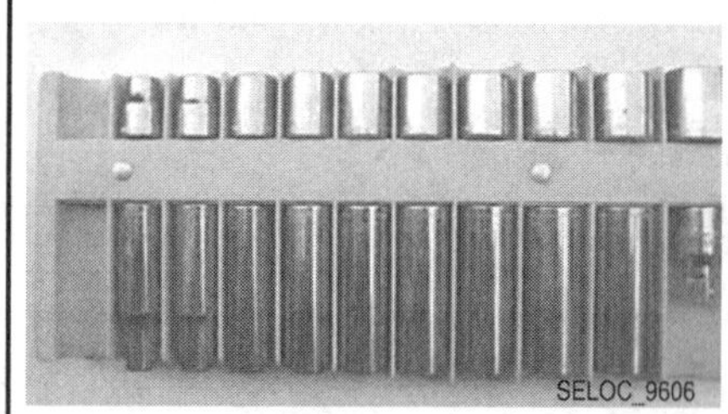

Fig. 40 Shallow sockets (top) are good for most jobs. But, some bolts require deep sockets (bottom)

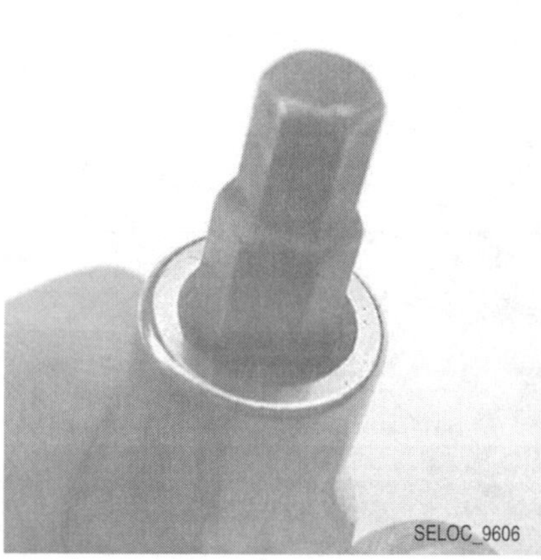

Fig. 41 Hex-head fasteners require a socket with a hex shaped driver

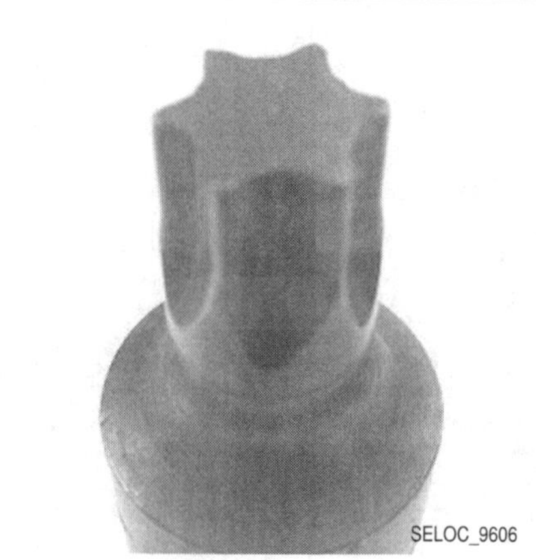

Fig. 42 Torx® drivers . . .

Fig. 43 . . . and tamper resistant drivers are required to remove special fasteners

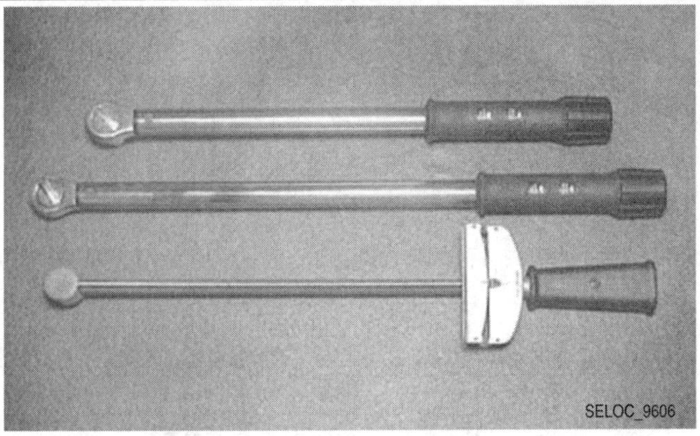

Fig. 44 Three types of torque wrenches. Top to bottom: a 3/8 in. drive beam type that reads in inch lbs., a 1/2 in. drive clicker type and a 1/2 in. drive beam type

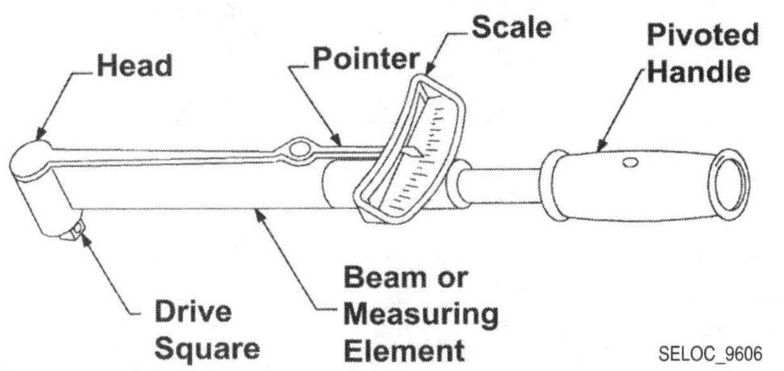

Fig. 45 Parts of a beam type torque wrench

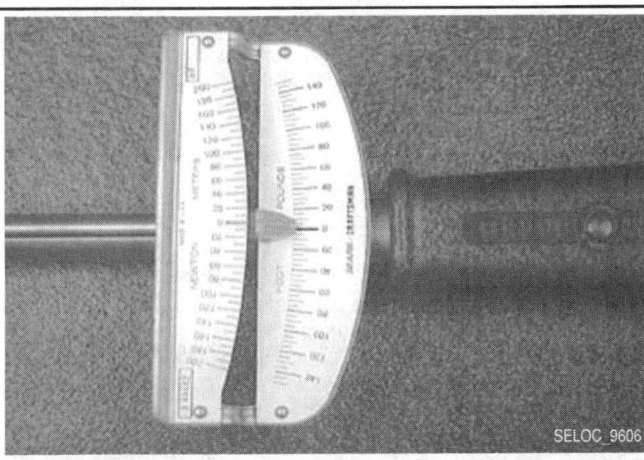

Fig. 46 A beam type torque wrench consists of a pointer attached to the head that runs the length of the flexible beam (shaft) to a scale located near the handle

wrench body, sending an impulse to the operator's hand. But be careful, as continuing the turn the wrench after the momentary release will increase torque on the fastener beyond the specified setting.

BREAKER BARS

◆ **See Figures 49**

Breaker bars are long handles with a drive lug. Their main purpose is to provide extra turning force when breaking loose tight bolts or nuts. They come in all drive sizes and lengths. Always take extra precautions and use the proper technique when using a breaker bar (pull on the bar, don't push, to prevent skinned knuckles).

WRENCHES

◆ **See Figures 50 thru 54**

Basically, there are 3 kinds of fixed wrenches: open end, box end, and combination.

Open-end wrenches have 2-jawed openings at each end of the wrench. These wrenches are able to fit onto just about any nut or bolt. They are extremely versatile but have one major drawback. They can slip on a worn or rounded bolt head or nut, causing bleeding knuckles and a useless fastener.

■ **Line wrenches are a special type of open-end wrench designed to fit onto more of the fastener than standard open-end wrenches, thus reducing the chance of rounding the corners of the fastener.**

Box-end wrenches have a 360° circular jaw at each end of the wrench. They come in both 6 and 12 point versions just like sockets and each type has some of the same advantages and disadvantages as sockets.

Combination wrenches have the best of both. They have a 2-jawed open end and a box end. These wrenches are probably the most versatile.

As for sizes, you'll probably need a range similar to that of the sockets, about 1/4 in. through 1 in. for standard fasteners, or 6mm through 19mm for metric fasteners. As for numbers, you'll need 2 of each size, since, in many instances, one wrench holds the nut while the other turns the bolt. On most fasteners, the nut and bolt are the same size so having two wrenches of the same size comes in handy.

■ **Although you will typically just need the sizes we specified, there are some exceptions. Occasionally you will find a nut that is larger. For these, you will need to buy ONE expensive wrench or a very large adjustable. Or you can always just convince the spouse that we are talking about safety here and buy a whole (read expensive) large wrench set.**

One extremely valuable type of wrench is the adjustable wrench. An adjustable wrench has a fixed upper jaw and a moveable lower jaw. The lower jaw is moved by turning a threaded drum. The advantage of an adjustable wrench is its ability to be adjusted to just about any size fastener.

The main drawback of an adjustable wrench is the lower jaw's tendency to move slightly under heavy pressure. This can cause the wrench to slip if it is not facing the right way. Pulling on an adjustable wrench in the proper direction will cause the jaws to lock in place. Adjustable wrenches come in a large range of sizes, measured by the wrench length.

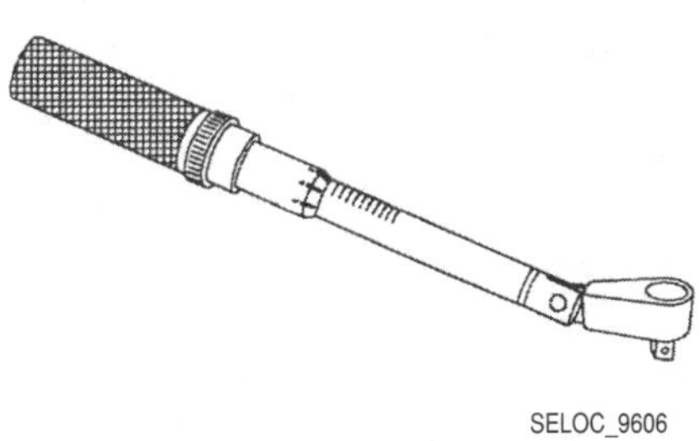

Fig. 47 A click type or breakaway torque wrench—note this one has a pivoting head

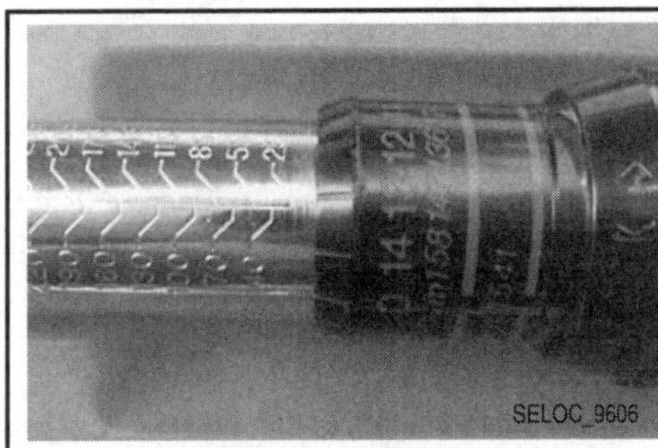

Fig. 48 Setting the torque on a click type wrench involves turning the handle until the specification appears on the dial

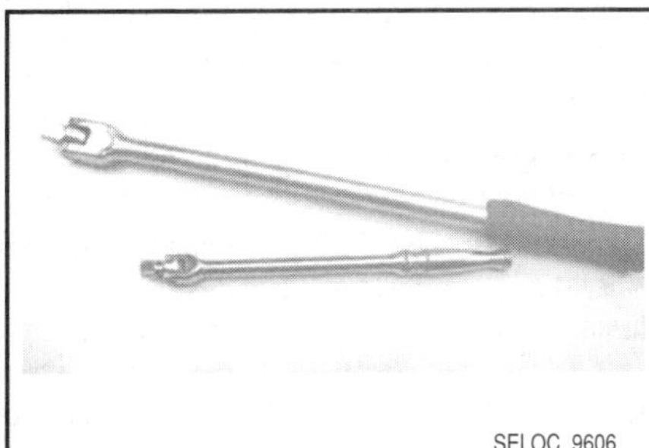

Fig. 49 Breaker bars are great for loosening large or stuck fasteners

INCHES	DECIMAL
1/8"	.125
3/16"	.187
1/4"	.250
5/16"	.312
3/8"	.375
7/16"	.437
1/2"	.500
9/16"	.562
5/8"	.625
11/16"	.687
3/4"	.750
13/16"	.812
7/8"	.875
15/16"	.937
1"	1.00

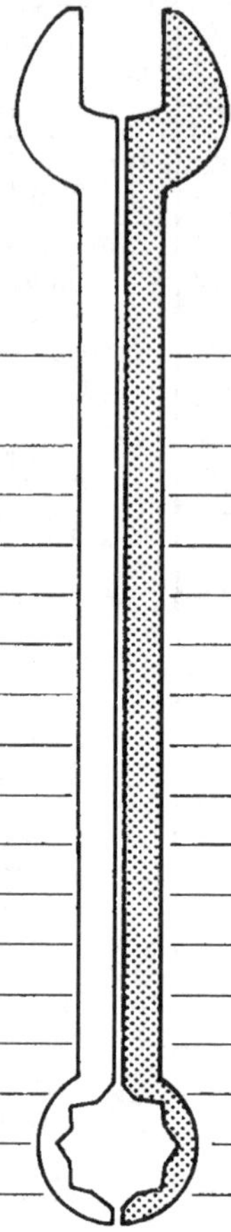

DECIMAL	MILLIMETERS
.118	3mm
.157	4mm
.236	6mm
.354	9mm
.394	10mm
.472	12mm
.512	13mm
.590	15mm
.630	16mm
.709	18mm
.748	19mm
.787	20mm
.866	22mm
.945	24mm
.984	25mm

SELOC_9606

Fig. 50 Comparison of U.S. measure and metric wrench sizes

PLIERS

◆ **See Figure 55**

Pliers are simply mechanical fingers. They are, more than anything, an extension of your hand. At least 3 pairs of pliers are an absolute necessity - standard, needle nose and slip joint.

In addition to standard pliers there are the slip-joint, multi-position pliers such as ChannelLock® pliers and locking pliers, such as Vise Grips(r).

Slip joint pliers are extremely valuable in grasping oddly sized parts and fasteners. Just make sure that you don't use them instead of a wrench too often since they can easily round off a bolt head or nut.

Locking pliers are usually used for gripping bolts or studs that can't be removed conventionally. You can get locking pliers in square jawed, needle-nosed and pipe-jawed. Locking pliers can rank right up behind duct tape as the handy-man's best friend.

SCREWDRIVERS

You can't have too many screwdrivers. They come in 2 basic flavors, either standard or Phillips. Standard blades come in various sizes and thickness for all types of slotted fasteners. Phillips screwdrivers come in sizes with number designations from 1 on up, with the lower number designating the smaller size. Screwdrivers can be purchased separately or in sets.

HAMMERS

◆ **See Figure 56**

You need a hammer for just about any kind of work. You need a ball-peen hammer for most metal work when using drivers and other like tools. A plastic hammer comes in handy for hitting things safely. A soft-faced dead-blow hammer is used for hitting things safely and hard. Hammers are also VERY useful with non air-powered impact drivers.

Other Common Tools

There are a lot of other tools that every DIYer will eventually need (though not all for basic maintenance). They include:

- Funnels
- Chisels
- Punches
- Files
- Hacksaw
- Portable Bench Vise
- Tap and Die Set
- Flashlight
- Magnetic Bolt Retriever
- Gasket scraper
- Putty Knife
- Screw/Bolt Extractors
- Prybars

Fig. 51 Always use a backup wrench to prevent rounding flare nut fittings

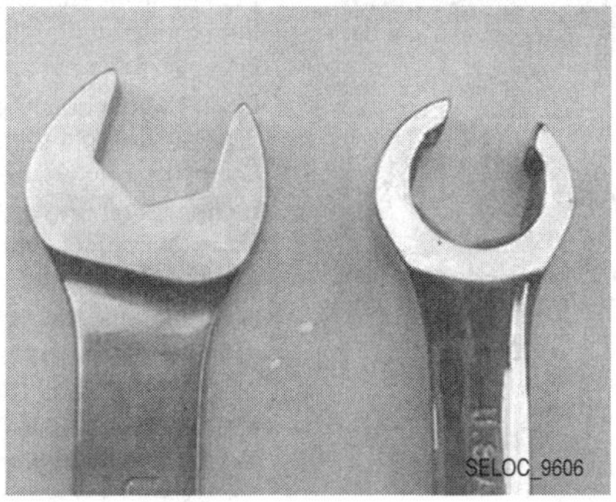

Fig. 52 Note how the flare wrench jaws are extended to grip the fitting tighter and prevent rounding

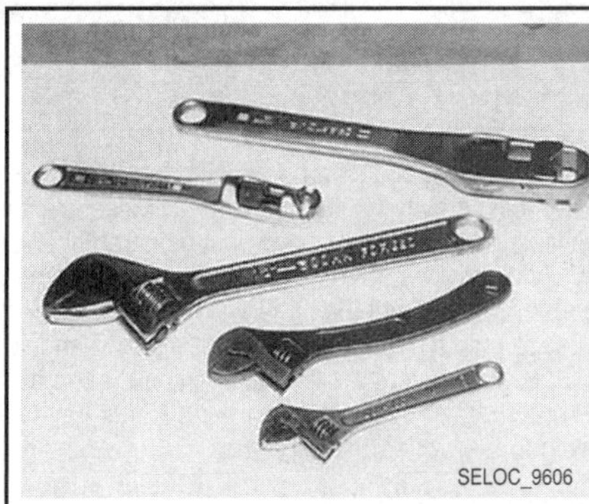

Fig. 53 Several types and sizes of adjustable wrenches

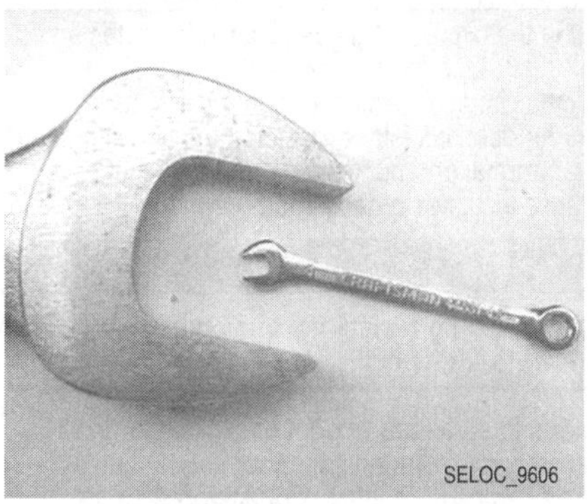

Fig. 54 You may find a nut that requires a particularly large or small wrench (that is usually available at your local tool store)

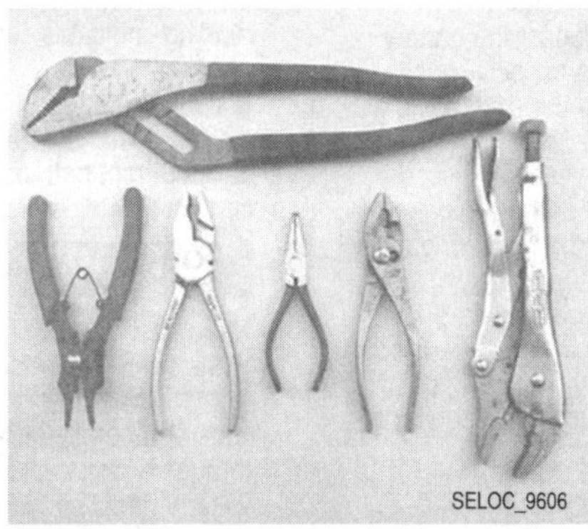

Fig. 55 Pliers come in many shapes and sizes. You should have an assortment on hand

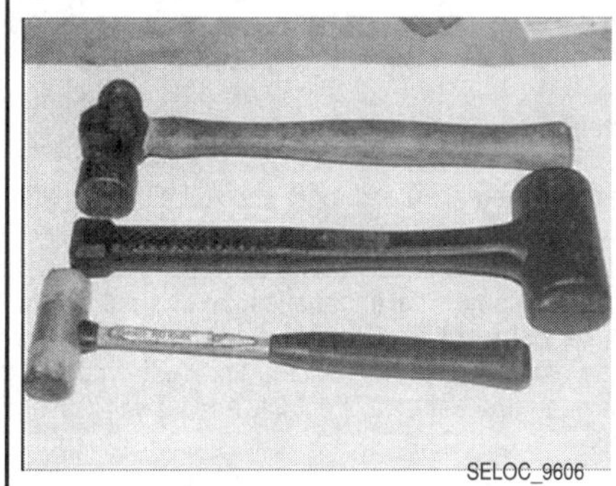

Fig. 56 Three types of hammers. Top to bottom: ball peen, rubber dead-blow, and plastic

Hacksaws have just one use - cutting things off. You may wonder why you'd need one for something as simple as maintenance or repair, but you never know. Among other things, guide studs to ease parts installation can be made from old bolts with their heads cut off.

A tap and die set might be something you've never needed, but you will eventually. It's a good rule, when everything is apart, to clean-up all threads, on bolts, screws or threaded holes. Also, you'll likely run across a situation in which you will encounter stripped threads. The tap and die set will handle that for you.

Gasket scrapers are just what you'd think, tools made for scraping old gasket material off of parts. You don't absolutely need one. Old gasket material can be removed with a putty knife or single edge razor blade. However, putty knives may not be sharp enough for some really stubborn gaskets and razor blades have a knack of breaking just when you don't want them to, inevitably slicing the nearest body part! As the old saying goes, "always use the proper tool for the job". If you're going to use a razor to scrape a gasket, be sure to always use a blade holder.

Putty knives really do have a use in a repair shop. Just because you remove all the bolts from a component sealed with a gasket doesn't mean it's going to come off. Most of the time, the gasket and sealer will hold it tightly. Lightly inserting a putty knife at various points between the two parts will break the seal without damage to the parts.

A small - 8-10 in. (20-25cm) long - prybar is extremely useful for removing stuck parts.

■ **Never use a screwdriver as a prybar! Screwdrivers are not meant for prying. Screwdrivers, used for prying, can break, sending the broken shaft flying!**

Screw/bolt extractors are used for removing broken bolts or studs that have broken off flush with the surface of the part.

Special Tools

◆ **See Figure 57**

Almost every marine engine around today requires at least one special tool to perform a certain task. In most cases, these tools are specially designed to overcome some unique problem or to fit on some oddly sized component.

When manufacturers go through the trouble of making a special tool, it is usually necessary to use it to ensure that the job will be done right. A special tool might be designed to make a job easier, or it might be used to keep you from damaging or breaking a part.

Don't worry, *most* maintenance procedures can either be performed without any special tools OR, because the tools must be used for such basic things, they are commonly available for a reasonable price. It is usually just the low production, highly specialized tools (like a super thin 7-point star-shaped socket capable of 150 ft. lbs. (203 Nm) of torque that is used only on the crankshaft nut of the limited production what-dya-callit engine) that tend to be outrageously expensive and hard to find. Hopefully, you will probably never need such a tool.

Special tools can be as inexpensive and simple as an adjustable strap wrench or as complicated as an ignition tester. A few common specialty tools are listed here, but check with your dealer or with other boaters for help in determining if there are any special tools for your particular engine. There is an added advantage in seeking advice from others, chances are they may have already found the special tool you will need, and know how to get it cheaper (or even let you borrow it).

Electronic Tools

BATTERY TESTERS

The best way to test a non-sealed battery is using a hydrometer to check the specific gravity of the acid. Luckily, these are usually inexpensive and are available at most parts stores. Just be careful because the larger testers are usually designed for larger batteries and may require more acid than you will be able to draw from the battery cell. Smaller testers (usually a short, squeeze bulb type) will require less acid and should work on most batteries.

Electronic testers are available and are often necessary to tell if a sealed battery is usable. Luckily, many parts stores have them on hand and are willing to test your battery for you.

BATTERY CHARGERS

◆ **See Figure 58**

If you are a weekend boater and take your boat out every week, then you will most likely want to buy a battery charger to keep your battery fresh. There are many types available, from low amperage trickle chargers to electronically controlled battery maintenance tools that monitor the battery voltage to prevent over or undercharging. This last type is especially useful if you store your boat for any length of time (such as during the severe winter months found in many Northern climates).

Even if you use your boat on a regular basis, you will eventually need a battery charger. The charger should be used anytime the boat is going to be in storage for more than a few weeks or so. Never leave the dock or loading ramp without a battery that is fully charged.

Also, some smaller batteries are shipped dry and in a partial charged state. Before placing a new battery of this type into service it must be filled and properly charged. Failure to properly charge a battery (which was shipped dry) before it is put into service will prevent it from ever reaching a fully charged state.

MULTI-METERS (DVOMS)

◆ **See Figure 59**

Multi-meters or Digital Volt Ohmmeter (DVOMs) are an extremely useful tool for troubleshooting electrical problems. They can be purchased in either analog or digital form and have a price range to suit any budget. A multi-meter is a voltmeter, ammeter and ohmmeter (along with other features) combined into one instrument. It is often used when testing solid state circuits because of its high input impedance (usually 10 mega-ohms or more). A brief description of the multi-meter main test functions follows:

- oltmeter - the voltmeter is used to measure voltage at any point in a circuit or to measure the voltage drop across any part of a circuit. Voltmeters usually have various scales and a selector switch to allow the reading of different voltage ranges. The voltmeter has a positive and a negative lead. To avoid the possibility of damage to the meter, whenever possible, connect the negative lead to the negative (-) side of the circuit (to ground or nearest the ground side of the circuit) and connect the positive lead to the positive (+) side of the circuit (to the power source or the nearest power source). Luckily, most quality DVOMs can adjust their own polarity internally and will indicate (without damage) if the leads are reversed. Note that the negative voltmeter lead will always be black and that the positive voltmeter will always be some color other than black (usually red).
- Ohmmeter - the ohmmeter is designed to read resistance (measured in ohms) in a circuit or component. Most ohmmeters will have a selector switch which permits the measurement of different ranges of resistance (usually the selector switch allows the multiplication of the meter reading by 10, 100, 1,000 and 10,000). Some ohmmeters are "auto-ranging" which means the meter itself will determine which scale to use. Since the meters are powered by an internal battery, the ohmmeter can be used like a self-powered test light. When the ohmmeter is connected, current from the ohmmeter flows through the circuit or component being tested. Since the ohmmeter's internal resistance and voltage are known values, the amount of current flow through the meter depends on the resistance of the circuit or component being tested. The ohmmeter can also be used to perform a continuity test for suspected open circuits. In using the meter for making continuity checks, do not be concerned with the actual resistance readings. Zero resistance, or any ohm reading, indicates continuity in the circuit. Infinite resistance indicates an opening in the circuit. A high resistance reading where there should be little or none indicates a problem in the circuit. Checks for short circuits are made in the same manner as checks for open circuits, except that the circuit must be isolated from both power and normal ground. Infinite resistance indicates no continuity, while zero resistance indicates a dead short.

**** WARNING**

Never use an ohmmeter to check the resistance of a component or wire while there is voltage applied to the circuit.

- Ammeter - an ammeter measures the amount of current flowing through a circuit in units called amperes or amps. At normal operating voltage, most circuits have a characteristic amount of amperes, called "current draw" which can be measured using an ammeter. By referring to a specified current draw rating, then measuring the amperes and comparing the two values; one can determine what is happening within the circuit to aid in diagnosis. An open circuit, for example, will not allow any current to flow, so the ammeter reading will be zero. A damaged component or circuit will have an increased current draw, so the reading will be high. The ammeter is always connected in series with the circuit being tested. All of the current that normally flows through the circuit must also flow through the ammeter; if there is any other path for the current to follow, the ammeter reading will not be accurate. The ammeter itself has very little resistance to current flow and, therefore, will not affect the circuit, but, it will measure current draw only when the circuit is closed and electricity is flowing. Excessive current draw can blow fuses and drain the battery, while a reduced current draw can cause motors to run slowly, lights to dim and other components to not operate properly.

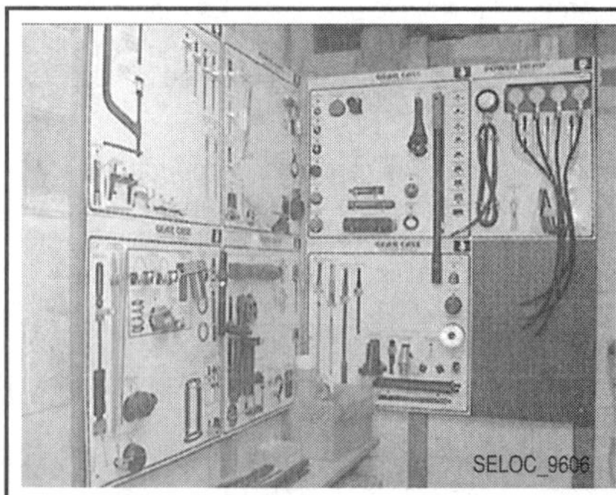

Fig. 57 Almost every marine engine around today requires at least one special tool to perform a certain task

Fig. 58 The Battery Tender® is more than just a battery charger, when left connected, it keeps your battery fully charged

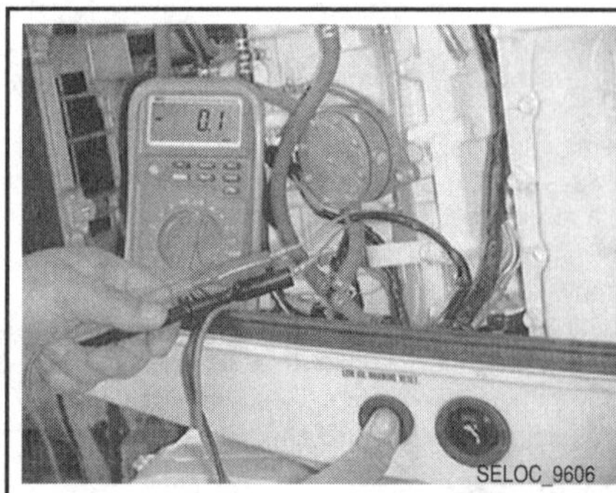

Fig. 59 Multimeters, such as this one from UEI, are an extremely useful tool for troubleshooting electrical problems

GAUGES

Compression Gauge

◆ See Figure 60

An important element in checking the overall condition of your engine is to check compression. This becomes increasingly more important on motors with high hours. Compression gauges are available as screw-in types and hold-in types. The screw-in type is slower to use, but eliminates the possibility of a faulty reading due to pressure escaping by the seal. A compression reading will uncover many problems that can cause rough running. Normally, these are not the sort of problems that can be cured by a tune-up.

Vacuum Gauge

◆ See Figures 61, 62 and 63

Vacuum gauges are handy for discovering air leaks, late ignition or valve timing, and a number of other problems.

Measuring Tools

Eventually, you are going to have to measure something. To do this, you will need at least a few precision tools.

MICROMETERS & CALIPERS

Micrometers and calipers are devices used to make extremely precise measurements. The simple truth is that you really won't have the need for many of these items just for routine maintenance. But, measuring tools, such as an outside caliper can be handy during repairs. And, if you decide to tackle a major overhaul, a micrometer will absolutely be necessary.

Should you decide on becoming more involved in boat engine mechanics, such as repair or rebuilding, then these tools will become very important. The success of any rebuild is dependent, to a great extent on the ability to check the size and fit of components as specified by the manufacturer. These measurements are often made in thousandths and ten-thousandths of an inch.

Micrometers

◆ See Figure 64 and 65

A micrometer is an instrument made up of a precisely machined spindle that is rotated in a fixed nut, opening and closing the distance between the end of the spindle and a fixed anvil. When measuring using a micrometer, don't over-tighten the tool on the part as either the component or tool may be damaged, and either way, an incorrect reading will result. Most micrometers are equipped with some form of thumbwheel on the spindle that is designed to freewheel over a certain light touch (automatically adjusting the spindle and preventing it from over-tightening).

Outside micrometers can be used to check the thickness of parts such shims or the outside diameter of components like the crankshaft journals. They are also used during many rebuild and repair procedures to measure the diameter of components such as the pistons. The most common type of micrometer reads in 1/1000 of an inch. Micrometers that use a vernier scale can estimate to 1/10 of an inch.

Inside micrometers are used to measure the distance between two parallel surfaces. For example, in powerhead rebuilding work, the "inside mike" measures cylinder bore wear and taper. Inside mikes are graduated the same way as outside mikes and are read the same way as well.

Remember that an inside mike must be absolutely perpendicular to the work being measured. When you measure with an inside mike, rock the mike gently from side to side and tip it back and forth slightly so that you span the widest part of the bore. Just to be on the safe side, take several readings. It takes a certain amount of experience to work any mike with confidence.

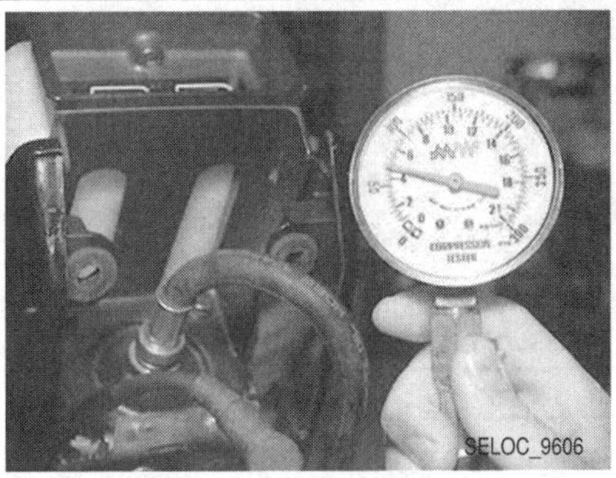

Fig. 60 Cylinder compression test results are extremely valuable indicators of internal engine condition

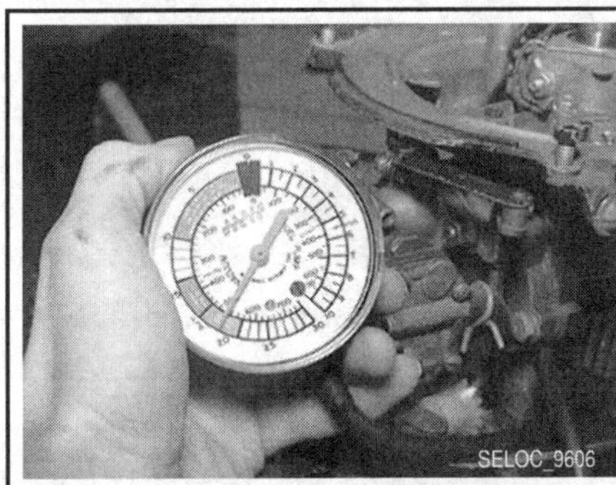

Fig. 61 Vacuum gauges are useful for troubleshooting including testing some fuel pumps

Fig. 62 You can also use the gauge on a hand-operated vacuum pump for tests

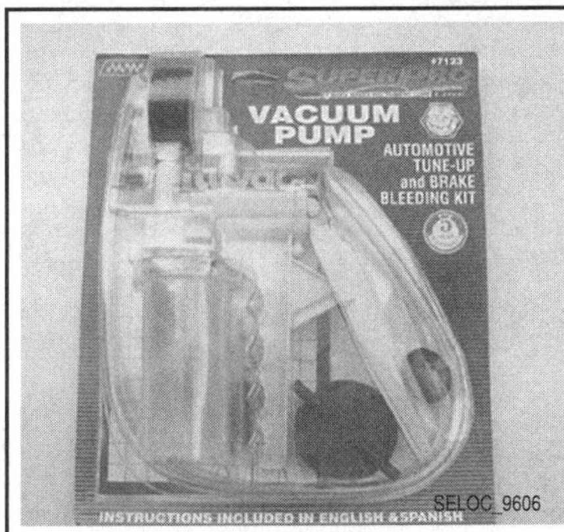

Fig. 63 Hand-held vacuum/pressure pumps are available at most parts stores

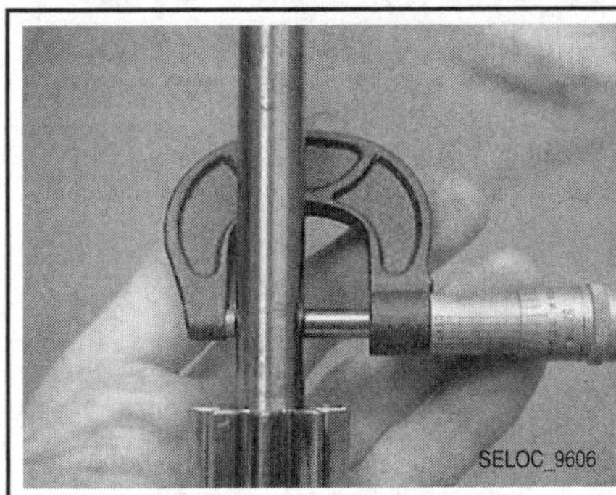

Fig. 64 Outside micrometers measure thickness, like shims or a shaft diameter

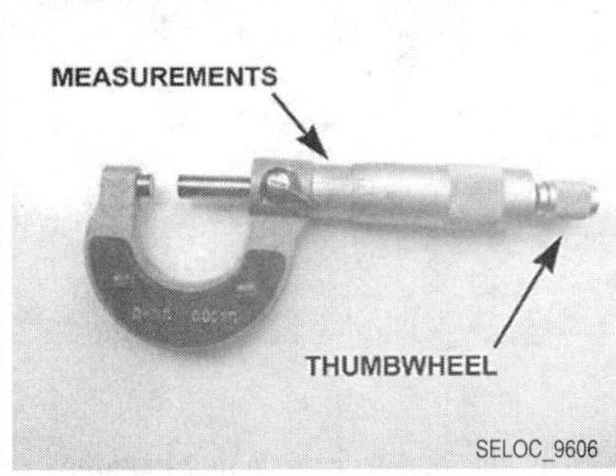

Fig. 65 Be careful not to over-tighten the micrometers always use the thumbwheel

Metric micrometers are read in the same way as inch micrometers, except that the measurements are in millimeters. Each line on the main scale equals 1mm. Each fifth line is stamped 5, 10, 15 and so on. Each line on the thimble scale equals 0.01mm. It will take a little practice, but if you can read an inch mike, you can read a metric mike.

Calipers

◆ See Figures 66 thru 68

Inside and outside calipers are useful devices to have if you need to measure something quickly and absolute precise measurement is not necessary. Simply take the reading and then hold the calipers on an accurate steel rule. Calipers, like micrometers, will often contain a thumbwheel to help ensure accurate measurement.

DIAL INDICATORS

◆ See Figure 69

A dial indicator is a gauge that utilizes a dial face and a needle to register measurements. There is a movable contact arm on the dial indicator. When the arm moves, the needle rotates on the dial. Dial indicators are calibrated to show readings in thousandths of an inch and typically, are used to measure end-play and run-out on various shafts and other components.

Dial indicators are quite easy to use, although they are relatively expensive. A variety of mounting devices are available so that the indicator can be used in a number of situations. Make certain that the contact arm is always parallel to the movement of the work being measured.

TELESCOPING GAUGES

◆ See Figure 70

A telescope gauge is really only used during rebuilding procedures (NOT during basic maintenance or routine repairs) to measure the inside of bores. It can take the place of an inside mike for some of these jobs. Simply insert the gauge in the hole to be measured and lock the plungers after they have contacted the walls. Remove the tool and measure across the plungers with an outside micrometer.

DEPTH GAUGES

◆ See Figure 71

A depth gauge can be inserted into a bore or other small hole to determine exactly how deep it is. One common use for a depth gauge is measuring the distance the piston sits below the deck of the block at top dead center. Some outside calipers contain a built-in depth gauge so you can save money and buy just one tool.

Fig. 66 Calipers are the fast and easy way to make precise measurements

Fig. 67 Calipers can also be used to measure depth . . .

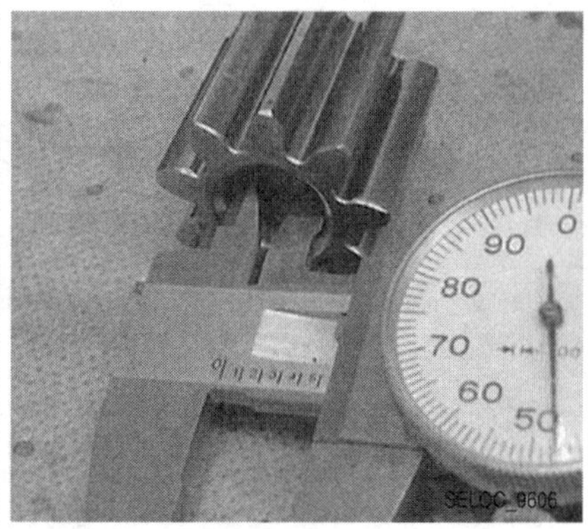

Fig. 68 . . . and inside diameter measurements, to 0.001 in. accuracy

Fig. 69 This dial indicator is measuring the end-play of a crankshaft during a powerhead rebuild

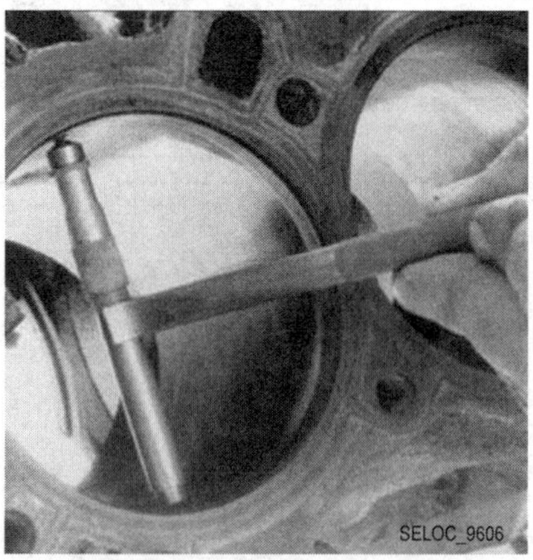

Fig. 70 Telescoping gauges are used during powerhead rebuilding procedures to measure the inside diameter of bores

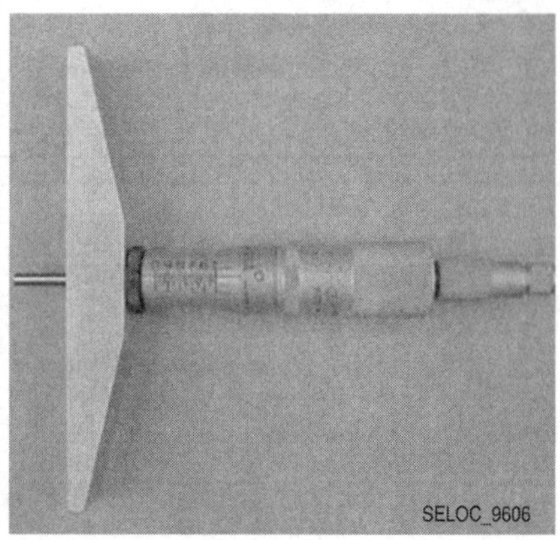

Fig. 71 Depth gauges are used to measure the depth of bore or other small holes

FASTENERS, MEASUREMENTS AND CONVERSIONS

Bolts, Nuts and Other Threaded Retainers

◆ **See Figures 72 and 73**

Although there are a great variety of fasteners found in the modern boat engine, the most commonly used retainer is the threaded fastener (nuts, bolts, screws, studs, etc). Most threaded retainers may be reused, provided that they are not damaged in use or during the repair.

■ **Some retainers (such as stretch bolts or torque prevailing nuts) are designed to deform when tightened or in use and should not be reused.**

Whenever possible, we will note any special retainers which should be replaced during a procedure. But you should always inspect the condition of a retainer when it is removed and you should replace any that show signs of damage. Check all threads for rust or corrosion that can increase the torque necessary to achieve the desired clamp load for which that fastener was originally selected. Additionally, be sure that the driver surface itself (on the fastener) is not compromised from rounding or other damage. In some cases a driver surface may become only partially rounded, allowing the driver to catch in only one direction. In many of these occurrences, a fastener may be installed and tightened, but the driver would not be able to grip and loosen the fastener again (this could lead to frustration down the line should that component ever need to be disassembled again).

If you must replace a fastener, whether due to design or damage, you must always be sure to use the proper replacement. In all cases, a retainer of the same design, material and strength should be used. Markings on the heads of most bolts will help determine the proper strength of the fastener. The same material, thread and pitch must be selected to assure proper installation and safe operation of the motor afterwards.

Thread gauges are available to help measure a bolt or stud's thread. Most part or hardware stores keep gauges available to help you select the proper size. In a pinch, you can use another nut or bolt for a thread gauge. If the bolt you are replacing is not too badly damaged, you can select a match by finding another bolt that will thread in its place. If you find a nut that will thread properly onto the damaged bolt, then use that nut as a gauge to help select the replacement bolt. If however, the bolt you are replacing is so badly damaged (broken or drilled out) that its threads cannot be used as a gauge, you might start by looking for another bolt (from the same assembly or a similar location) which will thread into the damaged bolt's mounting. If so, the other bolt can be used to select a nut; the nut can then be used to select the replacement bolt.

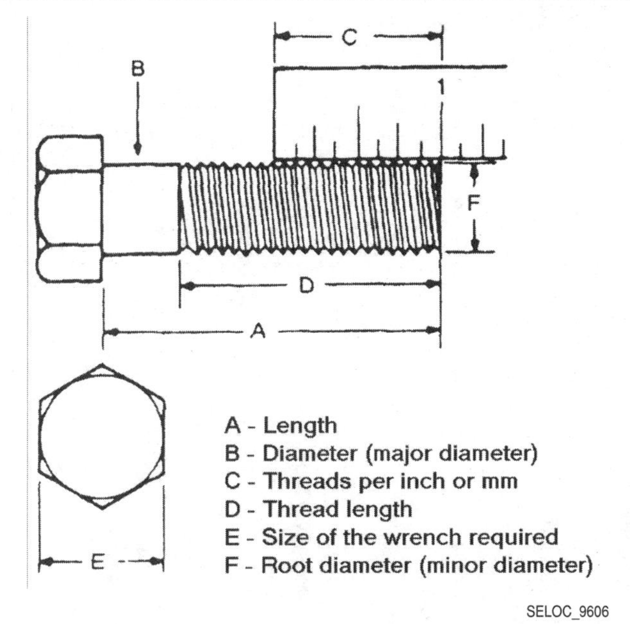

Fig. 72 Threaded retainer sizes are determined using these measurements

In all cases, be absolutely sure you have selected the proper replacement. Don't be shy, you can always ask the store clerk for help.

** WARNING

Be aware that when you find a bolt with damaged threads, you may also find the nut or tapped bore into which it was threaded has also been damaged. If this is the case, you may have to drill and tap the hole, replace the nut or otherwise repair the threads. Never try to force a replacement bolt to fit into the damaged threads.

Torque

Torque is defined as the measurement of resistance to turning or rotating. It tends to twist a body about an axis of rotation. A common example of this would be tightening a threaded retainer such as a nut, bolt or screw. Measuring torque is one of the most common ways to help assure that a threaded retainer has been properly fastened.

When tightening a threaded fastener, torque is applied in three distinct areas, the head, the bearing surface and the clamp load. About 50% of the measured torque is used in overcoming bearing friction. This is the friction between the bearing surface of the bolt head, screw head or nut face and the base material or washer (the surface on which the fastener is rotating). Approximately 40% of the applied torque is used in overcoming thread friction. This leaves only about 10% of the applied torque to develop a useful clamp load (the force that holds a joint together). This means that friction can account for as much as 90% of the applied torque on a fastener.

Standard & Metric Measurements

Specifications are often used to help you determine the condition of various components, or to assist you in their installation. Some of the most common measurements include length (in. or cm/mm), torque (ft. lbs., inch lbs. or Nm) and pressure (psi, in. Hg, kPa or mm Hg).

In some cases, that value may not be conveniently measured with what is available in your toolbox. Luckily, many of the measuring devices that are available today will have two scales so U.S. or Metric measurements may easily be taken. If any of the various measuring tools that are available to you do not contain the same scale as listed in your specifications, use the conversion factors that are provided in the Specifications section to determine the proper value.

The conversion factor chart is used by taking the given specification and multiplying it by the necessary conversion factor. For instance, looking at the first line, if you have a measurement in inches such as "free-play should be 2 in." but your ruler reads only in millimeters, multiply 2 in. by the conversion factor of 25.4 to get the metric equivalent of 50.8mm. Likewise, if a specification was given only in a Metric measurement, for example in Newton Meters (Nm), then look at the center column first. If the measurement is 100 Nm, multiply it by the conversion factor of 0.738 to get 73.8 ft. lbs.

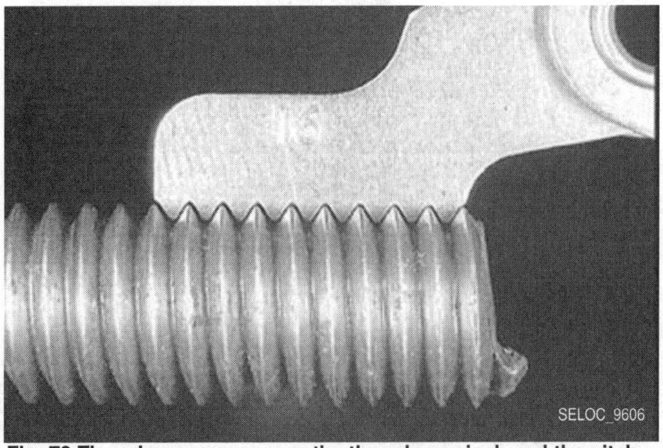

Fig. 73 Thread gauges measure the threads-per-inch and the pitch of a bolt or stud's threads

SPECIFICATIONS

CONVERSION FACTORS

LENGTH-DISTANCE

Inches (in.)	x 25.4	= Millimeters (mm)	x .0394	= Inches
Feet (ft.)	x .305	= Meters (m)	x 3.281	= Feet
Miles	x 1.609	= Kilometers (km)	x .0621	= Miles

VOLUME

Cubic Inches (in3)	x 16.387	= Cubic Centimeters	x .061	= in3
IMP Pints (IMP pt.)	x .568	= Liters (L)	x 1.76	= IMP pt.
IMP Quarts (IMP qt.)	x 1.137	= Liters (L)	x .88	= IMP qt.
IMP Gallons (IMP gal.)	x 4.546	= Liters (L)	x .22	= IMP gal.
IMP Quarts (IMP qt.)	x 1.201	= US Quarts (US qt.)	x .833	= IMP qt.
IMP Gallons (IMP gal.)	x 1.201	= US Gallons (US gal.)	x .833	= IMP gal.
Fl. Ounces	x 29.573	= Milliliters	x .034	= Ounces
US Pints (US pt.)	x .473	= Liters (L)	x 2.113	= Pints
US Quarts (US qt.)	x .946	= Liters (L)	x 1.057	= Quarts
US Gallons (US gal.)	x 3.785	= Liters (L)	x .264	= Gallons

MASS-WEIGHT

Ounces (oz.)	x 28.35	= Grams (g)	x .035	= Ounces
Pounds (lb.)	x .454	= Kilograms (kg)	x 2.205	= Pounds

PRESSURE

Pounds Per Sq. In. (psi)	x 6.895	= Kilopascals (kPa)	x .145	= psi
Inches of Mercury (Hg)	x .4912	= psi	x 2.036	= Hg
Inches of Mercury (Hg)	x 3.377	= Kilopascals (kPa)	x .2961	= Hg
Inches of Water (H_2O)	x .07355	= Inches of Mercury	x 13.783	= H_2O
Inches of Water (H_2O)	x .03613	= psi	x 27.684	= H_2O
Inches of Water (H_2O)	x .248	= Kilopascals (kPa)	x 4.026	= H_2O

TORQUE

Pounds-Force Inches (in-lb)	x .113	= Newton Meters (N·m)	x 8.85	= in-lb
Pounds-Force Feet (ft-lb)	x 1.356	= Newton Meters (N·m)	x .738	= ft-lb

VELOCITY

Miles Per Hour (MPH)	x 1.609	= Kilometers Per Hour (KPH)	x .621	= MPH

POWER

Horsepower (Hp)	x .745	= Kilowatts	x 1.34	= Horsepower

FUEL CONSUMPTION*

Miles Per Gallon IMP (MPG)	x .354	= Kilometers Per Liter (Km/L)
Kilometers Per Liter (Km/L)	x 2.352	= IMP MPG
Miles Per Gallon US (MPG)	x .425	= Kilometers Per Liter (Km/L)
Kilometers Per Liter (Km/L)	x 2.352	= US MPG

*It is common to covert from miles per gallon (mpg) to liters/100 kilometers (1/100 km), where mpg (IMP) x 1/100 km = 282 and mpg (US) x 1/100 km = 235.

TEMPERATURE

Degree Fahrenheit (°F)	= (°C x 1.8) + 32
Degree Celsius (°C)	= (°F – 32) x .56

SELOC_9606

Metric Bolts

Relative Strength Marking	4.6, 4.8			8.8		
Bolt Markings	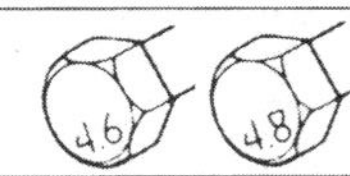			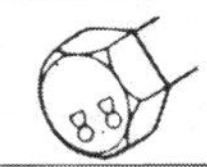		
Usage	Frequent			Infrequent		
Bolt Size	Maximum Torque			Maximum Torque		
Thread Size x Pitch (mm)	Ft-Lb	Kgm	Nm	Ft-Lb	Kgm	Nm
6 x 1.0	2–3	.2–.4	3–4	3–6	.4–.8	5–8
8 x 1.25	6–8	.8–1	8–12	9–14	1.2–1.9	13–19
10 x 1.25	12–17	1.5–2.3	16–23	20–29	2.7–4.0	27–39
12 x 1.25	21–32	2.9–4.4	29–43	35–53	4.8–7.3	47–72
14 x 1.5	35–52	4.8–7.1	48–70	57–85	7.8–11.7	77–110
16 x 1.5	51–77	7.0–10.6	67–100	90–120	12.4–16.5	130–160
18 x 1.5	74–110	10.2–15.1	100–150	130–170	17.9–23.4	180–230
20 x 1.5	110–140	15.1–19.3	150–190	190–240	26.2–46.9	160–320
22 x 1.5	150–190	22.0–26.2	200–260	250–320	34.5–44.1	340–430
24 x 1.5	190–240	26.2–46.9	260–320	310–410	42.7–56.5	420–550

SELOC_9606

SAE Bolts

SAE Grade Number	1 or 2			5			6 or 7		
Bolt Markings Manufacturers' marks may vary—number of lines always two less than the grade number.									
Usage	Frequent			Frequent			Infrequent		
Bolt Size (inches)—(Thread)	Maximum Torque			Maximum Torque			Maximum Torque		
	Ft-Lb	kgm	Nm	Ft-Lb	kgm	Nm	Ft-Lb	kgm	Nm
1/4—20	5	0.7	6.8	8	1.1	10.8	10	1.4	13.5
—28	6	0.8	8.1	10	1.4	13.6			
5/16—18	11	1.5	14.9	17	2.3	23.0	19	2.6	25.8
—24	13	1.8	17.6	19	2.6	25.7			
3/8—16	18	2.5	24.4	31	4.3	42.0	34	4.7	46.0
—24	20	2.75	27.1	35	4.8	47.5			
7/16—14	28	3.8	37.0	49	6.8	66.4	55	7.6	74.5
—20	30	4.2	40.7	55	7.6	74.5			
1/2—13	39	5.4	52.8	75	10.4	101.7	85	11.75	115.2
—20	41	5.7	55.6	85	11.7	115.2			
9/16—12	51	7.0	69.2	110	15.2	149.1	120	16.6	162.7
—18	55	7.6	74.5	120	16.6	162.7			
5/8—11	83	11.5	112.5	150	20.7	203.3	167	23.0	226.5
—18	95	13.1	128.8	170	23.5	230.5			
3/4—10	105	14.5	142.3	270	37.3	366.0	280	38.7	379.6
—16	115	15.9	155.9	295	40.8	400.0			
7/8— 9	160	22.1	216.9	395	54.6	535.5	440	60.9	596.5
—14	175	24.2	237.2	435	60.1	589.7			
1— 8	236	32.5	318.6	590	81.6	799.9	660	91.3	894.8
—14	250	34.6	338.9	660	91.3	849.8			

SELOC_9606

2

MAINTENANCE & TUNE-UP

ENGINE & DRIVE MAINTENANCE

At Seloc, we have estimated that some 75% of engine repair work can be directly or indirectly attributed to lack of proper care for the engine. This is especially true of care during the off-season period. There is no way on this green earth for a mechanical engine, particularly an watercraft motor, to be left sitting idle for an extended period of time, say for 6 months, and then be ready for instant satisfactory service.

Imagine, if you will, leaving your car or truck for 6 months, and then expecting to turn the key, having it roar to life, and being able to drive off in the same manner as a daily occurrence.

Therefore it is critical for a PWC engine to be run (at least once a month), preferably, in the water and properly maintained between uses; or for it to be specifically prepared for storage and serviced again immediately before the start of the season.

Only through a regular maintenance program can the owner expect to receive long life and satisfactory performance at minimum cost.

Many times, if an engine is not performing properly, the owner will "nurse" it through the season with good intentions of working on the unit once it is no longer being used. As with many New Year's resolutions, the good intentions are not completed and the craft may lie for many months before the work is begun or the unit is taken to the marine shop for repair.

Imagine, if you will, the cause of the problem being a blown head gasket. And let us assume water has found its way into a cylinder. This water, allowed to remain over a long period of time, will do considerably more damage than it would have if the unit had been disassembled and the repair work performed immediately. Therefore, if an outboard is not functioning properly, do not stow it away with promises to get at it when you get time, because the work and expense will only get worse, the longer corrective action is postponed. In the example of the blown head gasket, a relatively simple and inexpensive repair job could very well develop into major overhaul and rebuild work.

OK, perhaps no one thing that we do as boaters will protect us from risks involved with enjoying the wind and the water on a WaveRunner. But, each time we perform maintenance on our boat or motor, we increase the likelihood that we will find a potential hazard before it becomes a problem. Each time we inspect our boat and motor, we decrease the possibility that it could leave us stranded on the water.

In this way, performing boat and engine service is one of the most important ways that we, as boaters, can help protect ourselves, our boats, and the friends and family that we bring aboard.

Serial Numbers

Most manufacturers use 3 different identification numbers stamped, embossed or decaled somewhere on the craft. These numbers are the key to correct craft and engine identification; and also proper parts ordering.

ENGINE SERIAL NUMBER

◆ **See Figures 1 thru 5**

The engine serial number is the manufacturer's key to engine changes. This number identifies the year of manufacture, the qualified horsepower rating, and the parts book identification. If any correspondence or parts are required, the engine serial number MUST be used or proper identification is not possible. The serial number is stamped/embossed on a tag or plate located on or near the engine. Locations vary with models:

Fig. 1 The engine serial number plate is attached to the oil tank on 998cc and 1052cc FX models. . .

Fig. 2 . . .and also on VX models

Fig. 3 The engine serial number plate is attached to the forward end of the cylinder head cover (stbd side) on 1812cc FZ and FX/FX Cruiser SHO models. . .

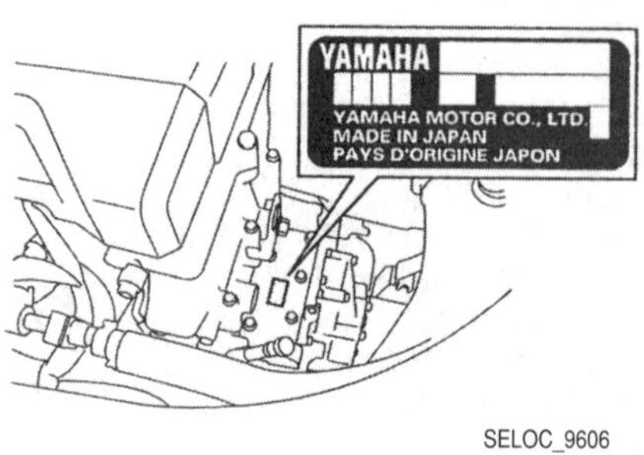

Fig. 4 . . .while on 1812cc FX/FX Cruiser HO and VXR/VZS models, it is attached to the plate on the forward end of the lower cylinder block

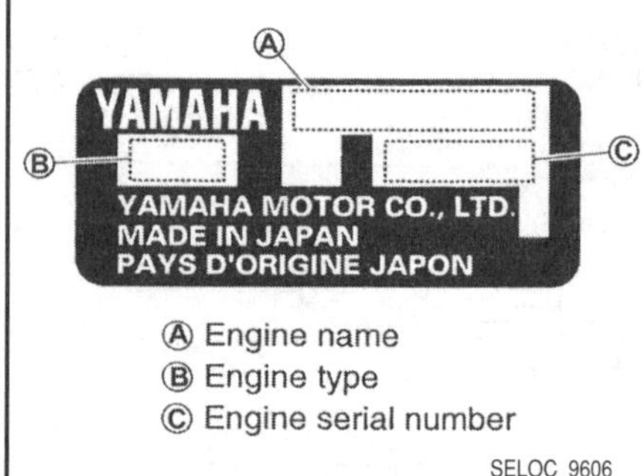

Fig. 5 A close look at the engine serial number tag

- 998cc and 1052cc FX/VX - a small plate attached to the oil tank
- 1812cc FX/FX Cruiser SHO and FZ - a small plate attached to the forward end of the cylinder head cover (stbd side). Access will require engine cover removal
- 1812cc FX/FX Cruiser HO and VXR/VXS - plate on the forward end of the lower cylinder block

PRIMARY I.D. NUMBER

◆ **See Figures 6 and 7**

This identification number, equally important to model identification and parts ordering as is the engine serial number, is normally found on the inside of the starboard hull.

HULL IDENTIFICATION NUMBER (HIN)

◆ **See Figure 8**

Federal regulations require all personal watercraft to have a Hull Identification Number (HIN) permanently attached to the craft in two separate locations: one on the transom on the starboard side, above the waterline and the other in an unexposed location, which is usually left to the discretion of the manufacturer. This HIN number can be most helpful when attempting to retrieve a stolen craft. The HIN number is required in most states in order to properly register the craft.

JET PUMP SERIAL NUMBER

◆ **See Figures 9 and 10**

You will also find a jet pump serial number plate stamped onto a label, usually attached to the intermediate housing.

EMISSION CONTROL LABEL

◆ **See Figures 11 and 12**

All models are also equipped with an Emissions Control Information label as well. When equipped this label gives basic tune-up or calibration information.

■ **You should ALWAYS check for an emission control label and go with the specifications on it before you believe anything in print or on the web. This label may contain running changes which never made it to published information from Yamaha or any other company.**

The emissions control information label states that the motor is in compliance with EPA emissions regulations for the model year of that engine. And, more importantly, it gives tune-up specifications that are vital to proper engine performance (that minimize harmful emissions). The specifications on this label may reflect changes that are made during production runs and are often not later reflected in a company's service literature. For this reason, specifications on the label always supersede those

Fig. 6 A good look at the primary I. D. number

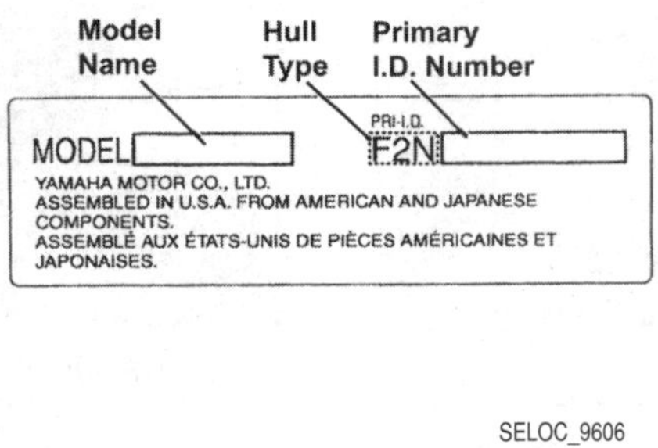

Fig. 7 The primary I. D. number can be found on the inside of the hull, in the engine compartment

Fig. 8 A good look at the HIN plate

Fig. 9 A good look at the jet pump serial number plate (VX models w/1052cc engines shown, others similar)

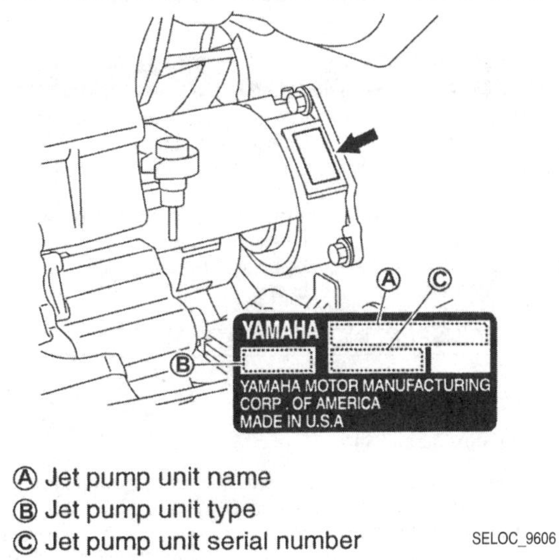

Fig. 10 A closer look at the tag itself

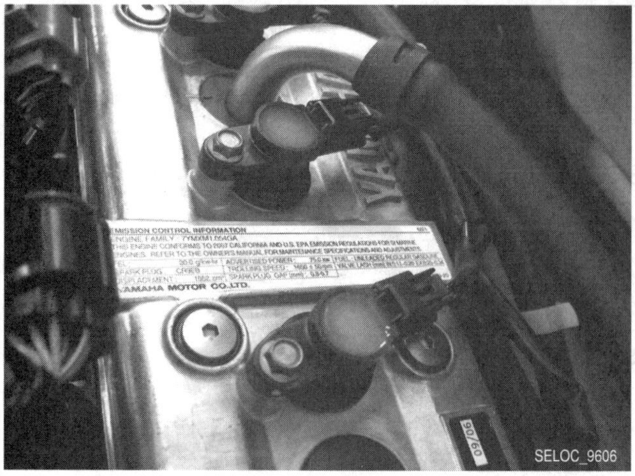

Fig. 11 Emission control labels may be found here. . .

Fig. 12 . . .or here

of a print manual. Typical specifications that are found on this label will include:

- Spark plug type and gap
- Power (kw)
- Trolling speed
- Fuel recommendations
- Engine displacement (in Cubic Inches or Cubic Centimeters, as noted on the label)

Before/After Each Use

As stated earlier, the best means of extending engine life and helping to protect yourself while on the water is to pay close attention to boat/engine maintenance. This starts with an inspection of systems and components before and after each time you use your boat.

A list of checks, inspections or required maintenance can be found in the Maintenance Intervals Chart at the end of this section. Some of these inspections or tasks are performed before the boat is launched, some only after it is retrieved and the rest, both times.

VISUALLY INSPECTING THE BOAT & MOTOR

1 EASY

◆ **See Figures 13 and 14**

Both before each launch and immediately after each retrieval, visually inspect the vessel and motor as follows:

1. **Check the fuel and oil levels** according to the procedures in this manual. Do NOT launch a vessel without properly topped off fuel and oil tanks. It is not worth the risk of getting stranded or of damage to the motor. Likewise, upon retrieval, check the oil and fuel levels while it is still fresh in your mind. This is a good way to track fuel consumption (one indication of engine performance). Oil consumption should be minimal, but all 4-stroke engines allow a small portion of oil to burn. Watch for sudden increases in the amount of oil burned and investigate further if found.

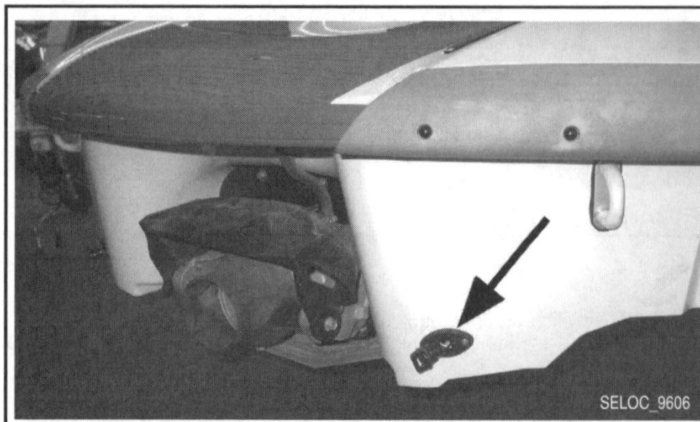

Fig. 13 Always make sure the transom plug is installed and tightened securely before a launch

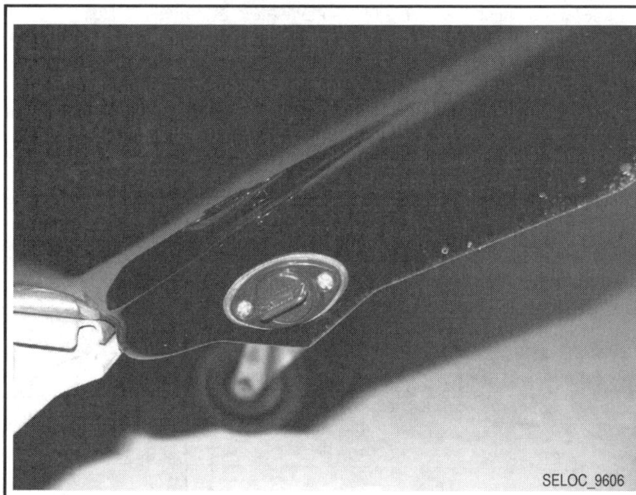

Fig. 14 A little closer look

2. **Check for signs of fuel or oil leakage.** Probably as important as making sure enough fuel and oil is onboard, is the need to make sure that no dangerous conditions might arise due to leaks. Thoroughly check all hoses, fittings and tanks for signs of leakage. Oil leaks may cause the vessel to become stranded, or worse, could destroy the motor if undetected for a significant amount of time. Fuel leaks can cause a fire hazard, or worse, an explosive condition. This check is not only about properly maintaining your motor, but about helping to protect your life.

3. **Inspect the hull** for signs of corrosion or damage. Don't launch a damaged craft. And don't surprise yourself dockside or at the launch ramp by discovering damage that went unnoticed last time the vessel was retrieved. Repair any hull or case damage now.

4. **Check the battery** connections to make sure they are clean and tight. A loose or corroded connection will cause charging problems (damaging the system or preventing charging). There's only one thing worse than a dead battery dockside/launch ramp and that's a dead battery in the middle of a bay, river or worse, the ocean. Whenever possible, make a quick visual check of battery electrolyte levels (keeping an eye on the level will give some warning of overcharging problems). This is especially true if the engine is operated at high speeds for extended periods of time.

5. **Check all accessible fasteners for tightness.** Make sure all easily accessible fasteners appear to be tight. This is especially true of any anode retaining bolts, all steering or throttle linkage fasteners and the engine clamps or mounting bolts. Don't risk losing control or becoming stranded due to loose fasteners. Perform these checks before heading out, and immediately after you return (so you'll know if anything needs to be serviced before you want to launch again.)

6. **Check operation of all controls including the throttle/shifter, steering and emergency stop/start switch and/or safety lanyard.** Before launching, make sure that all linkage and steering components operate properly and move smoothly through their range of motion. All electrical switches and especially the emergency stop system(s) must be in proper working order. While underway, watch for signs that a system is not working or has become damaged. With the steering, shifter or throttle, keep a watchful eye out for a change in resistance or the start of jerky/notchy movement.

7. **Check the pump intake grate.** The pump intake grate should be clean and undamaged before setting out. Remember that a damaged grate could allow debris into the system that could destroy the impeller or clog cooling passages.

8. **If used in salty, brackish or polluted waters,** thoroughly rinse the vessel, then flush the cooling system.

9. **For Pete's sake, make sure the plug is in!** We shouldn't have to say it, but unfortunately we do. If you've been boating for any length of time, you've seen or heard of someone who backed a trailer down a launch ramp, forgetting to check the drain plug before submerging (literally) the vessel. Always make sure the transom plug is installed and tight before a launch.

Air Filter

DESCRIPTION & OPERATION

In a marine engine compartment, the minimal amount of dust and dirt in the air mean that a marine air filter requires less maintenance than its counterpart in the automotive world. However, the maintenance of a marine air filter is equally important.

The marine filter prevents dirt from entering the engine and scoring the cylinder walls. This lessens oil consumption and extends the engine's life.

Over time, the air filter element will become clogged with dirt and oil, decreasing the amount of air entering the engine and lowering engine output. If an excessive amount of oil is clogging the filter, this could be an indication of worn cylinders or piston ring failure causing high pressure in the crankcase.

The maintenance intervals for filter element inspection/cleaning is at the end of the first boating season, and then every 100 hours of engine operation or once a year, whichever comes first. But, Yamaha has moved back and forth on this depending on the unit, so be sure to check the Maintenance Intervals chart for your engine.

REMOVAL & INSTALLATION

2002-08 FX Models With 998/1052cc Engines

EASY

◆ See Figures 15, 16 and 17

1. Disconnect the emergency lanyard at the handlebar to ensure the engine will not start accidentally.

2. Carefully pop the release and lift off the seat(s).

3. Remove the 2 mounting bolts in the center of the air filter case. Slide each of the case cover slide bars (8) out toward the nearest corner, and lift the cover off of the filter case.

4. Lift out the air filter element and then carefully cover the case with a lint-free towel to keep out any contaminants.

5. Inspect the element for cracks, tears, dirt or oil. Cleaning and/or replacing as necessary. Use only cold or lukewarm water to clean the element and leave enough time for it to air dry completely before installing.

■ **Do not use compressed air to speed the drying of the filter element.**

6. Remove the towel and then position the filter back into the housing making sure that the small tab on the filter fits into the slot in the case housing. Do not force it as it will only go in one way. If you have a problem, just spin it around 180° and attempt it again.

7. Position the cover over the case making sure everything lines up and then move the slide bars away from the corner until it hits the stop. Be sure that each of the slide bars is moved all the way to each corner *before* dropping the cover into position.

8. Coat the 2 mounting bolts with Loctite 572 and tighten them to 1.8 ft. lbs. (2.5 Nm).

9. Install the seat(s) and reconnect the lanyard.

2008-11 FX/FZ/VXR/VXS Models w/1812cc Engines

EASY

◆ See Figures 18 thru 21

1. Remove the seat(s), cargo container (if equipped) and then disconnect the battery cables.

2. Remove the 2 mounting bolts and lift off the engine cover. The cover is a press-fit over its support bracket ends, so carefully pop it off of the connectors by pulling back and upward at the same time.

3. Squeeze the pinch clamp on the breather hose where it attaches to the air intake pipe and wiggle the hose off of the connection. Carefully move the hose out of the way.

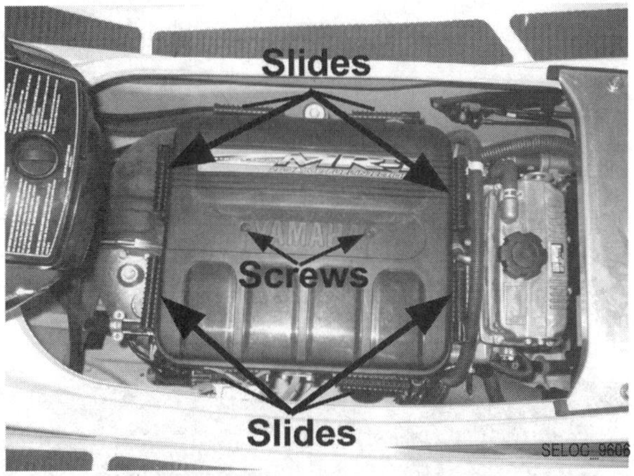

Fig. 15 Remove the 2 mounting bolts. . .

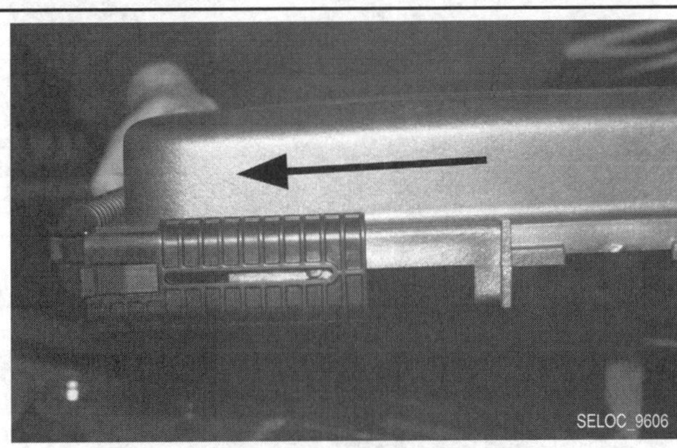

Fig. 16 . . .and then move the 8 slide bars to the nearest corner before lifting off the case cover

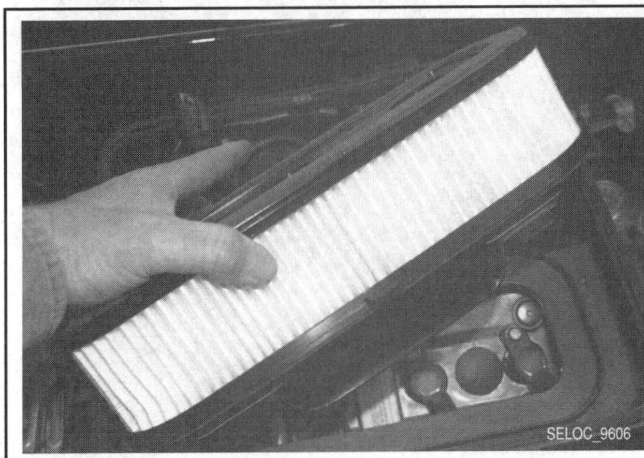

Fig. 17 Lift out the air filter element

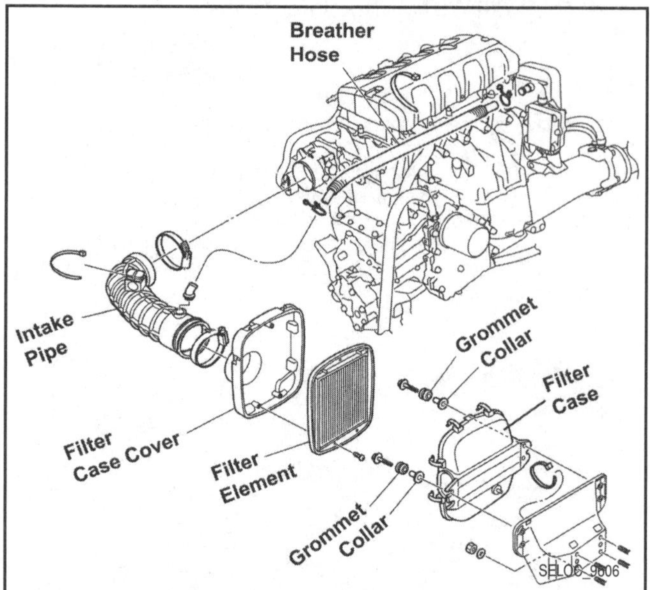

Fig. 18 Exploded view of the filter assembly on 1812cc FX/FZ models (HO shown, SHO and FZ similar)

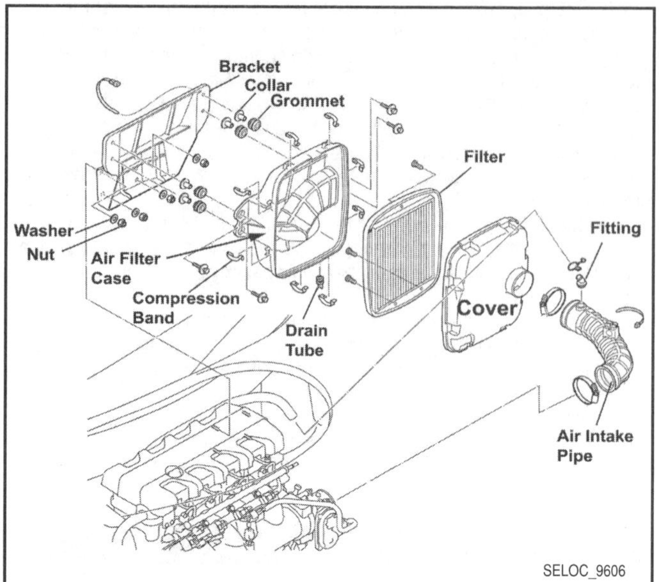

Fig. 19 Exploded view of the filter assembly on 1812cc VXR/VXS models

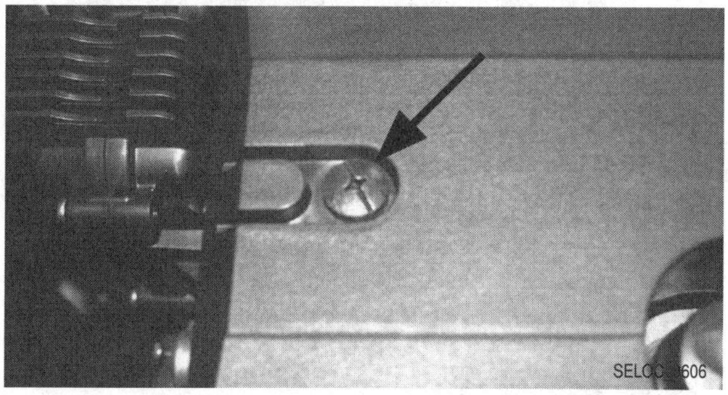

Fig. 20 Remove the 2 mounting bolts before popping the engine cover off of its brackets

Fig. 21 Pull back on the compression bands to release the cover latches

4. Take a careful look at the air filter case and tag/disconnect any other lines or hoses that might be in the way of access.
5. Loosen the hose clamp on the end of the intake pipe and slide it back on the pipe. Grasp the pipe where it connects to the filter case cover and wiggle it off of the lip.
6. Reach down and pop off the compression bands (should be 8, 2/side) holding the cover to the case.
7. Pull the cover away from the case and carefully lift out the cover and filter.
8. Remove the 3 mounting bolts (2 bottom, 1 top) and separate the filter from the cover.

To install:

9. Inspect, clean or replace the filter and then position it back into the cover. Tighten the 3 bolts securely, but not so tight as to warp the filter frame.
10. Slide the cover/filter back into position so it is sitting in the lip of the filter case and then pop the clamps back into place over the cover.
11. Slide the end of the intake pipe onto the cover. Move the clamp back into position inside the ribs on the pipe and tighten it to no more than 2.2 ft. lbs. (3 Nm).
12. Reconnect any hoses or line you chose to move for access.
13. Push the breather hose over the connector and pinch the clamp back into position.
14. Connect the battery cables.
15. Position the engine cover over the mounting bars and pop it into place. Tighten the 2 bolts to 3.7 ft. lbs. (5 Nm).
16. Connect the battery cables, and then install the container and/or seat.

VX Models w/1052cc Engines

① EASY

◆ See Figures 22 thru 25

1. Disconnect the emergency lanyard at the handlebar to ensure the engine will not start accidentally.
2. Carefully pop the release and lift off the seat(s).
3. Loosen the hose clamp on the air intake hose and carefully slide it off of the intake manifold.
4. Position some rags underneath the fuel line. Pinch the clamp and slide it down the line so you can wiggle the line off the nipple. Plug the line and move it out of the way.
5. Pinch the oil separator breather hose clamp and slide it off the nipple on the air intake hose. Position it out of the way.
6. Snip the 4 plastic ties over of near the filter case; remember where they came from.

■ **On 2005-09 models, Yamaha recommends moving the entire filter case from the engine compartment before removing the case cover and filter. Yet, in 2010 and later, they suggest removing the case cover first and then removing the case/filter. Take a look at the space in your engine compartment - you may wish to perform Steps 7 and 8 in the opposite order given depending on what see and what you are trying to do; there's no sense removing the entire assembly if you don't need to.**

7. Remove the 4 mounting bolts and lift out the entire air filter case. Remove the collars/grommets and inspect them for cracks or other damage.

8. Remove the 6 case cover mounting bolts (4 short and 2 long) and pop off the cover (with the air intake hose still attached). Remove the seal if it didn't come off with the cover. If you are doing this with the case still in the engine compartment, remove the water separator and swivel the air intake hose to the right before extracting the cover.

9. Pull the air filter up and out of its tracks in the case. Carefully check it for dirt, oil or other contaminants. Check for tears and other damage.

10. You can clean the filter if you like using cold water (no detergent!), but make sure that is dries *completely* before installation.

■ **Do not use compressed air to speed the drying of the filter element.**

To install:

11. Slide the air filter into the case making sure it stays in its tracks and the projections on the bottom of the filter are facing away (toward the port side) from the projections at the bottom of the case. We're sure you've figured this out already, but the filter is not completely square, so make sure the tapered end of the filter goes into the case.

12. Inspect the cover seal for cracks or other damage, replacing it as necessary. Press the seal into the groove and then install the case cover. Coat the 6 mounting bolts with Loctite 572 and thread them all in until finger tight. Tighten the long bolts to 11 inch lbs. (1.2 Nm) and the shorter ones to 13 inch lbs. (1.5 Nm) on 2005-09 models, or all bolts to 13 inch lbs. (1.5 Nm) on 2010-11 models.

13. Skip this step if you chose not to remove the entire filter case from the vessel. Install the collars/grommets and then position the case/cover into the engine compartment and install the mounting bolts, tightening them to 12 ft. lbs. (17 Nm).

14. Install the water separator if removed.

15. Reconnect the air intake hose to the throttle body and tighten the hose clamp securely.

16. Connect the fuel line (remember the plug) and breather hose, tightening their respective clamps securely.

17. Install 4 new plastic ties.

18. Install the seat.

Fig. 22 Disconnect the lines and hoses and move them out of the way. . .

Fuel Filter/Water Separator

◆ **See Figure 26**

A fuel filter is designed to keep particles of dirt and debris from entering the carburetors or the fuel injection system and clogging the tiny internal passages of either. A small speck of dirt or sand can drastically affect the ability of the fuel system to deliver the proper amount of air and fuel/oil to the engine. If a filter becomes clogged, the flow of gasoline will be impeded. This could cause lean fuel mixtures, hesitation and stumbling and idle problems in carburetors. Although a clogged fuel passage in a fuel injected engine could also cause lean symptoms and idle problems, dirt can also prevent a fuel injector from closing properly. A fuel injector that is stuck partially open by debris would likely cause the engine to run rich due to the unregulated fuel constantly spraying from the pressurized injector.

Regular cleaning or replacement of the fuel filter (depending on the type or types used) will decrease the risk of blocking the flow of fuel to the engine, which could leave you stranded on the water. It will also decrease the risk of damage to the small passages of a carburetor or fuel injector that could require more extensive and expensive replacement. Keep in mind that fuel filters are usually pretty inexpensive (at lease when compared to a tow) and replacement is a simple task. Service your fuel filter on a regular basis to avoid fuel delivery problems.

That being said, Yamaha does not utilize a conventional inline or canister filter on these units. There is a water separator situated directly above the fuel tank and it is recommended that this be checked after each use. If you notice water in the unit, simply position a container of sufficient size under the separator and then unscrew the drain cap on the bottom of the separator. Once drained, thread the cap back on.

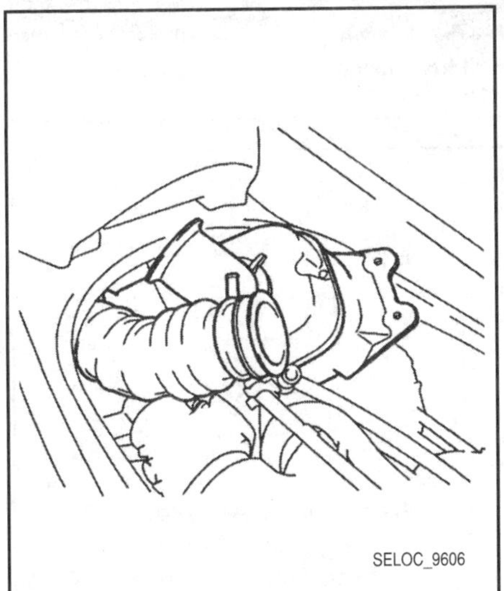

Fig. 23 . . .before lifting out the filter case

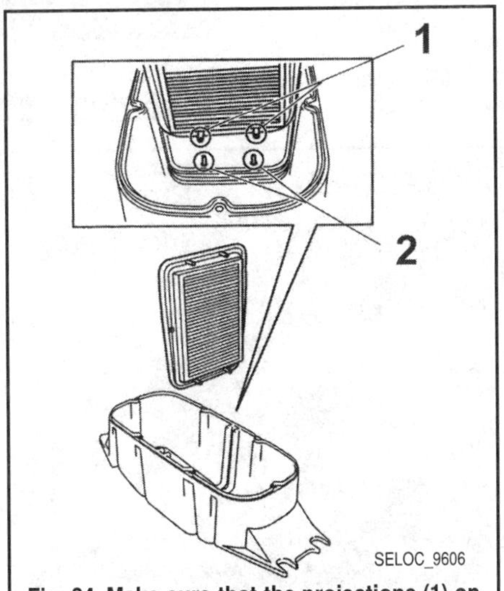

Fig. 24 Make sure that the projections (1) on the filter faces away from those on the filter case (2)

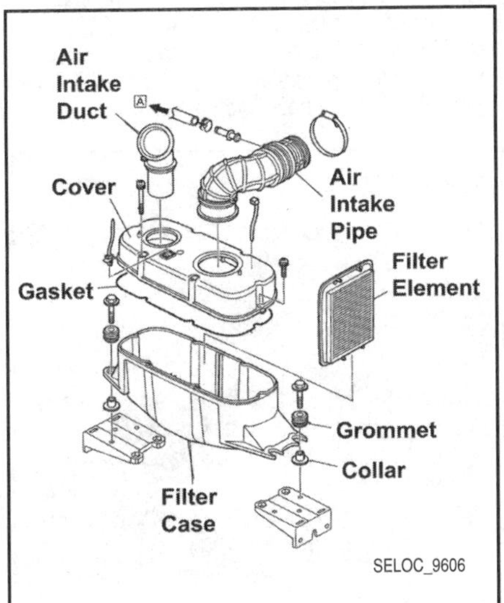

Fig. 25 A good look at the filter case

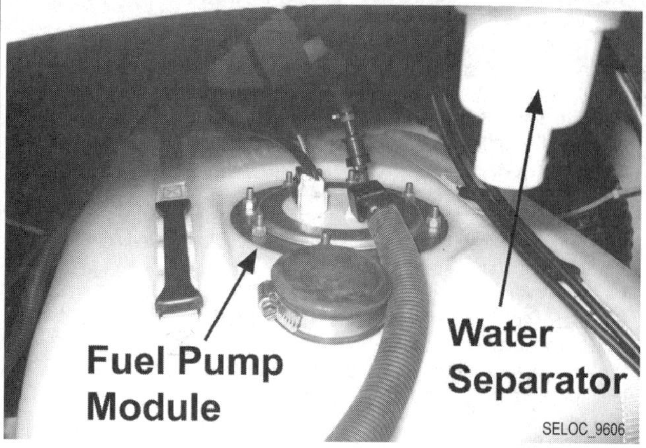

Fig. 26 If there is water in the separator, remove the drain cap and drain into a container.

In addition to the above there is fuel strainer in the tank where the fuel filler hose is connected. There is also a small screen filter in the bottom of the fuel pump module. Although there is no recommended maintenance on either of these, it wouldn't be a bad idea to check/clean them once a season.

■ **Although we see no reason why they have deleted it, Yamaha does not show the tank strainer on certain FX HO/FX Cruiser HO models.**

Cooling System

FLUSHING

① EASY

◆ **See Figures 27, 28 and 29**

Yamaha suggests in their Maintenance Intervals that the system be flushed after every use, particularly if used in salt water. To make things easier for the operator, all units covered here are equipped with an easily accessible flush fitting in the engine compartment.

Its quick, easy and will extend the life of your engine and drive, so try to do this whenever possible.

1. Ensure that the craft and trailer are horizontal.
2. Remove the seat. Remove the rear cargo container if equipped.
3. Locate the fitting, which looks like a long tube with a screw cap on it. FX models and VX models w/1812cc engines generally have the fitting attached to a cross-hull support behind the engine, while other VX models w/1952cc engines usually have it attached to the oil tank. Either way, it is easily identifiable after looking at the illustrations.
4. Unscrew the cap and insert the hose adapter into the fitting connector. Thread the end of a hose into the fitting. Do not turn the water on yet!

✳✳ CAUTION

You've got a double-edged sword here. . .never start the engine while out of the water without having a water source attached, and never turn on the water before starting the engine - this could allow water to flow back through the exhaust and into the crankcase, not good!

5. Unless you are right next to the spigot, have a friend give you hand here. Turn the engine on and immediately turn on the spigot. Adjust the water pressure until you see a continuous stream of water flowing from the jet nozzle.
6. Allow the engine to idle for approximately 3 minutes while keeping an eye on the temperature gauge. If you see the engine start to overheat or the engine stalls, turn the water off start from the beginning again making sure that all the connections are secure.
7. Turn the water off and then squeeze/release the throttle lever off and on for 10-15 seconds before turning the engine off. This will force any remaining water from the system.
8. Disconnect the hose and remove the adapter. Install the cap.
9. Install the cargo container and seat(s).

Thermostat

DESCRIPTION & OPERATION

◆ **See Figures 30 and 31**

■ **Thermostats are only used on FX, FZ and VXR/VXS models.**

The thermostat is a simple temperature sensitive valve that opens and closes to control cooling water flow through the engine. In operation, the thermostat hovers somewhere between open and closed. As engine load and temperature increase, the thermostat opens to allow more cooling water into the engine. As temperature and load decrease, the thermostat closes.

A sticking thermostat will either allow the temperature to rise well above the normal operating temperature before it opens, or, if stuck in the open position, will never allow the engine to reach operating temperature.

All thermostats are rated based on the temperature at which they open and this rating should always be stamped somewhere on the thermostat, usually on the flange area. All engines covered here utilize a 123-141°F thermostat, except the 1812cc HO and VX models which utilize 136-144 °F units and the 1812cc SHO/FZ models which use 118-126°F.

■ **On all engines covered here, the thermostat housing can be found on the rear side of the engine, often attached to the oil cooler - easily identifiable by the 3 hoses attached to it.**

✳✳ CAUTION

Serious damage may result from operating your engine without a thermostat! Don't even consider this!

Fig. 27 The flush fitting is either attached to a cross-hull support near the water lock (FX/FZ/VXR/VXS models). . .

Fig. 28 . . .or to the oil tank

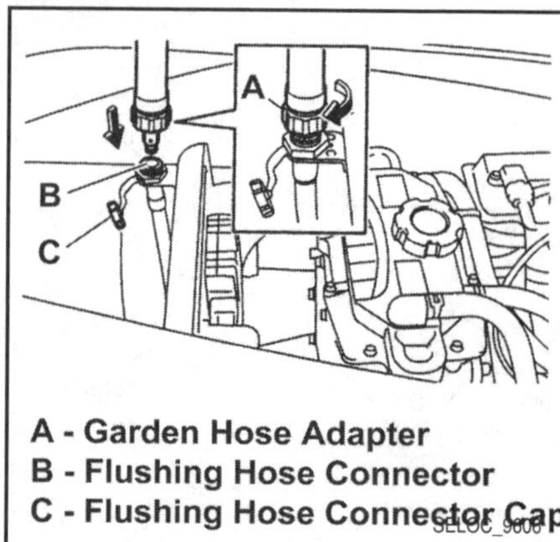

Fig. 29 Always use the adapter when connecting the hose to the fitting

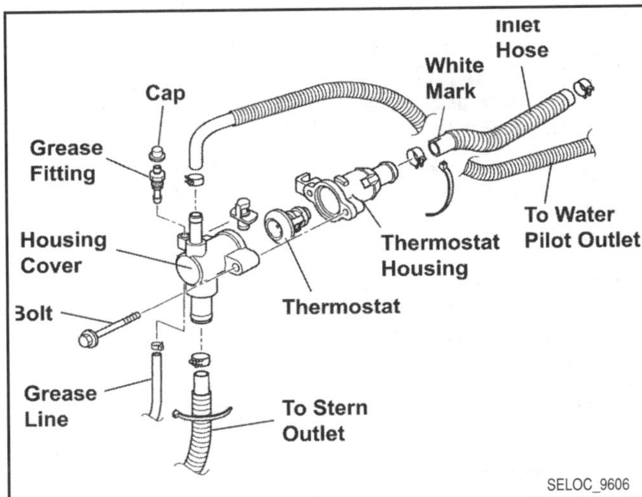

Fig. 30 Exploded view of the thermostat assembly - 998/1052cc FX models

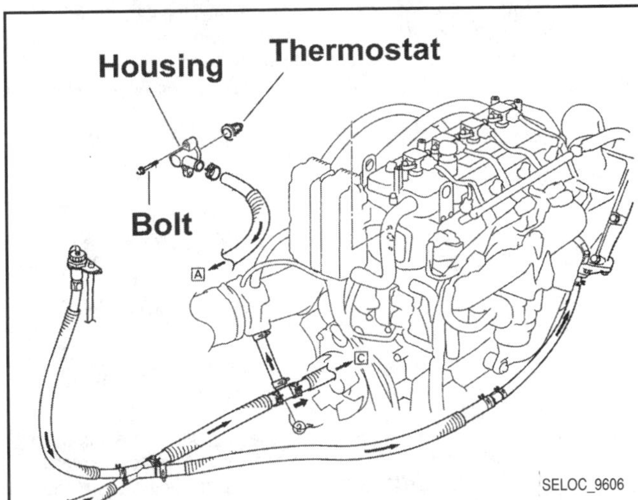

Fig. 31 Exploded view of the thermostat assembly - 1812cc FX, FZ, VXR and VXS models

✱✱ CAUTION

NEVER use an automotive/motorcycle thermostat in a marine engine. No matter how tempting this may seem, forget it!!

REMOVAL & INSTALLATION

998cc & 1052cc Engines

◆ **See Figures 30, 32 and 33**

1. Remove the seat(s) and disconnect the negative battery cable. You may want to remove the rear cargo container as well.

■ **Although not strictly necessary, you may find that removing the deck plate (above the tank) will make access much easier.**

2. Tag and disconnect the 3 water hoses connected to the housing; move them out of the way. You may have to remove a plastic tie as well. All 3 hoses should use conventional pinch clamps, so just squeeze the 2 tabs and slide the clamp back over the hose.
3. There is a small grease line on the side of the housing, remove the clamp and slide the line off the fitting. This line may have a full circle clamp on it so may require cutting. Lift the fitting and its cap out of the thermostat housing.
4. Slide the housing out of the mount. Loosen the 2 cover-to-housing bolts and lift off the cover. Remove the thermostat and gasket. Some later models may use 2 bolts attaching the housing to the mount instead. It is not necessary to remove the housing unless you need to.
5. Perform the testing procedures if necessary

To install:

6. Clean the housing and cover and check all passages for any blockage.
7. Position the thermostat, spring first, into the housing.
8. Position a new gasket onto the cover and then connect it to the housing. Coat the bolt threads with Loctite 572 and then install them and tighten to 2.7 ft. lbs. (32 inch lbs/3.7 Nm) in one pass and then again to 5.5 ft. lbs. (66 inch lbs./7.6 Nm). Move the assembly into the mount.
9. If you removed the housing attached with bolts, slide the collars into the holes, cot the bolts with Loctite 572 and then thread the 2 bolts through the housing and into the mount, tightening to 2.7 ft. lbs. (32 inch lbs/3.7 Nm) in one pass and then again to 5.5 ft. lbs. (66 inch lbs./7.6 Nm).
10. Press the grease fitting (if equipped) into the hole in the housing and then slide the line onto the bottom of the fitting. Slide the clamp up into place. If you have trouble with this, slide the line up through the hole and press the fitting into the line, and then press everything back into the hole - just make sure you install the clamp *after* you've pressed the assembly back down into the hole.

Fig. 32 The housing is attached to the oil tank on most 998/1052cc engines

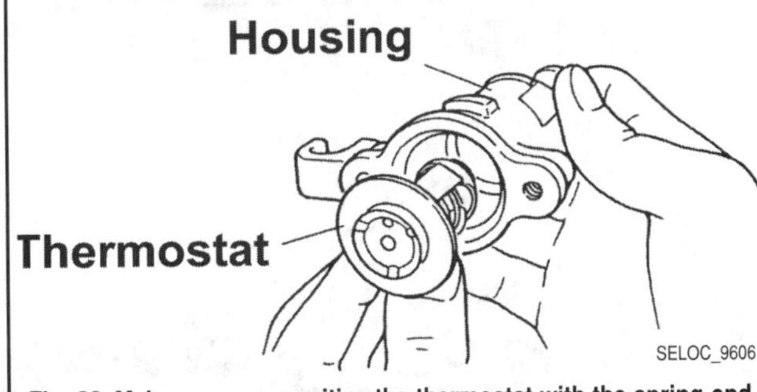

Fig. 33 Make sure you position the thermostat with the spring-end into the housing on 998/1052cc engines

■ **If you had to cut off a full circle clamp, make sure you use the correct clamp pliers to re-crimp the clamp.**

11. Reconnect the 3 water hoses and slide their clamps back up into position. There should be a white mark on the housing end of the inlet hose (the one connected to the bottom of the housing assembly) and this should be facing up when in place.

12. If you removed the deck plate for access, install the 4 bolts and tighten the nuts to 13 ft. lbs. (18 Nm). Remember there should be a washer on both sides - one with the nut and one under the bolts head.

13. Move the craft to the water and make certain you check everything carefully for leaks.

14. Connect the battery and install the cargo container and seat(s).

1812cc Engines

EASY

◆ See Figures 31, 34 and 35

1. Remove the seat(s) and disconnect the negative battery cable. You may want to remove the rear cargo container as well.

■ **Although not strictly necessary, you may find that removing the deck plate (above the tank) will make access much easier.**

2. Tag and disconnect the water hose connected to the housing; move it out of the way in order to drain the joint. You may have to remove a plastic tie as well. You may also wish to remove the other 2 hoses that are in the way of the thermostat housing. All hoses should use conventional pinch clamps, so just squeeze the 2 tabs and slide the clamp back over the hose.

3. Loosen the 2 thermostat housing mounting bolts and carefully pull off the housing.

4. Remove the thermostat and gasket.

5. Perform the testing procedures if necessary

Fig. 34 The housing is attached to back of the engine, just under the large water hose - 1812cc engines. . .

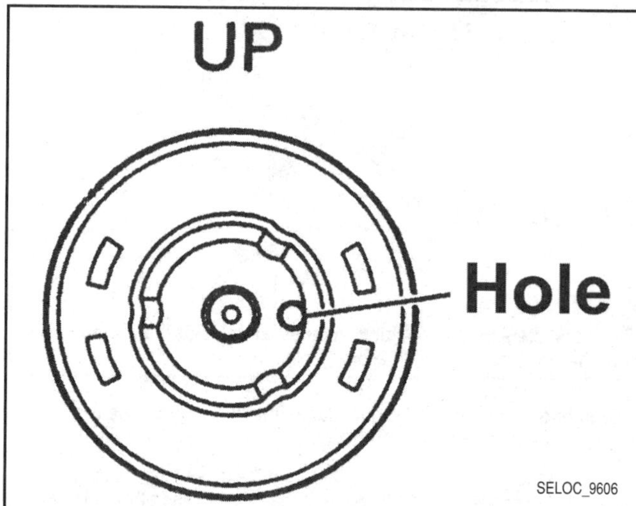

Fig. 35 Make sure the hole is positioned as shown when installed

To install:

6. Clean the housing and check all passages for any blockage.

7. Position the thermostat, spring first, into the engine.

■ **There is a small hole in the top of the thermostat - make sure it is positioned at roughly 3 o'clock when installed.**

8. Position a new gasket onto the cover and then position in place over the thermostat. Coat the bolt threads with Loctite 572 and then install them and tighten to 5.9 ft. lbs. (8 Nm).

9. Reconnect the all water hoses and slide their clamps back up into position.

10. If you removed the deck plate for access, install the 4 bolts and tighten the nuts to 13 ft. lbs. (18 Nm). Remember there should be a washer on both sides - one with the nut and one under the bolts head.

11. Move the craft to the water and make certain you check everything carefully for leaks.

12. Connect the battery and install the cargo container and seat(s).

TESTING

2002-07 Models

EASY

◆ See Figure 36

1. Inspect the thermostat at room temperature. If the thermostat is fully open, it is defective and must be replaced. Hold the thermostat up to the light and check it for leaks. A light leak around the perimeter indicates the thermostat is not closing, and therefore, it must be replaced.

2. Attach a length of thread to the thermostat. Now, suspend the thermostat and a thermometer inside a container filled with water (do not use distilled water or ethylene glycol!). Take care to be sure neither the thermostat or the thermometer touches the container. If either one does touch the container, the test will be unreliable. Stir the water occasionally to avoid direct heat being applied to the thermostat

3. Heat the water until the thermostat just begins to open - when this happens confirm that the temperature is the same as the thermostat rating (123°F/50°C). The thermometer reading must agree with the rating stamped on the thermostat. If the unit fails the test, it must be replaced.

4. Continue to heat the water until a temperature of 141°F/60°C is reached. At this time the thermostat should be completely open; if not, replace it.

5. Turn the heat off and allow the water to cool to a temperature 10° below the rating. The thermostat should now be completely closed; if not, replace it.

2008-11 Models

EASY

◆ See Figures 36 and 37

1. Inspect the thermostat at room temperature. If the thermostat is fully open, it is defective and must be replaced. Hold the thermostat up to the light and check it for leaks. A light leak around the perimeter indicates the thermostat is not closing, and therefore, it must be replaced.

2. Attach a length of thread to the thermostat. Now, suspend the thermostat and a thermometer inside a container filled with water (do not use distilled water or ethylene glycol!). Take care to be sure neither the thermostat or the thermometer touches the container. If either one touches the container, the test will be unreliable. Stir the water occasionally to avoid direct heat being applied to the thermostat

3. Heat the water to 118-126°F (48-52°C) on 998/1052cc engines, 136-144°F (58-62°C) on 1812cc HO and VX engines or, 118-126°F (48-52°C) on 1812cc SHO and FZ engines; until the thermostat just begins to open - when this happens, check that the valve opening is 0.002 in. (0.05mm). The thermometer reading must agree with the rating stamped on the thermostat. If the unit fails the test, it must be replaced.

4. Continue to heat the water until a temperature of 140°F (60°C) is reached on 998/1052cc engines and 1812cc SHO/FZ engines, or 158°F (70°C) for the 1812cc HO and VX engines. At this time the thermostat should be completely open and the valve clearance should be more that 0.17 in. (4.3mm); if not, replace it.

5. Turn the heat off and allow the water to cool to a temperature 10° below the initial rating. The thermostat should now be completely closed; if not, replace it.

Cylinder Compression

Cylinder compression test results are extremely valuable indicators of internal engine condition. The best marine mechanics automatically check an engine's compression as the first step in a comprehensive tune-up. A compression test will uncover many mechanical problems that can cause rough running or poor performance.

CHECKING COMPRESSION

OEM ② MODERATE

◆ **See Figures 38 and 39**

1. Make sure that the proper amount and viscosity of engine oil is in the crankcase, then ensure the battery is fully charged. Make sure that the valve clearance is properly adjusted as detailed in the Valve Lash Adjustment section.

2. Warm-up the engine to normal operating temperature, then shut the engine **OFF** and move the unit into a level position.

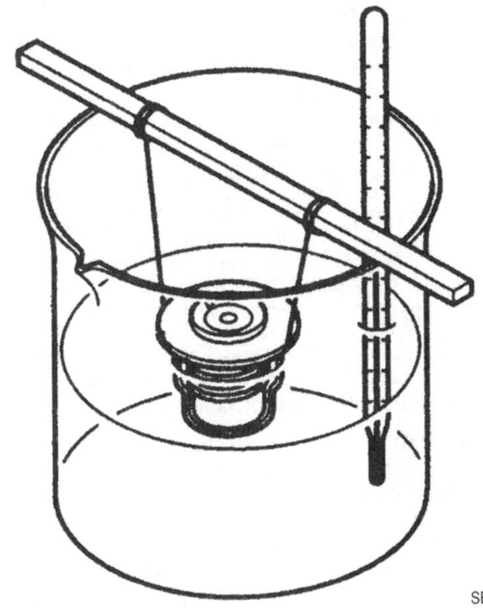

Fig. 36 Testing the thermostat

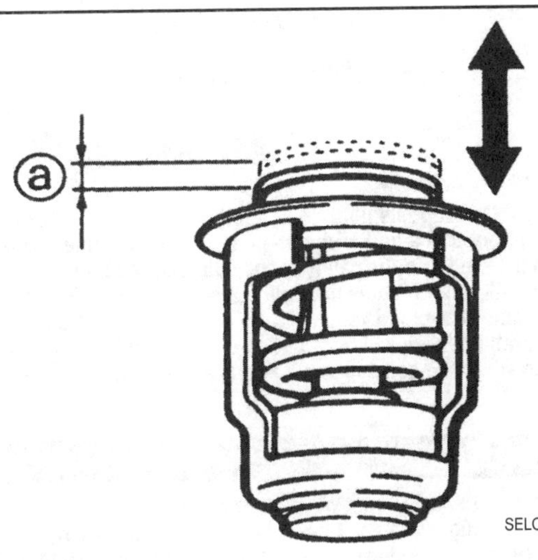

Fig. 37 Testing the thermostat

3. Remove the air filter case cover and then lift out the filter element on 2002-08 FX models with 998cc/1052cc engines. On all others, remove the engine cover.

4. Disconnect the spark plug caps/coils and then remove the spark plug(s). Make sure you use compressed air or something similar to remove any dirt or debris that may have collected in the recess.

5. Install a screw-in type compression gauge into the No. 1 cylinder spark plug hole until the fitting is snug. Please refer to the firing order illustrations for location of the No. 1 cylinder. You'll probably want to use an extension hose. When fitting the compression gauge adapter to the cylinder head, make sure the bleeder of the gauge (if equipped) is closed.

6. Make sure the throttle is open fully.

7. According to the tool manufacturer's instructions, connect a remote starting switch to the starting circuit.

8. With the ignition switch in the **OFF** position, use the remote starting switch to crank the engine through at least 5 compression strokes (approximately 5 seconds of cranking) and record the highest reading on the gauge.

9. Repeat the test on each cylinder, cranking the engine approximately the same number of compression strokes and/or time as the first.

10. Compare the highest readings from each cylinder to that of the others. The indicated compression pressures are considered within specifications if the lowest reading cylinder is within 14 psi of the pressure recorded for the highest reading cylinder. For example, if your highest reading cylinder pressure was 150 psi (1034 kPa), then the lowest reading cylinder should be no less than 136 psi. Yamaha suggests that no cylinder should be less than:

- 157 psi on 998cc engines
- 164 psi on 1052cc engines, exc. 2010-11 VX
- 115 psi on 2010-11 VX w/1052cc engines

Fig. 38 Remove all of the spark plug caps/coils (FX shown)

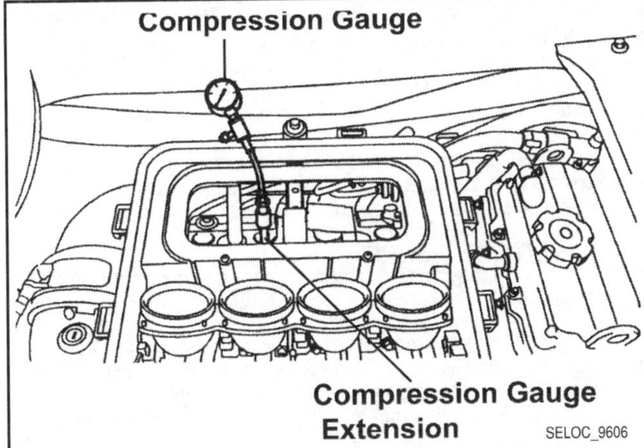

Fig. 39 You'll probably need an extension for your gauge (all models similar)

• 110.9 psi on 1812cc HO and VX units
• 86.8 psi on 1812cc SHO and FZ units

■ **Honestly, the minimum cylinder pressure on the 2010 VX w/1052cc engines and all1812cc models seems awfully low to us, but these are Yamaha's specifications. The key here is to use the 14 psi cylinder differential between low and high cylinders.**

11. Compression readings that are generally low indicate worn, broken, or sticking piston rings, scored pistons or worn cylinders.
12. If a cylinder exhibits an unusually low compression reading, squirt a tablespoon of clean engine oil into the cylinder through the injector hole and repeat the compression test. If the compression rises after adding oil, it means that the cylinder's piston rings and/or cylinder bore are damaged or worn. If the pressure remains low, the valves may not be seating properly (a valve job is needed), or the head gasket may be blown near that cylinder.
13. If compression in any two adjacent cylinders is low (with normal compression in the other cylinders), and if the addition of oil doesn't help raise compression, there is leakage past the head gasket. Oil and coolant in the combustion chamber, combined with blue or constant white smoke from the tailpipe, are symptoms of this problem. However, don't be alarmed by the normal white smoke emitted from the tailpipe during engine warm-up during cold weather. There may be evidence of water droplets on the engine oil dipstick and/or oil droplets in the cooling system if a head gasket is blown.
14. Remove the gauge and extension. Install the spark plugs and their plugs/coils.
15. Install the air filter, air filter case cover or the engine cover. Install the seat(s)

Spark Plugs

The spark plug performs four main functions:

• It fills a hole in the cylinder head.
• It acts as a dielectric insulator for the ignition system.
• It provides spark for the combustion process to occur.
• It removes heat from the combustion chamber.

It is important to remember that spark plugs do not create heat, they help remove it. Anything that prevents a spark plug from removing the proper amount of heat can lead to pre-ignition, detonation, premature spark plug failure and even internal engine damage.

In the simplest of terms, the spark plug acts as the thermometer of the engine. Much like a doctor examining a patient, this "thermometer" can be used to effectively diagnose the amount of heat present in each combustion chamber.

Spark plugs are valuable tuning tools, when interpreted correctly. They will show symptoms of other problems and can reveal a great deal about the engine's overall condition. By evaluating the appearance of the spark plug's firing tip, visual cues can be seen to accurately determine the engine's overall operating condition, get a feel for air/fuel ratios and even diagnose driveability problems.

As spark plugs grow older, they lose their sharp edges and material from the center and ground electrodes is slowly eroded away. As the gap between these two points grows, the voltage required to bridge this gap increases proportionately. The ignition system must work harder to compensate for this higher voltage requirement and hence there is a greater rate of misfires or incomplete combustion cycles. Each misfire means lost horsepower, reduced fuel economy and higher emissions. Replacing worn out spark plugs with new ones (with sharp new edges) effectively restores the ignition system's efficiency and reduces the percentage of misfires, restoring power, economy and reducing emissions.

How long spark plugs last will depend on a variety of factors, including engine compression, fuel used, gap, center/ground electrode material and the conditions in which the engine is operated.

SPARK PLUG HEAT RANGE

◆ **See Figure 40**

Spark plug heat range is the ability of the plug to dissipate heat from the combustion chamber. The longer the insulator (or the farther it extends into the engine), the hotter the plug will operate; the shorter the insulator (the closer the electrode is to the block's cooling passages) the cooler it will operate.

Selecting a spark plug with the proper heat range will ensure that the tip will maintain a temperature high enough to prevent fouling, yet be cool enough to prevent pre-ignition. A plug that absorbs little heat and remains too cool will quickly accumulate deposits of oil and carbon since it is not hot enough to burn them off. This leads to plug fouling and consequently to misfiring. A plug that absorbs too much heat will have no deposits but, due to the excessive heat, the electrodes will burn away quickly and might possibly lead to pre-ignition or other ignition problems.

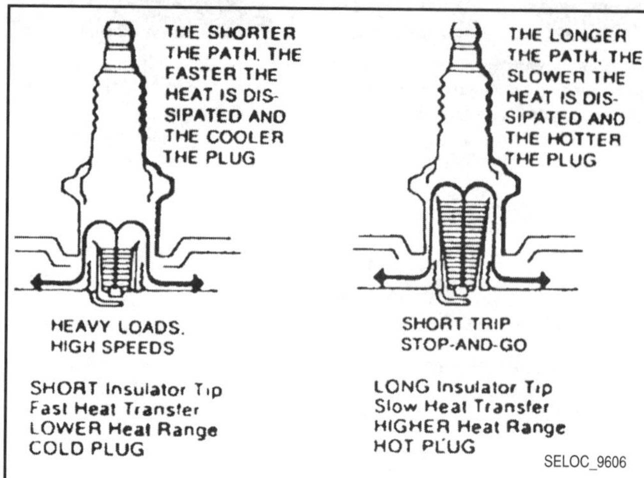

Fig. 40 Spark plug heat range

Pre-ignition takes place when plug tips get so hot that they glow sufficiently to ignite the air/fuel mixture before the actual spark occurs. This early ignition will usually cause a pinging during heavy loads and if not corrected will result in severe engine damage. While there are many other things that can cause pre-ignition, selecting the proper heat range spark plug will ensure that the spark plug itself is not a hot-spot source.

SPARK PLUG SERVICE

■ **New technologies in spark plug and ignition system design have pushed the recommended replacement interval higher and higher. However, this depends on usage and conditions.**

Spark plugs should only require replacement once a season; and at the very least need to be checked each season (refer to the Maintenance Intervals chart for your particular models). The electrode on a new spark plug has a sharp edge but with use, this edge becomes rounded by wear, causing the plug gap to increase. As the gap increases, the plug's voltage requirement also increases. It requires a greater voltage to jump the wider gap and about 2-3 times as much voltage to fire a plug at high speeds than at idle.

Tools needed for spark plug replacement include: a ratchet, short extension, spark plug socket (there are two types; either 13/16 in. or 5/8 in., depending upon the type of plug), a combination spark plug gauge and gapping tool and a can of anti-seize type compound.

REMOVAL & INSTALLATION

① EASY

◆ **See Figures 41 thru 46**

■ **2002-05 FX140 models utilize a 2-coil ignition system where the spark plug boot at the end of the spark plug wire is actually more of a plug, and pressed into the cavity in the cylinder head and over the spark plug. All other models (including a few 2005 FX140s) utilize individual coil-over-plug systems where there is no conventional spark plug wire; rather, there is a boot underneath the coil that fits over the spark plug when the coil assembly is pressed into the spark plug cavity.**

✻✻ CAUTION

When removing spark plugs, work on one at a time. Don't start by removing the plug wires all at once, because unless you number them, they may become mixed up. Take a minute before you begin and number the wires/coils with tape.

1. Disconnect the negative battery cable.

2. Remove the air filter case cover and then lift out the filter element on 2002-08 FX models with 998cc/1052cc engines. On all others, remove the engine cover.

3. If the engine has been run recently, allow the engine to thoroughly cool. Attempting to remove plugs from a hot cylinder head could cause the plugs to seize and damage the threads in the cylinder head, especially on aluminum heads!

4. On 2002-05 FX140/Cruiser models, carefully twist the spark plug wire boot plug to loosen it, then pull up on the plug using a twisting motion to remove it from the spark plug and out of the cavity. Be sure to pull on the boot/plug head and not on the wire.

5. On all other models, loosen the coil mounting bolt (2 bolts on 1812cc models) and carefully twist the coil pack/spark plug boot out of the spark plug recess. It is not necessary to disconnect the coil connector; but if you do, make certain to tag the wire bundle to the respective coil pack.

6. Using compressed air (and safety glasses), blow debris from the spark plug area to assure that no harmful contaminants are allowed to enter the combustion chamber when the spark plug is removed. If compressed air is not available, use a rag or a brush to clean the area. Compressed air is available from both an air compressor or from compressed air in cans available at photography stores.

■ **Remove the spark plugs when the engine is cold, if possible, to prevent damage to the threads. If plug removal is difficult, apply a few drops of penetrating oil to the area around the base of the plug and allow it a few minutes to work.**

7. Using a spark plug socket that is equipped with a rubber insert to properly hold the plug, turn the spark plug counterclockwise to loosen and remove the spark plug from the bore.

✳✳ WARNING

Avoid the use of a flexible extension on the socket. Use of a flexible extension may allow a shear force to be applied to the plug. A shear force could break the plug off in the cylinder head, leading to costly and frustrating repairs. In addition, be sure to support the ratchet with your other hand - this will also help prevent the socket from damaging the plug.

8. Evaluate each cylinder's performance by comparing the spark plug condition. Check each spark plug to be sure they are all made by the same manufacturer and have the same heat range rating. Inspect the threads in the spark plug opening of the block and clean the threads before installing the plug.

9. When purchasing new spark plugs, always ask the dealer if there has been a spark plug change for the engine being serviced. Many times manufacturers will update the type of spark plug used in an engine to offer better efficiency or performance.

10. Crank the engine through several revolutions to blow out any material that might have become dislodged during cleaning. Always use a new gasket (if applicable), but never use gaskets on taper seat plugs. The gasket must be fully compressed on clean seats to complete the heat transfer process and to provide a gas tight seal in the cylinder.

11. Inspect the spark plug boot plug for tears or damage. If a damaged one is found, the spark plug boot/plug/coils assy. will need replacement.

12. Check the spark plug gap, as detailed in the Tune-Up & Capacity specifications, prior to installing the plug. Most spark plugs do not come gapped to the proper specification.

13. Apply a thin coating of anti-seize on the thread of the plug. This is extremely important on aluminum head engines.

14. Carefully thread the plug into the bore by hand. If resistance is felt before the plug completely bottomed, back the plug out and begin threading again. In many applications here, you may need to use an extension with the socket on the end of it - be very careful not to strip the threads!

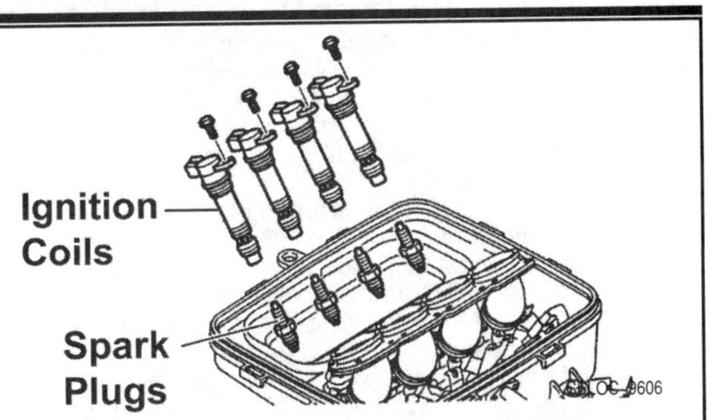

Fig. 41 A good look at a coil-over-plug set up on the 1052cc engine. . .

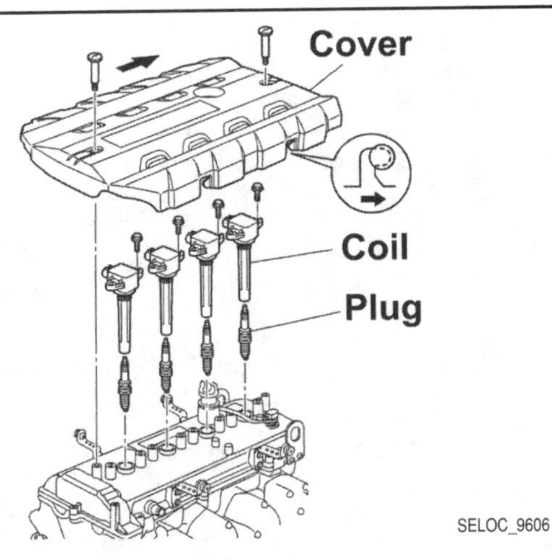

Fig. 42 . . .and on the 1812cc engines

Fig. 43 Coil-over-plug models are secured by 1 bolt. . .

Fig. 44 . . .or 2 bolts

15. Carefully tighten the spark plug. If the plug you are installing is equipped with a crush washer, tighten the plug until the washer seats, then turn it 1/4 turn to crush the washer. Whenever possible, spark plugs should be tightened to the factory torque specification:

- 998cc and 1052cc engines: 9.6 ft. lbs. (13 Nm)
- 1812cc engines: 18.4 ft. lbs. (25 Nm)

But, if you do not have a torque wrench, thread the plug in until it is finger-tight, and then turn it another 1/4 to 1/2 turn.

16. Apply a small amount of silicone dielectric grease to the end of the spark plug lead or inside the spark plug boot to prevent sticking, then press the boot/plug into the spark plug cavity and push until it clicks into place. The click may be felt or heard.

17. Install the mounting bolt(s) on models with coil-over-plug set-ups, tightening the ones with a single mounting bolt to 5.5 ft. lbs. (7.6 Nm) on all models but the 2010 VX where the figure is 5.9 ft. lbs. (8 Nm). Tighten the bolts on the 1812cc engines to 5.9 ft. lbs. (8 Nm) as well.

18. Replace the air filter element if removed and the install the filter case cover or engine cover, tightening to 1.8 ft. lbs. (2.5 Nm). On the 1812cc engine cover, tighten to 3.7 ft. lbs. (5 Nm).

19. Connect the negative battery cable, install the seat and then start the engine to insure proper operation.

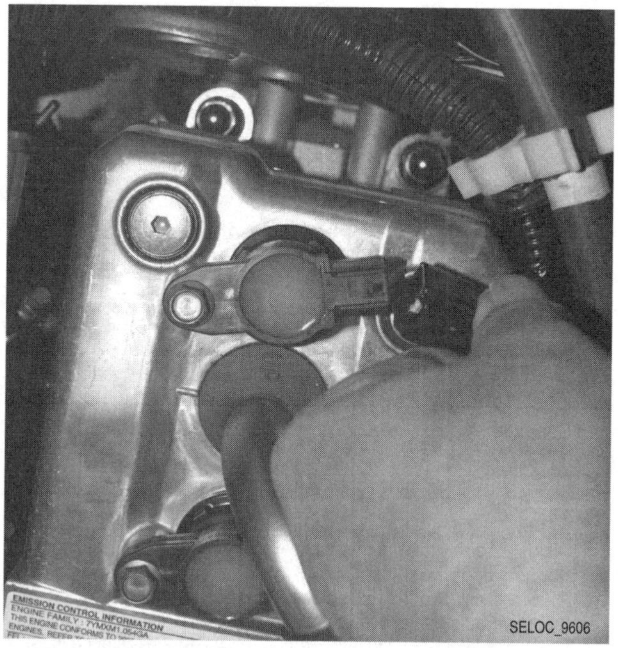

Fig. 45 It probably won't hurt to disconnect the electrical lead at the coil

Fig. 46 Although most coil leads should be marked to match the cylinder, it wouldn't hurt to do it yourself also

READING SPARK PLUGS

◆ **See Figures 47 thru 53**

Reading spark plugs can be a valuable tuning aid. By examining the insulator firing nose color, you can determine much about the engine's overall operating condition.

In general, a light tan/gray color tells you that the spark plug is at the optimum temperature and that the engine is in good operating condition.

Dark coloring, such as heavy black wet or dry deposits usually indicate a fouling problem. Heavy, dry deposits can indicate an overly rich condition, too cold a heat range spark plug, possible vacuum leak, low compression, overly retarded timing or too large a plug gap.

If the deposits are wet, it can be an indication of a breached head gasket, oil control from ring problems or an extremely rich condition, depending on what liquid is present at the firing tip.

Look for signs of detonation, such as silver specs, black specs or melting or breakage at the firing tip.

Compare your plugs to the illustrations shown to identify the most common plug conditions.

Fouled Spark Plugs

A spark plug is fouled when the insulator nose at the firing tip becomes coated with a foreign substance, such as fuel, oil or carbon. This coating makes it easier for the voltage to follow along the insulator nose and leach back down into the metal shell, grounding out, rather than bridging the gap normally.

Fuel, oil and carbon fouling can all be caused by different things but in any case, once a spark plug is fouled, it will not provide voltage to the firing tip and that cylinder will not fire properly. In many cases, the spark plug cannot be cleaned sufficiently to restore normal operation. It is therefore recommended that fouled plugs be replaced.

Signs of fouling or excessive heat must be traced quickly to prevent further deterioration of performance and to prevent possible engine damage.

Overheated Spark Plugs

When a spark plug tip shows signs of melting or is broken, it usually means that excessive heat and/or detonation was present in that particular combustion chamber or that the spark plug was suffering from thermal shock.

Since spark plugs do not create heat by themselves, one must use this visual clue to track down the root cause of the problem. In any case, damaged firing tips most often indicate that cylinder pressures or temperatures were too high. Left unresolved, this condition usually results in more serious engine damage.

Detonation refers to a type of abnormal combustion that is usually preceded by pre-ignition. It is most often caused by a hot spot formed in the combustion chamber.

As air and fuel is drawn into the combustion chamber during the intake stroke, this hot spot will "pre-ignite" the air fuel mixture without any spark from the spark plugs.

Detonation

Detonation exerts a great deal of downward force on the pistons as they are being forced upward by the mechanical action of the connecting rods. When this occurs, the resulting concussion, shock waves and heat can be severe. Spark plug tips can be broken or melted and other internal engine components such as the pistons or connecting rods themselves can be damaged.

Left unresolved, engine damage is almost certain to occur, with the spark plug usually suffering the first signs of damage.

■ **When signs of detonation or pre-ignition are observed, they are symptom of another problem. You must determine and correct the situation that caused the hot spot to form in the first place.**

INSPECTION & GAPPING

◆ **See Figures 54 and 55**

A particular spark plug might fit hundreds of engines and although the factory will typically set the gap to a pre-selected setting, this gap may not be the right one for your particular engine.

Insufficient spark plug gap can cause pre-ignition, detonation, even engine damage. Too much gap can result in a higher rate of misfires, noticeable loss of power, plug fouling and poor economy.

Check the spark plug gap before installation. The ground electrode (the L-shaped one connected to the body of the plug) must be parallel to the center electrode and the specified size wire gauge must pass between the electrodes with a slight drag.

Do not use a flat feeler gauge when measuring the gap on a used plug, because the reading may be inaccurate. A round wire-type gapping tool is the best way to check the gap. The correct gauge should pass through the electrode gap with a slight drag. If you're in doubt, try a wire that is one size smaller or larger. The smaller gauge should go through easily, while the larger one shouldn't go through at all.

Wire gapping tools usually have a bending tool attached. Use this tool to adjust the side electrode until the proper distance is obtained. Never attempt to bend the center electrode. Also, be careful not to bend the side electrode too far or too often as it may weaken and break off within the engine, requiring removal of the cylinder head to retrieve it.

Although spark plug gap specs are given in the Tune-Up chart for your specific model, for the most part, all 998cc/1052cc engines will be 0.028-0.031 in. (0.7-0.8mm) and all 1812cc engines will be 0.31-0.35 in. (0.8-0.9mm). But check the chart anyway; and more importantly, check the Emissions sticker on your engine to make sure there were no changes made during production.

Ignition Timing

Ignition timing is controlled by the ECM and not adjustable.

Idle Speed & Mixture

Idle speed is constantly monitored by the electronic control module (ECM) and/or an air bypass valve. Idle speed and mixture are not adjustable. Please refer to the Fuel System section for further information on the fuel injection system.

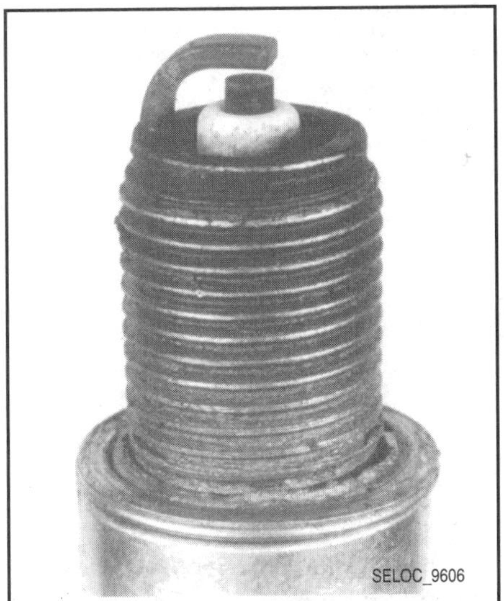

Fig. 47 A normally worn spark plug should have light tan or gray deposits on the firing tip (electrode)

Fig. 48 A carbon-fouled plug, identified by soft, sooty black deposits, may indicate an improperly tuned engine

Fig. 49 A physically damaged spark plug may be evidence of severe detonation in that cylinder. Watch the cylinder carefully between services, as a continued detonation will not only damage the plug but will most likely damage the engine

Fig. 50 An oil-fouled spark plug indicates an engine with worn piston rings

Fig. 51 This spark plug has been left in the engine too long, as evidenced by the extreme gap. Plugs with such an extreme gap can cause misfiring and stumbling accompanied by a noticeable lack of power

Fig. 52 A bridged or almost bridged spark plug, identified by the build-up between the electrodes caused by excessive carbon or oil build-up on the plug

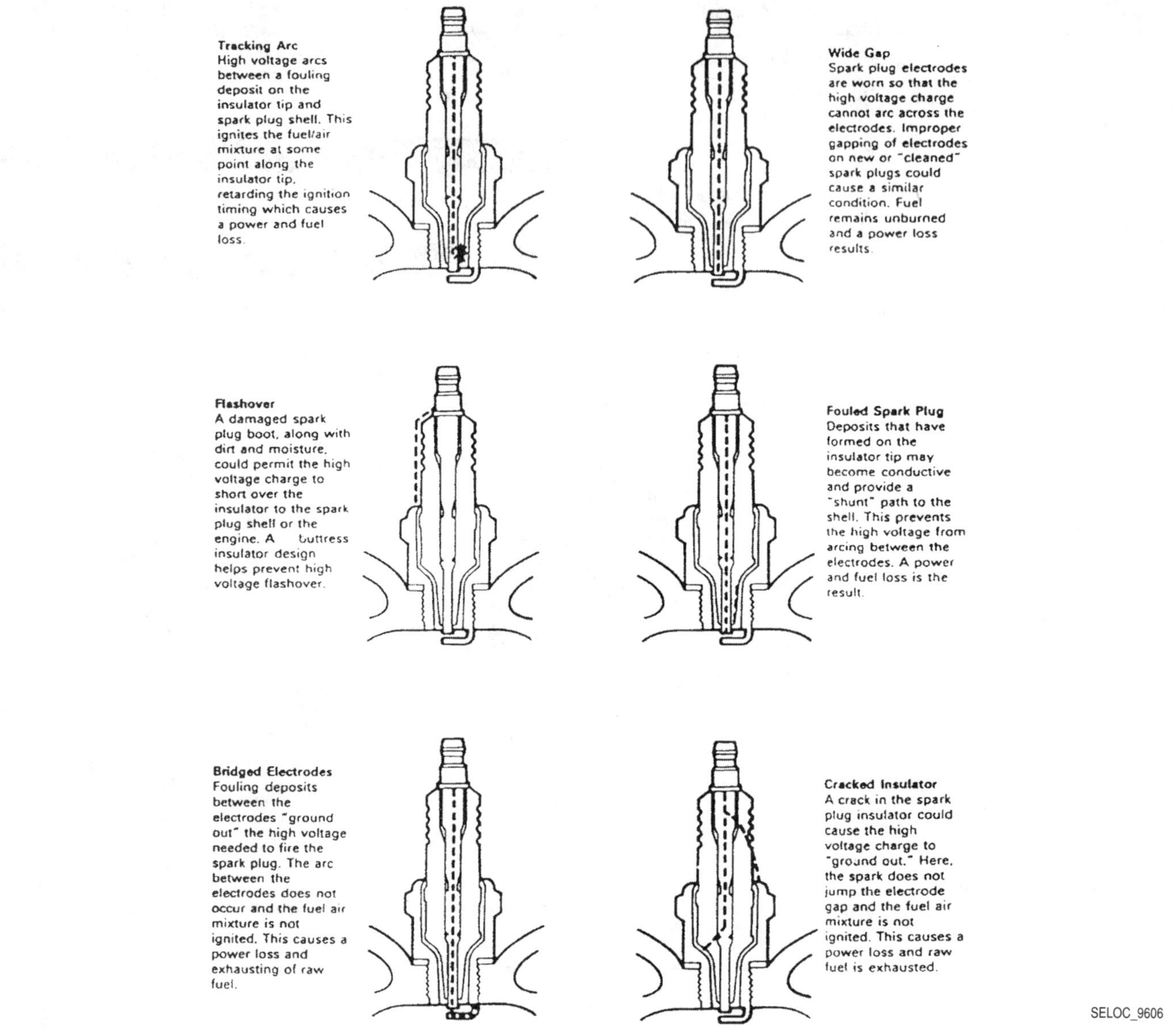

Fig. 53 Typical spark plug problems showing damage that may indicate engine problems

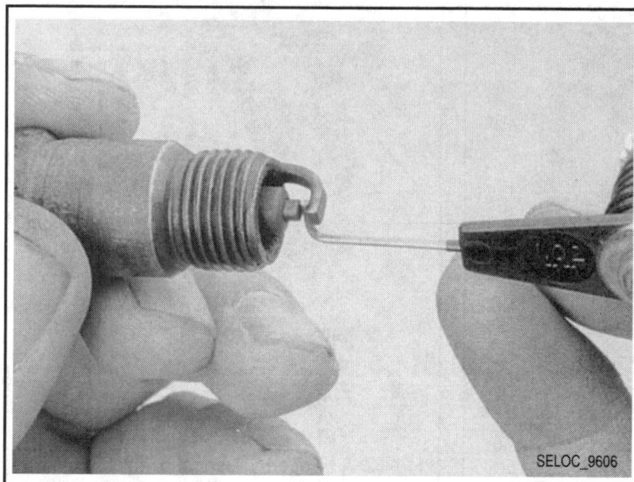

Fig. 54 Using a wire-type spark plug gapping tool to check the distance between center and ground electrodes

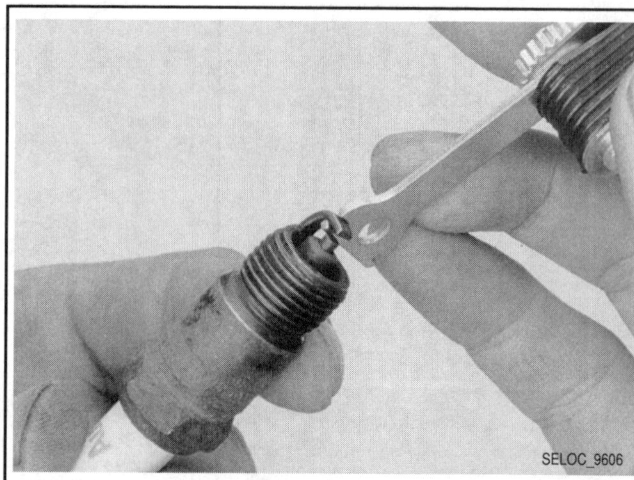

Fig. 55 Most spark plug gapping tools have an adjusting tool used to bend the ground electrode. USE IT! This tool greatly reduces the chance of breaking off the electrode and is much more accurate

Trolling Speed

ADJUSTMENT

◆ See Figure 56

1. With the craft in the water, secure all 4 corners and remove the seat.

2. Connect a shop tachometer as detailed in the manufacturer's instructions. If you do not have access to a tach, you can use the one on the dash, but we think it best to use a real tachometer.

3. Start the engine and allow it to idle in Neutral until the engine has reached normal operating temperature.

✱✱ CAUTION

Be sure you have access to the emergency lanyard in the following steps when the engine is running and the machine is in gear.

4. Now, with the machine in Forward gear, observe the tachometer. Confirm that the engine speed is:

- FX/FX Cruiser models w/998cc engines - 1600-1800 rpm
- 2004-07 FX/FX Cruiser models w/1052c engines: 1550-1750 rpm
- 2008 FX/FX Cruiser models w/1052cc engines: 1600-1700 rpm
- 2010 FX/FX Cruiser models w/1052cc engines: 1600-1700 rpm
- 2009-11 FX/FX Cruiser HO models w/1812cc engines: 1200-1400 rpm
- 2008-11 FX/FX Cruiser SHO models w/1812cc engines: 1150-1300 rpm
- 2009-11 FZ models: 1150-1350 rpm
- 2005-09 VX models w/1052c engines: 1550-1750 rpm
- 2010-11 VX models w/1052c engines: 1600-1700 rpm
- 2011 VX models w/1812cc engines: 1200-1400 rpm

If so, you're done. Shut the engine off, disconnect the tach and replace the seat. If the speed is outside the specification, move to the next step.

5. Turn off the engine and remove the air filter case cover or engine cover.

6. On FX/FX Cruiser models with 998/1952cc engines, start the engine again and engage Forward again. Find the star shaped throttle stop screw at the center of the throttle bodies, on the air filter side. Slowly turn the screw in (clockwise) to increase the engine speed, or out (counterclockwise) to decrease the speed so that it is within the correct range. Turn off the engine.

7. On VX models w/1052cc engines, and all models with 1812cc engines, if out of specification, you must check the Throttle cable adjustment, APS and throttle body as detailed in the Fuel System section.

8. Disconnect the tachometer and install the air filter case cover or engine cover.

9. Install the seat.

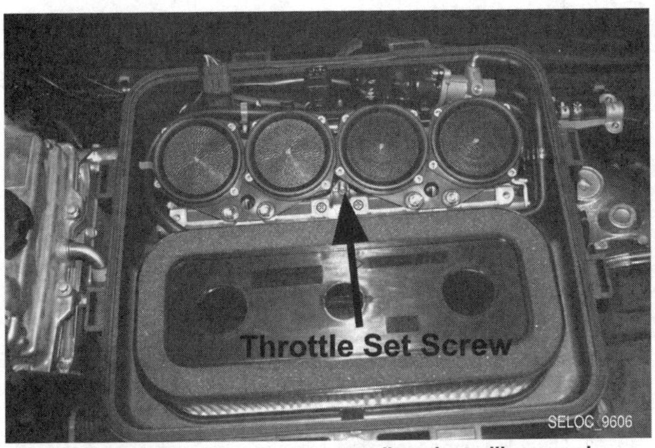

Fig. 56 Use the throttle stop screw to adjust the trolling speed (1052cc FX shown)

Valve Clearance

ADJUSTMENT

All 2002-08 FX & 2005-09 VX Models With 998cc/1052cc Engines

◆ See Figures 57 thru 64

■ **Cylinders are numbered front to back, 1 thru 4. The exhaust camshaft is on the right, starboard side of the engine and the intake is on the left, port side. You will see that there are 3 cam lobes/valves per cylinder on the intake side and 2 per cylinder on the exhaust side.**

Valve clearance adjustment should be performed every 200 hours or 2 years, whichever comes first. It will also be required anytime service work has been performed on the valve train.

The work should always be performed on a cold engine and the No. 1 piston should be at top dead center (TDC) of the compression stroke.

Valve clearance should be checked on each cylinder in the firing order: 1-2-4-3.

1. Remove the seat(s) and rear storage container. Disconnect the negative battery cable.

2. Remove the air filter case cover and then lift out the filter element on 2002-08 FX models with 998cc/1052cc engines. On all others, remove the engine cover. Both procedures are detailed in the Air Filter section.

3. Remove the spark plugs as detailed in this section.

4. Remove the cover on the drive coupler if equipped.

5. Remove the cylinder head cover (and gasket) as detailed in the Engine Mechanical section.

6. Position the No. 1 piston so it is at TDC of the compression stroke for the No. 1 cylinder (first cylinder at the front of the engine). To do this you could use a remote starter that has been hooked up according to the manufacturer's instructions, or you can simply rotate the engine coupler; either way, the object is to rotate the crankshaft so that the piston in the No.1 cylinder gets to TDC. We think it's easier, and certainly cheaper, to use the coupler.

With your finger over the No.1 cylinder's spark plug hole, have a friend *slowly* rotate the coupler counterclockwise until you detect pressure in the cylinder, indicating TDC on the compression stroke. You can also check the camshaft lobe positions - they will be pointing away from each other.

7. Thread a dial gauge stand (#90890-06583) into the spark plug hole on the No. 1 cylinder and then install the gauge into the stand. You may also

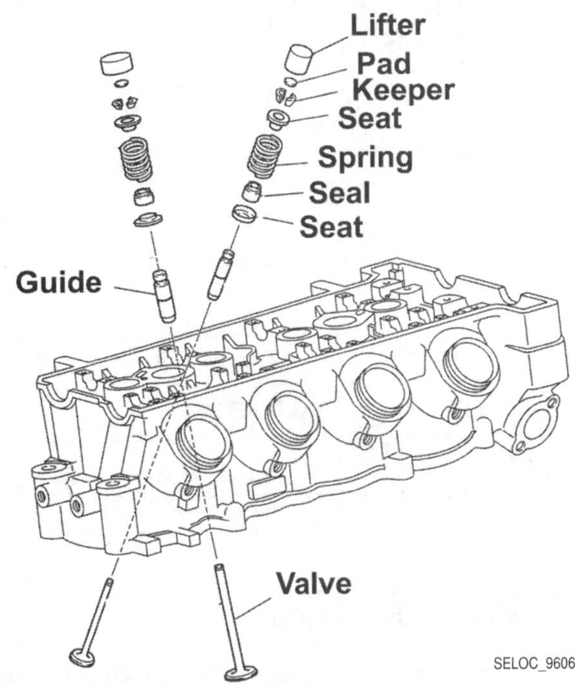

Fig. 57 Exploded view of the valve train

need a gauge needle (#90890-06584, or you can get the set (#90890-06585). Use the gauge to confirm your TDC positioning.

a. Keeping your eye on the dial gauge indicator, have your friend rotate, *very slowly*, the coupler counterclockwise again until you see the indicator just begin to move in the opposite direction.

b. Now rotate again in the opposite direction while watching the needle and stop as it just changes direction again.

c. Do this a few times until you are confident that you have pinpointed that point where the needle changes direction at the first turn of the coupler in either direction.

d. Make sure that the intake and exhaust cam lobes on the No. 1 cylinder are again pointing away from each other. If not, go back and start over again until you have achieved both confirmations.

e. Now you can be confident you have found TDC. Use a magic marker and place a small mark on the top of the coupler.

Fig. 58 You'll have to remove the mounting bolt and coupling cover. . .

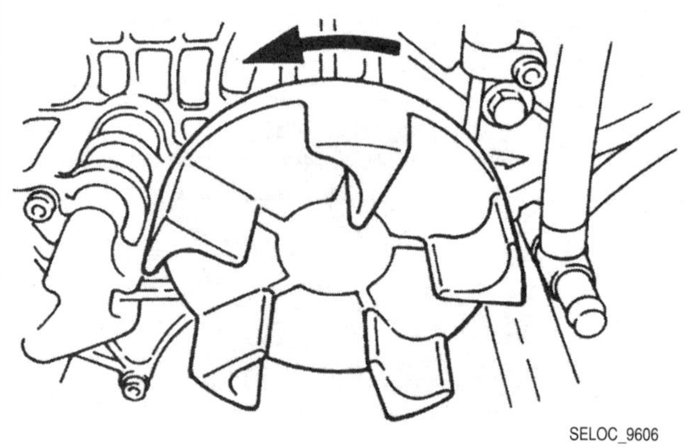

Fig. 59 . . .to gain access to the drive coupler

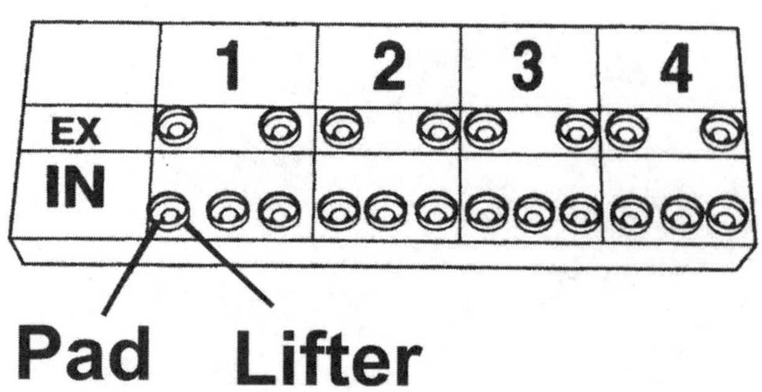

Fig. 61 We highly recommend taking the time to build an organizer for the lifters and pads

8. Insert a feeler gauge between the camshaft lobe and the top of the valve lifter until a very slight drag can be felt. Mark this measurement down to refer to later. Repeat this procedure on the other 4 valves for the No. 1 cylinder - There are 3 intake valves and lifters, and 2 exhaust valves/lifters. Remember?

■ **We suggest that you draw a picture of the valve train and then mark the measured clearance for each valve on its associated position in your picture.**

■ **We think the next 8 steps make checking the valves on each cylinder easier, but please remember that all they are doing is ensuring that each cylinder being checked is at TDC of the compression stroke. Obviously, you can confirm the same thing by performing the previous 3 steps for each cylinder you are checking. Your call, but we think the following is the way to go.**

9. Remove the dial gauge and hardware,

10. Now, **carefully and slowly**, rotate the coupler *counterclockwise* 180° so that the piston in the No. 2 cylinder is now at TDC on the compression stroke; this would mean that you have turned the coupler halfway around the complete circle from where you started. Check to make sure that the cam lobes on the No. 2 cylinder are facing away from one another as you did with the No. 1 cylinder, If not, repeat until they are.

11. Measure the clearance between each camshaft lobe and lifter as you did in Step 7, making sure that you mark each measurement down on the diagram that you drew.

12. Next, rotate the coupler counterclockwise an additional 180° (yep, bringing your little mark on the coupler back to its original position, 360° from the start, a complete circle) until the piston for No. 4 cylinder (*not* No. 3) is at TDC on its compression stroke. Make sure that you do this carefully and slowly and that the intake and exhaust lobes for this cylinder are facing away from each other.

13. Measure the clearance between each camshaft lobe and lifter as you did in Step 7, making sure that you mark each measurement down on the diagram that you drew.

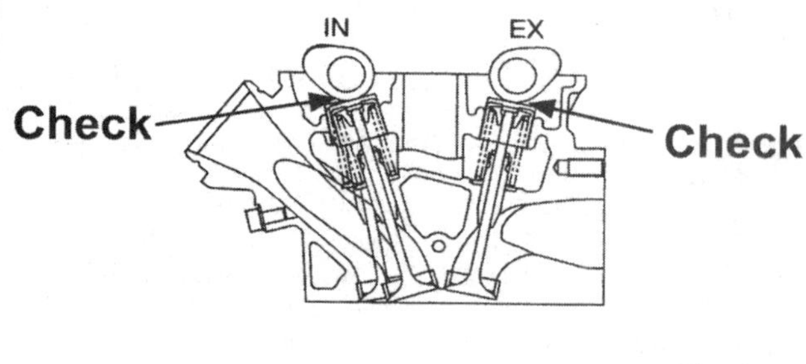

Fig. 60 Make sure that the camshaft lobes are facing away from each other before checking the clearance between the lifter and the cam lobe

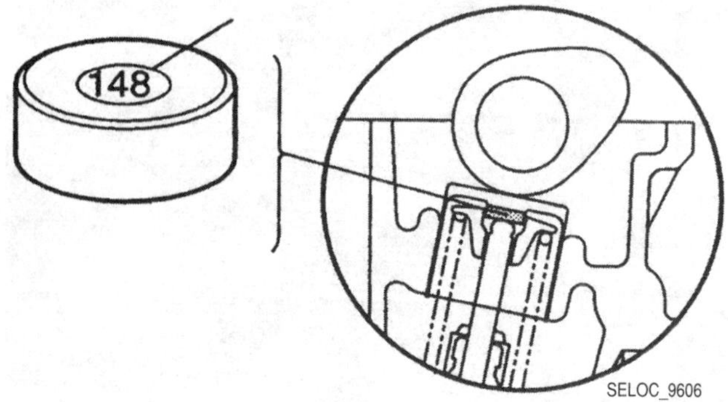

Fig. 62 All pads will be identified with a number corresponding to their thickness

Intake

MEASURED	ORIGINAL VALVE PAD NUMBER																							
CLEARANCE	120	125	130	135	140	145	150	155	160	165	170	175	180	185	190	195	200	205	210	215	220	225	230	235	240
0.00–0.02				120	125	130	135	140	145	150	155	160	165	170	175	180	185	190	195	200	205	210	215	220	225
0.03–0.07			120	125	130	135	140	145	150	155	160	165	170	175	180	185	190	195	200	205	210	215	220	225	230
0.08–0.10		120	125	130	135	140	145	150	155	160	165	170	175	180	185	190	195	200	205	210	215	220	225	230	235
0.11–0.20	STANDARD CLEARANCE																								
0.21–0.22	125	130	135	140	145	150	155	160	165	170	175	180	185	190	195	200	205	210	215	220	225	230	235	240	
0.23–0.27	130	135	140	145	150	155	160	165	170	175	180	185	190	195	200	205	210	215	220	225	230	235	240		
0.28–0.32	135	140	145	150	155	160	165	170	175	180	185	190	195	200	205	210	215	220	225	230	235	240			
0.33–0.37	140	145	150	155	160	165	170	175	180	185	190	195	200	205	210	215	220	225	230	235	240				
0.38–0.42	145	150	155	160	165	170	175	180	185	190	195	200	205	210	215	220	225	230	235	240					
0.43–0.47	150	155	160	165	170	175	180	185	190	195	200	205	210	215	220	225	230	235	240						
0.48–0.52	155	160	165	170	175	180	185	190	195	200	205	210	215	220	225	230	235	240							
0.53–0.57	160	165	170	175	180	185	190	195	200	205	210	215	220	225	230	235	240								
0.58–0.62	165	170	175	180	185	190	195	200	205	210	215	220	225	230	235	240									
0.63–0.67	170	175	180	185	190	195	200	205	210	215	220	225	230	235	240										
0.68–0.72	175	180	185	190	195	200	205	210	215	220	225	230	235	240											
0.73–0.77	180	185	190	195	200	205	210	215	220	225	230	235	240												
0.78–0.82	185	190	195	200	205	210	215	220	225	230	235	240													
0.83–0.87	190	195	200	205	210	215	220	225	230	235	240														
0.88–0.92	195	200	205	210	215	220	225	230	235	240															
0.93–0.97	200	205	210	215	220	225	230	235	240																
0.98–1.02	205	210	215	220	225	230	235	240																	
1.03–1.07	210	215	220	225	230	235	240																		
1.08–1.12	215	220	225	230	235	240																			
1.13–1.17	220	225	230	235	240																				
1.18–1.22	225	230	235	240																					
1.23–1.27	230	235	240																						
1.28–1.32	235	240																							
1.33–1.37	240																								

Example:
Intake valve clearance (cold)
0.11–0.20 mm (0.0043–0.0079 in)
Rounded value 150
Measured valve clearance is 0.24 mm (0.0094 in)
Replace pad 150 with pad 160
Pad No. 150 = 1.50 mm (0.0591 in)
Pad No. 160 = 1.60 mm (0.0630 in)

SELOC_9606

Fig. 63 Intake valve pad selection chart - 998/1052cc engines

Exhaust

MEASURED	ORIGINAL VALVE PAD NUMBER																							
CLEARANCE	120	125	130	135	140	145	150	155	160	165	170	175	180	185	190	195	200	205	210	215	220	225	230	235	240
0.00–0.01							120	125	130	135	140	145	150	155	160	165	170	175	180	185	190	195	200	205	210
0.02–0.06						120	125	130	135	140	145	150	155	160	165	170	175	180	185	190	195	200	205	210	215
0.07–0.11					120	125	130	135	140	145	150	155	160	165	170	175	180	185	190	195	200	205	210	215	220
0.12–0.16				120	125	130	135	140	145	150	155	160	165	170	175	180	185	190	195	200	205	210	215	220	225
0.17–0.21			120	125	130	135	140	145	150	155	160	165	170	175	180	185	190	195	200	205	210	215	220	225	230
0.22–0.24		120	125	130	135	140	145	150	155	160	165	170	175	180	185	190	195	200	205	210	215	220	225	230	235
0.25–0.34	STANDARD CLEARANCE																								
0.35–0.37	125	130	135	140	145	150	155	160	165	170	175	180	185	190	195	200	205	210	215	220	225	230	235	240	
0.38–0.42	130	135	140	145	150	155	160	165	170	175	180	185	190	195	200	205	210	215	220	225	230	235	240		
0.43–0.47	135	140	145	150	155	160	165	170	175	180	185	190	195	200	205	210	215	220	225	230	235	240			
0.48–0.52	140	145	150	155	160	165	170	175	180	185	190	195	200	205	210	215	220	225	230	235	240				
0.53–0.57	145	150	155	160	165	170	175	180	185	190	195	200	205	210	215	220	225	230	235	240					
0.58–0.62	150	155	160	165	170	175	180	185	190	195	200	205	210	215	220	225	230	235	240						
0.63–0.67	155	160	165	170	175	180	185	190	195	200	205	210	215	220	225	230	235	240							
0.68–0.72	160	165	170	175	180	185	190	195	200	205	210	215	220	225	230	235	240								
0.73–0.77	165	170	175	180	185	190	195	200	205	210	215	220	225	230	235	240									
0.78–0.82	170	175	180	185	190	195	200	205	210	215	220	225	230	235	240										
0.83–0.87	175	180	185	190	195	200	205	210	215	220	225	230	235	240											
0.88–0.92	180	185	190	195	200	205	210	215	220	225	230	235	240												
0.93–0.97	185	190	195	200	205	210	215	220	225	230	235	240													
0.98–1.02	190	195	200	205	210	215	220	225	230	235	240														
1.03–1.07	195	200	205	210	215	220	225	230	235	240															
1.08–1.12	200	205	210	215	220	225	230	235	240																
1.13–1.17	205	210	215	220	225	230	235	240																	
1.18–1.22	210	215	220	225	230	235	240																		
1.23–1.27	215	220	225	230	235	240																			
1.28–1.32	220	225	230	235	240																				
1.33–1.37	225	230	235	240																					
1.38–1.42	230	235	240																						
1.43–1.47	235	240																							
1.48–1.52	240																								

Example:
Exhaust valve clearance (cold)
0.25–0.34 mm (0.0098–0.0134 in)
Rounded value 170
Measured valve clearance is 0.44 mm (0.0173 in)
Replace pad 170 with pad 185
Pad No. 170 = 1.70 mm (0.0669 in)
Pad No. 185 = 1.85 mm (0.0728 in)

SELOC_9606

Fig. 64 Exhaust valve pad selection chart - 998/1052cc engines

14. Finally, rotate the coupler counterclockwise an additional 180° (now your mark will be in the same position as it was when working on the No. 2 cylinder, out of sight, 540° from where you originally started) until the piston for No. 3 cylinder is at TDC on its compression stroke. Make sure that you do this carefully and slowly and that the intake and exhaust lobes for this cylinder are facing away from each other.

15. Measure the clearance between each camshaft lobe and lifter as you did in Step 7, making sure that you mark each measurement down on the diagram that you drew.

16. Rotate the coupler counterclockwise 180° one more time, you're now back to TDC on the No. 1 cylinder.

17. Take a break. Sit down and take a look at all of your clearance measurements. Acceptable valve clearance on these engines is 0.0043-0.0079 in. (0.11-0.20mm) for all intake valves and 0.0098-0.0134 in. (0.25-0.34mm) on the exhaust side. Hopefully, all of your measurements have fallen into these ranges and you can skip to the end here and begin putting everything back together.

However, if any of your measurements were out of spec, you'll need to move to the next step.

18. Remove the front cover and then the camshafts as detailed in the Engine Mechanical section.

19. Run through the clearances on your drawing and mark any that are out of specification. Remove the valve lifter and valve pad (details in the Engine Mechanical section) from each of the valves that are out of spec; be sure to keep them together and identified as shown in the illustration.

■ **It's a good idea to cover the timing chain opening with a clean lint-free rag to prevent a lifter or pad from accidentally falling down into the crankcase.**

20. Original valve pads are numbered and range in thickness from 1.20-2.40mm with 25 available replacement pads in thicknesses of 0.05mm increments. When removing the pads from the engine, you will see that they are marked with a number corresponding to their thickness, *without* the decimal point or 'mm' designation - a pad marked **168** would have a measurement of 1.68mm, one marked **142** would be 1.42mm.

a. Because pads of various sizes were originally installed, you will need to round that original number to arrive at an available replacement size:

On all 2002-07 FX engines and 2005-09 VX engines, if the last digit in the pad's number is **0** or **2**, it will be rounded down - original number **132** will become **130**. If the original pad number ends with a **5**, it will stay the same - **145** stays as **145**. Finally, if the original number ended in an **8**, it will be rounded up - original number **168** will become **170**.

On all 2008 FX engines, if the last digit in the pad's number is **0**, **1** or **2**, it will be rounded down - original number **132** will become **130**. If the original pad number ends with a **4**, **5** or **6**, it will stay the same - **145** stays as **145**. Finally, if the original number ended in an **8** or **9**, it will be rounded up - original number **168** will become **170**.

a. Why are we telling you this? Because rather than having to perform more calculations, you're going to use the rounded number of the original pad, and the clearance measurement you noted previously of the same cylinder and plug them into the accompanying charts to arrive at the new replacement pad thickness. If you have a valve on the No. 2 cylinder that was out of spec, you'll go to the Intake or Exhaust chart and find your measurement in the left hand column (say it was 24mm so you'd go to the 0.23-0.27 range line).

b. Now you'll check the rounded valve pad number that you removed from this valve (say **150**), find it in the column across the top of the chart and then go to where the 2 positions intersect. In this case, you'll need a **160** replacement pad to bring your bad valve into spec. Make sense?

21. Once you have completed the previous step on each of the valves that measured out of specification, you can re-install the lifter and new pad. Making sure that everything goes back into the correct bore.

a. Coat the valve pad with molybdenum disulfide grease before positioning it in the bore.

b. Coat the valve lifter with molybdenum disulphide oil before positioning it into the bore. Make sure that rotates smoothly in the bore once it's in position.

22. Install the camshafts and front cover.

23. We know you're not going to like this next one. . .the new valve pad(s) you have installed was based on your measurement and a calculation. It is an approximation and may not have been sufficient to get the valve clearance back into spec. Soooo, you'll need to recheck the clearance on all valves again! If any of the valves are still out of specification, or have gone out of specification due to component removal, you will need to repeat this entire procedure again. Once you are certain that all valves fall inside the respective clearance allowances, you can start to wrap things up.

24. Install the cylinder head cover.

25. Install the coupler cover if removed.

26. Install the spark plugs and then the filter element/air filter case cover or the engine cover.

27. Reconnect the battery cable and install the seat(s) and/or storage container.

28. Phew! Did you have fun?

2010-11 VX Models w/1052cc Engines

OEM 3 DIFFICULT

◆ **See Figures 60 thru 70**

■ **Cylinders are numbered front to back, 1 thru 4. The exhaust camshaft is on the right, starboard side of the engine and the intake is on the left, port side. You will see that there are 3 cam lobes/valves per cylinder on the intake side and 2 per cylinder on the exhaust side.**

Valve clearance adjustment should be performed every 200 hours or 2 years, whichever comes first. It will also be required anytime service work has been performed on the valve train.

The work should always be performed on a cold engine and the No. 1 piston should be at top dead center (TDC) of the compression stroke.

1. Remove the seat(s) and rear storage container. Disconnect the negative battery cable.

2. Remove the engine cover (if equipped) as detailed in the Air Filter section.

3. Remove the spark plugs as detailed in this section.

4. Remove the cylinder head cover (and gasket) as detailed in the Engine Mechanical section.

5. Thread a dial gauge stand (#90890-06583) into the spark plug hole on the No. 1 cylinder and then install the gauge into the stand. You may also need a gauge needle (#90890-06584), or you can get the set (#90890-06585).

6. Position the No. 1 piston so it is at top dead center (TDC) of the compression stroke for the No. 1 cylinder (first cylinder at the front of the engine). To do this you will need a special camshaft wrench (#90890-06724). Connect the wrench to the exhaust camshaft sprocket and rotate it clockwise until the punch marks on the camshafts are all aligned with the corresponding marks on the camshaft bearing caps and the dial indicator indicates TDC.

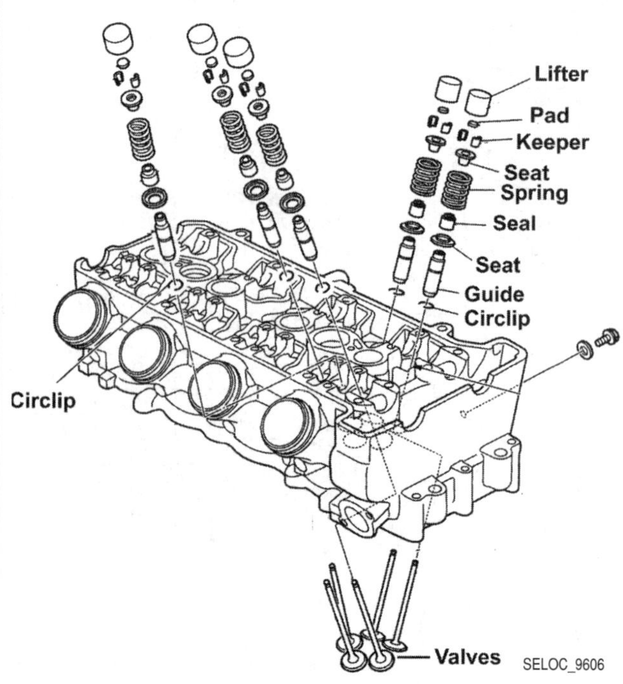

Fig. 65 Exploded view of the valve train - 2010-11 VX models

Make sure that the intake and exhaust cam lobes on the No. 1 cylinder are pointing away from each other. If not, go back and start over again until you have achieved both confirmations.

7. Insert a feeler gauge between the camshaft lobe and the top of the valve lifter until a very slight drag can be felt. Mark this measurement down to refer to later. Repeat this procedure on the other 4 valves for the No. 1 cylinder - there are 3 intake valves and lifters, and 2 exhaust valves/lifters. Remember?

8. Now, check the clearance on the exhaust valves for the No. 2 cylinder and the intake valves for the No. 3 cylinder. Remember to mark down the measurement for each valve.

■ **We suggest that you draw a picture of the valve train and then mark the measured clearance for each valve on its associated position in your picture.**

9. Remove the dial gauge from the No. 1 cylinder and re-install it into the spark plug hole for the No. 4 cylinder.

10. Using your camshaft wrench, rotate the camshaft 180° in the clockwise direction and this will put the No. 4 cylinder piston at TDC.

11. Measure the clearance of all 5 valves on the No. 4 cylinder, then the exhaust valves for the No. 3 cylinder and finally, the intake valves for the No. 2 cylinder.

12. Take a break. Sit down and take a look at all of your clearance measurements. Acceptable valve clearance on these engines is 0.0043-0.0079 in. (0.11-0.20mm) for all intake valves and 0.0098-0.0134 in. (0.25-0.34mm) on the exhaust side. Hopefully, all of your measurements have fallen into these ranges and you can skip to the end here and begin putting everything back together.

However, if any of your measurements were out of spec, you'll need to move to the next step.

13. Remove the front cover and then the camshafts as detailed in the Engine Mechanical section.

14. Run through the clearances on your drawing and mark any that are out of specification. Remove the valve lifter and valve pad (details in the Engine Mechanical section) from each of the valves that are out of spec; be sure to keep them together and identified as shown in the illustration.

■ **It's a good idea to cover the timing chain opening with a clean lint-free rag to prevent a lifter or pad from accidentally falling down into the crankcase.**

15. Original valve pads are numbered and range in thickness from 1.20-2.40mm with 25 available replacement pads in thicknesses of 0.05mm increments. When removing the pads from the engine, you will see that they are marked with a number corresponding to their thickness, *without* the decimal point or 'mm' designation - a pad marked **168** would have a measurement of 1.68mm, one marked **142** would be 1.42mm.

a. Because pads of various sizes were originally installed, you will need to round that original number to arrive at an available replacement size: if the last digit in the pad's number is **0**, **1** or **2**, it will be rounded down - original number **132** will become **130**. If the original pad number ends with a **4**, **5** or **6**, it will stay the same or round to the 5 spot - **145** stays as **145** or **147** rounds down to **145**. Finally, if the original number ended in an **8** or **9**, it will be rounded up - original number **168** will become **170**.

b. Why are we telling you this? Because rather than having to perform more calculations, you're going to use the rounded number of the original pad, and the clearance measurement you noted previously of the same cylinder and plug them into the accompanying charts to arrive at the new replacement pad thickness. If you have a valve on the No. 2 cylinder that was out of spec, you'll go to the Intake or Exhaust chart and find your measurement in the left hand column (say it was 24mm so you'd go to the 0.23-0.27 range line).

c. Now you'll check the rounded valve pad number that you removed from this valve (say **150**), find it in the column across the top of the chart and then go to where the 2 positions intersect. In this case, you'll need a **160** replacement pad to bring your bad valve into spec. Make sense?

16. Once you have completed the previous step on each of the valves that measured out of specification, you can re-install the lifter and new pad. Making sure that everything goes back into the correct bore.

a. Coat the valve pad with molybdenum disulfide grease before positioning it in the bore.

b. Coat the valve lifter with molybdenum disulphide oil before positioning it into the bore. Make sure that it rotates smoothly in the bore once it's in position.

17. Install the camshafts.

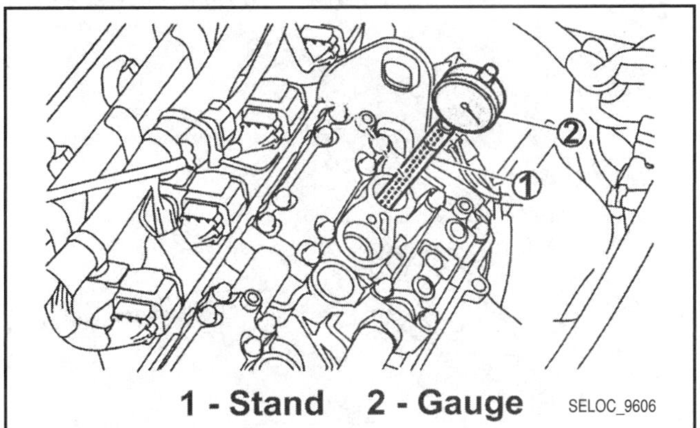

Fig. 66 You'll need a gauge stand when installing the dial gauge

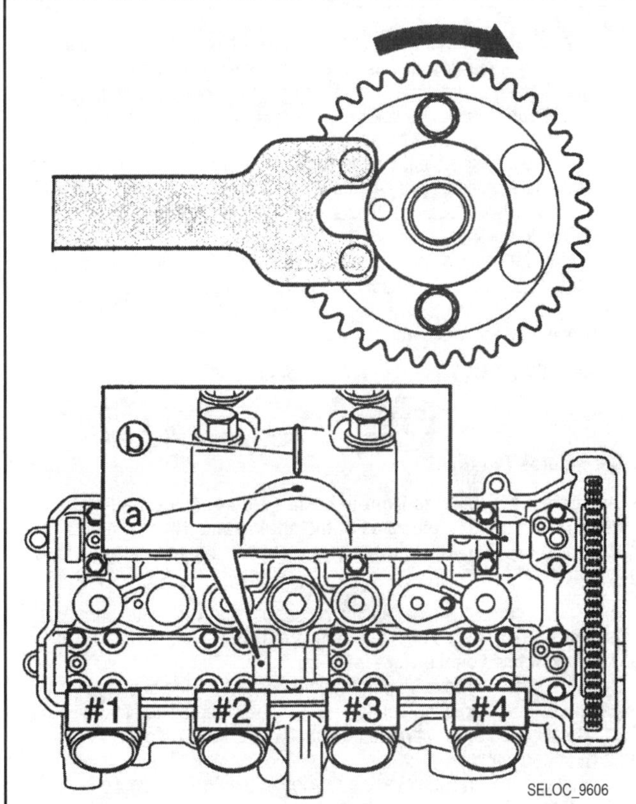

Fig. 67 Use the special tool to set the cylinder at TDC by rotating the camshaft sprocket until the marks align

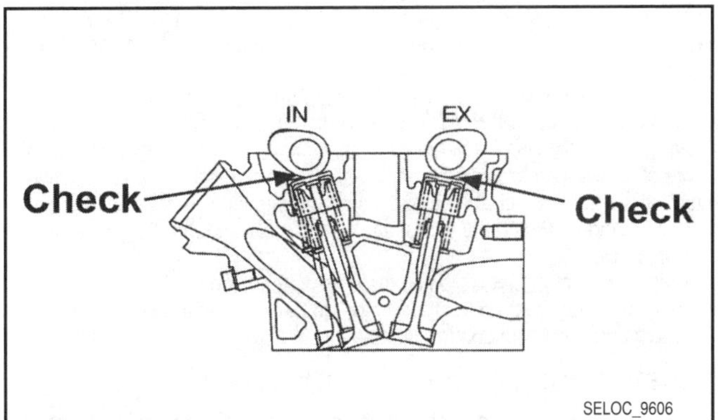

Fig. 68 Make sure that the camshaft lobes are facing away from each other before. . .

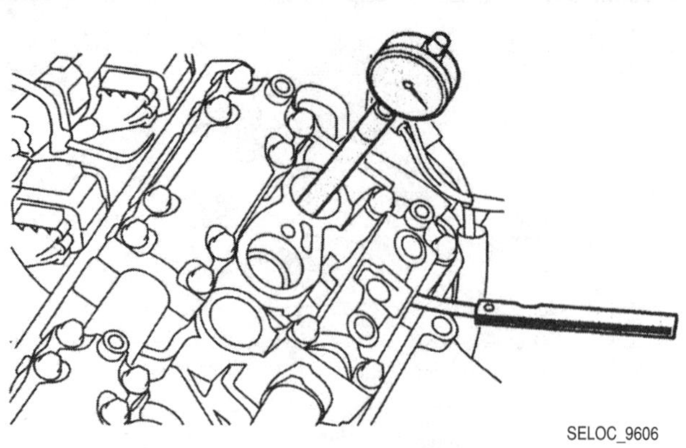

SELOC_9606

Fig. 69 . . .checking the clearance between the lifter and the cam lobe

18. We know you're not going to like this next one. . .the new valve pad(s) you have installed was based on your measurement and a calculation. It is an approximation and may not have been sufficient to get the valve clearance back into spec. Soooo, you'll need to recheck the clearance on all valves again! If any of the valves are still out of specification, or have gone out of specification due to component removal, you will need to repeat this entire procedure again. Once you are certain that all valves fall inside the respective clearance allowances, you can start to wrap things up.
19. Install the cylinder head cover.
20. Install the spark plugs and coils. Install the engine cover if removed.
21. Reconnect the battery cable and install the seat(s) and/or storage container.
22. Phew! Did you have fun?

All 2008-11 FX/FZ Models & 2011 VX Models w/1812cc Engines

OEM ③ DIFFICULT

◆ **See Figures 71 thru 82**

■ **Cylinders are numbered front to back, 1 thru 4. The exhaust camshaft is on the left, port side of the engine and the intake is on the right, starboard side. You will see that there are 2 cam lobes/4 valves per cylinder.**

Valve clearance adjustment should be performed every 200 hours or 2 years, whichever comes first. It will also be required anytime service work has been performed on the valve train.

The work should always be performed on a cold engine and the No. 1 piston should be at top dead center (TDC) of the compression stroke.

1. Remove the seat(s) and rear storage container. Disconnect the negative battery cable.
2. Remove the engine cover as detailed in the Air Filter section.
3. Remove the spark plugs as detailed in the Spark Plugs section.
4. Remove the cylinder head cover (and gasket) as detailed in the Engine Mechanical section.
5. Thread a dial gauge stand (#90890-06725) into the spark plug hole on the No. 1 cylinder and then install the gauge into the stand. You may also need a gauge needle for the (#90890-06584), or you can get the set (#YB-06585).
6. Position the No. 1 piston so it is at top dead center (TDC) of the compression stroke for the No. 1 cylinder (first cylinder at the front of the engine). To do this you will need a special camshaft wrench (#90890-06724). Connect the wrench to the exhaust camshaft sprocket and rotate it clockwise until the punch marks on the camshafts are all aligned with the corresponding marks on the camshaft bearing caps and the dial indicator indicates TDC.

Make sure that the intake and exhaust cam lobes on the No. 1 cylinder are pointing away from each other. If not, go back and start over again until you have achieved both confirmations.

On FX HO/FX Cruiser HO, FZ and VXR/VXS models:

7. Insert a feeler gauge between the camshaft lobe and the top of the valve lifter until a very slight drag can be felt. Mark this measurement down to refer to later. Repeat this procedure on the other valves for the No. 1 cylinder.

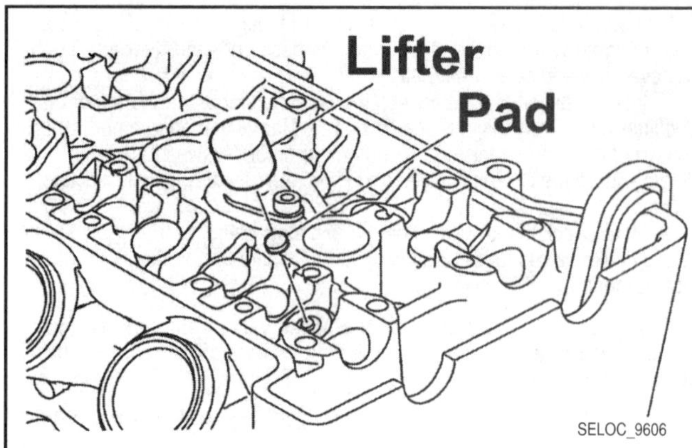

SELOC_9606

Fig. 70 Installing the new pad

8. Now, check the clearance on the exhaust valves for the No. 2 cylinder and the intake valves for the No. 3 cylinder. Remember to mark down the measurement for each valve.

■ **We suggest that you draw a picture of the valve train and then mark the measured clearance for each valve on its associated position in your picture.**

9. Remove the dial gauge from the No. 1 cylinder and re-install it into the spark plug hole for the No. 4 cylinder.
10. Using your camshaft wrench, rotate the camshaft 180° in the clockwise direction and this will put the No. 4 cylinder piston at TDC.
11. Measure the clearance of all valves on the No. 4 cylinder, then the exhaust valves for the No. 3 cylinder and finally, the intake valves for the No. 2 cylinder.

On FX SHO/FX Cruiser SHO models:

12. Insert a feeler gauge between the camshaft lobe and the top of the valve lifter until a very slight drag can be felt. Mark this measurement down to refer to later. Repeat this procedure on the other valves for the No. 1 cylinder.
13. Using your camshaft wrench, rotate the camshaft 90° in the clockwise direction and this will put the No. 2 cylinder piston at TDC. Measure the clearance at both the intake and exhaust valves for this cylinder, making sure that you mark them all down.

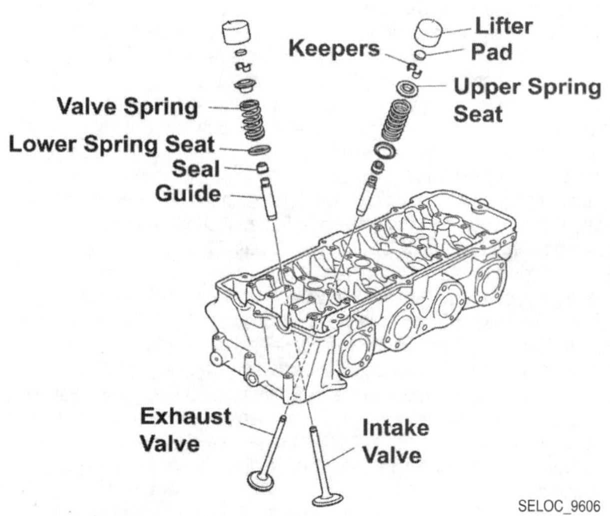

SELOC_9606

Fig. 71 Exploded view of the valve train - all 2008-11 FX/FZ/VXR/VXS models w/1812cc engines

Fig. 72 You'll need a gauge stand when installing the dial gauge

Fig. 73 Use the special tool to set the cylinder at TDC by rotating the camshaft sprocket. . .

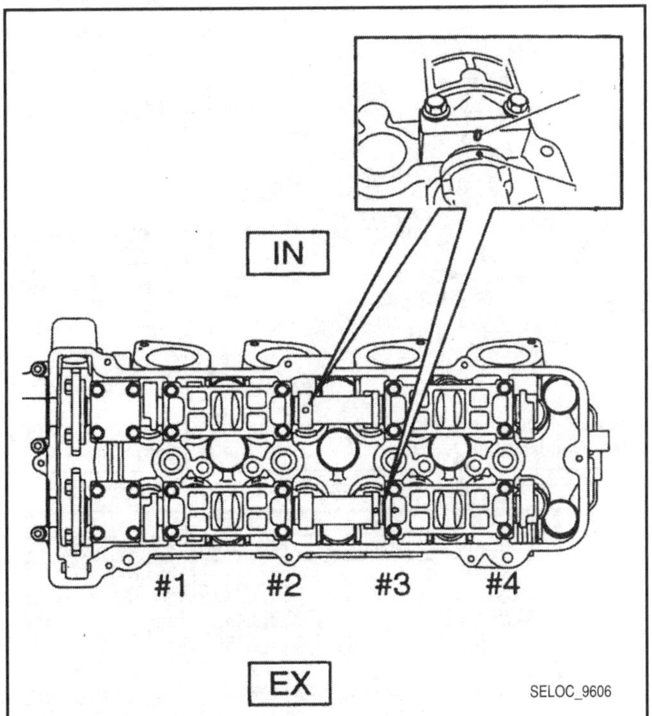

Fig. 74 . . .until the marks align

■ **We suggest that you draw a picture of the valve train and then mark the measured clearance for each valve on its associated position in your picture.**

14. Using your camshaft wrench, rotate the camshaft another 90° in the clockwise direction and this will put the No. 4 cylinder piston at TDC. Measure the clearance at both the intake and exhaust valves for this cylinder, making sure that you mark them all down.

15. Using your camshaft wrench again, rotate the camshaft a final 90° in the clockwise direction and this will put the No. 3 cylinder piston at TDC. Measure the clearance at both the intake and exhaust valves for this cylinder, making sure that you mark them all down.

On all models now:

16. Take a break. Sit down and take a look at all of your clearance measurements. Acceptable valve clearance on these engines is 0.006-0.009 in. (0.14-0.23mm) for all intake valves and 0.011-0.015 in. (0.28-0.37mm) on the exhaust side. Hopefully, all of your measurements have fallen into these ranges and you can skip to the end here and begin putting everything back together.

However, if any of your measurements were out of spec, you'll need to move to the next step.

17. Remove the front cover and then the camshafts as detailed in the Engine Mechanical section.

18. Run through the clearances on your drawing and mark any that are out of specification. Remove the valve lifter and valve pad (details in the Engine Mechanical section) from each of the valves that are out of spec; be sure to keep them together and identified as shown in the illustration.

■ **It's a good idea to cover the timing chain opening with a clean lint-free rag to prevent a lifter or pad from accidentally falling down into the crankcase.**

19. Original valve pads are numbered, and range in thickness from 1.20-2.30mm, with 23 available replacement pads in thicknesses of 0.05mm increments. When removing the pads from the engine, you will see that they are marked with a number corresponding to their thickness, *without* the decimal point or 'mm' designation - a pad marked **168** would have a measurement of 1.68mm, one marked **142** would be 1.42mm.

a. Because pads of various sizes were originally installed, you will need to round that original number to arrive at an available replacement size: if the last digit in the pad's number is **0**, **1** or **2**, it will be rounded down - original number **132** will become **130**. If the original pad number ends with a **4**, **5** or **6**, it will stay the same or be rounded to the 5 spot - **145** stays as **145** or **147** rounds down to **145**. Finally, if the original number ended in an **8** or **9**, it will be rounded up - original number **168** will become **170**.

b. Why are we telling you this? Because rather than having to perform more calculations, you're going to use the rounded number of the original pad, and the clearance measurement you noted previously of the same cylinder and plug them into the accompanying charts to arrive at the new replacement pad thickness. If you have a valve on the No. 2 cylinder that was out of spec, you'll go to the Intake or Exhaust chart and find your measurement in the left hand column (say it was 25mm so you'd go to the 0.24-0.26 range line).

c. Now you'll check the rounded valve pad number that you removed from this valve (say **150**), find it in the column across the top of the chart and then go to where the 2 positions intersect. In this case, you'll need a **155** replacement pad to bring your bad valve into spec. Make sense?

20. Once you have completed the previous step on each of the valves that measured out of specification, you can re-install the lifter and new pad; making sure that everything goes back into the correct bore.

a. Coat the valve pad with clean engine oil before positioning it in the bore.

b. Coat the valve lifter with clean engine oil before positioning it into the bore. Make sure that it rotates smoothly in the bore once it's in position.

21. Install the camshafts.

22. We know you're not going to like this next one. . .the new valve pad(s) you have installed were based on your measurement and a calculation. It is an *approximation* and may not have been sufficient to get the valve clearance back into spec. Soooo, you'll need to recheck the clearance on all valves again! If any of the valves are still out of specification, or have gone out of specification due to component removal, you will need to

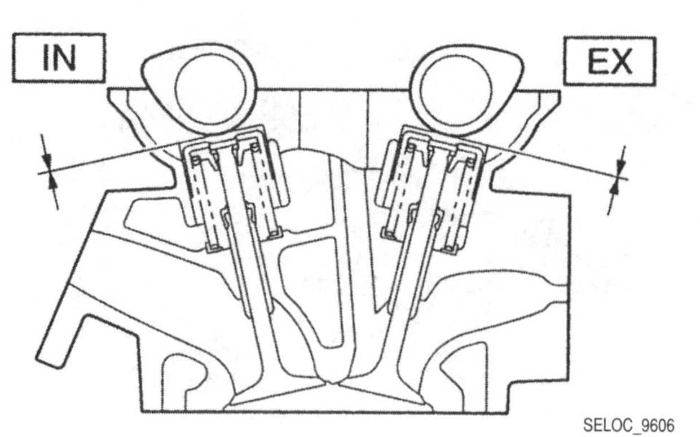

SELOC_9606

Fig. 75 Make sure that the camshaft lobes are facing away from each other before. . .

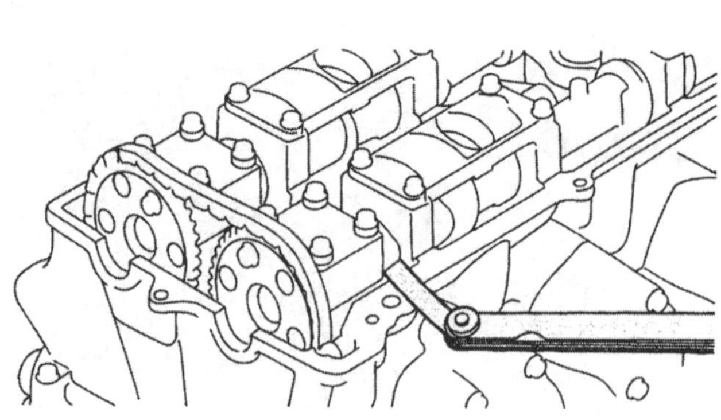

SELOC_9606

Fig. 76 . . .checking the clearance between the lifter and the cam lobe

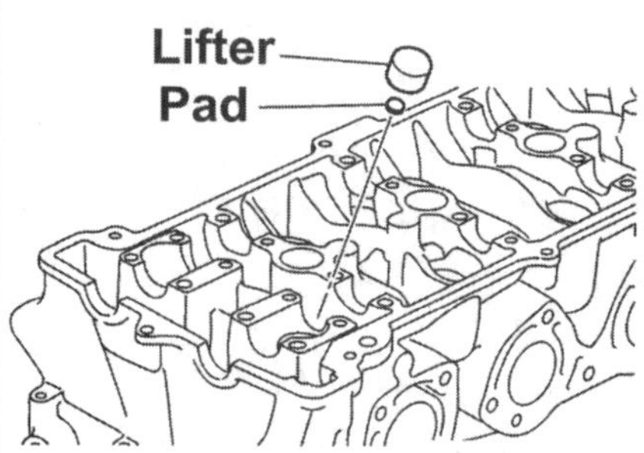

SELOC_9606

Fig. 77 Remove the lifters and pads from the cylinder head. . .

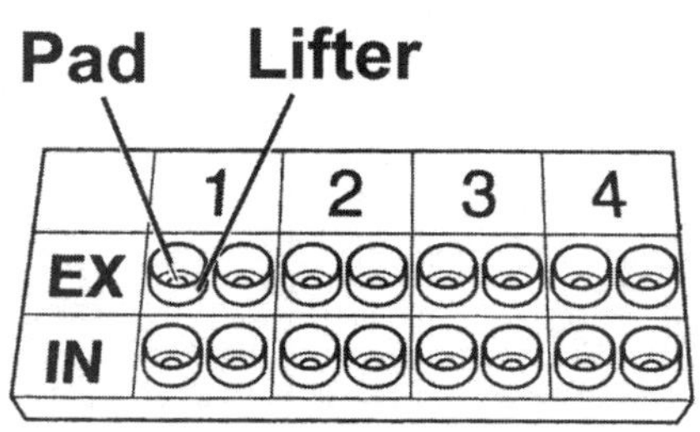

SELOC_9606

Fig. 78 . . .making sure that you store them correctly

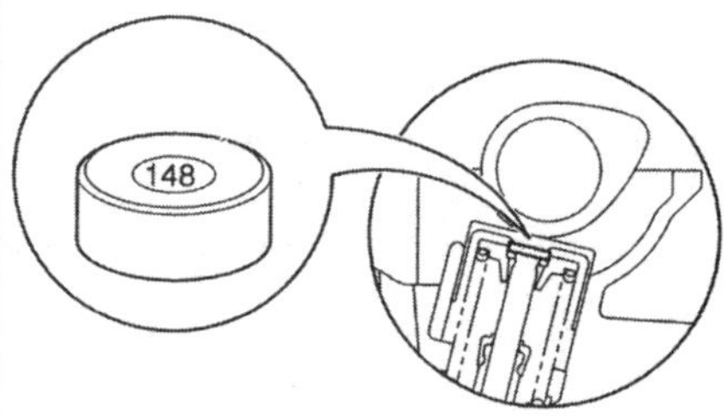

SELOC_9606

Fig. 79 All pads will be identified with a number corresponding to their thickness

MEASURED	ORIGINAL VALVE PAD NUMBER																					
CLEARANCE	120	125	130	135	140	145	150	155	160	165	170	175	180	185	190	195	200	205	210	215	220	225	230
0.00–0.02					120	125	130	135	140	145	150	155	160	165	170	175	180	185	190	195	200	205	210
0.03–0.06				120	125	130	135	140	145	150	155	160	165	170	175	180	185	190	195	200	205	210	215
0.07–0.10			120	125	130	135	140	145	150	155	160	165	170	175	180	185	190	195	200	205	210	215	220
0.11–0.13		120	125	130	135	140	145	150	155	160	165	170	175	180	185	190	195	200	205	210	215	220	225
0.14–0.23	STANDARD CLEARANCE																						
0.24–0.26	125	130	135	140	145	150	155	160	165	170	175	180	185	190	195	200	205	210	215	220	225	230	
0.27–0.31	130	135	140	145	150	155	160	165	170	175	180	185	190	195	200	205	210	215	220	225	230		
0.32–0.36	135	140	145	150	155	160	165	170	175	180	185	190	195	200	205	210	215	220	225	230			
0.37–0.41	140	145	150	155	160	165	170	175	180	185	190	195	200	205	210	215	220	225	230				
0.42–0.46	145	150	155	160	165	170	175	180	185	190	195	200	205	210	215	220	225	230					
0.47–0.51	150	155	160	165	170	175	180	185	190	195	200	205	210	215	220	225	230						
0.52–0.56	155	160	165	170	175	180	185	190	195	200	205	210	215	220	225	230							
0.57–0.61	160	165	170	175	180	185	190	195	200	205	210	215	220	225	230								
0.62–0.66	165	170	175	180	185	190	195	200	205	210	215	220	225	230									
0.67–0.71	170	175	180	185	190	195	200	205	210	215	220	225	230										
0.72–0.76	175	180	185	190	195	200	205	210	215	220	225	230											
0.77–0.81	180	185	190	195	200	205	210	215	220	225	230												
0.82–0.86	185	190	195	200	205	210	215	220	225	230													
0.87–0.91	190	195	200	205	210	215	220	225	230														
0.92–0.96	195	200	205	210	215	220	225	230															
0.97–1.01	200	205	210	215	220	225	230																
1.02–1.06	205	210	215	220	225	230																	
1.07–1.11	210	215	220	225	230																		
1.12–1.16	215	220	225	230																			
1.17–1.21	220	225	230																				
1.22–1.26	225	230																					
1.27–1.31	230																						

Example:

Measured valve clearance is 0.29 mm (0.0114 in)

Original valve pad number is 148 (thickness = 1.48 mm)

- Round off the original valve pad number 148 to 150 (thickness = 1.50 mm).
- Select the new valve pad number using the valve pad selection table.

New valve pad number is 160 (thickness = 1.60 mm)

SELOC_9606

Fig. 80 Intake valve pad selection chart - 1812cc engines

MEASURED	ORIGINAL VALVE PAD NUMBER																					
CLEARANCE	120	125	130	135	140	145	150	155	160	165	170	175	180	185	190	195	200	205	210	215	220	225	230
0.00–0.01								120	125	130	135	140	145	150	155	160	165	170	175	180	185	190	195
0.02–0.04							120	125	130	135	140	145	150	155	160	165	170	175	180	185	190	195	200
0.05–0.09						120	125	130	135	140	145	150	155	160	165	170	175	180	185	190	195	200	205
0.10–0.14					120	125	130	135	140	145	150	155	160	165	170	175	180	185	190	195	200	205	210
0.15–0.19				120	125	130	135	140	145	150	155	160	165	170	175	180	185	190	195	200	205	210	215
0.20–0.24			120	125	130	135	140	145	150	155	160	165	170	175	180	185	190	195	200	205	210	215	220
0.25–0.27		120	125	130	135	140	145	150	155	160	165	170	175	180	185	190	195	200	205	210	215	220	225
0.28–0.37	STANDARD CLEARANCE																						
0.38–0.40	125	130	135	140	145	150	155	160	165	170	175	180	185	190	195	200	205	210	215	220	225	230	
0.41–0.45	130	135	140	145	150	155	160	165	170	175	180	185	190	195	200	205	210	215	220	225	230		
0.46–0.50	135	140	145	150	155	160	165	170	175	180	185	190	195	200	205	210	215	220	225	230			
0.51–0.55	140	145	150	155	160	165	170	175	180	185	190	195	200	205	210	215	220	225	230				
0.56–0.60	145	150	155	160	165	170	175	180	185	190	195	200	205	210	215	220	225	230					
0.61–0.65	150	155	160	165	170	175	180	185	190	195	200	205	210	215	220	225	230						
0.66–0.70	155	160	165	170	175	180	185	190	195	200	205	210	215	220	225	230							
0.71–0.75	160	165	170	175	180	185	190	195	200	205	210	215	220	225	230								
0.76–0.80	165	170	175	180	185	190	195	200	205	210	215	220	225	230									
0.81–0.85	170	175	180	185	190	195	200	205	210	215	220	225	230										
0.86–0.90	175	180	185	190	195	200	205	210	215	220	225	230											
0.91–0.95	180	185	190	195	200	205	210	215	220	225	230												
0.96–1.00	185	190	195	200	205	210	215	220	225	230													
1.01–1.05	190	195	200	205	210	215	220	225	230														
1.06–1.10	195	200	205	210	215	220	225	230															
1.11–1.15	200	205	210	215	220	225	230																
1.16–1.20	205	210	215	220	225	230																	
1.21–1.25	210	215	220	225	230																		
1.26–1.30	215	220	225	230																			
1.31–1.35	220	225	230																				
1.36–1.40	225	230																					
1.41–1.45	230																						

Example:
Measured valve clearance is 0.47 mm (0.0185 in)
Original valve pad number is 168 (thickness = 1.68 mm)

- Round off the original valve pad number 168 to 170 (thickness = 1.70 mm).
- Select the new valve pad number using the valve pad selection table.

New valve pad number is 185 (thickness = 1.85 mm)

SELOC_9606

Fig. 81 Exhaust valve pad selection chart - 1812cc engines

Fig. 82 Use a coupler wrench when performing the final clearance checks

repeat this entire procedure again. Once you are certain that all valves fall inside the respective clearance allowances, you can start to wrap things up.

23. Reinstall the gauge and stand back into the spark plug hole for the No. 1 cylinder.
24. This time to set the No. 1 piston to TDC, you'll need to install a coupler wrench (#90890-06729) on the drive coupling and use it to rotate the engine counterclockwise until you get TDC.
25. Make sure that the marks on the cam and bearing caps are lined up again and recheck the clearance on all valves and confirm that they are all within specification. If not, ugh, start the adjustment process all over again. If they are, move to the next step.
26. Install the cylinder head cover.
27. Install the spark plugs and coils. Install the engine cover if removed.
28. Reconnect the battery cable and install the seat(s) and/or storage container.
29. Phew! Did you have fun?

Jet Pump

IMPELLER INSPECTION

2 MODERATE

◆ **See Figure 84**

1. Remove the seat and disconnect the negative battery cable.
2. Working underneath the unit, remove the intake grate, duct and ride plate as detailed in the Jet Drive section.

■ **Yamaha suggests removing the entire jet pump unit on 2010 and later VX models, but we don't see why you can't try to check with the pump in.**

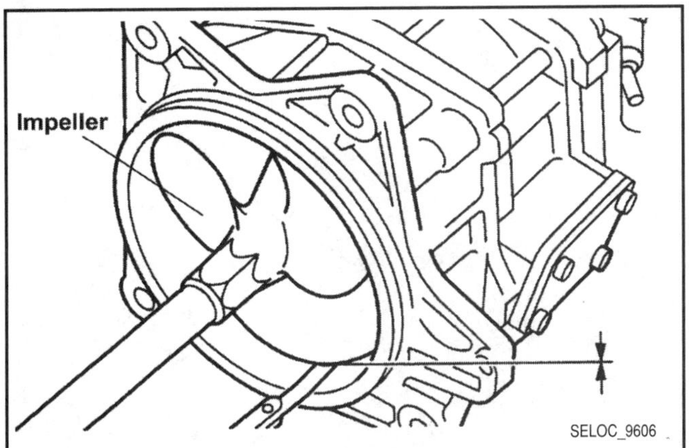

Fig. 84 Check the clearance between the housing and each of the impeller blades

3. Check the impeller for scratched, wear and other damage. Use a feeler gauge and measure the clearance between each of the 3 impeller blades and the pump housing; choose the gauge that slides in out with just a slight drag. Maximum clearance at any blade should be no more than 0.0236 in. (0.6mm) on models with 998cc engines, or 0.0138-0.0177 in. (0.35-0.45mm) on those with 1052cc/1812cc engines. Unfortunately, if clearance is too high, the impeller or housing will require replacement as detailed in the Jet Drive section. There is no adjustment possible.

Water Inlet Strainer

INSPECTION

EASY

◆ **See Figures 85 and 86**

The water inlet screen can be found on the left, port, side of the impeller housing assembly. Although it is not included in official maintenance intervals, we recommend inspecting it at least every season.

Fig. 85 The inlet strainer sits in the impeller housing on the left side

1. Remove the seat and disconnect the negative battery cable.
2. Remove the ride plate and the rubber plate as detailed in the Jet Drive section.
3. Remove the 4 mounting bolts and lift off the water inlet filter cover. Remove the old gasket and any remaining gasket/packing material from the cover or housing surfaces.
4. Lift out the inlet strainer and clean it thoroughly with compressed air. Inspect the filter mesh for tears or other damage, replacing if necessary.
5. Install a new filter (we figure why not replace it while you have it apart, but you can use the old one) and then new gasket/packing.
6. Carefully position the cover so the holes line up with those of the filter and housing. Coat the threads of the 4 bolts with Loctite 572 and thread them in until they are just snug. Now, in a criss-cross pattern, tighten them to 4.8 ft. lbs. (6.6 Nm) except on the 2010-11 VX w/1052cc engines and all 1812cc engines, where they should be tightened to 5.2 ft. lbs. (7 Nm).
7. Install the rubber plates and the ride plate.
8. Connect the battery cable and install the seat and/or storage container.

Bilge Pump

BILGE STRAINER

EASY

◆ **See Figures 87, 88 and 89**

The bilge strainer can be found inside the craft, under and slightly forward of the drive coupler. We recommend inspecting it at least every season or 100 hours.

1. Remove the seat(s) and/or storage container. Disconnect the negative battery cable.
2. Although not strictly necessary, you may find that removing the deck plate (above the tank) and/or the electrical box will make access much easier.
3. Remove the mounting bolt and then lift off the coupler cover.
4. Disconnect the strainer from the holder and then carefully slide it out of the bilge. Inspect it for cracks or other damage. If is appears dirty or clogged, disconnect it form the hose and clean it thoroughly. On 1052/1812cc FX/VX models, press in the hooks on the strainer and lift off the strainer case.
5. Slide the hose back onto the strainer (if removed) and reposition the strainer into the holder.
6. Position the coupler cover and tighten the bolt to 80 inch lbs. (7 Nm).

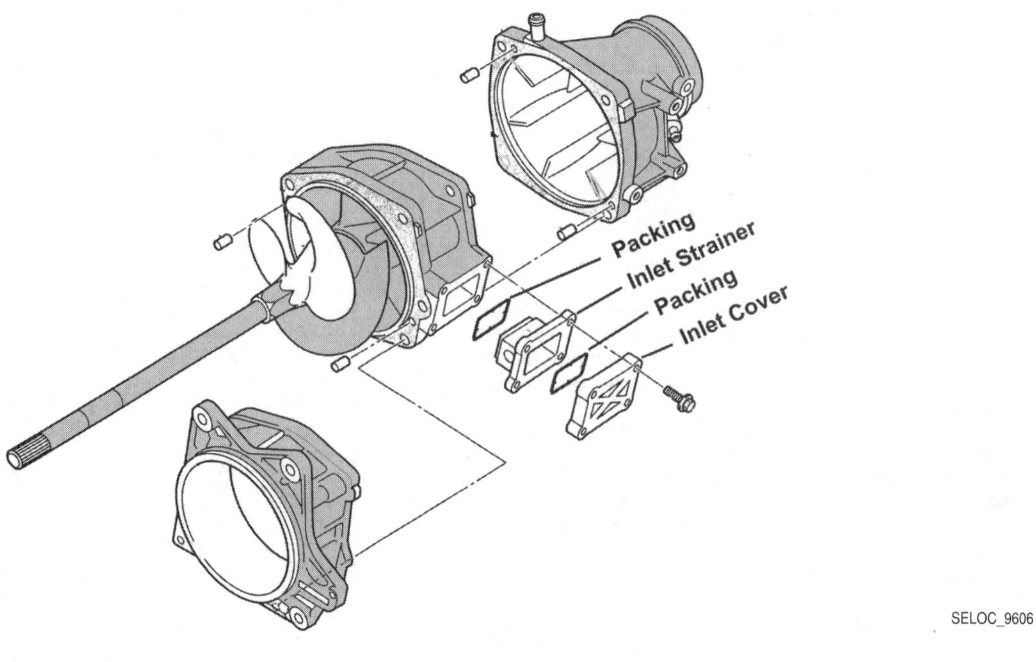

Fig. 86 Exploded view of the water inlet strainer

Fig. 87 The bilge strainer can be found under the drive coupler. . .

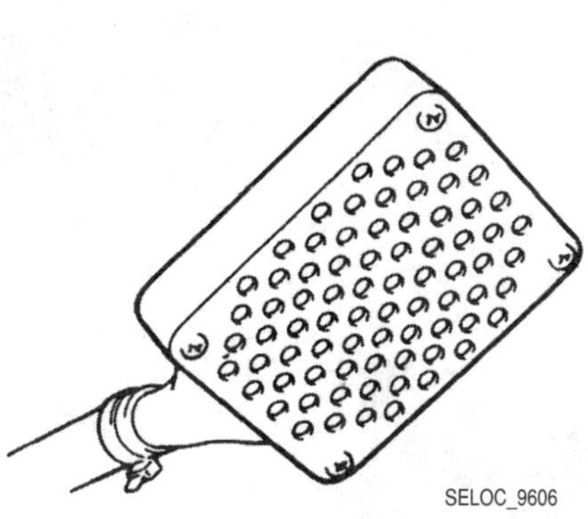

Fig. 88 . . .and looks like this on most 998cc models. . .

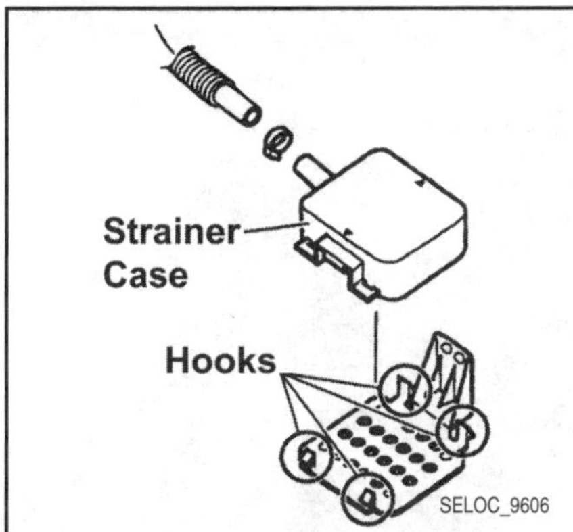

Fig. 89 . . .or this on most models with 1052cc and 1812cc engines

7. If you removed the deck plate for access, install the 4 bolts and tighten the nuts to 13 ft. lbs. (18 Nm). Remember there should be a washer on both sides - one with the nut and one under the bolt's head. Same thing with the electric box, reinstall the cover if removed.

8. Connect the battery cable and install the storage container/seat.

ELECTRIC BILGE PUMP STRAINER

① EASY

◆ See Figures 90, 91 and 92

The bilge strainer can be found inside the craft, near the battery; it should have a red cap with the words **500 gpm, Bilge Pump**. Although we're not sure why, we can find no evidence of one being used on the VX models w/1052cc engines.

We recommend inspecting it at least every season or 100 hours.

1. Remove the seat(s) and/or storage container. Disconnect the negative battery cable.
2. Reach in, carefully move the hoses aside and remove the band around the cap. Push in on the tabs and remove the pump cap/cover.
3. Carefully pry out the strainer screen out of the pump and inspect it for cracks, tears or other damage. Clean the strainer thoroughly.
4. Install the strainer back into the pump throat, rounded side facing up, until it seats on the inner flange.
5. Install the cap/cover, making sure that the tabs feed into their respective slots.
6. Install the battery cable and replace the seat/storage container.

Fuel, Water, Oil & Breather Lines

All motors are equipped with a number of breather, oil, water and/or fuel supply lines. As stated earlier under Before/After Each Use, all of these hoses should be checked visually each time you go out. However, at least once per year you should perform a more thorough check of these hoses. Get to know how hard or soft they are when new so that you can help spot when they have become dried out/brittle or softened and ready to perforate.

Many motors also contain breather hoses which interconnect parts of the crankcase and usually the intake tract. The function of a breather hose is to control the vacuum/pressure created inside the motor by the Otto cycle. When equipped, these hoses should be carefully checked at least once a year. Honestly, we think most people probably wait until they find the hose is brittle/cracking or is collapsing before they replace them, but they are mentioned in the manufacturer maintenance tables and thought we at least had to bring it up here. Use your best judgment when deciding when (or when not) to replace any hoses.

Drain Plugs

◆ See Figure 93

Yeah, yeah, yeah, we know; you always check the drain plugs! Well, you'd be surprised how many people don't, or forget. Just make it part of your pre- and post-ride routine. Check their condition regularly, including the O-rings.

Fuses

◆ See Figure 94

To check or replace a fuse, locate the electrical box inside the engine compartment and unthread the fuse cap. Use the fuse puller in the cap to remove the appropriate fuse. Please refer to your Owner's manual for a listing of fuses and their amperage.

On models with an 1812cc engine, you will need to slide back the lock-tabs on the electrical case and separate the halves. This will allow access to the fuse box cap.

Fig. 90 The bilge pump cap/cover is found in the engine compartment near the battery

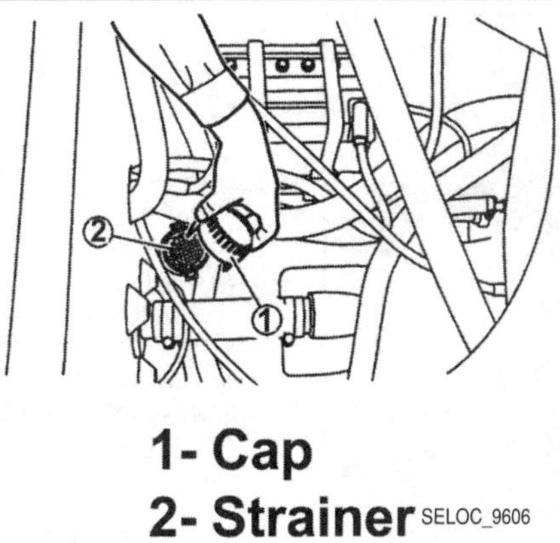

Fig. 91 Squeeze in the locking tabs to remove the cap. . .

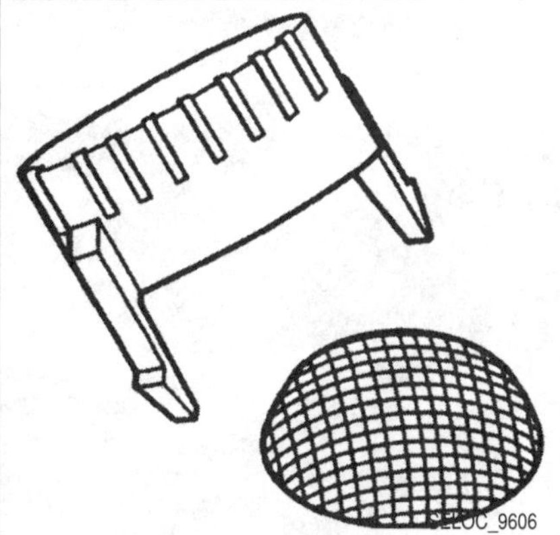

Fig. 92 . . .and then remove the strainer for inspection

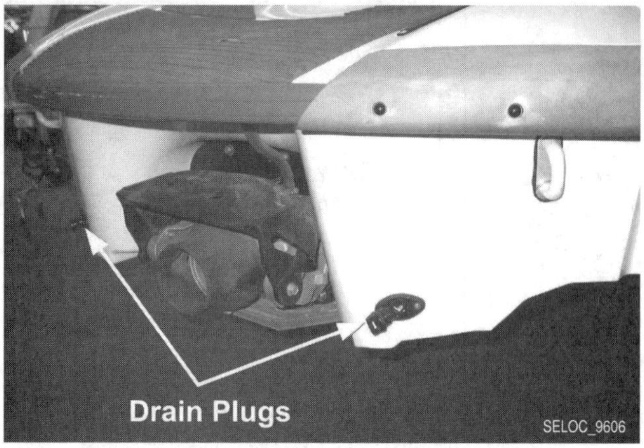

Fig. 93 Make sure they are threaded tightly before leaving the dock please

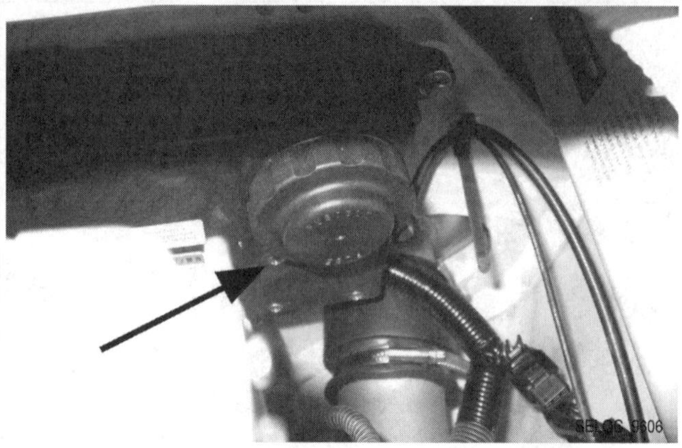

Fig. 94 The electrical box is attached to the inside of the engine compartment and the fuses are behind this cover, 1812cc engines similar

Checks & Adjustments

QSTS CABLE

■ **VX models do not utilize the QSTS system found on other models.**

2002-08 FX/FX Cruiser & 2004-07 FX/FX Cruiser HO Models

② MODERATE

◆ **See Figures 95 and 96**

1. Set the QSTS control grip to the Neutral position and check that the thrust nozzle is in the center position.

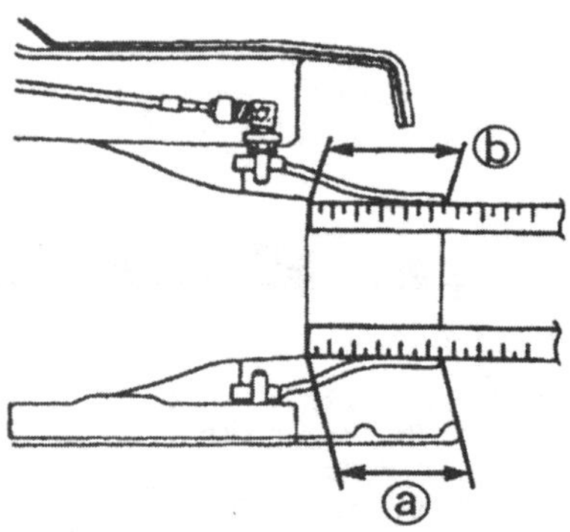

Fig. 95 Jet thrust nozzle set length. . .

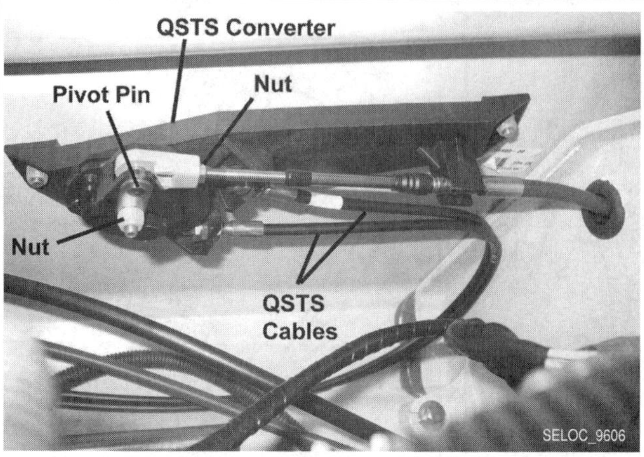

Fig. 96 . . .will require adjustment if the measurements are not the same

2. Measure the jet thrust nozzle set length at the top and bottom as shown in the illustration. Both measurements must be identical or adjustment will be required.
3. Locate the QSTS converter in the engine compartment and then loosen the locknut on the rear cable joint.
4. Remove the cable nut and pivot pin and pop out the rear cable/joint.
5. Make sure that the bars are still in the neutral position and the nozzle is in the center position. Rotating the cable joint inward will increase the lower set length measurement (A) and turning it outward will increase the upper measurement (B).
6. Reconnect the cable joint to the pivot pin and then tighten the nut to 2.7 ft. lbs. (3.8 Nm). Tighten the locknut to 2.2 ft. lbs. (3 Nm). Always make sure the cable joint is threaded onto the cable at least 0.31 in. (8mm).
7. Check the set length a final time. If the 2 measurements are still not identical, you will need to repeat the procedure.

■ **If you are unable to get the set length measurements adjusted correctly at the converter end, you will need to adjust the cable joint at the jet end as detailed in the Jet Drive section.**

2008-10 FX HO/SHO, FX Cruiser HO/SHO & FZ Models

② MODERATE

◆ **See Figures 97 thru 100**

1. Set the QSTS control grip to the Neutral position and check that the thrust nozzle is in the center position.
2. Measure the jet thrust nozzle set length at the top and bottom as shown in the illustration. Both measurements must be identical or adjustment will be required.
3. Move the shift lever to the Reverse position on FX/FX Cruiser HO and FZ models w/1812cc engines.
4. Loosen the locknut at the end of the QSTS rod above the drive unit. Pop the joint off of the ball joint.
5. Move the shift lever to the Forward position on FX/FX Cruiser HO and FZ models w/1812cc engines.
6. Adjust the angle set measurement. Rotating the joint outward will increase the set length of measurement A. Rotating the joint inward will increase the set length of measurement B.
7. On FX/FX Cruiser HO and FZ models w/1812cc engines, move the shift lever back to the Reverse position.
8. Reconnect the cable joint to the ball joint and then tighten the locknut to 2.7 ft. lbs. (3.8 Nm) on models w/1052cc engines, or 3 ft. lbs. (4 Nm) on models with 1812cc engines. Always make sure the rod joint is threaded onto the rod so at least 0.31 in. (8mm) of the threads is covered.
9. Check the set length a final time. If the 2 measurements are still not identical, you will need to repeat the procedure.

■ **If you are unable to get the set length measurements adjusted correctly at the converter end, you will need to adjust the cable joint at the jet end as detailed in the Jet Drive section.**

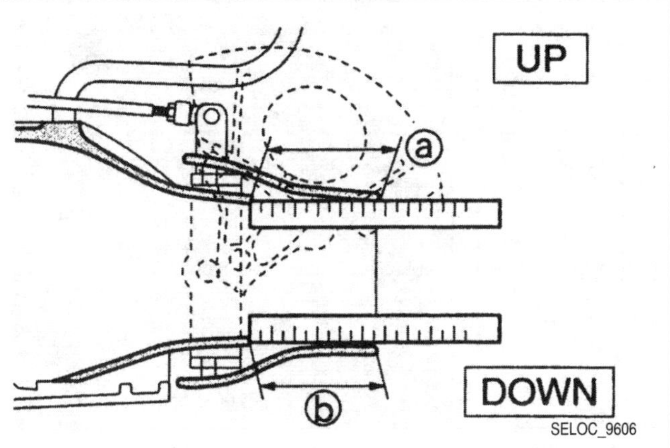

Fig. 97 Jet thrust nozzle set length. . .

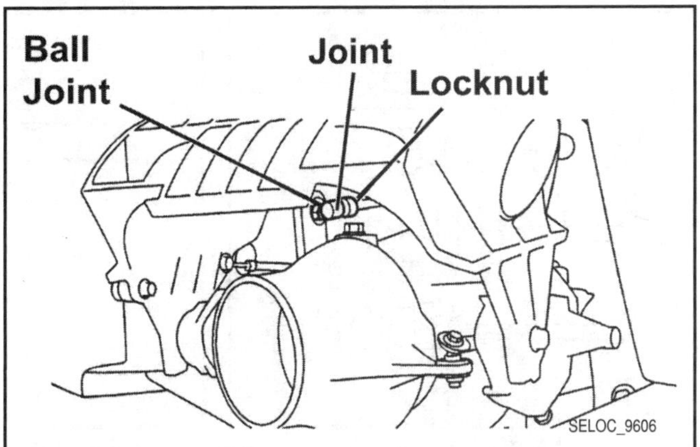

Fig. 98 . . .will require adjustment if the measurements are not the same

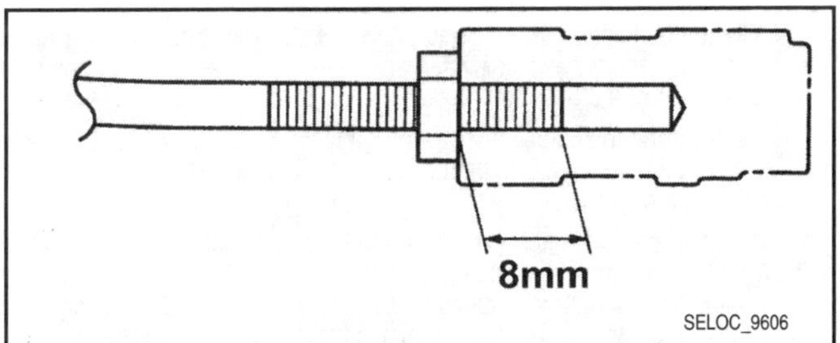

Fig. 99 Make sure the joint is covering at least 8mm of rod thread

Fig. 100 A good look at the QSTS cable and joint

SHIFT CABLE

2002-08 FX/FX Cruiser & 2004-07 FX HO/FX Cruiser HO Models

◆ See Figures 101 thru 106

2 MODERATE

1. Move the shift lever to the Reverse position and confirm that the reverse gate is in contact with the stopper on the bracket and that the lever is touching the gate.

2. Now move the shifter to the Forward position and check that the lever (at the drive) has been shifted over the bracket. If everything checks out, you're fine. Otherwise. . .

3. Loosen the locknut on the drive end of the cable and pop the joint off of the ball joint on the lever.

4. Move the reverse gate so it is against the stopper (on the bracket) and the lever is in contact with the gate.

5. Rotate the cable joint until it is lined up perfectly with the ball joint on the lever. Turning the joint in will shorten the length of the cable, while turning it out will lengthen the cable.

6. Once the joint socket is lined up with the ball joint, turn the joint outward 9 complete revolutions until the cable length has been extended 0.28 in. (7mm). Now connect the joint to the ball joint and tighten the locknut to 2.1 ft. lbs. (2.9 Nm).

■ **Always make sure that the cable joint is covering at least 0.31 in. (8mm) of the cable thread once the locknut is tightened.**

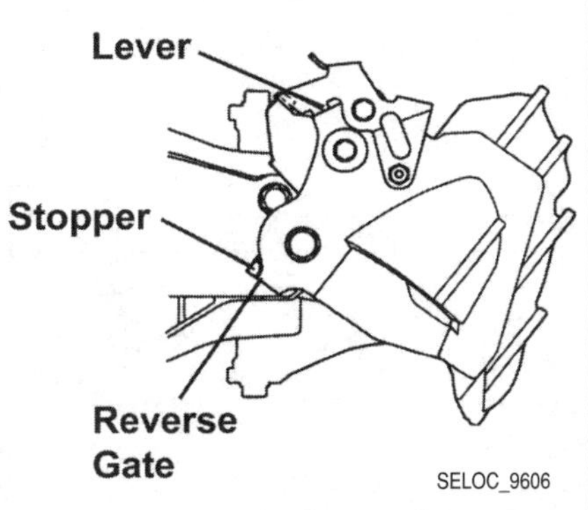

Fig. 101 With the shifter in Reverse, make sure the reverse gate is in contact with the stopper. . .

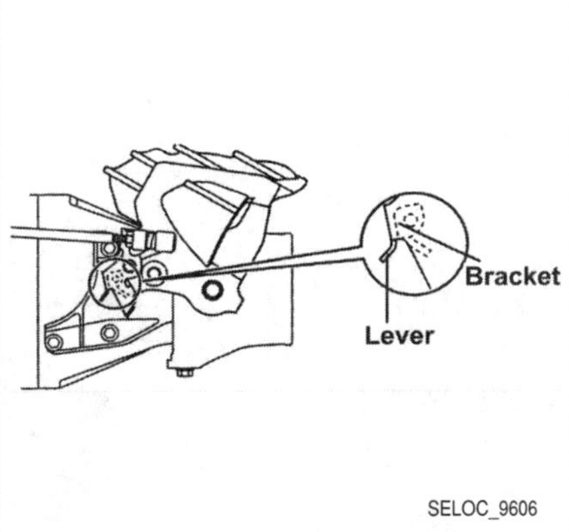

Fig. 102 . . .and that the lever is touching the gate

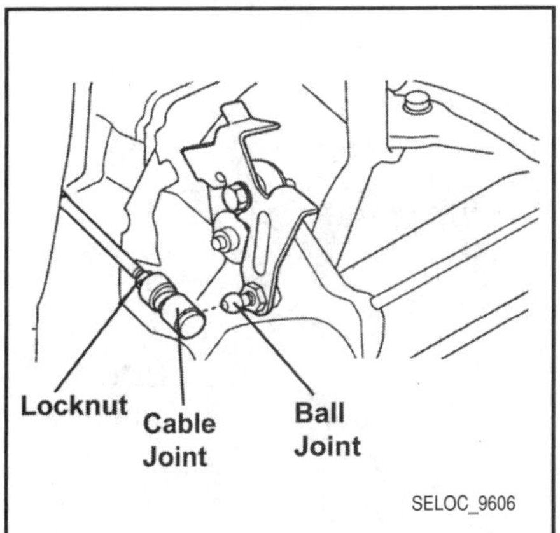

Fig. 103 Disconnect the cable joint to adjust the cable

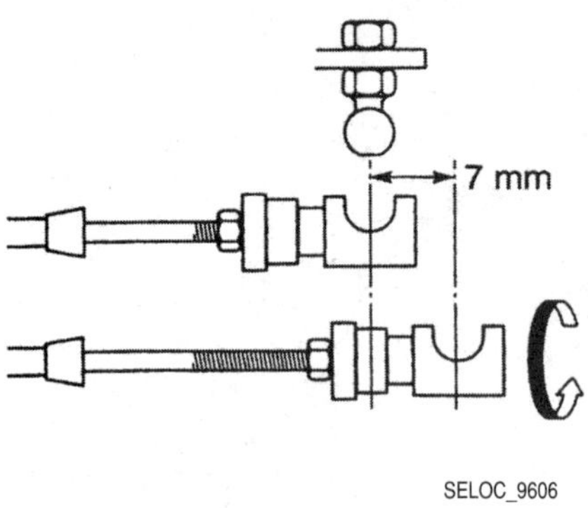

Fig. 104 Once everything is lined up correctly. . .

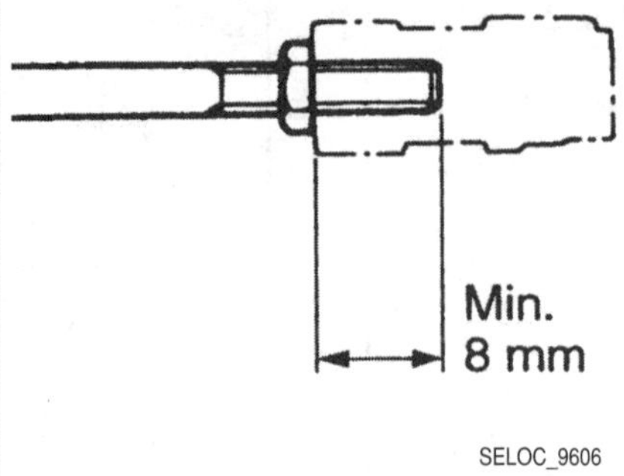

Fig. 105 . . .make sure that at least 8mm of the cable threads is covered by the cable join

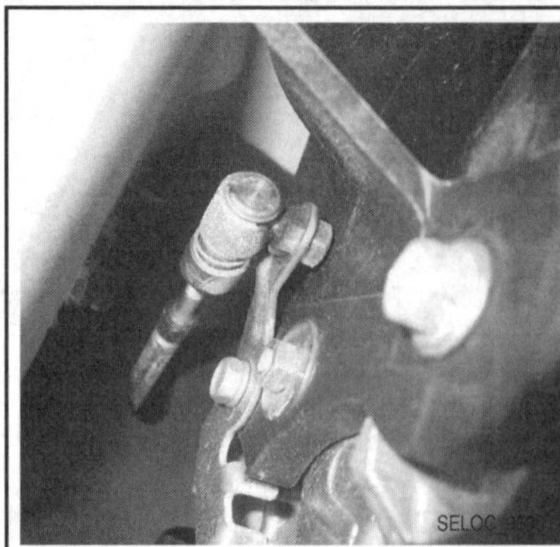

Fig. 106 A nice look at the shift cable, joint and lever

All 2005-11 VX models, 2008-11 FX Models & All FZ Models

◆ See Figures 106 thru 112

② MODERATE

1. Move the shift lever to the Reverse position and confirm that the reverse gate is in contact with the stopper on the bracket. If it does, you're OK; if not, move to the next step.

2. Loosen the locknut at the end of the shift cable joint and pop the joint off of the shift lever ball joint on the drive end.

3. Move the reverse gate so that it is resting against the stopper.

4. Rotate the joint on the cable to orient the cable joint directly over the ball joint/pivot. Turning the joint inward will decrease the length and turning it outward will increase the length.

5. Now extend the cable joint outward until the center of the joint is 0.28 in. (7mm) *past* the centerline of the ball joint (A). This should usually be 9 revolutions on most models.

6. Connect the joint back onto the ball joint and tighten the locknut to 2.1 ft. lbs. (2.9 Nm) on models w/1052cc engines (exc. 2010-11 VX), or 3 ft. lbs. (4 Nm) on models w/1812cc engines and the 2010-11 VX w/1052cc engine; using Loctite 572. Make sure that the joint is threaded onto the cable so that at least 0.31 in. (8mm) of thread has been covered.

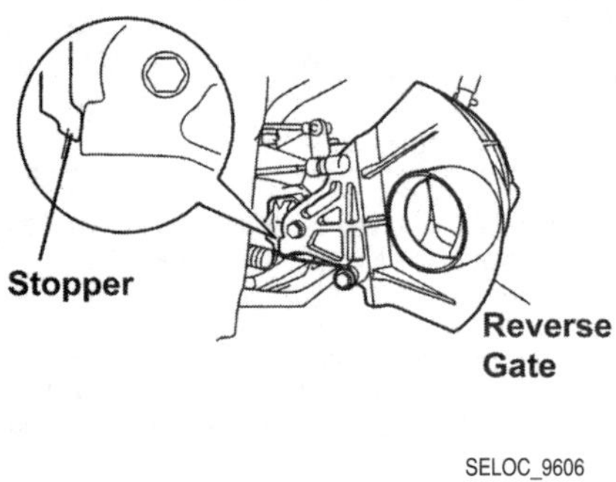

Fig. 107 Check to see if the reverse gate is touching the stopper. . .

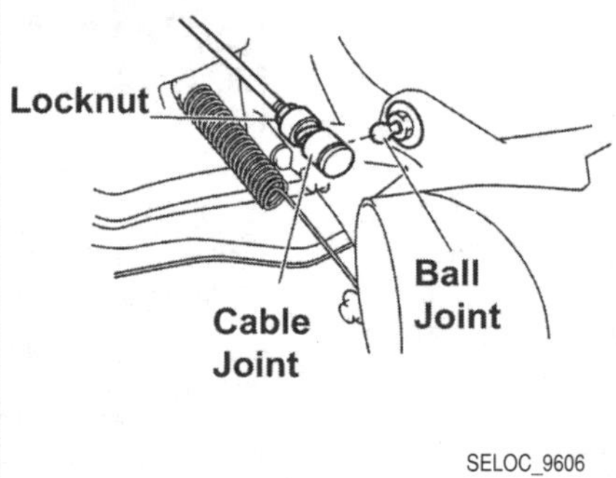

Fig. 108 . . .if not, disconnect the cable joint from the ball joint - FX/FZ models

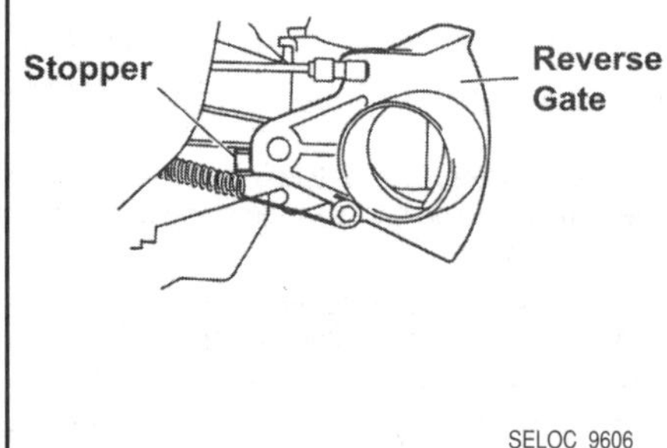

Fig. 109 Check to see if the reverse gate is touching the stopper. . .

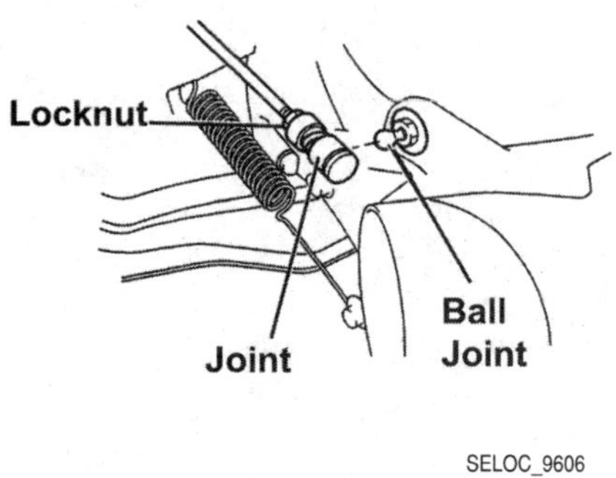

Fig. 110 . . .if not, disconnect the cable joint from the ball joint - VX models

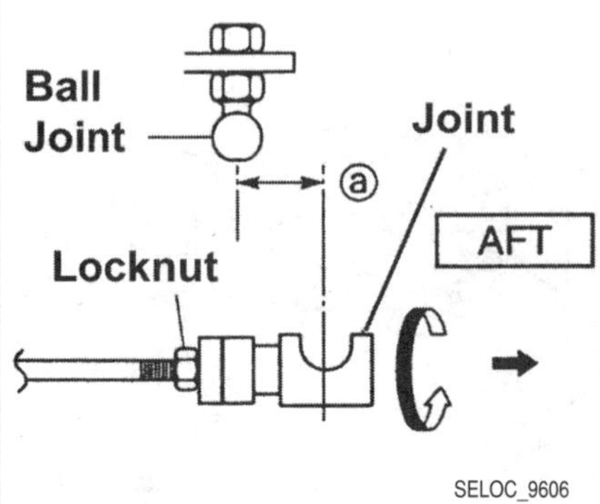

Fig. 111 Extend the joint even further before re-connecting it

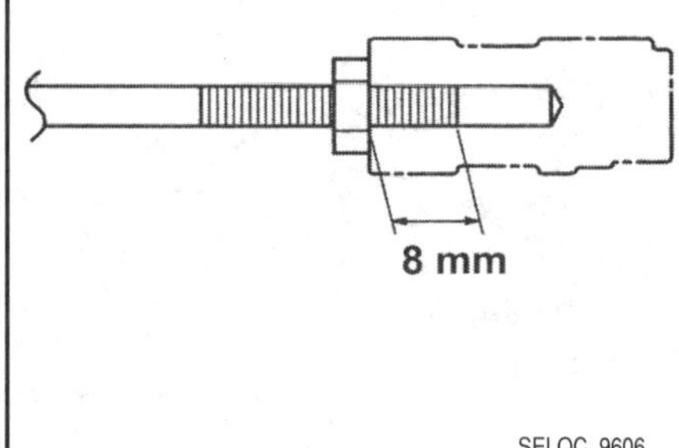

Fig. 112 Make sure that at least 8mm of cable thread is covered by the cable joint

STEERING CABLE

FX & FZ Models

MODERATE

◆ See Figures 113 thru 117

1. Set the QSTS control grip to the Neutral position.

2. On 2002-05 FX/FX Cruiser and FX/FX Cruiser HO models, move the handlebars from full lock to full lock. Measure the clearance between the edge of the jet thrust nozzle and the edge of the reverse gate stay. Move the handlebars to opposite lock and then measure again. Both measurements should be identical; if not, move to Step 4.

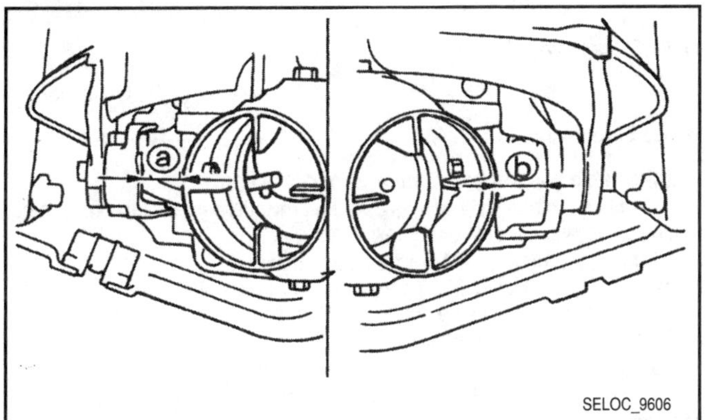

Fig. 113 Measuring the thrust nozzle clearance at opposite steering lock - 2002-05 FX/FX Cruiser & FX/FX Cruiser HO models

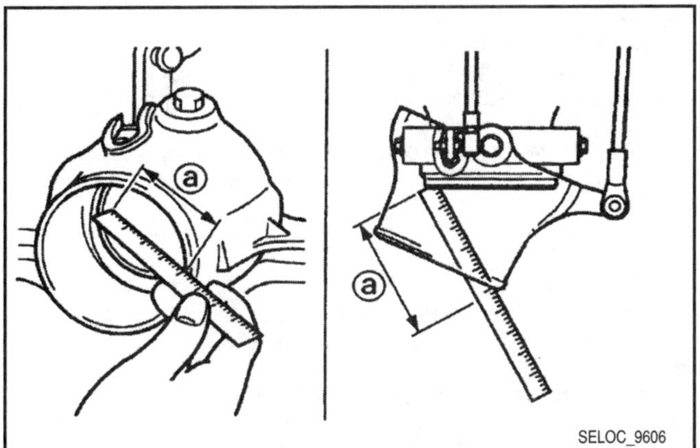

Fig. 114 Measuring the thrust nozzle clearance at steering lock. . .

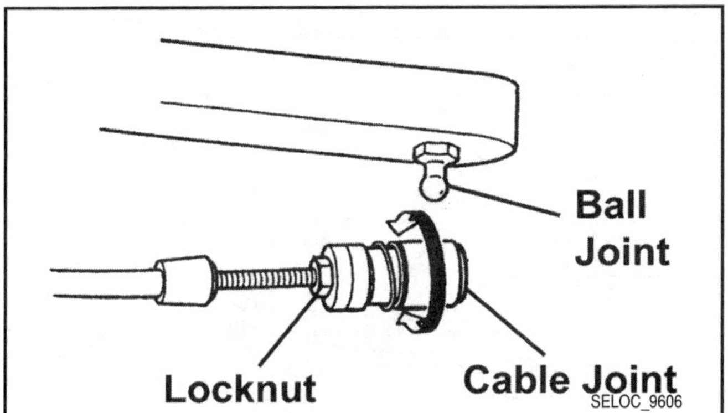

Fig. 116 Loosen the lock nut to remove the cable joint from the steering lever and then rotate it in or out to adjust clearance

3. On 2006-08 FX/FX Cruiser, 2006-10 FX/FX Cruiser HO/SHO and all FZ models, move the handlebars from full lock to full lock. Measure the clearance between the edge of the jet thrust nozzle and the opposite edge of the nozzle ring. Move the handlebars to opposite lock and then measure again. Both measurements should be within 0.2 in. (5mm) of each other; if not, move to the next step.

4. Move the QSTS control back to the Neutral position again and then find the forward steering joint where it connects to the steering lever in the hull (you'll probably have to remove the service lid).

5. Loosen the locknut on the cable joint and then pop it off the ball joint on the end of the lever.

6. Rotate the cable joint to adjust the nozzle clearance. Rotating the joint inward will increase the clearance on the (A) side of the nozzle. Rotating the joint outward will increase the clearance on the (B) side.

7. Install the joint back onto the steering lever and tighten the lock nut so it runs firmly up against the inner side of the joint - 4.6 ft. lbs. (6.4 Nm) on models w/1052cc engines, or 5.2 ft. lbs. (7 Nm) on models w/1812cc engines. Be very careful when adjusting the cable joint; it must always be at least 0.31 in. (8mm) down the threads of the cable.

8. Check the nozzle clearance again to confirm that both measurements are the same, If not, the cable joint will require further adjustment.

■ **If you are unable to equalize the nozzle clearances with the steering cable joint at the helm, you will need to move to the drive-end procedures in the Jet Drive section.**

VX Models

MODERATE

◆ See Figures 114, 115, 117

1. Set the control grip to the Neutral position.

2. Move the handlebars from full lock to full lock. Measure the clearance between the edge of the jet thrust nozzle and the opposite edge of the

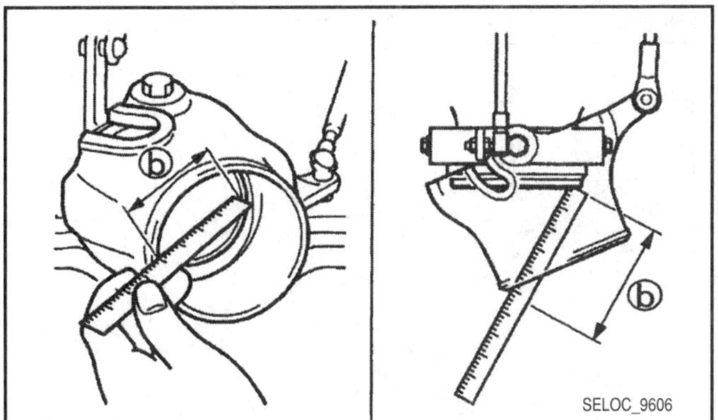

Fig. 115 . . .and opposite lock - VX, 2006-10 FX/FX Cruiser, FX/FX Cruiser HO/SHO and all FZ models

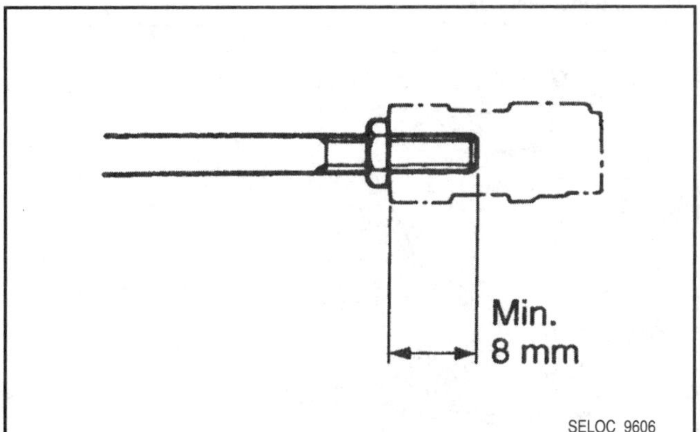

Fig. 117 There must be at least 0.31 in. (8mm) inside the joint

nozzle ring. Move the handlebars to opposite lock and then measure again. Both measurements should be within 0.2 in. (5mm) of each other; if not, move to the next step.

3. Move the control back to the Neutral position again and then find the forward steering joint where it connects to the steering lever in the hull (you'll probably have to remove the service lid).

4. Loosen the locknut on the cable joint and then pop it off the ball joint on the end of the lever.

5. Rotate the cable joint in or out to adjust the steering cable length until it is 8.61 in. (218.8mm) from the centerline of the ball joint at the lever to the top of the cable retaining bracket.

6. Working at the back of the vessel on the right side, measure the length of the cable from the hull to the centerline of the ball joint. Mark it down and then add it to the measurement taken in the previous step.

7. The sum of the measurements in the 2 previous steps should be 22.91 in. +/- 0.04 in. (581.8mm +/-1mm). If not, the adjustment is made at the drive end rather than at the lever in the vessel.

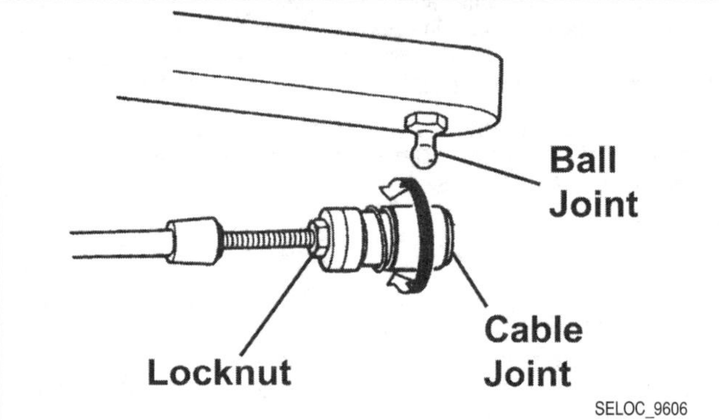

Fig. 118 Loosen the lock nut to remove the cable joint from the steering lever and then rotate it in or out to adjust clearance

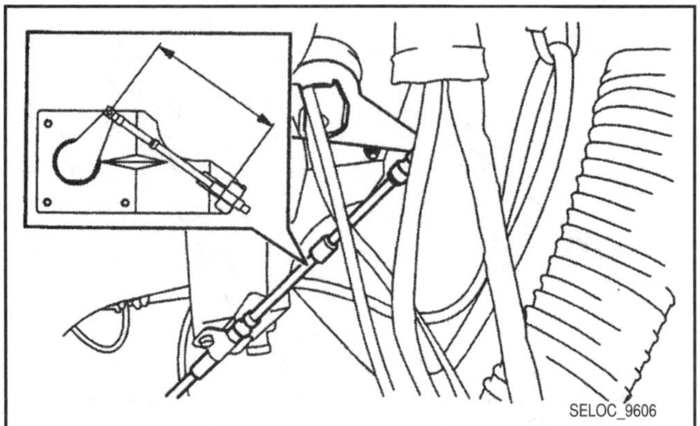

Fig. 119 Measure the cable length between the centerline of the joint and the top of the bracket. . .

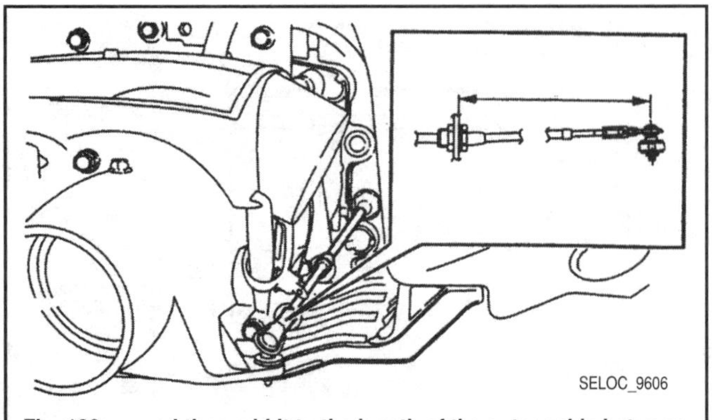

Fig. 120 . . .and then add it to the length of the outer cable between the hull and the centerline of the joint

8. Rotate the shift cable joint at the drive until the 2 measurements come into specification. This may take a couple times until you get it right. Once done, reconnect the cable joint to the ball joint and tighten the locknut to 4.9 ft. lbs. (6.8 Nm) on 2005-09 models, or 5.2 ft. lbs./63 inch lbs. (7 Nm) on 2010-11 models. Be very careful when adjusting the cable joint; it must always be at least 0.31 in. (8mm) down the threads of the cable.

9. Check the nozzle clearance again to confirm that both measurements are the same, If not, the cable joint will require further adjustment.

■ **If you are unable to equalize the nozzle clearances with the steering cable joint at the helm, you will need to move to the drive-end procedures in the Jet Drive section.**

THROTTLE CABLE

All FX Models With 998cc/1052cc Engines

② MODERATE

◆ **See Figures 121 thru 124**

If the cable has not been removed or replaced:

1. Check the trolling speed as detailed previously in this section. Once within specification, move to the next step.

2. Measure the throttle lever free-play - that is to say, the difference between the throttle lever when at rest and just before the cable starts to move when the lever is being squeezed. Clearance should be 0.16-0.28 in. (4-7mm), otherwise it will require adjustment.

3. Remove the handlebar cover as detailed in the appropriate section.

4. Loosen the cable locknut - the one butting up against the throttle lever assembly.

5. Turn the adjusting nut (the one against the cable housing) in to increase the free-play, or out to decrease the play.

6. Once free-play is within spec, tighten the locknut and install the cover, tightening the screws to 7 inch lbs. (1.1 Nm).

If the cable has been removed or replaced:

7. Remove the air filter case cover as detailed in the Air Filter section. Lift out the flame arrestor. Remove the ribbon sub-assembly on all but the 2002-05 FX models.

8. Pull back the cable end cover at the throttle body and loosen the large locknut on 2002-05 FX models. On all other models, just loosen the locknut.

9. Turn the smaller adjusting nut until you achieve a throttle cable adjusting length (a) of:

- 2002-05 FX/FX Cruiser - 0.72 in. +/- 0.04 in. (18.4mm +/- 1mm)
- 2004-05 FX/FX Cruiser HO - 0.96 in. +/- 0.04 in. (24.4mm +/- 1mm)
- 2006-08 FX/FX Cruiser and FX/FX Cruiser HO - 0.94 in. +/- 0.04 in. (23.9mm +/- 1mm)

This is the distance from the end of the cable housing to the outer edge of the adjusting nut on the 2002-05 FX/FX Cruiser models, or from the end of the cable cover to the edge of the block on all other models.

■ **When adjusting the cable at the throttle body, open the throttle fully and check that the butterfly valve in the throttle bore is fully open.**

10. Push the cover back into position (if removed) and then follow Steps 1-6.

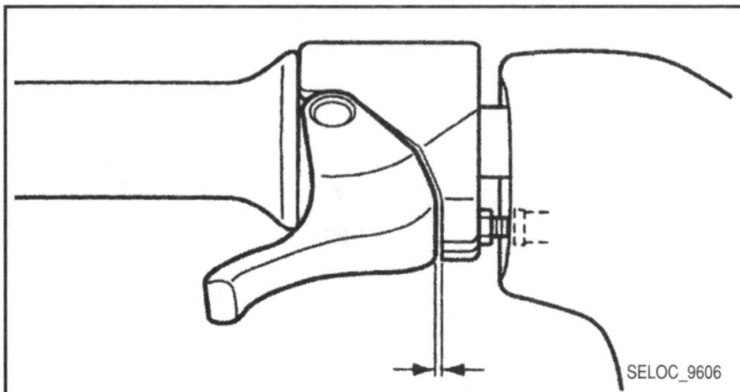

Fig. 121 Checking throttle cable free-play at the handlebar

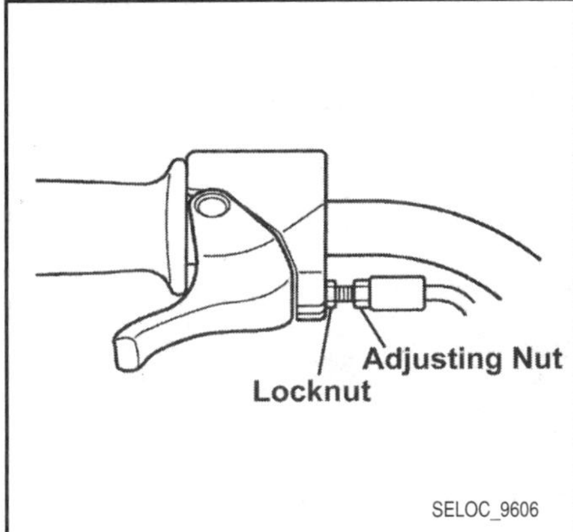

Fig. 122 A nice view of the locknut and adjusting nut

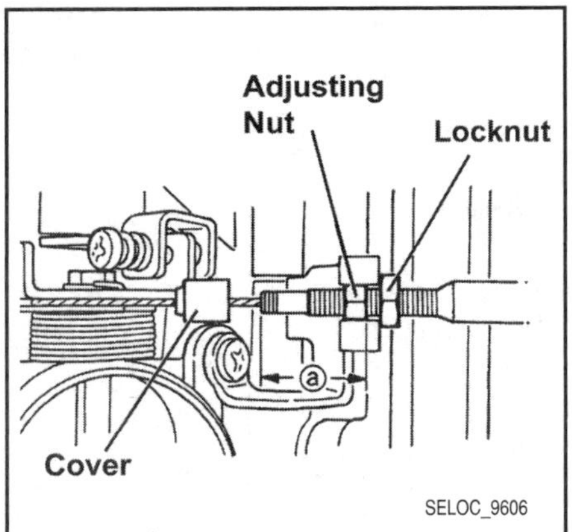

Fig. 123 Setting the throttle cable adjustment installation length - 2002-05 FX

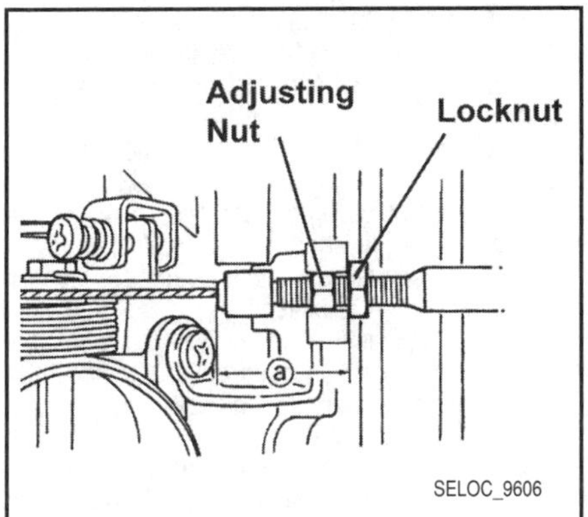

Fig. 124 Setting the throttle cable adjustment installation length - 2004-08 FX HO

In all instances now:

11. Install the handlebar cover and tighten the screws to 7 inch lbs. (1.1 Nm).

12. Install the air filter if removed. Install ribbon sub assembly (if applicable) and tighten to 4.7 ft. lbs. (6.5 Nm). Install the air filter case cover.

13. Once complete, start the engine and move the handlebars back and forth from lock to lock and make sure that the trolling speed does not change. Make sure you are hooked up to a suitable water supply when starting the engine.

■ **If you are not able to get the free-play within specifications, the entire cable will require replacement.**

All 2008-11 FX/FZ & 2010-11 VX Models With 1812cc Engines

2 MODERATE

◆ **See Figures 121 and 125 thru 128**

If the cable has not been removed or replaced:

1. Measure the throttle lever free-play - that is to say, the difference between the throttle lever when at rest and just before the cable starts to move when the lever is being squeezed. Clearance should be 0.08-0.20 in. (2-5mm) on all FX/FZ models 0.16-0.28 in. (4-7mm) on VX models; otherwise it will require adjustment.

2. Remove the handlebar cover as detailed in the Hull, Helm & Steering section.

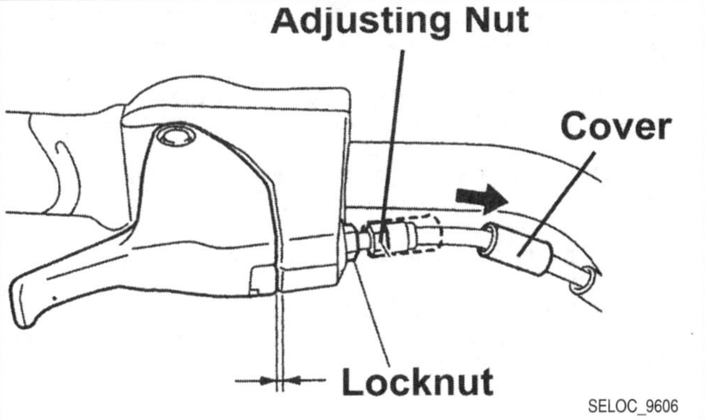

Fig. 125 Slide the rubber cover off of the adjusting nut

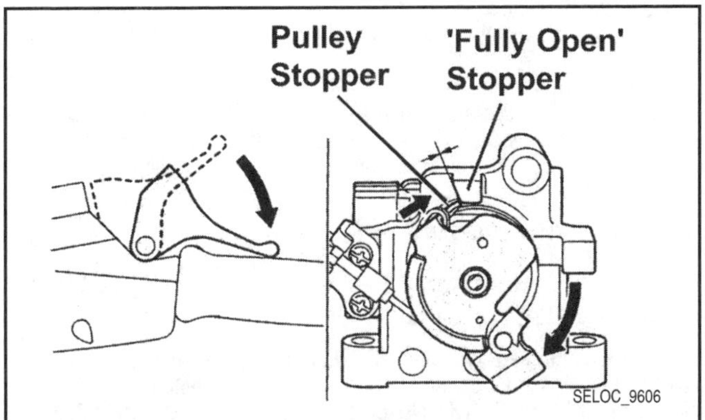

Fig. 126 Check the stopper on the APS unit when the throttle is fully opened. . .

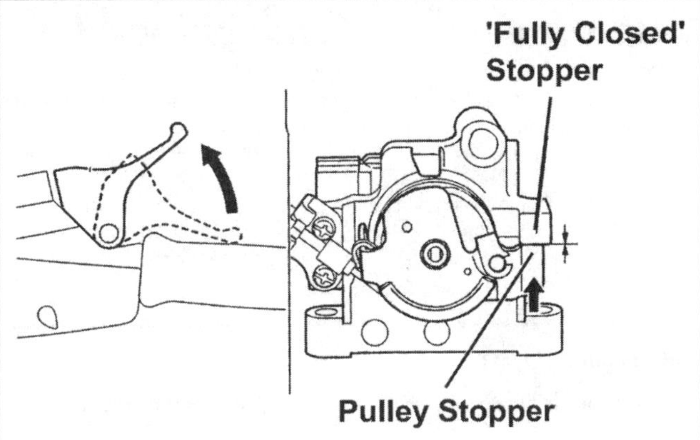

Fig. 127 . . .and fully closed

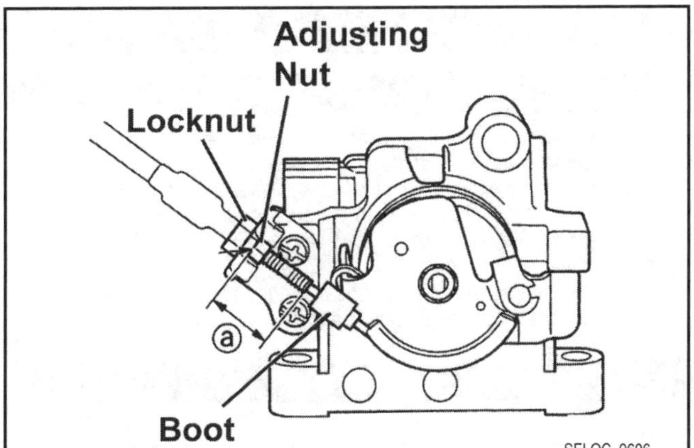

Fig. 128 Checking the throttle cable installation length at the APS

3. Move the handlebars to their lowest position on FZ models. Any other position will not allow the free-play to be within specifications.

4. Working at the handlebar, slide the rubber cover off of the locknut and adjusting nuts.

5. Loosen the cable locknut - the one butting up against the throttle lever assembly.

6. Turn the adjusting nut (the one against the cable housing) in to increase the free-play, or out to decrease the play.

7. Once free-play is within spec, tighten the locknut securely.

8. Locate the APS assembly (inside the service lid). Squeeze the throttle lever back into the fully open position and check that the APS pulley stopper is in contact with the 'fully open' stopper on the APS unit. Now release the throttle lever and confirm that the pulley stopper is in contact with the 'fully open' stopper on the APS unit.

9. On FZ models, move the handlebars back to the highest position and recheck the free-play again. If out of spec, try the adjustment procedure again.

■ **If you are still unable to get the free-play correctly adjusted and the throttle cable routing is correct, the cable itself will require replacement.**

If the cable has been removed or replaced:

10. Make sure the handlebars are facing straight ahead and then remove the front service lid.

11. Slide back the boot covering the lock nut and adjusting nut at the APS unit pulley side. Check that the throttle cable installation length (A) is 0.72 in. +/- 0.04 in. (18.4mm +/- 1mm) - this is the length from the top of the adjusting nut to the end of the threaded cable casing. If the measurement is within specifications, go back and perform Steps 1-9; if not, move to the next step.

12. Loosen the locknut and then turn the adjusting nut to bring the cable measurement into spec. Once done, coat the nut threads with Loctite 242 and then tighten the locknut to 5.2 ft. lbs. (7 Nm); also using the Loctite. Slide the boot back over the nuts.

13. To finish the procedure, you can now go back and follow Steps 1-9.

In all instances now:

14. Install the handlebar cover and tighten the screws to 7 inch lbs. (1.1 Nm).

15. Install the service lid.

16. Once complete, start the engine and move the handlebars back and forth from lock to lock and make sure that the trolling speed does not change. Make sure you are hooked up to a suitable water supply when starting the engine.

2005-11 VX Models w/1052cc Engines

② MODERATE

◆ **See Figures 121, 122 and 129**

■ **Yamaha has initiated a Factory Modification Campaign (TSB #WR2008-001R) for a possibly faulty outer throttle cable on all 2005-07 VX models, all 2008 VX/VX Deluxe/VX Sport models and certain 2008 VX Cruiser models (HIN #F1K-854101 thru 868200) that could corrode and become kinked or cracked. The campaign provides a new cable produced from a different material.**

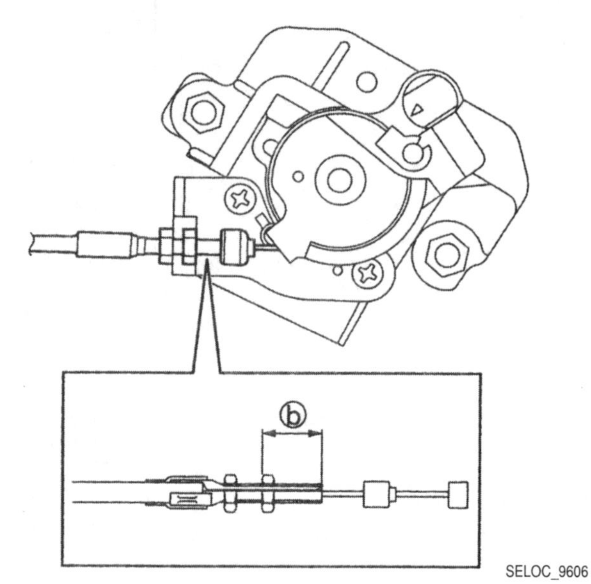

Fig. 129 Check the throttle cable length at the APS unit

1. Check the trolling speed as detailed previously in this section. Once within specification, move to the next step.

2. Measure the throttle lever free-play - that is to say, the difference between the throttle lever when at rest and just before the cable starts to move when the lever is being squeezed. Clearance should be 0.16-0.28 in. (4-7mm), otherwise it will require adjustment.

3. Remove the seat and then the engine cover.

4. Locate the APS unit in the engine compartment and check the throttle cable length between the outer edge of the adjusting nut and the end of the cable casing. Normal length should be 0.72 in. +/- 0.04 in. (18.4mm +/- 1mm)

5. Remove the handlebar cover as detailed in the appropriate section.

6. Loosen the cable locknut - the one butting up against the throttle lever assembly.

7. Turn the adjusting nut (the one against the cable housing) in to increase the free-play, or out to decrease the play.

8. Once free-play is within spec, tighten the locknut and install the cover, tightening the lower screw to 2.7 ft. lbs. (3.7 Nm), and the upper screw to 10 inch lbs. (1.1 Nm).

9. Install the engine cover and seat

10. Once complete, start the engine and move the handlebars back and forth from lock to lock and make sure that the trolling speed does not change. Make sure you are hooked up to a suitable water supply when starting the engine.

■ **If you are not able to get the free-play within specifications, the entire cable will require replacement.**

FLUIDS & LUBRICANTS

Fluid Disposal

Used fluids such as engine oil, gear oil, antifreeze and power steering fluid are hazardous wastes and must be disposed of properly and responsibly. Before draining any fluids it is always a good idea to check with your local authorities; in many areas there are recycling programs available for easy disposal. Service stations, Parts stores and Marinas also often will accept waste fluids for recycling.

Be sure of your local recyclers policies before draining any fluids, as many will not accept fluids that have been mixed together.

Fuel & Oil Recommendations

FUEL

All engines covered here are designed to run on unleaded fuel. Never use leaded fuel in your boat's engine. The minimum octane rating of fuel being used for your engine must be at least 86, which means regular unleaded, but some engines may require higher octane ratings. Fuel should be selected for the brand and octane that performs best with your engine.

The use of a fuel too low in octane (a measure of anti-knock quality) will result in spark knock. Newer systems have the capability to adjust the engine's ignition timing to compensate to some extent, but if persistent knocking occurs, it may be necessary to switch to a higher grade of fuel. Continuous or heavy knocking may result in engine damage.

OIL

◆ **See Figures 130 and 131**

Nothing affects the performance and durability of an engine more than the engine oil. If inferior oil is used, or if your engine oil is not changed regularly, the risk of piston seizure, piston ring sticking, accelerated wear of the cylinder walls or liners, bearings and other moving components increases significantly.

Maintaining the correct engine oil level is one of the most basic (and essential) forms of engine maintenance. Get into the habit of checking your oil on a regular basis; all engines naturally consume small amounts of oil, and if left neglected, can consume enough oil to damage the internal components of the engine. Assuming the oil level is correct because you "checked it the last time" can be a costly mistake.

If your engine has not been operated for more than 6 months it should be primed prior to starting.

When shutting the engine down, always let the engine idle a few minutes to bring engine temperature down to a normal level. Since the engine is, at least in part, cooled by engine oil, it is necessary to allow the engine oil temperature to stabilize prior to shutdown. Not allowing the temperature to stabilize can damage vital engine components.

Every container of engine oil for sale in the U.S. should have a label describing what standards it meets. Engine oil service classifications are designated by the American Petroleum Institute (API), based on the chemical composition of a given type of oil and testing of samples. The ratings include "S" (normal gasoline engine use) and "C" (commercial, fleet and diesel) applications.

Over the years, the S rating has been supplemented with various letters, each one representing the latest and greatest rating available at the time of its introduction. During recent years these ratings have changed and most recently (at the time of publication), the rating is SL or SJ. Each successive rating usually meets all of the standards of the previous alpha designation, but also meets some new criteria, meets higher standards and/or contains newer or different additives. Since oil is so important to the life of your engine, you should obviously NEVER use an oil of questionable quality. Oils that are labeled with modern API ratings, including the "energy conserving" donut symbol, have been proven to meet the API quality standards. Always use the highest grade of oil available. The better quality of the oil, the better it will lubricate the internals of your engine.

In addition to meeting the classification of the API, your oil should be of a viscosity suitable for the outside temperature in which your engine will be operating. Oil must be thin enough to get between the close-tolerance moving parts it must lubricate. Once there, it must be thick enough to

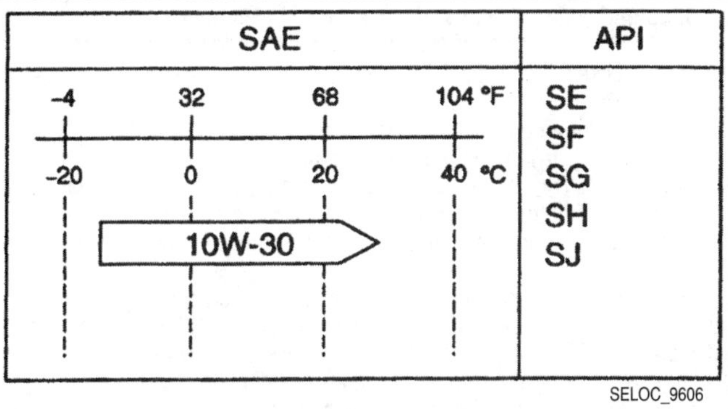

Fig. 130 Engine oil viscosity chart - 2002-08 FX models with 998/1052cc engines

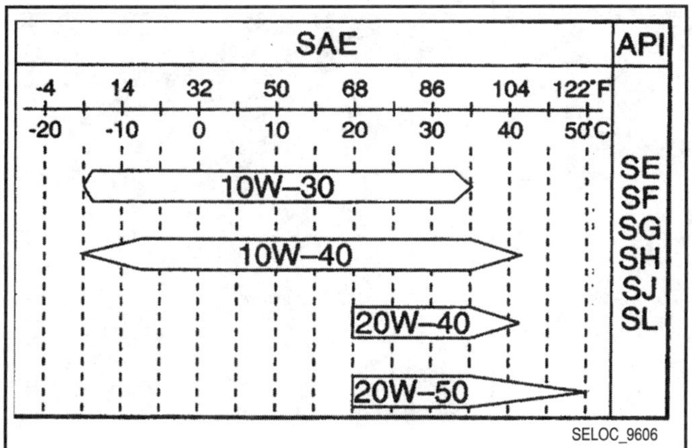

Fig. 131 Engine oil viscosity chart - most models, other than above

separate them with a slippery oil film. If the oil is too thin, it won't separate the parts; if it's too thick, it can't squeeze between them in the first place - either way, excess friction and wear takes place. To complicate matters, cold-morning starts require a thin oil to reduce engine resistance, while high speed boating requires a thick oil which can lubricate vital engine parts at temperatures.

According to the Society of Automotive Engineers' (SAE) viscosity classification system, an oil with a high viscosity number (such as SAE 40 or SAE 50) will be thicker than one with a lower number (SAE 10W). The "W" in 10W indicates that the oil is desirable for use in winter operation, and does not stand for "weight". Through the use of special additives, multiple-viscosity oils are available to combine easy starting at cold temperatures with engine protection at high speeds. For example, a 10W-40 oil is said to have the viscosity of a 10W oil when the engine is cold and that of a 40 oil when the engine is warm. The use of such an oil will decrease engine resistance and improve efficiency.

Yamaha has run through various recommendations with regard to multi-viscosity oils over the years so be certain to check your Owner's manual before purchasing your oil.

Most 2002-08 FX models with 998/1052cc engines come with the recommendation to use YamaLube Marine oil in a 10W-30 viscosity; or on 2004-09 VX models w/1052cc engines, 10W-30, 20W-40 or 20W-50. By 2008 or so on the FX/FZ models and the 2010-11 VX models (regardless of engine), Yamaha suggests that 10W-30, 10W-40, 20W-40 or 20W-50 are all acceptable. We were unable to find anything concrete on the use of synthetic oils, so suggest checking your Owner's manual or with your local dealer first.

Engine Oil

OIL LEVEL CHECK

Models With 998cc/1052cc Engines

◆ See Figures 132, 133 and 134

✱✱ CAUTION

The EPA warns that prolonged contact with used engine oil may cause a number of skin disorders, including cancer! You should make every effort to minimize your exposure to used engine oil. Protective gloves should be worn when changing the oil. Wash your hands and any other exposed skin areas as soon as possible after exposure to used engine oil. Soap and water, or waterless hand cleaner should be used.

When checking the oil level, it is best that the vessel be level and the oil be cool initially.

It is normal for an engine to naturally consume oil during the course of operation, particularly during break-in on a new engine. You should not be alarmed if the oil level in your engine drops slightly between inspections.

Also the color of the oil is usually a pitch black color. Smelling the oil is a better indicator of oil condition than the color. If the oil smells burned, it should be replaced immediately.

Over-filled crankcases can cause a fluctuation or drop in oil pressure. Take great care in checking and filling your engine with oil. Always maintain it between the **MIN** and **MAX** markings on the dipstick. Beware of false readings by checking the level too soon after adding oil.

Oil level should be checked each day the engine is operated.

1. Make sure the craft is level and then remove the seat(s). If on water, be careful, obviously, of wave action while the seat is off.

2. Clean off any debris that might be on the top of the oil tank, make sure you have a clean rag handy and then remove the cap and dipstick. Wipe the dipstick clean of any oil, reinsert it, screw the cap down firmly and then remove again.

3. Holding the dipstick level, check the oil level on the flat side of the stick. If there is no oil showing on the stick, add just enough oil through the filler hole until it shows on the tip of the stick (you may have to do this a couple times until is shows, because you don't want to put too much in. Thread the cap back into the tank.

4. **If you're on the water and have a 998cc engine:** pop the seat back on, start the engine and run it at 7000 rpm for about 5 minutes. Now throttle down and run it at trolling speed for another 2-3 minutes before shutting it off. Basically, you want to get the engine to its normal operating temperature, so adjust the amount of running time based on what you know that to be.

Fig. 132 The oil cap is on the top of the oil tank on most models. . .

Fig. 133 . . .remove it. . .

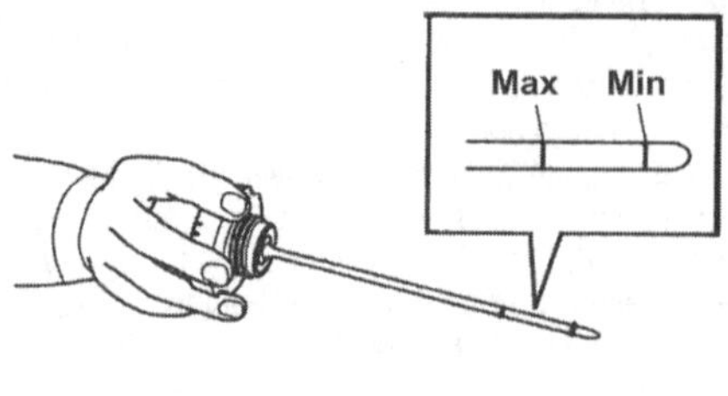

Fig. 134 . . .and check that the oil level is between the marks

5. **If you're on land and have a 998cc engine, or have a 1052cc engine:** connect a hose to the flush port in the engine compartment, turn on the water and then start the engine. Allow the engine to run at trolling speed for 6-8 minutes before turning off the water supply and the engine. Basically, you want to get the engine to its normal operating temperature, so adjust the amount of running time based on what you know that to be.

■ **If the air temperature is less than 59° F (15° C) on 2002-05 motors or 68° F (20° C) on 2006 and later motors, run the engine for at least another 5 minutes in both of the above situations.**

6. Now remove the dipstick, wipe it clean and reinsert it into the tank; threading the cap on all the way. Allow a few minutes for the oil to settle and then remove it once more and check the level on the dipstick again. Add or remove oil as necessary.

7. Add and/or remove oil slowly, a little at a time, rechecking the level frequently until you are satisfied the level is finally within the operating range.

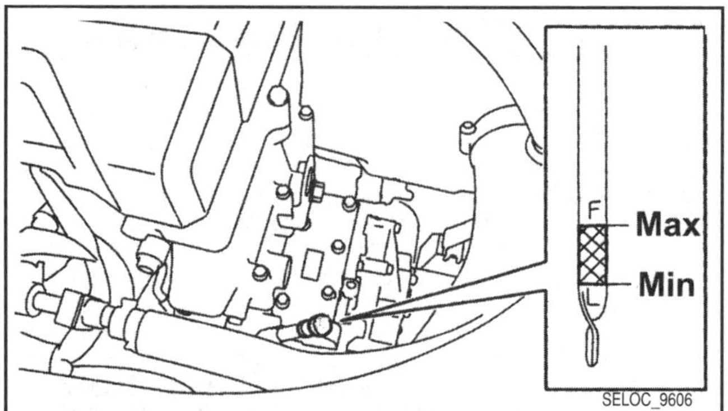

Fig. 135 Find the dipstick on the right side of the engine front and check to make sure the oil level is between the marks

8. Disconnect the hose if you were on land. Install the seat.

■ **Remember that if the engine does not get up to normal operating temperature before checking it in the 2nd stage, you will get a lower reading. Conversely, if the engine and oil get too hot before the check, you will notice a higher reading.**

Models w/1812cc Engines

◆ **See Figures 135 thru 138**

✳✳ CAUTION

The EPA warns that prolonged contact with used engine oil may cause a number of skin disorders, including cancer! You should make every effort to minimize your exposure to used engine oil. Protective gloves should be worn when changing the oil. Wash your hands and any other exposed skin areas as soon as possible after exposure to used engine oil. Soap and water, or waterless hand cleaner should be used.

When checking the oil level, it is best that the vessel be level and the oil be cool initially.

It is normal for an engine to naturally consume oil during the course of operation, particularly during break-in on a new engine. You should not be alarmed if the oil level in your engine drops slightly between inspections.

Also the color of the oil is usually a pitch black color. Smelling the oil is a better indicator of oil condition than the color. If the oil smells burned, it should be replaced immediately.

Over-filled crankcases can cause a fluctuation or drop in oil pressure. Take great care in checking and filling your engine with oil. Always maintain it between the **MIN** and **MAX** markings on the dipstick. Beware of false readings by checking the level too soon after adding oil.

Oil level should be checked each day the engine is operated.

1. Make sure the craft is level and then remove the seat(s). If on the water, be careful, obviously, of wave action while the seat is off.

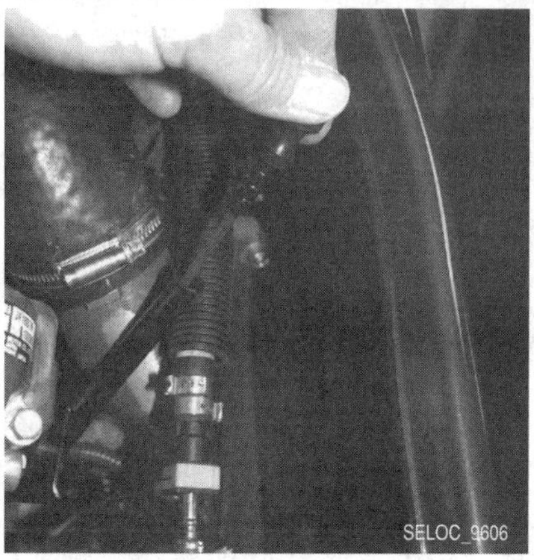

Fig. 136 Removing the dipstick

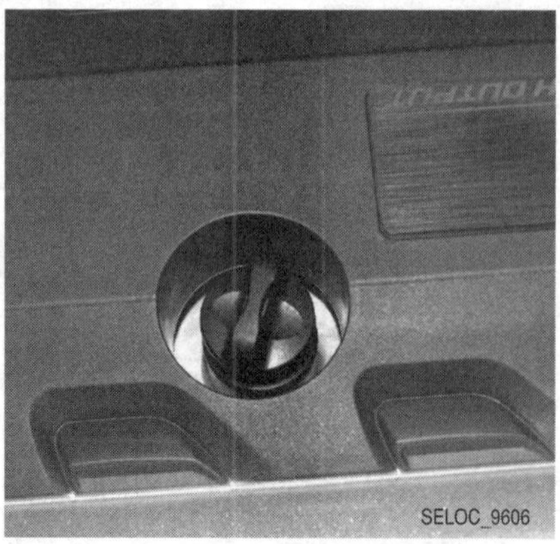

Fig. 137 You can access the oil cap through the engine cover. . .

Fig. 138 . . .and remove it to add oil

2. Locate the dipstick toward the front of the engine on the right hand side. Clean off any debris that might be on the top of the cap, make sure you have a clean rag handy and then remove the dipstick. Wipe the dipstick clean of any oil, reinsert it firmly and then remove again.

3. Holding the dipstick level, check the oil level on the flat side of the stick to be sure it is between the **Min** and **Max** marks. If the level is not between the lines, or there is no oil showing on the stick at all, you will need to add oil. Find the oil filler cap on the top of the engine, toward the rear - you do not have to remove the engine cover. Add just enough oil through the filler hole until it shows on the tip of the stick (you may have to do this a couple times until is shows, because you don't want to put too much in. Thread the cap back into place and install the dipstick.

4. Connect a hose to the flush port in the engine compartment, turn on the water and then start the engine. Allow the engine to run at idle for 6-8 minutes before turning off the water supply and the engine. Basically, you want to get the engine to its normal operating temperature, so adjust the amount of running time based on what you know that to be.

5. Allow a few minutes for the oil to settle and then remove the dipstick, wipe it clean and reinsert it; pressing the dipstick in all the way. Remove it once more and check the level on the dipstick again. Add or remove oil as necessary.

6. Add and/or remove oil slowly, a little at a time, rechecking the level frequently until you are satisfied the level is finally within the operating range.

7. Disconnect the flush hose. Install the seat.

■ **Remember that if the engine does not get up to normal operating temperature before checking it in the 2nd stage, you will get a lower reading. Conversely, if the engine and oil get too hot before the check, you will notice a higher reading.**

OIL & FILTER CHANGE

Models With 998cc/1052cc Engines

OEM ① EASY

◆ See Figures 139 thru 145

✻✻ CAUTION

The EPA warns that prolonged contact with used engine oil may cause a number of skin disorders, including cancer! You should make every effort to minimize your exposure to used engine oil. Protective gloves should be worn when changing the oil. Wash your hands and any other exposed skin areas as soon as possible after exposure to used engine oil. Soap and water, or waterless hand cleaner should be used.

A few precautions can make the messy job of oil and filter maintenance much easier. By placing oil absorbent pads, available at industrial supply stores, into the area below the engine, you can prevent oil spillage from reaching the bilge.

It is a good idea to warm the engine oil first so it will flow better; this will also ensure that many of the contaminates in the bottom of the pan will be suspended in the oil. This is accomplished by starting the engine and allowing it to reach normal operating temperature.

These engines do not have a conventional drain plug at the bottom of the engine so an evacuation pump must be used to remove the engine oil through the dipstick hole. Oil suction pumps come in a variety of configurations: motorized, hand-pumped or attached to an ordinary household drill.

The maintenance interval for oil and filter change on most models is every 100 hours of engine operation or annually, whichever comes first. There are a few models though that Yamaha suggests every 200 hours or 2 years - check the Maintenance Intervals chart for yours. This interval should be strictly kept; in fact we recommend cutting the interval in half!

■ **Since you will be hanging into the engine compartment, gather all the tools and spare parts necessary for the job. Don't forget plenty of rags to clean up any spills, and most importantly, remember a container or plastic milk jug to drain the oil into.**

1. Start the engine and run it until it reaches normal operating temperature. Turn the engine off.

✻✻ CAUTION

Make sure you have a flush kit attached.

2. Remove the seat.

3. Make sure you've got your pump, bucket and rags handy and then unscrew the oil tank cap/dipstick.

4. Insert the suction hose from the evacuation pump into the oil tank. There should be a baffle, so snake the hose down through the dipstick cut-out until you are sure the hose is at the bottom of the tank.

5. Crank, or turn on, the pump and begin removing the oil from the tank and into your container. When you notice no oil coming out of the expulsion hose, stop the pump. Wait a few minutes in case there's still residual oil in the tank and then run the pump a final time.

✻✻ CAUTION

Use caution around the hot oil; when at operating temperature, it is hot enough to cause a severe burn.

■ **Now is the time to change the oil filter as well. Although many folks change their filters every other oil change, we just don't understand it. Why take a chance?**

6. Locate the filter on the left side of the engine, near the back. Look hard, it's really there. As you can see, there's not a lot of room to work with; so make sure you've got some nice, absorbent, rags or towels spread out under the filter on the hull to absorb the inevitable oil that is going to spill.

7. The lack of room here will not allow you to use a conventional oil filter strap wrench so you'll need the kind that slips over the end of the filter. Yamaha offers one (#YU-38411/90890-01426) or you can purchase one at any auto parts retailer. Slide the wrench over the end of the filter and then insert a socket ratchet (should be 3/8 in.) into the opening on the filter wrench. Use the ratchet to turn the filter wrench and filter counterclockwise until it breaks loose. You can continue to use the wrench to remove the filter, but remember there's a decent amount of oil in that filter, so the faster you can get the opening turned upward, the less oil you'll spill. We recommend

Fig. 139 The oil filter may look easy to get to here, but it's a little tighter with the engine in the hull!

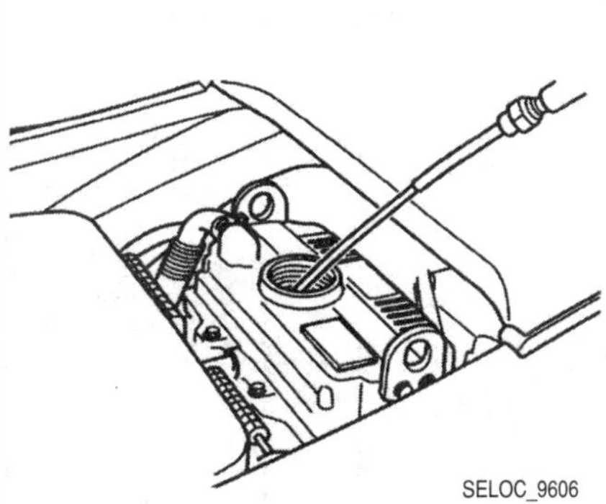

Fig. 140 Remove the dipstick cap on the oil tank and use a suction pump to remove the oil

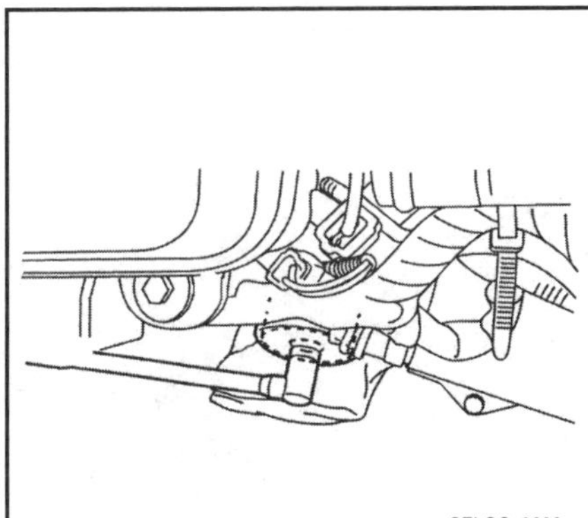

Fig. 141 Use a filter wrench and ratchet to loosen the filter. . .

removing the wrench/ratchet and unthreading the filter the remainder of the way with your hands. Once loose, get it turned upward quickly, being careful as you work it up and out of the engine compartment.

■ **The filter illustrations are of a typical application, although they may not look exactly like you engine, the procedure is the same.**

8. Coat the O-ring on your new filter lightly with *clean* engine oil and then reach down and thread the filter back onto the block until it is snug. Install the filter wrench and tighten the filter to 12 ft. lbs. (17 Nm) by using a torque wrench in place of the ratchet.

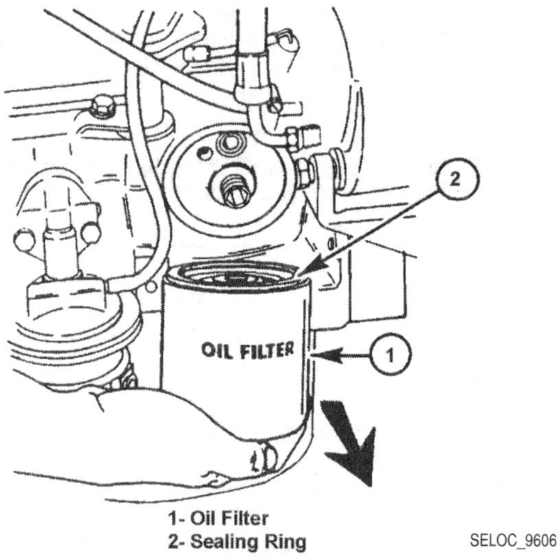

Fig. 142 . . .and then remove it the rest of the way with your hand

9. Add fresh engine oil through the dipstick opening in the oil tank until it is at the correct level on the dipstick. Remember that the design of these engines is such that you will not have drained all of the oil, so put the oil in slowly, checking frequently. It is unlikely you will use the full amount as shown in the Capacities chart.

10. Wipe up any spilled engine oil and remove the shop towels. Install the seat.

Models w/1812cc Engines

OEM ① EASY

◆ **See Figures 142 thru 148**

⁂ CAUTION

The EPA warns that prolonged contact with used engine oil may cause a number of skin disorders, including cancer! You should make every effort to minimize your exposure to used engine oil. Protective gloves should be worn when changing the oil. Wash your hands and any other exposed skin areas as soon as possible after exposure to used engine oil. Soap and water, or waterless hand cleaner should be used.

A few precautions can make the messy job of oil and filter maintenance much easier. By placing oil absorbent pads, available at industrial supply stores, into the area below the engine, you can prevent oil spillage from reaching the bilge.

It is a good idea to warm the engine oil first so it will flow better; this will also ensure that many of the contaminates in the bottom of the pan will be suspended in the oil. This is accomplished by starting the engine and allowing it to reach normal operating temperature.

These engines do not have a conventional drain plug at the bottom of the engine so an evacuation pump must be used to remove the engine oil through the dipstick hole. Oil suction pumps come in a variety of configurations: motorized, hand-pumped or attached to an ordinary household drill.

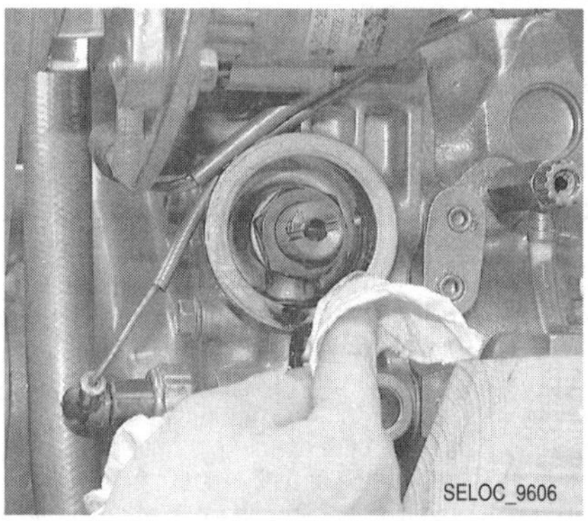

Fig. 143 Before installing the new filter, wipe the mating surface with a clean rag

Fig. 144 Dip your finger into a container of fresh oil and lubricate the O-ring on the new filter

Fig. 145 If you are unable to get your torque wrench on the new filter, tighten it 1/2 to 3/4 of turn more after the O-ring/gasket has come in contact with the mating surface

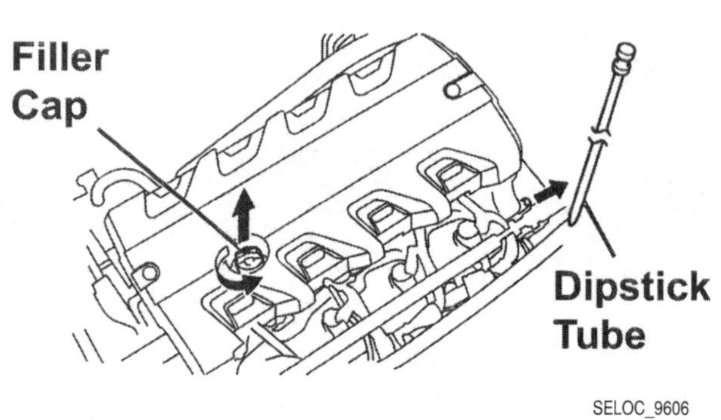

Fig. 146 Remove the dipstick and the oil filler cap. . .

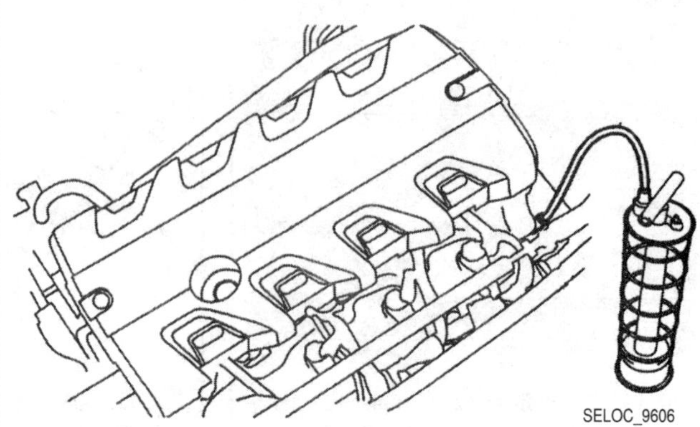

Fig. 147 . . .and use a suction pump to remove the oil

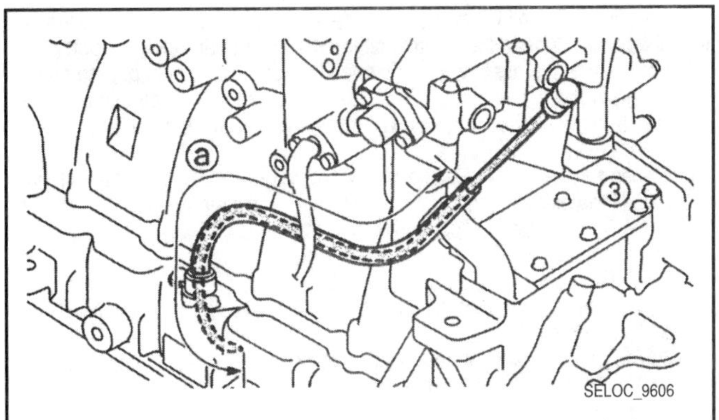

Fig. 148 Make sure that you insert the suction pump hose in far enough

The maintenance interval for oil change on these models is every 50 hours of engine operation or annually, whichever comes first. For the filter, Yamaha suggests every 100 hours or 1 year - check the Maintenance Intervals chart for yours. This interval should be strictly kept; in fact we recommend cutting the interval in half!

■ **Since you will be hanging into the engine compartment, gather all the tools and spare parts necessary for the job. Don't forget plenty of rags to clean up any spills, and most importantly, remember a container or plastic milk jug to drain the oil into.**

1. Start the engine and run it until it reaches normal operating temperature. Turn the engine off.

**** CAUTION**

Make sure you have a flush kit attached.

2. Remove the seat.
3. Make sure you've got your pump, bucket and rags handy and then unscrew the oil fill cap and remove the dipstick.
4. Insert the suction hose from the evacuation pump into the dipstick tube. Make sure that the hose goes into the tube at least 21.7 in. (550mm) on FX HO and VX models or 25.6 in. (650mm) on FX SHO and FZ models.
5. Crank, or turn on, the pump and begin removing the oil from the tank and into your container. When you notice no oil coming out of the expulsion hose, stop the pump. Wait a few minutes in case there's still residual oil in the tank and then run the pump a final time.

**** CAUTION**

Use caution around the hot oil; when at operating temperature, it is hot enough to cause a severe burn.

■ **Now is the time to change the oil filter as well. Although many folks change their filters every other oil change, we just don't understand it. Why take a chance?**

6. Locate the filter on the left side of the engine, near the front. Look hard, it's really there. As you can see, there's not a lot of room to work with; so make sure you've got some nice, absorbent, rags or towels spread out under the filter on the hull to absorb the inevitable oil that is going to spill.
7. The lack of room here will not allow you to use a conventional oil filter strap wrench so you'll need the kind that slips over the end of the filter. Yamaha offers one (#90890-06830) or you can purchase one at any auto parts retailer. Slide the wrench over the end of the filter and then insert a socket ratchet (should be 3/8 in.) into the opening on the filter wrench. Use the ratchet to turn the filter wrench and filter counterclockwise until it breaks loose. You can continue to use the wrench to remove the filter, but remember there's a decent amount of oil in that filter, so the faster you can get the opening turned upward, the less oil you'll spill. We recommend removing the wrench/ratchet and unthreading the filter the remainder of the way with your hands. Once loose, get it turned upward quickly, being careful as you work it up and out of the engine compartment.

■ **The filter illustrations are of a typical application, although they may not look exactly like your engine, the procedure is the same.**

8. Coat the O-ring on your new filter lightly with *clean* engine oil and then reach down and thread the filter back onto the block until it is snug. Install the filter wrench and tighten the filter to 13.3 ft. lbs. (18 Nm) by using a torque wrench in place of the ratchet.
9. Add fresh engine oil through the oil filler cap until it is at the correct level on the dipstick. Remember that the design of these engines is such that you will not have drained all of the oil, so put the oil in slowly, checking frequently. It is unlikely you will use the full amount as shown in the Capacities chart.
10. Wipe up any spilled engine oil and remove the shop towels. Install the seat.

Jet Pump/Intermediate Housing

LUBRICATION

① EASY

◆ **See Figures 149, 150 and 151**

The jet pump intermediate housing should be lubed with Yamaha A water resistant grease at 10 hrs. from start of service and then every 100 hrs. or 12 months, whichever comes first.

The initial service will require 1.11-1.18 oz, while subsequent services will only require 0.20-0.27 oz.

Locate the grease fitting next to the thermostat housing (on the oil tank) and remove the dust cap on FX models thru 2008. On other FX, VX and all FZ models, it will be found attached to the flush fitting retaining bracket. On 2010-11 VX models, you will find it on top of the housing itself.

Connect the grease gun and squeeze in the appropriate amount of grease.

Fig. 149 The intermediate housing grease fitting is located next to the thermostat housing on some models. . .

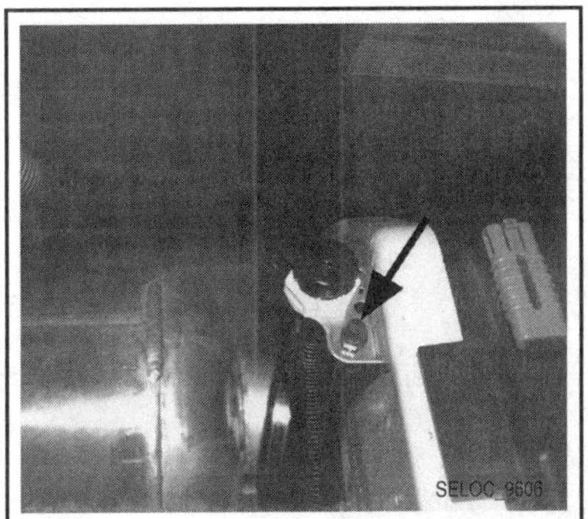

Fig. 150 . . .or in the flush fitting retaining bracket on others. . .

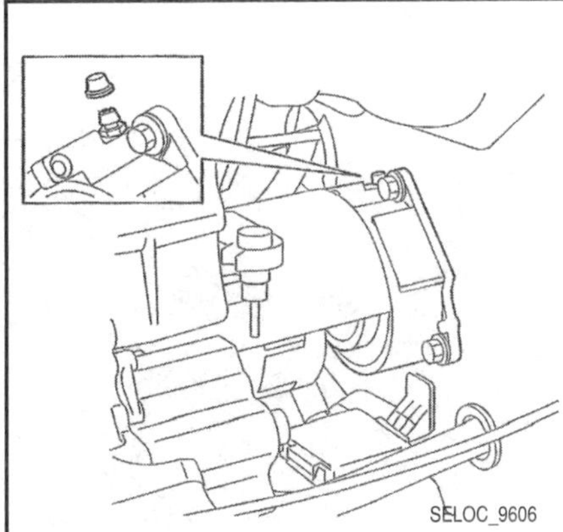

Fig. 151 . . .or on top of the housing itself

Lubrication Points

STEERING MASTER ASSEMBLY - FZ MODELS

◆ **See Figures 152 and 153**

1. Move the handlebars to the highest setting and then remove the handlebar covers.
2. Inject Yamaha Grease A into the grease nipple on the side of the handlebar bracket until it begins to seep out of the lower end of the bracket.
3. Squeeze a little grease onto the lower portion of the sliding surface and spread it around the tube.
4. Move the handlebars back to the lowest setting and repeat the previous step for the exposed portion of the sliding surface.
5. Now move the handlebars up and down through their range a few time to ensure the sliding surface is fully lubricated.
6. Remove any excess grease and install the handlebar covers. Tighten the longer screws to 3.7 ft. lbs. (5 Nm) and the shorter ones to 6 inch lbs. (1 Nm).

THROTTLE CABLE

◆ **See Figures 154 thru 157**

The throttle cable ends should be lubricated about every 100 hours of operation, or once a year, whichever comes first. The handle bar end will need a shot of rust inhibitor, while the other end takes Yamaha A water resistant grease.

1. On 2005-09 VX and 2002-05 FX models, squeeze the throttle lever all the way back to the handle bar and then pop out the little rubber seal where it runs into the mount - just slide it back on the exposed cable. Use the small straw on the can and squirt a little shot onto the cable and into the cavity.
2. Press the seal back into place and use a clean rag to wipe any spray that may have dripped.
3. 2006-11 FX, all FZ and 2010-11 (either engine) VX models do it a little differently - loosen the throttle cable-end lock nut and pull the outer cable out of the grip assembly so that you can spray into the opening of the thread end of the cable casing. Same idea as the other models but a little more labor intensive because you will probably need to remove the handlebar cover.
4. Insert the end of the cable back into the grip and tighten the lock nut securely. Hopefully you did not move the inner adjustment nut, but if you did, be sure you check the free-play adjustment detailed in the Adjustment section.
5. On FX models with 998/1052cc engines, remove the seat. Remove the air filter case cover as detailed in the Air Filter section. Or remove the engine cover on models so equipped.
6. The cable comes into the filter case on the port (left) side of the engine. Follow it in toward the throttle bodies and dab a small amount of grease on the exposed cable.
7. Install the case cover and replace the seat.
8. On FX/FZ models w/1812cc engines and all 2010-11 VX models, locate the accelerator position sensor (APS) unit toward the front of the craft and lubricate the cable and return spring with Yamaha A water repellent grease.

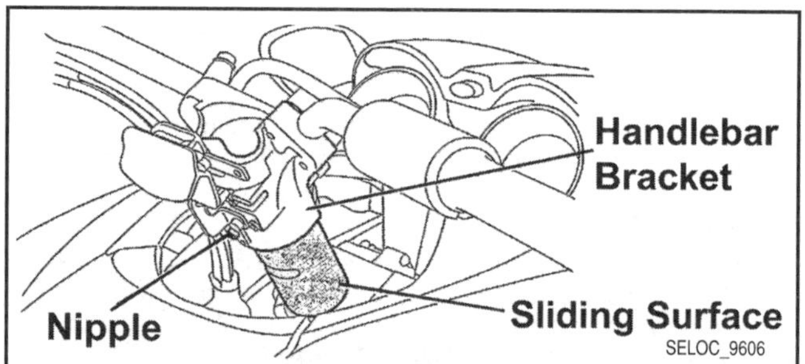

Fig. 152 Inject grease through the nipple until it comes out the bottom of the handlebar bracket and then lube the sliding surface

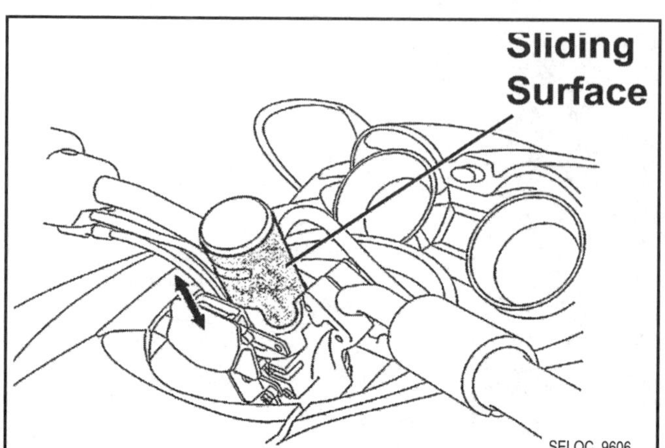

Fig. 153 Now lower the bars and lube the top portion of the surface

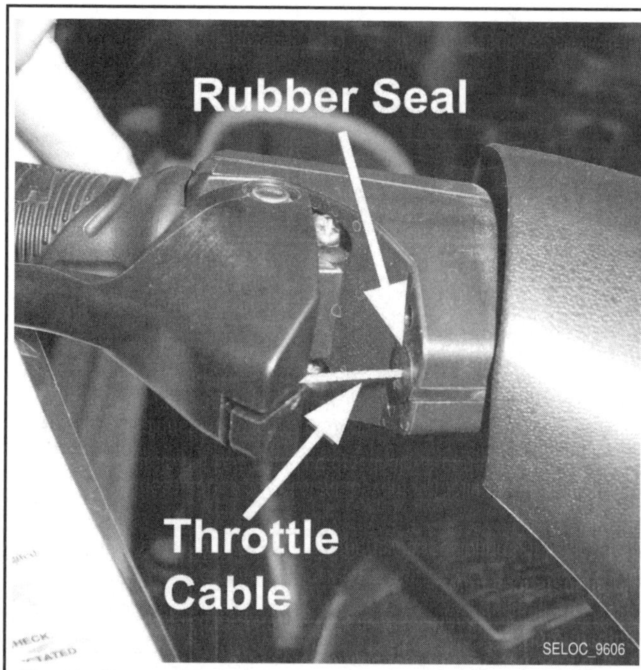

Fig. 154 Remove the rubber seal before spraying (2005-09 VX and 2002-05 FX)

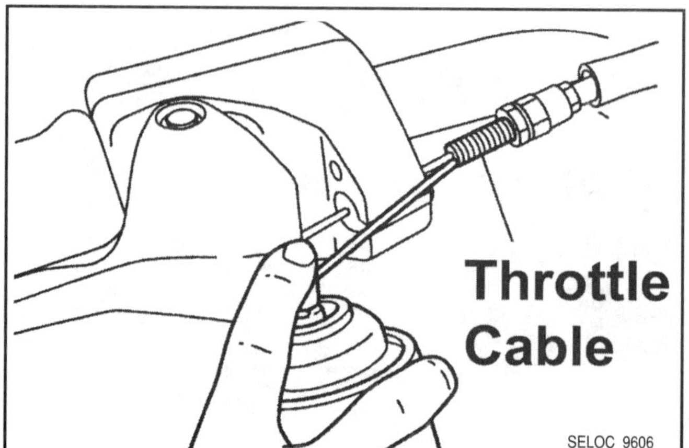

Fig. 155 On other models, you'll need to remove the handle bar cover and then the cable

Fig. 156 This end of the throttle cable gets a little bit of grease - FX models with 998/1052cc engines

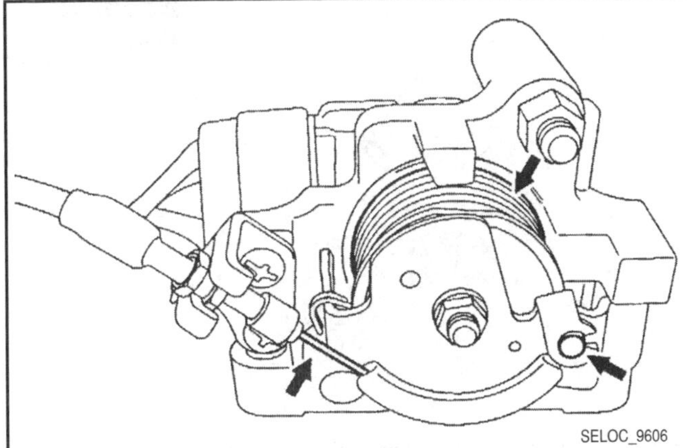

Fig. 157 This end of the throttle cable (APS) gets a little bit of grease - FX and FZ models w/1812cc engines and 2010-11 VX models

THROTTLE BODY

2006-08 FX Models w/1052cc Engines

MODERATE

◆ See Figures 158 and 159

Although Yamaha does provide a specific interval for lubricating the throttle body cable pulley, cam or return spring, it's probably a good idea to do it at least once a season. Always use Yamaha A water resistant grease.

1. Remove the seat and disconnect the negative battery cable.
2. Remove the air filter case cover as detailed in the Air Filter section.
3. Remove the 4 mounting bolts and lift off the flame arrestor(s); this will provide full access to the throttle bodies.
4. Carefully lubricate the cable pulley, cam and return springs for each throttle body.
5. Install the flame arrestor and tighten the bolts in 2 passes; once to 2.4 ft. lbs. (3.3 Nm) and then to 4.7 ft. lbs. (6.5 nm).
6. Install the air filter case cover.
7. Connect the battery cable and install the seat.

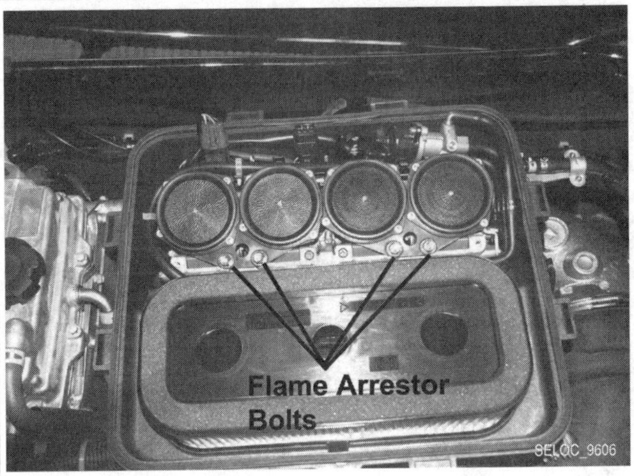

Fig. 158 You'll need to remove the flame arrestor. . .

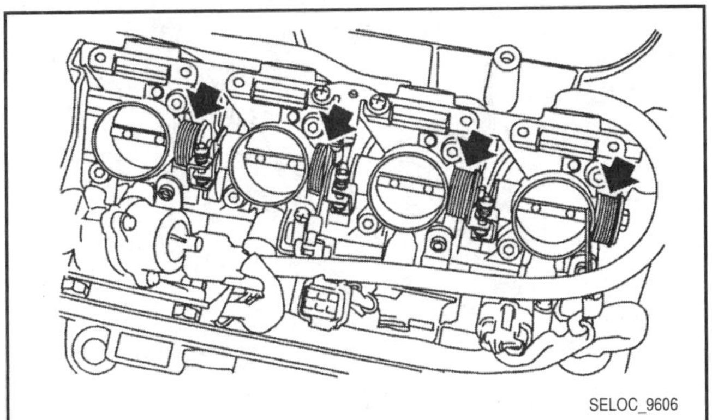

Fig. 159 . . .to gain access to the throttle bodies

FX HO Models w/1812cc Engines & All 2010-11 VX Models

MODERATE

◆ See Figures 160 and 161

Although Yamaha does provide a specific interval for lubricating the throttle body, it's probably a good idea to do it at least once a season. Always use Yamaha Rust Inhibiter.

1. Remove the seat and disconnect the negative battery cable.
2. Remove the engine cover if equipped.
3. Loosen the hose clamp and then wiggle the air intake hose off of the throttle body.
4. Spray a quick blast of Rust Inhibiter into the throttle bore.
5. Slide the hose back into position and tighten the hose clamp securely (2.2 ft. lbs./3 Nm).
6. Install the engine cover if removed.
7. Connect the battery cable and install the seat.

FZ Models

MODERATE

◆ See Figures 162 and 163

Although Yamaha does provide a specific interval for lubricating the throttle body, it's probably a good idea to do it at least once a season. Always use Yamaha Rust Inhibiter.

1. Remove the seat and disconnect the negative battery cable.
2. Remove the engine cover if equipped.
3. Loosen the hose clamp and then wiggle the breather hose off of the breather assembly.
4. Start the engine and then spray a blast of Rust Inhibiter into the hose for about 5 seconds. You may notice that the engine speed drops - this is normal.

Fig. 160 Loosen the hose clamp and slide the air intake hose off the throttle body. . .

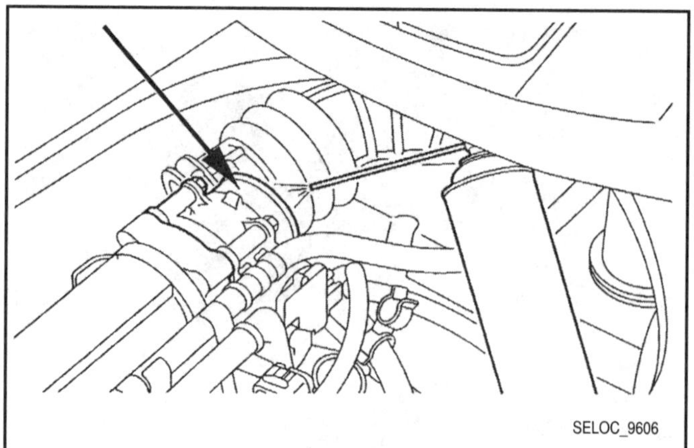

Fig. 161 . . .and then spray rust inhibiter into the bore - 1052cc VX shown, models w/1812cc engines similar

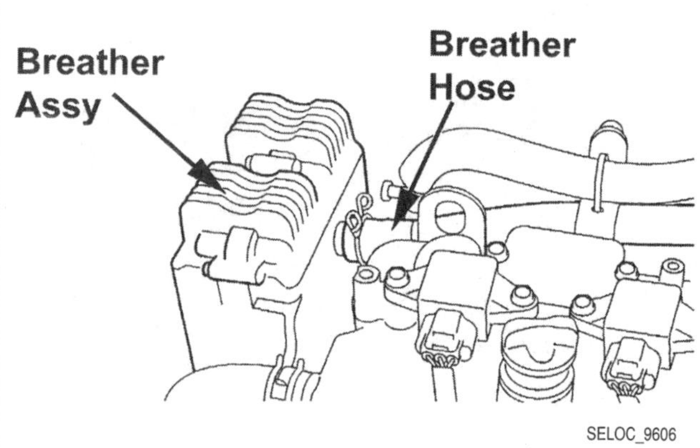

Fig. 162 Loosen the hose clamp and slide the breather hose off the breather assembly. . .

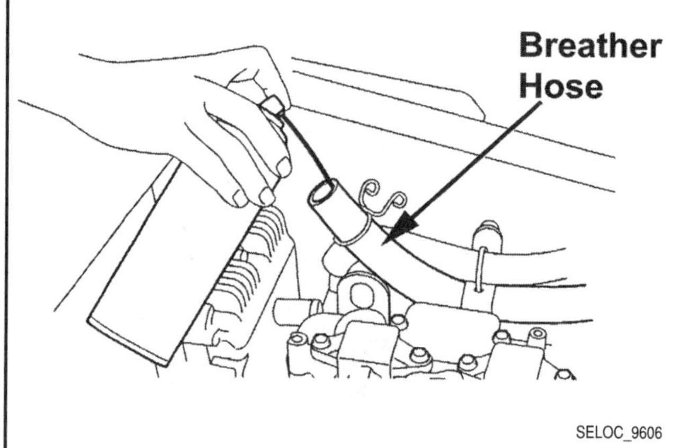

Fig. 163 . . .and then spray rust inhibiter into the hose

✳✳ CAUTION

Always make sure that you have the craft hooked up to a suitable water source.

5. Slide the hose back into position and squeeze the clamp back into position.
6. Install the engine cover if removed.
7. Install the seat.

QSTS CONTROL CABLE - FX & FZ MODELS

① EASY

◆ **See Figures 164 thru 167**

QSTS is an acronym for Quick Shift Trim Selector and it is responsible for changing the angle of the output nozzle on the jet pump. The handle bar cable ends and the cable ends at the cable housing in the hull should be lubricated with rust inhibitor every 100 hrs. of operation or once a year, whichever comes first. The inner cables inside the cable housing cover and also the jet drive cable end should be treated with Yamaha A water resistant grease.

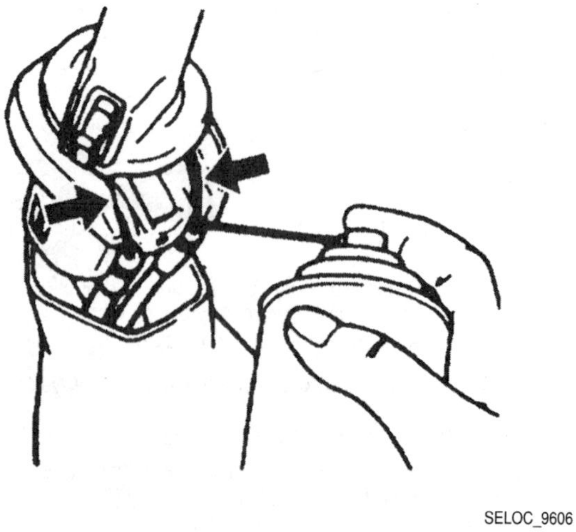

Fig. 164 Use your straw to get at the handle bar end of the cables

SHIFT & STEERING CABLES

◆ **See Figures 168 and 169**

All cable ends at the helm, joints and jet should be lubricated with Yamaha A water resistant grease every 100 hrs. or once a year, whichever comes first.

JET THRUST NOZZLE

◆ **See Figure 170**

The jet thrust nozzle pivot shaft should be lubricated with Yamaha A water resistant grease every 100 hrs. or once a year, whichever comes first.

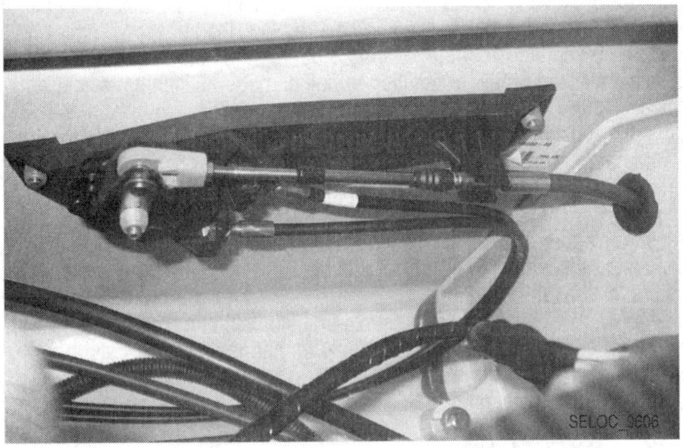

Fig. 165 The QSTS cable housing is inside the hull - 998/1052cc engines. . .

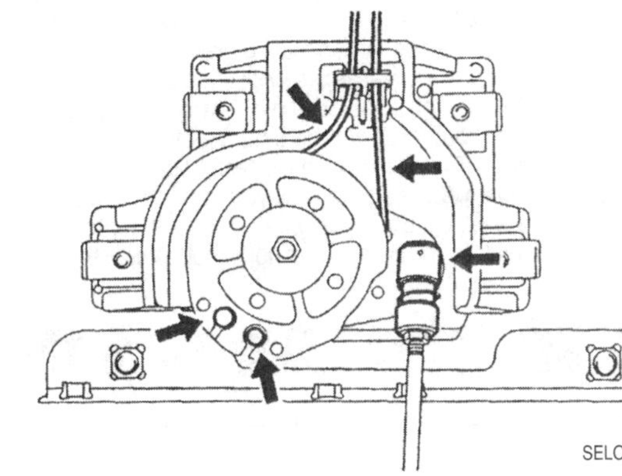

Fig. 166 . . .and 1812cc engines

Fig. 167 A good look at the jet end of the cable

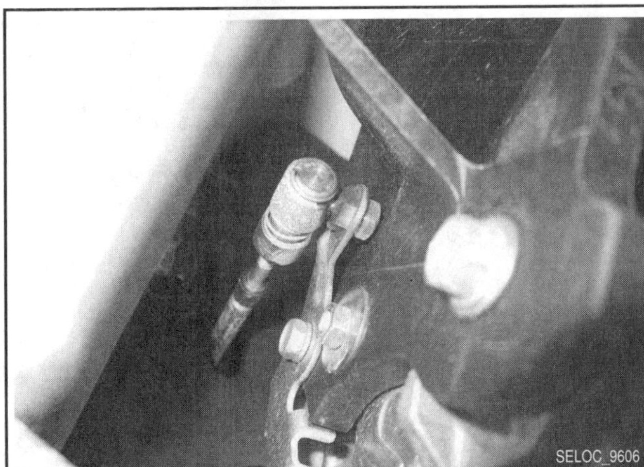

Fig. 168 A good look at the aft end of the shift cable

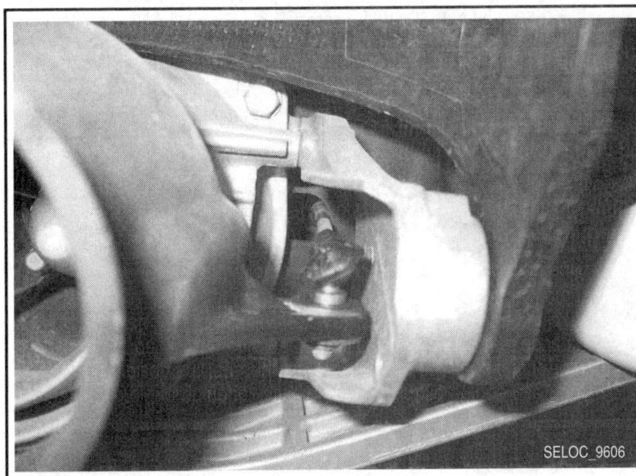

Fig. 169 A good look at the aft end of the steering cable

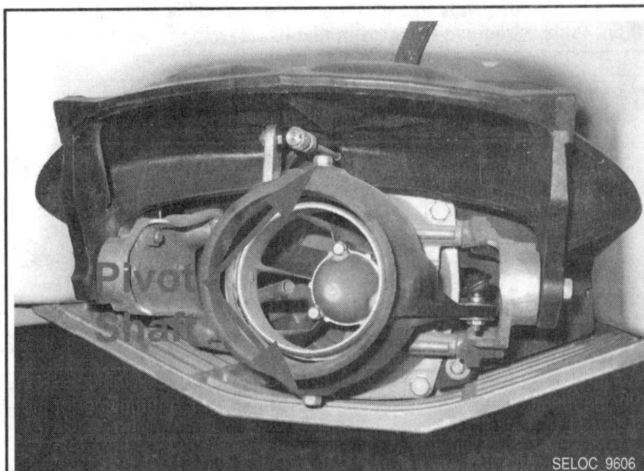

Fig. 170 A close-up of the thrust nozzle

PWC MAINTENANCE

Inside The Vessel

Probably the biggest surprise for owners is the extent to which mold and mildew develop in a PWC's interior. Preventing this growth is a two-fold process. First, the interior should be thoroughly (and regularly!) cleaned. Second, ventilation should be provided to allow adequate air circulation.

Properly cleaning the PWC includes removing as much as possible from it. The less there is in your engine bay and storage compartments, the less there is to attract mildew. Clothing, shoes, charts, paper goods, food stuffs, first aid supplies, odds and ends, etc., should be taken home. They're all fertile soil for mildew, as is dirt, grease, soap scum, etc.

The next step is to vacuum, scrub and polish every surface. Use a polish that leaves a protective coating on which it's hard for mold to get a foothold. The cleaner and more highly polished the insides of storage compartments are when you leave them, the less likely it is you'll find mold. Marine stores sell cleaners and polishes specifically for boats and for the marine environment. Many people find that regular household products are satisfactory. It's the elbow grease that counts.

The PWC's Exterior

HULL

◆ **See Figure 171**

Fiberglass reinforced plastic hulls are tough, durable, and highly resistant to impact. However, like any other material they can be damaged. One of the advantages of this type of construction is the relative ease with which it may be repaired. Because of its break characteristics, and the simple techniques used in restoration, these hulls have gained popularity throughout the world. From the most congested urban marina, to isolated lakes in wilderness areas, to the severe cold of northern seas, and in sunny tropic remote rivers of primitive islands or continents, fiberglass boats and PWCs can be found performing their daily task with a minimum of maintenance.

Fig. 171 The best way to care for a fiberglass hull is to wash it thoroughly to remove any growth that has developed on the bottom

A fiberglass hull has almost no internal stresses. Therefore, when the hull is broken or stove-in, it retains its true form. It will not dent to take an out-of-shape set. When the hull sustains a severe blow, the impact will either be absorbed by deflection of the laminated panel or the blow will result in a definite, localized break. In addition to hull damage, stiffening structures attached to the hull may also be affected and therefore, should be checked. Repairs are usually confined to the general area of the rupture.

The best way to care for a fiberglass hull is to first wash it thoroughly. Immediately after hauling the craft, while the bottom is still wet, is best, if possible. Remove any growth that has developed on the bottom. Use a pressure cleaner or a stiff brush to remove grass and slime. Pay particular attention to the waterline area. Pot scrubbers work well. Attend to any blisters; don't wait!

After washing the topsides, remove any stains that have developed with one of the fiberglass stain removers sold in marine stores. For stubborn stains, wet-sanding with 600-grit paper may be necessary. Remove oxidation and stains from metal parts. Apply a coat of wax to everything.

BELOW WATERLINE

◆ **See Figures 172 thru 175**

A foul bottom can seriously affect PWC performance. This is one reason why racers, large and small, PWC, powerboat or sail, are constantly giving attention to the condition of the hull below the waterline.

If growth has developed on the bottom, it can be removed with a solution of Muriatic acid applied with a brush or swab and then rinsed with clear water. Always use rubber gloves when working with Muriatic acid and take extra care to keep it away from your face and hands. The fumes are toxic. Therefore, work in a well-ventilated area, or if outside, keep your face on the windward side of the work.

Anodes (Zincs)

◆ **See Figures 176 and 177**

The idea behind anodes is simple: When dissimilar metals are dunked in water and a small current is leaked between or amongst them, the less-noble metal (galvanically speaking) is sacrificed.

The zinc alloy that most anodes are made of is designed to be less noble than the aluminum alloy your drive unit is made from. If there's any electrolysis and there almost always is, the inexpensive zinc anodes are consumed in lieu of the expensive drive.

This zinc needs a little attention in order to do its job. Make sure it's there, solidly attached to a clean mounting site and not covered with any kind of paint or wax.

Fig. 172 In areas where marine growth is prevalent, a coating of vinyl, anti-fouling bottom paint should be applied

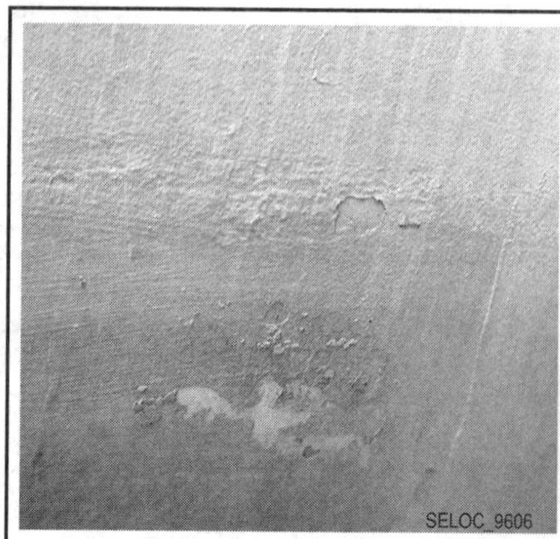

Fig. 173 This anti-foul paint has seen better days and must be replaced

Fig. 174 This hull is in even worse condition and should be sand blasted and repainted

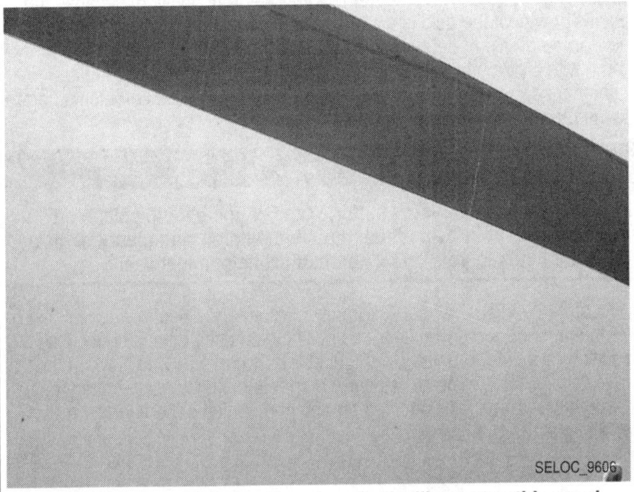

Fig. 175 This beautiful new fiberglass hull will not stay this good looking for long if it is not protected with anti-foul paint

Periodically inspect it to make sure it hasn't eroded too much. At a certain point in the erosion process, the mounting hole starts to enlarge, which is when the zinc might fall off. Obviously, once that happens, your motor no longer has any protection.

SERVICING

Engine

EASY

◆ **See Figures 176, 178 and 179**

The anode is hidden under a cover on the upper portion of the cylinder block on the starboard side, toward the rear of the engine.

1. Remove the seat.
2. Remove the air filter case cover or engine cover.
3. Locate the anode cover and clean it and the area around it thoroughly.

■ **On certain models, removing the intake manifold may provide better access.**

4. Remove the cover retaining bolt (the lower of the 2 bolts), but DO NOT remove the other bolt. Pry the cover loose from the block and remove the assembly - cover and anode.
5. Now you can remove the remaining bolt in the cover and pull the anode off. Inspect the anode for pitting or other damage. If the anode is more the half gone, replace it.
6. Pry the O-ring/grommet and gasket out of the cover carefully and inspect it for scuffs, cracks or other damage, replacing as necessary.

Fig. 178 Do not remove the anode bolt until you have removed the cover and anode from the block

Fig. 176 The main anode is toward the back of the engine

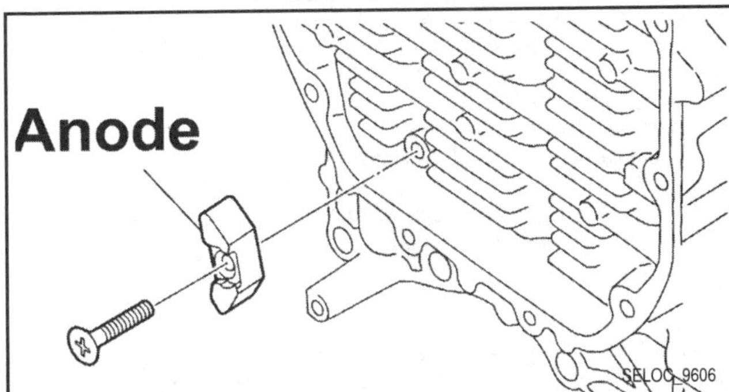

Fig. 177 The oil tank anode may be found on either the front or rear side of the oil cooler

■ **Although not strictly necessary, we at Seloc recommend replacing the O-ring/grommet each time you have the cover off - cheap insurance.**

7. Position the anode into the cover recess. Coat the bolt threads with Loctite 271 and thread the bolt in until it is finger-tight. Use a torque wrench and tighten it fully to 8.9 ft. lbs. (12 Nm) on all models but the 1812cc where it should be 5.9 ft. lbs. (8 Nm). Slide the O-ring/grommet over the anode and into position. Pat a little grease on the O-ring.
8. Insert the anode into the cylinder block so the cover mounting hole lines up with the bolt hole in the block. Thread the mounting bolt in and tighten it to 14.8 ft. lbs. (20 Nm).

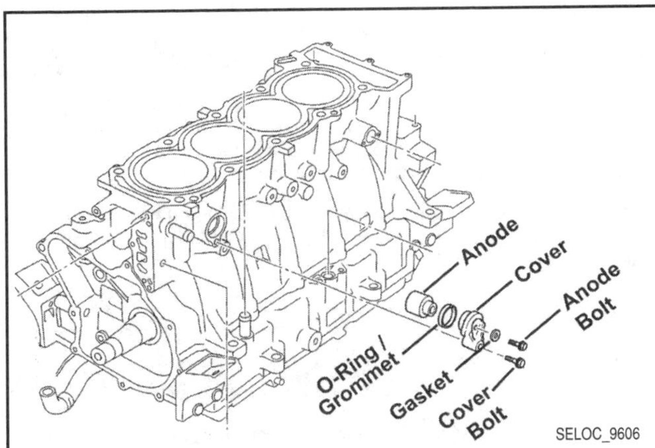

Fig. 179 Exploded view of the typical engine anode

9. Make sure that you check to see if the cover is leaking the first time you start the engine, especially if you chose not to change the O-ring. Reinstall the covers and seat.

Oil Tank - Models w/998cc and 1052cc Engines

② MODERATE

◆ See Figure 177

The anode is hidden behind the oil tank assembly's rear oil cooler cover on all models except the 2002 FX and the 2010-11 VX w/1052cc engine. On these models, the anode is found in the exact same place under the *front* cooler cover.

1. Remove the oil tank assembly and remove the appropriate cover as detailed in the Engine Mechanical section.
2. Loosen the anode bolt and remove the anode.
3. Install the anode and tighten the mounting bolt to 2.7 ft. lbs. (3.8 Nm) on the 2002 or 3 ft. lbs. (4 Nm) on the 2010-11 VX.
4. Install the cover and oil tank.

Battery

Difficulty in starting accounts for almost half of the service required on boats and watercraft each year. A survey by a major engine parts company indicated that roughly one third of all boat owners experienced a "won't start" condition in a given year. When an engine won't start, most people blame the battery when, in fact, it may be that the battery has run down in a futile attempt to start an engine with other problems.

Maintaining your battery in peak condition may be thought of as either tune-up or maintenance material. Most wise boaters will consider it to be both. A complete check up of the electrical system in your WaveRunner at the beginning of the boating season is a wise move. Continued regular maintenance of the battery will ensure trouble free starting on the water.

Complete battery service procedures are included in this section. The following are a list of basic electrical system service checks that should be performed.

- Check the battery for solid cable connections
- Check the battery and cables for signs of corrosion damage
- Check the battery case for damage or electrolyte leakage
- Check the electrolyte level in each cell
- Check to be sure the battery is fastened securely in position
- Check the battery's state of charge and charge as necessary
- Check battery voltage while cranking the starter. Voltage should remain above 9.5 volts
- Clean the battery, terminals and cables
- Coat the battery terminals with dielectric grease or terminal protector

Batteries that are not maintained on a regular basis can fall victim to parasitic loads (small current drains which are constantly drawing current from the battery, like clocks, small lights, etc.). Normal parasitic loads may drain a battery on boat that is in storage and not used frequently. Boats that have additional accessories with increased parasitic load may discharge a battery sooner. Storing a WaveRunner with the negative battery cable disconnected will minimize discharge due to parasitic loads.

CLEANING

① EASY

Keep the battery clean; as a film of dirt can help discharge a battery that is not used for long periods. A solution of baking soda and water mixed into a paste may be used for cleaning, but be careful to remove the battery from the hull first and to flush this off with clear water.

■ **Do not let any of the solution into the filler holes on non-sealed batteries. Baking soda neutralizes battery acid and will de-activate a battery cell.**

CHECKING SPECIFIC GRAVITY

OEM ① EASY

The electrolyte fluid (sulfuric acid solution) contained in the battery cells will tell you many things about the condition of the battery. Because the cell plates must be kept submerged below the fluid level in order to operate, maintaining the fluid level is extremely important. In addition, because the specific gravity of the acid is an indication of electrical charge, testing the fluid can be an aid in determining if the battery must be replaced. A battery in a boat with a properly operating charging system should require little maintenance, but careful, periodic inspection should reveal problems before they leave you stranded.

**** CAUTION**

Battery electrolyte contains sulfuric acid. If you should splash any on your skin or in your eyes, flush the affected area with plenty of clear water. If it lands in your eyes, get medical help immediately.

As stated earlier, the specific gravity of a battery's electrolyte level can be used as an indication of battery charge. At least once a year, check the specific gravity of the battery. It should be between 1.26 and 1.28 on the gravity scale. Most parts stores carry a variety of inexpensive battery testing hydrometers. These can be used on any non-sealed battery to test the specific gravity in each cell.

Conventional Battery

◆ See Figures 180 and 181

A hydrometer is required to check the specific gravity on all batteries that are not maintenance-free. The hydrometer has a squeeze bulb at one end and a nozzle at the other. Battery electrolyte is sucked into the hydrometer until the float or pointer is lifted from its seat. The specific gravity is then read by noting the position of the float/pointer. If gravity is low in one or more cells, the battery should be slowly charged and checked again to see if the gravity has come up. Generally, if after charging, the specific gravity of any 2 cells varies more than 50 points (0.50), the battery should be replaced, as it can no longer produce sufficient voltage to guarantee proper operation.

Check the battery electrolyte level at least once a month, or more often in hot weather or during periods of extended operation. Electrolyte level can be checked either through the case on translucent batteries or by removing the cell caps on opaque-case types. The electrolyte level in each cell should be kept filled to the split ring inside each cell, or between the 2 lines marked on the outside of the case - usually they will be marked **Upper Level** and **Lower Level**, or have a **-** and **+** in a small circle.

■ **Shining a flash light at the battery case will sometimes help in seeing the fluid level inside the case.**

■ **Never use mineral water or water obtained from a well. The iron content in these types of water is too high and will shorten the life of or damage the battery.**

If the level is low, remove the battery and add only distilled water through the opening until the level is correct and above the lower line on the case. Each cell is separate from the others, so each must be checked and filled individually. Distilled water should be used, because the chemicals and minerals found in most drinking water are harmful to the battery and could significantly shorten its life.

If water is added in freezing weather, the battery should be warmed to allow the water to mix with the electrolyte. Otherwise, the battery could freeze.

■ **Always allow the battery to sit for a few minutes after adding water and recheck the level a final time before installing the caps.**

Maintenance-Free Batteries

Although some maintenance-free batteries have removable cell caps for access to the electrolyte, the electrolyte condition and level is usually checked using the built-in hydrometer "eye." The exact type of eye varies between battery manufacturers, but most apply a sticker to the battery itself explaining the possible readings. When in doubt, refer to the battery manufacturer's instructions to interpret battery condition using the built-in hydrometer.

The readings from built-in hydrometers may vary, however a green eye usually indicates a properly charged battery with sufficient fluid level. A dark eye is normally an indicator of a battery with sufficient fluid, but one that may be low in charge. In addition, a light or yellow eye is usually an indication that electrolyte supply has dropped below the necessary level for battery (and hydrometer) operation. In this last case, sealed batteries with an insufficient electrolyte level must usually be discarded.

REMOVAL & INSTALLATION

EASY

◆ **See Figures 182 and 183**

Battery removal is pretty straightforward. . .pull back the cable-end boots and loosen the terminal screw to disconnect each cable. Position the cable out of the way, unhook the 2 rubber straps securing the battery to the case and then lift out the battery.

There are small nuts inside the battery terminals that will fall inside the terminal cage when you remove the screw - do not lose these nuts.

Do not lay the battery on its side; make sure it is sitting upright. Do not allow the battery to sit directly on a concrete floor for extended periods of time.

Fig. 180 On non-sealed batteries, you can use a hydrometer. . .

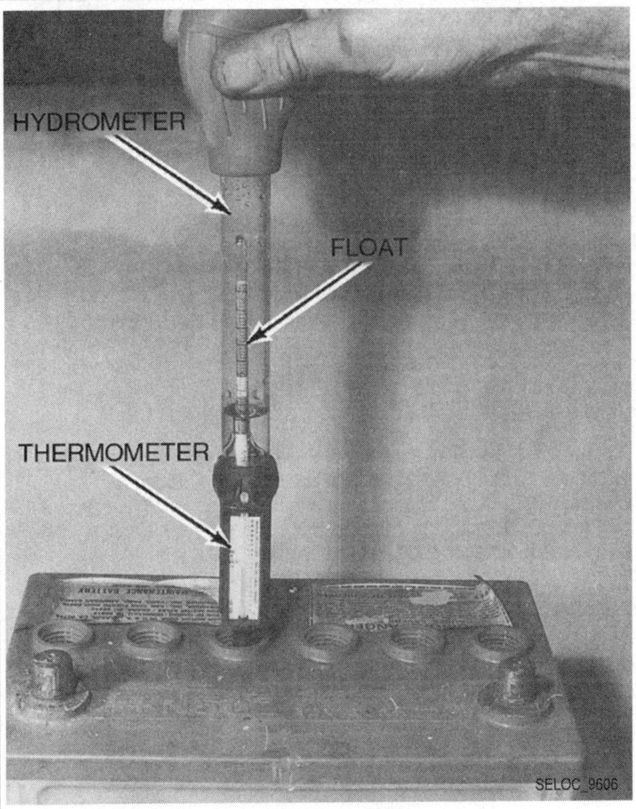

Fig. 181 . . .to check the battery charge/condition

■ **Remember, always disconnect the negative battery cable first during removal and then connect it last during installation.**

BATTERY TERMINALS

EASY

At least once a season, the battery terminals and cable clamps should be cleaned. Loosen the terminal screws and disconnect the cables, negative cable first.

Clean the cable-end clamps and the battery terminal with a wire brush, until all corrosion, grease, etc., is removed and the metal is shiny. It is especially important to clean the inside of the terminal cage thoroughly (a wire brush is useful here), since a small deposit of foreign material or oxidation there will prevent a sound electrical connection and inhibit either starting or charging. It is also a good idea to apply some dielectric grease to the terminal, as this will aid in the prevention of corrosion.

■ **Do not lose the small nut in the terminal cage while cleaning.**

After the cable clamps and terminals are clean, reinstall the cables, negative cable last; tightening the screws securely, but do not distort them. Slide the cable-end boots back over the connection and into position.

Check the cables themselves at the same time that the terminals are cleaned. If the insulation is cracked or broken, the cable should be replaced with a new one of the same length and gauge.

Fig. 182 Although your battery may use rubber straps like these. . .

Fig. 183 . . .or these, disconnecting the straps is what is necessary to lift out the battery

BATTERY & CHARGING SAFETY PRECAUTIONS

Always follow these safety precautions when charging or handling a battery.

- Wear eye protection when working around batteries. Batteries contain corrosive acid and produce explosive gas as a byproduct of their operation. Acid on the skin should be neutralized with a solution of baking soda and water made into a paste. In case acid contacts the eyes, flush with clear water and seek medical attention immediately.
- Avoid flame or sparks that could ignite the hydrogen gas produced by the battery and cause an explosion. Connection and disconnection of cables to battery terminals is one of the most common causes of sparks.
- Always turn a battery charger **OFF**, before connecting or disconnecting the leads. When connecting the leads, connect the positive lead first, then the negative lead, to avoid sparks.
- When lifting a battery, use a battery carrier or lift at opposite corners of the base.
- Ensure there is good ventilation in a room where the battery is being charged.
- Do not attempt to charge or load-test a maintenance-free battery when the charge indicator dot is indicating insufficient electrolyte.
- If you charge the battery while still in the hull, and you shouldn't, be sure the ignition switch is **OFF** before connecting or turning the charger **ON**. Sudden power surges can destroy electronic components.
- Use proper adapters to connect the charger leads to batteries with non-conventional terminals.

BATTERY CHARGERS

◆ **See Figure 184**

Before using any battery charger, consult the manufacturer's instructions for its use. Battery chargers are electrical devices that change Alternating Current (AC) to a lower voltage of Direct Current (DC) that can be used to charge a marine battery. There are two types of battery chargers - manual and automatic.

A manual battery charger must be physically disconnected when the battery has come to a full charge. If not, the battery can be overcharged, and possibly fail. Excess charging current at the end of the charging cycle will heat the electrolyte, resulting in loss of water and active material, substantially reducing battery life.

■ **As a rule, on manual chargers, when the ammeter on the charger registers half the rated amperage of the charger, the battery is fully charged. This can vary, and it is recommended to use a hydrometer to accurately measure state of charge.**

Fig. 184 When in storage, be sure to charge your battery periodically for maximum life

Automatic battery chargers have an important advantage - they can be left connected (for instance, overnight) without the possibility of overcharging the battery. Automatic chargers are equipped with a sensing device to allow the battery charge to taper off to near zero as the battery becomes fully charged. When charging a low or completely discharged battery, the meter will read close to full rated output. If only partially discharged, the initial reading may be less than full rated output, as the charger responds to the condition of the battery. As the battery continues to charge, the sensing device monitors the state of charge and reduces the charging rate. As the rate of charge tapers to zero amps, the charger will continue to supply a few milliamps of current - just enough to maintain a charged condition.

REPLACING BATTERY CABLES

① EASY

Battery cables don't go bad very often, but like anything else, they can wear out. If the cables on your WaveRunner are cracked, frayed or broken, they should be replaced.

When working on any electrical component, it is always a good idea to disconnect the negative (-) battery cable. This will prevent potential damage to many sensitive electrical components

Always replace the battery cables with one of the same length, or you will increase resistance and possibly cause hard starting. Smear the battery posts with a light film of dielectric grease, or a battery terminal protectant spray once you've installed the new cables. If you replace the cables one at a time, you won't mix them up.

■ **Any time you disconnect the battery cables, it is recommended that you disconnect the negative (-) battery cable first. This will prevent you from accidentally grounding the positive (+) terminal when disconnecting it, thereby preventing damage to the electrical system.**

Before you disconnect the cable(s), first turn the ignition to the **OFF** position. This will prevent a draw on the battery which could cause arcing. When the battery cable(s) are reconnected (negative cable last), be sure to check all electrical accessories are all working correctly.

STORAGE

If the craft is to be laid up for the winter or for more than a few weeks, special attention must be given to the battery to prevent complete discharge or possible damage to the terminals and wiring. Before putting the unit in storage, disconnect and remove the batteries. Clean them thoroughly of any dirt or corrosion and then charge them to full specific gravity reading. After they are fully charged, store them in a clean cool dry place where they will not be damaged or knocked over, preferably on a couple blocks of wood. Storing the battery up off the deck, will permit air to circulate freely around and under the battery and will help to prevent condensation.

Never store the battery with anything on top of it or cover the battery in such a manner as to prevent air from circulating around the filler caps. All batteries, both new and old, will discharge during periods of storage, more so if they are hot than if they remain cool. Therefore, the electrolyte level and the specific gravity should be checked at regular intervals. A drop in the specific gravity reading is cause to charge them back to a full reading.

In cold climates, care should be exercised in selecting the battery storage area. A fully-charged battery will freeze at about 60° below zero. A discharged battery, almost dead, will have ice forming at about 19° above zero.

WINTER STORAGE

General Information

Taking extra time to store your WaveRunner properly at the end of each season will increase the chances of satisfactory service at the next season. Remember, storage is the greatest enemy of a marine engine. In a perfect world the unit should be run on a monthly basis. The steering and shifting mechanism should also be worked through complete cycles several times each month. But who lives in a perfect world!

For most of us, if a small amount of time is spent in winterizing our beloved toys, the reward will be satisfactory performance, increased longevity and greatly reduced maintenance expenses. Remember that this is also the perfect time to review the Maintenance Interval charts and perform many of the procedures detailed in this section - do it now, because when it gets warm again, you want to go boating, not work on your engine!

Prepping For Storage

① EASY

WHERE TO STORE YOUR PWC

Ok, a well lit, locked, heated garage and work area is the best place to store you precious PWC, right? Well, we're probably not the only ones who wish we had access to a place like that, but if you're like most of us, we place our PWC wherever we can.

Of course, no matter what storage limitations are placed by where you live or how much space you have available, there are ways to maximize the storage site.

If possible, select an area that is dry. Covered is great, even if it is under a carport or sturdy portable structure designed for off-season storage. Many people utilize canvas and metal frame structures for such purposes. If you've got room in a garage or shed, that's even better. If you've got a heated garage, God bless you, when can we come over? If you do have a garage or shed that's not heated, an insulated area will help minimize the more extreme temperature variations and an attached garage is usually better than a detached for this reason. Just take extra care to make sure you've properly inspected the fuel system before leaving your vessel in an attached garage for any amount of time.

If a storage area contains large windows, mask them to keep sunlight off the hull and seats otherwise, use a high-quality, canvas cover if possible - the trailer too. A breathable cover is best to avoid the possible build-up of mold or mildew, but a heavy duty, non-breathable cover will work too. If using a non-breathable cover, place wooden blocks or length's of 2 x 4 under various reinforced spots in the cover to hold it up off the PWC's surface. This should provide enough room for air to circulate under the cover, allowing for moisture to evaporate and escape.

Whenever possible, avoid storing your vessel in industrial buildings or parks areas where corrosive emissions may be present. The same goes for storing your boat too close to large bodies of saltwater. Hey, on the other hand, if you live in the Florida Keys, we're jealous again, just enjoy it and service the boat often to prevent corrosion from causing damage.

Finally, when picking a place to store your PWC, consider the risk or damage from fire, vandalism or even theft. Check with your insurance agent regarding coverage while the vessel is stored.

THE GREAT ETHANOL DEBATE

As of this authoring, ethanol has been in use around the US and in various places in the world for quite some time. Truth is that decades ago manufacturers made changes to most fuel system components to make them less, or non-susceptible, to the affects of alcohol. However, in late 2005/early 2006 when Ethanol was suddenly introduced to large areas of the US where it had previously been absent, a few new problems were noted.

Ethanol can act like a solvent and in doing so it can clean away deposits in a fuel system that had previously hardened and stuck to components. The biggest concern would be for large fuel supplies (like fuel tanks) where a repeated pattern of evaporation over the years allowed large amounts of deposits to stick to component walls. Suddenly introducing ethanol to this environment, and allowing it to sit in the tank over a winter or longer, gives it time to reconstitute many of these deposits in a manner that they can be drawn into the fuel supply when the motor is taken out of storage, clogging filters; or worse, contaminating carburetors or injection systems. Ethanol hasn't seemed to cause this problem in markets where it has already been common for long periods of time and the thought is that it's constant use over the years has prevented deposits from sticking (literally) around. Basically, the Ethanol was essentially scrubbing the fuel systems clean in those markets, keeping them from developing enough deposits to clog the system all at once.

■ **There is a potential enzymatic solution to ethanol cleaning old fuel tanks and creating sludge/particulates in the form of a fuel treatment from Starbrite called StarTron. Purportedly StarTron was developed in Japan for use in extremely high-volume industrial equipment that could not economically be shut down in order to replace filters. As such, it uses enzymes to break down the particulates to a size small enough to not only pass through filters, but to harmlessly pass through carburetors or injectors for burning. Though we haven't personally tested this product yet, it comes with high recommendations from a company in which we have some faith, so we are optimistic.**

For aluminum tanks, we'd recommend storing the tank as empty to greatly reduce the amount of time that the tank is exposed to ethanol (basically, don't give it all winter to allow the ethanol to slowly attack the tank structure).

Another potential problem with Ethanol is that it has the unique ability (compared to other additives found in fuel) to bond with both fuel and water. The result is that where water would normally sit separate from the fuel in a contaminated system (where the water and fuel could be drawn off individually), water will instead mix with fuel containing ethanol which can lead to a greatly reduced ability to support combustion (to the point where an engine cannot run).

In the past we've recommended that it was a wash (six of one, half a dozen of the other) whether or not you stored your fuel tank FULL or EMPTY with regards to the amount of moisture that might enter the system from condensation. However, that was when the moisture might be drawn off from a full tank IF it got that bad. We don't think people are as willing today to risk 20, 50 or 100 gallons of fuel become contaminated with water and have changed our position. We think it is probably smarter to store the fuel tank as EMPTY as possible, and then in spring prior to refilling the tank to use a siphon or primer bulb to draw off any water which has condensed over the storage period.

Of course, these are the best current recommendations that we can offer, and this is likely a rapidly changing topic, so we also recommend asking around for other recommendations and solutions as time continues to go by.

Winter Storage Checklist

◆ **See Figures 185 and 186**

Proper winterizing involves adequate protection of the unit from physical damage, rust, corrosion, and dirt. The following steps provide guide to winterizing your marine engine at the end of a season.

Always keep a note or a checklist of just what maintenance needs to be done prior to starting the engine in the spring. In addition, make sure you mark down everything that was done before or during winter lay-up.

1. Always store your WaveRunner in the horizontal position on the trailer.
2. Flush the cooling system thoroughly as detailed in the Cooling System procedures found earlier in the section.
3. Fill the fuel tank with fresh fuel and then add 1 oz. of Yamaha Fuel Stabilizer & Conditioner to the tank for every gallon of fuel in the tank.
4. Remove the water separator cap as detailed in the Water Separator section and drain any accumulated water from the canister.
5. Replace the oil and oil filter as detailed elsewhere in this section. Normal combustion produces corrosive acids that are absorbed by the oil. Leaving dirty oil in the engine for an extended time allows these acids to attack and damage bearing surfaces.
6. Perform all applicable lubrication procedures detailed previously in this section using rust inhibiter or water repellant grease.

7. Rustproof the internal engine components:

a. On FX models with 998/1052cc engines, remove the air filter case cover and spray Yamaha Stor-Rite Engine Fogging oil into each intake opening for 3 seconds. Install the cover.

b. On VX models w/1052cc engines, loosen the air intake hose clamp, slide it down the hose and then pull the hose off of the throttle body. Spray Yamaha Stor-Rite Engine Fogging oil into the throttle body for 3 seconds. Reconnect the air intake hose at the throttle body.

c. Start the engine and let it run, at idle, for 15 seconds and then shut it off.

■ **We can find nothing in the service or owner's manuals detailing the fogging procedures on any of the 1812cc engine models. None-the-less, we think it a good idea to perform the procedure anyway - you be the judge.**

8. Fully charge the batteries. Disconnect all leads. Unattended, a battery naturally discharges over a period of several weeks. The electrolyte on a discharged battery can freeze at 20°F (-7°C), so keep the batteries fully charged or, better still, remove them to a warmer storage area. Small automatic trickle chargers work well.

9. Treat battery and cable terminals with petroleum jelly, silicone grease, or a heavy-duty corrosion inhibitor.

10. Protect external surfaces with a heavy-duty corrosion inhibitor such as Yamaha Silicone Protectant & Lubricant.

11. Make certain that the drain plugs have been removed.

Although not specifically recommended by the factory, you may also want to perform the following steps:

12. Inspect all hoses for signs of softening, cracking or bulging, especially those routinely exposed to high heat. Check hose clamps for tightness and corrosion.

13. Grease all greaseable points on the drivetrain.

14. Place a checklist in a handy spot to remind you of just what maintenance needs to be done prior to starting the engine in the spring - and also what you just completed!

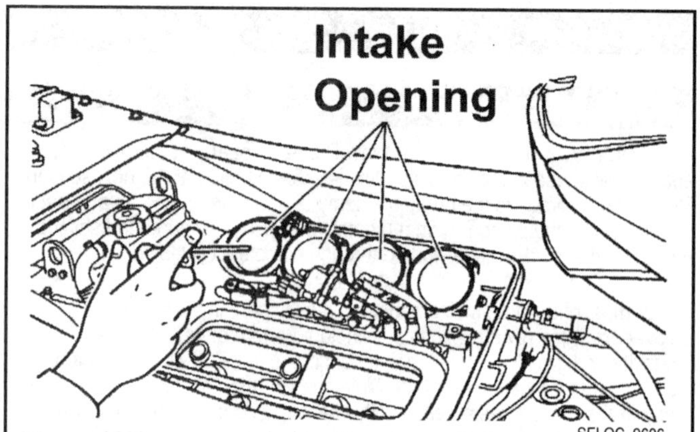

Fig. 185 Fogging the engine on certain FX models. . .

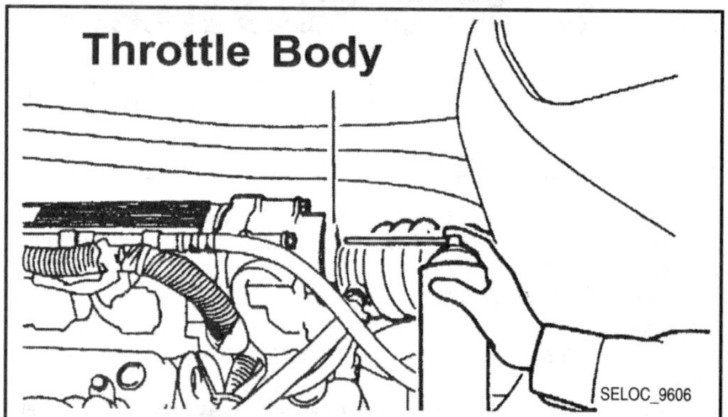

Fig. 186 . . .and VX models

SPRING COMMISSIONING

Satisfactory performance and maximum enjoyment can be realized if a little time is spent in commissioning your craft in the spring. Assuming you have followed the steps we recommended to winterize your vessel (in addition to any the manufacturer specifies) and the unit has been properly stored, a minimum amount of work should be required to prepare it for use.

After performing the spring commissioning and testing on the water, it is a good idea to perform a full tune-up. Remember, you are relying on your engine to get you where you want to go. Treat it good now and it will treat you good later.

Spring Commissioning Checklist

The following steps outline a logical sequence of tasks to be performed before starting your engine for the first time in a new season.

1. Pick up the checklist you made to remind yourself of just what maintenance needs to be done prior to starting the engine in the spring. You did remember to write yourself a checklist. . .right?

2. INSTALL THE DRAIN PLUG IF REMOVED!

3. Check and replace the anodes if necessary.

4. If you took our advice, you removed the battery for the winter. While it was in storage, you should have kept it fully charged and it should be ready to go. So install the battery and connect the battery cables. Treat battery and cable terminals with petroleum jelly, silicone grease, or a heavy-duty corrosion inhibitor. Capacity test the batteries.

5. As you did in the winter, inspect all hoses for signs of softening, cracking or bulging, especially those routinely exposed to high heat. Check hose clamps for tightness and corrosion.

6. Connect a water source and start the engine and allow it to reach operating temperature.

7. Once running, check the oil pressure, the raw water discharge and the engine for oil and water leaks.

8. Drain water from the separator bowl. If an excessive amount of water is noted in the fuel system, you may want to consider a fuel conditioner or replacing the old fuel with fresh fuel.

9. After testing the vessel on the water, it is a good idea to change the engine oil and filter. Yes, we know you probably changed it last winter, but do it again anyway! This eliminates the adverse effects of moisture in the oil.

10. Run through your winterizing check list and make sure that anything disconnected, closed or removed is put back in place.

TROUBLESHOOTING

Although all models covered here are set up to utilize the Yamaha Diagnostic System for all troubleshooting and diagnostic needs, we have provided some general troubleshooting charts here for those without access to the system. Generally speaking, these charts are appropriate when you experience a problem and the diagnostic system is not throwing a trouble code (details on trouble codes can be found in the Self-Diagnostic procedures covered in the Fuel System section.

Prior to checking any components in these charts, please make sure that the following conditions are correct:

a. The battery is fully charged and working properly

b. There are no broken, loose, incorrect or corroded wiring connections

c. The emergency shut-off lanyard is correctly installed in the shut-off switch

d. There is fuel of sufficient quality and quantity, and it is reaching the fuel injectors.2002-08 FX models w/998cc engines

Index

Trouble Mode **Check Elements**

ENGINE WILL NOT START	HARD STARTING	ROUGH IDLING	HIGH IDLING	ENGINE STALLS	POOR ACCELERATION	ENGINE WILL NOT STOP	POOR PERFORMANCE	LIMITED ENGINE SPEED	OVERHEATING	LOW OIL PRESSURE	LOOSE STEERING	BILGE INCREASE	IRREGULAR WARNING INDICATION	POOR BATTERY CHARGING	Check	Section
															FUEL SYSTEM	
O	O	O		O	O		O	O							Fuel tank	3
O	O	O		O	O		O	O							Fuel tank breather hose	3
O	O	O		O	O		O	O							Fuel hose	3
O	O	O		O	O		O	O							Fuel filter	2
O	O	O		O	O		O	O							Fuel pump	3
O	O	O		O	O		O	O							Fuel injectors	3
				O			O								Trolling speed	2
O	O			O			O	O							Air filter	2, 3

SELOC_9606

Troubleshooting chart - 2002-08 FX models w/998cc engines

ENGINE WILL NOT START	HARD STARTING	ROUGH IDLING	HIGH IDLING	ENGINE STALLS	POOR ACCELERATION	ENGINE WILL NOT STOP	POOR PERFORMANCE	LIMITED ENGINE SPEED	OVERHEATING	LOW OIL PRESSURE	LOOSE STEERING	BILGE INCREASE	IRREGULAR WARNING INDICATION	POOR BATTERY CHARGING	Check	Section
															POWER UNIT	
O	O	O		O	O		O	O							Compression	2
O	O	O		O	O		O	O	O						Cylinder head gaskets	6
O	O	O		O	O		O	O							Cylinder block	6
O	O	O		O	O		O	O							Crankcase	6
O	O	O		O	O		O	O							Piston rings	6
O	O	O		O	O		O	O							Pistons	6
O	O	O		O	O		O	O							Bearings	6
									O						Thermostat	2, 5
O		O		O	O		O								Valve(s) and valve seat(s)	6
O		O			O				O						Valve clearance adjusting pad(s)	2, 6
O		O			O										Camshaft(s)	6
O						O	O		O						Timing chain	6
									O	O					Oil pump	5
								O							Engine oil	2
										O					Oil filter	4,5
													O		Oil pressure switch	4.5
							O								Bearing housing	7
		O					O								Drive couplings	7
							O								Rubber coupling	7
									O			O			Pilot water hose	
									O			O			Water hose	
									O			O			Water passage	

Troubleshooting chart - 2002-08 FX models w/998cc engines (cont'd) SELOC_9606

ENGINE WILL NOT START	HARD STARTING	ROUGH IDLING	HIGH IDLING	ENGINE STALLS	POOR ACCELERATION	ENGINE WILL NOT STOP	POOR PERFORMANCE	LIMITED ENGINE SPEED	OVERHEATING	LOW OIL PRESSURE	LOOSE STEERING	BILGE INCREASE	IRREGULAR WARNING INDICATION	POOR BATTERY CHARGING	Check	Section
															JET PUMP UNIT	
							O		O			O			Duct	7
							O								Impeller	7
							O		O						Intake grate	7
		O					O								Bearings	7
							O		O						Intake duct	7
									O						Water inlet hose	7
											O				Bilge hose	7
											O				Bilge strainer	2
											O				Bilge hose joint	7
															ELECTRICAL	
															Ignition system, fuel control system	
O	O	O		O	O		O								Pulser coils	4
O			O	O		O	O	O							ECM	3
O	O	O		O	O		O								Ignition coils	4
O				O		O									Slant detector switch	4
O						O									Engine stop switch	4
O						O									Engine shut-off switch	4
O	O	O	O	O	O	O	O								Spark plugs	2
O						O									Main and fuel pump relay	3
													O		Thermoswitch	4
	O	O	O												Intake air pressure sensor	3
	O	O	O												Intake air temperature sensor	3
	O	O	O										O		Engine temperature sensor	3, 4
		O	O		O		O								Throttle position sensor	3
O	O	O		O	O		O	O							Cam position sensor	3

Troubleshooting chart - 2002-08 FX models w/998cc engines (cont'd) SELOC_9606

Trouble mode															Check elements	
ENGINE WILL NOT START	HARD STARTING	ROUGH IDLING	HIGH IDLING	ENGINE STALLS	POOR ACCELERATION	ENGINE WILL NOT STOP	POOR PERFORMANCE	LIMITED ENGINE SPEED	OVERHEATING	LOW OIL PRESSURE	LOOSE STEERING	BILGE INCREASE	IRREGULAR WARNING INDICATION	POOR BATTERY CHARGING	Check	Section
															Starting system	
O	O					O									• Start switch	4
O						O									• Starter relay	4
O															• Starter motor	4
															Charging system	
														O	• Lighting coil	4
														O	• Rectifier/regulator	4
O														O	• Fuses	4
O				O										O	• Battery leads	2, 4
O														O	• Battery	2, 4
															Electric bilge pump	
												O			• Electric bilge pump	8
															Hull & Hood	
											O				Steering master	8
							O					O			Water lock	8
				O			O					O			Exhaust hose	6
							O					O			Muffler	6
												O			Drain plugs	8

Troubleshooting chart - 2002-08 FX models w/998cc engines (cont'd)

SELOC_9606

Trouble mode															Check elements	
ENGINE WILL NOT START	HARD STARTING	ROUGH IDLING	HIGH IDLING	ENGINE STALLS	POOR ACCELERATION	ENGINE WILL NOT STOP	POOR PERFORMANCE	LIMITED ENGINE SPEED	OVERHEATING	LOW OIL PRESSURE	LOOSE STEERING	BILGE INCREASE	IRREGULAR WARNING INDICATION	POOR BATTERY CHARGING	Check	Section
															Fuel System	
O	O	O		O	O		O	O							Fuel tank	3
O	O	O		O	O		O	O							Fuel tank breather hose	3
O	O	O		O	O		O	O							Fuel hose	3
O	O	O		O	O		O	O							Fuel filter	2
O	O	O		O	O		O	O							Fuel pump	3
O	O	O		O	O		O	O							Fuel injectors	3
				O			O								Trolling speed	2
O	O	O		O	O		O	O							Air filter	2, 3

Troubleshooting chart - 2004-F07 FX HO models w/1052cc engines

SELOC_9606

ENGINE WILL NOT START	HARD STARTING	ROUGH IDLING	HIGH IDLING	ENGINE STALLS	POOR ACCELERATION	ENGINE WILL NOT STOP	POOR PERFORMANCE	LIMITED ENGINE SPEED	OVERHEATING	LOW OIL PRESSURE	LOOSE STEERING	BILGE INCREASE	IRREGULAR WARNING INDICATION	POOR BATTERY CHARGING	Check	Section
Trouble mode															Check elements	
															Power Unit	
O	O	O		O	O		O	O							Compression	2
O	O	O		O	O		O	O	O	O					Cylinder head gaskets	6
O	O	O		O	O		O	O	O						Cylinder block	6
O	O	O		O	O		O	O							Crankcase	6
O	O	O		O	O		O	O							Piston rings	6
O	O	O		O	O		O	O							Pistons	6
O	O	O		O	O		O	O							Bearings	4
									O						Thermostat	2, 5
O	O	O		O	O		O	O							Valve(s) and valve seat(s)	2, 6
O		O			O		O	O	O						Valve clearance adjusting pad(s)	2, 6
O		O		O	O		O								Camshaft(s)	6
O				O	O		O		O						Timing chain	6
									O	O					Oil pump	5
								O		O					Engine oil	2
										O					Oil filter	2
													O		Oil pressure switch	4, 5
							O								Bearing housing	7
		O					O								Drive couplings	7
							O								Rubber coupling	7
									O			O			Pilot water hose	8
									O			O			Water hose	8
									O			O			Water passage	8

Troubleshooting chart - 2004-07 FX HO models w/1052cc engines (cont'd) SELOC_9606

ENGINE WILL NOT START	HARD STARTING	ROUGH IDLING	HIGH IDLING	ENGINE STALLS	POOR ACCELERATION	ENGINE WILL NOT STOP	POOR PERFORMANCE	LIMITED ENGINE SPEED	OVERHEATING	LOW OIL PRESSURE	LOOSE STEERING	BILGE INCREASE	IRREGULAR WARNING INDICATION	POOR BATTERY CHARGING	Check	Section
Trouble mode															**Check elements**	
															Jet Pump Unit	
							O		O			O			Duct	7
							O								Impeller	7
							O		O						Intake grate	7
		O					O								Bearings	7
							O		O						Intake duct	7
									O						Water inlet hose	7
												O			Bilge hose	7
												O			Bilge strainer	2, 7
												O			Bilge hose joint	7
															ELECTRICAL	
															Ignition & fuel control systems	
O	O	O		O	O		O								• Pickup coils	4
O			O	O		O	O	O							• ECM	3
O	O	O		O	O		O								• Ignition coils	4
O				O		O									• Slant detector switch	4
O						O									• Engine stop switch	4
O						O									• Engine shut-off switch	4
O	O	O	O	O	O	O	O								• Spark plugs	2
O		O		O		O									• Main and fuel pump relay	3
													O		• Thermoswitch	4
	O	O	O		O		O	O							• Sensor assembly	4
	O	O	O										O		• Engine temperature sensor	4
		O	O		O		O	O							• Throttle position sensor	3
O	O	O		O	O		O	O							• Cam position sensor	3

Troubleshooting chart - 2004-07 FX HO models w/1052cc engines (cont'd) SELOC_9606

ENGINE WILL NOT START	HARD STARTING	ROUGH IDLING	HIGH IDLING	ENGINE STALLS	POOR ACCELERATION	ENGINE WILL NOT STOP	POOR PERFORMANCE	LIMITED ENGINE SPEED	OVERHEATING	LOW OIL PRESSURE	LOOSE STEERING	BILGE INCREASE	IRREGULAR WARNING INDICATION	POOR BATTERY CHARGING	Check	Section
															Starting system	
O	O					O									Start switch	4
O						O									Starter relay	4
O															Starter motor	4
															Charging system	
														O	Lighting coil	4
														O	Rectifier/regulator	4
O														O	Fuses	4
O				O										O	Battery leads	2, 4
O	O													O	Battery	2, 4
															Electric Bilge Pump	
												O			Electric bilge pump	2, 8
															Remote control system	8
O															Remote control unit	8
															Hull & Hood	
											O				Steering master	8
							O					O			Water lock	8
				O			O					O			Exhaust hose	8
							O					O			Muffler	6
												O			Drain plugs	2

Trouble mode / Check elements

Troubleshooting chart - 2004-07 FX HO models w/1052cc engines (cont'd)

SELOC_9606

ENGINE WILL NOT START	HARD STARTING	ROUGH IDLING	HIGH IDLING	ENGINE STALLS	POOR ACCELERATION	ENGINE WILL NOT STOP	POOR PERFORMANCE	LIMITED ENGINE SPEED	OVERHEATING	LOW OIL PRESSURE	LOOSE STEERING	BILGE INCREASE	IRREGULAR WARNING INDICATION	POOR BATTERY CHARGING	Check	Section
															Fuel System	
O	O	O		O	O		O	O				O			Fuel tank	3
O	O	O		O	O		O	O							Fuel tank breather hose	3
O	O	O		O	O		O	O				O			Fuel hose	3
O	O	O		O	O		O	O							Fuel filter	2
O	O	O		O	O		O	O							Fuel pump	3
O	O	O		O	O		O	O							Fuel injectors	3
			O	O			O								Trolling speed	2
O	O	O	O	O	O		O	O							Air filter	2, 3
													O		Fuel sender	3

Trouble mode / Check elements

Troubleshooting chart - 2008 FX HO models w/1052cc engines

SELOC_9606

Trouble mode															Check elements	
ENGINE WILL NOT START	HARD STARTING	ROUGH IDLING	HIGH IDLING	ENGINE STALLS	POOR ACCELERATION	ENGINE WILL NOT STOP	POOR PERFORMANCE	LIMITED ENGINE SPEED	OVERHEATING	LOW OIL PRESSURE	LOOSE STEERING	BILGE INCREASE	IRREGULAR WARNING INDICATION	POOR BATTERY CHARGING	Check	Section
															Power Unit	
O	O	O		O	O		O	O							Compression pressure	2
O	O	O		O	O		O	O							Cylinder head gaskets	6
O	O	O		O	O		O	O	O	O					Cylinder block	6
O	O	O		O	O		O	O							Crankcase	6
O	O	O		O	O		O	O							Crankshaft	6
O	O	O		O	O		O	O							Piston rings	6
O	O	O		O	O		O	O							Pistons	6
O	O	O		O	O		O	O							Bearings	6
		O			O				O						Thermostat	2, 5
O	O	O		O	O		O	O							Valves and valve seats	6
O	O	O			O		O	O							Valve clearance adjusting pads	2, 6
O	O	O		O	O		O	O							Camshafts	6
O	O	O		O	O		O	O							Timing chain	6
										O			O		Oil pump	5
	O	O		O	O		O	O		O			O		Engine oil	2
										O					Oil filter	2
													O		Oil pressure switch	4, 5
		O		O	O		O								Reduction drive gear	7
		O		O	O		O								Drive coupling	7
					O		O								Rubber coupling	7
												O			Pilot water hose	7
									O			O			Water hose	7
									O			O			Water passage	7

SELOC_9606

Troubleshooting chart - 2008 FX HO models w/1052cc engines (cont'd)

Trouble mode															Check elements	
ENGINE WILL NOT START	HARD STARTING	ROUGH IDLING	HIGH IDLING	ENGINE STALLS	POOR ACCELERATION	ENGINE WILL NOT STOP	POOR PERFORMANCE	LIMITED ENGINE SPEED	OVERHEATING	LOW OIL PRESSURE	LOOSE STEERING	BILGE INCREASE	IRREGULAR WARNING INDICATION	POOR BATTERY CHARGING	Check	Section
															Jet Pump Unit	
					O		O		O			O			Impeller duct	7
					O		O	O							Impeller	7
					O		O								Intake grate	7
	O	O			O		O								Bearings	7
									O						Water inlet hose	7
												O			Bilge hose	7
												O			Bilge strainer	2, 7
												O			Bilge hose joint	7
															Electrical	
O	O	O	O	O	O	O	O	O							ECM	3
O	O	O		O	O		O						O		Pickup coils	4
				O									O	O	Stator coil	4
				O									O	O	Rectifier/regulator	4
O	O	O		O	O		O	O							Ignition coils	4
	O	O	O		O		O	O					O		Sensor assembly (6B6 engine)	3
O	O	O	O		O		O	O					O		Intake air temperature sensor (6AA engine)	3
	O	O	O										O		Engine temperature sensor	4
O	O	O	O		O		O	O					O		Intake air pressure sensor (6AA engine)	3
													O		Thermoswitch (engine, exhaust) (6B6 engine)	4
				O									O		Thermoswitch (6AA engine)	4
				O									O		Thermo sensor (6AA engine)	4
				O									O		Exhaust temperature sensor (6AA engine)	4
	O	O	O		O		O	O					O		TPS	3
O	O	O		O	O		O	O					O		Cam position sensor	3
O				O		O							O		Slant detection switch	4

SELOC_9606

Troubleshooting chart - 2008 FX HO models w/1052cc engines (cont'd)

Trouble mode															Check elements	
ENGINE WILL NOT START	HARD STARTING	ROUGH IDLING	HIGH IDLING	ENGINE STALLS	POOR ACCELERATION	ENGINE WILL NOT STOP	POOR PERFORMANCE	LIMITED ENGINE SPEED	OVERHEATING	LOW OIL PRESSURE	LOOSE STEERING	BILGE INCREASE	IRREGULAR WARNING INDICATION	POOR BATTERY CHARGING	Check	Section
												O			Electric bilge pump	8
O				O									O	O	Fuses	4
O															Engine start switch	4
O						O									Engine stop switch	4
O						O									Engine shut-off switch	4
O		O		O		O									Main and fuel pump relay	3
O															Starter relay	4
O															Starter motor	4
O															Remote control receiver	4
O	O	O	O	O	O	O	O								Spark plugs	2
O	O			O									O	O	Battery	2, 4
O	O			O									O	O	Battery cables	2
															Hull & Hood	
											O				Steering master	8
O	O	O	O	O	O		O	O				O			Water lock	
O	O	O	O	O	O		O	O				O			Exhaust hose	8
												O			Drain plugs	2

SELOC_9606

Troubleshooting chart - 2008 FX HO models w/1052cc engines (cont'd)

Symptom 2	Cause	Checking step
Starter motor does not operate	Yamaha Security System set to lock mode	Check that the Yamaha Security System is set to the unlock mode.
	Discharged battery	Check the battery voltage and specific gravity.
	Loose connection of battery terminal	Check the battery cable and terminal for proper connection.
	Blown fuse	Check the fuse for continuity.
	Starter relay malfunction	Check the starter relay.
	Engine start switch malfunction	Check the engine start switch, (left handlebar switch)
	Short, open, or loose connection in starter motor cir-cuit	Measure the starter relay input voltage.
		Check the wiring harness.
	Starter motor malfunction	Disassemble and check the starter motor.
	Stuck piston or crankshaft	Disassemble and check the engine unit.
	Stuck impeller	Disassemble and check the jet pump unit.
Starter motor operates, but the engine does not crank	Idle gear or starter clutch malfunction	Check the idle gear, starter clutch, or crankshaft Woodruff key.

SELOC_9606

Troubleshooting chart - 2009-11 FX HO models w/1812cc engines

Symptom 2	Cause	Checking step
Throttle valve does not move properly.	Throttle lever squeezed	Check that the throttle lever is in the fully closed position.
	Throttle cable installed incorrectly	Adjust the throttle lever free play.
	APS mulfunction	Measure the APS 1 and 2 resistance.
	TPS mulfunction	Cfyeck the throttle valve opening.
	Throttle valve malfunction	Check the throttle body assembly.
Discharged battery	Battery performance decrement	Check the battery voltage and specific gravity.
	Stator coil assembly malfunction	Check the stator coil assembly.
	Short, open, or loose connection in charging circuit	Check the battery cable and terminal for proper connection.
		Check the wiring harness.
ECM does not operate	Blown fuse	Check the fuse for continuity.
	Main and fuel pump relay malfunction	Check the main and fuel pump relay.
	Short, open, or loose connection in ECM circuit	Measure the ECM input voltage.
		Check the ECM circuit for continuity.
	ECM malfunction	Replace the ECM.
Spark plug does not spark (all cylinders)	Engine shut-off cord (lanyard) clip not installed	Check that the engine shut-off cord (lanyard) clip is installed properly.
	Engine stop switch malfunction	Check the engine stop switch continuity, (left handlebar switch)
	Pick-up coil malfunction	Measure the pickup coil output peak voltage.
		Measure the pickup coil resistance.
	Short, open, or loose connection in pickup coil circuit	Check the white/black (W/B) and black/orange (B/O) pickup coil wiring harness leads for continuity.
	Short, open, or loose connection in ignition coil circuit	Measure the ignition coil input voltage (from the battery).
		Check the red/yellow (R/Y), black (B), black/red (B/R), black/white (B/ W), black/yellow (B/Y), and black/ green (B/G) ignition coil wiring harness leads for continuity.Check the red/yellow (R/Y), black (B), black/red (B/R), black/white (B/ W), black/yellow (B/Y), and black/ green (B/G) ignition coil wiring harness leads for continuity.
	ECM malfunction	Measure the ECM output peak voltage.

Troubleshooting chart - 2009-11 FX HO models w/1812cc (cont'd) SELOC_9606

Symptom 2	Cause	Checking step
Fuel not supplied (all cylinders)	Fuel leakage	Check the fuel hose.
	Blown fuse	Check the fuse for continuity.
	Clogged fuel pump filter	Clean the fuel pump filter.
	Fuel pump module malfunction	Check the fuel pump module operation sound using the YDIS.
		Measure the fuel pump module input voltage.
		Check the blue (L) and black (B) fuel pump module wiring harness leads for continuity.
	Main and fuel pump relay malfunction	Check the main and fuel pump relay.
	Fuel not supplied to the fuel rail	Measure the fuel pressure.
Compression pressure is low	Compression leakage	Measure the compression pressure.
		Measure the valve clearance.
		Check the camshaft for damage.
		Check the cylinder head gasket and cylinder head warpage.
		Check the valve and valve seat for wear.
		Check the piston and piston ring for damage.
		Check the cylinder for damage.
	Improper valve timing	Check the timing chain.

Troubleshooting chart - 2009-11 FX HO models w/1812cc (cont'd) SELOC_9606

Symptom 2	Cause	Checking step
Throttle valve does not move properly	Throttle cable installed incorrectly.	Adjust the throttle lever free play.
	Throttle valve malfunction	Check the throttle body assembly.
Spark plug does not spark (some cylinders)	Spark plug malfunction	Check the spark plug.
	Ignition coil malfunction	Check the ignition spark.
	Short, open, or loose connection in ignition coil circuit	Measure the ignition coil input voltage (from the battery).
		Check the red/yellow (R/Y), black (B), black/red (B/R), black/white (B/W), black/yellow (B/Y), and black/green (B/G) ignition coil wiring harness leads for continuity.
	ECM malfunction	Measure the ECM output peak voltage.
Fuel not supplied (some cylinders)	Fuel injector malfunction	Check the fuel injector operation sound using the YDIS.
		Measure the fuel injector resistance.
		Check the fuel injector O-ring.
	Short, open, or loose connection in fuel injector circuit	Measure the fuel injector input voltage.
		Check the red/yellow (R/Y), purple/red (Pu/R), purple/black (Pu/B), purple/yellow (Pu/Y), and purple/green (Pu/G) fuel injector wiring harness leads for continuity.
	Clogged fuel injector filter	Clean the fuel injector.
	ECM malfunction	Replace the ECM.
Compression pressure is low	Compression leakage	Measure the compression pressure.
		Measure the valve clearance.
		Check the camshaft for damage.
		Check the cylinder head gasket and cylinder head warpage.
		Check the valve and valve seat for wear.
		Check the piston and piston ring for damage.
		Check the cylinder for damage.
	Improper valve timing	Check the timing chain.

SELOC_9606

Troubleshooting chart - 2009-11 FX HO models w/1812cc (cont'd)

High engine idle speed

Symptom 2	Cause	Checking step
—	Throttle cable installed incorrectly.	Adjust the throttle lever free play and check the cable rooting.

Limited engine speed

Symptom 2	Cause	Checking step
• Buzzer sounds intermittently • Overheat warning indicator blinks • Check engine warning indicator blinks	Clogged cooling water passage	Check the cooling water pilot outlet for water discharge.
		Check the cooling water passage.
	Thermostat malfunction	Check the thermostat.
• Buzzer sounds intermittently Oil pressure warning indicator blinks	Insufficient engine oil	Check the engine oil level.
		Check for engine oil leakage.
		Check the oil passage.
	Oil pressure switch malfunction	Check the oil pressure switch continuity.
		Check that there is no short circuit in the pink/white (P/W) oil pressure switch wiring harness lead.
	APS malfunction	Check the APS.
	TPS malfunction	Check the throttle body assembly.
	Shift lever malfunction	Check the reverse sensor.

SELOC_9606

Troubleshooting chart - 2009-11 FX HO models w/1812cc (cont'd)

Discharged battery

Symptom 2	Cause	Checking step
	Battery performance decrease	Check the battery voltage and specific gravity.
	Loose connection of battery terminal	Check the battery cable and terminal for proper connection.
	Short, open, or loose connection in charging circuit	Check the charging circuit for wiring connection and damage.
	Stator coil malfunction	Measure the stator coil output peak voltage.
		Measure the stator coil resistance.
	Rectifier regulator malfunction	Measure the rectifier regulator output peak voltage.

Poor performance

Symptom 2	Cause	Checking step
Watercraft cannot reach high speeds	Jet pump unit malfunction	Check the impeller, impeller duct, and intake grate.
	Water entered hull	Check the drain plugs and O-rings for damage.
		Check the cooling water hoses for damage.
		Check the water lock, water tank, and rubber hoses for damage.
		Check the exterior of the hull for damage.

SELOC_9606

Troubleshooting chart - 2009-11 FX HO models w/1812cc (cont'd)

Trouble mode														Check items	
ENGINE WILL NOT START	HARD STARTING	ROUGH IDLING	HIGH IDLING	ENGINE STALLS	POOR ACCELERATION	ENGINE WILL NOT STOP	POOR PERFORMANCE	LIMITED ENGINE SPEED	OVERHEATING	LOW OIL PRESSURE	LOOSE STEERING	BILGE INCREASE	IRREGULAR WARNING INDICATION	Check	Section
														Fuel System	
O	O	O		O	O		O	O						Fuel tank breather hose	3
O	O	O		O	O		O	O						Check valve	3
O	O	O		O	O		O	O				O		Fuel hose	3
O	O	O		O	O		O	O						Fuel pump	3
O	O	O		O	O		O	O						Fuel filter	2
O	O	O		O	O		O	O				O		Fuel tank	3
													O	Fuel sender	3
O	O	O		O	O		O	O						Fuel rail	3
O	O	O		O	O		O	O						Fuel injectors	3

SELOC_9606

Troubleshooting chart - 2008-11 FX SHO models w/1812cc engines

Trouble mode														Check items	
ENGINE WILL NOT START	HARD STARTING	ROUGH IDLING	HIGH IDLING	ENGINE STALLS	POOR ACCELERATION	ENGINE WILL NOT STOP	POOR PERFORMANCE	LIMITED ENGINE SPEED	OVERHEATING	LOW OIL PRESSURE	LOOSE STEERING	BILGE INCREASE	IRREGULAR WARNING INDICATIO	Check	Section
														Power Unit	
									O			O		Water passage	
O	O	O		O	O		O	O						Compression pressure	2
	O	O		O	O		O	O		O			O	Engine oil	2
										O				Oil filter	2
O	O	O	O	O	O		O	O						Air filter	2, 3
					O		O	O				O		Air cooler	6
	O	O		O	O		O	O						Supercharger	6
					O		O		O	O		O		Oil cooler	5
										O			O	Oil pump	5
O	O	O		O	O		O	O						Valves and valve seats	6
O	O	O			O		O	O						Valve pads	6
O	O	O		O	O		O	O						Camshafts	6
O	O	O		O	O		O	O						Timing chain	6
		O		O	O		O							Drive coupling	6
O	O	O		O	O		O	O						Cylinder head gasket	6
O	O	O		O	O		O	O	O	O		O		Cylinder block	6
O	O	O		O	O		O	O						Crankcase	6
O	O	O		O	O		O	O						Crankshaft	6
O	O	O		O	O		O	O						Piston rings	6
O	O	O		O	O		O	O						Pistons	6
O	O	O		O	O		O	O						Bearings	6
		O			O				O					Thermostat	2, 5
												O		Pilot water hose	
									O			O		Water hose	

SELOC_9606

Troubleshooting chart - 2008-11 FX SHO models w/1812cc (cont'd)

ENGINE WILL NOT START	HARD STARTING	ROUGH IDLING	HIGH IDLING	ENGINE STALLS	POOR ACCELERATION	ENGINE WILL NOT STOP	POOR PERFORMANCE	LIMITED ENGINE SPEED	OVERHEATING	LOW OIL PRESSURE	LOOSE STEERING	BILGE INCREASE	IRREGULAR WARNING INDICATION	Check	Section
														Jet Pump Unit	
												O		Bilge strainer	2
					O		O		O			O		Impeller duct	7
					O		O	O						Impeller	7
					O		O							Intake grate	7
	O	O			O		O							Bearings	7
									O					Water inlet hose	7
												O		Bilge hose	7
												O		Bilge hose joint	7
					O		O							Rubber damper	7
														Electrical	
O	O	O		O	O		O							Spark plugs	2
O	O	O		O	O		O	O						Ignition coils	4
							O	O					O	Oil pressure switch	5
												O		Electric bilge pump	8
O														Engine start switch	4
O						O								Engine stop switch	4
O						O								Engine shut-off switch	4
O		O		O										Main and fuel pump relay	3
O														Starter relay	4
O														Starter motor	4
O														Remote control receiver	8
														Hull & Hood	
											O			Steering master	8
O	O	O		O	O		O	O				O		Water lock	8
O	O	O		O	O		O	O				O		Exhaust hose	8
												O		Drain plugs	2

(Trouble mode: first 14 columns; Check items: Check, Section)

SELOC_9606

Troubleshooting chart - 2008-11 FX SHO models w/1812cc (cont'd)

Engine does not crank.

Symptom 2	Cause	Checking step
Starter motor does not operate	Yamaha Security System set to lock mode	Check that the Yamaha Security System is set to the unlock mode.
	Discharged battery	Check the battery voltage and specific gravity.
	Loose connection of battery terminal	Check the battery cable and terminal for proper connection.
	Blown fuse	Check the fuse for continuity.
	Starter relay malfunction	Check the starter relay.
	Engine start switch malfunction	Check the engine start switch, (left handlebar switch)
	Short, open, or loose connection in starter motor circuit	Measure the starter relay input voltage.
		Check the wiring harness.
	Starter motor malfunction	Disassemble and check the starter motor.
	Stuck piston or crankshaft	Disassemble and check the engine unit.
	Stuck impeller	Disassemble and check the jet pump unit.
Starter motor operates, but the engine does not crank	Idle gear or starter clutch malfunction	Check the idle gear, starter clutch, or crankshaft Woodruff key.

SELOC_9606

Troubleshooting chart - 2009-11 FZX & FZR models w/1812cc engines

Engine will not start (engine cranks)

Symptom 2	Cause	Checking step
Throttle valve does not move properly.	Throttle lever squeezed	Check that the throttle lever is in the fully closed position.
	Throttle cable installed incorrectly	Adjust the throttle lever free play.
	APS mulfunction	Measure the APS 1 and 2 resistance.
	TPS mulfunction	Check the throttle valve opening.
	Throttle valve malfunction	Check the throttle body assembly.
Discharged battery	Battery performance decrement	Check the battery voltage and specific gravity.
	Stator coil assembly malfunction	Check the stator coil assembly.
	Short, open, or loose connection in charging circuit	Check the battery cable and terminal for proper connection.
		Check the wiring harness.
ECM does not operate	Blown fuse	Check the fuse for continuity.
	Main and fuel pump relay malfunction	Check the main and fuel pump relay.
	Short, open, or loose connection in ECM circuit	Measure the ECM input voltage.
		Check the ECM circuit for continuity.
	ECM malfunction	Replace the ECM.
Spark plug does not spark (all cylinders)	Engine shut-off cord (lanyard) clip not installed	Check that the engine shut-off cord (lanyard) clip is installed properly.
	Engine stop switch malfunction	Check the engine stop switch continuity, (left handlebar switch)
	Pickup coil malfunction	Measure the pickup coil output peak voltage.
		Measure the pickup coil resistance.
	Short, open, or loose connection in pickup coil circuit	Check the white/black (W/B) and black/orange (B/O) pickup coil wiring harness leads for continuity.
	Short, open, or loose connection in ignition coil circuit	Measure the ignition coil input voltage (from the battery).
		Check the red/yellow (R/Y), black (B), black/red (B/R), black/white (B/ W), black/yellow (B/Y), and black/ green (B/G) ignition coil wiring harness leads for continuity.
	ECM malfunction	Measure the ECM output peak voltage.

SELOC_9606

Troubleshooting chart - 2009-11 FZX & FZR models w/1812cc (cont'd)

Symptom 2	Cause	Checking step
Fuel not supplied (all cylinders)	Fuel leakage	Check the fuel hose.
	Blown fuse	Check the fuse for continuity.
	Clogged fuel pump filter	Clean the fuel pump filter.
	Fuel pump module malfunction	Check the fuel pump module operation sound using the YDIS.
		Measure the fuel pump module input voltage.
		Check the blue (L) and black (B) fuel pump module wiring harness leads for continuity.
	Main and fuel pump relay malfunction	Check the main and fuel pump relay.
	Fuel not supplied to the fuel rail	Measure the fuel pressure.
Compression pressure is low	Compression leakage	Measure the compression pressure.
		Measure the valve clearance.
		Check the camshaft for damage.
		Check the cylinder head gasket and cylinder head warpage.
		Check the valve and valve seat for wear.
		Check the piston and piston ring for damage.
		Check the cylinder for damage.
	Improper valve timing	Check the timing chain.

SELOC_9606

Troubleshooting chart - 2009-11 FZX & FZR models w/1812cc (cont'd)

Symptom 2	Cause	Checking step
Throttle valve does not move properly	Throttle cable installed incorrectly.	Adjust the throttle lever free play.
	Throttle valve malfunction	Check the throttle body assembly.
Spark plug does not spark (some cylinders)	Spark plug malfunction	Check the spark plug.
	Ignition coil malfunction	Check the ignition spark.
	Short, open, or loose connection in ignition coil circuit	Measure the ignition coil input voltage (from the battery).
		Check the red/yellow (R/Y), black (B), black/red (B/R), black/white (B/ W), black/yellow (B/Y), and black/ green (B/G) ignition coil wiring harness leads for continuity.
	ECM malfunction	Measure the ECM output peak voltage.
Fuel not supplied (some cylinders)	Fuel injector malfunction	Check the fuel injector operation sound using the YDIS.
		Measure the fuel injector resistance.
		Check the fuel injector Oring.
	Short, open, or loose connection in fuel injector circuit	Measure the fuel injector input voltage.
		Check the red/yellow (R/Y), purple/ red (Pu/R), purple/black (Pu/B), purple/yellow (Pu/Y), and purple/ green (Pu/G) fuel injector wiring harness leads for continuity.
	Clogged fuel injector filter	Clean the fuel injector.
	ECM malfunction	Replace the ECM.
Compression pressure is low	Compression leakage	Measure the compression pressure.
		Measure the valve clearance.
		Check the camshaft for damage.
		Check the cylinder head gasket and cylinder head warpage.
		Check the valve and valve seat for wear.
		Check the piston and piston ring for damage.
		Check the cylinder for damage.
	Improper valve timing	Check the timing chain.

SELOC_9606

Troubleshooting chart - 2009-11 FZX & FZR models w/1812cc (cont'd)

High engine idle speed

Symptom 2	Cause	Checking step
—	Throttle cable installed incorrectly.	Adjust the throttle lever free play and check the cable rooting.

Limited engine speed

Symptom 2	Cause	Checking step
Buzzer sounds intermittently Overheat warning indicator blinks Check engine warning indicator blinks	Clogged cooling water passage	Check the cooling water pilot outlet for water discharge.
		Check the cooling water passage.
	Thermostat malfunction	Check the thermostat.
Buzzer sounds intermittently Oil pressure warning indicator blinks	Insufficient engine oil	Check the engine oil level.
		Check for engine oil leakage.
		Check the oil passage.
	Oil pressure switch malfunction	Check the oil pressure switch continuity.
		Check that there is no short circuit in the pink/white (P/W) oil pressure switch wiring harness lead.
—	APS malfunction	Check the APS.
	TPS malfunction	Check the throttle body assembly.
	Shift lever malfunction	Check the reverse sensor.

SELOC_9606

Troubleshooting chart - 2009-11 FZX & FZR models w/1812cc (cont'd)

Discharged battery

Symptom 2	Cause	Checking step
-	Battery performance decrease	Check the battery voltage and specific gravity.
	Loose connection of battery terminal	Check the battery cable and terminal for proper connection.
	Short, open, or loose connection in charging circuit	Check the charging circuit for wiring connection and damage.
	Stator coil malfunction	Measure the stator coil output peak voltage.
		Measure the stator coil resistance.
	Rectifier regulator malfunction	Measure the rectifier regulator output peak voltage.

Poor performance

Symptom 2	Cause	Checking step
Watercraft cannot reach high speeds	Jet pump unit malfunction	Check the impeller, impeller duct, and intake grate.
	Water entered hull	Check the drain plugs and O-rings for damage.
		Check the cooling water hoses for damage.
		Check the water lock, water tank, and rubber hoses for damage.
		Check the exterior of the hull for damage.

SELOC_9606

Troubleshooting chart - 2009-11 FZX & FZR models w/1812cc (cont'd)

Trouble mode: ENGINE WILL NOT START	HARD STARTING	ROUGH IDLING	HIGH IDLING	ENGINE STALLS	POOR ACCELERATION	ENGINE WILL NOT STOP	POOR PERFORMANCE	LIMITED ENGINE SPEED	OVERHEATING	LOW OIL PRESSURE	LOOSE STEERING	BILGE INCREASE	IRREGULAR WARNING INDICATION	POOR BATTERY CHARGING	Check elements: Check	Section
															Fuel System	
O	O	O		O	O		O	O							Fuel tank	3
O	O	O		O	O		O	O							Fuel tank breather hose	3
O	O	O		O	O		O	O							Fuel hose	3
O	O	O		O	O		O	O							Fuel filter	2
O	O	O		O	O		O	O							Fuel pump	3
O	O	O		O	O		O	O							Fuel rail	3
O	O	O		O	O		O	O							Fuel injectors	3
				O			O								Trolling speed	2
O	O	O		O	O		O	O							Air filter	7

SELOC_9606

Troubleshooting chart - 2005-09 VX models w/1052cc engines

Trouble mode: ENGINE WILL NOT START	HARD STARTING	ROUGH IDLING	HIGH IDLING	ENGINE STALLS	POOR ACCELERATION	ENGINE WILL NOT STOP	POOR PERFORMANCE	LIMITED ENGINE SPEED	OVERHEATING	LOW OIL PRESSURE	LOOSE STEERING	BILGE INCREASE	IRREGULAR WARNING INDICATION	POOR BATTERY CHARGING	Check elements: Check	Section
															Power Unit	
O	O	O		O	O		O	O							Compression	2
O	O	O		O	O		O	O	O	O					Cylinder head gaskets	6
O	O	O		O	O		O	O	O						Cylinder block	6
O	O	O		O	O		O	O		O					Crankcase	6
O	O	O		O	O		O	O							Piston rings	6
O	O	O		O	O		O	O							Pistons	6
O	O	O		O	O		O	O							Bearings	6
O	O	O		O	O		O	O							Valve(s) and valve seat(s)	6
O		O			O		O	O							Valve clearance adjusting pad(s)	2, 6
O		O		O	O		O								Camshaft(s)	6
O				O	O		O								Timing chain	6
										O					Oil pump	5
								O		O					Engine oil	2
										O					Oil filter	2
										O			O		Oil pressure switch	5
							O								Bearing housing	7
		O					O								Drive couplings	7
							O								Rubber coupling	7
									O			O			Pilot water hose	8
									O			O			Water hose	
									O			O			Water passage	

SELOC_9606

Troubleshooting chart - 2005-09 VX models w/1052cc (cont'd)

Trouble mode															Check elements	
ENGINE WILL NOT START	HARD STARTING	ROUGH IDLING	HIGH IDLING	ENGINE STALLS	POOR ACCELERATION	ENGINE WILL NOT STOP	POOR PERFORMANCE	LIMITED ENGINE SPEED	OVERHEATING	LOW OIL PRESSURE	LOOSE STEERING	BILGE INCREASE	IRREGULAR WARNING INDICATION	POOR BATTERY CHARGING	Check	Section
															Jet Pump Unit	
							o		o			o			Duct	7
							o								Impeller	7
							o		o						Intake grate	7
		o					o								Bearings	7
									o						Water inlet hose	7
												o			Bilge hose	7
												o			Bilge strainer	2, 7
												o			Bilge hose joint	7
															Electrical **Ignition system, fuel control system**	
o	o	o		o	o		o						o		Pick-up coils	4
o			o	o		o	o	o							ECM	3
o	o	o		o	o		o								Ignition coils	4
o				o									o		Slant detection switch	4
o						o									Engine stop switch	4
o						o									Engine shut-off switch	4
o	o	o	o	o	o	o	o								Spark plugs	2
o		o		o		o	o								Main and fuel pump relay	3
o	o	o	o	o	o	o	o	o					o		Elect control throttle valve relay	
							o		o				o		Thermoswitch	4
	o	o	o		o		o	o					o		Sensor assembly	3
	o	o	o					o					o		Engine temperature sensor	4
o	o	o	o	o	o		o	o					o		Throttle body assembly (throttle position sensor)	3
o	o	o	o	o	o		o	o					o		Accelerator position sensor	3
o	o	o		o	o		o	o					o		Cam position sensor	3

SELOC_9606

Troubleshooting chart - 2005-09 VX models w/1052cc (cont'd)

Trouble mode															Check elements	
ENGINE WILL NOT START	HARD STARTING	ROUGH IDLING	HIGH IDLING	ENGINE STALLS	POOR ACCELERATION	ENGINE WILL NOT STOP	POOR PERFORMANCE	LIMITED ENGINE SPEED	OVERHEATING	LOW OIL PRESSURE	LOOSE STEERING	BILGE INCREASE	IRREGULAR WARNING INDICATION	POOR BATTERY CHARGING	Check	Section
															Starting System	
O	O					O									• Start switch	4
O						O									• Starter relay	4
O															• Starter motor	4
															Charging System	
														O	• Lighting coil	4
														O	• Rectifier/regulator	4
O		O	O	O										O	• Fuses	4
O		O		O										O	• Battery leads	2
O	O	O	O	O										O	• Battery	2, 4
															Remote Control System	
O															• Remote control unit	8
															Hull & Hood	
											O				Steering column	8
							O					O			Water lock	8
				O			O					O			Exhaust hose	8
							O					O			Muffler	6
												O			Drain plugs	2

SELOC_9606

Troubleshooting chart - 2005-09 VX models w/1052cc (cont'd)

Engine does not crank

Symptom 2	Cause	Checking step
Starter motor does not operate	Yamaha Security System set to lock mode (*1)	Check that the Yamaha Security System is set to the unlock mode.
	Discharged battery	Check the battery voltage and specific gravity.
	Loose connection of battery terminal	Check the battery cable and terminal for proper connection.
	Blown fuse	Check the fuse for continuity.
	Starter relay malfunction	Check the starter relay.
	Engine start switch malfunction	Check the engine start switch, (left handlebar switch)
	Short, open, or loose connection in starter motor circuit	Measure the starter relay input voltage.
		Check the wiring harness.
	Starter motor malfunction	Disassemble and check the starter motor.
	Stuck piston or crankshaft	Disassemble and check the engine unit.
	Stuck impeller	Disassemble and check the jet pump unit.
Starter motor operates, but the engine does not crank	Idle gear or starter clutch malfunction	Check the idle gear, starter clutch, or crankshaft Woodruff key.

SELOC_9606

Troubleshooting chart - 2010-11 VX models w/1052cc engines

Engine will not start (engine cranks).

Symptom 2	Cause	Checking step
Throttle valve does not move properly	Throttle lever squeezed	Check that the throttle lever is in the fully closed position.
	Throttle cable installed incorrectly	Adjust the throttle lever free play.
	APS mulfunction	Measure the APS 1 and 2 resistance.
	TPS mulfunction	Check the throttle valve opening.
	Throttle valve malfunction	Check the throttle body assy.
Discharged battery	Battery performance decrease	Check the battery voltage and specific gravity.
	Stator coil malfunction	Check the stator coil assy.
	Short, open, or loose connection in charging circuit	Check the battery cable and terminal for proper connection.
		Check the wiring harness.
ECM does not operate	Blown fuse	Check the fuse for continuity.
	Main and fuel pump relay malfunction	Check the main and fuel pump relay.
	Short, open, or loose connection in ECM circuit	Measure the ECM input voltage.
		Check the ECM circuit for continuity.
	ECM malfunction	Replace the ECM.
Spark plug does not spark (all cylinders)	Engine shut-off cord clip not installed	Check that the engine shut-off cord clip is installed properly.
	Engine stop switch malfunction	Check the engine stop switch continuity. (left handlebar switch)
	Pickup coil malfunction	Measure the pickup coil output peak voltage.
		Measure the pickup coil resistance.
	Short, open, or loose connection in pickup coil circuit	Check the white/black (W/B) and black/orange (B/O) pickup coil wiring harness leads for continuity.
	Short, open, or loose connection in ignition coil circuit	Measure the ignition coil input voltage (from the battery).
		Check the red/yellow (R/Y), black/red (B/R), black/white (B/W), black/yellow (B/Y), and black/green (B/G) ignition coil wiring harness leads for continuity.
	ECM malfunction	Measure the ECM output peak voltage.

SELOC_9606

Troubleshooting chart - 2010-11 VX models w/1052cc engines (cont'd)

Symptom 2	Cause	Checking step
Fuel not supplied (all cylinders)	Fuel leakage	Check the fuel hose.
	Blown fuse	Check the fuse for continuity.
	Clogged fuel pump filter	Clean the fuel pump filter.
	Fuel pump module malfunction	Check the fuel pump module operation sound using the YDIS.
		Measure the fuel pump module input voltage.
		Check the blue (L) and black (B) fuel pump module wiring harness leads for continuity.
		Measure the fuel pressure.
	Main and fuel pump relay malfunction	Check the main and fuel pump relay.
Compression pressure is low	Compression leakage	Measure the compression pressure.
		Measure the valve clearance.
		Check the camshaft for damage.
		Check the cylinder head gasket and cylinder head warpage.
		Check the valve and valve seat for wear.
		Check the piston and piston ring for damage.
		Check the cylinder for damage.
	Improper valve timing	Check the timing chain.

SELOC_9606

Troubleshooting chart - 2010-11 VX models w/1052cc engines (cont'd)

Symptom 2	Cause	Checking step
Throttle valve does not move properly	Throttle cable installed incorrectly	Adjust the throttle lever free play.
	Throttle valve malfunction	Check the throttle body assy.
Spark plug does not spark (some cylinders)	Spark plug malfunction	Check the spark plug.
	Ignition coil malfunction	Check the ignition spark.
	Short, open, or loose connection in ignition coil circuit	Measure the ignition coil input volt-age^from the battery).
		Check the red/yellow (R/Y), black/red (B/R), black/white (B/W), black/yellow (B/Y), and black/green (B/G) ignition coil wiring harness leads for continuity.
	ECM malfunction	Measure the ECM output peak voltage.
Fuel not supplied (some cylinders)	Fuel injector malfunction	Check the fuel injector operation sound using the YDIS.
		Measure the fuel injector resistance.
		Check the fuel injector O-ring.
	Short, open, or loose connection in fuel injector circuit	Measure the fuel injector input voltage.
		Check the red/yellow (R/Y), purple/red (Pu/R), purple/black (Pu/B), purple/yellow (Pu/Y), and purple/green (Pu/G) fuel injector wiring harness leads for continuity.
	Clogged fuel injector filter	Clean the fuel injector.
	ECM malfunction	Replace the ECM.
Compression pressure is low	Compression leakage	Measure the compression pressure.
		Measure the valve clearance.
		Check the camshaft for damage.
		Check the cylinder head gasket and cylinder head warpage.
		Check the valve and valve seat for wear.
		Check the piston and piston ring for damage.
		Check the cylinder for damage.
	Improper valve timing	Check the timing chain.

SELOC_9606

Troubleshooting chart - 2010-11 VX models w/1052cc engines (cont'd)

High engine idle speed

Symptom 2	Cause	Checking step
—	Throttle cable installed incorrectly	Adjust the throttle lever free play and check the cable rooting.

Limited engine speed

Symptom 2	Cause	Checking step
• Buzzer sounds intermittently • Overheat warning indicator blinks • Check engine warning indicator blinks	Clogged cooling water passage	Check the cooling water pilot outlet for water discharge.
		Check the cooling water passage.
• Buzzer sounds intermittently • Oil pressure warning indicator blinks	Insufficient engine oil	Check the engine oil level.
		Check for engine oil leakage.
		Check the oil passage.
	Oil pressure switch malfunction	Check the oil pressure switch continuity.
		Check that there is no short circuit in the pink/white (P/W) oil pressure switch wiring harness lead.
	APS malfunction	Check the APS.
	TPS malfunction	Check the throttle body assy.

SELOC_9606

Troubleshooting chart - 2010-11 VX models w/1052cc engines (cont'd)

Discharged battery

Symptom 2	Cause	Checking step
	Battery performance decrease	Check the battery voltage and specific gravity.
	Loose connection of battery terminal	Check the battery cable and terminal for proper connection.
	Short, open, or loose connection in charging circuit	Check the charging circuit for wiring connection and damage.
	Stator coil malfunction	Check the stator coil assy.
	Rectifier regulator malfunction	Measure the rectifier regulator output peak voltage.

Poor performance

Symptom 2	Cause	Checking step
Watercraft cannot reach high speeds	Jet pump unit malfunction	Check the impeller, impeller duct, and intake grate.
	Water entered hull	Check the drain plugs and O-rings for damage.
		Check the cooling water hoses for damage.
		Check the water lock, water tank, and rubber hoses for damage.
		Check the exterior of the hull for damage.

SELOC_9606

Troubleshooting chart - 2010-11 VX models w/1052cc engines (cont'd)

Symptom 2	Cause	Checking step
Starter motor does not operate	Yamaha Security System set to lock mode	Check that the Yamaha Security System is set to the unlock mode.
	Discharged battery	Check the battery voltage and specific gravity.
	Loose connection of battery terminal	Check the battery cable and terminal for proper connection.
	Blown fuse	Check the fuse for continuity.
	Starter relay malfunction	Check the starter relay.
	Engine start switch malfunction	Check the engine start switch, (left handlebar switch)
	Short, open, or loose connection in starter motor cir- cuit	Measure the starter relay input voltage.
		Check the wiring harness.
	Starter motor malfunction	Disassemble and check the starter motor.
	Stuck piston or crankshaft	Disassemble and check the engine unit.
	Stuck impeller	Disassemble and check the jet pump unit.
Starter motor operates, but the engine does not crank	Idle gear or starter clutch malfunction	Check the idle gear, starter clutch, or crankshaft Woodruff key.

SELOC_9606

Troubleshooting chart - 2011 VX models w/1812cc engines

Symptom 2	Cause	Checking step
Throttle valve does not move properly.	Throttle lever squeezed	Check that the throttle lever is in the fully closed position.
	Throttle cable installed incorrectly	Adjust the throttle lever free play.
	APS mulfunction	Measure the APS 1 and 2 resistance.
	TPS mulfunction	Cfyeck the throttle valve opening.
	Throttle valve malfunction	Check the throttle body assembly.
Discharged battery	Battery performance decrement	Check the battery voltage and specific gravity.
	Stator coil assembly malfunction	Check the stator coil assembly.
	Short, open, or loose connection in charging circuit	Check the battery cable and terminal for proper connection.
		Check the wiring harness.
ECM does not operate	Blown fuse	Check the fuse for continuity.
	Main and fuel pump relay malfunction	Check the main and fuel pump relay.
	Short, open, or loose connection in ECM circuit	Measure the ECM input voltage.
		Check the ECM circuit for continuity.
	ECM malfunction	Replace the ECM.
Spark plug does not spark (all cylinders)	Engine shut-off cord (lanyard) clip not installed	Check that the engine shut-off cord (lanyard) clip is installed properly.
	Engine stop switch malfunction	Check the engine stop switch continuity, (left handlebar switch)
	Pick-up coil malfunction	Measure the pickup coil output peak voltage.
		Measure the pickup coil resistance.
	Short, open, or loose connection in pickup coil circuit	Check the white/black (W/B) and black/orange (B/O) pickup coil wiring harness leads for continuity.
	Short, open, or loose connection in ignition coil circuit	Measure the ignition coil input voltage (from the battery).
		Check the red/yellow (R/Y), black (B), black/red (B/R), black/white (B/ W), black/yellow (B/Y), and black/ green (B/G) ignition coil wiring harness leads for continuity.Check the red/yellow (R/Y), black (B), black/red (B/R), black/white (B/ W), black/yellow (B/Y), and black/ green (B/G) ignition coil wiring harness leads for continuity.
	ECM malfunction	Measure the ECM output peak voltage.

Troubleshooting chart - 2011 VX models w/1812cc engines (cont'd)

SELOC_9606

Symptom 2	Cause	Checking step
Fuel not supplied (all cylinders)	Fuel leakage	Check the fuel hose.
	Blown fuse	Check the fuse for continuity.
	Clogged fuel pump filter	Clean the fuel pump filter.
	Fuel pump module malfunction	Check the fuel pump module operation sound using the YDIS.
		Measure the fuel pump module input voltage.
		Check the blue (L) and black (B) fuel pump module wiring harness leads for continuity.
	Main and fuel pump relay malfunction	Check the main and fuel pump relay.
	Fuel not supplied to the fuel rail	Measure the fuel pressure.
Compression pressure is low	Compression leakage	Measure the compression pressure.
		Measure the valve clearance.
		Check the camshaft for damage.
		Check the cylinder head gasket and cylinder head warpage.
		Check the valve and valve seat for wear.
		Check the piston and piston ring for damage.
		Check the cylinder for damage.
	Improper valve timing	Check the timing chain.

SELOC_9606

Troubleshooting chart - 2011 VX models w/1812cc engines (cont'd)

Symptom 2	Cause	Checking step
Throttle valve does not move properly	Throttle cable installed incorrectly.	Adjust the throttle lever free play.
	Throttle valve malfunction	Check the throttle body assembly.
Spark plug does not spark (some cylinders)	Spark plug malfunction	Check the spark plug.
	Ignition coil malfunction	Check the ignition spark.
	Short, open, or loose connection in ignition coil circuit	Measure the ignition coil input voltage (from the battery).
		Check the red/yellow (R/Y), black (B), black/red (B/R), black/white (B/W), black/yellow (B/Y), and black/green (B/G) ignition coil wiring harness leads for continuity.
	ECM malfunction	Measure the ECM output peak voltage.
Fuel not supplied (some cylinders)	Fuel injector malfunction	Check the fuel injector operation sound using the YDIS.
		Measure the fuel injector resistance.
		Check the fuel injector O-ring.
	Short, open, or loose connection in fuel injector circuit	Measure the fuel injector input voltage.
		Check the red/yellow (R/Y), purple/red (Pu/R), purple/black (Pu/B), purple/yellow (Pu/Y), and purple/green (Pu/G) fuel injector wiring harness leads for continuity.
	Clogged fuel injector filter	Clean the fuel injector.
	ECM malfunction	Replace the ECM.
Compression pressure is low	Compression leakage	Measure the compression pressure.
		Measure the valve clearance.
		Check the camshaft for damage.
		Check the cylinder head gasket and cylinder head warpage.
		Check the valve and valve seat for wear.
		Check the piston and piston ring for damage.
		Check the cylinder for damage.
	Improper valve timing	Check the timing chain.

SELOC_9606

Troubleshooting chart - 2011 VX models w/1812cc engines (cont'd)

High engine idle speed

Symptom 2	Cause	Checking step
—	Throttle cable installed incorrectly.	Adjust the throttle lever free play and check the cable rooting.

Limited engine speed

Symptom 2	Cause	Checking step
• Buzzer sounds intermittently • Overheat warning indicator blinks • Check engine warning indicator blinks	Clogged cooling water passage	Check the cooling water pilot outlet for water discharge.
		Check the cooling water passage.
	Thermostat malfunction	Check the thermostat.
• Buzzer sounds intermittently Oil pressure warning indicator blinks	Insufficient engine oil	Check the engine oil level.
		Check for engine oil leakage.
		Check the oil passage.
	Oil pressure switch malfunction	Check the oil pressure switch continuity.
		Check that there is no short circuit in the pink/white (P/W) oil pressure switch wiring harness lead.
	APS malfunction	Check the APS.
	TPS malfunction	Check the throttle body assembly.
	Shift lever malfunction	Check the reverse sensor.

SELOC_9606

Troubleshooting chart - 2011 VX models w/1812cc engines (cont'd)

Discharged battery

Symptom 2	Cause	Checking step
	Battery performance decrease	Check the battery voltage and specific gravity.
	Loose connection of battery terminal	Check the battery cable and terminal for proper connection.
	Short, open, or loose connection in charging circuit	Check the charging circuit for wiring connection and damage.
	Stator coil malfunction	Measure the stator coil output peak voltage.
		Measure the stator coil resistance.
	Rectifier regulator malfunction	Measure the rectifier regulator output peak voltage.

Poor performance

Symptom 2	Cause	Checking step
Watercraft cannot reach high speeds	Jet pump unit malfunction	Check the impeller, impeller duct, and intake grate.
	Water entered hull	Check the drain plugs and O-rings for damage.
		Check the cooling water hoses for damage.
		Check the water lock, water tank, and rubber hoses for damage.
		Check the exterior of the hull for damage.

SELOC_9606

Troubleshooting chart - 2011 VX models w/1812cc engines (cont'd)

FIRING ORDERS

■ **To avoid confusion, ALWAYS label the spark plug wires or coil packs with a piece of tape before disconnecting them from the plug or cylinder head cover. Always use the firing order you observe on your engine over the one depicted here if different!**

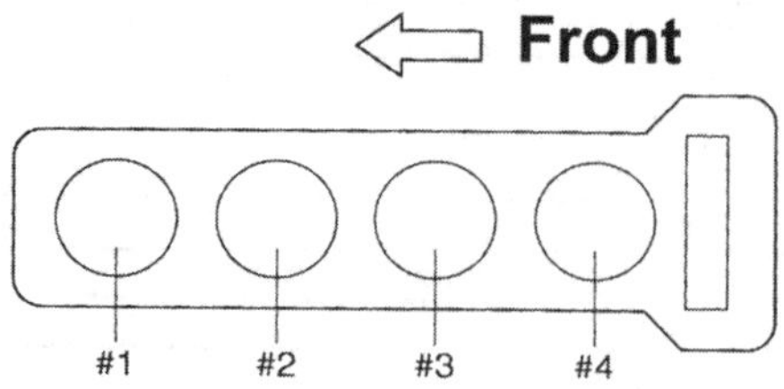

Firing Order - 1-2-4-3

Firing order - 998cc and 1052cc engines

SELOC_9606

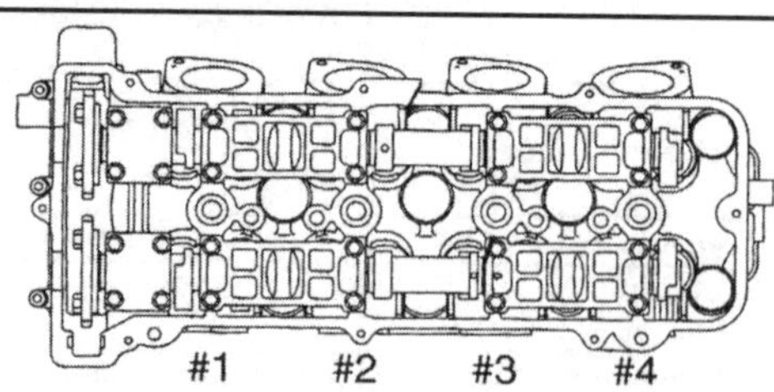

Firing Order - 1-2-4-3

Firing order - 1812cc engines

SELOC_9606

SPECIFICATIONS

Model Identification

Year	Model Name	Model Number	Model Code Hull/Engine	Length in. (mm)	Width in. (mm)	Height in. (mm)	Capacity Person/lb (kg)	Dry Weight lb (kg)
2002	FX140	FX1000A/AC	F1B/60E	131.5 (3340)	48.4 (1230)	45.7 (1160)	3/530 (240)	798 (362)
2003	FX Cruiser	FX1000A-B/AC-B	F1B/60E	131.5 (3340)	48.4 (1230)	45.7 (1160)	3/530 (240)	798 (362)
	FX140	FX1000B/C-B	F1B/60E	131.5 (3340)	48.4 (1230)	45.7 (1160)	3/530 (240)	798 (362)
2004	FX Cruiser	FX1000A-C/AC-C	F1B/60E	131.5 (3340)	48.4 (1230)	45.7 (1160)	3/530 (240)	798 (362)
	FX Cruiser High Output	FX1100A-C/A-C-C	F1S/6B6	131.5 (3340)	48.4 (1230)	45.7 (1160)	3/530 (240)	772 (350)
	FX High Output	FX1100-C/-C-C	F1S/6B6	131.5 (3340)	48.4 (1230)	45.7 (1160)	3/530 (240)	772 (350)
	FX140	FX1000-C	F1B/60E	131.5 (3340)	48.4 (1230)	45.7 (1160)	3/530 (240)	798 (362)
	FX140 (Calif.)	FX1000C-C	F1B/60E	131.5 (3340)	48.4 (1230)	45.7 (1160)	3/530 (240)	798 (362)
2005	FX	FX1000D/CD	F1B/60E	131.5 (3340)	48.4 (1230)	45.7 (1160)	3/530 (240)	798 (362)
	FX Cruiser	FX1000AD/ACD	F1B/60E	131.5 (3340)	48.4 (1230)	45.7 (1160)	3/530 (240)	798 (362)
	FX Cruiser High Output	FX1100AD/ACD	F1S/6B6	131.5 (3340)	48.4 (1230)	45.7 (1160)	3/530 (240)	772 (350)
	FX High Output	FX1100D	F1S/6B6	131.5 (3340)	48.4 (1230)	45.7 (1160)	3/530 (240)	772 (350)
	VX110 Deluxe	VX1100AD	F1K/6D3	126.8 (3220)	46.1 (1170)	45.3 (1150)	3/530 (240)	716 (325)
	VX110 Sport	VX1100D	F1K/6D3	126.8 (3220)	46.1 (1170)	45.3 (1150)	3/530 (240)	716 (325)
2006	FX	FX1000E	F1Y/60E	131.5 (3340)	48.4 (1230)	45.7 (1160)	3/530 (240)	87 (384)
	FX Cruiser	FX1000AE	F1Y/60E	131.5 (3340)	48.4 (1230)	48.8 (1240)	3/530 (240)	864 (392)
	FX Cruiser High Output	FX1100AE/BE	F1X/6B6	131.5 (3340)	48.4 (1230)	48.8 (1240)	3/530 (240)	838 (380)
	FX High Output	FX1100E/C-C	F1X/6B6	131.5 (3340)	48.4 (1230)	45.7 (1160)	3/530 (240)	820 (372)
	VX110 Deluxe	VX1100AE	F1K/6D3	126.8 (3220)	46.1 (1170)	45.3 (1150)	3/530 (240)	716 (325)
	VX110 Sport	VX1100E	F1K/6D3	126.8 (3220)	46.1 (1170)	45.3 (1150)	3/530 (240)	716 (325)
2007	FX	FX1000F	F1Y/60E	131.5 (3340)	48.4 (1230)	45.7 (1160)	3/530 (240)	87 (384)
	FX Cruiser	FX1000AF	F1Y/60E	131.5 (3340)	48.4 (1230)	48.8 (1240)	3/530 (240)	864 (392)
	FX Cruiser High Output	FX1100AF	F1X/6B6	131.5 (3340)	48.4 (1230)	48.8 (1240)	3/530 (240)	838 (380)
	FX High Output	FX1100F	F1X/6B6	131.5 (3340)	48.4 (1230)	45.7 (1160)	3/530 (240)	820 (372)
	VX	VX1100F	F1K/6D3	126.8 (3220)	46.1 (1170)	45.3 (1150)	3/530 (240)	716 (325)
	VX Crusier	VX1100AF	F1K/6D3	126.8 (3220)	46.1 (1170)	45.3 (1150)	3/530 (240)	716 (325)
	VX110 Deluxe	VX1100BF	F1K/6D3	126.8 (3220)	46.1 (1170)	45.3 (1150)	3/530 (240)	716 (325)
	VX110 Sport	VX1100CF	F1K/6D3	126.8 (3220)	46.1 (1170)	45.3 (1150)	3/530 (240)	716 (325)
2008	FX	FX1000G	F1Y/60E	131.5 (3340)	48.4 (1230)	45.7 (1160)	3/530 (240)	87 (384)
	FX Cruiser	FX1000AG	F1Y/60E	131.5 (3340)	48.4 (1230)	48.8 (1240)	3/530 (240)	864 (392)
	FX Cruiser HO	FX1100AG/ALG	F1X/6B6	131.5 (3340)	48.4 (1230)	48.8 (1240)	3/530 (240)	838 (380)
	FX Cruiser HO (Calif.)	FX1100AG/ALG	F1X/6AA	131.5 (3340)	48.4 (1230)	48.8 (1240)	3/530 (240)	844 (383)
	FX Cruiser SHO	FX1800AG	F1W/6S5	132.7 (3370)	48.4 (1230)	48.8 (1240)	3/530 (240)	829 (376)
	FX HO	FX1100G/LG	F1X/6B6	131.5 (3340)	48.4 (1230)	45.7 (1160)	3/530 (240)	827 (375)
	FX HO (Calif.)	FX1100G/LG	F1X/6AA	131.5 (3340)	48.4 (1230)	45.7 (1160)	3/530 (240)	833 (375)
	FX SHO	FX1800G	F1W/6S5	132.7 (3370)	48.4 (1230)	45.7 (1160)	3/530 (240)	829 (376)
	VX	VX1100G	F1K/6D3	126.8 (3220)	46.1 (1170)	45.3 (1150)	3/530 (240)	716 (325)

SELOC_9606

Model Identification

Year	Model Name	Model Number	Model Code Hull/Engine	Length in. (mm)	Width in. (mm)	Height in. (mm)	Capacity Person/lb (kg)	Dry Weight lb (kg)
2008	VX Crusier	VX1100AGW	F1K/6D3	126.8 (3220)	46.1 (1170)	45.3 (1150)	3/530 (240)	716 (325)
(cont'd)	VX Deluxe	VX1100BGB	F1K/6D3	126.8 (3220)	46.1 (1170)	45.3 (1150)	3/530 (240)	716 (325)
	VX110 Sport	VX1100CG	F1K/6D3	126.8 (3220)	46.1 (1170)	45.3 (1150)	3/530 (240)	716 (325)
2009	FX Cruiser HO		F2H/6BH	132.7 (3370)	48.4 (1230)	48.8 (1240)	3/530 (240)	805 (365)
	FX Cruiser SHO	FX1800AHR/AHZ	F1W/6S5	132.7 (3370)	48.4 (1230)	48.8 (1240)	3/530 (240)	829 (376)
	FX HO	FY1800AH/H	F2H/6BH	132.7 (3370)	48.4 (1230)	45.7 (1160)	3/530 (240)	794 (360)
	FX SHO	FX1800HB/HK	F1W/6S5	132.7 (3370)	48.4 (1230)	45.7 (1160)	3/530 (240)	829 (376)
	FZR		F2R/6AN	131.9 (3350)	48.4 (1230)	45.7 (1160)	2/353 (160)	807 (366)
	FZS	GX1800AHK/AHR	F2C/6AN	132.7 (3370)	48.4 (1230)	45.7 (1160)	3/530 (240)	813 (369)
	VX	VX1100F	F1K/6D3	126.8 (3220)	46.1 (1170)	45.3 (1150)	3/530 (240)	716 (325)
	VX Cruiser	VX1100AH	F1K/6D3	126.8 (3220)	46.1 (1170)	45.3 (1150)	3/530 (240)	716 (325)
	VX Deluxe	VX1100BH	F1K/6D3	126.8 (3220)	46.1 (1170)	45.3 (1150)	3/530 (240)	716 (325)
	VX Sport	VX1100CH	F1K/6D3	126.8 (3220)	46.1 (1170)	45.3 (1150)	3/530 (240)	716 (325)
2010	FX Cruiser HO	FY1800AJ	F2H/6BH	132.7 (3370)	48.4 (1230)	48.8 (1240)	3/530 (240)	805 (365)
	FX Cruiser SHO	FX1800AJA/AJB	F1W/6S5	132.7 (3370)	48.4 (1230)	48.8 (1240)	3/530 (240)	829 (376)
	FX HO	FY1800J	F2H/6BH	132.7 (3370)	48.4 (1230)	45.7 (1160)	3/530 (240)	794 (360)
	FX SHO	FX1800JA/JB	F1W/6S5	132.7 (3370)	48.4 (1230)	45.7 (1160)	3/530 (240)	829 (376)
	FZR	GX1800J	F2R/6AN	131.9 (3350)	48.4 (1230)	45.7 (1160)	2/353 (160)	807 (366)
	FZS	GX1800AJA/AJB	F2C/6AN	132.7 (3370)	48.4 (1230)	45.7 (1160)	3/530 (240)	813 (369)
	VX Cruiser	VX1100A/AJB	F2N/6BU	128.7 (3270)	46.1 (1170)	45.7 (1160)	3/530 (240)	750 (340)
	VX Deluxe	VX1100B/BJB	F2N/6BU	126.8 (3220)	46.1 (1170)	45.7 (1160)	3/530 (240)	743 (337)
	VX Sport	VX1100J	F2N/6BU	126.8 (3220)	46.1 (1170)	45.7 (1160)	3/530 (240)	736 (334)
2011	FX Cruiser HO	FY1800-AK	F2H/6BH	132.7 (3370)	48.4 (1230)	48.8 (1240)	3/530 (240)	805 (365)
	FX Cruiser SHO	FX1800A-K	F1W/6S5	132.7 (3370)	48.4 (1230)	48.8 (1240)	3/530 (240)	829 (376)
	FX HO	FY1800-K	F2H/6BH	132.7 (3370)	48.4 (1230)	45.7 (1160)	3/530 (240)	794 (360)
	FX SHO	FX1800-K	F1W/6S5	132.7 (3370)	48.4 (1230)	45.7 (1160)	3/530 (240)	829 (376)
	FZR	GX1800-K	F2R/6AN	131.9 (3350)	48.4 (1230)	45.7 (1160)	2/353 (160)	807 (366)
	FZS	GX1800-AK	F2C/6AN	132.7 (3370)	48.4 (1230)	45.7 (1160)	3/530 (240)	813 (369)
	VX Cruiser	VX1100A-K	F2N/6BU	128.7 (3270)	46.1 (1170)	45.7 (1160)	3/530 (240)	750 (340)
	VX Deluxe	VX1100B-K	F2N/6BU	126.8 (3220)	46.1 (1170)	45.7 (1160)	3/530 (240)	743 (337)
	VX Sport	VX1100C-K	F2N/6BU	126.8 (3220)	46.1 (1170)	45.7 (1160)	3/530 (240)	736 (334)
	VXR	VX1800A-K	F2M/6CN	128.7 (3270)	46.1 (1170)	45.7 (1160)	3/530 (240)	728 (330)
	VXS	VX1800-K	F2N/6BU	126.8 (3220)	46.1 (1170)	45.7 (1160)	3/530 (240)	721 (327)

SELOC_9606

General Engine Specifications

Year	Model Name	Engine Type	Diplacement cu. in. (cc)	Bore & Stroke in. (mm)	Max. Output (Hp@rpm)	Fuel System	Jet Pump
2002	FX140	4 cyl, 4-stroke	60.9 (998)	2.91 x 2.28 (74 x 58)	140 @ 10,000	MPI	Axial, Single Stage
2003	FX Cruiser	4 cyl, 4-stroke	60.9 (998)	2.91 x 2.28 (74 x 58)	140 @ 10,000	MPI	Axial, Single Stage
	FX140	4 cyl, 4-stroke	60.9 (998)	2.91 x 2.28 (74 x 58)	140 @ 10,000	MPI	Axial, Single Stage
2004	FX Cruiser	4 cyl, 4-stroke	60.9 (998)	2.91 x 2.28 (74 x 58)	140@10,000	MPI	Axial, Single Stage
	FX Cruiser High Output	4 cyl, 4-stroke	64.2 (1052)	2.99 x 2.28 (76 x 58)	160 @ 10,000	MPI	Axial, Single Stage
	FX High Output	4 cyl, 4-stroke	64.2 (1052)	2.99 x 2.28 (76 x 58)	160 @ 10,000	MPI	Axial, Single Stage
	FX140	4 cyl, 4-stroke	60.9 (998)	2.91 x 2.28 (74 x 58)	140 @ 10,000	MPI	Axial, Single Stage
	FX140 (Calif.)	4 cyl, 4-stroke	60.9 (998)	2.91 x 2.28 (74 x 58)	140 @ 10,000	MPI	Axial, Single Stage
2005	FX	4 cyl, 4-stroke	60.9 (998)	2.91 x 2.28 (74 x 58)	140 @ 10,000	MPI	Axial, Single Stage
	FX Cruiser	4 cyl, 4-stroke	60.9 (998)	2.91 x 2.28 (74 x 58)	140 @ 10,000	MPI	Axial, Single Stage
	FX Cruiser High Output	4 cyl, 4-stroke	64.2 (1052)	2.99 x 2.28 (76 x 58)	160 @ 10,000	MPI	Axial, Single Stage
	FX High Output	4 cyl, 4-stroke	64.2 (1052)	2.99 x 2.28 (76 x 58)	160 @ 10,000	MPI	Axial, Single Stage
	VX110 Deluxe	4 cyl, 4-stroke	64.2 (1052)	2.99 x 2.28 (76 x 58)	110 @ 8,000	MPI	Axial, Single Stage
	VX110 Sport	4 cyl, 4-stroke	64.2 (1052)	2.99 x 2.28 (76 x 58)	110 @ 8,000	MPI	Axial, Single Stage
2006	FX	4 cyl, 4-stroke	60.9 (998)	2.91 x 2.28 (74 x 58)	140 @ 10,000	MPI	Axial, Single Stage
	FX Cruiser	4 cyl, 4-stroke	60.9 (998)	2.91 x 2.28 (74 x 58)	140 @ 10,000	MPI	Axial, Single Stage
	FX Cruiser High Output	4 cyl, 4-stroke	64.2 (1052)	2.99 x 2.28 (76 x 58)	160 @ 10,000	MPI	Axial, Single Stage
	FX High Output	4 cyl, 4-stroke	64.2 (1052)	2.99 x 2.28 (76 x 58)	160 @ 10,000	MPI	Axial, Single Stage
	VX110 Deluxe	4 cyl, 4-stroke	64.2 (1052)	2.99 x 2.28 (76 x 58)	110 @ 8,000	MPI	Axial, Single Stage
	VX110 Sport	4 cyl, 4-stroke	64.2 (1052)	2.99 x 2.28 (76 x 58)	110 @ 8,000	MPI	Axial, Single Stage
2007	FX	4 cyl, 4-stroke	60.9 (998)	2.91 x 2.28 (74 x 58)	140 @ 10,000	MPI	Axial, Single Stage
	FX Cruiser	4 cyl, 4-stroke	60.9 (998)	2.91 x 2.28 (74 x 58)	140 @ 10,000	MPI	Axial, Single Stage
	FX Cruiser High Output	4 cyl, 4-stroke	64.2 (1052)	2.99 x 2.28 (76 x 58)	160 @ 10,000	MPI	Axial, Single Stage
	FX High Output	4 cyl, 4-stroke	64.2 (1052)	2.99 x 2.28 (76 x 58)	160 @ 10,000	MPI	Axial, Single Stage
	VX	4 cyl, 4-stroke	64.2 (1052)	2.99 x 2.28 (76 x 58)	110 @ 8,000	MPI	Axial, Single Stage
	VX Cruiser	4 cyl, 4-stroke	64.2 (1052)	2.99 x 2.28 (76 x 58)	110 @ 8,000	MPI	Axial, Single Stage
	VX110 Deluxe	4 cyl, 4-stroke	64.2 (1052)	2.99 x 2.28 (76 x 58)	110 @ 8,000	MPI	Axial, Single Stage
	VX110 Sport	4 cyl, 4-stroke	64.2 (1052)	2.99 x 2.28 (76 x 58)	110 @ 8,000	MPI	Axial, Single Stage
2008	FX	4 cyl, 4-stroke	60.9 (998)	2.91 x 2.28 (74 x 58)	140 @ 10,000	MPI	Axial, Single Stage
	FX Cruiser	4 cyl, 4-stroke	60.9 (998)	2.91 x 2.28 (74 x 58)	140 @ 10,000	MPI	Axial, Single Stage
	FX Cruiser HO	4 cyl, 4-stroke	64.2 (1052)	2.99 x 2.28 (76 x 58)	160 @ 10,000	MPI	Axial, Single Stage
	FX Cruiser HO (Calif.)	4 cyl, 4-stroke	64.2 (1052)	2.99 x 2.28 (76 x 58)	160 @ 10,000	MPI	Axial, Single Stage
	FX Cruiser SHO	4 cyl, 4-stroke	110.6 (1812)	3.39 x 3.07 (86 x 78)	210	MPI ①	Axial, Single Stage
	FX HO	4 cyl, 4-stroke	64.2 (1052)	2.99 x 2.28 (76 x 58)	160 @ 10,000	MPI	Axial, Single Stage
	FX HO (Calif.)	4 cyl, 4-stroke	64.2 (1052)	2.99 x 2.28 (76 x 58)	160 @ 10,000	MPI	Axial, Single Stage
	FX SHO	4 cyl, 4-stroke	110.6 (1812)	2.99 x 2.28 (76 x 58)	210	MPI ①	Axial, Single Stage
	VX	4 cyl, 4-stroke	64.2 (1052)	2.99 x 2.28 (76 x 58)	110 @ 8,000	MPI	Axial, Single Stage

SELOC_9606

General Engine Specifications

Year	Model Name	Engine Type	Diplacement cu. in. (cc)	Bore & Stroke in. (mm)	Max. Output (Hp@rpm)	Fuel System	Jet Pump
2008 (cont'd)	VX Cruiser	4 cyl, 4-stroke	64.2 (1052)	2.99 x 2.28 (76 x 58)	110 @ 8,000	MPI	Axial, Single Stage
	VX Deluxe	4 cyl, 4-stroke	64.2 (1052)	2.99 x 2.28 (76 x 58)	110 @ 8,000	MPI	Axial, Single Stage
	VX110 Sport	4 cyl, 4-stroke	64.2 (1052)	2.99 x 2.28 (76 x 58)	110 @ 8,000	MPI	Axial, Single Stage
2009	FX Cruiser HO	4 cyl, 4-stroke	110.6 (1812)	3.39 x 3.07 (86 x 78)	180	MPI	Axial, Single Stage
	FX Cruiser SHO	4 cyl, 4-stroke	110.6 (1812)	3.39 x 3.07 (86 x 78)	210	MPI ①	Axial, Single Stage
	FX HO	4 cyl, 4-stroke	110.6 (1812)	3.39 x 3.07 (86 x 78)	180	MPI	Axial, Single Stage
	FX SHO	4 cyl, 4-stroke	110.6 (1812)	3.39 x 3.07 (86 x 78)	210	MPI ①	Axial, Single Stage
	FZR	4 cyl, 4-stroke	110.6 (1812)	3.39 x 3.07 (86 x 78)	210	MPI ①	Axial, Single Stage
	FZS	4 cyl, 4-stroke	110.6 (1812)	3.39 x 3.07 (86 x 78)	210	MPI ①	Axial, Single Stage
	VX	4 cyl, 4-stroke	64.2 (1052)	2.99 x 2.28 (76 x 58)	110 @ 8,000	MPI	Axial, Single Stage
	VX Cruiser	4 cyl, 4-stroke	64.2 (1052)	2.99 x 2.28 (76 x 58)	110 @ 8,000	MPI	Axial, Single Stage
	VX Deluxe	4 cyl, 4-stroke	64.2 (1052)	2.99 x 2.28 (76 x 58)	110 @ 8,000	MPI	Axial, Single Stage
	VX Sport	4 cyl, 4-stroke	64.2 (1052)	2.99 x 2.28 (76 x 58)	110 @ 8,000	MPI	Axial, Single Stage
2010	FX Cruiser HO	4 cyl, 4-stroke	110.6 (1812)	3.39 x 3.07 (86 x 78)	180	MPI	Axial, Single Stage
	FX Cruiser SHO	4 cyl, 4-stroke	110.6 (1812)	3.39 x 3.07 (86 x 78)	210	MPI ①	Axial, Single Stage
	FX HO	4 cyl, 4-stroke	110.6 (1812)	3.39 x 3.07 (86 x 78)	180	MPI	Axial, Single Stage
	FX SHO	4 cyl, 4-stroke	110.6 (1812)	3.39 x 3.07 (86 x 78)	210	MPI ①	Axial, Single Stage
	FZR	4 cyl, 4-stroke	110.6 (1812)	3.39 x 3.07 (86 x 78)	210	MPI ①	Axial, Single Stage
	FZS	4 cyl, 4-stroke	110.6 (1812)	3.39 x 3.07 (86 x 78)	210	MPI ①	Axial, Single Stage
	VX Cruiser	4 cyl, 4-stroke	64.2 (1052)	2.99 x 2.28 (76 x 58)	110	MPI	Axial, Single Stage
	VX Deluxe	4 cyl, 4-stroke	64.2 (1052)	2.99 x 2.28 (76 x 58)	110	MPI	Axial, Single Stage
	VX Sport	4 cyl, 4-stroke	64.2 (1052)	2.99 x 2.28 (76 x 58)	110	MPI	Axial, Single Stage
2011	FX Cruiser HO	4 cyl, 4-stroke	110.6 (1812)	3.39 x 3.07 (86 x 78)	180 @ 7,600	MPI	Axial, Single Stage
	FX Cruiser SHO	4 cyl, 4-stroke	110.6 (1812)	3.39 x 3.07 (86 x 78)	210	MPI ①	Axial, Single Stage
	FX HO	4 cyl, 4-stroke	110.6 (1812)	3.39 x 3.07 (86 x 78)	180 @ 7,600	MPI	Axial, Single Stage
	FX SHO	4 cyl, 4-stroke	110.6 (1812)	3.39 x 3.07 (86 x 78)	210	MPI ①	Axial, Single Stage
	FZR	4 cyl, 4-stroke	110.6 (1812)	3.39 x 3.07 (86 x 78)	210	MPI ①	Axial, Single Stage
	FZS	4 cyl, 4-stroke	110.6 (1812)	3.39 x 3.07 (86 x 78)	210	MPI ①	Axial, Single Stage
	VX Cruiser	4 cyl, 4-stroke	64.2 (1052)	2.99 x 2.28 (76 x 58)	110	MPI	Axial, Single Stage
	VX Deluxe	4 cyl, 4-stroke	64.2 (1052)	2.99 x 2.28 (76 x 58)	110	MPI	Axial, Single Stage
	VX Sport	4 cyl, 4-stroke	64.2 (1052)	2.99 x 2.28 (76 x 58)	110	MPI	Axial, Single Stage
	VXR	4 cyl, 4-stroke	110.6 (1812)	3.39 x 3.07 (86 x 78)	180 @ 7,600	MPI	Axial, Single Stage
	VXS	4 cyl, 4-stroke	110.6 (1812)	3.39 x 3.07 (86 x 78)	180 @ 7,600	MPI	Axial, Single Stage

SELOC_9606

N/A Not Available

① Supercharged

Tune-Up Specifications & Capacities

Year	Model	Diplacement cu. in. (cc)	Spark Plug Make	Spark Plug Type	Spark Plug Gap in. (mm)	Ignition Timing Deg. Idle	Ignition Timing Deg. WOT	Valve Clearance ① Exhaust in. (mm)	Valve Clearance ① Intake in. (mm)	Fuel Tank gal. (L)	Engine Oil ② w/o Filter qt. (L)	Engine Oil ② w/filter qt. (L)
2002	FX140	60.9 (998)	NGK	CR9EB	0.028-0.031 (0.7-0.8)	35 BTDC	5 ATDC	0.0098-0.0134 (0.25-0.34)	0.0043-0.0079 (0.11-0.20)	18.5 (70)	2.1 (2.0)	2.3 (2.2)
2003	FX Cruiser	60.9 (998)	NGK	CR9EB	0.028-0.031 (0.7-0.8)	35 BTDC	5 ATDC	0.0098-0.0134 (0.25-0.34)	0.0043-0.0079 (0.11-0.20)	18.5 (70)	2.1 (2.0)	2.3 (2.2)
	FX140	60.9 (998)	NGK	CR9EB	0.028-0.031 (0.7-0.8)	35 BTDC	5 ATDC	0.0098-0.0134 (0.25-0.34)	0.0043-0.0079 (0.11-0.20)	18.5 (70)	2.1 (2.0)	2.3 (2.2)
2004	FX Cruiser	60.9 (998)	NGK	CR9EB	0.028-0.031 (0.7-0.8)	35 BTDC	5 ATDC	0.0098-0.0134 (0.25-0.34)	0.0043-0.0079 (0.11-0.20)	18.5 (70)	2.1 (2.0)	2.3 (2.2)
	FX Cruiser High Output	64.2 (1052)	NGK	CR9EB	0.028-0.031 (0.7-0.8)	N/A	N/A	0.0098-0.0134 (0.25-0.34)	0.0043-0.0079 (0.11-0.20)	18.5 (70)	2.2 (2.1)	2.4 (2.3)
	FX High Output	64.2 (1052)	NGK	CR9EB	0.028-0.031 (0.7-0.8)	N/A	N/A	0.0098-0.0134 (0.25-0.34)	0.0043-0.0079 (0.11-0.20)	18.5 (70)	2.2 (2.1)	2.4 (2.3)
	FX140	60.9 (998)	NGK	CR9EB	0.028-0.031 (0.7-0.8)	35 BTDC	5 ATDC	0.0098-0.0134 (0.25-0.34)	0.0043-0.0079 (0.11-0.20)	18.5 (70)	2.1 (2.0)	2.3 (2.2)
2005	FX	60.9 (998)	NGK	CR9EB	0.028-0.031 (0.7-0.8)	35 BTDC	5 ATDC	0.0098-0.0134 (0.25-0.34)	0.0043-0.0079 (0.11-0.20)	18.5 (70)	2.1 (2.0)	2.3 (2.2)
	FX Cruiser	60.9 (998)	NGK	CR9EB	0.028-0.031 (0.7-0.8)	35 BTDC	5 ATDC	0.0098-0.0134 (0.25-0.34)	0.0043-0.0079 (0.11-0.20)	18.5 (70)	2.1 (2.0)	2.3 (2.2)
	FX Cruiser High Output	64.2 (1052)	NGK	CR9EB	0.028-0.031 (0.7-0.8)	N/A	N/A	0.0098-0.0134 (0.25-0.34)	0.0043-0.0079 (0.11-0.20)	18.5 (70)	2.2 (2.1)	2.4 (2.3)
	FX High Output	64.2 (1052)	NGK	CR9EB	0.028-0.031 (0.7-0.8)	N/A	N/A	0.0098-0.0134 (0.25-0.34)	0.0043-0.0079 (0.11-0.20)	18.5 (70)	2.2 (2.1)	2.4 (2.3)
	VX110 Deluxe	64.2 (1052)	NGK	CR9EB	0.028-0.031 (0.7-0.8)	35 BTDC	5 BTDC	0.0098-0.0134 (0.25-0.34)	0.0043-0.0079 (0.11-0.20)	13.2 (60)	1.8 (2.0)	1.9 (2.2)
	VX110 Sport	64.2 (1052)	NGK	CR9EB	0.028-0.031 (0.7-0.8)	35 BTDC	5 BTDC	0.0098-0.0134 (0.25-0.34)	0.0043-0.0079 (0.11-0.20)	13.2 (60)	1.8 (2.0)	1.9 (2.2)
2006	FX	60.9 (998)	NGK	CR9EB	0.028-0.031 (0.7-0.8)	35 BTDC	5 ATDC	0.0098-0.0134 (0.25-0.34)	0.0043-0.0079 (0.11-0.20)	18.5 (70)	2.1 (2.0)	2.3 (2.2)
	FX Cruiser	60.9 (998)	NGK	CR9EB	0.028-0.031 (0.7-0.8)	35 BTDC	5 ATDC	0.0098-0.0134 (0.25-0.34)	0.0043-0.0079 (0.11-0.20)	18.5 (70)	2.1 (2.0)	2.3 (2.2)
	FX Cruiser High Output	64.2 (1052)	NGK	CR9EB	0.028-0.031 (0.7-0.8)	5BTDC	32 BTDC	0.0098-0.0134 (0.25-0.34)	0.0043-0.0079 (0.11-0.20)	18.5 (70)	2.2 (2.1)	2.4 (2.3)
	FX High Output	64.2 (1052)	NGK	CR9EB	0.028-0.031 (0.7-0.8)	5BTDC	32 BTDC	0.0098-0.0134 (0.25-0.34)	0.0043-0.0079 (0.11-0.20)	18.5 (70)	2.2 (2.1)	2.4 (2.3)
	VX110 Deluxe	64.2 (1052)	NGK	CR9EB	0.028-0.031 (0.7-0.8)	35 BTDC	5 BTDC	0.0098-0.0134 (0.25-0.34)	0.0043-0.0079 (0.11-0.20)	13.2 (60)	1.8 (2.0)	1.9 (2.2)
	VX110 Sport	64.2 (1052)	NGK	CR9EB	0.028-0.031 (0.7-0.8)	35 BTDC	5 BTDC	0.0098-0.0134 (0.25-0.34)	0.0043-0.0079 (0.11-0.20)	13.2 (60)	1.8 (2.0)	1.9 (2.2)
2007	FX	60.9 (998)	NGK	CR9EB	0.028-0.031 (0.7-0.8)	35 BTDC	5 ATDC	0.0098-0.0134 (0.25-0.34)	0.0043-0.0079 (0.11-0.20)	18.5 (70)	2.1 (2.0)	2.3 (2.2)

SELOC_9606

Tune-Up Specifications & Capacities

Year	Model	Diplacement	Spark Plug			Ignition Timing		Valve Clearance ①		Fuel	Engine Oil ②	
					Gap	Deg.		Exhaust	Intake	Tank	w/o Filter	w/filter
		cu. in. (cc)	Make	Type	in. (mm)	Idle	WOT	in. (mm)	in. (mm)	gal. (L)	qt. (L)	qt. (L)
2007 (cont'd)	FX Cruiser	60.9 (998)	NGK	CR9EB	0.028-0.031 (0.7-0.8)	35 BTDC	5 ATDC	0.0098-0.0134 (0.25-0.34)	0.0043-0.0079 (0.11-0.20)	18.5 (70)	2.1 (2.0)	2.3 (2.2)
	FX Cruiser High Output	64.2 (1052)	NGK	CR9EB	0.028-0.031 (0.7-0.8)	5BTDC	32 BTDC	0.0098-0.0134 (0.25-0.34)	0.0043-0.0079 (0.11-0.20)	18.5 (70)	2.2 (2.1)	2.4 (2.3)
	FX High Output	64.2 (1052)	NGK	CR9EB	0.028-0.031 (0.7-0.8)	5BTDC	32 BTDC	0.0098-0.0134 (0.25-0.34)	0.0043-0.0079 (0.11-0.20)	18.5 (70)	2.2 (2.1)	2.4 (2.3)
	VX	64.2 (1052)	NGK	CR9EB	0.028-0.031 (0.7-0.8)	35 BTDC	5 BTDC	0.0098-0.0134 (0.25-0.34)	0.0043-0.0079 (0.11-0.20)	13.2 (60)	1.8 (2.0)	1.9 (2.2)
	VX Crusier	64.2 (1052)	NGK	CR9EB	0.028-0.031 (0.7-0.8)	35 BTDC	5 BTDC	0.0098-0.0134 (0.25-0.34)	0.0043-0.0079 (0.11-0.20)	13.2 (60)	1.8 (2.0)	1.9 (2.2)
	VX110 Deluxe	64.2 (1052)	NGK	CR9EB	0.028-0.031 (0.7-0.8)	35 BTDC	5 BTDC	0.0098-0.0134 (0.25-0.34)	0.0043-0.0079 (0.11-0.20)	13.2 (60)	1.8 (2.0)	1.9 (2.2)
	VX110 Sport	64.2 (1052)	NGK	CR9EB	0.028-0.031 (0.7-0.8)	35 BTDC	5 BTDC	0.0098-0.0134 (0.25-0.34)	0.0043-0.0079 (0.11-0.20)	13.2 (60)	1.8 (2.0)	1.9 (2.2)
2008	FX	60.9 (998)	NGK	CR9EB	0.028-0.031 (0.7-0.8)	35 BTDC	5 ATDC	0.0098-0.0134 (0.25-0.34)	0.0043-0.0079 (0.11-0.20)	18.5 (70)	2.1 (2.0)	2.3 (2.2)
	FX Cruiser	60.9 (998)	NGK	CR9EB	0.028-0.031 (0.7-0.8)	35 BTDC	5 ATDC	0.0098-0.0134 (0.25-0.34)	0.0043-0.0079 (0.11-0.20)	18.5 (70)	2.1 (2.0)	2.3 (2.2)
	FX Cruiser HO	64.2 (1052)	NGK	CR9EB	0.028-0.031 (0.7-0.8)	31 BTDC ③	-	0.0098-0.0134 (0.25-0.34)	0.0043-0.0079 (0.11-0.20)	18.5 (70)	2.2 (2.1)	2.4 (2.3)
	FX Cruiser SHO	110.6 (1812)	NGK	LFR6A	0.031-0.035 (0.8-0.9)	20 BTDC ③	-	0.011-0.015 (0.28-0.37)	0.006-0.009 (0.14-0.23)	18.5 (70)	3.17 (3.0)	3.28 (3.1)
	FX HO	64.2 (1052)	NGK	CR9EB	0.028-0.031 (0.7-0.8)	31 BTDC ③	-	0.0098-0.0134 (0.25-0.34)	0.0043-0.0079 (0.11-0.20)	18.5 (70)	2.2 (2.1)	2.4 (2.3)
	FX SHO	110.6 (1812)	NGK	LFR6A	0.031-0.035 (0.8-0.9)	20 BTDC ③	-	0.011-0.015 (0.28-0.37)	0.006-0.009 (0.14-0.23)	18.5 (70)	3.17 (3.0)	3.28 (3.1)
	VX	64.2 (1052)	NGK	CR9EB	0.028-0.031 (0.7-0.8)	35 BTDC	5 BTDC	0.0098-0.0134 (0.25-0.34)	0.0043-0.0079 (0.11-0.20)	13.2 (60)	1.8 (2.0)	1.9 (2.2)
	VX Crusier	64.2 (1052)	NGK	CR9EB	0.028-0.031 (0.7-0.8)	35 BTDC	5 BTDC	0.0098-0.0134 (0.25-0.34)	0.0043-0.0079 (0.11-0.20)	13.2 (60)	1.8 (2.0)	1.9 (2.2)
	VX Deluxe	64.2 (1052)	NGK	CR9EB	0.028-0.031 (0.7-0.8)	35 BTDC	5 BTDC	0.0098-0.0134 (0.25-0.34)	0.0043-0.0079 (0.11-0.20)	13.2 (60)	1.8 (2.0)	1.9 (2.2)
	VX110 Sport	64.2 (1052)	NGK	CR9EB	0.028-0.031 (0.7-0.8)	35 BTDC	5 BTDC	0.0098-0.0134 (0.25-0.34)	0.0043-0.0079 (0.11-0.20)	13.2 (60)	1.8 (2.0)	1.9 (2.2)
2009	FX Cruiser HO	110.6 (1812)	NGK	LFR6A	0.031-0.035 (0.8-0.9)	25 BTDC ③	-	0.011-0.015 (0.28-0.37)	0.006-0.009 (0.14-0.23)	18.5 (70)	3.2 (3.0)	3.3 (3.1)
	FX Cruiser SHO	110.6 (1812)	NGK	LFR6A	0.031-0.035 (0.8-0.9)	20 BTDC ③	-	0.011-0.015 (0.28-0.37)	0.006-0.009 (0.14-0.23)	18.5 (70)	3.17 (3.0)	3.28 (3.1)
	FX HO	110.6 (1812)	NGK	LFR6A	0.031-0.035 (0.8-0.9)	25 BTDC ③	-	0.011-0.015 (0.28-0.37)	0.006-0.009 (0.14-0.23)	18.5 (70)	3.2 (3.0)	3.3 (3.1)
	FX SHO	110.6 (1812)	NGK	LFR6A	0.031-0.035 (0.8-0.9)	20 BTDC ③	-	0.011-0.015	0.006-0.009	18.5 (70)	3.17 (3.0)	3.28 (3.1)

SELOC_9606

Tune-Up Specifications & Capacities

Year	Model	Diplacement	Spark Plug			Ignition Timing		Valve Clearance ①		Fuel	Engine Oil ②	
					Gap	Deg.		Exhaust	Intake	Tank	w/o Filter	w/filter
		cu. in. (cc)	Make	Type	in. (mm)	Idle	WOT	in. (mm)	in. (mm)	gal. (L)	qt. (L)	qt. (L)
2009								(0.28-0.37)	(0.14-0.23)			
(cont'd)	FZR	110.6 (1812)	NGK	LFR6A	0.031-0.035 (0.8-0.9)	20 BTDC ③	-	0.011-0.015 (0.28-0.37)	0.006-0.009 (0.14-0.23)	18.5 (70)	3.2 (3.0)	3.3 (3.1)
	FZS	110.6 (1812)	NGK	LFR6A	0.031-0.035 (0.8-0.9)	20 BTDC ③	-	0.011-0.015 (0.28-0.37)	0.006-0.009 (0.14-0.23)	18.5 (70)	3.2 (3.0)	3.3 (3.1)
	VX	64.2 (1052)	NGK	CR9EB	0.028-0.031 (0.7-0.8)	35 BTDC	5 BTDC	0.0098-0.0134 (0.25-0.34)	0.0043-0.0079 (0.11-0.20)	13.2 (60)	1.8 (2.0)	1.9 (2.2)
	VX Cruiser	64.2 (1052)	NGK	CR9EB	0.028-0.031 (0.7-0.8)	35 BTDC	5 BTDC	0.0098-0.0134 (0.25-0.34)	0.0043-0.0079 (0.11-0.20)	13.2 (60)	1.8 (2.0)	1.9 (2.2)
	VX Deluxe	64.2 (1052)	NGK	CR9EB	0.028-0.031 (0.7-0.8)	35 BTDC	5 BTDC	0.0098-0.0134 (0.25-0.34)	0.0043-0.0079 (0.11-0.20)	13.2 (60)	1.8 (2.0)	1.9 (2.2)
	VX Sport	64.2 (1052)	NGK	CR9EB	0.028-0.031 (0.7-0.8)	35 BTDC	5 BTDC	0.0098-0.0134 (0.25-0.34)	0.0043-0.0079 (0.11-0.20)	13.2 (60)	1.8 (2.0)	1.9 (2.2)
2010	FX Cruiser HO	110.6 (1812)	NGK	LFR6A	0.031-0.035 (0.8-0.9)	25 BTDC ③	-	0.011-0.015 (0.28-0.37)	0.006-0.009 (0.14-0.23)	18.5 (70)	3.2 (3.0)	3.3 (3.1)
	FX Cruiser SHO	110.6 (1812)	NGK	LFR6A	0.031-0.035 (0.8-0.9)	20 BTDC ③	-	0.011-0.015 (0.28-0.37)	0.006-0.009 (0.14-0.23)	18.5 (70)	3.17 (3.0)	3.28 (3.1)
	FX HO	110.6 (1812)	NGK	LFR6A	0.031-0.035 (0.8-0.9)	25 BTDC ③	-	0.011-0.015 (0.28-0.37)	0.006-0.009 (0.14-0.23)	18.5 (70)	3.2 (3.0)	3.3 (3.1)
	FX SHO	110.6 (1812)	NGK	LFR6A	0.031-0.035 (0.8-0.9)	20 BTDC ③	-	0.011-0.015 (0.28-0.37)	0.006-0.009 (0.14-0.23)	18.5 (70)	3.17 (3.0)	3.28 (3.1)
	FZR	110.6 (1812)	NGK	LFR6A	0.031-0.035 (0.8-0.9)	20 BTDC ③	-	0.011-0.015 (0.28-0.37)	0.006-0.009 (0.14-0.23)	18.5 (70)	3.2 (3.0)	3.3 (3.1)
	FZS	110.6 (1812)	NGK	LFR6A	0.031-0.035 (0.8-0.9)	20 BTDC ③	-	0.011-0.015 (0.28-0.37)	0.006-0.009 (0.14-0.23)	18.5 (70)	3.2 (3.0)	3.3 (3.1)
	VX Cruiser	64.2 (1052)	NGK	CR9EB	0.028-0.031 (0.7-0.8)	32 BTDC ③	-	0.0098-0.0134 (0.25-0.34)	0.0043-0.0079 (0.11-0.20)	13.2 (60)	1.8 (2.0)	1.9 (2.2)
	VX Deluxe	64.2 (1052)	NGK	CR9EB	0.028-0.031 (0.7-0.8)	32 BTDC ③	-	0.0098-0.0134 (0.25-0.34)	0.0043-0.0079 (0.11-0.20)	13.2 (60)	1.8 (2.0)	1.9 (2.2)
	VX Sport	64.2 (1052)	NGK	CR9EB	0.028-0.031 (0.7-0.8)	32 BTDC ③	-	0.0098-0.0134 (0.25-0.34)	0.0043-0.0079 (0.11-0.20)	13.2 (60)	1.8 (2.0)	1.9 (2.2)
2011	FX Cruiser HO	110.6 (1812)	NGK	LFR6A	0.031-0.035 (0.8-0.9)	25 BTDC ③	-	0.011-0.015 (0.28-0.37)	0.006-0.009 (0.14-0.23)	18.5 (70)	3.2 (3.0)	3.3 (3.1)
	FX Cruiser SHO	110.6 (1812)	NGK	LFR6A	0.031-0.035 (0.8-0.9)	20 BTDC ③	-	0.011-0.015 (0.28-0.37)	0.006-0.009 (0.14-0.23)	18.5 (70)	3.17 (3.0)	3.28 (3.1)
	FX HO	110.6 (1812)	NGK	LFR6A	0.031-0.035 (0.8-0.9)	25 BTDC ③	-	0.011-0.015 (0.28-0.37)	0.006-0.009 (0.14-0.23)	18.5 (70)	3.2 (3.0)	3.3 (3.1)
	FX SHO	110.6 (1812)	NGK	LFR6A	0.031-0.035 (0.8-0.9)	20 BTDC ③	-	0.011-0.015 (0.28-0.37)	0.006-0.009 (0.14-0.23)	18.5 (70)	3.17 (3.0)	3.28 (3.1)
	FZR	110.6 (1812)	NGK	LFR6A	0.031-0.035 (0.8-0.9)	20 BTDC ③	-	0.011-0.015 (0.28-0.37)	0.006-0.009 (0.14-0.23)	18.5 (70)	3.2 (3.0)	3.3 (3.1)

SELOC_9606

Tune-Up Specifications & Capacities

			Spark Plug			Ignition Timing		Valve Clearance ①		Fuel	Engine Oil ②	
		Diplacement			Gap	Deg.		Exhaust	Intake	Tank	w/o Filter	w/filter
Year	Model	cu. in. (cc)	Make	Type	in. (mm)	Idle	WOT	in. (mm)	in. (mm)	gal. (L)	qt. (L)	qt. (L)
2011 (cont'd)	FZS	110.6 (1812)	NGK	LFR6A	0.031-0.035 (0.8-0.9)	20 BTDC ③	-	0.011-0.015 (0.28-0.37)	0.006-0.009 (0.14-0.23)	18.5 (70)	3.2 (3.0)	3.3 (3.1)
	VX Cruiser	64.2 (1052)	NGK	CR9EB	0.028-0.031 (0.7-0.8)	32 BTDC ③	-	0.0098-0.0134 (0.25-0.34)	0.0043-0.0079 (0.11-0.20)	13.2 (60)	1.8 (2.0)	1.9 (2.2)
	VX Deluxe	64.2 (1052)	NGK	CR9EB	0.028-0.031 (0.7-0.8)	32 BTDC ③	-	0.0098-0.0134 (0.25-0.34)	0.0043-0.0079 (0.11-0.20)	13.2 (60)	1.8 (2.0)	1.9 (2.2)
	VX Sport	64.2 (1052)	NGK	CR9EB	0.028-0.031 (0.7-0.8)	32 BTDC ③	-	0.0098-0.0134 (0.25-0.34)	0.0043-0.0079 (0.11-0.20)	13.2 (60)	1.8 (2.0)	1.9 (2.2)
	VXR	110.6 (1812)	NGK	LFR6A	0.031-0.035 (0.8-0.9)	25 BTDC ③	-	0.011-0.015 (0.28-0.37)	0.006-0.009 (0.14-0.23)	15.9 (60)	3.2 (3.0)	3.3 (3.1)
	VXS	110.6 (1812)	NGK	LFR6A	0.031-0.035 (0.8-0.9)	25 BTDC ③	-	0.011-0.015 (0.28-0.37)	0.006-0.009 (0.14-0.23)	15.9 (60)	3.2 (3.0)	3.3 (3.1)

SELOC_9606

① Adjusted cold

② After engine disassembly, add 4.5 qts. (4.3L) on 998cc & 1812cc, 4.8 qts. (4.5L) on 1052cc engines.

③ Maximum

Maintenance Intervals

Component	Procedure	Model	Year	Daily/ Before Use	Initial 10 hrs	Initial 50 hrs/ 6 Months	Initial 100 Hrs/ 12 Months	OnGoing, Every 100 Hrs/ 12 Months	OnGoing, Every 200 Hrs/ 24 Months
Air Filter	Inspect/Clean	998cc	2002						X
			2003-05		X				X
			2006-08				X	X	
		1052cc	2004-05		X				X
			2006-10				X	X	
	Check/Replace	1812cc					X	X	
Anodes	Inspect			X					
Battery	Check Fluid Level			X					
	Inspect						X	X	
Bilge	Bilge Strainer, Clean						X	X	
	Bilge Pump Strainer, Inspect/Clean	2003-10 FX					X	X	
	Stern Drain Plugs, Check			X					
	Stern Drain Plugs, Inspect/Replace	998cc	2002				X	X	
			2003-05		X		X	X	
			2006-08				X	X	
		1052cc	2004-05		X		X	X	
			2006-10				X	X	
		1812cc					X	X	
Cooling System	Flush Water Passages			X ①					
	Inspect/Clean Water Inlet Strainer						X	X	
Coupling, Rubber	Inspect	998cc, 1052cc							X
Damper, Rubber	Inspect	1812cc							X
Engine	Engine Mount								X
	Nuts & Bolts	998cc	2002	X			X	X	
			2003-08	X	X		X	X	
		1052cc, 1812cc		X	X		X	X	
	Oil, Check Level			X					
	Oil, Replace	998cc, 1052cc			X		X	X	
		1812cc				⑥			
	Oil Filter, Replace	998cc	2002-05						X ②
			2006-08				X	X	
		1052cc	2004-05						X ②
			2006-08				X	X	
		1812cc					X	X	
	Valve Clearance								X
Fuel System	Clean Fuel Tank						X	X	
	Inspect/Clean Filter	2002 FX ③					X	X	
	Inspect Fuel/Water Separator	2003-10 FX/FZ		X ①					
	Inspect System						X	X	
	Inspect Throttle Shaft						X	X	
Impeller	Inspect	998cc	2002			X	X	X	
			2003-05	X			X	X	
			2006-08				X	X	
		1052cc, 1812cc					X	X	
Idle Speed	Check/Adjust	2010 VZ					X	X	
Intermediate Housing	Lube	2003-10 FX, 2010 VX, 2011 VXR/VXS			X		X	X	
Internal Engine Components	Lubricate	1812cc					⑤		
Jet Thrust Nozzle Angle	Inspect	2003-10					X	X	
Lubrication Points	Lube						X	X	
QSTS Mechanism/Cables	Inspect/Adjust Mechanism	FX/FZ		X	X		X	X	
	Lubricate Cables	1812cc FX/FZ					X	X	
Shift Cable/Mechanism	Inspect/Adjust	998cc	2002				X	X	
			2003-05	X			X	X	
			2006-08				X	X	
		2004-09 1052cc ④					X	X	
Shift Cable/Reverse Gate	Inspect/Adjust	2010 1052cc ④					X	X	
		1812cc					X	X	
Spark Plugs/Wires	Check/Clean/Adjust	998cc	2002		X	X	X	X	
			2003-05		X		X	X	
			2006-08				X	X	
		1052, 1812cc			X		X	X	
Steering System	Inspect Cable	2002 FX				X	X	X	
	Inspect System	2003-10 FX			X		X	X	
Throttle Lever	Cable, Inspect/Adjust	998cc	2002-05				X	X	
			2006-08		X		X	X	
		1052cc, 1812cc			X		X	X	
	Lever, Check Operation			X					
Trolling Speed	Inspect	998cc	2003-05		X		X	X	
			2006-08				X	X	
		1052cc, 1812cc					X	X	

SELOC_9606

① After each use
② We actually recommend changing the filter every time you change the oil, cheap insurance
③ Although no interval is given for 2003 and later models, we suggest at least once a year
④ 2005-09 VZ Models: Deluxe only; 2010 VZ models: Cruiser and Deluxe only
⑤ Before long-term storage, 50 hrs or 12 months
⑥ Every 50 hrs or 12 months

3

FUEL SYSTEM

FUEL & COMBUSTION BASICS

General Information

**** CAUTION**

If equipped, disconnect the negative battery cable *anytime* work is performed on the engine, especially when working on the fuel system. This will help prevent the possibility of sparks during service (from accidentally grounding a hot lead or powered component). Sparks could ignite vapors or exposed fuel. Disconnecting the cable on electric start motors will also help prevent the possibility fuel spillage if an attempt is made to crank the engine while the fuel system is open.

**** CAUTION**

Fuel leaking from a loose, damaged, or incorrectly installed hose or fitting may cause a fire or an explosion. ALWAYS pressurize the fuel system and run the motor while inspecting for leaks after servicing any component of the fuel system.

To tune, troubleshoot or repair the fuel system of a PWC motor you must first understand the fuel injection and ignition principles of engine operation.

If you have any doubts concerning your understanding of engine operation, it would be best to study The Basic Operating Principles of an engine as detailed under Troubleshooting in Section 1, before tackling any work on the fuel system.

The fuel systems used on these engines utilize various forms of electronic fuel injection systems.

Fuel System Service Cautions

There is no way around it. Working with gasoline can provide for many different safety hazards and requires that extra caution is used during all steps of service. To protect yourself and others, you must take all necessary precautions against igniting the fuel or vapors (which will cause a fire at best or an explosion at worst).

**** CAUTION**

Take extreme care when working with the fuel system. NEVER smoke (it's bad for you anyhow, but smoking during fuel system service could kill you much faster!) or allow flames or sparks in the work area. Flames or sparks can ignite fuel, especially vapors, resulting in a fire at best or an explosion at worst.

For starters, disconnect the negative battery cable **EVERY** time a fuel system hose or fitting is going to be disconnected. It takes only one moment of forgetfulness for someone to crank the motor, possibly causing a dangerous spray of fuel from the opening. This is especially true on EFI systems that contain an electric fuel pump, where simply turning the key on (without cranking) will energize the pump and result in high-pressure fuel spray in certain fuel system lines/fittings.

Gasoline contains harmful additives and is quickly absorbed by exposed skin. As an additional precaution, always wear gloves and some form of eye protection (regular glasses help, but only safety glasses can really protect your eyes).

■ During service, pay attention to ensure that all components, hoses and fittings are installed them in the correct location and orientation to prevent the possibility of leakage. Matchmark components before they are removed as necessary.

Because of the dangerous conditions that result from working with gasoline and fuel vapors, always take extra care and be sure to follow these guidelines for safety:

- Keep a Coast Guard-approved fire extinguisher handy when working.
- Allow the engine to cool completely before opening a fuel fitting. Don't allow gasoline to drip on a hot engine.
- The first thing you must do after removing the engine cover is to check for the presence of gasoline fumes. If strong fumes are present, look for leaking or damage hoses, fittings or other fuel system components and repair.
- Do not repair the motor or any fuel system component near any sources of ignition; including sparks, open flames, or anyone smoking. Remember that fumes can pool near the ground and something as simple as a space or kerosene heater can ignite those vapors without warning.
- Clean up spilled gasoline right away using clean rags. Keep all fuel soaked rags in a metal container until they can be properly disposed of or cleaned. NEVER leave solvent, gasoline or oil soaked rags in the hull.
- Don't use electric powered tools in the hull or near the boat during fuel system service or after service, until the system is pressurized and checked for leaks.
- Fuel leaking from a loose, damaged or incorrectly installed hose or fitting may cause a fire or an explosion. ALWAYS pressurize the fuel system and run the motor while inspecting for leaks after servicing any component of the fuel system.

Fuel

GENERAL INFORMATION

◆ **See Figure 1**

Fuel recommendations have become more complex as the chemistry of modern gasoline changes. The major driving force behind the many of the changes in gasoline chemistry was the search for additives to replace lead as an octane booster and lubricant. These additives are governed by the types of emissions they produce in the combustion process. Also, the replacement additives do not always provide the same level of combustion stability, making a fuel's octane rating less meaningful.

In the 1960's and 1970's, leaded fuel was common. The lead served two functions. First, it served as an octane booster (combustion stabilizer) and second it served as a valve seat lubricant.

For decades now, all lead has been removed from the refining process. This means that the benefit of lead as an octane booster has been eliminated. Several substitute octane boosters have been introduced in the place of lead. While many are adequate in automobile engines, most do not perform nearly as well as lead did, even though the octane rating of the fuel is the same.

OCTANE RATING

◆ **See Figure 1**

A fuel's octane rating is a measurement of how stable the fuel is when heat is introduced. Octane rating is a major consideration when deciding whether a fuel is suitable for a particular application. For example, in an engine, we want the fuel to ignite when the spark plug fires and not before, even under high pressure and temperatures. Once the fuel is ignited, it must burn slowly and smoothly, even though heat and pressure are building up while the burn occurs. The unburned fuel should be ignited by the traveling

SELOC_9606

Fig. 1 Damaged piston, possibly caused by; using too-low an octane fuel or by insufficient oil

flame front, not by some other source of ignition, such as carbon deposits or the heat from the expanding gasses. A fuel's octane rating is known as a measurement of the fuel's anti-knock properties (ability to burn without exploding). Essentially, the octane rating is a measure of a fuel's stability.

Usually a fuel with a higher octane rating can be subjected to a more severe combustion environment before spontaneous or abnormal combustion occurs. To understand how 2 gasoline samples can be different, even though they have the same octane rating, we need to know how octane rating is determined.

The American Society of Testing and Materials (ASTM) has developed the method of determining the octane rating of a fuel sample which is used in the U.S. The octane rating we see on the pump at a gasoline station in the U.S. is known as the Pump Octane Number (PON). Look at the small print on the pump. The rating has a formula. The rating is determined by the R+M/2 method. This number is the average of the research octane reading and the motor octane rating.

- The Research Octane Rating is a measure of a fuel's anti-knock properties under a light load or part throttle conditions. During this test, combustion heat is easily dissipated.
- The Motor Octane Rating is a measure of a fuel's anti-knock properties under a heavy load or full throttle conditions, when heat buildup is at maximum.

All Yamaha motors covered here are designed to use fuel with a PON rating of 86 (a RON of about 90) which should translate as "Regular" grade fuel in most areas. Octane needs generally go down as altitude go up so it MAY be possible to run an even slightly lower octane at high altitude.

VAPOR PRESSURE

Fuel vapor pressure is a measure of how easily a fuel sample evaporates. Many additives used in gasoline contain aromatics. Aromatics are light hydrocarbons distilled off the top of a crude oil sample. They are effective at increasing the research octane of a fuel sample but can cause vapor lock (bubbles in the fuel line) on a very hot day. If you have an inconsistent running engine and you suspect vapor lock, use a piece of clear fuel line to look for bubbles, indicating that the fuel is vaporizing.

One negative side effect of aromatics is that they create additional combustion products such as carbon and varnish. If your engine requires high octane fuel to prevent detonation, de-carbon the engine more frequently with an internal engine cleaner to prevent ring sticking due to excessive varnish buildup.

ALCOHOL-BLENDED FUELS

When the Environmental Protection Agency mandated a phase-out of the leaded fuels in January of 1986, fuel suppliers needed an additive to improve the octane rating of their fuels. Although there are multiple methods currently employed, the addition of alcohol to gasoline seems to be favored because of its favorable results and low cost. Two types of alcohol are used in fuel today as octane boosters, methanol (wood alcohol) or ethanol (grain alcohol).

When used as a fuel additive, alcohol tends to raise the research octane of the fuel, so these additives will have limited benefit in an outboard motor. There are, however, some special considerations due to the effects of alcohol in fuel.

- Since alcohol contains oxygen, it replaces gasoline without oxygen content and tends to cause the air/fuel mixture to become leaner.
- On older units, the leaching affect of alcohol will, in time, cause fuel lines and plastic components to become brittle to the point of cracking. Unless replaced, these cracked lines could leak fuel, increasing the potential for hazardous situations. However, later model vessels (such as the models covered here) utilize fuel lines which are much more resistant to alcohol in the fuel mixture.

■ **Modern fuel lines and plastic fuel system components have been specially formulated to resist alcohol leaching effects.**

- When alcohol blended fuels become contaminated with water, the water combines with the alcohol then settles to the bottom of the tank. This leaves the gasoline on a top layer.

THE BOTTOM LINE WITH FUELS

If we could buy fuel of the correct octane rating, free of alcohol and aromatics, this would be our first choice.

Yamaha continues to recommend unleaded fuel. This is almost a redundant recommendation due to the near universal unavailability of any other type fuel.

According to the fuel recommendations that come with your engine, there is no engine in the product line that requires more than 86 octane (PON or 90 RON). An 86 (PON) octane rating generally means middle grade unleaded. Premium unleaded is more stable under severe conditions, but also produces more combustion products. Therefore, when using premium unleaded, more frequent de-carboning is often necessary.

Regardless of the fuel octane rating you choose, try to stay with a name brand fuel. You never know for sure what kinds of additives or how much is in off brand fuel.

HIGH ALTITUDE OPERATION

At elevated altitudes there is less oxygen in the atmosphere than at sea level. Less oxygen means lower combustion efficiency and less power output. Power output is reduced 3% for every thousand feet above sea level. At ten thousand feet, power is reduced 30% from that available at sea level.

Re-jetting carburetors for high altitude does not restore this lost power. It simply corrects the air-fuel ratio for the reduced air density, and makes the most of the remaining available power. If you re-jet an engine, you are locked into the higher elevation. You cannot operate at sea level until you re-jet for sea level. Understand that going below the elevation jetted for your motor will damage the engine with overly lean fuel:air ratios. As a general rule, jet for the lowest elevation anticipated. Spark plug insulator tip color is the best guide for high altitude jetting.

If you are in an area of known poor fuel quality, you may want to use fuel additives. Today's additives are mostly alcohol and aromatics, and their effectiveness may be limited. It is difficult to find additives without ethanol, methanol, or aromatics. If you use octane boosters frequent de-carboning may be necessary. If possible, the best policy is to use name brand pump fuel with no additional additives except Yamaha fuel conditioner and Ring-Free(tm).

CHECKING FOR STALE/CONTAMINATED FUEL

① EASY

◆ **See Figure 2**

PWC motors often sit for weeks at a time, making them the perfect candidate for fuel problems. Gasoline has a short life, as combustibles begin evaporating almost immediately. Even when stored properly, fuel starts to deteriorate within a few months, leaving behind a stale fuel mixture that can

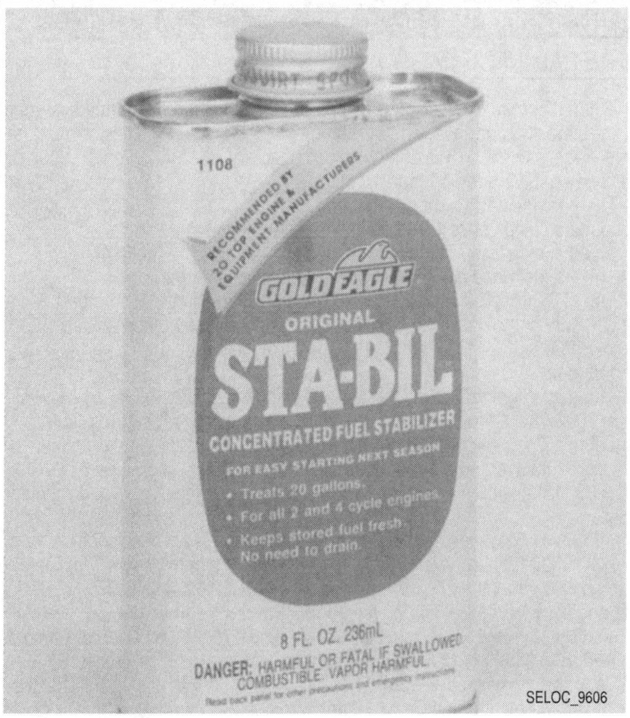

SELOC_9606

Fig. 2 Commercial additives, such as Sta-bil, may be used to help prevent "souring"

cause hard-starting, poor engine performance and even lead to possible engine damage.

Furthermore, as gasoline evaporates it leaves behind gum deposits that can clog filters, lines and small passages. Although the sealed high-pressure fuel system of an EFI motor is less susceptible to fuel evaporation, the low-pressure fuel systems of all engines can suffer the effects.

As mentioned under Alcohol-Blended fuels, modern fuels contain alcohol, which is hygroscopic (meaning it absorbs water). And, over time, fuel stored in a partially filled tank or a tank that is vented to the atmosphere will absorb water. The water/alcohol settles to the bottom of the tank, potentially promoting corrosion (in untreated metal tanks or older tanks where the treatment has peeled) and leaving a non-combustible mixture at the bottom of a tank that could leave a boater stranded.

One of the first steps to fuel system troubleshooting is to make sure the fuel source is not at fault for engine performance problems. Check the fuel if the engine will not start and there is no ignition problem. Stale or contaminated fuels will often exhibit an unusual or even unpleasant unusual odor.

■ **The best method of disposing stale fuel is through a local waste pickup service, automotive repair facility or marine dealership. But, this can be a hassle. If fuel is not too stale or too badly contaminated, it may be mixed with greater amounts of fresh fuel and used to power lawn/yard equipment or (as long as we're not talking about disposing pre-mix here) even an automobile (if greatly diluted so as to prevent misfiring, unstable idle or damage to the automotive engine). But we feel that it is much less of a risk to have a lawn mower stop running because of the fuel problem than it is to have your boat motor quit or refuse to start.**

■ **If a sample cannot be obtained from the fuel filter or pump supply hose, there is a problem with the fuel tank-to-motor fuel circuit. Check the tank, fuel hose, fuel pump or fitting.**

1. Check the appearance and odor of the fuel. An unusual smell, signs of visible debris or a cloudy appearance (or even the obvious presence of water) points to a fuel that should be replaced.
2. If contaminated fuel is found, drain the fuel system and dispose of the fuel in a responsible manner, then clean the entire fuel system.

■ **If debris is found in the fuel system, clean and/or replace all fuel filters.**

3. When finished, reconnect the negative battery cable, then properly pressurize the fuel system and check for leaks.

Fuel System Pressurization

GENERAL INFORMATION

When it comes to safety and outboards, the condition of the fuel system is of the utmost importance. The system must be checked for signs of damage or leakage with every use and checked especially carefully when portions of the system have been opened for service.

The best method to check the fuel system is to visually inspect the lines, hoses and fittings once the system has been properly pressurized.

All engines utilize a circuit which delivers fuel from the fuel tank to the engine via fuel hoses and the action of an electric fuel pump.

The high-pressure circuit of all motors works by using a submerged electric fuel pump located within the fuel tank to achieve fuel pressure somewhere in the 45-53 psi (310-370 kPa) range depending upon the year and model.

The actual fuel test pressure and test fittings varies by model as follows:

- 2002-08 FX models w/998cc and 1052cc motors require an adaptor (such as #YW-06842/90890-06842) which connects between the high-pressure fuel line and the top of the fuel tank. Pressure should be 45-47 psi (310-330 kPa) with the engine at normal operating temperature and running at idle
- 2008-11 FX models w/1812cc motors require an adaptor (such as #YW-06842/90890-06842) which connects between the high-pressure fuel line and the top of the fuel tank. Pressure should be 49.1-52.6 psi (345-370 kPa) with the engine at normal operating temperature and running at idle
- 2009-11 FZR/FZS models w/1812cc motors require an adaptor (such as #YW-06842/90890-06842) which connects between the high-pressure fuel line and the top of the fuel tank. Pressure should be 49.1-52.6 psi (345-370 kPa) with the engine at normal operating temperature and running at idle
- 2005-09 VX models w/1052cc motors require an adaptor (such as #YW-06842/90890-06842) which connects between the high-pressure fuel line and the top of the fuel tank. Pressure should be 46-47 psi (320-327 kPa) with the engine at normal operating temperature and running at idle
- 2010-11 VX models w/1052cc motors require an adaptor (such as #YW-06842/90890-06842) which connects between the high-pressure fuel line and the top of the fuel tank. Pressure should be 45.4-48 psi (319-340 kPa) with the engine at normal operating temperature and running at idle
- 2011 VX models w/1812cc motors require an adaptor (such as #YW-06842/90890-06842) which connects between the high-pressure fuel line and the top of the fuel tank. Pressure should be 49.1-52.6 psi (345-370 kPa) with the engine at normal operating temperature and running at idle

RELIEVING FUEL SYSTEM PRESSURE (EFI MOTORS)

② MODERATE

Interestingly, Yamaha does not provide a fuel pressure relief procedure like most other manufacturers. Instead, they suggest that you remove the fuel tank filler cap and then *slowly* disconnect the fuel line quick connector while covering it with a clean rag. That's it!

Combustion

GENERAL INFORMATION

A piston from a 2 cycle (stroke) engine has a power stroke every revolution of the crankshaft. A piston from a 4 cycle engine has a power stroke every other revolution of the crankshaft. Therefore, the 2 cycle engine has twice as many power strokes for any given engine speed. If the displacement of the 2 types of engines is identical, then the 2 cycle engine has to dissipate twice as much heat as the 4 cycle engine. In such a high heat environment, the fuel must be very stable to avoid detonation. If any parameters affecting combustion change suddenly (the engine runs lean for example), uncontrolled heat build-up occurs very rapidly in a 2 cycle engine.

Similarly, if a 4-stroke engine is tuned for high output, either through higher compression, advanced ignition timing or other factors, like through variable valve timing, it may benefit from a higher octane fuel as well.

ABNORMAL COMBUSTION

There are 2 types of abnormal combustion:

- Pre-ignition - Occurs when the air-fuel mixture is ignited by some other incandescent source other than the correctly timed spark from the spark plug.
- Detonation - Occurs when excessive heat and or pressure ignites the air/fuel mixture rather than the spark plug. The burn becomes explosive.

FACTORS AFFECTING COMBUSTION

The combustion process is affected by several interrelated factors. This means that when one factor is changed, the other factors also must be changed to maintain the same controlled burn and level of combustion stability.

Compression

Determines the level of heat buildup in the cylinder when the air-fuel mixture is compressed. As compression increases, so does the potential for heat buildup. Over time, the build-up of carbon deposits can increase compression (by decreasing the available size of the combustion chamber). Similarly, machining the cylinder head (to correct for warpage) can increase compression. Increases of compression often bring with them the need to increase fuel octane to prevent abnormal combustion.

Ignition Timing

Determines when the ignited gasses will start to expand in relation to the motion of the piston. If the gasses begin to expand too soon, such as they would during pre-ignition or in an overly advanced ignition timing, the motion of the piston opposes the expansion of the gasses, resulting in extremely high combustion chamber pressures and heat.

As ignition timing is retarded, the burn occurs later in relation to piston position. This means that the piston has less distance to travel under power to the bottom of the cylinder, resulting in less usable power.

Fuel Mixture

Determines how efficient the burn will be. A rich mixture burns cooler and slower than a lean one. However, if the mixture is too lean, it can't become explosive. The slower the burn, the cooler the combustion chamber, because pressure buildup is gradual.

Fuel Quality (Octane Rating)

Determines how much heat is necessary to ignite the mixture. Once the burn is in progress, heat is on the rise. The unburned poor quality fuel is ignited all at once by the rising heat instead of burning gradually as a flame front of the burn passing by. This action results in detonation (pinging).

Other Factors

In general, anything that can cause abnormal heat buildup can be enough to push an engine over the edge to abnormal combustion, if any of the 4 basic factors previously discussed are already near the danger point, for example, excessive carbon buildup raises the compression and retains heat as glowing embers.

FUEL TANK & LINES

General Information

** CAUTION

Disconnect the negative battery cable ANYTIME work is performed on the engine, especially when working on the fuel system. This will help prevent the possibility of sparks during service (from accidentally grounding a hot lead or powered component). Sparks could ignite vapors or exposed fuel. Disconnecting the cable on will also help prevent the possibility fuel spillage if an attempt is made to crank the engine while the fuel system is open.

** CAUTION

Fuel leaking from a loose, damaged or incorrectly installed hose or fitting may cause a fire or an explosion. ALWAYS pressurize the fuel system and run the motor while inspecting for leaks after servicing any component of the fuel system.

** CAUTION

Interestingly, Yamaha does not provide a fuel pressure relief procedure like most other manufacturers. Instead, they suggest that you remove the fuel tank filler cap and then *slowly* disconnect the fuel line quick connector while covering it with a clean rag. That's it!

Fuel Filter/Water Separator

■ **Please refer to the Fuel Filter/Water Separator procedures in the Maintenance & Tune-up section for all details on the fuel filter.**

Fuel Tank

REMOVAL & INSTALLATION

FX Models

3 DIFFICULT

◆ **See Figures 3 thru 7**

1. Remove the seat(s) and then disconnect the battery cables.
2. Release the hood and lift it to its fully open position.
3. On 2002-08 models, remove the 4 bolts (rivets, 2006-08) and lift out the square service lid 1. Reach in, remove the 4 bolts (8 rivets, 2006-08) and remove the small round service lid 2.

■ **To remove a rivet, push the center pin in until you here it click and then remove the rivet.**

4. Drain the fuel tank.
5. Disconnect the ventilation hose and position it out of the way.
6. On 2002-08 models and 2008-10 FX SHO/FX Cruiser SHO models with 1812cc engines, remove the engine as detailed in the Engine Mechanical section.

■ **For some reason, Yamaha does not indicate that engine removal is necessary on 2009-11 FX HO/FX Cruiser HO models. This may have been an accidental omission, so we suggest you take a good look at your set up and determine if you'll need to remove the engine as well.**

7. Remove the hose clamp and disconnect the fuel breather hose at the fuel pump module (2002-08) or tank (2009 and later). Position the hose (along with the check valve and rollover valve) out of the way.
8. Tag and disconnect the fuel sender coupler (black, front) and the pump module coupler (white, rear). Position them out of the way.
9. On 2002-08 models pop off the fuel line connector cover and then remove the fuel line. Plug the line and rest it aside. On 2009 and later models, pull out the ends of the fuel line quick connector while pulling it up at the same time and then remove the fuel line. Plug the line and rest it aside. Details on the quick connect can be found in the Fuel Lines section.

** CAUTION

Interestingly, Yamaha does not provide a fuel pressure relief procedure like most other manufacturers. Instead, they suggest that you remove the fuel tank cap and then slowly disconnect the fuel line quick connector while covering it with a clean rag. That's it!

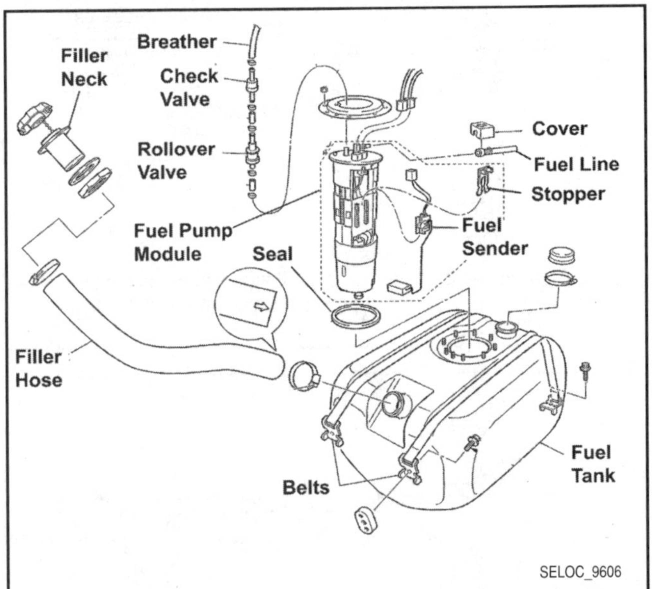

Fig. 3 Exploded view of the fuel tank assembly. Certain models may not use the fuel tank strainer; while on 2009 and later models, the breather hose is connected to the fuel tank and not the module as shown - FX Models

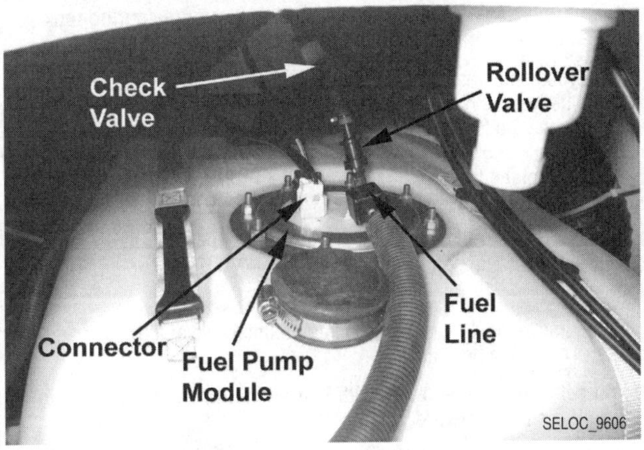

Fig. 4 A good look at the fuel pump module and tank

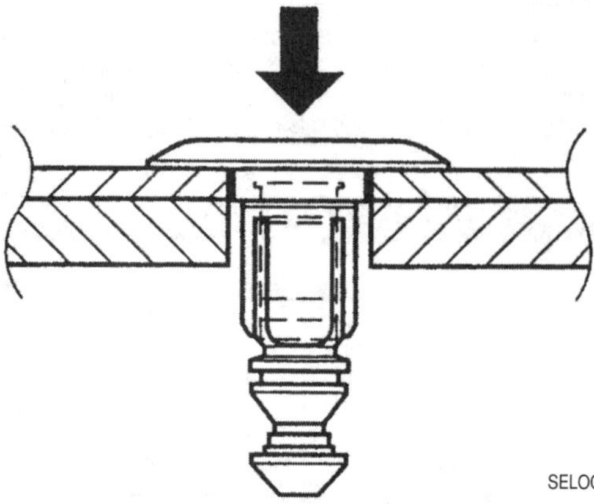

Fig. 5 To remove a rivet, press the center pin in until it clicks and then pull out the rivet

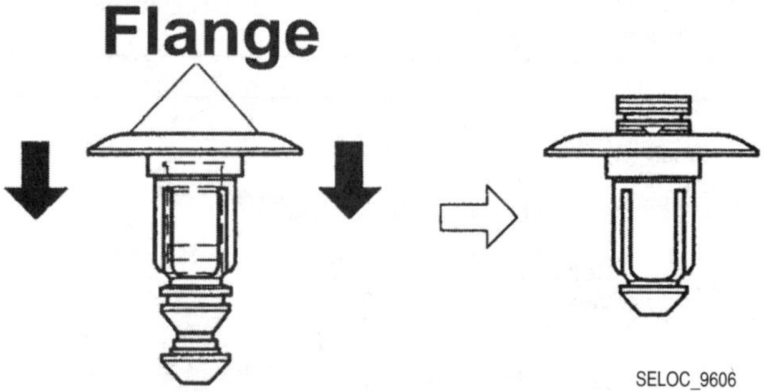

Fig. 6 To re-install that rivet, hold the flange and press the pin against a hard surface until it pops back up and protrudes from the head of the rivet. . .

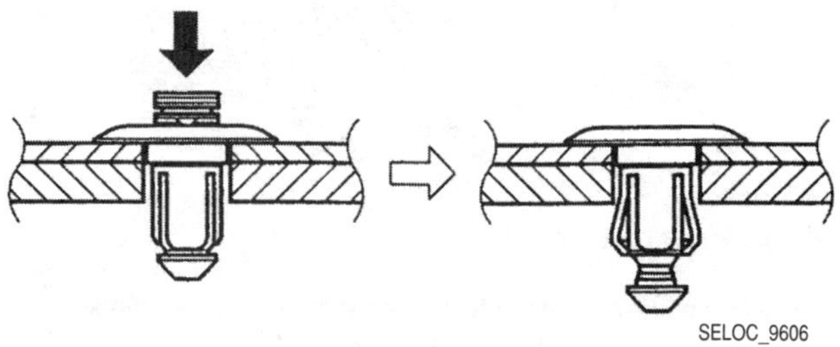

Fig. 7 . . .then insert it into the hole and push the pin inward until it clicks into place, flush with the rivet head

■ **It is not necessary to remove the fuel pump module during tank removal. Details can be found in the Fuel Pump Module section.**

10. Loosen the hose clamps at each end of the fuel filler hose, slide them back on the hose and wiggle the hose off of the filler neck and tank. If equipped, and the fuel strainer comes out of the tank with the hose, be sure to insert it back into place. Most 2008 and later models will have an arrow on the tank end of the filler hose - take note of its orientation to the tank neck before disconnecting it.

■ **Yamaha does not show a fuel strainer on certain FX HO/FX Cruiser HO models, but check just to make sure even if you have one of these units.**

11. Remove the 4 fuel tank belt/strap mounting bolts, 2 per belt, front and rear, and move the belts aside.

■ **You can probably get away with just removing the 2 rear bolts and pulling the straps forward, over the tank.**

12. Lift the fuel tank out through the engine compartment.

To install:

13. Position the tank back in the hull and guide the belts/straps into their depressions in the tank. Install the mounting bolts and tighten them to 11.8 ft. lbs. (16 Nm).

14. If equipped, make sure that the fuel strainer is in place and then wiggle the filler hose over the inlet neck. Do the same at the filler neck and then slide the clamps back into position, tightening them to no more than 3 ft. lbs./36 inch lbs. (4 Nm). If your hose was equipped with the arrow on the tank end, make sure the hose goes back on with the arrow in the same position.

■ **Although not strictly necessary, we recommend replacing the strainer, if equipped, while you have access to it.**

15. Reconnect the fuel line to the module nozzle and install the cover on 2002-08 models. Details on the quick connector are found in the Fuel Lines section.
16. Reconnect the 2 sender connectors.
17. Reconnect the breather and ventilation hoses.
18. Install the engine if removed.
19. Install the service lids on the early models. Tighten the round lid screws to 13 inch lbs. (1.5 Nm) and the square lid screws to 3.9 ft. lbs./45 inch lbs. (5.4 Nm). On models using rivets, hold the flange with both hands and with the pin against a hard flat surface, press down until the pin protrudes from the top. Insert the rivet into the hole until the lip is flush with the surface and then press in the pin until it is flush with the surface - it will click.
20. Close the hood and reconnect the battery cables.
21. Fill the tank with fuel and carefully check that there are no leaks.

FZR/FZS Models

③ DIFFICULT

◆ **See Figures 4 and 8**

1. Remove the seat(s) and then disconnect the battery cables.
2. Drain the fuel tank.
3. Disconnect the ventilation hose and position it out of the way.
4. On FX SHO/FX Cruiser SHO models, remove the engine as detailed in the Engine Mechanical section.

■ **For some reason, Yamaha does not indicate that engine removal is necessary on the FZR/FZS models. This may have been an accidental omission, so we suggest you take a good look at your set up and determine if you'll need to remove the engine as well.**

5. Remove the bottom hose clamp or plastic tie and disconnect the fuel breather hose at the tank, just forward of the module. Position the hose (along with the check valve and rollover valve) out of the way.
6. Tag and disconnect the fuel sender coupler (black, front) and the fuel pump module coupler (white, rear) where they connect to the module. Position them out of the way.
7. Locate the fuel line where it attaches to the fuel pump module. Pull out the ends of the fuel line quick connector while pulling it up at the same time and then remove the fuel line. Plug the line and rest it aside. Details on the quick connect can be found in the Fuel Lines section.

✻✻ CAUTION

Interestingly, Yamaha does not provide a fuel pressure relief procedure like most other manufacturers. Instead, they suggest that you remove the fuel tank cap and then slowly disconnect the fuel line quick connector while covering it with a clean rag. That's it!

■ **It is not necessary to remove the fuel pump module during tank removal. Details can be found in the Fuel Pump Module section.**

8. Loosen the hose clamps at each end of the fuel filler hose, slide them back on the hose and wiggle the hose off of the filler neck and tank. Most 2008 and later models will have an arrow on the tank end of the filler hose - take note of its orientation to the tank neck before disconnecting it if yours has one.
9. Remove the 2 fuel tank belt/strap mounting bolts at the front of the tank and then unclip them at the back of the tank.
10. Lift the fuel tank out through the engine compartment.

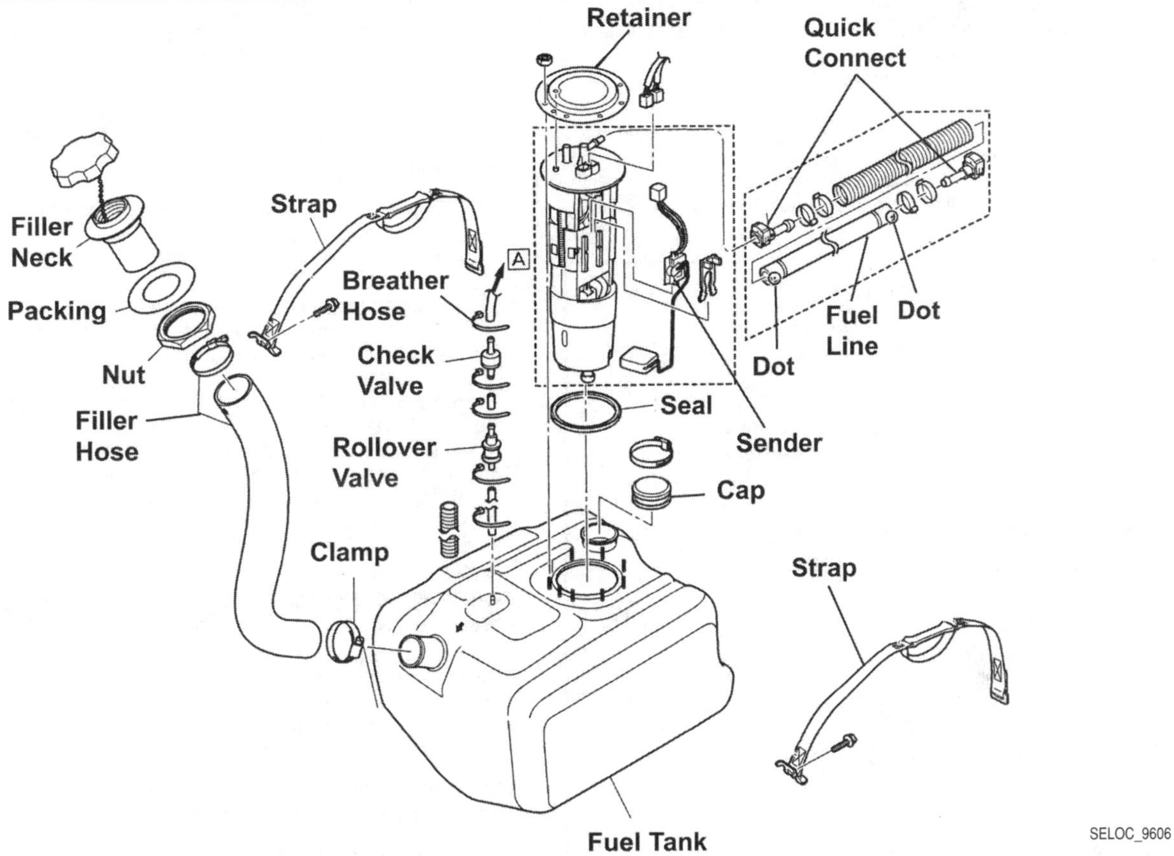

Fig. 8 Exploded view of the fuel tank assembly - FZR and FZS models

To install:

11. Position the tank back in the hull and guide the belts/straps into their depressions in the tank. Clip the rear ends over the lip on the tank and then install the mounting bolts and tighten them to 11.8 ft. lbs. (16 Nm).
12. Wiggle the filler hose over the inlet neck. Do the same at the filler neck and then slide the clamps back into position, tightening them to no more than 3 ft. lbs./36 inch lbs. (4 Nm). If your hose was equipped with the arrow on the tank end, make sure the hose goes back on with the arrow in the same position.
13. Reconnect the fuel line to the module nozzle. Details on the quick connector are found in the Fuel Lines section.
14. Reconnect the 2 sender connectors.
15. Reconnect the breather and ventilation hoses.
16. Install the engine if removed.
17. Reconnect the battery cables.
18. Fill the tank with fuel and carefully check that there are no leaks.
19. Install the seat.

2005-09 VX Models

③ DIFFICULT

◆ **See Figures 4 thru 7 and 9**

1. Remove the seat(s) and then disconnect the battery cables.
2. Release the hood and lift it to its fully open position.
3. Remove the 4 rivets and lift out the storage compartment panel.

■ **To remove a rivet, push the center pin in until you here it click and then remove the rivet.**

4. Drain the fuel tank.
5. Disconnect the ventilation hose and position it out of the way.
6. Remove the engine as detailed in the Engine Mechanical section.
7. Remove the hose clamp and disconnect the fuel breather hose at top of the tank. Position the hose (along with the check valve) out of the way.
8. Tag and disconnect the fuel sender and the pump module couplers from the top of the pump module. Position them out of the way.
9. Pop off the fuel line connector cover and then remove the fuel line. Plug the line and rest it aside. Details on the quick connect can be found in the Fuel Lines section.

**** CAUTION**

Interestingly, Yamaha does not provide a fuel pressure relief procedure like most other manufacturers. Instead, they suggest that you remove the fuel tank cap and then slowly disconnect the fuel line quick connector while covering it with a clean rag. That's it!

■ **It is not necessary to remove the fuel pump module during tank removal. Details can be found in the Fuel Pump Module section.**

10. Loosen the hose clamp at the lower end of the fuel filler hose where it attaches to the tank, slide it back on the hose and wiggle the hose off of the tank. Position the hose out of the way. Or, you can loosen the clamp at the filler neck and simply remove the hose all altogether.
11. Remove the 2 fuel tank belts and move the belts aside.

■ **You can probably get away with just disconnecting the belts at the rear and pulling the straps forward, over the tank.**

12. Lift the fuel tank out through the engine compartment.

To install:

13. Position the tank back in the hull and guide the belts into their depressions in the tank, clipping the ends into the hooks.
14. Wiggle the filler hose over the inlet neck. Do the same at the filler neck and then slide the clamps back into position, tightening them to no more than 2.7 ft. lbs./33 inch lbs. (3.7 Nm).
15. Reconnect the fuel line to the module nozzle and install the cover. For details on connecting the line, you can always refer to the Fuel Lines procedures.
16. Reconnect the 2 sender connectors. Make sure that the inner seal on the sender coupler is correctly installed.
17. Reconnect the breather and ventilation hoses.

■ **Note that the check valve has TANK written on it with an arrow - make sure that the arrow is facing TOWARD the tank. Yes, that means the word will be upside down and the arrow pointing down!**

18. Install the engine.

19. Hold the rivet flange with both hands and, with the pin against a hard flat surface, press down until the pin protrudes from the top. Position the storage compartment panel so the holes line up and then insert the rivet into the hole until the lip is flush with the surface. Press in the pin until it is flush with the surface - it will click.

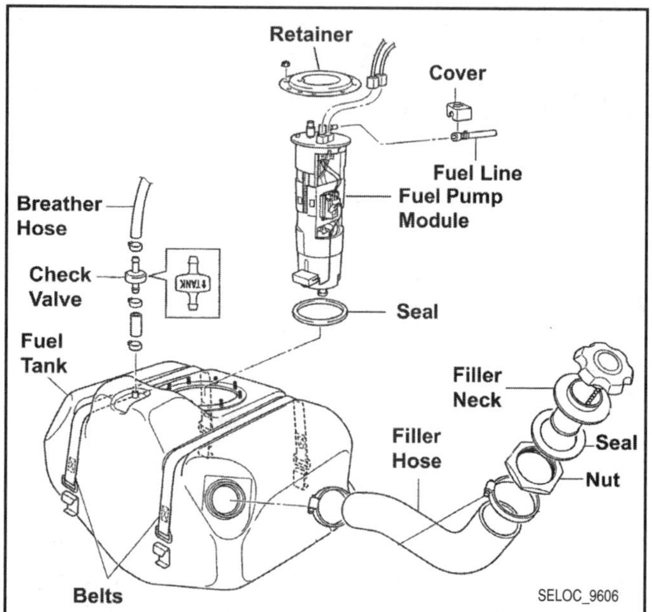

Fig. 9 Exploded view of the fuel tank assembly - 2005-09 VX models

20. Reconnect the battery cables.
21. Fill the tank with fuel and carefully check that there are no leaks.
22. Install the seat(s) and close the hood.

2010-11 VX Models

③ DIFFICULT

◆ **See Figures 10, 11 and 12**

1. Remove the seat(s) and then disconnect the battery cables.
2. Drain the fuel tank.
3. Remove the air filter case as detailed in the Air Filter Case, Housing & Cover section.
4. Remove the engine as detailed in the Engine Mechanical section.
5. Remove the fuse box as detailed in the Electrical section and then remove the 2 nuts/washers to lift out the fuse box bracket. Remove the air filter case brackets.
6. Remove the plastic tie and disconnect the fuel breather hose at top of the tank. Position the hose (along with the check valve) out of the way.
7. Tag and disconnect the fuel sender and the pump module couplers from the top of the pump module. Position them out of the way.
8. Pull out the ends of the fuel line quick connector, while pulling it up at the same time and then remove the fuel line as detailed in the Fuel Lines section. Plug the line and rest it aside. You may already have done this when removing the engine.

**** CAUTION**

Interestingly, Yamaha does not provide a fuel pressure relief procedure like most other manufacturers. Instead, they suggest that you remove the fuel tank cap and then slowly disconnect the fuel line quick connector while covering it with a clean rag. That's it!

■ **It is not necessary to remove the fuel pump module during tank removal. Details can be found in the Fuel Pump Module section.**

9. Loosen the hose clamp at the lower end of the fuel filler hose where it attaches to the tank, slide it back on the hose and wiggle the hose off of the tank. Position the hose out of the way. Or, you can loosen the clamp at the filler neck and simply remove the hose all altogether.

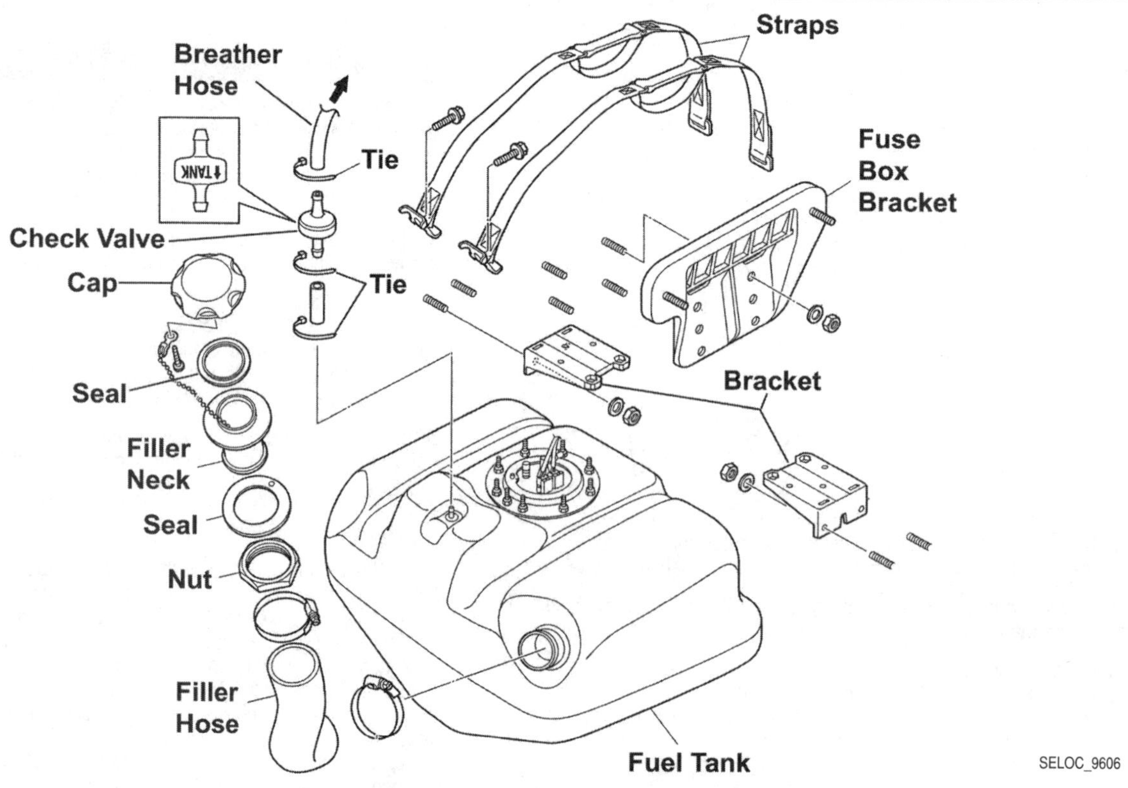

Fig. 10 Exploded view of the fuel tank assembly - 2010-11 VX models w/1052cc engines

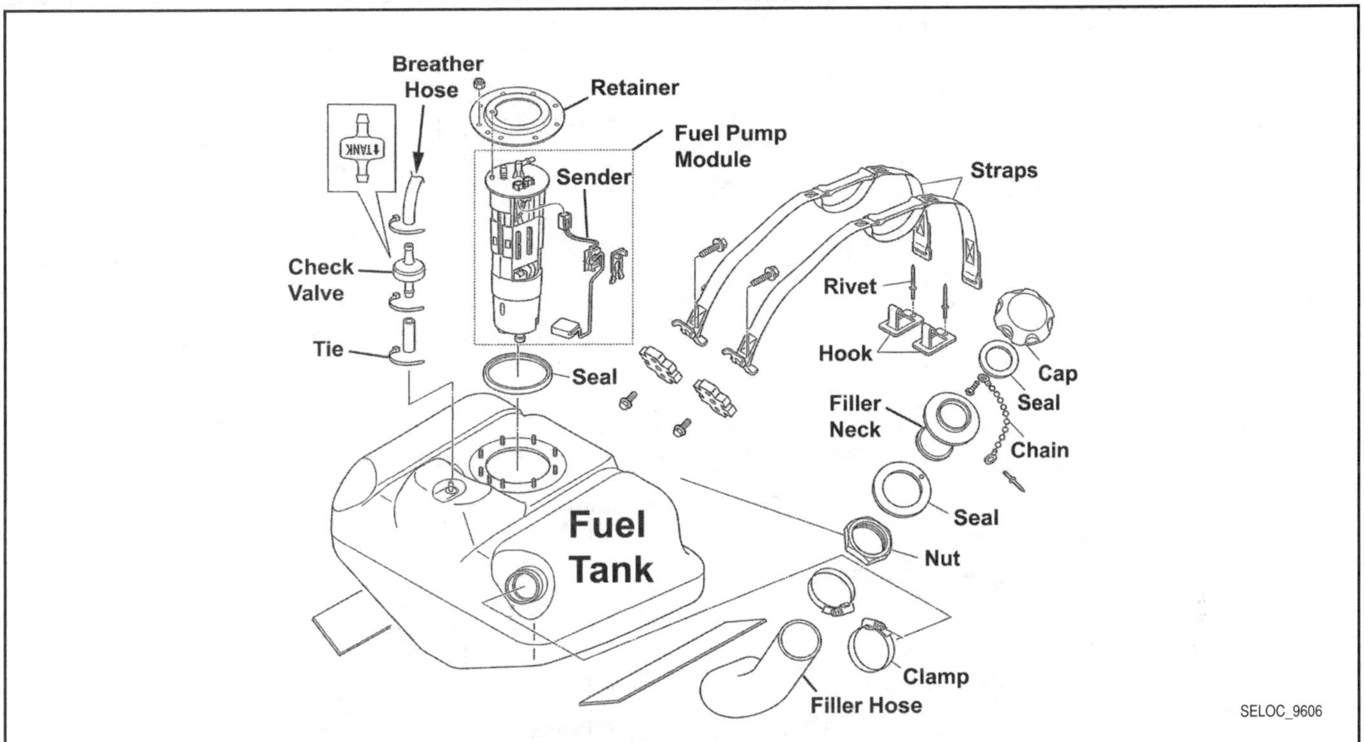

Fig. 11 Exploded view of the fuel tank assembly - 2011 VX models w/1812cc engines

10. Remove the 2 fuel tank belts/straps and move the belts aside. The forward side of each strap will be bolted down and the rear side uses a loop clip going over a stationary hook.

■ **You can probably get away with just disconnecting the belts at the front and pulling the straps rearward, over the tank.**

11. Lift the fuel tank out through the engine compartment.

To install:

12. Check the tank carefully for cracks or other signs of damage.

13. Position the tank back in the hull and guide the belts into their depressions in the tank, clipping the loop ends into the hooks and threading in the bolts on the forward end. Tighten the 2 bolts to 11.8 ft. lbs. (16 Nm).

14. Wiggle the filler hose over the inlet neck. Do the same at the filler neck and then slide the clamps back into position, tightening them to no more than 2.7 ft. lbs./33 inch lbs. (3.7 Nm); 3 ft. lbs./36 inch lbs. (4 Nm) on the models w/1812cc engines. Before tightening the filler neck clamp, make sure that there is 1.2-1.6 in. (30-40mm) between the bottom of the filler neck nut and the center line of the clamp (the bolt) and the arrow on the hose lines up with the projection on the filler neck; adjust the clamp and hose as necessary.

15. Reconnect the fuel line to the module nozzle and install the cover. For details on connecting the line, you can always refer to the Fuel Lines procedures.

16. Reconnect the 2 sender connectors. Make sure that the inner seal on the sender coupler is correctly installed.

17. Reconnect the breather and ventilation hoses.

■ **Note that the check valve has TANK written on it with an arrow - make sure that the arrow is facing TOWARD the tank. Yes, that means the word will be upside down and the arrow pointing down!**

18. Install the filter case brackets and fuse box bracket, tightening all nuts to 12.5 ft. lbs. (17 Nm). Install the fuse box.

19. Install the engine.

20. Install the air filter case

21. Reconnect the battery cables.

22. Fill the tank with fuel and carefully check that there are no leaks.

23. Install the seat(s).

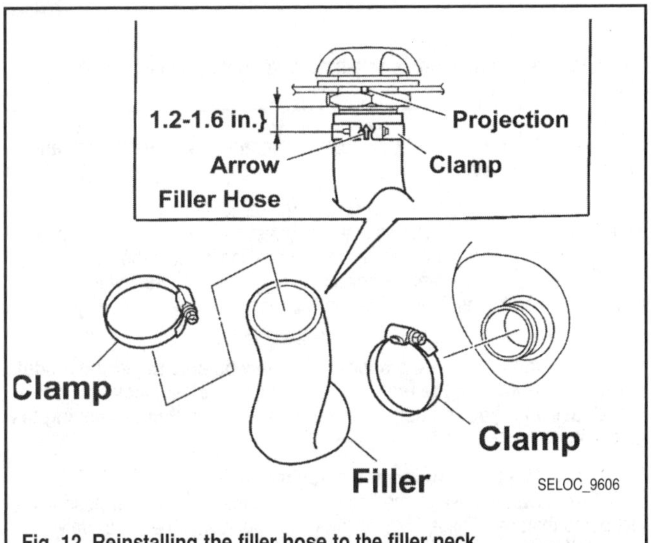

Fig. 12 Reinstalling the filler hose to the filler neck

Fuel Pump Module

REMOVAL & INSTALLATION

FX Models & 2005-09 VX Models

③ DIFFICULT

◆ See Figures 3 thru 7, 9, and 13 thru 16

1. Remove the seat(s) and then disconnect the battery cables.
2. Release the hood and lift it to its fully open position (certain models).
3. On 2002-08 FX models, remove the 4 bolts/rivets and lift out the square service lid 1. Reach in, remove the 4 bolts or 8 rivets, and remove the small round service lid 2.
4. On VX models, remove the 4 rivets and lift out the storage compartment panel.

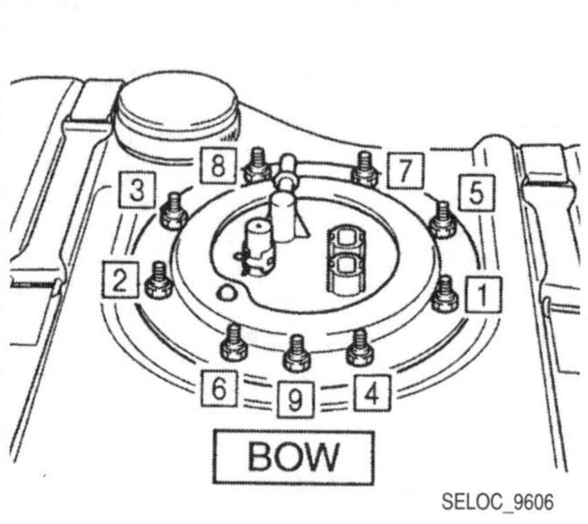

Fig. 13 Fuel pump module retaining ring loosening sequence - FX shown, VX similar (cap in different place)

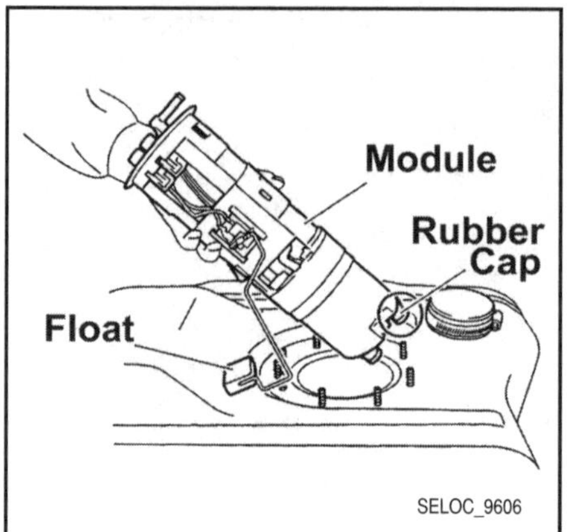

Fig. 14 Always tilt the module during remove removal so as not to bend the float

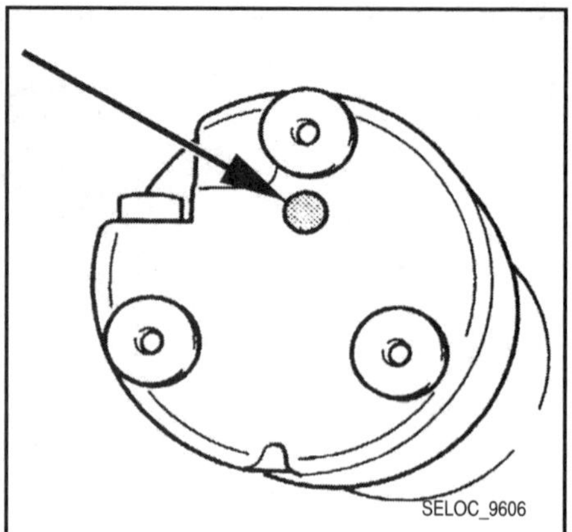

Fig. 15 Check the fuel pump filter

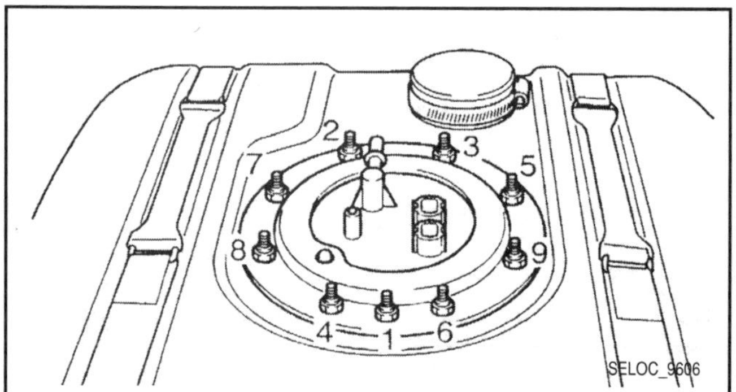

Fig. 16 Fuel pump module retaining ring tightening sequence - FX and 2005-09 VX models

■ **To remove a rivet, push the center pin in until you here it click and then remove the rivet.**

5. Drain the fuel tank. Strictly speaking, you could probably get away without draining the tank, but we highly suggest draining for safety's sake.
6. Disconnect the ventilation hose and position it out of the way.
7. Remove the fuel tank as detailed in the Fuel Tank Removal & Installation procedures. This, of course will require removing the engine as well.

■ **On most models, *if you are lucky*, you may be able to get the module out without removing the fuel tank. Take a good, close, look and if this is the case with your model, skip to the next step without removing the fuel tank.**

On models where you are not removing the fuel tank:

8. Remove the hose clamp and disconnect the fuel breather hose at the fuel pump module or tank. Position the hose (along with the check valve, FX/FZ/VX and rollover valve, FX/FZ) out of the way.
9. Tag and disconnect the fuel sender coupler (black, front) and the pump module coupler (white, rear). Position them out of the way.
10. Remove the fuel line quick connector and then remove the fuel line as detailed in the Fuel Lines section. Plug the line and rest it aside.

**** CAUTION**

Interestingly, Yamaha does not provide a fuel pressure relief procedure like most other manufacturers. Instead, they suggest that you remove the fuel tank cap and then slowly disconnect the fuel line quick connector while covering it with a clean rag. That's it!

On all models:

11. Loosen the module retainer nuts (9) in the order shown in the illustration and then lift off the retainer.
12. Carefully remove the module, making sure you tilt it *toward* the float arm (forward) so as not to snag the float on the lip of the tank. Be careful not to knock off the small rubber cap(s) on the bottom of the module.
13. Remove the packing ring and throw it away.

To install:

14. Check the fuel pump filter, cleaning it as necessary.
15. Install a new module packing ring and then carefully slide the module into the tank and then position the retainer ring over the studs. Tighten the ring nuts in the order shown in the illustration to:
    - 2002-08 FX models w/998 and 1052cc engine - a 1st pass of 2.3 ft. lbs./24 inch lbs. (3.2 Nm). Now tighten them a final time (in order) to 4.6 ft. lbs./55 inch lbs. (6.4 Nm).
    - 2008 and later FX models w/1812cc engine - a 1st pass of 2.2 ft. lbs./26 inch lbs. (3 Nm). Now tighten them a final time (in order) to 4.4 ft. lbs./53 inch lbs. (6 Nm).
    - 2009 and later FZ models w/1812cc engine - a 1st pass of 2.2 ft. lbs./26 inch lbs. (3 Nm). Now tighten them a final time (in order) to 4.4 ft. lbs./53 inch lbs. (6 Nm).
    - 2005-09 VX - a 1st pass of 2.3 ft. lbs./24 inch lbs. (3.2 Nm). Now tighten them a final time (in order) to 4.6 ft. lbs./55 inch lbs. (6.4 Nm).
16. Install the tank if you had to remove it to get at the module.
17. If you didn't remove the tank:
    a. Coat the fuel pipe lightly with clean engine oil and then reconnect the fuel line as detailed in the Fuel Lines section.
    b. Reconnect the 2 sender connectors. Make sure that the inner seal on the sender coupler is correctly installed.
    c. Reconnect the breather and ventilation hoses.
18. Install the engine if removed.
19. Install the service lids on all models but the 2008-11 FX and FZ models w/1812cc engines. On models w/o rivets, tighten the round lid screws to 13 inch lbs. (1.5 Nm) and the square lid screws to 3.9 ft. lbs./47 inch lbs. (5.4 Nm). On late models, hold the rivet flange with both hands and, with the pin against a hard flat surface, press down until the pin protrudes from the top. Position the storage compartment panel so the holes line up and then insert the rivet into the hole until the lip is flush with the surface. Press in the pin until it is flush with the surface - it will click.
20. Close the hood and reconnect the battery cables.
21. Fill the tank with fuel and carefully check that there are no leaks.

2010-11 VX Models

3 DIFFICULT

◆ **See Figures 5, 6, 7, 10, 11, 14 and 17 thru 19**

1. Remove the seat(s) and then disconnect the battery cables.
2. Release the hood and lift it to its fully open position.
3. Remove the 5 rivets (see the illustrations) and lift out the storage compartment panel (service lid).

■ **To remove a rivet, push the center pin in until you hear it click and then remove the rivet.**

4. Drain the fuel tank. Strictly speaking, you could probably get away without draining the tank, but we highly suggest draining for safety's sake.
5. Remove the hose clamp and disconnect the fuel breather hose at the fuel pump module. Position the hose (along with the check valve) out of the way.
6. Tag and disconnect the fuel sender coupler (usually black, front) and the pump module coupler (usually white, rear). Position them out of the way.

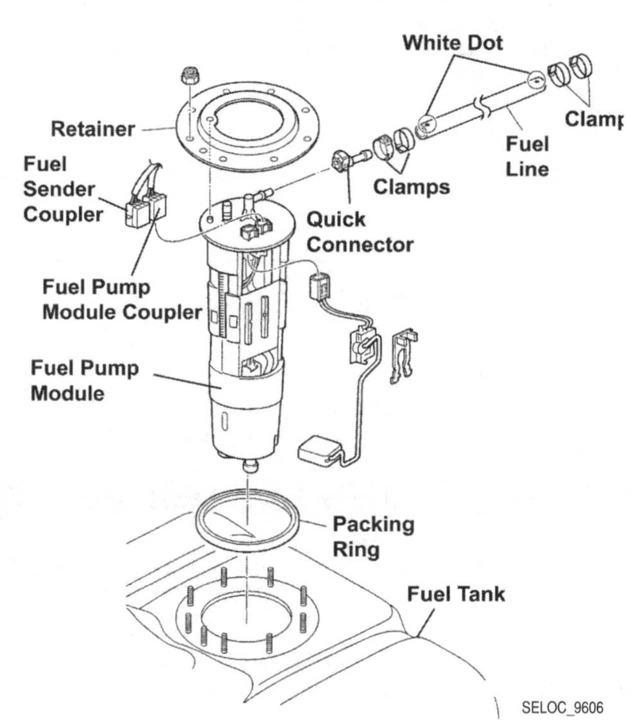

Fig. 17 Exploded view of the fuel pump module - 2010-11 VX models

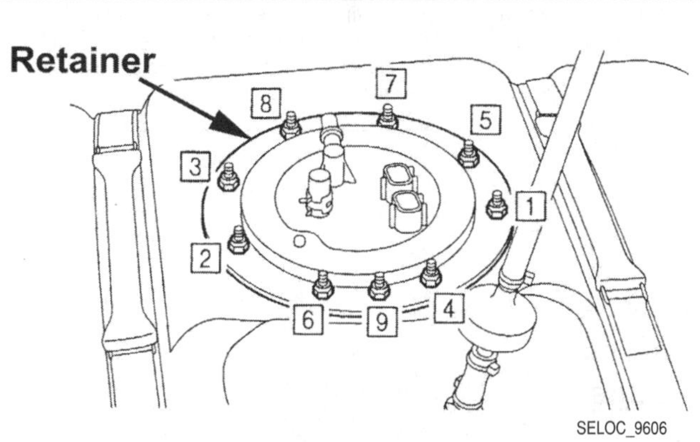

Fig. 18 Fuel pump module retaining ring loosening sequence - 2010-11 VX models

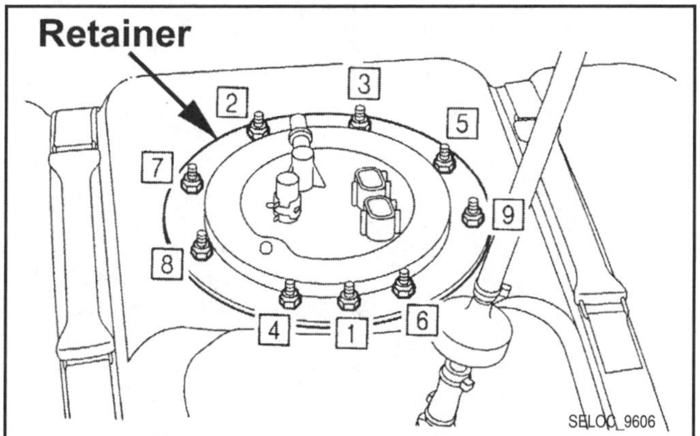

Fig. 19 Fuel pump module retaining ring tightening sequence - 2010-11 VX models

7. Pull out the ends of the fuel line quick connector, while pulling it up at the same time and then remove the fuel line as detailed in the Fuel Lines section. Plug the line and rest it aside.

✳✳ CAUTION

Interestingly, Yamaha does not provide a fuel pressure relief procedure like most other manufacturers. Instead, they suggest that you remove the fuel tank cap and then *slowly* disconnect the fuel line quick connector while covering it with a clean rag. That's it!

8. Loosen the module retainer nuts (9) in the order shown in the illustration and then lift off the retainer.

9. Carefully remove the module, making sure you tilt it *toward* the float arm (forward) so as not to snag the float on the lip of the tank. Do not lose the rubber caps on the bottom of the module.

10. Remove the packing ring and throw it away.

To install:

11. Install a new module packing ring and then carefully slide the module into the tank and then position the retainer ring over the studs. Tighten the ring nuts in the order shown in the illustration to a 1st pass of 2.2 ft. lbs./26 inch lbs. (3 Nm). Now tighten them a final time (in order) to 4.4 ft. lbs./53 inch lbs. (6 Nm).

12. Install the tank if you had to remove it to get at the module.

13. Coat the fuel pipe lightly with clean engine oil and then reconnect the fuel line. Press the quick connector down until it clicks into position.

■ If the connector is not installed correctly, the retainer will not be able to 'click' into place.

14. Reconnect the 2 sender connectors. Make sure that the inner seal on the sender coupler is correctly installed.

15. Reconnect the breather and ventilation hoses.

16. Install the service lid. Hold the rivet flange with both hands and, with the pin against a hard flat surface, press down until the pin protrudes from the top. Position the storage compartment panel so the holes line up and then insert the rivet into the hole until the lip is flush with the surface. Press in the pin until it is flush with the surface - it will click.

17. Close the hood and reconnect the battery cables.

18. Fill the tank with fuel and carefully check that there are no leaks.

19. Install the seat.

FUEL SENDER

Removal & Installation

DIFFICULT

◆ See Figures 20 and 21

■ For some reason, Yamaha does not show a fuel sender removal procedure on the FX HO/FX Cruiser HO models prior to 2009 or the 2005-09 VX models, even though the units appear to be identical *and* they have a checking procedure (which calls for replacement if the check fails). So, we suspect this procedure will work for these models as well.

1. Remove the fuel pump module as detailed in the Fuel Pump Module Removal & Installation procedure.

2. Disconnect the sender coupler.

3. Locate the sender on the side of the module and squeeze the sender retainer tabs (B) inward (toward each other) and then slide out the retainer in the direction of (A).

4. Now press in on the sender tabs and remove it in the same direction (B).

5. Installation is essentially the reverse of removal - slide the sender in until the tabs snap into position and then do the same with the retainer.

6. Connect the coupler and install the module.

Checking

DIFFICULT

◆ See Figure 22

1. Remove the fuel pump module from the fuel tank and support it upright on a clean, flat surface.

2. Disconnect the forward (black) coupler on top of the module and connect a multi-meter as per the manufacturer's instructions.

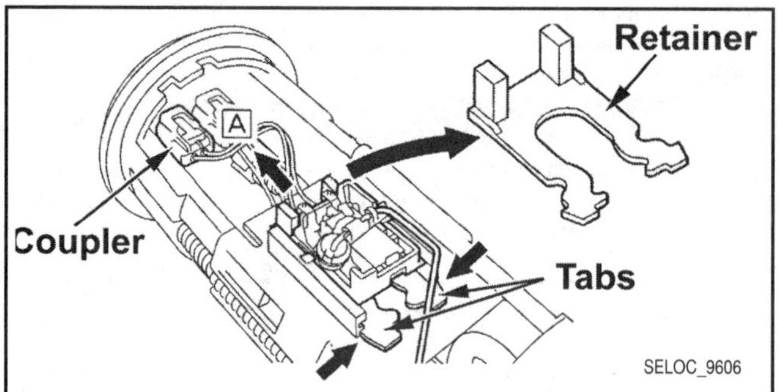

Fig. 20 Remove the sender retainer. . .

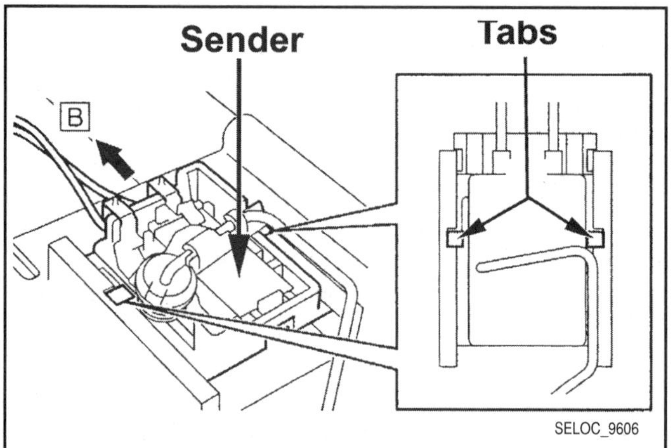

Fig. 21 . . .and then the sender itself

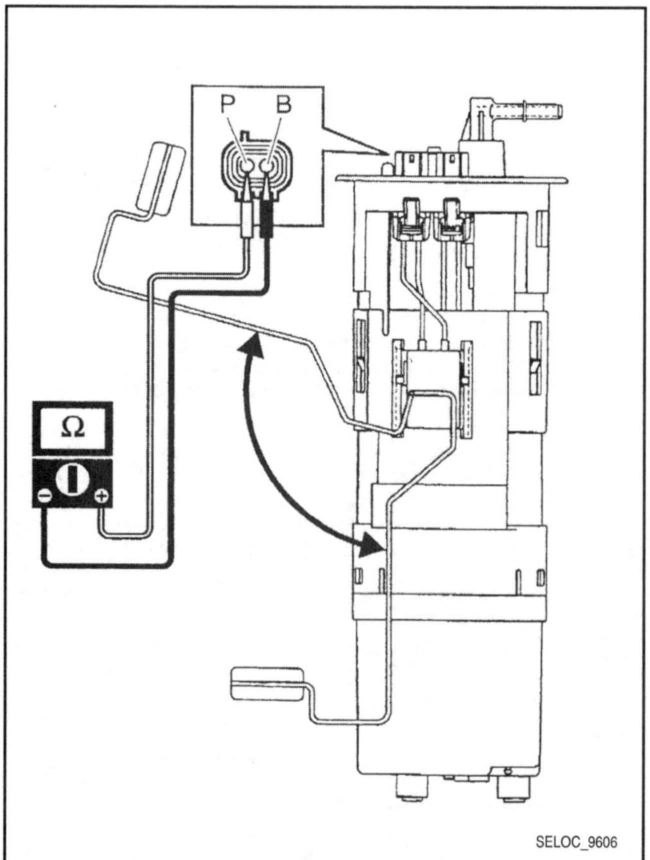

Fig. 22 Checking the fuel sender, some models may use a yellow or red for the left terminal

3. The meter should read 91-93 ohms with the float in the lowered position and 6-8 ohms in the raised position on FX/FX140 and FX Cruiser/FX140 Cruiser models; or 133.5-136.5 ohms in the lower position and 5-7 ohms in the raised position on FX HO/SHO, FX Cruiser HO/SHO, FZR/FZS and VX models.

■ **Do not raise the float more than 105°.**

4. Any reading outside of the specified ranges will indicate the send unit needs to be replaced.

CHECKING FUEL PRESSURE

OEM ② MODERATE

◆ **See Figures 4 and 23**

1. Remove the seat and locate the fuel line where it connects to the fuel pump module. Remove the connection cover and then carefully disconnect the fuel line from the elbow.

✳✳ CAUTION

Interestingly, Yamaha does not provide a fuel pressure relief procedure like most other manufacturers. Instead, they suggest that you remove the fuel tank cap and then slowly disconnect the fuel line quick connector while covering it with a clean rag. That's it!

2. Install a pressure gauge adapter (#YW-06842 or 90890-06842) between the elbow and the fuel line. Once connected, install the pressure gauge to the adapter.

3. Start the engine and allow it to reach normal operating temperature. Make sure that you have the craft connected to a sufficient water source. While the engine is warming up, keep an eye on the adapter connections to make sure there is no fuel leaking anywhere.

4. Once the engine is warm, confirm that the gauge is showing:

• 2002-08 FX models w/998cc and 1052cc motors - pressure should be 45-47 psi (310-330 kPa) with the engine at normal operating temperature and running at idle

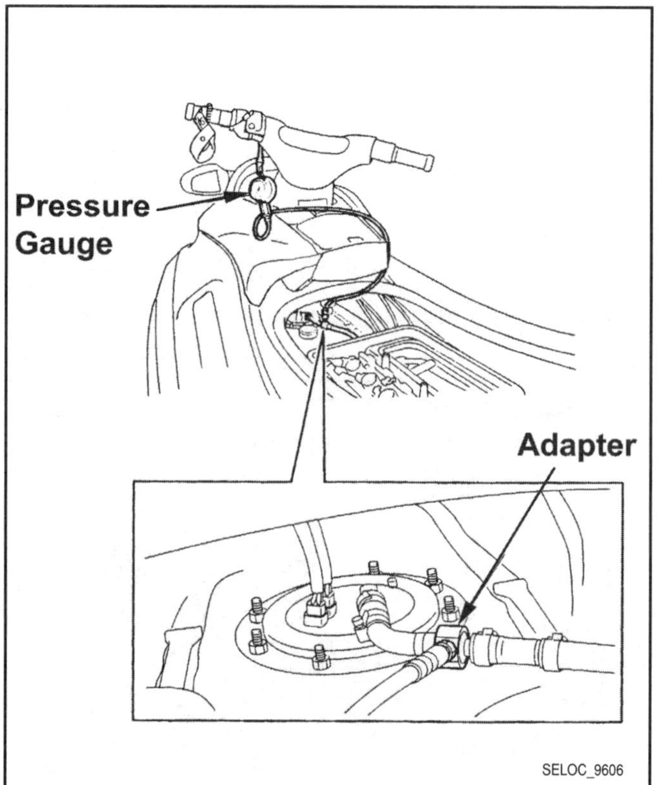

Fig. 23 Always check the fuel pressure at the fuel line connection to the module

• 2008-11 FX models w/1812cc motors: pressure should be 49.1-52.6 psi (345-370 kPa) with the engine at normal operating temperature and running at idle

• 2009-11 FZ models w/1812cc motors: pressure should be 49.1-52.6 psi (345-370 kPa) with the engine at normal operating temperature and running at idle

• 2005-09 VX models w/1052cc motors - pressure should be 46-47 psi (320-327 kPa) with the engine at normal operating temperature and running at idle

• 2010-11 VX models w/1052cc motors - pressure should be 45.4-48 psi (319-340 kPa) with the engine at normal operating temperature and running at idle

• 2011 VX models w/1812cc motors: pressure should be 49.1-52.6 psi (345-370 kPa) with the engine at normal operating temperature and running at idle

5. Turn the engine off and allow it to cool before disconnecting the gauge and adapter. Once cool, slowly disconnect the adapter and reconnect the fuel line to the elbow.

Main & Fuel Pump Relay

REMOVAL & INSTALLATION

EASY

◆ See Figures 24 thru 29

1. Remove the seat(s) and disconnect the battery cables. Remove the rear storage tray if equipped.

2. Locate the electrical box behind the oil tank (2002-08 FX), behind the engine (2009 and later FX, FZ and VX w/1812cc engines) or on the starboard side hull (VX w/1052cc engines) and remove the cover as detailed in the Electrical Box procedures found in the Ignition & Electrical section.

3. Unplug the 8 pin connector, remove the relay mounting bolt and lift out the relay on 2002-08 FX models. On VX models and 2008-11 FX and FZ models, simply pull the relay out of the rubber mount.

4. Position the relay on the wall of the box and tighten the bolt to 34 inch lbs./2.8 ft. lbs. (3.9 Nm) on 2002-08 FX models. On VX and later FX/FZ models, press it back into the mount.

5. Reconnect the 8 pin connector.

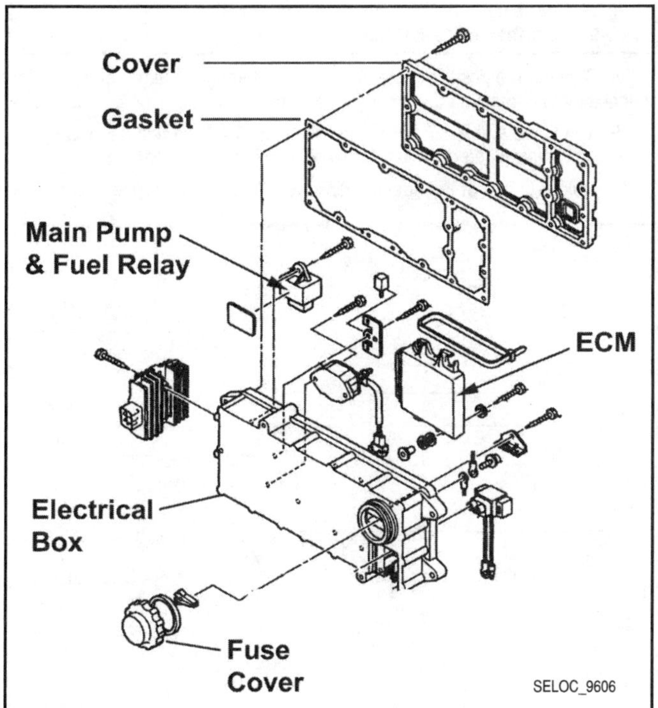

Fig. 24 Exploded view of the electrical box showing the fuel pump relay - FX/FX Cruiser and FX140/FX140 Cruiser

6. Install the cover and tighten the 18 screws to 43 inch lbs./3.5 ft. lbs.(4.9 Nm). On 2008 and later FX, FZ and VX models w/1812cc engines, simply slide the latches back into position.

7. Connect the cables and install the tray/seats.

TESTING

2002-08 FX With 998/1052cc Engines & 2005-09 VX Models

MODERATE

◆ **See Figures 24, 25 and 27 thru 31**

1. Remove the seat(s). Remove the rear storage tray if equipped.

2. On FX models, locate the electrical box behind the oil tank and remove the cover as detailed in the Electrical Box procedures found in the

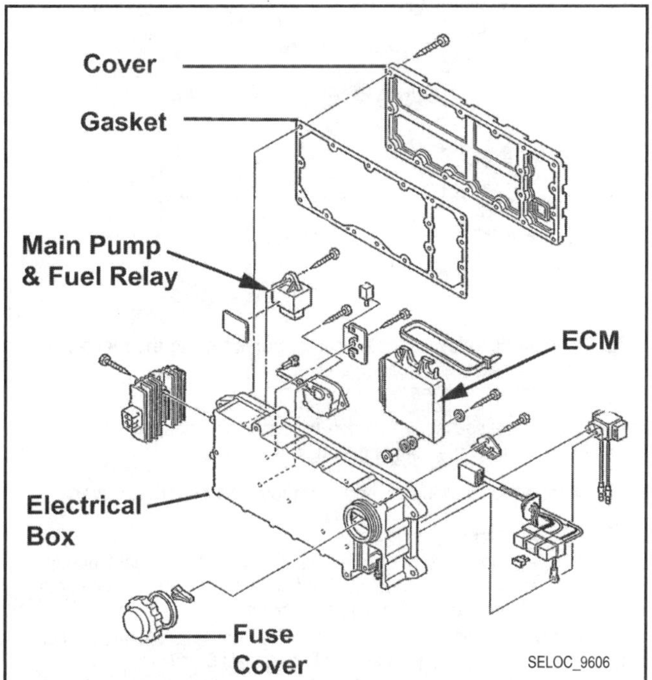

Fig. 25 Exploded view of the electrical box showing the fuel pump relay - 2004-08 FX HO/FX Cruiser HO (2006 shown)

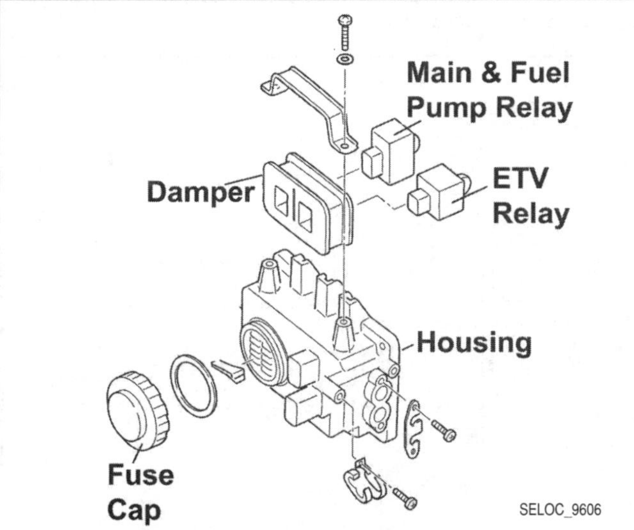

Fig. 26 Exploded view of the electrical box showing the fuel pump relay - 2008 and later FX and FZ models w/1812cc engines, 2011 VXR/VXS models similar

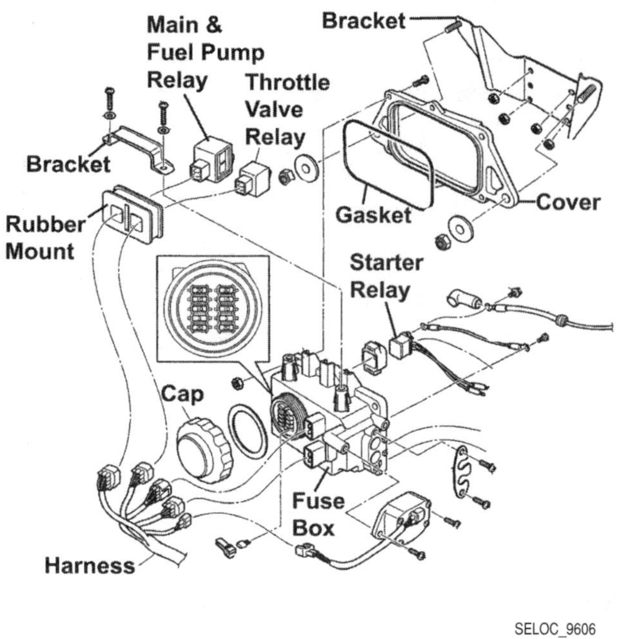

Fig. 27 Exploded view of the electrical box showing the fuel pump relay - VX models w/1052cc engines

Ignition & Electrical section. VX models will have the box attached to the starboard hull, toward the front of the engine.

■ **On VX models, there is no need to remove the electrical cover.**

3. Unplug the 8 pin connector.

4. Using a jumper wire of sufficient length, connect terminal **1** on the relay to the negative battery terminal. Connect a jumper wire between relay terminals **2** or **3** and the positive battery terminal.

5. Connect the positive meter lead to relay terminal **5** and then check for continuity when touching the other lead to terminals **6** or **7**.

6. Now remove the 2 jumpers from the battery (at the relay end) and confirm that there is no continuity between terminal **5** and terminals **6** or **7**.

7. Now, with the jumper wire, connect terminal **4** on the relay to the negative battery terminal. Connect a jumper wire between relay terminal **6** and the positive battery terminal.

Fig. 28 A better look at the VX box (1052cc engines)

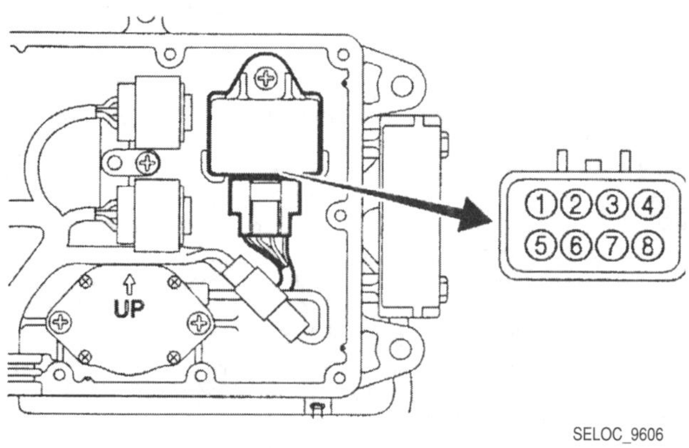

Fig. 29 The relay is in the upper right corner of the box when viewed from the stern - FX models

8. Connect the positive meter lead to relay terminal **7** and then check for continuity when touching the other lead to terminal **8**.

9. Now remove the 2 jumpers from the battery (at the relay end) and confirm that there is no continuity between terminal **7** and terminal **8**.

10. Failure of any of the above checks will require relay replacement.

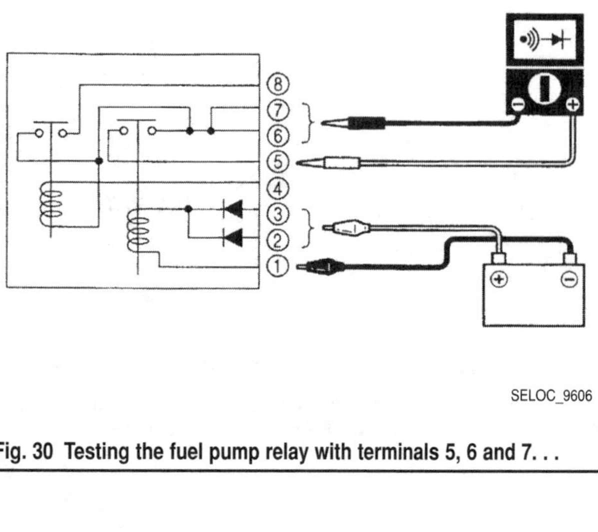

Fig. 30 Testing the fuel pump relay with terminals 5, 6 and 7. . .

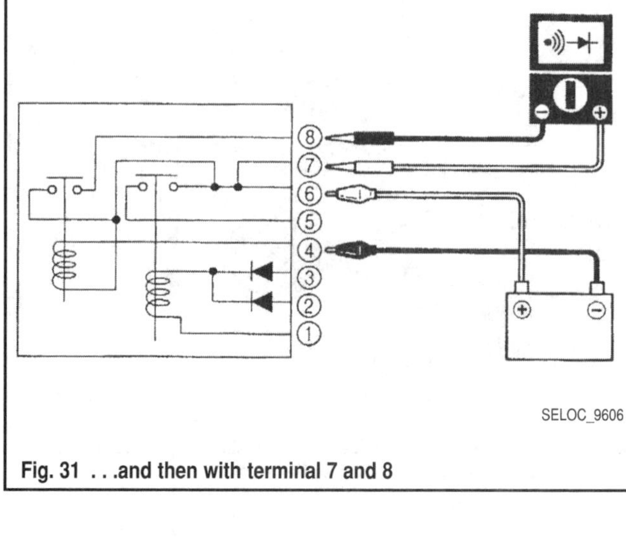

Fig. 31 . . .and then with terminal 7 and 8

2008-11 FX/FZ Models w/1812cc Engines & 2010-11 VX Models

② MODERATE

◆ See Figures 27, 28 and 32 thru 35

1. Remove the seat(s). Remove the rear storage tray if equipped.
2. On FZ models, remove the 7 bolts and lift off the deck beam assembly. Two bolts will have nuts and a washer under the bolt head and the nut, two will have a nut with a washer under it and 3 will have no nuts or washers at all.
3. Slide back the slide locks and remove the cover to the electrical box.
4. Unplug the 8 pin connector.
5. Connect a DVOM and measure voltage between terminals **3** and ground. Repeat with terminal **5**. Battery voltage (12V) should be present at both terminals. FZR and FZS models use the same terminals as in the illustration, but designate them as **2** (top) and **3** (bottom).

■ Remember that the connector terminals will be the reverse of the ones shown in the relay terminals picture. The important thing is that you are checking the 2 red (R) wire terminals in the connector/coupler.

6. Remove the relay.
7. Main relay continuity. Using a jumper wire of sufficient length, connect terminal **1** on the relay to the negative battery terminal. Connect a jumper wire between relay terminals **2** or **3** and the positive battery terminal.
8. Connect the positive meter lead to relay terminal **5** and then check for continuity when touching the other lead to terminals **6** or **7**.
9. Now remove the 2 jumpers from the battery (at the relay end) and confirm that there is no continuity between terminal **5** and terminals **6** or **7**.
10. Fuel pump relay continuity. Now, with the jumper wire, connect terminal **4** on the relay to the negative battery terminal. Connect a jumper wire between relay terminal **6** and the positive battery terminal.
11. Connect the positive meter lead to relay terminal **7** and then check for continuity when touching the other lead to terminal **8**.
12. Now remove the 2 jumpers from the battery (at the relay end) and confirm that there is no continuity between terminal **7** and terminal **8**.
13. Failure of any of the above checks will require relay replacement.
14. When installing the deck plate on FZ models, tighten the 7 bolts to 13.3 ft. lbs. (18 Nm).

Fuel Lines

REMOVAL & INSTALLATION

OEM ① EASY

◆ See Figures 36 thru 48

■ Yamaha has initiated a Factory Modification Campaign (TSB #WR2008-004R) for a faulty fuel line connector on 2008 FX SHO and FX Cruiser SHO models and certain 2009 FX SHO and FX Cruiser SHO (HIN #F1W-809301 thru 815100) that could cause the line to come loose at the fuel rail. The campaign provides a new fuel line assembly with a different connector and an additional clamp.

**** CAUTION**

Always make sure that the battery is disconnected before performing fuel line work.

Aside from the conventional screw clamps and scissors clamps, fuel lines on these models may also be equipped with a different style of crimp (circle) clamp and/or a quick connector clamp.

Crimp clamps require that you actually cut the joint off before removal and then on installation you physically pinch in the clamp protrusion to secure the clamp.

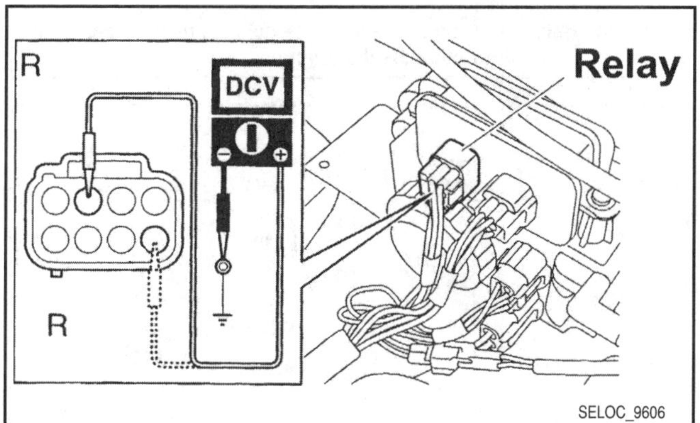

Fig. 32 Testing relay input voltage

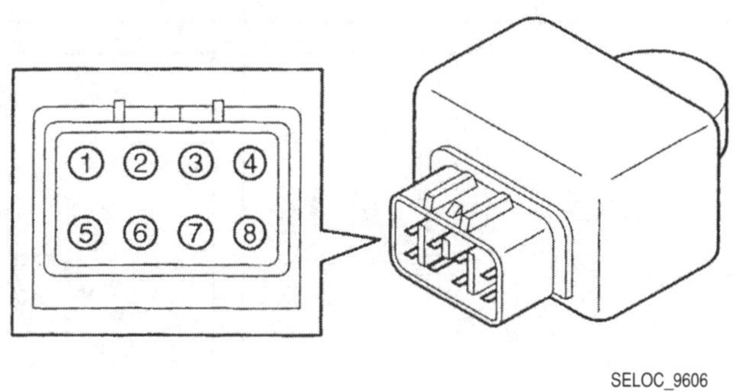

Fig. 33 Relay terminals

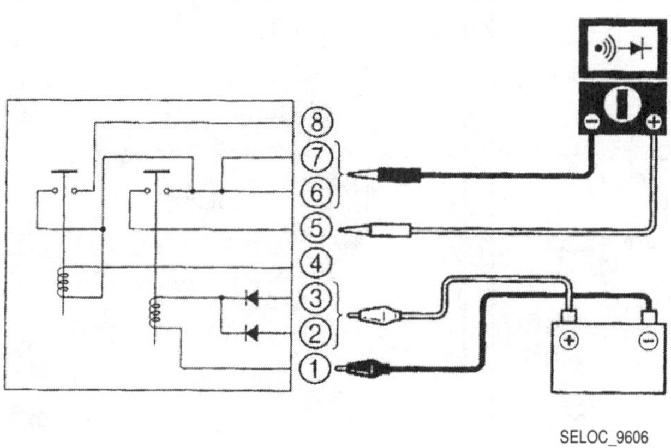

Fig. 34 Testing main relay continuity

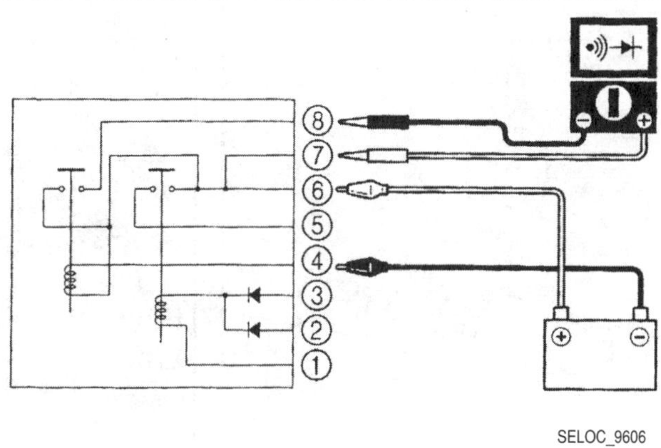

Fig. 35 Testing fuel pump relay continuity

■ **There are special pliers that will facilitate both cutting the clamp and pinching the new one correctly.**

** CAUTION

Interestingly, Yamaha does not provide a fuel pressure relief procedure like most other manufacturers. Instead, they suggest that you remove the fuel tank cap and then slowly disconnect the fuel line quick connector at the pump module while covering it with a clean rag. That's it!

On 2002-08 FX and 2005-09 VX models:

Wrap a shop towel around the connector and then SLOWLY turn the connector tab until it meets the stopper position. You may wish to use a pair of pliers for this - be careful. Remove the retainer from the connector quickly and be sure not to lose it.

** CAUTION

Be certain that you do not rotate the connector past the stopper or the assembly will be damaged.

Cover the fuel pipe/elbow and the connector end of the fuel line with a clean plastic bag. Always check the condition of the connector O-ring, replacing if necessary.

■ **When installing the fuel line onto the module fitting (or fuel rail for that matter) always make sure that the small white dot is facing upward.**

When installing a new quick connect, slide the fuel line onto the connector so that the small white mark on the fuel line is facing upward and lined up with the checker tab on the new connector.

Coat the fuel elbow/pipe lightly with engine oil and then slide the quick connect onto the pipe until you hear it click into position.

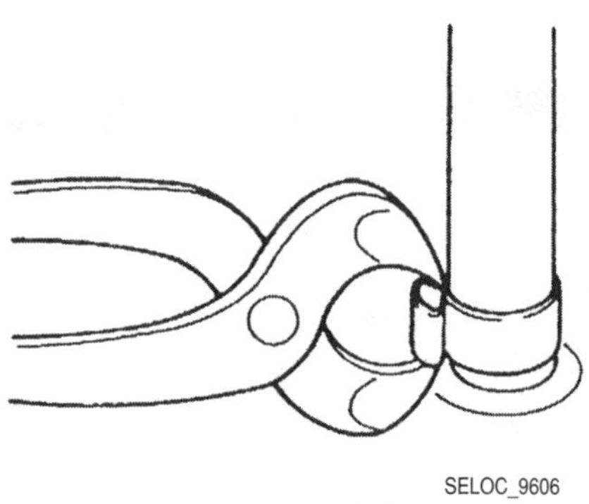

Fig. 36 Use the special pliers to cut the crimp (circle) clamp at the joint

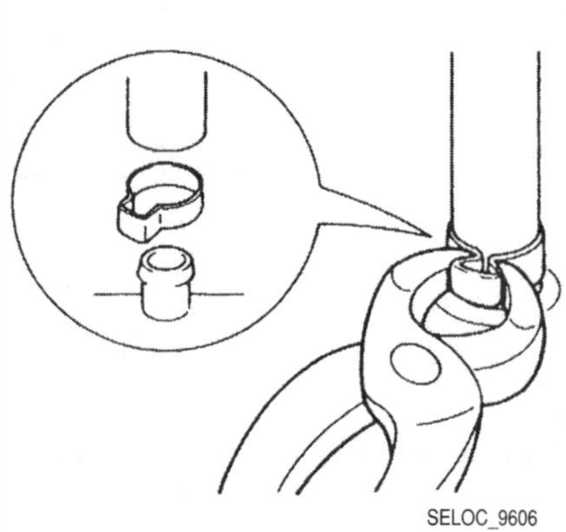

Fig. 37 You can use the same tool to crimp the new clamp

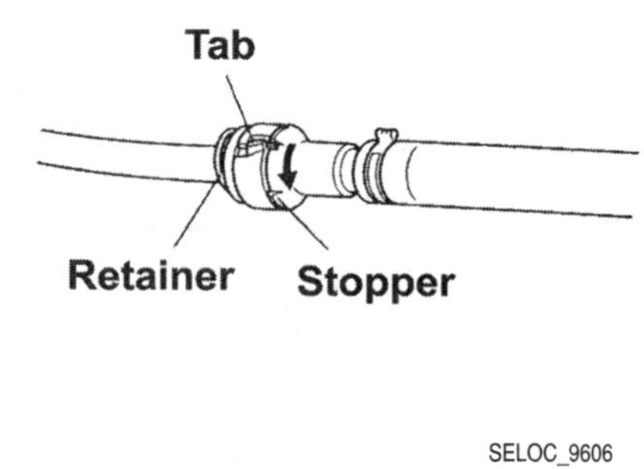

Fig. 38 Rotate the quick connector tab until it meets the stopper. . .

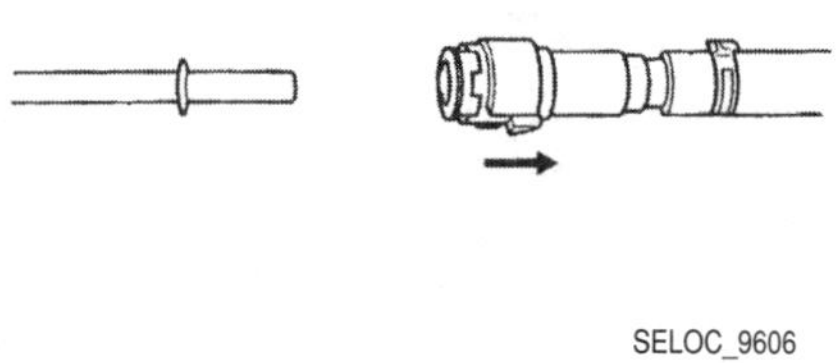

Fig. 39 . . .and then pull the connector off the fuel pipe - 2002-08 models

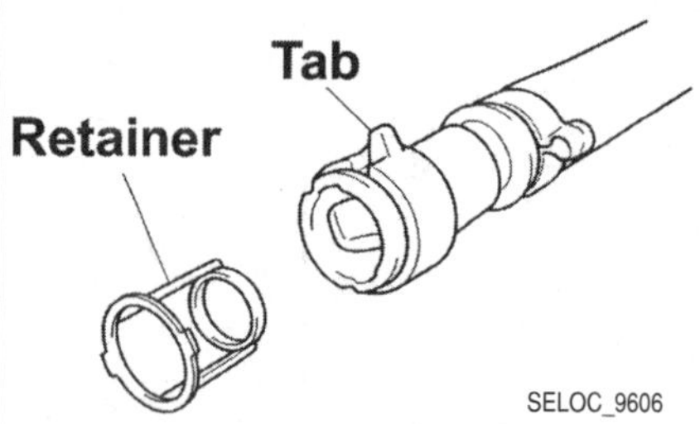

Fig. 40 Once you've disconnected the line, remove the retainer so it does not get lost - 2002-08 models

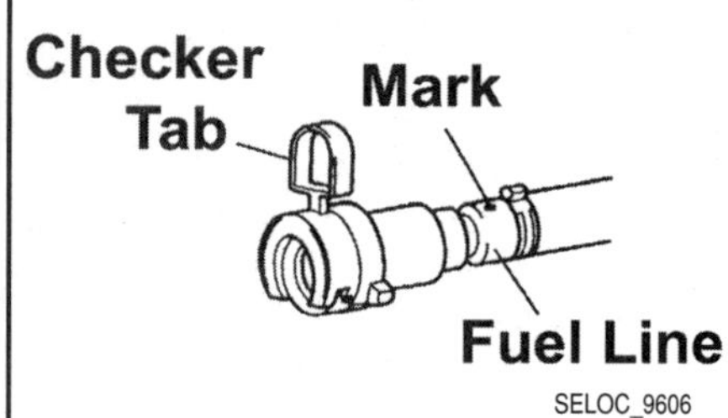

Fig. 41 New quick connects are equipped with a checker tab. . .

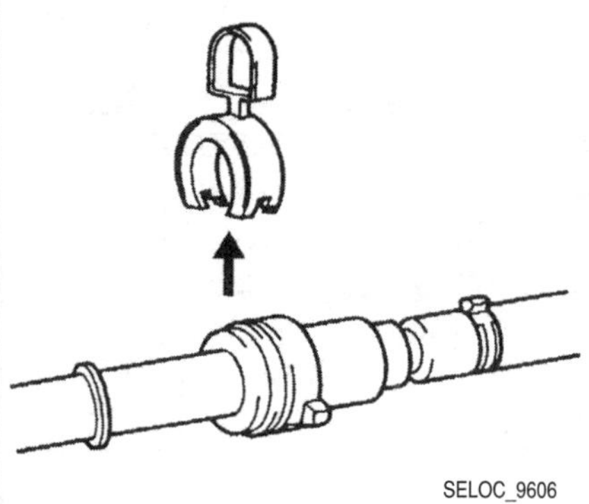

Fig. 42 . . .which can be removed once the connector is secure - 2002-08 models

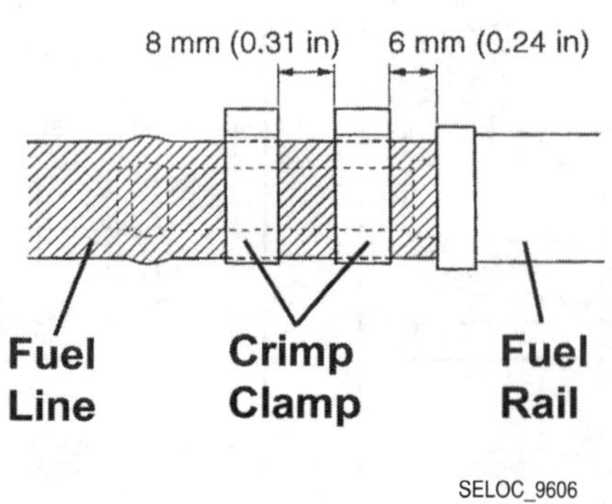

Fig. 43 Clamp positioning at the fuel rail end - 2005-09 VX models

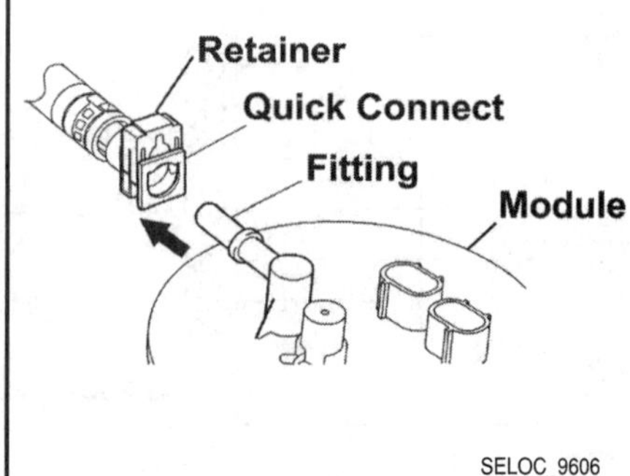

Fig. 44 A view of the new style connector used on most 2008/9 and later models

All quick connects should have 0.08-0.12 in. (2-3mm) of free-play, so carefully grab the connector and the fuel elbow/pipe and wiggle them (push/pull, not side-to-side) until you can see that the free-play is within spec.

If you are unable to get the free-play into spec, disconnect everything and check the O-ring for damage or improper positioning.

Once done with everything, you can pull off the checker tab.

Although Yamaha does not provide any advice for connecting the fuel line to the fuel rail on FX models, they are pretty specific when it comes to the VX models.

Slide 2 new crimp clamps over the fuel rail end of the fuel line and back along the hose. Slide the end of the line over the fuel rail fitting until it is on as far as it will go. Make sure the white dot is still facing upward. Slide the clamps toward the fuel rail until you are sure they are both over the rib in the fitting and then evenly space them between the rib and the rail. The 1st clamp should be 0.24 in. (6mm) back from the fitting flange on the rail, while the 2nd clamp should be 0.31 in. (8mm) outboard (behind) that one. Use the special crimping pliers and crimp them until they are tight.

On 2008-11 FX/FZ models w/1812cc engines and all 2010-11 VX models:

Lay a shop towel over the connection and then pull out on the retainer ends (tabs) while pulling up at the same time to release the retainer and then pull the connector off of the fitting. Do this slowly please!

Cover the fuel pipe/elbow and the connector end of the fuel line with a clean plastic bag. Always check the condition of the connector O-ring, replacing if necessary.

■ **When installing the fuel line onto the module fitting (or fuel rail for that matter) always make sure that the small white dot is facing upward on VX models or sideways, lined up with the notch or projection on the quick connectors.**

When installing a new quick connect on the fuel line of VX models, slide the fuel line onto the connector so that the small white mark on the fuel line is facing upward. Slide 2 new crimp clamps over the end of the fuel line and back along the hose. Slide the ends of the line over the quick connect fitting until it is on as far as it will go. Make sure the white dot is still facing upward. Slide the clamps toward the fuel rail until you are sure they are both over the rib in the fitting and then evenly space them between the rib and the connect. The inside edge of the 1st clamp should be 0.67 in. (17mm) back from the fitting flange on the rail, while the *outside* edge of the 2nd clamp should be 1.36 in. (34.5mm) from the flange. Use the special crimping pliers and crimp them until they are tight (0.04 in./1mm or less) and 180° apart.

On the fuel rail end of the line for VX models w/1052cc engines, simply follow the instructions given in the 2002-08 procedures.

When installing a new connector on the fuel line of FX and FZ models, Yamaha is not quite as specific; simply stating that the crimps should be 180° apart and the gap between the crimp should be no more than 0.4 in. (1mm).

Coat the fuel pump module fitting lightly with engine oil and then slide the quick connect onto the pipe. Push the retainer down into position until it clicks.

Although there is no mention of it on FX or FZ models, on VX models, all quick connects should have 0.08-0.12 in. (2-3mm) of free-play, so carefully grab the connect and the fuel elbow/pipe and wiggle them (push/pull, not side-to-side) until you can see that the free-play is within spec.

If you are unable to get the free-play into spec, disconnect everything and check the O-ring for damage or improper positioning.

■ **Although VX models w/1052cc engines still use the crimp clamps on the fuel rail end of the fuel line; FX, FZ and VX models w/1812cc engines utilize the new style quick connector on the fuel rail end as well as the module end.**

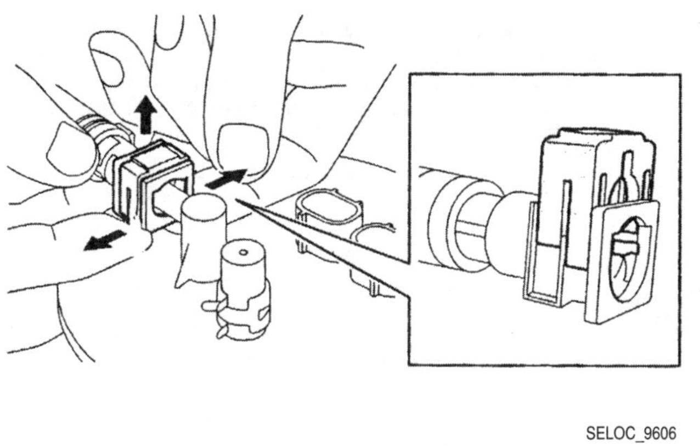

Fig. 45 Pull out the retainer tabs while pulling up at the same time

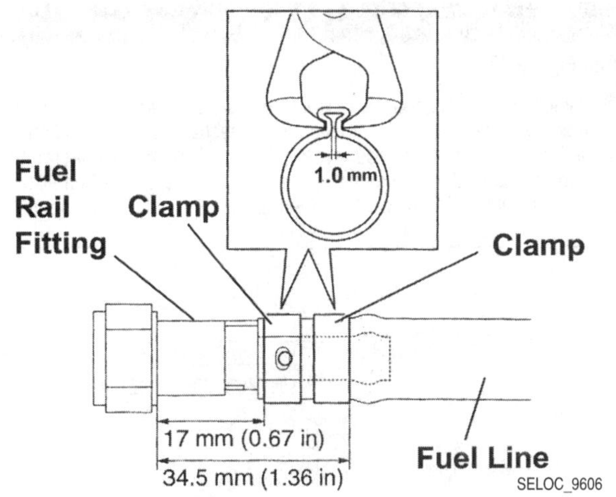

Fig. 46 Clamp positioning at the fuel pump module end - 2010-11 VX models

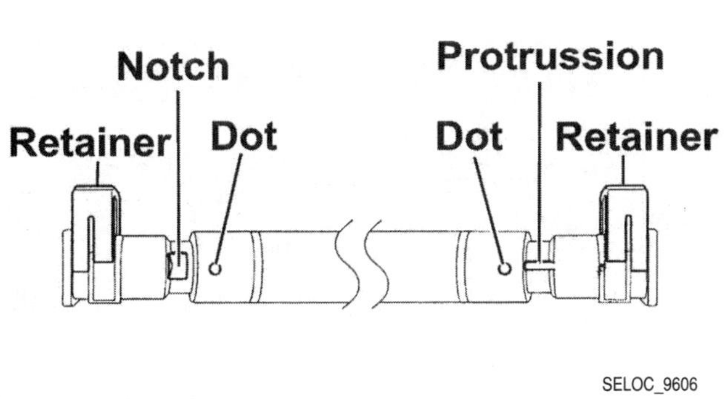

Fig. 47 2008/9 and later FX/FZ models use a quick connect at both ends of the fuel line

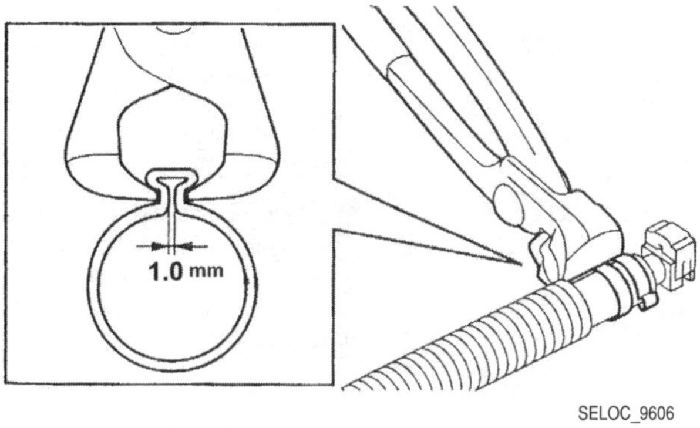

Fig. 48 On 2008/9 and later FX/FZ models, make sure the crimp gap is no more than 0.4 in. (1mm)

ELECTRONIC FUEL INJECTION SYSTEM

General Information

Beginning in 2002, Yamaha introduced the first of their EFI equipped 4-stroke PWCs.

The systems used on these motors are essentially all versions of a manifold-injection system with minor differences from model-to-model. In all cases, these systems improve low speed smoothness, reduce maintenance, improve durability, and increase fuel efficiency while lowering harmful engine emissions.

On all motors an engine control computer known as the Engine or Electronic Control Module (ECM) is capable of adjusting fuel delivery and ignition timing over a wide range to match engine operation. Unlike most carburetor-equipped engines, fuel-injected engines will automatically adjust for changes in altitude, air temperature, and barometric pressure. This precise control over engine operation results in crisp throttle response and maximum efficiency over a broad range of conditions.

■ **Yamaha *sometimes* calls the ECM a TCI unit. However, since this unit does more than just control ignition, we think that is a little misleading and will refer to it as the ECM (the more commonly used term for fuel and ignition control units/models) throughout this repair guide. However, knowing that Yamaha MIGHT call it that on some motors could be handy when dealing with a parts counterperson.**

■ **Although we have included the normal fuel system-related switches and sensors within this Fuel System section, there are many more covered under Switches & Sensors within the Ignition & Electrical section.**

Fuel Injection Basics

◆ **See Figure 49**

Fuel injection is not a new invention. Even as early as the 1950s, various automobile manufacturers experimented with mechanical-type injection systems. There was even a vacuum tube-equipped control unit offered for one system! This might have been the first "electronic fuel injection system." Early problems with fuel injection revolved around the control components. The electronics were not very smart or reliable. These systems have steadily improved since. Today's fuel injection technology, responding to the need for better economy and emission control, has become amazingly reliable and efficient. Computerized engine management, the brain of fuel injection, continues to get more reliable and more precise.

Components needed for a basic computer-controlled system are as follows:

• A computer-controlled engine manager, which is the Electronic Control Module (ECM), with a set of internal maps to follow (you know, "if this, then do that" type of information for the fuel and ignition systems).

• A set of input devices to inform the ECM of engine performance parameters.

• A set of output devices. Each device is controlled by the ECM. These devices modify fuel delivery and timing. Changes to fuel injection and timing are based on input information matched to the map programs.

This list gets a little more complicated when you start to look at specific components. Some fuel injection systems may have twenty or more input devices. On many systems, output control can extend beyond fuel and timing. The Yamaha Fuel Injection Systems provide more than just the basic functions, but they are still straight forward in layout. There are generally somewhere between 7 and 20 input devices and generally between eight and ten output controls (depending upon the system, year and model). The accompanying diagrams show the typical input and output devices of the various fuel injection systems, but the Wiring Diagrams found in the Ignition & Electrical System section will confirm the specific devices that you should expect to find on given models.

There are several fuel injection delivery technologies in wide use around the world and the manufacturers have chosen one major form of them for these motors. A brief discussion of the various types (even those not used by Yamaha) would be helpful in understanding these systems.

Throttle body injection is relatively inexpensive and was used widely in early automotive systems. This is usually a low pressure system running at 15 PSI or less. Often an engine with a single carburetor was selected for throttle body injection. The carburetor was recast to hold a single injector and the original manifold was retained. Throttle body injection is not as precise or efficient as port injection.

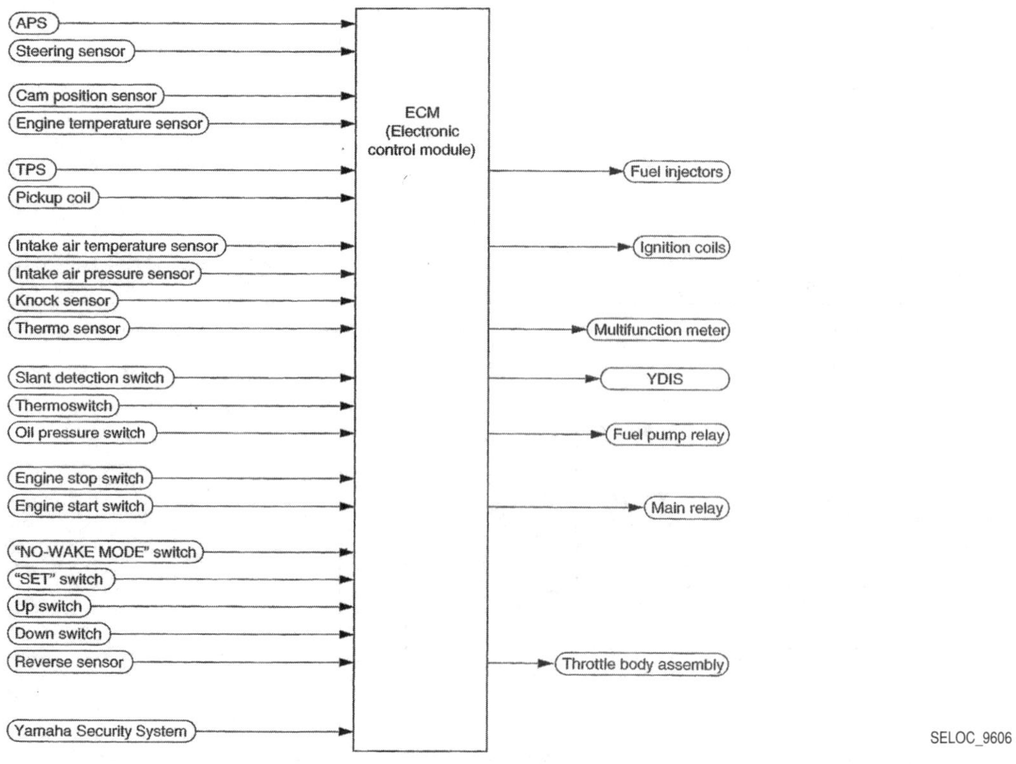

Fig. 49 Schematic of inputs and outputs for a typical Yamaha EFI system

Multi-port fuel injection is defined as an engine that uses one or more electrically activated solenoid injectors. Multi-port injection generally operates at higher pressures than throttle body systems. The Yamaha EFI systems operate with a system pressure of somewhere in the 45-53 psi range depending upon the year and model.

Another type of Multi-port fuel injection used predominantly on 2-stroke marine engines (but also found on automotive and marine diesels) is Direct Injection, which is called so because fuel is introduced from the injector DIRECTLY into the combustion chamber (as opposed to somewhere on the intake tract).

No matter whether the fuel is injected to the manifolds, behind valves, or directly into the combustion chamber, the real advantage of a fuel injection system is the precise control afforded the ECM by the use of high response electronic components. This gives electronic fuel injectors are great advantage over mechanical injectors when it comes to emissions and operating efficiency. Port injectors can be triggered two ways. One system uses simultaneous injection. All injectors are triggered at once. The fuel "hangs around" until the pressure drop in the cylinder pulls the fuel into the combustion chamber. This type of system loses some of the advantage that is possible with an electronic fuel injector. The second type is more precise and follows the firing order or SEQUENCE of the engine. Each cylinder gets a squirt of fuel precisely when needed. If you haven't already guessed, this second type of fuel injection is generally known as sequential multi-port fuel injection.

The Yamaha EFI 4-stroke system is a tuned port, sequential method of fuel injection. The injectors are mounted to a common fuel rail and either the intake manifold or cylinder head.

Like most marine fuel injection systems made during this timeframe, the 4-stroke EFI system does not contain an oxygen sensor, therefore all engine fuel mapping decisions are based upon a combination of factors including engine speed from the pulser/pick-up coil (which provides crankshaft position information), throttle opening from the throttle position sensor (TPS) and intake air from the intake air temperature (IAT) and intake air pressure (IAP) sensors. In this way the ECM keeps track of just how much air is entering the engine (between the TPS and the temperature/pressure of the air) and can adjust air/fuel mixture ratios to suit operating conditions. The lack of an oxygen sensor makes this form of injection open-loop only (meaning the ECM receives no feedback on the combustion process).

In fuel injection terms this type of system is known as a speed-density injection system. This is because the basic engine fuel mapping decisions are made by the ECM based on a comparison between the engine rpm (speed) and the air pressure (air density). The basic, pre-programmed, fuel mapping is then modified based upon input from the remaining sensors. Specifically, engine and air temperatures are taken into account.

Like most fuel injection systems, the Yamaha EFI 4-stroke system normally incorporates a return-to-port capability or limp-home mode. If there is a major sensor failure that prevents the ECM from processing one or more signals, the engine is run on a minimum performance map. In most instances the ECM will substitute a fixed value for the missing sensor signal. Usually the only input that will shut the engine down completely is battery voltage. If the battery is disconnected or the battery voltage falls below a certain point, the fuel pump quits pumping so the engine stops.

Major components of this system include the ECM and sensors along with the fuel delivery circuits. The electronic control sensors include:

- Throttle position sensor (TPS)
- Thermo sensor (sometimes called engine cooling temp/ECT or engine temp/ET)
- Intake air temperature (IAT) sensor
- Intake air pressure (IAP) sensor
- Water temperature sensor (WTS)
- Oil pressure sensor or switch
- Pulser/pick-up coil
- Camshaft position sensor
- And a host of other sensors/switches depending on the year and model

Air Filter Case, Cover Or Housing

REMOVAL & INSTALLATION

2002-08 FX Models With 998/1052cc Engines

② MODERATE

◆ **See Figures 50 thru 57**

1. Carefully pop the release and lift off the seat(s). Disconnect the battery cables.
2. Remove the 2 mounting bolts in the center of the air filter case. Slide each of the case cover slide bars (8) out toward the nearest corner, and lift the cover off of the filter case.
3. Lift out the air filter element.
4. Remove the throttle body assembly as detailed in the Throttle Body Assembly procedures.
5. Reach under the back (port) corner of the case and back the breather hose clamp off before pulling the hose from its connection. Secure the clamp to the hose (or store somewhere you won't forget) and position it out of the way.
6. Loosen the mounting bolt and remove the plastic band attached to the bottom of the case on the port side.
7. Trace the fuel line to where it attaches to the fitting on the front side of the housing and then disconnect the lighting coil, pulser coil and engine thermo-switch connectors. MAKE SURE you take the time to tag each connector for installation.
8. Now move to the back side of the cover and disconnect the exhaust thermo-switch (thermo-sensor on 2008 Calif. HOs), camshaft position sensor and wiring harness connectors. Yep, you remember, MAKE SURE you take the time to tag each connector for installation.
9. Remove the 3 case mounting bolts; 1 on the starboard side and 2 on the port.
10. Swivel the case slightly to gain access to the harness bracket on the rear corner. Remove the 2 mounting bolts and the pull out the bracket and sub-wiring harness.
11. If you haven't already done so, disconnect the fuel line at the front of the case. Remove the 2 bolts and lift out the fuel line bracket, before removing the wiring harness bracket bolt and bracket. Finally, remove the grommet if it stayed with the housing when you removed the harness bracket.
12. Pop (carefully) the 4 joint seals out of the bottom of the filter case. Loosen the hose clamps holding the 4 throttle body joints to the intake manifold and lift out the joints.
13. Remove the plastic band and then the mounting bolts (2) at the port side case brackets. If the sleeve(s) and grommet(s) do not come with the bracket, make sure they do not fall out of the bolt holes in the case.

■ **Yamaha does not show the grommets on the HO models, but we suspect you will find one there.**

14. Repeat the previous step for the starboard case bracket and then lift out the air filter case.
15. Remove the 2 cover bolts in the case and lift out the cover along with the gasket. Throw the gasket away.

To install:

16. Clean the filter case and allow it to dry thoroughly.
17. If equipped, and/or removed, position the cover gasket (*new*) in the bottom of the case after it has dried and then install the cover, tightening the bolts to 16 inch lbs. (1.7 Nm).
18. If the grommets came out of the case cover mounting bolt holes when you removed the support brackets, pop them back into their respective holes (a little soapy water will help if you have any problems) and then press in the sleeves from the bottom until the lip seats on the bottom of the grommet.
19. Position the starboard bracket under the case mounting bolt hole (recessed side toward the case) and then thread in the bolt until it is just snug - do not tighten the bolt yet.
20. Do the same with the 2 port brackets, flat side upward and the recessed side facing away from the case.
21. Press each of the throttle body joints onto their respective intake runners until they bottom out and then slide the clamps into position, being careful of the screw positioning so you will be able to get a screwdriver on them when tightening.
22. Position the filter case over the engine and mounting bracket. Align the starboard mounting hole grommet over the bracket and thread the mounting bolt in finger tight.
23. Reach down and swivel the throttle body joints so that the bolt holes and alignment pins are all lined up with the case and then tighten each of the clamps securely.
24. Now you can thread in the case mounting bolts on the 2 port brackets until they are finger tight.
25. Remove the 3 *bracket* mounting bolts and coat the threads with Loctite 271. Thread each bolt back into place and tighten them to a 1st pass of 11 ft. lbs. (15 Nm). Once all three have been tightened, go back and tighten again to 28 ft. lbs. (39 Nm).

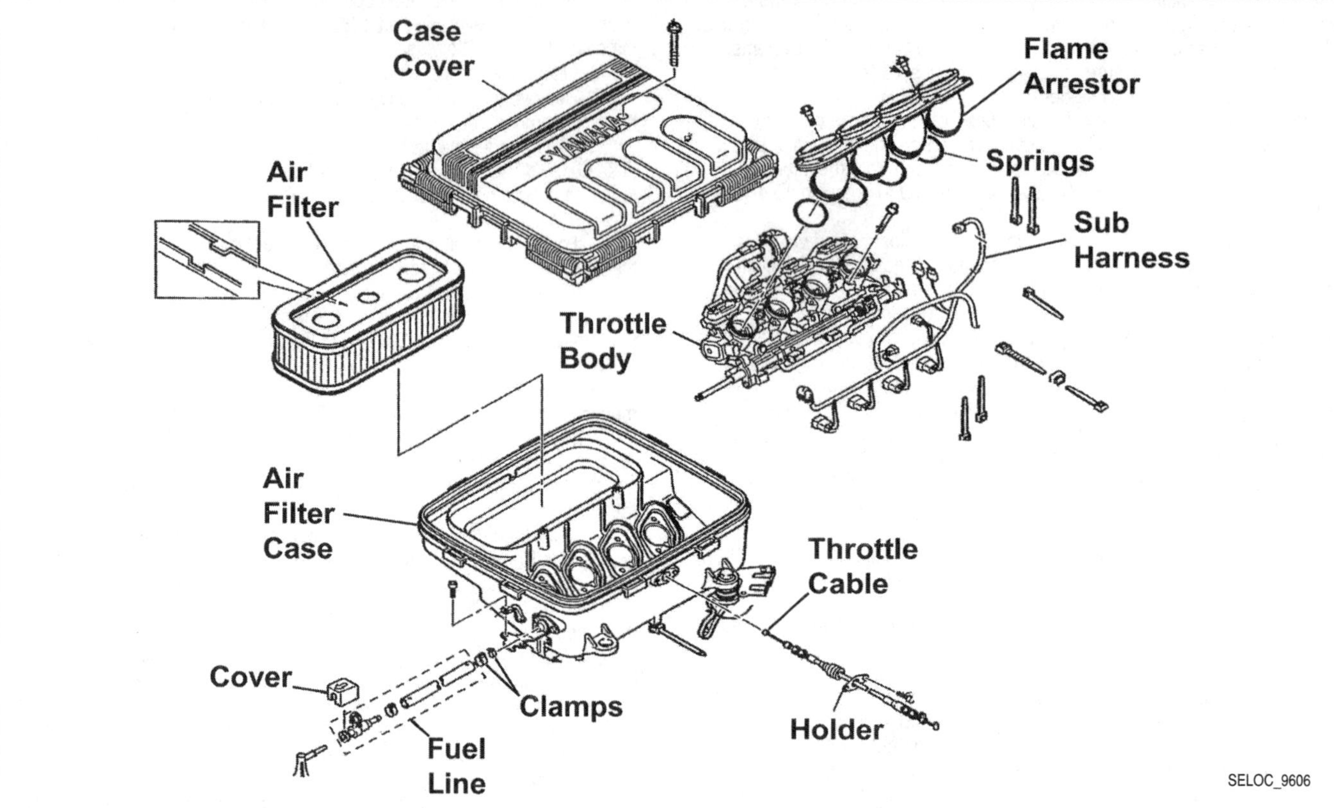

Fig. 50 Exploded view of the air filter case and cover assembly - FX/FX14, FX Cruiser/FX140 Cruiser and 2008 California FX HO/FX Cruiser HO models

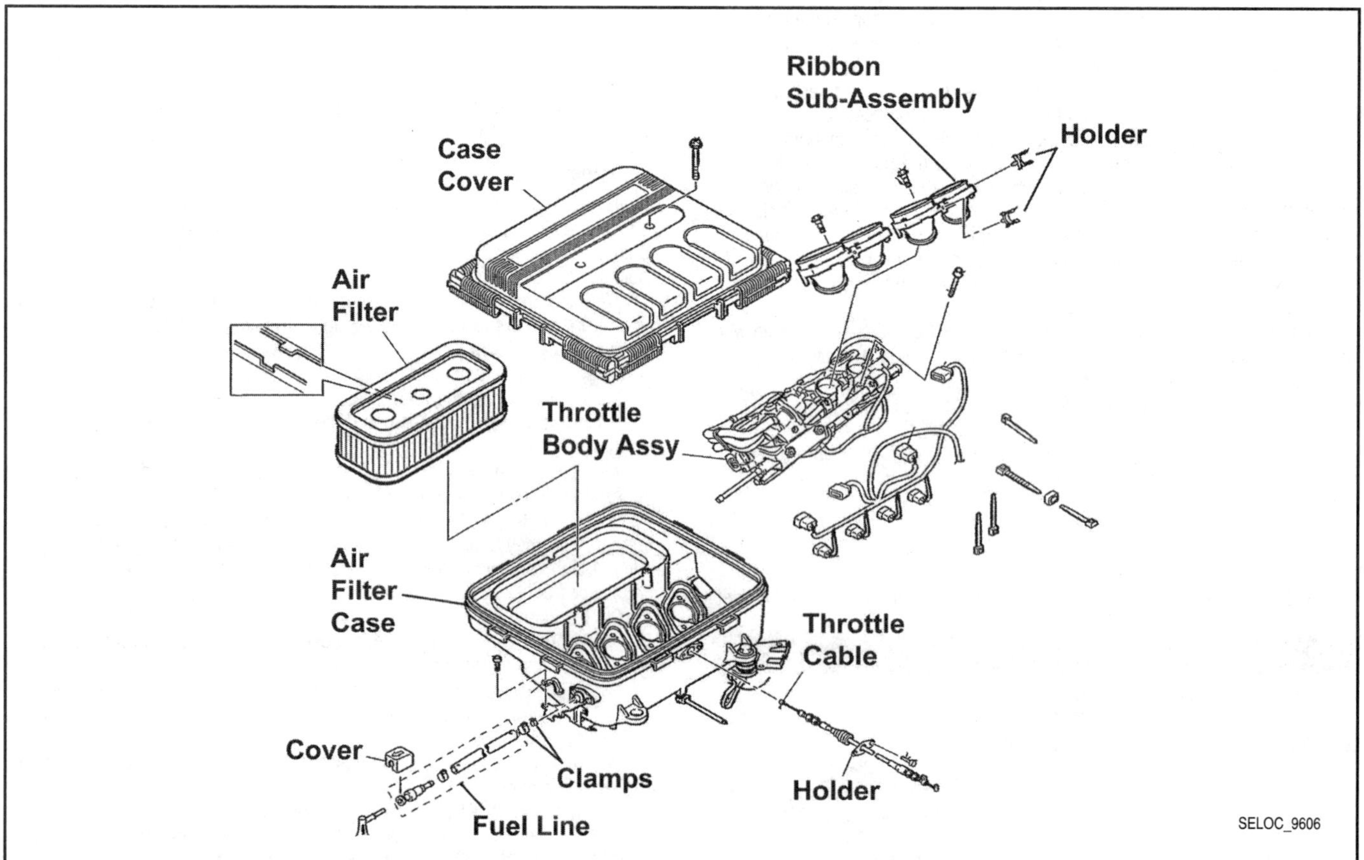

Fig. 51 Exploded view of the air filter case and cover assembly - 2004-08 FX HO/FX Cruiser HO models (exc. California engines)

26. Loosen the case mounting bolts and lift off the case. Position it upside down on a clean surface and then press the upper lip of a seal into the groove in the bottom of the case, making sure the tab in the seal fits into the notch in the case. Repeat this for the other 3 seals and then lay the case back in position over the engine.

27. Feed the sub harness into the hole in the bottom of the case (at the back) so the retainer holes line up and then position the harness bracket, coat the bolt threads with Loctite 572 (271 on 2008 models, or equivalent thread locking compound), tightening the 2 bolts to a 1st pass of 2.7 ft. lbs./32 inch lbs. (3.8 Nm) and then to a final pass of 5.5 ft. lbs./66 inch lbs. (7.6 Nm). On 2008 models, skip the 1st pass and just tighten them to 5.5 ft. lbs.

28. Reconnect the connectors/couplers, remember? You tagged them? Exhaust thermo-switch/thermo-sensor, cam position sensor and harness connector. Press any connectors into their holders.

29. Back at the front of the case, press the fuel line grommet into position in the recess in the case. Slide on the harness bracket and then the fuel line bracket. Coat the bolt threads with Loctite 242 or equivalent, (271 on 2008 models) and tighten the 2 bolts to a 1st pass of 2.7 ft. lbs./32 inch lbs. (3.8 Nm) and to a final pass of 5.5 ft. lbs./66 inch lbs. (7.6 Nm). On HO models, just tighten them to the final figure without the 1st pass.

30. Feed the fuel line in, connect it to the grommet and then reconnect the connectors for the lighting/stator coil, pulser/pick-up coil and engine thermo-sensor connectors. Press any connectors into their holders.

31. By now, you should have most everything hooked up and lined up right? Remove the 3 case mounting bolts and coat their threads with Loctite 242 or equivalent. Thread them back in and tighten each bolt to a 1st pass of 6.4 ft. lbs./77 inch lbs. (8.8 Nm) and then a final pass of 13 ft. lbs. (18 Nm).

32. Connect the plastic band to the bottom of the case on the port side and tighten the bolt to 2.5 ft. lbs./30 inch lbs. (3.4 Nm).

33. Push the breather hose back onto the nipple and slide the clamp into position.

34. Install the throttle body assembly as detailed in the appropriate section.

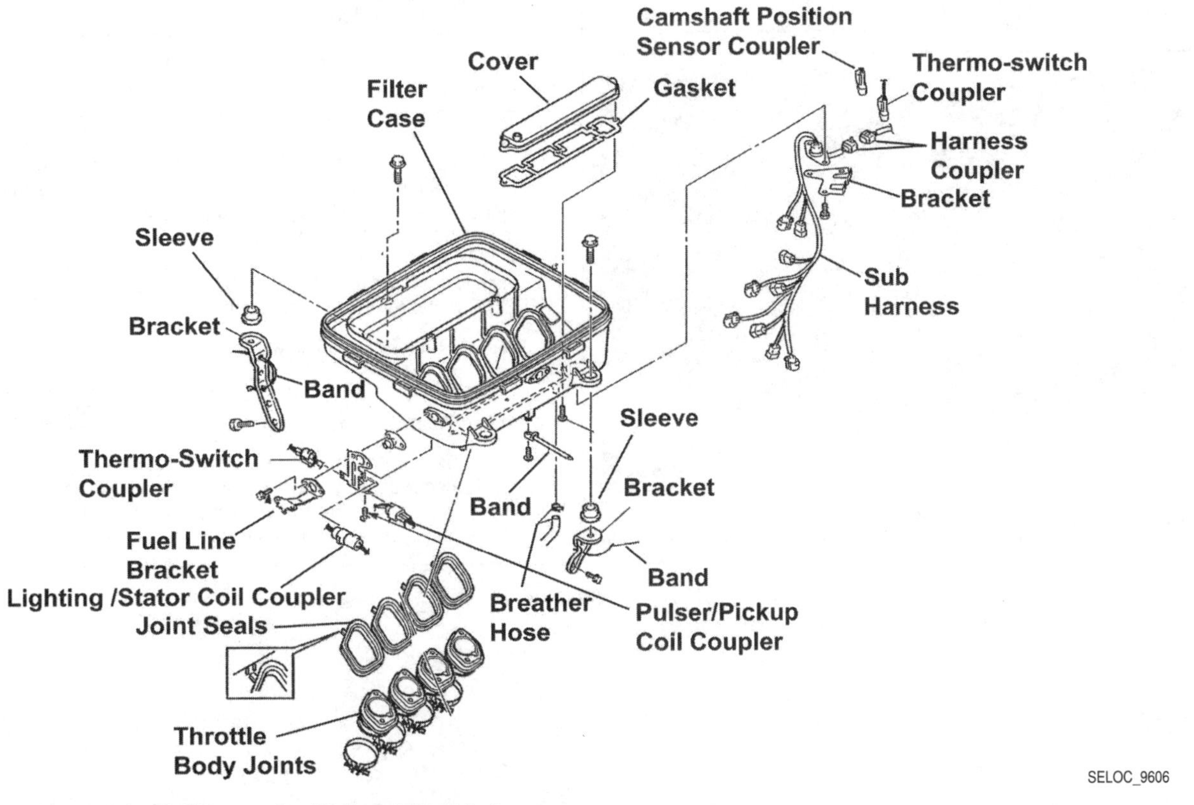

Fig. 52 Exploded view of the air filter case itself (all models similar)

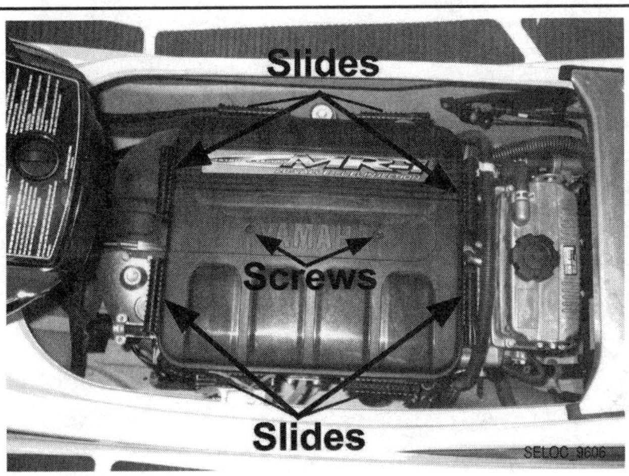

Fig. 53 Remove the 2 mounting bolts. . .

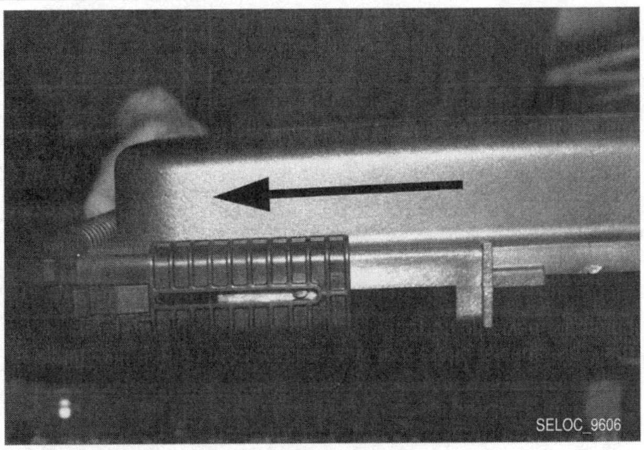

Fig. 54 . . .and then move the 8 slide bars to the nearest corner before lifting off the case cover

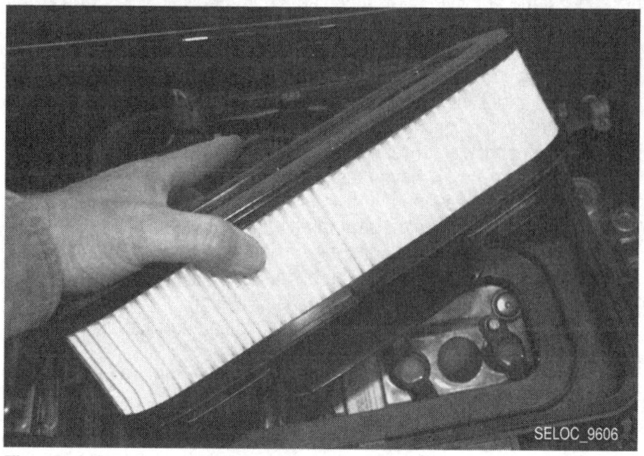

Fig. 55 Lift out the air filter element

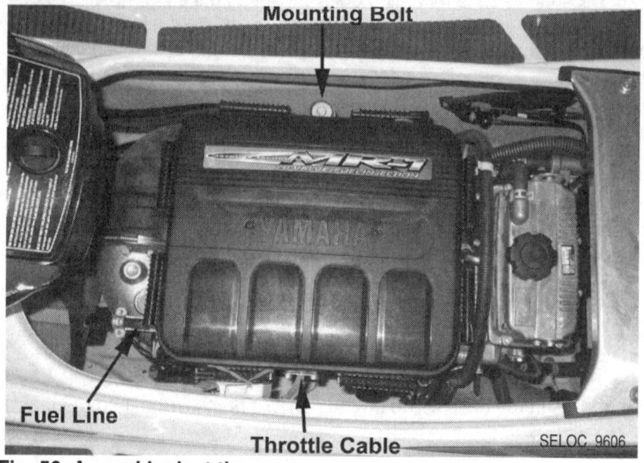

Fig. 56 A good look at the case

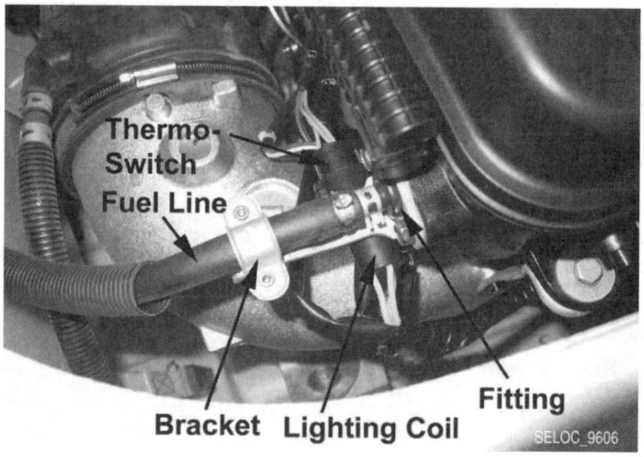

Fig. 57 Fuel line connection and couplers

35. Position the air filter back into the housing making sure that the small tab on the filter fits into the slot in the case housing. Do not force it as it will only go in one way. If you have a problem, just spin it around 180° and attempt it again.

36. Position the cover over the case making sure everything lines up and then move the slide bars away from the corner until it hits the stop. Be sure that each of the slide bars is moved all the way to each corner *before* dropping the cover into position.

37. Coat the 2 mounting bolts with Loctite 572, or equivalent, and tighten them to 1.8 ft. lbs./22 inch lbs. (2.5 Nm).

38. Install the seat(s) and reconnect the battery cables.

2008-11 FX/FZ/VX Models w/1812cc Engines

EASY

◆ See Figures 58, 59 and 60

1. Remove the seat and then disconnect the battery cables.

2. Remove the 2 mounting bolts and lift off the engine cover. The cover is a press-fit over its support bracket ends, so carefully pop it off of the connectors by pulling back and upward at the same time.

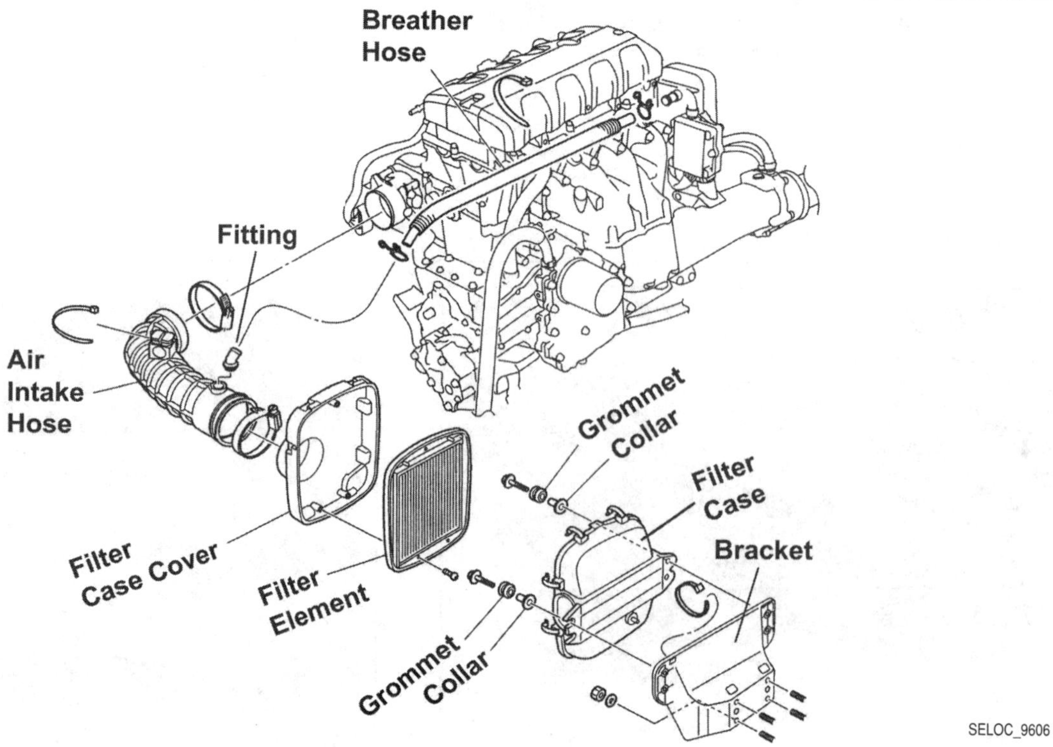

Fig. 58 Exploded view of the filter assembly on 1812cc FX/FZ models (HO shown, SHO, FZ and VX similar)

3. Remove the fuel filler cap and release the quick connect at each end of the fuel line as detailed in the Fuel Lines section and remove the line. You will have to disconnect it from the clamp on the air intake duct.

4. Squeeze the pinch clamp on the breather hose where it attaches to the air intake pipe and wiggle the hose off of the connection. Carefully move the hose out of the way.

5. Take a careful look at the air filter case and tag/disconnect any other lines or hoses that might be in the way of access.

6. Loosen the hose clamp on the end of the intake pipe and slide it back on the pipe. Grasp the pipe where it connects to the filter case cover and wiggle it off of the lip.

7. Reach down and pop off the compression bands holding the cover to the case.

8. Pull the cover away from the case and carefully lift out the cover and filter.

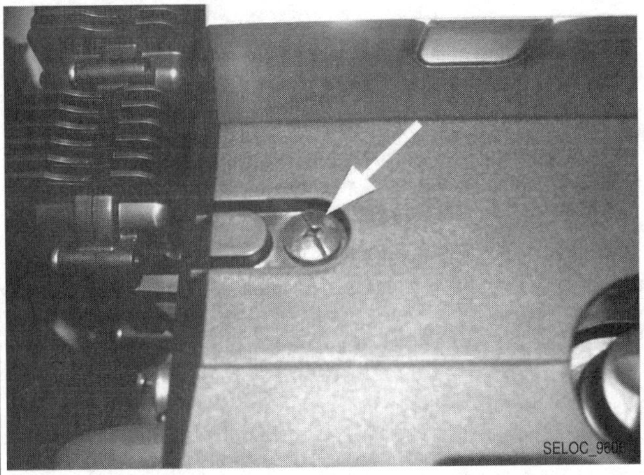

Fig. 59 Remove the 2 mounting bolts before popping the cover off of its brackets

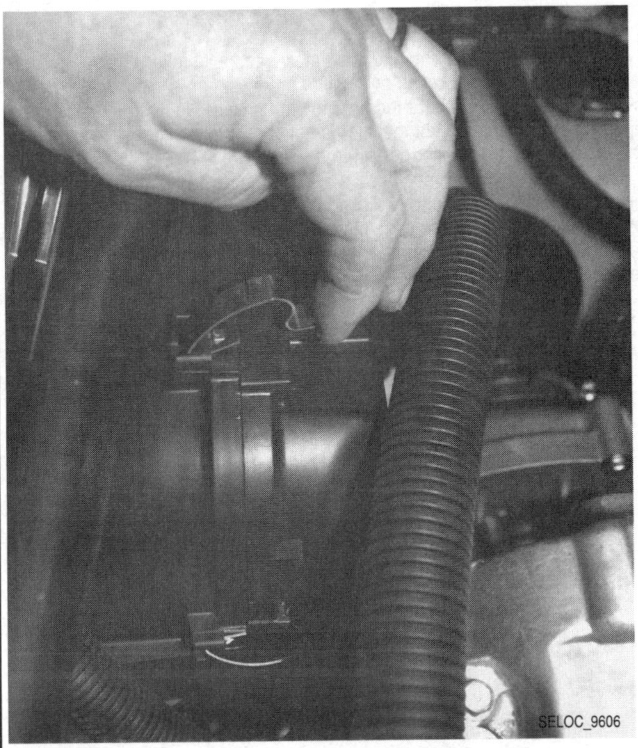

Fig. 60 Pull back on the compression bands to release the cover latches

9. Remove the 3 mounting bolts (2 bottom, 1 top) and separate the filter from the cover.

10. Remove the 4 mounting bolts and lift out the filter case. Be careful not to lose the collars and grommets.

To install:

11. Position the filter case on the bracket with the collar and grommet between the two. Thread in the bolts and tighten them to 8.9 ft. lbs./106 inch lbs. (12 Nm) on FX HO, FZ and VX models, or 9.6 ft. lbs./115 inch lbs. (13 Nm) on FX SHO models.

12. Inspect, clean or replace the filter and then position it back into the cover. Tighten the 3 bolts securely, but not so tight as to warp the filter frame.

13. Slide the cover/filter back into position so it is sitting in the lip of the filter case and then pop the clamps back into place over the cover.

14. Slide the end of the intake pipe onto the cover. Move the clamp back into position inside the ribs on the pipe and tighten it to no more than 2.2 ft. lbs. (3 Nm).

15. Reconnect any hoses or line you chose to move for access.

16. Push the breather hose over the connector and pinch the clamp back into position.

17. Install the fuel line, routing it through the holder on the air hose.

18. Connect the battery cables.

19. Position the engine cover over the mounting bars and pop it into place. Tighten the 2 bolts to 3.7 ft. lbs. (5 Nm).

20. Install the seat.

VX Models w/1052cc Engines

① EASY

◆ **See Figures 61 thru 66**

■ **You do not have to remove the entire air filter case to remove the filter.**

1. Carefully pop the release and lift off the seat(s).

2. Remove the water separator.

3. Loosen the fuel filler cap to relieve pressure and then position some rags underneath the fuel line where it connects to the fuel rail. Cut the 2 pinch clamps so you can wiggle the line off the nipple. Plug the line and move it out of the way. You can find more information on this in the Fuel Lines section.

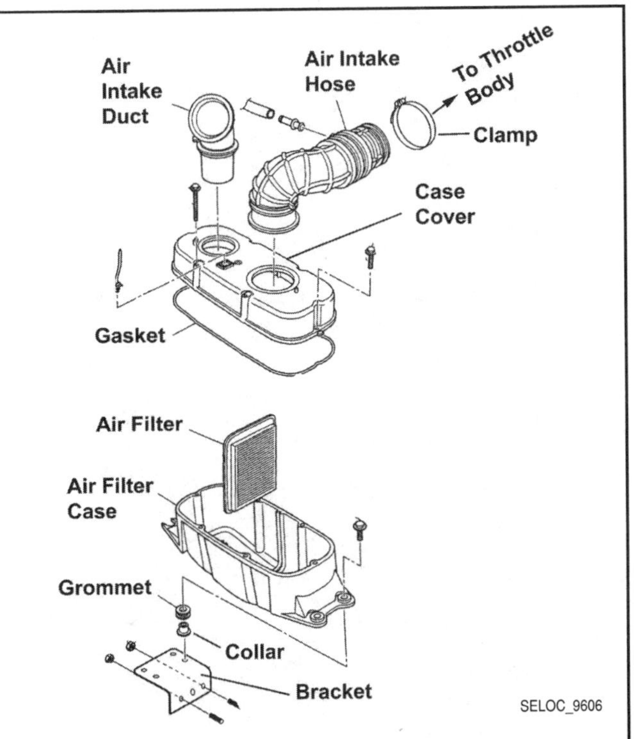

Fig. 61 Exploded view of the air filter case and cover - 2002-09 VX models, 2010-11 w/1052cc engines similar

Fig. 62 Disconnect the lines and hoses and move them out of the way. . .

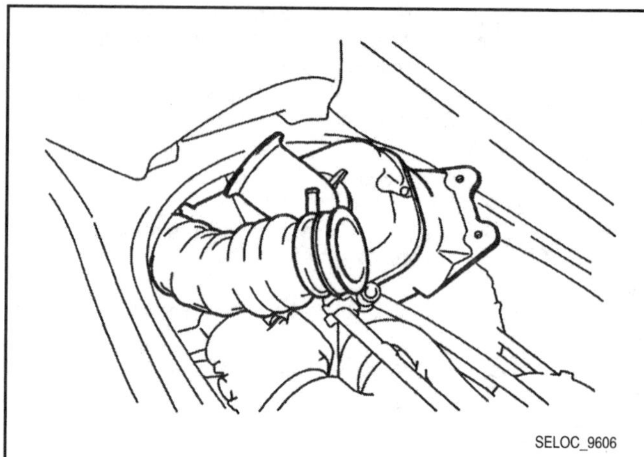

Fig. 63 . . .before lifting out the filter case

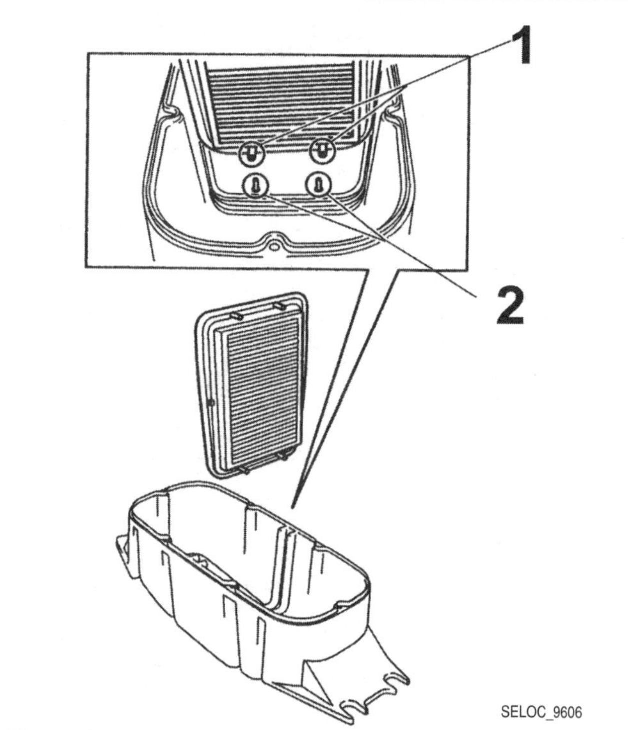

Fig. 64 Make sure that the projections (1) on the filter faces away from those on the filter case (2)

4. Pinch the oil separator breather hose clamp and slide it off the nipple on the air intake hose. Position it out of the way.

5. Loosen the hose clamp on the air intake hose and carefully slide it off of the throttle body.

6. Snip the 4 plastic ties over of near the filter case; remember where they came from.

7. Remove the 4 hex head mounting bolts and lift out the entire air filter case. Remove the collars/grommets and inspect them for cracks or other damage. You will probably have to swivel the intake hose toward the starboard side to ease removal.

8. Remove the 6 case cover mounting bolts (4 short and 2 long) and pop off the cover (with the air intake hose and duct still attached). Remove the seal if it didn't come off with the cover.

9. Pull the air filter up and out of its tracks in the case. Carefully check it for dirt, oil or other contaminants. Check for tears and other damage.

10. You can clean the filter if you like using cold water (no detergent!), but make sure that is dries *completely* before installation.

■ **Do not use compressed air to speed the drying of the filter element.**

To install:

11. Slide the air filter into the case making sure it stays in its tracks and the projections on the bottom of the filter are facing away (toward the port side) from the projections at the bottom of the case. We're sure you've figured this out already, but the filter is not completely square, so make sure the tapered end of the filter goes into the case.

12. Inspect the cover seal for cracks or other damage, replacing it as necessary - we recommend replacing each time. Press the seal into the groove and then install the case cover. Coat the 6 mounting bolts with Loctite 572 and thread them all in until finger tight. Tighten the long bolts to 0.9 ft. lbs/11 inch lbs. (1.2 Nm) and the shorter ones to 1.1 ft. lbs./13 inch lbs. (1.5 Nm) on 2005-09 models, or all bolts to 13 inch lbs. (1.5 Nm) on 2010-11 models.

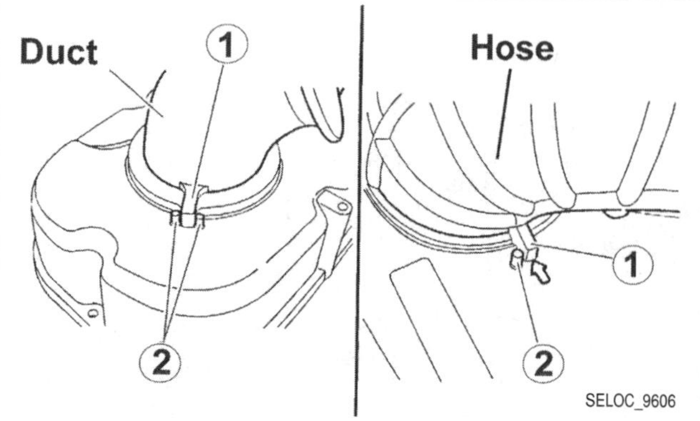

Fig. 65 Make sure that the projections (1) on the air hose and duct align with those on the filter case (2)

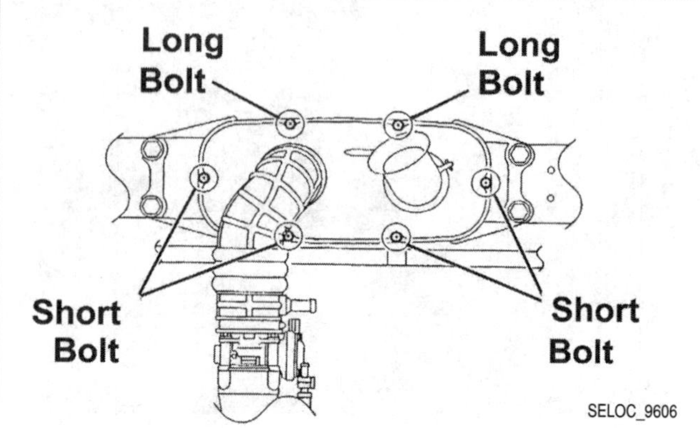

Fig. 66 The case cover has 2 different size bolts

13. Install the collars/grommets and then position the case/cover into the engine compartment and install the mounting bolts, tightening them to 12.5 ft. lbs. (17 Nm).
14. Install the water separator if removed.
15. Reconnect the air intake hose to the throttle body and tighten the hose clamp securely.
16. In the event that the air intake hose or duct came disconnected from the housing cover, make sure that the projections on each are aligned with the projections on the cover when reconnecting them.
17. Connect the fuel line (remember the plug) and breather hose, tightening their respective clamps securely. Clamp placement details can be found in the Fuel Lines section
18. Install 4 new plastic ties.
19. Install the seat.

Fuel Rail & Injectors

REMOVAL, DISASSEMBLY & INSTALLATION

FX/FX140 & FX Cruiser/FX140 Cruiser Models With 998/1052cc Engines

② MODERATE

◆ **See Figure 67**

1. Remove the air filter case cover and filter as detailed in the Maintenance & Tune-Up section or in the Air Filter Case, Cover Or Housing procedures.
2. Remove the throttle body assembly as detailed in the appropriate section. Set the assembly carefully on a suitable, clean work surface.
3. Although not necessary, you may wish to remove the bypass valve motor for extra room.
4. Although this should already have been done when removing the assembly, tag and disconnect any vacuum lines still connected to the fuel rail.
5. Remove the 3 rail mounting bolts and then pull out their spacers from under the fuel rail.
6. Carefully grab the fuel rail with both hands and then slowly wiggle it back and forth (side to side, not front to back) until you feel the injectors come free from their recesses in the throttle body.
7. Each injector rides in a grommet that should have stayed on the end of the injector but may have been left in the throttle body; remove each one and throw them away.
8. Grab each injector and wiggle it back and forth while pulling on it until it pops out of the rail. Repeat for the other 3 injectors.
9. *Carefully* remove the O-rings from the end of each injector. Once off, throw them away, *never* re-use them. Check the O-ring groove for signs of damage.

■ **Be very careful when removing the injector O-rings - scratching or damaging the groove will necessitate injector replacement.**

10. Remove the mounting bolt from the bracket carrying the intake air pressure (IAP) and intake air temperature (IAT) sensors and lift it off of the fuel rail along with the sensors.
11. Working on the forward end of the rail, remove the 2 mounting bolts and pull off the fuel pipe. If the O-ring doesn't come off with the fitting, pull it out of the rail and discard it.
12. Remove the 2 harness bracket mounting bolts at the other end of the fuel rail and pull off the bracket.

To install:

13. Clean the fuel rail thoroughly with solvent and allow it to dry thoroughly - use compressed air if at all possible. Inspect the fuel rail for cracks and other damage, replacing as necessary.
14. Position the harness bracket on the aft end of the fuel rail so the tabs are facing away from the injector bores and then thread in the mounting bolts, tightening to 2.5 ft. lbs./30 inch lbs. (3.5 Nm).
15. Slide a new O-ring over the inner portion of the fuel pipe until it seats against flange and then slide the pipe into the forward end of the rail. Tighten the mounting bolts to 2.5 ft. lbs./30 inch lbs. (3.5 Nm).
16. Position the IAT/IAP bracket on the rail and thread in the lower mounting bolt just enough to hold up the bracket.

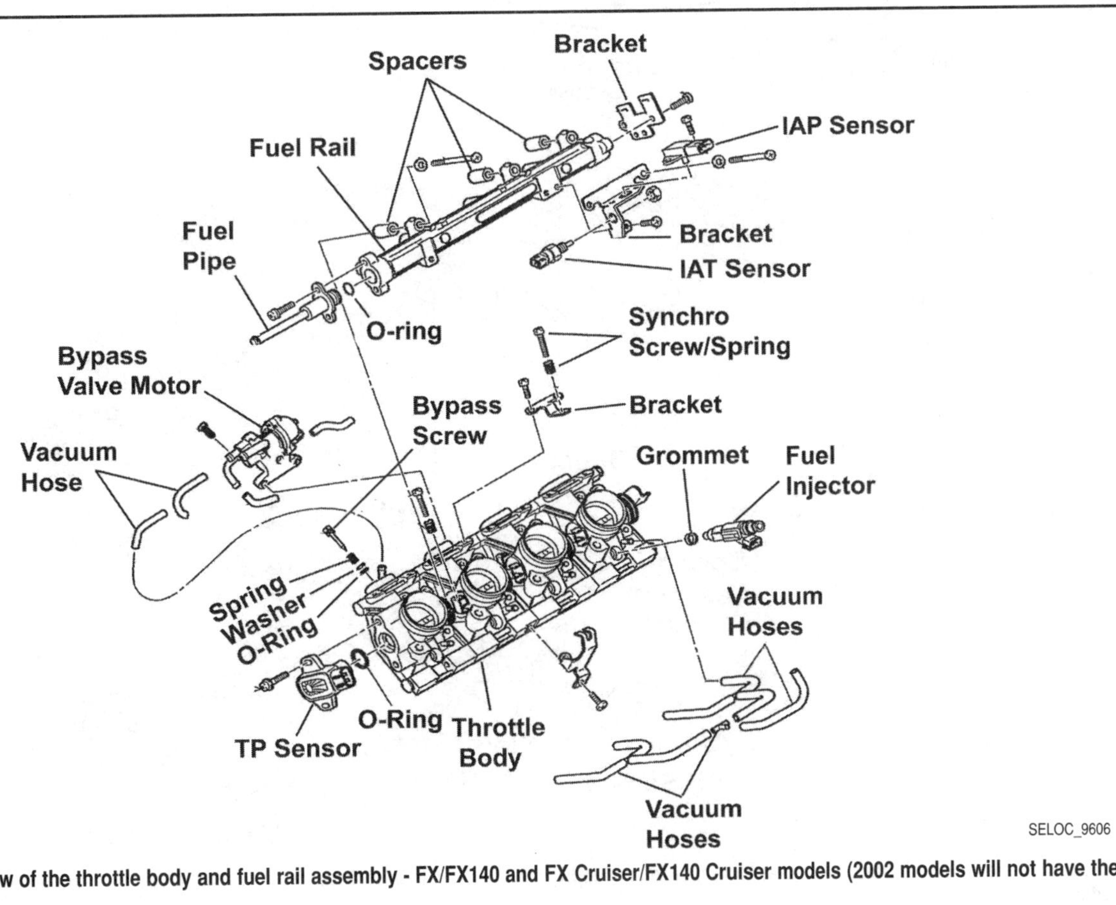

Fig. 67 Exploded view of the throttle body and fuel rail assembly - FX/FX140 and FX Cruiser/FX140 Cruiser models (2002 models will not have the bypass valve motor)

17. Coat your *new* injector O-rings lightly with clean engine oil and then carefully slide them into position in their groove on the top side of the injector.

18. Position each injector into its bore in the fuel rail and wiggle it carefully and slowly into position, making sure that the electrical connector is facing up and toward the Port side. Be very careful not to nick or twist the O-rings during assembly.

19. Now coat your new grommets lightly with clean engine oil and press them into position on the bottom of each injector.

20. Position the fuel rail and injectors over the throttle body so that the ends of each injector line up with their respective bores in the throttle body. Carefully press the rail/injectors into the bores until they are fully seated.

■ **Always be very careful to not bump or damage the injector during assembly.**

21. Slide the spacers in between the rail mounting bolt bosses and the throttle body - you may have to move the fuel rail assembly slightly to fit them in. Move the sensor bracket (remember you haven't tightened the bottom bolt yet?) so the holes line up over the rear 2 mounting bolt holes on the rail and then thread in the 2 bolts with their washers. Install the forward mounting bolt.

22. Confirm that the injectors are all still seated correctly and then tighten the 3 fuel rail mounting bolts to 3.6 ft. lbs. (5 Nm). Now you can tighten the lower bracket bolt to the same torque.

23. If you had removed either of the 2 sensors, tighten the IAT sensor nut to 11 ft. lbs. (15 Nm) and the IAP sensor bolt to 2.5 ft. lbs./30 inch lbs. (3.5 Nm).

24. Install the throttle body assembly.

25. Install the air filter case, filter and case.

2004-08 FX HO/FX Cruiser HO Models With 1052cc Engines

◆ See Figures 68, 69 and 70

② MODERATE

1. Remove the air filter case cover and filter as detailed in the Maintenance & Tune-Up section or in the Air Filter Case, Cover Or Housing procedures.

2. Remove the throttle body assembly as detailed in the appropriate section. Set the assembly carefully on a suitable, clean work surface.

3. Although not necessary, you may wish to remove the bypass valve motor for extra room.

4. Although this should already have been done when removing the assembly, tag and disconnect any vacuum lines still connected to the fuel rail.

5. Remove the 4 rail mounting bolts.

6. Carefully grab the fuel rail with both hands and then slowly wiggle it back and forth (side to side, not front to back) until you feel the injectors come free from their recesses in the throttle body.

7. Each injector rides in a grommet that should have stayed on the end of the injector but may have been left in the throttle body; remove each one and throw them away.

8. Grab each injector and wiggle it back and forth while pulling on it until it pops out of the rail. Repeat for the other 3 injectors.

9. *Carefully* remove the O-rings from the end of each injector. Once off, throw them away, *never* re-use them. Check the O-ring groove for signs of damage.

■ **Be very careful when removing the injector O-rings - scratching or damaging the groove will necessitate injector replacement.**

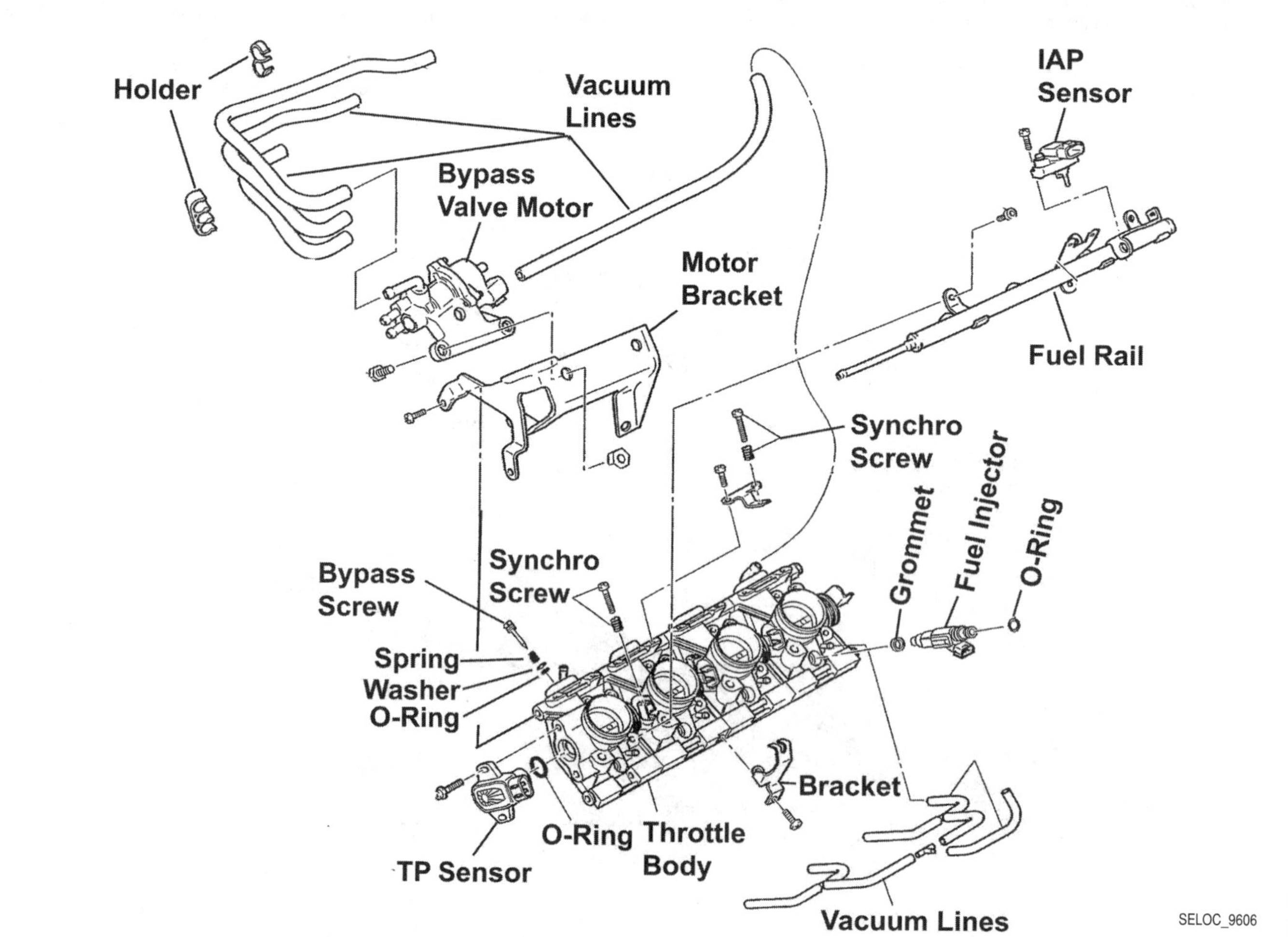

Fig. 68 Exploded view of the throttle body and fuel rail assembly - 2004-08 FX HO/FX Cruiser HO models (2008 IAP sensor looks different on Calif. Models)

10. Remove the 2 mounting bolts from the intake air pressure (IAP) sensor and lift it off of the fuel rail.

11. Remove the nut and pull the IAT sensor from the flange on the rail.

To install:

12. Clean the fuel rail thoroughly with solvent and allow it to dry thoroughly - use compressed air if at all possible. Inspect the fuel rail for cracks and other damage, replacing as necessary.

13. Position the IAT sensor through the flange and tighten the nut to 11 ft. lbs. (15 Nm).

14. Position the IAP on the back of the rail and thread in the mounting bolt, tightening it to 2.5 ft. lbs./30 inch lbs. (3.5 Nm).

15. Coat your *new* injector O-rings lightly with clean engine oil and then carefully slide them into position in their groove on the top side of the injector.

16. Position each injector into its bore in the fuel rail and wiggle it carefully and slowly into position, making sure that the electrical connector is facing up and toward the Port side. Be very careful not to nick or twist the O-rings during assembly.

17. Now coat your new grommets lightly with clean engine oil and press them into position on the bottom of each injector.

18. Position the fuel rail and injectors over the throttle body so that the ends of each injector line up with their respective bores in the throttle body. Carefully press the rail/injectors into the bores until they are fully seated.

■ **Always be very careful to not bump or damage the injector during assembly.**

19. Confirm that the injectors are all still seated correctly and then tighten the 4 fuel rail mounting bolts to 3.6 ft. lbs./44 inch lbs. (5 Nm).

20. Install the throttle body assembly.

21. Install the air filter case, filter and case.

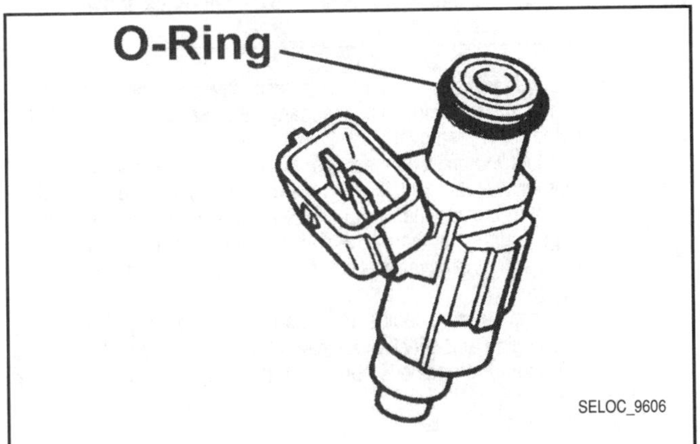

Fig. 69 Make sure you coat the O-ring with clean engine oil

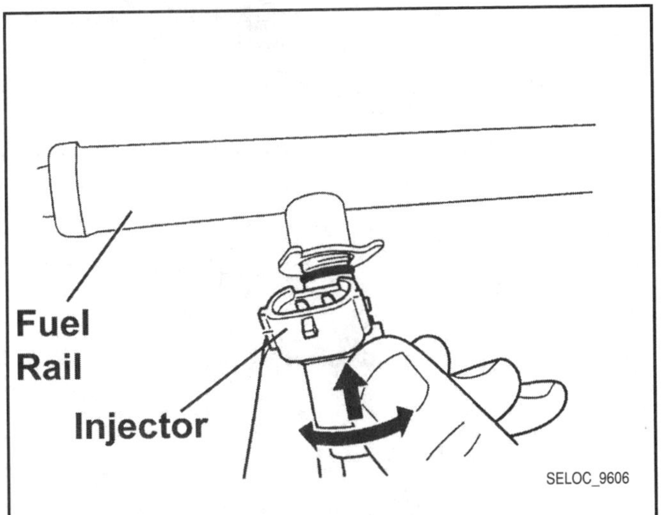

Fig. 70 Wiggle the injector when seating it into the rail

2008 & Later FX/FZ/VX Models w/1812cc Engines

MODERATE

◆ **See Figures 71 thru 75**

■ **Yamaha has initiated a Factory Modification Campaign (TSB #WR2008-004R) for a faulty fuel line connector on 2008 FX SHO and FX Cruiser SHO models and certain 2009 FX SHO and FX Cruiser SHO (HIN #F1W-809301 thru 815100) that could cause the line to come loose at the fuel rail. The campaign provides a new fuel line assembly with a different connector and an additional clamp.**

1. Remove the seat(s) and then disconnect the battery cables - removing the cargo tray if necessary.

2. Remove the 2 mounting bolts and lift off the engine cover.

3. Reach into the front of the compartment and pull the quick connect retainer apart and upward, releasing the connector at the fuel pump module as detailed in the Fuel Lines section. Make sure you have a container in position because there will most likely be some residual fuel in the line.

**** CAUTION**

Interestingly, Yamaha does not provide a fuel pressure relief procedure like most other manufacturers. Instead, they suggest that you remove the fuel tank cap and then slowly disconnect the fuel line quick connector while covering it with a clean rag. That's it!

4. Trace the fuel line back to the rail and then remove the quick connector here as you just did at the module. Have some rags handy - the fuel should already have been drained, but you never know. Remove the fuel line after disconnecting it from the holder on the air intake hose.

5. Tag each of the fuel injector connectors, press the clip in and disconnect each coupler from the injector. Do this for each of the 4 injectors. Carefully grasp the harness and pop each clamp out of the fuel rail bracket, carefully moving the harness out of the way.

6. Remove the 3 mounting bolts and then gently grab the rail, pulling upward while slowly rocking it back and forth (side-to-side, not fore-to-aft). The injectors should come out with the rail, but may stick in the manifold.

7. Each injector rides in a grommet/gasket that should have stayed on the end of the injector but may have been left in the manifold; remove each one and throw them away.

8. Grab each injector and wiggle it back and forth while pulling on it until it pops out of the rail (or manifold). Repeat for the other 3 injectors, keeping track of which injector goes where.

9. *Carefully* remove the O-rings from the rail end of each injector. Once off, throw them away, *never* re-use them. Check the O-ring groove for signs of damage.

■ **Be very careful when removing the injector O-rings - scratching or damaging the groove will necessitate injector replacement.**

To install:

10. Clean the fuel rail thoroughly with solvent and allow it to dry thoroughly - use compressed air if at all possible. Inspect the fuel rail for cracks and other damage, replacing as necessary. Do the same with the fuel line and the injectors.

11. Check the small filter screen in the top of the injector, cleaning *very* carefully if necessary.

12. Coat your *new* injector O-rings lightly with clean engine oil and then carefully slide them into position in their groove on the top side of the injector.

13. Position each injector into its bore in the fuel rail and wiggle it carefully and slowly into position, making sure that the electrical connector is facing up. Be very careful not to nick or twist the O-rings during assembly.

14. Now coat your *new* grommets lightly with clean engine oil and press them into position on the bottom of each injector.

15. Position the fuel rail and injectors over the intake manifold so that the ends of each injector line up with their respective bores. Carefully press the rail/injectors into the bores until they are fully seated.

■ **Always be very careful to not bump or damage the injector during assembly.**

16. Confirm that the injectors are all still seated correctly and then tighten the 3 fuel rail mounting bolts to 9.6 ft. lbs./116 inch lbs. (13 Nm).

17. Press each harness clamp back into their respective slots in the

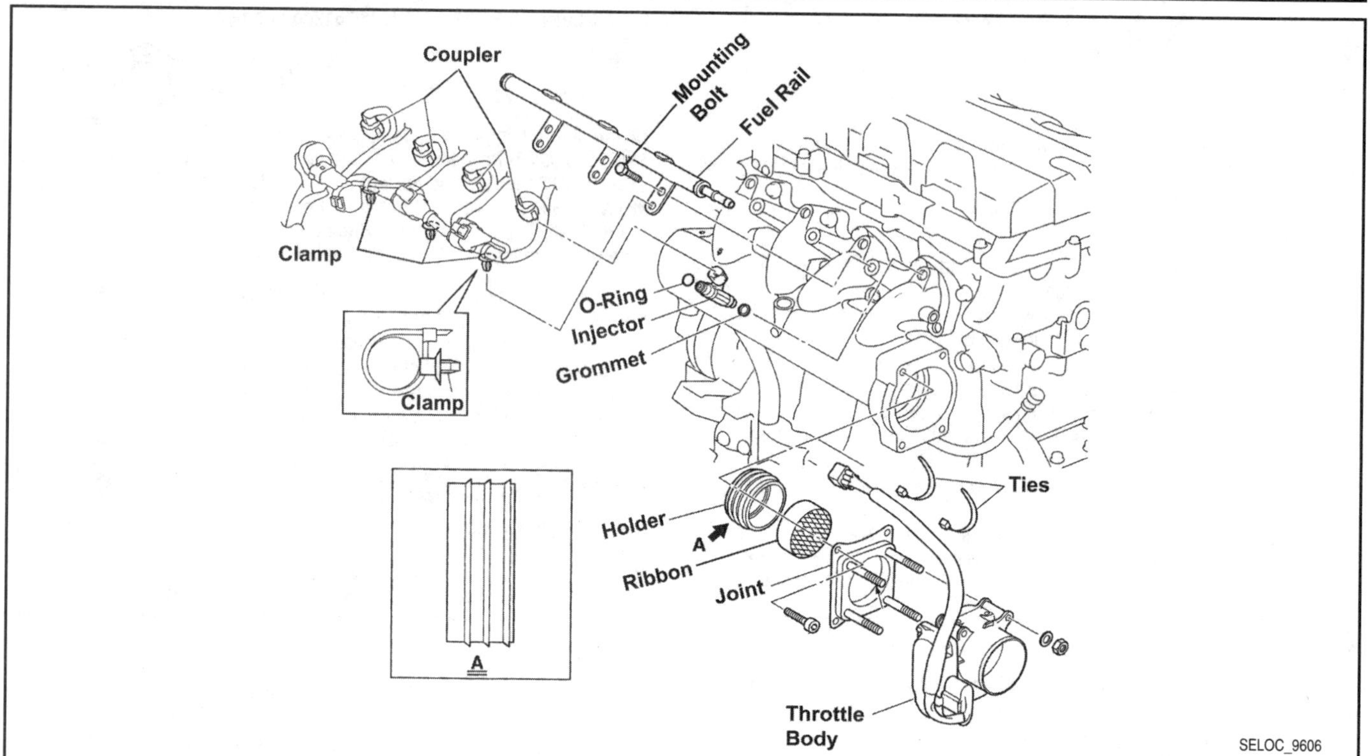

Fig. 71 Exploded view of the throttle body and fuel rail assembly - 2009 and later FX HO/Cruiser HO and VX models w/1812cc engines shown (throttle body will be different on SHO and FZ models)

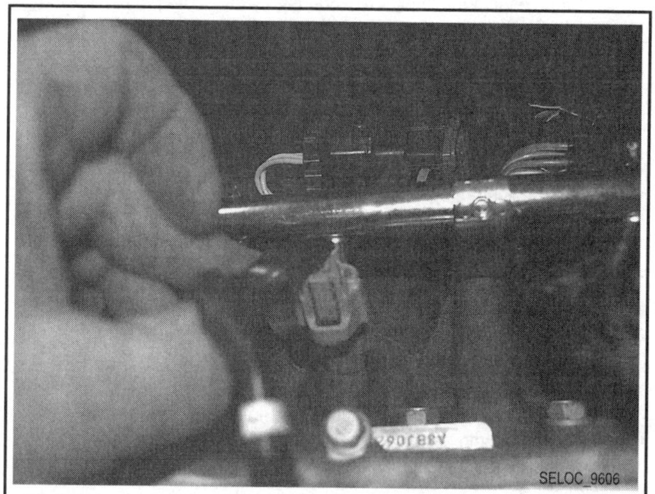

Fig. 72 A good look at the fuel injector coupler

bracket and then press each coupler into its respective injector connector. They should 'click' into position. Reconnect any other sensor couplers that you disconnected during removal.

18. Slide the fuel line quick connect onto the fuel rail fitting and click the retainer back into place. Repeat this with other end at the fuel module, making sure you get the fuel line into the clamp on the air intake hose. Please refer to the Fuel Lines procedures for more information on the quick connects.

19. Position the engine cover and tighten the 2 mounting bolts to 3.7 ft. lbs./44 inch lbs. (5 Nm).

20. Reconnect the battery cables and insert the storage tray (if removed).

21. Start the engine and check everything carefully for fuel leaks. Obviously, make sure the engine is hooked up to a water source if the unit is not in the water!

22. Replace the seat.

2005-11 VX Models w/1052cc Engines

◆ See Figures 76 thru 82

② MODERATE

1. Remove the seat(s) and then disconnect the battery cables - removing the cargo tray if necessary.

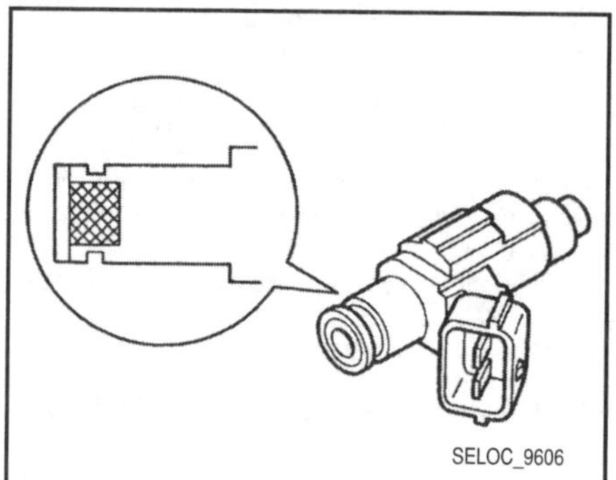

Fig. 73 Don't forget to clean the filter screen. . .

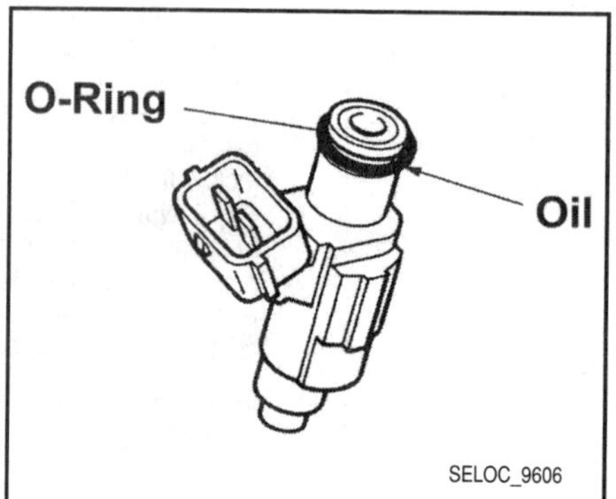

Fig. 74 . . .or coat the O-ring with clean oil

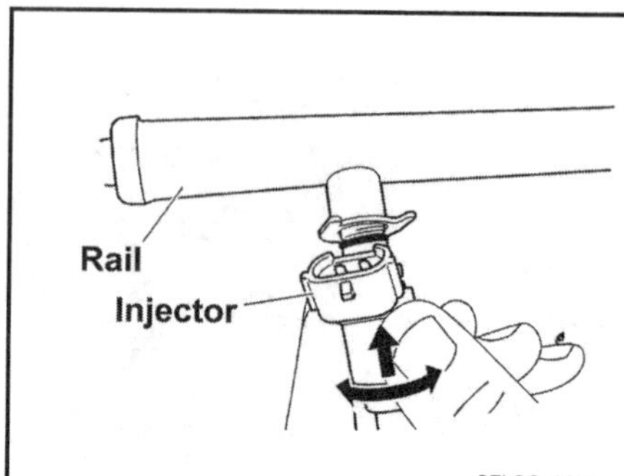

Fig. 75 You need to wiggle the injector as you press it back into the fuel rail

2. Reach into the front of the compartment and pop off the connector cover where the fuel line connects to the fuel module (2002-09 models). Disconnect the fuel line quick connector as detailed in the Fuel Lines section. Make sure you have a container in position, because there will most likely be some residual fuel in the line.

**** CAUTION**

Interestingly, Yamaha does not provide a fuel pressure relief procedure like most other manufacturers. Instead, they suggest that you remove the fuel tank cap and then slowly disconnect the fuel line quick connector while covering it with a clean rag. That's it!

3. Trace the fuel line back to the rail and then remove (cut) the 2 clamps attaching the line to the rail end. Wiggle the line from the rail and over the ridge. Have some rags handy - the fuel should already have been drained, but you never know.

4. Tag and disconnect the camshaft position sensor, sensor, oil pressure switch and exhaust thermo-sensor couplers. Move them out of the way.

5. Tag each of the fuel injector connectors, press the clip in and disconnect the coupler from the injector. Do this for each of the 4 injectors, carefully moving each harness out of the way.

6. Remove the 2 mounting bolts and then gently grab the rail, pulling upward while slowly rocking it back and forth (side-to-side, not fore-to-aft). The injectors should come out with the rail, but may stick in the manifold.

7. Each injector rides in a grommet/gasket that should have stayed on the end of the injector but may have been left in the manifold; remove each one and throw them away.

8. Grab each injector and wiggle it back and forth while pulling on it until it pops out of the rail (or manifold). Repeat for the other 3 injectors.

9. *Carefully* remove the O-rings from the rail end of each injector. Once off, throw them away, *never* re-use them. Check the O-ring groove for signs of damage.

■ **Be very careful when removing the injector O-rings - scratching or damaging the groove will necessitate injector replacement.**

To install:

10. Clean the fuel rail thoroughly with solvent and allow it to dry thoroughly - use compressed air if at all possible. Inspect the fuel rail for cracks and other damage, replacing as necessary. Do the same with the fuel line and the injectors.

11. Check the small filter screen in the top of the injector, cleaning *very* carefully if necessary.

12. Coat your *new* injector O-rings lightly with clean engine oil and then carefully slide them into position in their groove on the top side of the injector.

13. Position each injector into its bore in the fuel rail and wiggle it carefully and slowly into position, making sure that the electrical connector is facing up and toward the Port side. Be very careful not to nick or twist the O-rings during assembly.

14. Now coat your *new* grommets lightly with clean engine oil and press them into position on the bottom of each injector.

15. Position the fuel rail and injectors over the intake manifold so that the ends of each injector line up with their respective bores. Carefully press the rail/injectors into the bores until they are fully seated.

■ **Always be very careful to not bump or damage the injector during assembly.**

16. Confirm that the injectors are all still seated correctly and then tighten the 2 fuel rail mounting bolts to 9.6 ft. lbs./116 inch lbs. (13 Nm).

17. Press each harness coupler into its respective injector connector. They should 'click' into position. Reconnect the other 4 sensor couplers that you disconnected during removal.

18. Slide 2 new crimp clamps over the fuel rail end of the fuel line and back along the hose. Slide the end of the line over the fuel rail fitting until it is on as far as it will go. Make sure the white dot is still facing upward. Slide the clamps toward the fuel rail until you are sure they are both over the rib in the fitting and then evenly space them between the rib and the rail. The 1st clamp should be 0.24 in. (6mm) back from the fitting flange on the rail, while the 2nd clamp should be 0.31 in. (8mm) outboard (behind) that one. Use the special crimping pliers and crimp them until they are tight. You can see more information on this process in the Fuel Lines section.

19. Now slide the quick connector on the other end over the pump module fitting until it 'clicks' into place. Again, you can find details and illustrations in the Fuel Lines section - remember that most 2009 and later models use a different connector, so check out the appropriate section. As with the fuel rail end, make sure that the white dot on the fuel line is facing upward. Flip the cover into position.

20. Reconnect the battery cables and insert the storage tray (if removed).

21. Start the engine and check everything carefully for fuel leaks. Obviously, make sure the engine is hooked up to a water source if the unit is not in the water!

22. Replace the seat.

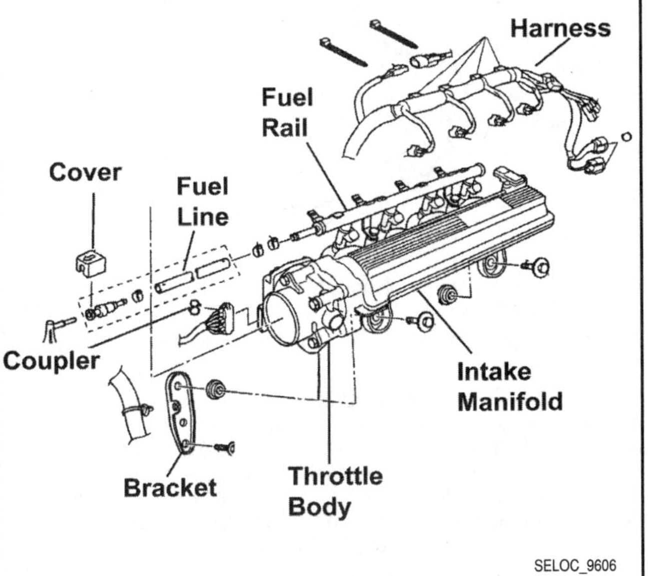

Fig. 76 Exploded view of the fuel injection system components - VX models w/1052cc engines

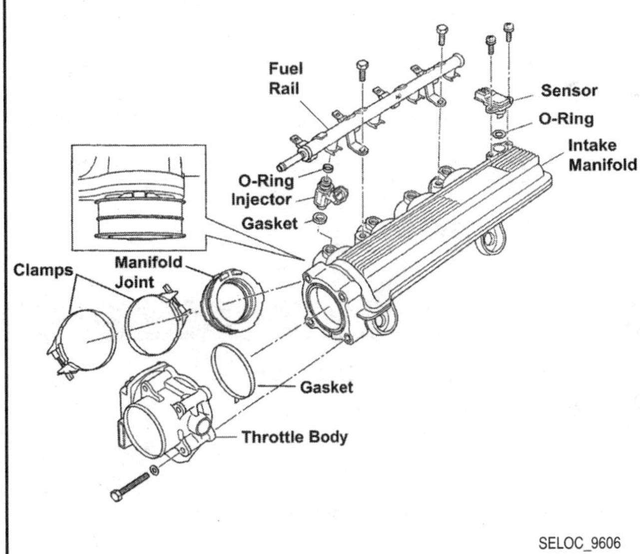

Fig. 77 Exploded view of the throttle body, intake manifold and fuel rail assembly - VX models w/1052cc engines

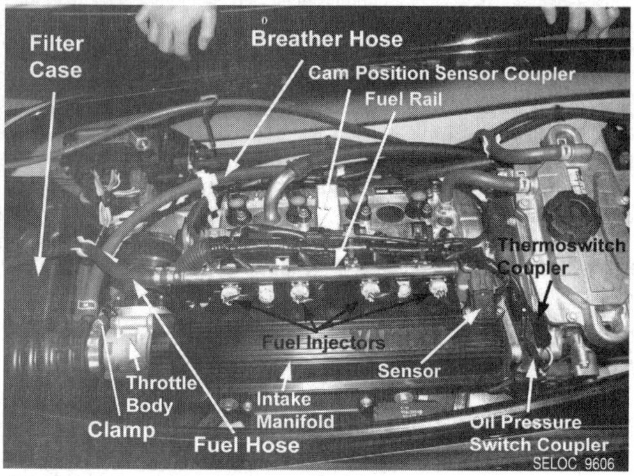

Fig. 78 A good look at the fuel injection components

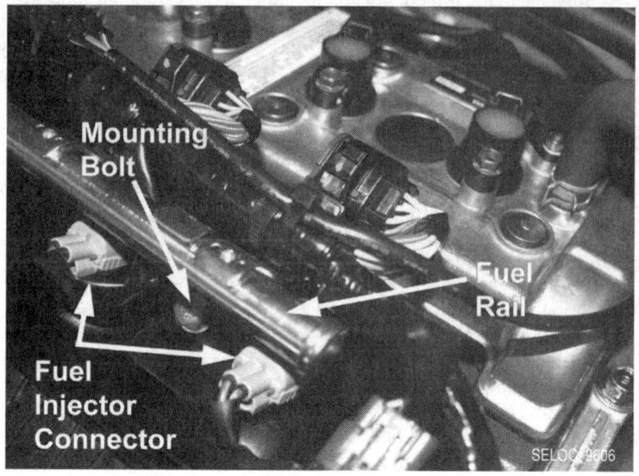

Fig. 79 Removing the fuel rail is pretty simple. . .

Fig. 80 . . .after you've disconnected the fuel line

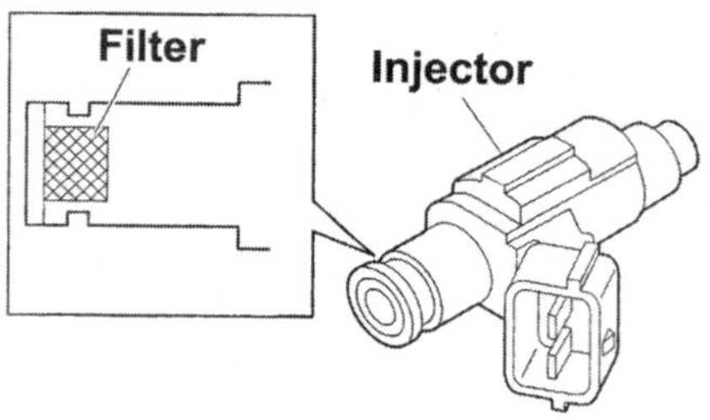

Fig. 81 Don't forget to check the injector filter

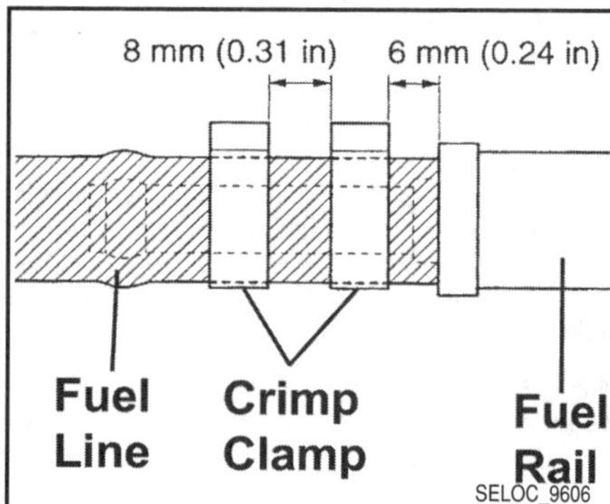

Fig. 82 Correct crimp clamp spacing on the fuel rail end of the fuel line - VX models w/1052cc engines

TESTING

MODERATE

◆ See Figure 83

Yamaha suggests checking the injectors by using the Stationary Test from the Yamaha Diagnostic system. But a quick check can be performed also if one is not available.

Remove the fuel injectors and connect a multi-meter as shown in the accompanying illustration. Resistance should be:

- FX/FX140 and FX Cruiser/FX140 Cruiser models - 14-15 ohms at 68°F (20°C)
- 2004-11 FX HO/FX Cruiser HO models - 11.5-12.5 ohms at 68°F (20°C)
- 2008-11 FX SHO/FX Cruiser SHO models - 11.5-12.5 ohms at 68°F (20°C)
- 2009-11 FZR/FZS models - 11.5-12.5 ohms at 68°F (20°C)
- 2005-11 VX models - 11.5-12.5 ohms at 68°F (20°C)

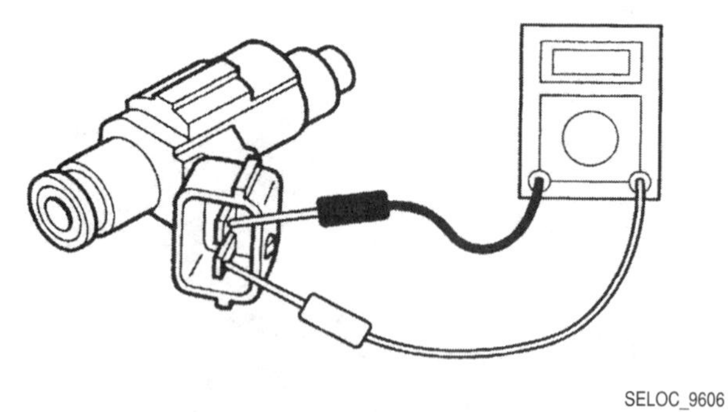

Fig. 83 Testing the fuel injector

Throttle Body Assembly

REMOVAL, DISASSEMBLY & INSTALLATION

FX/FX140 & FX Cruiser/FX140 Cruiser Models

DIFFICULT

◆ See Figures 50, 57, 67, and 84 thru 88

1. Disconnect the battery cables.
2. Remove the air filter case cover and filter as detailed in the Air Filter Case, Cover Or Housing or Maintenance & Tune-Up sections.
3. Remove the mounting bolts (4) and then lift off the flame arrestor. The springs may come off with the arrestor or stay with the throttle body; remove them either way.
4. Disconnect the throttle cable at the throttle shaft lever. While squeezing the throttle lever to hold open the throttle valves, insert a small, clean piece of wood into the bore to hold the valves open after you release the throttle lever. Wiggle the cable end out of the throttle shaft lever. Loosen the cable locknut, carefully remove the piece of wood and then lift the cable out of the bracket. Remove the 2 cable holder bolts on the side of the filter case and then pull the cable out of the filter case.
5. Remove the fuel filler cap to relieve fuel pressure and then pop off the fuel line quick connect cover at the fuel tank. Position a small container, or rag, beneath the section of fuel line running from the quick connect to the filter case to catch any residual fuel that will leak out of the line and

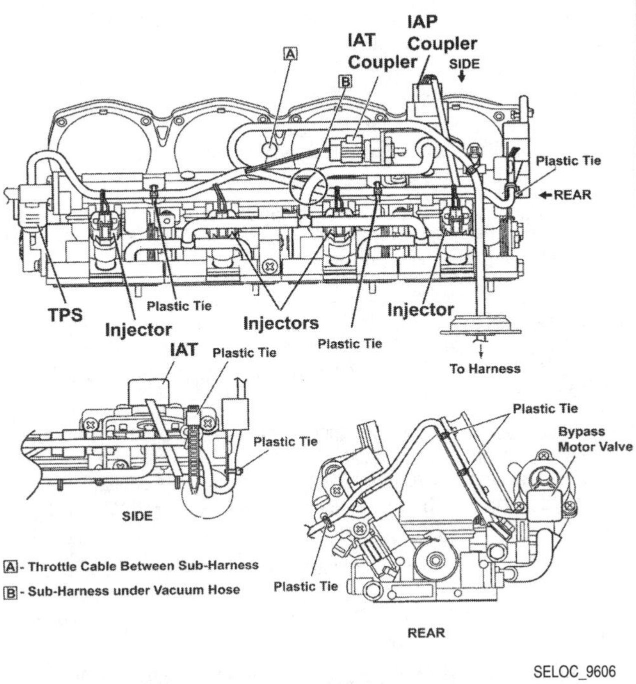

Fig. 84 A good look at the assembly with its wires and lines

connections and then disconnect the line with the quick connect as detailed in the Fuel Lines and/or Fuel Tank sections.

6. Trace the fuel line back to where it runs into the front of the filter housing. Remove the 2 bolts and lift out the fuel line bracket, before removing the wiring harness bracket bolt and bracket. You will need to cut off the outer crimp clamp (sometimes called a circle clamp) as detailed in the Fuel Lines section and then slide back the small inner pinch clamp before wiggling the line of the connecter grommet.

7. Finally, remove the grommet if it stayed with the housing when you removed the harness bracket.

■ **The previous 3 steps are recommended by Yamaha, but it seems to us that it would be a lot easier to simple relieve the fuel pressure and then disconnect the fuel line at the grommet.**

8. Tag and disconnect all electrical connectors connected to, or near, the throttle body assembly that you can reach (IAP, IAT, TP sensor couplers and also the injector and joint couplers). Tag and connect the breather hose at the MAP sensor.

9. Loosen and remove the 8 throttle body mounting bolts. Carefully grab the assembly and tilt the back end up enough to allow the fuel hose fitting on the end of the fuel rail to clear the hole in the filter case while pulling the assembly toward the back of the case. Check that there are no more connectors or lines that you may have missed in the previous step and remove the assembly. Carefully set the unit down on a clean, sturdy work bench or appropriate surface.

■ **The remaining disassembly procedures should only be performed if necessary. Otherwise, you can move to the installation procedures.**

10. Remove the fuel rail and injectors as detailed in the Fuel Rail & Injectors procedures.

Fig.85 A good look at the bypass valve motor (2007 shown)

Fig. 86 Remove the 2 mounting bolts before pulling out the throttle cable holder

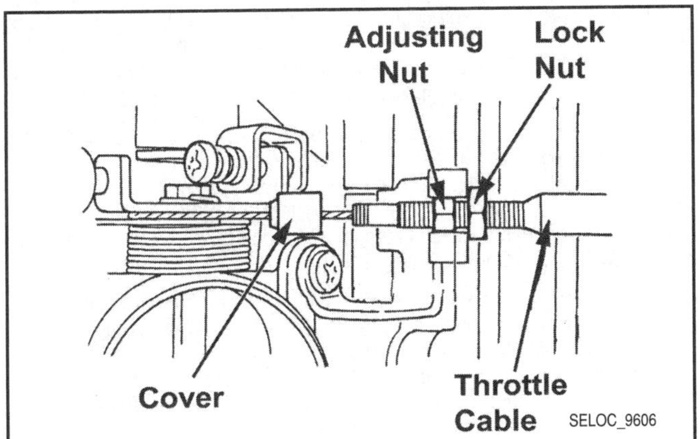

Fig. 87 A closer view of the throttle cable end

Fig. 88 Another good look at the fuel line and clamps

11. Tag and disconnect the coupler and vacuum lines (should be 4) at the bypass valve motor and then remove the 2 mounting bolts. Lift off the motor.

12. Loosen the mounting screw and then remove the throttle stop guide from the side of the assembly.

13. Do the same with the 2 screws and the throttle stop screw bracket on the opposite side.

14. Remove the 4 bypass air screws (preset at the factory) and 4 synchro screws along with their associated hardware (springs, washers and O-rings) only if *absolutely* necessary and only *after* referring to the Throttle Body Synchronization procedures first.

** WARNING

We will say it again - DO NOT REMOVE THESE SCREWS UNLESS YOU HAVE TO! Before loosening any of the screws, make a small mark between the screw head and throttle body and then count the *exact* number of turns it takes to unscrew it. In the event that you somehow did not bother to follow these instructions, use 2.5 turns as the general starting point when threading in the screws.

15. Remove the 2 mounting bolts and lift off the throttle position (TP) sensor at the front of the assembly. There will be an O-ring and this must be discarded and replaced with a new one on assembly.

To assemble:

16. Thoroughly inspect the throttle body assembly and all associated components for wear, cracks, loose connections or any other damage. Check all fuel passages for obstructions. You may wash the assembly in a petroleum-based cleansing solvent, but DO NOT use carburetor cleaner. Use compressed air to blow out all passages and bores, making certain that everything is completely dry before installation.

17. Install the TP sensor as detailed in the Throttle Position (TP) Sensor procedures and tighten the 2 bolts to 18 inch lbs. (2 Nm).

18. If you removed any (or all) of the bypass or synchro screws, refer to the Synchronization procedures before doing anything!

19. Install the throttle stop screw bracket and tighten the 2 bolts to 44 inch lbs. (5 Nm). Do the same for the throttle stop guide on the other side.

20. Install the fuel rail and injectors.

21. Install the bypass valve motor and tighten the 2 bolts to 9.4 ft. lbs. (13 Nm). Reconnect the coupler and each of the vacuum lines.

To install:

22. If you are jumping to this step from the removal procedures please perform this step. If you disassembled the unit then you have already performed this step and can skip to the next one. Thoroughly inspect the throttle body assembly and all associated components for wear, cracks, loose connections or any other damage. Check all fuel passages for obstructions. You may wash the assembly in a petroleum-based cleansing solvent, but DO NOT use carburetor cleaner. Use compressed air to blow out all passages and bores, making certain that everything is completely dry before installation.

23. Carefully lift the throttle body assembly and move it over to the filter case. There may have been some sensor couplings that could only be disconnected once the assembly was raised out of the box, if so, connect them now and make sure that locking tabs are completely engaged.

24. Now you can pick it up again and, while tilting the assembly slightly at the forward end, slowly move it forward while feeding the fuel pipe through the hole in the side of the filter case and through the grommet. Once through, position the assembly in place so the mounting bolt holes are all lined up.

25. Coat the threads of the 8 mounting bolts lightly with clean engine oil and then thread them each in by hand until they are finger-tight. Now, tighten the bolts, in a crisscross pattern, from the outside corners, to 96 inch lbs. (11 Nm). Once complete, run through them again, tightening to a final torque of 16 ft. lbs. (22 Nm).

26. Reconnect all remaining couplers, making certain that their locking tabs engage the boss completely and the connections are secure.

27. Back at the front of the case, press the fuel line grommet into position in the recess in the case and over the fuel pipe if you had removed it.

28. Slide on the harness bracket and then the fuel line bracket. Coat the bolt threads with Loctite 242 or equivalent, and tighten the 2 bolts to a 1st pass of 2.7 ft. lbs./32 inch lbs. (3.8 Nm) and to a final pass of 5.5 ft. lbs./66 inch lbs. (7.6 Nm).

29. Slide a new crimp clamp and pinch clamp over the end of the line, feed the line in and over the fitting until it is against the grommet and then slide on the 2 clamps. Remember that the pinch clamp is closest to the air filter case (over the end of the grommet) and the crimp clamp is behind it. Crimp the clamp using the special pliers (as detailed in the Fuel Lines section), making sure the clamp is far enough on the line that it has cleared the small ridge on the end of the fuel pipe.

30. Reconnect the connectors for the lighting coil, pulser coil and engine thermo-sensor connectors if you disconnected them for access. Press any connectors into their holders.

31. Referring to the Fuel Lines section again, reconnect the quick connect at the fuel tank (if you removed it) and pop down the cover. Remember that the dot on the line must be facing up.

32. Position the throttle cable holder on the side of the filter case and tighten the 2 bolts to 2.7 ft. lbs./34 inch lbs. (3.8 Nm); and then retighten them again to a final torque of 5.5 ft. lbs./67 inch lbs. (7.6 Nm). Feed the throttle cable through and connect it to the throttle shaft lever. Press the cable down into the slot on the bracket.

33. Slide the 4 flame arrestor springs up and onto the lower tubes and then position the arrestor tubes over the throttle body bores so they seat on the flanges. Roll the springs down the tubes until they are covering the joint where the bottom of each tube comes into contact with the throttle body flange.

34. Coat the threads of the flame arrestor bolts with Loctite 242 and thread them into position, tightening the centers and then the outers to an initial pass of 2.4 ft. lbs./21 inch lbs. (3.3 Nm), and then a final pass of 16 ft. lbs. (22 Nm).

35. Install the air filter and case cover. Connect the battery cables.

36. Adjust the throttle free-play as detailed in the Maintenance & Tune-Up section.

37. Adjust the throttle body synchronization as detailed in this section. You'll probably need to adjust the idle on 2002 models as well.

38. Make sure that the fuel filler cap is tightened - remember you loosened it to relieve system pressure?

2004-08 FX HO/FX Cruiser HO Models With 998/1052cc Engines

3 DIFFICULT

◆ See Figures 51, 57, 68 and 85 thru 89

1. Disconnect the battery cables.

2. Remove the air filter case cover and filter as detailed in the Air Filter Case, Cover Or Housing or Maintenance & Tune-Up sections.

3. Remove the mounting bolts (4) and then lift off the ribbon sub-assembly. The holders may come off with the assembly or stay with the throttle body; remove them either way and keep with the ribbon.

4. Disconnect the throttle cable at the throttle shaft lever. While squeezing the throttle lever to hold open the throttle valves, insert a small, clean piece of wood into the bore to hold the valves open after you release the throttle lever. Wiggle the cable end out of the throttle shaft lever. Loosen the cable locknut, carefully remove the piece of wood and then lift the cable out of the bracket. Remove the 2 cable holder bolts on the side of the filter case and then pull the cable out of the filter case.

5. Remove the fuel filler cap to relieve fuel pressure and then pop off the fuel line quick connect cover at the fuel tank. Position a small container, or rag, beneath the section of fuel line running from the quick connect to the filter case to catch any residual fuel that will leak out of the line and connections and then disconnect the line with the quick connect as detailed in the Fuel Lines and/or Fuel Tank sections.

6. Trace the fuel line back to where it runs into the front of the filter housing. Remove the 2 bolts and lift out the fuel line bracket, before removing the wiring harness bracket bolt and bracket. You will need to cut off the outer crimp clamp (sometimes called a circle clamp) as detailed in the Fuel Lines section and then slide back the small inner pinch clamp before wiggling the line of the connecter grommet.

7. Finally, remove the grommet if it stayed with the housing when you removed the harness bracket.

■ The previous 3 steps are recommended by Yamaha, but it seems to us that it would be a lot easier to simple relieve the fuel pressure and then disconnect the fuel line at the grommet.

8. Tag and disconnect all electrical connectors connected to, or near, the throttle body assembly that you can reach (IAP, IAT, TP sensor couplers and also the bypass valve motor, injector and joint couplers). Tag and connect the breather hose at the MAP sensor.

9. Loosen and remove the 8 throttle body mounting bolts. Carefully grab the assembly and tilt the back end up enough to allow the fuel hose fitting on the end of the fuel rail to clear the hole in the filter case while pulling the assembly toward the back of the case. Check that there are no more connectors or lines that you may have missed in the previous step and remove the assembly. Carefully set the unit down on a clean, sturdy work bench or appropriate surface.

■ The remaining disassembly procedures should only be performed if necessary. Otherwise, you can move to the installation procedures.

10. Remove the fuel rail and injectors as detailed in the Fuel Rail & Injectors procedures.

11. Tag and disconnect the coupler and vacuum lines (should be 4) at the bypass valve motor. Be careful not to lose the 2 hose holders. Remove the 2 mounting bolts and their nuts. Lift off the motor.

12. Remove the 3 mounting bolts and lift off the bypass motor bracket.

■ On 2008 6AA California engines, follow the bypass valve motor removal steps as detailed in the FX/FX140 & FX Cruiser/FX140 Cruiser Models section. . .basically, remove the 2 bolts and lift off the motor

13. Loosen the mounting screw and then remove the throttle stop guide from the side of the assembly.

14. Do the same with the 2 screws and the throttle stop screw bracket on the opposite side.

15. Remove the 4 bypass air screws (preset at the factory) and 4 synchro screws along with their associated hardware (springs, washers and O-rings) only if absolutely</e> necessary and only *after* referring to the Throttle Body Synchronization procedures first.

✻✻ WARNING

We will say it again - DO NOT REMOVE THESE SCREWS UNLESS YOU HAVE TO! Before loosening any of the screws, make a small mark between the screw head and throttle body and then count the *exact* number of turns it takes to unscrew it. In the event that you somehow did not bother to follow these instructions, use 2.5 turns as the general starting point when threading in the screws.

16. Remove the 2 mounting bolts and lift off the throttle position (TP) sensor at the front of the assembly. There will be an O-ring and this must be discarded and replaced with a new one on assembly.

To assemble:

17. Thoroughly inspect the throttle body assembly and all associated components for wear, cracks, loose connections or any other damage. Check all fuel passages for obstructions. You may wash the assembly in a petroleum-based cleansing solvent, but DO NOT use carburetor cleaner. Use

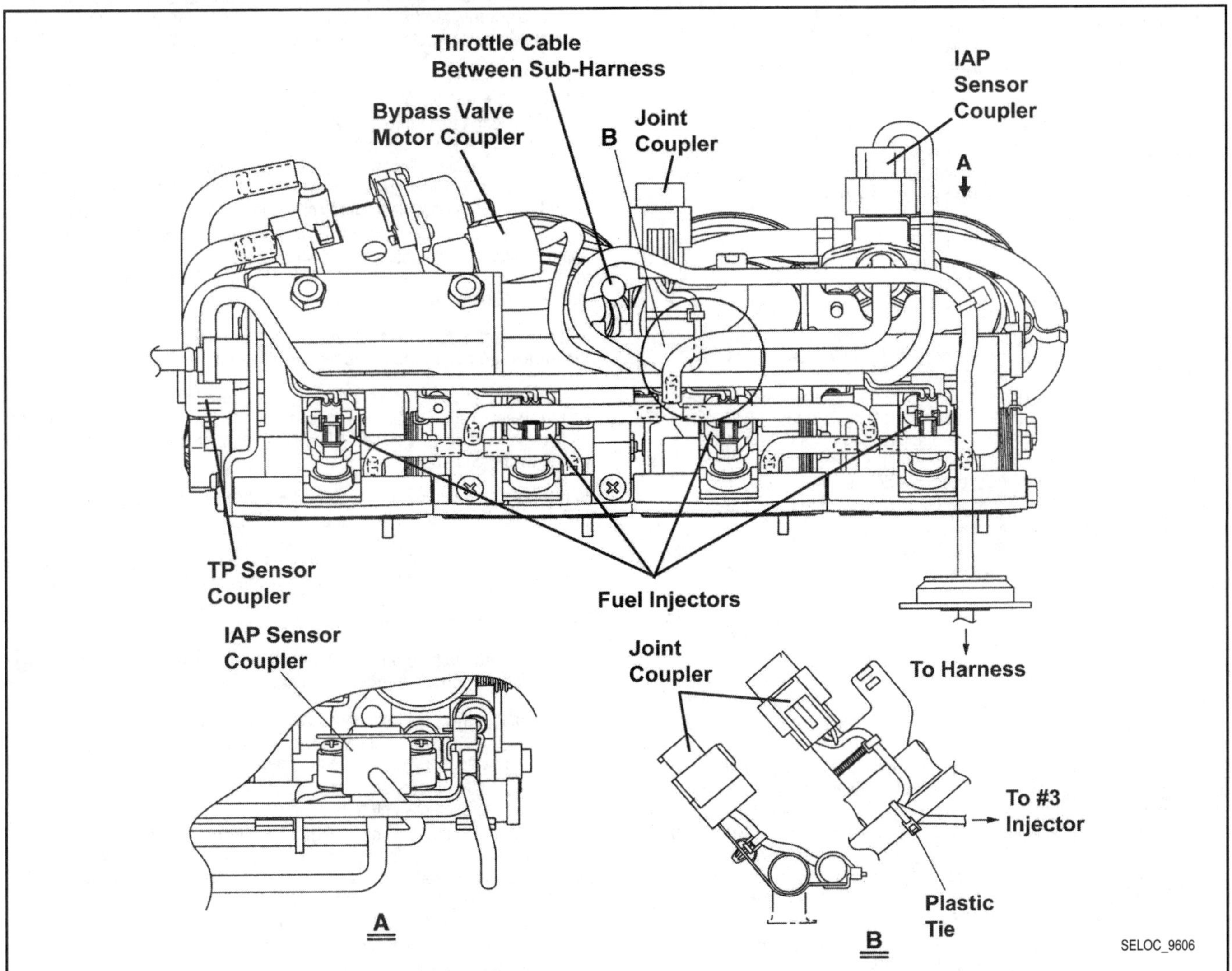

Fig. 89 A good look at the assembly with its wires and lines

compressed air to blow out all passages and bores, making certain that everything is completely dry before installation.

18. Install the TP sensor as detailed in the Throttle Position (TP) Sensor procedures and tighten the 2 bolts to 18 inch lbs. (2 Nm).

19. If you removed any (or all) of the bypass or synchro screws, refer to the Synchronization procedures before doing anything!

20. Install the throttle stop screw bracket and tighten the 2 bolts to 44 inch lbs. (5 Nm). Do the same for the throttle stop guide on the other side.

21. Position the bypass valve motor bracket and tighten the 4 bolts to 44 inch lbs. (5 Nm).

22. Position the motor onto the bracket and tighten the bolts (2) and nuts to 9.4 ft. lbs. (13 Nm). Make sure the bolts come in from the bottom and the nuts are on top of the motor boss. Reconnect the vacuum lines and coupler. On the 2008 California motor, tighten the bolts to 9.4 ft. lbs. (13 Nm).

23. Install the fuel rail and injectors.

To install:

24. If you are jumping to this step from the removal procedures please perform this step. If you disassembled the unit then you have already performed this step and can skip to the next one. Thoroughly inspect the throttle body assembly and all associated components for wear, cracks, loose connections or any other damage. Check all fuel passages for obstructions. You may wash the assembly in a petroleum-based cleansing solvent, but DO NOT use carburetor cleaner. Use compressed air to blow out all passages and bores, making certain that everything is completely dry before installation.

25. Carefully lift the throttle body assembly and move it over to the filter case. There may have been some sensor couplings that could only be disconnected once the assembly was raised out of the box, if so, connect them now and make sure that the locking tabs are completely engaged.

26. Now you can pick it up again and, while tilting the assembly slightly at the forward end, slowly move it forward while feeding the fuel pipe through the hole in the side of the filter case and through the grommet. Once through, position the assembly in place so the mounting bolt holes are all lined up.

27. Coat the threads of the 8 mounting bolts lightly with clean engine oil and then thread them each in by hand until they are finger-tight. Now, tighten the bolts, in a crisscross pattern, from the outside corners, to 16 ft. lbs. (22 Nm).

28. Reconnect all remaining couplers, making certain that their locking tabs engage the boss completely and the connections are secure.

29. Back at the front of the case, press the fuel line grommet into position in the recess in the case and over the fuel pipe if you had removed it.

30. Slide on the harness bracket and then the fuel line bracket. Coat the bolt threads with Loctite 572 or equivalent, and tighten the 2 bolts to 2.4 ft. lbs./30 inch lbs. (3.3 Nm).

31. Slide a new crimp clamp and pinch clamp over the end of the line, feed the line in and over the fitting until it is against the grommet and then slide on the 2 clamps. Remember that the pinch clamp is closest to the air filter case (over the end of the grommet) and the crimp clamp is behind it. Crimp the clamp using the special pliers (as detailed in the Fuel Lines section), making sure the clamp is far enough on the line that it has cleared the small ridge on the end of the fuel pipe.

32. Reconnect the connectors for the lighting/stator coil, pulser/pick-up coil and engine thermo-sensor connectors if you disconnected them for access. Press any connectors into their holders.

33. Referring to the Fuel Lines section again, reconnect the quick connect at the fuel tank (if you removed it) and pop down the cover. Remember that the dot on the line must be facing up.

34. Position the throttle cable holder on the side of the filter case and tighten the 2 bolts to 5.5 ft. lbs./67 inch lbs. (7.6 Nm). Feed the throttle cable through and connect it to the throttle shaft lever. Press the cable down into the slot on the bracket.

35. Pop the holders into the ribbon sub-assembly and position the assembly so the holes line up and the holders pop into place. Coat the threads bolts with Loctite 242 (271 on 2008 models) and thread them into position, tightening the centers and then the outers to 4.7 ft. lbs./42 inch lbs. (6.5 Nm).

36. Install the air filter and case cover. Connect the battery cables.

37. Adjust the throttle free-play as detailed in the Maintenance & Tune-Up section.

38. Adjust the throttle body synchronization as detailed in this section.

39. Make sure that the fuel filler cap is tightened - remember you loosened it to relieve system pressure?

SYNCHRONIZATION

■ **Throttle body synchronization is only performed on 2002-08 FX models with 998cc and 1052cc engines. VX models and 2008 and later models with the 1812cc engine utilize a single throttle body and there is no need for synchronization.**

2002-07 FX Models With 998/1052cc Engines

OEM ③ DIFFICULT

◆ **See Figures 50 and 90 thru 99**

■ **Throttle body synchronization is only performed on 2002-08 FX models with 99cc and 1052cc engines. VX models and 2008 and later models with the 1812cc engine utilize a single throttle body and there is no need for synchronization.**

■ **The 4 bypass air screws (side of each throttle body, preset at the factory) should not normally require adjustment or removal. If it becomes necessary, make sure that you COUNT THE NUMBER of turns it takes to remove the screw(s) - we recommend that you use a marker and make a mark across the screw head and onto the boss. In the event that the screws are removed without counting the turns, thread the screw in until it seats completely and then back it out 2.5 turns.**

■ **In addition to a number of other special tools, you will need a Yamaha Diagnostic System scan tool to complete these procedures. This is a complicated procedure. . .we suggest you read through it thoroughly before determining you wish to proceed.**

1. Remove the air filter case cover, throttle body assembly and case as detailed in the appropriate sections.

2. Working on the throttle body assembly, loosen the throttle stop screw and 3 synchro screws until they are released from their levers. The throttle (butterfly) valves on cylinders #1, 3 and 4 should be closed half way, while the valve on #2 cylinder should be fully closed. Verify the valve is completely closed by using a good flashlight to check for light leaks on the edges.

3. Now, tighten (clockwise) the synchro screw on the #2 throttle body approximately 7 turns until it just comes in contact with the lever. Check that the #2 and #3 valves are now fully closed - if turned more than 7 times, the #2 valve will begin to open.

■ **If the throttle valves on #2 and #3 are *not* fully closed, simply turn the screw (on #2) a little at a time until they are.**

4. Now, tighten (clockwise) the synchro screw on the #1 throttle body approximately 7 turns until it just comes in contact with the lever. Check that the #1 valve is now fully closed - if turned more than 7 times, the #2 and #3 valves will begin to open.

■ **If the throttle valves on #1, #2 and #3 are *not* fully closed, simply turn the screw (on #1) a little at a time until they are.**

5. Finally, tighten (clockwise) the synchro screw on the #3 throttle body approximately 7 turns until it just comes in contact with the lever. Check that the #4 valve is now fully closed - if turned more than 7 times, the #1, #2 and #3 valves will begin to open.

■ **If the throttle valves on #1, #2 and #3, #4 are *not* fully closed, simply turn the screw (on #3) a little at a time until they are.**

6. Confirm that all 4 butterfly valves are now closed fully and they open at the same time (in sync!). If they are not all closed, or if they do not all open at the same time, you will need to repeat the previous steps.

7. Tighten the throttle stop screw (clockwise) about 1 1/2 turns until it just comes into contact with the throttle lever.

8. Working in the engine compartment now, remove the plugs in each intake runner, just under the throttle body joint. Install a carburetor synchronizer (#YU-08030) or vacuum gauge (preferably one with 4 separate hoses), connecting the lines to each runner.

9. Install the throttle body assembly back onto the joints *without* installing the air filter case. Reconnect the fuel line to the rail.

10. Connect the Yamaha Diagnostic System scan tool as per the tool's instructions.

11. If you neglected to remove the sub wire harness from the filter case during removal, disconnect it now and reconnect it between the throttle bodies and main harness.

12. Install a 3-pin test harness (#YB-06793 / 90890-06793) between the throttle position sensor and its coupling.

13. Start the ECM using the scan tool. If this does not work, install a test connector (#YB-06862 / 90890-06862).

**** CAUTION**

Energizing the ECM may cause some fuel spray - make sure you have rags handy.

14. Measure the throttle position sensor output voltage: Pink (P) - Black/Orange (B/O), 0.8 +/- 0.1 V. If the sensor does not fall within this range, loosen the screw(s) and rotate the sensor ever-so-slightly until the voltage comes into range. Clockwise to decrease the voltage and counterclockwise to increase. Once in range, tighten the mounting screw finger-tight.

■ **HO models are a little more specific here. . .check that the sensor voltage is 0.756 +/- 0.016 V.**

15. With the vessel in the water and secured properly, start the engine and allow it run at trolling speed for about 20 minutes, making sure it gets to 122° F (50° C).

16. Once at temperature, adjust the throttle stop screw until the engine speed is at the following specifications based on the conditions as noted:

- 1650 rpm, summer with the temperature at 86° F (30° C) or above
- 1720 rpm, winter with the temperature at 50° F (10° C) or below
- 1680 rpm, spring or fall with the temperature at 68° F (20° C) or below

■ **Obviously, go by the *temperatures* if the seasons do not match your location.**

17. Now you need to check the vacuum pressure differences of each cylinder, always using cylinder #2 as the standard and the synchro screws as the means of adjustment. The allowable differences are as follows on non-HO models:

- Cylinder #1: 1.2 +/- 0.4 in. Hg. (4 +/- 1.33 kPa) *less than* cylinder #2's standard
- Cylinder #2: this is your standard, take this reading and apply the parameters shown for the other cylinders to it
- Cylinder #3: should be at cylinder #2's standard +/- 0.4 in. Hg (1.33 kPa)
- Cylinder #4: 0.8 +/- 0.4 in. Hg. (2.67 +/- 1.33 kPa) *less than* cylinder #2's standard

While on HO models:

- Cylinder #1: 0.8 +/- 0.4 in. Hg. (2.4 +/- 1.33 kPa) *less than* cylinder #2's standard
- Cylinder #2: this is your standard, take this reading and apply the parameters shown for the other cylinders to it
- Cylinder #3: 0.5 +/- 0.4 in. Hg (1.53 +/-1.33 kPa) *less than* cylinder #2's standard
- Cylinder #4: should be at cylinder #2's standard +/- 0.4 in. Hg (1.33 kPa)

In example, if readings for cylinder #2 indicate -7.1 in. Hg (-23.99 kPa), then cylinder #1 readings should indicate -8.3 +/- 0.4 in. Hg (-27.99 +/- 1.33 kPa). Understand? -7.1 in. Hg (cyl #2) *plus* -1.2 +/- 0.4 in. Hg (cyl #1's difference value) *equals* -8.3 +/- 0.4 in. Hg (cyl #1's vacuum range.

Or, using the same value for cylinder #2 (-7.1 in. Hg), cylinder #3's range should be 7.1 in. Hg +/- 0.4 in. Hg.

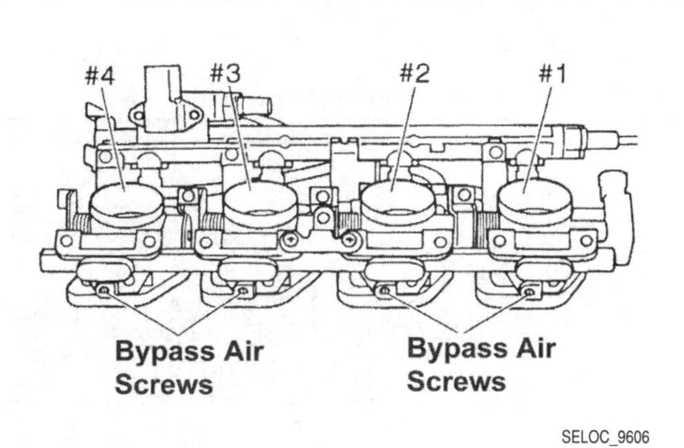

Fig. 90 Bypass air screws are on the side of each throttle body - non-HO models. . .

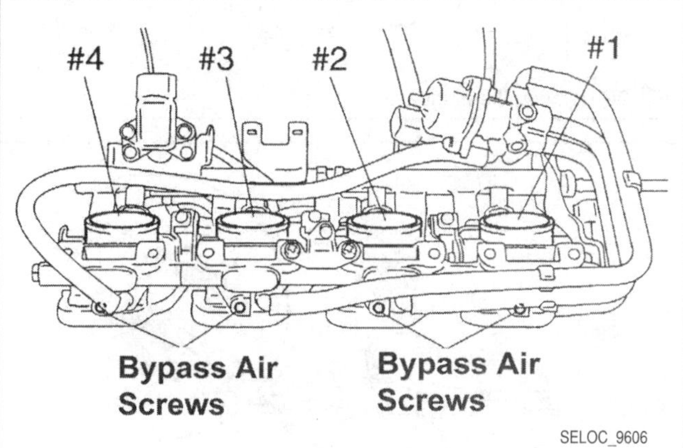

Fig. 91 . . .and HO models. . .

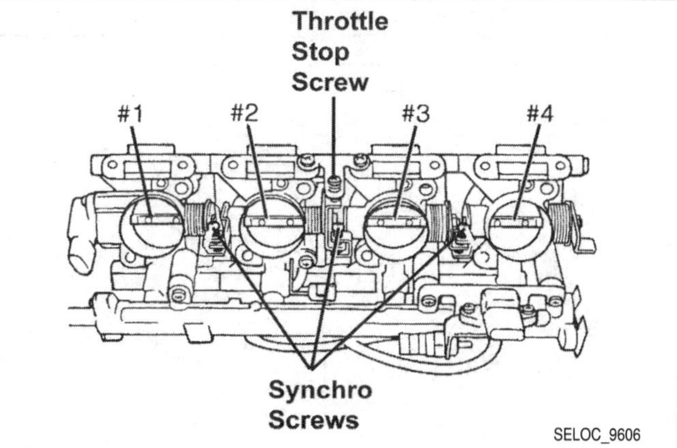

Fig. 92 While synchro and throttle stop screws are on top - non-HO models. . .

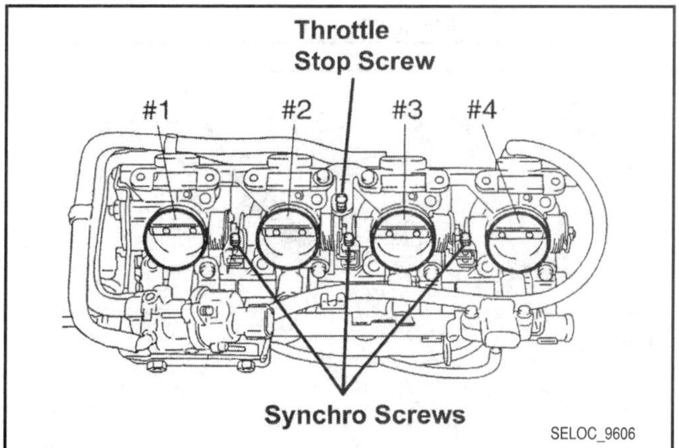

Fig. 93 . . . and HO models

Adjust each cylinder with the synchro screw until cylinders 1, 3 and 4 are within the correct range.

■ **The engine should be running at the previously indicated trolling speed thru-out the previous step.**

18. Finally, check the throttle position sensor output voltage again and confirm that it now reads 0.760 +/- 0.016 V - again, HO models will be 0.756 +/- 0.016 V. Adjust it as you did the first time around if necessary and then tighten the screws to 18 inch lbs. (2 Nm).
19. Turn off the machine and remove the scan tool, vacuum gauge and either of the connectors you may have used.
20. Carefully remove the throttle body assembly (don't forget to disconnect the sub harness) and set it aside in a clean, dry place.
21. Install the plugs for the synchronizer screw.
22. Install the air filter case, and then re-install the throttle body assembly again. Both of these procedures are detailed completely in their respective sections.
23. Install the air filter and case cover. Connect the battery cables and test drive the vessel as soon as possible.

2008 FX HO/FX Cruiser HO Models w/1052cc Engines

◆ **See Figures 50, 95, 96 and 100 thru 108**

■ **Throttle body synchronization is only performed on 2002-08 FX models with 99cc and 1052cc engines. VX models and 2008 and later models with the 1812cc engine utilize a single throttle body and there is no need for synchronization.**

■ **The 4 bypass air screws (side of each throttle body, preset at the factory) should not normally require adjustment or removal. If it becomes necessary, make sure that you COUNT THE NUMBER of turns it takes to remove the screw(s) - we recommend that you use a marker and make a mark across the screw head and onto the boss. In the event that the screws are removed without counting the turns, thread the screw in until it seats completely and then back it out 2.5 turns.**

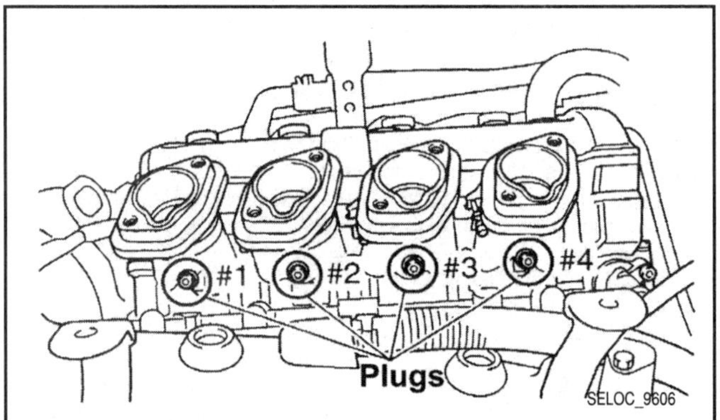

Fig. 94 Remove the plugs. . .

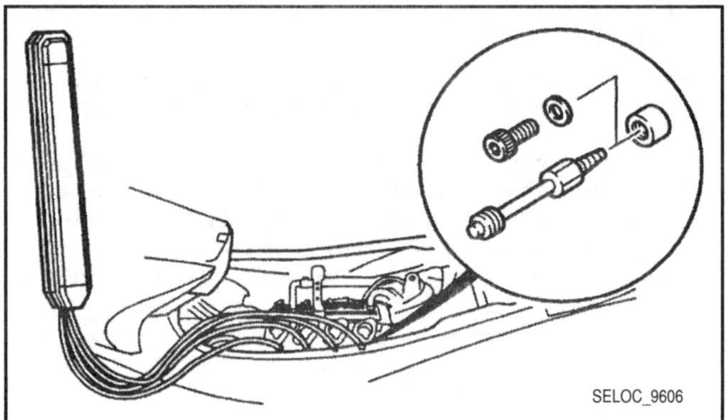

Fig. 95 . . .and then install the synchronizer. . .

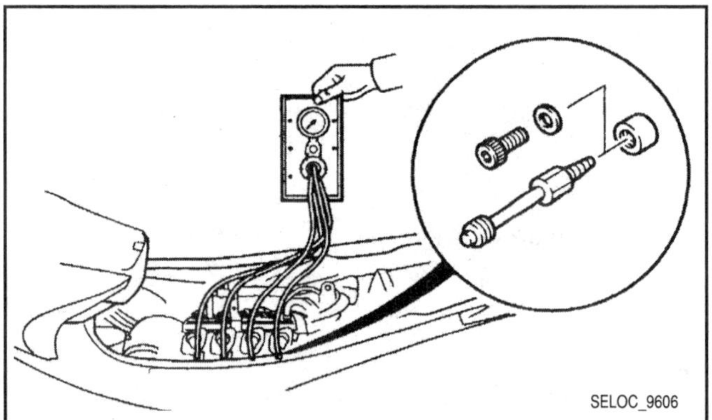

Fig. 96 . . .or the vacuum gauge

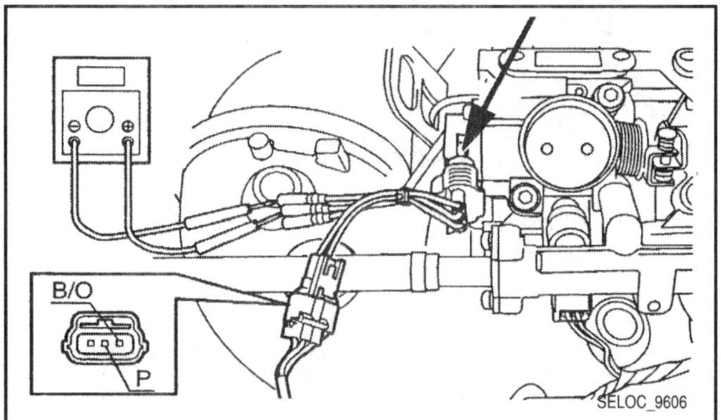

Fig. 97 Installing the test harness on non-HO models. . .

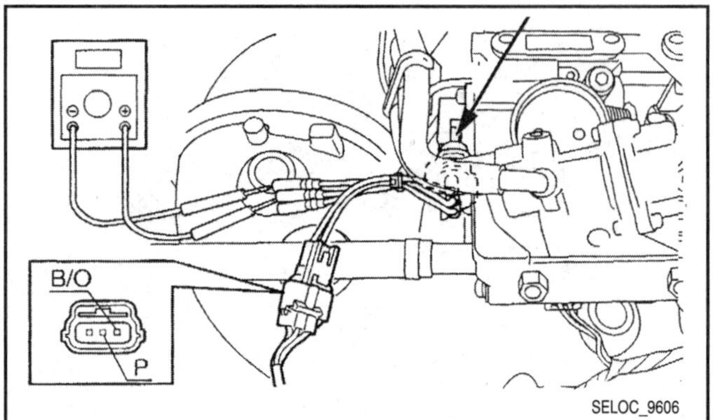

Fig. 98 . . .and HO models. . .

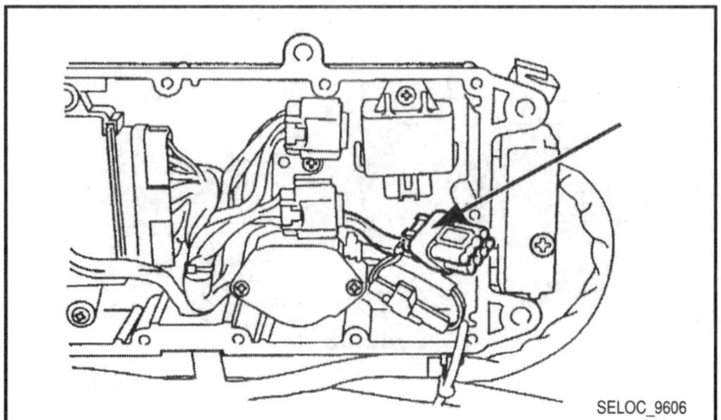

Fig. 99 Installing the test connector

■ **In addition to a number of other special tools, you will need a Yamaha Diagnostic System scan tool to complete these procedures. This is a complicated procedure. . .we suggest you read through it thoroughly before determining you wish to proceed.**

1. Remove the air filter case cover, throttle body assembly and case as detailed in the appropriate sections.
2. Working on the throttle body assembly, loosen the throttle stop screw and 3 synchro screws until they are released from their levers. The throttle (butterfly) valves on cylinders #1, 3 and 4 should be closed half way, while the valve on #2 cylinder should be fully closed. Verify the valve is completely closed by using a good flashlight to check for light leaks on the edges.
3. Now, tighten (clockwise) the synchro screw on the #2 throttle body approximately 7 turns *after* it comes in contact with the lever. Check that the #3 valve is now fully closed.
4. Now, tighten (clockwise) the synchro screw on the #1 throttle body approximately 7 turns *after* it comes in contact with the lever. Check that the #1 valve is now fully closed.
5. Finally, tighten (clockwise) the synchro screw on the #3 throttle body approximately 7 *after* it comes in contact with the lever. Check that the #4 valve is now fully closed.
6. Confirm that all 4 butterfly valves are now closed fully and they open at the same time (in sync!). If they are not all closed, or if they do not all open at the same time, you will need to repeat the previous steps.
7. Tighten the throttle stop screw (clockwise) about 1 1/2 turns until it just comes into contact with the throttle lever.
8. On California models, remove the 2 muffler bolts and 2 muffler stay bolts and then remove the muffler stay. Now, reinstall the 2 muffler bolts using 0.24 in. (6mm) washers and tighten them in 3 passes. 1st to 1.4 ft. lbs./18 inch lbs. (2 Nm), next to 11 ft. lbs. (15 Nm), and finally to 25 ft. lbs. (34 Nm).
9. On all models now. Working in the engine compartment now, remove the plugs in each intake runner, just under the throttle body joint. Install a carburetor synchronizer (#YU-08030) or vacuum gauge (preferably one with 4 separate hoses), connecting the lines to each runner.
10. Install the throttle body assembly back onto the joints *without* installing the air filter case. Reconnect the fuel line to the rail.
11. Connect the Yamaha Diagnostic System (YDIS) scan tool as per the tool's instructions.
12. If you neglected to remove the sub wire harness from the filter case during removal, disconnect it now and reconnect it between the throttle bodies and main harness.
13. Start the ECM using the scan tool.

✻✻ CAUTION

Energizing the ECM may cause some fuel spray - make sure you have rags handy.

14. Measure the throttle position sensor output voltage: Pink (P) - Black/Orange (B/O), 0.756 +/- 0.01g V. If the sensor does not fall within this range, loosen the screw(s) and rotate the sensor ever-so-slightly until the voltage comes into range. Clockwise to decrease the voltage and counterclockwise to increase. Once in range, tighten the mounting screw finger-tight.
15. With the vessel in the water and secured properly, start the engine and allow it run at trolling speed until the engine reaches normal operating temperature. Once at temperature, adjust the throttle stop screw until the engine speed is 1600-1700 rpm.
16. Now you need to check the vacuum pressure differences of each cylinder, always using cylinder #2 as the standard and the synchro screws as the means of adjustment. The allowable differences are as follows:

- Cylinder #1: 0.7 in. Hg. (2.30 kPa) *less than* cylinder #2's standard
- Cylinder #2: this is your standard, take this reading and apply the parameters shown for the other cylinders to it
- Cylinder #3: 1.2 in. Hg. (4.0 kPa) *less than* cylinder #2's standard
- Cylinder #4: 0.9 in. Hg. (3.0 kPa) *less than* cylinder #2's standard

In example, if readings for cylinder #2 indicate -7.1 in. Hg (-23.99 kPa), then cylinder #1 readings should indicate -7.8 in. Hg (-30.29 kPa). Understand? -7.1 in. Hg (cyl #2) *plus* -0.7 in. Hg (cyl #1's difference value) *equals* -7.8 in. Hg (-30.29 kPa) (cyl #1's vacuum range).

Or, using the same value for cylinder #2 (-7.1 in. Hg), cylinder #3's range should be 8.3 in. Hg.

Adjust each cylinder with the synchro screw until cylinders 1, 3 and 4 are within the correct range.

■ **The engine should be running at the previously indicated trolling speed thru-out the previous step.**

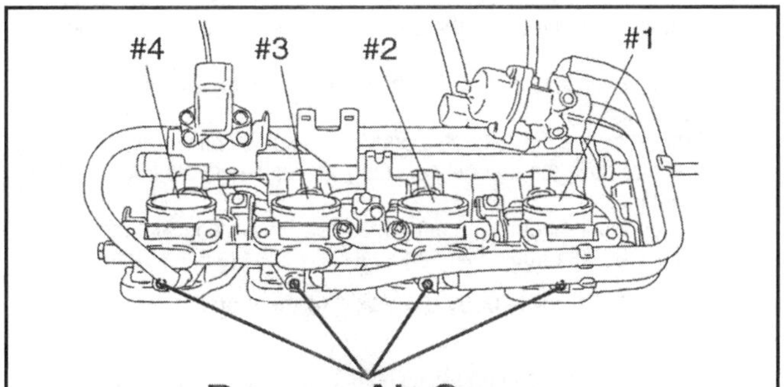

Fig. 100 Bypass air screws are on the side of each throttle body - non-California models. . .

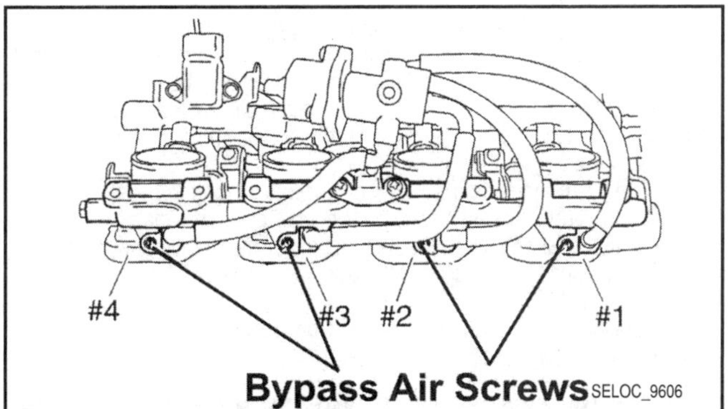

Fig. 101 . . . and California models

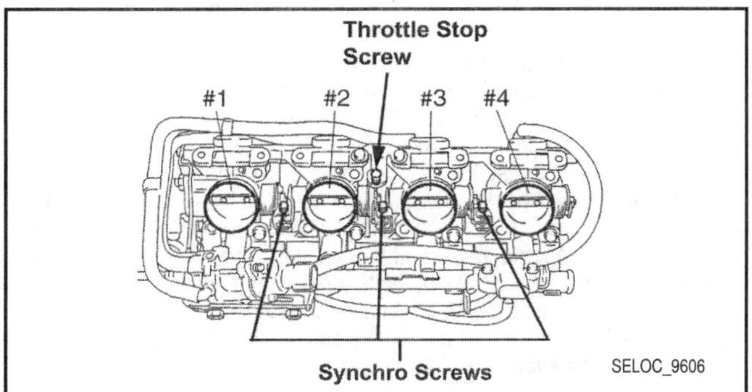

Fig. 102 While synchro and throttle stop screws are on top - non-California models. . .

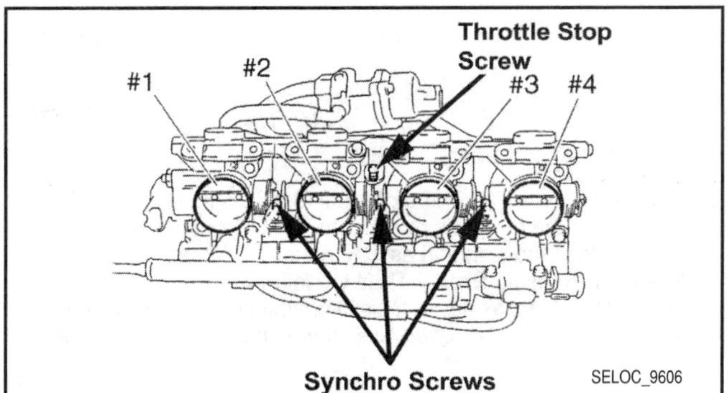

Fig. 103 . . . and California models

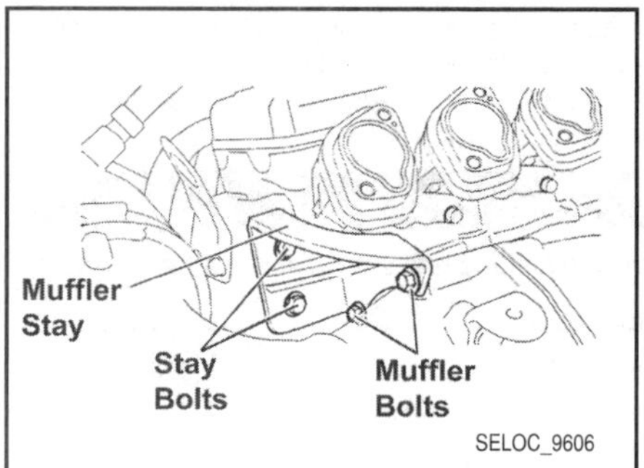

Fig. 104 Remove the muffler bolts and stay...

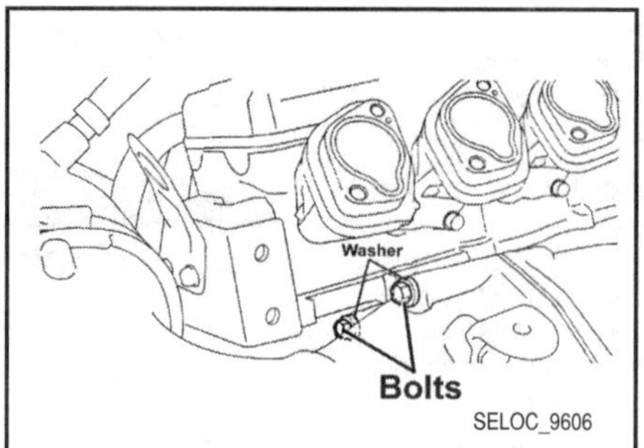

Fig. 105 . . .and then re-install the 2 bolts with a washer

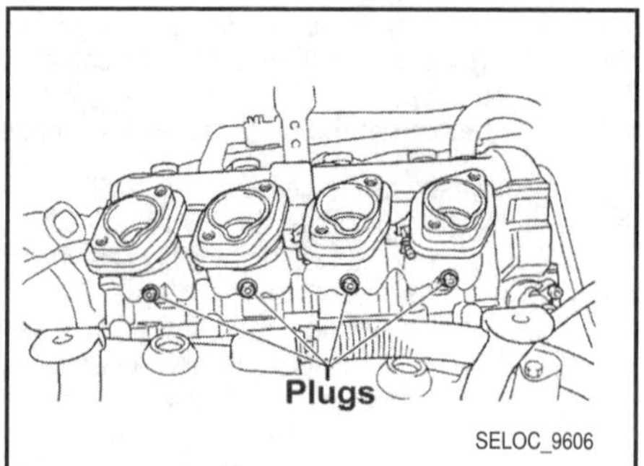

Fig. 106 Remove the plugs

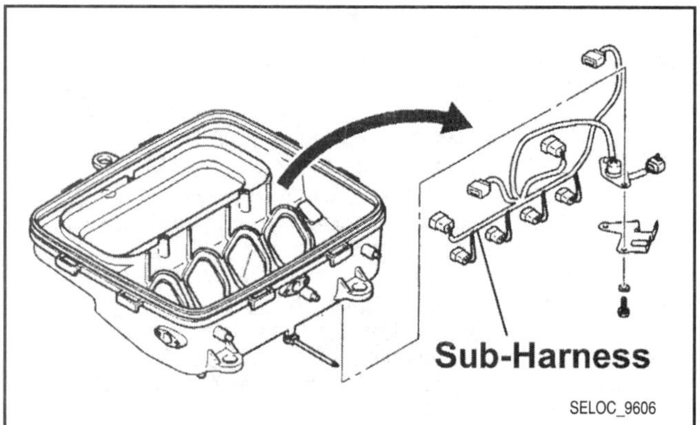

Fig. 107 Installing the test harness on non-HO models...

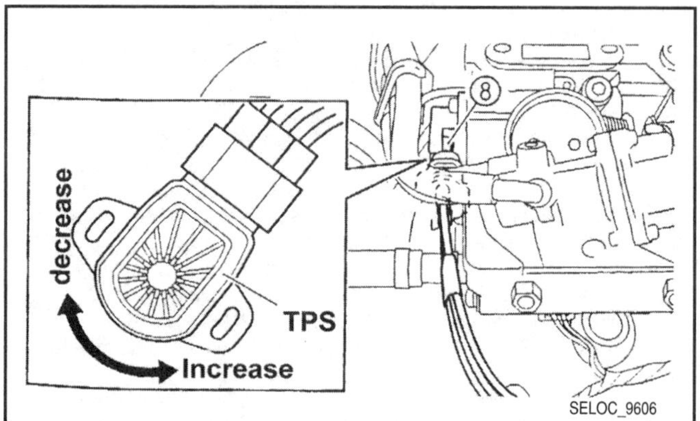

Fig. 108 . . .and HO models. . .

17. Finally, check the throttle position sensor output voltage again and confirm that it now reads 0.768 +/- 0.016 V. Adjust it as you did the first time around if necessary and then tighten the screws to 18 inch lbs. (2 Nm).

18. Turn off the machine and remove the scan tool, vacuum gauge and either of the connectors you may have used.

19. Install the plugs and tighten them to 7.2 ft. lbs. (10 Nm).

20. On California models, remove the muffler bolts and slide their original washers on. Coat the threads of these bolts and the stay bolts with Loctite 572, position the muffler stay and then tighten the bolts in 3 passes: 1st to 1.4 ft. lbs./18 inch lbs. (2 Nm), next to 11 ft. lbs. (15 Nm), and finally to 25 ft. lbs. (34 Nm)

21. Carefully remove the throttle body assembly (don't forget to disconnect the sub harness) and set it aside in a clean, dry place.

22. Install the air filter case, and then re-install the throttle body assembly again. Both of these procedures are detailed completely in their respective sections.

23. Install the air filter and case cover. Connect the battery cables and test drive the vessel as soon as possible.

Throttle Body/Intake Manifold Assembly

REMOVAL & INSTALLATION

2009 & Later FX HO/FX Cruiser HO Models & 2011 VX Models w/1812cc Engines

2 MODERATE

◆ See Figures 71 and 109 thru 112

■ **Throttle body removal, in and of itself, is quite simple and will not entail removing the fuel lines or other major components. If your intent is to also remove the manifold as well though, fuel line and fuel rail removal will also be necessary. Read the following steps *prior* to starting your work so as not to remove components unnecessarily, but know ahead that there is a good likelihood that you will need to remove the engine.**

1. Remove the seat(s) and cargo tray and then disconnect the battery cables.

2. Loosen the clamp on the breather hose and pull it off of the air intake hose; moving it out of the way.

3. Loosen the hose clamp at the throttle body end of the air intake hose (coming from the air filter case) and wiggle the hose off of the flange on the throttle body. Swivel it out of the way as much as you are able to - you may have to take the fuel line out of the clamp to provide more leeway to get the hose out of the way.

4. Press the tab and disconnect the connector at the side of the throttle body assembly.

5. If removing the entire throttle body/manifold assembly, disconnect the fuel lines and remove the fuel rail assembly as detailed in the Fuel Lines or Fuel Rail & Injectors sections.

6. Loosen the 4 mounting nuts and remove them along with their washers. Support the throttle body and carefully remove it from the joint.

■ **If removing the throttle body only, skip to the installation procedures.**

7. Remove the 4 Allen head bolts holding the joint to the intake manifold and lift off the joint. Remove the ribbon and holder from the end of the manifold if they don't come out with the joint.

8. If not already done, remove the fuel rail as detailed in the Fuel Rail & Injectors section.

9. Tag and disconnect the intake air temperature and air pressure sensor connectors, moving each of them out of the way.

10. If replacing the manifold, remove the intake air temperature sensor and its washer. Pop the small hose off of the bottom of the intake air pressure sensor and remove it and the grommet.

11. Under the manifold are 2 stays. Remove the 2 stay-to-manifold bolts and then remove the 2 stay-to-cylinder block bolts. Remove the stays.

12. The manifold is held in position by 2 nuts (top side of the flange, at each end) and 6 bolts, loosen them from the center outward (just reverse the tightening sequence!) and then lift off the manifold. There are 2 dowel pins (bottom, front and back) and 2 stud bolts (upper, front and back), check to make sure that they stayed with the engine and did not come out with the manifold. The 2 mounting nuts are what were attached to the stud bolts.

13. Remove the intake manifold gasket and discard.

To install:

14. If you removed the manifold:

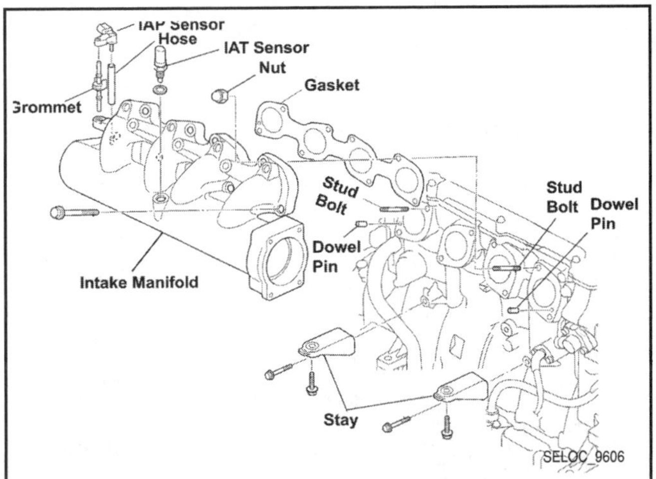

Fig. 109 Exploded view of the intake manifold assembly - 2009 and later FX HO/Cruiser HO models and 2011 VX models w/1812cc engines

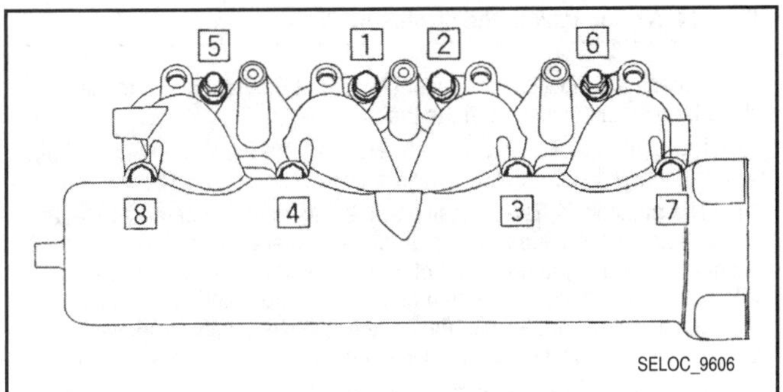

Fig. 110 Intake manifold tightening sequence

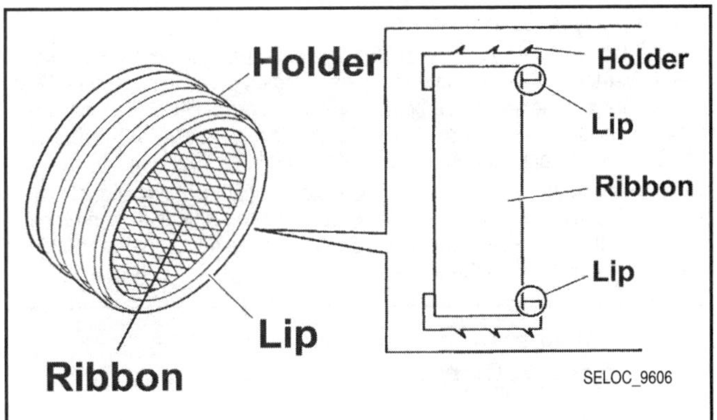

Fig. 111 Make sure that the holder lip hooks into the ribbon

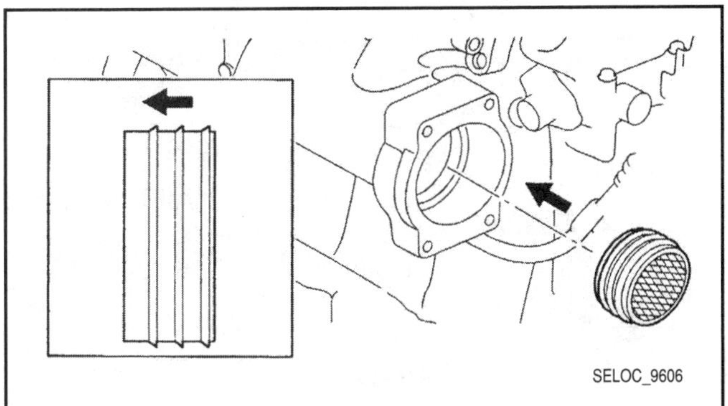

Fig. 112 Installing the ribbon assembly

a. Clean and inspect the manifold assembly. Position a *new* manifold gasket over the stud bolts and dowel pins and then install the manifold. Coat the bolt threads and the 2 stud bolt threads with Loctite 242 and then tighten them in 2 passes, in the order shown. Tighten the 1st pass to 7.4 ft. lbs./88 inch lbs. (10 Nm) and then the final pass to 14.8 ft. lbs. (20 Nm).

15. Position the stays under the manifold and tighten the cylinder block bolts to 11.1 ft. lbs./97 inch lbs. (15 Nm) and then again to 22.1 ft. lbs. (30 Nm). Now install the stay-to-manifold bolts and tighten them to the same specs.

16. If removed, install the air pressure sensor grommet and then the sensor itself. Reconnect the hose. Slide the washer over the air temperature sensor threads and then screw it in, tightening to 11.1 ft. lbs./133 inch lbs. (15 Nm).

17. Install the fuel rail and lines if removed.

18. Install the ribbon into the holder making sure that the lip of the holder hooks into the ribbon. Now install the assembly into the manifold as shown.

19. Install the joint and tighten the 4 bolts in a criss-cross pattern to 17.7 ft. lbs. (24 Nm). Don't forget to coat the bolt threads with Loctite 242.

Whether you removed the manifold or just the throttle body:

20. Install the throttle body over the joint studs and tighten the nuts, in a criss-cross pattern, to 9.6 ft. lbs./116 inch lbs. (13 Nm).

21. Connect the electrical coupler, pushing it in until it clicks.

22. Slide the air intake hose onto the throttle body flange and tighten the clamp to 22 inch lbs. (2.5 Nm). Reconnect the fuel line to the clamp if you had disconnected it.

23. Connect the battery cables and install the cargo try and/or seat.

2008 & Later FX SHO/Cruiser SHO And FZR/FXS Models w/1812cc Engines

3 DIFFICULT

◆ **See Figures 111 thru 117**

■ **Throttle body removal, in and of itself, is quite simple and should not entail removing the fuel lines or other major components. If your intent is to also remove the manifold as well though, fuel line, fuel rail, air cooler and supercharger removal will also be necessary. Read the following steps *prior* to starting your work so as not to remove components unnecessarily, but know ahead that there is a good likelihood that you will need to remove the engine.**

1. Remove the seat(s) and cargo tray and then disconnect the battery cables.

2. Remove the 2 bolts and lift off the engine cover.

If removing only the throttle body:

3. Loosen the hose clamp at the air cooler hose end of the boost pipe.

4. Remove the boost pipe mounting bolts at the bottom bend on the pipe.

5. Loosen the 2 hose clamps and wiggle the boost pipe/joint a little while pulling down on it until it separates from the throttle body. Pull the upper end out of the cooler hose and remove the pipe.

If removing the entire throttle body/manifold assembly:

6. Disconnect the fuel lines and remove the fuel rail assembly as detailed in the Fuel Lines or Fuel Rail & Injectors sections.

7. Remove the supercharger and air cooler as detailed in the Engine Mechanical section - this means you are going to have to remove the engine.

With either choice now:

8. Press the tab and disconnect the connector at the side of the throttle body assembly.

9. Loosen the 4 mounting nuts and remove them along with their washers. Support the throttle body and carefully remove it from the joint.

■ **If removing the throttle body only, skip to the installation procedures.**

10. Remove the 4 Allen head bolts holding the joint to the intake manifold and lift off the joint. Remove the ribbon and holder from the end of the manifold if they don't come out with the joint. Remove the O-ring and throw it away.

11. Tag and disconnect the intake air temperature and air pressure sensor connectors, moving each of them out of the way.

12. If replacing the manifold, remove the intake air temperature sensor and its washer. Pop the small hose off of the bottom of the intake air pressure sensor and remove it. Remove the bracket along with the collar and grommets.

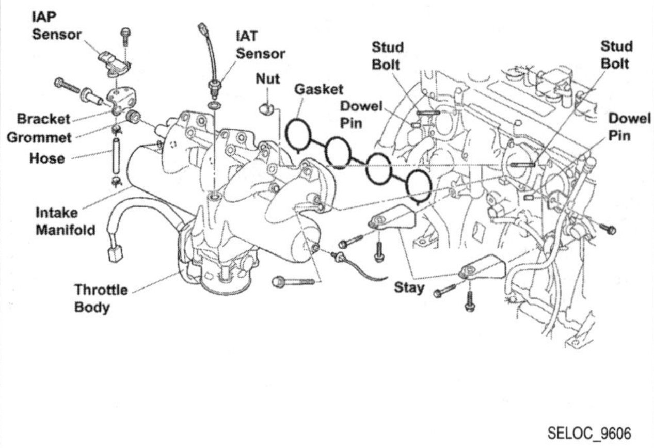

Fig. 113 Exploded view of the intake manifold assembly - 2008 and later FX SHO/Cruiser SHO and FZ models

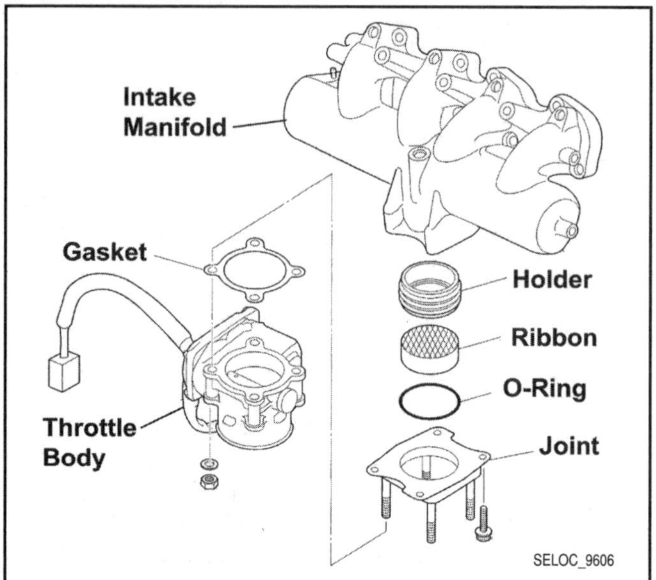

Fig. 114 A good look at the throttle body assembly

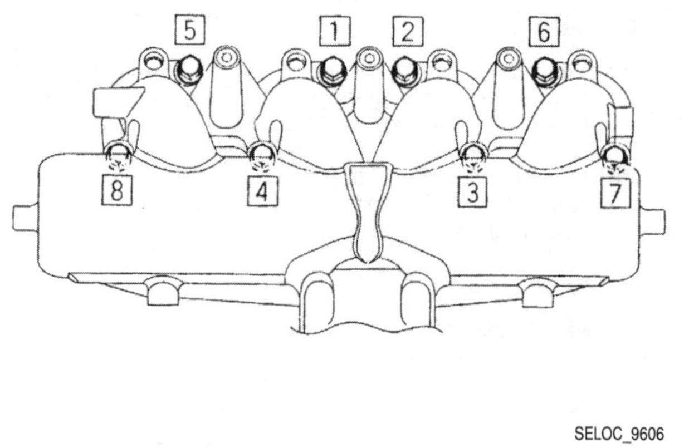

Fig. 115 Intake manifold tightening sequence

13. Loosen the dipstick tube mounting bolt at the front of the manifold and pull the tube/bracket away from the manifold.

14. Under the manifold are 2 stays. Remove the 2 stay-to-manifold bolts and then remove the 2 stay-to-cylinder block bolts. Remove the stays.

15. The manifold is held in position by 2 nuts (top side of the flange, at each end) and 6 bolts, loosen them from the center outward (just reverse the tightening sequence!) and then lift off the manifold. There are 2 dowel pins (bottom, front and back) and 2 stud bolts (upper, front and back), check to make sure that they stayed with the engine and did not come out with the manifold. The 2 mounting nuts are what were attached to the stud bolts.

16. Remove the intake manifold gasket and discard.

To install:

17. If you removed the manifold:

a. Clean and inspect the manifold assembly. Position a *new* manifold gasket over the stud bolts and dowel pins and then install the manifold. Coat the bolt threads and the 2 stud bolt threads with Loctite 242 and then tighten them in 2 passes, in the order shown. Tighten the 1st pass to 7.4 ft. lbs./88 inch lbs. (10 Nm) and then the final pass to 14.8 ft. lbs. (20 Nm).

b. Position the stays under the manifold and tighten the cylinder block bolts to 17 ft. lbs. (23 Nm), or 16.2 ft. lbs. (22 Nm) on FZ models, and then again to 31 ft. lbs. (42 Nm) on all models. Now install the stay-to-manifold bolts and tighten them to the same specs.

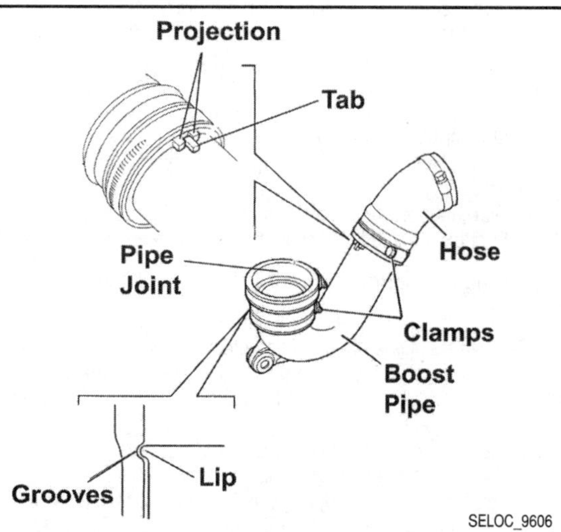

Fig. 116 Connecting the boost pipe to the air hose. . .

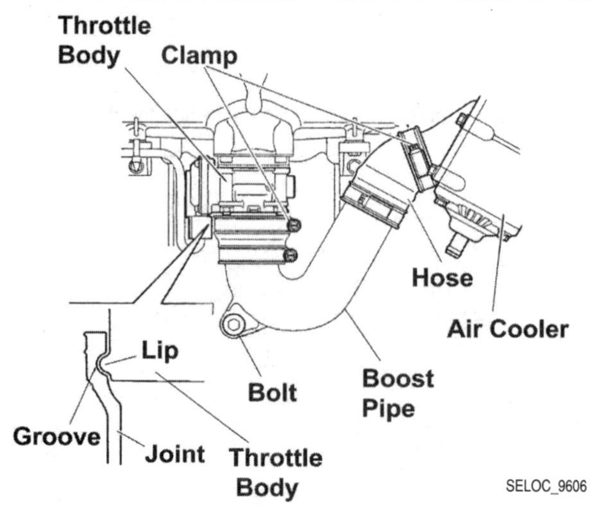

Fig. 117 . . .and then to the throttle body

c. If removed, install the air pressure sensor bracket with the collar and grommet, and then the sensor itself. Tightening the bracket bolts to 3.7 ft. lbs./44 inch lbs. (5 Nm) and the sensor bolts to 2.6 ft. lbs./31 inch lbs. (3.5 Nm).Reconnect the hose.

d. Slide the washer over the air temperature sensor threads and then screw it in, tightening to 11.1 ft. lbs./133 inch lbs. (15 Nm).

e. Install the ribbon into the holder making sure that the lip of the holder hooks into the ribbon. Now install the assembly into the manifold as shown.

f. Install the joint and tighten the 4 bolts in a criss-cross pattern to 9.6 ft.lbs./115 inch lbs. (13 Nm). Don't forget to coat the bolt threads with Loctite 242.

Whether you removed the manifold or not:

18. Install the throttle body over the joint studs and tighten the nuts, in a criss-cross pattern, to 9.6 ft. lbs./116 inch lbs. (13 Nm).

19. Connect the electrical coupler, pushing it in until it clicks.

20. Slide the boost pipe up into the cooler hose so that the tab on the pipe nests in between the 2 projections on the hose; making sure that pipe goes in far enough for the lip to fit into the groove in the hose.

■ **Obviously, if you removed the supercharger and air cooler, you will not be able to perform the previous step until you install the other components later on so just skip to the next step.**

21. Slide the pipe joint over the throttle body throat until the groove in the joint slide over the lip on the throat.

22. Tighten the 3 hose clamps (2 on the pipe joint, 1 at the cooler hose) to 2.2 ft. lbs./27 inch lbs (3 Nm). Tighten the boost pipe mounting bolt to 14.8 ft. lbs. (20 Nm).

23. Install the air cooler, supercharger and fuel rail/lines if removed.

24. Install the engine cover and tighten the bolts securely

25. Connect the battery cables and install the cargo try and/or seat.

2005-11 VX Models w/1052cc Engines

② MODERATE

◆ **See Figures 46, 76 thru 80, 82, and 118 thru 123**

■ **Throttle body removal, in and of itself, is quite simple and will not entail removing the fuel lines or other major components. If your intent is to also remove the manifold as well though, fuel line and fuel rail removal will also be necessary. Read the following steps *prior* to starting your work so as not to remove components unnecessarily.**

1. Remove the seat(s) and cargo tray and then disconnect the battery cables.

2. Loosen the clamp on the breather hose and pull it off of the air intake hose; moving it out of the way.

3. Loosen the hose clamp at the throttle body end of the air intake hose (coming from the air filter case) and wiggle the hose off of the flange on the throttle body. Swivel it out of the way as much as you are able to.

Fig. 118 The throttle position sensor is attached to the throttle body

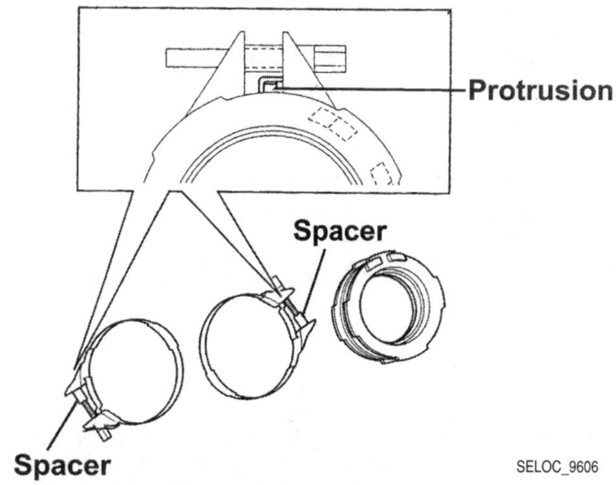

Fig. 119 Installing the manifold joint and clamps - 2005-09 VX models

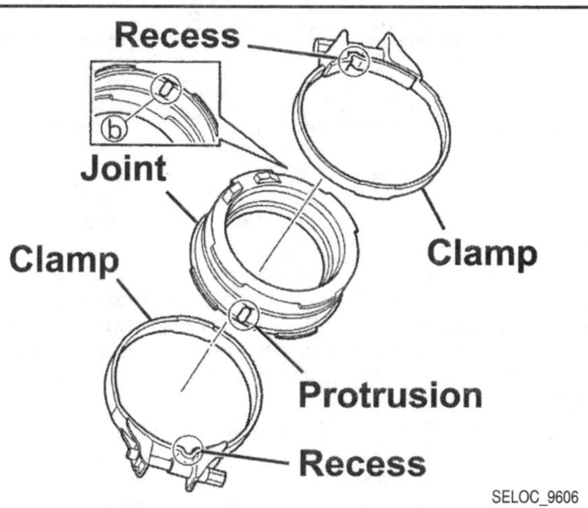

Fig. 120 Installing the manifold joint and clamps - 2010-11 VX models w/1052cc engines

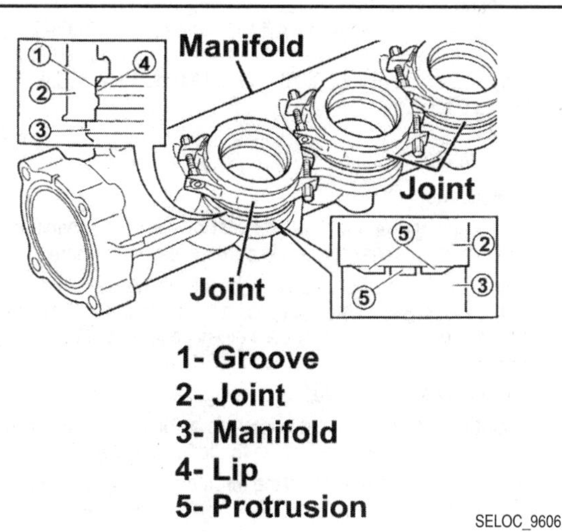

Fig. 121 Installing the manifold joint and clamps to the manifold

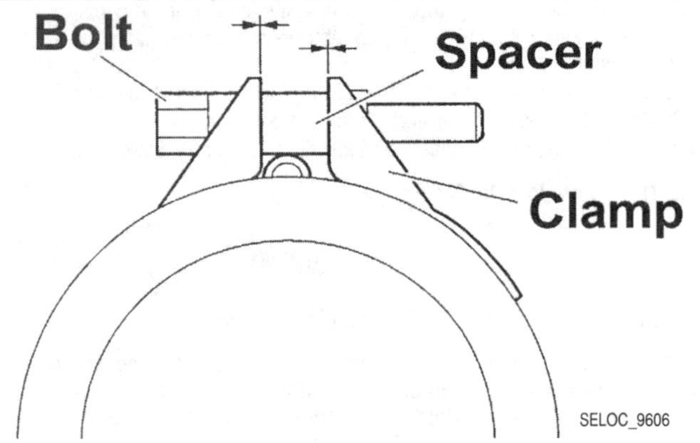

Fig. 122 Both clamp ends should be against the spacer

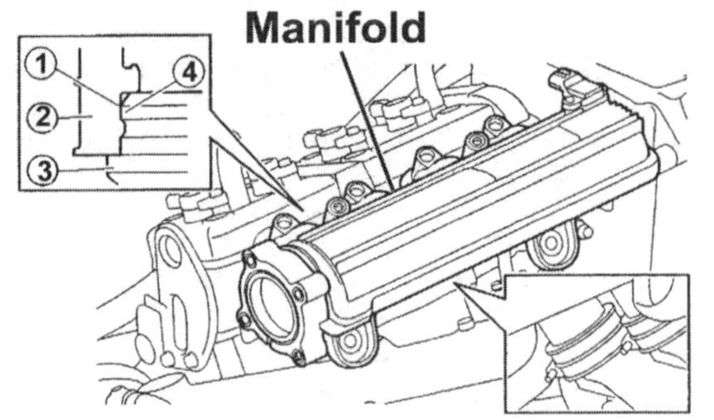

Fig. 123 Installing the manifold and joints to the cylinder head

4. Press the tab and disconnect the connector at the side of the throttle body assembly.

5. If removing the entire throttle body/manifold assembly, disconnect the fuel lines and remove the fuel rail assembly as detailed in the Fuel Lines or Fuel Rail & Injectors sections.

6. Loosen the 4 mounting bolts and remove them along with their washers. Support the throttle body and carefully remove it from the intake manifold. Remove the gasket and throw it away.

■ **If removing the throttle body only, skip to the installation procedures.**

7. If not already done, remove the fuel rail as detailed in the Fuel Rail & Injectors section.

8. Loosen the 2 clamps on each of the 4 intake manifold joints.

9. Remove the 2 mounting bolts and carefully lift off the intake manifold. You should be able to get these bolts off cleanly, but if necessary, you can remove the ECU and rectifier and their support bracket. The manifold may come off with the joints attached, but if they stay with the engine, simply lift them off as well.

To install:

10. Inspect the manifold joint for cracks or other damage, replacing as necessary. Install the clamps over the joints so the protrusion in the joint lines up with the recess in the clamp. Make sure that the clamp bolts are opposing and then slide everything down into position on the manifold. Make sure that the protrusion on the manifold runner fits between the protrusions on the joint. Fit the lip on the manifold into the groove in the joint. Tighten the clamp bolts to 2.2 ft. lbs./27 inch lbs. (3 Nm) - the spacer should be in contact with each side of the clamp.

11. Clean and inspect the manifold assembly and then position it on the engine so the lip on the cylinder head fits into the grooves in the joints. Thread in the mounting bolts and tighten them, in 2 passes, to 6.4 ft. lbs. (8.8 Nm) on 2005-09 models or 6.6 ft. lbs. (9 Nm) on later models. Finally, tighten them to 13.3 ft. lbs. (18 Nm).

12. Recheck the joint clamp bolts to make sure they are still torque correctly.

13. Install the fuel rail.

14. Install a new gasket into the throttle body end of the manifold, making sure that the tab on the gasket feeds into the recess in the bore.

15. Position the throttle body assembly over the manifold end so the holes line up and the gasket stays in position. Thread in the 4 mounting bolts and then tighten them, in and 'X' or criss-cross pattern, to 9.6 ft. lbs. (13 Nm).

16. Connect the electrical coupler, pushing it in until it clicks.

17. Check the throttle valve opening angle as detailed by the Throttle Position Sensor Checking procedures in the Ignition & Electrical section.

18. Slide the air intake hose onto the throttle body flange and tighten the clamp to 22 inch lbs. (2.5 Nm).

19. Install the fuel line if disconnected.

20. Connect the battery cables and install the cargo try and/or seat.

Bypass Valve Motor

REMOVAL & INSTALLATION

2003-07 FX Models With 998/1052cc Engines

② MODERATE

◆ **See Figure 85**

■ **Depending on how nimble you are, you should be able to remove the motor without removing the throttle body assembly; otherwise refer to the throttle body disassembly procedures.**

1. Remove the seat and disconnect the battery cables.
2. Remove the air filter case cover as detailed in the Air Filter Case, Cover Or Housing section.
3. Tag and disconnect the coupler and vacuum lines (should be 4) at the bypass valve motor.
4. Reach into the case, remove the 2 mounting bolts and their nuts and lift off the motor. California models will use 2 bolts only, no nuts.
5. Install the bypass valve motor and tighten the 2 bolts to 9.4 ft. lbs. (13 Nm). On non-California models, install the bolts from the bottom and tighten the nuts (on top of the boss).

■ **Bracket removal, if necessary, is covered in the throttle body disassembly procedures.**

6. Reconnect the coupler and each of the vacuum lines.
7. Install the air filter case cover.
8. Connect the battery cables and install the seat.

Sensor & Circuit Resistance/Output Tests

TESTING ELECTRONIC COMPONENTS

③ DIFFICULT

◆ **See Figure 124**

Before conducting resistance or voltage checks on a component/circuit, make sure all electrical connections in the system are clean and tight. Visually check the wiring for obvious breaks, defects or other problems.

Keep in mind that although a haphazard approach might find the cause of problems, only a systematic approach will prevent wasted time and the possibility of unnecessary component replacement. In some cases, installing an electronic component into a faulty circuit that damaged or destroyed the previous component, will instantly destroy the replacement. For various reasons, including this possibility, most parts suppliers do not accept returns on electrical components.

Make sure you perform each test procedure and especially, make all test connections as described. If a component tests out of range (faulty), but the reading is close to the service limit, bring the component to your marine dealer or parts supplier so they can verify your result before purchasing the replacement. In some cases, your dealer may be willing to check a sensor against the reading on a new one to verify whether or not your sensor is actually faulty.

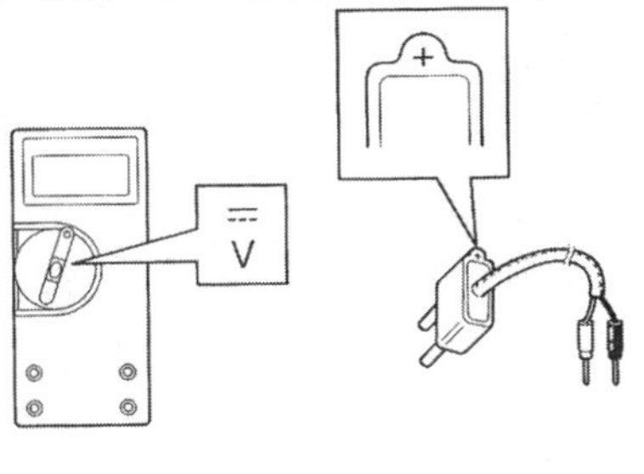

Fig. 124 Yamaha recommends a high quality DVOM and a peak reading voltage adaptor for testing on their motors

Although resistance checks are relatively easy to conduct, their results can be difficult to interpret. Remember that resistance in any electrical component or circuit will vary with temperature. Also, ohmmeters will vary with quality, meaning that readings can vary on the same component between different meters. The specifications provided here are based upon the use of a high-quality digital multi-meter applied to a component that is currently at about 68°F (20°) unless otherwise noted.

■ **More information on electrical test equipment is can be found in the General Information section, while additional information on basic electrical theory and electrical troubleshooting can be found in the Ignition & Electrical System section.**

As a general rule, the resistance of a circuit or component will increase (rise) as temperature increases. Although this is true for most resistor and windings, some manufacturers use negative temperature coefficient sensors for certain functions. A negative temperature coefficient sensor reacts in an opposite manner, meaning that its resistance decreases (lowers) as temperature increases (rises) and vice-versa.

Most components of the EFI systems are checked using resistance or voltage checks. This means that a voltmeter or ohmmeter is connected to the component or circuit under the proper circumstances. Yamaha recommends the use of a high quality Digital Volt Ohm Meter (DVOM) such as #YU-34899-A or equivalent and a Peak Voltage Adaptor such as #YU-39991.

For resistance checks, no voltage must be applied to the circuit (except that as provided by the meter to perform the check). Typically, the circuit or component wiring is isolated, by disconnecting the wiring from the component itself or somewhere else in the circuit, then applying the meter probes across 2 terminals that connect through the component. When conducting resistance checks, you *must* isolate the ECM from the meter (unless specifically directed otherwise) in order to prevent the possibility of damage to the control module. By identifying the proper circuit you can conduct tests of the entire component circuit right from the harness connectors that attach to the ECM. Otherwise, use the wiring diagrams and the test procedures to determine other points to disconnect the wiring and test the circuit/component.

We recommend you work in 1 of 2 directions. Either test the component first, then work your way back through the component wiring toward the ECM, or disconnect the harness connectors from the ECM, testing the entire circuit first, then working your way toward the component. A bad test reading at the ECM harness connector must be verified by making sure it is not the result of a bad wiring harness or connection along the way to the component. Likewise, a good reading at a component (when there is a bad reading on the circuit as noted by the presence of a trouble code or by a reading at the ECM harness) must be verified by checking the circuit wiring for trouble.

When checking the circuit wiring, isolate the harness by disconnecting it from both the component and the ECM, then using an ohmmeter to check resistance from 1 connector to the next on each wire. Wiggle the wire and connectors while conducting the test to see if readings fluctuate. Although no specifications are provided for wiring resistance, all sensor/switch wiring for these motors should have very little resistance. Also, be sure to check wiring for shorts to ground (by checking resistance between 1 terminal and a good engine ground), and for shorts to power (by checking a terminal using a voltmeter and connecting the other probe to a good engine ground or the negative battery cable).

**** WARNING**

Be VERY careful when checking circuits for electronic engine control components. Always isolate the ECM by disconnecting the wiring from it before using an ohmmeter on the wires. Keep in mind that making the wrong connections across the terminals of an ECM can fry the unit, requiring an expensive replacement.

When conducting voltage tests, the circuit must be complete in order to receive a reading. In most cases, this means the wiring must be left connected and the meter probes should be carefully inserted through the back of the connector (called back-probing) in order to get a reading without opening and disrupting circuit operation. However, back-probing should NOT be done on waterproof connectors as you will compromise the seal, and jumper wires should instead be used to complete the circuit. The best method for manufacturing a jumper wire is to obtain 2 identical size/type connectors as the portion of the harness to which you will be inserting the jumper wires. Wire these jumper connectors together along with a second wire connected to EACH terminal. Each second wire from a terminal should route off to a female bullet connector which can be safely probed without the fear of accidental shorting by connecting 2 or more wires together.

Engine Control Module (ECM)

GENERAL INFORMATION

The engine control module (ECM) controls all functions of the EFI and ignition systems. Probably as a holdover from the carbureted 2-stroke motors, some of the earliest of the Yamaha fuel injection systems refer to this as the TCI Unit. However, since it also controls the fuel injection system we think this is a little misleading and will refer to it as the ECM throughout this guide.

Problems with the ECM are rare, but when they occur, can cause a no-start, stumbling, misfire, hesitation, incorrect engine timing, rough idle or incorrect speed limiting through improper control of the ignition and/or fuel injection systems.

Unfortunately, solid state components like the ECM cannot be directly tested in many ways. In most cases, ECM testing involves a process of elimination, testing all other possible causes of a symptom. Condemn the ECM only if all other components that could cause a problem have been eliminated. Remember that many of the circuits used by the ECM for information or for direct control of the motor are sensitive to changes in resistance. Simple problems such as loose, dirty or corroded connectors, even pinched wires can cause symptoms making an otherwise good ECM seem bad.

REMOVAL & INSTALLATION

FX, FZR/FZX & VXR/VXS Models

(2) MODERATE

◆ See Figures 125 and 126

1. Remove the seat and rear cargo box. Disconnect the battery cables.
2. Locate the electrical box at the rear of the engine, attached to the bulkhead and remove the cover as detailed in the Ignition & Electrical section. You will probably have to remove the entire box to get at the cover bolts on most models thru 2007. On 1812cc models, move the grey slide locks to the outside, tilt the cover back and lift it off.
3. Once accessed, ECM removal and installation is a relatively simple matter of disconnecting the wiring (tag it unless each of the connectors are different) and removing the 4 fasteners which secure it to the box. There is a good chance that the grommets and spacers will come away with the ECM.
4. *Carefully* clean the body of the ECM and the connector.
5. Check the grommets, replacing if necessary and slide them back into the corresponding slots in the ECM if they were removed. Slide in the spacers and then position the ECM, tightening the bolts to 2.8 ft. lbs/34 inch lbs. (3.9 Nm) on 998cc and 1052cc models; or 3.7 ft. lbs./45 inch lbs. (5 Nm) on 1812cc models. On all models, tighten the bolts in a criss-cross pattern. Always use Loctite 572 on the bolt threads prior to threading them in.
6. Reconnect the connector and then install the cover and/or electrical box.
7. Connect the battery cables and install the seat.

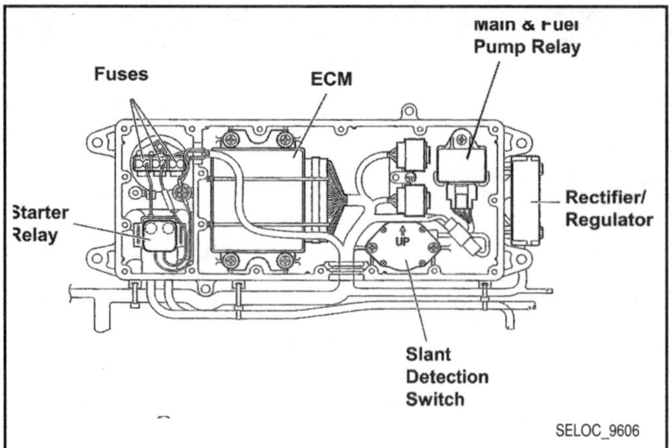

Fig. 125 The ECM is located in the electrical box - 998cc and 1052cc engines

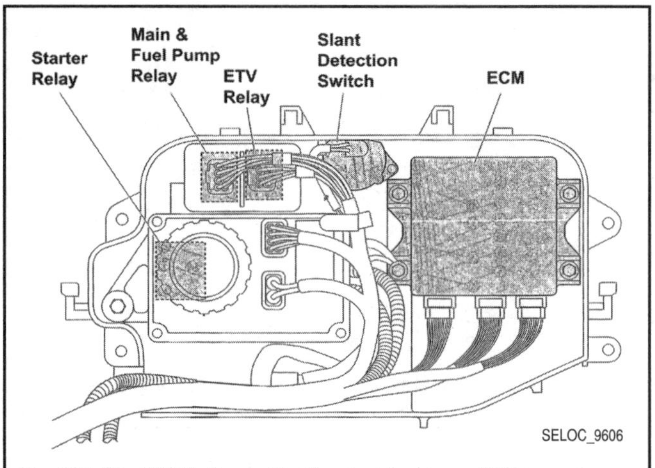

Fig. 126 The ECM is located in the electrical box - 1812cc engines

VX Models w/1052cc Engines

MODERATE

◆ See Figures 127, 128 and 129

1. Remove the seat and rear cargo box. Disconnect the battery cables.

2. On 2005-09 models, locate the ECM on the port side of the engine, below the intake manifold. It, and the rectifier/regulator, are attached to a mounting bracket.

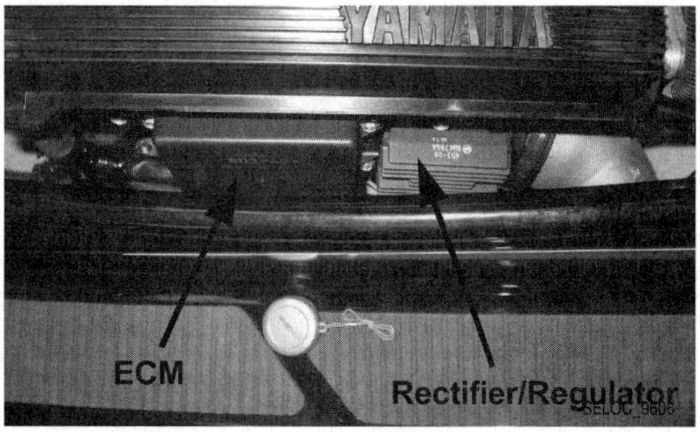

Fig. 127 The ECM is located on the port side of the engine, next to the rectifier/regulator (2005-09 models)

a. Tag and disconnect the 3 harness connectors at the bottom of the ECM.

3. On 2010 and later models, locate the unit at the rear of the engine, near the oil tank. Unlike earlier models, it utilizes a cover.

b. Remove the 7 mounting bolts and tilt the cover back enough so you can tag and disconnect the 3 harness connectors. We recommend leaving the harness attached to the cover; so just move the cover, with the wires still attached, out of the way.

c. Remove the cover gasket and throw it away.

4. On all models now, remove the 4 mounting bolts and lift out the unit. Be careful not to lose the grommets and collars. Note that 2010 and later models also have nuts on the back of the mounting plate!

5. *Carefully* clean the body of the ECM, the bracket or mounting plate and the connectors.

6. Check the grommets, replacing if necessary, and slide them back into the corresponding slots in the ears of the ECM, if they were removed. Slide in the spacers and then position the ECM onto the bracket. Coat the threads lightly with Loctite 572 and then tighten the bolts to 2.9 ft. lbs/35 inch lbs. (3.9 Nm) on 2005-09 models, or 3 ft. lbs/36 inch lbs. (4 Nm) on 2010-11 models. Always in a criss-cross pattern.

7. Reconnect the connectors, clicking them into place. On the 2010-11 models, you'll want to position the cover over the unit so you have enough room to plug in the connectors.

8. On the 2010-11 models, once the harness couplers are connected, move the cover back into position (don't forget the NEW gasket!) and tighten the bolts to 3 ft. lbs/36 inch lbs. (4 Nm).

9. Connect the battery cables and install the seat.

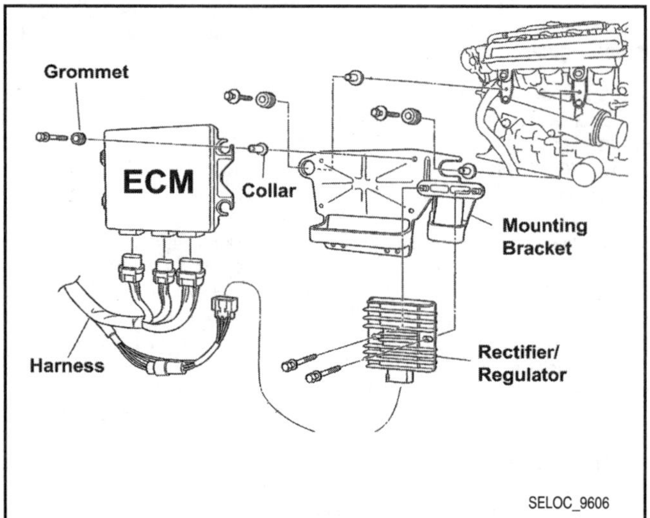

Fig. 128 A good look at the ECM mounting (2005-09 models). . .

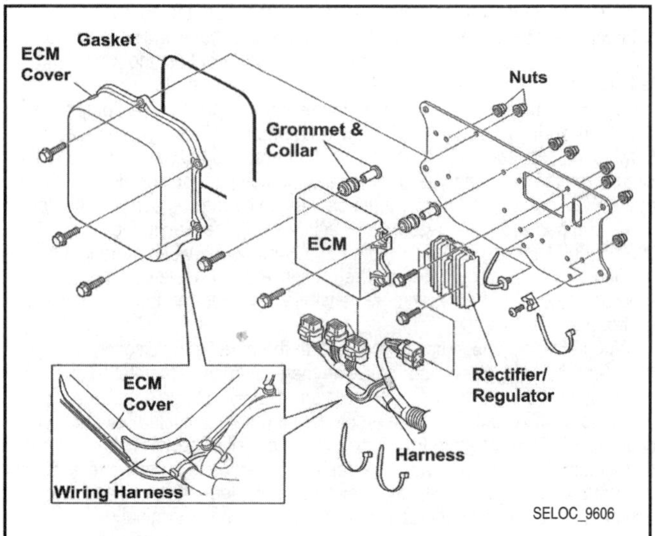

Fig. 129 . . .and on 2010-11 models (w/1052cc engines)

ECM PIN-OUTS & CIRCUIT DIAGRAMS

◆ **See Figures 130 thru 142**

■ **Unfortunately, pin-outs and circuit diagrams are not available for 2002-07 models.**

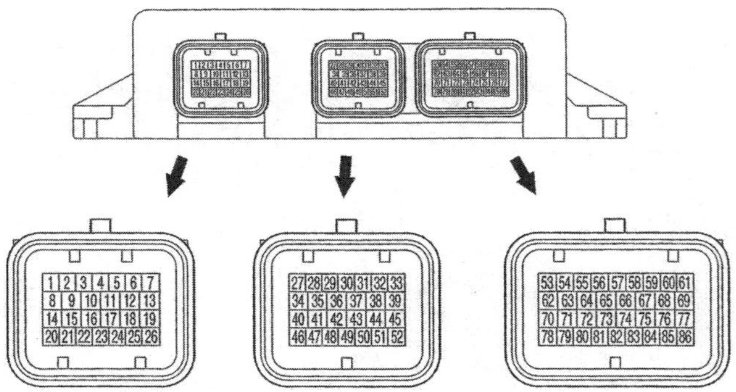

No.	Lead color	External connection lead	No.	Lead color	External connection lead	No.	Lead color	External connection lead
1	—	—	15	B/R	Intake air temperature sensor	29	—	—
2	—	—	16	P/G	Intake air pressure sensor	30	—	—
3	G/O	Cam position sensor	17	W/L	Steering sensor	31	—	—
4	—	—	18	G/W	Reverse sensor	32	—	—
5	L/B	Slant detection switch	19	Y/B	Cruise assist "SET" switch	33	—	—
6	—	—	20	Y/G	Main and fuel pump relay (Main relay)	34	B/R	APS 1 ground
7	R/Y	Battery power source	21	Y	Engine start switch	35	B/W	APS 2 ground
8	W/B	Pickup coil	22	—	—	36	P/R	APS 1
9	—	—	23	B/Y	Engine temperature sensor	37	P/W	APS 2
10	—	—	24	L/R	"NO-WAKE MODE" switch	38	—	—
11	—	—	25	O/R	Steering sensor power source	39	R/Y	ETV relay power source (contact)
12	—	—	26	O	Sensor power source	40	B/O	Sensor ground
13	B/L	Thermo sensor	27	R/Y	Battery power source	41	B/O	TPS ground
14	Y/W	Main relay drive signal	28	W	Immobilizer serial communication	42	P	TPS 1

SELOC_9606

Fig. 130 ECM pin-out and connectors - 2009-11 FX HO models w/1812cc engines

No.	Lead color	External connection lead	No.	Lead color	External connection lead	No.	Lead color	External connection lead
43	P/B	TPS 2	58	Pu/R	Fuel injector #1	73	—	—
44	O	TPS power source	59	Pu/B	Fuel injector #2	74	—	—
45	R/Y	ETV relay power source (contact)	60	B/G	Ignition coil #4	75	—	—
46	B	ECM ground	61	B/Y	Ignition coil #3	76	—	—
47	W/B	Self-diagnosis lead (communication), meter unit communication	62	B	ETV ground	77	—	—
48	—	—	63	Y/G	Cruise assist Down switch	78	G	ETV motor (positive)
49	O/R	APS 1 power source	64	P	Thermoswitch	79	P/W	Oil pressure switch
50	O/W	APS 2 power source	65	—	—	80	Y/W	ETV relay power source (coil)
51	—	—	66	—	—	81	L/R	Main and fuel pump relay (fuel pump relay)
52	L	ETV motor (negative)	67	B	Ground	82	—	—
53	—	—	68	—	—	83	Pu/G	Fuel injector #4
54	Y/R	Cruise assist Up switch	69	—	—	84	Pu/Y	Fuel injector #3
55	—	—	70	B	ETV ground	85	B/R	Ignition coil #1
56	—	—	71	—	—	86	B/W	Ignition coil #2
57	—	—	72	W	Engine shut-off switch			

Color code

B	: Black	B/Y	: Black/yellow	Pu/G	: Purple/green
G	: Green	G/O	: Green/orange	Pu/R	: Purple/red
L	: Blue	G/W	: Green/white	Pu/Y	: Purple/yellow
O	: Orange	L/B	: Blue/black	R/Y	: Red/yellow
P	: Pink	L/R	: Blue/red	W/B	: White/black
W	: White	O/R	: Orange/red	W/L	: White/blue
Y	: Yellow	O/W	: Orange/white	Y/B	: Yellow/black
B/G	: Black/green	P/B	: Pink/black	Y/G	: Yellow/green
B/L	: Black/blue	P/G	: Pink/green	Y/R	: Yellow/red
B/O	: Black/orange	P/R	: Pink/red	Y/W	: Yellow/white
B/R	: Black/red	P/W	: Pink/white		
B/W	: Black/white	Pu/B	: Purple/black		

SELOC_9606

Fig. 131 ECM pin-out and connectors - 2009-11 FX HO models w/1812cc engines (cont'd)

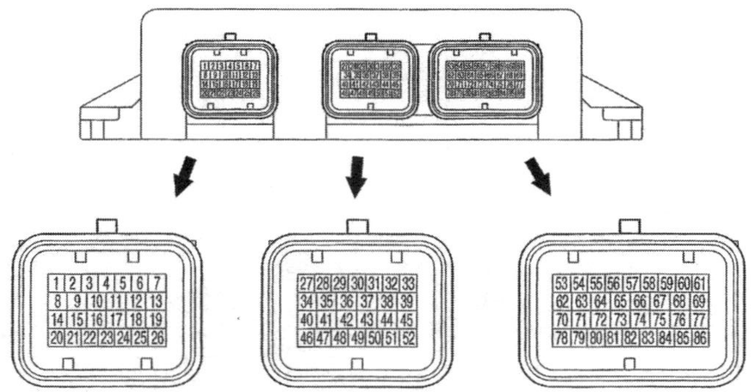

No.	Lead color	External connection lead	No.	Lead color	External connection lead	No.	Lead color	External connection lead
1	G	Knock sensor	15	B/R	Intake air temperature sensor	29	–	–
2	–	–	16	P/G	Intake air pressure sensor	30	–	–
3	G/O	Cam position sensor	17	W/L	Steering sensor	31	–	–
4	–	–	18	G/W	Reverse sensor	32	–	–
5	L/B	Slant detection switch	19	–	–	33	–	–
6	–	–	20	Y/G	Main and fuel pump relay (main relay)	34	B/R	APS 1 ground
7	R/Y	Battery power source	21	Y	Engine start switch	35	B/W	APS 2 ground
8	W/B	Pickup coil	22	–	–	36	P/R	APS 1
9	–	–	23	B/Y	Engine temperature sensor	37	P/W	APS 2
10	–	–	24	–	–	38	–	–
11	–	–	25	O/R	Steering sensor power source	39	R/Y	ETV relay power source (contact)
12	–	–	26	O	Sensor power source	40	B/O	Sensor ground
13	B/L	Thermo sensor	27	R/Y	Battery power source	41	B/O	TPS ground
14	Y/W	Main relay drive signal	28	W	Immobilizer serial communication	42	P	TPS 1

SELOC_9606

Fig. 132 ECM pin-out and connectors - 2008-11 FX SHO & FZR/FZX models w/1812cc engines

No.	Lead color	External connection lead	No.	Lead color	External connection lead	No.	Lead color	External connection lead
43	P/B	TPS 2	58	Pu/R	Fuel injector #1	73	–	–
44	O	TPS power source	59	Pu/B	Fuel injector #2	74	–	–
45	R/Y	ETV relay power source (contact)	60	B/G	Ignition coil #4	75	–	–
46	B	ECM ground	61	B/Y	Ignition coil #3	76	–	–
47	W/B	Self-diagnosis lead (communication), dual analog meter unit communication	62	B	ETV ground	77	–	–
48	–	–	63	–	–	78	G	ETV motor (positive)
49	O/R	APS 1 power source	64	P	Thermoswitch	79	P/W	Oil pressure switch
50	O/W	APS 2 power source	65	–	–	80	Y/W	ETV relay power source (coil)
51	–	–	66	–	–	81	L/R	Main and fuel pump relay (fuel pump relay)
52	L	ETV motor (negative)	67	B	Ground	82	–	–
53	–	–	68	–	–	83	Pu/G	Fuel injector #4
54	–	–	69	–	–	84	Pu/Y	Fuel injector #3
55	–	–	70	B	ETV ground	85	B/R	Ignition coil #1
56	–	–	71	–	–	86	B/W	Ignition coil #2
57	–	–	72	W	Engine shut-off switch			

SELOC_9606

Color code

B	: Black	B/R	: Black/red	P/G	: Pink/green
Br	: Brown	B/W	: Black/white	P/R	: Pink/red
G	: Green	B/Y	: Black/yellow	P/W	: Pink/white
Gy	: Gray	G/L	: Green/blue	Pu/B	: Purple/black
L	: Blue	G/O	: Green/orange	Pu/G	: Purple/green
O	: Orange	G/W	: Green/white	Pu/R	: Purple/red
P	: Pink	L/B	: Blue/black	Pu/Y	: Purple/yellow
R	: Red	L/R	: Blue/red	R/Y	: Red/yellow
W	: White	L/Y	: Blue/yellow	W/B	: White/black
Y	: Yellow	O/R	: Orange/red	W/L	: White/blue
B/G	: Black/green	O/W	: Orange/white	Y/G	: Yellow/green
B/L	: Black/blue	P/B	: Pink/black	Y/W	: Yellow/white
B/O	: Black/orange				

Fig. 133 ECM pin-out and connectors - 2008-11 FX SHO & FZR/FZX models w/1812cc engines (cont'd)

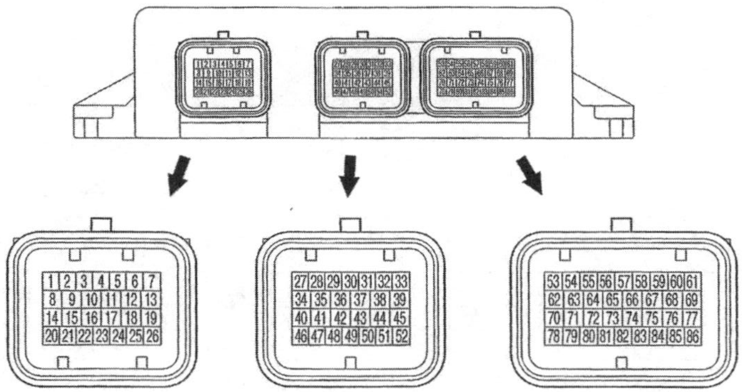

No.	Lead color	External connection lead	No.	Lead color	External connection lead	No.	Lead color	External connection lead
1	—	—	15	B/Y	Intake air temperature sensor	29	—	—
2	—	—	16	P/G	Intake air pressure sensor	30	—	—
3	G/O	Cam position sensor	17	W/L	Steering sensor	31	—	—
4	—	—	18	—	—	32	—	—
5	L/B	Slant detection switch	19	—	—	33	—	—
6	—	—	20	Y/G	Main and fuel pump relay (main relay)	34	B/R	APS 1 ground
7	R/Y	Battery power source	21	Y	Engine start switch	35	B/W	APS 2 ground
8	W/B	Pickup coil	22	—	—	36	P/R	APS 1
9	—	—	23	B/Y	Engine temperature sensor	37	P/W	APS 2
10	—	—	24	—	—	38	—	—
11	—	—	25	O/R	Steering sensor power source	39	R/Y	ETV relay power source (contact)
12	—	—	26	O	Sensor power source	40	B/O	Sensor ground
13	—	—	27	R/Y	Battery power source	41	B/O	TPS ground
14	Y	Main relay drive signal	28	W	Immobilizer serial communication (*1)	42	P	TPS 1

SELOC_9606

Fig. 134 ECM pin-out and connectors - 2010-11 VX models w/1052cc engines

No.	Lead color	External connection lead	No.	Lead color	External connection lead	No.	Lead color	External connection lead
43	P/B	TPS 2	58	Pu/R	Fuel injector #1	73	—	—
44	O	TPS power source	59	Pu/B	Fuel injector #2	74	—	—
45	R/Y	ETV relay power source (contact)	60	B/G	Ignition coil #4	75	—	—
46	B	ECM ground	61	B/Y	Ignition coil #3	76	—	—
47	W/B	Self-diagnosis lead (communication), meter unit communication	62	B	ETV ground	77	B	Ground
48	—	—	63	—	—	78	G	ETV motor (positive)
49	O/R	APS 1 power source	64	P	Thermoswitch	79	P/W	Oil pressure switch
50	O/W	APS 2 power source	65	—	—	80	Y/W	ETV relay power source (coil)
51	—	—	66	—	—	81	L/R	Main and fuel pump relay (fuel pump relay)
52	L	ETV motor (negative)	67	B	Ground	82	—	—
53	—	—	68	—	—	83	Pu/G	Fuel injector #4
54	—	—	69	B	Ground	84	Pu/Y	Fuel injector #3
55	—	—	70	B	ETV ground	85	B/R	Ignition coil #1
56	—	—	71	—	—	86	B/W	Ignition coil #2
57	—	—	72	W	Engine shut-off switch			

(*1) VX Cruiser and VX Deluxe

Color code

B : Black
G : Green
L : Blue
O : Orange
P : Pink
W : White
Y : Yellow
B/G : Black/Green
B/O : Black/Orange
B/R : Black/Red
B/W : Black/White
B/Y : Black/Yellow
G/O : Green/Orange
L/B : Blue/Black
L/R : Blue/Red
O/R : Orange/Red
O/W : Orange/White
P/B : Pink/Black
P/G : Pink/Green
P/R : Pink/Red
P/W : Pink/White
Pu/B : Purple/Black
Pu/G : Purple/Green
Pu/R : Purple/Red
Pu/Y : Purple/Yellow
R/Y : Red/Yellow
W/B : White/Black
W/L : White/Blue
Y/G : Yellow/Green
Y/W : Yellow/White

SELOC_9606

Fig. 135 ECM pin-out and connectors - 2010-11 VX models w/1052cc engines (cont'd)

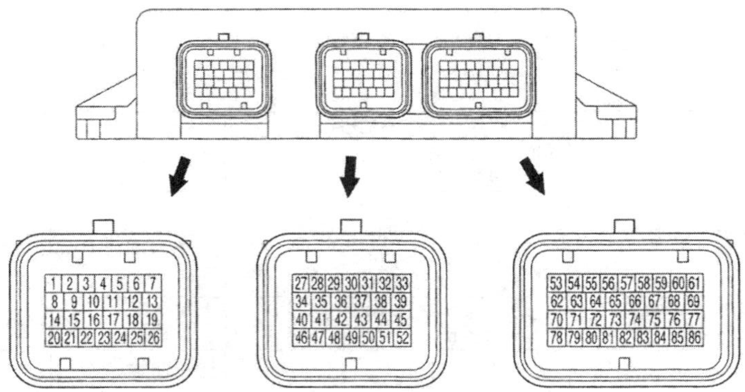

No.	Lead color	External connection lead	No.	Lead color	External connection lead	No.	Lead color	External connection lead
1	—	—	15	B/R	Intake air temperature sensor	29	—	—
2	—	—	16	P/G	Intake air pressure sensor	30	—	—
3	G/O	Cam position sensor	17	W/L	Steering sensor	31	—	—
4	—	—	18	—	—	32	—	—
5	L/B	Slant detection switch	19	—	—	33	—	—
6	—	—	20	Y/G	Main and fuel pump relay (main relay)	34	B/R	APS 1 ground
7	R/Y	Battery power source	21	Y	Engine start switch	35	B/W	APS 2 ground
8	W/B	Pickup coil	22	—	—	36	P/R	APS 1
9	—	—	23	B/Y	Engine temperature sensor	37	P/W	APS 2
10	—	—	24	—	—	38	—	—
11	—	—	25	O/R	Steering sensor power source	39	R/Y	ETV relay power source (contact)
12	—	—	26	O	Sensor power source	40	B/O	Sensor ground
13	B/L	Thermo sensor	27	R/Y	Battery power source	41	B/O	TPS ground
14	Y/W	Main relay drive signal	28	W	Immobilizer serial communication	42	P	TPS 1

SELOC_9606

Fig. 136 ECM pin-out and connectors - 2011 VX models w/1812cc engines

No.	Lead color	External connection lead	No.	Lead color	External connection lead	No.	Lead color	External connection lead
43	P/B	TPS 2	58	Pu/R	Fuel injector #1	73	—	—
44	O	TPS power source	59	Pu/B	Fuel injector #2	74	—	—
45	R/Y	ETV relay power source (contact)	60	B/G	Ignition coil #4	75	—	—
46	B	ECM ground	61	B/Y	Ignition coil #3	76	—	—
47	W/B	Self-diagnosis lead (communication), meter unit communication	62	B	ETV ground	77	—	—
48	—	—	63	—	—	78	G	ETV motor (positive)
49	O/R	APS 1 power source	64	P	Thermoswitch	79	P/W	Oil pressure switch
50	O/W	APS 2 power source	65	—	—	80	Y/W	ETV relay power source (coil)
51	—	—	66	—	—	81	L/R	Main and fuel pump relay (fuel pump relay)
52	L	ETV motor (negative)	67	B	Ground	82	G/L	—
53	—	—	68	—	—	83	Pu/G	Fuel injector #4
54	—	—	69	—	—	84	Pu/Y	Fuel injector #3
55	—	—	70	B	ETV ground	85	B/R	Ignition coil #1
56	—	—	71	—	—	86	B/W	Ignition coil #2
57	—	—	72	W	Engine shut-off switch			

Color code

B	: Black	L/R	: Blue/Red	Pu/R	: Purple/Red
B/G	: Black/Green	O	: Orange	Pu/Y	: Purple/Yellow
B/O	: Black/Orange	O/R	: Orange/Red	R/Y	: Red/Yellow
B/R	: Black/Red	O/W	: Orange/White	W	: White
B/W	: Black/White	P	: Pink	W/B	: White/Black
B/Y	: Black/Yellow	P/B	: Pink/Black	W/L	: White/Blue
G	: Green	P/G	: Pink/Green	Y	: Yellow
G/L	: Green/Blue	P/R	: Pink/Red	Y/G	: Yellow/Green
G/O	: Green/Orange	P/W	: Pink/White	Y/W	: Yellow/White
L	: Blue	Pu/B	: Purple/Black		
L/B	: Blue/Black	Pu/G	: Purple/Green		

SELOC_9606

Fig. 137 ECM pin-out and connectors - 2011 VX models w/1812cc engines (cont'd)

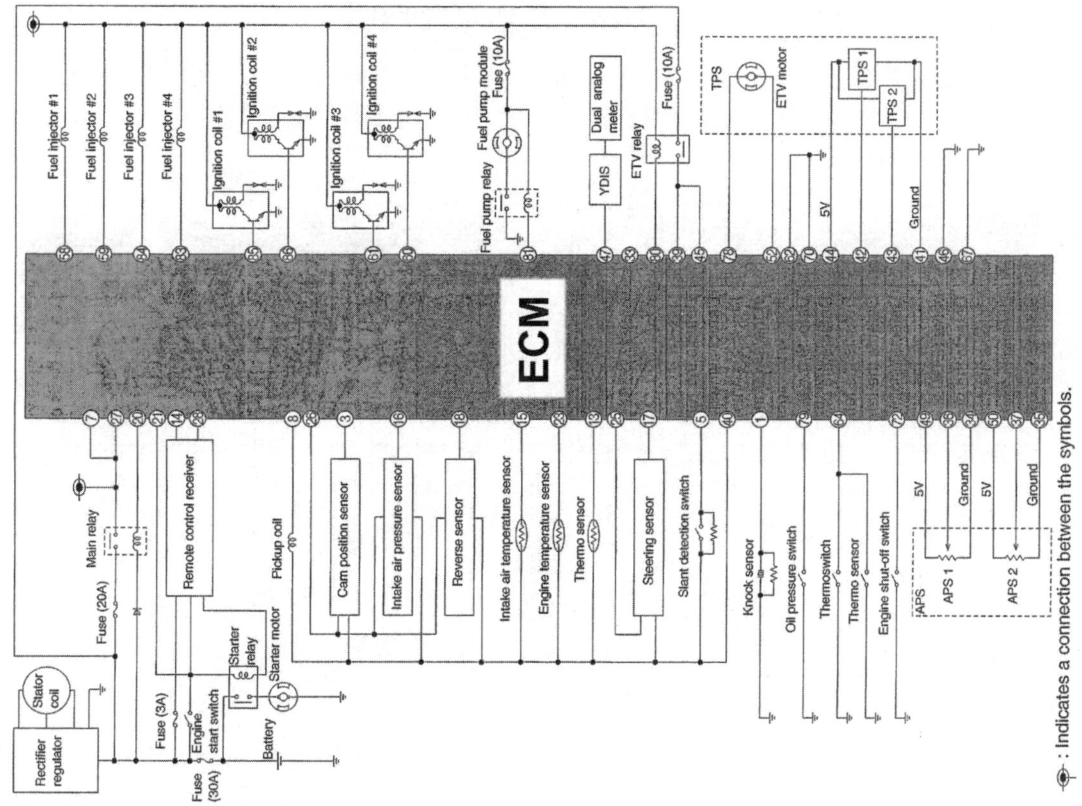

Fig. 139 ECM circuit diagram - 2008-11 FX SHO & FZR/FZX models w/1812cc engines

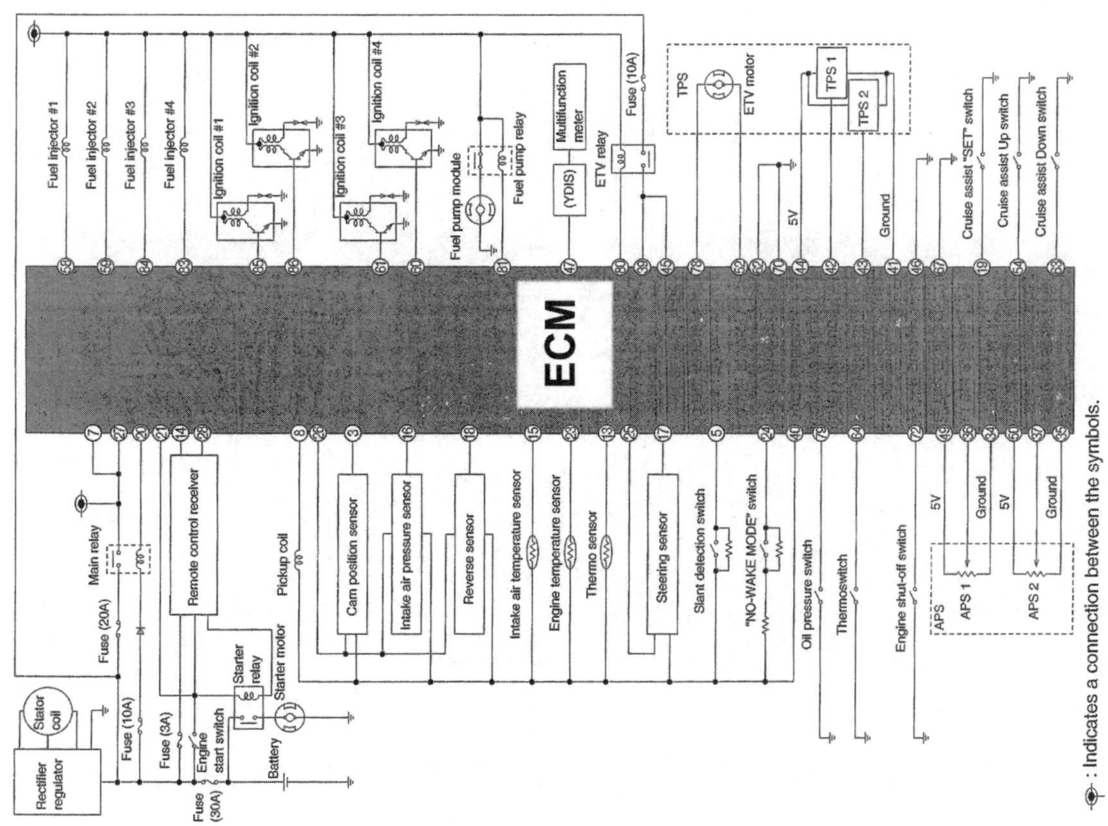

Fig. 138 ECM circuit diagram - 2009-11 FX HO models w/1812cc engines

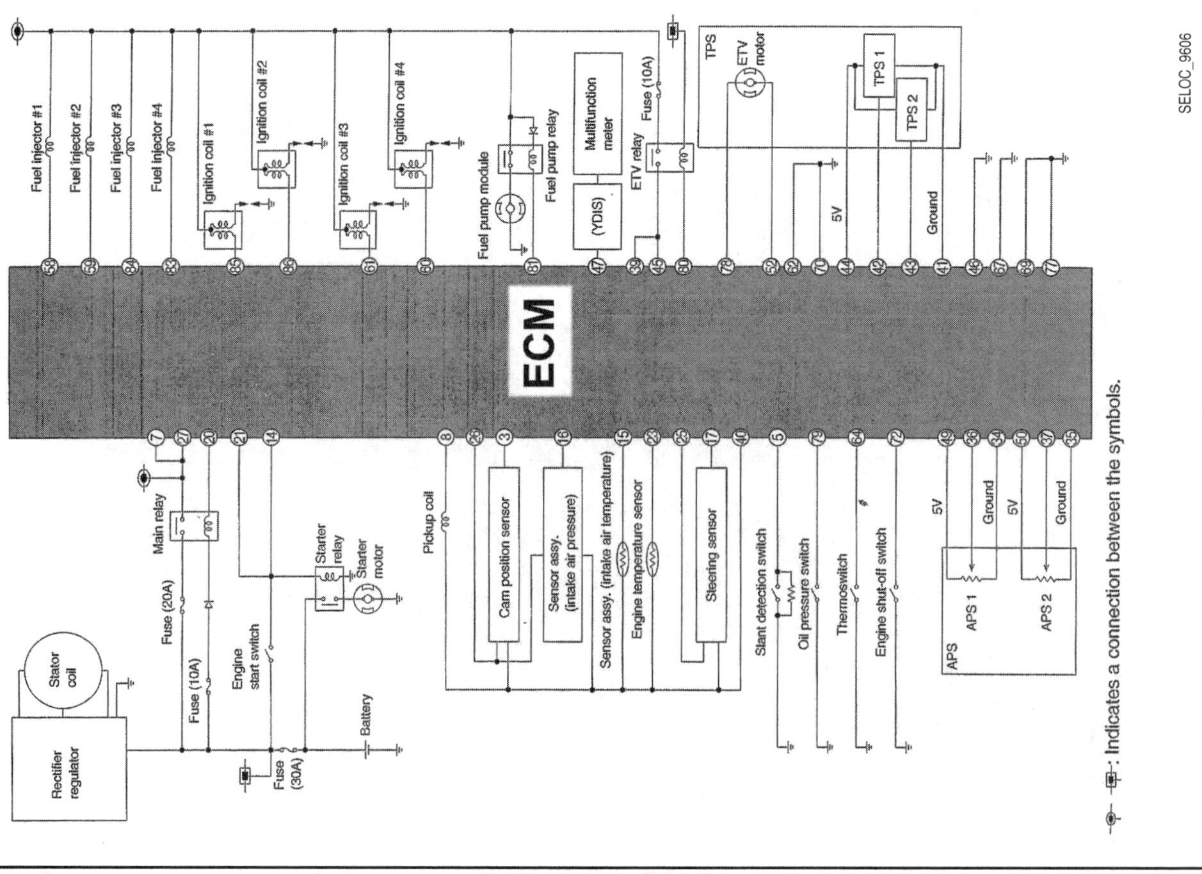

Fig. 140 ECM circuit diagram - 2010-11 VX Sport models w/1052cc engines

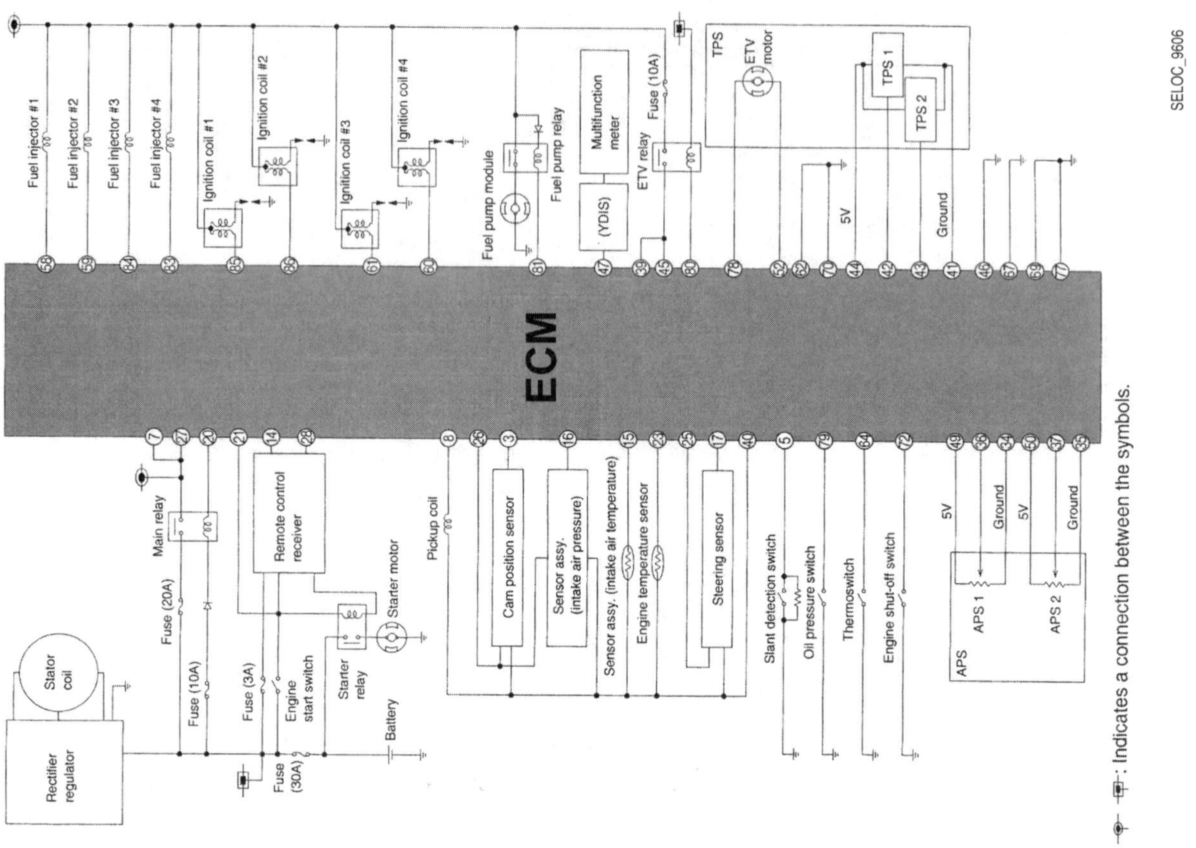

Fig. 141 ECM circuit diagram - 2010-11 VX Cruiser & Deluxe models w/1052cc engines

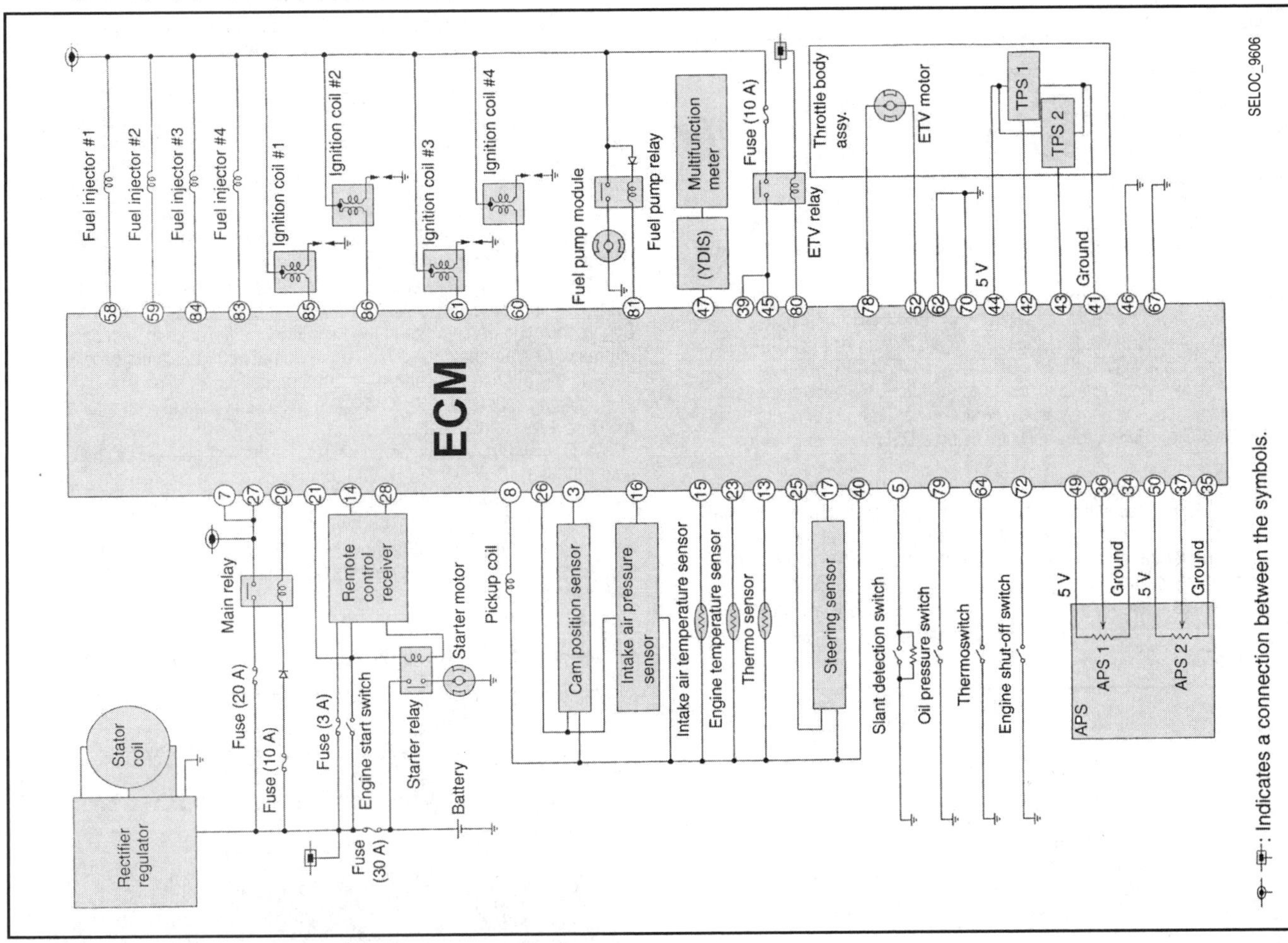

Fig. 142 ECM circuit diagram - 2011 VX models w/1812cc engines

Throttle Position Sensor (TPS)

GENERAL INFORMATION

The throttle position sensor (TPS) is a rotary potentiometer, meaning it sends a variable signal to the ECM based on physical throttle (and therefore sensor) positioning.

The sensor itself receives a reference voltage (usually about 5 volts) from the ECM, and as the throttle lever is rotated, the ECM receives a return voltage signal through a separate wire. This signal will change with throttle position. Generally they are designed so the voltage will increase as the throttle shaft opens, or decrease as the shaft closes. A third wire is used to complete the ground circuit back to the ECM.

Sensor location varies slightly from model-to-model, but on most FX motors, the TPS sensor is located on the top or side of the throttle body assembly, in a position where it can be directly, mechanically actuated by a throttle valve.

REMOVAL & INSTALLATION

② MODERATE

◆ See Figure 143

■ **The throttle position sensor on VX models and all models w/1812cc engines is an integral part of the throttle body/accelerator position sensor (APS)/electronic throttle valve (ETV) and cannot be removed separately. The APS/ETV unit is located under the hood on the port side of the inner hull.**

The TPS is located at the forward edge of the throttle body assembly. Removing the air filter case cover and filter should provide sufficient access to get at the sensor. If not, you will need to remove the flame arrestor assembly as well. Both components are covered in their respective procedures in this section.

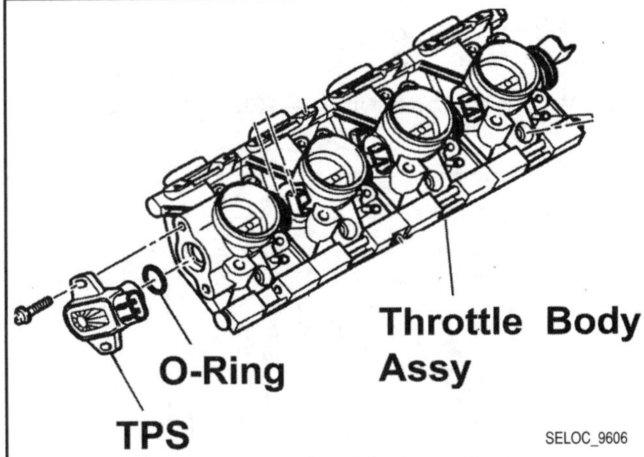

Fig. 143 A good look at the throttle position sensor - 2002-08 FX models w/998cc and 1052cc engines

1. Disconnect the battery cables.
2. Once you have access, press the tab and unlock the sensor connector.
3. Remove the 2 mounting bolts and lift out the sensor. There is an O-ring that may come off with the sensor or stay in the throttle body assy - either way, remove it and throw it away.
4. Position a *new* O-ring onto the sensor and then install them to the throttle body, swiveling the sensor until the bolt holes line up. Thread in the mounting bolts and tighten them to 1.4 ft. lbs./18 inch lbs. (2 Nm). Press the connector back in until it clicks.
5. Install the flame arrestor if it had been removed and the install the filter and case cover. Connect the battery cables.

TESTING

2002-08 FX Models w/998cc & 1052cc Engines

OEM MODERATE

◆ See Figures 144 thru 147

Since Yamaha first started producing 4-stoke EFI motors, they normally recommend checking the TPS output voltage as part of each tune-up (or at least annually).

Since the sensor does not generate a voltage, but instead acts upon a reference voltage from the ECM, all testing must occur with the circuit complete. Yamaha recommends using a 3-pin jumper harness (#YB-06793 / 90890-06793) which is installed inline between the engine ECM wiring harness and the TPS itself to allow access to the circuit voltage readings while the circuit is closed/connected. They also supply a test connector (#YW-06862 / 90890-06862) for use when using a DVOM rather than the YDIS.

However, Yamaha strongly recommends using the YDIS software and connectors. Basically they want you to plug in a laptop and use it as a scan tool to monitor TPS voltages. Honestly, if you can get a hold of the proper software and connectors that really is the best way to go since you don't risk shorting a connection and possible irrevocably damaging the ECM.

So you'll have to make up your own mind how to proceed when testing the TPS, but just be aware of the risks, as an incorrect test connection/short could damage the ECM.

1. If using the YDIS software and wiring, connect the laptop with the appropriate software to the motor diagnostic connector and skip ahead to the voltage reading portions of this procedure.

2. If you've decided to use a test harness to check the TPS, disconnect the wiring from the TPS and connect a test harness (electrical box) or jumper wires (at sensor) to complete the circuit. Disconnect the harness directly at the switch itself.

3. Either use the YDIS system or connect a DVOM set to read DC volts to the appropriate terminals (Pink and Black/Orange) and turn the ignition switch **ON** without starting the motor to check for the following results depending upon motor and throttle setting:

• 2002-08 FX models w/988cc and 1052cc engines - sensor output voltage should be 0.760 V +/- 0.016 V. If out of specification, refer to the Synchronization procedures in the Maintenance & Tune-Up section. Interestingly, Yamaha suggests that the output voltage should be 0.8 V +/- 0.1 V in the Synchro procedure for 2002 models.

Beginning with 2008 FX HO models w/1052cc engines, Yamaha also provides a resistance test.

4. Remove the sensor and connect a 3-pin test harness (#YB-06793 / 90890-06793). Connect the DVOM and check that resistance across the Orange and Black/Orange terminals is 4-6 kilo-ohms at an ambient temperature of 68° F (20° C. If out of specification, replace the sensor.

5. Now connect the DVOM between the Pink and Black/Orange terminals. Slowly rotate the projections on the underside of the TPS in a clockwise direction checking that the resistance changes continuously. If not, replace the sensor.

■ **Always recheck the output voltage after replacing the TP sensor.**

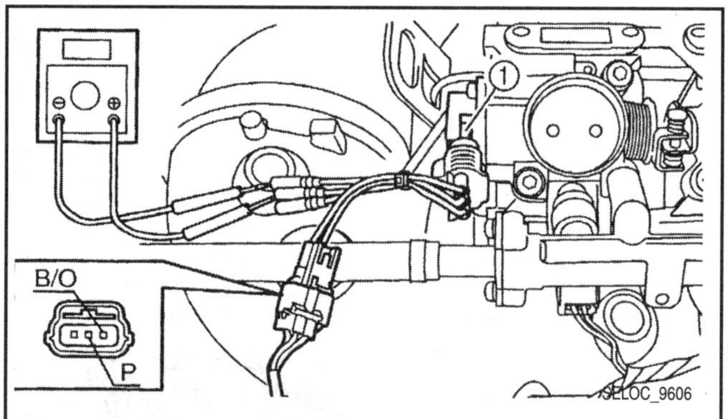

Fig. 144 Using the 3 pin test harness - 2002-08 FX models w/998cc and 1052cc engines

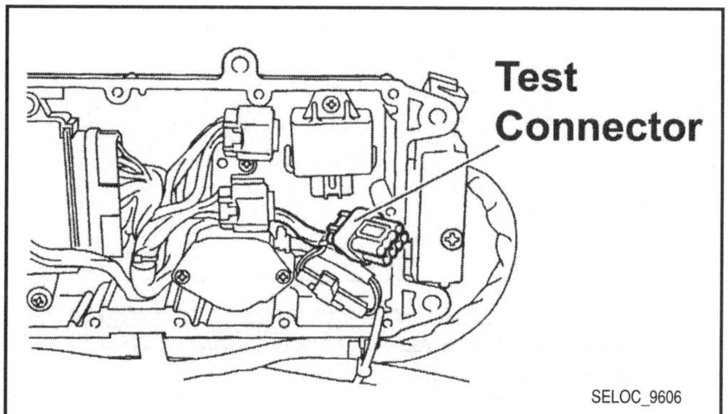

Fig. 145 The test connector is connected in the electrical box - 2002-08 FX models w/998cc and 1052cc engines

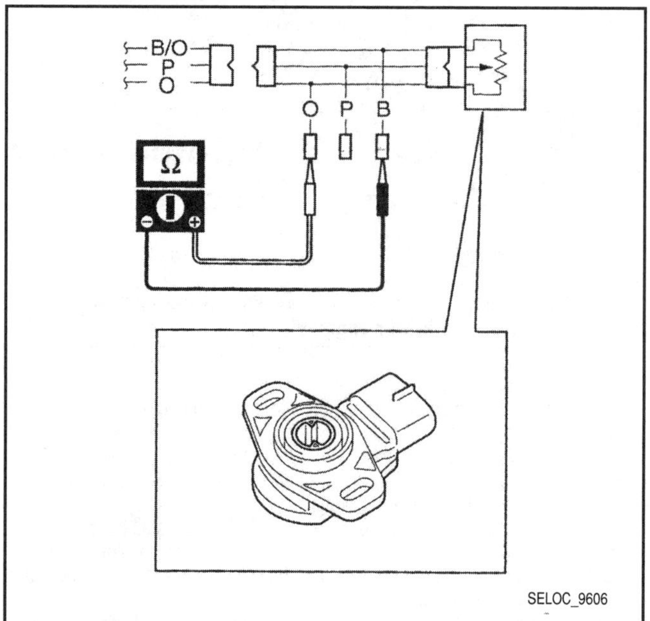

Fig. 146 Check the TPS resistance across these terminals. . .

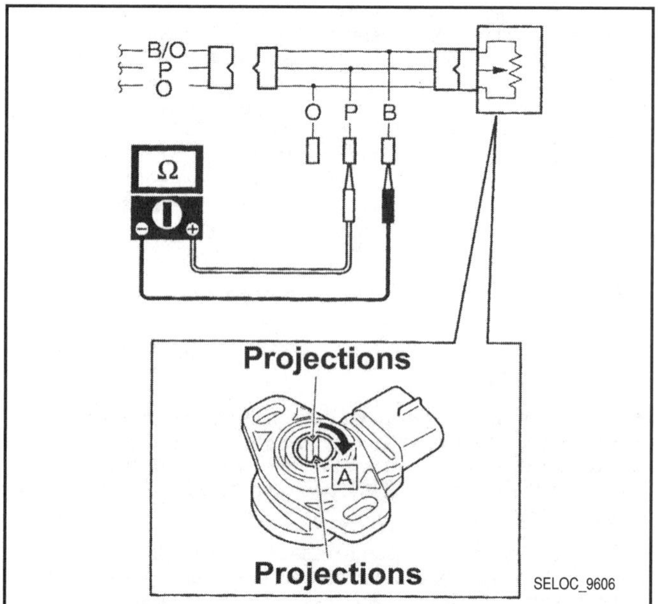

Fig. 147 . . and then that is changes continuously across these while rotating the projections - 2008 FX HO w/1052cc engines

VX Models & All Models w/1812cc Engines

◆ See Figures 148 thru 156

■ **Unfortunately, Yamaha requires the use of their Yamaha Diagnostic Information System (YDIS) software for throttle position sensor checking. They do not provide any test procedures using a conventional DVOM.**

✱✱ CAUTION

Never loosen the throttle stop screw nut at the throttle body and never turn the throttle stop screw itself.

■ **Do not start the engine when using the YDIS software!**

1. Connect your laptop to the diagnostic connector under the hood and then instruct the YDIS software to display the TPS 1 output voltage, the throttle valve opening angle and then the TPS 2 output voltage.
2. Make sure the throttle lever at the handlebar is in the fully closed position, which is to say the lever is all the way out.
3. On 2010-11 VX models and all models w/1812cc engines, squeeze the throttle lever slowly and confirm that the throttle valve opening angle increases as you open the throttle. Also, make sure that the pulley stopper on the APS is in contact with the other fully closed stopper when the throttle lever is fully closed (open on the handlebar). If not, adjust the cable as detailed in the Maintenance & Tune-Up section.

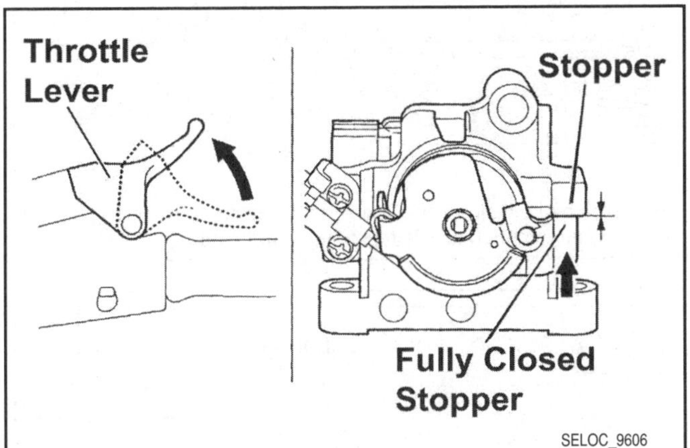

Fig. 148 Check that the cable is adjusted correctly on 2010-11 VX models and all FX/FZ models w/1812cc engines

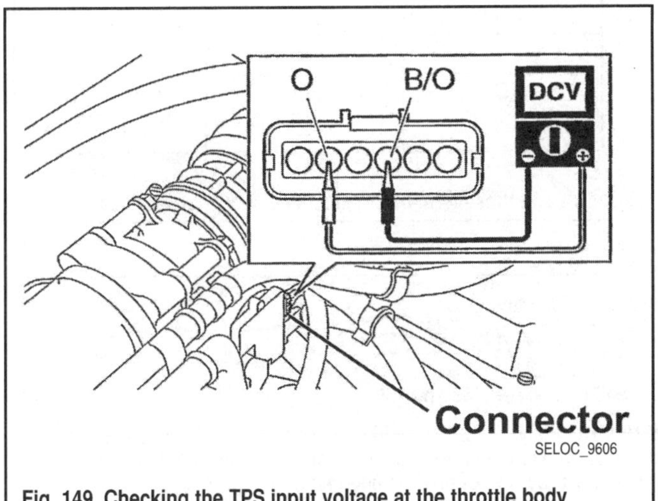

Fig. 149 Checking the TPS input voltage at the throttle body connector on 2010-11 VX models w/1052cc engines. . .

■ **Sensor output voltage and valve opening angle may, and will, vary according to local environmental conditions.**

4. Confirm that the TPS 1 output voltage is:
   - 2005-09 VX models - 0.45-0.95 V
   - 2010-11 VX models (all engines) - 0.6-0.9 V
   - 2008-11 FX and FZ models w/1812cc engines - 0.6-0.9 V

   The throttle lever must be fully closed when checking output voltage. Now check also that the throttle valve opening angle is:
   - 2005-09 VX models - 1.1-9.4°
   - 2010-11 VX models (all engines) - 2-8°
   - 2008-11 FX and FZ models w/1812cc engines - 2-8°

   Also with the lever fully closed.
5. Now squeeze the throttle lever so that it is in the fully opened position (all the way in) and check that the TPS 2 output voltage is 4.6-4.7 V (4.5-4.8 V on 2011 VX w/1812cc engine) and the valve opening angle is above 80° (2005-09 VX) or above 70° (2010-11 VX and all models w/1812cc engines).
6. Slowly ease off on the throttle lever while watching the screen and hold the lever when the TPS 2 voltage is less than 4.5V and the valve opening angle is more than 10°.
7. Subtract the TPS 1 voltage reading from the TPS 2 reading and make sure that it is within 1.9-2.1 V (2.0 V on 2011 VX w/1812cc engine).
8. Squeeze the throttle lever in and out making sure that the values for both sensors changes continuously.

■ **Maximum output for TPS 2 is 4.65 V +/- 0.05 V at the half throttle position.**

There are few more checks available:

9. Move to the throttle body and disconnect the side connector. Turn the ignition key to the **On** position, without starting the engine, and then use a DVOM to check the input voltage at the Orange and Black/Orange connector terminals. Allowable range is 4.75-5.25 V - if out of range, move to the next step.
10. Remove the ECM cover as detailed in the ECM section and then disconnect the throttle body coupler at the throttle body and 2 of the ECM couplers at the bottom of the unit. These would be the port and center connectors. Check for continuity, if none, replace the harness. If there is continuity, replace the throttle body.

Accelerator Position Sensor (APS)

The accelerator position sensor (APS) is used on all VX models and all FX/FZ models w/1812cc engines. The throttle position sensor is also incorporated and they must be replaced as a complete unit - there are no disassembly procedures. In fact, there are actually 2 TP sensors and 2 AP sensors incorporated into the assembly

Sensor location varies slightly from model-to-model, but on all models, the APS sensor is located under the hood, attached to the hull

REMOVAL & INSTALLATION

◆ See Figures 157 and 158

1. Remove the seat and luggage tray (if necessary) and disconnect the battery cables.

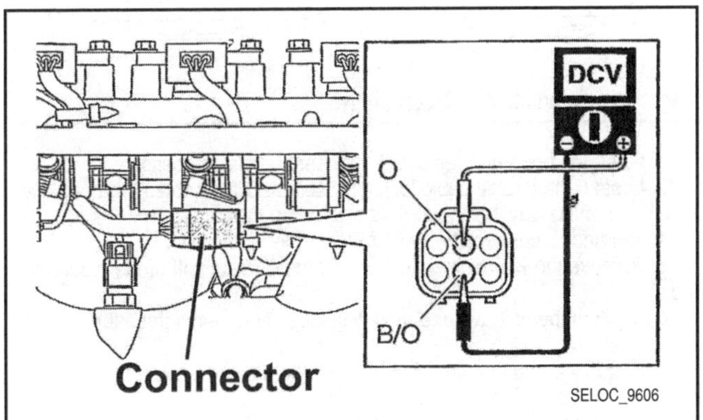

Fig. 150 . . .and all models w/1812cc engines

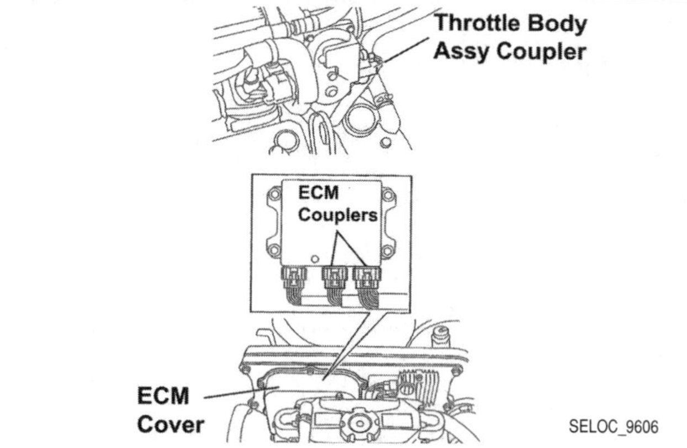

Fig. 151 When checking circuit continuity, disconnect the 3 harness connectors. . .

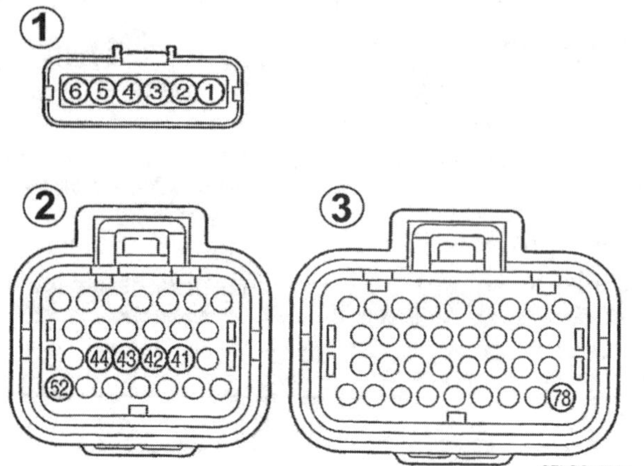

Fig. 152 . . .and check for continuity. . .

TPS Circuit Continuity:

Terminal No.		Color
Coupler ①	Coupler ②, ③	
①	78	G
②	52	L
③	41	B/O
④	42	P
⑤	44	O
⑥	43	P/B

SELOC_9606

Fig. 153 . . .at these terminals - 2010-11 VX models w/1052cc engines

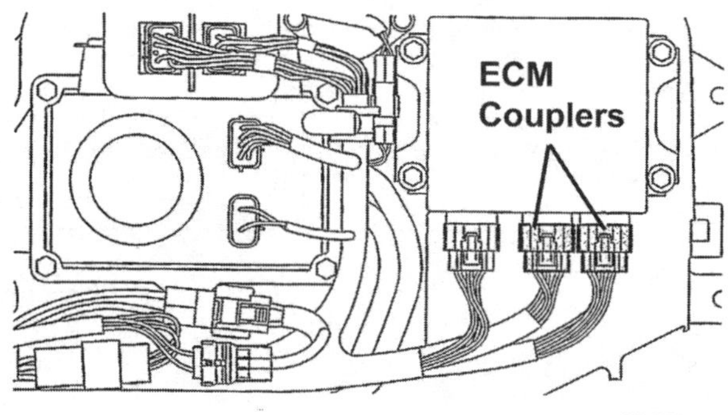

Fig. 154 When checking circuit continuity, disconnect the 2 harness connectors (and the one at the throttle body). . .

Fig. 155 . . .and check for continuity. . .

Wiring harness continuity:

Terminal No.		Color
Coupler ①	Coupler ②, ③	
1	42	P
2	44	O
3	78	G
4	43	P/B
5	41	B/O
6	52	L

SELOC_9606

Fig. 156 . . .at these terminals - all models w/1812cc engines

2. Open the hood and remove the access panels if necessary.
3. Loosen the throttle cable locknut, disconnect the cable at the unit and move it out of the way, being careful not to crimp the cable.
4. Tag and disconnect the 8-pin APS coupler at the unit.
5. Remove the 2 mounting nuts and lift the APS unit off of the mounting studs.
6. Position the unit over the mounting studs and then tighten the nuts to 12 ft. lbs. (17 Nm).
7. Reconnect the throttle cable and then tighten the cable casing lock nuts to 4.7 ft. lbs./58 inch lbs. (6.5 Nm).
8. Reconnect the electrical coupler.
9. Install the access panel(s) if removed and then close the hood.
10. Connect the battery cables and install the seat.

TESTING

2005-09 VX Models

OEM ② MODERATE

◆ See Figures 157, 158 and 159

■ **Unfortunately, Yamaha requires the use of their Yamaha Diagnostic Information System (YDIS) software for accelerator position sensor checking. They do not provide any test procedures using a conventional DVOM as the only means.**

■ **Do not start the engine when using the YDIS software!**

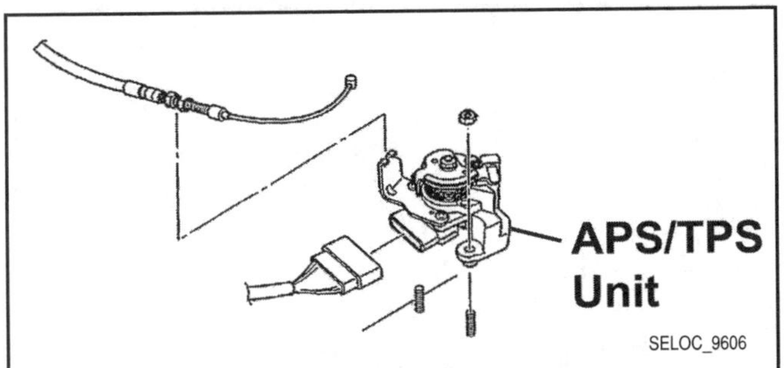

Fig. 157 A good look at the accelerator position sensor, all models similar

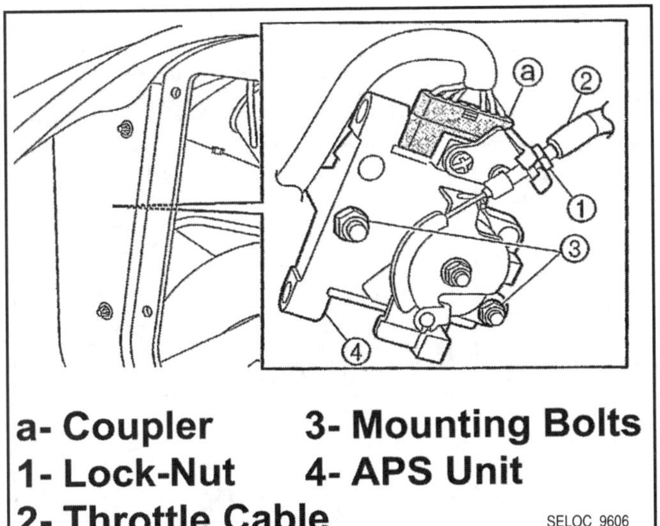

Fig. 158 Removing the APS unit

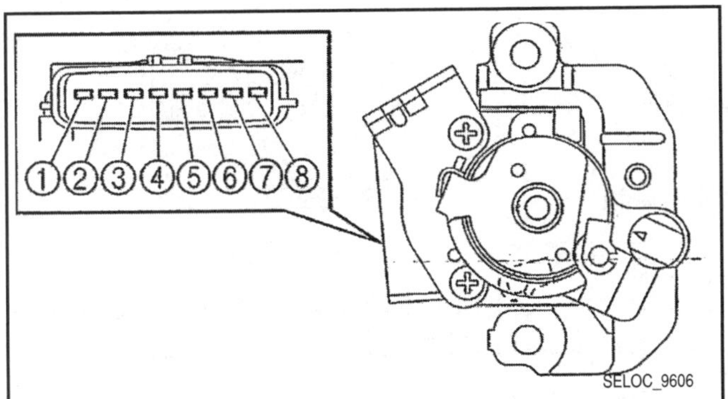

Fig. 159 Testing APS terminal resistance at the unit connector

1. Connect your laptop to the diagnostic connector under the hood and then instruct the YDIS software to display the APS 1 output voltage, the APS 2 output voltage and the throttle valve opening angle.
2. Make sure the throttle lever at the handlebar is in the fully closed position, which is to say the lever is all the way *out*.

■ **Sensor output voltage and valve opening angle may, and will, vary according to local environmental conditions.**

3. Check the output voltage and throttle valve opening angle with the throttle fully closed.
   a. Confirm that the APS 1 output voltage is 0.50-0.90 V
   b. Confirm that the APS 2 output voltage is 0.35-1.05 V
   c. Now, check also that the throttle valve opening angle is 2.7-4.8°
4. Now squeeze the throttle lever so that it is in the fully opened position (all the way in).
   a. Confirm that the APS 1 output voltage is 3.75-4.35 V
   b. Confirm that the APS 2 output voltage is 3.50-4.50 V
   c. Now, check also that the throttle valve opening angle is above 61°
5. Subtract the figure you got for APS 2 while the throttle was fully open from the APS 1 figure (also w/throttle open). The resulting figure must be less than 0.75 V. If not, the unit will require replacement.
6. Squeeze the throttle lever in and out making sure that the values for both sensors change continuously.
7. Connect a DVOM and measure the resistance at terminals **1** and **3** while the throttle lever is in the fully closed position (lever all the way out). APS 1 should be 0.50-0.90 K ohms.
8. Move the tester leads to terminals **6** and **7**. Resistance for APS 2 should be 0.35-1.05 K ohms with the throttle fully closed.
9. Now move the APS cam to the fully opened position and move the meter leads back to terminals **1** and **3** and confirm that resistance for APS 1 is 3.75-4.35 K ohms.
10. Do the same for APS 2, checking that resistance between terminals **6** and **7** is 3.60-4.50 K ohms.
11. Failure to fall within any of these resistance readings will necessitate replacing the APS unit.
12. Squeeze the throttle lever in and out making sure that the values for both sensors change continuously.

2010-11 VX Models & All FX/FZ Models w/1812cc Engines

◆ **See Figures 157, 158 and 160 thru 166**

■ **Unfortunately, Yamaha requires the use of their Yamaha Diagnostic Information System (YDIS) software for accelerator position sensor (APS) checking. They do not provide any test procedures using a conventional DVOM only.**

■ **Do not start the engine when using the YDIS software!**

1. Connect your laptop to the diagnostic connector under the hood and then instruct the YDIS software to display APS 1 and APS 2.
2. Make sure the throttle lever at the handlebar is in the fully closed position, which is to say the lever is all the way *out*.
3. Squeeze the throttle lever slowly and confirm that the throttle valve opening angle increases as you open the throttle. Also, make sure that the pulley stopper on the APS is in contact with the other fully closed stopper when the throttle lever is fully closed (open on the handlebar). If not, adjust the cable as detailed in the Maintenance & Tune-Up section.

■ **Sensor output voltage and valve opening angle may, and will, vary according to local environmental conditions.**

4. Check the output voltage with the throttle fully closed.
   a. Confirm that the APS 1 output voltage is 0.50-0.90 V
   b. Confirm that the APS 2 output voltage is 0.35-1.05 V
5. Now squeeze the throttle lever so that it is in the fully opened position (all the way in).
   a. Confirm that the APS 1 output voltage is 3.75-4.35 V
   b. Confirm that the APS 2 output voltage is 3.60-4.50 V
6. Subtract the figure you got for APS 2 while the throttle was fully open from the APS 1 figure (also w/throttle open). The resulting figure must be less than 0.75 V. If not, the unit will require replacement.
7. Squeeze the throttle lever in and out making sure that the values for both sensors change continuously.
8. Disconnect the APS coupler at the unit and then press the unlock button. Connect the DVOM leads to the Orange/Red and Black/Red terminals in the harness end of the coupler and check that input voltage for APS 1 is 4.75-5.25 V.
9. Move the leads to the Orange/White and Black/White terminals and check that input voltage for APS 2 is 4.75-5.25 V. If either measurement is outside the spread, check the APS circuit.
10. On VX models w/1052cc engines, remove the APS unit as detailed in the Removal & Installation procedures. No other source suggests this for the other models, so it may or may not be necessary.
11. Confirm that the fully closed stopper (fixed) is in contact the APS pulley stopper when the pulley is in the fully closed position.
12. Connect a DVOM and measure the resistance at terminals **1** and **3** while the pulley is in the fully closed position. APS 1 should be 0.50-0.90 K ohms.

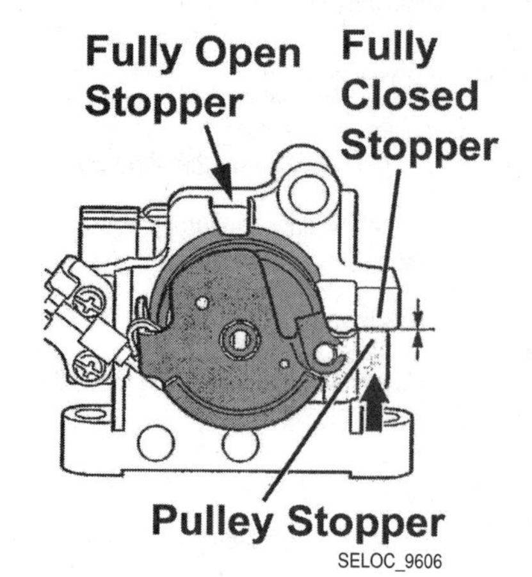

Fig. 160 Testing APS output voltage. . .

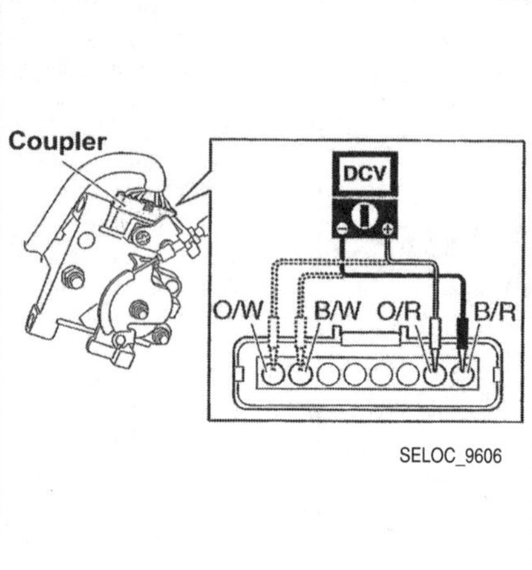

Fig. 161 . . .and input voltage

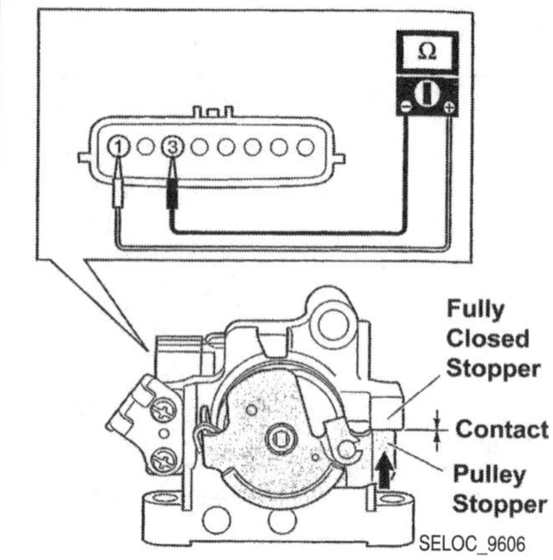

Fig. 162 Testing resistance for APS 1 (shown with pulley fully closed). . .

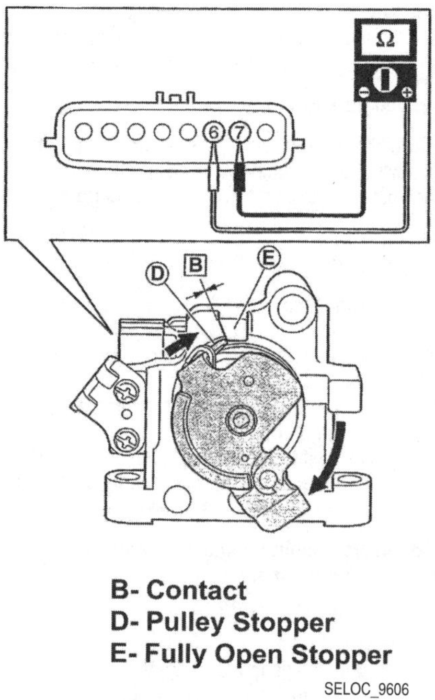

Fig. 163 . . .and APS 2 (shown with pulley fully open)

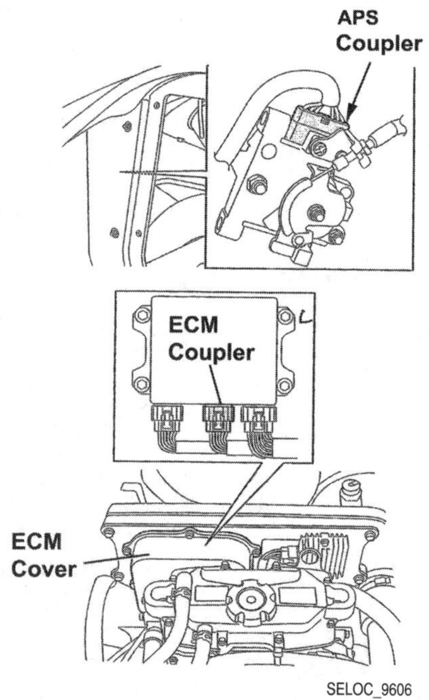

Fig. 164 When checking APS circuit continuity disconnect the harness at each end

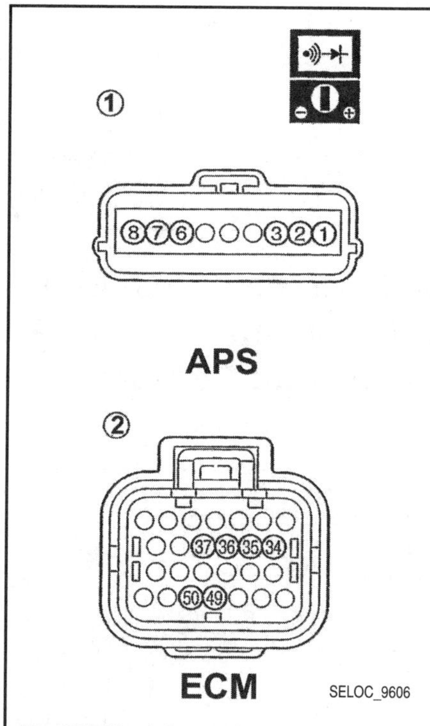

Fig. 165 Check for continuity between these terminals. . .

APS circuit continuity:		
Terminal No.		Color
Coupler ①	Coupler ②	
①	㉞	B/R
②	㊾	O/R
③	㊱	P/R
⑥	㊲	P/W
⑦	㉟	B/W
⑧	㊿	O/W

SELOC_9606

Fig. 166 . . .as suggested in the chart

13. Move the tester leads to terminals **6** and **7**. Resistance for APS 2 should be 0.35-1.05 K ohms with the pulley fully closed.

14. Now move the APS pulley to the fully opened position and the pulley stopper is in contact with the fixed fully open stopper. Move the meter leads back to terminals **1** and **3** and confirm that resistance for APS 1 is 3.75-4.35 K ohms.

15. Do the same for APS 2, checking that resistance between terminals **6** and **7** is 3.60-4.50 K ohms.

16. Failure to fall within any of these resistance readings will necessitate replacing the APS unit.

17. Install the unit, making sure to check the throttle lever free-play after the cable had been reconnected. Details on free-play adjustment can be found in the Maintenance & Tune-Up section.

18. And finally, let's check the APS circuit continuity! Remove the ECM cover and disconnect the middle (ECM) coupler. Disconnect the coupler at the APS unit. Reconnect the DVOM and test for continuity at the terminals as shown in the accompanying table. If continuity is not found at any of them, replace the harness.

19. Reconnect the harness ends and install the ECM cover.

Camshaft Position Sensor (CMP)

REMOVAL & INSTALLATION

◆ See Figure 167

Locate the camshaft position sensor on the cylinder head cover, toward the rear end. Tag and disconnect the sensor coupler, remove the mounting bolt and then lift out the sensor.

■ **Depending on the model/engine, you may have to move a number of components to gain access to the sensor. Take a good look at your set-up and then refer to the appropriate component removal and installation procedures**

When installing the sensor, tighten the bolt to 7.4 ft. lbs./89 inch lbs. (10 Nm) and connect the coupler.

TESTING & SERVICE

◆ See Figures 168 thru 171

Like most sensors, you can check for the proper ECM reference voltage easily enough using a DVOM if you know the correct terminals to check. However to test the sensor itself you'll need to make up a jumper harness so that you can see the output voltage with the circuits intact. Yamaha supplies a 3-pin test harness (#YB-06777 / 90890-06777).

■ **Yamaha suggests using the YDIS system while testing the cam sensor as it can energize (supply power) to the sensor during testing; but we imagine a very careful person without the benefit of this system could probably accomplish the same thing by turning the ignition key to the ON position, but not starting the engine.**

To check the sensor itself, loosen the retaining bolt and remove it from the cylinder cover. Keep in mind that they are sealed to the cover with an O-ring which Yamaha states you should replace once removed (though you might get away with reusing it *if* it's in good condition, but why take the chance?).

Install the test connector between the terminals of the sensor and the wiring harness connector so the circuit will be complete for testing. Connect a DVOM (or use the YDIS) to take readings across the sensor output circuit for the sensor being tested - Green/Orange and Black/Orange.

■ **It is extremely important that the screwdriver passes only in the direction noted and with the orientation shown, or you can't trust the readings.**

Now while watching the DVOM, slowly pass the tip of a screwdriver in front of the sensor moving from the side of the bolt mount past/under the tip. Readings on the sensor output circuit should vary with the position of the screwdriver as follows depending upon the model:

- 2002-05 FX models w/998cc engines - less than 0.8 V when moving the screwdriver toward the edge of the bottom of the sensor. More than 4.8 V when moving it away from the edge

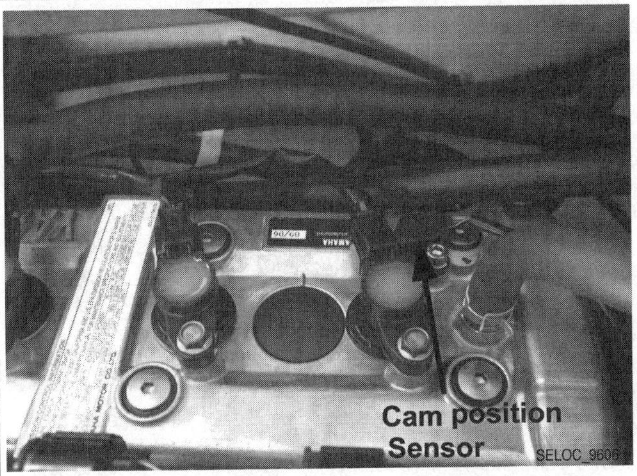

Fig. 167 The camshaft sensor is found in the cylinder head cover, toward the rear - 1052cc engine shown, others similar

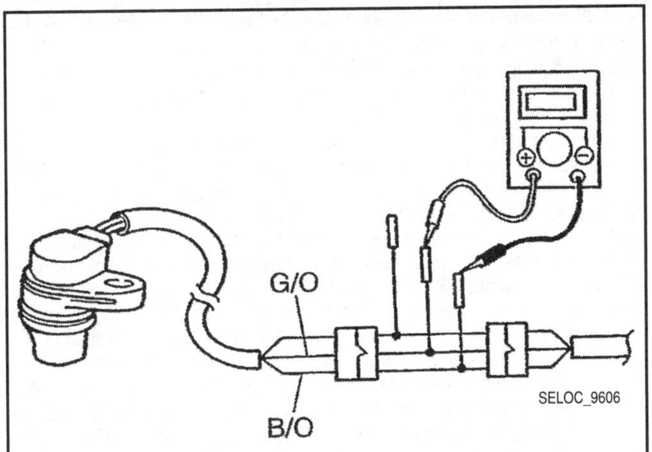

Fig. 168 You'll need the test harness to test the sensor properly

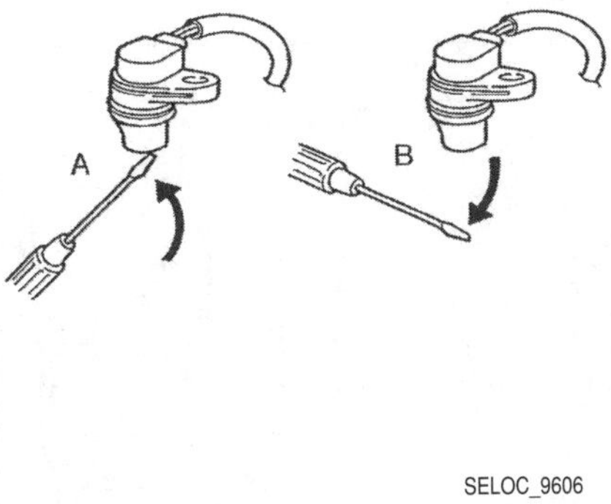

Fig. 169 Use a screwdriver to check output voltage - 2002-08 FX models w/998cc engines

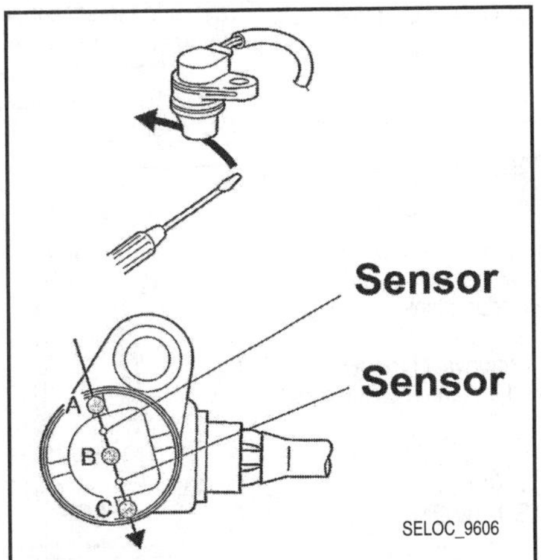

Fig. 170 Use a screwdriver to check output voltage - 2004-07 FX HO models w/1052cc engines and 2005-09 VX models

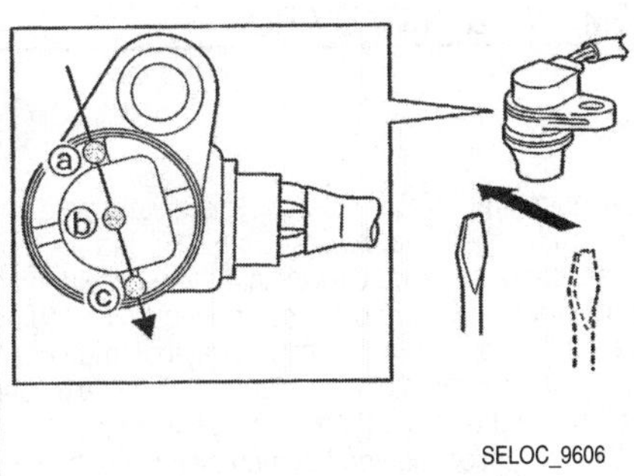

Fig. 171 Use a screwdriver to check output voltage - 2008 FX HO models w/1052cc engines, 2008-11 FX/FZ models w/1812cc engines and 2010-11 VX models (both engines). Note the different orientation of the screwdriver

• 2004-08 FX models w/1052cc engines - with the sensor in the upright position pass a screwdriver underneath the sensor, from the mounting bolt flange side to the other side. As you first come under the sensor (**a**) the meter should read more the 4.8 V. It should drop to less than 0.8 V as you move across the middle of the sensor (**b**), and then go back up to above 4.8 V as you move the screwdriver toward the far edge (**c**). Note the different orientation of the screwdriver on 2008 models.

■ There are actually 2 individual sensors incorporated into the CMP on 2004-07 models w/1052cc engines.

• 2008-11 FX/FZ models and 2011 VX models w/1812cc engines - with the sensor in the upright position pass a screwdriver underneath the sensor, from the mounting bolt flange side to the other side. As you first come under the sensor (**a**) the meter should read more the 4.8 V. It should drop to less than 0.8 V as you move across the middle of the sensor (**b**), and then go back up to above 4.8 V as you move the screwdriver toward the far edge (**c**).

• 2005-11 VX models w/1052cc engines - with the sensor in the upright position pass a screwdriver underneath the sensor, from the mounting bolt flange side to the other side. As you first come under the sensor (**a**) the meter should read more the 4.8 V. It should drop to less than 0.8 V as you move across the middle of the sensor (**b**), and then go back up to above 4.8 V as you move the screwdriver toward the far edge (**c**). Note the different orientation of the screwdriver on 2010-11 models.

Once tests are completed, remove the test harness and make sure a freshly oiled O-ring is installed on the sensor before bolting it back in place on the valve cover. Reinstall any other components removed for access.

Intake Air Temperature Sensor (IAT)

Signals from the intake air temperature sensor (IAT) are used by the EFI system to manage engine operation. As the name suggests, the sensor is used to monitor air temperature. The air temperature sensor is also used by the ECM to help determine air/fuel ratios.

Temperature sensors for modern fuel injection systems are normally thermistors, meaning that they are variable resistors or electrical components that change their resistance value with changes in temperature. From the test data provided by the manufacturer it appears that most sensors are usually Negative Temperature Coefficient (NTC) thermistors. Whereas the resistance of most thermistors (and most electrical circuits) increases with temperature increases (or lowers as the temperature goes down), an NTC sensor operates in an opposite manner. The resistance of an NTC thermistor goes down as temperature rises (or goes up when temperature goes down).

Sensor locations vary slightly by engine and are as follows:

• 2002-08 FX models w/998cc engine and 2008 FX HO models w/1052cc 6AA engines - attached to a bracket riding on the rear side of the fuel rail

• 2004-08 FX HO models w/1052cc 6B6 engines - incorporated with the IAP sensor and installed in the rear of the fuel rail

• 2008-11 FX/FZ models w/1812cc engines - threaded into the intake manifold at the center

• 2005-11 VX models w/1052cc engines - incorporated with the IAP sensor and installed in the rear of the intake manifold

• 2011 VX models w/1812cc engines - threaded into the intake manifold at the center

REMOVAL & INSTALLATION

① EASY

Although removal of the sensor is quite simple. . .disconnect the coupler and remove the nut or bolt(s). On FX/FZ and VX models w/1812cc engines, the sensor is threaded directly into the manifold; no bolt or nut. Many applications may have other components blocking access to the sensor so take a moment and look over your lay-out to determine if removal of the air filter case or some other component will be necessary and then refer to the appropriate procedure for that component(s).

On installation, tighten the mounting nut/bolt to:

• 2002-08 FX models w/998cc engines - 11 ft. lbs. (15 Nm)

• 2004-08 FX HO models w/1052cc 6B6 engines - 2.5 ft. lbs./31 inch lbs (3.5 Nm). Use a new O-ring

• 2008 FX HO models w/1052cc 6AA engines - 11 ft. lbs. (15 Nm)

• 2008-11 FX/FZ models w/1812cc engines - 11.1 ft. lbs. (15 Nm)

• 2005-09 VX models w/1052cc engines - 2.5 ft. lbs./31 inch lbs (3.5 Nm). Use a new O-ring

• 2010-11 VX models w/1052cc engines - 3 ft. lbs./36 inch lbs (4 Nm). Use a new O-ring

• 2011 VX models w/1812cc engines - 11.1 ft. lbs. (15 Nm)

TESTING

2002-08 FX Models w/998cc Engines & 2008 FX HO Models w/1052cc 6AA Engines

② MODERATE

◆ See Figure 172

It is important to EFI operation that the temperature sensors provide accurate signals across the entire operating range and not just when fully hot or fully cold. For this reason, when resistance specifications are available, it is best to test the sensor by watching resistance constantly as the sensor is heated from a cold temperature to the upper end of the engine's operating range. The most accurate way to do this is to suspend the sensor in a container of water, connect a DVOM and slowly heat the liquid while watching resistance on the meter.

To perform this check, you will need a high quality (accurate) DVOM, a thermometer, a length of wire, a metal or laboratory grade glass container and a heat source (such as a hot plate or camp stove). A DVOM with alligator clip style probes will make this test a lot easier; otherwise, alligator clip adaptors can be used, but check before testing to make sure they do not add significant additional resistance to the circuit. This check is performed by connecting the 2 alligator clips together and checking for a very low or 0 resistance reading. If readings are higher than 0, record the value to subtract from the sensor resistance readings that are taken with the clips in order to compensate for the use of the alligator clips.

1. Remove the temperature switch as detailed elsewhere in this section.
2. Suspend the sensor and the thermometer in a container of cool water.

■ To ensure accurate readings make sure the temperature sensor and the thermometer are suspended in the liquid and are not touching the bottom or sides of the container (as the temperature of the container may vary somewhat from the liquid contained within and sensor or thermometer held in suspension).

3. Set the DVOM to the resistance scale, then attach the probes to the sensor or sensor pigtail terminals, as applicable.
4. Allow the temperature of the sensor and thermometer to stabilize, then note the temperature and the resistance reading. If a resistance specification is provided for low-temperatures, you may wish to add ice to the water in order to cool it down and start the test well below ambient temperatures.
5. Use the hot plate or camp stove to slowly raise the temperature. Watch the display on the DVOM closely (in case there are any sudden dips or spikes in the reading which could indicate a problem). Continue to note resistance readings as the temperature rises to 176°F (80°C). Up to that point, the meter should show a steady decrease in resistance that is proportional to the rate at which the liquid is heated. Extreme peaks or valleys in the sensor signal should be rechecked to see if they are results of sudden temperature increases or a possible problem with the sensor.
6. Compare the readings to the sensor test specifications:

• 2002-08 FX models w/998cc engine - 32°F (0°C): 5.4-6.6 K ohms and 176°F (80°C): 0.29-0.39 K ohms.

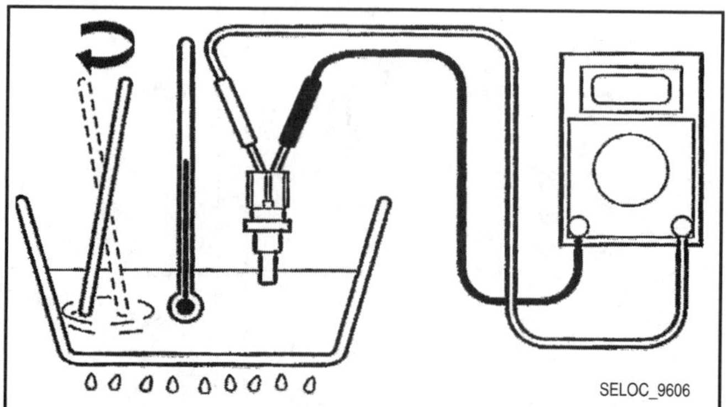

Fig. 172 A direct way of testing the sensors is to use water to change their temperature

• 2008 FX HO models w/1052cc engine - 32°F (0°C): 5.4-6.6 K ohms and 176°F (80°C): 0.29-0.39 K ohms.

• 2008-11 FX HO models w/1812cc engine - 32°F (0°C): 5.4-6.6 K ohms and 176°F (80°C): 0.29-0.39 K ohms.

• 2010-11 VX models w/1052cc engine - 32°F (0°C): 5.4-6.6 K ohms and 176°F (80°C): 0.29-0.39 K ohms.

7. If the sensor falls outside these parameters, it will require replacement.

2004-08 FX HO Models w/1052cc 6B6 Engines, 2008-11 FX/FZ Models w/1812cc Engines & 2005-11 VX Models

OEM ② MODERATE

◆ **See Figures 173 thru 177**

Testing can only be handled with the Yamaha Diagnostic system.

Connect a laptop to the YDIS system connector under the hood and set to display the IAT sensor.

Use a quality thermometer to measure the ambient temperature and check it against the value on the computer. If the displayed intake air temperature differs from ambient by +/- 9°F (5°C), replace the sensor.

■ **Interestingly, the 2008 FX HO, 2010-11 VX and FX/FZ models w/1812cc engines also show a standard resistance test as detailed in the 2002-08 FX Models w/998cc Engines & 2008 FX HO Models w/1052cc 6AA Engines section, so please refer to this procedure as well for the final test. We suspect that this would also apply to earlier VX models, but there is no mention of it what-so-ever.**

On models w/1812cc engines, you can also check the input voltage. Disconnect the connector at the sensor and probe the connector terminals as shown. Voltage should be 4.5-5.5 V (4.75-5.25 V on the VXR/VXS), if not, check and/or replace the harness. Although this test is with the YDIS still connected, we suspect that without the YDIS, you could carefully turn the ignition key **On**, being very careful to not start the engine. This would energize the system as the YDIS would have done.

Fig. 173 VX (1052cc) and certain FX HO models have incorporated the IAT and IAP sensors into one unit

Intake Air Pressure Sensor (IAP)

◆ **See Figures 173, 178 and 179**

All fuel injection systems are equipped with some form of an air pressure sensor. Yamaha usually call it an intake air pressure (IAP) sensor.

Although the mounting point differs slightly from motor-to-motor and may

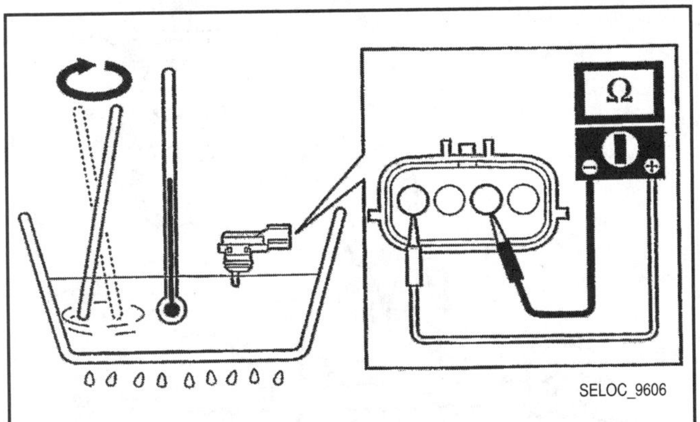

Fig. 174 2010-11 VX models w/1052cc engines provide a resistance test as do other models, but you'll need to be careful how you support the sensor in the water

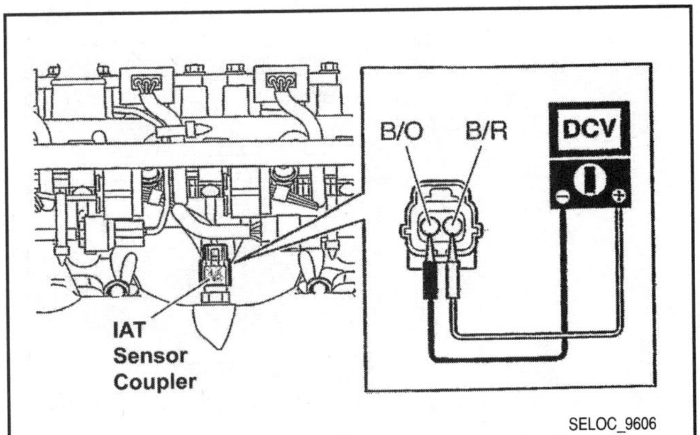

Fig. 175 Checking the input voltage on all 2008-11 models w/1812cc engines - FX HO shown, others similar

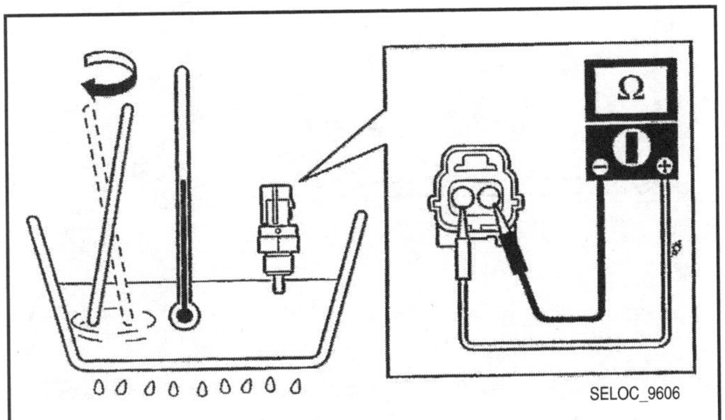

Fig. 176 2008-11 FX HO models and 2011 VX models w/1812cc engines provide a resistance test, as do other models, but you'll need to be careful how you support the sensor in the water

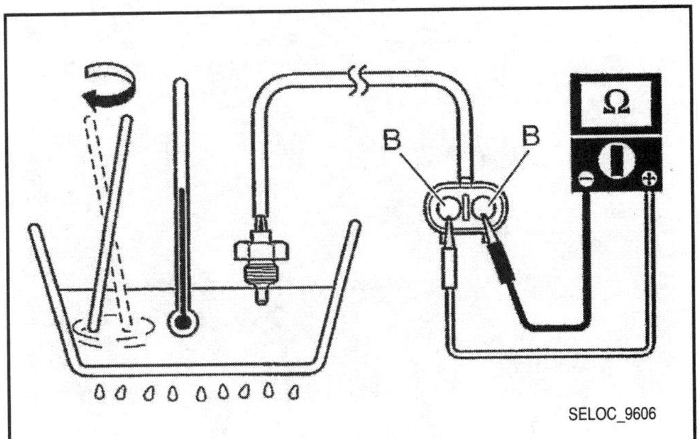

Fig. 177 2009-10 FX SHO and FZ models w/1812cc engines provide a resistance test as do other models

have an effect on exactly what the sensor is measuring (and for what the ECM is calibrated) the basic function performed by these sensors is the same. The sensor works like a barometer and reports barometric air pressure. This sensor allows the computer to compensate for barometric pressure changes at sea level and the normal reduction in atmospheric pressure found at high altitudes.

During normal EFI operation, a large portion of fuel mapping decisions (injector on-time strategy) is made based upon signals from the sensor.

You should also note that most of these motors utilize oil pressure sensors to monitor internal engine lubrication functions (however for more information on Oil Pressure sensors, please refer to the Lubrication & Cooling section as it is not a direct component of the fuel injection system).

Sensor locations vary slightly by engine and are as follows:

- 2002-08 FX models w/998cc engine and 2008 FX HO models w/1052cc 6AA engines - attached to a bracket riding on the rear side of the fuel rail
- 2004-08 FX HO models w/1052cc 6B6 engines - incorporated with the IAP sensor and installed in the rear of the fuel rail
- 2008-11 FX/FZ models w/1812cc engines - connected to a small bracket, just aft of the fuel rail and injectors
- 2005-11 VX models w/1052cc engines - incorporated with the IAP sensor and installed in the rear of the intake manifold
- 2011 VX models w/1812cc engines - connected to a small bracket, just aft of the fuel rail and injectors

REMOVAL & INSTALLATION

① EASY

Although removal of the sensor is quite simple. . .disconnect the coupler and remove the bolt(s); or on others, lift the sensor off of the grommet. Many applications may have other components blocking access to the sensor so take a moment and look over your lay-out to determine if removal of the air filter case or some other component will be necessary and then refer to the appropriate procedure for that component(s).

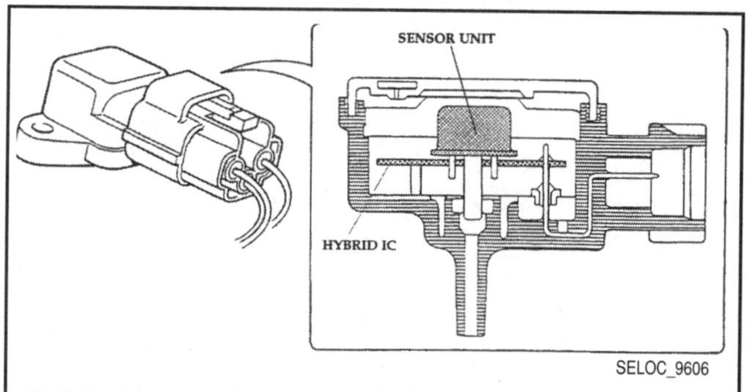

Fig. 178 Cross-section of typical IAP sensor

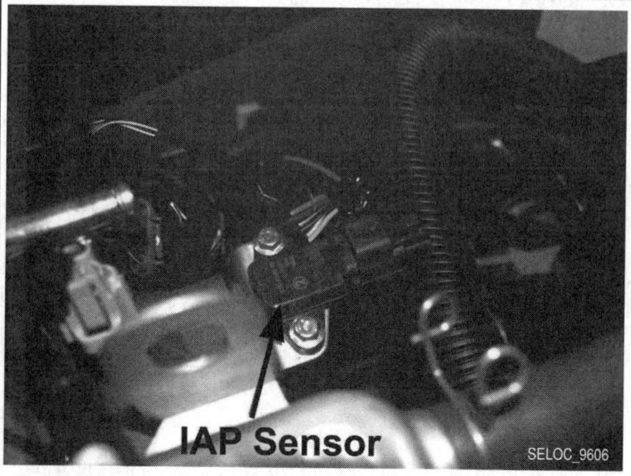

Fig. 179 Typical IAP sensor - 2009 SHO shown, others similar

On installation, tighten the mounting bolts to 2.5 ft. lbs./31 inch lbs (3.5 Nm) on all models but the 2010-11 VX where it is 3 ft. lbs./36 inch lbs (4 Nm). Use a new O-ring if yours came out with one (not all models utilize an O-ring).

TESTING

2002-08 FX Models w/998cc Engines

OEM ③ DIFFICULT

◆ **See Figure 180**

Pressure sensors operate using a reference voltage from the ECM, as a result, there is no way to test them unless the circuit is complete and operating. For the IAP sensor, this means with the ignition keyswitch **ON** and the engine not running, unless you have access to the YDIS, which will supply power to the system on its own.

The best method for testing the sensors themselves is to use the YDIS software from Yamaha.

On all models, you're basically going to remove the sensor from the motor and attach a hand vacuum pump, then use the YDIS software to monitor the signal as vacuum is applied to the sensor nipple. Use the YDIS software to make sure the displayed IAP values change when vacuum is applied.

Yamaha suggests that you use a 3-pin test harness (such as #YB-06769 / 90890-06769, or equivalent) to monitor voltage readings on the sensor while a vacuum pump applies some negative pressure to the sensor. Testing across the Pink/Green and Black/Orange harness wires should show 0.79 volts at 2.8 psi (20 kPa), 1.97 V at 7.1 psi (50 kPa), and finally, 4 V at 14.4 psi (101.3 kPa).

Failure to meet these parameters will necessitate replacing the sensor.

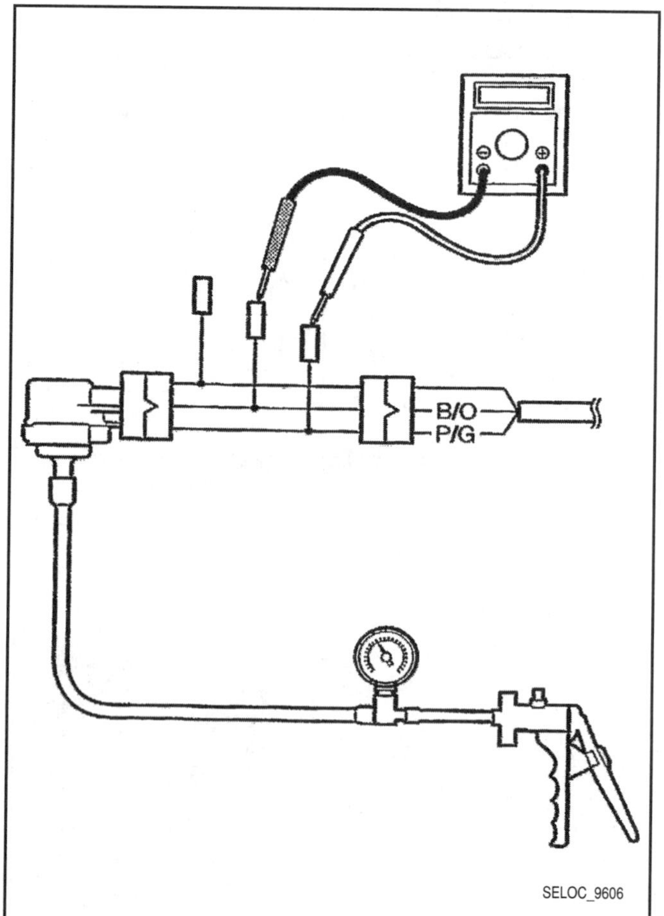

Fig. 180 Testing the IAP sensor - 2002-08 FX models w/998cc engines

2004-08 FX HO Models w/1052cc 6B6 Engines & 2005-09 VX Models

■ As you may have already learned, these models have combined the IAT and IAP sensors into a single unit and Yamaha has not provided any testing procedures for the IAP portion of the combined sensor.

2010-11 VX Models w/1052cc Engines

OEM ② MODERATE

◆ **See Figure 181**

Well, the good news is that unlike its earlier models, it is possible to test the sensor on the 2010 and later models. The bad news is that you can't do it without the YDIS system.

Connect the laptop to the vessel and set the system to display Intake Pressure. Start the engine and then slowly run through the throttle's range, confirming that the air pressure increase or decreases as the throttle does. If not, replace the sensor.

You are also able to check input voltage on this model, so disconnect the coupler at the sensor. Use the YDIS to energize the system, or turn the ignition switch to the **ON** position without starting the engine. Check that the reference signal across the Orange and Black/Orange terminals of the coupler is 4.75-5.25 V. . .if not, inspect the wiring harness.

2008 FX HO Models w/1052cc 6AA Engines

OEM ③ DIFFICULT

◆ **See Figure 182**

Pressure sensors operate using a reference voltage from the ECM, as a result, there is no way to test them unless the circuit is complete and operating. For the IAP sensor, this means with the ignition keyswitch **ON** and the engine not running, unless you have access to the YDIS, which will supply power to the system on its own.

The best method for testing the sensor itself, well ok, the only way, is to use the YDIS software from Yamaha.

On all models, you're basically going to remove the sensor from the motor and attach a hand vacuum pump, then use the YDIS software to monitor the signal as vacuum is applied to the sensor nipple. Use the YDIS software to make sure the displayed IAP values change when vacuum is applied.

Failure to meet these parameters will necessitate replacing the sensor.

All FX/FZ & VX Models w/1812cc Engines

OEM ③ DIFFICULT

◆ **See Figures 183 and 184**

Pressure sensors operate using a reference voltage from the ECM, as a result, there is no way to test them unless the circuit is complete and operating. For the IAP sensor, this means with the ignition keyswitch **ON** and the engine not running, unless you have access to the YDIS, which will supply power to the system on its own.

The best method for testing the sensor itself, well ok, the only way, is to use the YDIS software from Yamaha.

On all models, you're basically going to remove the sensor from the motor and attach a hand vacuum pump, then use the YDIS software to monitor the signal as vacuum is applied to the sensor nipple. Use the YDIS software to make sure the displayed IAP values change when vacuum is applied.

Failure to meet these parameters will necessitate replacing the sensor.

You are also able to check input voltage on these models, so disconnect the coupler at the sensor. Use the YDIS to energize the system, or turn the ignition switch to the **ON** position without starting the engine. Check that the reference signal across the Orange and Black/Orange terminals of the coupler is 4.75-5.25 V. . .if not, inspect the wiring harness.

Self-Diagnostic System

When the electronic engine control system detects a problem with one of its circuits, the ECM will store and display diagnostic trouble codes either

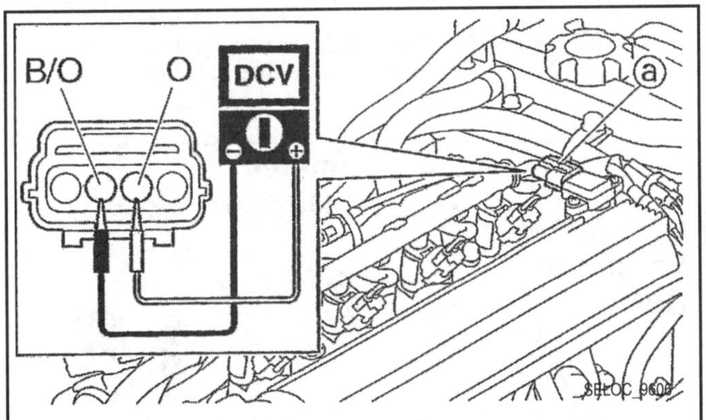

Fig. 181 Testing the IAP sensor input voltage on the 2010-11 VX models w/1052cc - yes we know the coupler should be disconnected!

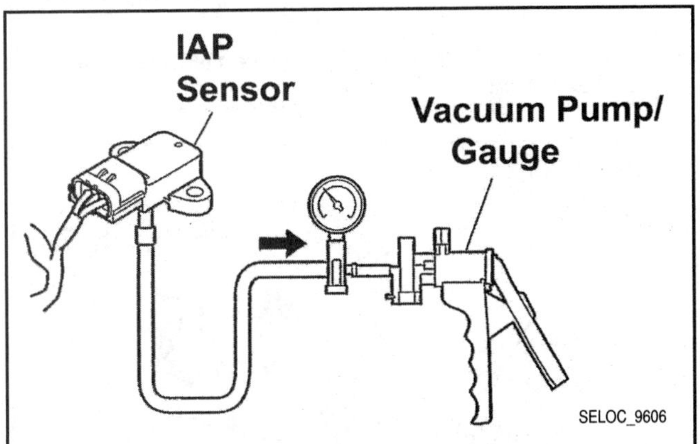

Fig. 182 Testing the IAP sensor - 2008 FX HO models w/1052cc 6AA engines

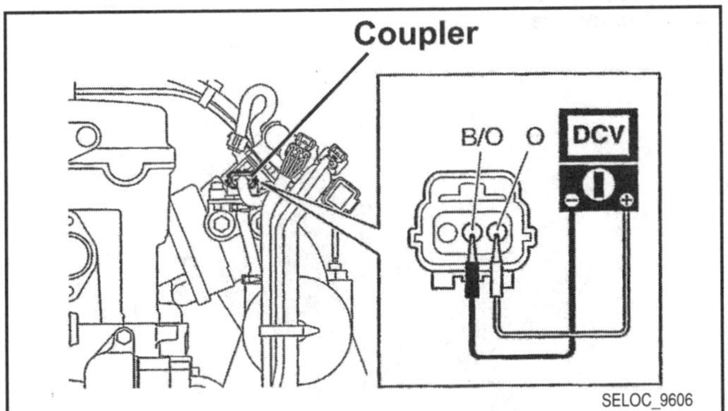

Fig. 183 Testing the IAP sensor input voltage - models w/1812cc engines

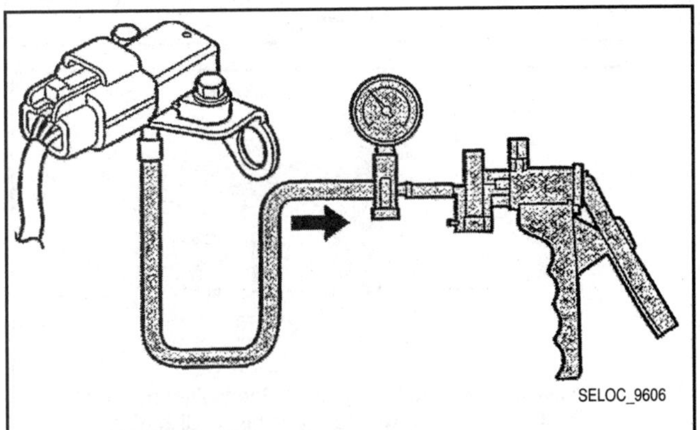

Fig. 184 Testing the IAP sensor

when the appropriate buttons are used on the multi-function display or the Yamaha Diagnostic Information System (YDIS) is connected to the wiring harness.

Yamaha also sells software and the appropriate connectors so that a technician can use a Windows® based computer (usually Windows® 2000, XP® or Vista®) into a diagnostic scan tool. If available this method is usually much more convenient than a diagnostic flash adaptor as it is much easier to read, but also it gives the user the ability to datalog (monitor and record powerhead operating conditions).

As a result of most faults, the ECM will ignore the circuit signal and enter a fail-safe mode designed to keep the vessel from becoming stranded. During fail-safe operation the ECM will provide a fixed substitute value for the faulty circuit. During fail-safe operation the engine will run, but usually with reduced performance (power and economy).

Once a malfunction ceases (has been fixed or the circuit signal returns to something within the anticipated normal range), engine operation will return to normal and the diagnostic flash indicator will cease to display the trouble code for that circuit.

Although the lion's share of diagnostics is accomplished via the YDIS/software, Yamaha has made it very easy to access the trouble codes - simply pushing a button, or buttons, on the multifunction display will call and display a code from memory. Nice! Once done, use the accompanying trouble code charts to determine the affected component or circuit - most can be diagnosed by using the test procedures detailed in the appropriate section here for that component.

READING TROUBLE CODES

② MODERATE

◆ **See Figures 185 thru 189**

As mentioned previously, the bad news is that Yamaha has set up their diagnostics so that they are best handled using the Yamaha Diagnostic Information System (YDIS); bad news for those not having access to this tool that is. . .we certainly recommend purchasing the tool or software if you are able as it will make everything much easier.

Conversely though, the good news is that they have also made it very easy to access stored trouble codes without the special tool(s) and most of the symptoms associated with each code are easily checked by following the testing procedures for the specific sensor or switch that are detailed elsewhere in this section or in the Ignition & Electrical section.

■ **When using the multi-function display to access a stored trouble code, the system can only show one code at a time, so always remember to check it again after you have rectified the problem with the first code. Another reason to spring for the YDIS!**

Proceed as follows to display stored codes:

Start the engine and allow it to run at idle, making sure that you have an adequate supply of cooling water if the vessel is not in the water. Once running:

- 2002-08 FX models w/998cc engines - press the Hour Meter/Voltmeter button at the same time as the Speedometer Display button and hold them both down for 8 seconds until the code comes up on the display screen
- 2004-05 FX HO models w/1052cc engines - press the Select button and hold it down for 8 seconds until the code comes up on the display screen
- 2006-08 FX HO models w/1052cc engines - press the Hour Meter/Voltmeter button at the same time as the Speedometer Display button and hold them both down for 8 seconds until the code comes up on the display screen
- 2008-11 FX HO/SHO models w/1812cc engines - press the Select button and hold it down for 8 seconds until the code comes up on the display screen

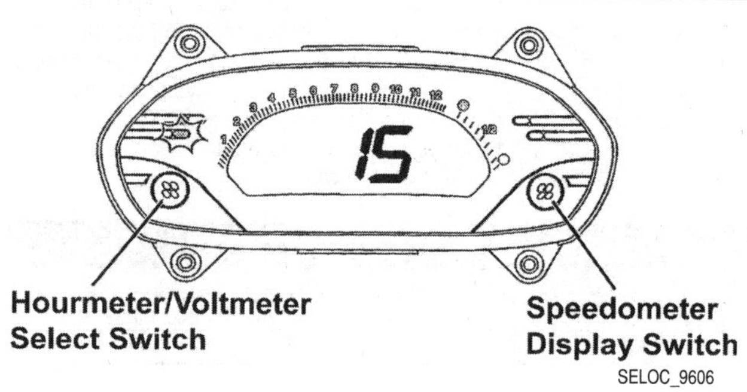

Fig. 185 Accessing the trouble code using the multi-function meter - 2002-05 FX models w/998cc engines

Fig. 186 Accessing the trouble code using the multi-function meter - 2004-05 FX HO models w/1052cc engines

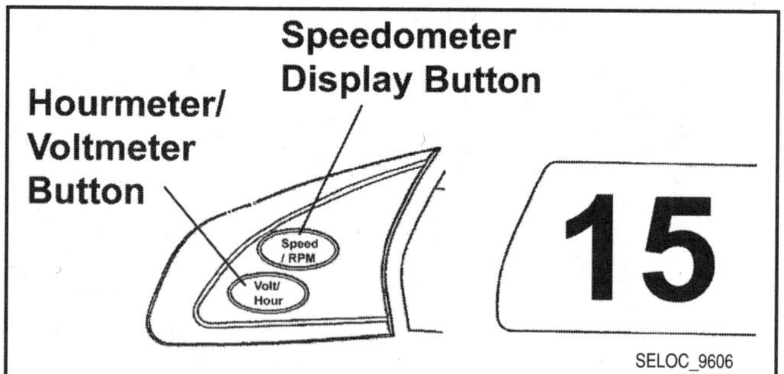

Fig. 187 Accessing the trouble code using the multi-function meter - 2006-08 FX models w/998cc engines & 2006-08 FX HO models w/1052cc engines

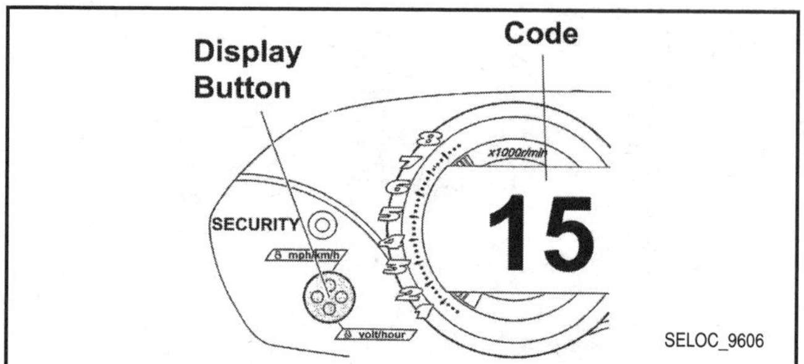

Fig. 188 Accessing the trouble code using the multi-function meter - 2008-11 FX HO/SHO models w/1812cc engines & all 2005-11 VX models

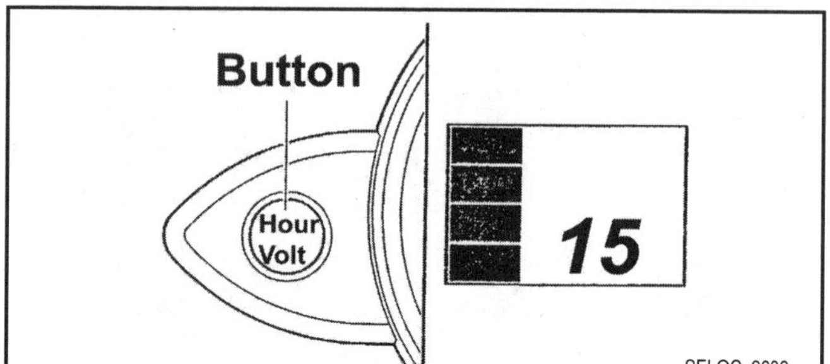

Fig. 189 Accessing the trouble code using the multi-function meter - 2009-11 FZR/FZX models w/1812cc engines

• 2009-11 FZR/FZX models w/1812cc engines - press the Hour Volt button and hold it down for 8 seconds until the code comes up on the display screen

• 2005-11 VX models (both engines) - press the Select button and hold it down for 8 seconds until the code comes up on the display screen

Once you have pulled the trouble code, refer to the appropriate Trouble Code & Symptom chart for you engine. In most instances, the code will indicate a malfunction or voltage problem with a particular sensor or switch. If so, simply find the sensor/switch in its pertinent section, probably either Fuel System or Ignition & Electrical, and follow the testing procedures to determine if the unit is operating out of specifications.

Once you have read the code and fixed the problem, please remember to read the codes through the display a final time to make sure there were no other codes that had been set. Remember, without the YDIS/software, you can only display a single trouble code at a time.

TROUBLE CODE & SYMPTOM CHARTS

◆ **See Figures 190 thru 198**

Please use these charts to determine the symptoms related to each trouble code for your engine.

In most instances, the code will indicate a malfunction or voltage problem with a particular sensor or switch. If so, simply find the sensor/switch in its pertinent section, probably either Fuel System or Ignition & Electrical, and follow the testing procedures to determine if the unit is operating out of specifications.

Once you have read the code and fixed the problem, please remember to read the codes through the display a final time to make sure there were no other codes that had been set. Remember, without the YDIS/software, you can only display a single trouble code at a time.

Code	Symptom
13	Incorrect pulser coil signal
15	Incorrect engine temperature sensor signal
18	Incorrect TPS signal
19	Incorrect battery voltage
23	Incorrect IAT sensor signal
24	Incorrect cam position sensor signal
29	Incorrect IAP sensor signal
39	Incorrect oil pressure sensor signal
47	Incorrect slant detection switch signal
48	Incorrect data transmission

SELOC_9606

Fig. 190 Diagnostic trouble code chart - 2002-05 FX models w/998cc engines

Code	Symptom
01	Normal
13	Incorrect pick-up coil signal
15	Incorrect engine temperature sensor signal
18	Incorrect TPS signal
19	Incorrect battery voltage
23	Incorrect IAT sensor signal
24	Incorrect cam position sensor signal
29	Incorrect IAP sensor signal
47	Incorrect slant detection switch signal
48	Incorrect data transmission
54	Bypass valve motor malfunction
55	Steering switch malfunction
63	Intake air system malfunction

SELOC_9606

Fig. 191 Diagnostic trouble code chart - 2006-08 FX models w/998cc engines & 2004-07 FX HO models w/1052cc engines

Code	Symptom
01	Normal
13	Incorrect pick-up coil signal
15	Incorrect engine temperature sensor signal
18	Incorrect TPS signal
19	Incorrect battery voltage
23	Incorrect IAT sensor signal
24	Incorrect cam position sensor signal
29	Incorrect IAP sensor signal
47	Incorrect slant detection switch signal
48	Incorrect data transmission
53	Exhaust temp sensor malfunction (6AA)
54	Bypass valve motor malfunction
55	Steering switch malfunction
65	Thermo sensor malfunction (6AA engine)

SELOC_9606

Fig. 192 Diagnostic trouble code chart - 2008 FX HO models w/1052cc engines

Code	Symptom
01	Normal
13	Incorrect pick-up coil signal
15	Incorrect engine temperature sensor signal
19	Incorrect battery voltage
23	Incorrect IAT sensor signal
24	Incorrect cam position sensor signal
29	Incorrect IAP sensor signal
47	Incorrect slant detection switch signal
48	Incorrect data transmission
55	Steering switch malfunction
112-123	Electronic control throttle system malfunction
124-128	TPS malfunction
129	Electronic control throttle system malfunction
131-135	Accelerator position sensor malfunction
136-139	Electronic control throttle system malfunction
141-145	Electronic control throttle system malfunction

SELOC_9606

Fig. 193 Diagnostic trouble code chart - 2005-09 VX models w/1052cc engines

Code	Symptom
01	Normal
13	Incorrect pick-up coil signal
15	Incorrect engine temperature sensor signal
19	Incorrect battery voltage
23	Incorrect IAT sensor signal
24	Incorrect cam position sensor signal
29	Incorrect IAP sensor signal
47	Incorrect slant detection switch signal
55	Steering sensor malfunction
112-123	Electronic control throttle system malfunction
124-128	TPS malfunction
129	Electronic control throttle system malfunction
131-135	Accelerator position sensor malfunction
136-139	Electronic control throttle system malfunction
141-145	Electronic control throttle system malfunction
252	Overheat warning
253	Low oil pressure warning

SELOC_9606

Fig. 194 Diagnostic trouble code chart - 2010-11 VX models w/1052cc engines

Code	Symptom
01	Normal
13	Pick-up coil malfunction
15	Engine temperature sensor malfunction
19	Battery voltage malfunction
23	IAT sensor malfunction
24	Camshaft position sensor malfunction
29	IAP sensor malfunction
47	Slant detection switch malfunction
55	Steering sensor malfunction
112-123	Electronic throttle system malfunction
124-128	TPS malfunction
129	Electronic throttle system malfunction
131-135	Accelerator position sensor malfunction
136-139	Electronic control throttle system malfunction
141-145	Electronic control throttle system malfunction
252	Overheat warning
253	Low oil pressure warning

SELOC_9606

Fig. 195 Diagnostic trouble code chart - 2011 VX models w/1812cc engines

Code	Symptom
01	Normal
13	Incorrect pick-up coil signal
15	Incorrect engine temperature sensor signal
17	Knock sensor malfunction
19	Incorrect battery voltage
23	Incorrect IAT sensor signal
24	Incorrect cam position sensor signal
29	Incorrect IAP sensor signal
47	Incorrect slant detection switch signal
55	Steering sensor malfunction
65	Ex. manifold water temp sensor malfunction
68	No-Wake mode switch malfunction
78	Reverse switch malfunction
112-123	Electronic control throttle system malfunction
124-128	TPS malfunction
129	Electronic control throttle system malfunction
131-135	Accelerator position sensor malfunction
136-139	Electronic control throttle system malfunction
141-145	Electronic control throttle system malfunction

SELOC_9606

Fig. 196 Diagnostic trouble code chart - 2008-11 FX SHO models w/1812cc engines

Code	Symptom
01	Normal
13	Incorrect pick-up coil signal
15	Incorrect engine temperature sensor signal
19	Incorrect battery voltage
23	Incorrect IAT sensor signal
24	Incorrect cam position sensor signal
29	Incorrect IAP sensor signal
47	Incorrect slant detection switch signal
55	Steering sensor malfunction
65	Ex. Manifold water temp sensor malfunction
68	No-Wake mode switch malfunction
78	Reverse switch malfunction
112-123	Electronic throttle system malfunction
124-128	TPS malfunction
129	Electronic throttle system malfunction
131-135	Accelerator position sensor malfunction
136-139	Electronic control throttle system malfunction
141-145	Electronic control throttle system malfunction
252	Overheat warning
253	Low oil pressure warning

SELOC_9606

Fig. 197 Diagnostic trouble code chart - 2009-11 FX HO models w/1812cc engines

Code	Symptom
01	Normal
13	Incorrect pick-up coil signal
15	Incorrect engine temperature sensor signal
17	Knock sensor malfunction
19	Incorrect battery voltage
23	Incorrect IAT sensor signal
24	Incorrect cam position sensor signal
29	Incorrect IAP sensor signal
47	Incorrect slant detection switch signal
55	Steering sensor malfunction
65	Ex. Manifold water temp sensor malfunction
78	Reverse switch malfunction
112-123	Electronic throttle system malfunction
124-128	TPS malfunction
129	Electronic throttle system malfunction
131-135	Accelerator position sensor malfunction
136-139	Electronic control throttle system malfunction
141-145	Electronic control throttle system malfunction
252	Overheat warning
253	Low oil pressure warning

SELOC_9606

Fig. 198 Diagnostic trouble code chart - 2009-11 FZR/FZX models w/1812cc engines

TROUBLE CODE SYMPTOM TEST CHARTS

Please use these charts to determine the symptoms tests related to each trouble code for your engine.

These charts have been created for use with the YDIS tool, but in many instances, the code will indicate a malfunction or voltage problem with a particular sensor or switch. If so, simply find the sensor/switch in its pertinent section, probably either Fuel System or Ignition & Electrical, and follow the testing procedures to determine if the unit is operating out of specifications.

■ **Symptom test charts were not available for earlier models than what we have provided here.**

Trouble Code Symptom Test Chart

2009-10 FX HO models w/1812cc engines

Diagnostic Code	System/ Component	Symptom	Check
13	Pulser coil malfunction	Engine will not start	1. Measure the pick-up coil output peak voltage.
			2. Measure the pick-up coil resistance.
			3. Check the white/black (W/B) and black/orange (B/O) pick-up coil wiring harness leads for continuity.
15	Engine temp sensor malfunction	Trolling speed is unstable ①	1. Check the engine temperature using the YDIS.
			2. Measure the engine temperature sensor input voltage.
			3. Check the black/yellow (B/Y) and black/orange (B/O) engine temperature sensor wiring harness leads for continuity.
			4. Measure the engine temperature sensor resistance.
19	Battery voltage malfunction	Trolling speed is unstable ①	1. Check the battery voltage using the YDIS.
			2. Check the fuse for continuity.
			3. Measure the stator coil output peak voltage.
			4. Measure the rectifier regulator output peak voltage.
			5. Check the battery cables and terminals for proper connection.
23	IAT sensor malfunction	Trolling speed is unstable ①	1. Check the intake air temperature using the YDIS.
			2. Measure the intake air temperature sensor input voltage.
			3. Check the black/red (B/R) and black/ orange (B/O) intake air temperature sensor wiring harness leads for continuity.
			4. Measure the intake air temperature sensor resistance.
24	Cam position sensor malfunction	Engine speed is limited	1. Measure the cam position sensor input voltage.
			2. Check the orange (O), green/orange (G/O), and black/orange (B/O) cam position sensor wiring harness leads for continuity.
			3. Measure the cam position sensor output voltage.
29	IAP sensor malfunction	Engine stalls Trolling speed is unstable ①	1. Check the intake air pressure using the YDIS.
			2. Measure the intake air pressure sensor input voltage.
			3. Check the orange (O), pink/green (P/ G), and black/orange (B/O) intake air pressure sensor wiring harness leads for continuity.
			4. Check the intake air pressure sensor operation.
47	Slant detection switch malfunction	Normal operation	1. Check the slant detection switch operation using the YDIS.
			2. Measure the slant detection switch input voltage.
			3. Check the blue/black (L/B) and black/ orange (B/O) slant detection switch wiring harness leads for continuity.
			4. Check the slant detection switch continuity.
55	Steering sensor malfunction	Normal operation	1. Check the steering sensor operation using the YDIS.
			2. Measure the steering sensor input voltage.
			3. Check the orange/red (O/R), white/ blue (W/L), and black/orange (B/O) steering sensor wiring harness leads for continuity.
65	Ex. manifold water temp. sensor malfunction ②	Normal operation	1. Check the thermo sensor operation using the YDIS.
			2. Measure the thermo sensor input voltage.
			3. Check the black/blue (B/L) and black/ orange (B/O) thermo sensor wiring harness leads for continuity.
			4. Measure the thermo sensor resistance.

① This symptom may only be exhibited in certain environmental conditions.

② Thermo sensor

SELOC_9606

Trouble Code Symptom Test Chart

2009-10 FX HO models w/1812cc engines

Diagnostic Code	System/ Component	Symptom	Check
68	No-Wake mode switch malfunction	Normal operation	1. Check the right handlebar switch operation using the YDIS.
			2. Measure the right handlebar switch input voltage.
			3. Check the blue/red (L7R), yellow/black (Y/B), yellow/red (Y/R), yellow/green (Y/G), and black (B) right handlebar switch assembly wiring harness leads for continuity.
78	Reverse switch malfunction ③	Normal operation	1. Check the reverse sensor operation using the YDIS.
			2. Measure the reverse sensor input voltage.
			3. Check the orange (O), green/white (G/ W), and black/orange (B/O) reverse sensor wiring harness leads for continuity.
112 113 114	Electronic throttle system malfunction ④	Engine speed is limited	1. Check the TPS output voltage and throttle valve opening using the YDIS.
			2. Check the ECM circuit for continuity.
			3. Check the ETV relay for continuity.
			4. Check the charging system. • Stator coil output peak voltage • Stator coil resistance • Rectifier regulator output peak voltage • Rectifier regulator continuity
			5. Check the fuse for continuity.
			6. Check the battery voltage and specific gravity.
			7, Check the red (R) and black (B) power supply circuit cables and leads for continuity.
115 116	Electronic throttle system malfunction ④	Engine speed is limited	1. Check the pink (P), orange (O), pink/ black (P/B), and black/orange (B/O) throttle body assembly wiring harness leads for continuity.
			2. Check the TPS output voltage and throttle valve opening using the YDIS.
			3. Check the throttle shaft and throttle inner surface for wear or damage.
117 118 119	Electronic throttle system malfunction ④	Engine speed is limited	1. Check the throttle body assembly coupler terminals for rust and corrosion.
			2. Check the green (G) and blue (L) throttle body assembly wiring harness leads for continuity.
			3. Check the ECM circuit for continuity.
121	Electronic throttle system malfunction ④	Engine speed is limited	1. Cross-check the ECM or replace.
122	Electronic throttle system malfunction ④	Engine speed is limited	1. Cross-check the throttle body assembly.
123	Electronic throttle system malfunction ④	Engine speed is limited	1. Check the ETV relay and fuse (ETV relay) for continuity.
			2. Check the ECM circuit for continuity.
124 125 127 128	Throttle position sensor malfunction	Engine speed is limited	1. Check the pink (P), orange (O), pink/ black (P/B), and black/orange (B/O) throttle body assembly wiring harness leads for continuity.
			2. Check the ECM circuit for continuity.

③ Reverse sensor
④ Throttle body assembly

SELOC_9606

Trouble Code Symptom Test Chart

2009-10 FX HO models w/1812cc engines

Diagnostic Code	System/ Component	Symptom	Check
126	Throttle position sensor malfunction	Engine speed is limited	1. Check the TPS output voltage using the YDIS, 2. Check the ECM circuit for continuity.
129	Electronic throttle system malfunction ④	Engine speed is limited	1. Check the TPS output voltage using the YDIS. 2. Check the ETV relay and fuse (ETV relay) for continuity.
131 132 133 134	Accelerator position sensor malfunction	Engine speed is limited	1. Check the APS output voltage using the YDIS. 2. Check the APS circuit for continuity. 3. Measure the APS 1 and 2 resistance.
135	Accelerator position sensor malfunction	Engine speed is limited	1. Measure the APS 1 and 2 resistance. 2. Check the APS circuit for continuity.
136 137 138 139	Electronic throttle system malfunction ④	Engine speed is limited	1. Check the ECM circuit for continuity. 2. Check the ETV relay for continuity. 3. Check the charging system. • Stator coil output peak voltage • Stator coil resistance • Rectifier regulator output peak voltage • Rectifier regulator continuity 4. Check the fuse for continuity. 5. Check the battery voltage and specific gravity. 6. Check the red (R) and black (B) power supply circuit cables and leads for continuity.
141	Electronic throttle system malfunction ④	Engine speed is limited	1. Check the ECM circuit for continuity. 2. Check the ETV relay and fuse (ETV relay) for continuity.
142	Electronic throttle system malfunction ④	Engine speed is limited	1. Check the TPS output voltage and throttle valve opening using the YDIS. 2. Check the throttle shaft and throttle inner surface for wear or damage.
143 144	Electronic throttle system malfunction ④	Engine speed is limited	1. Check the ECM circuit for continuity. 2. Check the ETV relay for continuity. 3. Check the charging system. • Stator coil output peak voltage • Stator coil resistance • Rectifier regulator output peak voltage • Rectifier regulator continuity 4. Check the fuse for continuity. 5. Check the battery voltage and specific gravity. 6. Check the red (R) and black (B) power supply circuit cables and leads for continuity.
145	Electronic throttle system malfunction ④	Engine speed is limited	1. Check the ETV relay and fuse (ETV relay) for continuity. 2. Check the throttle shaft and throttle inner surface for wear or damage.

SELOC_9606

Trouble Code Symptom Test Chart

2009-10 FX HO models w/1812cc engines

Diagnostic Code	System/ Component	Symptom	Check
252	Overheat warning	Engine speed is limited	1. Check the cooling water passages for obstructions.
			2. Measure the engine temperature sensor resistance.
			3. Check that there is no short circuit between the black/yellow (B/Y) and black/orange (B/O) engine temperature sensor wiring harness leads. ⑤
			4. Check the thermoswitch for continuity.
			5. Check that there is no short circuit between the pink (P) and black/orange (B/O) thermoswitch wiring harness leads. ⑤
			6. Measure the thermo sensor resistance.
			7. Check that there is no short circuit between the black/blue (B/L) and black/orange (B/O) thermo sensor wiring harness leads. ⑤
253	Low oil pressure warning	Engine speed is limited	1. Check the oil pressure switch for continuity.
			2. Check that there is no short circuit in the pink/white (P/W) oil pressure switch wiring harness lead. ⑤
			3. Check the oil passages for obstructions.

① This symptom may only be exhibited in certain environmental conditions.

SELOC_9606

② Thermo sensor

③ Reverse sensor

④ Throttle body assembly

⑤ Disconnect the sensor coupler or switch coupler, and then check that there is no short circuit at the coupler (wiring harness end)

Trouble Code Symptom Test Chart

2008-10 FX SHO models w/1812cc engines

Diagnostic Code	System/ Component	Symptom	Check
13	Pulser coil malfunction	Engine will not start	1. Measure the pick-up coil output peak voltage.
			2. Measure the pick-up coil resistance.
			3. Check the white/black and black/ orange (W/B and B/O) pick-up coil wiring harness leads for continuity.
15	Engine temp sensor malfunction	Trolling speed is unstable ①	1. Check the engine temperature using the YDIS.
			2. Measure the engine temperature sensor input voltage.
			3. Check the black/yellow and black/ orange (B/Y and B/O) engine temperature sensor wiring harness leads for continuity.
			4. Measure the engine temperature sensor resistance.
17	Knock sensor malfunction	Normal operation	1. Check that the knock sensor is installed correctly.
			2. Measure the knock sensor resistance.
			3. Check the green (G) knock sensor wiring harness lead for continuity.
19	Battery voltage malfunction	Trolling speed is unstable ①	1. Check the battery voltage using the YDIS.
			2. Check the fuse for continuity.
			3. Measure the stator coil output peak voltage.
			4. Measure the rectifier regulator output peak voltage.
			5. Check the battery cables and terminals for proper connection.
23	IAT sensor malfunction	Trolling speed is unstable ①	1. Check the intake air temperature using the YDIS.
			2. Measure the intake air temperature sensor input voltage.
			3. Check the black/red and black/orange (B/R and B/O) intake air temperature sensor wiring harness leads for continuity.
			4. Measure the intake air temperature sensor resistance.
24	Cam position sensor malfunction	Engine speed is limited	1. Measure the cam position sensor input voltage.
			2. Check the orange, green/orange, and black/orange (O, G/O, and B/O) cam position sensor wiring harness leads for continuity.
			3. Measure the cam position sensor output voltage.
29	IAP sensor malfunction	Engine stalls Trolling speed is unstable ①	1, Check the intake air pressure using the YDIS.
			2. Measure the intake air pressure sensor input voltage.
			3. Check the orange, pink/green, and black/orange (O, P/G, and B/O) intake air pressure sensor wiring harness leads for continuity.
			4. Check the intake air pressure sensor operation.
47	Slant detection switch malfunction	Normal operation	1. Check the slant detection switch operation using the YDIS.
			2. Measure the slant detection switch input voltage.
			3. Check the biue/black and black/ orange (L/B and B/O) slant detection switch wiring harness leads for continuity.
			4. Check the slant detection switch continuity.
55	Steering sensor malfunction	Normal operation	1. Check the steering sensor operation using the YDIS.
			2. Measure the steering sensor input voltage.
			3. Check the orange/red, white/blue, and black/orange (O/R, W/L, and B/O) steering sensor wiring harness leads for continuity.

SELOC_9606

① This symptom may only be exhibited in certain environmental conditions.

Trouble Code Symptom Test Chart
2008-10 FX SHO models w/1812cc engines

Diagnostic Code	System/ Component	Symptom	Check
65	Ex. manifold water temp. sensor malfunction ②	Normal operation	1. Check the thermo sensor operation using the YDIS.
			2. Measure the thermo sensor input voltage.
			3. Check the black/blue and black/ orange (B/L and B/O) thermo sensor wiring harness leads for continuity.
			4. Measure the thermo sensor resistance.
68	No-Wake mode switch malfunction	Normal operation	1. Check the right handlebar switch operation using the YDIS.
			2. Measure the right handlebar switch input voltage.
			3. Check the blue/red, yellow/black, yellow/red, yellow/green, and black (L/R, Y/B, Y/R, Y/G, and B) right handlebar switch assembly wiring harness leads for continuity.
78	Reverse switch malfunction ③	Normal operation	1. Check the reverse sensor operation using the YDIS.
			2. Measure the reverse sensor input voltage.
			3. Check the orange, green/white, and black/orange (O, G/W, and B/O) reverse sensor wiring harness leads for continuity.
112 113 114	Electronic throttle system malfunction ④	Engine speed is limited	1. Check the TPS output voltage and throttle valve opening using the YDIS.
			2. Check the ECM circuit for continuity.
			3. Check the ETV relay for continuity.
			4. Check the charging system. • Stator coil output peak voltage • Stator coil resistance • Rectifier regulator output peak voltage • Rectifier regulator continuity
			5. Check the fuse for continuity.
			6, Check the battery voltage and specific gravity.
			7. Check the red and black (R and B) power supply circuit cables and leads for continuity.
115 116	Electronic throttle system malfunction ④	Engine speed is limited	1. Check the pink, orange, pink/black, and black/orange (P, O, P/B, and B/O) throttle body assembly wiring harness leads for continuity.
			2. Check the TPS output voltage and throttle valve opening using the YDIS.
			3. Check the throttle shaft and throttle inner surface for wear or damage.
117 118 119	Electronic throttle system malfunction ④	Engine speed is limited	1. Check the throttle body assembly coupler terminals for rust and corrosion.
			2. Check the green and blue (G and L) throttle body assembly wiring harness leads for continuity.
			3. Check the ECM circuit for continuity.
121	Electronic throttle system malfunction ④	Engine speed is limited	1. Cross-check the ECM or replace.
122	Electronic throttle system malfunction ④	Engine speed is limited	1. Cross-check the throttle body assembly.

SELOC_9606

② Thermosensor

③ Reverse sensor

④ Throttle body assembly

Trouble Code Symptom Test Chart

2008-10 FX SHO models w/1812cc engines

Diagnostic Code	System/ Component	Symptom	Check
123	Electronic throttle system malfunction ④	Engine speed is limited	1. Check the ETV relay and fuse (ETV relay) for continuity. 2. Check the ECM circuit for continuity.
124 125 127 128	Throttle Position Sensor malfunction	Engine speed is limited	1. Check the pink, orange, pink/black, and black/orange (P, O, P/B, and B/O) throttle body assembly wiring harness leads f&r continuity. 2, Check the ECM circuit for continuity.
126	Throttle Position Sensor malfunction	Engine speed is limited	1. Check the TPS output voltage using the YDIS. 2. Check the ECM circuit for continuity.
129	Electronic throttle system malfunction ④	Engine speed is limited	1. Check the TPS output voltage using the YDIS. 2. Check the ETV relay and fuse (ETV relay) for continuity.
131 132 133 134	Accelerator position sensor malfunction	Engine speed is limited	1. Check the APS output voltage using the YDIS. 2. Check the APS circuit for continuity. 3. Measure the APS 1 and 2 resistance.
135	Accelerator position sensor malfunction	Engine speed is limited	1. Measure the APS 1 and 2 resistance. 2. Check the APS circuit for continuity.
136 137 138 139	Electronic throttle system malfunction ④	Engine speed is limited	1. Check the ECM circuit for continuity. 2. Check the ETV relay for continuity. 3. Check the charging system. • Stator coil output peak voltage • Stator coil resistance • Rectifier regulator output peak voltage • Rectifier regulator continuity 4. Check the fuse for continuity. 5. Check the battery voltage and specific gravity. 6. Check the red and black (R and B) power supply circuit cables and leads for continuity.
141	Electronic throttle system malfunction ④	Engine speed is limited	1. Check the ECM circuit for continuity. 2. Check the ETV relay and fuse (ETV relay) for continuity.
142	Electronic throttle system malfunction ④	Engine speed is limited	1. Check the TPS output voltage and throttle valve opening using the YDIS. 2. Check the throttle shaft and throttle inner surface for wear or damage.

SELOC_9606

① This symptom may only be exhibited in certain environmental conditions.
② Thermosensor
③ Reverse sensor
④ Throttle body assembly

Trouble Code Symptom Test Chart

2008-10 FX SHO models w/1812cc engines

Diagnostic Code	System/ Component	Symptom	Check
143 144	Electronic throttle system malfunction ④	Engine speed is limited	1. Check the ECM circuit for continuity. 2. Check the ETV relay for continuity. 3. Check the charging system. • Stator coil output peak voltage • Stator coil resistance • Rectifier regulator output peak voltage • Rectifier regulator continuity 4. Check the fuse for continuity. 5. Check the battery voltage and specific gravity. 6. Check the red and black (R and B) power supply circuit cables and leads for continuity.

Diagnostic Code	System/ Component	Symptom	Check
145	Electronic throttle system malfunction ④	Engine speed is limited	1. Check the ETV relay and fuse (ETV relay) for continuity. 2. Check the throttle shaft and throttle inner surface for wear or damage.

SELOC_9606

① This symptom may only be exhibited in certain environmental conditions.

② Thermosensor

③ Reverse sensor

④ Throttle body assembly

Trouble Code Symptom Test Chart

2009-10 FZR/FZX models w/1812cc engines

Diagnostic Code	System/ Component	Symptom	Check
13	Pulser coil malfunction	Engine will not start	1. Measure the pick-up coil output peak voltage.
			2. Measure the pick-up coil resistance.
			3. Check the white/black (W/B) and black/orange (B/O) pick-up coil wiring harness leads for continuity.
15	Engine temp sensor malfunction	Trolling speed is unstable ①	1. Check the engine temperature using the YDIS.
			2. Measure the engine temperature sensor input voltage.
			3. Check the black/yellow (B/Y) and black/orange (B/O) engine temperature sensor wiring harness leads for continuity.
			4. Measure the engine temperature sensor resistance.
17	Knock sensor malfunction	Normal operation	1. Check that the knock sensor is installed correctly.
			2. Measure the knock sensor resistance.
			3. Check the green (G) knock sensor wiring harness lead for continuity.
19	Battery voltage malfunction	Trolling speed is unstable ①	1. Check the battery voltage using the YDIS.
			2. Check the fuse for continuity.
			3. Measure the stator coil output peak voltage.
			4. Measure the rectifier regulator output peak voltage.
			5. Check the battery cables and terminals for proper connection.
23	IAT sensor malfunction	Trolling speed is unstable ①	1. Check the intake air temperature using the YDIS.
			2. Measure the intake air temperature sensor input voltage.
			3. Check the black/red (B/R) and black/ orange (B/O) intake air temperature sensor wiring harness leads for continuity.
			4. Measure the intake air temperature sensor resistance.
24	Cam position sensor malfunction	Engine speed is limited	1. Measure the cam position sensor input voltage,
			2. Check the orange (O), green/orange (G/O), and black/orange (B/O) cam position sensor wiring harness leads for continuity.
			3. Measure the cam position sensor output voltage.
29	IAP sensor malfunction	Engine stalls Trolling speed is unstable ①	1. Check the intake air pressure using the YDIS.
			2. Measure the intake air pressure sensor input voltage.
			3. Check the orange (O), pink/green (P/ G), and black/orange (B/O) intake air pressure sensor wiring harness leads for continuity.
			4. Check the intake air pressure sensor operation.
47	Slant detection switch malfunction	Normal operation	1. Check the slant detection switch operation using the YDIS.
			2. Measure the slant detection switch input voltage.
			3. Check the blue/black (L/B) and black/ orange (B/O) slant detection switch wiring harness leads for continuity.
			4. Check the slant detection switch continuity.
55	Steering sensor malfunction	Normal operation	1. Check the steering sensor operation using the YDIS.
			2. Measure the steering sensor input voltage.
			3. Check the orange/red (O/R), white/ blue (W/L), and black/orange (B/O) steering sensor wiring harness leads for continuity.

SELOC_9606

① This symptom may only be exhibited in certain environmental conditions.

Trouble Code Symptom Test Chart

2009-10 FZR/FZX models w/1812cc engines

Diagnostic Code	System/ Component	Symptom	Check
65	Ex. manifold water temp. sensor malfunction ②	Normal operation	1. Check the thermo sensor operation using the YDIS. 2. Measure the thermo sensor input voltage. 3. Check the black/blue (B/L) and black/ orange (B/O) thermo sensor wiring harness leads for continuity. 4. Measure the thermo sensor resistance.
78	Reverse switch malfunction ③	Normal operation	1. Check the reverse sensor operation using the YDIS. 2. Measure the reverse sensor input voltage. 3. Check the orange (O), green/white (G/ W), and black/orange (B/O) reverse sensor wiring harness leads for continuity.
112 113 114	Electronic throttle system malfunction ④	Engine speed is limited	1. Check the TPS output voltage and throttle valve opening using the YDIS, 2. Check the ECM circuit for continuity. 3. Check the ETV relay for continuity. 4. Check the charging system. • Stator coil output peak voltage • Stator coil resistance • Rectifier regulator output peak voltage • Rectifier regulator continuity 5. Check the fuse for continuity. 6. Check the battery voltage and specific gravity. 7. Check the red (R) and black (B) power supply circuit cables and leads for continuity.
115 116	Electronic throttle system malfunction ④	Engine speed is limited	1. Check the pink (P), orange (O), pink/ black (P/B), and black/orange (B/O) throttle body assembly wiring harness leads for continuity. 2. Check the TPS output voltage and throttle valve opening using the YDIS. 3. Check the throttle shaft and throttle inner surface for wear or damage.
117 118 119	Electronic throttle system malfunction ④	Engine speed is limited	1. Check the throttle body assembly coupler terminals for rust and corrosion. 2. Check the green (G) and blue (L) throttle body assembly wiring harness leads for continuity. 3. Check the ECM circuit for continuity.
121	Electronic throttle system malfunction ④	Engine speed is limited	1. Cross-check the ECM or replace.
122	Electronic throttle system malfunction ④	Engine speed is limited	1. Cross-check the throttle body assembly.
123	Electronic throttle system malfunction ④	Engine speed is limited	1. Check the ETV relay and fuse (ETV relay) for continuity. 2. Check the ECM circuit for continuity.

② Thermosensor
③ Reverse sensor
④ Throttle body assembly

SELOC_9606

Trouble Code Symptom Test Chart

2009-10 FZR/FZX models w/1812cc engines

Diagnostic Code	System/ Component	Symptom	Check
124 125 127 128	Throttle position sensor malfunction	Engine speed is limited	1. Check the pink (P), orange (O), pink/ black (P/B), and black/orange (B/O) throttle body assembly wiring harness leads for continuity. 2. Check the ECM circuit for continuity.
126	Throttle position sensor malfunction	Engine speed is limited	1. Check the TPS output voltage using the YDIS. 2. Check the ECM circuit for continuity.
129	Electronic throttle system malfunction ④	Engine speed is limited	1. Check the TPS output voltage using the YDIS. 2. Check the ETV relay and fuse (ETV relay) for continuity.
131 132 133 134	Accelerator position sensor malfunction	Engine speed is limited	1. Check the APS output voltage using the YDIS. 2. Check the APS circuit for continuity. 3. Measure the APS 1 and 2 resistance.
135	Accelerator position sensor malfunction	Engine speed is limited	1. Measure the APS 1 and 2 resistance. 2. Check the APS circuit for continuity.
136 137 138 139	Electronic throttle system malfunction ④	Engine speed is limited	1. Check the ECM circuit for continuity. 2. Check the ETV relay for continuity. 3. Check the charging system. • Stator coil output peak voltage • Stator coil resistance • Rectifier regulator output peak voltage • Rectifier regulator continuity 4. Check the fuse for continuity. 5. Check the battery voltage and specific gravity. 6. Check the red (R) and black (B) power supply circuit cables and leads for continuity.
141	Electronic throttle system malfunction ④	Engine speed is limited	1. Check the ECM circuit for continuity. 2. Check the ETV relay and fuse (ETV relay) for continuity.
142	Electronic throttle system malfunction ④	Engine speed is limited	1. Check the TPS output voltage and throttle valve opening using the YDIS. 2. Check the throttle shaft and throttle inner surface for wear or damage.

SELOC_9606

Trouble Code Symptom Test Chart
2009-10 FZR/FZX models w/1812cc engines

Diagnostic Code	System/ Component	Symptom	Check
143 144	Electronic throttle system malfunction ④	Engine speed is limited	1. Check the ECM circuit for continuity.
			2. Check the ETV relay for continuity.
			3. Check the charging system. • Stator coil output peak voltage • Stator coil resistance • Rectifier regulator output peak voltage • Rectifier regulator continuity
			4. Check the fuse for continuity.
			5. Check the battery voltage and specific gravity.
			6. Check the red (R) and black (B) power supply circuit cables and leads for continuity.
145	Electronic throttle system malfunction ④	Engine speed is limited	1. Check the ETV relay and fuse (ETV relay) for continuity.
			2. Check the throttle shaft and throttle inner surface for wear or damage.
252	Overheat warning	Engine speed is limited	1. Check the cooling water passages for obstructions.
			2. Measure the engine temperature sensor resistance.
			3. Check that there is no short circuit between the black/yellow (B/Y) and black/orange (B/O) engine temperature sensor wiring harness leads. ⑤
			4. Check the thermoswitch for continuity.
			5. Check that there is no short circuit between the pink (P) and black/orange (B/O) thermoswitch wiring harness leads.⑤
			6. Measure the thermo sensor resistance.
			7. Check that there is no short circuit between the black/blue (B/L) and black/orange (B/O) thermo sensor wiring harness leads. ⑤
253	Low oil pressure	Engine speed is	1. Check the oil pressure switch for continuity.
			2. Check that there is no short circuit in the pink/white (P/W) oil pressure switch wiring harness lead*.
			3. Check the oil passages for obstructions.

SELOC_9606

① This symptom may only be exhibited in certain environmental conditions.

② Thermosensor

③ Reverse sensor

④ Throttle body assembly

⑤ Disconnect the sensor coupler or switch coupler, and then check that there is no short circuit at the coupler (wiring harness end).

Trouble Code Symptom Test Chart

2010 VX models w/1052cc engines

Diagnostic Code	System/ Component	Symptom	Check
13	Pulser coil malfunction	Engine will not start	1. Measure the pick-up coil output peak voltage.
			2. Measure the pick-up coil resistance.
			3. Check the white/black (W/B) and black/orange (B/O) pick-up coil wiring harness leads for continuity.
15	Engine temp sensor malfunction	Trolling speed is unstable ①	1. Check the engine temperature using the YDIS.
			2. Measure the engine temperature sensor input voltage.
			3. Check the black/yellow (B/Y) and black/orange (B/O) engine temperature sensor wiring harness leads for continuity.
			4. Measure the engine temperature sensor resistance.
19	Battery voltage malfunction	Trolling speed is unstable ①	1. Check the battery voltage using the YDIS.
			2. Check the fuse for continuity.
			3. Measure the stator coil output peak voltage.
			4. Measure the rectifier regulator output peak voltage.
			5. Check the battery cables and terminals for proper connection.
23	IAT sensor malfunction	Trolling speed is unstable ①	1. Check the intake air temperature using the YDIS.
			2. Measure the sensor assy, input voltage.
			3. Check the black/yellow (B/Y) and black/orange (B/O) sensor assy, wiring harness leads for continuity.
			4. Measure the sensor assy, resistance.
24	Cam position sensor malfunction	Engine speed is limited	1. Measure the cam position sensor input voltage.
			2. Check the orange (0), green/orange (G/O), and black/orange (B/O) cam position sensor wiring harness leads for continuity.
			3. Measure the cam position sensor output voltage.
29	IAP sensor malfunction	Engine stalls Trolling speed is unstable ①	1. Check the intake air pressure using the YDIS.
			2. Measure the sensor assy, input voltage.
			3. Check the orange (O), pink/green (P/G), and black/orange (B/O) sensor assy, wiring harness leads for continuity.
47	Slant detection switch malfunction	Normal operation	1. Check the slant detection switch operation using the YDIS.
			2. Measure the slant detection switch input voltage.
			3. Check the blue/black (L/B) and black/orange (B/O) slant detection switch wiring harness leads for continuity.
			4. Check the slant detection switch continuity.
55	Steering sensor malfunction	Normal operation	1. Check the steering sensor operation using the YDIS.
			2. Measure the steering sensor input voltage.
			3. Check the orange/red (O/R), white/blue (W/L), and black/orange (B/O) steering sensor wiring harnessleads for continuity.

SELOC_9606

① This symptom may only be exhibited in certain environmental conditions.

Trouble Code Symptom Test Chart

2010 VX models w/1052cc engines

Diagnostic Code	System/ Component	Symptom	Check
112 113 114	Electronic throttle system malfunction ②	Engine speed is limited	1. Check the TPS output voltage and throttle valve opening using the YDIS. 2. Check the ECM circuit for continuity. 3. Check the ETV relay for continuity. 4. Check the charging system. • Stator coil output peak voltage • Stator coil resistance • Rectifier regulator output peak voltage • Rectifier regulator continuity 5. Check the fuse for continuity. 6. Check the battery voltage and specific gravity. 7. Check the red (R) and black (B) power supply circuit cables and leads for continuity.
115 116	Electronic throttle system malfunction ②	Engine speed is limited	1. Check the pink (P), orange (O), pink/black (P/B), and black/orange (B/O) throttle body assy, wiring harness leads for continuity. 2. Check the TPS output voltage and throttle valve opening using the YDIS. 3. Check the throttle shaft and throttle inner surface for wear or damage.
117 118 119	Electronic throttle system malfunction ②	Engine speed is limited	1. Check the throttle body assy, coupler terminals for rust and corrosion. 2, Check the green (G) and blue (L) throttle body assy, wiring harness leads for continuity. 3. Check the ECM circuit for continuity.
121	Electronic throttle system malfunction ②	Engine speed is limited	1. Cross-check the ECM or replace.
122	Electronic throttle system malfunction ②	Engine speed is limited	1. Cross-check the throttle body assy.
123	Electronic throttle system malfunction ②	Engine speed is limited	1. Check the ETV relay and fuse (ETV relay) for continuity. 2. Check the ECM circuit for continuity.
124 125 127 128	Throttle position sensor malfunction	Engine speed is limited	1. Check the pink (P), orange (O), pink/black (P/B), and black/orange (B/O) throttle body assy, wiring harness leads for continuity. 2. Check the ECM circuit for continuity.
126	Throttle position sensor malfunction	Engine speed is limited	1. Check the TPS output voltage using the YDIS. 2. Check the ECM circuit for continuity.
129	Electronic throttle system malfunction ②	Engine speed is limited	1. Check the TPS output voltage using the YDIS. 2. Check the ETV relay and fuse (ETV relay) for continuity.

② Throttle body assembly

SELOC_9606

Trouble Code Symptom Test Chart

2010 VX models w/1052cc engines

Diagnostic Code	System/ Component	Symptom	Check
131 132 133 134	Accelerator position sensor malfunction	Engine speed is limited	1. Check the APS output voltage using the YDIS. 2. Check the APS circuit for continuity. 3. Measure the APS 1 and 2 resistance.
135	Accelerator position sensor malfunction	Engine speed is limited	1. Measure the APS 1 and 2 resistance. 2. Check the APS circuit for continuity.
136 137 138 139	Electronic throttle system malfunction ②	Engine speed is limited	1. Check the ECM circuit for continuity. 2. Check the ETV relay for continuity. 3. Check the charging system. • Stator coil output peak voltage • Stator coil resistance • Rectifier regulator output peak voltage • Rectifier regulator continuity 4. Check the fuse for continuity. 5. Check the battery voltage and specific gravity. 6. Check the red (R) and black (B) power supply circuit cables and leads for continuity.
141	Electronic throttle system malfunction ②	Engine speed is limited	1. Check the ECM circuit for continuity. 2. Check the ETV relay and fuse (ETV relay) for continuity.
142	Electronic throttle system malfunction ②	Engine speed is limited	1. Check the TPS output voltage and throttle valve opening using the YDIS. 2. Check the throttle shaft and throttle inner surface for wear or damage.
143 144	Electronic throttle system malfunction ②	Engine speed is limited	1. Check the ECM circuit for continuity. 2. Check the ETV relay for continuity. 3. Check the charging system. • Stator coil output peak voltage • Stator coil resistance • Rectifier regulator output peak voltage • Rectifier regulator continuity 4. Check the fuse for continuity. 5. Check |he battery voltage and specific gravity. 6. Check the red (R) and black (B) power supply circuit cables and leads for continuity.
145	Electronic throttle system malfunction ②	Engine speed is limited	1. Check the ETV relay and fuse (ETV relay) for continuity. 2. Check the throttle shaft and throttle inner surface for wear or damage.

③ Disconnect the sensor coupler or switch coupler, and then check that there is no short circuit at the coupler (wiring harness end).

SELOC_9606

Trouble Code Symptom Test Chart
2010 VX models w/1052cc engines

Diagnostic Code	System/ Component	Symptom	Check
252	Overheat warning	Engine speed is limited	1. Check the cooling water passages for obstructions.
			2. Measure the engine temperature sensor resistance.
			3. Check that there is no short circuit between the black/yellow (B/Y) and black/orange (B/O) engine temperature sensor wiring harness leads. ③
			4. Check the thermoswitch (engine) for continuity.
			5. Check that there is no short circuit between the pink (P) and black (B) thermoswitch (engine) wiring harness leads.③
			6. Check the thermoswitch (exhaust) for continuity.
			7. Check that there is no short circuit between the pink (P) and black (B) thermoswitch (exhaust) wiring harness leads.③
253	Low oil pressure warning	Engine speed is limited	1. Check the oil pressure switch for continuity.
			2. Check that there is no short circuit in the pink/white (P/W) oil pressure switch wiring harness lead. ③
			3. Check the oil passages for obstructions.

SELOC_9606

① This symptom may only be exhibited in certain environmental conditions.
② Throttle body assembly
③ Disconnect the sensor coupler or switch coupler, and then check that there is no short circuit at the coupler (wiring harness end).

Trouble Code Symptom Test Chart

2011 VX models w/1812cc engines

Diagnostic Code	System/ Component	Symptom	Check
13	Pulser coil malfunction	Engine will not start	1. Measure the pick-up coil output peak voltage.
			2. Measure the pick-up coil resistance.
			3. Check the white/black (W/B) and black/orange (B/O) pick-up coil wiring harness leads for continuity.
15	Engine temp sensor malfunction	Trolling speed is unstable ①	1. Check the engine temperature using the YDIS.
			2. Measure the engine temperature sensor input voltage.
			3. Check the black/yellow (B/Y) and black/orange (B/O) engine temperature sensor wiring harness leads for continuity.
			4. Measure the engine temperature sensor resistance.
19	Battery voltage malfunction	Trolling speed is unstable ①	1. Check the battery voltage using the YDIS.
			2. Check the fuse for continuity.
			3. Measure the stator coil output peak voltage.
			4. Measure the rectifier regulator output peak voltage.
			5. Check the battery cables and terminals for proper connection.
23	IAT sensor malfunction	Trolling speed is unstable ①	1. Check the intake air temperature using the YDIS.
			2. Measure the intake air temperature sensor input voltage.
			3. Check the black/red (B/R) and black/ orange (B/O) intake air temperature sensor wiring harness leads for continuity.
			4. Measure the intake air temperature sensor resistance.
24	Cam position sensor malfunction	Engine speed is limited	1. Measure the cam position sensor input voltage.
			2. Check the orange (O), green/orange (G/O), and black/orange (B/O) cam position sensor wiring harness leads for continuity.
			3. Measure the cam position sensor output voltage.
29	IAP sensor malfunction	Engine stalls Trolling speed is unstable ①	1. Check the intake air pressure using the YDIS.
			2. Measure the intake air pressure sensor input voltage.
			3. Check the orange (O), pink/green (P/ G), and black/orange (B/O) intake air pressure sensor wiring harness leads for continuity.
			4. Check the intake air pressure sensor operation.
47	Slant detection switch malfunction	Normal operation	1. Check the slant detection switch operation using the YDIS.
			2. Measure the slant detection switch input voltage.
			3. Check the blue/black (L/B) and black/ orange (B/O) slant detection switch wiring harness leads for continuity.
			4. Check the slant detection switch continuity.
55	Steering sensor malfunction	Normal operation	1. Check the steering sensor operation using the YDIS.
			2. Measure the steering sensor input voltage.
			3. Check the orange/red (O/R), white/ blue (W/L), and black/orange (B/O) steering sensor wiring harness leads for continuity.
65	Ex. manifold water temp. sensor malfunction ②	Normal operation	1. Check the thermo sensor operation using the YDIS.
			2. Measure the thermo sensor input voltage.
			3. Check the black/blue (B/L) and black/ orange (B/O) thermo sensor wiring harness leads for continuity.
			4. Measure the thermo sensor resistance.

① This symptom may only be exhibited in certain environmental conditions.

② Thermo sensor

SELOC_9606

Trouble Code Symptom Test Chart

2011 VX models w/1812cc engines

Diagnostic Code	System/ Component	Symptom	Check
112 113 114	Electronic throttle system malfunction ③	Engine speed is limited	1. Check the TPS output voltage and throttle valve opening using the YDIS.
			2. Check the ECM circuit for continuity.
			3. Check the ETV relay for continuity.
			4. Check the charging system. • Stator coil output peak voltage • Stator coil resistance • Rectifier regulator output peak voltage • Rectifier regulator continuity
			5. Check the fuse for continuity.
			6. Check the battery voltage and specific gravity.
			7, Check the red (R) and black (B) power supply circuit cables and leads for continuity.
115 116	Electronic throttle system malfunction ③	Engine speed is limited	1. Check the pink (P), orange (O), pink/ black (P/B), and black/orange (B/O) throttle body assembly wiring harness leads for continuity.
			2. Check the TPS output voltage and throttle valve opening using the YDIS.
			3. Check the throttle shaft and throttle inner surface for wear or damage.
117 118 119	Electronic throttle system malfunction ③	Engine speed is limited	1. Check the throttle body assembly coupler terminals for rust and corrosion.
			2. Check the green (G) and blue (L) throttle body assembly wiring harness leads for continuity.
			3. Check the ECM circuit for continuity.
121	Electronic throttle system malfunction ③	Engine speed is limited	1. Cross-check the ECM or replace.
122	Electronic throttle system malfunction ③	Engine speed is limited	1. Cross-check the throttle body assembly.
123	Electronic throttle system malfunction ③	Engine speed is limited	1. Check the ETV relay and fuse (ETV relay) for continuity.
			2. Check the ECM circuit for continuity.
124 125 127 128	Throttle position sensor malfunction	Engine speed is limited	1. Check the pink (P), orange (O), pink/ black (P/B), and black/orange (B/O) throttle body assembly wiring harness leads for continuity.
			2. Check the ECM circuit for continuity.
126	Throttle position sensor malfunction	Engine speed is limited	1. Check the TPS output voltage using the YDIS,
			2. Check the ECM circuit for continuity.
129	Electronic throttle system malfunction ③	Engine speed is limited	1. Check the TPS output voltage using the YDIS.
			2. Check the ETV relay and fuse (ETV relay) for continuity.
131 132 133 134	Accelerator position sensor malfunction	Engine speed is limited	1. Check the APS output voltage using the YDIS.
			2. Check the APS circuit for continuity.
			3. Measure the APS 1 and 2 resistance.

③ Throttle body assembly

SELOC_9606

Trouble Code Symptom Test Chart

2011 VX models w/1812cc engines

Diagnostic Code	System/ Component	Symptom	Check
135	Accel position sensor malfunction	Engine speed is limited	1. Measure the APS 1 and 2 resistance.
			2. Check the APS circuit for continuity.
136 137 138 139	Electronic throttle system malfunction ③	Engine speed is limited	1. Check the ECM circuit for continuity.
			2. Check the ETV relay for continuity.
			3. Check the charging system. • Stator coil output peak voltage • Stator coil resistance • Rectifier regulator output peak voltage • Rectifier regulator continuity
			4. Check the fuse for continuity.
			5. Check the battery voltage and specific gravity.
			6. Check the red (R) and black (B) power supply circuit cables and leads for continuity.
141	Electronic throttle system malfunction ③	Engine speed is limited	1. Check the ECM circuit for continuity.
			2. Check the ETV relay and fuse (ETV relay) for continuity.
142	Electronic throttle system malfunction ③	Engine speed is limited	1. Check the TPS output voltage and throttle valve opening using the YDIS.
			2. Check the throttle shaft and throttle inner surface for wear or damage.
143 144	Electronic throttle system malfunction ③	Engine speed is limited	1. Check the ECM circuit for continuity.
			2. Check the ETV relay for continuity.
			3. Check the charging system.
			4. Check the fuse for continuity.
			5. Check the battery voltage and specific gravity.
			6. Check the red (R) and black (B) power supply circuit cables and leads for continuity.
145	Electronic throttle system malfunction ③	Engine speed is limited	1. Check the ETV relay and fuse (ETV relay) for continuity.
			2. Check the throttle shaft and throttle inner surface for wear or damage.
252	Overheat warning	Engine speed is limited	1. Check the cooling water passages for obstructions.
			2. Measure the engine temperature sensor resistance.
			3. Check that there is no short circuit between the black/yellow (B/Y) and black/orange (B/O) engine temperature sensor wiring harness leads. ⑤
			4. Check the thermoswitch for continuity.
			5. Check that there is no short circuit between the pink (P) and black/orange (B/O) thermoswitch wiring harness leads. ⑤
			6. Measure the thermo sensor resistance.
			7. Check that there is no short circuit between the black/blue (B/L) and black/orange (B/O) thermo sensor wiring harness leads. ⑤
253	Low oil pressure warning	Engine speed is limited	1. Check the oil pressure switch for continuity.
			2. Check that there is no short circuit in the pink/white (P/W) oil pressure switch wiring harness lead. ④
			3. Check the oil passages for obstructions.

① This symptom may only be exhibited in certain environmental conditions.

SELOC_9606

② Thermo sensor

③ Throttle body assembly

④ Disconnect the sensor coupler or switch coupler, and then check that there is no short circuit at the coupler (wiring harness end)

COMPONENT LOCATIONS

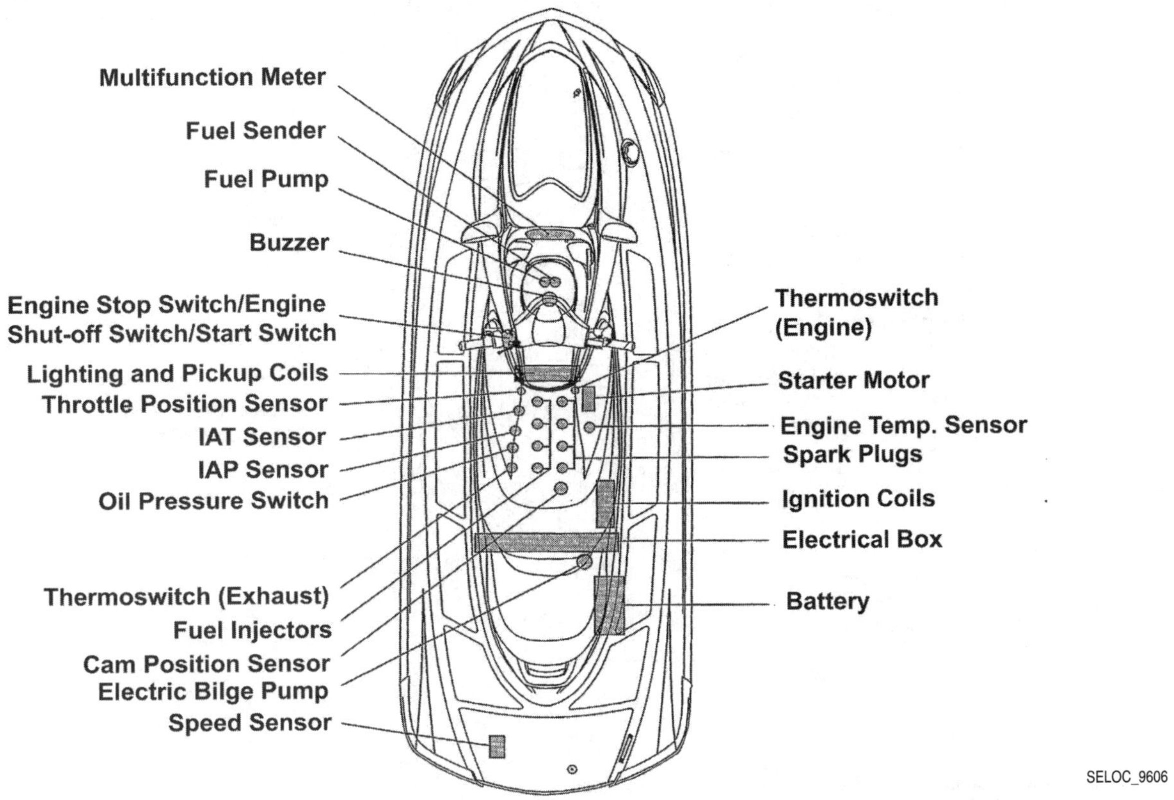

Electrical component locations - 2002 FX models w/998cc engines

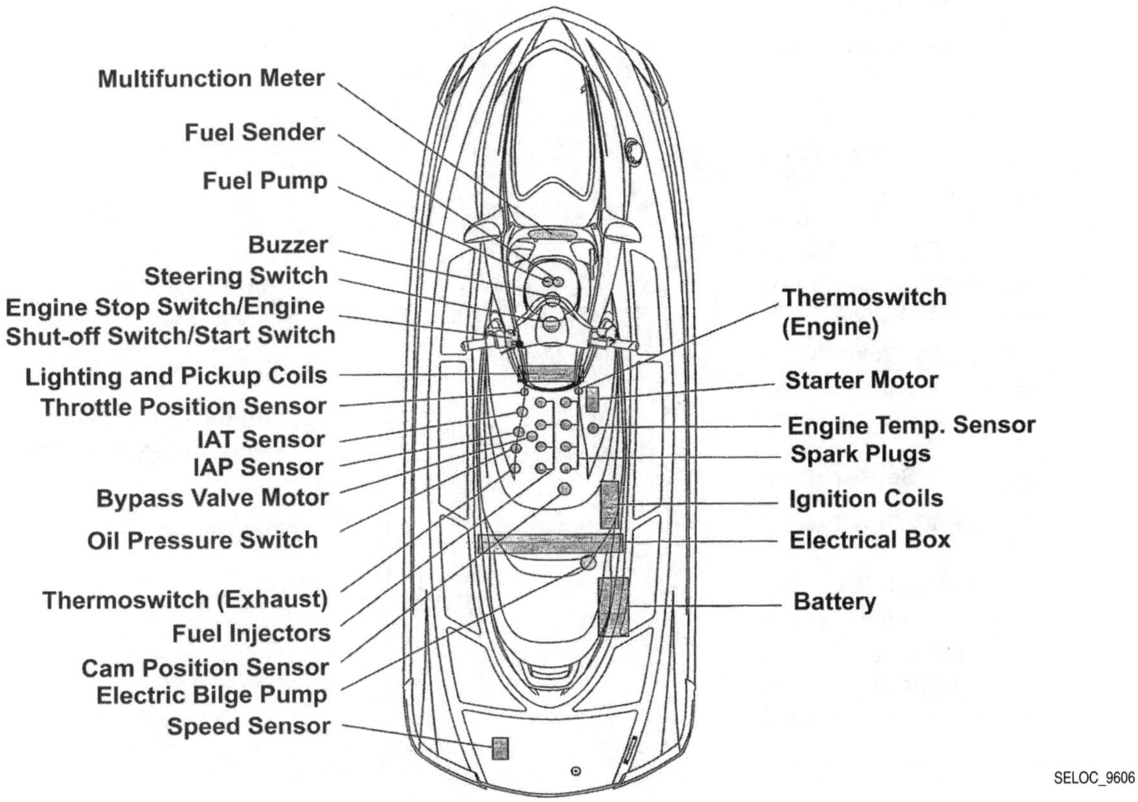

Electrical component locations - 2003-05 FX models w/998cc engines

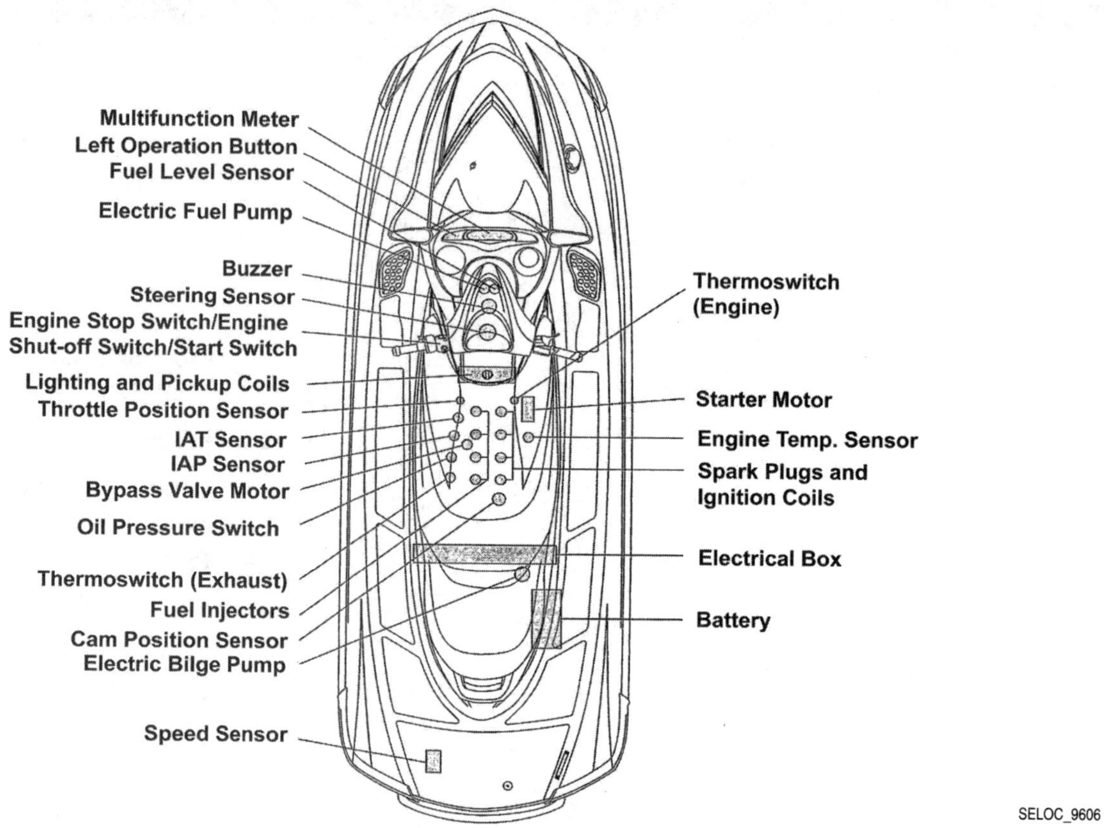

Electrical component locations - 2006-07 FX models w/998cc engines

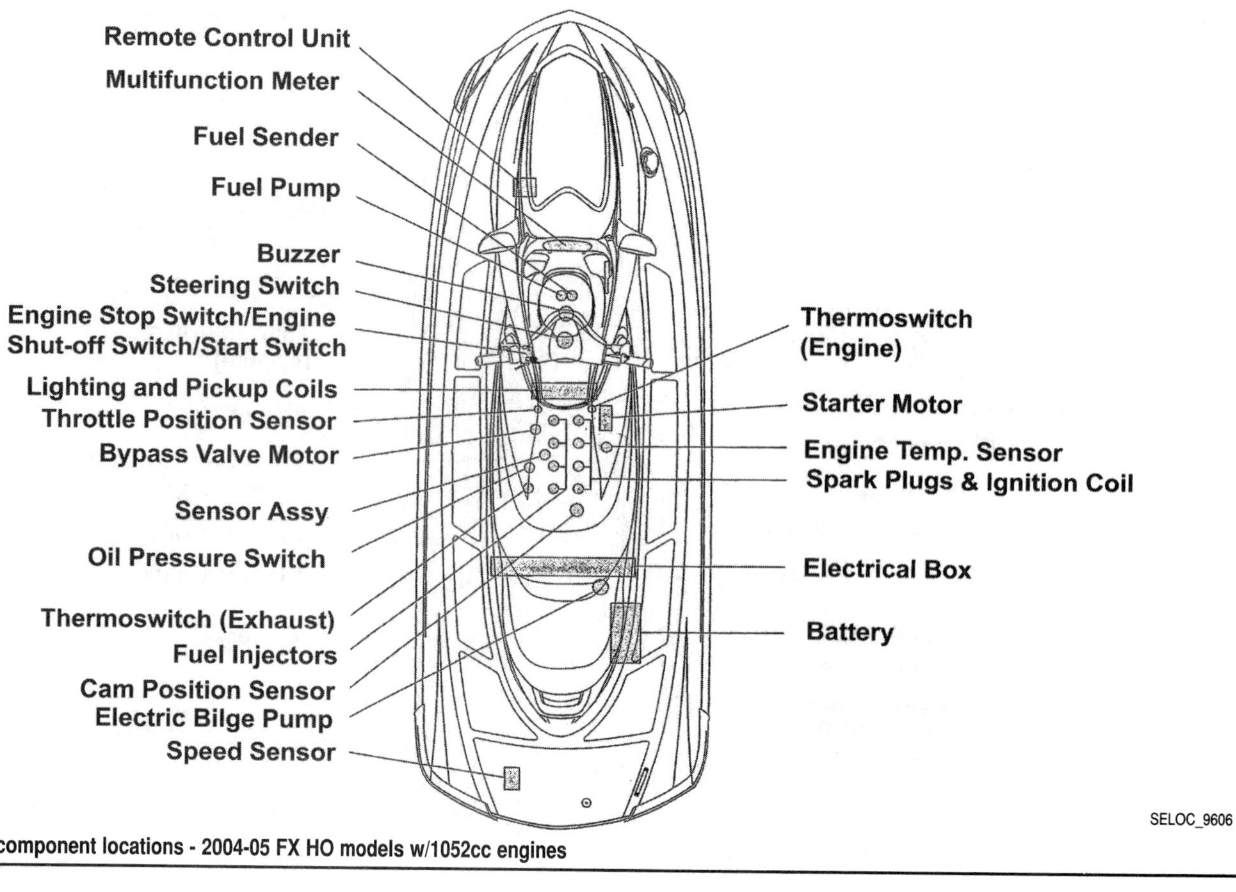

Electrical component locations - 2004-05 FX HO models w/1052cc engines

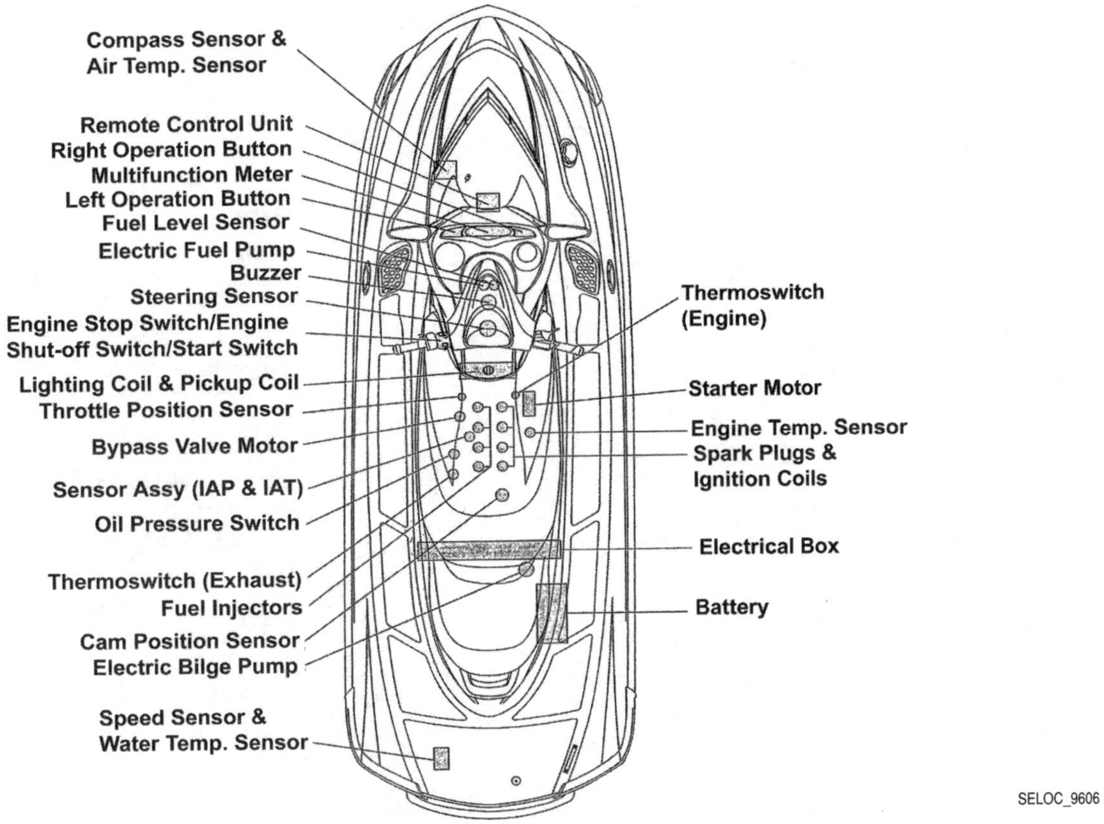

Electrical component locations - 2006-07 FX HO models w/1052cc engines

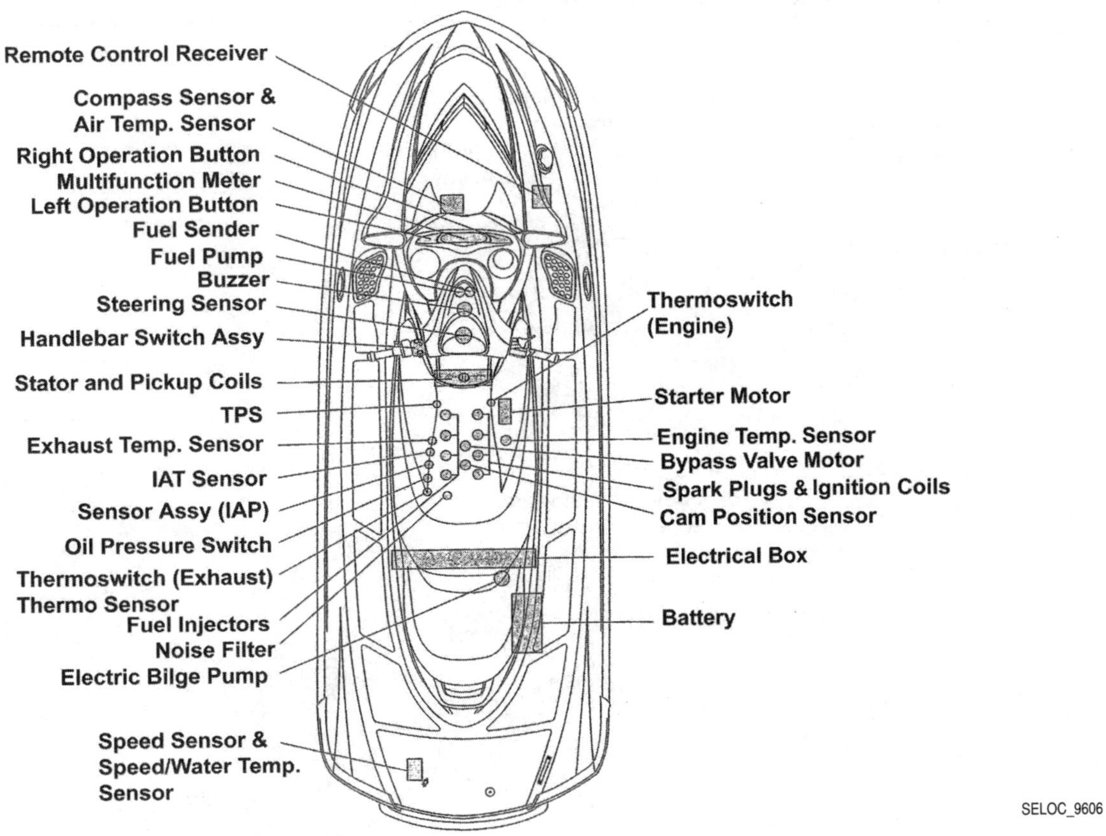

Electrical component locations - 2008 FX HO models w/1052cc engines

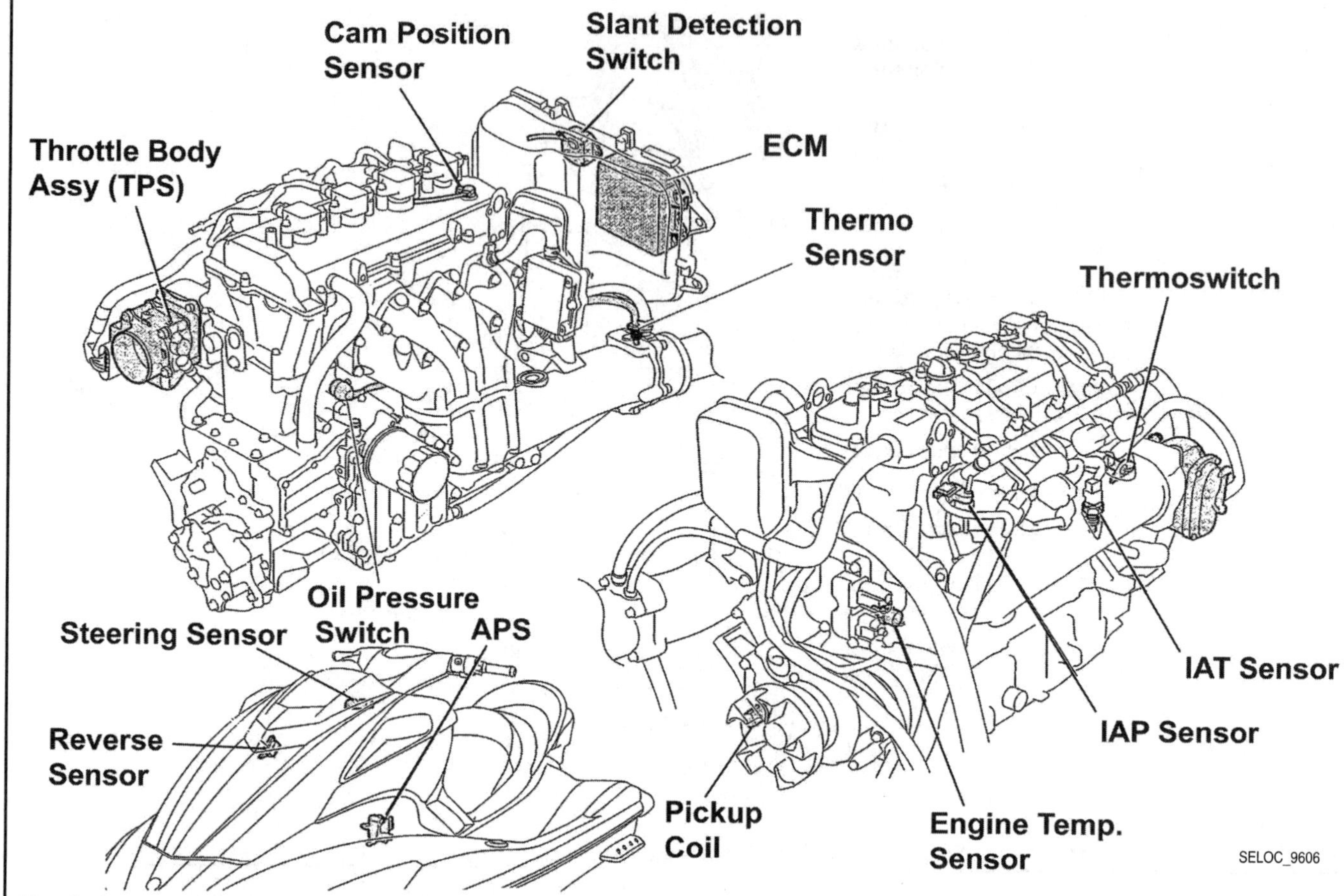

Electrical component locations - 2009-11 FX HO models w/1812cc engines

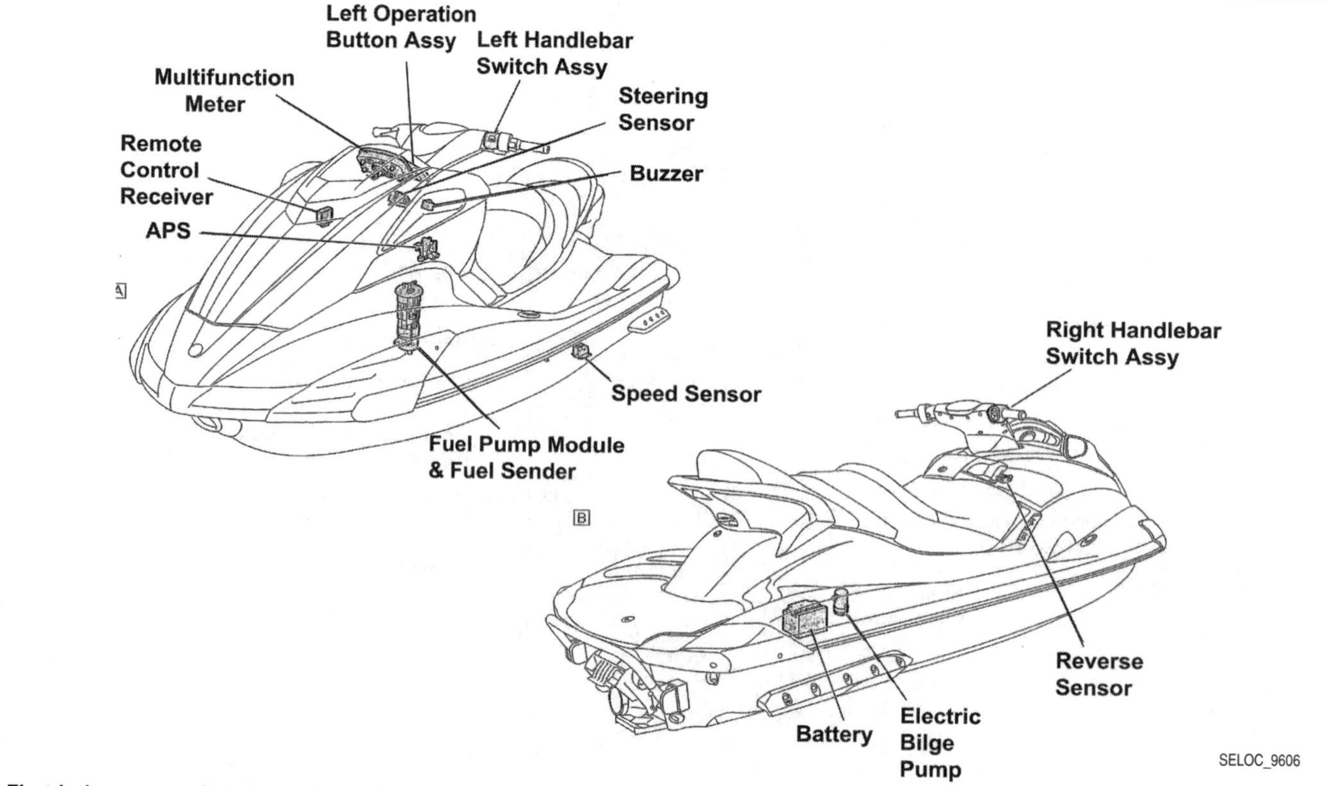

Electrical component locations - 2009-11 FX HO models w/1812cc engines (cont'd)

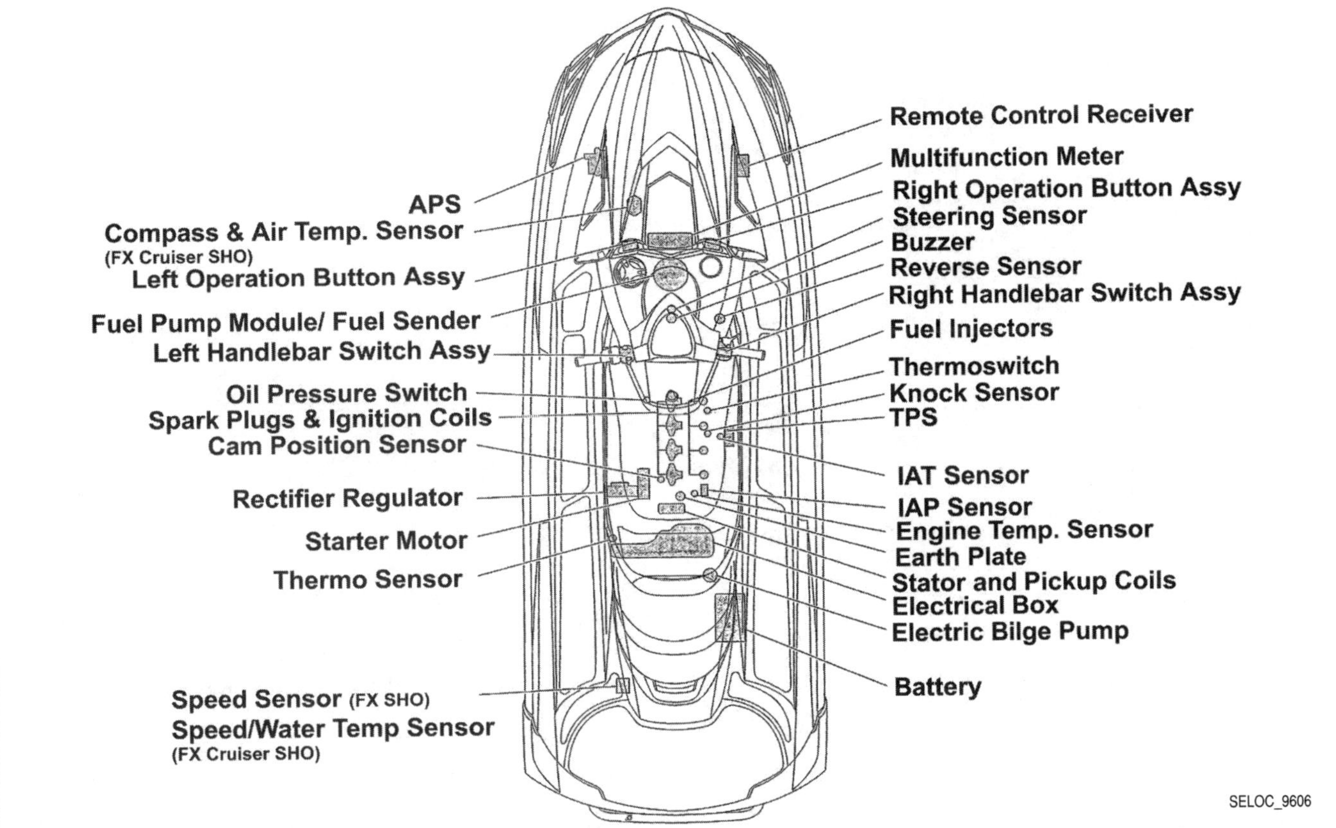

Electrical component locations - 2008-11 FX SHO models w/1812cc engines

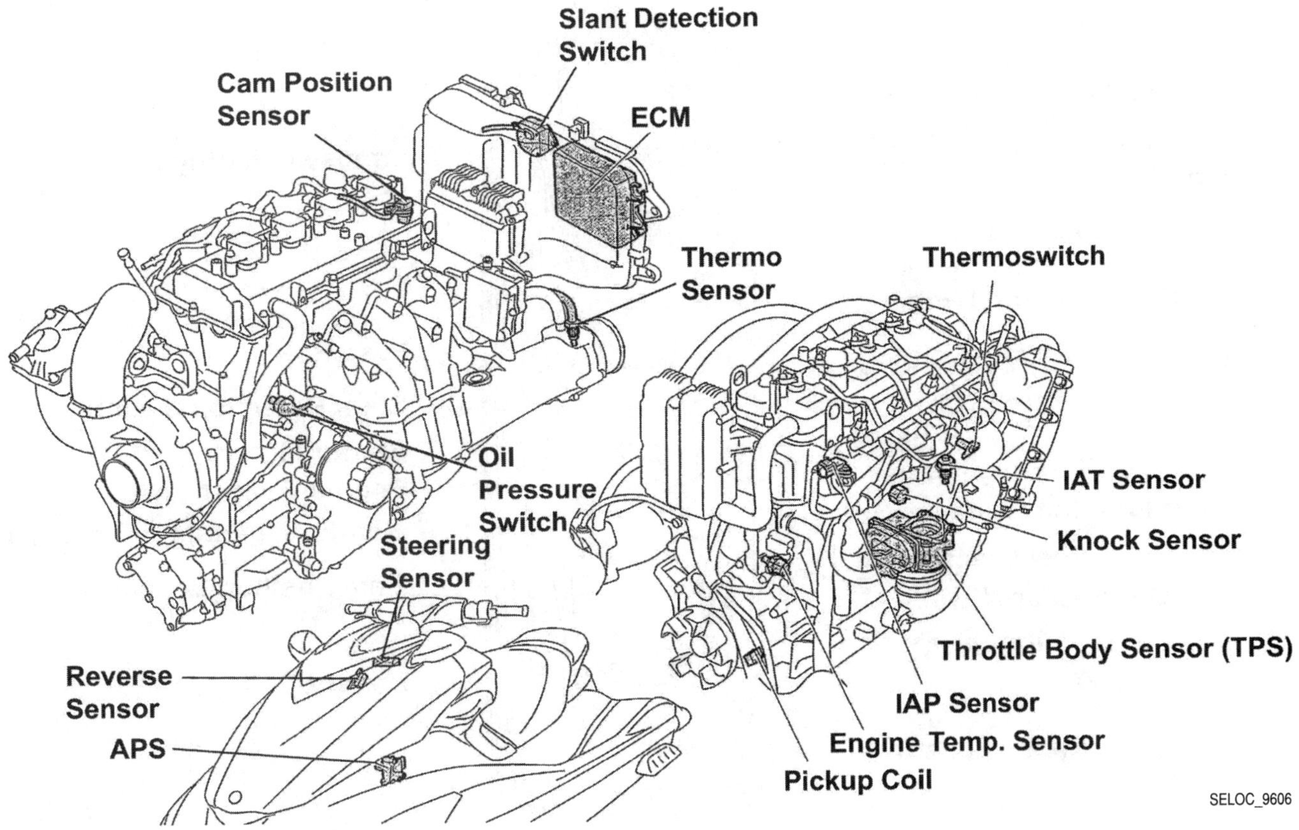

Electrical component locations - 2009-11 FX SHO models w/1812cc engines

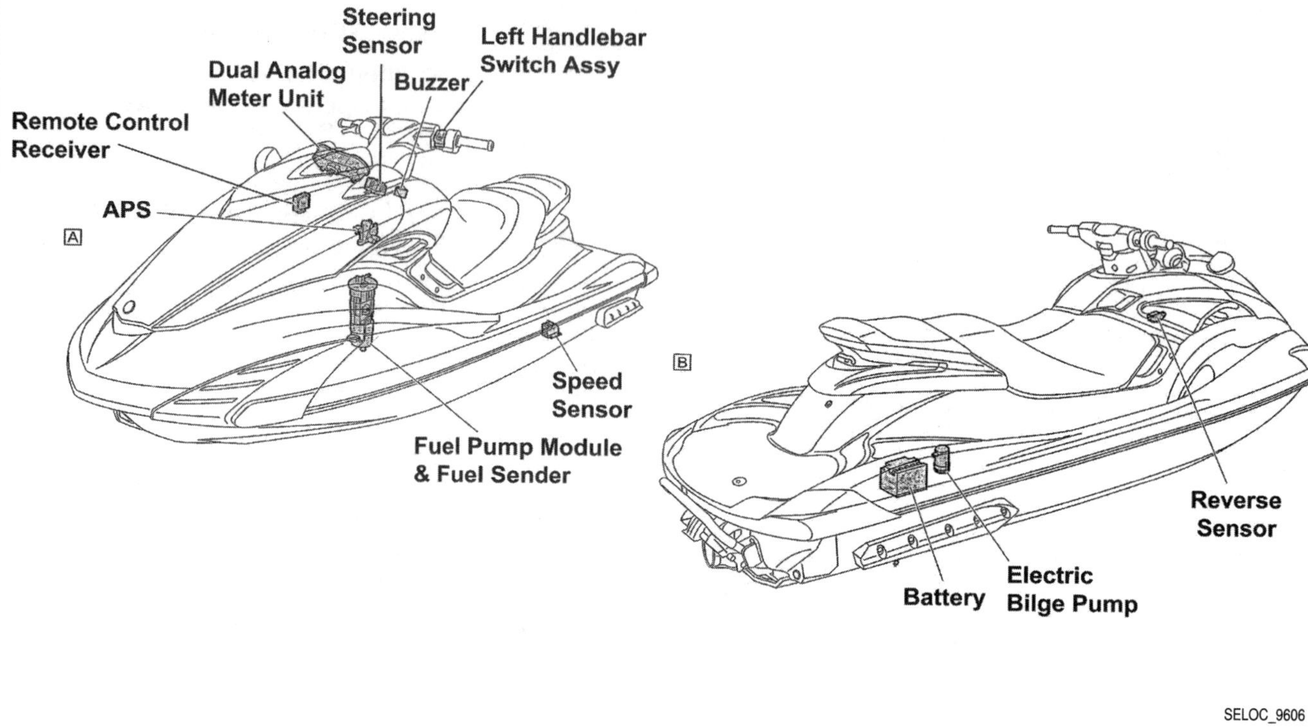

Electrical component locations - 2009-11 FX SHO models w/1812cc engines (cont'd)

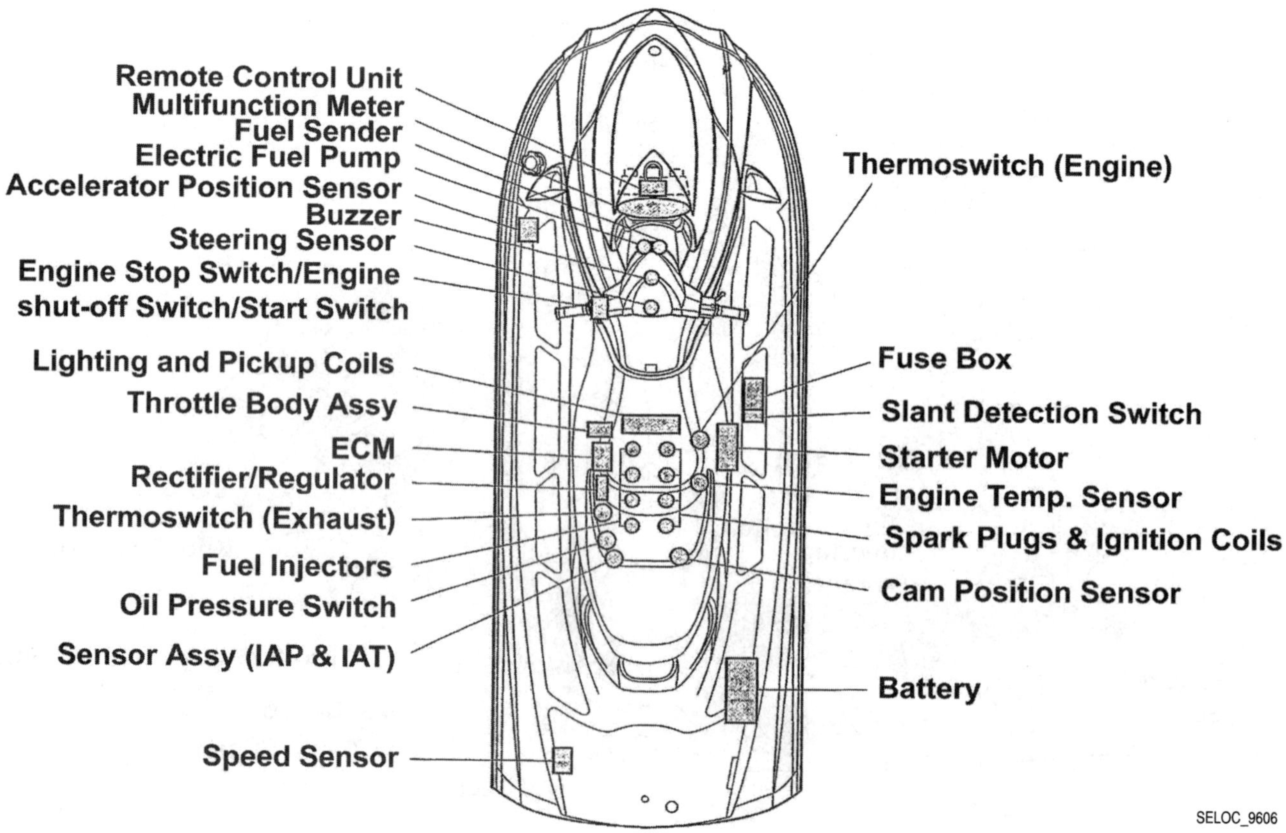

Electrical component locations - 2005-09 VX models w/1052cc engines

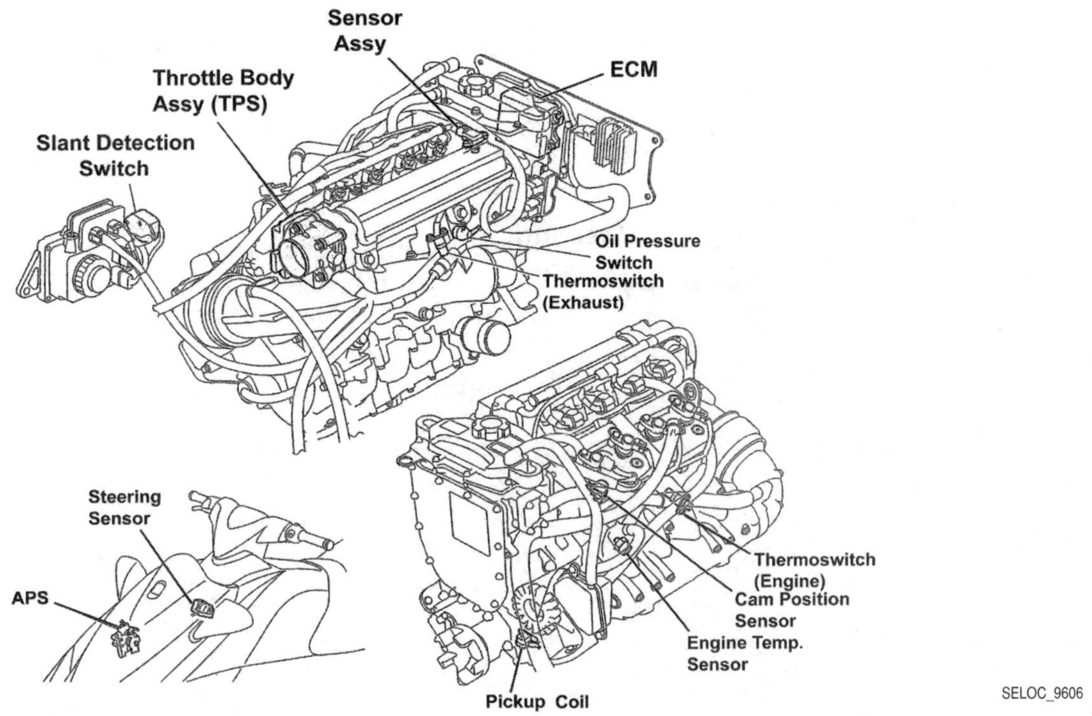

Electrical component locations - 2010-11 VX models w/1052cc engines

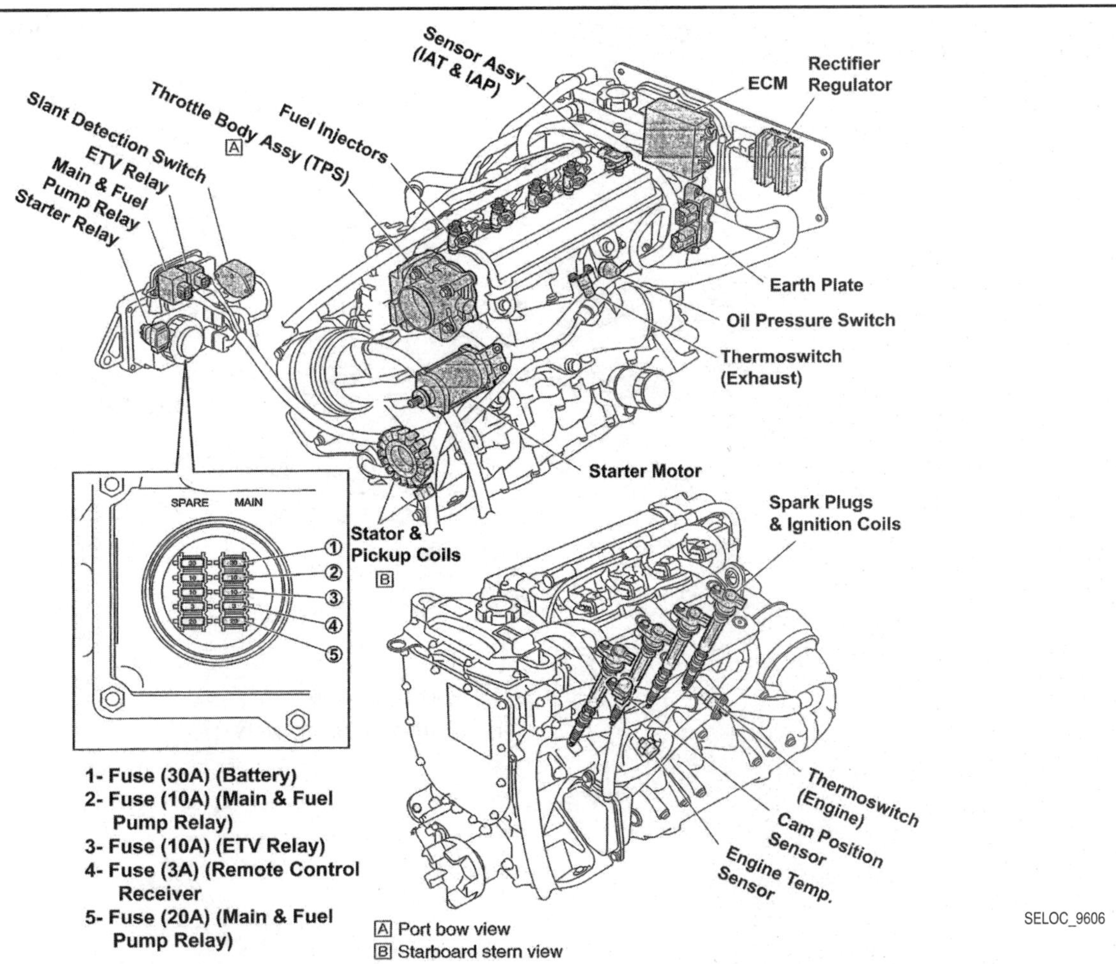

Electrical component locations - 2010-11 VX models w/1052cc engines (cont'd)

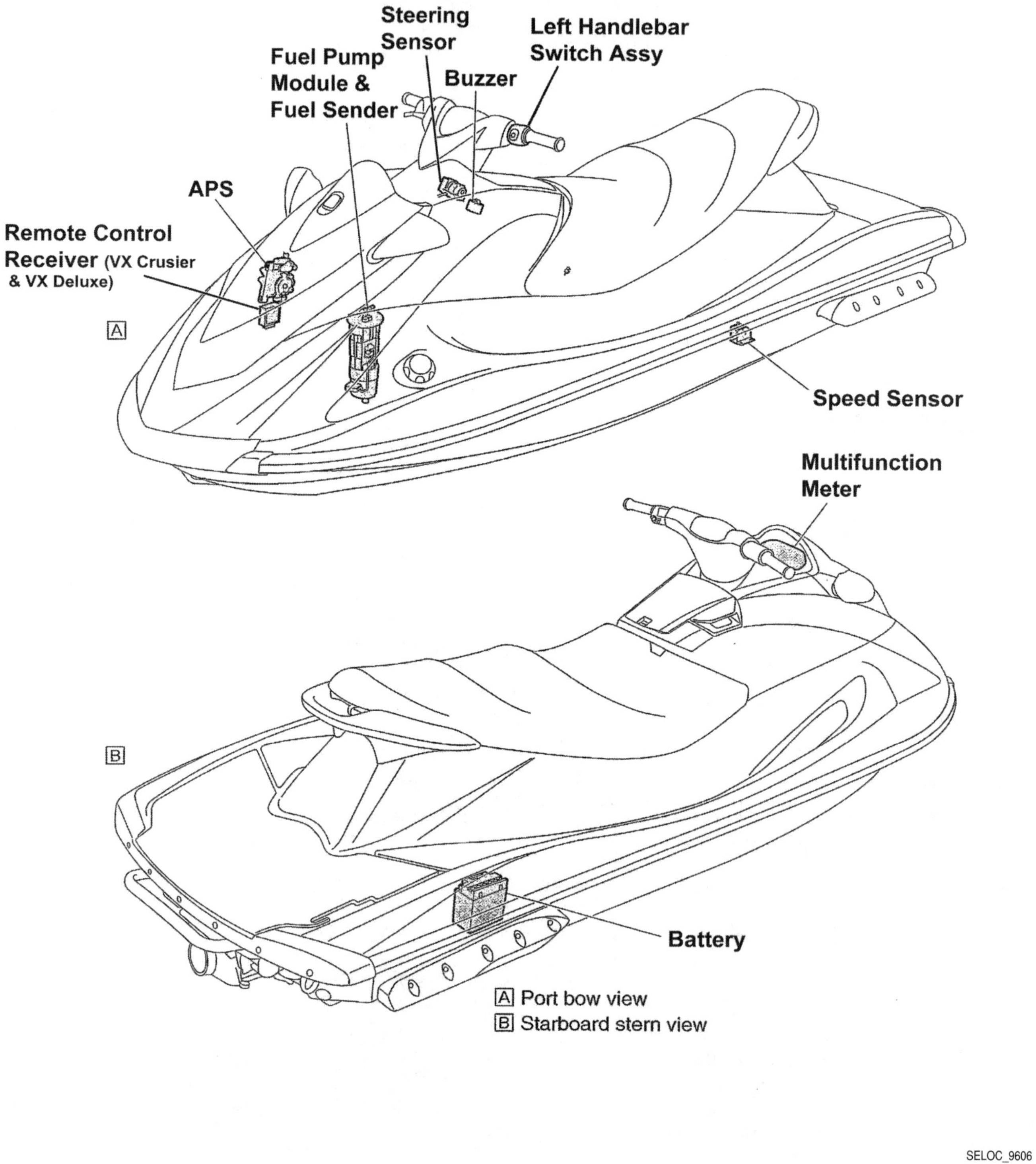

Electrical component locations - 2010-11 VX models w/1052cc engines (cont'd)

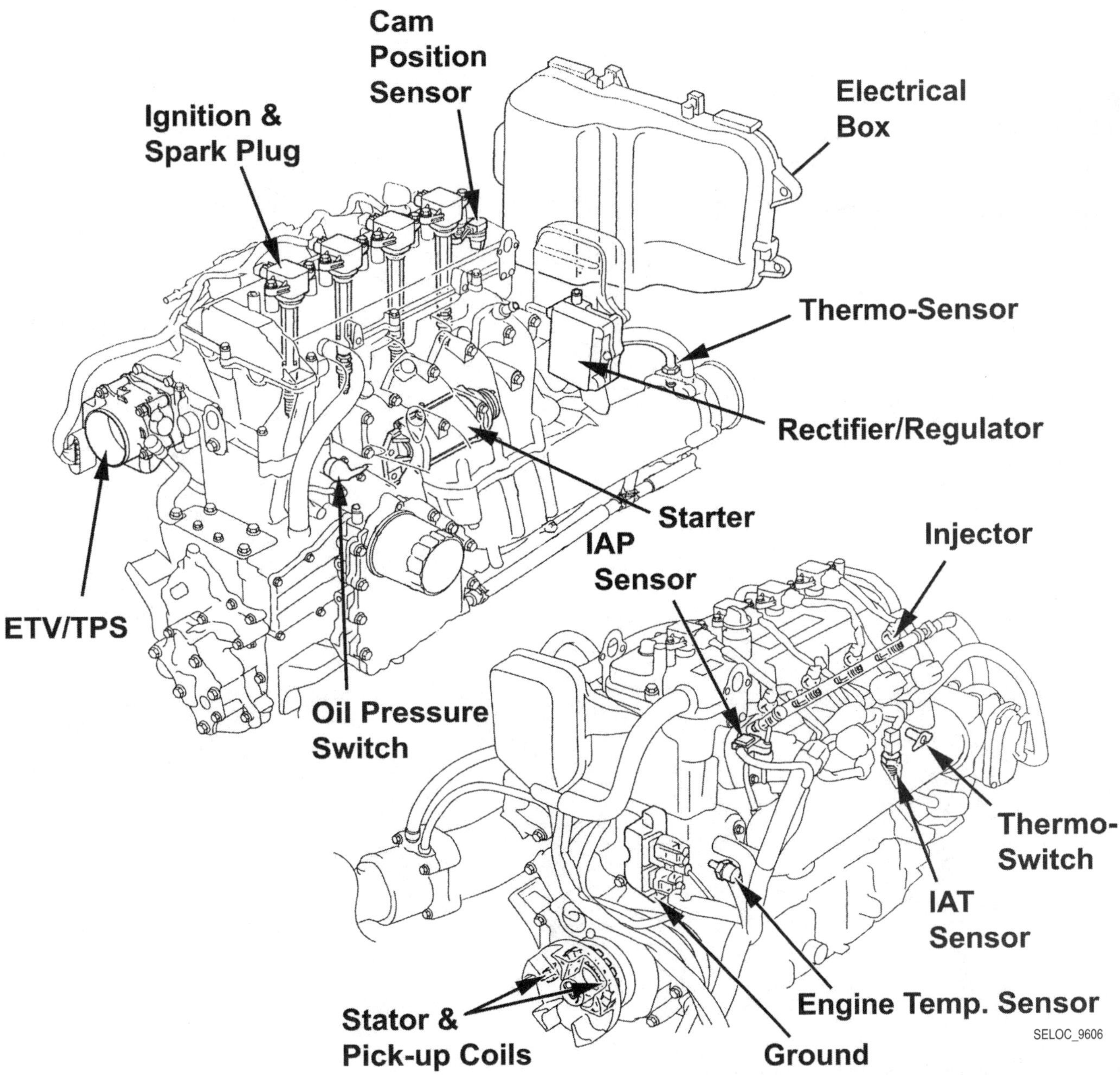

Electrical component locations - 2011 VX models w/1812cc engines

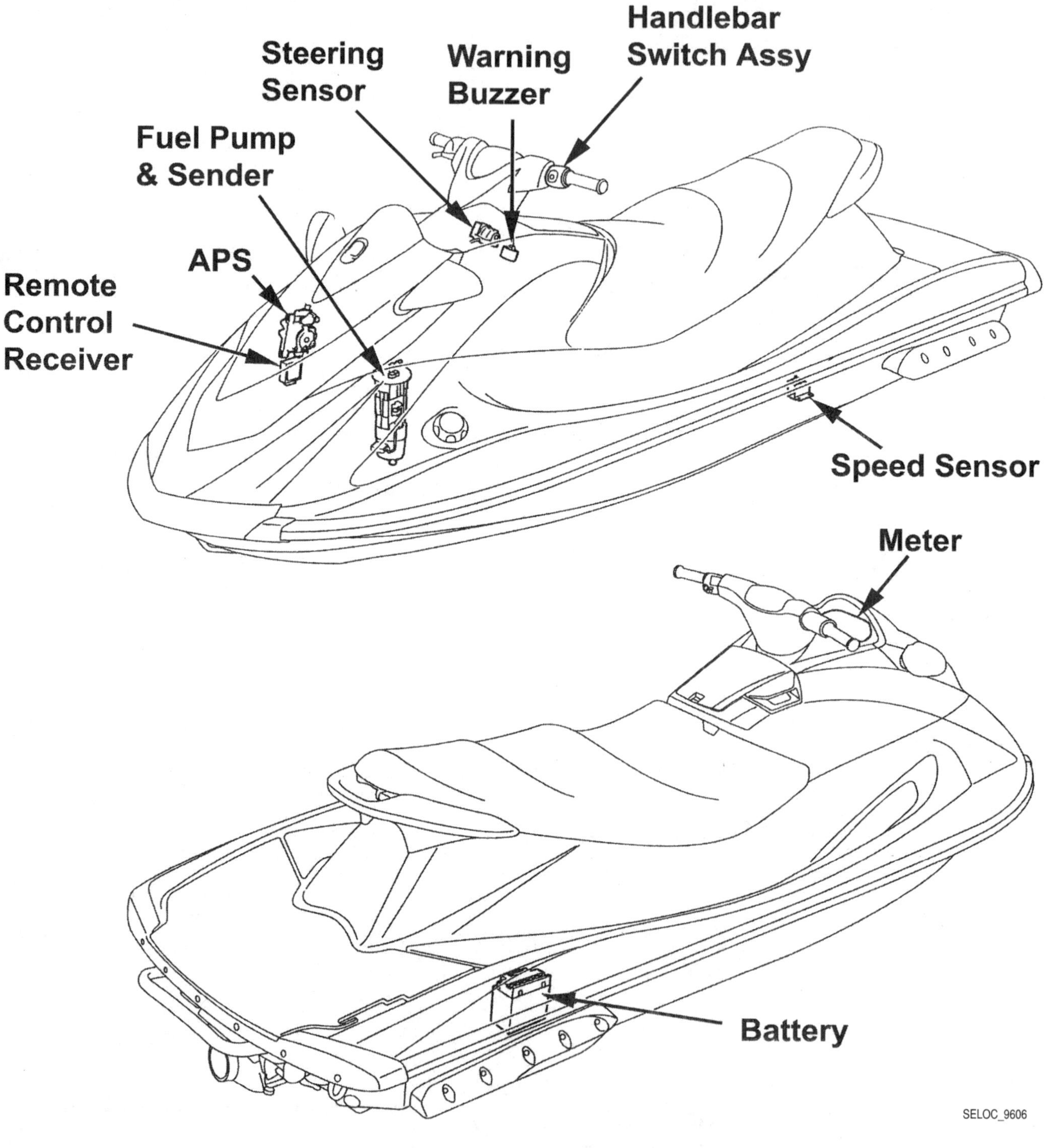

Electrical component locations - 2011 VX models w/1812cc engines (cont'd)

4

IGNITION & ELECTRICAL SYSTEMS

UNDERSTANDING & TROUBLESHOOTING ELECTRICAL SYSTEMS

Basic Electrical Theory

◆ **See Figure 1**

For any 12-volt, negative ground, electrical system to operate, the electricity must travel in a complete circuit. This simply means that current (power) from the positive terminal (+) of the battery must eventually return to the negative terminal (-) of the battery. Along the way, this current will travel through wires, fuses, switches and components. If, for any reason, the flow of current through the circuit is interrupted, the component fed by that circuit would cease to function properly.

Perhaps the easiest way to visualize a circuit is to think of connecting a light bulb (with 2 wires attached to it) to the battery - 1 wire attached to the negative (-) terminal of the battery and the other wire to the positive (+) terminal. With the 2 wires touching the battery terminals, the circuit would be complete and the light bulb would illuminate. Electricity would follow a path from the battery to the bulb and back to the battery. It's easy to see that with wires of sufficient length, our light bulb could be mounted nearly anywhere on the boat. Further, q wire could be fitted with a switch inline so that the light could be turned on and off without having to physically remove the wire(s) from the battery.

The normal marine circuit differs from this simple example in two ways. First, instead of having a return wire from each bulb to the battery, the current travels through a single ground wire that handles all the grounds for a specific circuit. Secondly, most marine circuits contain multiple components that receive power from a single circuit. This lessens the overall amount of wire needed to power components.

HOW DOES ELECTRICITY WORK: THE WATER ANALOGY

Electricity is the flow of electrons - the sub-atomic particles that constitute the outer shell of an atom. Electrons spin in an orbit around the center core of an atom. The center core is comprised of protons (positive charge) and neutrons (neutral charge). Electrons have a negative charge and balance out the positive charge of the protons. When an outside force causes the number of electrons to unbalance the charge of the protons, the electrons will split off the atom and look for another atom to balance out. If this imbalance is kept up, electrons will continue to move and an electrical flow will exist.

Many people find electrical theory easier to understand when using an analogy with water. In a comparison with water flowing through a pipe, the electrons would be the water and the wire is the pipe.

The flow of electricity can be measured much like the flow of water through a pipe. The unit of measurement used is amperes, frequently abbreviated as amps (a). You can compare amperage to the volume of water flowing through a pipe (for water that would mean a measurement of mass usually measured in units delivered over a set amount of time such as gallons or liters per minute). When connected to a circuit, an ammeter will measure the actual amount of current flowing through the circuit. When relatively few electrons flow through a circuit, the amperage is low. When many electrons flow, the amperage is high.

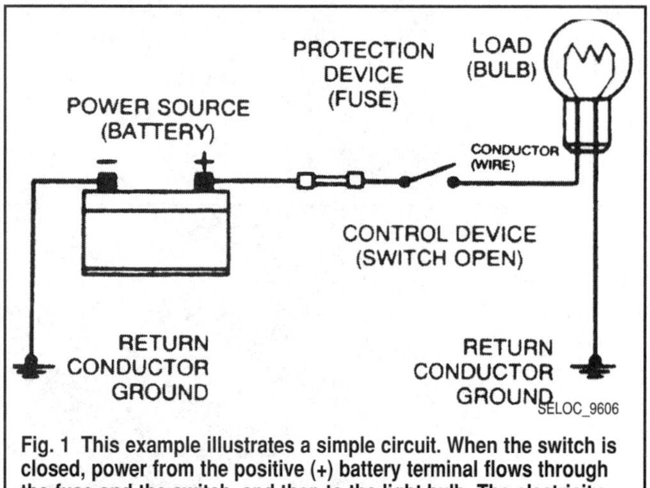

Fig. 1 This example illustrates a simple circuit. When the switch is closed, power from the positive (+) battery terminal flows through the fuse and the switch, and then to the light bulb. The electricity illuminates the bulb and the circuit is completed through the ground wire back to the negative (-) battery terminal.

Water pressure is measured in units such as pounds per square inch (psi). The electrical pressure is measured in units called volts (v). When a voltmeter is connected to a circuit, it is measuring the electrical pressure.

The actual flow of electricity depends not only on voltage and amperage, but also on the resistance of the circuit. The higher the resistance, the higher the force necessary to push the current through the circuit. The standard unit for measuring resistance is an ohm (omega). Resistance in a circuit varies depending on the amount and type of components used in the circuit. The main factors that determine resistance are:

- Material - some materials have more resistance than others. Those with high resistance are said to be insulators. Rubber materials (or rubber-like plastics) are some of the most common insulators used, as they have a very high resistance to electricity. Very low resistance materials are said to be conductors. Copper wire is among the best conductors. Silver is actually a superior conductor to copper and is used in some relay contacts, but its high cost prohibits its use as common wiring. Most marine wiring is made of copper.
- Size - the larger the wire size being used, the less resistance the wire will have (just as a large diameter pipe will allow small amounts of water to just trickle through). This is why components that use large amounts of electricity usually have large wires supplying current to them.
- Length - for a given thickness of wire, the longer the wire, the greater the resistance. The shorter the wire, the less the resistance. When determining the proper wire for a circuit, both size and length must be considered to design a circuit that can handle the current needs of the component.
- Temperature - with many materials, the higher the temperature, the greater the resistance (positive temperature coefficient). Some materials exhibit the opposite trait of lower resistance with higher temperatures (these are said to have a negative temperature coefficient). These principles are used in many engine control sensors (especially those found on microcomputer controlled ignition and fuel injection systems).

OHM'S LAW

There is a direct relationship between current, voltage and resistance. The relationship between current, voltage and resistance can be summed up by a statement known as Ohm's law.

Voltage (E) is equal to amperage (I) times resistance (R): $E=I \times R$

Other forms of the formula are $R=E/I$ and $I=E/R$

In each of these formulas, E is the voltage in volts, I is the current in amps and R is the resistance in ohms. The basic point to remember is that if the voltage of a circuit remains the same, as the resistance of that circuit goes up, the amount of current that flows in the circuit will go down.

The amount of work that electricity can perform is expressed as power. The unit of power is the watt (w). The relationship between power, voltage and current is expressed as:

Power (W) is equal to amperage (I) times voltage (E): $W=I \times E$

This is only true for direct current (DC) circuits; the alternating current formula is a tad different, but since the electrical circuits in most vessels are DC type, we need not get into AC circuit theory.

Electrical Components

POWER SOURCE

◆ **See Figure 2**

Typically, power is supplied to a vessel by 2 devices: The battery and the stator (or lighting coil). The stator supplies electrical current anytime the engine is running in order to recharge the battery and in order to operate electrical devices of the vessel. The battery supplies electrical power during starting or during periods when the current demand of the vessel's electrical system exceeds stator output capacity (which includes times when the motor is shut off and stator output is zero).

The Battery

In most modern vessels, the battery is a lead/acid electrochemical device consisting of six 2-volt subsections (cells) connected in series, so that the

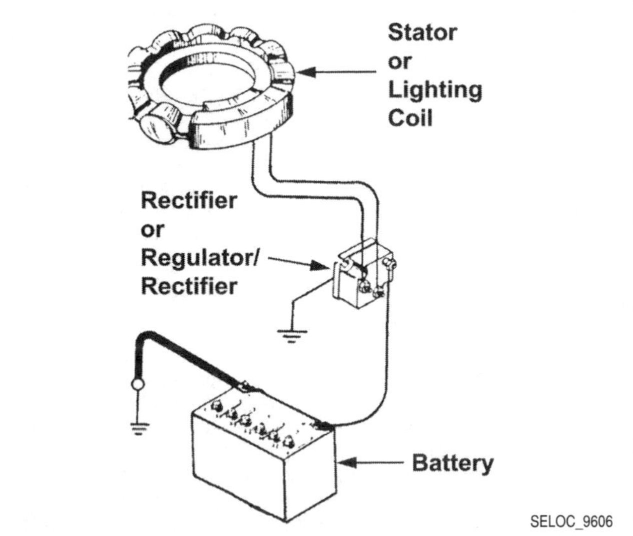

Fig. 2 Functional diagram of a typical charging circuit showing the relationship between the stator (lighting coil), rectifier (or regulator/rectifier) and battery

unit is capable of producing approximately 12 volts of electrical pressure. Each subsection consists of a series of positive and negative plates held a short distance apart in a solution of sulfuric acid and water.

The 2 types of plates in each battery cell are of dissimilar metals. This sets up a chemical reaction, and it is this reaction which produces current flow from the battery when its positive and negative terminals are connected to an electrical load. Power removed from the battery in use is replaced by current from the stator and restores the battery to its original chemical state.

The Stator (Lighting Coil)

Alternators and generators are devices that consist of coils of wires wound together making big electromagnets. The coil is normally referred to as a stator or lighting coil. Either, one group of coils spins within another set (or a set of permanently charged magnets, usually attached to the flywheel, are spun around a set of coils) and the interaction of the magnetic fields generates an electrical current. This current is then drawn off the coils and fed into the vessel's electrical system.

■ **Most vessels utilize a generator instead of an alternator. Although the terms are often misused and interchanged, the main difference is that an alternator supplies alternating current that is changed to direct current for use on the vessel, while a generator produces direct current. Alternators tend to be more efficient and that is why they are used on almost all modern engines.**

GROUND

Two types of grounds are used in marine electric circuits. Direct ground components are grounded to the electrically conductive metal through their mounting points. All other components use some sort of ground wire that leads back to the battery. The electrical current runs through the ground wire and returns to the battery through the ground or negative (-) cable; if you look, you'll see that the battery ground cable connects between the battery and a heavy gauge ground wire.

■ **A large percentage of electrical problems can be traced to bad grounds.**

If you refer back to the basic explanation of a circuit, you'll see that the ground portion of the circuit is just as important as the power feed. The wires delivering power to a component can have perfectly good, clean connections, but the circuit would fail to operate if there was a damaged ground connection. Since many components ground through their mounting or through wires that are connected to an engine surface, contamination from dirt or corrosion can raise resistance in a circuit to a point where it cannot operate.

PROTECTIVE DEVICES

◆ See Figures 3 and 4

Problems can occur in the electrical system that will cause large surges of current to pass through the electrical system of your vessel. These problems can be the fault of the charging circuit, but more likely would be a problem with the operating electrical components that causes an excessively high load. An unusually high load can occur in a circuit from problems such as a seized electric motor (like a damaged starter) or the excessive resistance caused by a bad ground (from loose or damaged wires or connections). A short to ground that bypasses the load and allows the battery to quickly discharge through a wire can also cause current surges.

■ **Inline or plug-in fuses are the most common forms of circuit protection found on these Yamaha motors. For more details, check the wiring diagram on the model which you are servicing.**

If this surge of current were to reach the load in the circuit, the surge could burn it out or severely damage it. It can also overload the wiring, causing the harness to get hot and melt the insulation. To prevent this, fuses, circuit breakers and/or fusible links are connected into the supply wires of the electrical system. These items are nothing more than a built-in weak spot in the system. When an abnormal amount of current flows through the system, these protective devices work as follows to protect the circuit:

- Fuse - when an excessive electrical current passes through a fuse, the fuse blows (the conductor melts) and opens the circuit, preventing current flow.
- Circuit Breaker - a circuit breaker is basically a self-repairing fuse. It will open the circuit in the same fashion as a fuse, but when the surge subsides, the circuit breaker can be reset and does not need replacement. Most circuit

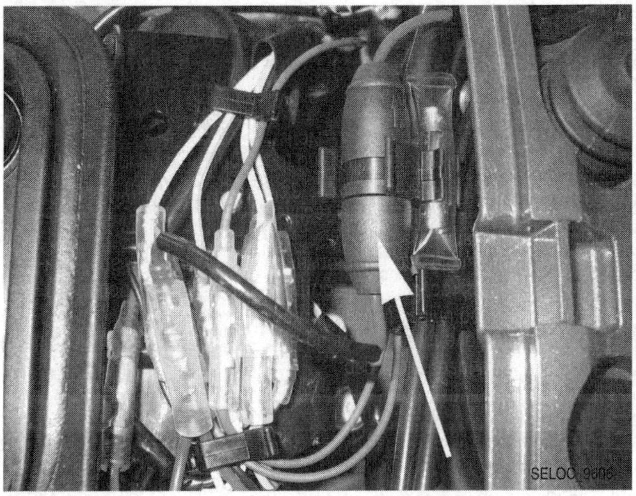

Fig. 3 Some motors generally have one or more inline fuses. . .

Fig. 4 All motors contain an electrical junction box of fuses (and relays), VX Cruiser shown

breakers on marine engine applications are self-resetting, but some that operate accessories (such as on larger vessels with a circuit breaker panel) must be reset manually (just like the circuit breaker panels in most homes). However, keep in mind these aren't common on Yamahas.

• Fusible Link - a fusible link (fuse link or main link) is a short length of special, high temperature insulated wire that acts as a fuse. When an excessive electrical current passes through a fusible link, the thin gauge wire inside the link melts, creating an intentional open to protect the circuit. To repair the circuit, the link must be replaced. Some newer type fusible links are housed in plug-in modules, which are simply replaced like a fuse, while older type fusible links must be cut and spliced if they melt. Since this link is very early in the electrical path, it's the first place to look if nothing on the vessel works, yet the battery seems to be charged and is otherwise properly connected. However, keep in mind these aren't common on Yamahas.

** CAUTION

Always replace fuses, circuit breakers and fusible links with identically rated components. Under no circumstances should a component of higher or lower amperage rating be substituted. A lower rated component will disable the circuit sooner than necessary (possibly during normal operation), while a higher rated component can allow dangerous amounts of current that could damage the circuit or component (or even melt insulation causing sparks or a fire).

SWITCHES & RELAYS

◆ **See Figure 5**

Switches are used in electrical circuits to control the passage of current. The most common use is to open and close circuits between the battery and the various electric devices in the system. Switches are rated according to the amount of amperage they can handle. If a sufficient amperage rated switch is not used in a circuit, the switch could overload and cause damage.

Some electrical components that require a large amount of current to operate use a special switch called a relay. Since these circuits carry a large amount of current, the thickness of the wire in the circuit is also greater. If this large wire were connected from the load to the control switch, the switch would have to carry the high amperage load and the space needed for wiring in the vessel would be twice as big to accommodate the increased size of the wiring harness. A relay is used to prevent these problems.

Think of relays as essentially "remote controlled switches." They allow a smaller current to throw the switch that operates higher amperages devices. Relays are composed of a coil and a set of contacts. When current is passed through the coil, a magnetic field is formed that causes the contacts to move together, closing the circuit. Most relays are normally open, preventing current from passing through the main circuit until power is applied to the coil. But, relays can take various electrical forms depending on the job for which they are intended. Some common circuits that may use relays are horns, lights, starters, electric fuel pumps and other potentially high draw circuits.

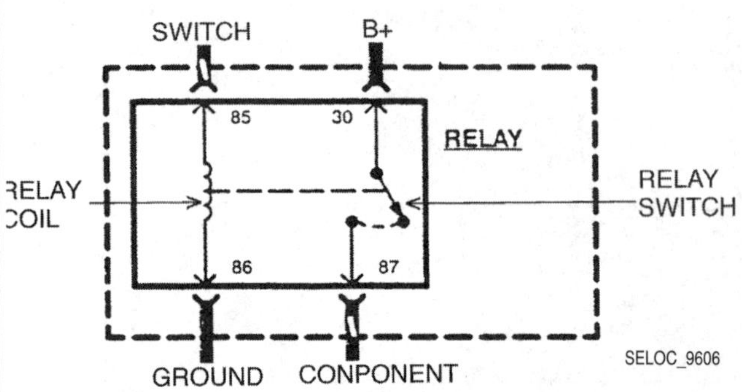

Fig. 5 Relays are composed of a coil and a switch. These two components are linked together so that when one is operated it actuates the other. The large wires in the circuit are connected from the battery to one side of the relay switch (B+) and from the opposite side of the relay switch to the load (component). Smaller wires are connected from the relay coil to the control switch for the circuit and from the opposite side of the relay coil to ground

LOAD

Every electrical circuit must include a load (something to use the electricity coming from the source). Without this load, the battery would attempt to deliver its entire power supply from one pole to another. This is called a short circuit. All this electricity would take a short cut to ground and cause a great amount of damage to other components in the circuit (including the battery) by developing a tremendous amount of heat. This condition could develop sufficient heat to melt the insulation on all the surrounding wires and reduce a multiple wire cable to a lump of plastic and copper. A short can allow sparks that could ignite fuel vapors or other combustible materials in the vessel, causing an extremely hazardous condition.

WIRING & HARNESSES

The average vessel contains miles of wiring, with hundreds of individual connections. To protect the many wires from damage and to keep them from becoming a confusing tangle, they are organized into bundles, enclosed in plastic or taped together and called wiring harnesses. Different harnesses serve different parts of the vessel. Individual wires are color coded to help trace them through a harness where sections are hidden from view.

Marine wiring or circuit conductors can be either a single strand wire, multi-strand wire or printed circuitry. Single strand wire has a solid metal core and is usually used inside such components as stator coil windings, motors, relays and other devices. Multi-strand wire has a core made of many small strands of wire twisted together into a single conductor. Most of the wiring in a marine electrical system is made up of multi-strand wire, either as a single conductor or grouped together in a harness. All wiring is color coded on the insulator, either as a solid color or as a colored wire with an identification stripe. A printed circuit is a thin film of copper or other conductor that is printed on an insulator backing. Occasionally, a printed circuit is sandwiched between 2 sheets of plastic for more protection and flexibility. A complete printed circuit, consisting of conductors, insulating material and connectors is called a printed circuit board. Printed circuitry is used in place of individual wires or harnesses in places where space is limited, such as behind 1-piece instrument clusters.

Since marine electrical systems are very sensitive to changes in resistance, the selection of properly sized wires is critical when systems are repaired. A loose or corroded connection or a replacement wire that is too small for the circuit will add extra resistance and an additional voltage drop to the circuit.

The wire gauge number is an expression of the cross-section area of the conductor. Vessels from countries that use the metric system will typically describe the wire size as its cross-sectional area in square millimeters. In this method, the larger the wire, the greater the number. Another common system for expressing wire size is the American Wire Gauge (AWG) system. As gauge number increases, area decreases and the wire becomes smaller. Using the AWG system, an 18 gauge wire is smaller than a 4 gauge wire. A wire with a higher gauge number will carry less current than a wire with a lower gauge number. Gauge wire size refers to the size of the strands of the conductor, not the size of the complete wire with insulator. It is possible, therefore, to have 2 wires of the same gauge with different diameters because one may have thicker insulation than the other.

It is essential to understand how a circuit works before trying to figure out why it doesn't. An electrical schematic shows the electrical current paths when a circuit is operating properly. Schematics break the entire electrical system down into individual circuits. In most schematics no attempt is made to represent wiring and components as they physically appear on the vessel; switches and other components are shown as simply as possible. But, this is **not** always the case on Yamaha schematics and some of the wiring diagrams provided here. So, when using a Yamaha schematic if the component in question is represented by something more than a small square or rectangle with a label, it is likely a intended to be a representation of the actual component's shape. On most schematics, the face views of harness connectors show the cavity or terminal locations in all multi-pin connectors to help locate test points.

Test Equipment

Pinpointing the exact cause of trouble in an electrical circuit is usually accomplished by the use of special test equipment, but the equipment does not always have to be expensive. The following sections describe different types of commonly used test equipment and briefly explains how to use them in diagnosis. In addition to the information covered below, be sure to read and understand the tool manufacturer's instruction manual (provided with most tools) before attempting any test procedures.

JUMPER WIRES

◆ See Figure 6

✻✻ CAUTION

Never use jumper wires made from a thinner gauge wire than the circuit being tested. If the jumper wire is of too small a gauge, it may overheat and possibly melt. Never use jumpers to bypass high resistance loads in a circuit. Bypassing resistances, in effect, creates a short circuit. This may, in turn, cause damage and fire. Jumper wires should only be used to bypass lengths of wire or to simulate switches.

Jumper wires are simple, yet extremely valuable, pieces of test equipment. They are basically test wires that are used to bypass sections of a circuit. Although jumper wires can be purchased, they are usually fabricated from lengths of standard marine wire and whatever type of connector (alligator clip, spade connector or pin connector) that is required for the particular application being tested. In cramped, hard-to-reach areas, it is advisable to have insulated boots over the jumper wire terminals in order to prevent accidental grounding. It is also advisable to include a standard marine fuse in any jumper wire. This is commonly referred to as a fused jumper. By inserting an in-line fuse holder between a set of test leads, a fused jumper wire is created for bypassing open circuits. Use a 5-amp fuse to provide protection against voltage spikes.

Jumper wires are used primarily to locate open electrical circuits, on either the ground (-) side of the circuit or on the power (+) side. If an electrical component fails to operate, connect the jumper wire between the component and a good ground. If the component operates only with the jumper installed, the ground circuit is open. If the ground circuit is good, but the component does not operate, the circuit between the power feed and component may be open. By moving the jumper wire successively back from the component toward the power source, you can isolate the area of the circuit where the open is located. When the component stops functioning, or the power is cut off, the open is in the segment of wire between the jumper and the point previously tested.

You can sometimes connect the jumper wire directly from the battery to the hot terminal of the component, but first make sure the component uses a full 12 volts in operation. Some electrical components, such as sensors, are designed to operate on smaller voltages like 4 or 5 volts, and running 12 volts directly to these components can damage or destroy them.

■ **Yamaha sells 2-pin, 3-pin and 6-pin test harnesses for use with all of their motors. These harnesses are designed to connect into place between the component and engine harness keeping the circuit intact, but are T'd so you can test the completed circuit under normal operating conditions.**

TEST LIGHTS

◆ See Figure 7

The test light is used to check circuits and components while electrical current is flowing through them. It is used for voltage and ground tests. To use a 12-volt test light, connect the ground clip to a good ground and probe connectors the pick where you are wondering if voltage is present. The test light will illuminate when voltage is detected. This does not necessarily mean that 12 volts (or any particular amount of voltage) is present; it only means that some voltage is present. It is advisable before using the test light to touch its ground clip and probe across the battery posts or terminals to make sure the light is operating properly and to note how brightly the light glows when 12 volts is present.

✻✻ WARNING

Do not use a test light to probe electronic ignition, spark plug or coil wires, as the circuit is much, much higher than 12 volts. Also, never use a pick-type test light to probe wiring on electronically controlled systems unless specifically instructed to do so. Whenever possible, avoid piercing insulation with the test light pick, as you are inviting shorts or corrosion and excessive resistance. But, any wire insulation that is pierced by necessity, must be sealed with silicone and taped after testing.

Like the jumper wire, the 12-volt test light is used to isolate opens in circuits. But, whereas the jumper wire is used to bypass the open to operate the load, the 12-volt test light is used to locate the presence or lack of voltage in a circuit. If the test light illuminates, there is power up to that point in the circuit; if the test light does not illuminate, there is an open circuit (no power). Move the test light in successive steps back toward the power source until the light in the handle illuminates. The open is between the probe and the point that was previously probed.

The self-powered test light is similar in design to the 12-volt test light, but contains a 1.5 volt penlight battery in the handle. It is most often used in place of a multi-meter to check for open or short circuits when power is isolated from the circuit (thereby performing a continuity test).

The battery in a self-powered test light does not provide much current. A weak battery may not provide enough power to illuminate the test light even when a complete circuit is made (especially if there is high resistance in the circuit). Always make sure that the test battery is strong. To check the battery, briefly touch the ground clip to the probe; if the light glows brightly, the battery is strong enough for testing.

■ **A self-powered test light should not be used on any electronically controlled system or component. Even the small amount of electricity transmitted by the test light is enough to damage many electronic components.**

MULTI-METERS

◆ See Figure 8

■ **Yamaha recommends the use of the #YU-34899-A multi-meter to best approximate their testing specifications.**

Multi-meters are extremely useful for troubleshooting electrical problems. They can be purchased in both analog or digital form and have a price range to suit nearly any budget. A multi-meter is a voltmeter, ammeter and ohmmeter (along with other features) combined into one instrument. It is often used when testing solid state circuits because of its high input impedance (usually 10 mega ohms or more). A high-quality digital multi-

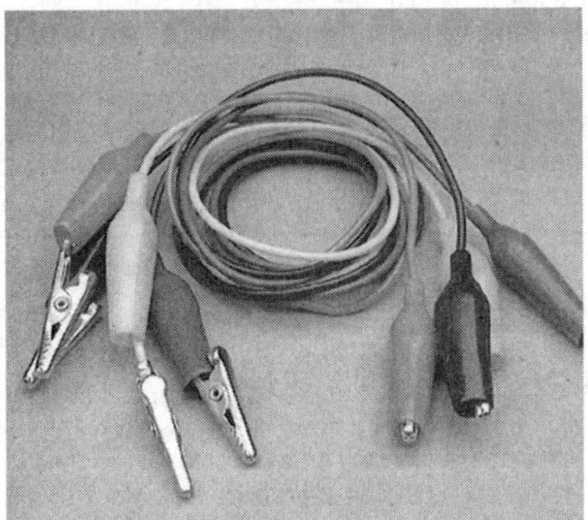

Fig. 6 Jumper wires are simple, but valuable pieces of test equipment

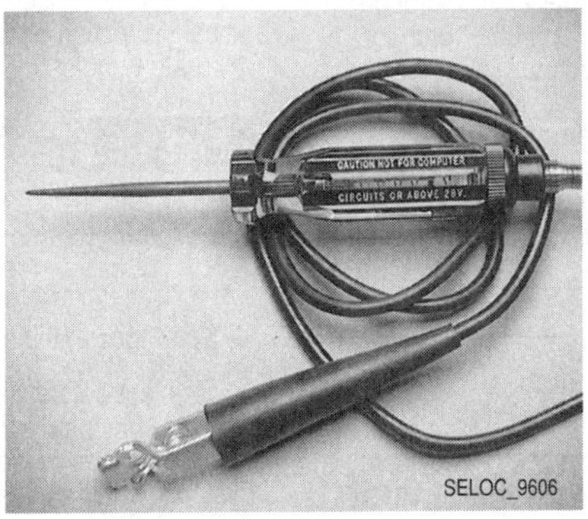

Fig. 7 A 12-volt test light is used to detect the presence of voltage in a circuit

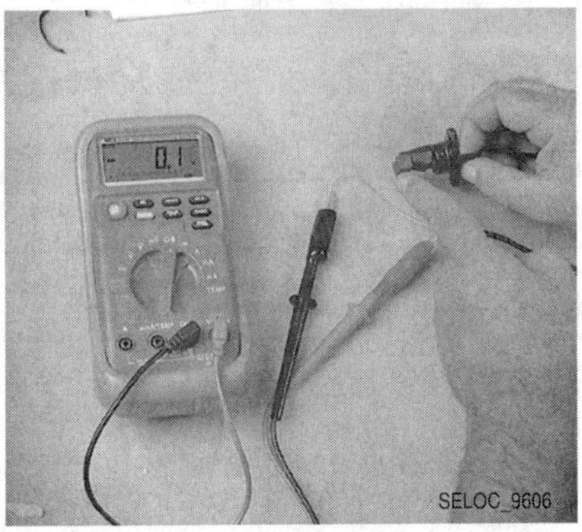

Fig. 8 Multi-meters are probably the most versatile and handy tools for diagnosing faulty electrical components or circuits

meter or Digital Volt Ohm Meter (DVOM) helps to ensure the most accurate test results and, although not absolutely necessary for electronic components such as computer controlled ignition systems and charging systems, is highly recommended. A brief description of the main test functions of a multi-meter follows:

- Voltmeter - the voltmeter is used to measure voltage at any point in a circuit or to measure the voltage drop across any part of a circuit. Voltmeters usually have various scales and a selector switch to allow metering and display of different voltage ranges. The voltmeter has a positive and a negative lead. To avoid damage to the meter, connect the negative lead to the negative (-) side of the circuit (to ground or nearest the ground side of the circuit) and connect the positive lead to the positive (+) side of the circuit (to the power source or the nearest power source). This is mostly a concern on analog meters, as DVOMs are not normally adversely affected (as they are usually designed to take readings even with reverse polarity and display accordingly). Note that the negative voltmeter lead will always be black and that the positive voltmeter will always be some color other than black (usually red).
- Ohmmeter - the ohmmeter is designed to read resistance (measured in ohms) in a circuit or component. Most ohmmeters will have a selector switch which permits the measurement of different ranges of resistance (usually the selector switch allows the multiplication of the meter reading by 10, 100, 1,000 and 10,000). Most modern ohmmeters (especially DVOMs) are auto-ranging which means the meter itself will determine which scale to use. Since ohmmeters are powered by an internal battery, the ohmmeter can be used like a self-powered test light. When the ohmmeter is connected, current from the ohmmeter flows through the circuit or component being tested. Since the ohmmeter's internal resistance and voltage are known values, the amount of current flow through the meter depends on the resistance of the circuit or component being tested. The ohmmeter can also be used to perform a continuity test for suspected open circuits. When using the meter for continuity checks, do not be concerned with the actual resistance readings. Zero resistance, or any ohm reading, indicates continuity in the circuit. Infinite resistance indicates an opening in the circuit. A high resistance reading where there should be none indicates a problem in the circuit. Checks for short circuits are made in the same manner as checks for open circuits, except that the circuit must be isolated from both power and normal ground. Infinite resistance indicates no continuity, while zero resistance indicates a dead short.

** WARNING

Never use an ohmmeter to check the resistance of a component or wire while there is voltage applied to the circuit. Voltage in the circuit can damage or destroy the meter.

- Ammeter - an ammeter measures the amount of current flowing through a circuit in units called amperes or amps. At normal operating voltage, most circuits have a characteristic amount of amperes, called current draw that can be measured using an ammeter. By referring to a specified current draw rating, and then measuring the amperes and comparing the two values, you can determine what is happening within the circuit to aid in diagnosis. An open circuit, for example, will not allow any current to flow, so the ammeter reading will be zero. A damaged component or circuit will have an increased current draw, so the reading will be high. The ammeter is always connected in series with the tested circuit. All of the current that normally flows through the circuit must also flow through the ammeter; if there is any other path for the current to follow, the ammeter reading will not be accurate. The ammeter itself has very little resistance to current flow and, therefore, will not affect the circuit, but it will measure current draw only when the circuit is closed and electricity is flowing. Excessive current draw can blow fuses and drain the battery, while a reduced current draw can cause motors to run slowly, lights to dim and other components to not operate properly.

PEAK VOLTAGE READING ADAPTOR

◆ See Figure 9

Some of the best testing procedures for the ignition and charging system involve capturing the peak voltage output of a particular coil assembly during a dynamic test. Some of the better multi-meters will have a peak reading setting and can be used for these tests. Other multi-meters, such as the #YU-34889-A which Yamaha recommends, have an available peak reading adaptor #YU-39991 that can be used for this purpose.

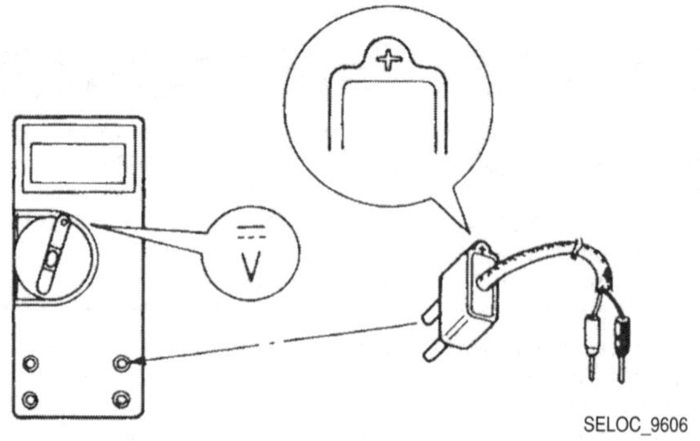

Fig. 9 A peak voltage reading multi-meter or a meter with a suitable peak reading adaptor is necessary for some ignition and charge coil testing

Troubleshooting Electrical Systems

When diagnosing a specific problem, organized troubleshooting is a must. The complexity of a modern marine vessel demands that you approach any problem in a logical, organized manner. There are certain troubleshooting techniques, however, which are standard:

- **Establish when the problem occurs**. Does the problem appear only under certain conditions? Were there any noises, odors or other unusual symptoms? Isolate the problem area. To do this, make some simple tests and observations, and then eliminate the systems that are working properly. Check for obvious problems, such as broken wires and loose or dirty connections. Always check the obvious before assuming something complicated is the cause.
- **Test for problems systematically to determine the cause once the problem area is isolated**. Are all the components functioning properly? Is there power going to electrical switches and motors? Performing careful, systematic checks will often turn up most causes on the first inspection, without wasting time checking components that have little or no relationship to the problem.
- **Test all repairs after the work is done to make sure that the problem is fixed**. Some causes can be traced to more than one component, so a careful verification of repair work is important in order to pick up additional malfunctions that may cause a problem to reappear or a different problem to arise. A blown fuse, for example, is a simple problem that may require more than another fuse to repair. If you don't look for a problem that caused a fuse to blow, a shorted wire (for example) may go undetected and cause the new fuse to blow right away (if the short is still present) or during subsequent operation (as soon as the short returns if it is intermittent).

Experience shows that most problems tend to be the result of a fairly simple and obvious cause, such as loose or corroded connectors, bad grounds or damaged wire insulation that causes a short. This makes careful visual inspection of components during testing essential to quick and accurate troubleshooting.

Electrical Testing

VOLTAGE

◆ See Figure 10

This test determines the voltage available from the battery and should be the first step in any electrical troubleshooting procedure after visual inspection. Many electrical problems, especially on electronically controlled systems, can be caused by a low state of charge in the battery. Many circuits cannot function correctly if the battery voltage drops below normal operating levels.

Loose or corroded battery cable terminals can cause poor contact that will prevent proper charging and full battery current flow.

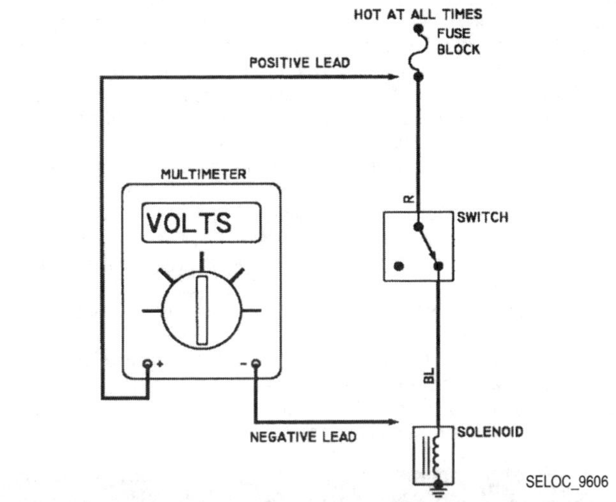

Fig. 10 A voltage check determines the amount of battery voltage available and, as such, should be the first step in any troubleshooting procedure

1. Set the voltmeter selector switch to the 20V position.
2. Connect the meter negative lead to the battery's negative (-) post or terminal and the positive lead to the battery's positive (+) post or terminal.
3. Turn the ignition switch **ON** to provide a small load.
4. A well charged battery should register over 12 volts. If the meter reads below 11.5 volts, the battery power may be insufficient to operate the electrical system properly. Check and charge or replace the battery as detailed under Engine Maintenance before further tests are conducted on the electrical system.

VOLTAGE DROP

◆ See Figure 11

When current flows through a load, the voltage beyond the load drops. This voltage drop is due to the resistance created by the load and also by small resistances created by corrosion at the connectors (or by damaged insulation on the wires). Since all voltage drops are cumulative, the maximum allowable voltage drop under load is critical, especially if there is more than one load in the circuit.

1. Set the voltmeter selector switch to the 20 volts position.
2. Connect the multi-meter negative lead to a good ground.
3. Operate the circuit and check the voltage prior to the first component (load).
4. There should be little or no voltage drop in the circuit prior to the first component. If a voltage drop exists, the wire or connectors in the circuit are suspect.
5. While operating the first component in the circuit, probe the ground side of the component with the positive meter lead and observe the voltage readings. A small voltage drop should be noticed. This voltage drop is caused by the resistance of the component.
6. Repeat the test for each component (load) down the circuit.
7. If an excessively large voltage drop is noticed, the preceding component, wire or connector is suspect.

RESISTANCE

◆ See Figure 12

✲✲ WARNING

Never use an ohmmeter with power applied to the circuit. The ohmmeter is designed to operate on its own power supply. The normal 12-volt electrical system voltage will damage or destroy many meters!

1. Isolate the circuit from the vessel's power source.
2. Ensure that the ignition key is **OFF** when disconnecting any components or the battery.
3. Where necessary, also isolate at least one side of the circuit to be checked, in order to avoid reading parallel resistances. Parallel circuit resistances will always give a lower reading than the actual resistance of either of the branches.
4. Connect the meter leads to both sides of the circuit (wire or component) and read the actual measured ohms on the meter scale. Make sure the selector switch is set to the proper ohm scale for the circuit being tested, to avoid misreading the ohmmeter test value.

■ **The resistance reading of most electrical components will vary with temperature. Unless otherwise noted, specifications given are for testing under ambient conditions of 68°F (20°C). If the component is tested at higher or lower temperatures, expect the readings to vary slightly. When testing engine control sensors or coil windings with smaller resistance specifications (less than 1000 ohms) it is best to use a high quality DVOM and be especially careful of your test results. Whenever possible, double-check your results against a known good part before purchasing the replacement. If necessary, bring the old part to the marine parts dealer and have them compare the readings to prevent possibly replacing a good component.**

OPEN CIRCUITS

◆ See Figure 13

This test already assumes the existence of an open in the circuit and it is used to help locate position of the open.

1. Isolate the circuit from power and ground.
2. Connect the self-powered test light or ohmmeter ground clip to the ground side of the circuit and probe sections of the circuit sequentially.

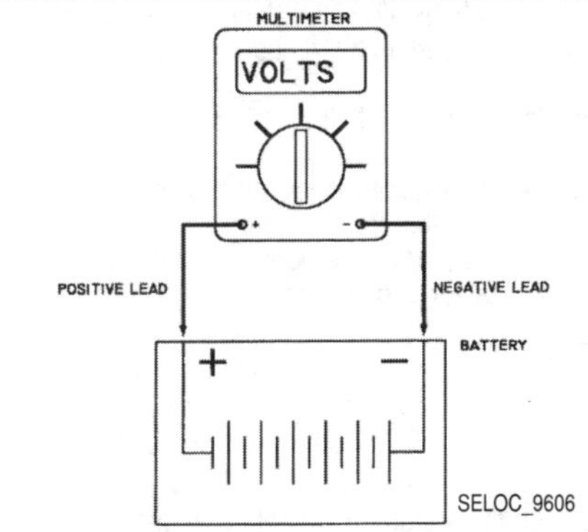

Fig. 11 Voltage drops are due to resistance in the circuit, from the load or from problems with the wiring

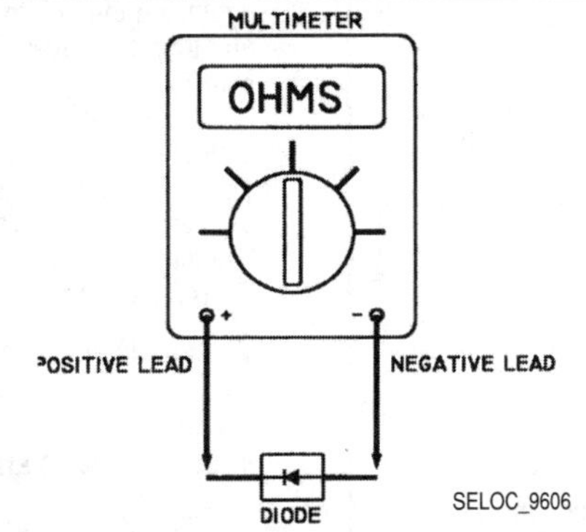

Fig. 12 Resistance tests must be conducted on portions of the circuit, isolated from battery power

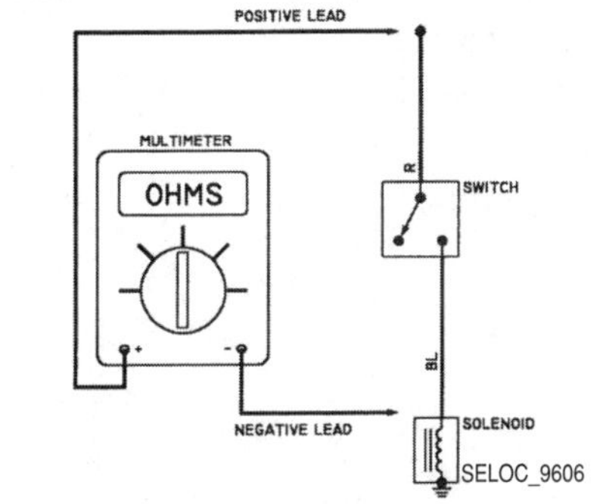

Fig. 13 The easiest way to illustrate an open circuit is to picture a circuit in which the switch is turned OFF (creating an opening in the circuit) that prevents power from reaching the load

3. If the light is out or there is infinite resistance, the open is between the probe and the circuit ground.

4. If the light is on or the meter shows continuity, the open is between the probe and the end of the circuit toward the power source.

SHORT CIRCUITS

◆ See Figure 14

■ **Never use a self-powered test light to perform checks for opens or shorts when power is applied to the circuit under test. The test light can be damaged by outside power.**

1. Isolate the circuit from power and ground.
2. Connect the self-powered test light or ohmmeter ground clip to a good ground and probe any easy-to-reach point in the circuit.
3. If the light comes on or there is continuity, there is a short somewhere in the circuit.
4. To isolate the short, probe a test point at either end of the isolated circuit (the light should be on or the meter should indicate continuity).
5. Leave the test light probe engaged and sequentially open connectors or switches, remove parts, etc. until the light goes out or continuity is broken.
6. When the light goes out, the short is between the last 2 circuit components that were opened.

Wire & Connector Repair

Almost anyone can replace damaged wires, as long as the proper tools and parts are available. Wire and terminals are available to fit almost any need. Even the specialized weatherproof, molded and hard shell connectors used by many marine manufacturers.

Be sure the ends of all the wires are fitted with the proper terminal hardware and connectors. Wrapping a wire around a stud is not a permanent solution and will only cause trouble later. Replace wires one at a time to avoid confusion. Always route wires in the same manner of the manufacturer.

When replacing connections, make absolutely certain that the connectors are certified for marine use. Automotive wire connectors may not meet United States Coast Guard (USCG) specifications.

■ **If connector repair is necessary, only attempt it if you have the proper tools. Weatherproof and hard shell connectors may require special tools to release the pins inside the connector. Attempting to repair these connectors with conventional hand tools will damage them.**

Electrical System Precautions

- Wear safety glasses when working on or near the battery.
- Don't wear a watch with a metal band when servicing the battery or starter. Serious burns can result if the band completes the circuit between the positive battery terminal (or a hot wire) and ground.
- Be absolutely sure of the polarity of a booster battery before making connections. Remember that even momentary connection of a booster battery with the polarity reversed will damage charging system diodes. Connect the cables positive-to-positive, and negative (of the good battery)-to-a good ground on the engine (away from the battery to prevent the possibility of an explosion if hydrogen vapors are present from the electrolyte in the discharged battery). Connect positive cables first (starting with the discharged battery), and then make the last connection to ground on the body of the booster vessel so that arcing cannot ignite hydrogen gas that may have accumulated near the battery. • Disconnect both vessel battery cables before attempting to charge a battery.
- Never ground the alternator or generator output or battery terminal. Be cautious when using metal tools around a battery to avoid creating a short circuit between the terminals.
- When installing a battery, make sure that the positive and negative cables are not reversed.
- Always disconnect the battery (negative cable first) when charging.
- Never smoke or expose an open flame around the battery. Hydrogen gas is released from battery electrolyte during use and accumulates near the battery. Hydrogen gas is **highly** explosive.

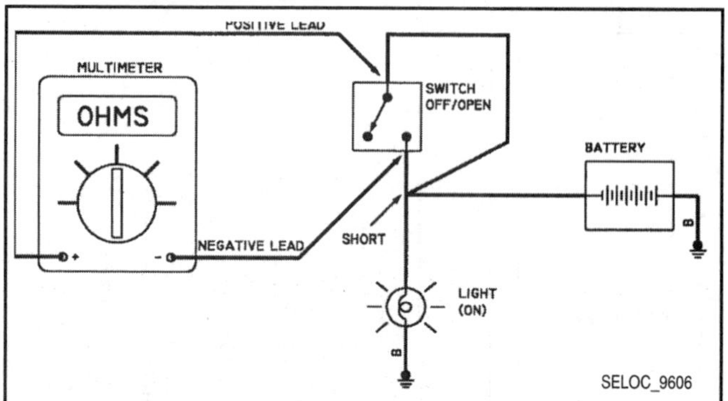

Fig. 14 In this illustration a load (the light) is powered when it should not be (since the switch should be creating an open condition), but a short to power (battery) is powering the circuit. Shorts like this can be caused by chaffed wires with worn or broken insulation

Electrical/Fuse Box

REMOVAL & INSTALLATION

2002-08 FX Models With 998cc/1052cc Engines

MODERATE

◆ See Figure 15

The electrical box is located behind the oil tank, toward the rear of the engine compartment.

1. Remove the seat(s) and rear cargo container (if equipped). Remove the battery as detailed in the Maintenance & Tune-Up section.
2. Lift out the damper and then remove the 4 battery box mounting bolts. Lift out the box and spacers.
3. Remove the 4 box mounting bolts from the corners.
4. Remove the 4 terminal cover mounting screws and pull off the cover and gasket. Loosen the bolt and remove the starter motor lead and battery lead. Tag them and position out of the way.
5. Locate the ground leads cover on the port side of the oil tank. Remove the 3 screws and lift off the cover and gasket. Disconnect the 2 ground leads, tag them and move them out of the way.
6. Trace all electrical leads coming from the box (should be approximately 15-22 depending on the model), tag them on both sides of the couplers and disconnect them.
7. Lift out the electrical box (along with the rectifier/regulator).

To install:

8. Position the box in place and run in the mounting screws just enough to hold it in place.
9. Reconnect all electrical leads disconnected during removal.
10. Reconnect the 2 ground leads at the oil tank. Install a *new* gasket and then the cover, tightening the 3 screws to 5.5 ft. lbs./66 inch lbs. (7.6 Nm). Be sure to coat the screw threads with Loctite 572.
11. Reconnect the starter and battery leads at the chamber in the box, tightening the bolt to 5.5 ft. lbs./66 inch lbs. (7.6 Nm). Install a *new* gasket and then the cover, tightening the 4 screws to 3.5 ft. lbs./41 inch lbs. (7.6 Nm).
12. Tighten the 4 box mounting bolts to 12 ft. lbs. (17 Nm). Be sure to coat the screw threads with Loctite 572.
13. Position the battery box dampers and then lay the box in position. Tighten the 4 bolts to 3.9 ft. lbs./46 inch lbs. Be sure to coat the 2 starboard bolt threads with Loctite 572 and the 2 port bolt threads with Loctite 271.
14. Install the battery, cargo container and seat(s).

VX Models w/1052cc Engines

MODERATE

◆ See Figures 4 and 16

The electrical box is located on the starboard side of the engine compartment, just forward of the engine.

1. Remove the seat(s) and rear cargo container (if equipped). Remove the battery as detailed in the Maintenance & Tune-Up section.
2. Lift out the damper and then remove the 4 battery box mounting bolts.

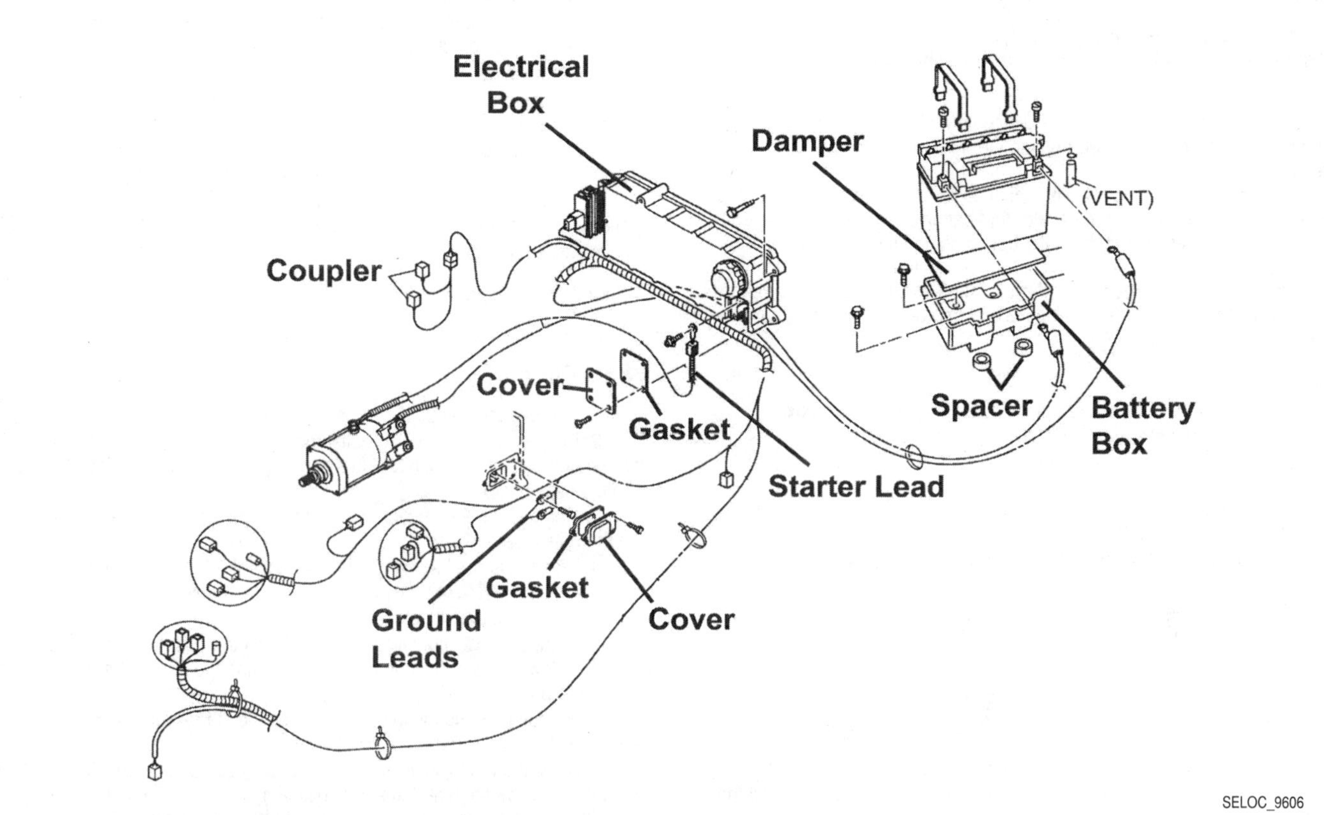

Fig. 15 A good look at the electrical box - 2002-08 FX models with 998cc/1052cc engines

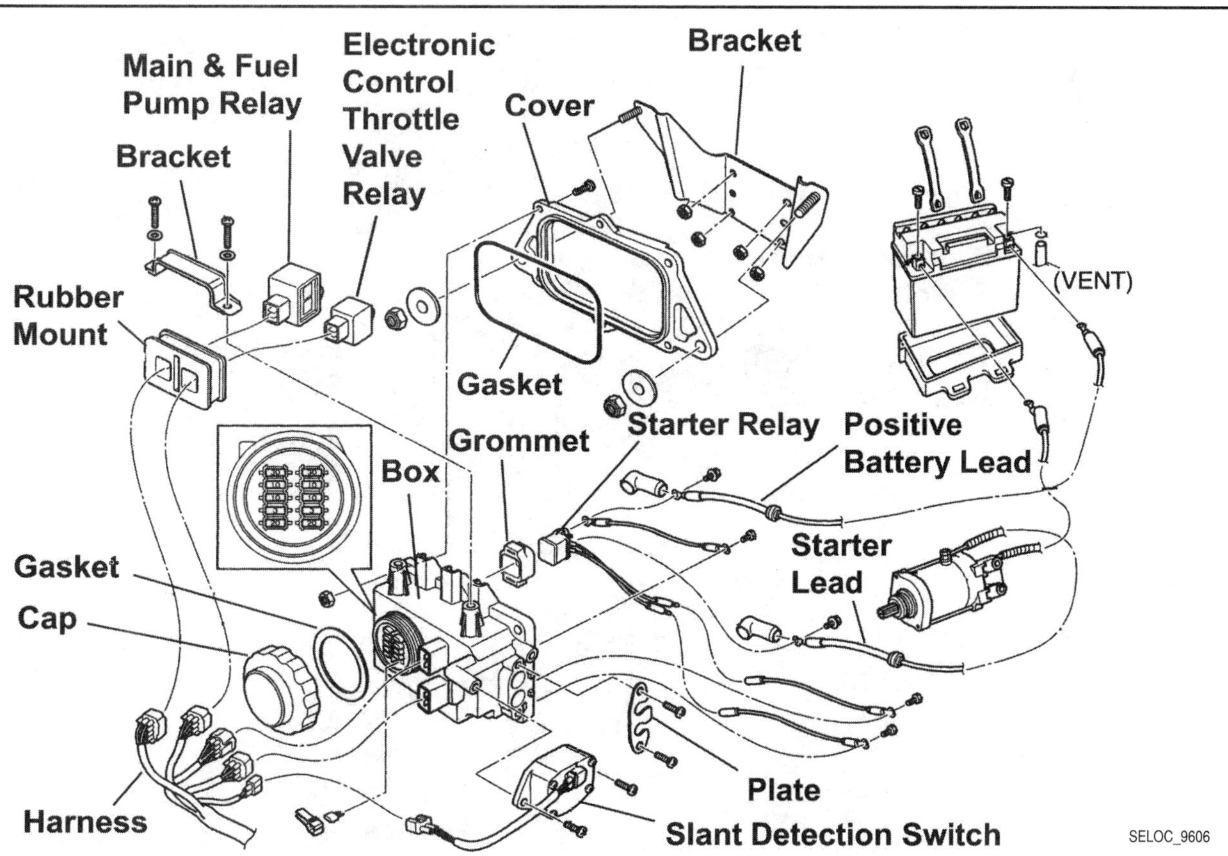

Fig. 16 Exploded view of the electrical/fuse box - 2005-09 VX models (2010-11 similar)

Lift out the box and spacers.

■ **Yamaha suggests removing the battery and battery box, but we're not sure that we would simply disconnect the battery cables at the battery.**

3. Tag and disconnect all electrical lead couplers coming from the box or in the way of removal - be smart and tag the wires on each side of the coupler.
4. Remove the 2 box-to-bracket mounting bolts and their washers. Lift out the box. You'll notice that the starter and battery lead is still connected to the box, if you are simply removing the box to get it out of the way we'd suggest disconnecting these 2 leads at their respective components and lifting it out with them still attached. If this is not feasible, you need to remove the cover and disconnect them inside the box in order to completely remove the box from the vessel.

To install:

5. If the starter and battery leads were disconnected at the box, reconnect them, tightening the bolts securely.
6. Position the box into the bracket and tighten the 2 bolts to 12 ft. lbs. (17 Nm).
7. Reconnect all disconnected electrical couplers - you did take our advice and tag them, right?
8. Install the battery (if removed), cargo container and seat(s).

DISASSEMBLY & ASSEMBLY

2002-08 FX Models With 998cc/1052cc Engines

② MODERATE

◆ **See Figures 17 and 18**

1. Remove the electrical box as detailed previously.
2. Working at the back of the box, remove the 18 screws and lift off the cover and gasket.
3. Remove the fuse holder cap.
4. You can remove the entire wiring harness at this time if inclined, but there's really no need unless there is a problem.
5. Remove the 2 rectifier/regulator mounting bolts and lift off the unit.
6. Lay the unit on its back (the forward side) and remove any components necessary.
7. All components, including the rectifier/regulator, should have their mounting screws tightened to 2.8 ft. lbs./34 inch lbs. (3/9 Nm).
8. Thread in all 18 cover screws and then tighten them, in a criss-cross pattern, to 3.5 ft. lbs./42 inch lbs. (4.9 Nm).
9. Install the box.

VX Models w/1052cc Engines

② MODERATE

◆ **See Figure 16**

1. Remove the electrical box as detailed previously.
2. Remove the 2 mounting bolts and lift off the upper bracket and grommet. Pop out the main and fuel pump and throttle valve relays from the grommet.
3. Remove the 2 mounting bolts and take off the slant detection switch on the right side of the box.
4. Working at the back of the box, remove the 6 screws and lift off the cover and gasket. Make sure you do not lose the nuts.
5. Remove the fuse holder cap.
6. Remove the 2 mounting bolts and lift off the plate on the right side of the box. Now you can disconnect the starter motor lead and the battery lead inside the box.
7. Remove the mounting bolt and lift off the starter relay. Pull out the grommet if it did not come out with the relay.
8. All components, including the slant detection switch, should have their mounting screws tightened to 2.8 ft. lbs./34 inch lbs. (3/9 Nm). On 2010 models, the detection switch and plate should be tightened to 2.2 ft. lbs/27

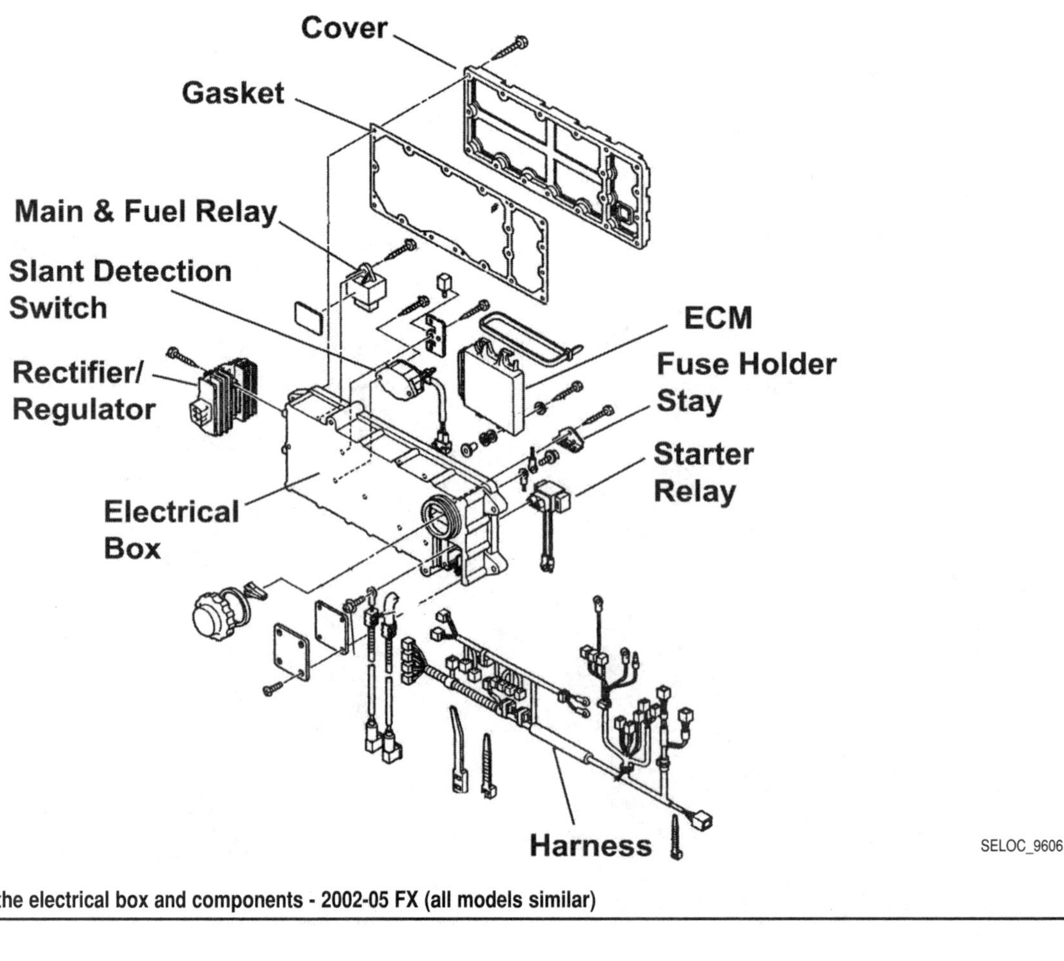

Fig. 17 Exploded view of the electrical box and components - 2002-05 FX (all models similar)

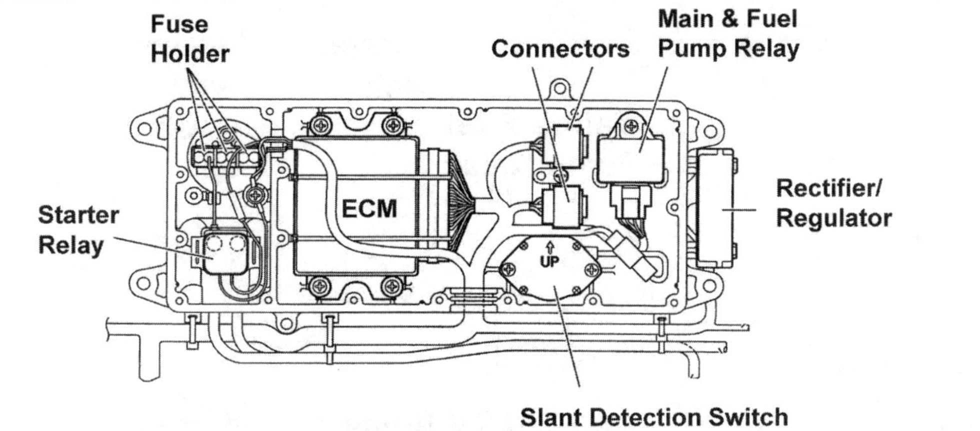

Fig. 18 A good look at the back of the box and component positioning (all models similar)

inch lbs. (3 Nm), while the starter and battery lead bolts should be 3 ft. lbs./36 inch lbs. (4 Nm).

9. Press a *new* gasket into the cover and then position it over the box. Install the 6 bolts and then thread on the nuts and then tighten them, in a criss-cross pattern, to 3.5 ft. lbs./42 inch lbs. (4.9 Nm).

10. Install the box.

REMOVAL, DISASSEMBLY & INSTALLATION

FX, FZ & VX Models With 1812cc Engines

◆ See Figures 19 thru 22

② MODERATE

The electrical box is located behind the rear of the engine compartment. These models are interesting in that they have a conventional FX electrical box with a VX (1052cc) fuse box inside of it.

1. Remove the seat(s) and rear cargo container (if equipped). Disconnect the battery.

2. Slide the 2 slide-locks on each upper corner of the box outward to release and then lift out the electrical box cover.

3. Tag and disconnect the following electrical lead couplers:

- ECM (3)
- Fuse box (2)
- Slant detection switch (1)
- ETV relay (1)
- Main and fuel pump relay (1)

Move the wiring harness out of the way.

If you are simply removing the electrical box to gain access to something else, skip to Step 7. Otherwise, move to the next step.

4. Remove the 4 mounting bolts and lift out the ECM. Be careful not to lose the grommets and collars.

5. Remove the 2 screws and take out the slant detection switch.

6. On FX HO, FZ and VX models, the fuse box has no cover and bolts directly to the electrical box. Remove the 6 mounting bolts and lift out the fuse box along with the O-ring/seal. On FX SHO models, the fuse box has a cover and then bolts to the back of the electrical box. Remove the 2 mounting bolts and remove the fuse box from the electrical box. Now you

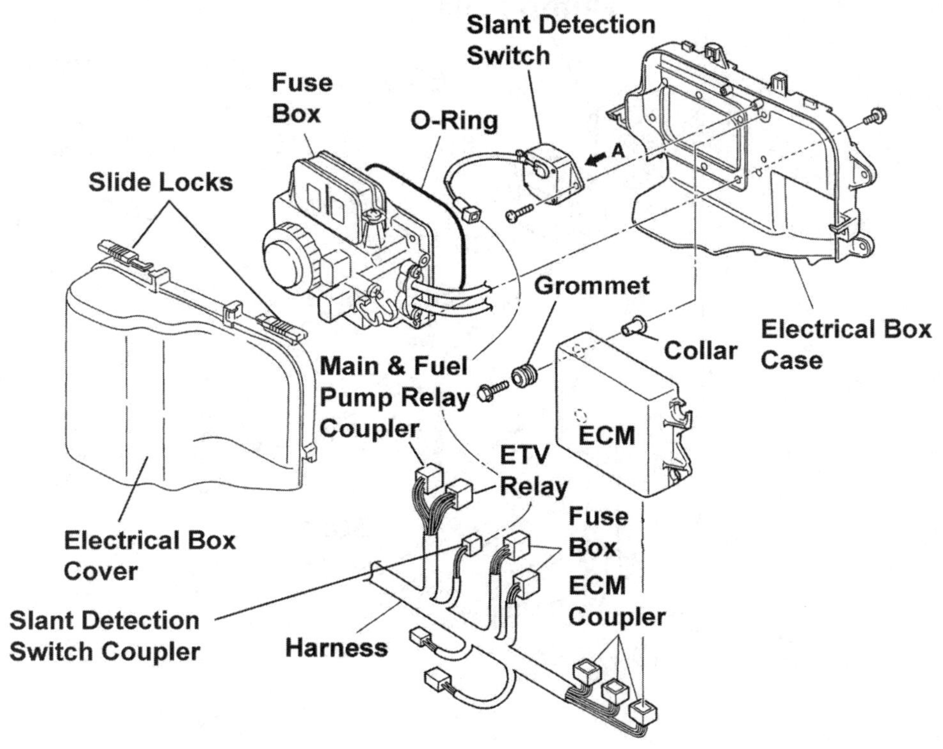

Fig. 19 A good look at the electrical box - FX, FZ & VX models w/1812cc engines (HO shown, SHO, VX and FZ similar)

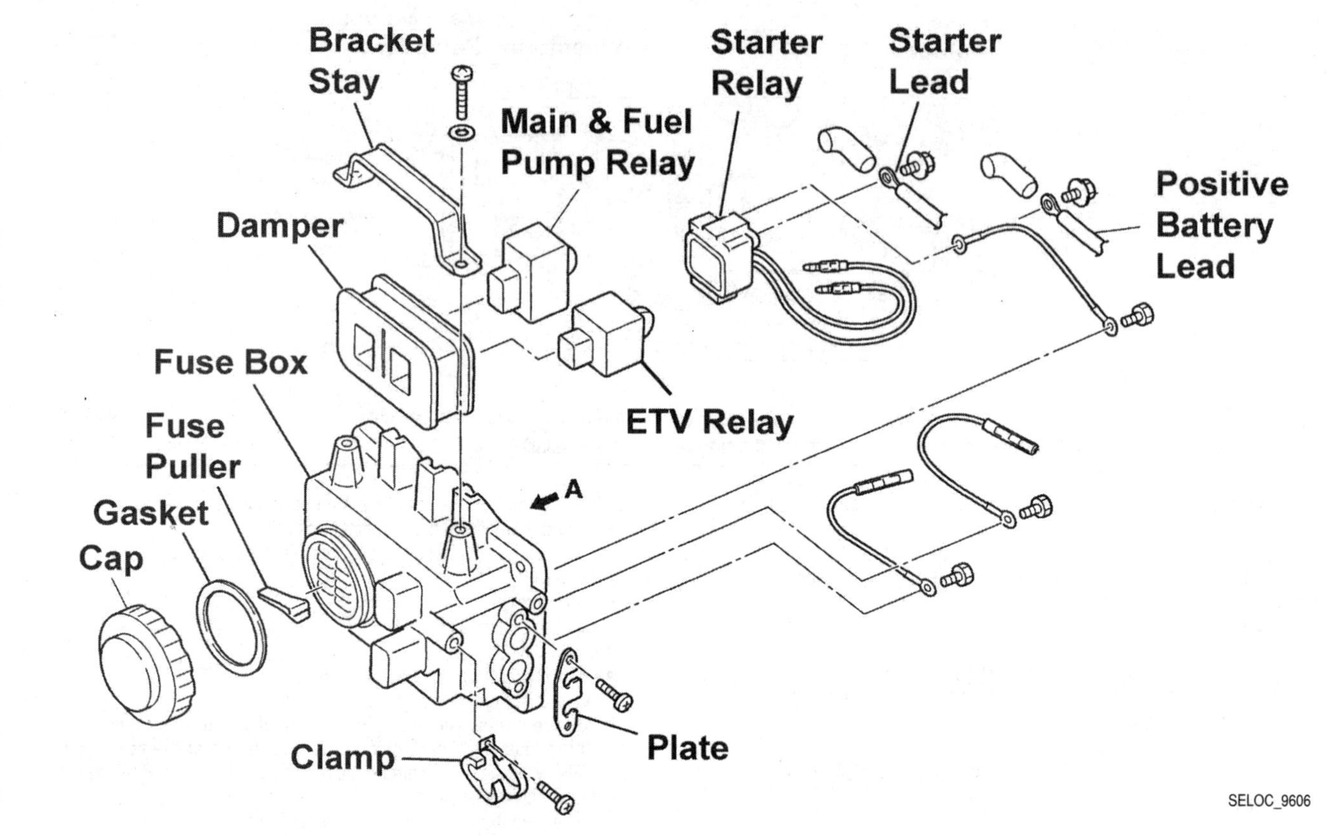

Fig. 20 A good look at the fuse box - FX HO, FZ and VX models w/1812cc engines

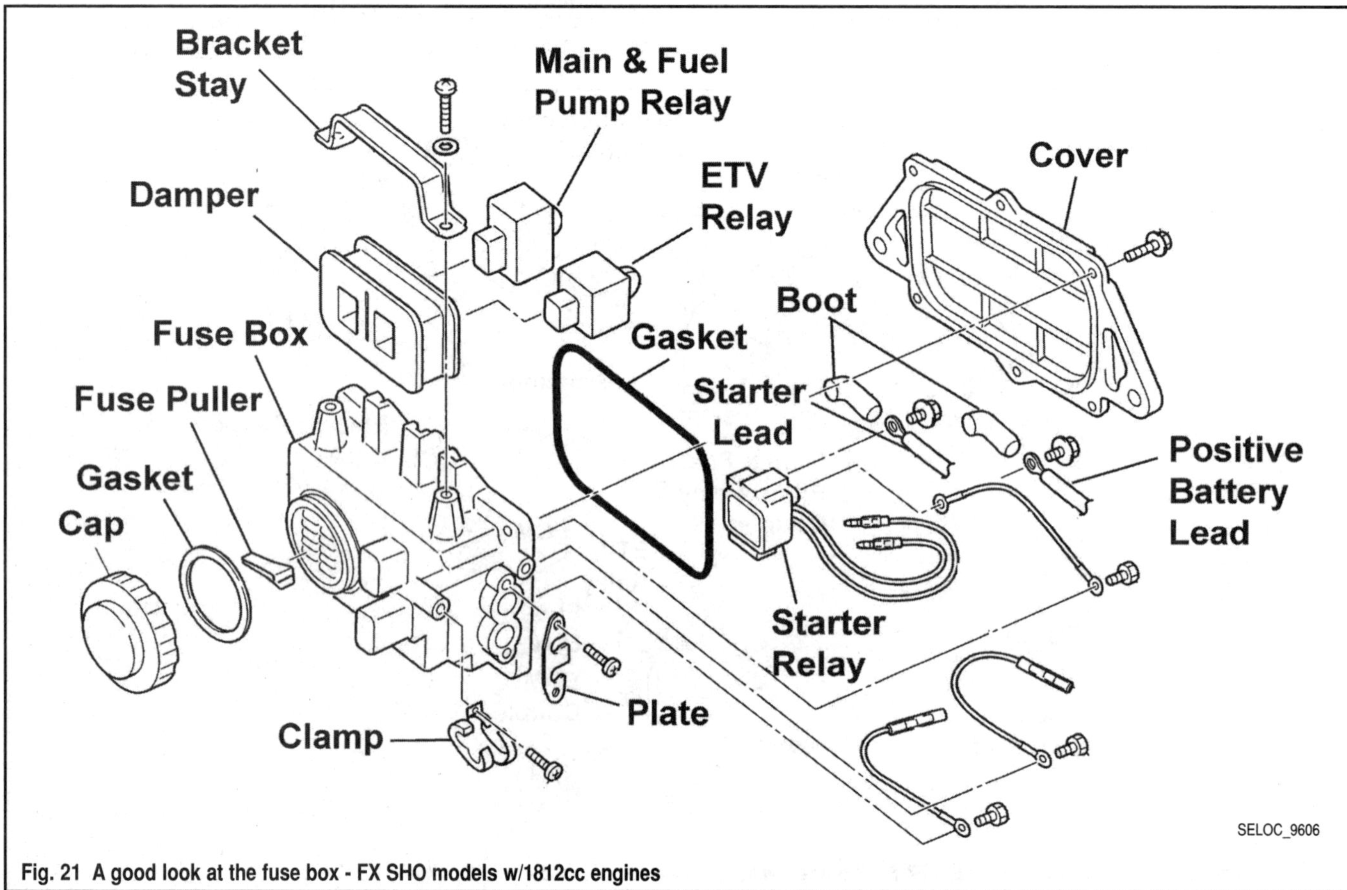

Fig. 21 A good look at the fuse box - FX SHO models w/1812cc engines

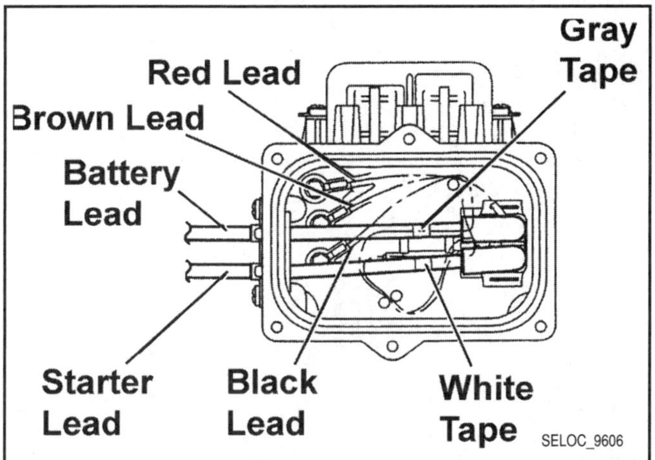

Fig. 22 Fuse box leads - FX shown, others similar

can remove the 6 cover mounting bolts and lift off the fuse box cover along with the O-ring.

7. Remove the 4 mounting bolts and remove the electrical box case from the mounting bracket.

If necessary, fuse box disassembly is as follows:

8. Remove the 2 mounting bolts and lift off the upper bracket stay and grommet. Pop out the main and fuel pump and throttle valve (ETV) relays from the grommet.

9. Remove the fuse holder cap and gasket.

10. Remove the 2 mounting bolts and lift off the plate (ring on VX) on the right side of the box as you face it. Now you can disconnect the starter motor lead (bottom) and the battery lead (top) inside the box and pull them through the side of the box.

11. Lift out the starter relay (you already removed the bolts when you disconnected the starter and battery leads). Pull out the grommet if it did not come out with the relay.

To install:

If you had simply removed the electrical box, perform the next step and then skip to the last two for installation.

12. Position the electrical box case on the bracket and tighten the 4 bolts securely.

13. Position the starter relay into the fuse box.

14. Feed the starter and battery leads through the side of the box and then connect them to the starter relay, tightening them to 3 ft. lbs./36 inch lbs. (4 Nm). Slide the boots over the bolt and ring terminal.

15. Position the plate/ring around the starter and battery leads on the side of the box and tighten the 2 bolts to 3 ft. lbs./36 inch lbs. (4 Nm).

16. Pop the ETV and main/fuel pump relays into their damper and position them on top of the box. Install the bracket stay and tighten the 2 bolts securely.

17. Install the cap.

18. On FX HO, FZ and VX models, press a *new* O-ring/seal into the box and then position it into the electrical box case. Install the 6 bolts and then tighten them, in a criss-cross pattern, to 5.9 ft. lbs./70 inch lbs. (8 Nm). Coat the bolt threads with Loctite 242.

19. On FX SHO models, press a *new* gasket into the box and then position the cover on the back of the fuse box. Install the 6 bolts and then tighten them, in a criss-cross pattern, to 5.9 ft. lbs./70 inch lbs. (8 Nm). Coat the bolt threads with Loctite 242. Now position the unit on the back of the electrical box and tighten the 2 mounting bolts to 9.6 ft. lbs./110 inch lbs. (13 Nm).

20. Install the slant detection switch and tighten the bolts to 3 ft. lbs./36 inch lbs. (4 Nm). Make sure that the word **UP** is, you guessed it, pointed up, in the direction of the arrow.

21. Press the 4 grommets into their slots on the sides of the ECM. Install the 4 bolts and collars and then tighten them to 3.7 ft. lbs./44 inch lbs. (5 Nm). Coat the bolt threads with Loctite 242.

22. Position the electrical box cover into the latches on the case and then slide the locks inward until they click into place.

23. Move the harness back into position and reconnect each of the couplers.

24. Connect the battery and installs the seat(s) and cargo containers.

ELECTRONIC IGNITION & CHARGING SYSTEMS

General Information

The less a PWC engine is operated, the more care it needs. Allowing an engine to remain idle will do more harm than if it is used regularly. To maintain the engine in top shape and always ready for efficient operation at any time, the engine ideally should be operated every 3-4 weeks throughout the year.

The fuel and ignition principles of 4-stroke engine operation must be understood in order to perform a proper tune-up on a motor. If you have any doubts concerning your understanding of engine operation, it would be best to study the operation theory before tackling any work on the ignition system.

Description & Operation

TCI ELECTRONIC IGNITION SYSTEM

◆ **See Figures 23 thru 32**

All Yamaha 4-stroke motors are equipped with some form of a Transistor Controlled (TCI) electronic ignition system. Some models may be equipped with a micro-computer controlled version of the system (TCI Micro) which means the TCI unit or ECM (as the unit is known on fuel injected motors) utilizes additional input (sensor input such as a crank position sensor, temperature sensor or even oil pressure sensor on some motors) to help make spark timing decisions, however, the basic function of the various Yamaha ignition systems are all similar.

To understand basic TCI ignition operation, it is important to understand the basic theory of induction. Induction theory states that if we move a magnet (magnetic field) past a coil of wire (or the coil by the magnet), AC current will be generated in the coil.

The amount of current produced depends on several factors:

- How fast the magnet moves past the coil
- The size of the magnet (strength)
- How close the magnet is to the coil
- Number of turns of wire and the size of the windings

The first thing these ignitions have in common is that a flywheel/magneto/rotor assembly is used to generate signals by rotating over/past 1 (VX and 1812cc FX and FZ models) or 2 (998cc/1052cc FX models) pulser or pick-up coil(s) and power over/past a single lighting coil or stator. In all cases, the power and signals are fed to a control unit which is used to power and "trigger" the ignition coils and spark plugs.

Although the basic function of the components is the same across most systems, their actual jobs may be performed by different physical components or their functions may be combined into a single unit on some motors.

Generally speaking, lighting/stator coils are used to generate system power. Similarly, pulser/pick-up coils are normally used to generate signals triggering the TCI/ECM unit to fire the ignition coils.

The TCI/ECM unit takes in power from the lighting coil (or stator coil) and signals from the pulser/pick-up coil(s) and then uses the power and signals to control function of the ignition coils. As mentioned previously, fuel injected motors combine the function of the TCI unit into the electronic control module (ECM), which also controls the fuel injection system (details of which can be found in the Fuel System section.

■ **Throughout this section we'll refer to the TCI unit as the ECM.**

The ignition system used on 998cc motors utilizes 2 pulser coils, 1 stator-mounted lighting coil winding, a TCI unit (the ECM) and 2 dual-lead ignition coils.

All ignition and timing functions are controlled by the ECM based on input from the pulser coil and voltage from the lighting coil. These models contain dual-lead ignition coils, so the spark plugs are fired in pairs (one cylinder at TDC of the compression stroke while the other is on TDC of the exhaust stroke) using a traditional "waste-spark" configuration. Typically, 1 dual-lead

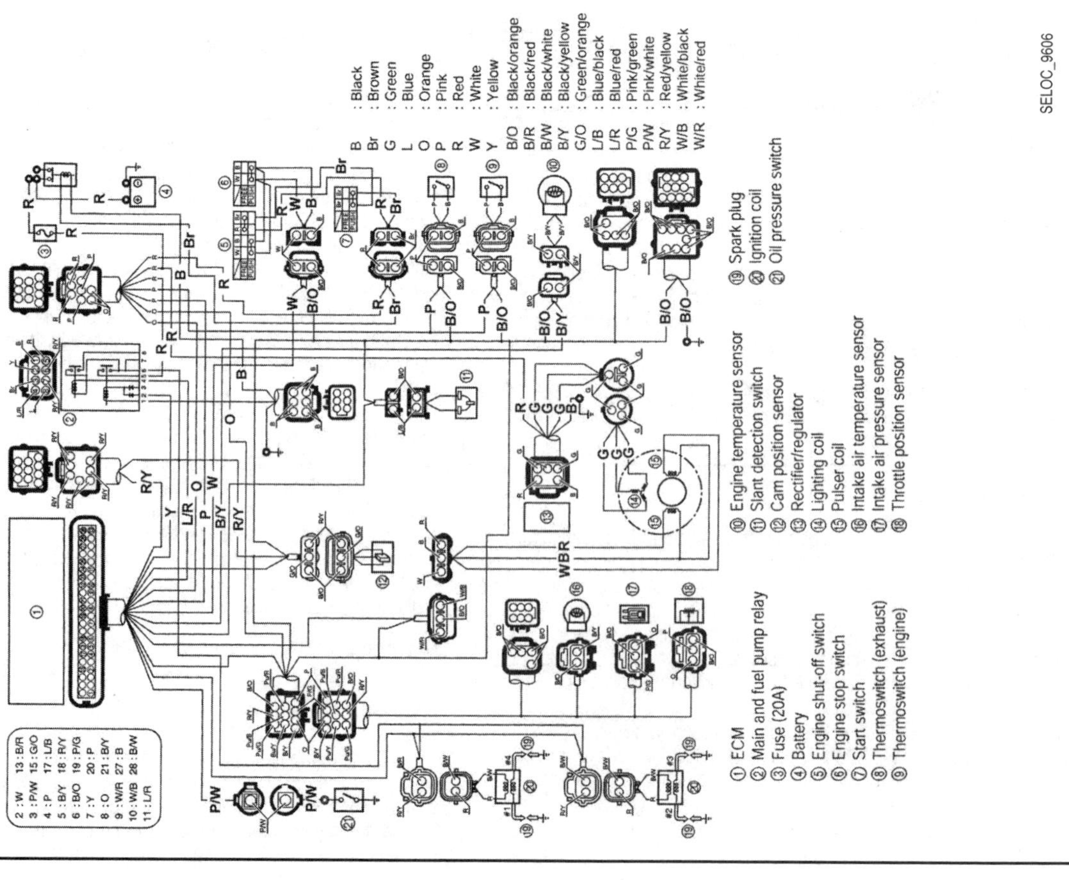

Fig. 23 Functional diagram of the TCI ignition system - FX models w/998cc engines

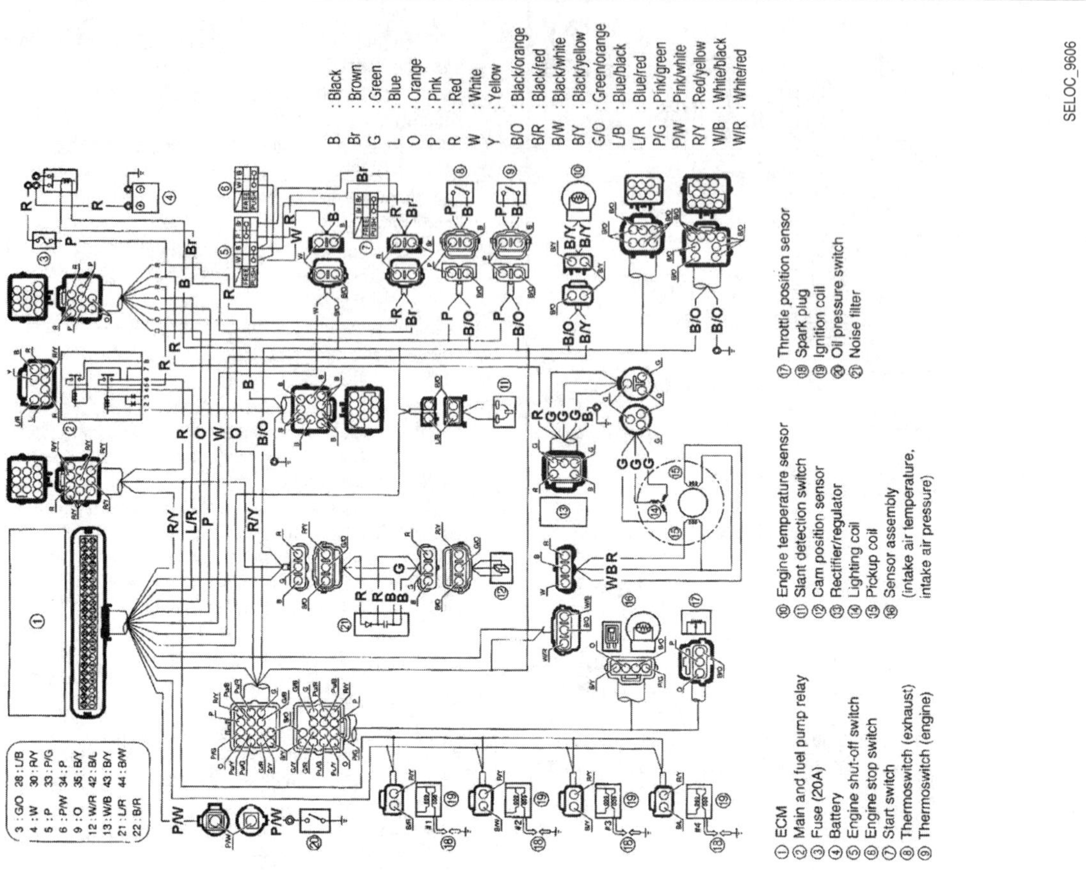

Fig. 24 Functional diagram of the TCI ignition system - FX models w/1052cc engines

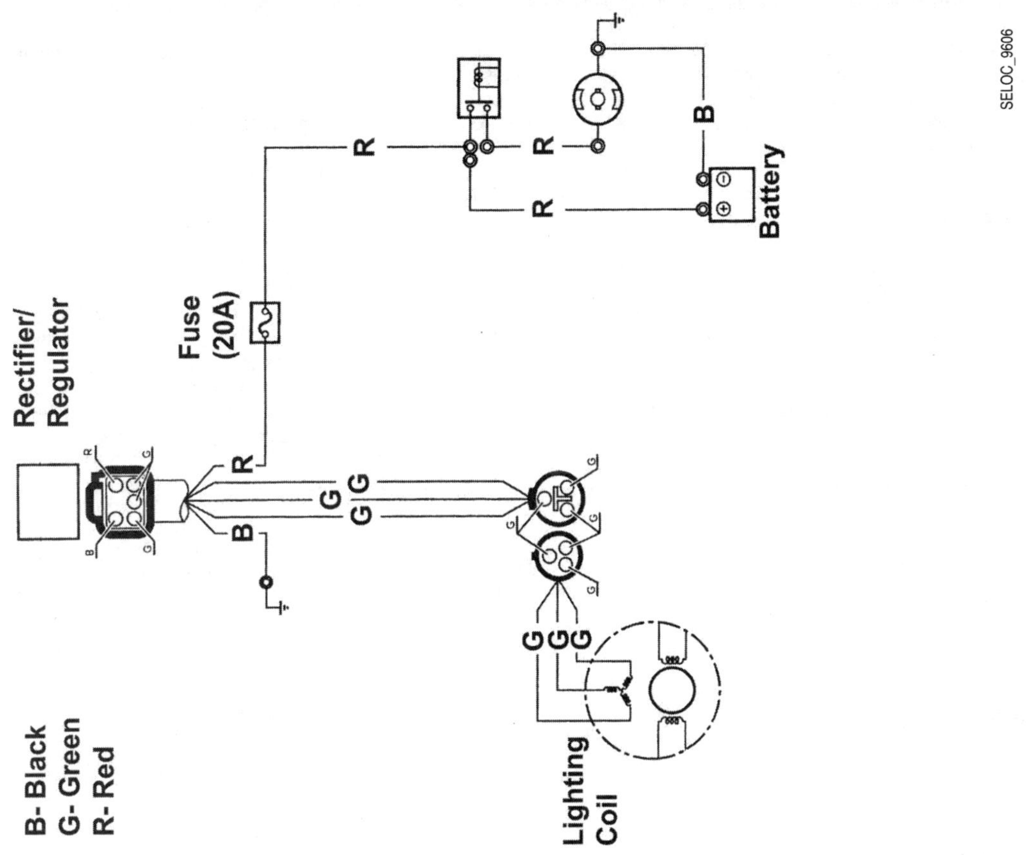

Fig. 26 Functional diagram of the charging system - 2002-05 FX models w/998cc engines

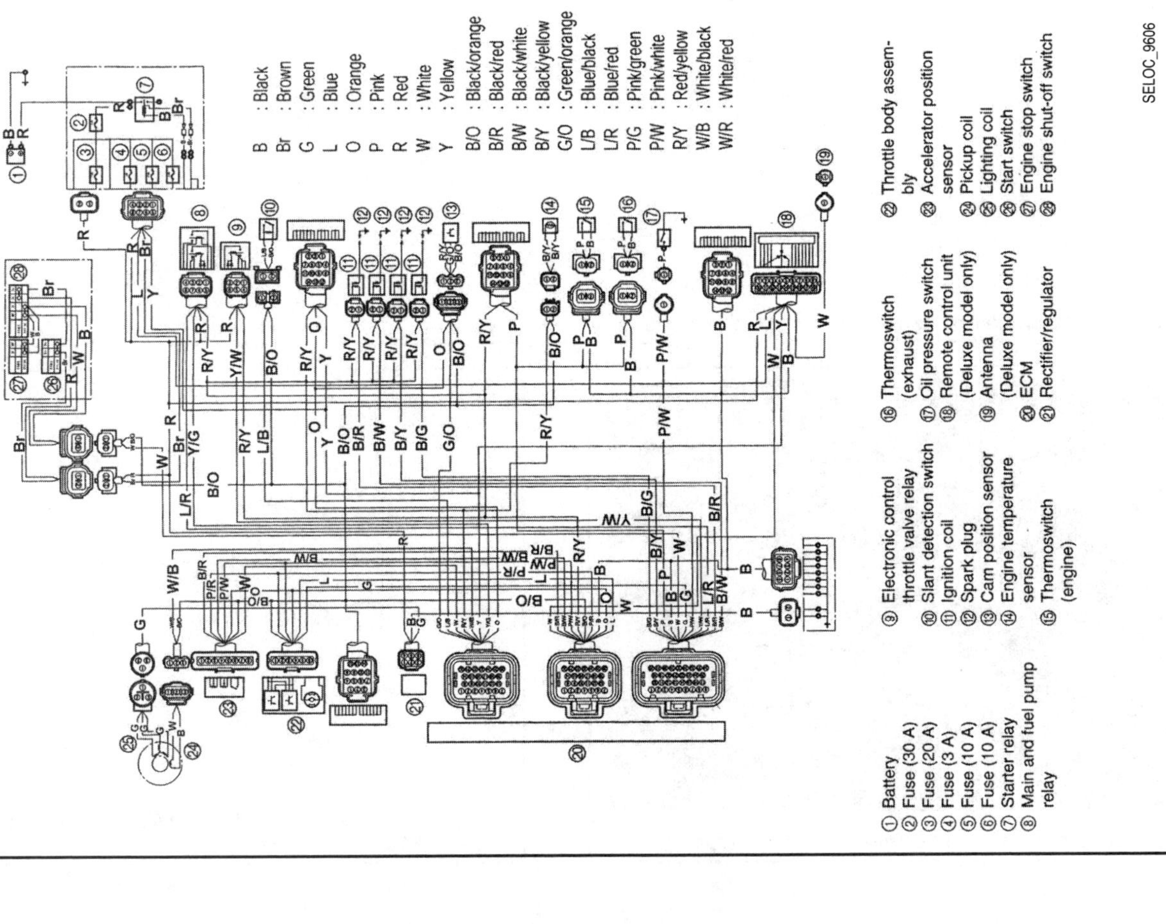

Fig. 25 Functional diagram of the TCI ignition system - VX models w/1052cc engines

ignition coil is used for the #1 and #4 cylinders, while the other is used to fire the #2 and #3 cylinders.

These models contain dual-lead ignition coils, so the spark plugs are fired in pairs (one cylinder at TDC of the compression stroke while the other is on TDC of the exhaust stroke) using a traditional "waste-spark" configuration. Typically, 1 dual-lead ignition coil is used for the #1 and #4 cylinders, while the other is used to fire the #2 and #3 cylinders.

The ignition systems used on 1052/1812cc motors utilize 2 pulser or pick-up coils on the 1052cc FX, or a single pulser or pick-up coil on the VX (both engines) and 1812cc FX and FZ, 1 stator or lighting coil, a TCI unit (the ECM) and 4 ignition coils.

On all models, ignition and timing functions are controlled by the ECM based on input from the pulser/pick-up coil(s) and voltage from the stator/lighting coil.

In addition to thermo-switch and oil pressure switch input, microcomputer control on these models also uses input from the throttle position sensor (TPS) and other engine control sensors.

Yamaha ignition coils boost the DC voltage instantly to over 20,000 volts (and potentially as much as 50,000+ volts) for spark. This completes the primary side of the ignition circuit.

Once the voltage is discharged from the ignition coil, the secondary circuit begins and only stretches from the ignition coil to the spark plugs via extremely large high tension leads (or on most motors, via the direct coil mounting). At the spark plug, the voltage arcs in the form of a spark, across from the center electrode to the outer electrode, and then to ground via the spark plug threads. This completes the ignition circuit.

CHARGING SYSTEM

◆ **See Figures 30, 31 and 32**

For many years, single-phase, full-wave charging systems were the dominant design. Most manufacturers used single-phase systems because they were simple and reliable. The drawback to these systems is low output. A typical single-phase lighting coil system is capable of only 10-15 amps.

In response to higher electrical demands, multi-phase (usually 3-phase) systems were introduced by the early 1990s and since they've found their way onto many motors, especially EFI 4-strokes. At the heart of a 3-phase charging system is an interconnected 3 coil winding. By using 3 coils instead of one, output is more than doubled. The multi-phase rectifier/regulator works

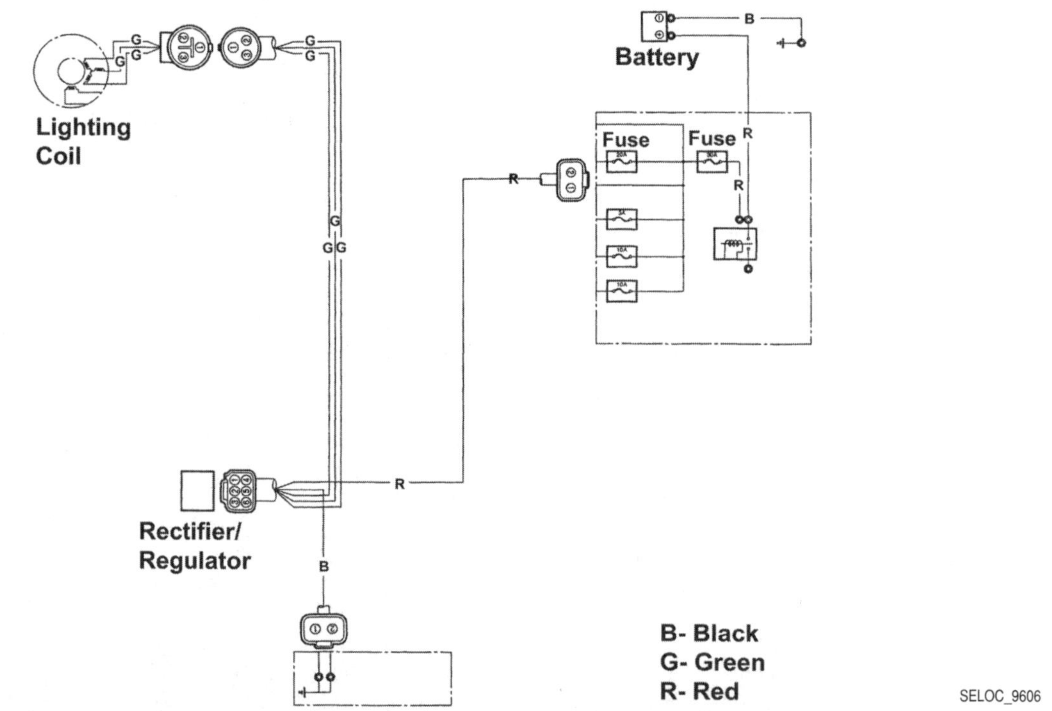

Fig. 27 Functional diagram of the charging system - VX models

Fig. 28 Most engines utilize an individual coil over each spark plug. . .

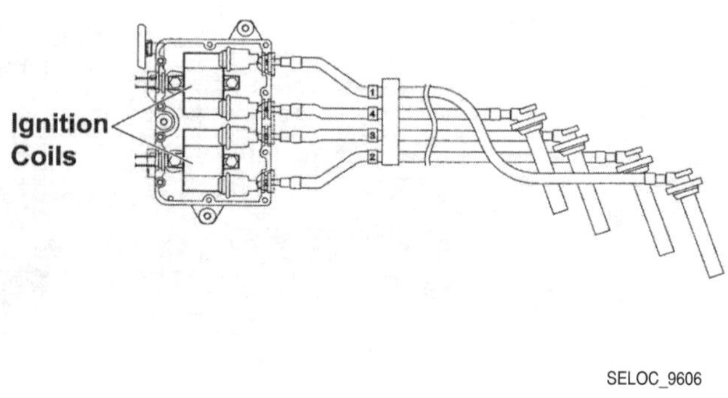

Fig. 29 . . .although some early models use 2 dual-lead coils (2002-05 FX w/998cc engine shown)

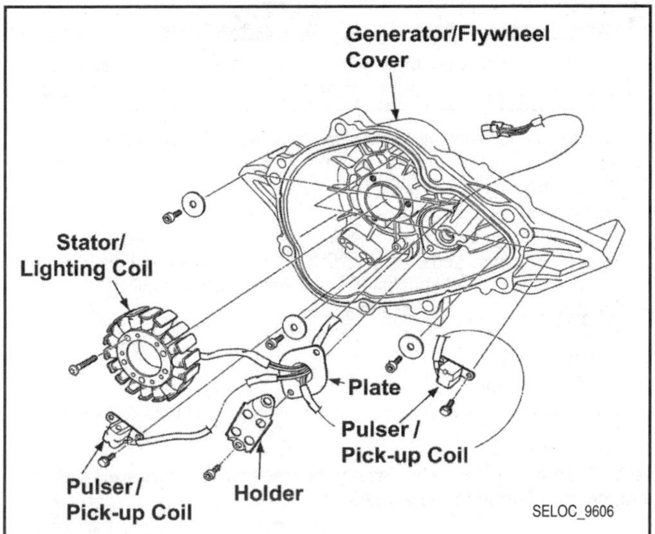

Fig. 30 Typical ignition and charging system components (FX models with 998cc/1052cc engines)

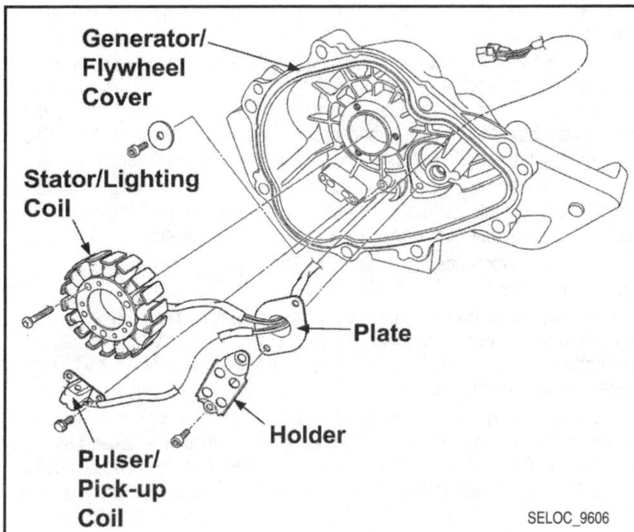

Fig. 31 Typical ignition and charging system components (VX models w/1052cc engines)

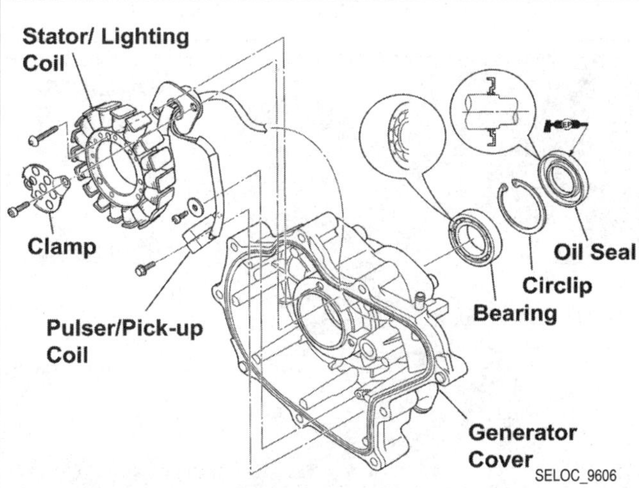

Fig. 32 Typical ignition and charging system components (FX, FZ and VX models w/1812cc engines)

in a similar way to the single-phase system. The difference is the addition of more diodes.

■ **The lighting coils are part of a 1-piece stator unit which is also an integral part of the ignition system.**

This charging system produces electricity by moving a magnet past a fixed coil. Alternating current is produced by this method. Since a battery cannot be charged by AC (alternating current), the AC current produced by the lighting coil is normally rectified or changed into DC (direct current) to charge the battery.

In order to control the charging rate, an additional device called a regulator is used. When the battery voltage reaches approximately 14.6 volts, the regulator sends the excess current to ground. This prevents the battery from overcharging and boiling away the electrolyte.

The charging system consists of the following components:

- A flywheel/rotor containing magnets
- The lighting coil or stator coil
- The battery, fuse assembly and wiring
- A regulator/rectifier

These days, most 4-strokes, including all EFI motors, are equipped with a 3-phase charging system. The charging system operates in essentially the same manner as the single-phase system. The differences in the system come in output and testing, as it uses 3 coils wired to each other in a Y-configuration (and 3 stator coil wires for the charging system circuit that various combinations of which must be tested for shorts to ground and for proper output).

Servicing charging systems is not difficult if you follow a few basic rules. Always start by verifying the problem. If the complaint is that the battery will not stay charged, do not automatically assume that the charging system is at fault. Something as simple as an accessory that draws current with the key off will convince anyone they have a bad charging system. Another culprit is the battery or wiring. Remember to check, clean and service your battery and connections regularly. Battery abuse is the number one charging system problem.

The regulator/rectifier is the brains of the charging system. This assembly controls current flow in the charging system. If battery voltage is below about 14.5-14.6 volts the regulator sends the available current from the rectifier to the battery. If the battery is fully charged (is at or above 14.5-14.6 volts) the regulator diverts most of the current from the rectifier back to the lighting coil through ground.

Do not expect the regulator/rectifier to send current to a fully charged battery. You may find that you must pull down the battery voltage below 12.5 volts to test charging system output.

In the charging system, the regulator/rectifier is both the hardest and easiest item to troubleshoot. This is normally accomplished by checking around it. Check the battery and charge or replace it as needed. Check the AC voltage output of the lighting coil.

Ignition & Charging System Troubleshooting

GENERAL

Accurate troubleshooting of an ignition system is viewed by many as mission impossible. But this really does not have to be the case. Having a systematic approach to troubleshooting is the key. Your first priority is to understand how the system works.

Get yourself mentally prepared for ignition troubleshooting before you touch the motor. Make notes about how the motor behaves. Is it totally dead? Misfiring? If it is dead, what were the circumstances that led to its demise? Did it run poorly before it died? If it is running, but not well, is this a permanent condition? Only when cold? Only when warm? Under load? Ask yourself a lot of questions. The more information you can gather the faster the repair can be made.

It is easy to forget to ask yourself if the root cause of the failure is truly the ignition? Could it be something else? Look at the plug(s) and check the spark. Once you have done the basics, the hands-on troubleshooting begins.

The key to successful troubleshooting is a systematic approach. Do not skip around. On new or unfamiliar equipment, try writing up a check list on a 3 x 5 note card. This makes a handy reference for future troubleshooting.

Use a spark tester to observe spark quality. Does it jump the gap? Is it bright and blue? Orange spark usually means trouble in the ignition system.

Do not forget to check the ignition controlling components such as the main or stop switches. These can be disconnected from the system. These switches can cause failures that checks of the individual ignition components may not reveal.

Motors with microcomputer controlled ignition systems utilize other sensor inputs to help determine spark timing. A thermo-switch or oil pressure sensor giving an incorrect reading may prevent spark timing advance. Don't forget these and other ignition control sensors. Be sure to check for trouble codes to see if any related EFI components may be causing the symptoms.

Always attempt to proceed with the troubleshooting in an orderly manner. The shot in the dark approach will only result in wasted time, incorrect diagnosis, replacement of unnecessary parts, and frustration.

There are several conditions you can find when troubleshooting a seemingly problematic charging system:

- Normal - In a normal system the charge rate is correct. There could be a problem elsewhere, such as an electrical short, too many accessories or a bad battery. Whatever the problem, there is no need to work on the charging system. It is working correctly!
- Overcharging - The charging system continues to supply current to the battery even after the battery is fully charged. Usually this is caused by a bad regulator portion of the regulator/rectifier. However, in some cases it could also potentially be due to the wiring the regulator uses to determine battery voltage.
- Undercharging - The charging system is not producing current at all or is producing it at a reduced rate. In this circumstance the question is whether or not the lighting coils are producing what they should and, if they ARE, the fault is in the wiring or the regulator.
- Battery drain during non-use - Isolate the battery draw to the motor or to any accessories by removing loads one at a time. Although it is rare, one of the diodes in the rectifier can short and slowly drain the battery.

Keep the above possibilities in mind when troubleshooting. By simplifying your troubleshooting approach you may get to the solution more quickly.

The first step in testing a charging system is to confirm there is a charging problem.

■ **The absolute number one charging system problem is a faulty battery!**

The real problem may be a draw with the key off. The charging system may really be OK, but if the draw on the system exceeds the charging output, the charging system will appear faulty.

There are 2 methods for checking 3-phase charging systems, the DC amp check and the DC voltage check. The DC amp check method is usually preferred, but both methods are useful.

CHECKING FOR KEY-OFF AMPERAGE DRAW

Eliminate the battery as the problem by using a known good one. Now check for any amperage drain on the battery with an ammeter. Connect the ammeter between the positive battery cable and the positive post on the battery. Take your first reading with the key switch in the **OFF** position. Your reading should be at, or near, zero. Now take a reading with the key switch in the **ON** position. This reading will vary according to how many accessories are connected to key-on power. Record the amperage reading.

EXCESSIVE CHARGING

There is really only 1 cause for this type of failure, the regulator is not working (is not controlling voltage supplied to the battery), either because it is not getting a proper battery voltage signal or because it is broken. If there are no wiring problems, since there is no repair of the regulator/rectifier, replace it.

IGNITION SYSTEM PEAK VOLTAGE CHECK

FX Models w/998cc Engines

OEM ② MODERATE

◆ See Figures 33 thru 37

■ **If no spark or weak spark, perform this test. If spark is good, there is no need for the test as the problem is not with the ignition system. Check the spark plugs.**

✻✻ CAUTION

Be sure the vessel is secured in a body of water or connected to an adequate water supply before performing any of the following tests.

1. Find the ignition coil box on the starboard hull, just aft of the engine, disconnect the 2 electrical couplers on the left side of the box and connect a 2-pin test harness (#YB-06792 / 90890-06792).
2. Connect a DVOM as per the manufacturer's instructions.
3. Check for peak voltage on cylinder #1 and #4, between the Black/Red and ground terminals and then on cylinder #2 and #3 between the Black/White and ground terminals. Check voltage while cranking, at 2000 rpm and at 3500 rpm. Voltage should be:
   - Cranking - 7 V
   - 2000 rpm - 258 V
   - 3500 rpm - 258 V

   If not within specifications, check the pulser coil and then replace the ECM.

■ **You can crank the engine easily by pressing the start switch and engine stop button at the same time.**

4. Locate the pulser coupler near the front of the air filter housing, just under the fuel line (not the black one, but the blue one underneath). Disconnect it and install a 3-pin test harness (#YB-06791 / 90890-06791).
5. Using the DVOM, test the pulser coil peak voltage while unloaded, cranking, at 2000 rpm and at 3500 rpm. Test between the White and Black

Fig. 33 The pulser/pick-up and stator/lighting coil couplers are under the fuel line on FX models w/998cc and 1052cc engines

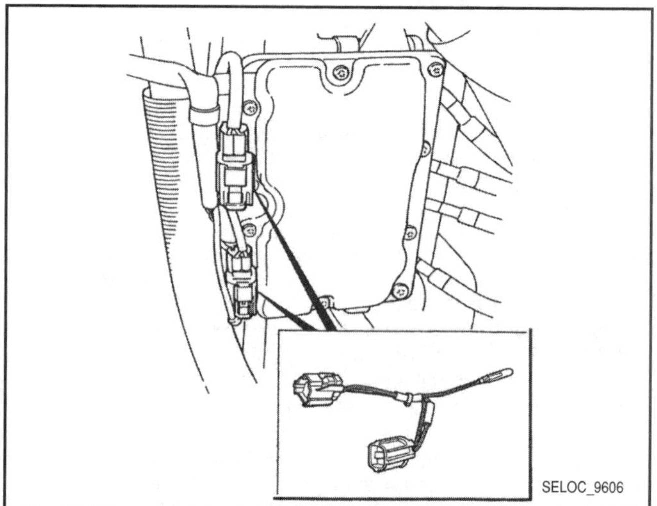

Fig. 34 When checking the ECM voltage on 998cc engines, install a test harness between these couplers at the ignition coil box

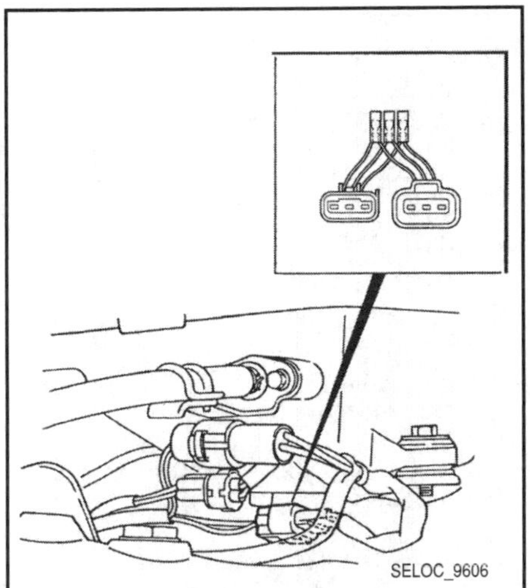

Fig. 35 Checking the pulser coil output peak voltage. . .

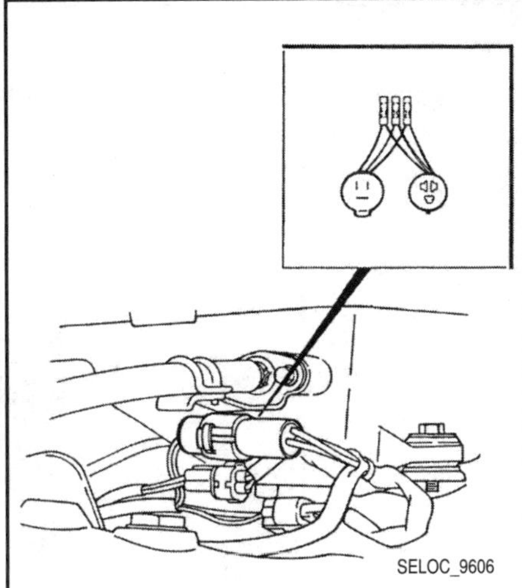

Fig. 36 . . .and the lighting coil

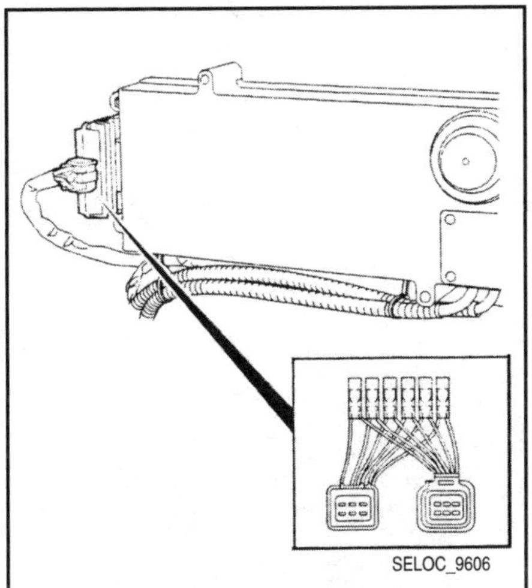

Fig. 37 The rectifier/regulator is attached to the end of the electrical box and tested like this on the 998cc engines

terminals and then between the Red and Black terminals. Peak voltage should be:

- Unloaded - 4 V
- Cranking - 4 V
- 2000 rpm - 23 V
- 3500 rpm - 38 V

If it falls below any of these levels, replace the pulser coil.

6. Now you can disconnect the lighting coil coupler and install a different 3-pin test harness (#YB-06770 / 90890-06770). The coupler is just above the pulser coil coupler, a little larger, and black.

7. Using the DVOM again, test the lighting coil peak voltage while unloaded, cranking, at 2000 rpm and at 3500 rpm. Test between the Green and Green terminals. Peak voltage should be:

- Unloaded - 9 V
- Cranking - 8 V
- 2000 rpm - 11 V
- 3500 rpm - 12 V

If it falls below any of these levels, replace the lighting coil.

8. Finally, you can disconnect the rectifier/regulator coupler and install a 6-pin test harness (#YB-06790 / 90890-06790). The rectifier/regulator is connected to the end of the electrical box behind the engine.

9. Using the DVOM again, test the peak voltage while unloaded at 3500 rpm. Test between the Red and Black terminals. Output voltage should be 14.5 V at 3500 rpm, but voltage will increase slightly after starting so make sure you let it run for a minute or so before checking.

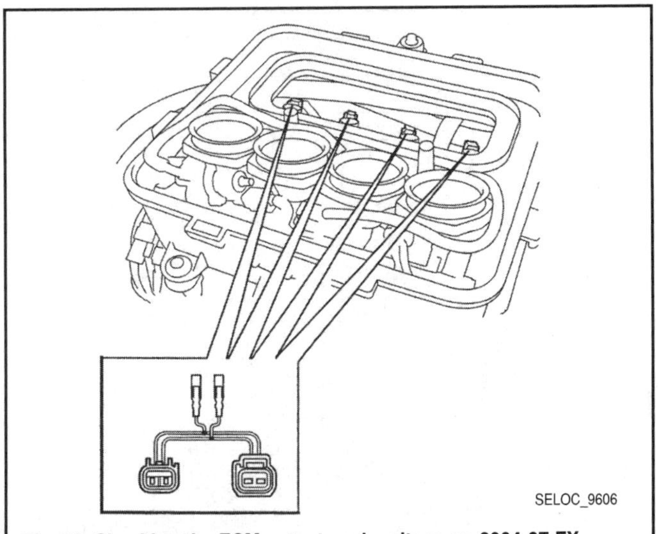

Fig. 38 Checking the ECM output peak voltage on 2004-07 FX models w/1052cc engines. . .

■ **The battery must be fully charged in order to perform this test correctly, an undercharged battery will affect the test results.**

If it falls below this level, replace the rectifier/regulator.

FX Models w/1052cc Engines

OEM ② MODERATE

◆ **See Figures 33 and 38 thru 45**

■ **If no spark or weak spark, perform this test. If spark is good, there is no need for the test as the problem is not with the ignition system. Check the spark plugs.**

⁂ CAUTION

Be sure the vessel is secured in a body of water or connected to an adequate water supply before performing any of the following tests.

1. Remove the air filter cover and the filter. Connect a 2-pin test harness (#YB-06867 / 90890-06867 on 2004-07, or #YB-06877 / 90890-06877) as shown in the illustrations.

2. Connect a DVOM as per the manufacturer's instructions.

3. Check for peak voltage on each cylinder between:

- Cyl. #1 - Black/Red and Red/Yellow

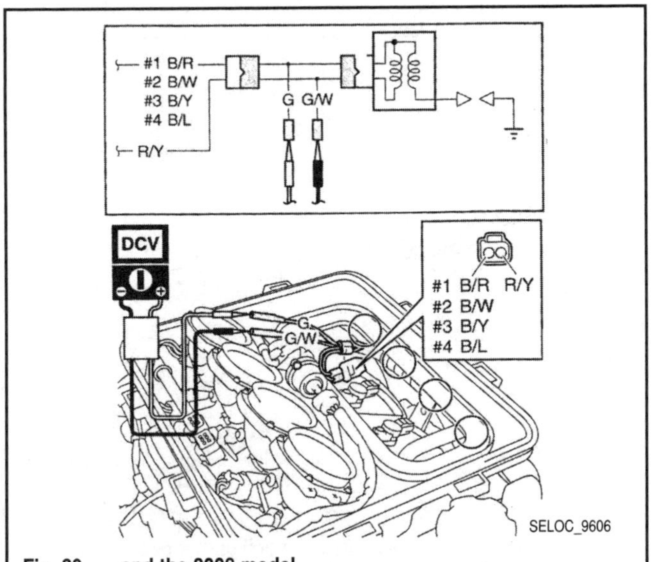

Fig. 39 . . .and the 2008 model

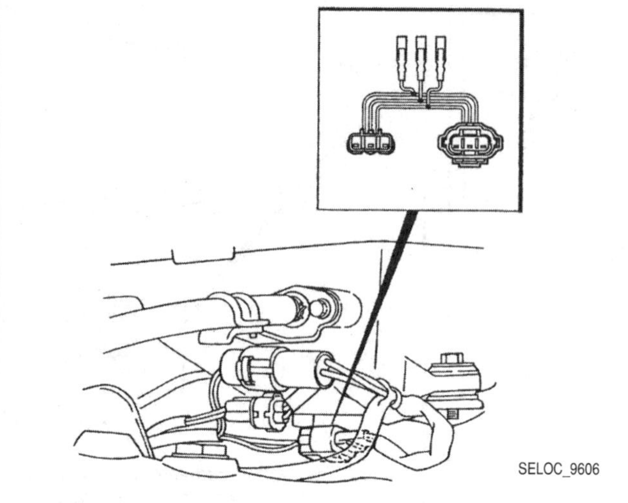

Fig. 40 Checking the pick-up coil output peak voltage on 2004-07 FX models w/1052cc engines. . .

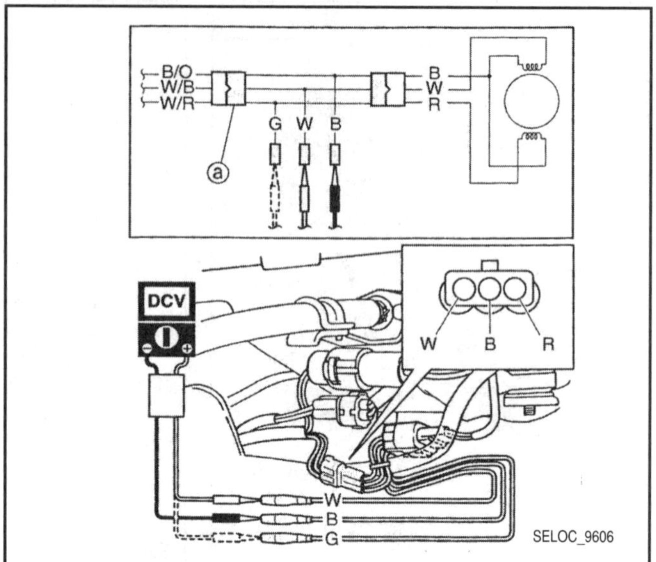

Fig. 41 . . .and on 2008 engines

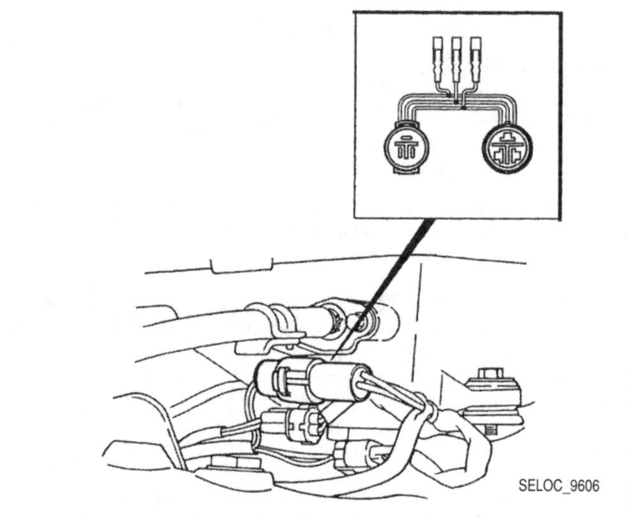

Fig. 42 Checking the stator/lighting coil output peak voltage on 2004-07 FX models w/1052cc engines. . .

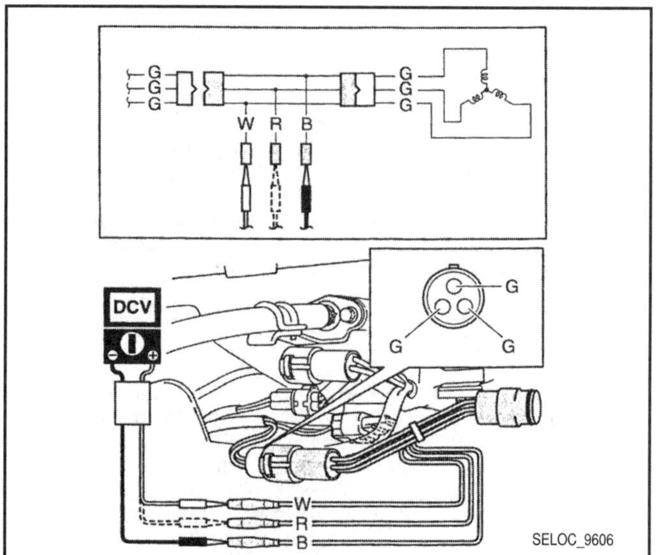

Fig. 43 . . .and on 2008 engines

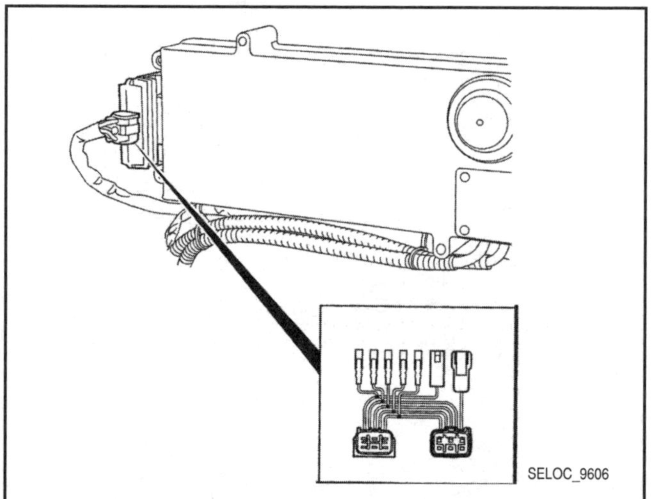

Fig. 44 Checking the regulator/rectifier coil output peak voltage on 2004-07 FX models w/1052cc engines. . .

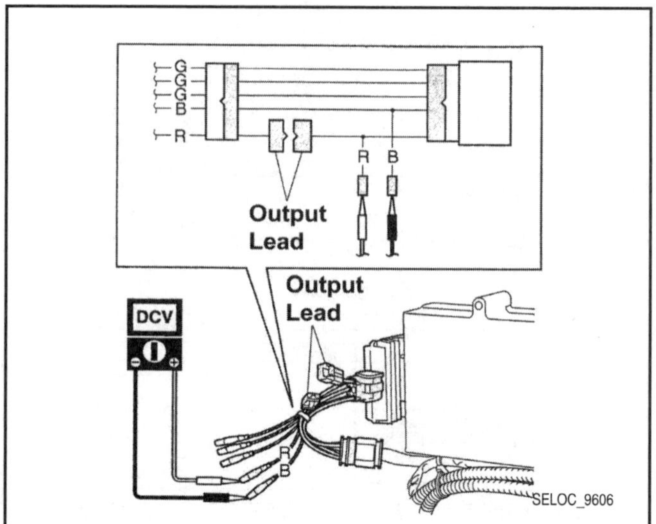

Fig. 45 . . .and on 2008 engines

- Cyl. #2 - Black/White and Red/Yellow
- Cyl. #3 - Black/Yellow and Red/Yellow
- Cyl. #4 - Black/Blue and Red/Yellow

Disconnect the ignition coil coupler for each cylinder except the one being tested. On 2008 models, disconnect the fuel pump module coupler as well. Check voltage while cranking, at 2000 rpm and at 3500 rpm. Voltage should be:

- Cranking - 2004-07: 87 V, 2008: 90 V
- 2000 rpm - 2004-07: 38.5 V, 2008: 110 V
- 3500 rpm - 2004-07: 63.2 V, 2008: 130 V

If not within specifications, check the pulser coil and then replace the ECM.

■ **You can crank the engine easily by connecting the engine shut-off lanyard to the shut-off switch and then pressing the start switch.**

4. Locate the pulser/pick-up coil coupler near the front of the air filter housing, just under the fuel line (not the black one, but the blue one underneath it). Disconnect it and install a 3-pin test harness (#YB-06877 / 90890-06877).

5. Using the DVOM, test the pulser/pick-up coil peak voltage while unloaded, cranking, at 2000 rpm and at 3500 rpm. Test between the White and Black terminals and then between the Red and Black terminals. Peak voltage should be:

- Unloaded - 5.2 V
- Cranking - 4.7 V
- 2000 rpm - 26.1 V
- 3500 rpm - 41.3 V

If it falls *below* any of these levels, replace the pulser/pick-up coil. If it tests *above* any of these levels, replace the ECM

6. Now you can disconnect the stator/lighting coil coupler and install a different 3-pin test harness (#YB-06770 / 90890-06770). The coupler is just above the pick-up coil coupler, a little larger, and black.

7. Using the DVOM again, test the lighting coil peak voltage while cranking, at 2000 rpm and at 3500 rpm. Test between the Green and Green terminals. Peak voltage should be:

- Cranking - 9.2 V
- 2000 rpm - 38.5 V
- 3500 rpm - 63.2 V

If it falls below any of these levels, replace the stator/lighting coil.

8. Finally, you can disconnect the rectifier/regulator coupler and install a 6-pin test harness (#YB-06848 / 90890-06848). Be sure to disconnect the output lead on the harness with this test. The rectifier/regulator is connected to the end of the electrical box behind the engine.

9. Using the DVOM again, test the peak voltage while unloaded, at 3500 rpm. Test between the Red and Black terminals. Output voltage should be 15 V.

■ **Do not use the peak voltage adapter when checking output voltage at the rectifier/regulator**

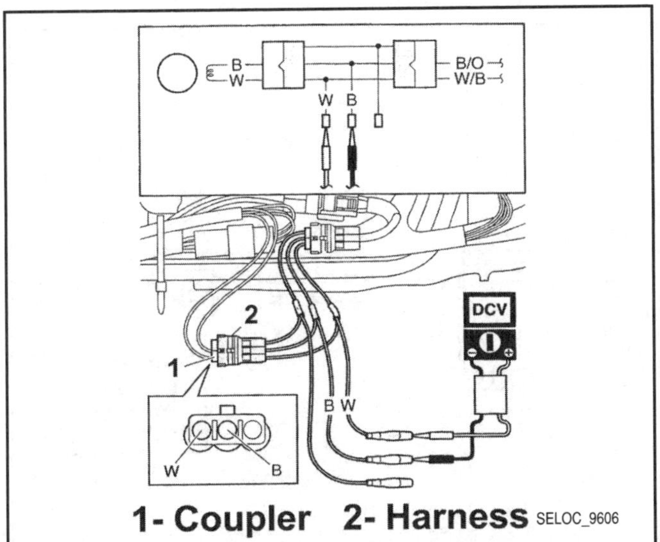

Fig. 47 Checking the pick-up coil output peak voltage on models w/1812cc engines. .

■ **The battery must be fully charged in order to perform this test correctly, an undercharged battery will affect the test results.**

If it falls below this level, replace the rectifier/regulator.

FX, FZ & VX Models w/1812cc Engines

OEM — ② MODERATE

◆ **See Figures 46 thru 54 and 66**

■ **If no spark or weak spark, perform this test. If spark is good, there is no need for the test as the problem is not with the ignition system. Check the spark plugs.**

✻✻ CAUTION

Be sure the vessel is secured in a body of water or connected to an adequate water supply before performing any of the following tests.

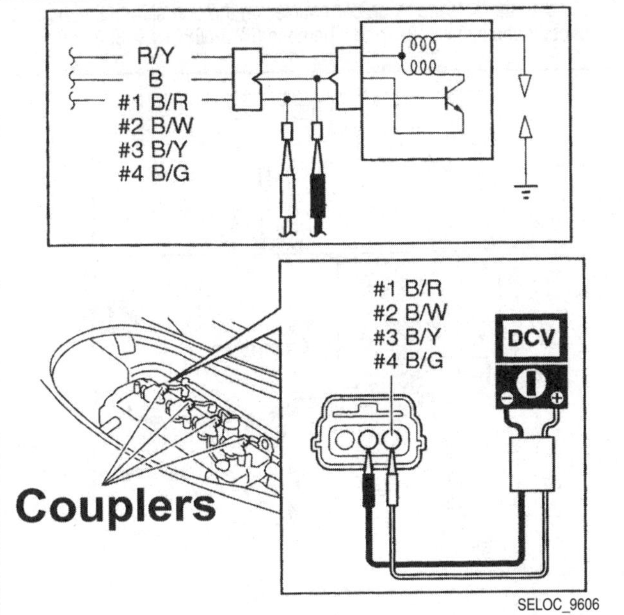

Fig. 46 Checking the ECM output peak voltage on models w/1812cc engines

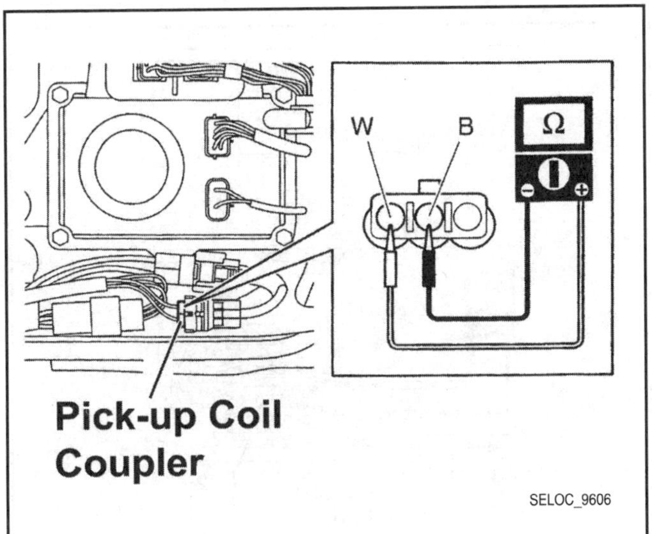

Fig. 48 . . .and then the resistance

1. Remove the engine cover as detailed in the Maintenance & Tune-Up section.
2. Connect a DVOM as per the manufacturer's instructions.
3. Check for ECM output peak voltage on each cylinder between:
   - Cyl. #1 - Black/Red and Black
   - Cyl. #2 - Black/White and Black
   - Cyl. #3 - Black/Yellow and Black
   - Cyl. #4 - Black/Green and Black

Disconnect the ignition coil coupler for each cylinder except the one being tested. When checking in the cranking condition, make sure you disconnect the fuel injector couplers for all 4 cylinders. Check voltage while cranking, at 2000 rpm and at 3500 rpm. Voltage should be:

- Cranking - 2.5 V
- 2000 rpm - 2.6 V
- 3500 rpm - 2.6 V

If not within specifications, check the pulser/pick-up coil and then replace the ECM.

■ **You can crank the engine easily by connecting the engine shut-off lanyard to the shut-off switch and then pressing the start switch.**

4. Locate the pulser/pick-up coil coupler on the port side of the engine. Disconnect it and install a 3-pin test harness (#YB-06877 / 90890-06877).
5. Using the DVOM, test the pulser/pick-up coil output peak voltage while unloaded, cranking, at 2000 rpm and at 3500 rpm. Test between the White and Black terminals. When checking under unloaded cranking, disconnect the coupler (3). Peak voltage should be:
   - Unloaded - 5.5 V for FX HO and VX, 5.4 V for FX SHO and FZ
   - Cranking - 4.7 V for FX HO and VX, 4.6 V for FX SHO and FZ
   - 2000 rpm - 16.4 V for FX HO and VX, 18.1 V for FX SHO and FZ
   - 3500 rpm - 21.5 V for FX HO and VX, 23.9 V for FX SHO and FZ

   If it falls *below* any of these levels, check the pulser/pick-up resistance.
6. Disconnect the 3-pin harness and check the resistance between the White and Black terminals. Resistance should be 459-561 ohms at 68° F (20° C); if not within range, replace the stator/lighting coil.
7. Now you can disconnect the stator/lighting coil coupler (right next to the pulser/pick-up coil coupler) and install a different 3-pin test harness ((#YB-06870 / 90890-06870).
8. Using the DVOM again, test the stator/lighting coil peak output voltage while cranking, at 2000 rpm and at 3500 rpm. Test between the Green and Green terminals. Peak voltage should be:
   - Cranking - 8.3 V for FX HO and VX, 8.4 V for FX SHO and FZ
   - 2000 rpm - 36.4 V for FX HO and VX, 42.7 V for FX SHO and FZ
   - 3500 rpm - 64.7 V for FX HO and VX, 74 V for FX SHO and FZ

   If it falls below any of these levels, check the stator resistance.

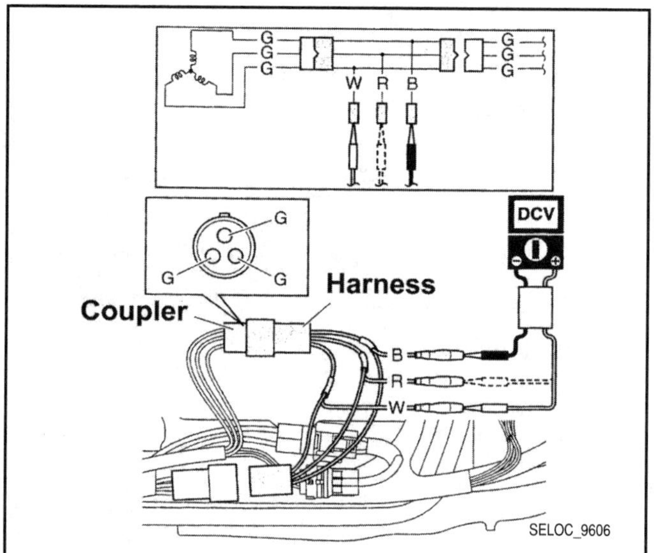

Fig. 49 Checking the stator/lighting coil output peak voltage on models w/1812cc engines. . .

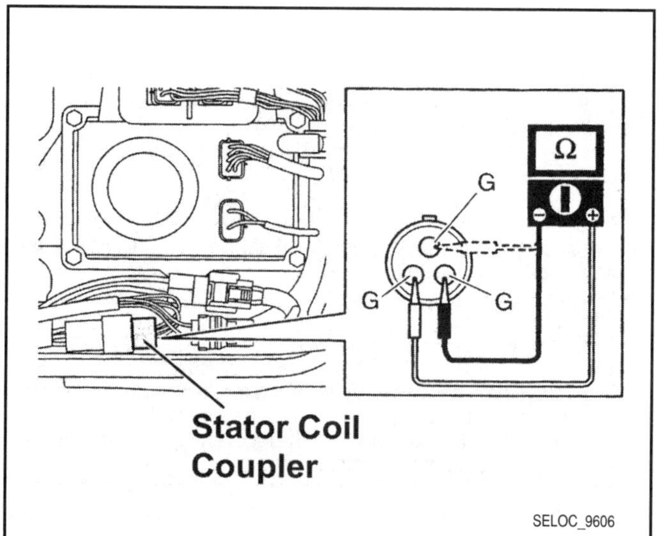

Fig. 50 . . .and then the resistance

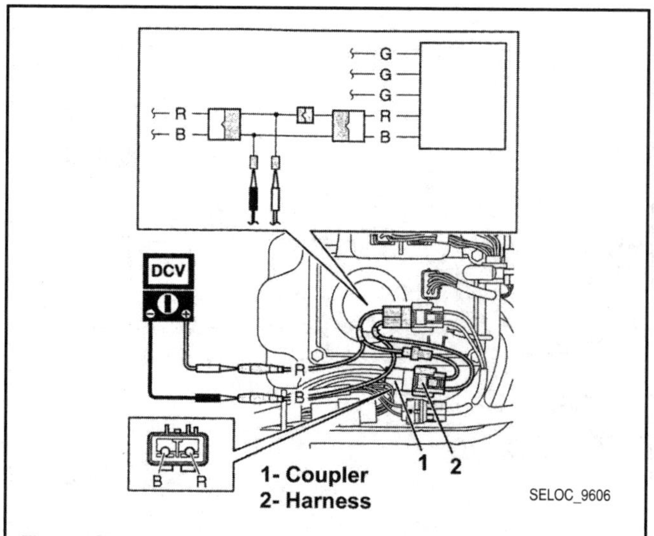

Fig. 51 Checking the regulator/rectifier coil output peak voltage on models w/1812cc engines

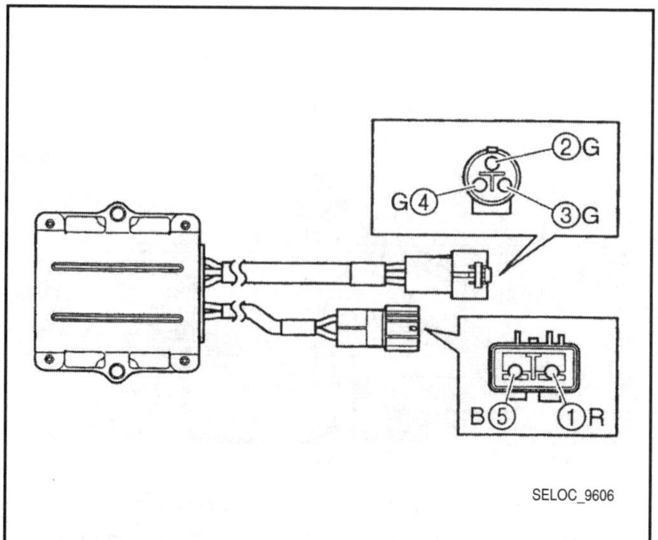

Fig. 52 Rectifier/regulator terminal identification

Tester lead ⊕	Tester lead ⊖	Display value (V) (reference data)
① R	② G	OL
① R	③ G	
① R	④ G	
① R	⑤ B	
⑤ B	② G	0.43–0.51
⑤ B	③ G	0.43–0.51
⑤ B	④ G	0.43–0.51
⑤ B	① R	0.66–0.74
② G	① R	0.43–0.51
③ G	① R	0.43–0.51
④ G	① R	0.43–0.51
② G	⑤ B	OL
③ G	⑤ B	
④ G	⑤ B	

OL: overload

SELOC_9606

Fig. 53 Check the regulator/rectifier terminal continuity as detailed here. Use the terminal identification illustration for terminal locations - FX and FZ models

Tester lead ⊕	Tester lead ⊖	Display value (V) (reference data)
①	②	OL
	③	
	④	
	⑤	
②	①	0.43–0.51
	③	OL
	④	
	⑤	
③	①	0.43–0.51
	②	OL
	④	
	⑤	
④	①	0.43–0.51
	②	OL
	③	
	⑤	
⑤	①	0.66–0.74
	②	0.43–0.51
	③	0.43–0.51
	④	0.43–0.51

OL: Indicates an overload

SELOC_9606

Fig. 54 Check the regulator/rectifier terminal continuity as detailed here. Use the terminal identification illustration for terminal locations - VX models

9. With the harness disconnected, check the resistance between the Green and Green terminals. Resistance should be 0.31-0.38 ohms at 68° F (20° C); if not within range, replace the stator/lighting coil.

10. Finally, you can disconnect the rectifier/regulator coupler and install a 2-pin test harness (#90890-06850). The rectifier/regulator is on the port side of the engine, at the rear.

11. Using the DVOM again, test the peak voltage while loaded, at 2000 and 3500 rpm. Test between the Red and Black terminals. Output peak voltage should be 13 V at both speeds.

■ **Do not use the peak voltage adapter when checking output voltage at the rectifier/regulator.**

■ **The battery must be fully charged in order to perform this test correctly, an undercharged battery will affect the test results.**

If it falls below this level, check the rectifier/regulator for continuity.

12. Remove the rectifier/regulator and check for continuity as shown in the accompanying illustrations. If out of range, it will require replacement. If in range, install the unit and reconnect it.

2005-09 VX Models

◆ **See Figures 55 thru 58**

■ **If no spark or weak spark, perform this test. If spark is good, there is no need for the test as the problem is not with the ignition system. Check the spark plugs.**

**** CAUTION**

Be sure the vessel is secured in a body of water or connected to an adequate water supply before performing any of the following tests.

1. Connect a 2-pin test harness (#YB-06867 / 90890-06867 or #YB-06767 / 90890-06767) as shown in the illustrations.

2. Connect a DVOM as per the manufacturer's instructions.

3. Check for peak voltage on each cylinder between:

- Cyl. #1 - Black/Red and Red/Yellow
- Cyl. #2 - Black/White and Red/Yellow
- Cyl. #3 - Black/Yellow and Red/Yellow
- Cyl. #4 - Black/Green and Red/Yellow

Disconnect the ignition coil coupler for each cylinder except the one being tested. Check voltage while cranking, at 2000 rpm and at 3500 rpm. Voltage on should be:

- Cranking - 90 V
- 2000 rpm - 130 V
- 3500 rpm - 140 V

If not within specifications, check the pulser/pick-up coil and then replace the ECM.

■ **You can crank the engine easily by connecting the engine shut-off lanyard to the shut-off switch and then pressing the start switch.**

4. Locate the pulser/pick-up coil coupler near ECM and regulator/rectifier on the port side of the engine. Disconnect it and install a 3-pin test harness (#YB-06877 / 90890-06877 or #YB-06777 / 90890-06777).

5. Using the DVOM, test the pulser/pick-up coil peak voltage while unloaded, cranking, at 2000 rpm and at 3500 rpm. Test between the White and Black terminals and Black and Orange terminals. Peak voltage should be:

- Unloaded - 8 V
- Cranking - 6.9 V
- 2000 rpm - 19.4 V
- 3500 rpm - 25.1 V

If it falls *below* any of these levels, replace the pulser/pick-up coil. If it tests *above* any of these levels, replace the ECM.

6. Now you can disconnect the stator/lighting coil coupler (right next to the pulser/pick-up coil coupler) and install a different 3-pin test harness ((#YB-06870 / 90890-06870 or #YB-06770 / 90890-06770).

7. Using the DVOM again, test the stator/lighting coil peak voltage while cranking, at 2000 rpm and at 3500 rpm. Test between the Green and Green terminals. Peak voltage should be:

- Cranking - 7.8 V

- 2000 rpm - 24.5 V
- 3500 rpm - 41.6 V

If it falls below any of these levels, replace the stator/lighting coil.

8. Finally, you can disconnect the rectifier/regulator coupler and install a 6-pin test harness (#YB-06848 / 90890-06848). Be sure to disconnect the output lead on the harness with this test. The rectifier/regulator is on the port side of the engine, next to the ECM.

9. Using the DVOM again, test the peak voltage while unloaded, at 3500 rpm. Test between the Red and Black terminals. Output voltage should be 13 V.

■ **Do not use the peak voltage adapter when checking output voltage at the rectifier/regulator.**

■ **The battery must be fully charged in order to perform this test correctly, an undercharged battery will affect the test results.**

If it falls below this level, replace the rectifier/regulator.

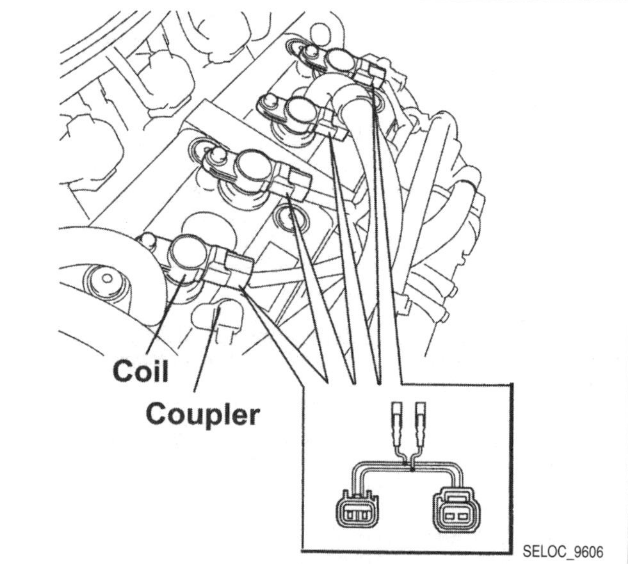

Fig. 55 Checking the ECM output peak voltage on 2005-09 VX models

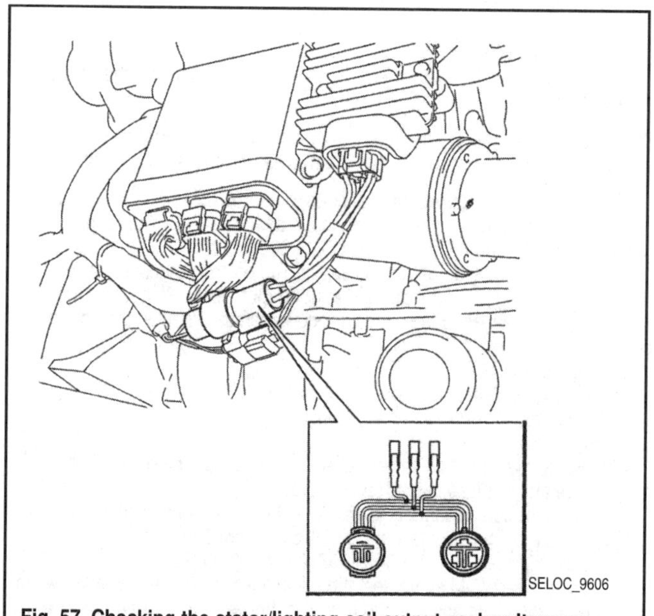

Fig. 57 Checking the stator/lighting coil output peak voltage on 2005-09 VX models

2010-11 VX Models w/1052cc Engines

◆ **See Figures 59 thru 67**

■ **If no spark or weak spark, perform this test. If spark is good, there is no need for the test as the problem is not with the ignition system. Check the spark plugs.**

**** CAUTION**

Be sure the vessel is secured in a body of water or connected to an adequate water supply before performing any of the following tests.

1. Connect a 2-pin test harness (#YB-06867 / 90890-06867) as shown in the illustrations.

2. Connect a DVOM as per the manufacturer's instructions.

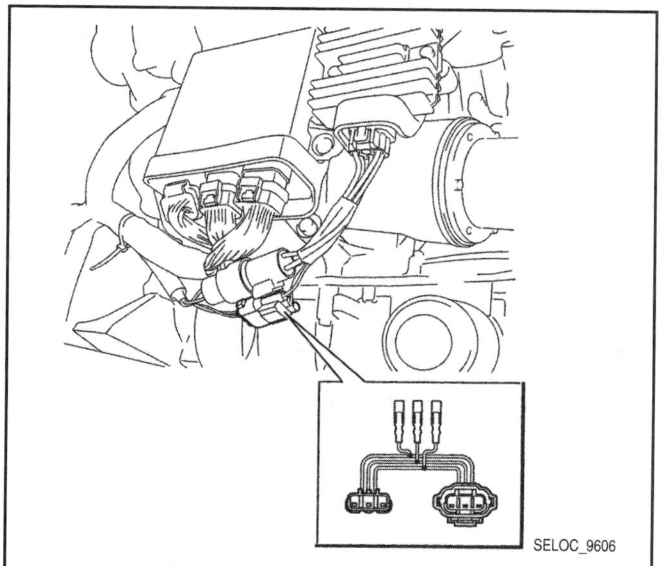

Fig. 56 Checking the pick-up coil output peak voltage on 2005-09 VX models

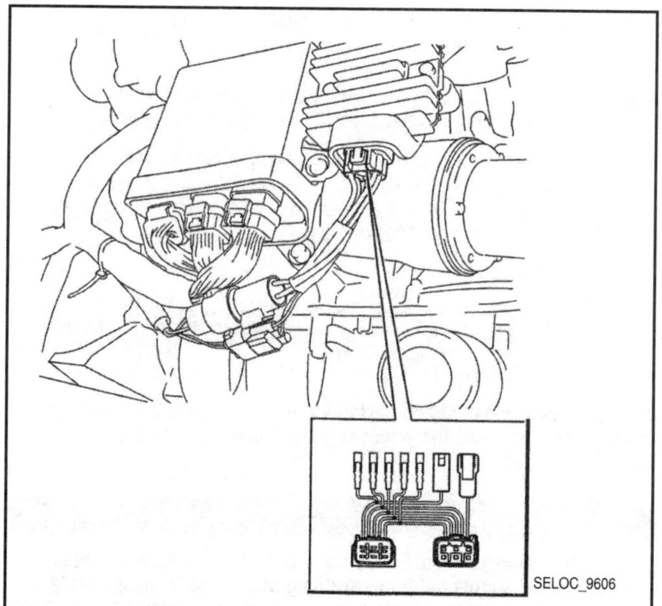

Fig. 58 Checking the regulator/rectifier coil output peak voltage on 2005-09 VX models

3. Check for ECM peak output voltage on each cylinder between:

- Cyl. #1 - Black/Red and Red/Yellow
- Cyl. #2 - Black/White and Red/Yellow
- Cyl. #3 - Black/Yellow and Red/Yellow
- Cyl. #4 - Black/Green and Red/Yellow

Disconnect the ignition coil coupler for each cylinder except the one being tested. Check voltage while cranking, at 2000 rpm and at 3500 rpm. Voltage should be:

- Cranking - 94 V
- 2000 rpm - 120 V
- 3500 rpm - 148 V

If not within specifications, check the pulser/pick-up coil and then replace the ECM.

■ **You can crank the engine easily by connecting the engine shut-off lanyard to the shut-off switch and then pressing the start switch.**

4. Locate the pulser/pick-up coil coupler near ECM and regulator/rectifier on the port side of the engine. Disconnect it and install a 3-pin test harness (#YB-06877 / 90890-06877).

5. Using the DVOM, test the pulser/pick-up coil peak voltage while unloaded, cranking, at 2000 rpm and at 3500 rpm. Test between the White and Black terminals. Peak voltage should be:

- Unloaded - 6 V
- Cranking - 5 V
- 2000 rpm - 17 V
- 3500 rpm - 23 V

If it falls *below* any of these levels, check the pulser/pick-up resistance.

6. With the harness still connected, check the resistance between the White and Black terminals. Resistance should be 459-561 ohms at 68° F (20° C); if not within range, replace the stator/lighting coil.

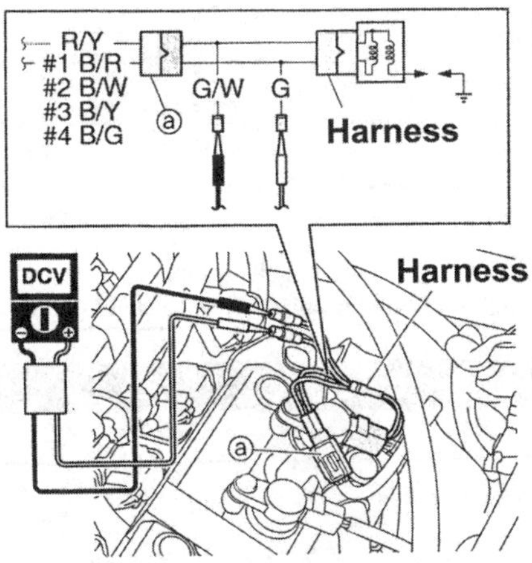

a - Ignition Coil Coupler

SELOC_9606

Fig. 59 Checking the ECM output peak voltage on 2010-11 VX models w/1052cc engines

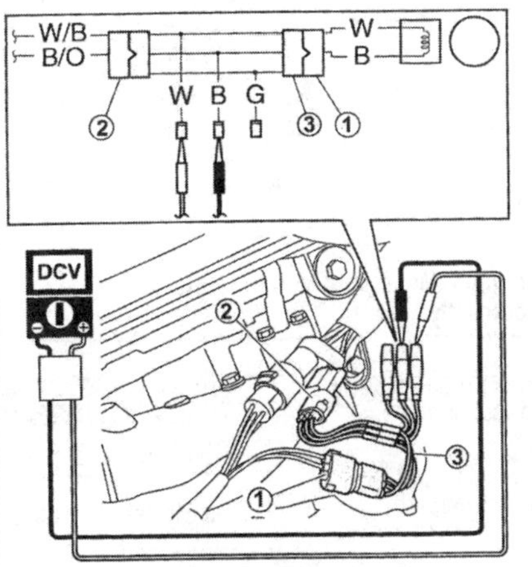

1 - Pick-up Coil Coupler
2 - Pick-up Coil Coupler (Harness End)
3 - Test Harness

SELOC_9606

Fig. 60 Checking the pick-up coil output peak voltage on 2010-11 VX models w/1052cc engines

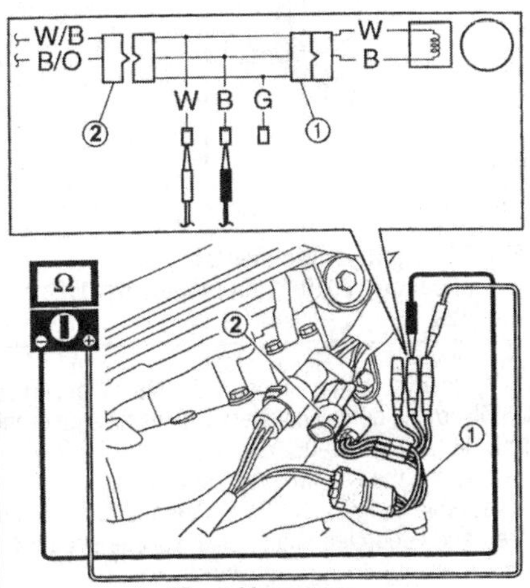

1 - Test Harness
2 - Pick-up Coil Coupler

SELOC_9606

Fig. 61 Checking the pick-up coil resistance on 2010-11 VX models w/1052cc engines

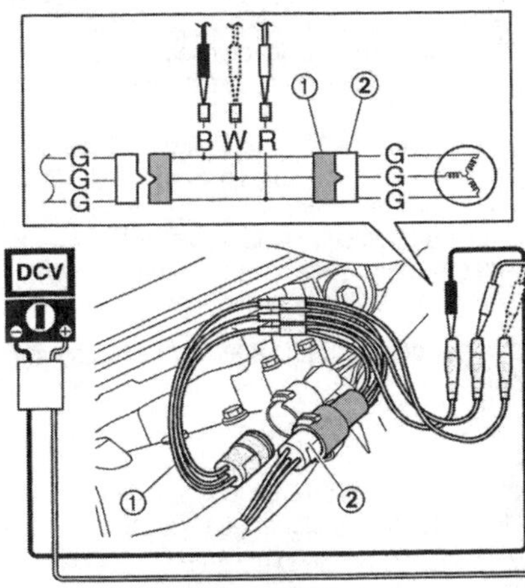

1 - Test Harness
2 - Stator Coil Coupler

SELOC_9606

Fig. 62 Checking the stator/lighting coil output peak voltage on 2010-11 VX models w/1052cc engines

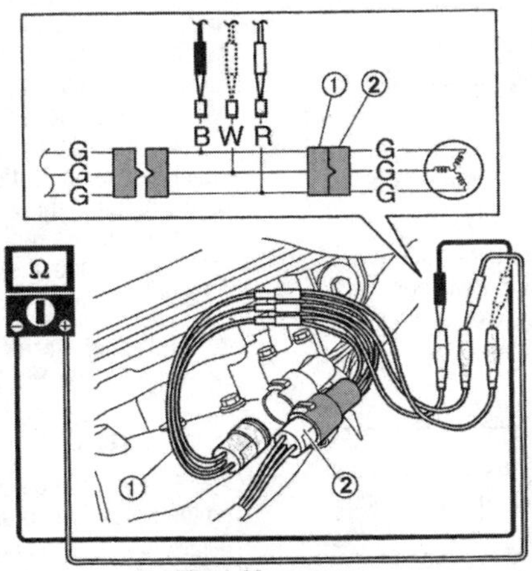

1 - Test Harness
2 - Stator Coil Coupler

SELOC_9606

Fig. 63 Checking the stator/lighting coil resistance on 2010-11 VX models w/1052cc engines

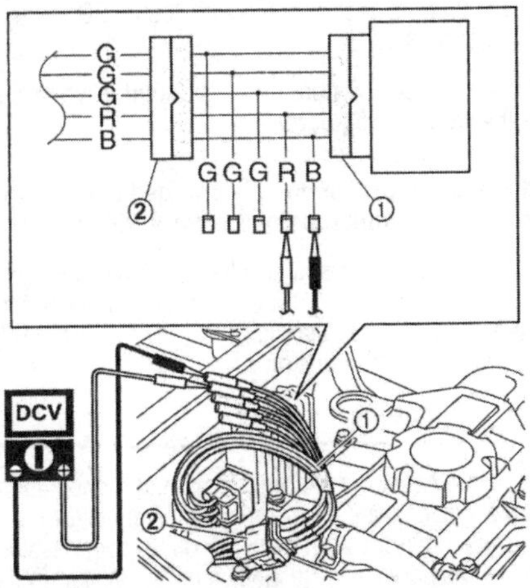

1 - Test Coupler
2 - Rectifier/Regulator Coupler

SELOC_9606

Fig. 64 Checking the regulator/rectifier coil output peak voltage on 2010-11 VX models w/1052cc engines

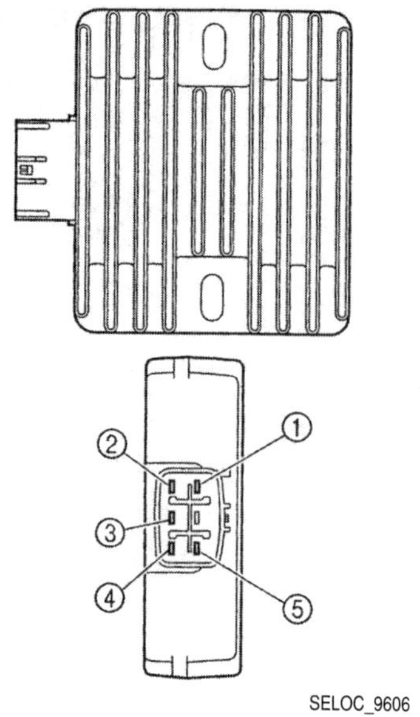

Fig. 65 Regulator/rectifier terminal identification - 2010-11 VX models w/1052cc engines

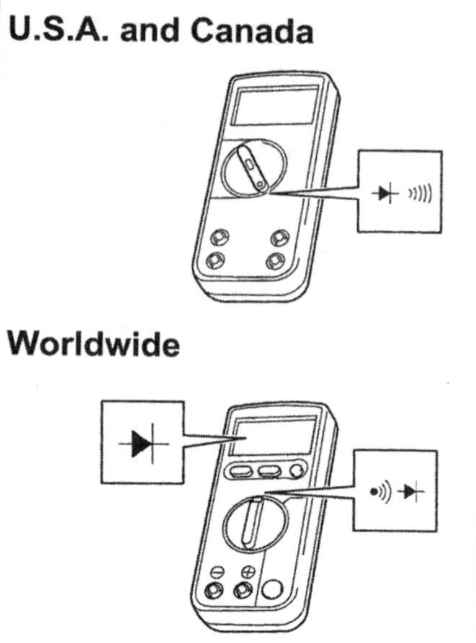

Fig. 66 Make sure you set your meter correctly. . .

Tester lead ⊕	Tester lead ⊖	Display value (V) (reference data)
①	②	OL
	③	
	④	
	⑤	
②	①	0.41–0.49
	③	OL
	④	
	⑤	
③	①	0.41–0.49
	②	OL
	④	
	⑤	
④	①	0.41–0.49
	②	OL
	③	
	⑤	
⑤	①	0.77–0.85
	②	0.41–0.49
	③	0.41–0.49
	④	0.41–0.49

OL: Overload

SELOC_9606

Fig. 67 . . .before checking the regulator/rectifier terminal as detailed here.

7. Now you can disconnect the stator/lighting coil coupler (right next to the pulser/pick-up coil coupler) and install a different 3-pin test harness ((#YB-06870 / 90890-06870).

8. Using the DVOM again, test the stator/lighting coil peak output voltage while cranking, at 2000 rpm and at 3500 rpm. Test between the Green and Green terminals. Peak voltage should be:

- Cranking - 6 V
- 2000 rpm - 26 V
- 3500 rpm - 45 V

If it falls below any of these levels, check the resistance.

9. With the harness still connected, check the resistance between the Green and Green terminals. Resistance should be 0.23-0.29 ohms at 68° F (20° C); if not within range, replace the stator/lighting coil.

10. Finally, you can disconnect the rectifier/regulator coupler and install a 6-pin test harness (#YB-06790 / 90890-06790). The rectifier/regulator is on the port side of the engine, next to the ECM.

11. Using the DVOM again, test the peak voltage while loaded, at 2000 and 3500 rpm. Test between the Red and Black terminals. Output voltage should be 13 V.

■ **Do not use the peak voltage adapter when checking output voltage at the rectifier/regulator.**

■ **The battery must be fully charged in order to perform this test correctly, an undercharged battery will affect the test results.**

If it falls below this level, check the rectifier/regulator for continuity.

12. Remove the rectifier/regulator and check for continuity as shown in the accompanying illustrations. If out of range, it will require replacement. If in range, install the unit and reconnect it.

COMPRESSION

Before spending too much time and money attempting to trace a problem to the ignition system, a compression check of the cylinder should be made. If the cylinder does not have adequate compression, troubleshooting and attempted service of the ignition or fuel system will fail to give the desired results of satisfactory engine performance.

For details, please refer to Compression Testing in the Maintenance & Tune-Up section.

Electronic Control Module (ECM)

■ **Please refer to the Fuel System section for ECM removal and installation procedures.**

Flywheel/Flywheel Magneto/Rotor

REMOVAL & INSTALLATION

All Models With 998cc/1052cc Engines

OEM ③ DIFFICULT

◆ **See Figures 68 thru 74**

1. Remove the generator/flywheel cover as detailed in the appropriate section.

2. Install a flywheel holding tool (#YS-01880-A / 90890-01701) over the flywheel and have someone hold the tool as you loosen and remove the mounting bolt/washer. Throw them both away as you cannot reuse either one of them.

3. Now thread a flywheel puller (#YS-01082 / 90890-01080) into the bolt hole in the crankshaft and carefully tighten the bolt, while your friend is still holding the flywheel from rotating, until you are able to unseat the assembly from the crankshaft. Remove the tools and set the assembly down where it will not get damaged.

4. The starter gear may or may not come off with the flywheel/starter clutch assembly, but either way remove the woodruff key from the slot in the crankshaft and throw it away - you can't re-use it either.

5. If you are replacing the rotor (remember, Yamaha calls the flywheel a rotor, or at times, the flywheel magneto), hold the assembly and remove the 6 bolts attaching the starter clutch to the back of the flywheel.

■ **Additional starter clutch, starter gear and idle gear procedures can be found in the Starter section.**

To install:

6. If you removed the starter clutch, position it on the back off the rotor so the holes line up and the roller is facing the starter gear. Install the 6 bolts finger tight, making sure everything lines up; and then tighten the bolts, in a star pattern, to 17 ft. lbs. (24 Nm).

■ **On later models, be sure that the correct side of the clutch is facing the rotor.**

7. Install the starter gear onto the clutch assembly by spinning it clockwise slightly. Spinning it counterclockwise should engage the starter gear.

8. Clean the tapered end of the crankshaft thoroughly, position the NEW woodruff key and then slide the rotor assembly into position on the shaft.

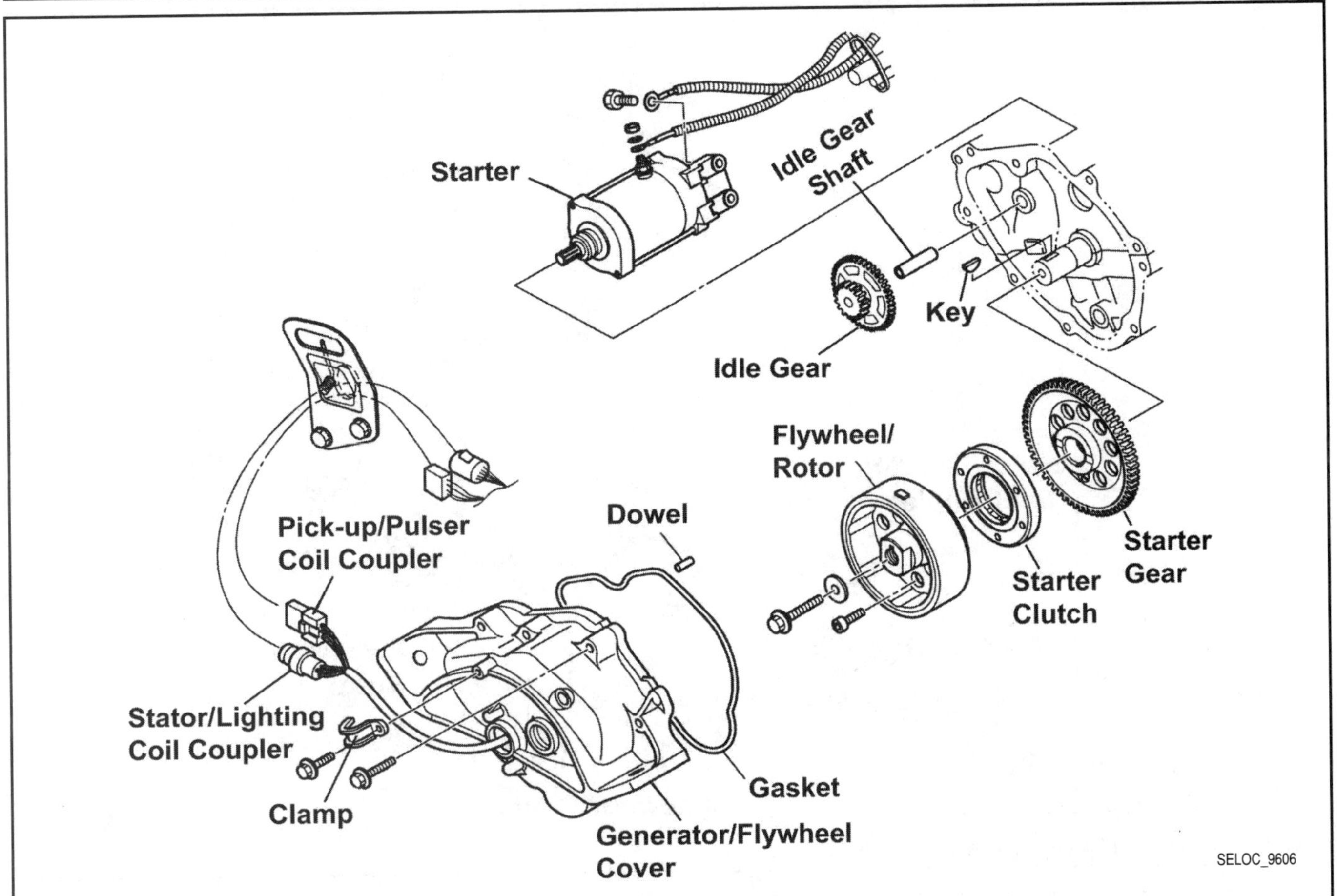

Fig. 68 Exploded view of the flywheel/rotor assembly - FX models w/998cc engines

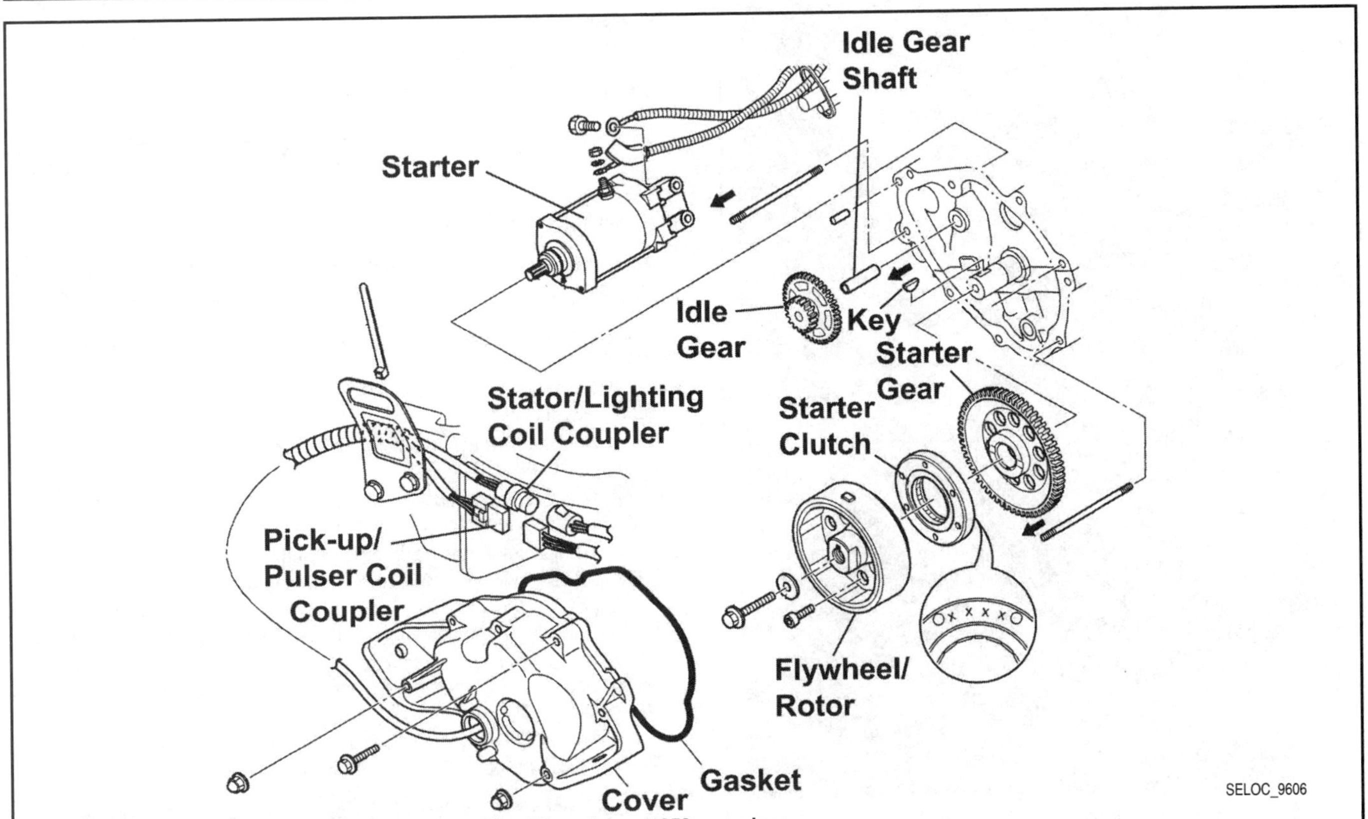

Fig. 69 Exploded view of the flywheel/rotor assembly - FX models w/1052cc engines

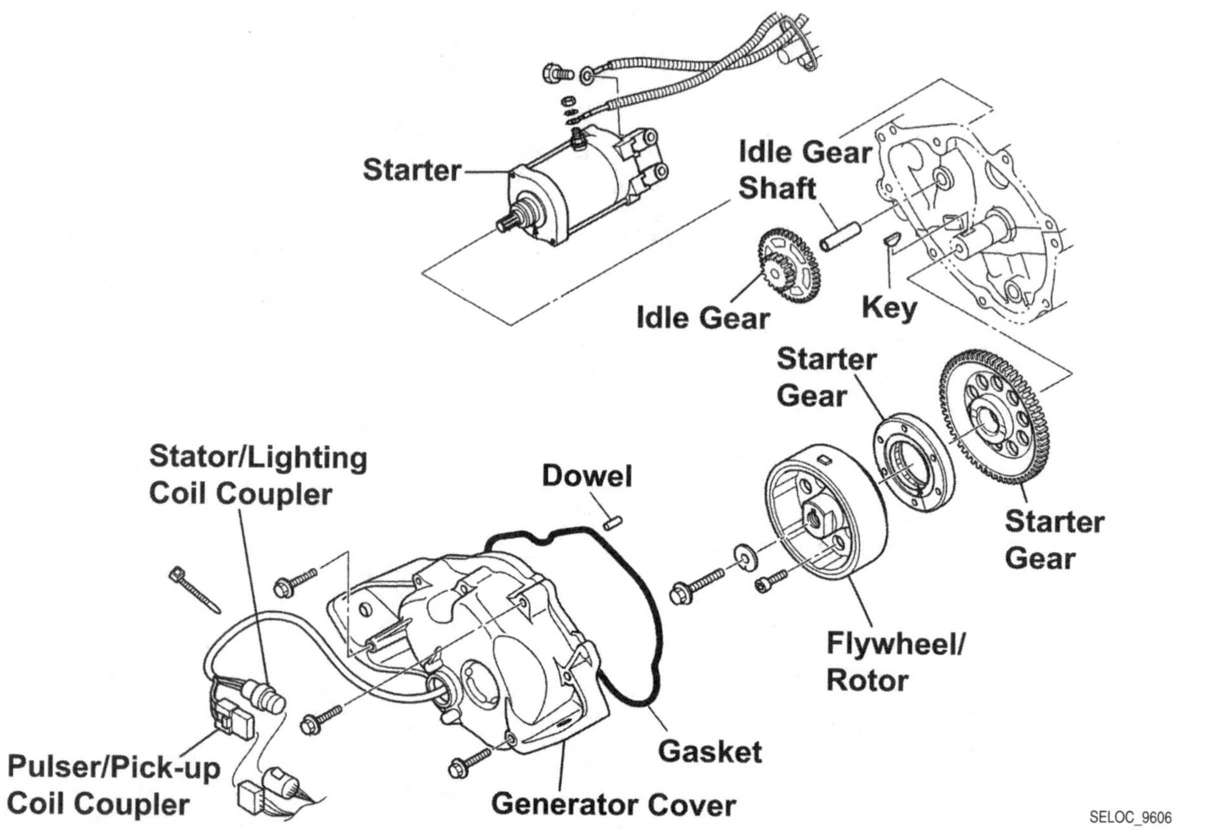

Fig. 70 Exploded view of the flywheel/rotor assembly - VX models

Fig. 71 A good look at the flywheel/rotor assembly

Coat the NEW mounting bolt and washer with clean engine oil and thread it into the crankshaft. Tighten the bolt to 54 ft. lbs. (75 Nm) while holding the flywheel with the holding tool.

9. Install the generator cover.

All Models w/1812cc Engines

OEM ③ DIFFICULT

◆ **See Figures 75 thru 87**

■ **You may want to read through this procedure before beginning - it is not going to be fun! Particularly if working on the FX HO, FZ or VX.**

1. Remove the generator/flywheel cover as detailed in the appropriate section.
2. Use a heat gun and carefully heat the transfer shaft assembly for a good 20-30 minutes.
3. Hold the flywheel by installing a shaft holder (#90890-06721) and driver handle (#90890-06722).
4. Use a 12mm hexagon bit socket that is at least 150mm (5.91 in.) long and loosen the special nut until it can no longer be turned.
5. Install a crankshaft holder (#90890-06732) and then remove the transfer shaft assembly. The assembly is not reusable, so discard it,
6. Install a flywheel puller (#90890-06723) and rotor puller (#90890-01080) onto the flywheel and remove the flywheel magneto and starter gear.
7. The starter gear may or may not come off with the flywheel/starter clutch assembly, but either way remove the woodruff key from the slot in the crankshaft and discard it as well.
8. Remove the 6 flywheel bolts and then separate the starter clutch from the flywheel. You can throw the flywheel away - you can't re-use it either.

■ **Additional starter clutch, starter gear and idle gear procedures can be found in the Starter section.**

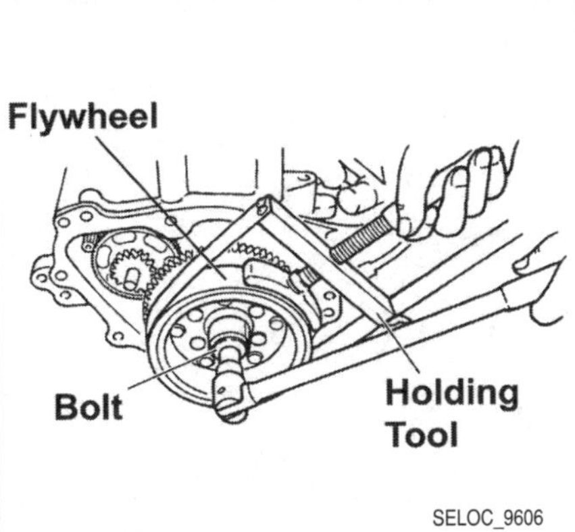

Fig. 72 Loosen the mounting bolt while using a holding tool. . .

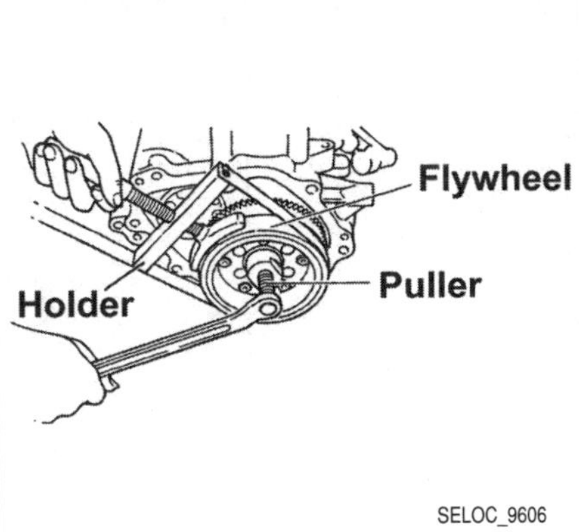

Fig. 73 . . .and then use a puller to get the assembly off of the crankshaft

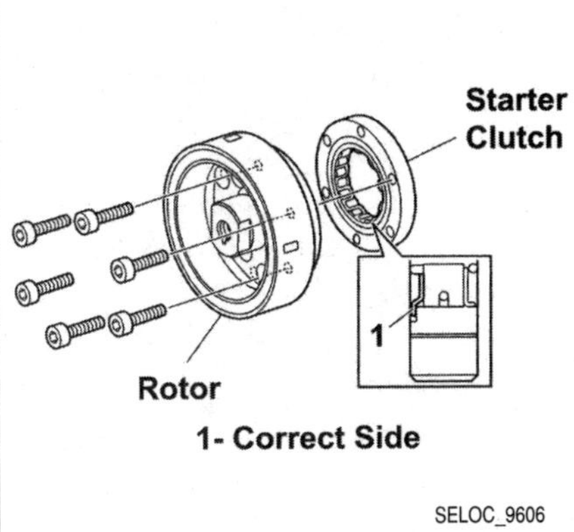

Fig. 74 Make sure the this side of the starter clutch is facing the rotor on later models

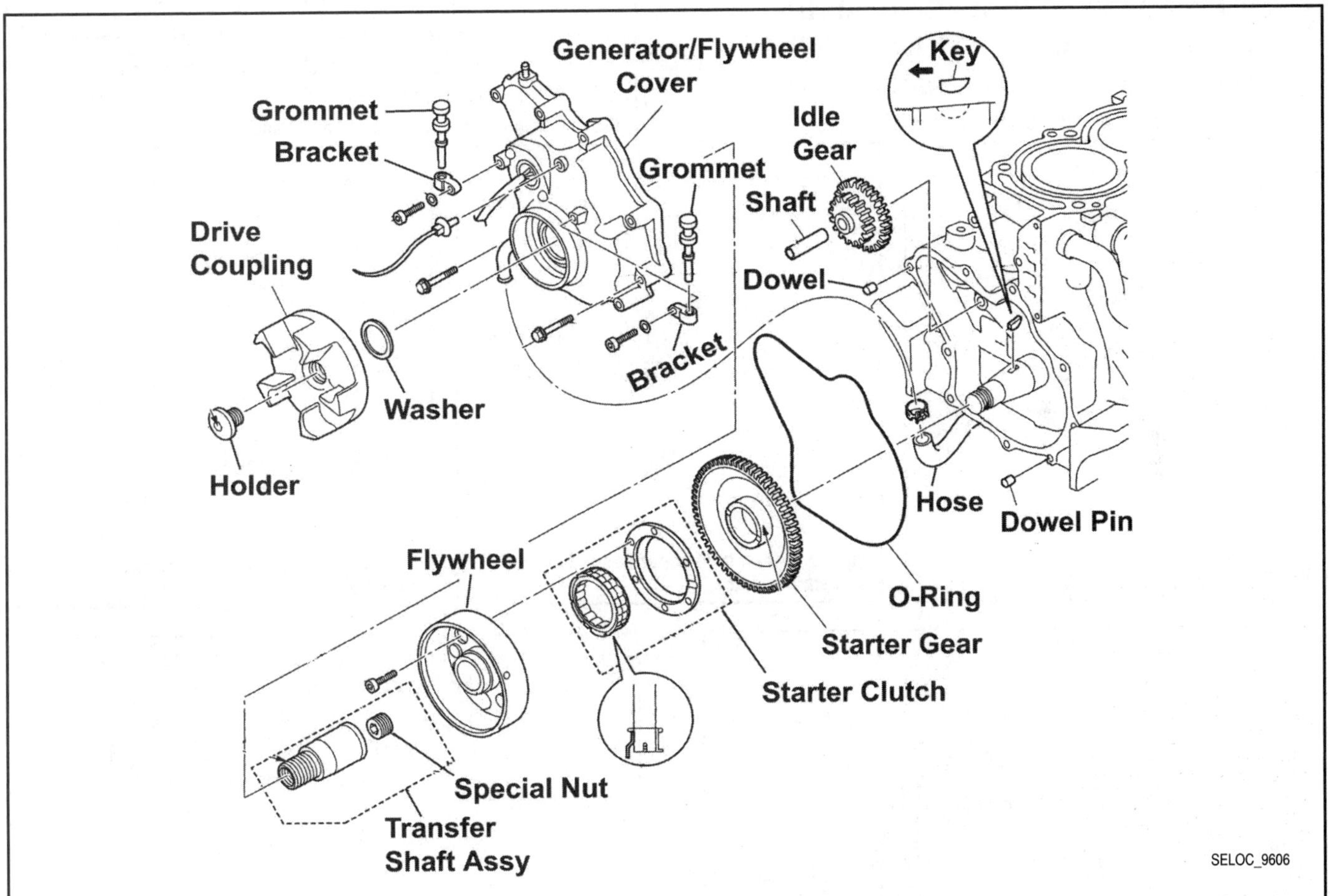

Fig. 75 Exploded view of the flywheel/rotor assembly - models w/1812cc engines

To install:

9. If you removed the starter clutch, position it on the back of the *new* flywheel/rotor so the holes line up as shown in the illustration. Install the 6 *new* bolts finger tight, making sure everything lines up; and then tighten the bolts, in a star pattern, to 17.7 ft. lbs. (24 Nm).

10. Install the starter gear onto the clutch assembly by spinning it clockwise slightly. Spinning it counterclockwise should engage the starter gear.

11. Check the *new* transfer shaft and special nut carefully to make sure the threads are clean and free of any grease. Thread the special nut into the end of the shaft until it cannot be turned any more.

12. Clean the tapered end of the crankshaft thoroughly - there can be no grease on the threads or inside the flywheel, position the *new* woodruff key and then slide the assembly into position on the shaft.

13. Apply a short line of Loctite 648 to the end of the crankshaft and then coat the mating surfaces between the flywheel assembly and transfer shaft assembly with Yamaha Grease A.

On FX HO, FZ and VX models:

14. Position the transfer shaft assembly and then install the shaft holder and driver handle again. Install the crankshaft holder and tighten the transfer shaft assembly to 36.9 ft. lbs. (50 Nm).

15. Now, use the hexagonal portion of the crankshaft holder (still installed) as a reference, mark the shaft holder tool at 0°, 60° and 120°. Once done, add another mark onto the transfer shaft assembly where the shaft is in alignment with the 0° mark.

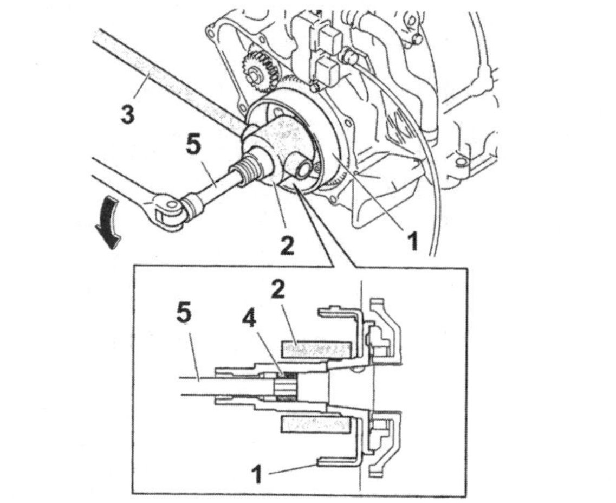

1- Flywheel
2- Shaft Holder
3- Driver Handle
4 - Special Nut
5 - Hexagon Bit Socket

SELOC_9606

Fig. 76 Loosening the transfer shaft special nut

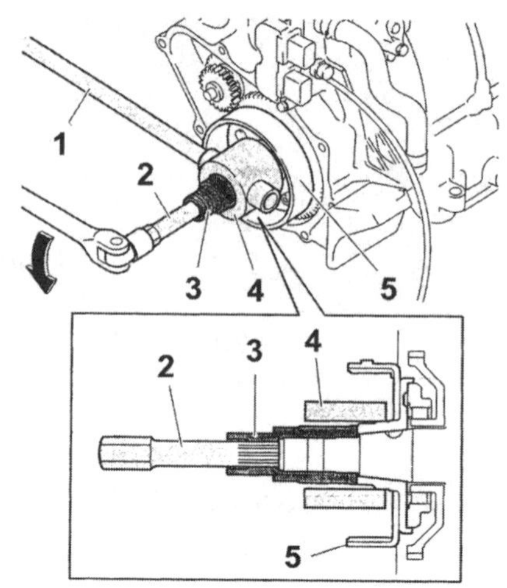

1- Driver Handle
2- Crankshaft Holder
3- Transfer Shaft
4- Shaft Holder
5- Flywheel

SELOC_9606

Fig. 77 Remove the transfer shaft assembly. . .

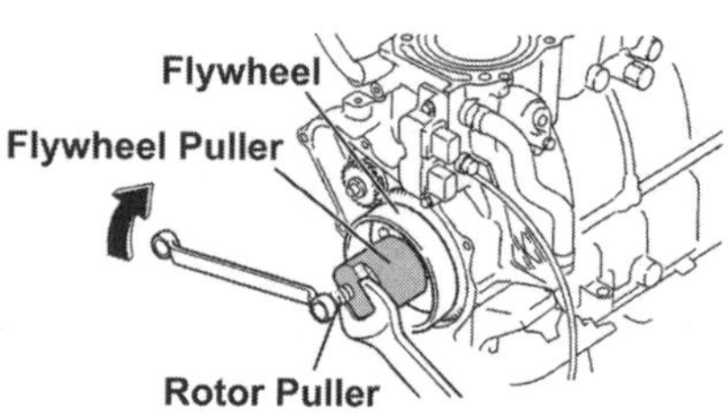

SELOC_9606

Fig. 78 . . .the flywheel. . .

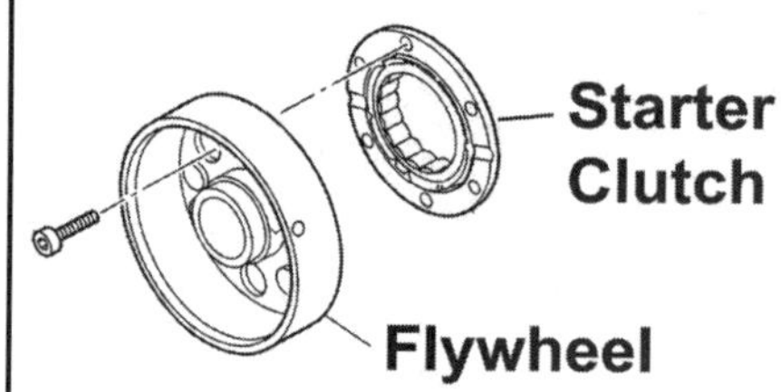

SELOC_9606

Fig. 79 . . .and then separate the flywheel and starter clutch

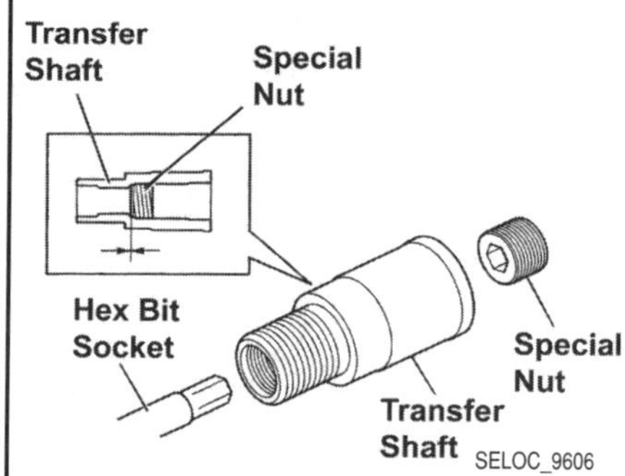

SELOC_9606

Fig. 80 A good look at the transfer shaft assembly

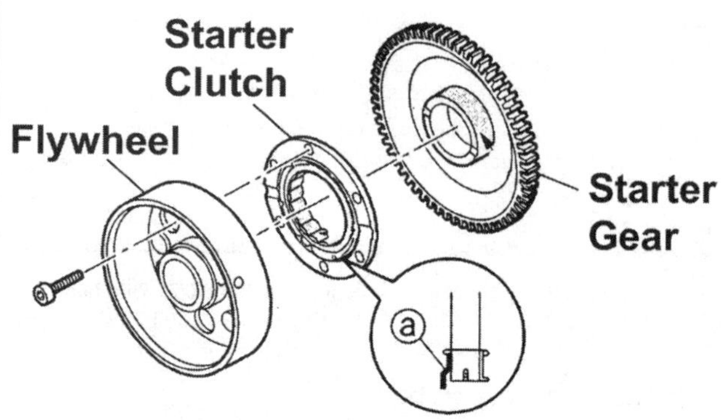

a - Faces Flywheel

SELOC_9606

Fig. 81 Reassembling the flywheel/starter clutch

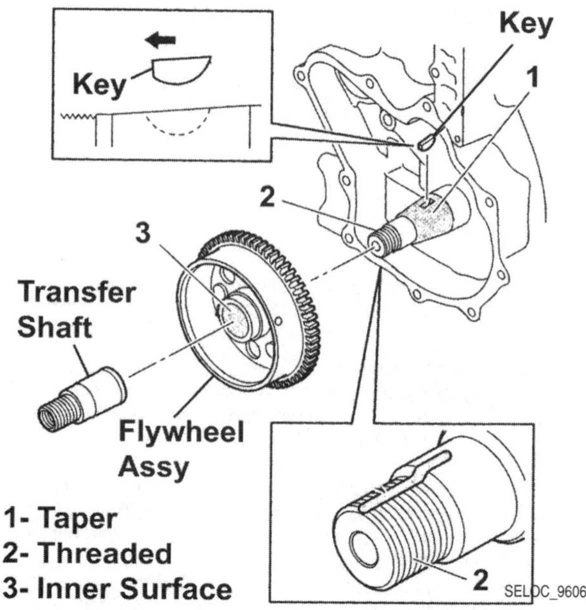

1- Taper
2- Threaded
3- Inner Surface

SELOC_9606

Fig. 82 Setting up the flywheel assembly for installation

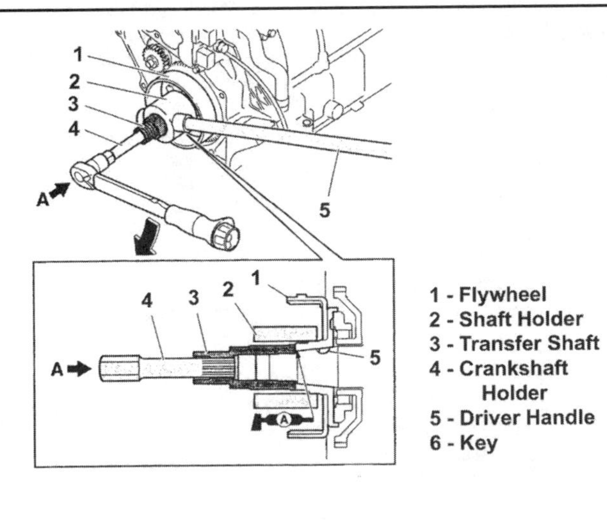

Fig. 83 Install the transfer shaft assembly

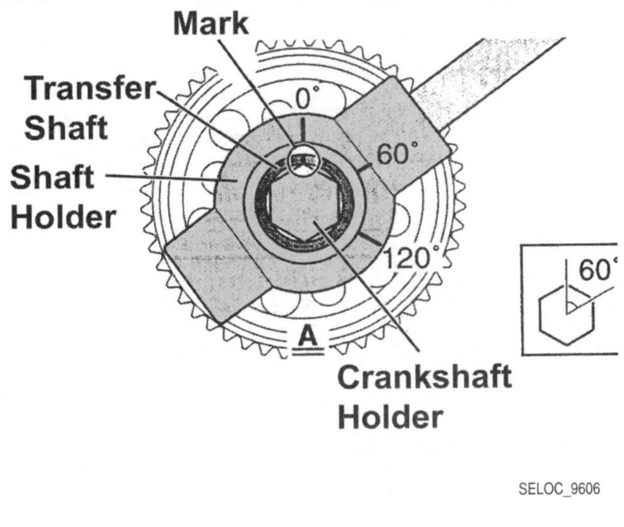

Fig. 84 Mark the shaft holder as shown (FX HO, FZ and VX models only)

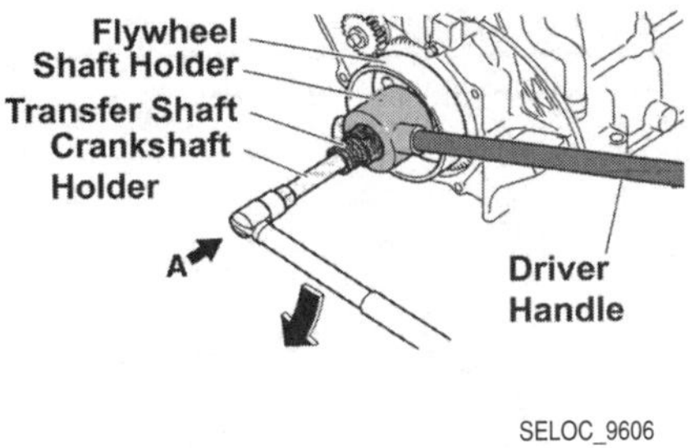

Fig. 85 Tighten the shaft assembly to its initial setting. . .

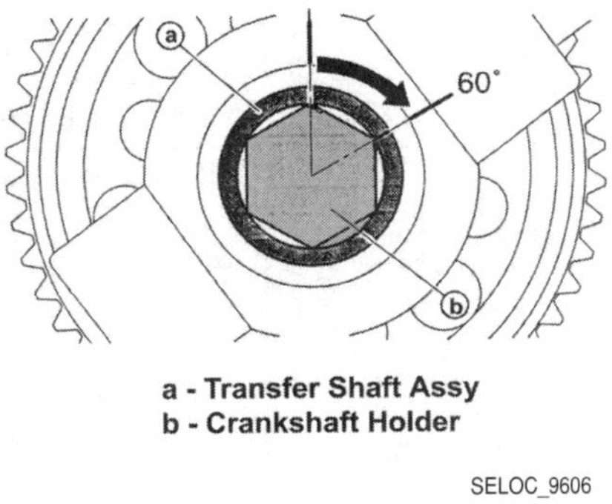

Fig. 86 . . .and then tighten it an additional 60°. . .

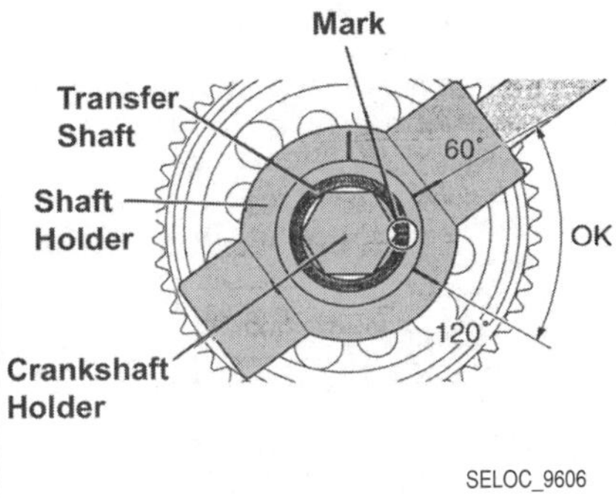

Fig. 87 . . .and then to its final torque (FX HO, FZ and VX models only)

■ **The angle from one point/corner of the hexagon to another is 60°.**

16. While still holding the flywheel assembly with the 2 tools, use the crankshaft holder to tighten the transfer shaft assembly an additional 60°. Once done, you can tighten the assembly to a final torque of 184.4 ft. lbs. (250 Nm).

■ **If tightening the transfer shaft to the final torque figure means that your initial 0° mark on the transfer shaft has now moved past the 120° mark on the shaft holder tool, you will need to undo everything and go back to Step 11 and start over.**

17. Remove the crankshaft holder tool, but leave the shaft holder and drive handle in place. Reinstall the hexagon bit socket from Step 3 and now you can tighten the special nut to a final torque of 184.4 ft. lbs. (250 Nm) as well.

18. Skip to the last step.

On FX SHO models:

19. Position the transfer shaft assembly and then install the shaft holder and driver handle again. Install the crankshaft holder and tighten the transfer shaft assembly to 184.4 ft. lbs. (250 Nm).

20. Remove the crankshaft holder tool, but leave the shaft holder and drive handle in place. Reinstall the hexagon bit socket from Step 3 and now you can tighten the assembly to a final torque of 184.4 ft. lbs. (250 Nm) as well.

On all models now:

21. Install the flywheel/generator cover. Allow at least an hour before operation so that the Loctite 648 is allowed to set properly. Phew, we told you this was gonna be a fun one!

Generator/Flywheel Cover

REMOVAL & INSTALLATION

All Models With 998cc/1052cc Engines

3 DIFFICULT

◆ **See Figures 68 thru 71**

1. Remove the seat(s) and disconnect the battery cables.

2. Remove the engine as detailed in the Engine Mechanical section. We have heard of people who are able to accomplish this task by removing the exhaust manifold and *not* the engine - you take a look and decide for yourself.

3. Yamaha also suggests disconnecting the starter and removing it.

4. Tag and disconnect the pulser/pick-up and stator/lighting coil couplers.

5. Remove the 8 cover mounting bolts (on FX models w/1052cc engines there are also 2 cap nuts) and carefully pull off the generator cover (many people call this the flywheel cover). Be careful, as there are 2 dowel pins that may come off with the cover, or stay in the cylinder block; remove them both either way. Remember that the stator and coils are mounted inside the cover, so store it carefully and make sure you cover it.

■ **On VX models, there are 2 bolts that are a different length. . .they should be the 2 outermost bolts on the bottom of the cover, but take note of which holes they came out of to be sure!**

6. Remove the cover gasket/ring and discard it.

To install:

7. After cleaning them, insert the 2 studs back into their bore on the cylinder block.

8. Install a *new* gasket into the cover and then position the assembly so that the stator/lighting coil will slide into the flywheel correctly as you press the cover into position. Make sure that the studs line up perfectly and then press the cover assembly all of the way in. Make sure that the 2 long bolts on the FX models w/1052cc engines feed through their cover holes.

9. Coat the bolt threads with Loctite 572 and then thread them all in finger-tight to hold the cover in position. Use a criss-cross pattern to tighten the bolts to 11 ft. lbs. (15 Nm) and then follow with a 2nd pass of 36 ft. lbs. (50 Nm) in the same pattern. On FX models w/1052cc engines, tighten the 2 cap nuts to the same torque.

10. Feed the coil harness through the clamp and position it on the cover without pinching the harness cover. Tighten the bolt securely.

11. Reconnect the stator/lighting and pick-up/pulser coil couplers to the main harness.

12. Install the starter and engine if removed.

All Models w/1812cc Engines

OEM ③ DIFFICULT

◆ **See Figure 75**

■ **Unlike other models, the generator cover, and its associated components, is found on the *back* of the engine and the oil pump is at the front.**

1. Remove the seat(s) and cargo container. Disconnect the battery cables.

2. Remove the drive coupling (back of the engine, connecting to the intermediate housing) as detailed in the Engine Mechanical section.

3. Pry out the washer left in the generator cover bore if it doesn't come out with the coupling.

4. Cut the plastic tie around the stator and pick-up coil leads coming out of the cover.

5. If needed, remove the grommet and bolt to lift off the small bracket found on each side of the cover.

6. Loosen the small hose clamp and pull the hose off of the fitting at the bottom of the cover.

7. Loosen the 9 cover bolts and carefully pry the cover off of the engine. If the large O-ring does not come off with it, remove it and discard.

8. Make sure the 2 dowel pins are in the block and not in the cover.

9. If necessary, carefully pry out the oil seal.

To install:

10. Press a new oil seal into position if removed.

11. Carefully position the cover onto the cylinder block and thread in the 9 bolts. Alternately, tighten the bolts to 19.2 ft. lbs. (26 Nm).

12. Slide the hose clamp back on the hose (actually, we'd replace it but you don't have to), wiggle the hose onto the neck so that you have enough room to install the clamp over the hose upstream of the ridge on the end of the neck. Tighten the clamp screw securely.

13. Reinstall the 2 brackets if removed and tighten the bolts to 5.9 ft. lbs./71 inch lbs. (8 Nm). Press the grommets into each bracket.

14. Use a new plastic tie to secure the 2 leads coming out of the cover.

15. Install a *new* washer into the cover bore and then install the drive coupling.

16. Connect the battery cables and install the seat(s) and cargo container.

Ignition Coils

REMOVAL & INSTALLATION

FX Models w/998cc Engines

② MODERATE

◆ **See Figures 88, 89 and 90**

1. Remove the seat(s) and disconnect the battery cable.

2. Remove the air filter case cover and lift out the air filter.

3. Locate the ignition coil box on the aft starboard side and disconnect the 2 couplers on the right side of the box. Make sure you tag each one.

4. If removing the ignition coil only, skip to Step 5. If removing the entire ignition coil box:

a. Wiggle each spark plug cap while pulling up on it and remove all 4 caps. The plug wires should already be marked with their cylinder number, but if not, make sure that you mark them yourself.

b. Remove the 3 mounting bolts and lift out the coil box. Be careful not to lose the bolt collars and grommets.

5. Remove the 10 mounting bolts securing the cover to the coil box and pull off the cover. You may have to pop the plug wires out of the holder.

6. Tag each plug wire where it comes from the ignition coil and then remove the 2 wires from the coil you need to remove.

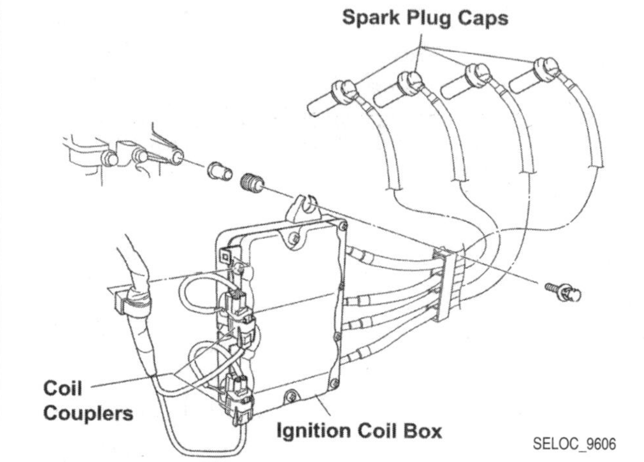

Fig. 88 A good look at the ignition coil box. . .

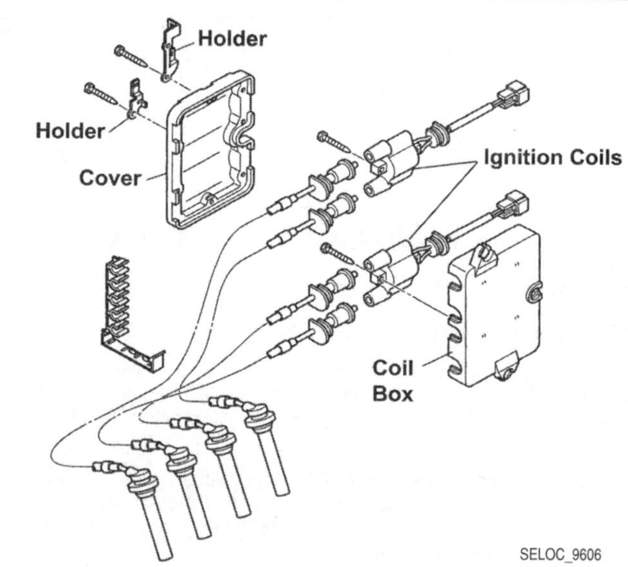

Fig. 89 . . .and the coils themselves

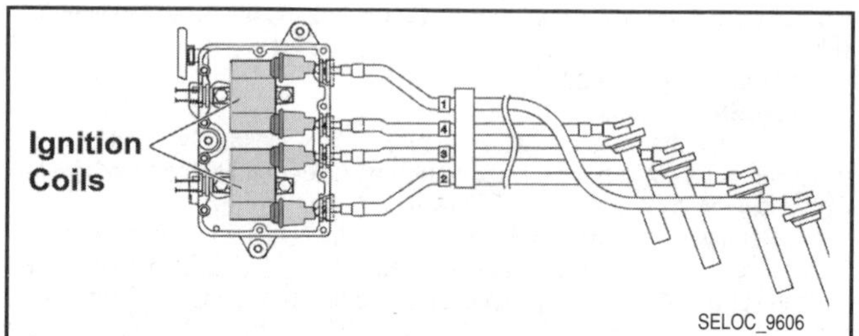

Fig. 90 The spark plug wires should be numbered but it's a good idea to tag them anyway. Remember that they are listed in the firing order at the coils and in the cylinder order at the plugs

7. Remove the 2 mounting bolts and lift the coil out of the box.

To install:

8. Position the coil over the mounting bosses and thread in the mounting bolts, tightening to 3.5 ft. lbs./43 inch lbs. (4.9 Nm).
9. Press the plug wires back into their respective towers on the coil - it is imperative that they are fully seated.
10. Feed the coil leads and plug wires through their openings in the sides of the box, making sure their seal grommets are seated correctly in the case cut-outs.
11. Position the cover so the bolt holes all line up and then thread in all of the mounting bolts - don't forget the 2 holders that are held in position by 2 of the bolts. Tighten each bolt, in a criss-cross pattern, to 3.5 ft. lbs./43 inch lbs. (4.9 Nm).
12. If the entire box was removed, position it now and thread in the 3 bolts with their collars and grommets. Tighten the bolts to 5.5 ft. lbs./67 inch lbs (7.6 Nm).
13. Reconnect each of the couplers.
14. Press each of the spark plug caps back into their plug bores and then install the air filter and cover.
15. Connect the battery cables and install the seat.

FX Models With 1052/1812cc Engines & All FZ And VX Models

2 MODERATE

◆ **See Figures 91 thru 96**

These engines utilize a single coil per cylinder.

1. Remove the seat(s) and disconnect the battery cable.
2. Remove the air filter case cover and lift out the air filter on FX models w/1052cc engines. Remove the engine cover on all models w/1812cc engines.
3. Reach in and disconnect the ignition coil coupler and position it out of the way. If you are removing all 4 coils, we recommend tagging each coupler so you know what cylinder it came from.
4. Loosen and remove the coil mounting bolt(s). Wiggle the coil while pulling up on it and remove it.
5. Repeat the procedure for the remaining 3 coils if necessary.

To install:

6. Position the coil over the spark plug bore in the head and tighten the mounting bolt to 5.5 ft. lbs./65 inch lbs. (7.6 Nm) on all models but the 2010-11 VX (both engines) and the FX/FZ w/1812cc engines, which should be tightened to 5.9 ft. lbs./70 inch lbs. (8 Nm).
7. Reconnect each of the couplers that may have been disconnected.
8. Install the air filter and cover if removed. Install the engine cover if removed.
9. Connect the battery cables and install the seat.

TESTING

FX Models w/998cc Engines

2 MODERATE

◆ **See Figures 97 and 98**

1. With the coil removed, install a DVOM as per the manufacturer's instructions.
2. Check resistance between the Black/White and Red terminals in the coil coupler; this will show the coil primary resistance spec. Any reading outside of 1.53-2.07 ohms (at 68°F/20°C) will necessitate coil replacement.
3. Now check between the 2 spark plug wire towers (inside, on the connector) for the secondary resistance spec. Readings outside of the 12.5-16.9 kila-ohm (at 68°F/20°C) range will necessitate replacement.

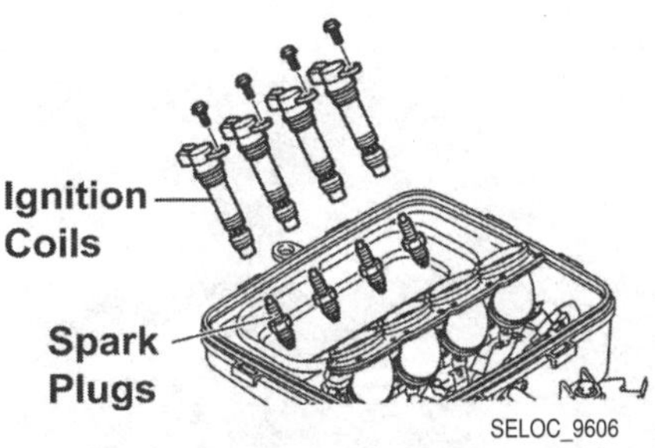

Fig. 91 The ignition coils on these engines pop into the cylinder head, over the spark plugs (FX w/1052cc shown, VX w/1052cc similar)

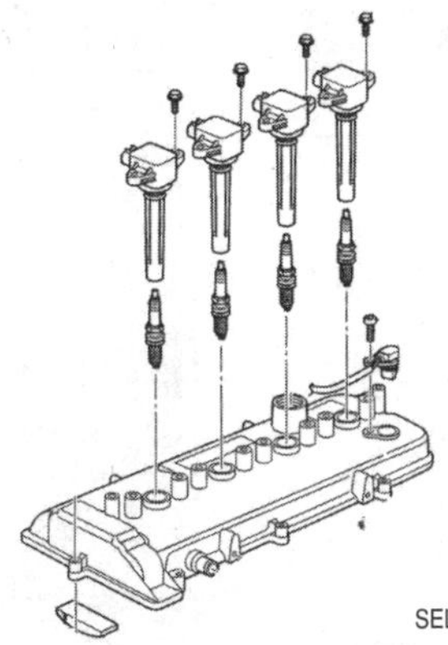

Fig. 92 Here they are on the 1812cc engines

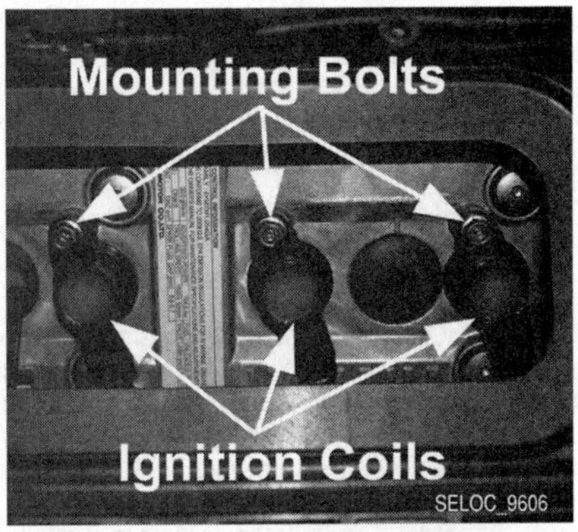

Fig. 93 The coils are held in position by 1 bolt on FX models w/1052cc engines. . .

Fig. 94 . . .and also on VX models w/1052cc engines. . .

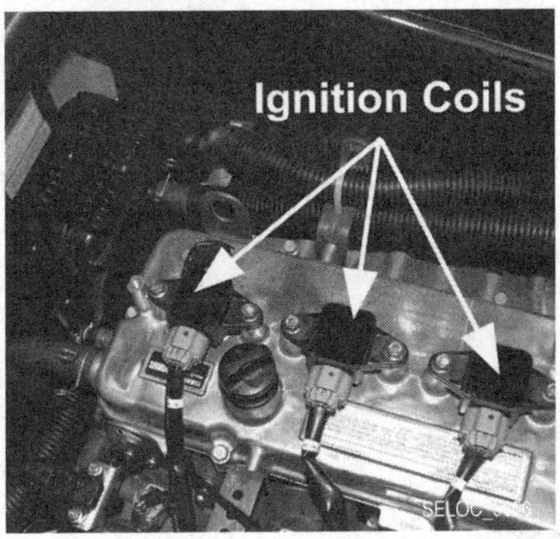

Fig. 95 . . .but use 2 bolts on all models w/1812cc engines

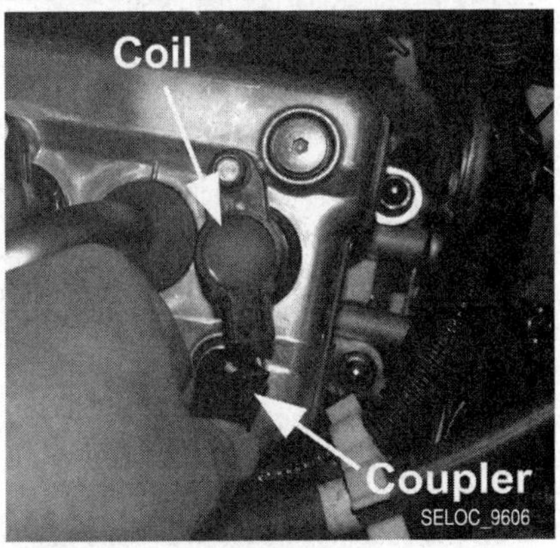

Fig. 96 Disconnect the coupler before removing the coil (FX shown, VX similar)

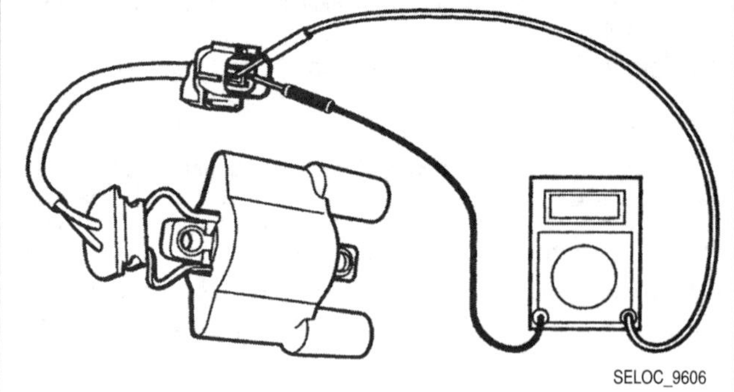

Fig. 97 Testing the primary coil resistance on 998cc engines. . .

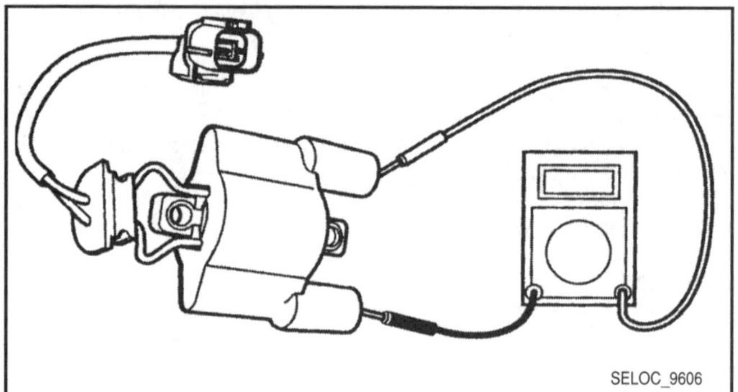

Fig. 98 . . .and the secondary resistance

All Models w/1052cc Engines

◆ See Figure 99

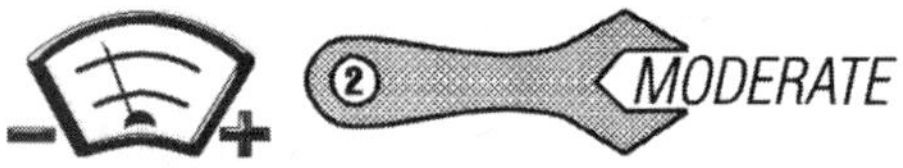

1. With the coil removed, install a DVOM as per the manufacturer's instructions.

2. Check resistance between the **A** and **B** terminals in the coil coupler; this will show the coil primary resistance spec. Any reading outside of 1.19-1.61 ohms (at 68°F/20°C) will necessitate coil replacement.

3. Now check between the **B** terminal and the end of the coil (**C**) for the secondary resistance spec. Readings outside of the 8.5-11.5 kila-ohm (at 68°F/20°C) range will necessitate replacement.

All Models w/1812cc Engines

■ **Yamaha does not provide a specific ignition coil resistance test for these models.**

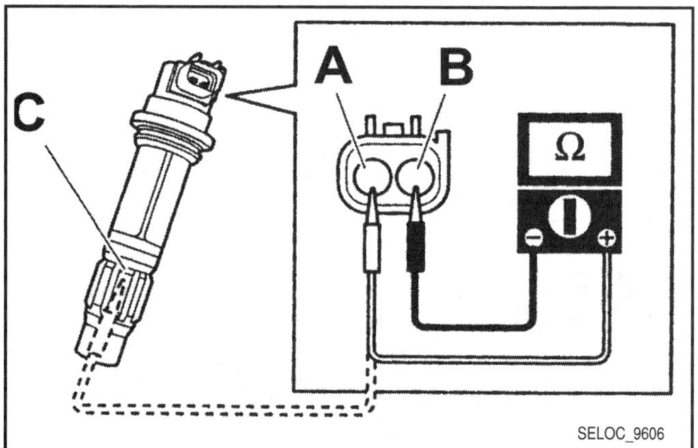

Fig. 99 On 1052cc engines, check the primary resistance between A and B. Check the secondary resistance between B and C

Stator/Lighting Coil & Pulser/Pick-Up Coils

REMOVAL & INSTALLATION

FX Models With 998cc/1052cc Engines

◆ See Figures 30 and 100

■ **As previously mentioned, the lighting coil is also commonly known as the stator. Following in that vein, the pulser coil is frequently referred to as a pick-up coil or trigger coil. Although separate components, they must be removed and replaced as an assembly because they share common wiring harness.**

1. Remove the seat(s) and disconnect the battery cable.

2. Remove the generator cover as detailed in the Generator/Flywheel Cover section (yep, this means you're probably going to have to remove the engine!).

3. Lay the cover on a clean work surface with the inside facing upward.

4. There are 3 Allen head bolts with large washers that are used to route the coil leads through the cover - remove them, but pay close attention to the routing of the harness (take a picture if you can).

5. Remove the pulser/pick-up coil mounting bolts (2/coil) and pull off the coils -

6. Remove the Allen head bolts securing the holder and lift it out before removing the plate and seal from the cover.

7. Remove the 3 stator/lighting coil mounting bolts and then remove the assembly, carefully pulling the harness through the opening in the cover.

To install:

8. Carefully feed the harness and connectors back through the harness plate and seal. Now feed the connectors through the hole in the cover. Press the plate into position in the cover making sure that the harness is not pinched.

■ **Be sure to coat the threads of all bolts with Loctite 242.**

9. Position the pulser and lighting coils and thread in the mounting bolts without tightening them. Now carefully feed the wires into position as shown in the illustration (or your picture), carefully feeding any excess through the retaining plate. Once done, install the 3 bolts with the large washers and tighten each bolt to 3.5 ft. lbs./43 inch lbs. (4.9 Nm).

Fig. 100 A good look at the generator components (2002-08 FX shown, FX w/1812cc engines and all VX similar but with only a single pulser/pick-up coil)

10. Now that you are confident that the wire routing is correct, install the holder over the wires and harness plate, tightening the 2 bolts to 10 ft. lbs. (14 Nm).

11. Tighten the stator/lighting coil bolts to 10 ft. lbs. (14 Nm).

12. Tighten the pulser/pick-up coil bolts to 3.5 ft. lbs./43 inch lbs. (4.9 Nm).

13. Install the generator cover and then the engine.

14. Connect the battery cables and install the seat(s).

VX Models & All FX Models w/1812cc Engines

◆ See Figures 31, 32, 100 and 101

(3) DIFFICULT

■ **As previously mentioned, the lighting coil is also commonly known as the stator. Following in that vein, the pulser coil is frequently referred to as a pick-up coil or trigger coil. Although separate components, they must be removed and replaced as an assembly because they share common wiring harness.**

1. Remove the seat(s) and disconnect the battery cable.
2. Remove the generator cover as detailed in the Generator/Flywheel Cover section (yep, this means you're probably going to have to remove the engine!).
3. Lay the cover on a clean work surface with the inside facing upward.
4. There is an Allen head bolt with a large washer that is used to route the coil leads through the cover - remove them, but pay close attention to the routing of the harness (take a picture if you can).
5. Remove the pulser/pick-up coil mounting bolts (2/coil) and pull off the coil.
6. Remove the Allen head bolts securing the holder and lift it out before removing the plate and seal from the cover.
7. Remove the 3 stator/lighting coil mounting bolts and then remove the assembly, carefully pulling the harness through the opening in the cover.

To install:

8. Carefully feed the harness and connectors back through the harness plate and seal. Now feed the connectors through the hole in the cover. Press the plate into position in the cover making sure that the harness is not pinched.

■ **Be sure to coat the threads of all bolts with Loctite 242.**

9. Position the pulser and lighting coils and thread in the mounting bolts without tightening them. Now carefully feed the wires into position as shown in the illustration (or your picture), carefully feeding any excess through the retaining plate. Once done, install the bolt with the large washer and tighten it to 3.7 ft. lbs./46 inch lbs. (5 Nm).
10. Now that you are confident that the wire routing is correct, install the holder over the wires and harness plate, tightening the 2 bolts to 10.3 ft. lbs. (14 Nm) on the FX HO, FZ, 2005-09 VX and 2011 VX w/1812cc engine; or 11.1 ft. lbs. (15 Nm) on the FX SHO and 2010-11 VX w/1052cc engine.
11. Tighten the stator/lighting coil bolts to 10.3 ft. lbs. (14 Nm) on the FX HO, FZ, 2005-09 VX and 2011 VX w/1812cc engine, or 11.1 ft. lbs. (15 Nm) on the FX SHO and 2010-11 VX w/1052cc engine.
12. Tighten the pulser/pick-up coil bolts to 3.7 ft. lbs./46 inch lbs. (5 Nm).
13. Install the generator cover and then the engine if removed.
14. Connect the battery cables and install the seat(s).

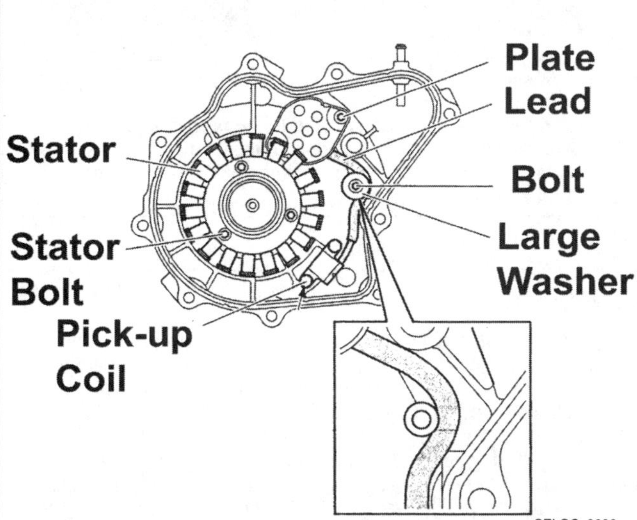

Fig. 101 A good look at the coil lead routing - models w/1812cc engines

Rectifier/Regulator

REMOVAL & INSTALLATION

FX Models With 998cc/1052cc Engines

(2) MODERATE

The rectifier/regulator assembly is attached to the starboard side of the electrical box, just behind the oil tank. Disconnect the electrical coupler on the forward edge and remove the 2 mounting bolts to take out the unit.

When installing, make sure that the connector is facing forward and tighten the bolts to 2.8 ft. lbs./35 inch lbs. (3.9 Nm). Connect the coupler.

All Models w/1812cc Engines

◆ See Figure 102

(2) MODERATE

The rectifier/regulator is found on the port side of the engine, at the back and just forward of the breather assembly.

1. Remove the seat(s) and cargo container. Disconnect the battery cables.
2. Loosen the hose clamps and slide the 3 hoses off of their nipples and position out of the way.
3. Disconnect the electrical lead coupler and move it out of the way.
4. Loosen the 2 mounting bolts (side and bottom) and pull them out without losing the grommets or collars.
5. Remove the rectifier/regulator.
6. When installing, coat the bolt threads with Loctite 572 and tighten to 19.2 ft. lbs. (26 Nm). Tighten the hose clamps securely, making sure that the clamp and hose are completely over the ridge on the end. Notice that the there is a mark on the upper hose and on the top of the unit - make sure that they are in alignment when tightening the clamp on the top hose.
7. Connect the coupler. Connect the battery cables and install the seat and cargo container.

VX Models w/1052cc Engines

◆ See Figures 103 and 104

(2) MODERATE

■ **Yamaha has initiated a Factory Modification Campaign (TSB #WR2005-009R) for a possibly faulty rectifier/regulator connector terminals on all 2005 VX models that could cause excessive electrical resistance. The campaign provides a new rectifier/regulator with improved watertight characteristics, along with a new mounting bracket. In addition, a new APS assembly will be installed as well.**

The rectifier/regulator is located on the port side of the engine, next to the ECM on 2005-09 models. On 2010-11 models, it is attached to a hull plate at the rear of the engine, still next to the engine.

1. Remove the seat and disconnect the battery.
2. Disconnect the electrical lead at the bottom of the rectifier/regulator.
3. Remove the 2 mounting bolts and lift the unit from the bracket.

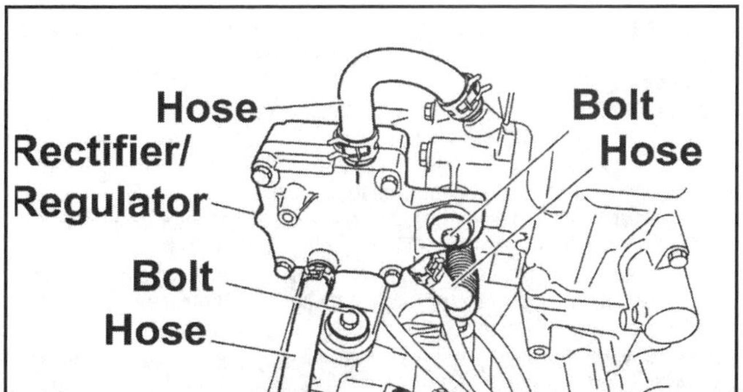

Fig. 102 A good shot of the rectifier/regulator on all models w/1812cc engines

Fig. 103 The rectifier/regulator on 2005-09 VX models is located next to the ECM on the side of the engine. . .

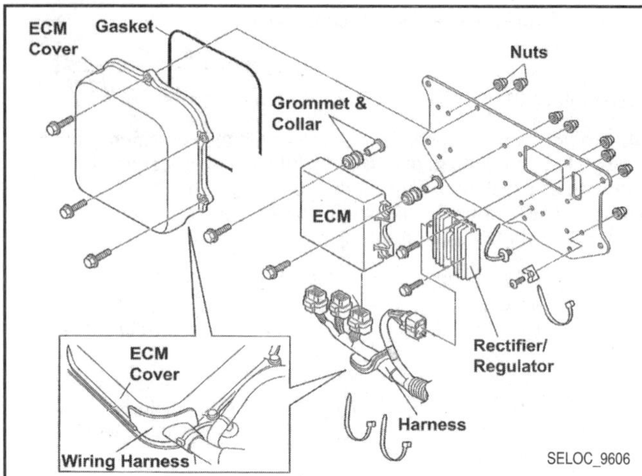

Fig. 104 . . .or behind the engine on 2010-11 VX models w/1052cc engines

To install:

4. Position the rectifier/regulator on the bracket and thread in the mounting bolts, tightening to 2.9 ft. lbs./34 inch lbs (4 Nm).
5. Connect the coupler at the bottom (or side) of the unit.
6. Connect the battery cables and install the seat.

Battery

■ **Additional information on the battery may be found in the Maintenance & Tune-Up section.**

The battery is one of the most important (and vulnerable) components of the electrical system. In addition to providing electrical power to start the engine, it also provides power for operation of the running lights, radio, and electrical accessories.

Because of its job and the potential consequences of a failure (especially in an emergency), the best advice is to purchase a well-known brand, with a sufficient warranty period, from a reputable dealer.

The usual warranty covers a pro-rated replacement policy, which means the purchaser would be entitled to a consideration for the time left on the warranty period if the battery should prove defective before its time.

Do not consider a battery of less than the rating provided in the General Engine System Specifications chart from the Maintenance & Tune-Up section. For most Yamaha motors, this means a battery of 70- amp/hour or 124-minute reserve capacity. But it never hurts to buy a battery of larger capacity. Especially as the boat and motor's electrical system ages, cables crack internally and resistance increases.

MARINE BATTERIES

◆ **See Figure 105**

Because marine batteries are required to perform under much more rigorous conditions than automotive batteries, they are constructed differently than those used in automobiles or trucks. Therefore, a marine battery should always be the No. 1 unit for the boat and other types of batteries used only in an emergency.

Marine batteries have a much heavier exterior case to withstand the violent pounding and shocks imposed on it as the boat moves through rough water and in extremely tight turns. The plates are thicker and each plate is securely anchored within the battery case to ensure extended life. The caps are spill proof to prevent acid from spilling into the bilges when the boat heels to one side in a tight turn, or is moving through rough water. Because of these features, the marine battery will recover from a low charge condition and give satisfactory service over a much longer period of time than any type intended for automotive use.

✻✻ WARNING

Avoid the use an automotive type batteries with an outboard unit. It is not built for the pounding or deep cycling to which this use will subject it and it may be quickly damaged.

Fig. 105 A fully charged battery, filled to the proper level with electrolyte, is the heart of the ignition and electrical systems. Engine cranking and efficient performance of electrical items depend on a full rated battery

BATTERY CONSTRUCTION

◆ **See Figure 106**

A battery consists of a number of positive and negative plates immersed in a solution of diluted sulfuric acid. The plates contain dissimilar active materials and are kept apart by separators. The plates are grouped into elements. Plate straps on top of each element connect all of the positive plates and all of the negative plates into groups.

The battery is divided into cells holding a number of the elements apart from the others. The entire arrangement is contained within a hard plastic case. The top is a 1-piece cover and contains the filler caps for each cell. The terminal posts protrude through the top where the battery connections for the boat are made. Each of the cells is connected to its neighbor in a positive-to-negative manner with a heavy strap called the cell connector.

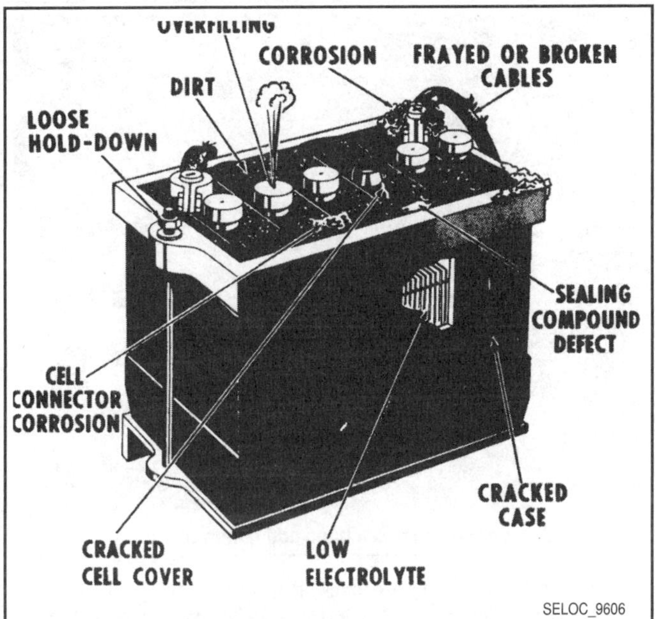

Fig. 106 A visual inspection of the battery should be made each time the boat is used. Such a quick check may reveal a potential problem in its early stages. A dead battery in a busy waterway or far from assistance could have serious consequences

BATTERY RATINGS

◆ **See Figure 107**

Typically, at least 3 different methods are used to measure and indicate battery electrical capacity:

- Amp/hour rating
- Cold cranking performance
- Reserve capacity

The amp/hour rating of a battery refers to the battery's ability to provide a set amount of amps for a given amount of time under test conditions at a constant temperature. Therefore, if the battery is capable of supplying 4 amps of current for 20 consecutive hours, the battery is rated as an 80 amp/hour battery. The amp/hour rating is useful for some service operations, such as slow charging or battery testing.

Cold cranking performance is measured by cooling a fully charged battery to 0°F (-17°C) and then testing it for 30 seconds to determine the maximum current flow. In this manner the cold cranking amp rating is the number of amps available to be drawn from the battery before the voltage drops below 7.2 volts.

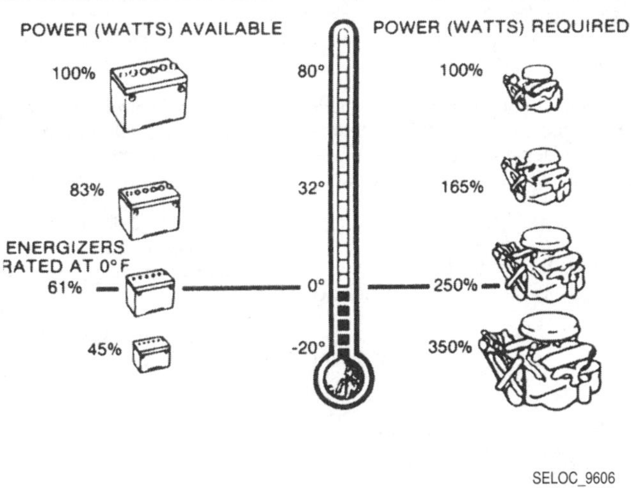

Fig. 107 Comparison of battery efficiency and engine demands at various temperatures

The illustration depicts the amount of power in watts available from a battery at different temperatures and the amount of power in watts required of the engine at the same temperature. It becomes quite obvious - the colder the climate, the more necessary for the battery to be fully charged.

Reserve capacity of a battery is considered the length of time, in minutes, at 80°F (27°C), a 25 amp current can be maintained before the voltage drops below 10.5 volts. This test is intended to provide an approximation of how long the engine, including electrical accessories, could operate satisfactorily if the stator assembly or lighting coil did not produce sufficient current. A typical rating is 100 minutes.

If possible, the new battery should always have a power rating equal to or higher than the original unit.

BATTERY LOCATION

◆ **See Figure 108**

Every battery must be secured in a well protected, ventilated area. If the battery area lacks adequate ventilation, hydrogen gas, which is given off during charging, is very explosive. This is especially true if the gas is concentrated and confined.

Remember, the vessel is going to take a pounding at some point and it is critical that the battery is secured so that wire terminals are not subjected to stresses which could break the cables or battery (and could provide for other dangerous conditions, including the potential for shorts, sparks and even an explosion of that hydrogen gas we just mentioned). There are many types of battery tray mounting assemblies available and you will normally find the perfect one for your application on the shelves of boat supply stores. Make sure the tie down bracket or strap is snug, but not over-tightened to the point of distorting or cracking the case.

BATTERY & CHARGING SAFETY PRECAUTIONS

◆ **See Figure 109**

Always follow these safety precautions when charging or handling a battery:

- Wear eye protection when working around batteries. Batteries contain corrosive acid and produce explosive gas as a byproduct of their operation. Acid on the skin should be neutralized with a solution of baking soda and water made into a paste. In case acid contacts the eyes, flush with clear water and seek medical attention immediately.
- Avoid flame or sparks that could ignite the hydrogen gas produced by the battery and cause an explosion. Connection and disconnection of cables to battery terminals is one of the most common potential causes of sparks. Use extreme caution with jumper cables and make sure all switches are **OFF** before making any other connections.

Fig. 108 It is imperative that the battery is secured properly

Fig. 109 Explosive hydrogen gas is released from the batteries in a discharged state. This one exploded when something ignited the gas. Explosions can be caused by a spark from the battery terminals or jumper cables

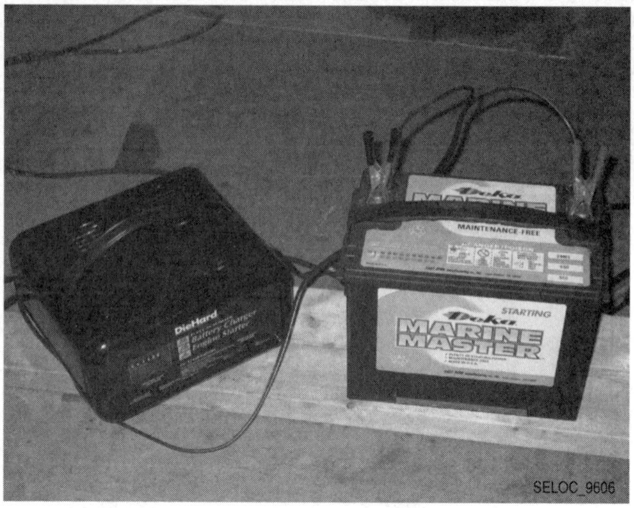

Fig. 110 A battery charger is a good addition to the tool kit!

- Always turn a battery charger **OFF**, before connecting or disconnecting the leads. When connecting the leads, connect the positive lead first, then the negative lead, to avoid sparks.
- When lifting a battery, use a battery carrier or lift at opposite corners of the base.
- Ensure there is good ventilation in a room where the battery is being charged.
- Do not attempt to charge or load-test a maintenance-free battery when the charge indicator dot is indicating insufficient electrolyte.
- Disconnect the negative battery cable if the battery is to remain in the boat during the charging process.
- Be sure the ignition switch is **OFF** before connecting or turning the charger **ON**. Sudden power surges can destroy electronic components.
- Use proper adapters to connect charger leads to batteries with non-conventional terminals.

BATTERY CHARGERS

◆ See Figure 110

Before using any battery charger, consult the manufacturer's instructions for its use. Battery chargers are electrical devices that change Alternating Current (AC) to a lower voltage of Direct Current (DC) that can be used to charge a marine battery. There are 2 types of battery chargers - manual and automatic.

A manual battery charger must be physically disconnected when the battery has come to a full charge. If not, the battery can be overcharged, and possibly fail. Excess charging current at the end of the charging cycle will heat the electrolyte, resulting in loss of water and active material, substantially reducing battery life.

■ As a rule, on manual chargers, when the ammeter on the charger registers half the rated amperage of the charger, the battery is fully charged. This can vary, and it is recommended to use a hydrometer to accurately measure state of charge.

Automatic battery chargers have an important advantage - they can be left connected (for instance, overnight) without the possibility of overcharging the battery. Automatic chargers are equipped with a sensing device to allow the battery charge to taper off to near zero as the battery becomes fully charged. When charging a low or completely discharged battery, the meter will read close to full rated output. If only partially discharged, the initial reading may be less than full rated output, as the charger responds to the condition of the battery. As the battery continues to charge, the sensing device monitors the state of charge and reduces the charging rate. As the rate of charge tapers to zero amps, the charger will continue to supply a few milliamps of current - just enough to maintain a charged condition.

✻✻ WARNING

Even with automatic storage chargers, don't assume the charger is working properly. When a battery is connected for any time longer than a day, check it frequently (at least daily for the first week and weekly thereafter). Feel the battery case for excessive heat, listen and look for gassing, check the electrolyte levels.

REPLACING BATTERY CABLES

Battery cables don't go bad very often, but like anything else, they can wear out. If the cables on your boat are cracked, frayed or broken, they should be replaced.

When working on any electrical component, it is always a good idea to disconnect the negative (-) battery cable. This will prevent potential damage to many sensitive electrical components.

Always replace the battery cables with one of the same length, or you will increase resistance and possibly cause hard starting. Smear the battery posts with a light film of dielectric grease, or a battery terminal protectant spray once you've installed the new cables. If you replace the cables one at a time, you won't mix them up.

■ Any time you disconnect the battery cables, it is recommended that you disconnect the negative (-) battery cable first. This will prevent you from accidentally grounding the positive (+) terminal when disconnecting it, thereby preventing damage to the electrical system.

Before you disconnect the cable(s), first turn the ignition to the **OFF** position. This will prevent a draw on the battery which could cause arcing. When the battery cable(s) are reconnected (negative cable last), be sure to check all electrical accessories are all working correctly.

STARTING SYSTEM

Description & Operation

◆ See Figures 111, 112 and 113

The system utilizes an electric starting motor coupled with a mechanical gear mesh between the motor and the engine crankshaft, similar in some ways to the method used to crank an automobile engine. On these engines though, the starter shaft is connected to an idle gear which meshes with a starter gear and clutch mounted on the crankshaft behind the flywheel.

As the name implies, the sole purpose of the starting motor circuit is to control operation of the starter motor to turn or "crank" the engine until it is operating. The circuit includes a solenoid or magnetic switch to connect or disconnect the motor from the battery. The operator controls the solenoid with a key switch or button.

On models with Reverse, a neutral safety switch might be installed into the circuit to permit operation of the cranking motor only when the shift control lever is in Neutral. This switch is a safety device to prevent accidental engine start when the engine is in gear.

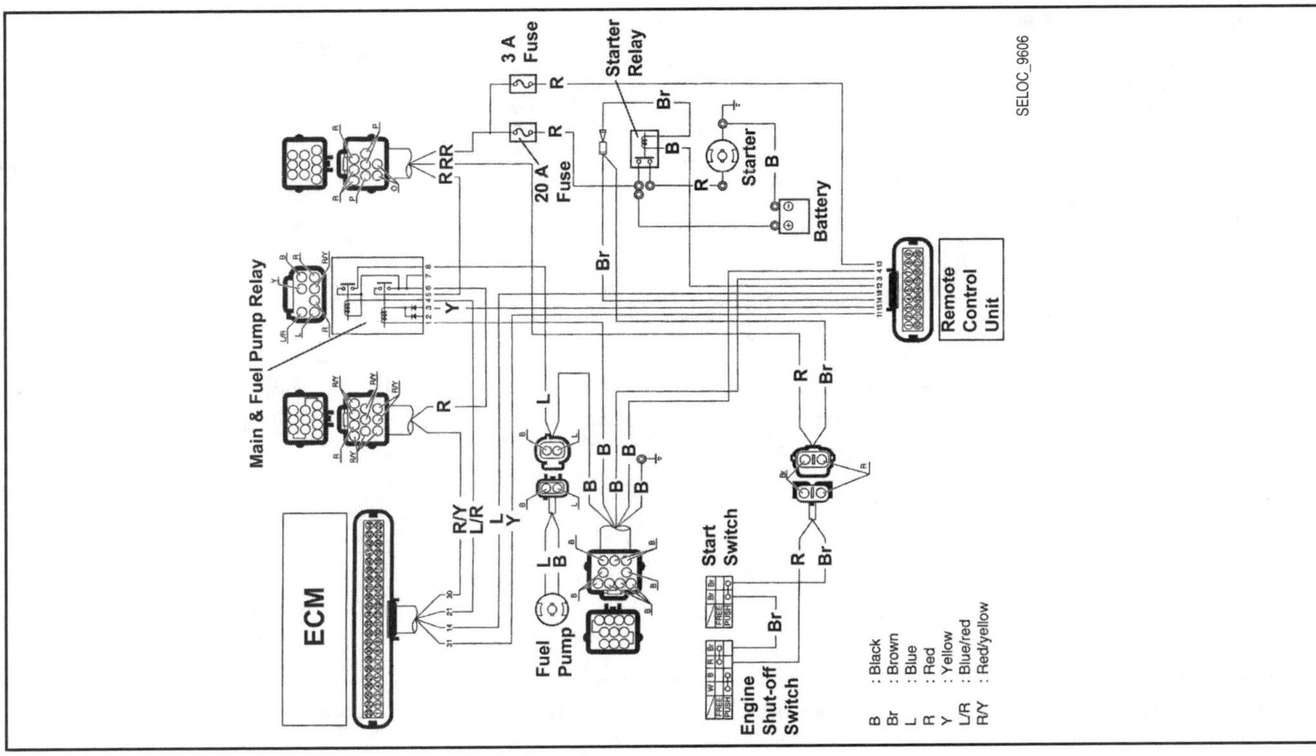

Fig. 112 Starting system schematic - 2004-07 FX models w/1052cc engines

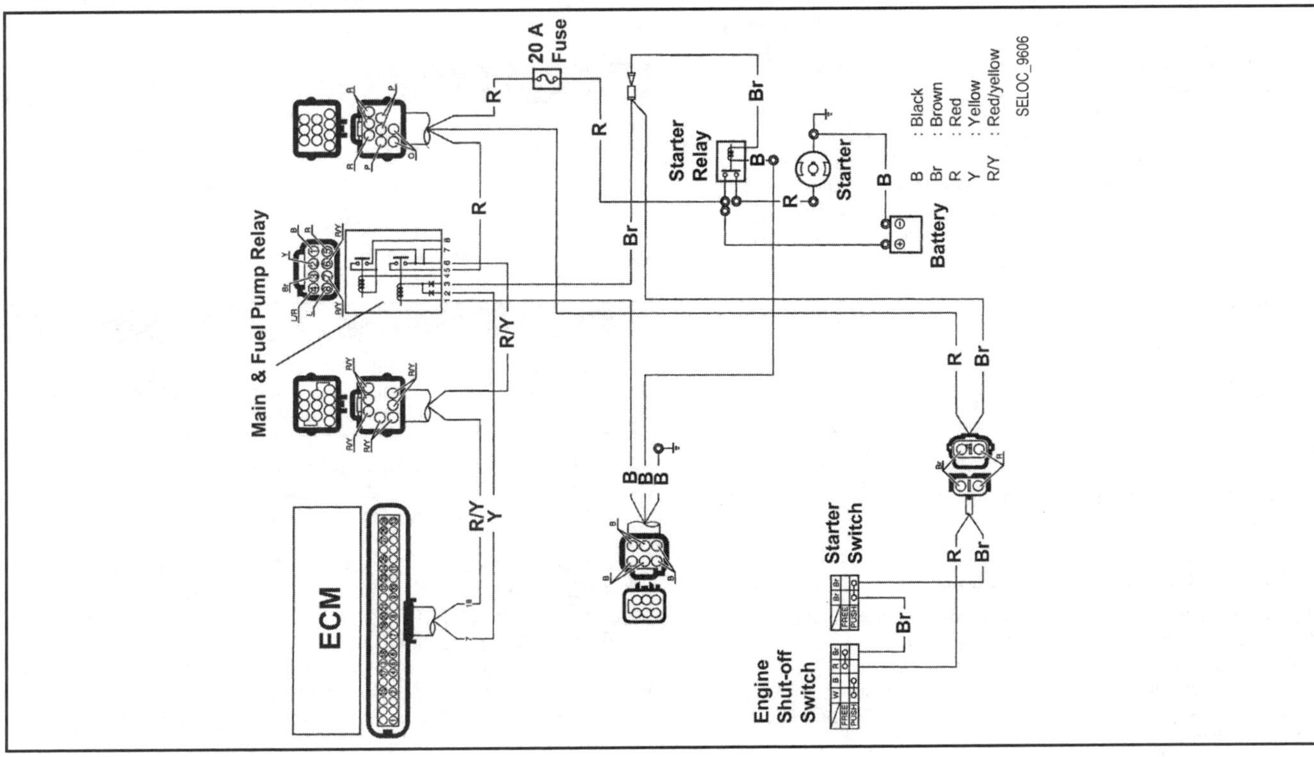

Fig. 111 Starting system schematic - 998cc Engines

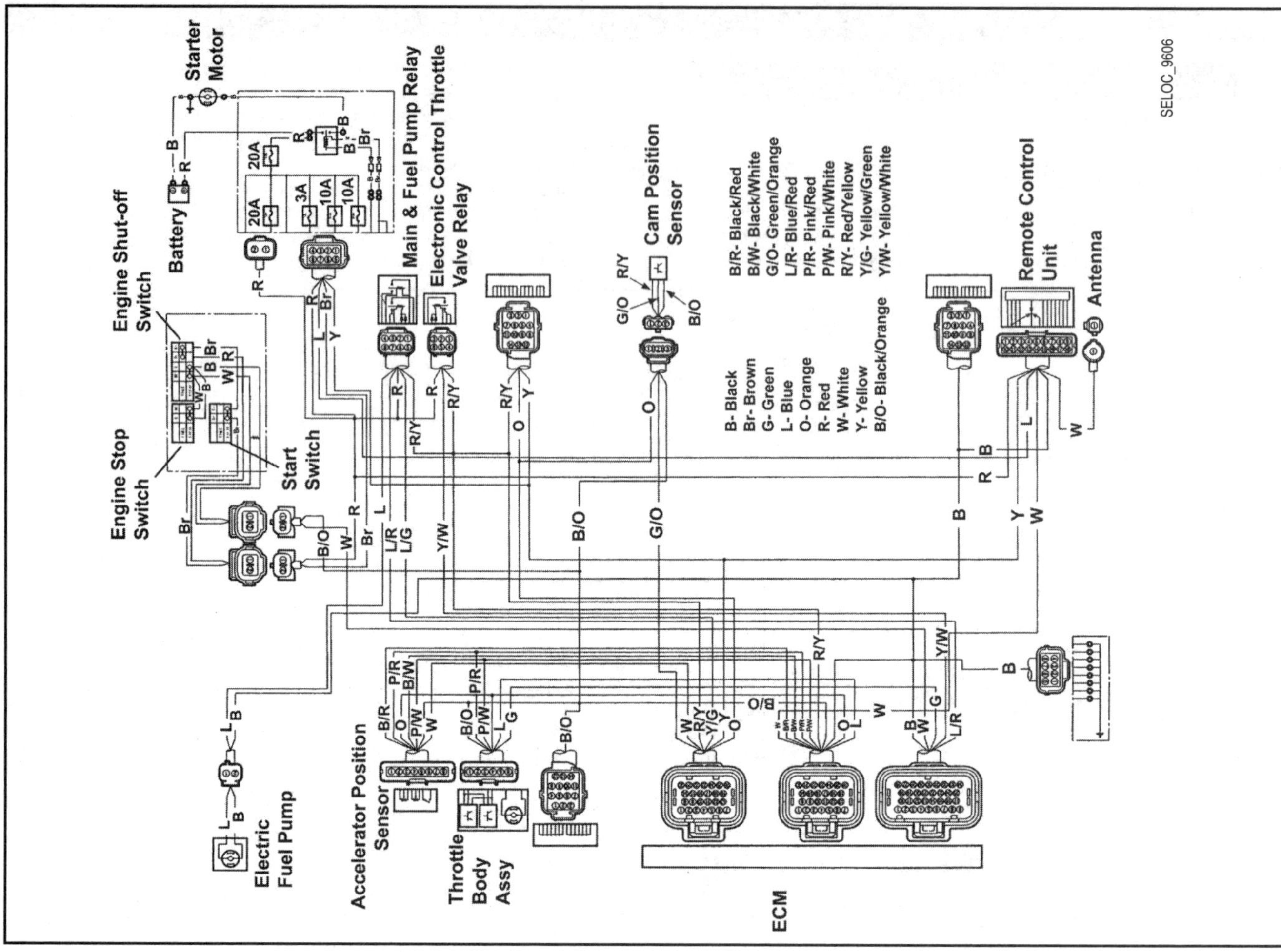

Fig. 113 Starting system schematic - VX models w/1052cc engines

The starter motor is a series-wound electric motor which draws a heavy current from the battery. It is designed to be used only for short periods of time to crank the engine for starting. To prevent overheating the motor, cranking should not be continued for more than 30 seconds without allowing the motor to cool for at least 3 minutes. Actually, this time can be spent in making preliminary checks to determine why the engine fails to start (such as checking the safety lanyard, the fuel supply, the fuel line primer bulb, etc).

The sound of the motor during cranking is a good indication of whether the starter motor is operating properly or not. Naturally, temperature conditions will affect the speed at which the motor is able to crank the engine. The speed when cranking a cold engine will be much slower than when the same starter and battery are used to crank a warm engine. An experienced operator will learn to recognize the favorable sounds of the powerhead cranking under various conditions.

MAINTENANCE

Generally speaking, the starter motor does not require periodic maintenance or lubrication. If the motor fails to perform properly, the checks outlined in the previous paragraph should be performed. The frequency of starts governs how often the motor should be removed and reconditioned. On late-model motors Yamaha usually recommends removal and reconditioning (or replacement) about every 1000 hours. Naturally, the motor will have to be replaced if corrective actions do not restore the motor to satisfactory operation.

FAULTY SYMPTOMS

If the starter motor cranks the engine too slowly, the following are possible causes and the corrective actions that may be taken:

- Battery charge is low. Charge the battery to full capacity.
- High resistance connections at the battery, solenoid, or motor. Clean and tighten all connections (a loose or bad ground/battery negative cable is often the culprit here).
- Undersize battery cables (creating too much resistance). Replace cables with sufficient size.
- Battery cables too long (creating too much resistance). Relocate the battery to shorten the run to the solenoid.

Starter Relay

TESTING

MODERATE

◆ See Figures 114, 115 and 116

** WARNING

The following test must be conducted with the relay wiring harness disconnected (to prevent accidental starting and to prevent potential damage to the meter) and the relay removed.

1. Remove the relay from the electrical box.
2. Hook up a DVOM as per the manufacturer's instructions.
3. Connect the meter leads to the relay terminals as shown in the illustration.
4. Use a jumper wire and connect the ring terminal on the Brown lead to the positive (+) battery terminal. Connect the Black lead terminal to the negative (-) battery terminal.
5. The meter should show continuity between the two relay terminals.
6. Now disconnect one of the jumper wires at the battery (doesn't matter which one) and check that there is no continuity now.
7. Failure to meet either of these conditions will call for relay replacement.

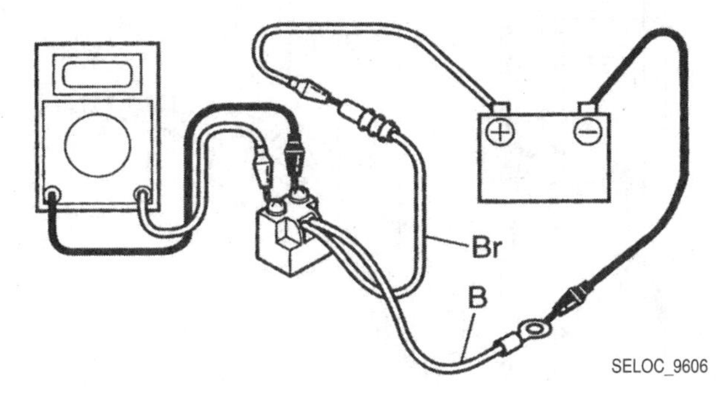

Fig. 114 Testing the starter relay

REMOVAL & INSTALLATION

FX Models With 998cc/1052cc Engines

◆ See Figure 115

■ **The wire colors provided in this procedure are true for most years and models, but if yours differs, check the wiring diagrams in this section. In some cases wires may not agree with either the procedure or the diagram as all or part of a harness may have been replaced previously. In these cases, trace the wires to determine which wire fits each function.**

The starter relay is contained within the electrical components box.

1. Remove the seat, disconnect the battery cables and then remove the electrical box as detailed elsewhere in this section.
2. Once the box is free, lean it out and remove the 4 terminal cover screws. Lift off the cover and gasket. This is on the back of the box, directly under the fuse cap.
3. Carefully disconnect the starter and battery leads, tag them and move them out of the way. Generally speaking, the starter cable is on the left (as you face the 2 terminals) and the battery cable is on the left.
4. Working inside the box now, find the relay (left side, next to the ECM) and disconnect the Brown and Black relay leads, making sure that you tag them if planning to reuse.
5. Grab the relay and pop it out of the mount.

To install:

6. Press the relay back into position in the holder.
7. Reconnect the 2 relay leads.
8. Position the battery and starter lead ring terminals on the relay and then tighten each screw to 5.5 ft. lbs./76 inch lbs. (7.6 Nm). Coat the terminal liberally with Yamaha Marine grease (or equivalent).
9. Position the gasket and terminal cover over the recess and then tighten the screws to 3.5 ft. lbs./43 inch lbs. (4.9 Nm). We highly suggest using a new gasket here, but it is not always necessary if the old one looks to be in good shape.

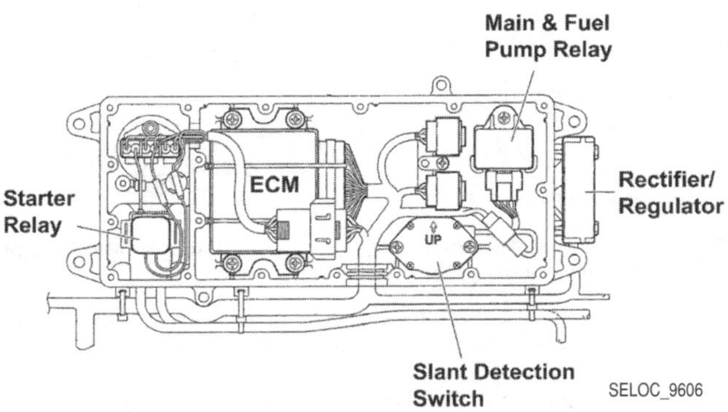

Fig. 115 The relay is located in the electrical box - FX models (1812cc models similar)

10. Reinstall the box cover with a new gasket and then install the assembly. Make sure you get all 18 of those cover screws!
11. Connect the battery cables and install the seat.

All Models w/1812cc Engines

◆ See Figure 115

■ **The wire colors provided in this procedure are true for most years and models, but if yours differs, check the wiring diagrams in this section. In some cases wires may not agree with either the procedure or the diagram as all or part of a harness may have been replaced previously. In these cases, trace the wires to determine which wire fits each function.**

The starter relay is contained in the fuse box within the electrical components box.

1. Remove the seat, disconnect the battery cables and then remove the electrical box cover as detailed elsewhere in this section.
2. Remove the fuse box as detailed elsewhere as well.
3. Working in the open side of the fuse box, pull back the lead boots and carefully disconnect the starter and battery leads, tag them and move them out of the way. Generally speaking, the starter cable is on the bottom and will have a piece of white tape around the casing near the ring terminal. The battery cable is on the top and will have a piece of gray tape around it.
4. Working inside the box now, find the relay (right side looking into the open end) and disconnect the Brown and Black relay leads, making sure that you tag them if planning to reuse.
5. Grab the relay and pop it out of the mount.

To install:

6. Press the relay back into position in the holder.
7. Reconnect the 2 relay leads.
8. Position the battery and starter lead ring terminals on the relay and then tighten each screw to 3 ft. lbs./36 inch lbs. (4 Nm). Coat the terminal liberally with Yamaha Marine grease (or equivalent). Press the boots back into position.
9. Reinstall the fuse box and electrical box cover with a new gasket.
10. Connect the battery cables and install the seat.

VX Models w/1052cc Engines

◆ See Figure 116

■ **The wire colors provided in this procedure are true for most years and models, but if yours differs, check the wiring diagrams in this section. In some cases wires may not agree with either the procedure or the diagram as all or part of a harness may have been replaced previously. In these cases, trace the wires to determine which wire fits each function.**

Fig. 116 The relay is located in the electrical box, VX w/1052cc engines shown

The starter relay is contained within the electrical components/fuse box.

1. Remove the seat, disconnect the battery cables and then remove the electrical box as detailed later in this section.
2. Once the box is free, lean it out and remove the 6 fuse box cover screws. Lift off the cover and gasket. This is on the back of the box.
3. Working inside the box now, find the relay (left side) and disconnect the Brown and Black relay leads, making sure that you tag them if planning to reuse.
4. Grab the relay and pop it out of the grommet mount.

To install:

5. Press the relay back into position in the grommet.
6. Reconnect the 2 relay leads.
7. Position the gasket and box cover over the recess and then tighten the screws securely. We highly suggest using a new gasket here, but it is not always necessary if the old one looks to be in good shape.
8. Reinstall the fuse box.
9. Connect the battery cables and install the seat.

Start Switch

TESTING

2 MODERATE

◆ **See Figure 117**

✳✳ WARNING

The following test must be conducted with the relay wiring harness disconnected (to prevent accidental starting and to prevent potential damage to the meter) and the relay removed.

1. Follow the switch harness down from the handlebar and disconnect the harness coupler.
2. Hook up a DVOM as per the manufacturer's instructions.
3. Connect the meter leads to the relay terminals as shown in the illustration. If no continuity is found, replace the switch

Starter Motor

DESCRIPTION

◆ **See Figure 118**

The starter motor used on all Yamaha motors is of a pretty standard design. The housing/mounting and exterior appearance of the motor itself will vary slightly from model-to-model and with it there may be subtle differences in the servicing procedures. Therefore, the accompanying illustrations for the removal and installation procedures may differ slightly from the unit being serviced, but the procedures and maintenance instructions are valid. We've attempted to put as many exploded views as possible for overhaul and you should be able to find a detailed match to your motor if you decide to take it apart for inspection, cleaning and/or component replacement.

Marine starter motors are very similar in construction and operation to the units used in the automotive industry with the exception that they don't generally have the small Bendix gear on the end of the armature shaft.

■ **The exception to the exception is the 1812cc engine, which *does* utilize a more conventional auto-style gear on the end of the shaft.**

Instead, the splined shaft mates/meshes with a small idle gear once it has been inserted through the mounting bore. This idle gear actually has an even smaller gear attached to it which in turn meshes with the actual starter gear connected to the end of the crankshaft. A starter clutch is also used between the gear and the rotor.

As with most starter motors, the housing is not designed to provide airflow which would be necessary for continuous or repeated operation. Therefore, never operate a cranking motor for more than 30 seconds without allowing it to cool for at least 3 minutes. Continuous operation without the cooling period can cause serious damage to or destroy the cranking motor.

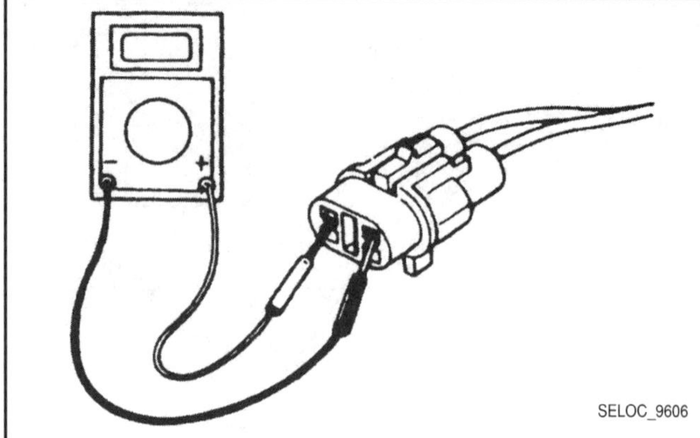

Fig. 117 Testing the starter switch

Fig. 118 The starter motor shaft mates with the idle gear, which in turn drives the starter gear. The starter clutch (not visible here) is found between the rotor and the starter gear (1812cc engines similar)

REMOVAL & INSTALLATION

All Except 1812cc Engines

3 DIFFICULT

◆ **See Figures 119 and 120**

■ **Unfortunately, starter motor removal will require removing the entire motor, and probably the exhaust manifold as well.**

1. Remove the engine assembly as detailed in the Engine Mechanical section. Remove any exhaust system components that may block access to the starter motor as well.
2. The starter is mounted to the leading edge of the engine, on the starboard side. Remove the rubber boot and then tag and disconnect the starter lead (should be Red) from the stud on top of the motor case. The starter lead will use a nut and washer to secure the ring terminal.

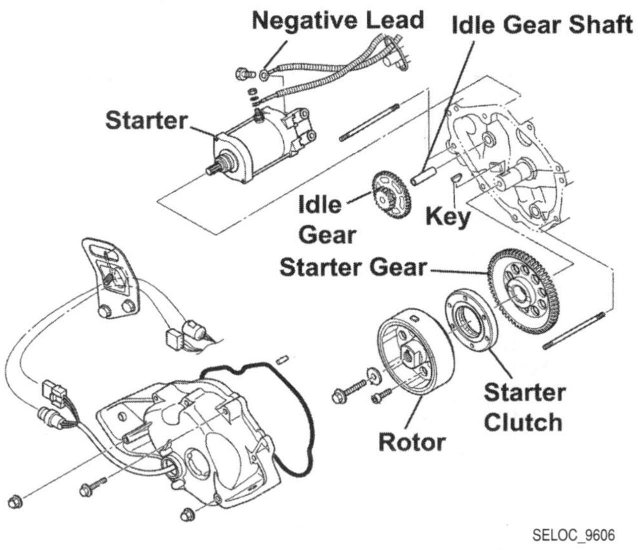

Fig. 119 The starter is mounted to the block on the starboard side

Fig. 120 A good shot of the starter motor, the mounting bolts are just barely visible behind the coiled starter leads

3. Loosen the 2 starter mounting bolts and then carefully remove the starter motor from the cylinder block. Note that the battery ground lead will be connected to one of the mounting bolts (usually the top one). Note also that the mounting bolts are at the *back* of the starter - unlike automotive starters where the mounting bolts connect the shaft end of the starter directly through a flange on the end of the motor.

To install:

4. Coat the armature shaft splines and mating surface with a good all-purpose marine grease and then feed the starter through the shaft hole so that the splines engage the teeth of the idle gear. You may have to swivel the starter body slightly to get the splines and teeth to mesh.

5. The mounting bolt holes should be lined up with the holes in the block, but you can also move the starter housing slightly as well. Thread in the bolts and tighten to 13 ft. lbs. (18 Nm). Make sure you slid the battery cable ring terminal over the upper bolt before threading it in. You may have to hold the lead while tightening the bolt as it will have a tendency to 'follow' the bolt as it is being turned. Slide the rubber boot over the end of the lead and bolt.

6. Slide the starter lead ring over the stud on top of the housing, followed by the washer and nut. Tighten the nut to 3.5 ft. lbs./43 inch lbs. (4.9 Nm) - you'll probably have to hold the lead from spinning when tightening. Coat the ring terminal and nut liberally with a good marine all-purpose grease and then slide the rubber boot back into position.

7. Install the engine and any other components removed for access.

1812cc Engines

③ DIFFICULT

◆ See Figure 121

■ **Unfortunately, starter motor removal will require removing the entire exhaust manifold and muffler.**

1. Remove the exhaust manifold and muffler as detailed in the Engine Mechanical section.

2. The starter is mounted to the port side of the engine, at the rear. Remove the rubber boot and then tag and disconnect the starter lead (should be Red) from the stud on top of the motor case. The starter lead will use a nut and washer to secure the ring terminal.

3. Loosen the 2 starter mounting bolts and then carefully remove the starter motor from the cylinder block. Note that the battery ground lead will be connected to one of the mounting bolts (usually the bottom one). Note also that the mounting bolts are at the *back* of the starter - unlike automotive starters where the mounting bolts connect the shaft end of the starter directly through a flange on the end of the motor.

To install:

4. Coat front cover O-ring with a good all-purpose marine grease (Yamaha Grease A) and then feed the starter through the shaft hole so that the teeth of the drive fear engage the teeth of the idle gear inside the housing. You may have to swivel the starter body slightly to get the splines and teeth to mesh.

■ **Why not just replace the O-ring while you're at it??**

5. The mounting bolt holes should be lined up with the holes in the block, but you can also move the starter housing slightly as well. Thread in the bolts and tighten to 13.3 ft. lbs. (18 Nm). Make sure you slid the battery cable ring terminal over the lower bolt before threading it in. You may have to hold the lead while tightening the bolt as it will have a tendency to 'follow' the bolt as it is being turned. Slide the rubber boot over the end of the lead and bolt.

6. Slide the starter lead ring over the stud on top of the housing, followed by the washer and nut. Tighten the nut to 3.7 ft. lbs./47 inch lbs. (5 Nm) - you'll probably have to hold the lead from spinning when tightening. Coat the ring terminal and nut liberally with a good marine all-purpose grease and then slide the rubber boot back into position.

7. Install the exhaust assembly and any other components removed for access.

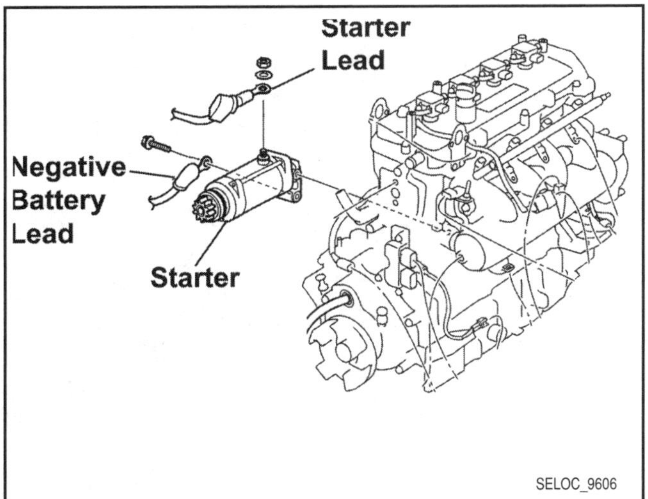

Fig. 121 The starter is mounted to the block on the port side

STARTER MOTOR EXPLODED VIEWS

◆ See Figures 122 thru 125

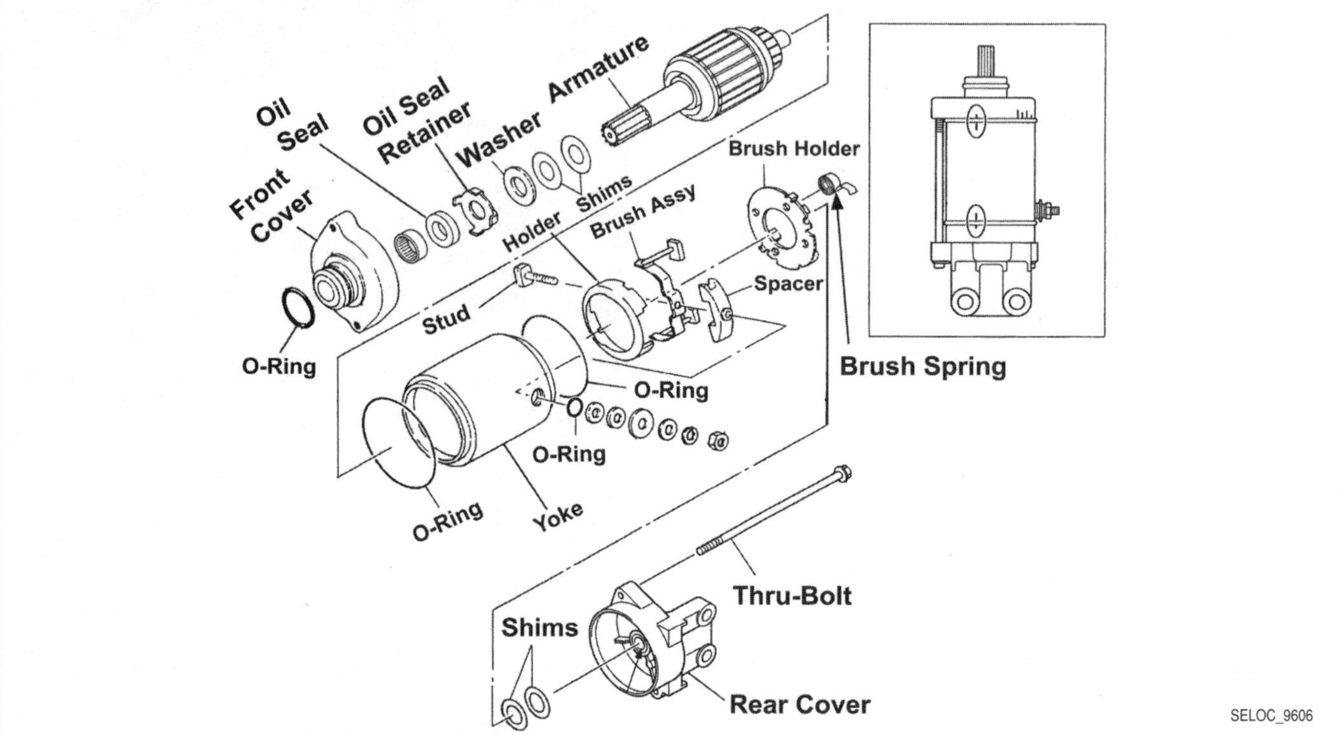

Fig. 122 Exploded view of the starter assembly - 2002-08 FX/FX Cruiser and FX140/FX140 Cruiser models w/998cc engine

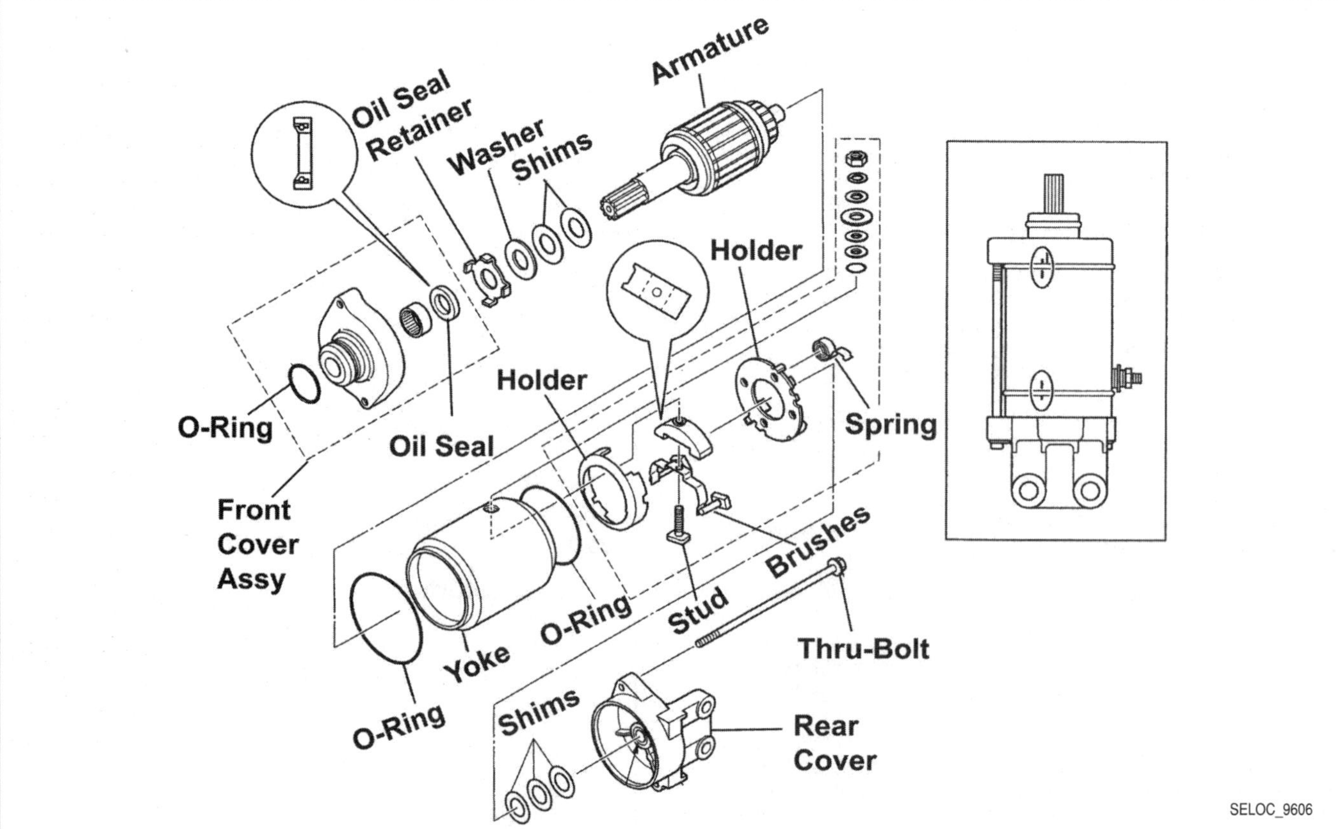

Fig. 123 Exploded view of the starter assembly - 2004-08 FX HO /FX Cruiser HO models w/1052cc engines and all VX models w/1052cc engines

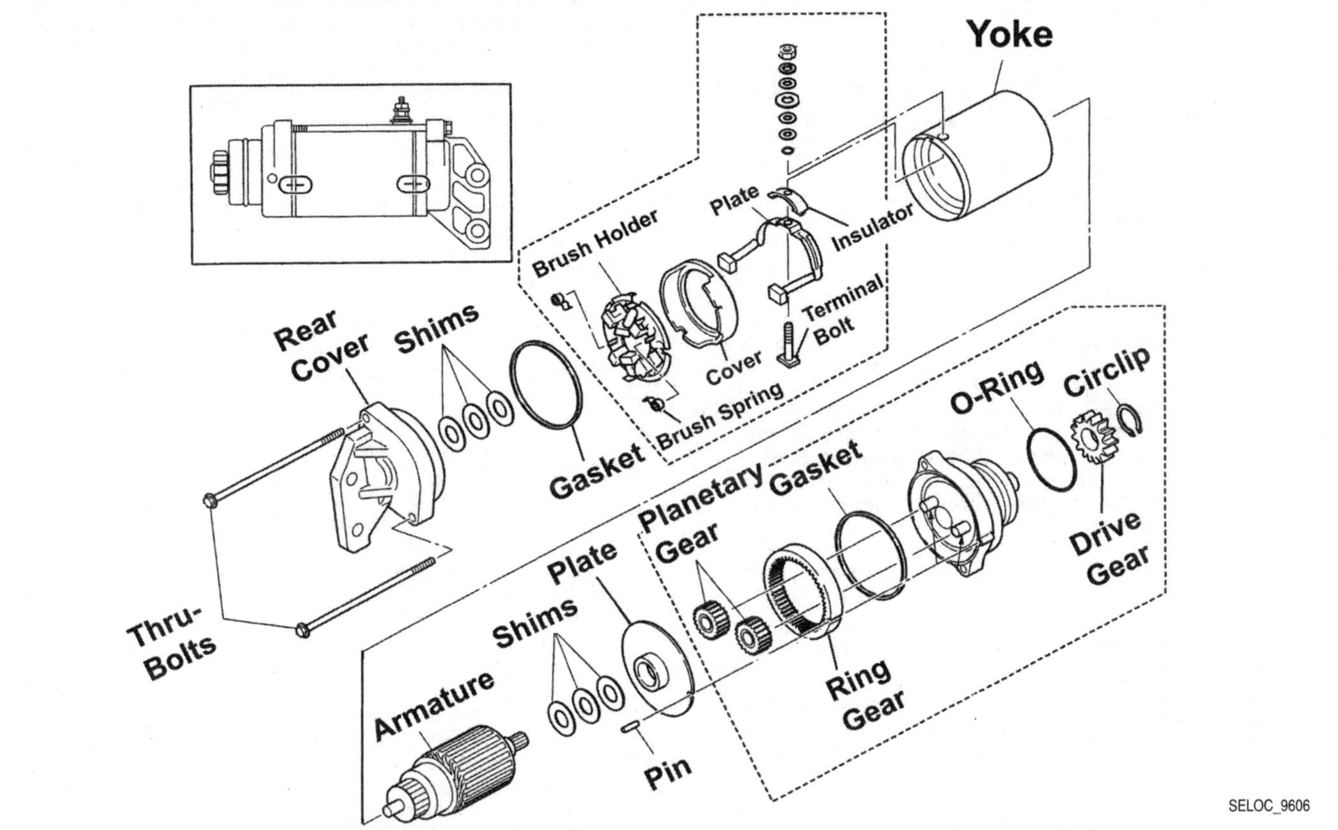

Fig. 124 Exploded view of the starter assembly - FX/FZ models w/1812cc engines

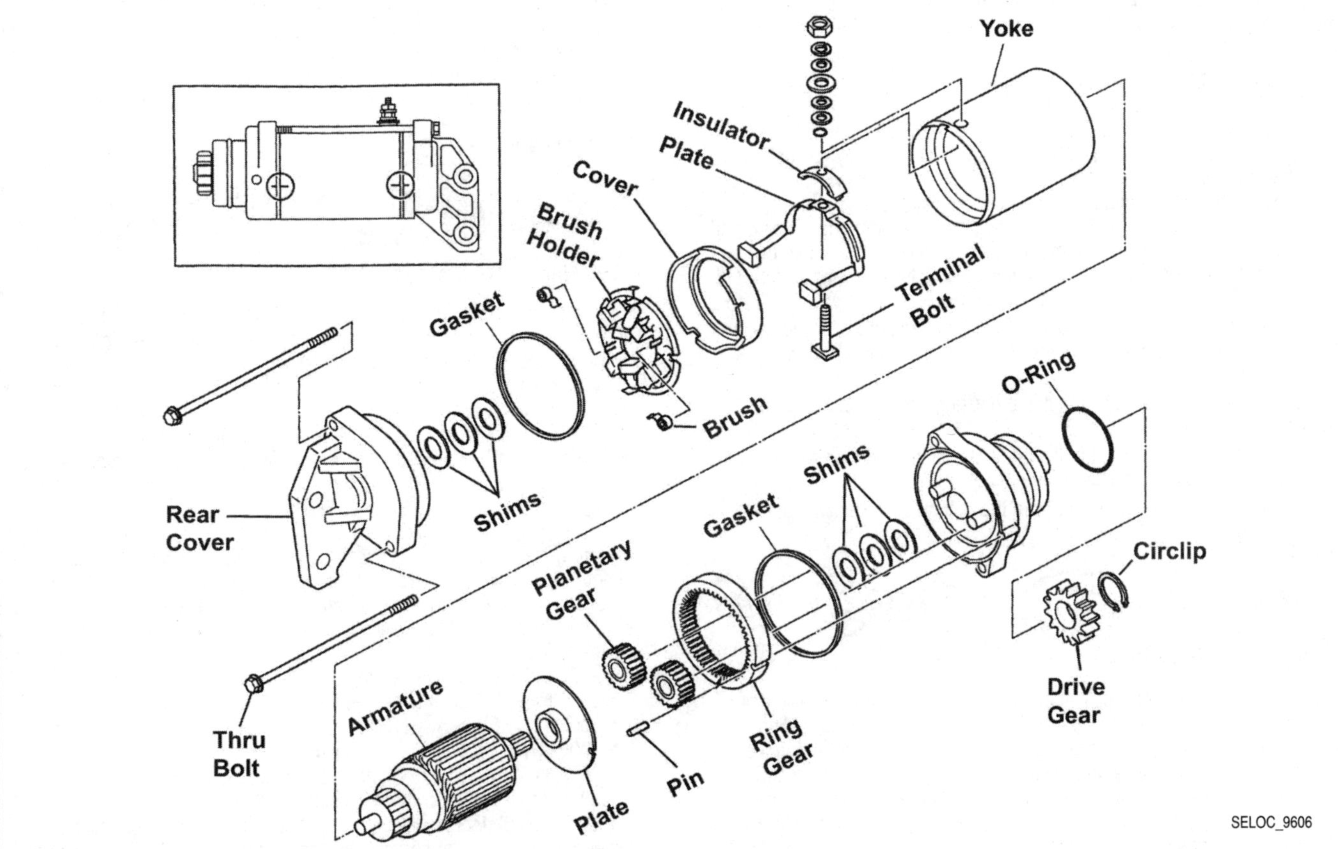

Fig. 125 Exploded view of the starter assembly - VX models w/1812cc engines

DISASSEMBLY

All Exc. 1812cc Engines

◆ See Figures 122 and 123

The wide availability of rebuilt starters on the marine parts market today dissuades most technicians and many DIYers from actually overhauling their starter; but, rebuild kits are normally available. If you decide to undertake this, keep close track of all starter components as you disassemble it. Digital cameras are relatively cheap these days, so if you've got access to one, work slowly taking pics as you go. They can be a great reference when it comes time for assembly. Be sure to keep track of all washers and shims so that they are placed in the same positions upon assembly.

■ **Yamaha reuses what is essentially the same starter design assembly on multiple engines. Besides small differences in the top or bottom housing, differences will occur with the thickness of shims used between armature washers from year-to-year or model-to-model. For this reason, it is critical that you keep track of all washers and shims for installation in the same positions.**

1. Remove the starter motor as detailed in the Removal & Installation section.
2. Scribe matchmarks between the upper/lower (front/rear) covers and the starter housing (also sometimes known as the armature housing or yoke). This will ensure easy alignment during assembly.
3. Remove the 2 long thru-bolts connecting the front and back covers and then remove the front cover. You may have to 'persuade' the cover with a small rubber mallet, so be very careful if you resort to this. Remove the O-ring on the outside of the cover and the one inside the cover - throw them both away.
4. Turn the cover over and remove the oil seal and its retainer.
5. Reach into the housing and slide the large washer and any shims up and off of the armature shaft.
6. Remove the rear cover. You may have to 'persuade' the cover with a small rubber mallet, so be very careful if you resort to this. Remove the O-ring and throw it away.
7. Reach into the rear cover and remove the shim(s) where the end of the armature rests. Keep them in order.
8. Hold the armature housing (some people call this a starter frame or yoke) grab the splined end of the armature shaft and carefully pull the assembly out of the housing. Keep in mind that the magnets which make up a large part of the starter motor are quite strong. Once the thru-bolts and covers are removed it can be difficult to separate the field frame/housing and the armature assembly. It is even more important to keep this in mind when assembling the starter as it is easy to badly pinch a finger during that part of the process. So take your time and watch the piggies.
9. Remove the starter lead stud nut and then remove the spring washer along with the 4 washers. Remove the small O-ring from the stud hole and discard it.
10. Remove the terminal bolt from the housing and then slide out the brush assembly.
11. Pull the brush holder off of the brush assembly and then remove the spacer from the brush assembly.
12. Remove the 4 brush springs from the bottom of the brush plate.

1812cc Engines

◆ See Figures 124 and 125

The wide availability of rebuilt starters on the marine parts market today dissuades most technicians and many DIYers from actually overhauling their starter; but, rebuild kits are normally available. If you decide to undertake this, keep close track of all starter components as you disassemble it. Digital cameras are relatively cheap these days, so if you've got access to one, work slowly taking pics as you go. They can be a great reference when it comes time for assembly. Be sure to keep track of all washers and shims so that they are placed in the same positions upon assembly.

■ **Yamaha reuses what is essentially the same starter design assembly on multiple engines/models. Besides small differences in the top or bottom housing, differences may occur with the thickness of shims used between armature washers from year-to-year or model-to-model. For this reason, it is critical that you keep track of all washers and shims for installation in the same positions.**

1. Remove the starter motor as detailed in the Removal & Installation section.
2. Scribe matchmarks between the front/rear covers and the starter housing (also sometimes known as the armature housing or yoke). This will ensure easy alignment during assembly.
3. Remove the circlip at the end of the armature shaft and then work the drive gear off of the shaft.
4. Remove the 2 long thru-bolts connecting the front and rear covers and then remove the front cover. You may have to 'persuade' the cover with a small rubber mallet, so be very careful if you resort to this. Remove the O-ring on the outside of the cover and the gasket/seal inside the cover - throw them both away.
5. Turn the cover over and remove the 2 small planetary gears along with the ring gear. Pull out the dowel pin.
6. Reach into the housing and slide the large plate and any shims up and off of the armature shaft. The shims may have stayed with the cover, either way, keep them together and in order.
7. Remove the rear cover. You may have to 'persuade' the cover with a small rubber mallet, so be very careful if you resort to this. Remove the inner gasket/seal and throw it away.
8. Reach into the rear cover and remove the shim(s) where the end of the armature rests. Keep them together and in order.
9. Hold the armature housing/yoke, grab the splined end of the armature shaft and carefully pull the assembly out of the housing. Keep in mind that the magnets which make up a large part of the starter motor are quite strong. Once the thru-bolts and covers are removed it can be difficult to separate the housing and the armature assembly. It is even more important to keep this in mind when assembling the starter as it is easy to badly pinch a finger during that part of the process. So take your time and watch the piggies.
10. Remove the starter lead stud nut and then remove the spring washer along with the 4 washers. Remove the small O-ring from the stud hole and discard it.
11. Remove the terminal bolt from the housing and then slide out the lead plate assembly and insulator.
12. Pull the brush holder out of the cover assembly.
13. Remove the 4 brush springs from the bottom of the brush holder.

CLEANING & INSPECTION

◆ See Figures 126 thru 133

Once disassembled, carefully clean all metal components with a mild solvent and either blow dry with compressed air or allow them to air dry. Thoroughly inspect all components as follows:

1. Check the armature shaft for nicks, scratches, wear or other damage, replacing as necessary.
2. A dirty commutator may be cleaned using #600 grit sandpaper, but should be thoroughly cleaned afterwards using solvent and compressed air.
3. Be sure to inspect the housing and end caps for signs of wear or damage such as cracks.
4. Using a sliding caliper, measure the outer diameter of the commutator (armature) and, if specs make sure it is not smaller than the minimum diameter as follows: standard - 1.10 in. (28mm), limit - 1.06 in. (27mm)
5. Carefully exam the mica undercut (depth of the grooves) on the end of the commutator. Build specifications vary greatly, however service limits are generally:
   - 2002-007 FX - 0.01 in. (0.2mm)
   - 2008 FX w/1052cc engines - 0.006-0.028 in. (0.15-0.7mm)
   - 2008-10-11 FX/FZ w/1812cc engines - 0.008-0.028 in. (0.2-0.7mm)
   - 2005-09 VX - 0.006-0.028 in. (0.15-0.7mm)
   - 2010-11 VX w/1052cc engines - 0.008-0.028 in. (0.2-0.7mm)
   - 2011 VX w/1812cc engines - 0.008-0.028 in. (0.2-0.7mm)

   If the undercut depth is less than this spec, remove some metal from between the commutator segments using a hacksaw blade, then remove all metal and mica particles using compressed air (and safety goggles!).

Specifications, where available, are by model are as follows:

6. Inspect the brush lengths (height of each brush from top to bottom, if the point where the wire attaches is considered the side). Again, build specifications vary greatly, however minimum brush length is 0.26 in. (6.5mm). If you are unsure as to whether or not the brushes are serviceable, check with a local parts supplier to see if you can measure the height of a replacement set (or just spring for the replacements, since you've come this far). Specifications, where available, are by model as follows:

- All exc. 1812cc engines - standard 0.49 in. (12.5mm), limit 0.26 in. (6.5mm)
- 1812cc engines - standard 0.49 in. (12.5mm), limit 0.20 in. (5mm)

7. Use an ohmmeter to inspect armature continuity. First check for continuity between each of the segments on the commutator, there should be continuity. Next, check between the commutator segments and the armature core (the thick main shaft of the armature), there should be NO continuity. Finally check between the commutator segments and the

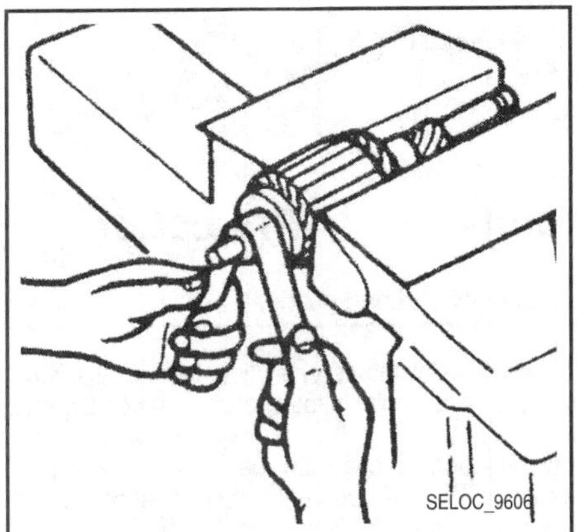

Fig. 126 If necessary, clean the commutator with sandpaper

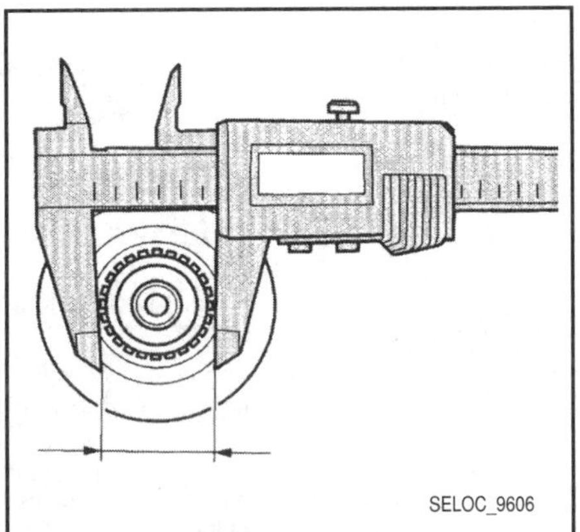

Fig. 127 Use a sliding caliper to check the commutator outer-diameter

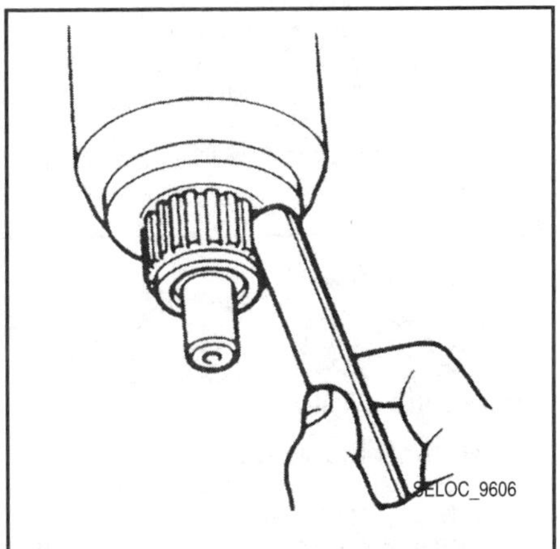

Fig. 128 Check the mica undercut depth

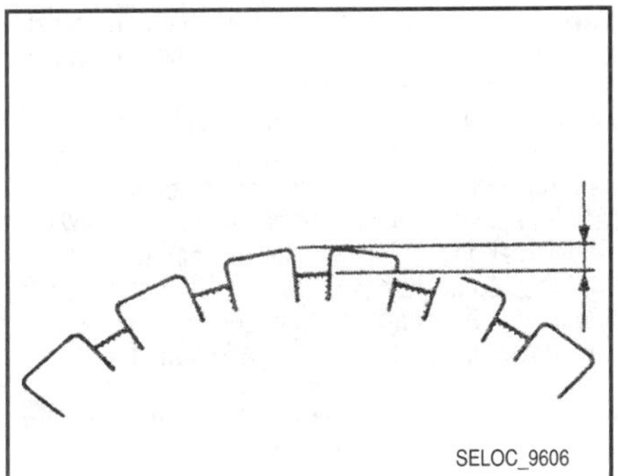

Fig. 129 If necessary, increase undercut depth using a hacksaw blade or file

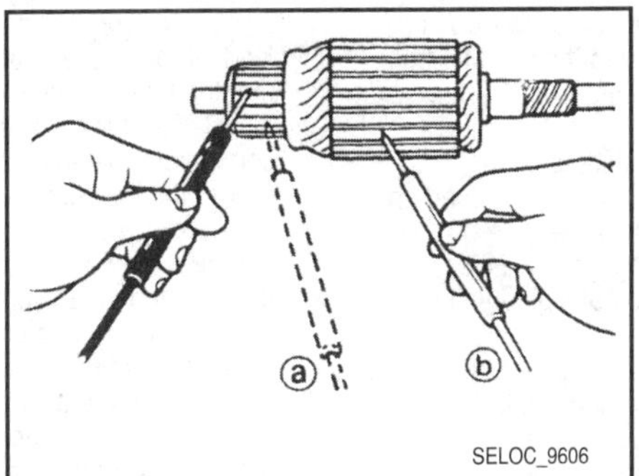

Fig. 130 Use an ohmmeter to check for continuity between the commutator segments (a), the segments and the armature core (b) and lastly, between the segments and armature shaft (c)

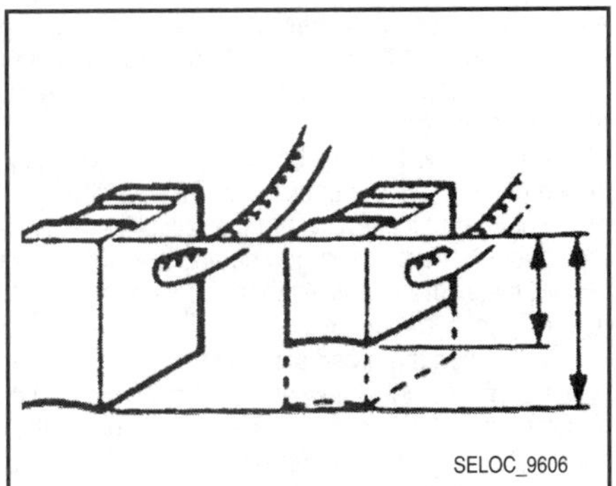

Fig. 131 Brushes wear out, so check the length (A)

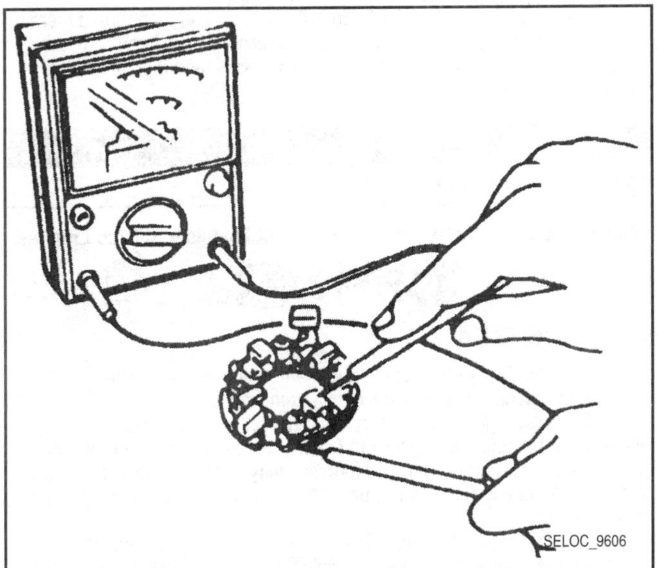

Fig. 132 Checking brush holder continuity - all exc. 1812cc engines

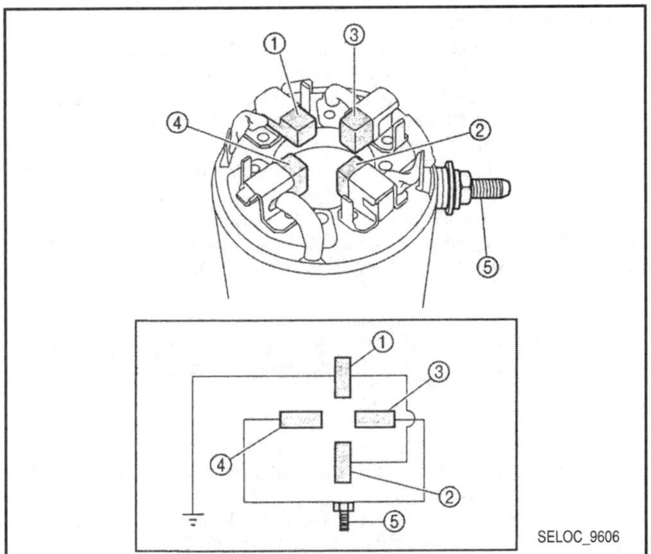

Fig. 133 Checking brush holder continuity - 1812cc engines

armature shaft (the long, thin shaft at the other end of the armature from the commutator segments), again there should be NO continuity. Replace the armature if it fails this armature continuity test.

8. Use an ohmmeter to inspect brush set continuity. There should be NO continuity between the brush holder and the base plate on all engines exc. the 1812cc. On the 1812cc engines check as follows:

- Brush (1) to brush (2) - continuity
- Brush (3) to brush (4) - continuity
- Brush (3) to terminal (5) - continuity
- Brush (4) to terminal (5) - continuity
- there should be *no continuity* when checking any other combinations (for instance: brush (1) to brush (3) or brush (2) to terminal (5). . .no continuity

Replace any components which do not appear to be serviceable.

ASSEMBLY

All Exc. 1812cc Engines

◆ See Figures 122 and 123

1. Grab the armature assembly by the shaft and then carefully slide it into the housing from the top side (forward when installed) toward the bottom (back) side, but not all the way.

2. Line up the brush assembly so the terminal bolt hole matches the larger opening in the brush holder and press it into position.

3. Now line up the hole in the spacer with the bolt hole in the brush assembly and carefully press it down into position so the holes line up and the curved side is facing outward. Install the assembly into the housing from the bottom/back side so that the holes you've already aligned match the stud hole in the housing.

4. Install the terminal bolt through the assembly making sure that it is not contacting any of the brushes. Press a *new* O-ring over the exposed end of the bolt and all the way into the recess in the housing. Now slide on 2 of the small washers, the large one and then the remaining small washer. Finally, install the spring washer and then thread on the nut, tightening to 6.4 ft. lbs./77 inch lbs. (8.8 Nm).

5. Position the brush plate over the rear housing opening, feed the brush leads through the openings in the plate and then slide in into the housing so that the tabs line up with the slots in the brush holder.

6. Carefully slide the armature further down into the housing until the commutator passes through the assembly. Now you can push the brushes down into their slots on the brush plate and then install the 4 springs.

7. Install the 2 shims back into position in the rear cover. Install a *new* O-ring over the back of the housing and roll it into its recess. Coat the end of the armature shaft lightly with a good all-purpose marine grease and then press the cover into position so that the shaft feeds into the bushing inside the cover and then rotate the cover so that the marks made during disassembly are aligned.

8. At the other end now, slide the 2 shims and the washer over the armature shaft and work them down the shaft until fully seated.

9. Coat the needle bearings (in the cover) and the oil seal with a good marine grease and then press the oil seal and its retainer into the cover.

10. Position another *new* O-ring onto the housing and then guide the cover over the armature shaft and press it fully onto the housing, rotating it until the marks line up.

11. Slide the thru-bolts through the bottom cover and then thread them into the top cover, tightening to 4.6 ft. lbs./56 inch lbs. (6.4 Nm) an all models but the 2010 VX which is 4.4 ft. lbs./52 inch lbs. (6 Nm).

1812cc Engines

◆ See Figures 124, 125 and 134

1. Grab the armature assembly by the shaft and then carefully slide it into the housing from the top side (forward when installed) toward the bottom (back) side, but not all the way.

2. Line up the brush assembly so the terminal bolt hole matches the larger opening in the brush holder and press it into position.

3. Now line up the hole in the insulator with the bolt hole in the brush assembly and carefully press it down into position so the holes line up and the curved side is facing outward. Install the assembly into the housing from the bottom/back side so that the holes you've already aligned match the stud hole in the housing.

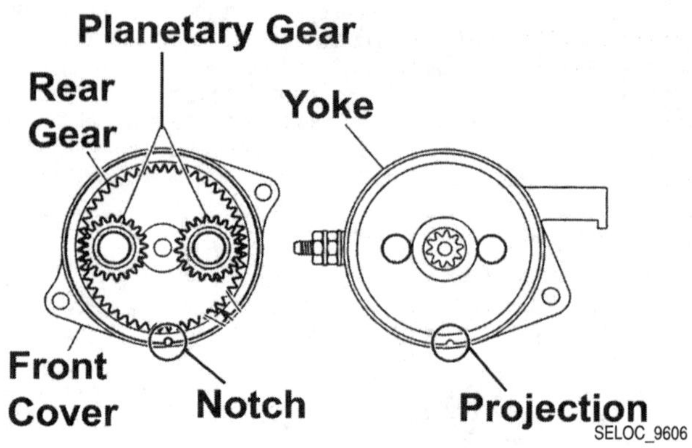

Fig. 134 Installing the planetary and ring gears

4. Install the terminal bolt through the assembly making sure that it is not contacting any of the brushes. Press a *new* O-ring over the exposed end of the bolt and all the way into the recess in the housing. Now slide on 2 of the small washers, the large one and then the remaining small washer. Finally, install the spring washer and then thread on the nut, tightening to 6.6 ft. lbs./79 inch lbs. (9 Nm).

5. Position the lead plate over the rear housing opening, feed the brush leads through the openings in the plate and then slide in into the housing so that the tabs line up with the slots in the brush holder.

6. Carefully slide the armature further down into the housing until the commutator passes through the assembly. Now you can push the brushes down into their slots on the brush plate and then install the 4 springs.

7. Install the shims back into position in the rear cover. Install a *new* O-ring over the back of the housing and roll it into its recess. Coat the end of the armature shaft lightly with a good all-purpose marine grease and then press the cover into position so that the shaft feeds into the bushing inside the cover and then rotate the cover so that the marks made during disassembly are aligned.

8. At the other end now, slide the shims and the plate over the armature shaft and work them down the shaft until fully seated.

9. Coat the planetary gears and ring gear with a good marine grease and then install the large ring gear. Press the 2 small gears onto their studs in the cover so that they mesh with the ring gear.

10. Position another *new* O-ring onto the housing and then guide the cover over the armature shaft and press it fully onto the housing, rotating it until the marks line up. Make sure that the notch on the ring gear lines up with the small projection on the edge of the yoke.

11. Slide the thru-bolts through the bottom cover and then thread them into the top cover, tightening to 5.2 ft. lbs./56 inch lbs. (6.4 Nm).

12. Press the drive gear onto the end of the shaft and then install the circlip.

Starter Clutch, Starter Gear & Idle Gear

REMOVAL & INSTALLATION

FX Models With 998cc/1052cc Engines & VX Models w/1052cc Engines

◆ See Figures 68 thru 73

1. Remove the flywheel cover and flywheel as detailed in the appropriate procedures in the Ignition System section.

2. The starter gear may or may not have come off with the flywheel/starter clutch assembly, but either way remove the woodruff key from the slot in the crankshaft and throw it away - you can't re-use it. If still on the crankshaft, carefully pull it off - it may require a little finesse, but do not use a mallet.

3. The starter clutch is actually attached to the flywheel. Hold the assembly and remove the 6 bolts attaching the starter clutch to the back of the flywheel.

4. Slide the idle gear off of the shaft.

To install:

5. Position the idle gear over the shaft and slide it on so that the inner teeth mesh with starter armature teeth.

6. If you removed the starter clutch, position it on the back of the rotor so the holes line up and the roller is facing the starter gear. Install the 6 bolts finger tight, making sure everything lines up; and then tighten the bolts, in a star pattern, to 17 ft. lbs. (24 Nm).

7. Clean the tapered end of the crankshaft thoroughly, position the NEW woodruff key and then slide the starter gear into position, followed by the rotor assembly. Coat the NEW mounting bolt and washer with clean engine oil and thread it into the crankshaft. Tighten the bolt to 54 ft. lbs. (75 Nm) while holding the flywheel with the holding tool.

8. Install the generator cover.

All Models w/1812cc Engines

OEM 3 DIFFICULT

◆ **See Figures 75 thru 87**

1. Remove the flywheel cover and flywheel as detailed in the appropriate procedures in the Ignition System section.

2. The starter gear may or may not have come off with the flywheel/starter clutch assembly, but either way remove the woodruff key from the slot in the crankshaft and throw it away - you can't re-use it. If still on the crankshaft, carefully pull it off - it may require a little finesse, but do not use a mallet.

3. The starter clutch is actually attached to the flywheel and you have probably already separated it. If not, hold the assembly and remove the 6 bolts attaching the starter clutch to the back of the flywheel.

4. Slide the idle gear off of the shaft.

To install:

5. Position the idle gear over the shaft and slide it on so that the inner teeth mesh with starter drive gear teeth.

6. If you removed the starter clutch, position it on the back of the rotor so the holes line up and the roller is facing the starter gear. Install the 6 bolts finger tight, making sure everything lines up; and then tighten the bolts, in a star pattern, to 17 ft. lbs. (24 Nm).

7. Clean the tapered end of the crankshaft thoroughly, position the NEW woodruff key and then slide the starter gear into position, followed by the flywheel/rotor assembly.

8. Install the flywheel and cover.

ELECTRICAL SWITCH/SOLENOID SERVICE

This short section provides testing procedures for other electrical parts installed on the engine or instrument panel (the later ASSUMING that the craft was rigged with the expected Yamaha original parts). If a unit fails the testing, the faulty part must be replaced. In most cases, removal and installation is through attaching hardware.

Sensor & Circuit Resistance/Output Tests

TESTING ELECTRONIC COMPONENTS

3 DIFFICULT

◆ **See Figure 135**

Before conducting resistance or voltage checks on a component/circuit, make sure all electrical connections in the system are clean and tight. Visually check the wiring for obvious breaks, defects or other problems.

Keep in mind that although a haphazard approach might find the cause of problems, only a systematic approach will prevent wasted time and the possibility of unnecessary component replacement. In some cases, installing an electronic component into a faulty circuit that damaged or destroyed the previous component, will instantly destroy the replacement. For various reasons, including this possibility, most parts suppliers do not accept returns on electrical components.

Make sure you perform each test procedure and especially, make all test connections as described. If a component tests out of range (faulty), but the reading is close to the service limit, bring the component to your marine dealer or parts supplier so they can verify your result before purchasing the replacement. In some cases, your dealer may be willing to check a sensor against the reading on a new one to verify whether or not your sensor is actually faulty.

Although resistance checks are relatively easy to conduct, their results can be difficult to interpret. Remember that resistance in any electrical component or circuit will vary with temperature. Also, ohmmeters will vary with quality, meaning that readings can vary on the same component between different meters. The specifications provided here are based upon the use of a high-quality digital multi-meter applied to a component that is currently at about 68°F (20°) unless otherwise noted.

■ **More information on electrical test equipment is can be found in the General Information section, while additional information on basic electrical theory and electrical troubleshooting can be found elsewhere in this section.**

As a general rule, the resistance of a circuit or component will increase (rise) as temperature increases. Although this is true for most resistor and windings, some manufacturers use negative temperature coefficient sensors for certain functions. A negative temperature coefficient sensor reacts in an opposite manner, meaning that its resistance decreases (lowers) as temperature increases (rises) and vice-versa.

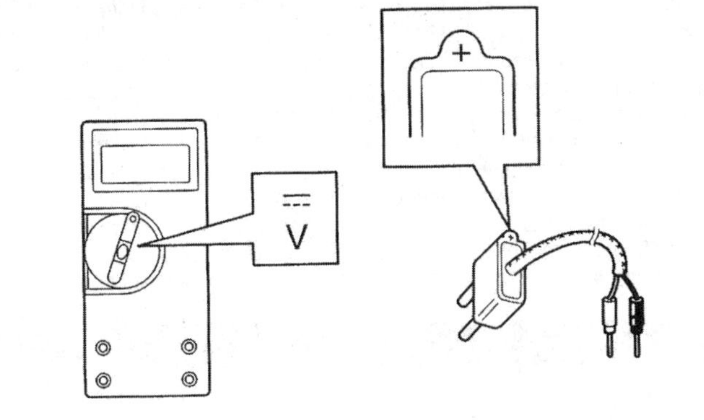

Fig. 135 Yamaha recommends a high quality DVOM and a peak reading voltage adaptor for testing on their motors

Most components of the EFI systems are checked using resistance or voltage checks. This means that a voltmeter or ohmmeter is connected to the component or circuit under the proper circumstances. Yamaha recommends the use of a high quality Digital Volt Ohm Meter (DVOM) such as #YU-34899-A or equivalent and a Peak Voltage Adaptor such as #YU-39991.

For resistance checks, no voltage must be applied to the circuit (except that as provided by the meter to perform the check). Typically, the circuit or component wiring is isolated, by disconnecting the wiring from the component itself or somewhere else in the circuit, then applying the meter probes across 2 terminals that connect through the component. When conducting resistance checks, you *must* isolate the ECM from the meter (unless specifically directed otherwise) in order to prevent the possibility of damage to the control module. By identifying the proper circuit you can conduct tests of the entire component circuit right from the harness connectors that attach to the ECM. Otherwise, use the wiring diagrams and the test procedures to determine other points to disconnect the wiring and test the circuit/component.

We recommend you work in 1 of 2 directions. Either test the component first, then work your way back through the component wiring toward the ECM, or disconnect the harness connectors from the ECM, testing the entire circuit first, then working your way toward the component. A bad test reading at the ECM harness connector must be verified by making sure it is not the result of a bad wiring harness or connection along the way to the component. Likewise, a good reading at a component (when there is a bad reading on the circuit as noted by the presence of a trouble code or by a reading at the ECM harness) must be verified by checking the circuit wiring for trouble.

When checking the circuit wiring, isolate the harness by disconnecting it from both the component and the ECM, then using an ohmmeter to check resistance from 1 connector to the next on each wire. Wiggle the wire and connectors while conducting the test to see if readings fluctuate. Although no specifications are provided for wiring resistance, all sensor/switch wiring for these motors should have very little resistance. Also, be sure to check wiring for shorts to ground (by checking resistance between 1 terminal and a good engine ground), and for shorts to power (by checking a terminal using a voltmeter and connecting the other probe to a good engine ground or the negative battery cable).

**** WARNING**

Be VERY careful when checking circuits for electronic engine control components. Always isolate the ECM by disconnecting the wiring from it before using an ohmmeter on the wires. Keep in mind that making the wrong connections across the terminals of an ECM can fry the unit, requiring an expensive replacement.

When conducting voltage tests, the circuit must be complete in order to receive a reading. In most cases, this means the wiring must be left connected and the meter probes should be carefully inserted through the back of the connector (called back-probing) in order to get a reading without opening and disrupting circuit operation. However, back-probing should NOT be done on waterproof connectors as you will compromise the seal, and jumper wires should instead be used to complete the circuit. The best method for manufacturing a jumper wire is to obtain 2 identical size/type connectors as the portion of the harness to which you will be inserting the jumper wires. Wire these jumper connectors together along with a second wire connected to EACH terminal. Each second wire from a terminal should route off to a female bullet connector which can be safely probed without the fear of accidental shorting by connecting 2 or more wires together.

Engine Stop Switch

TESTING

② MODERATE

◆ **See Figure 136**

The kill switch is normally located on the front left handlebar assembly. It may be called the left handle bar switch assembly elsewhere in this manual.

Trace the kill switch button harness containing 2 wires from the switch to their nearest coupler. The colors may vary for different models, but it shouldn't make a difference, because it is only a 2 wire switch and your tests will be with 1 meter lead attached to each wire. For more details, refer to the wiring diagrams in this section but they will usually be Black and White.

Disengage the 2 leads, then connect an ohmmeter or DVOM set to read resistance across the disconnected leads.

Verify the emergency tether is in place behind the kill switch button. Check the meter, with the button released and the tether installed there should be NO continuity (continuity would shut down the powerhead if the switch was connected and the powerhead was operating).

Now, depress the kill button. The meter should indicate continuity.

Release the button (making sure the tether is still in place). The meter should now indicate NO continuity.

Remove the emergency tether, and the meter should again indicate continuity.

All tests must be successful. If the switch fails any one test, the switch is defective and must be replaced. The switch is a 1 piece sealed unit and cannot be serviced.

Slant Detection Switch

TESTING

① EASY

◆ **See Figures 137 and 138**

The slant detection switch can be found *in* (FX/FZ and VX w/1812cc engine) or attached *to* (VX w/1052cc engine) the electrical box.

Connect a DVOM as detailed by the manufacturer.

Remove the switch and connect the positive meter lead to the Black/Orange terminal and the negative lead to the Blue/Black terminal.

■ **For some reason, on models w/1812cc engines, Yamaha shows the terminals reversed from what is shown for the other models and they also suggest connecting the positive lead to the Blue/Black terminal and the negative to the Black/Orange. Seems odd to us, but that is what they show so we suggest so go with the connections as shown in the illustration.**

Holding the switch in its normal upright position should indicate no continuity on all FX, FZ and 2010-11 VX models and 4.18-5.10 kili-ohms on 2005-09 VX models. Turning the switch to either side, or upside down should show continuity.

■ **Make sure that you test the switch in all 4 positions as shown.**

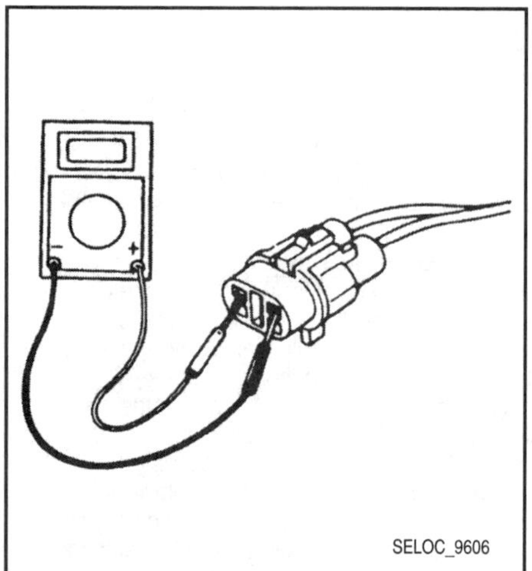

Fig. 136 Using an ohmmeter to test the engine stop switch with the emergency tether in place

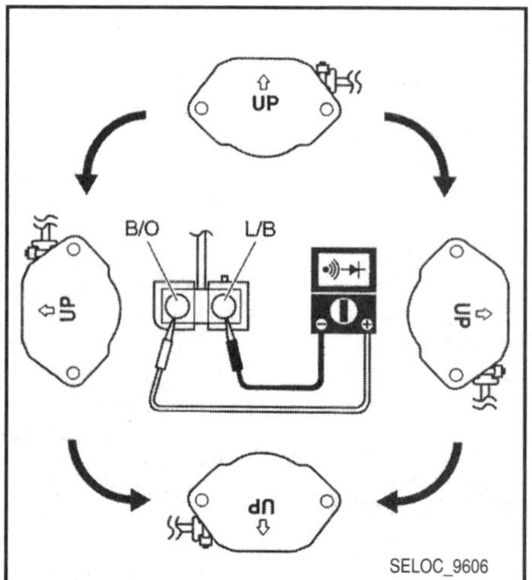

Fig. 137 Testing the slant detection switch - all except 1812cc engines

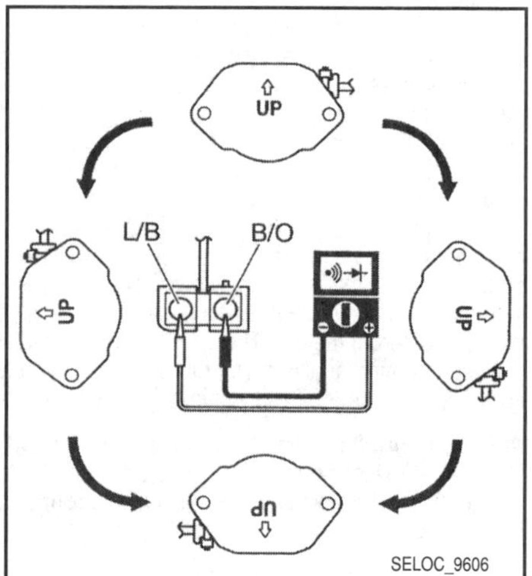

Fig. 138 Testing the slant detection switch - 1812cc engines

Warning Buzzer/Horn

TESTING

◆ See Figure 139

The buzzer or horn is a warning device to indicate low oil pressure, an over rev. condition, or overheating, etc. The buzzer is usually located under the handlebar console on the upper left side.

Remove the control box cover and identify the 2 leads from the buzzer, one is normally Red and the other is normally Black. Of course, it shouldn't make a difference, because it is only a 2 wire component and your tests will be with 1 meter lead attached to each wire. For more details, refer to the Wiring Diagrams section.

Obtain a 12-volt battery. Momentarily connect the buzzer leads to the battery (generally the Black buzzer lead would connect to the negative battery terminal, while the Red buzzer lead would connect to the positive terminal). As soon as the leads are connected, the buzzer should sound. If the buzzer is silent, or the sound emitted does not capture your attention immediately, the buzzer should be replaced. Service or adjustment is not possible.

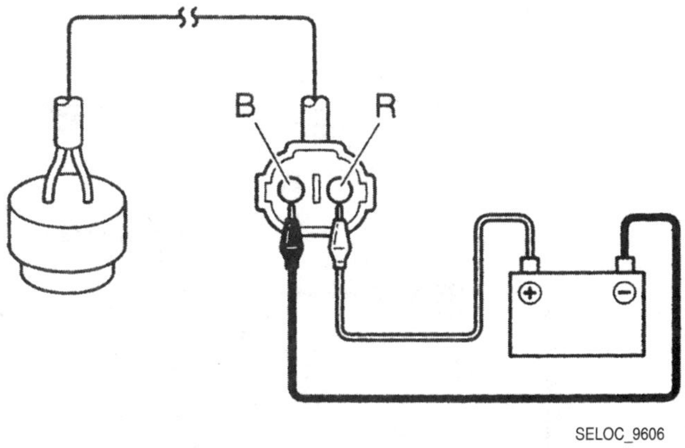

Fig. 139 The warning buzzer/horn is tested by applying 12 volts and listening for activation

COMPONENT LOCATIONS

■ **Please refer to Component Locations at the end of the Fuel System section.**

WIRING DIAGRAMS

The following wiring diagrams represent the most common models of outboard engines covered in this manual. Models with accessory options may not be depicted here.

Index

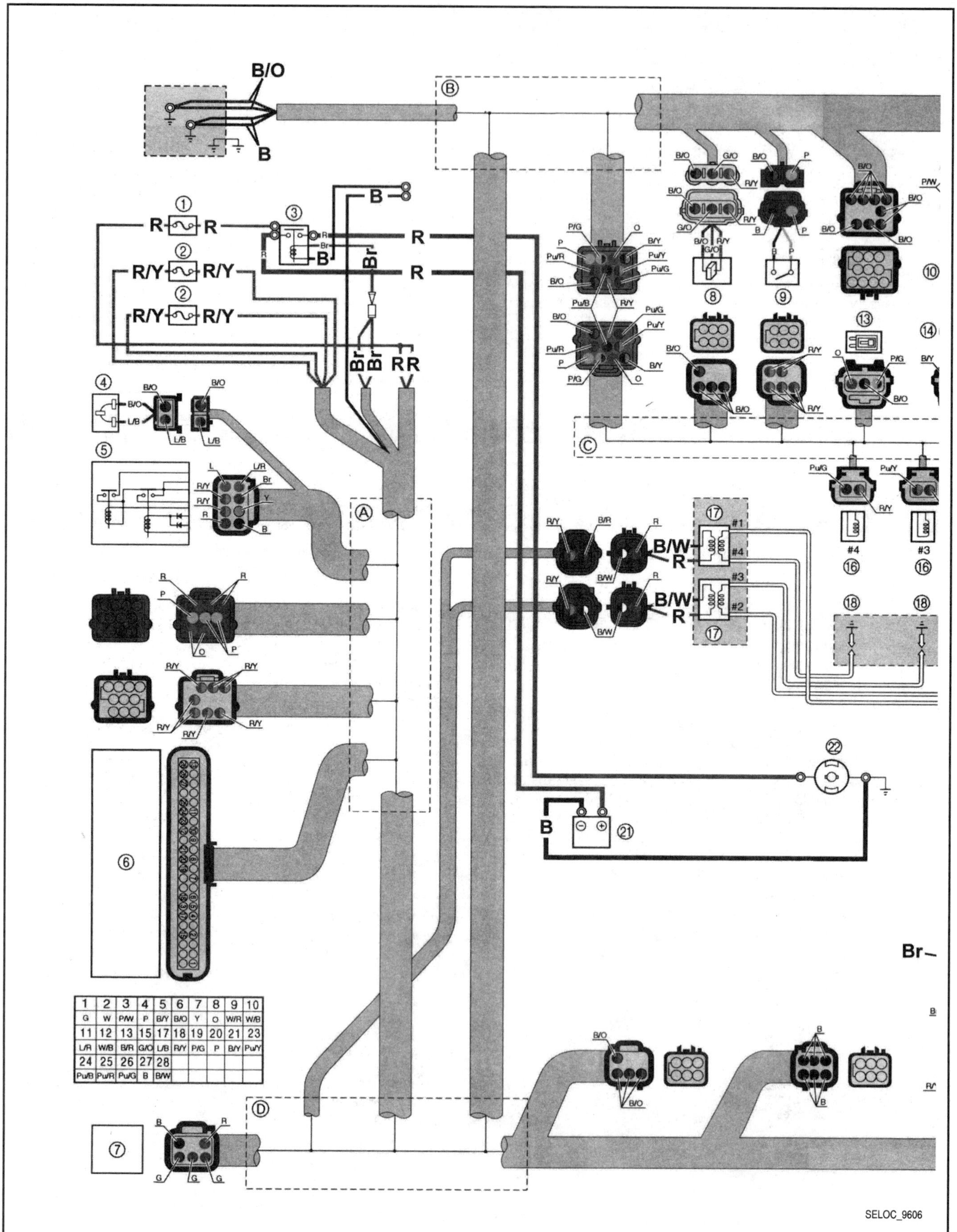

1	2	3	4	5	6	7	8	9	10
G	W	P/W	P	B/Y	B/O	Y	O	W/R	W/B
11	12	13	15	17	18	19	20	21	23
L/R	W/B	B/R	G/O	L/B	R/Y	P/G	P	B/Y	Pu/Y
24	25	26	27	28					
Pu/B	Pu/R	Pu/G	B	B/W					

Wiring diagram - 2002 FX140 Models w/998cc Engines (Left Side)

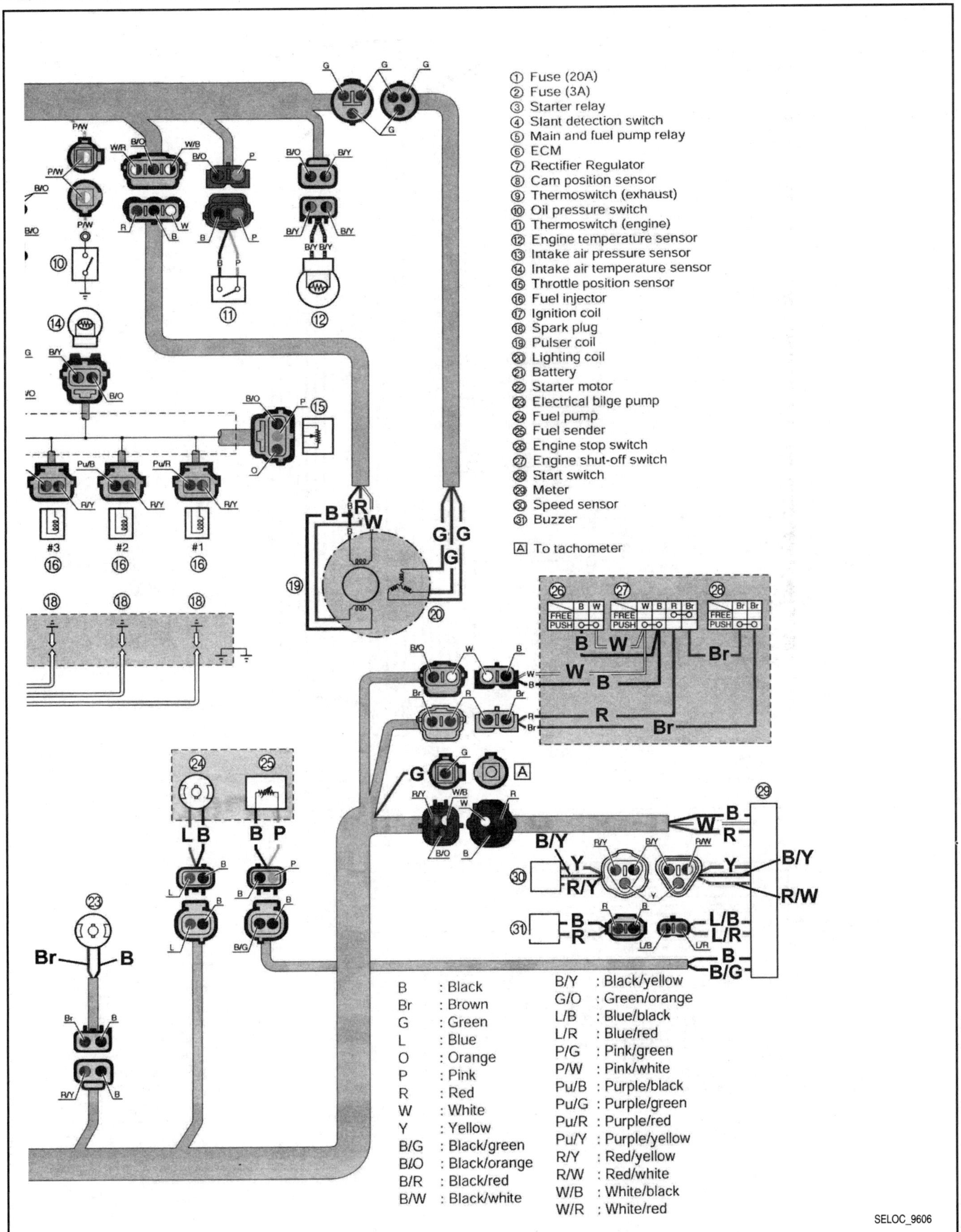

Wiring diagram - 2002 FX140 Models w/998cc Engines (Right Side)

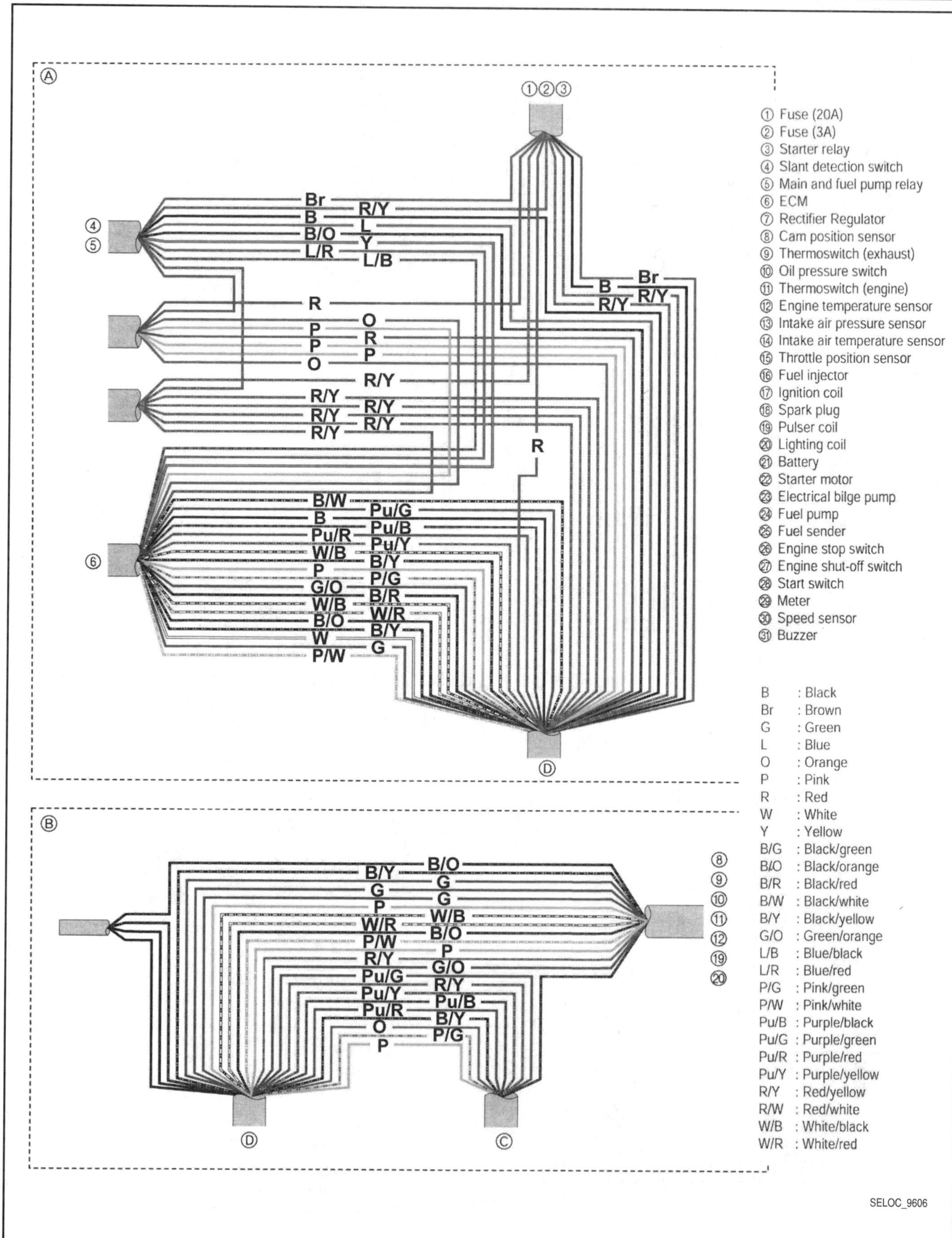

Wiring diagram harness wires - 2002 FX140 Models w/998cc Engines (Left Side)

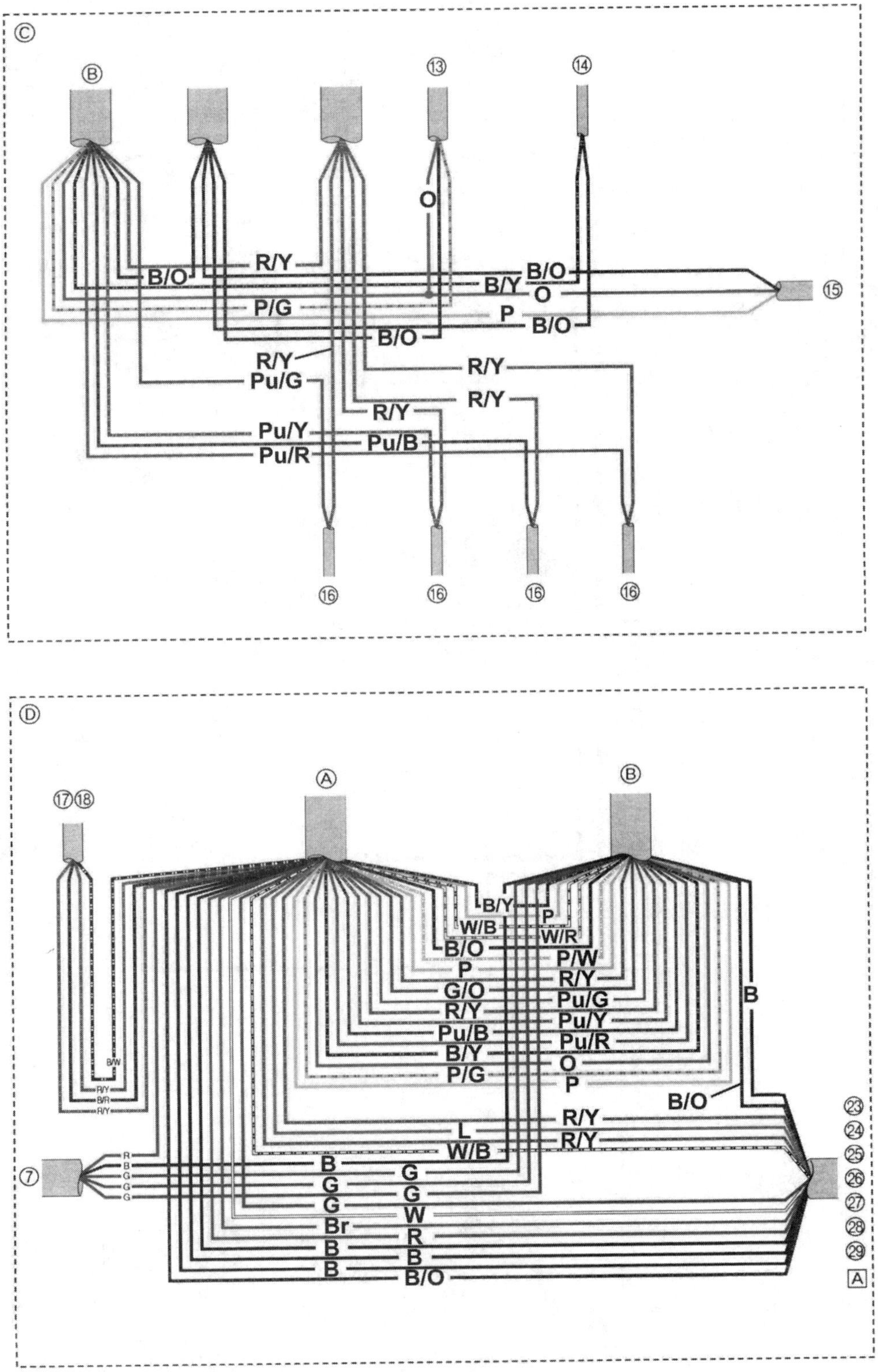

1. Fuse (20A)
2. Fuse (3A)
3. Starter relay
4. Slant detection switch
5. Main and fuel pump relay
6. ECM
7. Rectifier Regulator
8. Cam position sensor
9. Thermoswitch (exhaust)
10. Oil pressure switch
11. Thermoswitch (engine)
12. Engine temperature sensor
13. Intake air pressure sensor
14. Intake air temperature sensor
15. Throttle position sensor
16. Fuel injector
17. Ignition coil
18. Spark plug
19. Pulser coil
20. Lighting coil
21. Battery
22. Starter motor
23. Electrical bilge pump
24. Fuel pump
25. Fuel sender
26. Engine stop switch
27. Engine shut-off switch
28. Start switch
29. Meter
30. Speed sensor
31. Buzzer

B : Black
Br : Brown
G : Green
L : Blue
O : Orange
P : Pink
R : Red
W : White
Y : Yellow
B/G : Black/green
B/O : Black/orange
B/R : Black/red
B/W : Black/white
B/Y : Black/yellow
G/O : Green/orange
L/B : Blue/black
L/R : Blue/red
P/G : Pink/green
P/W : Pink/white
Pu/B : Purple/black
Pu/G : Purple/green
Pu/R : Purple/red
Pu/Y : Purple/yellow
R/Y : Red/yellow
R/W : Red/white
W/B : White/black
W/R : White/red

SELOC_9606

Wiring diagram harness wires - 2002 FX140 Models w/998cc Engines (Right Side)

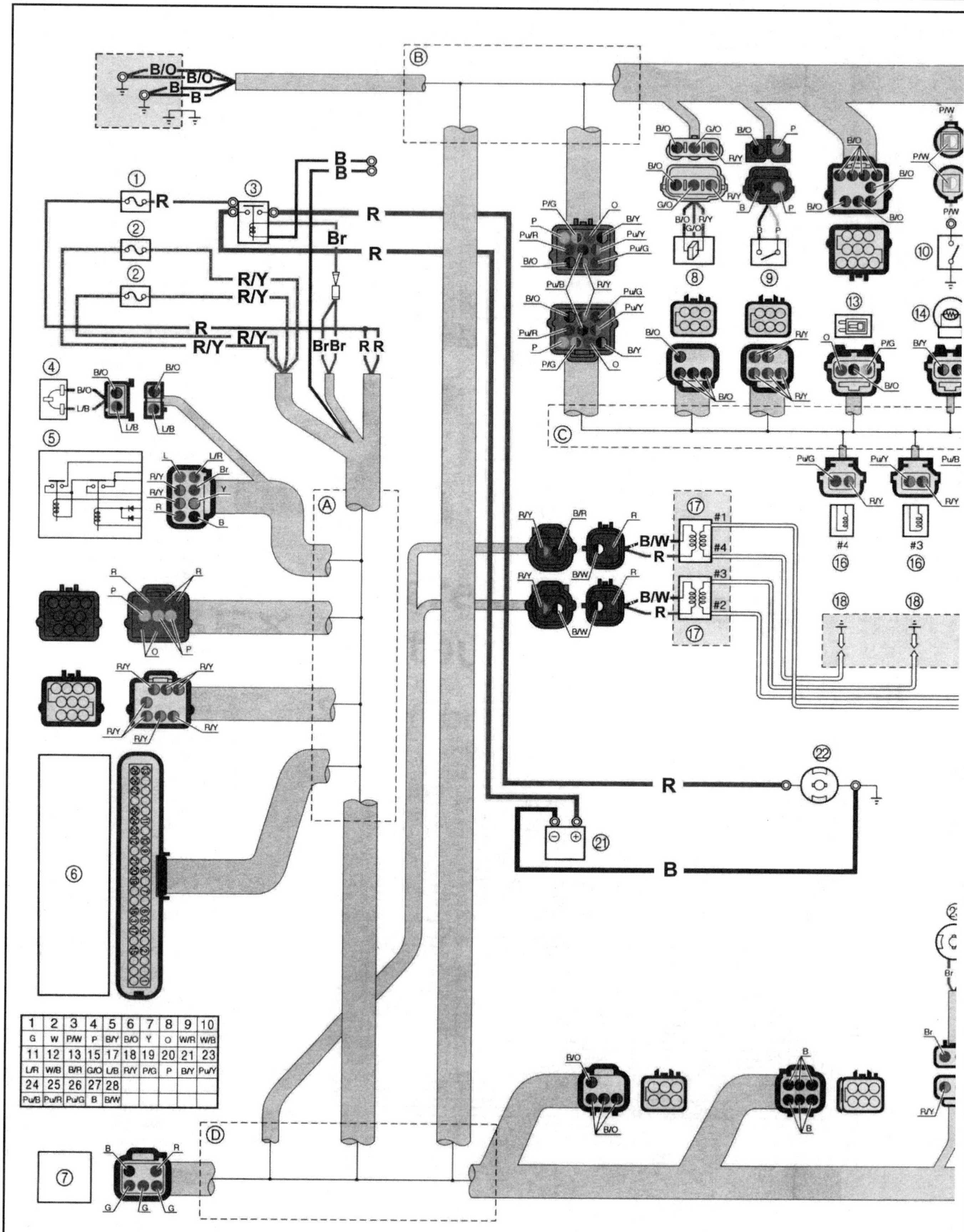

1	2	3	4	5	6	7	8	9	10
G	W	P/W	P	B/Y	B/O	Y	O	W/R	W/B
11	12	13	15	17	18	19	20	21	23
L/R	W/B	B/R	G/O	L/B	R/Y	P/G	P	B/Y	Pu/Y
24	25	26	27	28					
Pu/B	Pu/R	Pu/G	B	B/W					

SELOC_9606

Wiring diagram - 2003-04 FX140/FX140 Cruiser Models w/998cc Engines (Left Side)

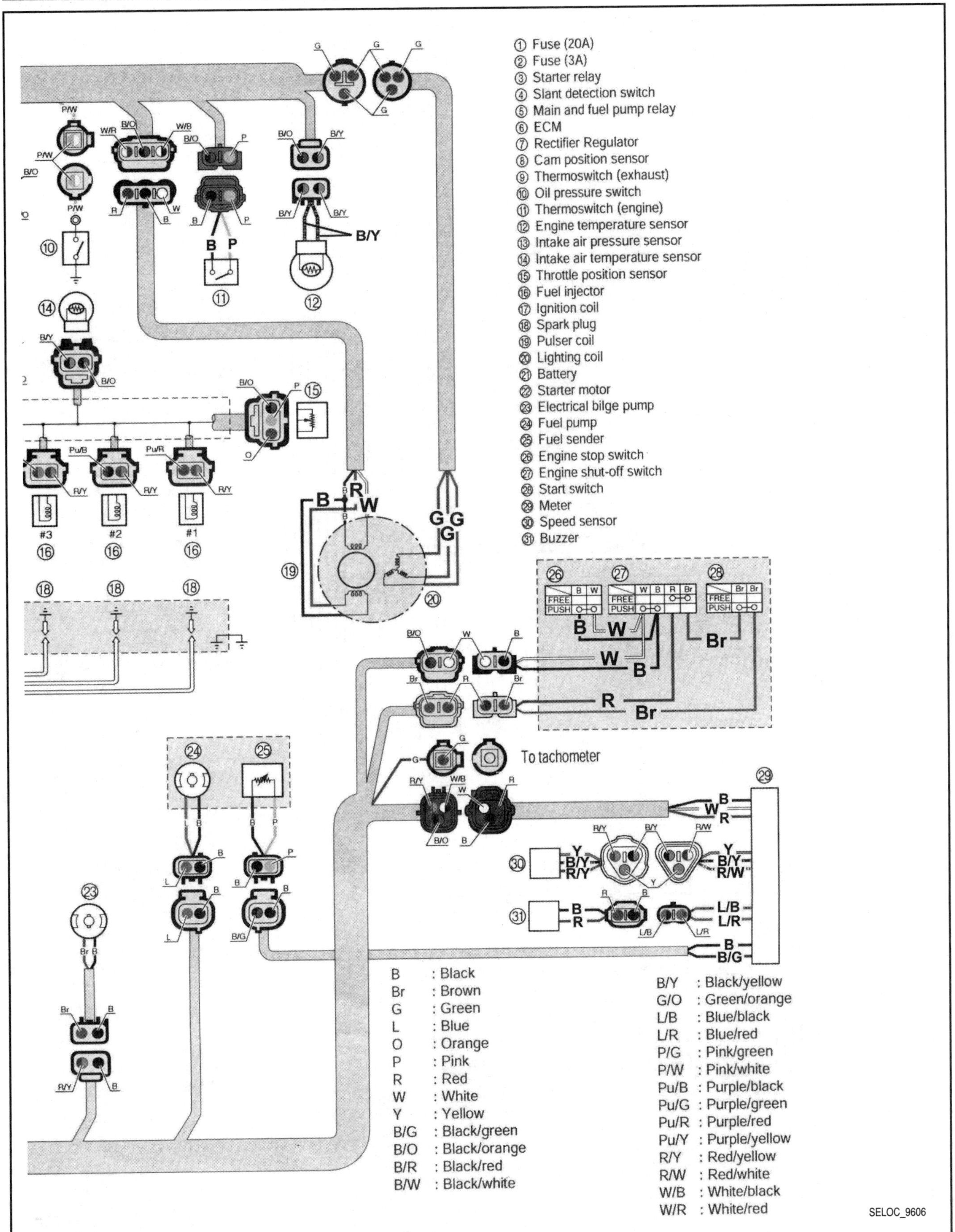

Wiring diagram - 2003-04 FX140/FX140 Cruiser Models w/998cc Engines (Right Side)

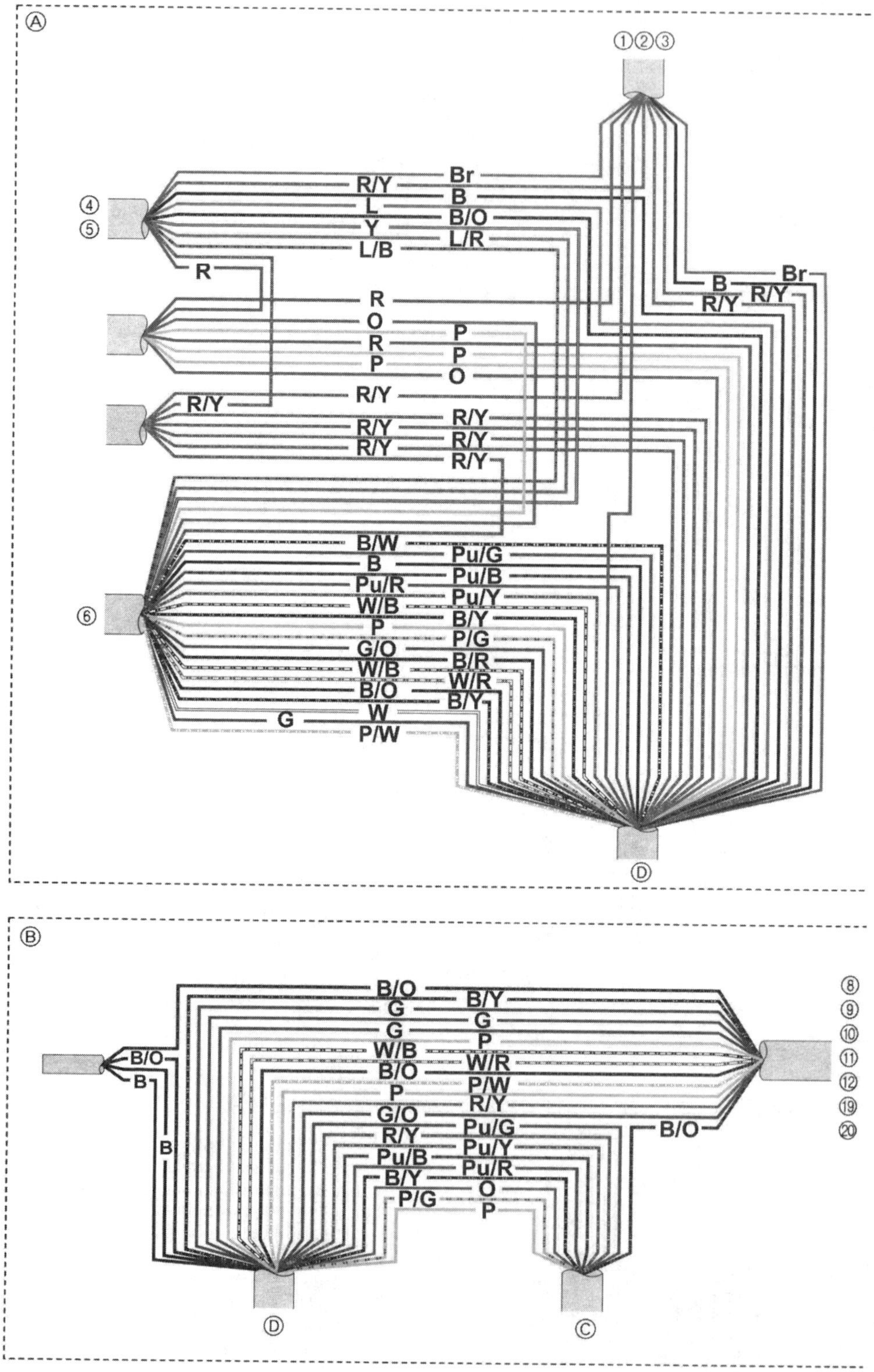

① Fuse (20A)
② Fuse (3A)
③ Starter relay
④ Slant detection switch
⑤ Main and fuel pump relay
⑥ ECM
⑦ Rectifier Regulator
⑧ Cam position sensor
⑨ Thermoswitch (exhaust)
⑩ Oil pressure switch
⑪ Thermoswitch (engine)
⑫ Engine temperature sensor
⑬ Intake air pressure sensor
⑭ Intake air temperature sensor
⑮ Throttle position sensor
⑯ Fuel injector
⑰ Ignition coil
⑱ Spark plug
⑲ Pulser coil
⑳ Lighting coil
㉑ Battery
㉒ Starter motor
㉓ Electrical bilge pump
㉔ Fuel pump
㉕ Fuel sender
㉖ Engine stop switch
㉗ Engine shut-off switch
㉘ Start switch
㉙ Meter
㉚ Speed sensor
㉛ Buzzer

[A] To tachometer

Color code

B	: Black
Br	: Brown
G	: Green
L	: Blue
O	: Orange
P	: Pink
R	: Red
W	: White
Y	: Yellow
B/G	: Black/green
B/O	: Black/orange
B/R	: Black/red
B/W	: Black/white
B/Y	: Black/yellow
G/O	: Green/orange
L/B	: Blue/black
L/R	: Blue/red
P/G	: Pink/green
P/W	: Pink/white
Pu/B	: Purple/black
Pu/G	: Purple/green
Pu/R	: Purple/red
Pu/Y	: Purple/yellow
R/Y	: Red/yellow
R/W	: Red/white
W/B	: White/black
W/R	: White/red

SELOC_9606

Wiring diagram harness wires - 2003-04 FX140/FX140 Cruiser Models w/998cc Engines (Left Side)

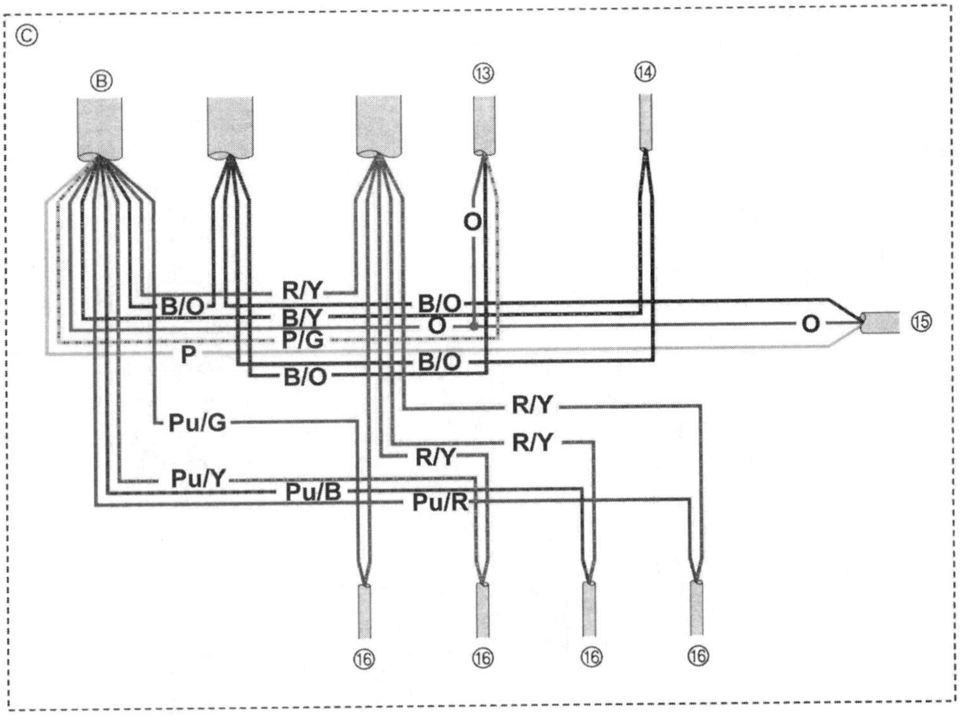

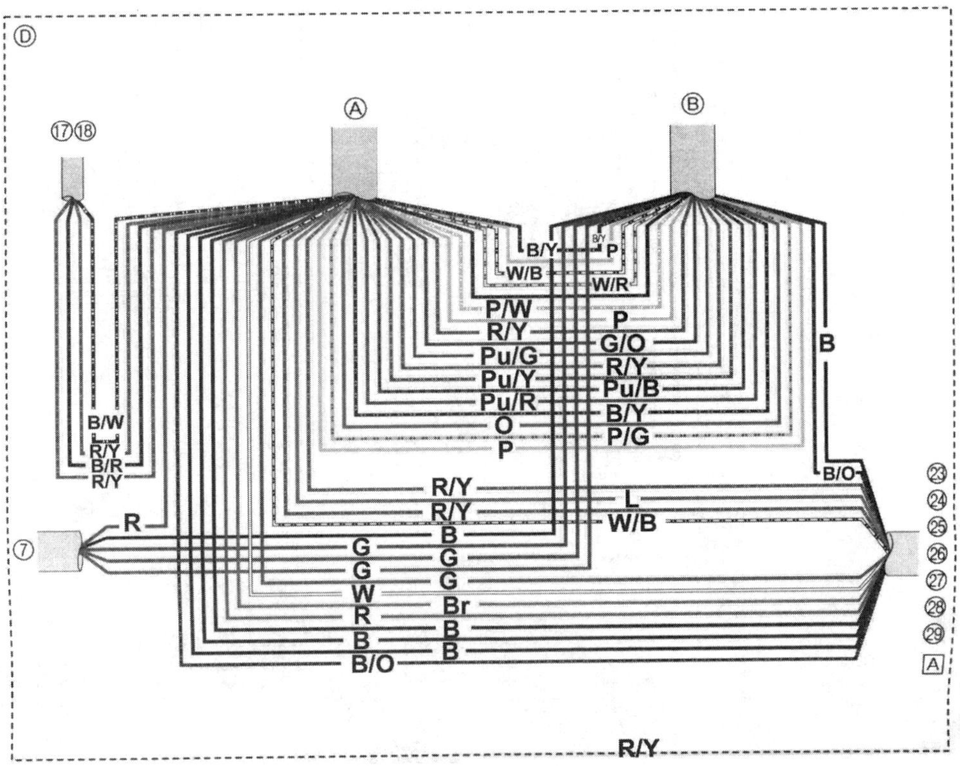

1. Fuse (20A)
2. Fuse (3A)
3. Starter relay
4. Slant detection switch
5. Main and fuel pump relay
6. ECM
7. Rectifier Regulator
8. Cam position sensor
9. Thermoswitch (exhaust)
10. Oil pressure switch
11. Thermoswitch (engine)
12. Engine temperature sensor
13. Intake air pressure sensor
14. Intake air temperature sensor
15. Throttle position sensor
16. Fuel injector
17. Ignition coil
18. Spark plug
19. Pulser coil
20. Lighting coil
21. Battery
22. Starter motor
23. Electrical bilge pump
24. Fuel pump
25. Fuel sender
26. Engine stop switch
27. Engine shut-off switch
28. Start switch
29. Meter
30. Speed sensor
31. Buzzer

[A] To tachometer

Color code

B	: Black
Br	: Brown
G	: Green
L	: Blue
O	: Orange
P	: Pink
R	: Red
W	: White
Y	: Yellow
B/G	: Black/green
B/O	: Black/orange
B/R	: Black/red
B/W	: Black/white
B/Y	: Black/yellow
G/O	: Green/orange
L/B	: Blue/black
L/R	: Blue/red
P/G	: Pink/green
P/W	: Pink/white
Pu/B	: Purple/black
Pu/G	: Purple/green
Pu/R	: Purple/red
Pu/Y	: Purple/yellow
R/Y	: Red/yellow
R/W	: Red/white
W/B	: White/black
W/R	: White/red

SELOC_9606

Wiring diagram harness wires - 2003-04 FX140/FX140 Cruiser Models w/998cc Engines (Right Side)

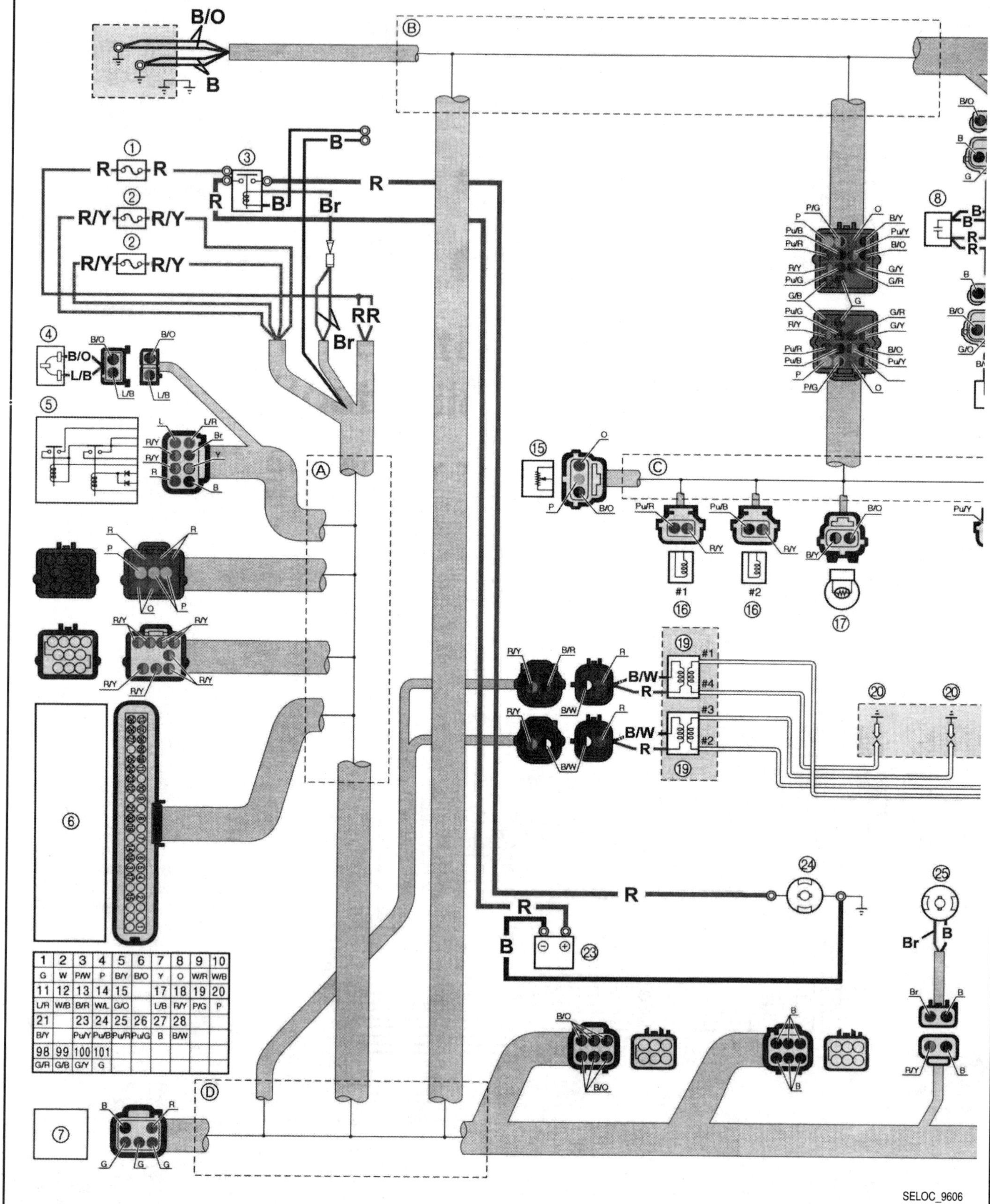

Wiring diagram - 2005 FX/FX Cruiser Models w/998cc Engines (Left Side)

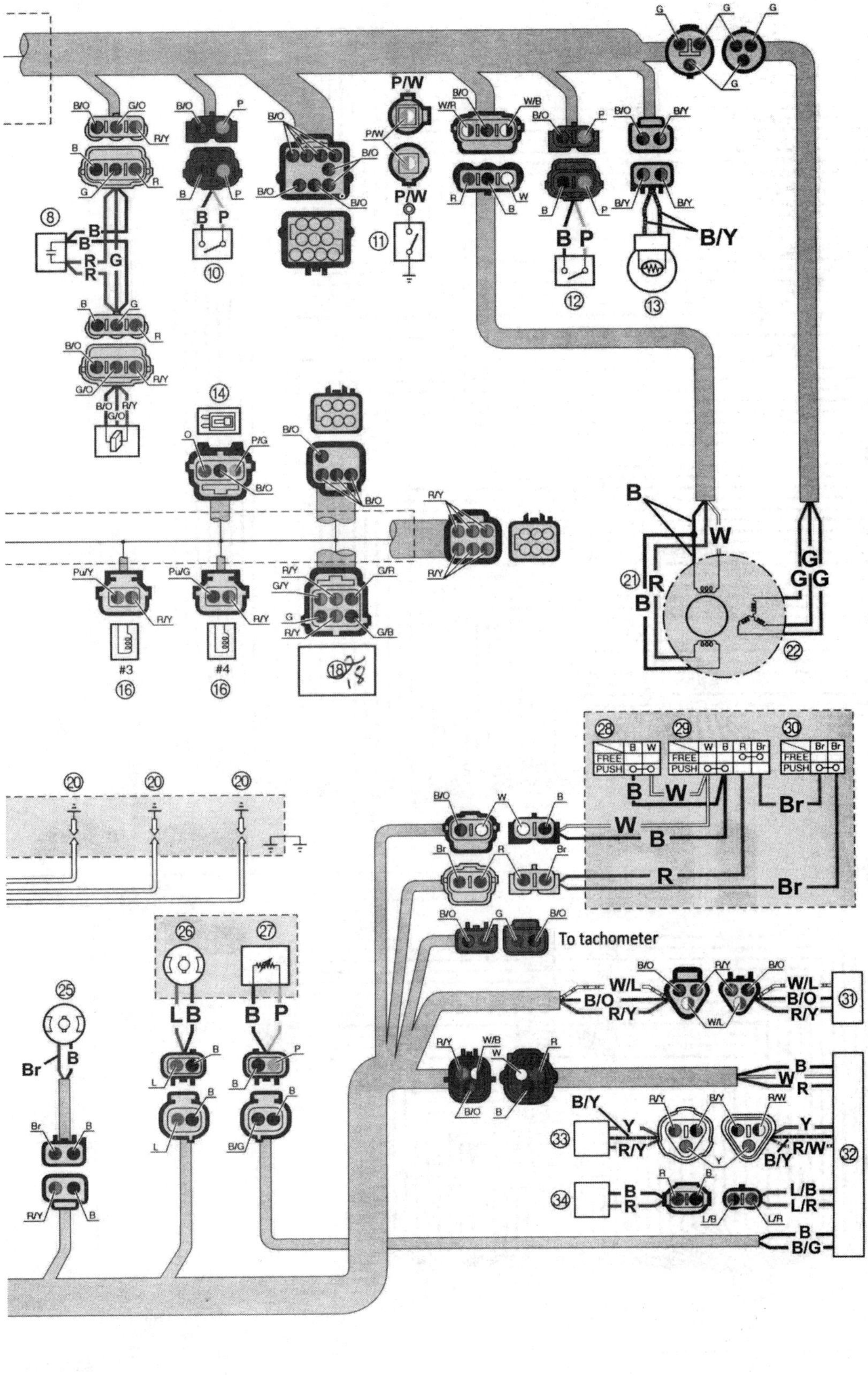

1 – Fuse (20A)
2 – Fuse (3A)
3 – Starter relay
4 – Slant detection switch
5 – Main and fuel pump relay
6 – ECM
7 – Rectifier/regulator
8 – Noise filter
9 – Cam position sensor
10 – Thermoswitch (exhaust)
11 – Oil pressure switch
12 – Thermoswitch (engine)
13 – Engine temperature sensor
14 – IAP sensor
15 – TP sensor
16 – Fuel injector
17 – IAT sensor
18 – Bypass valve motor
19 – Ignition coil
20 – Spark plug
21 – Pulser coil
22 – Lighting coil
23 – Battery
24 – Starter motor
25 – Electric bilge pump
26 – Fuel Pump
27 – Fuel sender
28 – Engine stop switch
29 – Engine shut-off switch
30 – Start switch
31 – Steering switch
32 – Meter
33 – Speed sensor
34 – Warning buzzer

B	: Black
Br	: Brown
G	: Green
L	: Blue
O	: Orange
P	: Pink
R	: Red
W	: White
Y	: Yellow
B/G	: Black/green
B/O	: Black/orange
B/R	: Black/red
B/W	: Black/white
B/Y	: Black/yellow
G/O	: Green/orange
L/B	: Blue/black
L/R	: Blue/red
P/G	: Pink/green
P/W	: Pink/white
Pu/B	: Purple/black
Pu/G	: Purple/green
Pu/R	: Purple/red
Pu/Y	: Purple/yellow
R/Y	: Red/yellow
R/W	: Red/white
W/B	: White/black
W/R	: White/red

SELOC_9606

Wiring diagram - 2005 FX/FX Cruiser Models w/998cc Engines (Left Side)

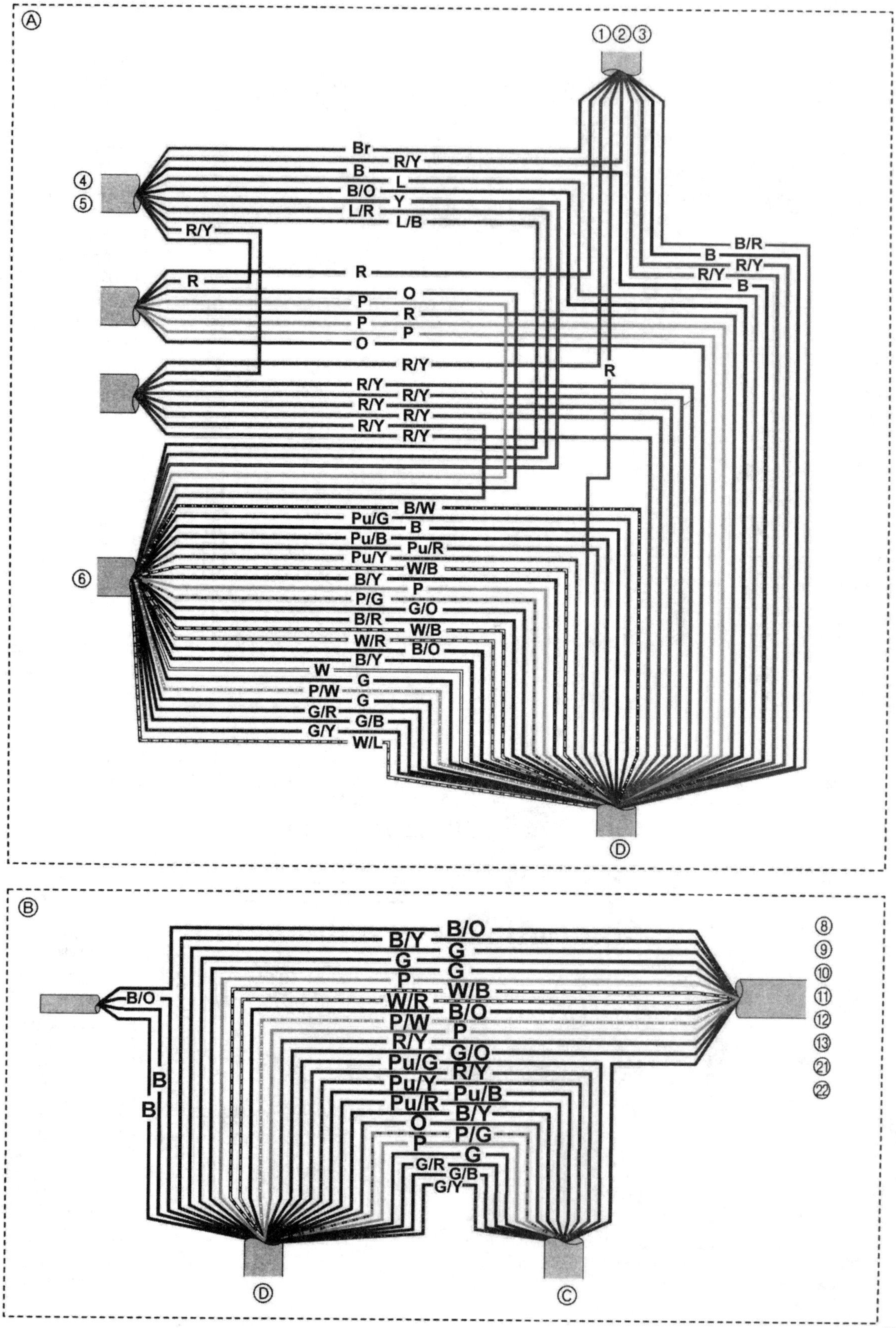

SELOC_9606

Wiring diagram harness wires - 2005 FX/FX Cruiser Models w/998cc Engines (Left Side)

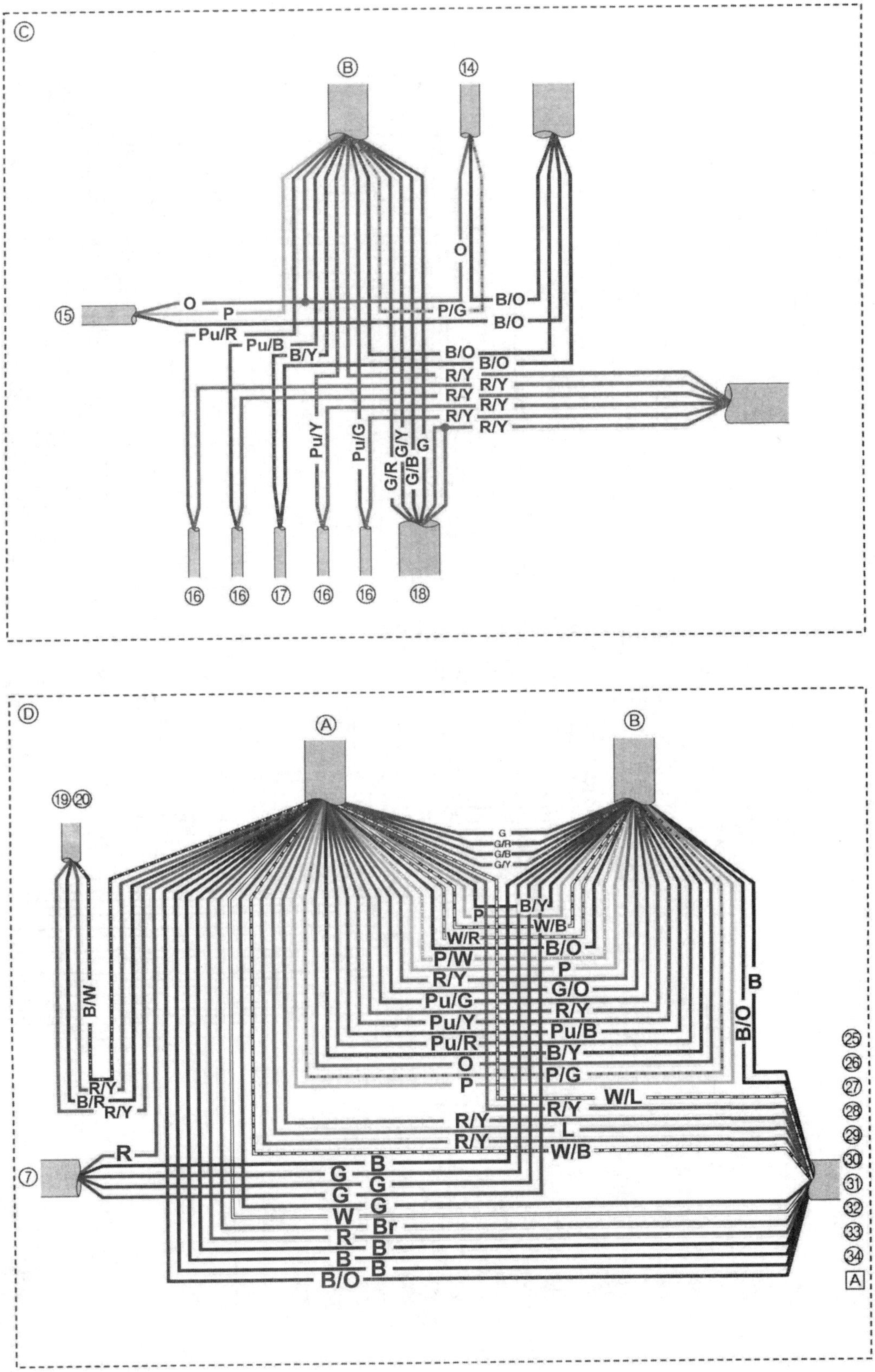

1 – Fuse (20A)
2 – Fuse (3A)
3 – Starter relay
4 – Slant detection switch
5 – Main and fuel pump relay
6 – ECM
7 – Rectifier/regulator
8 – Noise filter
9 – Cam position sensor
10 – Thermoswitch (exhaust)
11 – Oil pressure switch
12 – Thermoswitch (engine)
13 – Engine temperature sensor
14 – IAP sensor
15 – TP sensor
16 – Fuel injector
17 – IAT sensor
18 – Bypass valve motor
19 – Ignition coil
20 – Spark plug
21 – Pulser coil
22 – Lighting coil
23 – Battery
24 – Starter motor
25 – Electric bilge pump
26 – Fuel Pump
27 – Fuel sender
28 – Engine stop switch
29 – Engine shut-off switch
30 – Start switch
31 – Steering switch
32 – Meter
33 – Speed sensor
34 – Warning buzzer
A – To tachometer

Color code

B : Black
Br : Brown
G : Green
L : Blue
O : Orange
P : Pink
R : Red
W : White
Y : Yellow
B/G : Black/green
B/O : Black/orange
B/R : Black/red
B/W : Black/white
B/Y : Black/yellow
G/O : Green/orange
L/B : Blue/black
L/R : Blue/red
P/G : Pink/green
P/W : Pink/white
Pu/B : Purple/black
Pu/G : Purple/green
Pu/R : Purple/red
Pu/Y : Purple/yellow
R/Y : Red/yellow
R/W : Red/white
W/B : White/black
W/R : White/red

SELOC_9606

Wiring diagram harness wires - 2005 FX/FX Cruiser Models w/998cc Engines (Right Side)

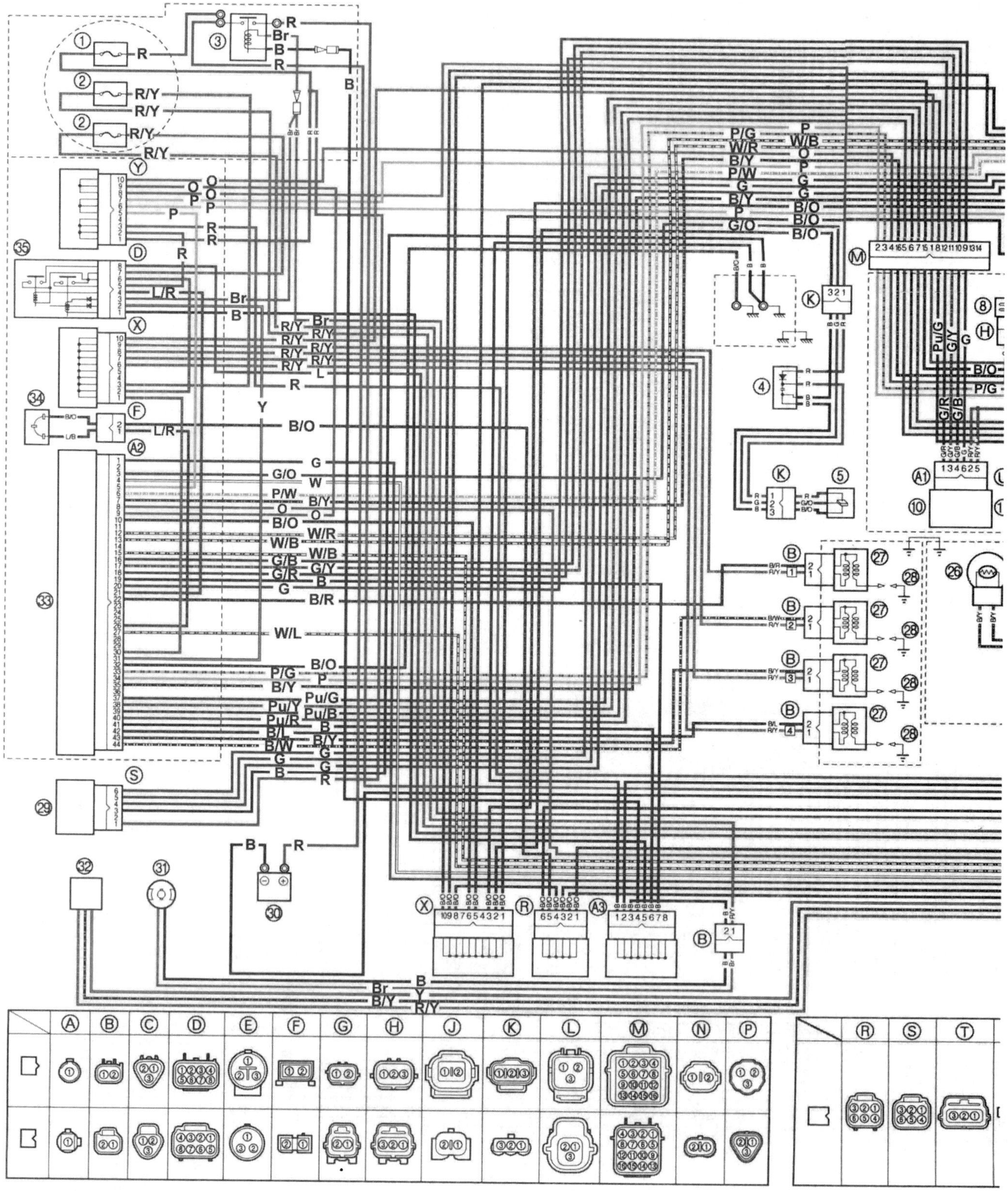

SELOC_9606

Wiring diagram - 2006-07 FX/FX Cruiser Models w/998cc Engines (Left Side)

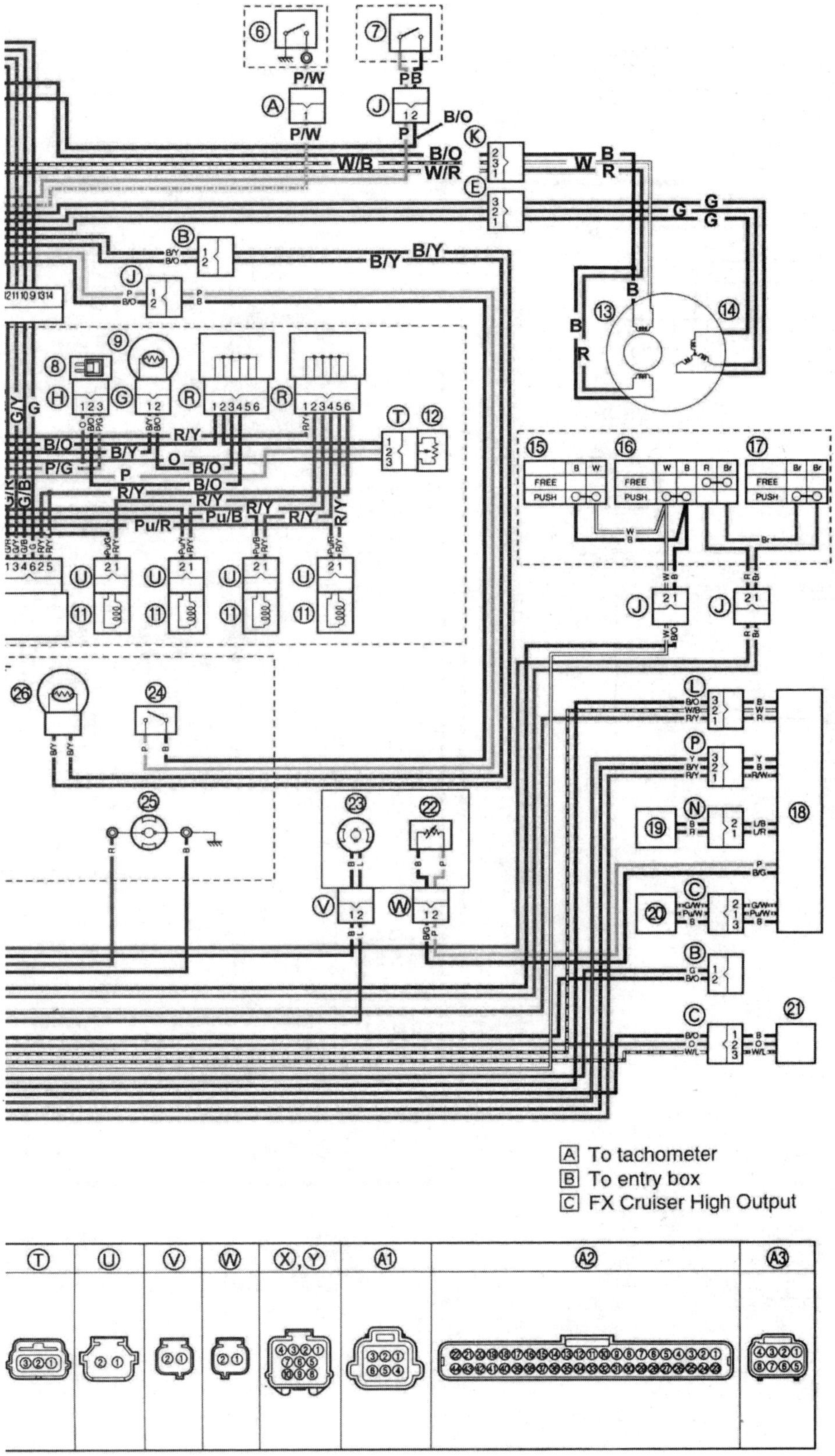

1 – Fuse (20A)
2 – Fuse (3A)
3 – Starter relay
4 – Noise filter
5 – Cam position sensor
6 – Oil pressure switch
7 – Thermoswitch (exhaust)
8 – IAT sensor
9 – IAP sensor
10 – Bypass valve motor
11 – Fuel injector
12 – TP sensor
13 – Pick-up coil
14 – Stator coil
15 – Engine stop switch
16 – Engine shut-off switch
17 – Start switch
18 – Meter
19 – Warning buzzer
20 – Left operation button assembly
21 – Steering sensor
22 – Fuel sender
23 – Fuel Pump
24 – Thermoswitch (engine)
25 – Starter motor
26 – Engine temperature sensor
27 – Ignition coil
28 – Spark plug
29 – Rectifier/Regulator
30 – Battery
31 – Electric bilge pump
32 – Speed sensor
33 – ECM
34 – Slant detection switch
35 – Main and fuel pump relay

B : Black
Br : Brown
G : Green
Gy : Gray
L : Blue
O : Orange
P : Pink
R : Red
W : White
Y : Yellow
B/G : Black/green
B/L : Black/blue
B/O : Black/orange
B/R : Black/red
B/W : Black/white
B/Y : Black/yellow
G/B : Green/black
G/O : Green/orange
G/R : Green/red
G/W : Green/white
G/Y : Green/yellow
L/B : Blue/black
L/R : Blue/red
L/Y : Blue/yellow
P/G : Pink/green
P/W : Pink/white
Pu/B : Purple/black
Pu/G : Purple/green
Pu/R : Purple/red
Pu/W : Purple/white
Pu/Y : Purple/yellow
R/Y : Red/yellow
R/W : Red/white
W/B : White/black
W/L : White/blue
W/R : White/red
Y/W : Yellow/white

SELOC_9606

Wiring diagram - 2006-07 FX/FX Cruiser Models w/998cc Engines (Right Side)

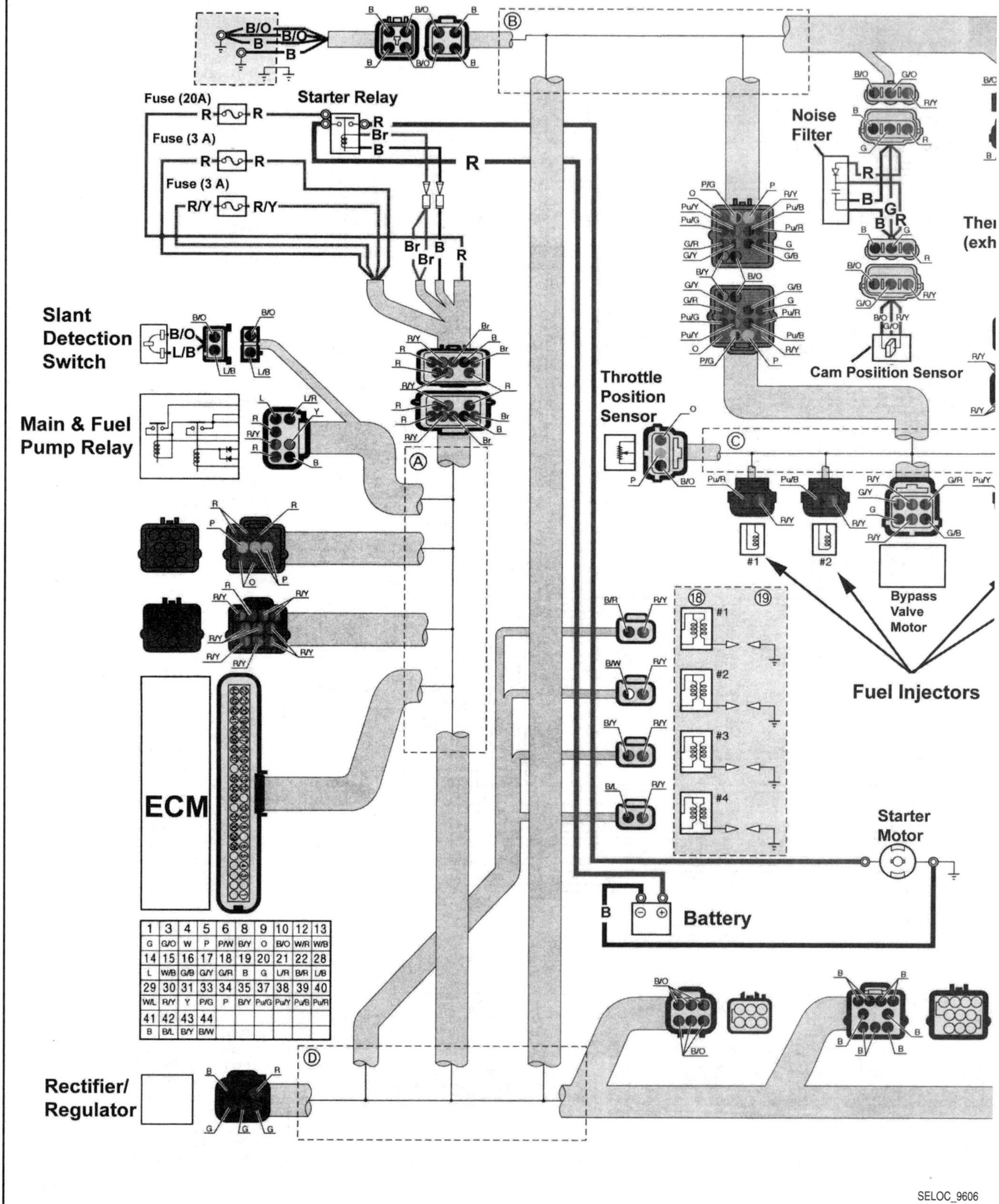

1	3	4	5	6	8	9	10	12	13
G	G/O	W	P	P/W	B/Y	O	B/O	W/R	W/B
14	15	16	17	18	19	20	21	22	28
L	W/B	G/B	G/Y	G/R	B	G	L/R	B/R	L/B
29	30	31	33	34	35	37	38	39	40
W/L	R/Y	Y	P/G	P	B/Y	Pu/G	Pu/Y	Pu/B	Pu/R
41	42	43	44						
B	B/L	B/Y	B/W						

Wiring diagram - 2004-05 FX HO and FX Cruiser HO Models (Left Side)

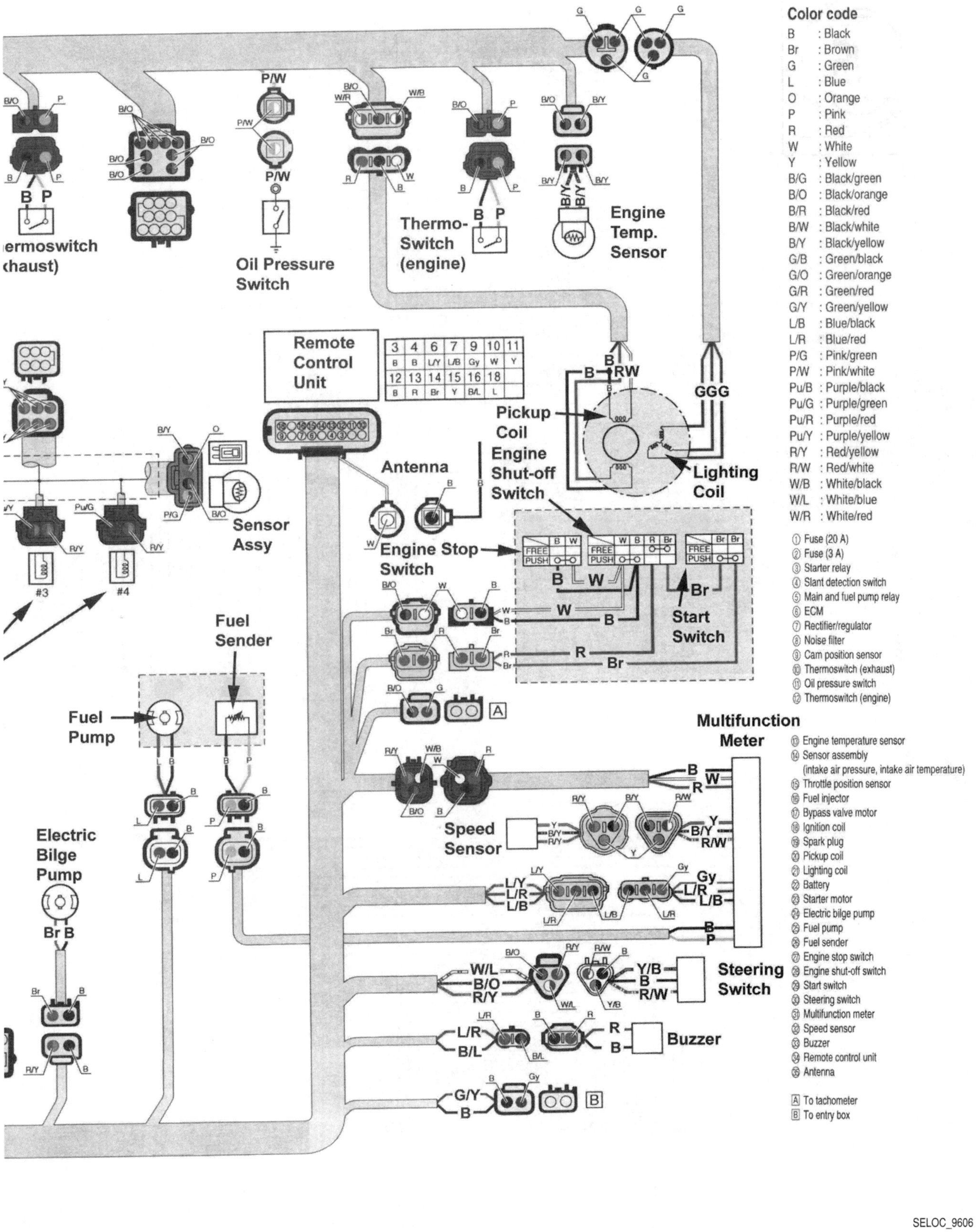

Wiring diagram - 2004-05 FX HO and FX Cruiser HO Models (Right Side)

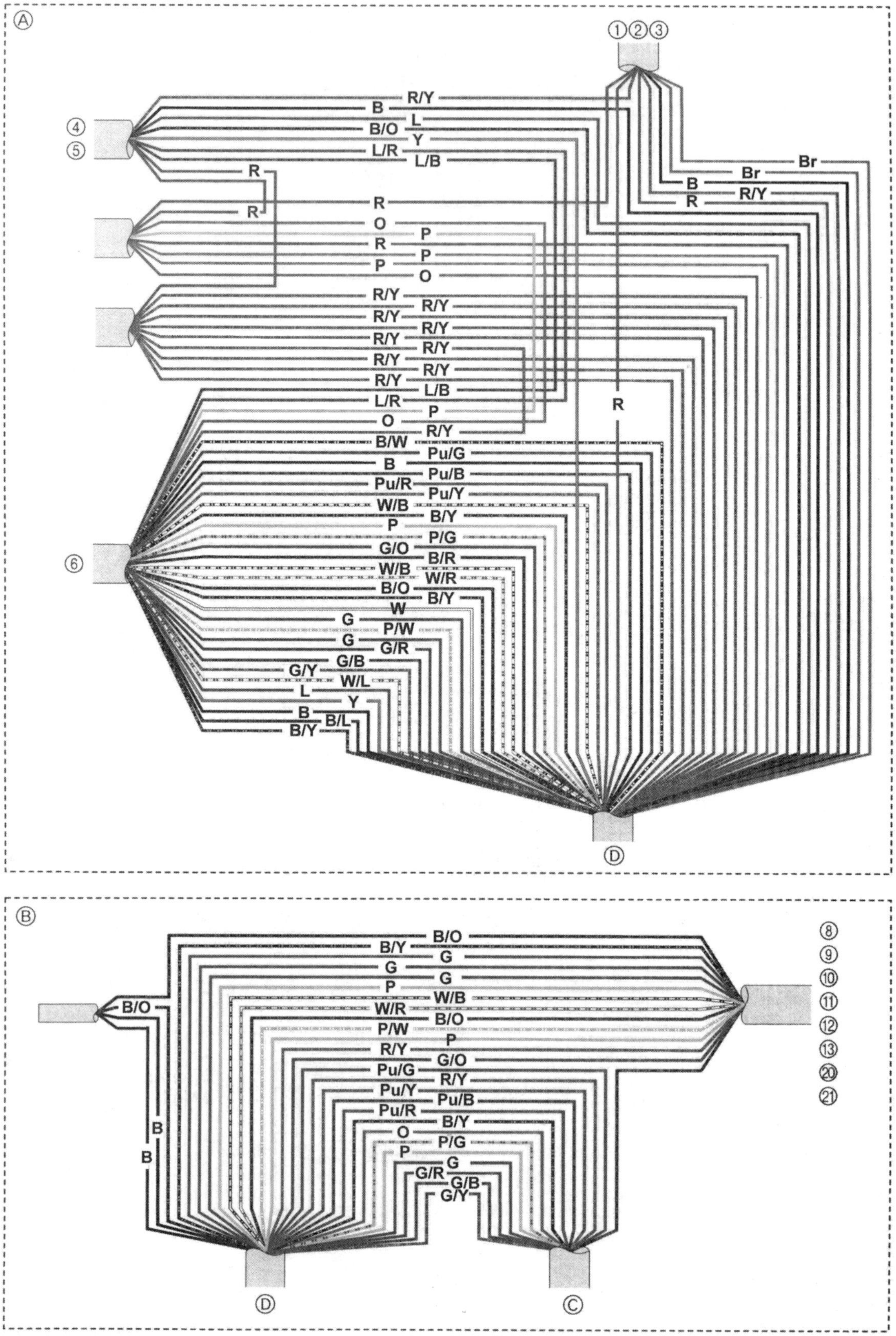

SELOC_9606

Wiring diagram harness wires - 2004-05 FX HO and FX Cruiser HO Models (Left Side)

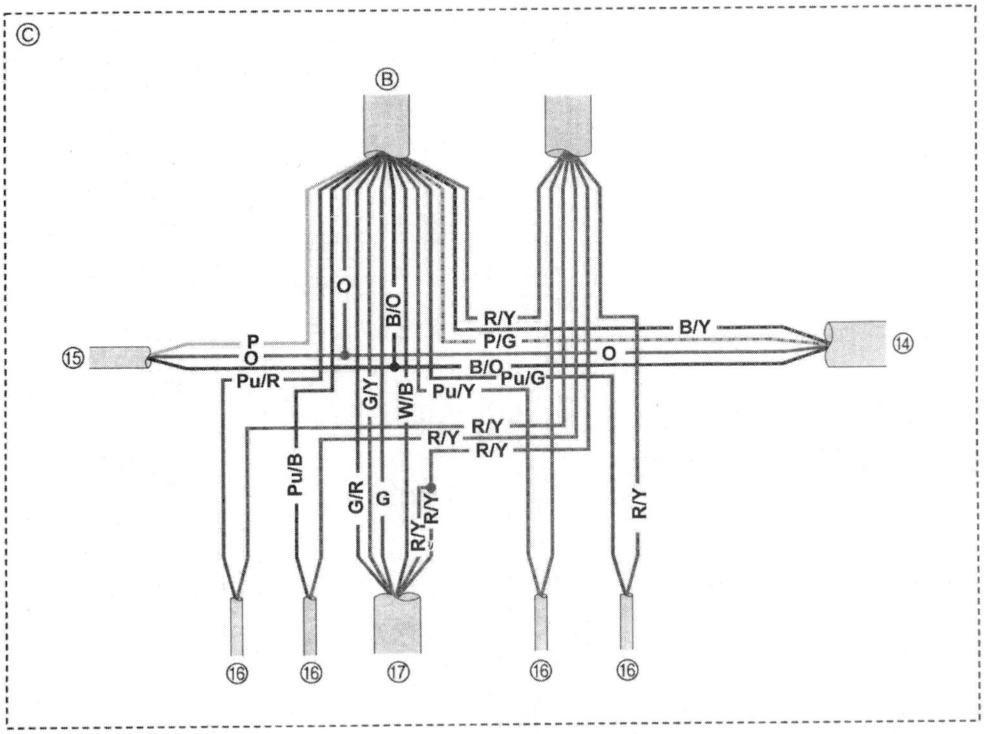

1. Fuse (20 A)
2. Fuse (3 A)
3. Starter relay
4. Slant detection switch
5. Main and fuel pump relay
6. ECM
7. Rectifier/regulator
8. Noise filter
9. Cam position sensor
10. Thermoswitch (exhaust)
11. Oil pressure switch
12. Thermoswitch (engine)
13. Engine temperature sensor
14. Sensor assembly (intake air pressure, intake air temperature)
15. Throttle position sensor
16. Fuel injector
17. Bypass valve motor
18. Ignition coil
19. Spark plug
20. Pickup coil
21. Lighting coil
22. Battery
23. Starter motor
24. Electric bilge pump
25. Fuel pump
26. Fuel sender
27. Engine stop switch
28. Engine shut-off switch
29. Start switch
30. Steering switch
31. Multifunction meter
32. Speed sensor
33. Buzzer
34. Remote control unit
35. Antenna

A To tachometer
B To entry box

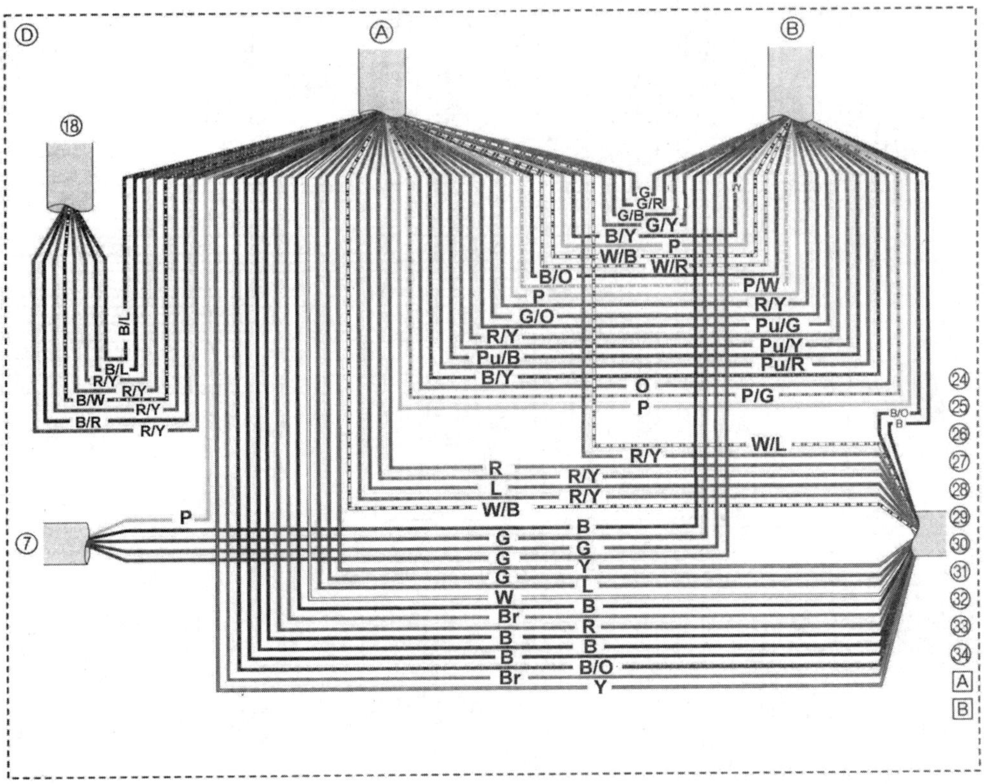

Color code

B	: Black
Br	: Brown
G	: Green
L	: Blue
O	: Orange
P	: Pink
R	: Red
W	: White
Y	: Yellow
B/G	: Black/green
B/O	: Black/orange
B/R	: Black/red
B/W	: Black/white
B/Y	: Black/yellow
G/B	: Green/black
G/O	: Green/orange
G/R	: Green/red
G/Y	: Green/yellow
L/B	: Blue/black
L/R	: Blue/red
P/G	: Pink/green
P/W	: Pink/white
Pu/B	: Purple/black
Pu/G	: Purple/green
Pu/R	: Purple/red
Pu/Y	: Purple/yellow
R/Y	: Red/yellow
R/W	: Red/white
W/B	: White/black
W/L	: White/blue
W/R	: White/red

SELOC_9606

Wiring diagram harness wires - 2004-05 FX HO and FX Cruiser HO Models (Right Side)

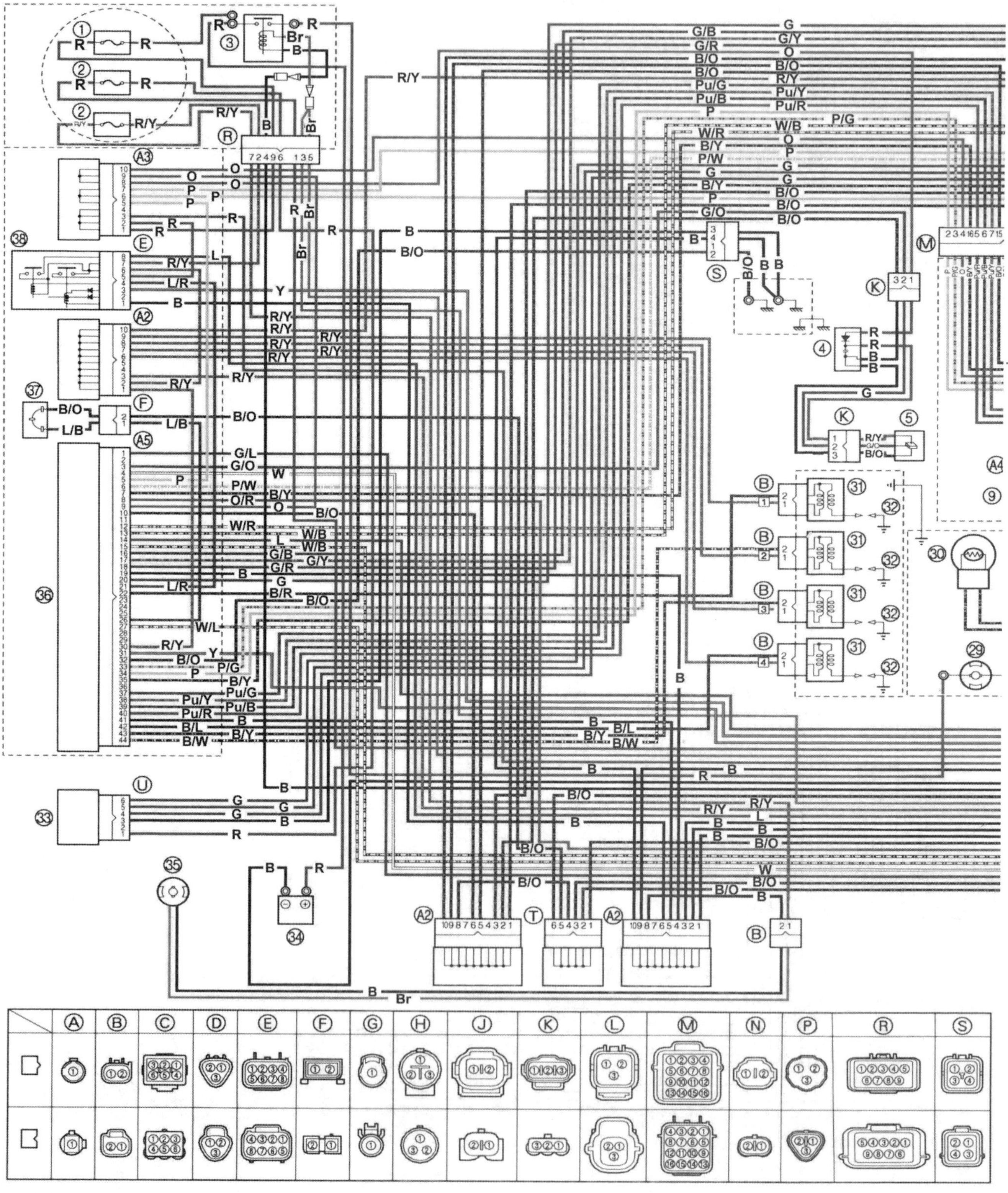

SELOC_9606

Wiring diagram - 2006-08 FX HO and FX Cruiser HO Models w/6B6 Engines (Left Side)

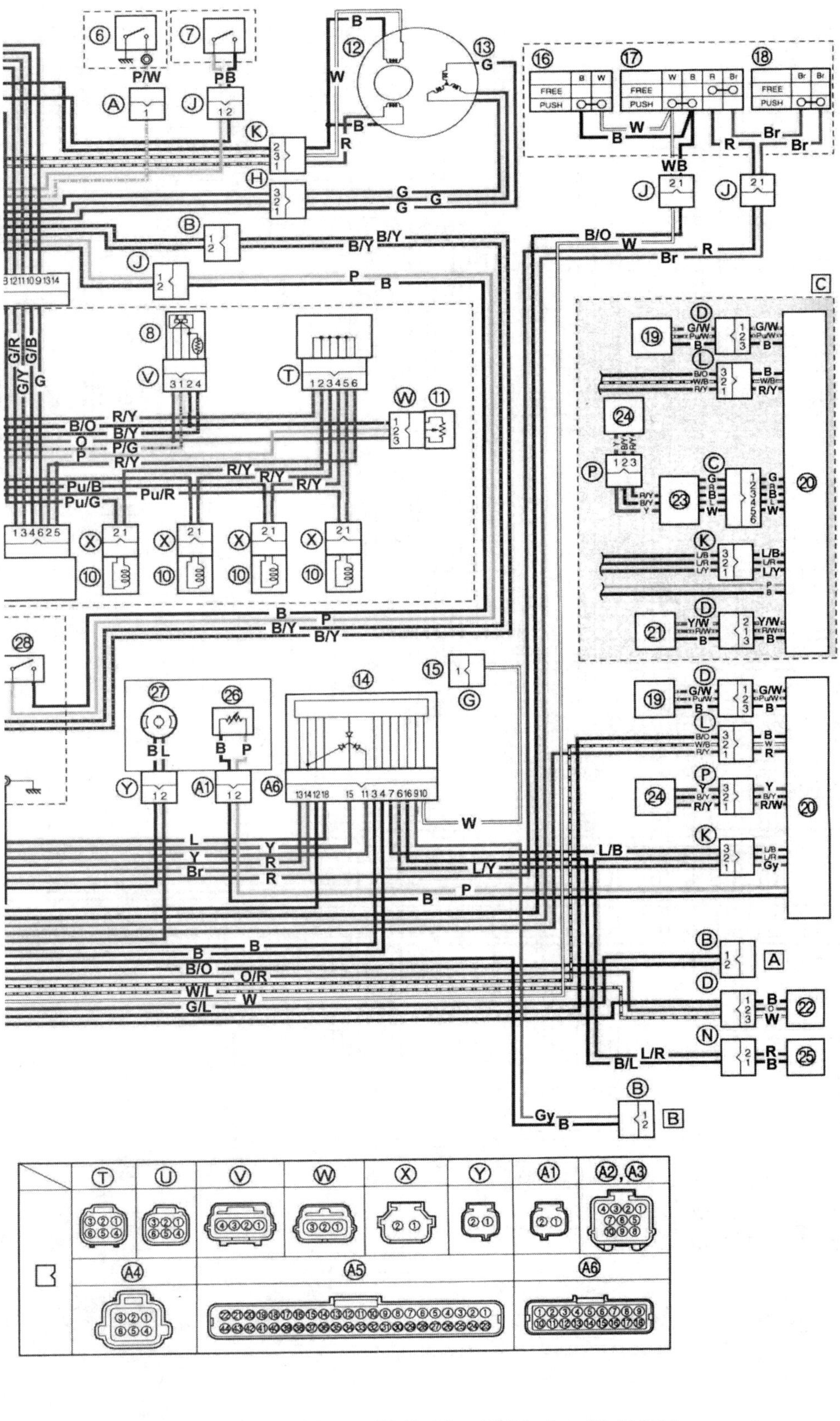

① Fuse (20 A)
② Fuse (3 A)
③ Starter relay
④ Noise filter
⑤ Cam position sensor
⑥ Oil pressure switch
⑦ Thermoswitch (exhaust)
⑧ Sensor assembly (intake air pressure, intake air temperature)
⑨ Bypass valve motor
⑩ Fuel injector
⑪ TPS
⑫ Pickup coil
⑬ Stator coil
⑭ Remote control receiver
⑮ Antenna
⑯ Engine stop switch
⑰ Engine shut-off switch
⑱ Engine start switch
⑲ Left operation button assembly
⑳ Multifunction meter
㉑ Right operation button assembly (FX Cruiser High Output)
㉒ Steering sensor
㉓ Compass/air temperature sensor (FX Cruiser High Output)
㉔ Speed sensor (FX High Output) Speed/water temperature sensor (FX Cruiser High Output)
㉕ Buzzer
㉖ Fuel sender
㉗ Fuel pump
㉘ Thermoswitch (engine)
㉙ Starter motor
㉚ Engine temperature sensor
㉛ Ignition coil
㉜ Spark plug
㉝ Rectifier/regulator
㉞ Battery
㉟ Electric bilge pump
㊱ ECM
㊲ Slant detection switch
㊳ Main and fuel pump relay

[A] To tachometer
[B] To entry box
[C] FX Cruiser High Output

Color code

B	Black	G/W	Green/white
Br	Brown	G/Y	Green/yellow
G	Green	L/B	Blue/black
Gy	Gray	L/R	Blue/red
L	Blue	L/Y	Blue/yellow
O	Orange	P/G	Pink/green
P	Pink	P/W	Pink/white
R	Red	Pu/B	Purple/black
W	White	Pu/G	Purple/green
Y	Yellow	Pu/R	Purple/red
B/G	Black/green	Pu/W	Purple/white
B/L	Black/blue	Pu/Y	Purple/yellow
B/O	Black/orange	R/Y	Red/yellow
B/R	Black/red	R/W	Red/white
B/W	Black/white	W/B	White/black
B/Y	Black/yellow	W/L	White/blue
G/B	Green/black	W/R	White/red
G/O	Green/orange	Y/W	Yellow/white
G/R	Green/red		

SELOC_9606

Wiring diagram - 2006-08 FX HO and FX Cruiser HO Models w/6B6 Engines (Right Side)

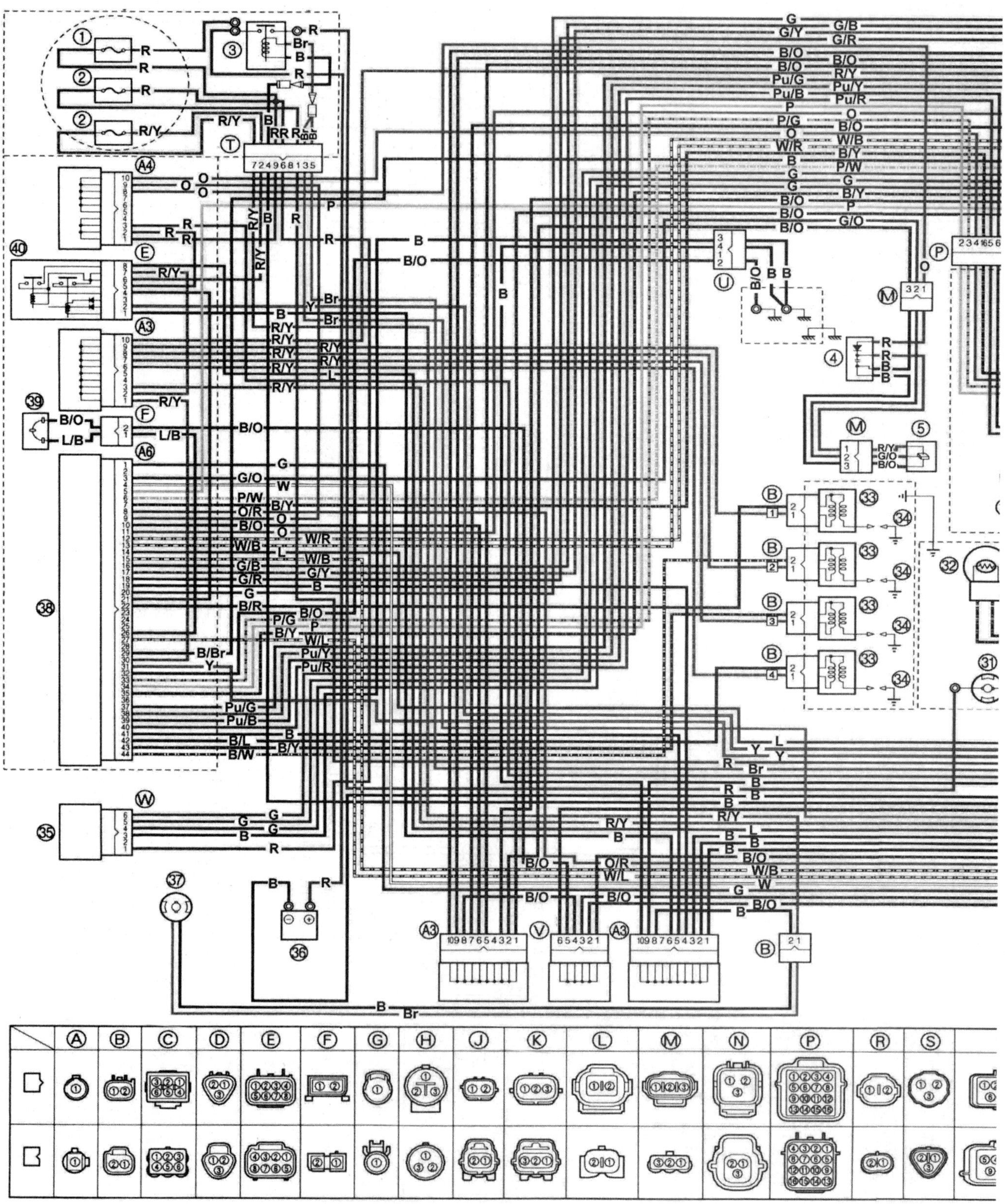

SELOC_9606

Wiring diagram - 2008 FX HO and FX Cruiser HO Models w/6AA Engines (Left Side)

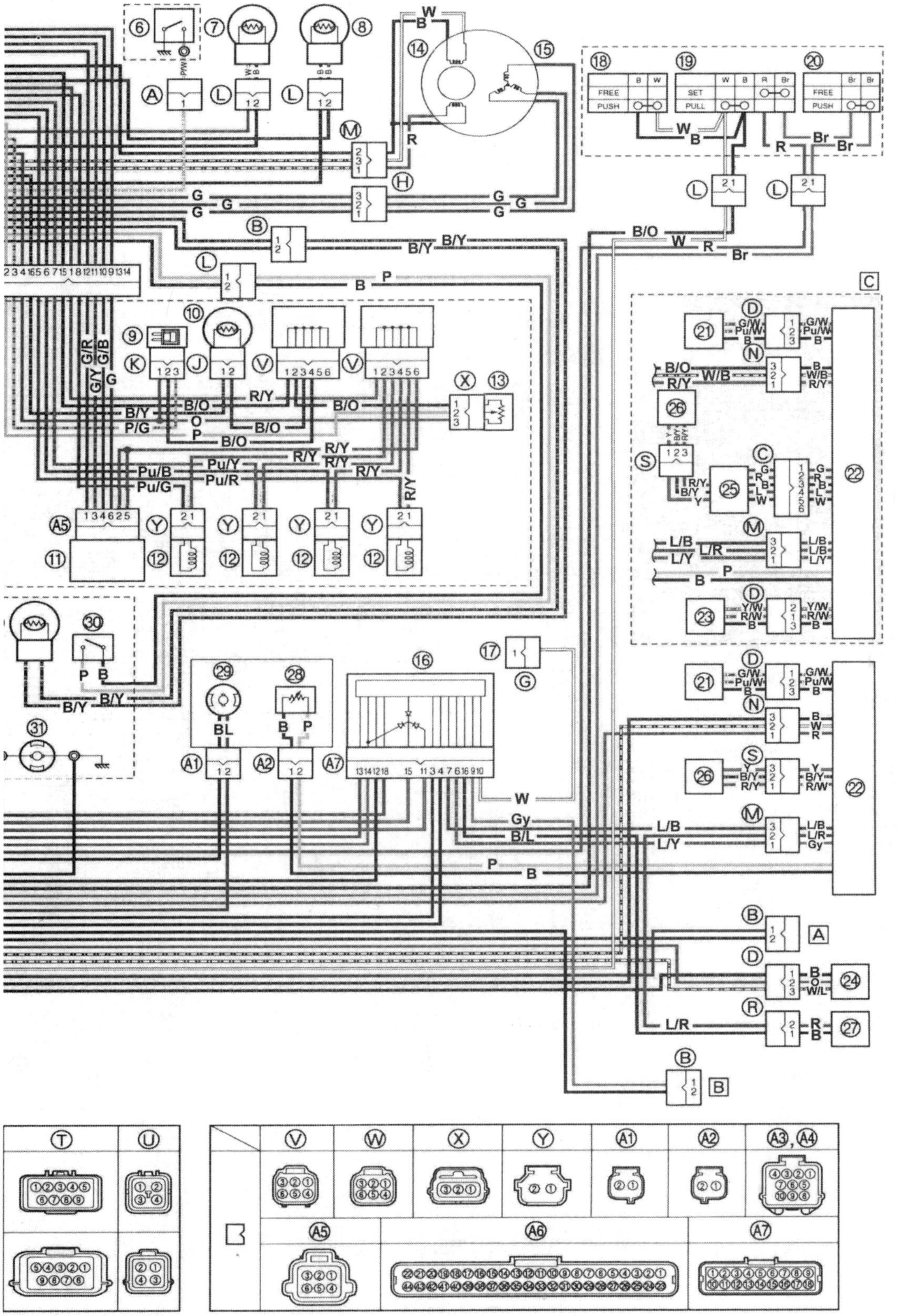

1. Fuse (20 A)
2. Fuse (3 A)
3. Starter relay
4. Noise filter
5. Cam position sensor
6. Oil pressure switch
7. Exhaust temperature sensor
8. Thermo sensor
9. Intake air pressure sensor
10. Intake air temperature sensor
11. Bypass valve motor
12. Fuel injector
13. TPS
14. Pickup coil
15. Stator coil
16. Remote control receiver
17. Antenna
18. Engine stop switch
19. Engine shut-off switch
20. Engine start switch
21. Left operation button assembly
22. Multifunction meter
23. Right operation button assembly (FX Cruiser High Output)
24. Steering sensor
25. Compass/air temperature sensor (FX Cruiser High Output)
26. Speed sensor (FX High Output) Speed/water temperature sensor (FX Cruiser High Output)
27. Buzzer
28. Fuel sender
29. Fuel pump
30. Thermoswitch (engine)
31. Starter motor
32. Engine temperature sensor
33. Ignition coil
34. Spark plug
35. Rectifier/regulator
36. Battery
37. Electric bilge pump
38. ECM
39. Slant detection switch
40. Main and fuel pump relay

A To tachometer
B To entry box
C FX Cruiser High Output

Color code

B : Black
Br : Brown
G : Green
Gy : Gray
L : Blue
O : Orange
P : Pink
R : Red
W : White
Y : Yellow
B/Br : Black/brown
B/L : Black/blue
B/O : Black/orange
B/R : Black/red
B/W : Black/white
B/Y : Black/yellow
G/B : Green/black
G/O : Green/orange
G/R : Green/red
G/W : Green/white
G/Y : Green/yellow
L/B : Blue/black
L/R : Blue/red
L/Y : Blue/yellow
O/R : Orange/red
P/G : Pink/green
P/W : Pink/white
Pu/B : Purple/black
Pu/G : Purple/green
Pu/R : Purple/red
Pu/W : Purple/white
Pu/Y : Purple/yellow
R/W : Red/white
R/Y : Red/yellow
W/B : White/black
W/L : White/blue
W/R : White/red
Y/W : Yellow/white

SELOC_9606

Wiring diagram - 2008 FX HO and FX Cruiser HO Models w/6AA Engines (Right Side)

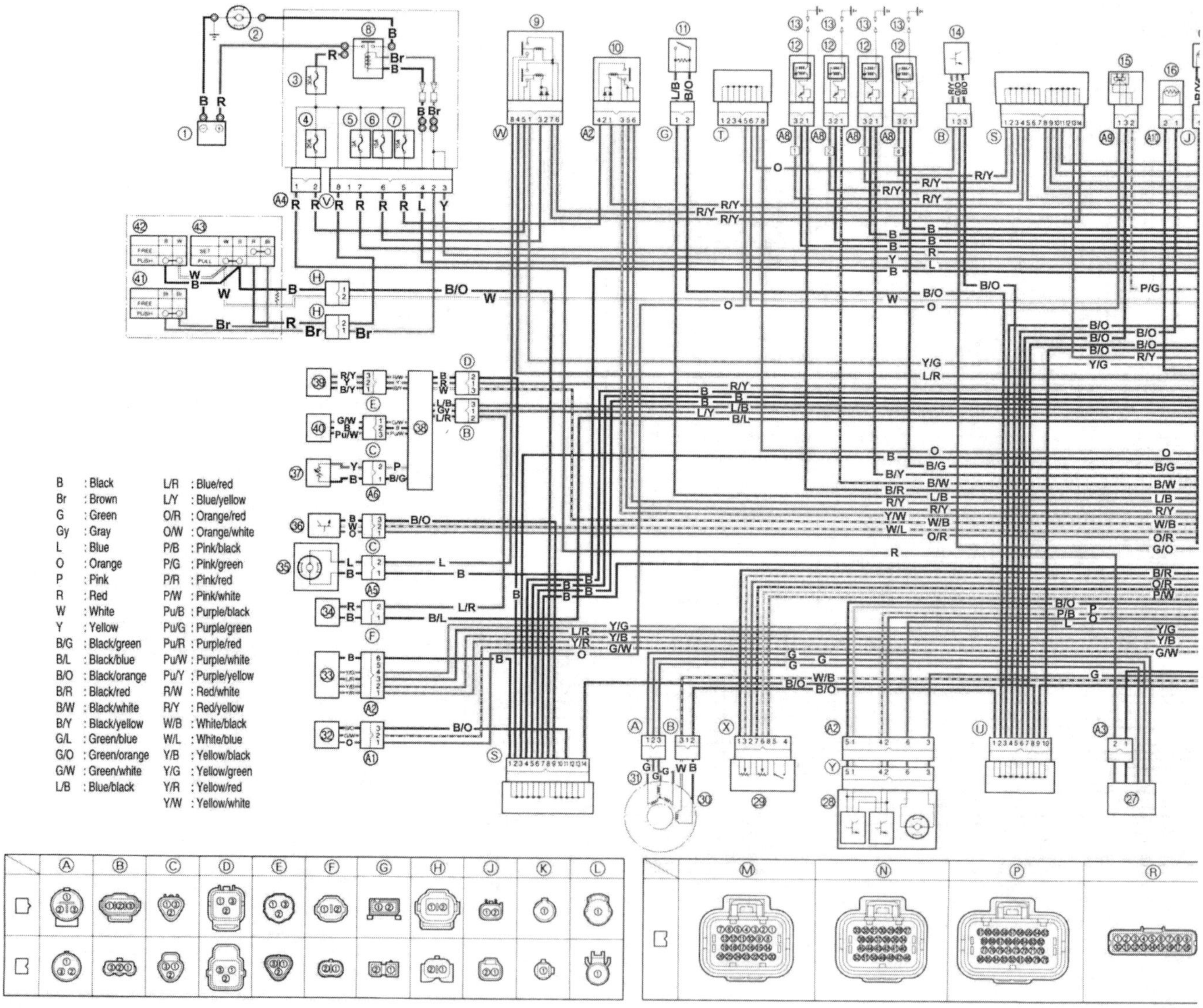

Wiring diagram - 2009-11 FX HO and FX Cruiser HO Models w/1812cc Engines (Left Side)

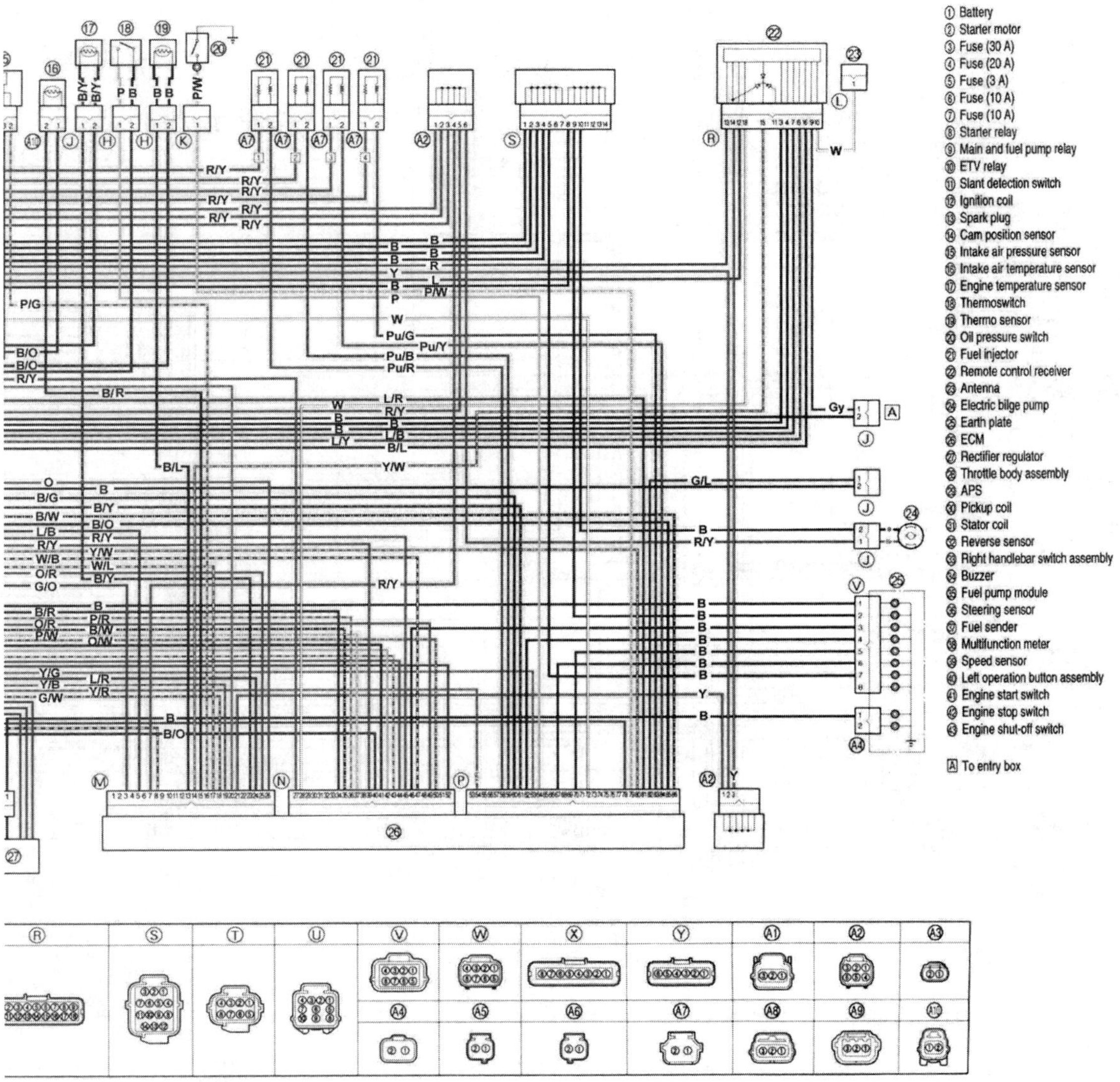

SELOC_9606

Wiring diagram - 2009-11 FX HO and FX Cruiser HO Models w/1812cc Engines (Right Side)

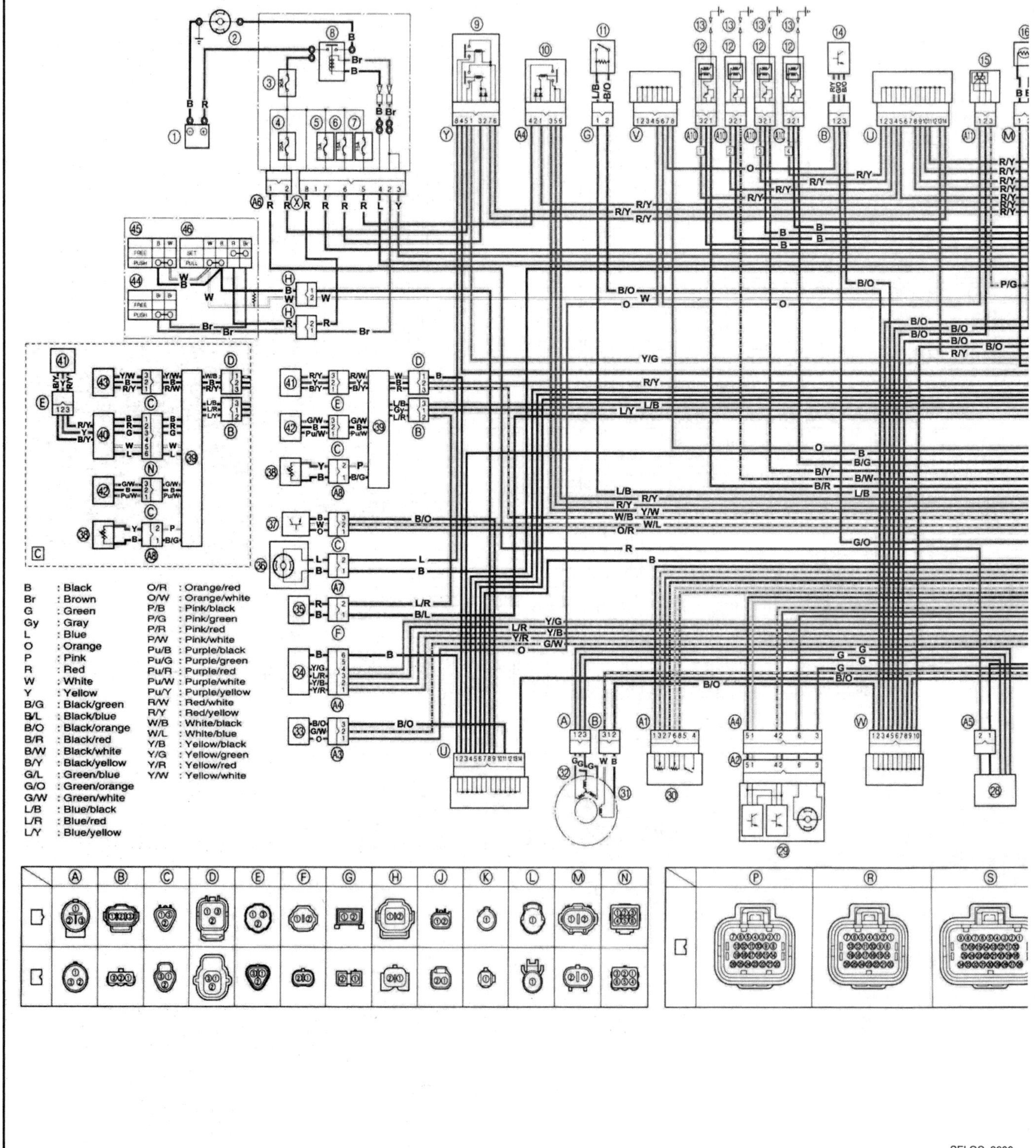

Wiring diagram - 2008-11 FX SHO and FX Cruiser SHO Models w/1812cc Engines (Left Side)

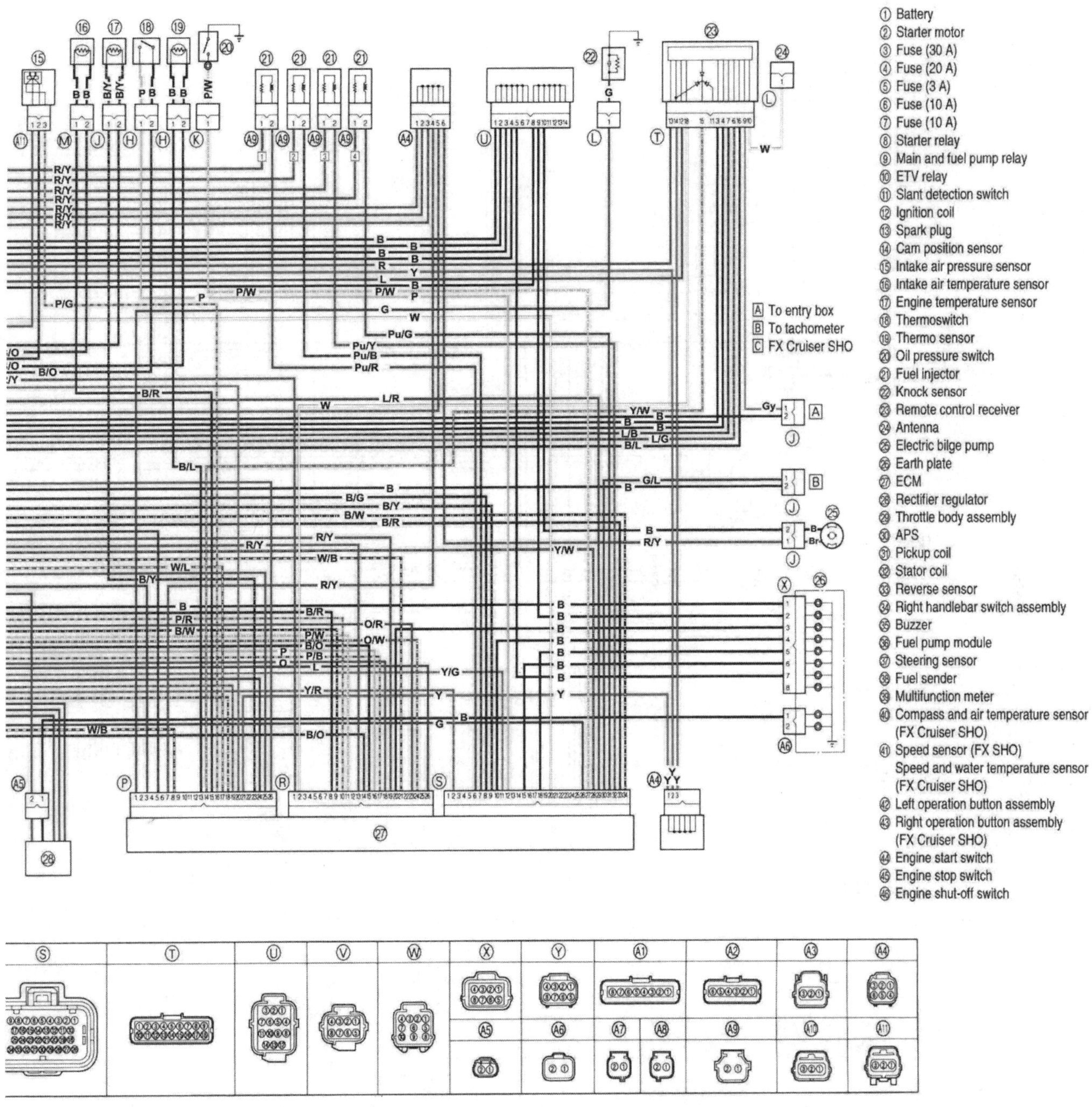

SELOC_9606

Wiring diagram - 2008-11 FX SHO and FX Cruiser SHO Models w/1812cc Engines (Right Side)

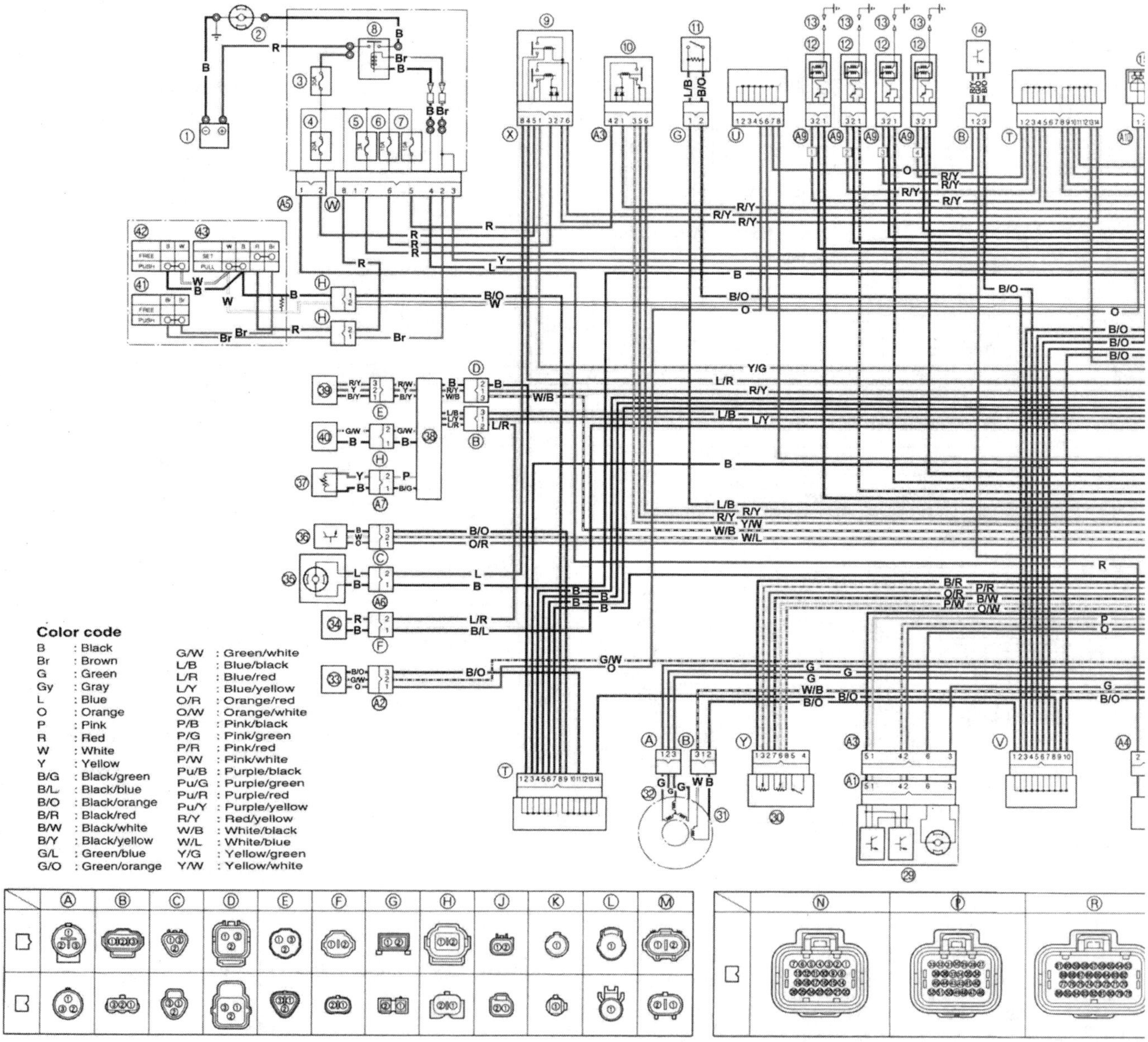

Color code

B	Black	G/W	Green/white
Br	Brown	L/B	Blue/black
G	Green	L/R	Blue/red
Gy	Gray	L/Y	Blue/yellow
L	Blue	O/R	Orange/red
O	Orange	O/W	Orange/white
P	Pink	P/B	Pink/black
R	Red	P/G	Pink/green
W	White	P/R	Pink/red
Y	Yellow	P/W	Pink/white
B/G	Black/green	Pu/B	Purple/black
B/L	Black/blue	Pu/G	Purple/green
B/O	Black/orange	Pu/R	Purple/red
B/R	Black/red	Pu/Y	Purple/yellow
B/W	Black/white	R/Y	Red/yellow
B/Y	Black/yellow	W/B	White/black
G/L	Green/blue	W/L	White/blue
G/O	Green/orange	Y/G	Yellow/green
		Y/W	Yellow/white

SELOC_9606

Wiring diagram - 2009-11 FZR and FZS Models w/1812cc Engines (Left Side)

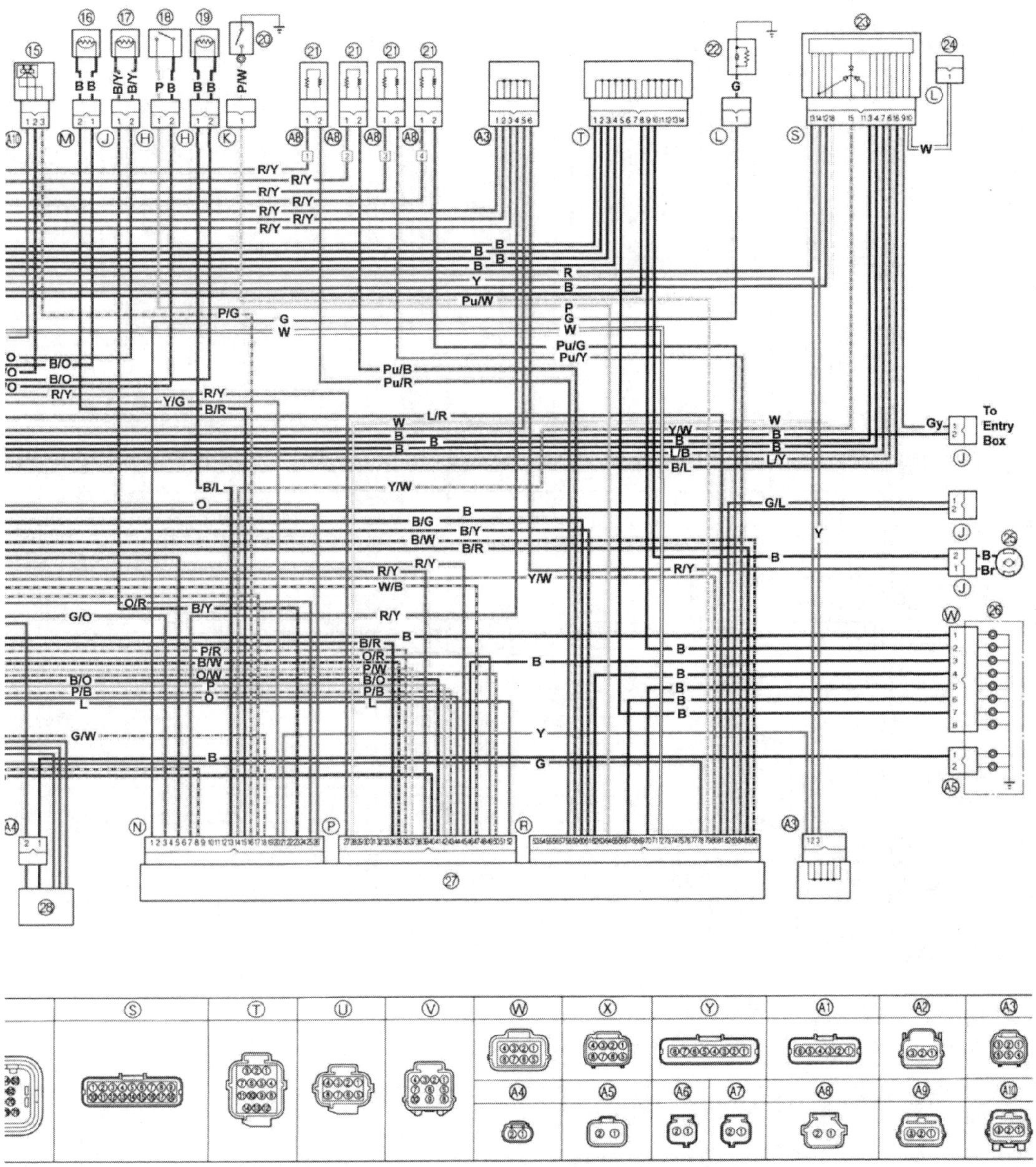

① Battery
② Starter motor
③ Fuse (30 A)
④ Fuse (20 A)
⑤ Fuse (3 A)
⑥ Fuse (10 A)
⑦ Fuse (10 A)
⑧ Starter relay
⑨ Main and fuel pump relay
⑩ ETV relay
⑪ Slant detection switch
⑫ Ignition coil
⑬ Spark plug
⑭ Cam position sensor
⑮ Intake air pressure sensor
⑯ Intake air temperature sensor
⑰ Engine temperature sensor
⑱ Thermoswitch
⑲ Thermo sensor
⑳ Oil pressure switch
㉑ Fuel injector
㉒ Knock sensor
㉓ Remote control receiver
㉔ Antenna
㉕ Electric bilge pump
㉖ Earth plate
㉗ ECM
㉘ Rectifier regulator
㉙ Throttle body assembly
㉚ APS
㉛ Pickup coil
㉜ Stator coil
㉝ Reverse sensor
㉞ Buzzer
㉟ Fuel pump module
㊱ Steering sensor
㊲ Fuel sender
㊳ Dual analog meter unit
㊴ Speed sensor
㊵ Switch assembly
㊶ Engine start switch
㊷ Engine stop switch
㊸ Engine shut-off switch

SELOC_9606

Wiring diagram - 2009-11 FZR and FZS Models w/1812cc Engines (Right Side)

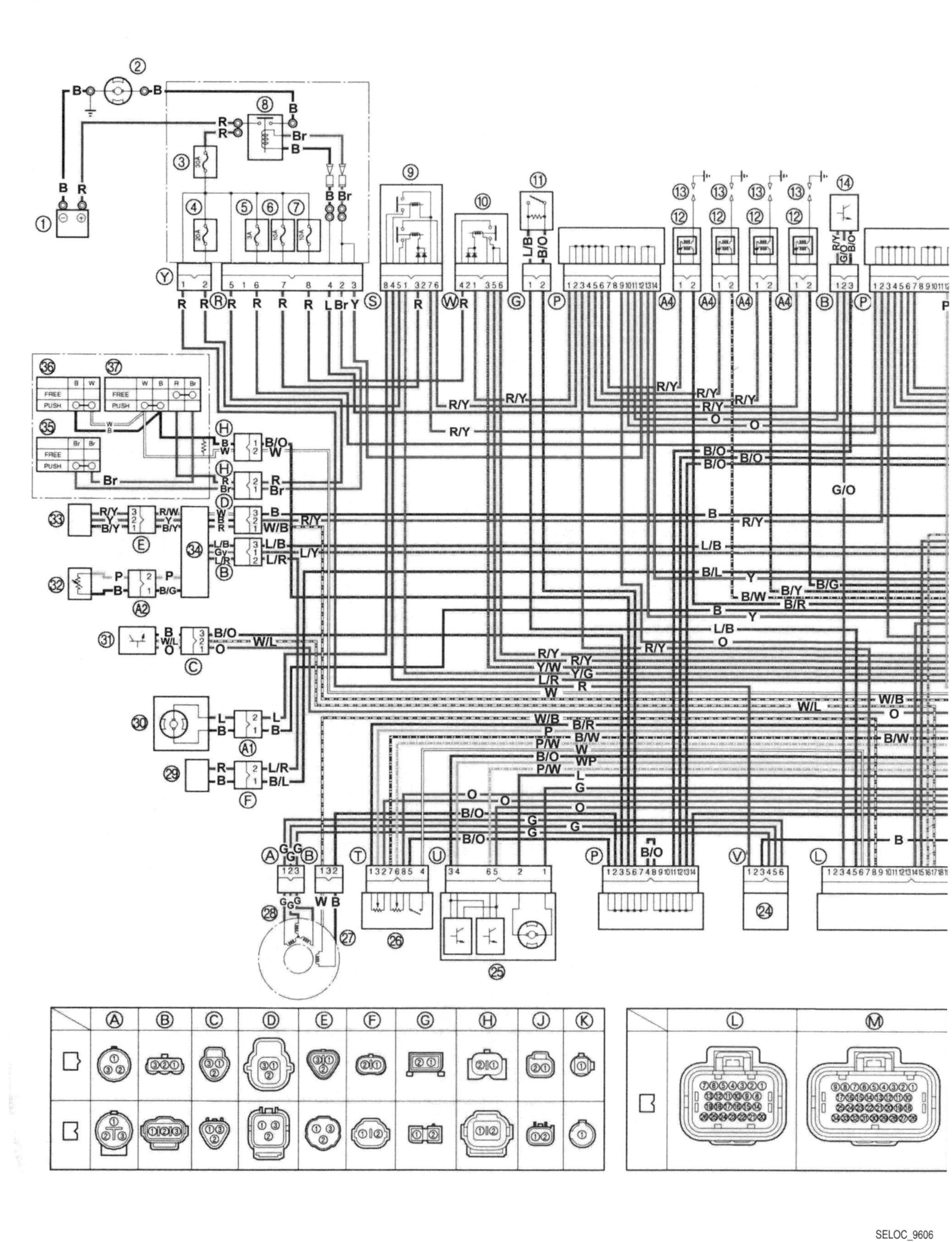

Wiring diagram - 2005-09 VX Cruiser and VX Deluxe Models (Left Side)

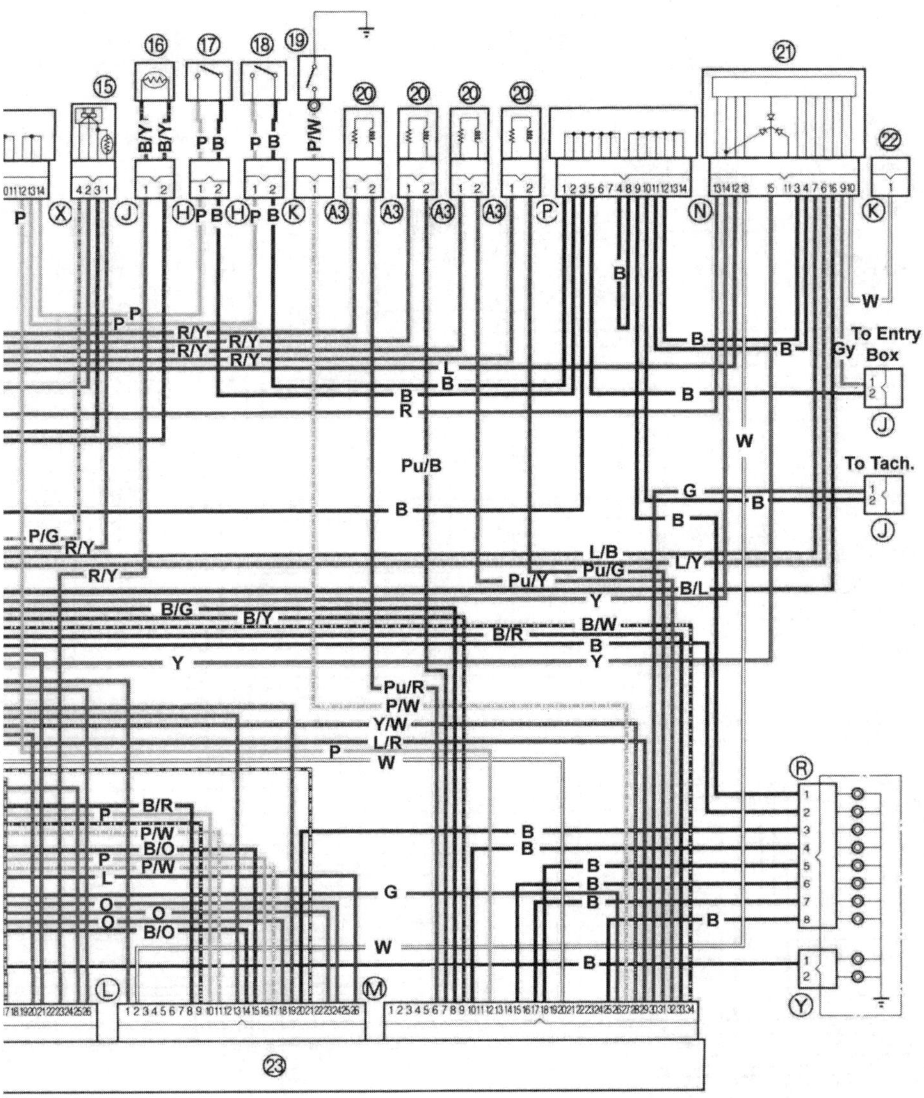

1. Battery
2. Starter motor
3. Fuse (30 A)
4. Fuse (20 A)
5. Fuse (3 A)
6. Fuse (10 A)
7. Fuse (10 A)
8. Starter relay
9. Main and fuel pump relay
10. Electronic control throttle valve relay
11. Slant detection switch
12. Ignition coil
13. Spark plug
14. Cam position sensor
15. Sensor assembly (intake air pressure and intake air temperature)
16. Engine temperature sensor
17. Thermoswitch (engine)
18. Thermoswitch (exhaust)
19. Oil pressure switch
20. Fuel injector
21. Remote control unit
22. Antenna
23. ECM
24. Rectifier/regulator
25. Throttle body assembly
26. Accelerator position sensor
27. Pickup coil
28. Lighting coil
29. Buzzer
30. Electric fuel pump
31. Steering sensor
32. Fuel sender
33. Speed sensor
34. Multifunction meter
35. Start switch
36. Engine stop switch
37. Engine shut-off switch

Color code

B : Black
Br : Brown
G : Green
Gy : Gray
L : Blue
O : Orange
P : Pink
R : Red
W : White
Y : Yellow
B/G : Black/green
B/O : Black/orange
B/R : Black/red
B/W : Black/white
B/Y : Black/yellow
G/B : Green/black
G/O : Green/orange
G/R : Green/red
G/Y : Green/yellow
L/B : Blue/black
L/R : Blue/red
L/Y : Blue/yellow
P/G : Pink/green
P/W : Pink/white
Pu/B : Purple/black
Pu/G : Purple/green
Pu/R : Purple/red
Pu/Y : Purple/yellow
R/W : Red/white
R/Y : Red/yellow
W/B : White/black
W/L : White/blue
W/R : White/red
Y/G : Yellow/green
Y/W : Yellow/white

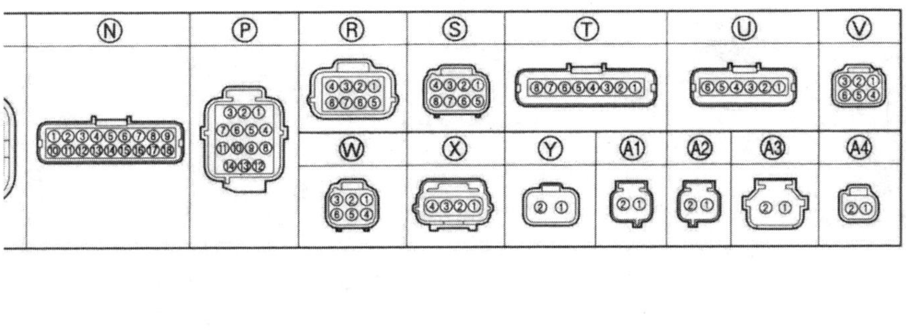

SELOC_9606

Wiring diagram - 2005-09 VX Cruiser and VX Deluxe Models (Right Side)

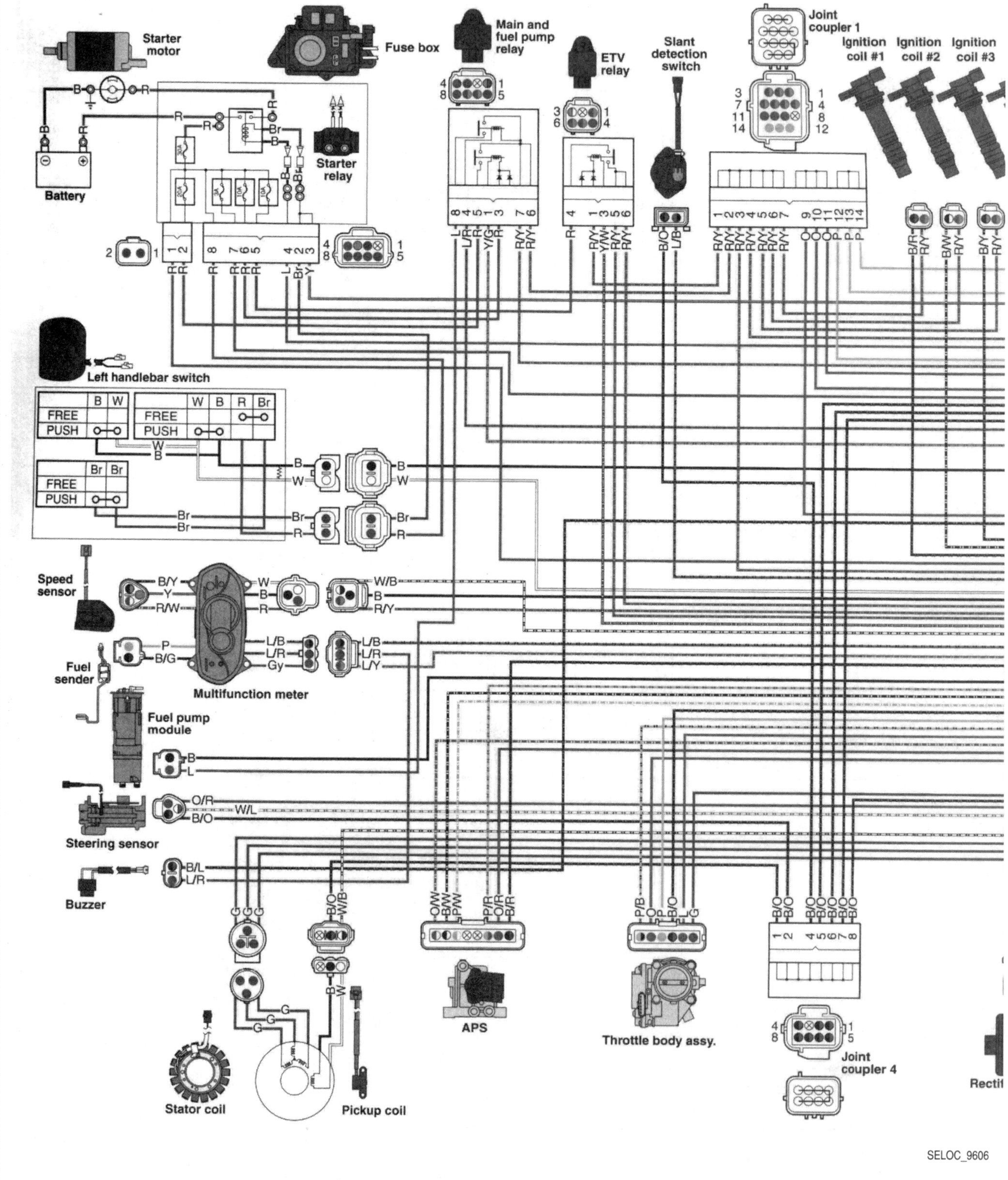

Wiring diagram - 2010-11 VX Cruiser and VX Deluxe Models (Left Side)

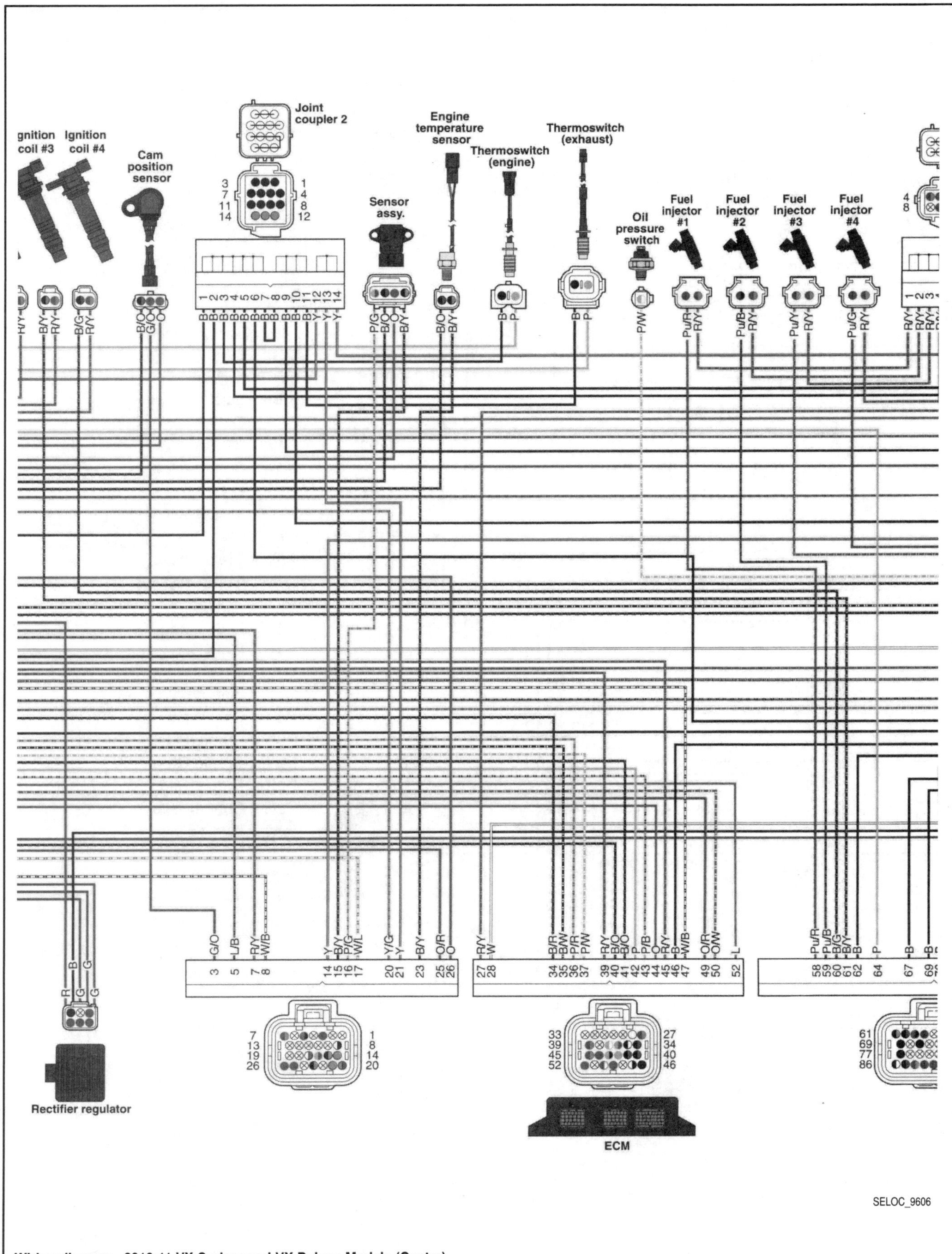

Wiring diagram - 2010-11 VX Cruiser and VX Deluxe Models (Center)

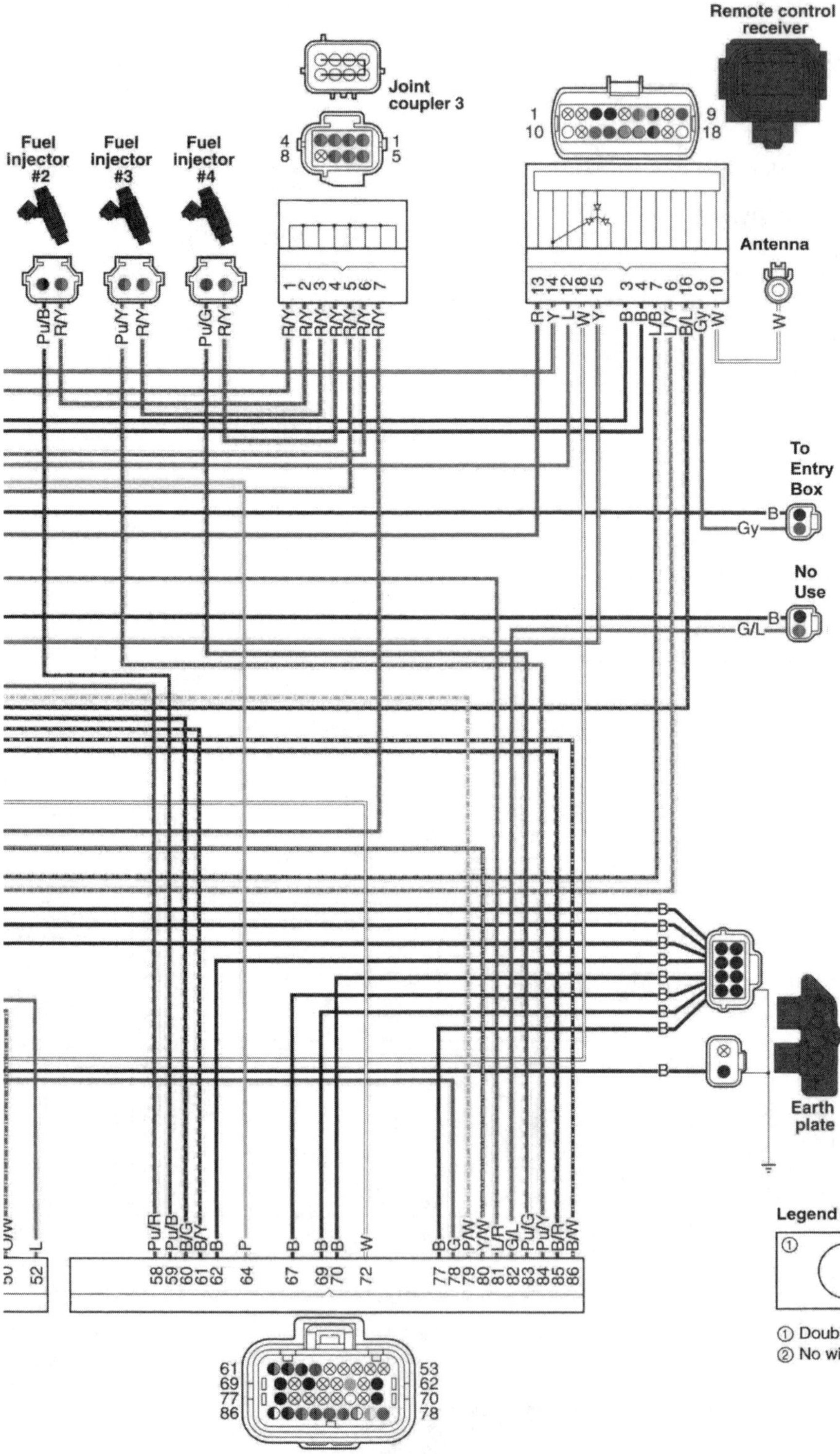

Color code

B	: Black
Br	: Brown
G	: Green
L	: Blue
O	: Orange
P	: Pink
R	: Red
W	: White
Y	: Yellow
B/G	: Black/Green
B/O	: Black/Orange
B/R	: Black/Red
B/L	: Black/Blue
Gy	: Gray
B/W	: Black/White
B/Y	: Black/Yellow
G/O	: Green/Orange
L/B	: Blue/Black
L/R	: Blue/Red
O/R	: Orange/Red
O/W	: Orange/White
P/B	: Pink/Black
P/G	: Pink/Green
P/R	: Pink/Red
L/Y	: Blue/Yellow
G/L	: Green/Blue
P/W	: Pink/White
Pu/B	: Purple/Black
Pu/G	: Purple/Green
Pu/R	: Purple/Red
Pu/Y	: Purple/Yellow
R/Y	: Red/Yellow
W/B	: White/Black
W/L	: White/Blue
Y/G	: Yellow/Green
Y/W	: Yellow/White

Legend symbols in the wiring diagrams

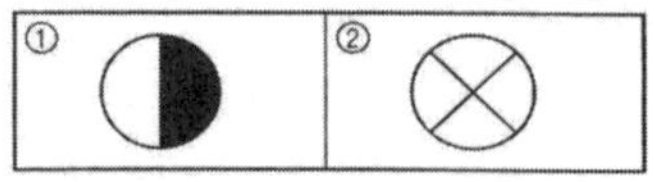

① Double colors wire
② No wire connector

SELOC_9606

Wiring diagram - 2010-11 VX Cruiser and VX Deluxe Models (Right Side)

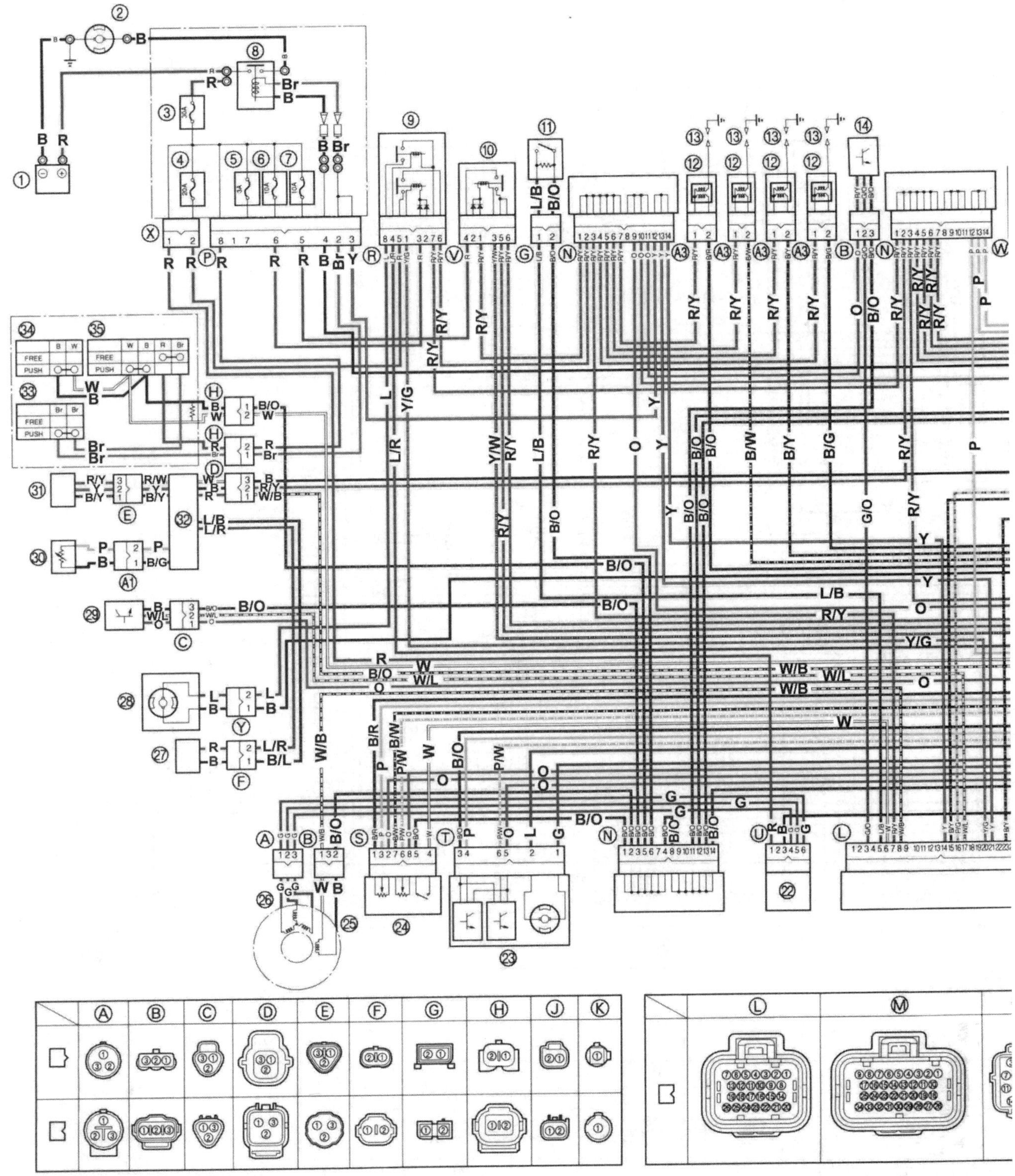

SELOC_9606

Wiring diagram - 2005-09 VX Sport Models (Left Side)

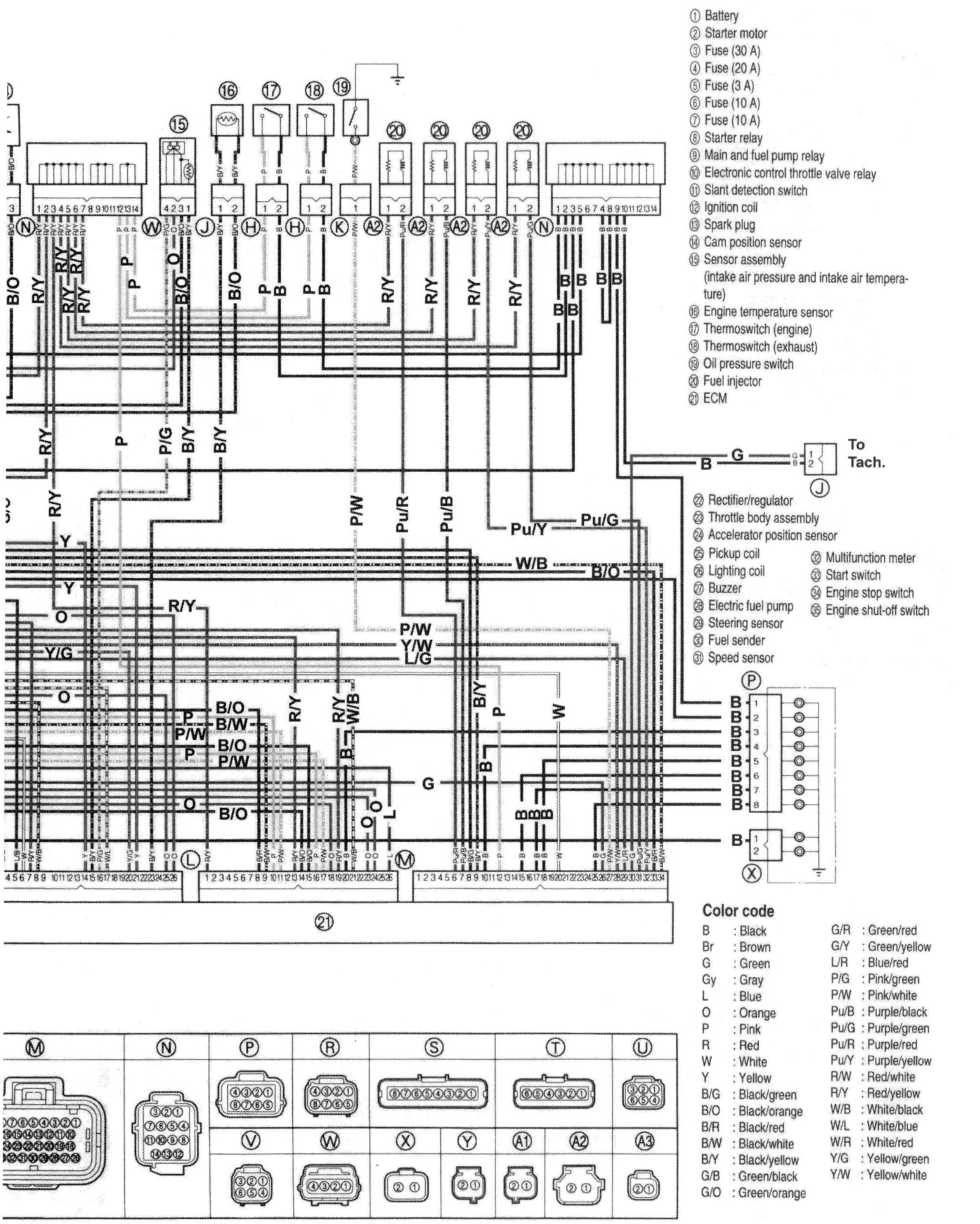

Wiring diagram - 2005-09 VX Sport Models (Right Side)

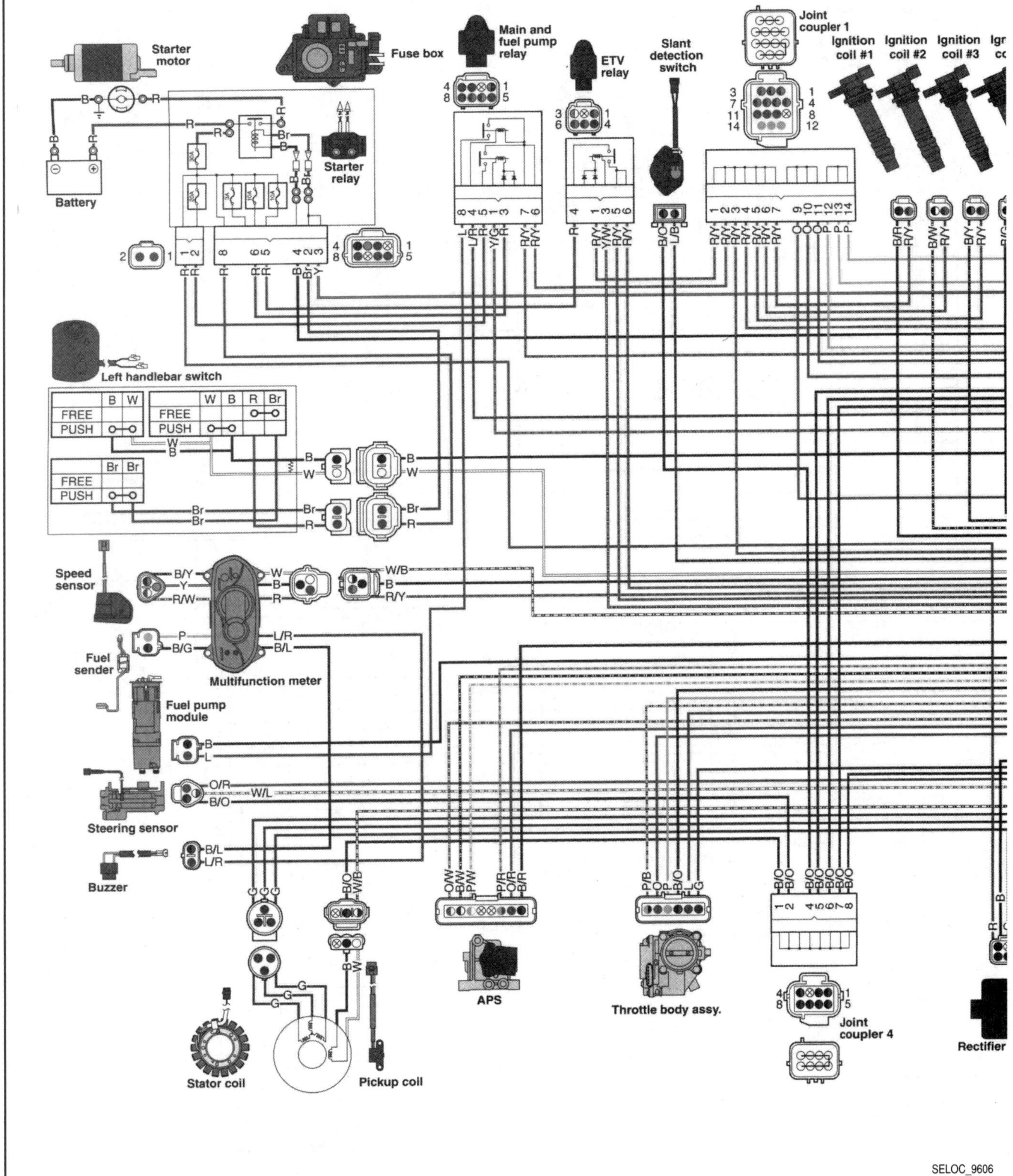

Wiring diagram - 2010-11 VX Sport Models (Left Side)

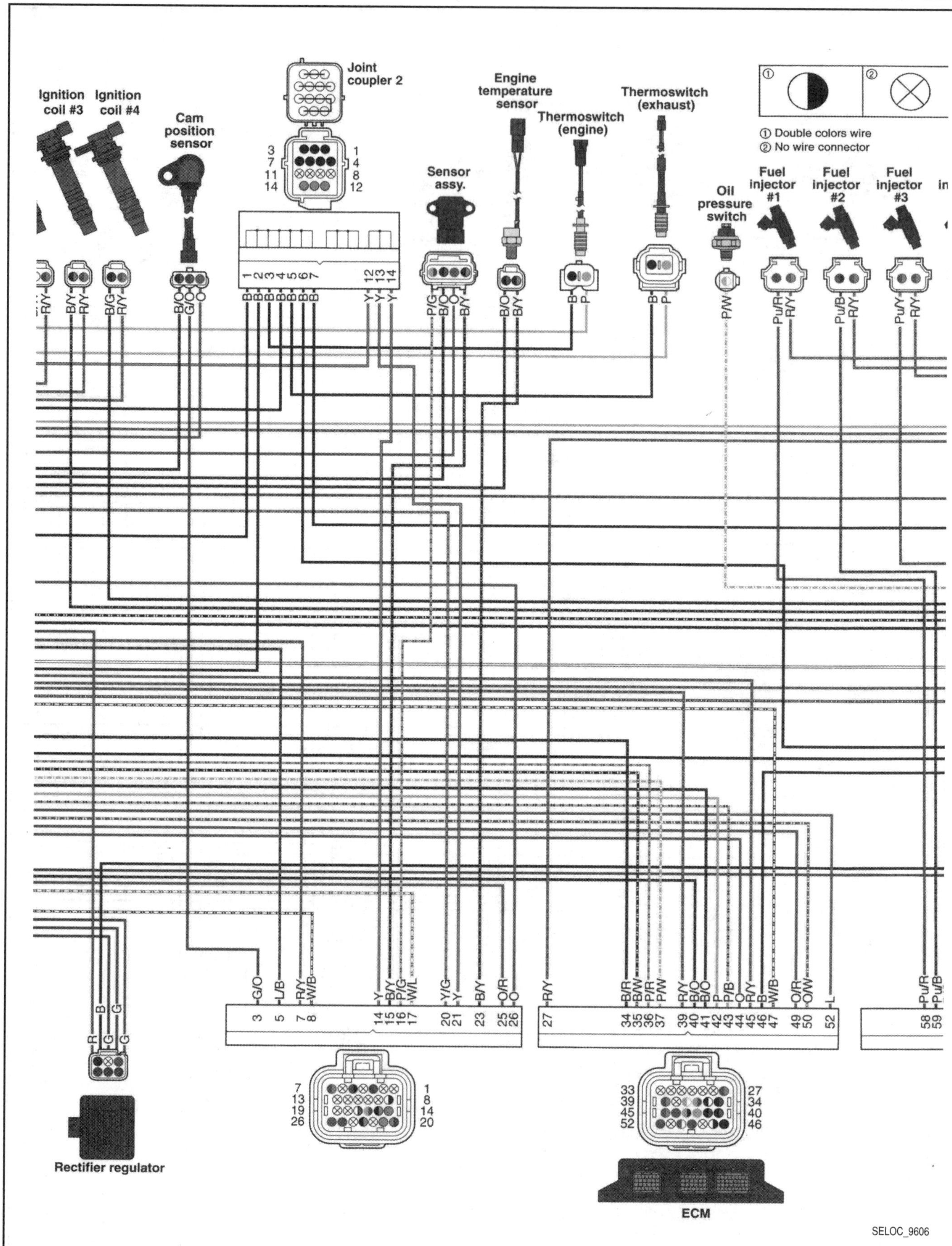

Wiring diagram - 2010-11 VX Sport Models (Center)

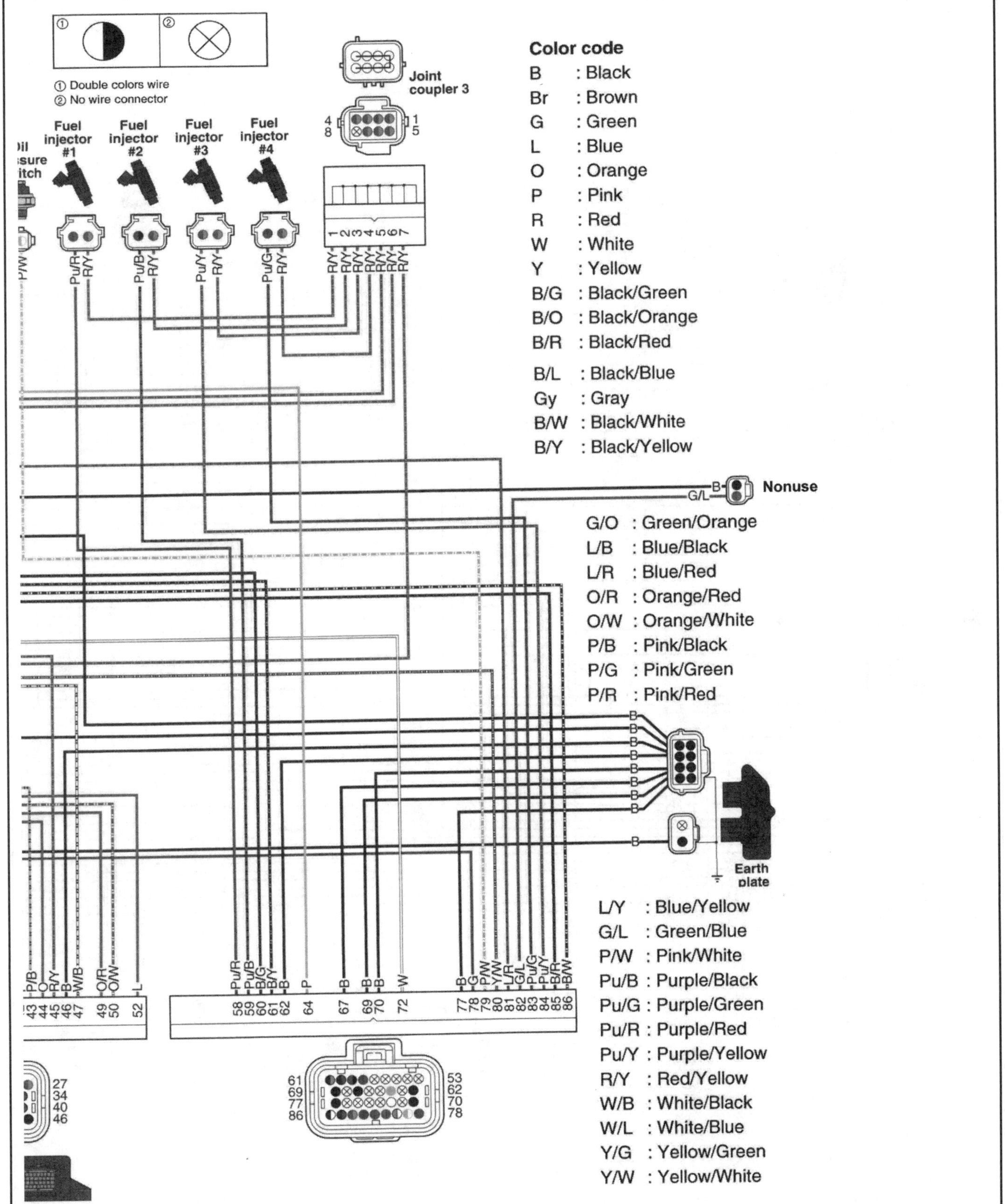

SELOC_9606

Wiring diagram - 2010-11 VX Sport Models (Right Side)

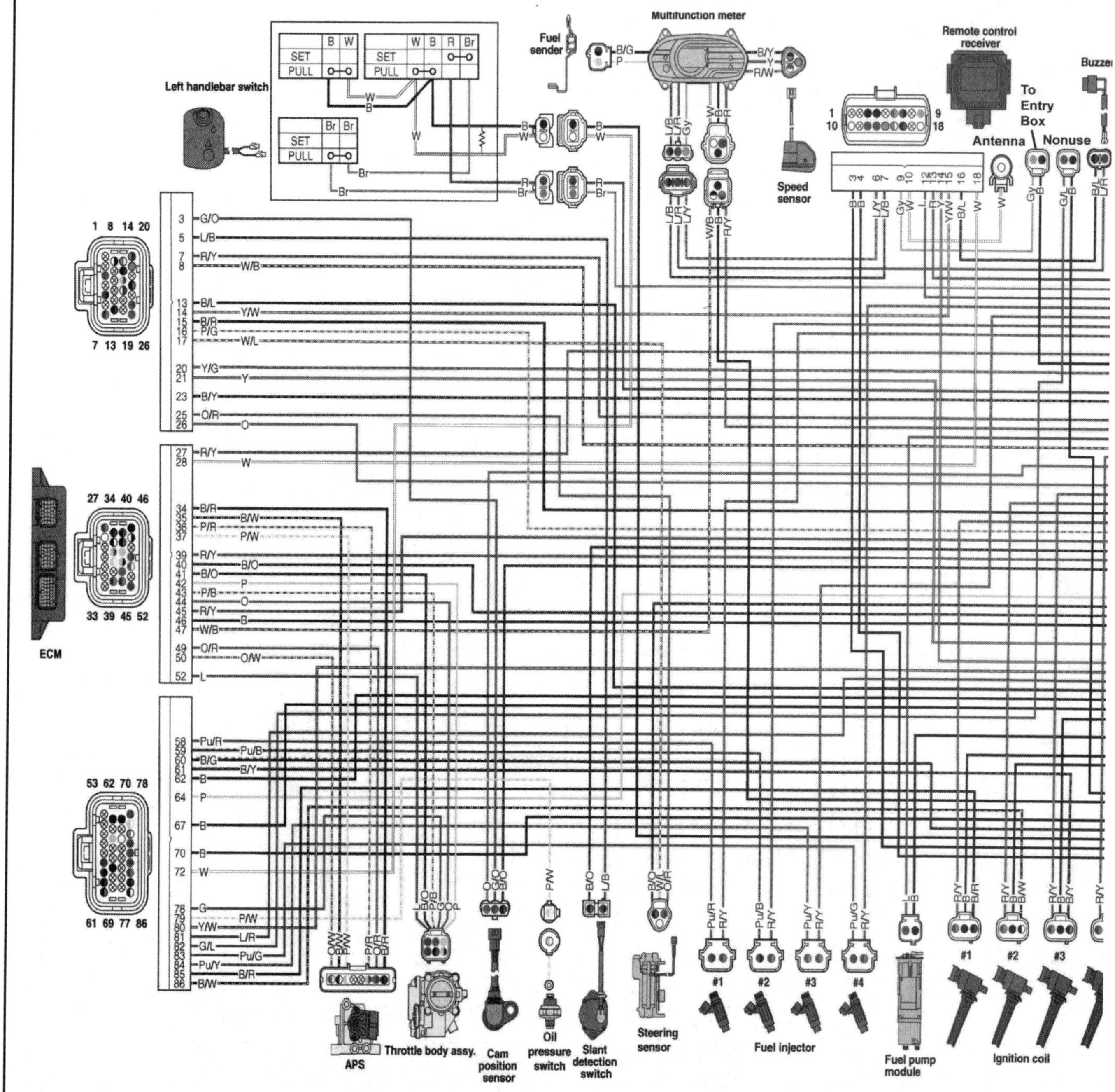

Wiring diagram - 2011 VXR and VXS Models (Left Side)

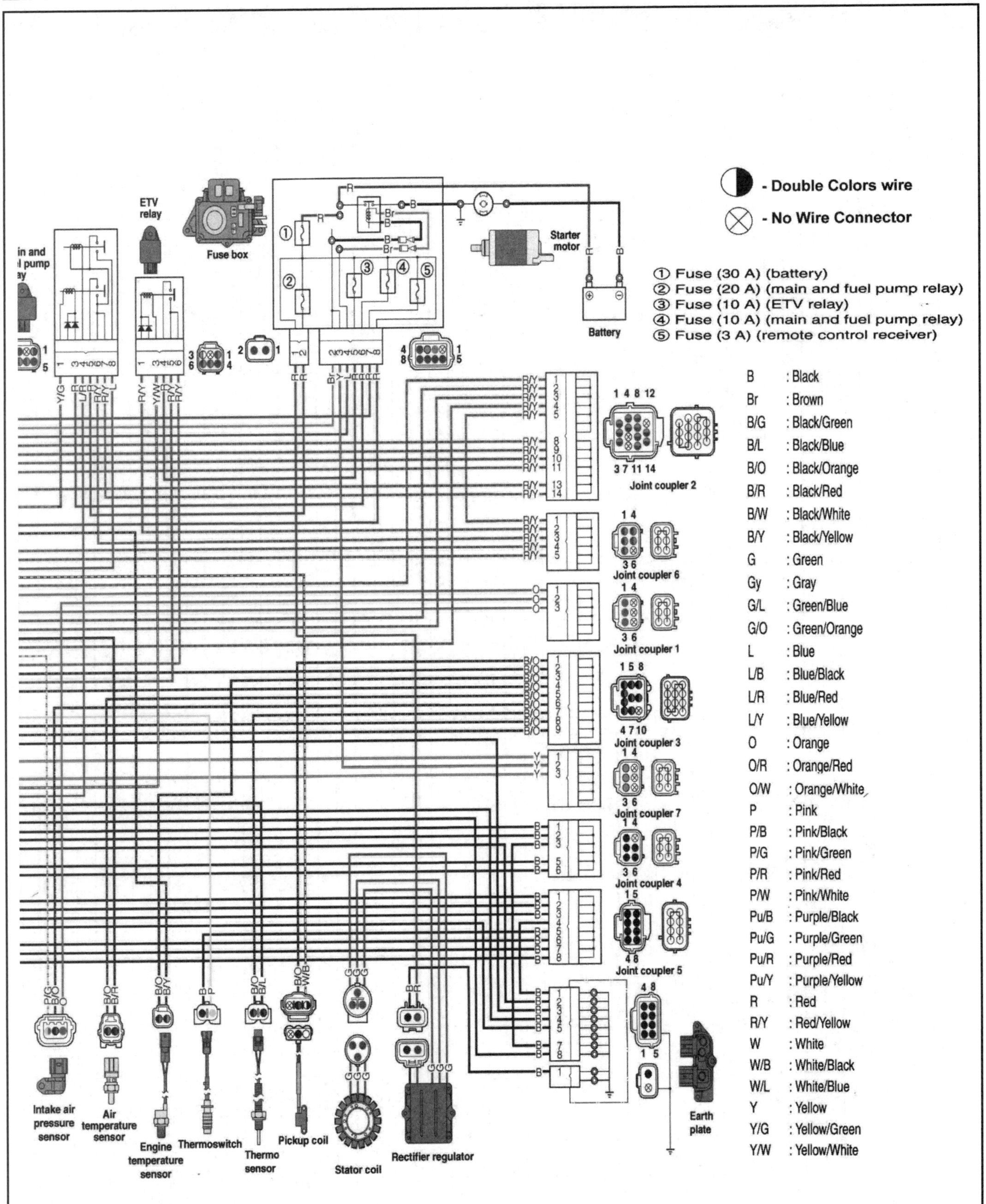

Wiring diagram - 2011 VXR and VXS Models (Right Side)

Electronic Systems Testing Specifications

		ECM Peak Output			Pulser/Pick-Up Coil Peak Peak Output			Lighting/Stator Coil Peak Output			Ignition Coil Resistance			Rectifier/ Regulator
Year	Model Name	V @ Cranking	V @ 2000 rpm	V @ 3500 rpm	V @ Cranking	V @ 2000 rpm	V @ 3500 rpm	V @ Cranking	V @ 2000 rpm	V @ 3500 rpm	Min. Spark Gap in. (mm)	Primary (ohm)	Secondary (Kohm)	Peak Output V @ 3500 rpm
2002	FX140	7	258	258	①	23	38	②	11	12	0.28-0.31 (7-8)	1.53-2.07	12.5-16.9	14.5
2003	FX Cruiser	7	258	258	①	23	38	②	11	12	0.28-0.31 (7-8)	1.53-2.07	12.5-16.9	14.5
	FX140	7	258	258	①	23	38	②	11	12	0.28-0.31 (7-8)	1.53-2.07	12.5-16.9	14.5
2004	FX Cruiser	7	258	258	①	23	38	②	11	12	0.28-0.31 (7-8)	1.53-2.07	12.5-16.9	14.5
	FX Cruiser HO	87	86	85	③	26.1	41.3	9.2	38.5	63.2	-	1.19-1.61	8.5-11.5	15
	FX HO	87	86	85	③	26.1	41.3	9.2	38.5	63.2	-	1.19-1.61	8.5-11.5	15
	FX140	7	258	258	①	23	38	②	11	12	0.28-0.31 (7-8)	1.53-2.07	12.5-16.9	14.5
2005	FX	7	258	258	①	23	38	②	11	12	0.28-0.31 (7-8)	1.53-2.07	12.5-16.9	14.5
	FX Cruiser	7	258	258	①	23	38	②	11	12	0.28-0.31 (7-8)	1.53-2.07	12.5-16.9	14.5
	FX Cruiser HO	87	86	85	③	26.1	41.3	9.2	38.5	63.2	-	1.19-1.61	8.5-11.5	15
	FX HO	87	86	85	③	26.1	41.3	9.2	38.5	63.2	-	1.19-1.61	8.5-11.5	15
	VX110 Deluxe	90	130	140	④	19.4	25.1	7.8	24.5	41.6	-	1.19-1.61	8.5-11.5	13
	VX110 Sport	90	130	140	④	19.4	25.1	7.8	24.5	41.6	-	1.19-1.61	8.5-11.5	13
2006	FX	87	86	85	①	23	38	9	11	12	-	1.19-1.61	8.5-11.5	15
	FX Cruiser	87	86	85	①	23	38	9	11	12	-	1.19-1.61	8.5-11.5	15
	FX Cruiser HO	87	86	85	③	26.1	41.3	9.2	38.5	63.2	-	1.19-1.61	8.5-11.5	15
	FX HO	87	86	85	③	26.1	41.3	9.2	38.5	63.2	-	1.19-1.61	8.5-11.5	15
	VX110 Deluxe	90	130	140	④	19.4	25.1	7.8	24.5	41.6	-	1.19-1.61	8.5-11.5	13
	VX110 Sport	90	130	140	④	19.4	25.1	7.8	24.5	41.6	-	1.19-1.61	8.5-11.5	13
2007	FX	87	86	85	①	23	38	9	11	12	-	1.19-1.61	8.5-11.5	15
	FX Cruiser	87	86	85	①	23	38	9	11	12	-	1.19-1.61	8.5-11.5	15
	FX Cruiser HO	87	86	85	③	26.1	41.3	9.2	38.5	63.2	-	1.19-1.61	8.5-11.5	15
	FX HO	87	86	85	③	26.1	41.3	9.2	38.5	63.2	-	1.19-1.61	8.5-11.5	15
	VX	90	130	140	④	19.4	25.1	7.8	24.5	41.6	-	1.19-1.61	8.5-11.5	13
	VX Crusier	90	130	140	④	19.4	25.1	7.8	24.5	41.6	-	1.19-1.61	8.5-11.5	13
	VX110 Deluxe	90	130	140	④	19.4	25.1	7.8	24.5	41.6	-	1.19-1.61	8.5-11.5	13
	VX110 Sport	90	130	140	④	19.4	25.1	7.8	24.5	41.6	-	1.19-1.61	8.5-11.5	13
2008	FX	87	86	85	①	23	38	9	11	12	-	1.19-1.61	8.5-11.5	15
	FX Cruiser	87	86	85	①	23	38	9	11	12	-	1.19-1.61	8.5-11.5	15
	FX Cruiser HO	90	110	130	③	26.1	41.3	9.2	38.5	63.2	-	1.19-1.61	8.5-11.5	15
	FX Cruiser SHO	2.5	2.6	2.6	⑤	18.1	23.9	8.4	42.7	74	-	-	-	13
	FX HO	90	110	130	③	26.1	41.3	9.2	38.5	63.2	-	1.19-1.61	8.5-11.5	15
	FX SHO	2.5	2.6	2.6	⑤	18.1	23.9	8.4	42.7	74	-	-	-	13
	VX	90	130	140	④	19.4	25.1	7.8	24.5	41.6	-	1.19-1.61	8.5-11.5	13
	VX Crusier	90	130	140	④	19.4	25.1	7.8	24.5	41.6	-	1.19-1.61	8.5-11.5	13
	VX Deluxe	90	130	140	④	19.4	25.1	7.8	24.5	41.6	-	1.19-1.61	8.5-11.5	13
	VX110 Sport	90	130	140	④	19.4	25.1	7.8	24.5	41.6	-	1.19-1.61	8.5-11.5	13
2009	FX Cruiser HO	2.5	2.6	2.6	⑥	16.4	21.5	8.3	36.4	64.7	-	-	-	13
	FX Cruiser SHO	2.5	2.6	2.6	⑤	18.1	23.9	8.4	42.7	74	-	-	-	13
	FX HO	2.5	2.6	2.6	⑥	16.4	21.5	8.3	36.4	64.7	-	-	-	13
	FX SHO	2.5	2.6	2.6	⑤	18.1	23.9	8.4	42.7	74	-	-	-	13
	FZR	2.5	2.6	2.6	⑤	18.1	23.9	8.4	42.7	74	-	-	-	13
	FZS	2.5	2.6	2.6	⑤	18.1	23.9	8.4	42.7	74	-	-	-	13
	VX	90	130	140	④	19.4	25.1	7.8	24.5	41.6	-	1.19-1.61	8.5-11.5	13

SELOC_9606

Electronic Systems Testing Specifications

Year	Model Name	ECM Peak Output V @ Cranking	ECM Peak Output V @ 2000 rpm	ECM Peak Output V @ 3500 rpm	Pulser/Pick-Up Coil Peak Peak Output V @ Cranking	Pulser/Pick-Up Coil Peak Peak Output V @ 2000 rpm	Pulser/Pick-Up Coil Peak Peak Output V @ 3500 rpm	Lighting/Stator Coil Peak Output V @ Cranking	Lighting/Stator Coil Peak Output V @ 2000 rpm	Lighting/Stator Coil Peak Output V @ 3500 rpm	Ignition Coil Resistance Min. Spark Gap in. (mm)	Ignition Coil Resistance Primary (ohm)	Ignition Coil Resistance Secondary (Kohm)	Rectifier/ Regulator Peak Output V @ 3500 rpm
2009	VX Cruiser	90	130	140	④	19.4	25.1	7.8	24.5	41.6	-	1.19-1.61	8.5-11.5	13
(cont'd)	VX Deluxe	90	130	140	④	19.4	25.1	7.8	24.5	41.6	-	1.19-1.61	8.5-11.5	13
	VX Sport	90	130	140	④	19.4	25.1	7.8	24.5	41.6	-	1.19-1.61	8.5-11.5	13
2010	FX Cruiser HO	2.5	2.6	2.6	⑥	16.4	21.5	8.3	36.4	64.7	-	-	-	13
	FX Cruiser SHO	2.5	2.6	2.6	⑤	18.1	23.9	8.4	42.7	74	-	-	-	13
	FX HO	2.5	2.6	2.6	⑥	16.4	21.5	8.3	36.4	64.7	-	-	-	13
	FX SHO	2.5	2.6	2.6	⑤	18.1	23.9	8.4	42.7	74	-	-	-	13
	FZR	2.5	2.6	2.6	⑤	18.1	23.9	8.4	42.7	74	-	-	-	13
	FZS	2.5	2.6	2.6	⑤	18.1	23.9	8.4	42.7	74	-	-	-	13
	VX Cruiser	90	130	140	④	19.4	25.1	7.8	24.5	41.6	-	1.19-1.61	8.5-11.5	13
	VX Deluxe	90	130	140	④	19.4	25.1	7.8	24.5	41.6	-	1.19-1.61	8.5-11.5	13
	VX Sport	90	130	140	④	19.4	25.1	7.8	24.5	41.6	-	1.19-1.61	8.5-11.5	13
2011	FX Cruiser HO	2.5	2.6	2.6	⑥	16.4	21.5	8.3	36.4	64.7	-	-	-	13
	FX Cruiser SHO	2.5	2.6	2.6	⑤	18.1	23.9	8.4	42.7	74	-	-	-	13
	FX HO	2.5	2.6	2.6	⑥	16.4	21.5	8.3	36.4	64.7	-	-	-	13
	FX SHO	2.5	2.6	2.6	⑤	18.1	23.9	8.4	42.7	74	-	-	-	13
	FZR	2.5	2.6	2.6	⑤	18.1	23.9	8.4	42.7	74	-	-	-	13
	FZS	2.5	2.6	2.6	⑤	18.1	23.9	8.4	42.7	74	-	-	-	13
	VX Cruiser	90	130	140	④	19.4	25.1	7.8	24.5	41.6	-	1.19-1.61	8.5-11.5	13
	VX Deluxe	90	130	140	④	19.4	25.1	7.8	24.5	41.6	-	1.19-1.61	8.5-11.5	13
	VX Sport	90	130	140	④	19.4	25.1	7.8	24.5	41.6	-	1.19-1.61	8.5-11.5	13
	VXR	2.5	2.6	2.6	⑤	16.4	21.5	8.3	36.4	64.7	-	-	-	13
	VXS	2.5	2.6	2.6	⑤	16.4	21.5	8.3	36.4	64.7	-	-	-	13

SELOC_9606

① Cranking 1: 4V, Cranking 2: 4V
② Cranking 1: 9V, Cranking 2: 8V
③ Cranking unloaded: 5.2V, Cranking loaded: 4.7V
④ Cranking unloaded: 8V, Cranking loaded: 6.9V
⑤ Cranking unloaded: 5.4V, Cranking loaded: 4.6V
⑥ Cranking unloaded: 5.5V, Cranking loaded: 4.7V

Electronic Component Test Specifications

Year	Model Name	Thermoswitch On/Eng (°F)	Thermoswitch Off/Eng (°F)	Thermoswitch On/Exh (°F)	Thermoswitch Off/Exh (°F)	Eng Temp Sensor Resistance @ 68°F (k ohms)	Eng Temp Sensor Resistance @ 212°F (k ohms)	Intake Air Temp Sensor Resistance @ 32°F (k ohms)	Intake Air Temp Sensor Resistance @ 176°F (k ohms)	Speed Sensor Output (V)	TPS Output @ Trolling (V)	Cam Position Sensor Output @ A (V)	Cam Position Sensor Output @ B (V)	Cam Position Sensor Output @ C (V)	Fuel Sender Output @ A (ohm)	Fuel Sender Output @ B (ohm)	Fuel Injector Resistance (ohm)
2002	FX140	183-194	158-183	210-212	176-201	54.2-69.0	3.12-3.48	5.4-6.6	0.29-0.39	11.6	0.76 +/- 0.016	>0.8	<4.8	-	91-93	6-8	14-15
2003	FX Cruiser	183-194	158-183	210-212	176-201	54.2-69.0	3.12-3.48	5.4-6.6	0.29-0.39	11.6	0.76 +/- 0.016	>0.8	<4.8	-	91-93	6-8	14-15
	FX140	183-194	158-183	210-212	176-201	54.2-69.0	3.12-3.48	5.4-6.6	0.29-0.39	11.6	0.76 +/- 0.016	>0.8	<4.8	-	91-93	6-8	14-15
2004	FX Cruiser	183-194	158-183	210-212	176-201	54.2-69.0	3.12-3.48	5.4-6.6	0.29-0.39	11.6	0.76 +/- 0.016	>0.8	<4.8	-	91-93	6-8	14-15
	FX Cruiser HO	183-194	158-183	210-212	176-201	54.2-69.0	3.12-3.48	-	-	11.6	0.76 +/- 0.016	>4.8	<0.8	>4.8	133.5-136.5	5-7	11.5-12.5
	FX HO	183-194	158-183	210-212	176-201	54.2-69.0	3.12-3.48	-	-	11.6	0.76 +/- 0.016	>4.8	<0.8	>4.8	133.5-136.5	5-7	11.5-12.5
	FX140	183-194	158-183	210-212	176-201	54.2-69.0	3.12-3.48	5.4-6.6	0.29-0.39	11.6	0.76 +/- 0.016	>0.8	<4.8	-	91-93	6-8	14-15
2005	FX	183-194	158-183	210-212	176-201	54.2-69.0	3.12-3.48	5.4-6.6	0.29-0.39	11.6	0.76 +/- 0.016	>0.8	<4.8	-	91-93	6-8	14-15
	FX Cruiser	183-194	158-183	210-212	176-201	54.2-69.0	3.12-3.48	5.4-6.6	0.29-0.39	11.6	0.76 +/- 0.016	>0.8	<4.8	-	91-93	6-8	14-15
	FX Cruiser HO	183-194	158-183	210-212	176-201	54.2-69.0	3.12-3.48	-	-	11.6	0.76 +/- 0.016	>4.8	<0.8	>4.8	133.5-136.5	5-7	11.5-12.5
	FX HO	183-194	158-183	210-212	176-201	54.2-69.0	3.12-3.48	-	-	11.6	0.76 +/- 0.016	>4.8	<0.8	>4.8	133.5-136.5	5-7	11.5-12.5
	VX110 Deluxe	183-194	158-183	-	-	54.2-69.0	3.12-3.48	-	-	11.6	④	>4.8	<0.8	>4.8	133.5-136.5	5-7	11.5-12.5
	VX110 Sport	183-194	158-183	-	-	54.2-69.0	3.12-3.48	-	-	11.6	④	>4.8	<0.8	>4.8	133.5-136.5	5-7	11.5-12.5
2006	FX	183-194	158-183	210-212	176-201	54.2-69.0	3.12-3.48	5.4-6.6	0.29-0.39	11.6	0.76 +/- 0.016	>4.8	<0.8	>4.8	133.5-136.5	5-7	11.5-12.5
	FX Cruiser	183-194	158-183	210-212	176-201	54.2-69.0	3.12-3.48	5.4-6.6	0.29-0.39	11.6	0.76 +/- 0.016	>4.8	<0.8	>4.8	133.5-136.5	5-7	11.5-12.5
	FX Cruiser HO	183-194	158-183	210-212	176-201	54.2-69.0	3.12-3.48	-	-	5	0.76 +/- 0.016	>4.8	<0.8	>4.8	133.5-136.5	5-7	11.5-12.5
	FX HO	183-194	158-183	210-212	176-201	54.2-69.0	3.12-3.48	-	-	11.6	0.76 +/- 0.016	>4.8	<0.8	>4.8	133.5-136.5	5-7	11.5-12.5
	VX110 Deluxe	183-194	158-183	-	-	54.2-69.0	3.12-3.48	-	-	11.6	④	>4.8	<0.8	>4.8	133.5-136.5	5-7	11.5-12.5
	VX110 Sport	183-194	158-183	-	-	54.2-69.0	3.12-3.48	-	-	11.6	④	>4.8	<0.8	>4.8	133.5-136.5	5-7	11.5-12.5
2007	FX	183-194	158-183	210-212	176-201	54.2-69.0	3.12-3.48	5.4-6.6	0.29-0.39	11.6	0.76 +/- 0.016	>4.8	<0.8	>4.8	133.5-136.5	5-7	11.5-12.5
	FX Cruiser	183-194	158-183	210-212	176-201	54.2-69.0	3.12-3.48	5.4-6.6	0.29-0.39	11.6	0.76 +/- 0.016	>4.8	<0.8	>4.8	133.5-136.5	5-7	11.5-12.5
	FX Cruiser HO	183-194	158-183	210-212	176-201	54.2-69.0	3.12-3.48	-	-	5	0.76 +/- 0.016	>4.8	<0.8	>4.8	133.5-136.5	5-7	11.5-12.5
	FX HO	183-194	158-183	210-212	176-201	54.2-69.0	3.12-3.48	-	-	11.6	0.76 +/- 0.016	>4.8	<0.8	>4.8	133.5-136.5	5-7	11.5-12.5
	VX	183-194	158-183	-	-	54.2-69.0	3.12-3.48	-	-	11.6	④	>4.8	<0.8	>4.8	133.5-136.5	5-7	11.5-12.5
	VX Crusier	183-194	158-183	-	-	54.2-69.0	3.12-3.48	-	-	11.6	④	>4.8	<0.8	>4.8	133.5-136.5	5-7	11.5-12.5
	VX110 Deluxe	183-194	158-183	-	-	54.2-69.0	3.12-3.48	-	-	11.6	④	>4.8	<0.8	>4.8	133.5-136.5	5-7	11.5-12.5
	VX110 Sport	183-194	158-183	-	-	54.2-69.0	3.12-3.48	-	-	11.6	④	>4.8	<0.8	>4.8	133.5-136.5	5-7	11.5-12.5
2008	FX	183-194	158-183	210-212	176-201	54.2-69.0	3.12-3.48	5.4-6.6	0.29-0.39	11.6	0.76 +/- 0.016	>4.8	<0.8	>4.8	133.5-136.5	5-7	11.5-12.5
	FX Cruiser	183-194	158-183	210-212	176-201	54.2-69.0	3.12-3.48	5.4-6.6	0.29-0.39	11.6	0.76 +/- 0.016	>4.8	<0.8	>4.8	133.5-136.5	5-7	11.5-12.5
	FX Cruiser HO	183-194	158-183	-	-	54.2-69.0	3.12-3.48	-	-	-	0.76 +/- 0.016	>4.8	<0.8	>4.8	133.5-136.5	5-7	11.5-12.5
	FX Cruiser HO (	183-194	158-183	-	-	54.2-69.0	3.12-3.48	5.4-6.6	0.29-0.39	-	0.76 +/- 0.016	>4.8	<0.8	>4.8	133.5-136.5	5-7	11.5-12.5
	FX Cruiser SHO	-	-	210-212	176-201	54.2-69.0	3.12-3.48	5.4-6.6	0.29-0.39	-	③	>4.8	<0.8	>4.8	133.5-136.5	5-7	11.5-12.5
	FX HO	183-194	158-183	-	-	54.2-69.0	3.12-3.48	-	-	-	0.76 +/- 0.016	>4.8	<0.8	>4.8	133.5-136.5	5-7	11.5-12.5
	FX HO (Calif.)	183-194	158-183	-	-	54.2-69.0	3.12-3.48	5.4-6.6	0.29-0.39	-	0.76 +/- 0.016	>4.8	<0.8	>4.8	133.5-136.5	5-7	11.5-12.5
	FX SHO	-	-	210-212	176-201	54.2-69.0	3.12-3.48	5.4-6.6	0.29-0.39	-	③	>4.8	<0.8	>4.8	133.5-136.5	5-7	11.5-12.5
	VX	183-194	158-183	-	-	54.2-69.0	3.12-3.48	-	-	11.6	④	>4.8	<0.8	>4.8	133.5-136.5	5-7	11.5-12.5
	VX Crusier	183-194	158-183	-	-	54.2-69.0	3.12-3.48	-	-	11.6	④	>4.8	<0.8	>4.8	133.5-136.5	5-7	11.5-12.5
	VX Deluxe	183-194	158-183	-	-	54.2-69.0	3.12-3.48	-	-	11.6	④	>4.8	<0.8	>4.8	133.5-136.5	5-7	11.5-12.5
	VX110 Sport	183-194	158-183	-	-	54.2-69.0	3.12-3.48	-	-	11.6	④	>4.8	<0.8	>4.8	133.5-136.5	5-7	11.5-12.5

SELOC_9606

Electronic Component Test Specifications

		Thermoswitch				Eng Temp Sensor Resistance		Intake Air Temp Sensor Resistance		Speed Sensor	TPS Output	Cam Position Sensor Output			Fuel Sender Output		Fuel Injector
Year	Model Name	On/Eng (°F)	Off/Eng (°F)	On/Exh (°F)	Off/Exh (°F)	@ 68°F (k ohms)	@ 212°F (k ohms)	@ 32°F (k ohms)	@ 176°F (k ohms)	Output (V)	@ Trolling (V)	@ A (V)	@ B (V)	@ C (V)	@ A (ohm)	@ B (ohm)	Resistance (ohm)
2009	FX Cruiser HO	-	-	210-212	176-201	54.2-69.0	3.12-3.48	5.4-6.6	0.29-0.39	-	③	>4.8	<0.8	>4.8	133.5-136.5	5-7	11.5-12.5
	FX Cruiser SHO	-	-	210-212	176-201	54.2-69.0	3.12-3.48	5.4-6.6	0.29-0.39	-	③	>4.8	<0.8	>4.8	133.5-136.5	5-7	11.5-12.5
	FX HO	-	-	210-212	176-201	54.2-69.0	3.12-3.48	5.4-6.6	0.29-0.39	-	③	>4.8	<0.8	>4.8	133.5-136.5	5-7	11.5-12.5
	FX SHO	-	-	210-212	176-201	54.2-69.0	3.12-3.48	5.4-6.6	0.29-0.39	-	③	>4.8	<0.8	>4.8	133.5-136.5	5-7	11.5-12.5
	FZR	-	-	210-212	176-201	54.2-69.0	3.12-3.48	5.4-6.6	0.29-0.39	-	③	>4.8	<0.8	>4.8	133.5-136.5	5-7	11.5-12.5
	FZS	-	-	210-212	176-201	54.2-69.0	3.12-3.48	5.4-6.6	0.29-0.39	-	③	>4.8	<0.8	>4.8	133.5-136.5	5-7	11.5-12.5
	VX	183-194	158-183	-	-	54.2-69.0	3.12-3.48	-	-	11.6	④	>4.8	<0.8	>4.8	133.5-136.5	5-7	11.5-12.5
	VX Cruiser	183-194	158-183	-	-	54.2-69.0	3.12-3.48	-	-	11.6	④	>4.8	<0.8	>4.8	133.5-136.5	5-7	11.5-12.5
	VX Deluxe	183-194	158-183	-	-	54.2-69.0	3.12-3.48	-	-	11.6	④	>4.8	<0.8	>4.8	133.5-136.5	5-7	11.5-12.5
	VX Sport	183-194	158-183	-	-	54.2-69.0	3.12-3.48	-	-	11.6	④	>4.8	<0.8	>4.8	133.5-136.5	5-7	11.5-12.5
2010	FX Cruiser HO	-	-	210-212	176-201	54.2-69.0	3.12-3.48	5.4-6.6	0.29-0.39	-	③	>4.8	<0.8	>4.8	133.5-136.5	5-7	11.5-12.5
	FX Cruiser SHO	-	-	210-212	176-201	54.2-69.0	3.12-3.48	5.4-6.6	0.29-0.39	-	③	>4.8	<0.8	>4.8	133.5-136.5	5-7	11.5-12.5
	FX HO	-	-	210-212	176-201	54.2-69.0	3.12-3.48	5.4-6.6	0.29-0.39	-	③	>4.8	<0.8	>4.8	133.5-136.5	5-7	11.5-12.5
	FX SHO	-	-	210-212	176-201	54.2-69.0	3.12-3.48	5.4-6.6	0.29-0.39	-	③	>4.8	<0.8	>4.8	133.5-136.5	5-7	11.5-12.5
	FZR	-	-	210-212	176-201	54.2-69.0	3.12-3.48	5.4-6.6	0.29-0.39	-	③	>4.8	<0.8	>4.8	133.5-136.5	5-7	11.5-12.5
	FZS	-	-	210-212	176-201	54.2-69.0	3.12-3.48	5.4-6.6	0.29-0.39	-	③	>4.8	<0.8	>4.8	133.5-136.5	5-7	11.5-12.5
	VX Cruiser	183-194	158-183	-	-	54.2-69.0	3.12-3.48	-	-	11.6	④	>4.8	<0.8	>4.8	133.5-136.5	5-7	11.5-12.5
	VX Deluxe	183-194	158-183	-	-	54.2-69.0	3.12-3.48	-	-	11.6	④	>4.8	<0.8	>4.8	133.5-136.5	5-7	11.5-12.5
	VX Sport	183-194	158-183	-	-	54.2-69.0	3.12-3.48	-	-	11.6	④	>4.8	<0.8	>4.8	133.5-136.5	5-7	11.5-12.5
2011	FX Cruiser HO	-	-	210-212	176-201	54.2-69.0	3.12-3.48	5.4-6.6	0.29-0.39	-	③	>4.8	<0.8	>4.8	133.5-136.5	5-7	11.5-12.5
	FX Cruiser SHO	-	-	210-212	176-201	54.2-69.0	3.12-3.48	5.4-6.6	0.29-0.39	-	③	>4.8	<0.8	>4.8	133.5-136.5	5-7	11.5-12.5
	FX HO	-	-	210-212	176-201	54.2-69.0	3.12-3.48	5.4-6.6	0.29-0.39	-	③	>4.8	<0.8	>4.8	133.5-136.5	5-7	11.5-12.5
	FX SHO	-	-	210-212	176-201	54.2-69.0	3.12-3.48	5.4-6.6	0.29-0.39	-	③	>4.8	<0.8	>4.8	133.5-136.5	5-7	11.5-12.5
	FZR	-	-	210-212	176-201	54.2-69.0	3.12-3.48	5.4-6.6	0.29-0.39	-	③	>4.8	<0.8	>4.8	133.5-136.5	5-7	11.5-12.5
	FZS	-	-	210-212	176-201	54.2-69.0	3.12-3.48	5.4-6.6	0.29-0.39	-	③	>4.8	<0.8	>4.8	133.5-136.5	5-7	11.5-12.5
	VX Cruiser	183-194	158-183	-	-	54.2-69.0	3.12-3.48	-	-	11.6	④	>4.8	<0.8	>4.8	133.5-136.5	5-7	11.5-12.5
	VX Deluxe	183-194	158-183	-	-	54.2-69.0	3.12-3.48	-	-	11.6	④	>4.8	<0.8	>4.8	133.5-136.5	5-7	11.5-12.5
	VX Sport	183-194	158-183	-	-	54.2-69.0	3.12-3.48	-	-	11.6	④	>4.8	<0.8	>4.8	133.5-136.5	5-7	11.5-12.5
	VXR	-	-	210-212	176-201	54.2-69.0	3.12-3.48	5.4-6.6	0.29-0.39	-	⑤	>4.8	<0.8	>4.8	133.5-136.5	5-7	11.5-12.5
	VXS	-	-	210-212	176-201	54.2-69.0	3.12-3.48	5.4-6.6	0.29-0.39	-	⑤	>4.8	<0.8	>4.8	133.5-136.5	5-7	11.5-12.5

SELOC_9606

① Cranking 1: 4V, Cranking 2: 4V
② Cranking 1: 9V, Cranking 2: 8V
③ Throttle fully closed: 0.6-0.9, Throttle fully open: 4.6-4.7
④ Throttle fully closed: 0.45-0.95, Throttle fully open: 4.6-4.7
⑤ Throttle fully closed: 0.6-0.9, Throttle fully open: 4.5-4.8

SELOC LABOR GUID MANUAL
Stern Drives
Outboards
EASY
MODERATE
DIFFICULT
SKILLED
OEM
WELCOME TO SELOC PRO

5

EXHAUST & COOLING

EXHAUST SYSTEM

Description & Operation

The exhaust system on these models is made up of a number of different components, starting at the exhaust manifold and ending at the flapper valve where the exhaust/water exits the craft.

On 2002-08 FX models with 998cc/1052cc engines, a 2-piece manifold is connected to the engine on the starboard side. From there, the system uses a series of 3 additional joints, pipes and fittings to route everything across the front of the engine and back along the port side of the engine.

On 2008-11 FX, FZ and VX models w/1812cc engines, a 1-piece manifold is connected to the engine on the port side. From there, the system uses a series of 3 additional pipes and fittings to route everything toward the rear of the craft.

On VX models w/1052cc engines, a 1-piece manifold is connected to the engine on the starboard side. From here, the system uses a series of 2 additional joints, pipes and fittings to route everything across the front of the engine and back along the port side.

At the rear of the engine on both models, the #2 or #3 exhaust pipe connects to an exhaust tube/joint via an adapter (exc. 1812cc engines). This tube/joint runs into the water lock (which helps silence the exhaust note like a muffler). From the water lock, we continue through another exhaust tube to the water tank where it then runs through another tube to the outlet/flapper, where it exits the craft. 2008 and later models w/1812cc engines also use a resonator which is attached to the center exhaust tube in the rear.

Since component removal or disconnection pretty much starts with the #2 (VX) or #3 (FX) pipe and moves forward or backward from there, we will begin at the rear and work forward with our procedures because the rear components require a little less effort.

Rear Exhaust Components

REMOVAL & INSTALLATION

FX Models With 998cc/1052cc Engines & VX Models w/1052cc Engines

◆ See Figures 1 thru 15

③ DIFFICULT

The rear exhaust components are comprised of 3 exhaust hoses, the water lock, water tank and an exhaust valve (sometimes called a flapper). The forward and middle exhaust tubes and the water lock *should* be able to be removed without removing any other major components, but water tank removal and lower hose removal will definitely require removal of the jet drive unit on all models except the 2010-11 VX, where you'll need to remove the ride plate.

■ **The following procedure should be performed with the craft out of the water.**

1. Remove the seat(s) and disconnect the battery cables. Remove the storage compartment if equipped.
2. On FX models, remove the 4 deck beam mounting bolts and nuts and then lift it out of the craft. You may have to swivel it slightly, and make sure you don't lose the 8 washers.
3. Remove the jet pump unit or ride plate as detailed in the Drive Unit section. Remember, if you are only removing the forward/middle exhaust tubes or the water lock, this is not necessary.
4. On early FX models w/998cc engines, loosen the forward exhaust hose clamps at the exhaust pipe adaptor and water lock. Slide them back down the hose and remove it. You may have to do some careful friendly persuasion to get it off.
5. On later FX models with 998cc/1052cc engines, the forward hose is connected to the #3 exhaust pipe adaptor and the water lock with exhaust joints. Same with the VX models where it is connected to the #2 pipe. Loosen the forward hose clamp (around the adaptor) and wiggle (may have to pry it) the forward joint off of it. At the other end, loosen the rear clamp and wiggle/pry the rear joint off of the water lock. If replacing the forward hose, simply loosen the 2 remaining clamps and pull the joints out of the pipe.

■ **The water tank and water lock should have marks on their fittings showing how far on the hose must go. If not, make sure that you mark each fitting at the end of the hose before removal.**

■ **On 2008 models with 6AA engines, loosen the clamp and disconnect the 2-hose joint inserted into the forward exhaust joint.**

a. If necessary on 2002-05 FX models w/998cc engines, remove the 6 exhaust hose protector bolts and pull off the 2 halves of the outer casing.

6. Loosen the clamps on the middle exhaust hose and pry it off of the water lock neck.
7. Unhook the retaining strap and lift out the water lock.
8. Loosen the lower exhaust hose clamp at the water tank.
9. Remove the 6 nuts and washers securing the lower hose and flapper valve to the hull. Pull the retainer back and away from the hose flange and then remove the water tank and 2 hoses as an assembly. Disconnect the lower and center exhaust hose from the tank.

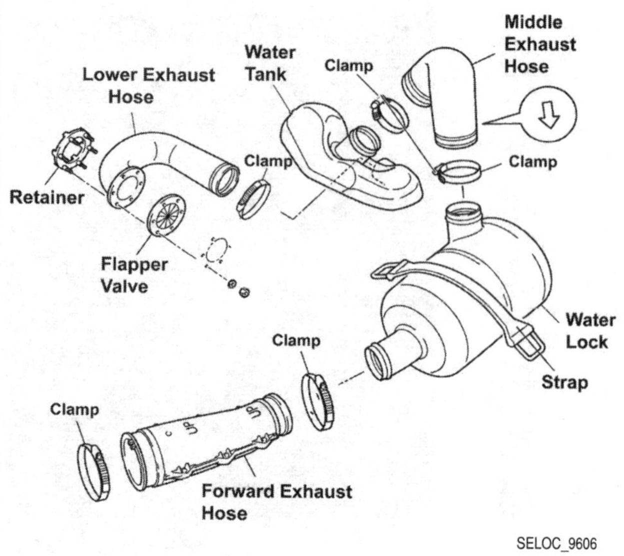

Fig. 1 Exploded view of the rear exhaust components - early FX models w/998cc engines

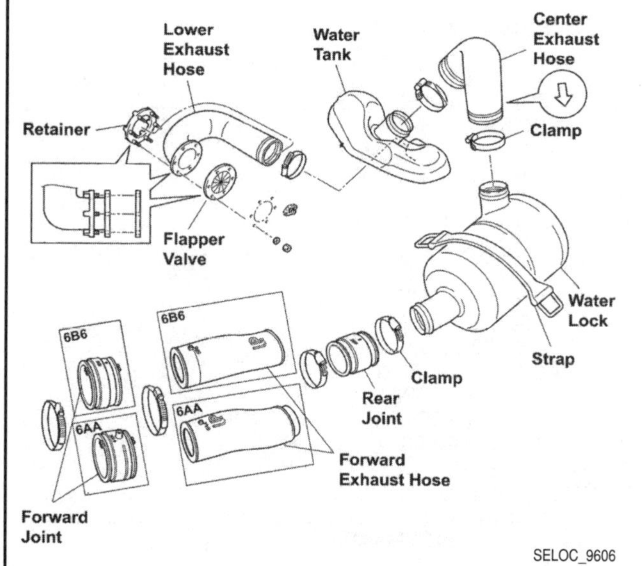

Fig. 2 Exploded view of the rear exhaust components - all other FX models with 998cc/1052cc engines

10. Remove the flapper valve if it didn't come off with the retainer.

To install:

11. Clean and inspect all hoses and clamps. Look carefully for cracks in all hoses, particularly in the vicinity of the hose clamps.

12. Clean and inspect the water tank and water lock for cracks or other obvious wear or damage.

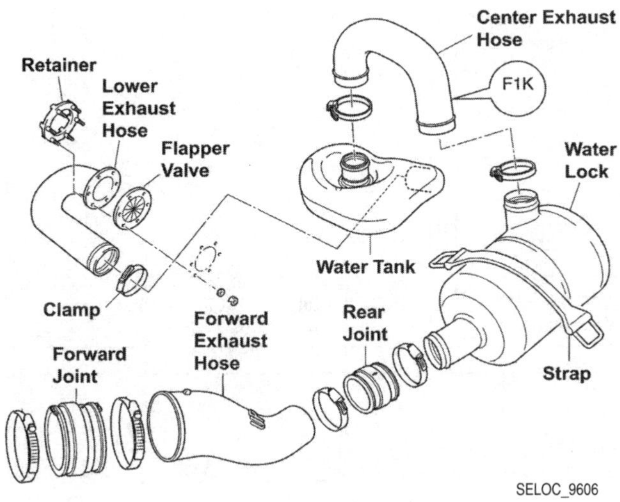

Fig. 3 Exploded view of the rear exhaust components - 2005-09 VX models (2010-11 similar) w/1052cc engines

13. Inspect the flapper valve for damaged fins - we suggest replacing it regardless since you already have it out.

14. Spin the retainer so that the nibs on the lower hose flange line up with the notches in the retainer. Slide the retainer bolts through the flange and then slide on the flapper valve. Coat the surface of the valve lightly with a good silicone marine sealant and then position the assembly through the hull. Thread on the washers and nuts, tightening them to 3.9 ft. lbs./46 inch lbs. (5.4 Nm), in a criss-cross pattern. On 2010-11 VX models, tighten to 3.7 ft. lbs./43 inch lbs. (5 Nm).

■ **When installing the remaining exhaust hoses to the water tank and water lock, you may find that you'll want to leave the hose clamps untightened until you are confident that all hoses and fittings line up correctly when their components are in position.**

15. Slide the clamp over the lower hose and then coat the inner mating surface with a little soapy water and slide it onto the lower fitting on the water tank. Make sure that the hose slides on until it reaches the mark on the fitting (or the one you made). Rotate the tank in the hose to ensure that the molding lines on both the hose and tank are at least 1.9 in. (47mm) apart on FX models. Interestingly, on 2005-09 VX models the molding lines should be lined up! On 2010-11 VX models, make sure that the line on the end of the hose lines up with the line on the water tank fitting. Ensure that the tank is oriented correctly in the hull and then tighten the clamp to 2.7 ft. lbs./30 inch lbs. (3.7 Nm).

16. Slide a clamp over the lower end of the middle hose and then coat the inner mating surface with a little soapy water. Slide the hose onto the upper water tank fitting until it meets the positioning line on the fitting and so that the molding lines are in alignment. On 2010-11 VX models, make sure that the small nib on the hose lines up with the line on the fitting. Tighten the clamp to 2.7 ft. lbs./30 inch lbs. (3.7 Nm).

17. Install the water lock carefully and then connect the retaining strap. Slide a clamp over the upper end of the middle hose (later models have an arrow that should point at the lock) and then slide the hose onto the upper

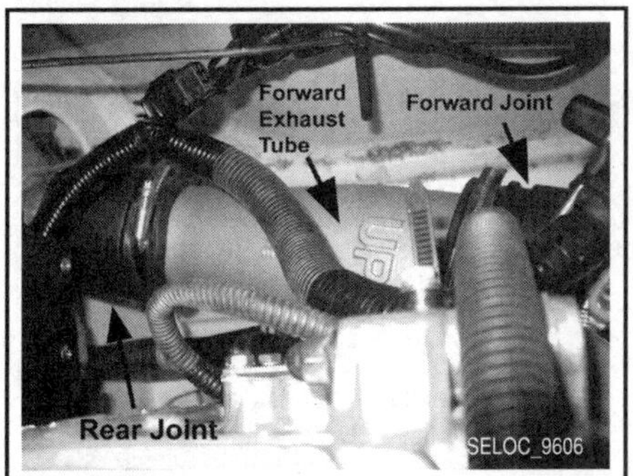

Fig. 4 A good look at the forward exhaust tube on later FX models with 998cc/1052cc engines

Fig. 5 The water lock acts like a muffler

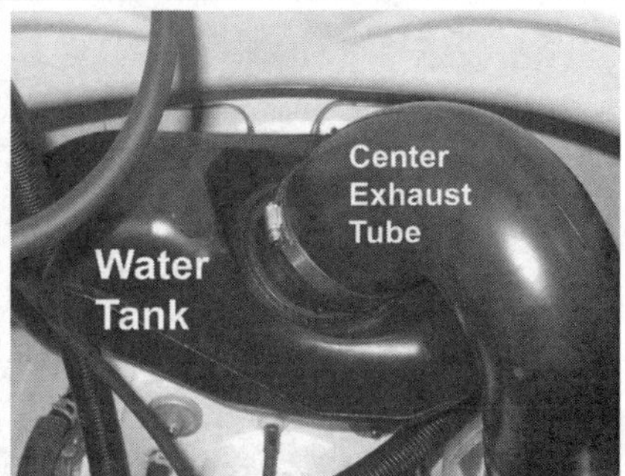

Fig. 6 A good look at the water tank

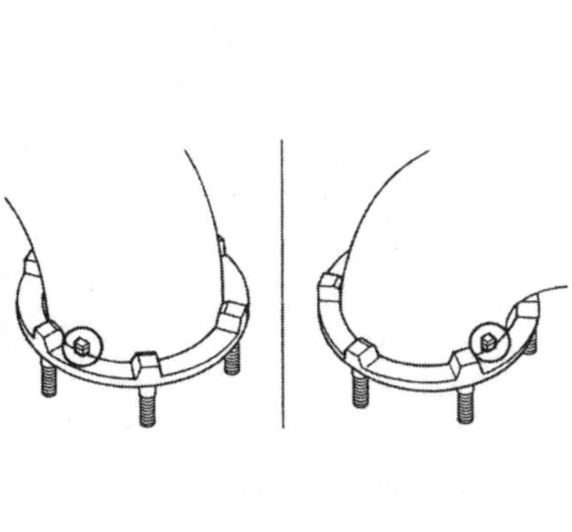

Fig. 7 Make sure that the nibs on the lower hose flange match the indent on the retainer

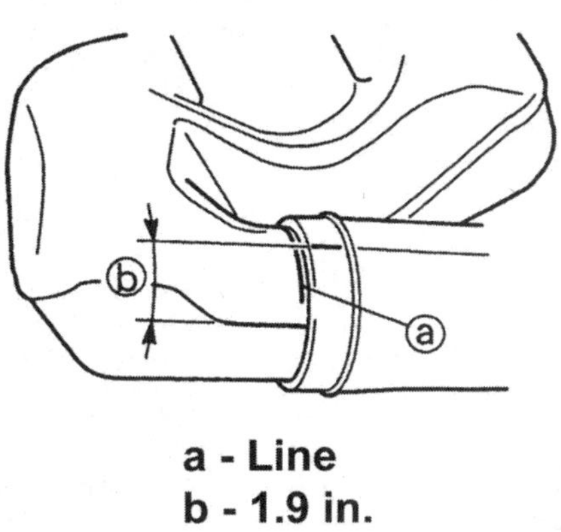

Fig. 8 On the lower hose-to-water tank fitting, the molding lines must be separated (exc. on VX models). . .

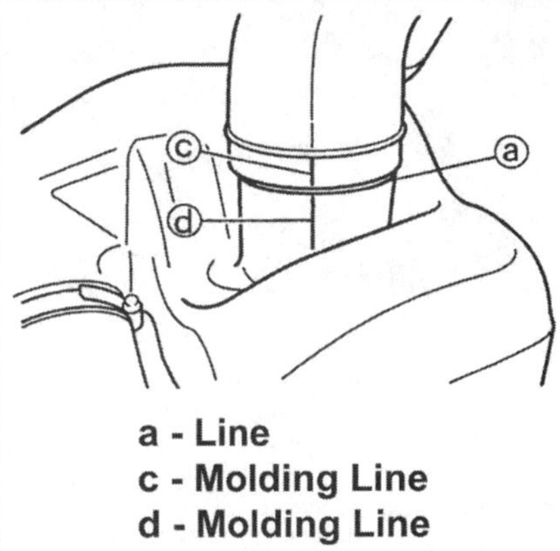

Fig. 9 . . .while on the middle hose-to-tank fitting they should be aligned

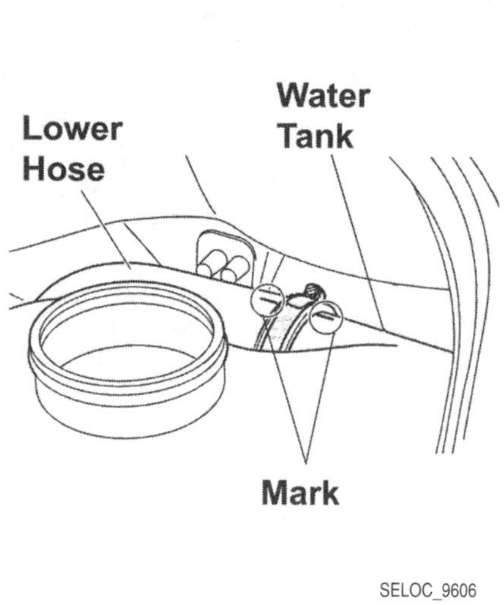

Fig. 10 Line the marks up on 2010-11 VX models

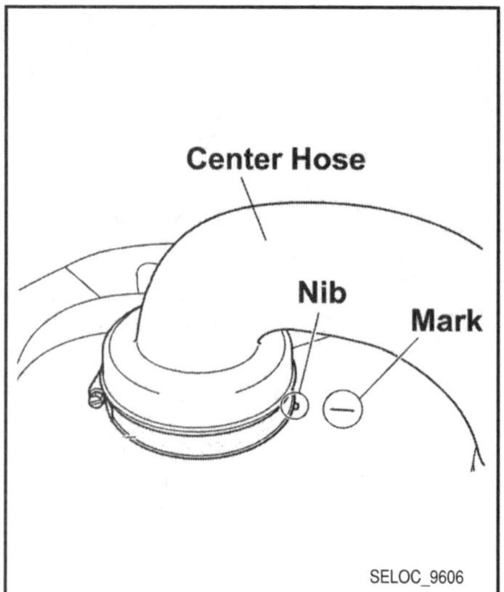

Fig. 11 Line the small nib up with the line on the water lock - 2010-11 VX models

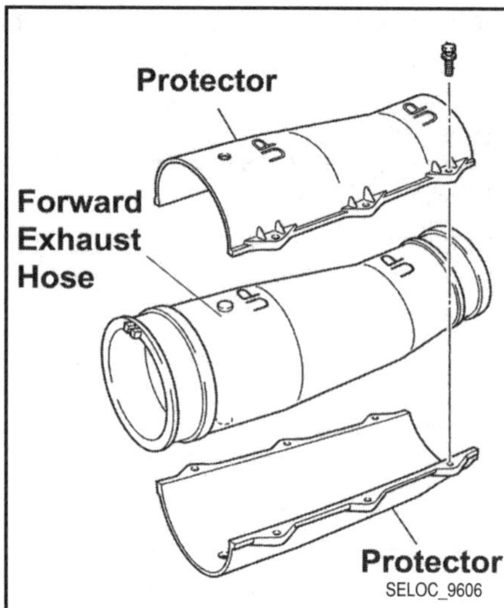

Fig. 12 Early models (2002-05 w/998cc engines) utilized a protector over the forward exhaust hose

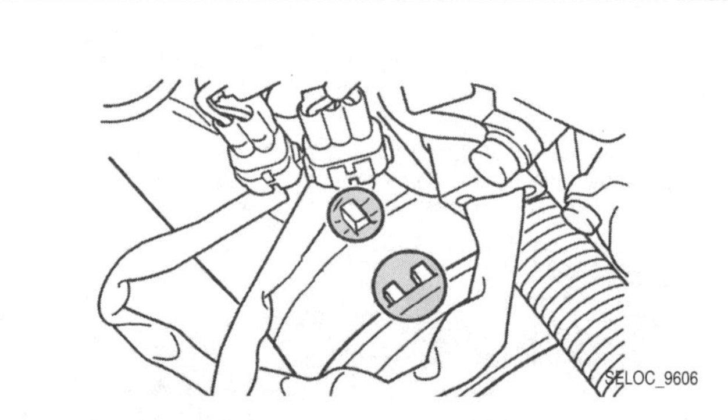

Fig. 13 Lining up the forward exhaust hose with the #3 exhaust pipe - FX models w/998cc engines

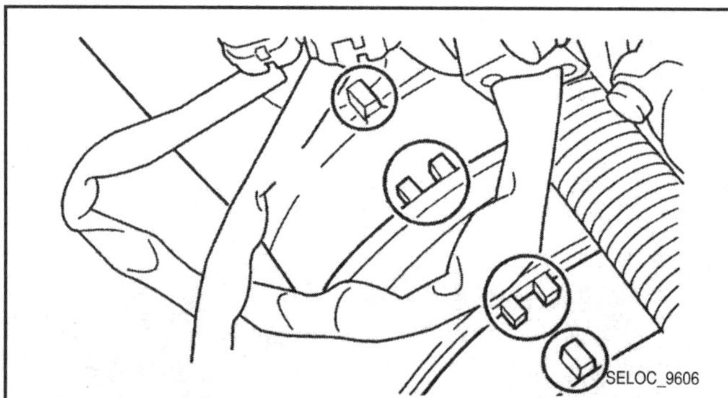

Fig. 14 Lining up the exhaust joint with forward exhaust hose and the #3 exhaust pipe adaptor - FX models w/1052cc engines (VX models similar)

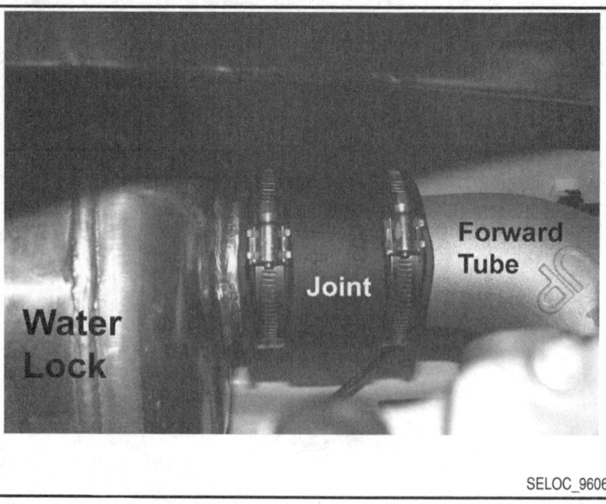

water lock fitting until it meets the positioning line on the fitting. Tighten the clamp to 2.7 ft. lbs./30 inch lbs. (3.7 Nm).

18. If you have a 2002-05 FX w/998cc engine with the hose protectors, position them over the inner hose so that the words **UP** are, you guessed it, up and then tighten the 6 bolts to a first pass of 2.7 ft. lbs./30 inch lbs. (3.7 Nm), working in a criss-cross pattern from the center bolts outward. Now tighten them again to a final pass of 8.8 ft. lbs. (12 Nm).

19. On 2002-05 FX models w/998cc engines, carefully squeeze the forward tube into position between the water lock and exhaust pipe adaptor - you may need to swivel the water lock a bit to get everything in position. Make sure that the nib on the end of the exhaust pipe fits in between the 2 on the exhaust hose and then tighten both clamps to 2.7 ft. lbs./30 inch lbs. (3.7 Nm).

20. On 2004-08 FX models w/1052cc engines and all VX models, coat the 2 joints with a little soapy water (on the inside) and slide them over their respective ends on the forward tube. Make sure that the arrow on the rear joint is pointing to the water lock and that the arrow on the forward joint is pointing back to the tube. Ensure that the 2 nibs on the forward joint fit perfectly around the single nib on the tube. Tighten the clamps to:

- 2008 FX HO: 1.7 ft. lbs./20 inch lbs. (2.4 Nm)
- All other FX models: 1.8 ft. lbs./21 inch lbs. (2.5 Nm)
- 2005-09 VX: 1.7 ft. lbs./20 inch lbs. (2.4 Nm)
- 2010-11 VX: 1.5 ft. lbs./18 inch lbs. (2 Nm)

21. Position the assembly between the #3 pipe (or #2 on VX) and water lock and then slide the joints into position on the adaptor and lock. Make sure that the 2 nibs on the forward end of the joint line up with the single one on the adaptor. You may have to wiggle the tube or water lock to get everything lined up. Tighten both clamps to 2.7 ft. lbs./30 inch lbs. (3.7 Nm).

22. Now go back and check the inner clamps on the 2008 FX HO and VX models - make sure they're still at 1.7 ft. lbs./20 inch lbs. (2.4 Nm).

23. Press the 2-hose joint into the exhaust joint on 2008 models w/6AA engines.

24. Install the jet pump or ride plate if they were removed.

25. If removed, position the deck plate on the hull and tighten the 4 bolts to 3.8 ft. lbs./43 inch lbs. (5.2 Nm) on 998cc models, 4.6 ft. lbs./55 inch lbs. (6.4 Nm) on the 2004-07 1052cc models, or 13 ft. lbs. (18 Nm) on the 2008 units.

26. Connect the battery cables, install the cargo container and then install the seat(s).

All Models w/1812cc Engines

3 DIFFICULT

◆ See Figures 16, 17 and 18

The rear exhaust components are comprised of 3 exhaust hoses (well, the forward hose is actually a joint), the water lock, water tank, resonator and an exhaust valve (sometimes called a flapper). These components *should* be able to be removed without removing any other major components.

■ **The following procedure should be performed with the craft out of the water.**

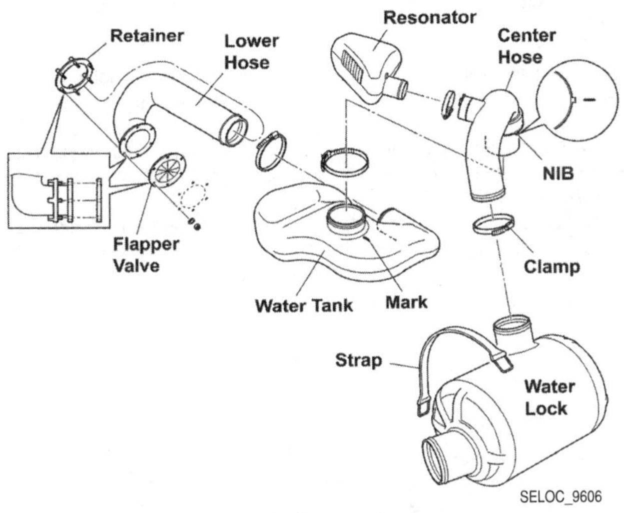

Fig. 16 Exploded view of the rear exhaust components - 1812cc engines

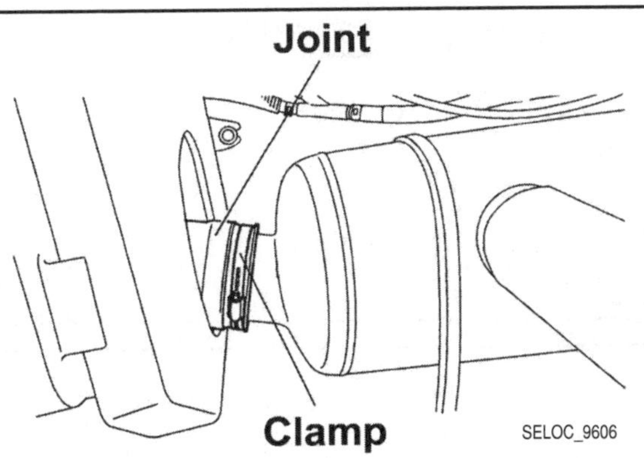

Fig. 17 The water lock is connected to the #3 exhaust pipe by a joint - 1812cc engines

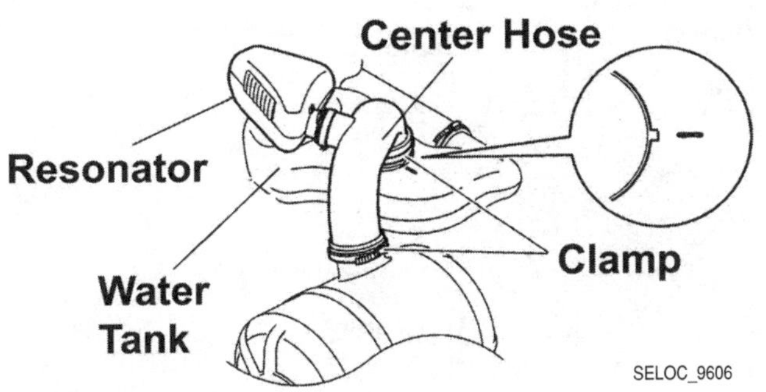

Fig. 18 Line the small nib up with the line on the water tank - 1812cc engines

1. Remove the seat(s) and disconnect the battery cables. Remove the storage compartment if equipped.
2. If equipped, remove the 4 deck beam mounting bolts and nuts and then lift it out of the craft. You may have to swivel it slightly, and make sure you don't lose the 8 washers.
3. Remove the rubber plate as detailed in the Jet Pump Unit procedures found in the Drive Unit section. Remember, if you are only removing the forward/middle exhaust tubes or the water lock, this is probably not necessary.
4. Loosen the rear hose clamp (around the joint) and wiggle (may have to pry it) the forward joint off of the water lock flange.

■ **The water tank and water lock should have marks on their fittings showing how far on the hose must go. If not, make sure that you mark each fitting at the end of the hose before removal to be safe.**

5. Loosen the hose clamp and then pull the resonator off of the fitting on the center exhaust hose/tube.
6. Loosen the clamps on the center exhaust hose and pry it off of the water lock neck.
7. Unhook the retaining strap and lift out the water lock.
8. Loosen the lower exhaust hose clamp at the water tank.
9. Remove the 6 nuts and washers securing the lower hose and flapper valve to the hull. Pull the retainer back and away from the hose flange and then remove the water tank and 2 hoses as an assembly. Disconnect the lower and center exhaust hose from the tank.
10. Remove the flapper valve if it didn't come off with the retainer.

To install:

11. Clean and inspect all hoses and clamps. Look carefully for cracks in all hoses, particularly in the vicinity of the hose clamps.
12. Clean and inspect the water tank, resonator and water lock for cracks or other obvious wear or damage.
13. Inspect the flapper valve for damaged fins - we suggest replacing it regardless since you already have it out.
14. Spin the retainer so that the holes on the lower hose flange line up with those in the retainer. Slide the retainer bolts through the flange and then slide on the flapper valve. Coat the surface of the valve lightly with a good silicone marine sealant and then position the assembly through the hull. Thread on the washers and nuts, tightening them to 3.7 ft. lbs./44 inch lbs. (5 Nm), in a criss-cross pattern.

■ **When installing the remaining exhaust hoses to the water tank and water lock, you may find that you'll want to leave the hose clamps untightened until you are confident that all hoses and fittings line up correctly when their components are in position.**

15. Slide the clamp over the lower hose and then coat the inner mating surface with a little soapy water and slide it onto the lower fitting on the water tank. Make sure that the hose slides on until it reaches the mark on the fitting (or the one you made). Ensure that the tank is oriented correctly in the hull and then tighten the clamp to 2.7 ft. lbs./30 inch lbs. (3.7 Nm).
16. Slide a clamp over the lower end of the middle hose and then coat the inner mating surface with a little soapy water. Slide the hose onto the upper water tank fitting making sure that the small nib on the hose lines up with the line on the fitting. Tighten the clamp to 2.7 ft. lbs./30 inch lbs. (3.7 Nm).
17. Support the back of the lower hose and then press the resonator into place on the fitting. Tighten the clamp to 2.7 ft. lbs./30 inch lbs. (3.7 Nm).
18. Install the water lock carefully while feeding the fitting around the #3 exhaust pipe and into the forward joint coming off of the #3 exhaust pipe and then connect the retaining strap. Slide a clamp over the upper end of the center hose, slather some soapy water on the inner surface and then slide the hose onto the upper water lock fitting until it meets the positioning line on the fitting. Tighten the clamp to 2.7 ft. lbs./30 inch lbs. (3.7 Nm).
19. Tighten the forward joint clamp to 1.5 ft. lbs./18 inch lbs. (2 Nm).
20. Install the rubber plate if removed.
21. If removed, position the deck plate on the hull and tighten the 4 bolts/nuts to 13.3 ft. lbs. (18 Nm). Remember there is a washer under the bolt head and also under the nut.
22. Connect the battery cables, install the cargo container and then install the seat(s).

Forward Exhaust Components

REMOVAL

FX Models With 998/1052cc Engines

◆ **See Figures 19 thru 24**

DIFFICULT

The forward exhaust components are comprised of 3 exhaust pipes/tubes/mufflers (port side, center and starboard side), an inner and outer exhaust joint, and a 2-piece exhaust manifold. All component services will require removal of the engine.

■ **The following procedure should be performed with the craft out of the water.**

1. Remove the seat(s) and disconnect the battery cables. Remove the storage compartment.
2. Remove the engine as detailed in the Engine Mechanical section.
3. Remove the air filter and case as detailed in the Fuel System section.

#3 Exhaust Pipe (Port Side) - All Exc. 2008 Models w/6AA Engines

■ **In 2008, Yamaha refers to this pipe as a muffler.**

4. Squeeze the water hose clamp and wiggle the hose off of its fitting at the rear of the #3 exhaust pipe. Position the hose out of the way.
5. Do the same to the 2 water hoses connected to the outer exhaust joint between the #3 exhaust pipe and the #2 pipe. Position them out of the way. Notice that each hose it actually connected to a small inner tube fitting seated in the actual cast fitting and also held via a hose clamp. You can remove the clamp holding the hose to the inner fitting or the clamp holding the inner fitting to the outer fitting - your choice.
6. Loosen the clamp holding the forward exhaust tube joint to the rear of the pipe and then work the joint off of the adaptor on the end of the #3 pipe.
7. Locate the exhaust thermo-switch at the rear of the #3 pipe (top), remove the 2 mounting bolts and then the thermo-switch. Although not strictly necessary unless replacing the pipe, this will protect the switch from damage if the pipe were to be dropped or roll over.
8. Loosen the large clamp(s) securing the outer exhaust joint to the inner joint and carefully move the outer joint toward the starboard side until it separates from the inner joint.
9. Remove the 2 lower exhaust pipe mounting bolts. Remove the 2 upper mounting bolts and then lift off the #3 pipe. Be careful not to lose the collar and grommet on the forward upper bolt. Lay the pipe on a clean, secure surface.
10. Remove the 3 mounting bolts securing the rear adaptor to the back end of the exhaust pipe and lift off the adaptor and gasket (throw away the gasket).
11. Loosen the inner joint clamp(s) and then remove the joint and seal. You have probably already done this, but if not, loosen the clamp and pull the outer exhaust joint off of the #2 exhaust pipe.

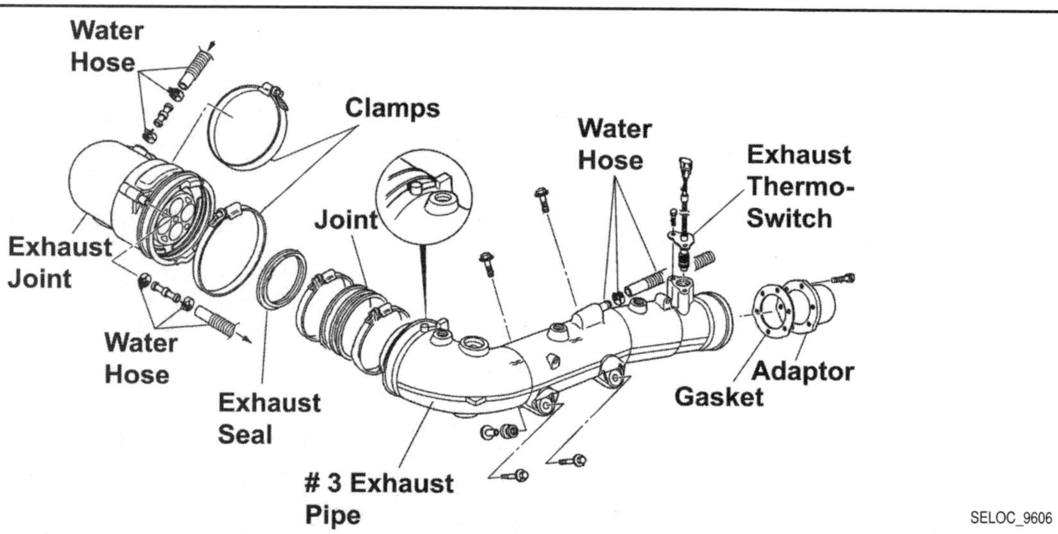

Fig. 19 Exploded view of the #3 exhaust pipe components (exc. 2008 6AA engines)

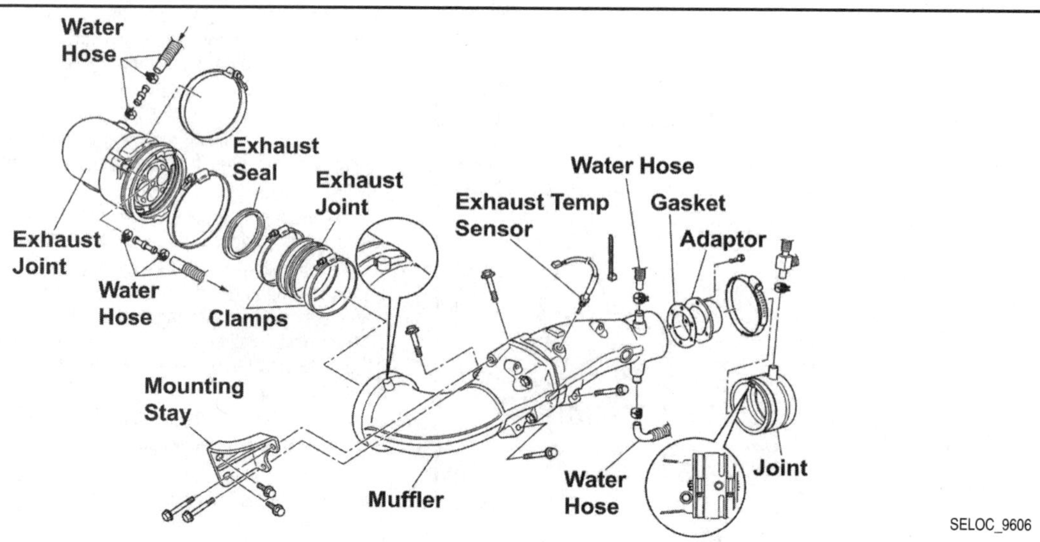

Fig. 20 Exploded view of the muffler components (2008 6AA engines only)

#3 Muffler/Catalyst Converter (Port Side) - 2008 Models w/6AA Engines

12. Squeeze the water hose clamp and wiggle the hose off of its fitting at the rear (top) of the muffler. Position the hose out of the way. Do the same with the lower water hose on the underside of the muffler.

13. Disconnect the 2 water hoses connected to the outer exhaust joint between the muffler and the #2 pipe/ring joint. Position them out of the way. Notice that each hose it actually connected to a small inner tube fitting seated in the actual cast fitting and also held via a hose clamp. You can remove the clamp holding the hose to the inner fitting or the clamp holding the inner fitting to the outer fitting - your choice.

14. Loosen the clamp holding the forward exhaust tube joint to the rear of the pipe and then work the joint off of the adaptor on the end of the #3 pipe.

15. Locate the exhaust temperature sensor at the center of the muffler (top, just past where the two halves meet), disconnect the coupler and then unthread the thermo-switch. Although not strictly necessary unless replacing the pipe, this will protect the switch from damage if the pipe were to be dropped or roll over.

16. Loosen the large clamp(s) securing the outer exhaust joint to the inner joint and carefully move the outer joint toward the starboard side until it separates from the inner joint.

17. Remove the 2 bolts holding the unit to the forward mounting stay.

18. Remove the 2 lower muffler mounting bolts. Remove the 2 upper mounting bolts and then lift off the assembly. Lay the pipe on a clean, secure surface.

19. Remove the 3 mounting bolts securing the rear adaptor to the back end of the exhaust pipe and lift off the adaptor and gasket (throw away the gasket).

20. Loosen the inner joint clamp(s) and then remove the joint and seal. You have probably already done this, but if not, loosen the clamp and pull the outer exhaust joint off of the #2 exhaust pipe/ring joint.

21. If necessary, loosen the 3 mounting bolts and separate the outer exhaust cover (back half of the assembly) from the muffler (front half). Make sure that you do not lose the 2 dowel pins.

22. Remove the gaskets and catalyst converter.

#2 Exhaust Pipe (Center)

■ **Access to the 2 inner mounting bolts may be limited and you may wish to remove the #1 and #2 pipes from the manifold 1st and then separate them.**

■ **In 2008, Yamaha refers to this pipe as a ring joint.**

23. Remove the #3 exhaust pipe/muffler if not already done.

24. Remove the generator cover as detailed in the Electrical & Ignition section.

25. Reach under the pipe where they mate and remove the mounting stud nut from the bottom (this will be the one next to the engine). Remove 4 upper mounting nuts holding the #2 pipe to the #1 pipe and then lift off the # 2 exhaust pipe. The mounting studs should stay in place, but give a quick look to make sure that one hasn't come out.

26. Remove the gasket from one of the pipe mating surfaces and throw it away.

27. There are 2 small dowel pins used between the two pipes for alignment and one, or both, may have come out with the #2 pipe - make sure that they are both in the flange of the #1 pipe.

#1 Exhaust Pipe (Starboard)

28. Remove the #2 exhaust pipe if you haven't already.

29. Remove the bolt attaching the #1 pipe to the small mounting stay. Now remove the 2 bolts attaching the stay to the block and lift out the stay. Of course, you could simply stop at the pipe-to-stay bolt and leave the stay attached to the engine block.

30. There are 2 water hoses connected to the rear of the pipe. Loosen their clamps and wiggle them off of the fittings.

31. Working under the bottom of the pipe, remove the 10 bolts that secure the pipe to the manifold. There 6 long bolts and 4 short bolts - take careful note of which bolts came out of which holes. Loosen the bolts a little at a time, starting at the corners and working inward.

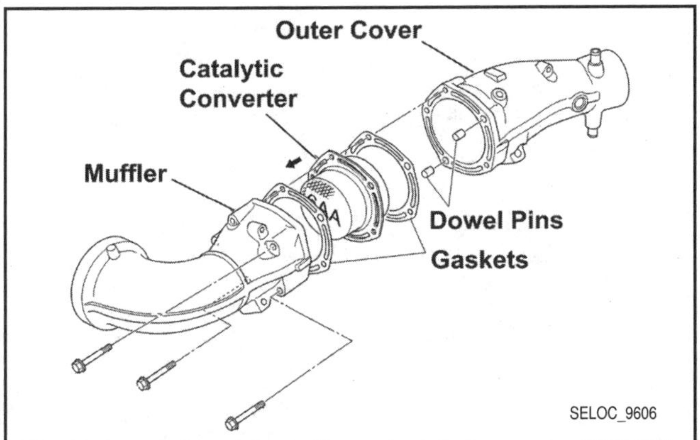

Fig. 21 A good shot of the muffler and catalyst converter (2008 6AA engines only)

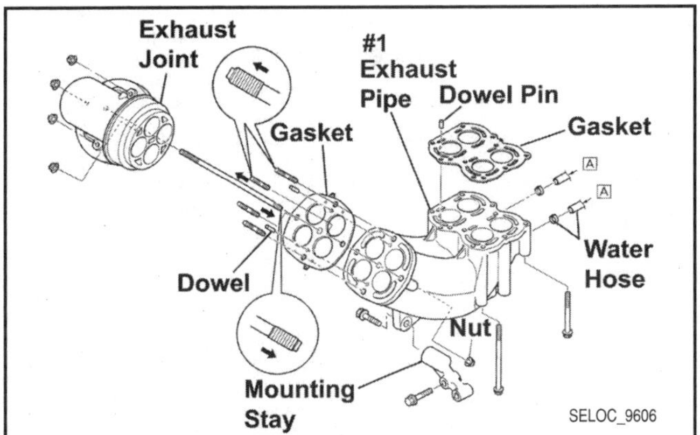

Fig. 22 Exploded view of the #1 and #2 exhaust pipe components - FX with 998cc/1052cc engines

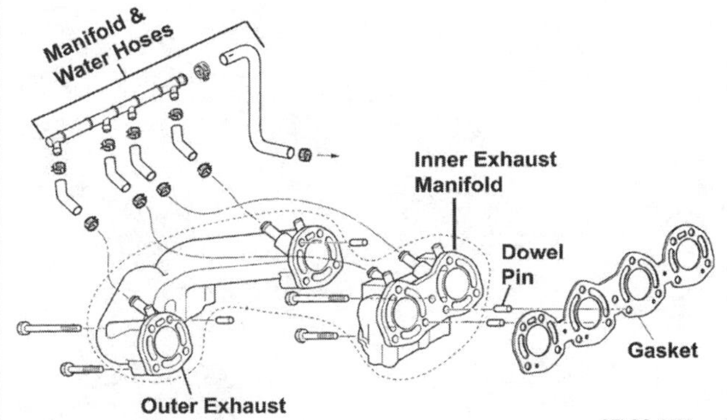

Fig. 23 Exploded view of the exhaust manifold - FX with 998cc/1052cc engines

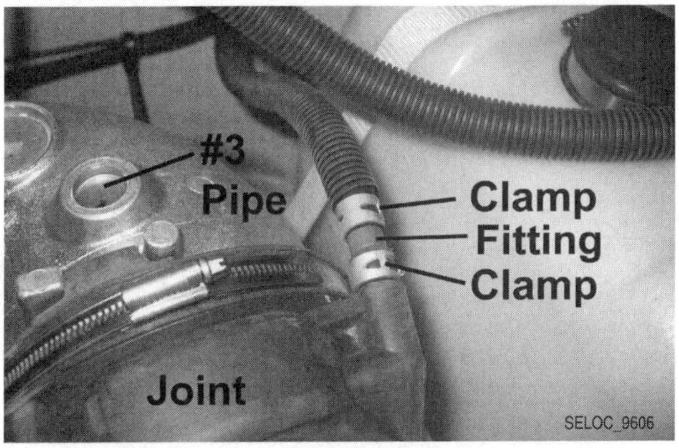

Fig. 24 A good look at the small fitting between the hose and the joint fitting - FX with 998cc/1052cc engines

32. Throw away the gasket and make sure you do not lose the 2 dowel pins.

Exhaust Manifold

■ **The exhaust manifold is comprised of 2 pieces.**

33. Remove the 3 exhaust pipes if not already done.

34. Squeeze the hose clamp on each of the water hoses and slide it back on the hose until it is off of the fitting. Tag each of the 4 hoses and then wiggle them off of their respective fittings. Move out of the way.

35. Remove the 6 bolts securing the outer manifold - remember these models use a 2-piece manifold, and the outer one covers the front and rear cylinder (#1 and #4). There are 2 long bolts and 4 shorter ones so make sure you mark which one went where. Remove the manifold and make sure that the small dowel pin stays in the flange for each cylinder.

■ **You may have to give the manifolds a slight tap with a rubber mallet in order to break them loose from the cylinder head.**

36. Now remove the 5 bolts (1 long, 4 short) securing the inner manifold (cyl. # 2 and #3) and lift it off the cylinder head. Remove the gasket and discard it. Take note that although the manifold assembly is a 2-piece set-up, they still utilize a single gasket. Check that the 2 dowels are still in the manifold.

All Models w/1812cc Engines

◆ **See Figures 25 thru 28**

The forward exhaust components are comprised of a 1-piece exhaust manifold, 3 exhaust pipe and a forward exhaust joint (actually part of the rear system. It may be possible to get a few components off without removing the engine, but we recommend removing it.

1. Remove the seats(s) and the cargo container. Disconnect the battery cables.

2. You may want to remove the engine cover for access.

3. Drain the oil and remove the oil filter.

4. Locate the thermo-sensor toward the back of the exhaust pipe and disconnect the electrical lead at the coupler. Tag the lead.

5. Tag and disconnect the water hose at the top of the rectifier/regulator. Do the same with the other hose where it attaches to the exhaust pipe.

6. Remove the rectifier/regulator as detailed in the Ignition & Electrical section. Make sure that the 2 collars and grommets stay with the unit.

7. Remove the 2 bolts and lift of the exhaust pipe mounting stay.

8. Loosen the clamp connecting the forward exhaust joint to the water lock.

9. Tag and disconnect any other hoses, leads or lines that may impede removal of the exhaust system.

10. Remove the 11 (8 long and 3 short) exhaust manifold bolts in the reverse order of the tightening sequence shown here. Pull the manifold out from the cylinder head (you may have to nudge it carefully with a rubber mallet!) and then slide the entire assembly forward until it clears the water lock. Lift the assembly out of the engine compartment and carefully lay it on a clean work surface.

■ **Do not lose the 2 dowel pins.**

11. Loosen the inner hose clamp and slide the forward exhaust joint off the back of the #3 pipe.

12. Unthread the thermo-sensor and remove it from the #2 pipe. Be careful, as the sensor unit is quite fragile.

13. The #3 exhaust pipe is threaded into the #2 pipe and will need to be heated to a temperature of approximately 392° F (200° C) on FX HO and VX models. Acquire a heat gun and carefully heat the entire circumference of pipe just behind where it threads into the #2 pipe until it reaches the suggested temperature. Once at the temperature on HO models or without

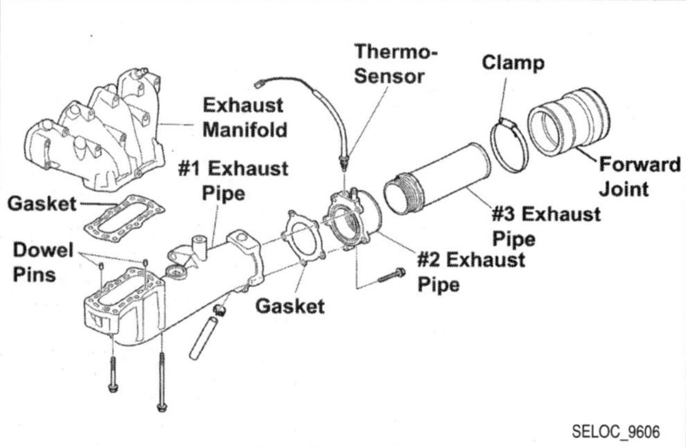

Fig. 25 Exploded view of the forward exhaust components - FX HO and VX models w/1812cc engines

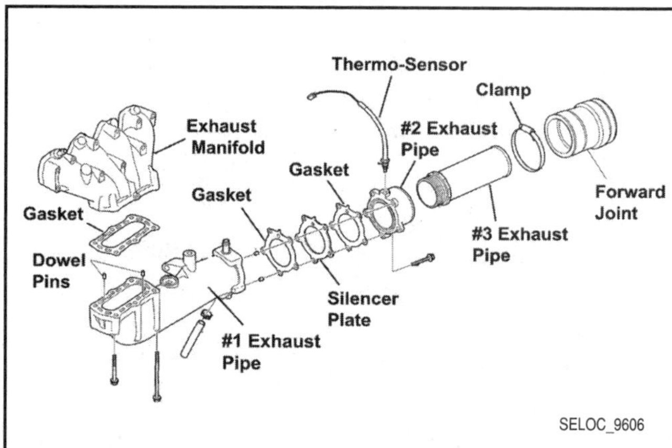

Fig. 26 Exploded view of the forward exhaust components - FX SHO & FZ models w/1812cc engines

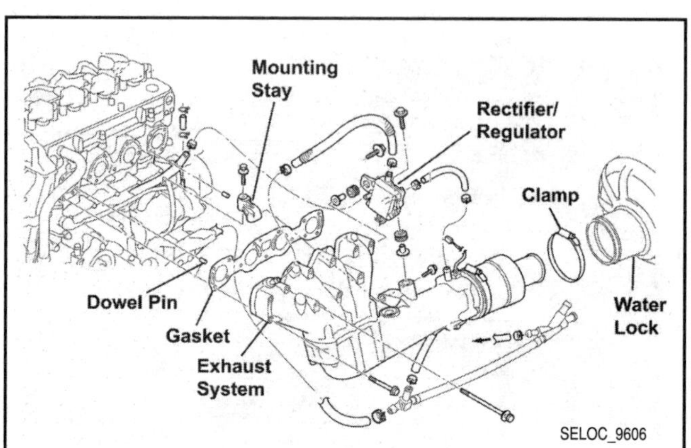

Fig. 27 You should be able to remove the entire assembly without removing the engine on 1812cc engines

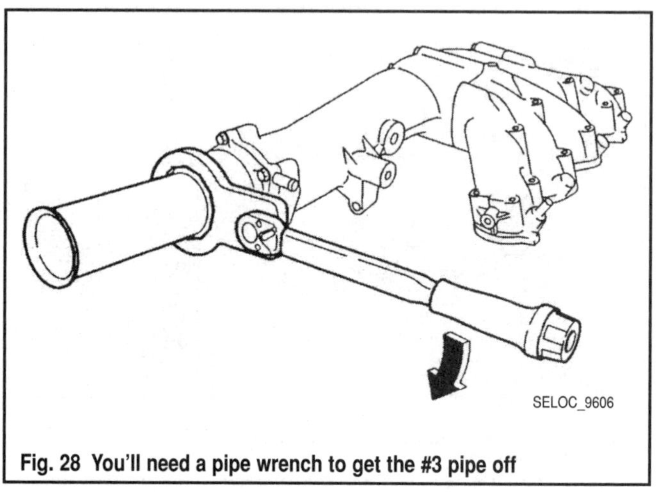

Fig. 28 You'll need a pipe wrench to get the #3 pipe off

heating on SHO and FZ models, use an exhaust pipe wrench (#90890-06726) to loosen the pipe and then thread it out of the #2 pipe. Be careful here if you're working on an HO model, as both pipes are going to be quite hot. WEAR HEAVY GLOVES.

■ **You may want to have someone hold down the assembly as you are loosening the pipe.**

13. Loosen the 5 bolts securing the #2 pipe to the #1 pipe and separate the two. Discard the gasket and check both mounting surfaces for any residual gasket material that may have stuck to them. SHO models utilize a silencer plate sandwiched between 2 gaskets.
14. Remove the 6 bolts (2 short and 4 long) and separate the exhaust manifold from the #1 pipe. Discard the gasket and check the mounting surfaces for any residual gasket material that may have stuck to them.

VX Models w/1052cc Engines

③ DIFFICULT

◆ **See Figures 29 thru 33**

The forward exhaust components are comprised of 2 exhaust pipes (port side and starboard side), an inner and outer exhaust joint, and a 1-piece exhaust manifold. All component services will require removal of the engine.

■ **The following procedure should be performed with the craft out of the water.**

1. Remove the seat and disconnect the battery cables
2. Remove the engine as detailed in the Engine Mechanical section.

#2 Exhaust Pipe (Port Side)

3. Locate the water hose at the forward bend in the #2 exhaust pipe. Pinch the clamp, slide it back on the hose and then wiggle the hose off of the fitting (you probably already did this when you pulled the engine).
4. Loosen the clamp holding the forward exhaust tube joint to the rear of the pipe and then work the joint off of the adaptor on the end of the #3 pipe.
5. Remove the 2 mounting bolts and then remove the exhaust thermo-switch at the rear end of the #2 pipe. Position it out of the way where it will not get dropped or crushed.
6. Loosen the 2 outer exhaust joint clamps. Slide the #2 pipe side of the hose back toward the #1 pipe so you can get at the hose clamp securing the inner joint to the #2 pipe. Loosen the clamp and pry the joint off of the #2 pipe flange.
7. Remove the 2 upper and 2 lower pipe mounting bolts and lift off the #2 pipe. There is a collar and grommet in the forward lower bolt hole that may fall out so keep an eye on them.
8. Remove the 3 adaptor mounting bolts and pull out the adaptor with the 2 gaskets and silencer screen. Check that there is no residual gasket material on the mating surfaces.

#1 Exhaust Pipe (Starboard Side)

9. Remove the #2 exhaust pipe if you haven't already.
10. If necessary, loosen the remaining clamp on the outer joint and then pull/pry it off of the #1 pipe. Do the same with the remaining clamp on the inner joint and remove it as well. Obviously, you could remove the inner first and then the outer if you like!
11. Locate the water hose attached to the pipe, loosen the clamp and then wiggle the hose off of the fitting.
12. Remove the 4 pipe mounting bolts and lift the pipe from the manifold. Remove the gasket, carefully scraping off any residual material that may be on the mounting flanges of the pipe and manifold. Note the 2 dowel pins and make sure you do not lose them.

Exhaust Manifold

13. Remove the #1 and #2 exhaust pipes if you haven't already.
14. Pull back the pinch clamp and wiggle the water hose off of the fitting on the manifold. Repeat with the remaining hose.
15. Remove the 11 mounting bolts and lift the manifold from the engine - you may have to use a little friendly persuasion but be careful. There are 5 short bolts and 6 long ones, so please make a note of which ones went in which holes.

■ **Although no mention is made of this, we would remove the bolts in the reverse order of the tightening sequence.**

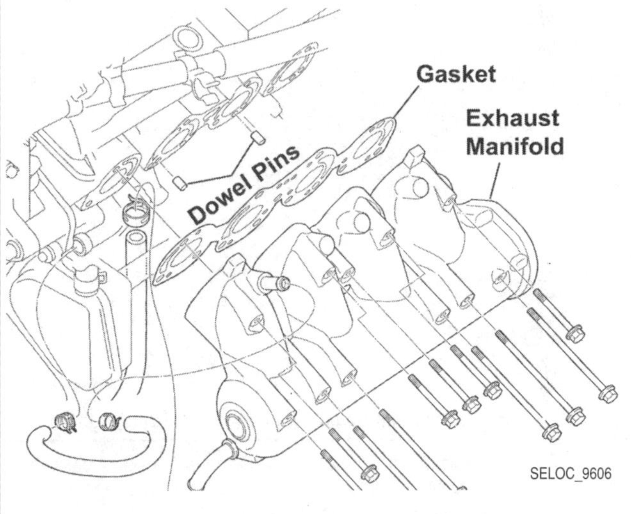

Fig. 30 VX models utilize a 1-piece exhaust manifold

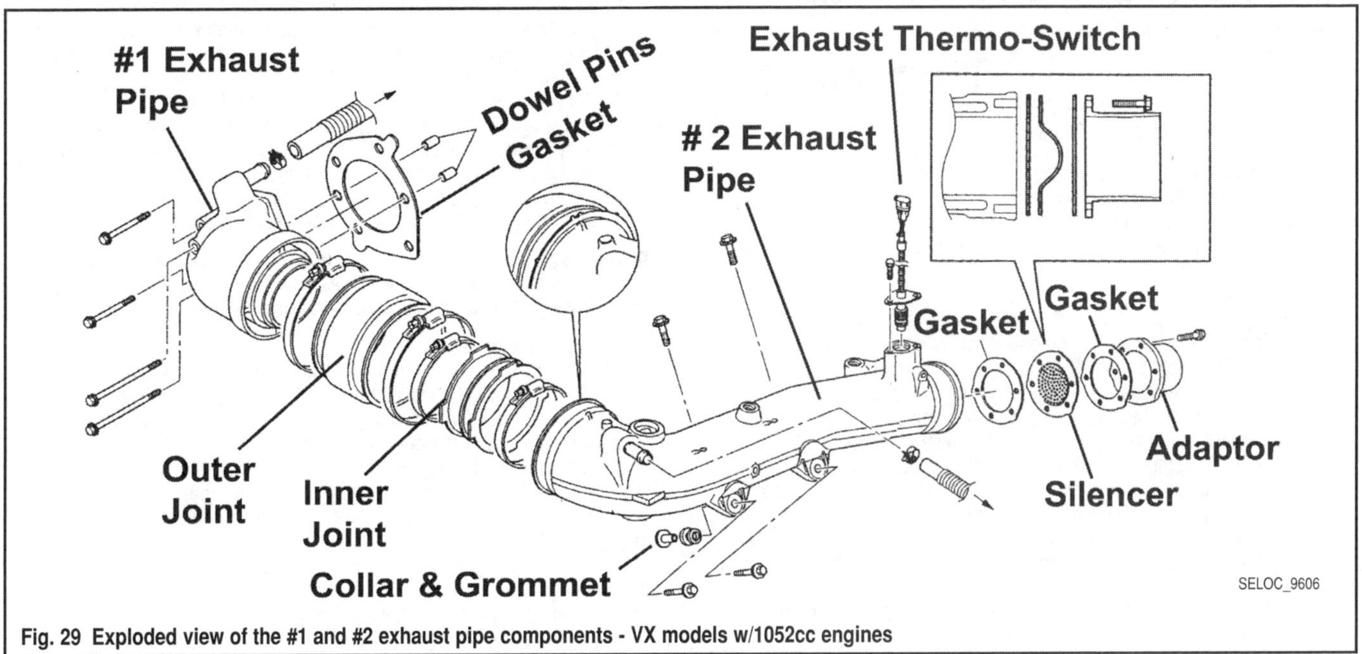

Fig. 29 Exploded view of the #1 and #2 exhaust pipe components - VX models w/1052cc engines

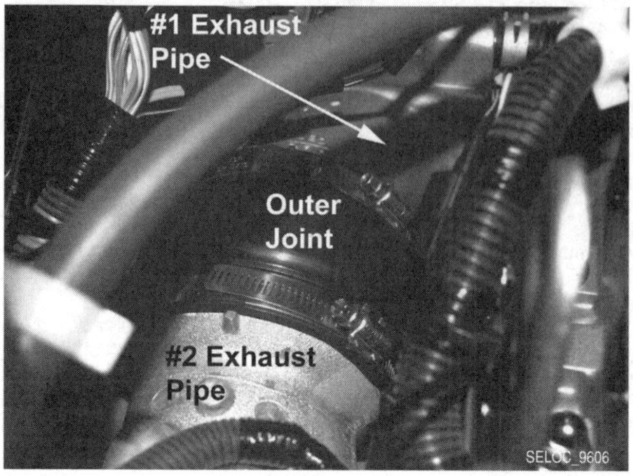

Fig. 31 A good look at the outer exhaust joint between the #1 and #2 pipes - VX models w/1052cc engines

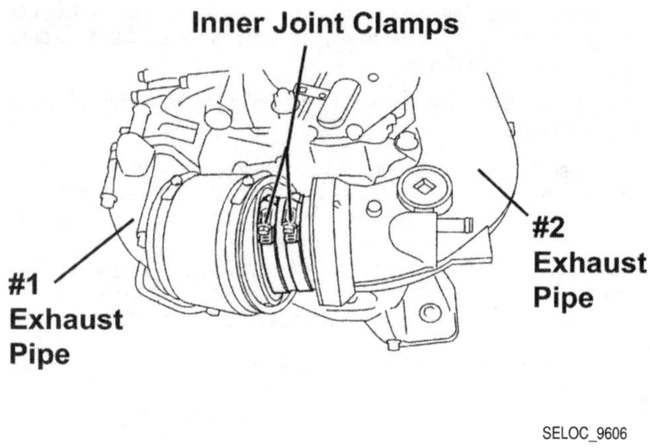

Fig. 32 Pull back the outer joint to get at the inner joint clamps - VX models w/1052cc engines

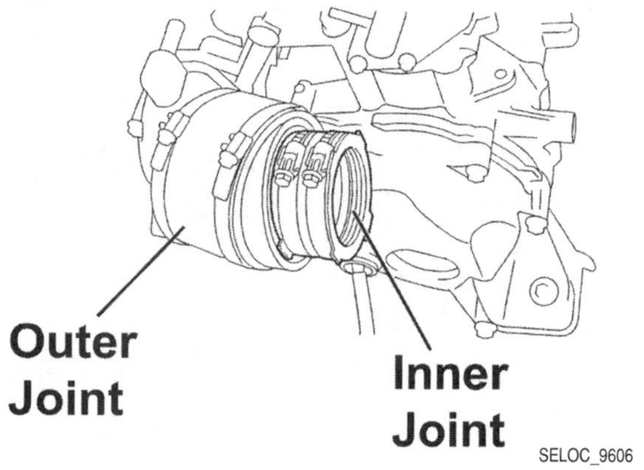

Fig. 33 Removing the inner joint - VX models w/1052cc engines

16. Remove the gasket and throw it away. Clean any residual gasket material from the 2 mating surfaces. Look for the 2 dowel pins and make sure that they are seated correctly in their bores.

INSTALLATION

FX Models With 998/1052cc Engines

DIFFICULT

◆ See Figures 19 thru 23 and 34 thru 38

1. Clean all components thoroughly and allow to dry completely. If available, use compressed air to dry them.

2. Inspect all components for cracks, pitting and any other form of damage; paying particular attention to the flange mating surfaces of each unit.

Exhaust Manifold

3. Position a *new* exhaust manifold gasket onto the cylinder head making sure that all cylinder, bolt and dowel holes line up correctly and then install the inner manifold so that the pins fit all of the way into the head.

4. Coat the bolt threads with Loctite 242 and then screw them in until just finger tight. Tighten the bolts, in a criss-cross pattern, to a 1st pass of 16 ft. lbs. (22 Nm). Once done, tighten them a 2nd time to the same torque. Now you can make a final pass on all bolts to 25 ft. lbs. (35 Nm).

5. Install the outer manifold, making sure that the holes line up and the pins press completely into the head. Tighten the bolts, in a criss-cross pattern, to a 1st pass of 16 ft. lbs. (22 Nm). Once done, tighten them a 2nd time to the same torque. Now you can make a final pass on all bolts to 25 ft. lbs. (35 Nm).

6. Reconnect all 4 water hoses, pinch the clamps and slide them up the hose and over the rib on the fitting. If the main hose running to the central tube came off, make sure that the small white mark on the end is facing UP when you reconnect it.

#1 Exhaust Pipe (Starboard)

■ **Please read ahead to the #2 pipe procedure. You may find that access to a mounting nut in that procedure will make you decide to connect the #1 and #2 pipes first and then connect the assembly to the manifold. Read both procedures, check it out and then make your call.**

7. Make sure that the 2 dowel pins are seated correctly in the pipe-to-manifold mating surface and then position a *new* gasket onto the pipe.

8. Coat the threads of the bolts with Loctite 242. Position the pipe under the exhaust manifolds and thread in the mounting bolts until they are finger tight. Make sure you have the right bolt in the right hole. Tighten the bolts, in a criss-cross pattern, to a 1st pass of 16 ft. lbs. (22 Nm). Once done, tighten them a 2nd time to the same torque. Now you can make a final pass on all bolts to 25 ft. lbs. (35 Nm).

9. If you removed the support stay, position it on the engine, coat the bolt threads with Loctite 242 and then tighten them to 11 ft. lbs. (15 Nm). Now tighten them again, but to 30 ft. lbs. (42 Nm) this time.

10. Coat the threads of the pipe-to-stay bolt with Loctite 242 and tighten it to a first pass of 11 ft. lbs. (15 Nm). Now tighten it again, but to 30 ft. lbs. (42 Nm) this time.

11. Connect the 2 water hoses on the rear of the pipe. Pinch each clamp and slide it up the hose until it is positioned inside of the rib on each fitting.

#2 Exhaust Pipe (Center)

12. Position a *new* gasket onto the #1 pipe's mating surface, making sure everything lines up correctly and the small tabs on the gasket are at the top and bottom (*not* the sides).

13. Move the #2 pipe into position over the studs and dowels. Coat the stud threads lightly with clean engine oil and thread on each of the 5 nuts. Tighten the lower nut to 28 ft. lbs. (39 Nm). Now tighten the 4 upper nuts to the same torque, starting with the bottom one, then the top, the inner and finally the outer nut. Finally, repeat this procedure one more time, in the same order and confirming that they are all tightened to 28 ft. lbs. (39 Nm).

■ **You may find it tricky tightening the 2 inner nuts. Some people actually connect the #1 and #2 pipes first before installing the #1 pipe to the manifold. Your call, but you may find this necessary.**

14. Install the generator cover.

#3 Exhaust Pipe (Port Side) - All Exc. 2008 Models w/6AA Engines

15. Position the outer exhaust joint onto the #2 pipe and tighten the clamp(s) to 3.2 ft.lbs./38 inch lbs. (4.4 Nm). Check them again after a little bit to make sure they are at the same torque.

16. Coat the threads of the 3 exhaust adapter bolts with Loctite 572. Position a *new* gasket, install the adaptor and then tighten the bolts to a 1st pass of 2.7 ft. lbs./33 inch lbs. (3.7 Nm). Tighten one more time to 5.5 ft. lbs./66 inch lbs. (7.6 Nm).

17. Press the grommet and collar into the forward bolt hole. Coat the bolt threads with Loctite 242. Position the pipe over the mounting bolt bosses on the block and then thread in the top 2 bolts until they are finger tight. Now thread in the 2 lower bolts.

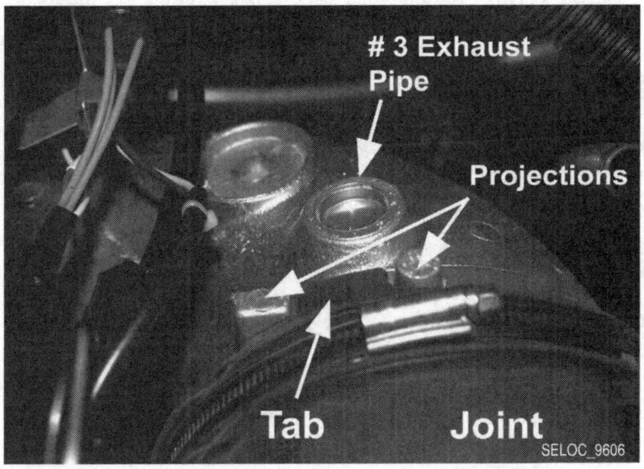

Fig. 34 The tab on the joint must fit between the 2 projections on the #3 pipe

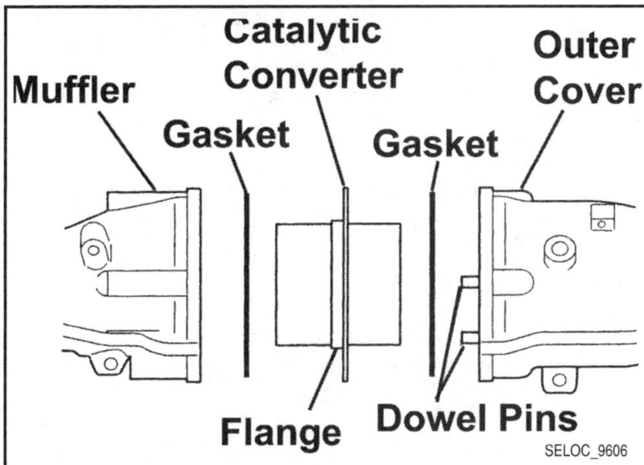

Fig. 35 Install the catalyst converter so the flange is toward the muffler (6AA engines)

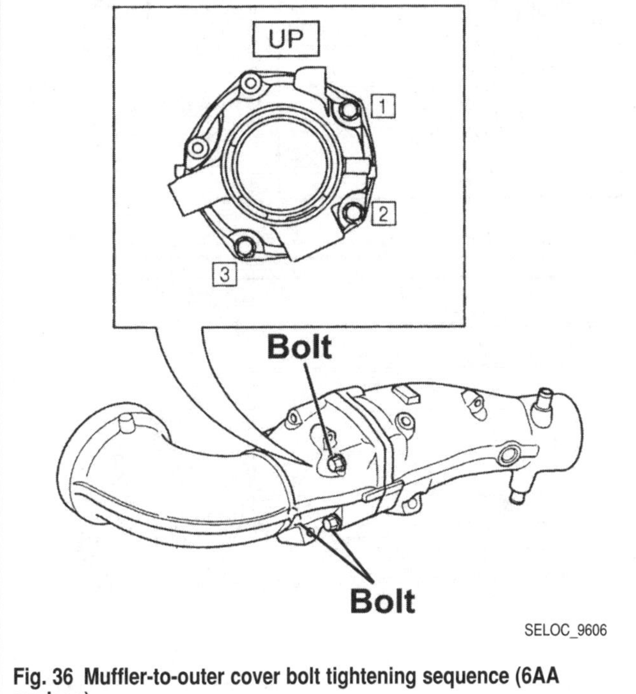

Fig. 36 Muffler-to-outer cover bolt tightening sequence (6AA engines)

18. Tighten all 4 bolts to 1.4 ft. lbs./17 inch lbs. (2 Nm). Start with the forward upper side bolt, move to the rear upper side bolt, then the rear lower side bolt and finally the forward lower side bolt. Make another pass, in the same order, tightening to 11 ft. lbs. (15 Nm) this time. Now make a final pass to 28 ft. lbs. (39 Nm).
19. Install a *new* joint seal.
20. Coat the inner exhaust joint with a little soapy water and slide it into position with the outer joint. Make sure that the clamp(s) is seated in the recess in the joint and tighten it to 3.2 ft.lbs./38 inch lbs. (4.4 Nm). Repeat this process on the other side of the inner joint attaching it to the #3 pipe flange.
21. Install the exhaust thermo-switch into its recess at the back of the pipe and tighten the 2 bolts to 5.5 ft. lbs./66 inch lbs. (7.6 Nm).
22. Reconnect the 3 water hoses, making sure that their pinch clamps are positioned *inside* the rib on the hose fitting. Check the hoses to make sure there are no cracks around the clamp.
23. Reconnect the forward exhaust hose joint to the adaptor at the end of the pipe (details can be found in the Rear Exhaust Components section.
24. Go back and check the torque on all of the joint hose clamps. Install the air filter case and then the engine.
25. Connect the battery cables and install the cargo container and seat(s).

#3 Muffler/Catalyst Converter (Port Side) - 2008 Models w/6AA Engines

26. Insert the catalyst converter into the muffler as shown using a *new* forward gasket.
27. Position a *new* gasket over the 2 dowel pins and onto the outer exhaust cover mating surface.
28. Connect the two pipe halves and then thread in the 3 mounting bolts after coating their threads with Loctite 271. Tighten the bolts, in the order shown, to 11 ft. lbs. (15 Nm) and then to a 2nd pass of 25 ft. lbs. (34 Nm). Finally, go back one more time and retighten to 25 ft. lbs. (34 Nm) again.

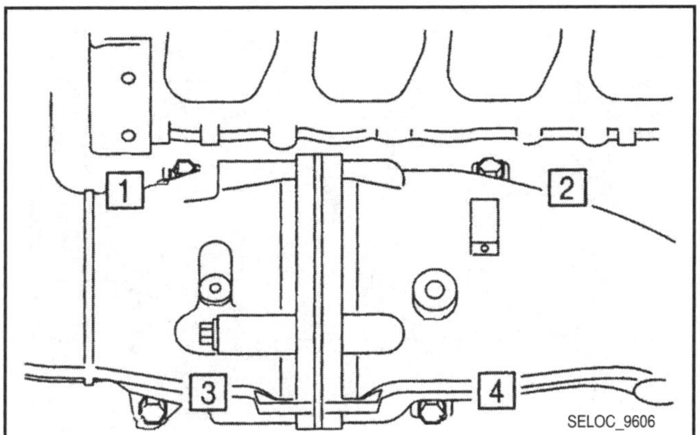

Fig. 37 Muffler pipe mounting bolt sequence (6AA engines)

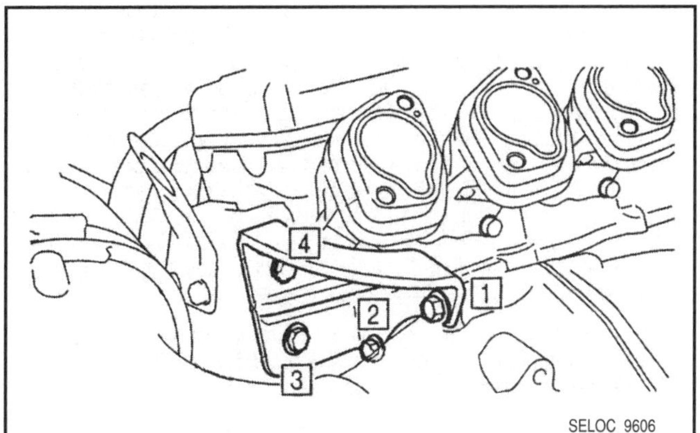

Fig. 38 Tighten the mounting stay bolts in this order (6AA engines)

29. Position the outer exhaust joint onto the #2 pipe/ring joint and tighten the clamp(s) to 3.2 ft.lbs./38 inch lbs. (4.4 Nm). Check them again after a little bit to make sure they are at the dame torque.

30. Coat the threads of the 3 exhaust adapter bolts with Loctite 572. Position a *new* gasket, install the adaptor and then tighten the bolts to a 1st pass of 2.7 ft. lbs./33 inch lbs. (3.7 Nm). Tighten one more time to 5.5 ft. lbs./66 inch lbs. (7.6 Nm).

31. Coat the muffler mounting bolt threads with Loctite 271. Position the pipe over the mounting bolt bosses on the block and then thread in the top 2 bolts until they are finger tight. Now thread in the 2 lower bolts.

32. Tighten all 4 bolts to 1.4 ft. lbs./17 inch lbs. (2 Nm) in the sequence shown. Make another pass, in the same order, tightening to 11 ft. lbs. (15 Nm) this time. Now make a final pass to 28 ft. lbs. (39 Nm).

33. Coat the mounting stay bolt threads with Loctite 271, screw them in and tighten, in the sequence shown, to 1.4 ft. lbs./17 inch lbs. (2 Nm), then 11 ft. lbs. (15 Nm) and then make a final pass to 28 ft. lbs. (39 Nm).

34. Install a *new* joint seal.

35. Coat the inner exhaust joint with a little soapy water and slide it into position with the outer joint. Make sure that the clamp(s) is seated in the recess in the joint and tighten it to 3.2 ft.lbs./38 inch lbs. (4.4 Nm). Repeat this process on the other side of the inner joint attaching it to the muffler flange.

36. Install the exhaust temperature sensor into its recess at the back of the pipe and tighten it to 28 ft. lbs. (39 Nm).

37. Reconnect the 4 water hoses, making sure that their pinch clamps are positioned *inside* the rib on the hose fitting. Check the hoses to make sure there are no cracks around the clamp.

38. Reconnect the forward exhaust hose joint to the adaptor at the end of the pipe (details can be found in the Rear Exhaust Components section.

39. Go back and check the torque on all of the joint hose clamps. Install the air filter case and then the engine.

40. Connect the battery cables and install the cargo container and seat(s).

41. Install the engine.

42. Connect the battery cables and install the seat(s) and cargo container.

All Models w/1812cc Engines

OEM 3 DIFFICULT

◆ **See Figures 25 thru 28, 39, 40, 41**

1. Carefully inspect all components for cracks or other damage. Check the thermo-sensor bends in the sensor tube and carbon deposits.

2. Position a *new* gasket on the #1 pipe mounting surface and then move the exhaust manifold into place. Coat the threads of the mounting bolts with Loctite 242 and screw them all the way in. Remember that the 2 short ones go in the forward holes and the long ones go in the rear holes. Tighten the bolts in a criss-cross pattern to 14.8 ft. lbs. (20 Nm). Wait a few minutes and then go back and re-tighten them to 25.8 ft. lbs. (35 Nm).

3. On HO and VX models, use a *new* gasket and position the #2 pipe on the back of the #1 pipe. On SHO and FZ models, position a *new* gasket, then the silencer plate and then another *new* gasket. Make certain that all the holes line up before installing the bolts. Coat the bolt threads with Loctite 242 and then tighten them, in a criss-cross pattern, to 11.1 ft. lbs. (15 Nm). Wait a few minutes and then re-tighten them to 21.4 ft. lbs. (29 Nm).

4. On HO and VX models, apply 2 short lines of Loctite 648 across the threads at the bottom of the inner #2 pipe, making sure they are each about 15° off of the vertical centerline of the end of the pipe. Now do the same thing across the top of the #3 pipe threads. On SHO models, simply coat the threads of the #3 pipe with Loctite 242, or do the same with Loctite 648 on FZ models. Thread the #3 pipe into the #2 pipe and then use the pipe wrench to tighten it 44.3 ft. lbs (60 Nm) on FX models, 147.5 ft. lbs. (200 Nm) on VX models or 46.5 ft. lbs. (63 Nm) on FZ models. The 648 sealant needs to set entirely, so allow the assembly to sit for about an hour before continuing if working on an HO, VX or FZ model.

■ **We know what you're thinking, and so are we! We have no idea why the sealant procedure and torque is different between the 3 models because they are essentially the same part.**

5. Slide the forward exhaust joint into position over the #3 pipe until it butts up against the flange. Tighten the inner (forward) clamp to 1.5 ft. lbs./18 inch lbs. (2 Nm) except on the VX where it should be 2.6 ft. lbs./25 inch lbs. (3.5 Nm).

6. Thread in the thermo-sensor and tighten in to 14.8 ft. lbs. (20 Nm).

7. Carefully move the assembly into the engine compartment on an angle with the back end lower. Feed it into the water lock as you lower the forward end until it is positioned correctly with the cylinder head and over the 2 dowel pins.

8. Coat the threads of the manifold bolts with Loctite 242 and then thread them in until finger-tight. Now tighten the bolts, in the order shown, to a 1st pass of 14.8 ft. lbs. (20 Nm). Wait a little bit and then re-tighten them to a final pass of 25.8 ft. lbs. (35 Nm).

On VX models, tighten to a 1st pass of 14.8 ft. lbs. (20 Nm). Wait a little bit and then re-tighten them to another pass of 14.8 ft. lbs. (20 Nm) and then finally a 3rd time to 25.8 ft. lbs. (35 Nm).

9. Make sure that the pipe is still inserted fully into the water lock and then tighten the rear clamp on the forward joint to 1.5 ft. lbs./18 inch lbs. (2 Nm) except on the VX where it should be 2.6 ft. lbs./25 inch lbs. (3.5 Nm).

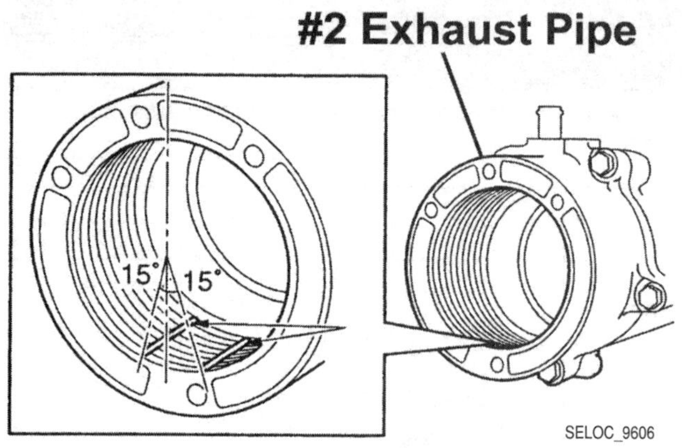

Fig. 39 Apply 2 lines of sealant to the inner threads of the #2 pipe. . .

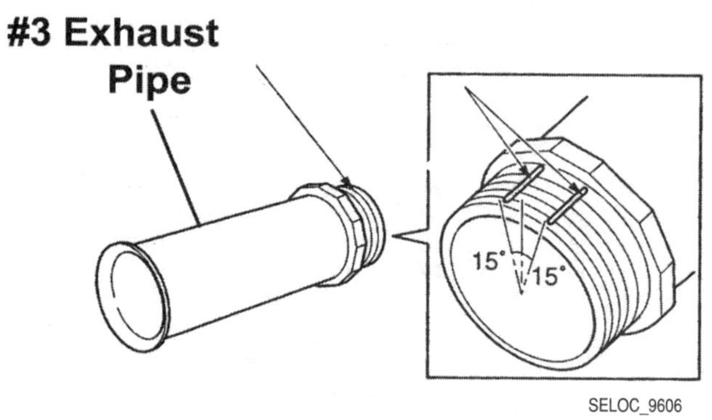

Fig. 40 . . .and then do the same with the threads on the #3 pipe - HO and VX models w/1812cc engines

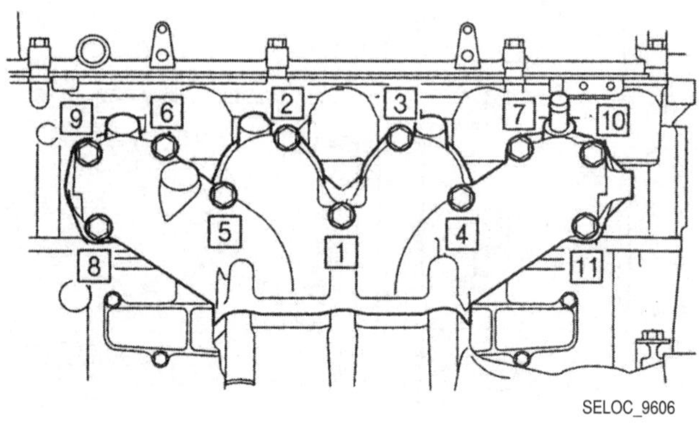

Fig. 41 Exhaust manifold tightening sequence - 1812cc engines

10. Position the mounting stay on the pipe and block, coat the bolt threads with Loctite 242 and then tighten the bolts to 16.2 ft. lbs. (22 Nm). Wait a few minutes and then re-tighten them to 31 ft. lbs. (42 Nm).

11. Install the rectifier/regulator and then connect all water hoses that were removed originally. Make sure they slide onto their fittings fully and check the hose ends around the clamping area.

12. Install the oil filter and fill with oil. Reconnect the electrical lead to the thermo-sensor at the coupler.

13. Install the engine cover if removed.

14. Connect the battery cables. Install the cargo container and seat(s).

VX Models w/1052cc Engines

3 DIFFICULT

◆ **See Figures 29, 30, 31, 42 and 43**

1. Clean all components thoroughly and allow to dry completely. If available, use compressed air to dry them.

2. Inspect all components for cracks, pitting and any other form of damage; paying particular attention to the flange mating surfaces of each unit.

Exhaust Manifold

3. Inspect the manifold thoroughly for cracks, blockage or any other visible damage.

4. Position a *new* gasket over the dowel pins and onto the head.

5. Coat the threads of all bolts with Loctite 242, position the manifold so that all holes line up and then thread the bolts in until finger tight. Hopefully you remembered to mark down which ones went where, but if not, remember that the longer ones (6) will go in the holes with the longer bosses. Tighten the bolts, in the sequence shown, to 16.2 ft. lbs. (22 Nm). Let everything sit for a bit and then go back and retighten them to 16 ft. lbs. (22 Nm) again. Now you can make a final pass to 25 ft. lbs. (35 Nm). On 2010-11 models, Yamaha has skipped the 2nd step where you retighten to 16 ft. lbs.

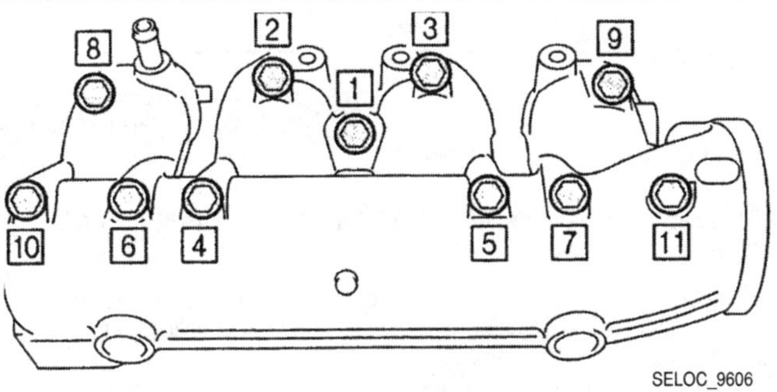

Fig. 42 Exhaust manifold tightening sequence - VX models w/1052cc engines

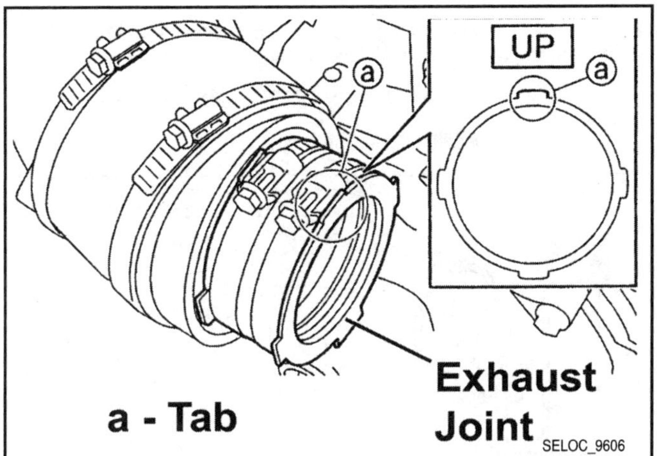

Fig. 43 Make sure the tabs on the inner joint are as shown - VX models w/1052cc engines

#1 Exhaust Pipe (Starboard Side)

6. Clean the pipe and manifold mating surfaces thoroughly.

7. Install a *new* gasket over the dowel pins and then carefully seat the pipe on the manifold, making sure that all of the holes line up. Coat the bolt threads with Loctite 242 and thread them in until finger-tight. Remember that the 2 longer bolts are for the inner holes. Tighten the bolts in criss-cross pattern to an initial torque of 16.2 ft. lbs, (22 Nm). Wait a few minutes and then go back and retighten them to the same torque a 2nd time. Make a final pass at 25 ft. lbs. (35 Nm). On 2010-11 models, Yamaha has skipped the 2nd step where you retighten to 16 ft. lbs.

8. Reconnect the water hoses to their fittings on the pipe, making sure that the hose and pinch clamp are pressed far enough on to the fitting.

9. Slide on the outer joint making sure it goes all the way into position.

10. Now slide the inner joint onto the pipe flange so that the flat tabs are position at the top, bottom and each side. Move the clamp (#1 pipe side) into the recess on the joint and tighten it to 2.5 ft. lbs./30 inch lbs. (3.4 Nm). Wait a while and then do it again.

11. Move the clamp into the recess on the outer joint and tighten it to 2.5 ft. lbs./30 inch lbs. (3.4 Nm). Wait a while and then do it again.

12. Install the #2 exhaust pipe.

#2 Exhaust Pipe (Port Side)

13. Position a *new* gasket on the end of the pipe. Follow this with the silencer screen and another *new* gasket and then the adaptor. Coat the bolt threads with Loctite 572 and tighten to 2.7 ft. lbs./31 inch lbs. (3.7 Nm). Wait a little bit and then tighten the 3 bolts to 5.5 ft. lbs./66 inch lbs. (7.6 Nm).

■ **Make sure that the cupped side of the silencer screen is facing toward the rear.**

14. Check the exhaust thermo-switch for carbon deposits and then press it back into the exhaust pipe - *carefully*. Coat the bolt threads with Loctite 572 and tighten them to 5.5 ft. lbs./66 inch lbs. (7.6 Nm) on 2005-09 models or 5.9 ft.lbs./70 inch lbs. (8 Nm).

15. Insert the forward end of the pipe into the inner joint pressing hard enough that it fully seats itself. Make sure that joint is situated so that the flat tabs are still on the top, bottom and both sides and then slide the clamp into position in the recess and tighten it to 2.5 ft. lbs./30 inch lbs. (3.4 Nm).

16. Coat the 4 pipe mounting bolt threads with Loctite 242, position the pipe over the bolts bosses and thread the bolt in finger-tight. Don't forget the collar and grommet on the lower forward bolt. Tighten the bolts to an initial torque of 1.5 ft. lbs./18 inch lbs. (2 Nm) starting with the forward top bolt, the rear top bolt, the rear lower bolt and then the forward lower bolt. Make a 2nd pass in the same order to 11.1 ft. lbs. (15 Nm). Now make a final pass, still in the same order, to 28 ft. lbs. (39 Nm). On 2010-11 models, tighten to 29.5 ft. lbs. (40 Nm) on the final pass.

17. Check the inner joint clamp again and retighten it to 2.5 ft. lbs./30 inch lbs. (3.4 Nm) again.

18. Slide the outer joint over the #2 pipe and position the clamp in the recess over the pipe flange. Tighten it to 2.5 ft. lbs./30 inch lbs. (3.4 Nm). Wait a while and then do it again.

19. Reconnect the water hose to the fitting at the forward end of the pipe, making sure that the hose and pinch clamp are pressed far enough on to the fitting.

20. Reconnect the forward exhaust hose joint to the adaptor at the end of the pipe (details can be found in the Rear Exhaust Components section.

21. Connect the battery cables and install the cargo container and seat(s).

22. Install the engine.

23. Connect the battery cables and install the seat.

Exhaust Thermo-Switch

TESTING

2 MODERATE

◆ **See Figure 44**

■ **Models w/1812cc engines and the 2008 1052cc 6AA engines do not use an exhaust thermo-switch in the pipe. Instead, they utilize an exhaust thermo-sensor.**

A thermo-switch is a simple device that is basically an on/off switch which is activated by temperature; and usually used in conjunction with the warning system.

The exhaust thermo-switch is located at the back of the #2 (VX w/1052cc engines) or #3 (FX with 998cc/1052cc engines) exhaust pipe on the port side.

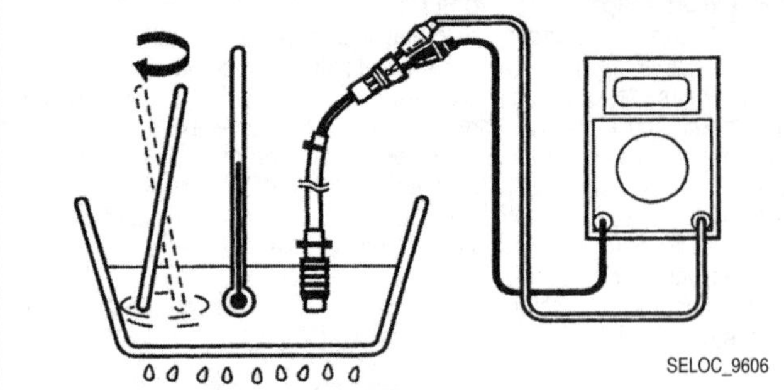

Fig. 44 Testing the thermo-switch

Locate and remove the thermo-switch from the exhaust pipe. The thermo-switch resembles a condenser, usually with Pink and Black leads (but check the Wiring Diagrams from the Ignition & Electrical System section for the motor on which you are working). The thermo-switch can normally be removed easily, by pulling it out of its recess. However, on a few models the thermo-switch may be secured by a cover or tab that is bolted in place. So, if present, make sure you remove any covers or tab before attempting to full the switch free. The thermo-switch is a very fragile switch and should never be dropped or handled in a rough manner.

Disconnect the 2 leads at their quick disconnect fittings. Since the switch should have no continuity until it is heated past a point which could damage the powerhead, there are no resistance tests to be performed on this switch while it is installed.

Testing is a simple matter of watching for continuity across the switch contacts (using an ohmmeter or DVOM set to read resistance across the switch wiring) as the switch is slowly heated to activation temperature. The best method for this is to suspend the switch in water which is slowly heated to temperature while you watch the meter. Obtain a thermometer, an ohmmeter, and a suitable pan in which to boil water.

Place the container of water, at room temperature, on a stove. Secure the thermometer in the water in such a manner to prevent the thermometer bulb from contacting the bottom or sides of the pan (this ensures the thermometer will read the temperature of the water and NOT the pan, which could be considerably higher).

Immerse the sensing end of the thermo-switch into the water up to the shoulder and suspend it, also NOT touching the pan for the same reason as the thermometer.

Make contact with the Red meter lead to the Pink thermo-switch lead, and the Black meter lead to the Black thermo-switch lead. The meter should indicate no continuity,

Observe at what temperature the switch begins to show continuity as you heat the water. Turn off the heat source and then observe at what temperature it reverts back to no continuity as the water cools.

The thermo-switch should begin to show continuity when the water temperature reaches:

- 2008 HO - 183-194°F (84-90°C)
- All other FX models - 201-212°F (94-100°C)
- VX models - 176-187°F (80-86°C)

As the water cools down, the meter should stop showing continuity as the temperature reaches

- 2008 HO - 158-183°F (70-84°C)
- All other FX models - 176-201°F (80-94°C)
- VX models - 151-176°F (66-80°C)

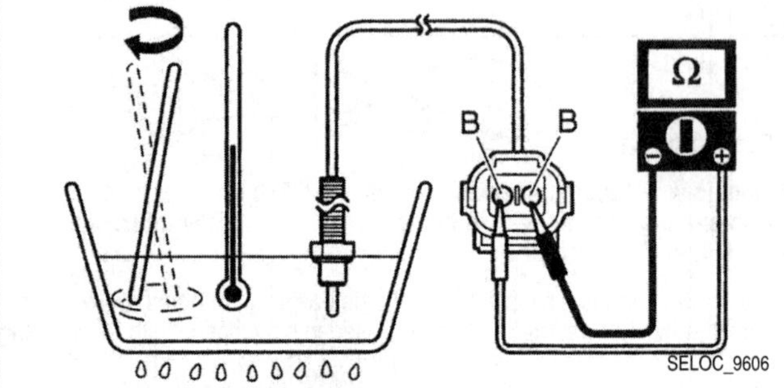

Fig. 45 Testing the thermo-sensor

■ **A slight variation is allowable above or below the stated temperatures listed. But keep in mind that too great a variation could either allow damage to the powerhead by operating it under conditions that are too hot or could allow activation of the warning system when the powerhead is not in danger.**

If the meter reading is acceptable, the thermo-switch is functioning correctly. If the meter readings are not acceptable, the thermo-switch can be easily replaced with a new unit.

Exhaust Thermo-Sensor

TESTING

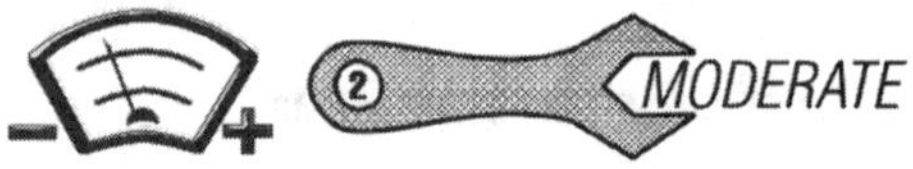

◆ **See Figure 45**

■ **2008 models w/1052cc 6AA engines and all models w/1812cc engines use an exhaust thermo-sensor, while all other models utilize an exhaust thermo-switch.**

1. Disconnect the electrical lead at the coupler, tag it and position it out of the way.
2. Unthread the sensor.
3. Connect a DVOM as per the manufacturer's instructions and then insert the probes into the sensor coupler.
4. Suspend the switch in a container of very cold water and slowly heat the water while observing the resistance at:
   - 32°F (0°C) - 24-37.1 kili-ohms
   - 212°F (100°C) - 0.87-1.18 kili-ohms

   Failure to meet these criteria will require replacement of the thermo-sensor.
5. Install the sensor and tighten securely.

Exhaust Temperature Sensor

TESTING

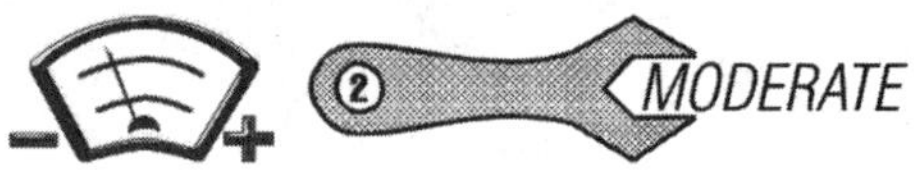

◆ **See Figure 46**

■ **This sensor is used only by 2008 FX models with a 1052cc 6AA engine.**

1. Locate the sensor in the muffler, disconnect the coupler and remove it.
2. Connect a DVOM as per the manufacturer's instructions and then insert the probes into the sensor coupler.
3. Heat the sensor using an electric furnace and measure the resistance at:
   - 572°F (300°C) - 73-241 kili-ohms
   - 1112°F (600°C) - 0.86-1.58 kili-ohms
   - 1652°F (900°C) - 64-90 ohms

■ **Personally, if you don't already own, or have access to, this type of heating device, it's probably cheaper to simply replace the sensor.**

Any reading outside of the above ranges will necessitate sensor replacement.

4. Install the sensor and tighten securely.

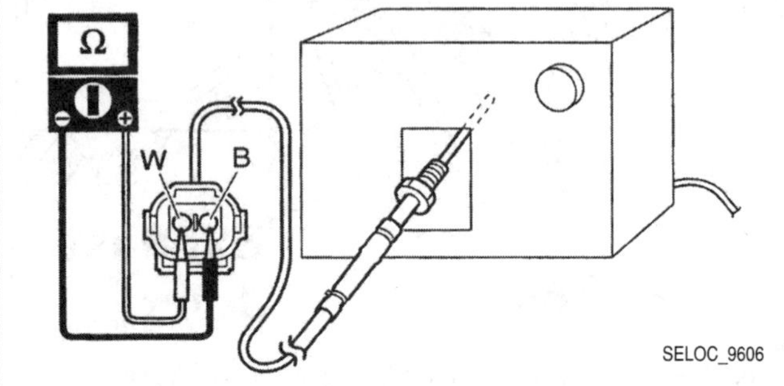

Fig. 46 Testing the exhaust temperature sensor (6AA engines)

COOLING SYSTEM

Description & Operation

All models covered here utilize the jet pump as a water pump, similar to what a water pump would do on other vessels. The pump takes raw water (fresh or salt) to both the oil cooler and the water distribution manifold (FX with 998cc/1052cc engines only).

From here the cooling water moves on to the engine or back to the oil cooler depending on conditions. FX models (exc. HO/SHO) utilize a water pressure control valve between the distribution manifold and the engine. On VX models the water goes right to the cylinder block or exhaust manifold. A series of passages in the cylinder block and head carry cooling water thru-out the engine until it exits at the exhaust manifold. Once in the manifold, water is either directed back to the distribution manifold (oil cooler on VX models) or exits through the exhaust system. Water can also be directed from the oil cooler to the exhaust system as well.

On models w/1812cc engines, water from the jet pump runs directly to the oil pump, but uses a 'T' along the way to direct some to the exhaust manifold (and/or air cooler on SHO or FZ models). From the oil pump, water moves through the passages in the cylinder block and head, though the rectifier/regulator, through the thermostat, to the exhaust pipe and then back on out from the thermostat to the drain joint.

All models also use a fitting on the #3 exhaust pipe or muffler (FX models with 998cc/1052cc engines), oil pump (all models w/1812cc engines) or #2 exhaust pipe (VX w/1052cc engine) that directs water to a water indicator which allows a small stream of water to be visible to the operator indicating that the cooling system is operating.

All FX models and also VX models w/1812cc engines are equipped with a thermostat that restricts the amount of cooling water allowed into the engine until it reaches normal operating temperature. The purpose of the thermostat is to increase engine performance and reduce emissions by making sure the engine warms as quickly as possible to operating temperature and remains at the proper temperature during use under all conditions.

Running a motor without a thermostat may prevent it from fully warming, not only increasing emissions and reducing fuel economy, but it will likely lead to carbon fouling, stumbling and poor performance in general. It can even damage the motor, especially if the motor is then run under load (such as full-throttle operation without allowing it to thoroughly warm). A restricted thermostat can promote engine overheating. The good news is that should you be caught on the water with a restricted thermostat, you should be able to easily remove it and get back to shore, just make sure you replace it before the next outing.

Cooling passages have the tendency to become clogged gradually over time by debris and corrosion. The best way to prevent this is to flush the cooling system *after each use* regardless of where you are operating (salt or freshwater). But obviously, this form of maintenance is even more important on vessels used in salt, brackish or polluted waters that will promote internal corrosion of the cooling passages.

Troubleshooting

■ **When troubleshooting the cooling system, especially for overheat conditions, the motor should be run on a launched vessel (to simulate normal running conditions.). Running the motor on a flushing device may provide both a higher volume of water than the system would deliver (and often times with water that is much colder than would be found in your local lake or bay).**

A water-cooled engine has a lot of potential problems to consider when talking about overheating. The most overlooked tends to be the simplest, clogged cooling passages or water intake grate. Although a visual inspection of the intake grate will go a long way, the cooling passage condition can really only be completely checked by operating the motor or disassembling it to observe the passages.

Damaged or worn cooling system components tend to cause most other problems with the system. And since there are relatively few components, they are easy to discuss. The most obvious is a thermostat that is damaged or corroded, which will often cause the motor to run hot or cold (depending on the position in which the thermostat is stuck, closed or open). The other most obvious component is the jet pump impeller which is really the heart of the cooling system and it is easy to check, easy that is once it is accessed. But of course, since the impeller, unlike on an outboard engine, is the source of propulsion, you will probably know you have a problem there already!

Overheating at high RPM, but not under light load, may indicate a leaking head gasket. If a head gasket it leaking, water can go into the cylinder, or hot exhaust gases may go into the water jacket, creating exhaust bubbles and excessive heat. Remember that aluminum heads have a tendency to warp, and need to be checked (if not re-surfaced) each time they are removed. If necessary, they can be resurfaced by using emery paper and a surface block moving in a figure-eight motion. If this is done, you should also inspect the cylinders and pistons for damage. Other areas to consider are the exhaust manifold gasket. Look for corrosion pin holes. This is rare, but if the unit has been operated in salt water over the years, there may just be a problem.

Whenever the motor has been run in polluted, brackish or saltwater, the cooling system should be flushed. Follow the instructions provided under Flushing the Cooling System in the Engine Maintenance section for more details. But in most cases, the motor should be flushed for at least 5 minutes. This will wash the salt from the castings and reduce internal corrosion.

TESTING COOLING SYSTEM EFFICIENCY

OEM ② MODERATE

◆ See Figure 47

If trouble is suspected, cooling system efficiency can be checked by running the motor (while an assistant navigates) and monitoring cylinder head temperatures. There are 2 common methods available to monitor cylinder head temperature, the use of a heat sensitive marker or an electronic pyrometer.

The Stevens Instrument company markets a product known as the Markal Thermomelt Stik®. This is a physical marker that can be purchased to check different heat ranges. The marker is designed to leave a chalky mark behind on a part of the motor that will remain solid and chalky until it is warmed to a specific temperature, at which point the mark will melt appearing liquid and glossy.

Alternately, and more easily, an electronic pyrometer may be used. Many DVOMs are available with thermo-sensor adapters that can be touched to the cylinder head in order to get a reading. Also, a number of instrument companies are now producing relatively inexpensive infra-red pyrometers (such as the Raytek® MiniTemp®) of a point-and-shoot design. These units are simply pointed toward the cylinder head while holding down the trigger and the electronic display will give cylinder head temperature. For ease of use and relative accuracy of information, it is hard to beat these infra-red pyrometers. Be sure to follow the tool manufacturer's instructions closely when using any pyrometer to ensure accurate readings.

■ **The infra-red pyrometers are EXTREMELY useful when attempting to find a cooling system blockage. You can, in a matter of seconds, take temperature readings of all different parts/locations on the outboard. Once you've used a pyrometer, you won't go back to the markers.**

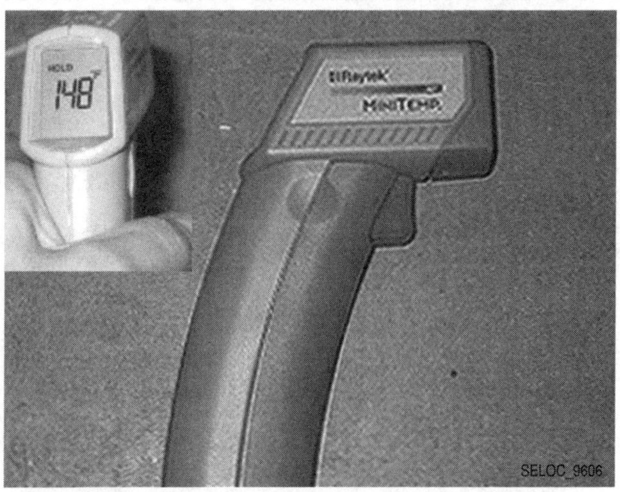

Fig. 47 By far the easiest way to check cylinder head temperature is with a hand-held pyrometer like the MiniTemp® from Raytek® pictured here

Thermostat

DESCRIPTION & OPERATION

◆ See Figures 48 and 49

■ **Remember, thermostats are only used on VX models w/1812cc engines, FX and FZ models.**

The thermostat is a simple temperature sensitive valve that opens and closes to control cooling water flow through the engine. In operation, the thermostat hovers somewhere between open and closed. As engine load and temperature increase, the thermostat opens to allow more cooling water into the engine. As temperature and load decrease, the thermostat closes.

A sticking thermostat will either allow the temperature to rise well above the normal operating temperature before it opens, or, if stuck in the open position, will never allow the engine to reach operating temperature.

All thermostats are rated based on the temperature at which they open and this rating should always be stamped somewhere on the thermostat, usually on the flange area. All engines covered here utilize a 123-141°F thermostat, except the 1812cc engines where FX HO and VX models utilize 136-144 °F units and the FX SHO/FZ models which use a 118-126°F.

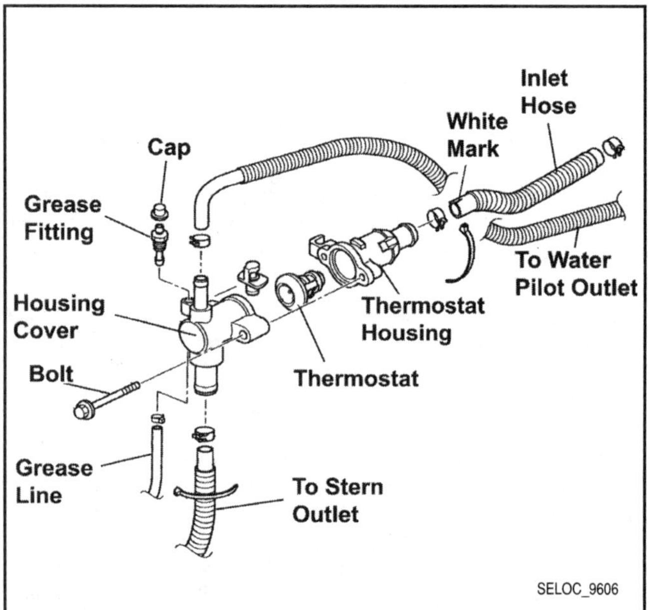

Fig. 48 Exploded view of the thermostat assembly - 998/1052cc FX models

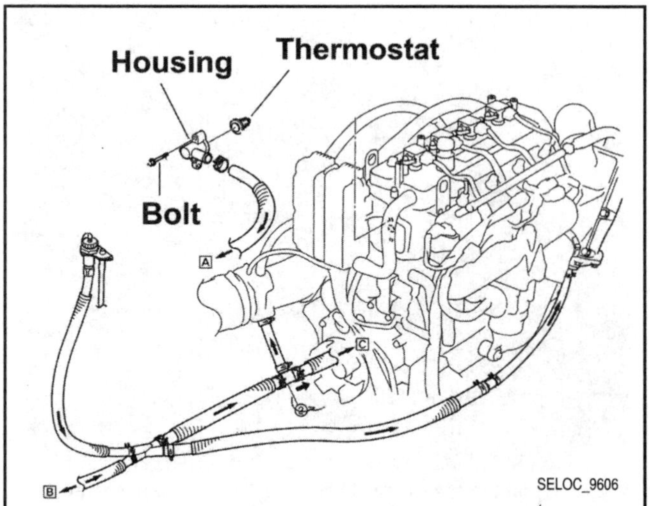

Fig. 49 Exploded view of the thermostat assembly - 1812cc models

■ **The thermostat housing can be found on the rear side of the engine, often attached to the oil cooler - easily identifiable by the 3 hoses attached to it.**

✻✻ CAUTION

Serious damage may result from operating your engine without a thermostat! Don't even consider this!

✻✻ CAUTION

NEVER use an automotive/motorcycle thermostat in a marine engine. No matter how tempting this may seem, forget it!!

REMOVAL & INSTALLATION

FX Models With 998cc/1052cc Engines

① EASY

◆ See Figures 48, 50 and 51

1. Remove the seat(s) and disconnect the negative battery cable. You may want to remove the rear cargo container as well.

■ **Although not strictly necessary, you may find that removing the deck plate (above the tank) will make access much easier.**

2. Tag and disconnect the 3 water hoses connected to the housing; move them out of the way. You may have to remove a plastic tie as well. All

Fig. 50 The housing is attached to the oil tank on most 998/1052cc engines

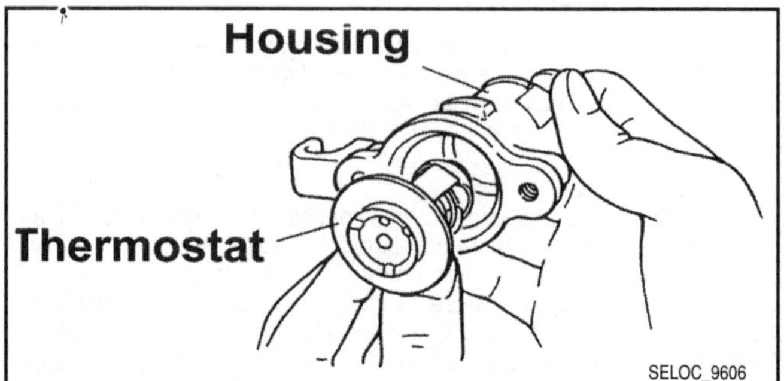

Fig. 51 Make sure you position the thermostat with the spring-end into the housing on 998/1052cc engines

3 hoses should use conventional pinch clamps, so just squeeze the 2 tabs and slide the clamp back over the hose.

3. There is a small grease line on the side of the housing, remove the clamp and slide the line off the fitting. This line may have a full circle clamp on it so may require cutting. Lift the fitting and its cap out of the thermostat housing.

4. Slide the housing out of the mount. Loosen the 2 cover-to-housing bolts and lift off the cover. Remove the thermostat and gasket. Some later models may use 2 bolts attaching the housing to the mount instead. It is not necessary to remove the housing unless you need to.

5. Perform the testing procedures if necessary.

To install:

6. Clean the housing and cover and check all passages for any blockage.

7. Position the thermostat, spring first, into the housing.

8. Position a new gasket onto the cover and then connect it to the housing. Coat the bolt threads with Loctite 572 and then install them and tighten to 2.7 ft. lbs. (32 inch lbs/3.7 Nm) in one pass and then again to 5.5 ft. lbs. (66 inch lbs./7.6 Nm). Move the assembly into the mount.

9. If you removed the housing attached with bolts, slide the collars into the holes, coat the bolts with Loctite 572 and then thread the 2 bolts through the housing and into the mount, tightening to 2.7 ft. lbs. (32 inch lbs/3.7 Nm) in one pass and then again to 5.5 ft. lbs. (66 inch lbs./7.6 Nm).

10. Press the grease fitting (if equipped) into the hole in the housing and then slide the line onto the bottom of the fitting. Slide the clamp up into place. If you have trouble with this, slide the line up through the hole and press the fitting into the line, and then press everything back into the hole - just make sure you install the clamp *after* you've pressed the assembly back down into the hole.

■ **If you had to cut off a full circle clamp, make sure you use the correct clamp pliers to re-crimp the clamp.**

11. Reconnect the 3 water hoses and slide their clamps back up into position. There should be a white mark on the housing end of the inlet hose (the one connected to the bottom of the housing assembly) and this should be facing up when in place.

12. If you removed the deck plate for access, install the 4 bolts and tighten the nuts to 13 ft. lbs. (18 Nm). Remember there should be a washer on both sides - one with the nut and one under the bolts head.

13. Move the craft to the water and make certain you check everything carefully for leaks.

14. Connect the battery and install the cargo container and seat(s).

All Models w/1812cc Engines

① EASY

◆ **See Figures 49, 52 and 53**

1. Remove the seat(s) and disconnect the negative battery cable. You may want to remove the rear cargo container as well.

■ **Although not strictly necessary, you may find that removing the deck plate (above the tank) will make access much easier.**

2. Tag and disconnect the water hose connected to the housing; move it out of the way in order to drain the joint. You may have to remove a plastic tie as well. You may also wish to remove the other 2 hoses that are in the way of the thermostat housing. All hoses should use conventional pinch clamps, so just squeeze the 2 tabs and slide the clamp back over the hose.

3. Loosen the 2 thermostat housing mounting bolts and carefully pull off the housing.

4. Remove the thermostat and gasket.

5. Perform the testing procedures if necessary

To install:

6. Clean the housing and check all passages for any blockage.

7. Position the thermostat, spring first, into the engine.

■ **There is a small hole in the top of the thermostat - make sure it is positioned at roughly 3 o'clock when installed.**

8. Position a new gasket onto the cover and then position in place over the thermostat. Coat the bolt threads with Loctite 572 and then install them and tighten to 5.9 ft. lbs. (8 Nm).

9. Reconnect the all water hoses and slide their clamps back up into position.

Fig. 52 The housing is attached to back of the engine, just under the large water hose - 1812cc engines. . .

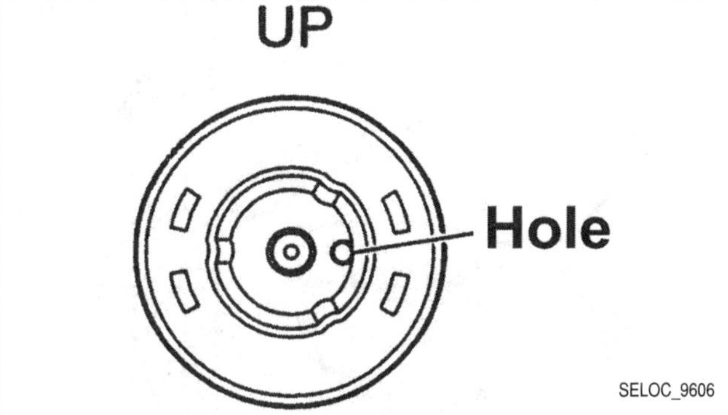

Fig. 53 Make sure the hole is positioned as shown when installed

10. If you removed the deck plate for access, install the 4 bolts and tighten the nuts to 13 ft. lbs. (18 Nm). Remember there should be a washer on both sides - one with the nut and one under the bolts head.

11. Move the craft to the water and make certain you check everything carefully for leaks.

12. Connect the battery and install the cargo container and seat(s).

TESTING

2002-07 FX Models

① EASY

◆ **See Figure 54**

1. Inspect the thermostat at room temperature. If the thermostat is fully open, it is defective and must be replaced. Hold the thermostat up to the light and check it for leaks. A light leak around the perimeter indicates the thermostat is not closing, and therefore, it must be replaced.

2. Attach a length of thread to the thermostat. Now, suspend the thermostat and a thermometer inside a container filled with water (do not use distilled water or ethylene glycol!). Take care to be sure neither the thermostat or the thermometer touches the container. If either one does touch the container, the test will be unreliable. Stir the water occasionally to avoid direct heat being applied to the thermostat

3. Heat the water until the thermostat just begins to open - when this happens confirm that the temperature is the same as the thermostat rating (123° F/50° C). The thermometer reading must agree with the rating stamped on the thermostat. If the unit fails the test, it must be replaced.

4. Continue to heat the water until a temperature of 141° F/60° C is reached. At this time the thermostat should be completely open; if not, replace it.

5. Turn the heat off and allow the water to cool to a temperature 10° below the rating. The thermostat should now be completely closed; if not, replace it.

All Models w/1812cc Engines

EASY

◆ **See Figures 54 and 55**

1. Inspect the thermostat at room temperature. If the thermostat is fully open, it is defective and must be replaced. Hold the thermostat up to the light and check it for leaks. A light leak around the perimeter indicates the thermostat is not closing, and therefore, it must be replaced.

2. Attach a length of thread to the thermostat. Now, suspend the thermostat and a thermometer inside a container filled with water (do not use distilled water or ethylene glycol!). Take care to be sure neither the thermostat or the thermometer touches the container. If either one touches the container, the test will be unreliable. Stir the water occasionally to avoid direct heat being applied to the thermostat

3. Heat the water to 118-126°F (48-52°C) on 998/1052cc engines, 136-144°F (58-62°C) on FX HO and VX w/1812cc engines or, 118-126°F (48-52°C) on FX SHO and FZ w/1812cc engines; until the thermostat just begins to open - when this happens, check that the valve opening is 0.002 in. (0.05mm). The thermometer reading must agree with the rating stamped on the thermostat. If the unit fails the test, it must be replaced.

4. Continue to heat the water until a temperature of 140°F (60°C) is reached on 998/1052cc engines and FX SHO/FZ models w/1812cc engines, or 158°F (70°C) for the FX HO and VX w/1812cc engines. At this time the thermostat should be completely open and the valve clearance should be more that 0.17 in. (4.3mm); if not, replace it.

5. Turn the heat off and allow the water to cool to a temperature 10° below the initial rating. The thermostat should now be completely closed; if not, replace it.

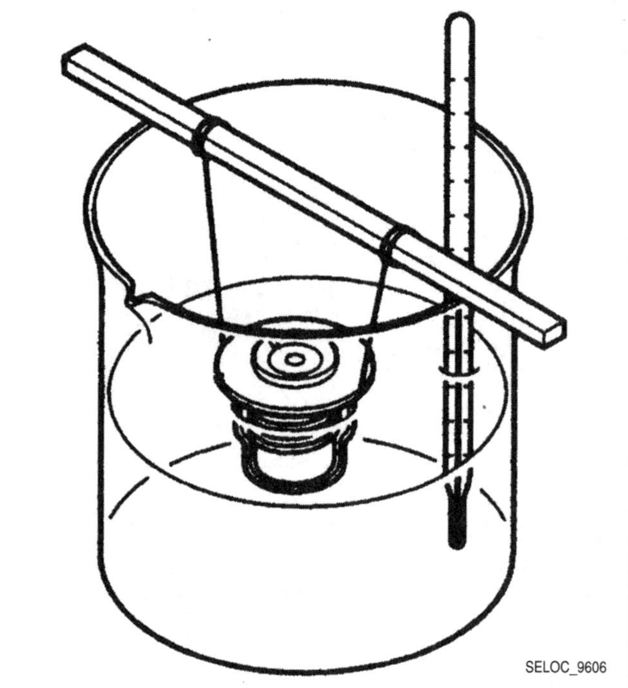

Fig. 54 Testing the thermostat

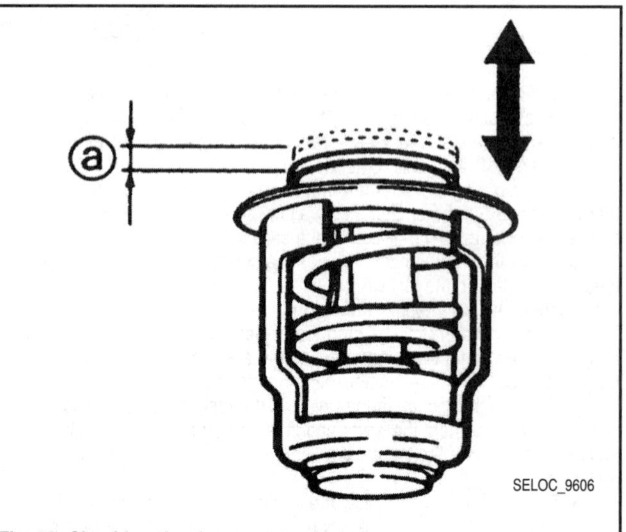

Fig. 55 Checking the thermostat opening

Engine Thermo-Switch

TESTING

MODERATE

◆ **See Figure 44**

A thermo-switch is a simple device that is basically an on/off switch which is activated by temperature; and usually used in conjunction with the warning system.

The engine thermo-switch is located at the front starboard side of the engine, near the manifold.

Locate and remove the thermo-switch from the motor. The thermo-switch resembles a condenser, usually with Pink and Black leads (but check the Wiring Diagrams from the Ignition & Electrical System section for the motor on which you are working). The thermo-switch can normally be removed easily, by pulling it out of its recess. However, on many models the thermo-switch may be secured by a cover or tab that is bolted in place. So, if present, make sure you remove any covers or tab before attempting to full the switch free. The thermo-switch is a very fragile switch and should never be dropped or handled in a rough manner.

Disconnect the 2 leads at their quick disconnect fittings. Since the switch should have no continuity until it is heated past a point which could damage the powerhead, there are no resistance tests to be performed on this switch while it is installed.

Testing is a simple matter of watching for continuity across the switch contacts (using an ohmmeter or DVOM set to read resistance across the switch wiring) as the switch is slowly heated to activation temperature. The best method for this is to suspend the switch in water which is slowly heated to temperature while you watch the meter. Obtain a thermometer, an ohmmeter, and a suitable pan in which to boil water.

Place the container of water, at room temperature, on a stove. Secure the thermometer in the water in such a manner to prevent the thermometer bulb from contacting the bottom or sides of the pan (this ensures the thermometer will read the temperature of the water and NOT the pan, which could be considerably higher).

Immerse the sensing end of the thermo-switch into the water up to the shoulder and suspend it, also NOT touching the pan for the same reason as the thermometer.

Make contact with the Red meter lead to the Pink thermo-switch lead, and the Black meter lead to the Black thermo-switch lead. The meter should indicate no continuity,

Observe at what temperature the switch begins to show continuity as you heat the water. Turn off the heat source and then observe at what temperature it reverts back to no continuity as the water cools.

On 998cc and 1052cc engines, the engine thermo-switch should begin to show continuity when the water temperature reaches 183-194°F (84-90°C). As the water cools down, the meter should stop showing continuity as the temperature reaches 158-183°F (70-84°C). While on 1812cc engines, the switch should begin to show continuity when the water temperature reaches 201-212°F (94-100°C). As the water cools down, the meter should stop showing continuity as the temperature reaches 176-201°F (80-94°C).

■ **A slight variation is allowable above or below the stated temperatures listed. But keep in mind that too great a variation could either allow damage to the powerhead by operating it under conditions that are too hot or could allow activation of the warning system when the powerhead is not in danger.**

If the meter reading is acceptable, the thermo-switch is functioning correctly. If the meter readings are not acceptable, the thermo-switch can be easily replaced with a new unit.

Engine Temperature Sensor

TESTING

MODERATE

◆ See Figure 56

1. Locate the sensor on the starboard side of the engine, disconnect the coupler and remove it.
2. Connect a DVOM as per the manufacturer's instructions and then insert the probes into the sensor coupler.

Testing is a simple matter of checking resistance across the switch contacts (using an ohmmeter or DVOM set to read resistance across the switch wiring, usually Black/Yellow) as the switch is slowly heated to activation temperature. The best method for this is to suspend the switch in water which is slowly heated to temperature while you watch the meter. Obtain a thermometer, an ohmmeter, and a suitable pan in which to boil water.

Place the container of water, at room temperature, on a stove. Secure the thermometer in the water in such a manner to prevent the thermometer bulb from contacting the bottom or sides of the pan (this ensures the thermometer will read the temperature of the water and NOT the pan, which could be considerably higher).

Immerse the sensing end of the sensor into the water up to the shoulder and suspend it, also NOT touching the pan for the same reason as the thermometer.

Slowly heat the water, while watching the thermometer and meter. Resistance reading should be:

- 68°F (20°C) - 54.2-69 kili-ohms
- 212°F (100°C) - 3.12-13.48 kili-ohms

Resistance readings outside of these ranges will necessitate sensor replacement.

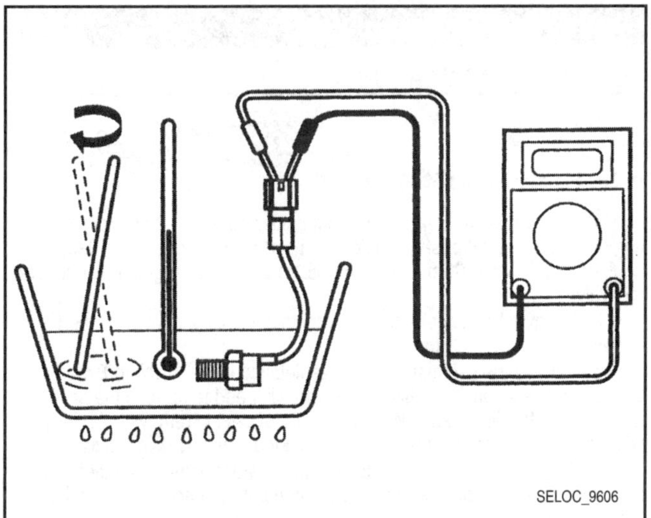

Fig. 56 Testing the engine temperature sensor

Water Inlet Strainer

■ **Beginning in 2005, all models are equipped with an inlet filter located on the side of the jet unit. Inspection and removal procedures can be found in the Maintenance & Tune-up section.**

Water Distribution Manifold

REMOVAL & INSTALLATION

MODERATE

◆ See Figure 57

■ **VX models and models w/1812cc engines do not utilize a water distribution manifold.**

The water distribution manifold is hung from the underside of the hull and can be removed by cutting the small plastic its connecting to the hanger. Once done, simply pinch the clamps and move them back down the hose and then work each of the 3 hoses off their respective fittings. Make sure you tag each hose for reconnection.

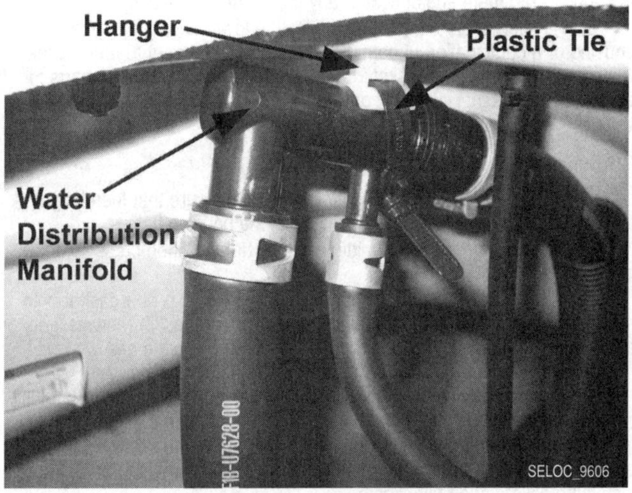

Fig. 57 A good look at the water distribution manifold

Water Pilot Outlet

REMOVAL & INSTALLATION

MODERATE

◆ See Figure 58

With the craft out of water, unscrew the outlet nut on the hull. Grab the water hose inside the engine compartment and carefully pull the outlet through the hull. Remove the grommet and then disconnect the hose. When replacing, make sure that the projection on the outlet goes through the notch in the grommet and hull. Tighten the nut securely.

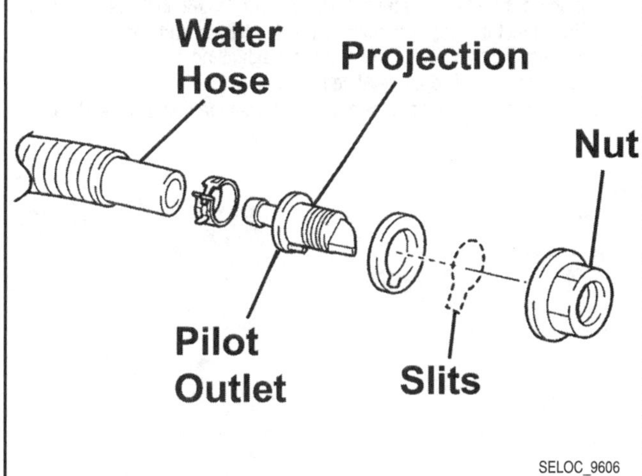

Fig. 58 A good look at the water pilot outlet

Water Pressure Control Valve

REMOVAL & INSTALLATION

MODERATE

◆ See Figures 59 and 60

The water pressure control valve is mounted to the right side of the engine, under the exhaust manifold. It is used on FX models with 998cc engines. FX/FZ and VX models with 1052cc or 1812cc engines do not utilize this valve,

1. Remove the seat(s) and cargo container. Disconnect the battery cables.
2. Remove the exhaust manifold as detailed in the Forward Exhaust Components section. Yep, this means you will need to remove the engine just to get at this little valve! We have heard of people removing the valve without doing this so you may want take a long look and give it a try.
3. Make sure that you tag each of the 3 water hoses, pinch their clamps and work clamps back down the hose. Carefully pull each hose from its fitting on the valve.
4. Remove the 3 mounting bolts and lift off the valve and the old gasket.
5. If necessary, remove the 2 bolts on the bottom of the valve housing cover and separate the 2 halves. Each bolt runs in a collar - if they didn't come out, grab them and check for any cracks or other wear.
6. The internal parts may all stay with the lower cover, the upper housing or a combination of both. Either way, you will want to remove the valve itself, the spring, 2 gaskets a plate and grommet. Check all parts for signs of wear of other damage, paying particular attention to the spring and valve.

To install:

7. If you disassembled the control valve, make sure that there is no gasket material left on the mating flanges of the housing halves.
8. Coat the grommet with a little soapy water and then press it into the housing.
9. Hold the upper housing upside down and install a *new* gasket onto the flange. Position the plate and then the remaining gasket (*new* as well) on top. Notice from the illustration that the flat side of the plate and gaskets must face to the side of the housing with the hose fittings.
10. Press the longer side of the valve into the grommet. Slide the spring over the other (shorter) end and then carefully compress the spring while installing the lower cover making sure that the end of the spring fits into its bore in the cover. Coat the mounting bolt threads with Loctite 572 and then install the collars and bolts, tightening to 2.7 ft. lbs./31 inch lbs. (3.7 Nm). After a couple minutes, go back and retighten the bolts to 5.5 ft. lbs./66 inch lbs. (7.6 Nm).
11. Position a *new* gasket onto the control valve mounting flange and install it on the engine. Coat the bolts threads with Loctite 572 and then thread them in, tightening to 2.7 ft. lbs./31 inch lbs. (3.7 Nm). After a couple minutes, go back and retighten the bolts to 5.5 ft. lbs./66 inch lbs. (7.6 Nm).
12. Reconnect the hoses to their respective fittings and slide the pinch clamps back into position over the ridge on each fitting.
13. Install the manifold and engine if removed.
14. Connect the battery cables and then install the container and seat(s)

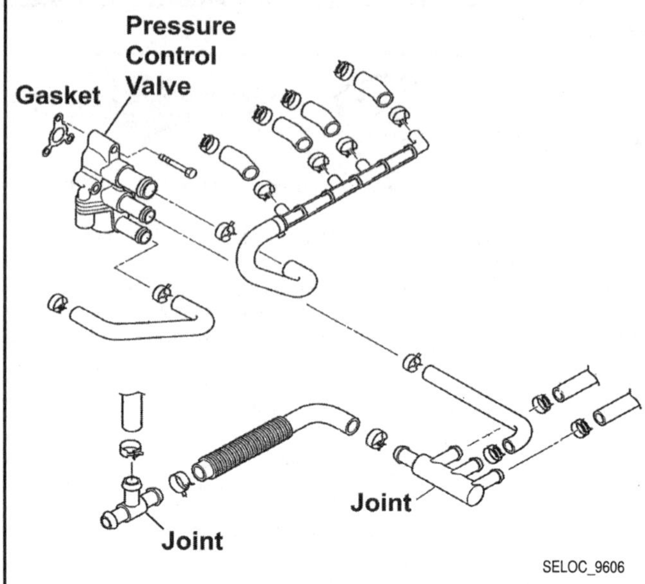

Fig. 59 The water pressure control valve and its associated components

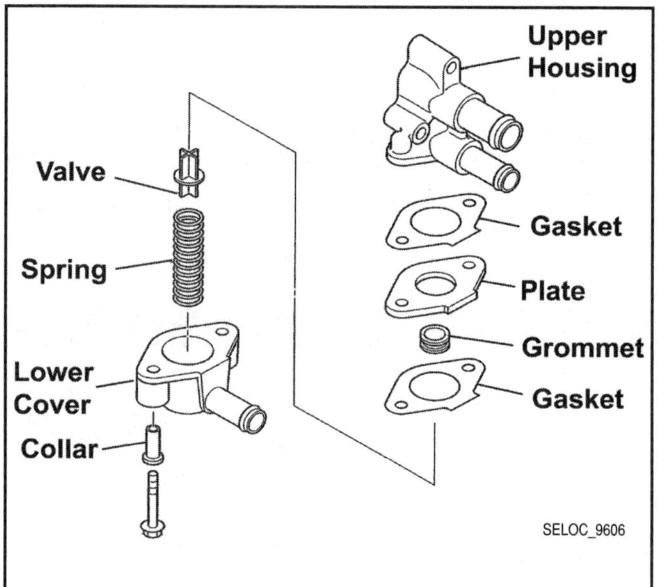

Fig. 60 Exploded view of the control valve

Water Hose Diagrams

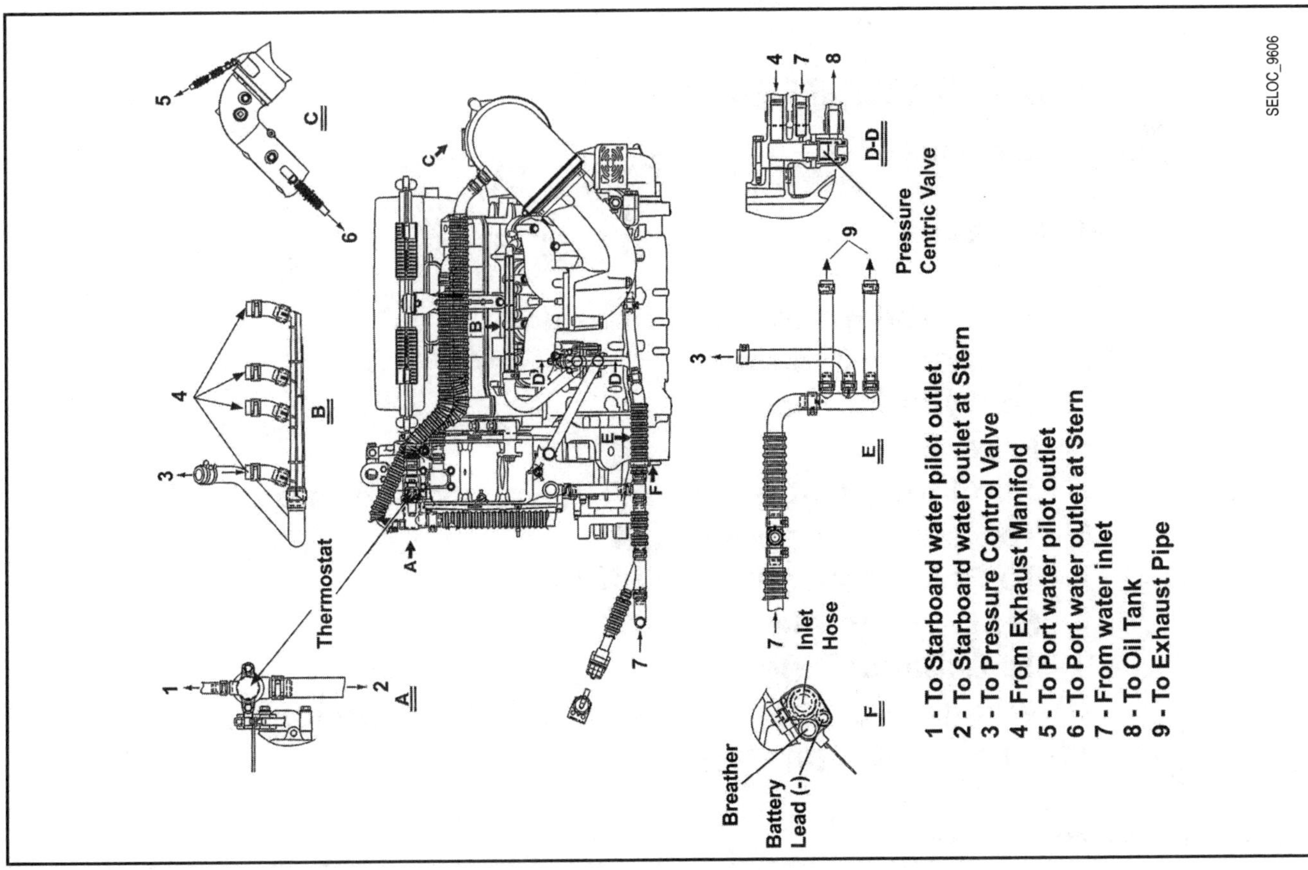

Water hose diagram - 2002 FX w/998cc engines

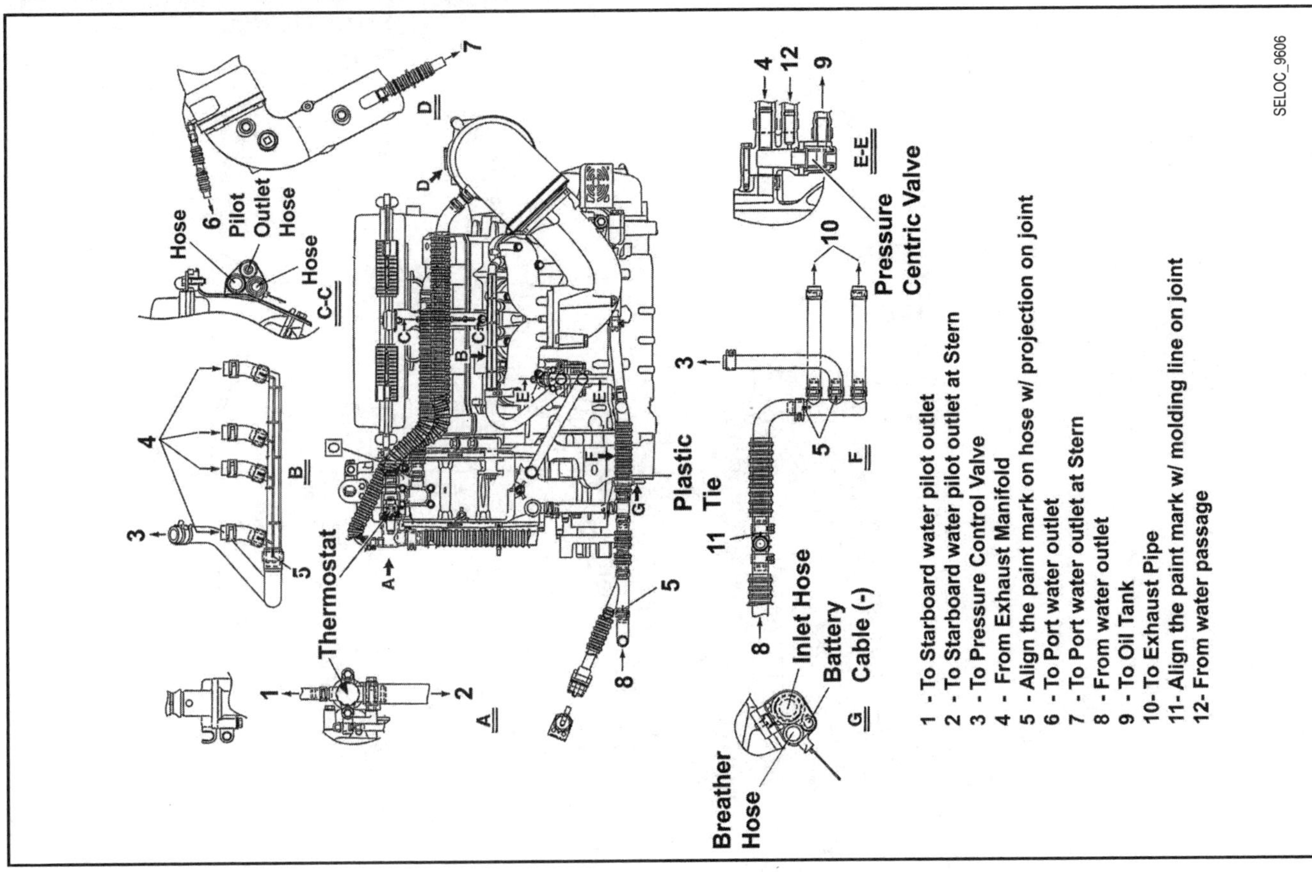

Water hose diagram - 2003-08 FX w/998cc engines

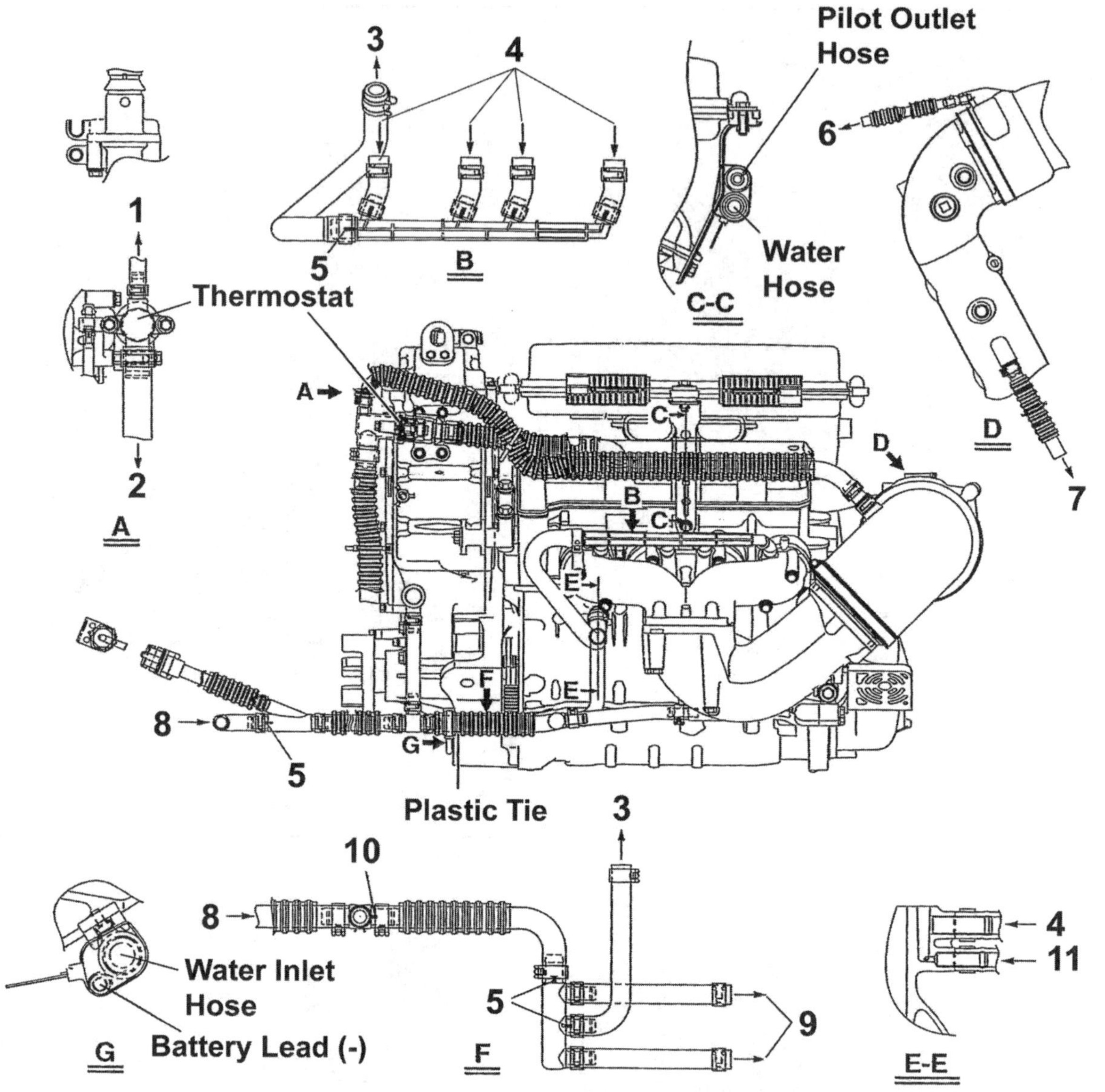

1 - To Starboard water pilot outlet
2 - To Starboard water outlet at Stern
3 - To Cylinder Block
4 - From Exhaust Manifold
5 - Align the paint mark on hose w/projection on joint
6 - To Port water pilot outlet
7 - To Port water outlet at Stern
8 - From water inlet
9 - To Exhaust Pipe
10- Align the paint mark w/ molding line on joint
11- From water passage

SELOC_9606

Water hose diagram - 2004-07 FX w/1052cc engines

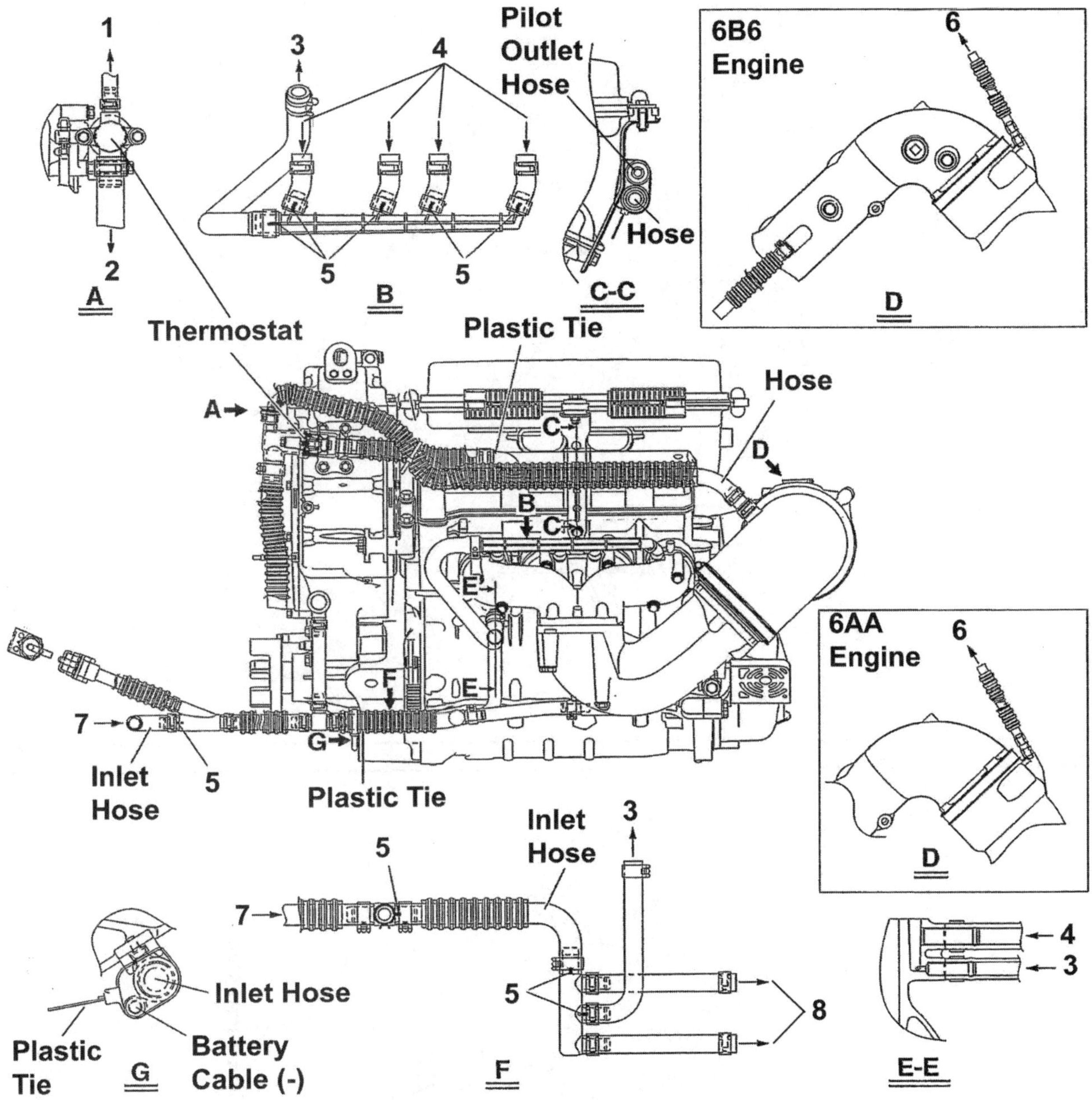

1 - To Starboard Water Pilot outlet
2 - To Starboard water outlet at Stern
3 - To Cylinder Block
4 - From Exhaust Manifold
5 - Align the paint mark on hose w/projection on joint
6 - To Port Water Pilot Outlet
7 - From Water Inlet
8 - To Exhaust Pipe
9 - Align the paint mark w/ molding line on joint

SELOC_9606

Water hose diagram - 2008 FX w/1052cc engines

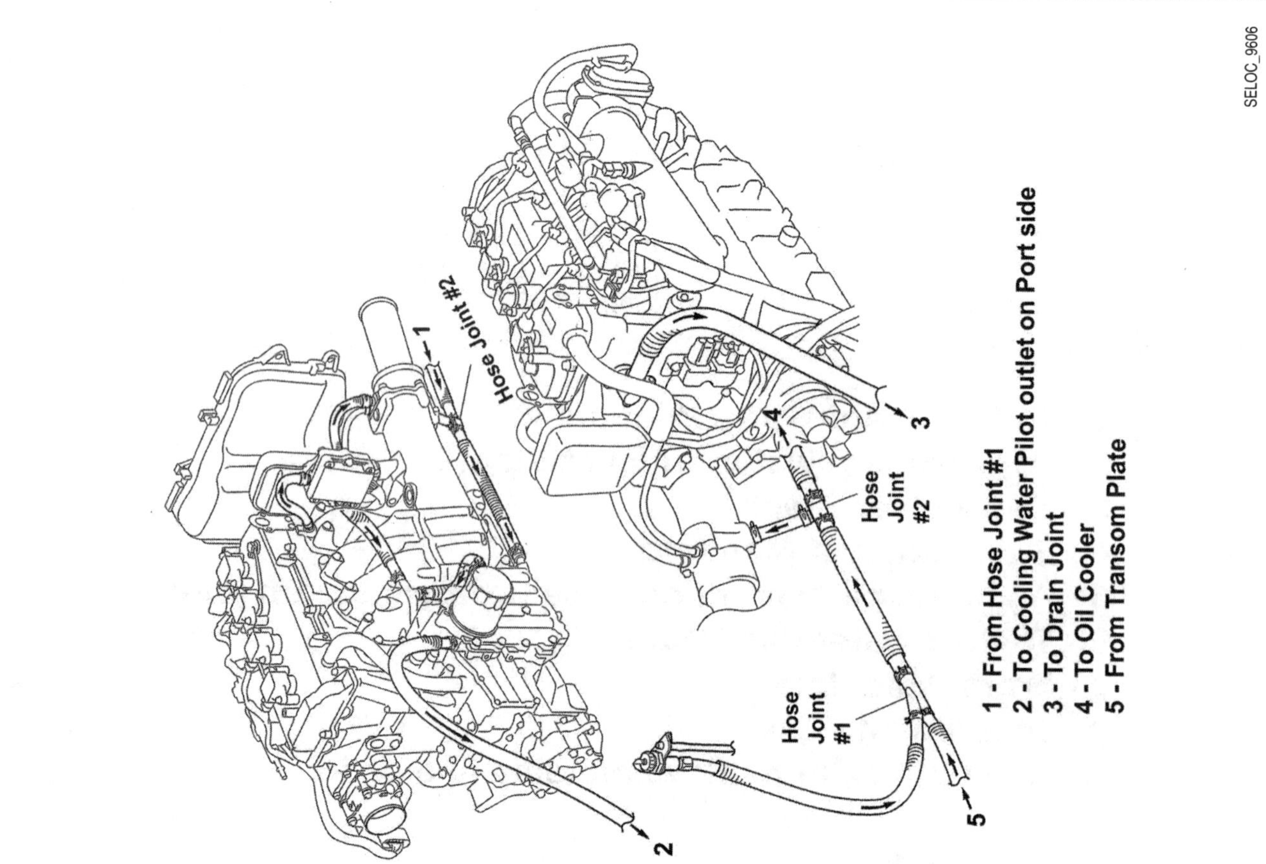

Water flow diagram - 2008-11 FX HO w/1812cc engines

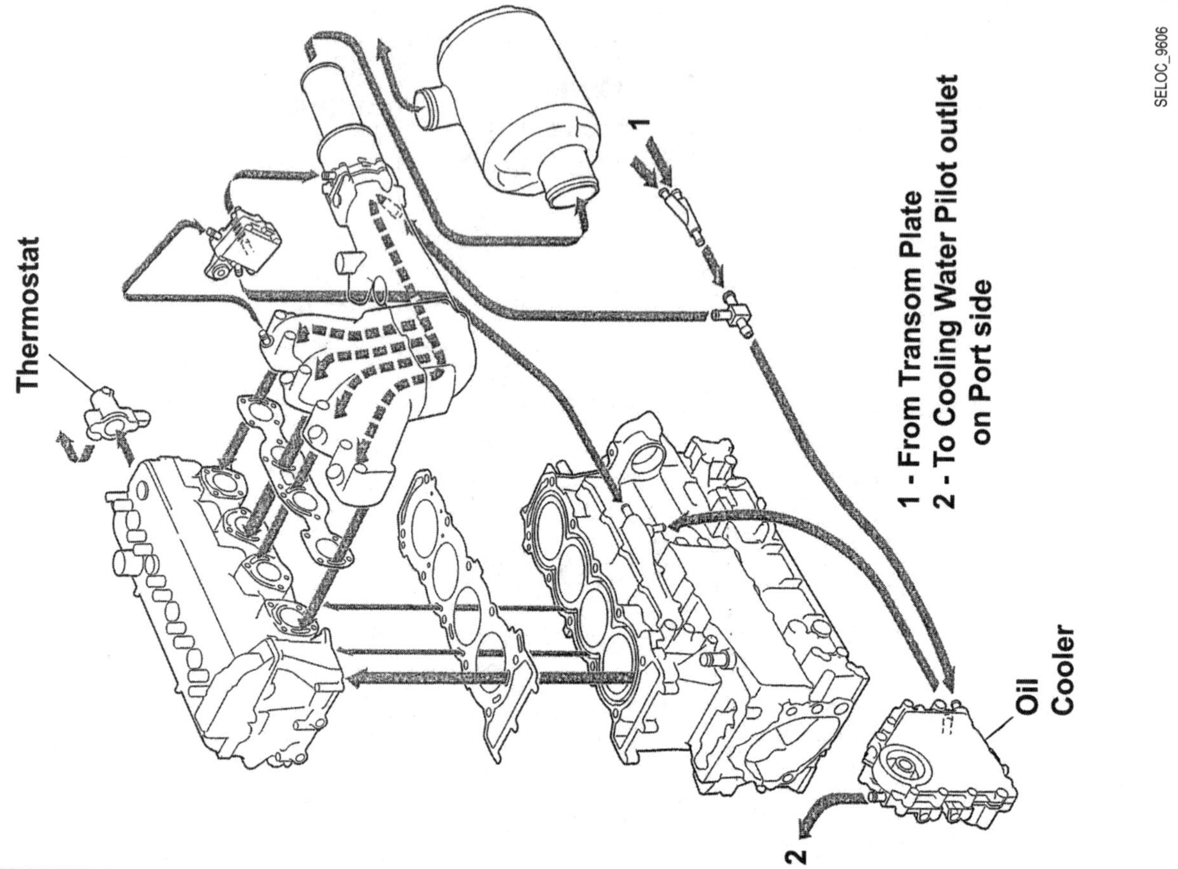

Water flow diagram - 2008-11 FX HO w/1812cc engines (cont'd)

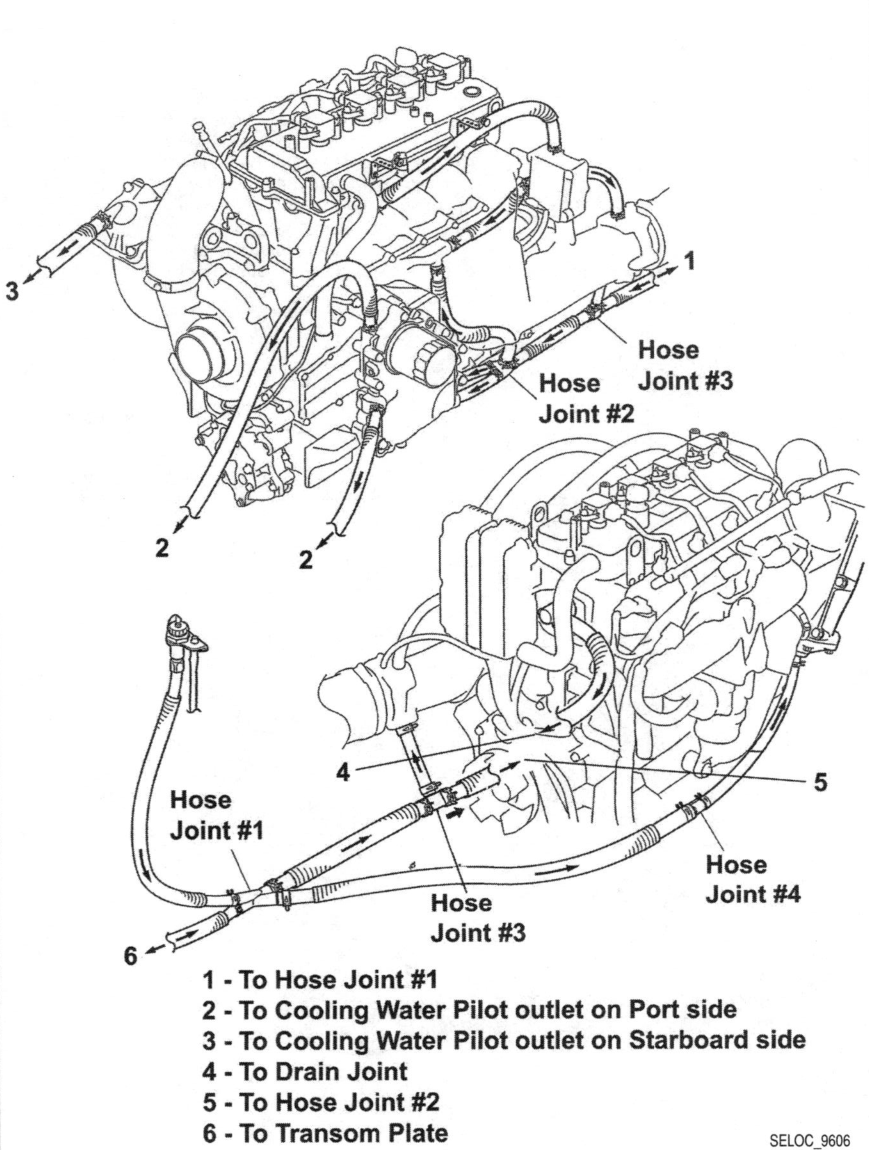

Water flow diagram - 2008-11 FX SHO & FZ w/1812cc engines

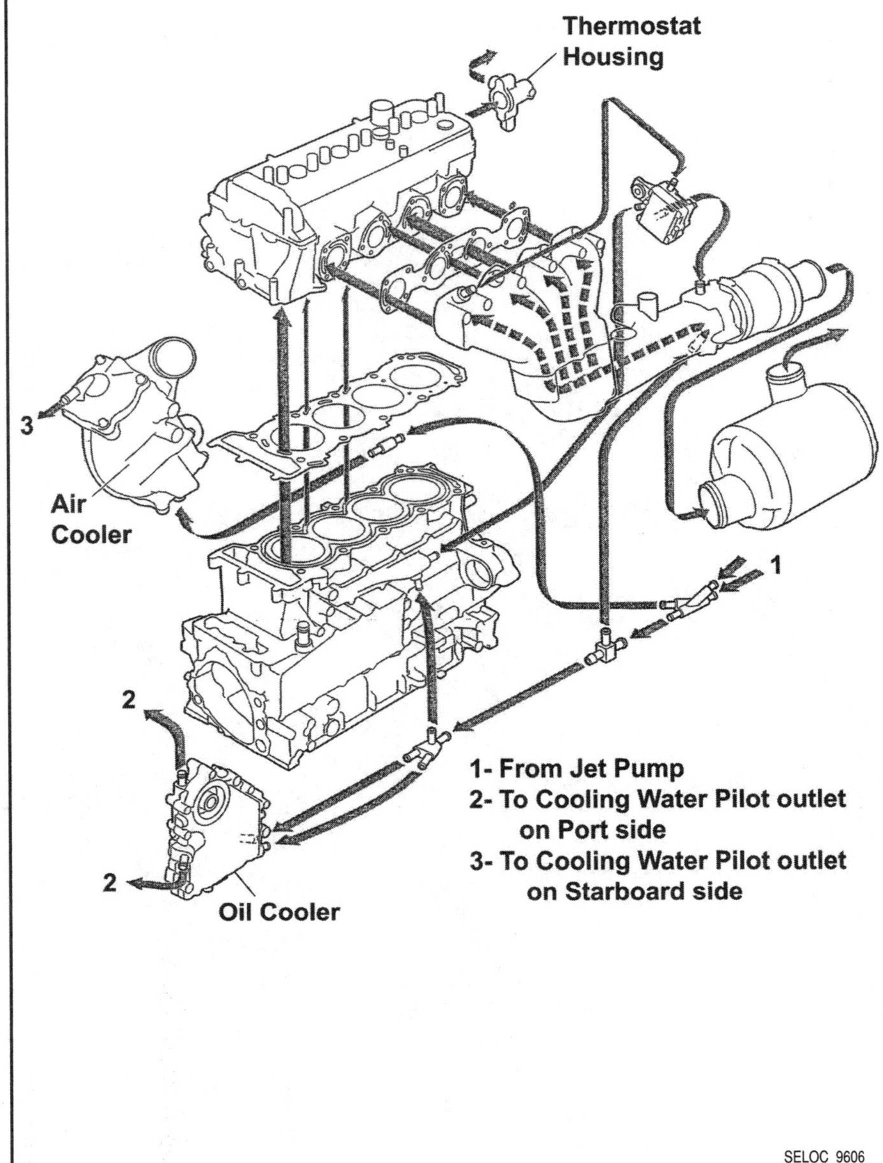

Water flow diagram - 2008-11 FX SHO & FZ w/1812cc engines (cont'd)

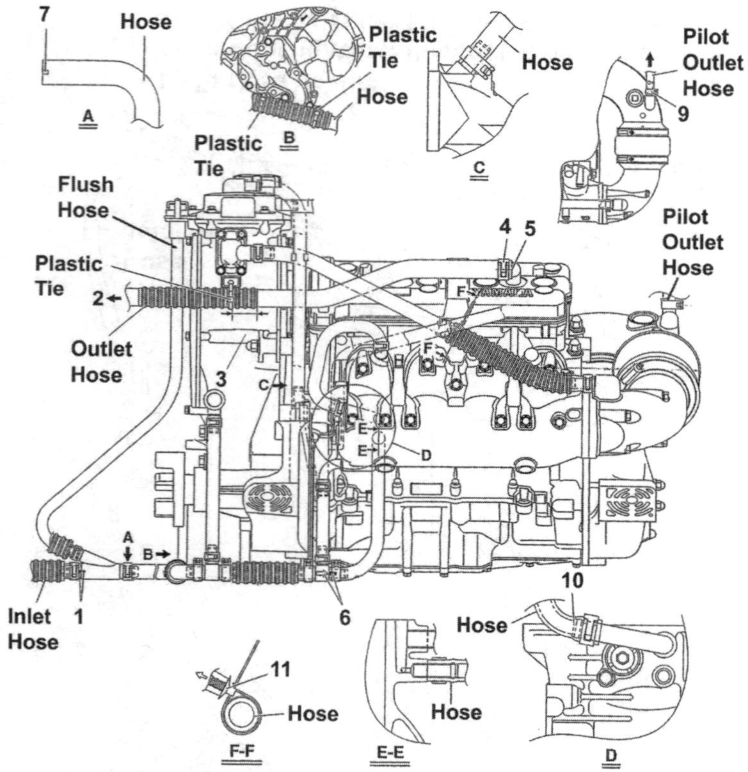

1 - Align the white mark with the projection on the joint
2 - To Transom Plate
3 - 0.98-1.77 in. (25-45mm)
4 - Face the Hose Clamp ends toward the Starboard side
5 - Insert hose to the paint mark
6 - Align the white mark with the arrow on the joint
7 - White mark facing up
8 - To Water Pilot outlet
9 - Face the Hose Clamp ends toward the Bow
10- Install hose until pipe reaches the bend in hose
11- Insert the plastic tie into the hole in the boss on the block

SELOC_9606

Water hose diagram - 2005-09 VX w/1052cc engines

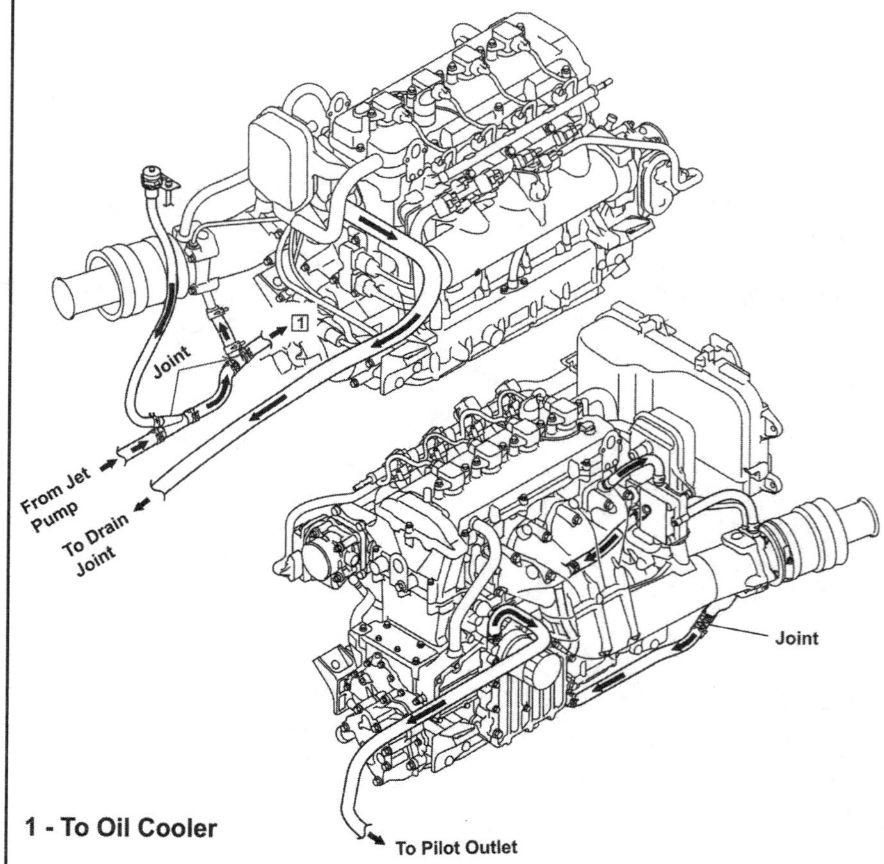

1 - To Oil Cooler

SELOC_9606

Water flow diagram - 2010-11 VX w/1052cc engines

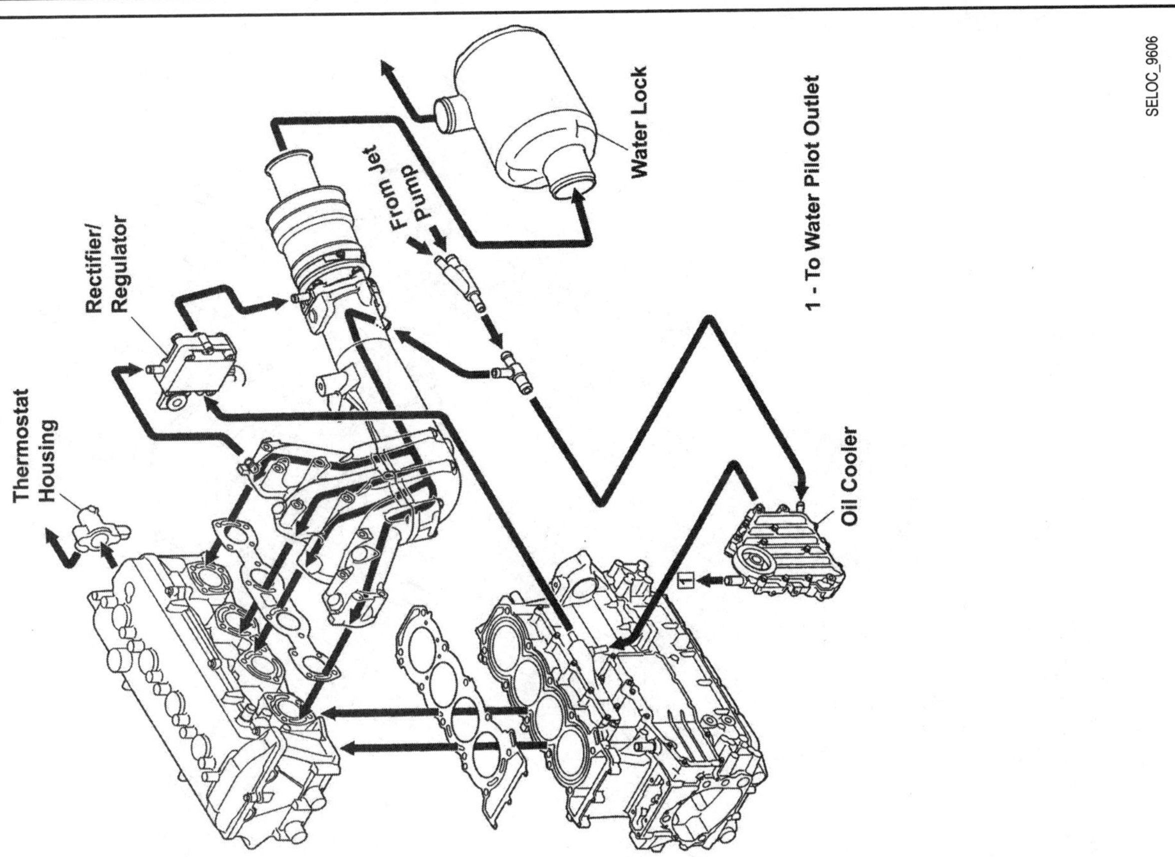

Water flow diagram - 2011 VX w/1812cc engines

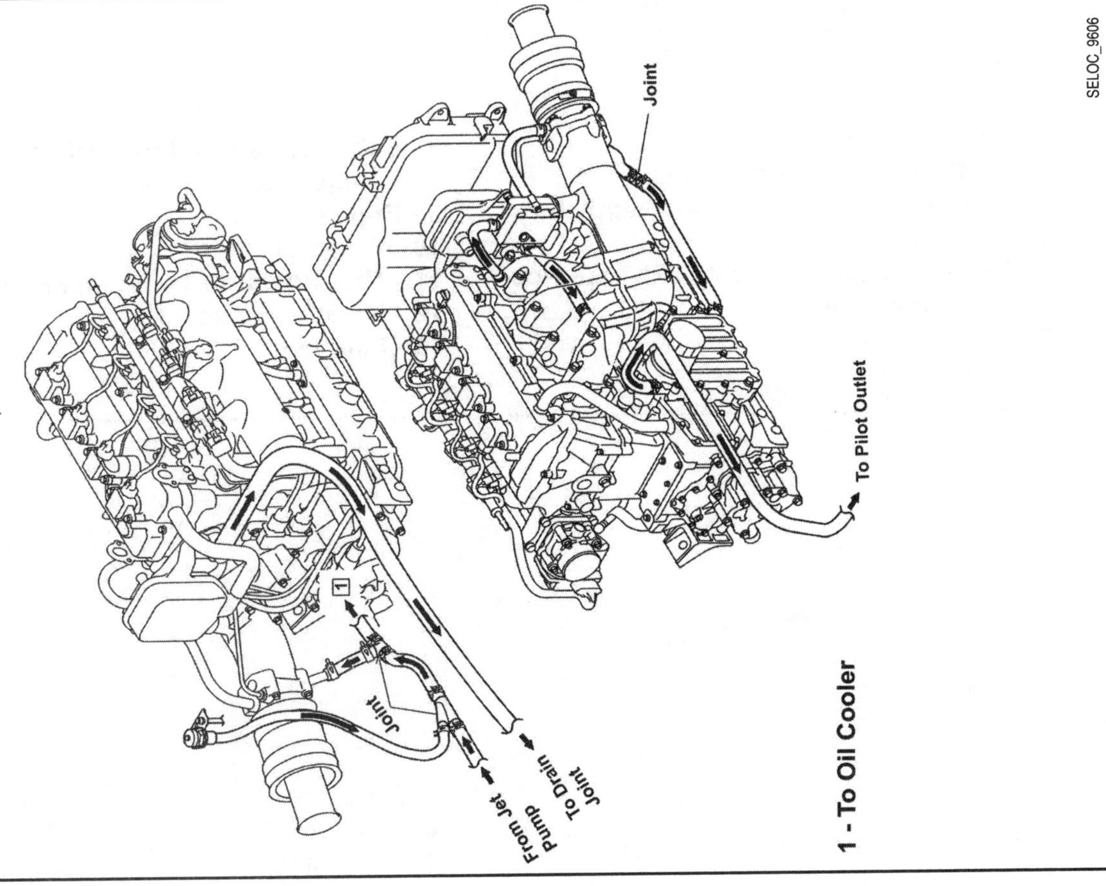

Water flow diagram - 2011 VX w/1812cc engines

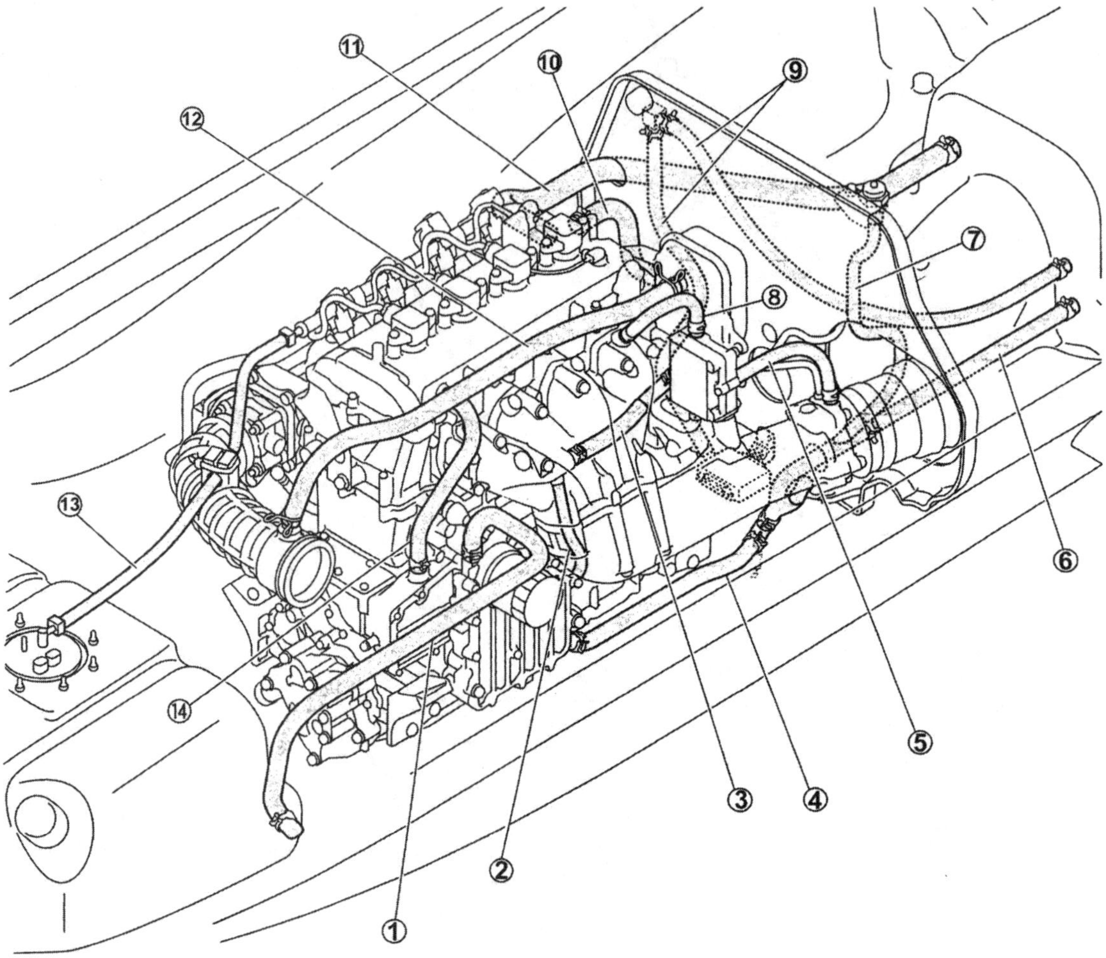

1 - To Water Pilot Outlet
2 - Oil Cooler to Cylinder Block
3 - Cylinder Block to Rectifier/Regulator
4 - Joint to Oil Cooler
5 - Rectifier/Regulator to #2 Exhaust Pipe
6 - Jet Pump to Joint
7 - Flush Hose
8 - Exhaust Manifold to Rectifier/ Regulator
9 - Bilge Hose
10 - Breather Hose
11 - Thermostat to Transom Plate
12 - Breather Hose
13 - Fuel Line

SELOC_9606

Water hose diagram - 2011 VX w/1812cc engines

6

ENGINE MECHANICAL

ENGINE MECHANICAL

General Information

BASICS

You can compare the major components of a PWC engine with the engine and drivetrain of your car or truck. In doing so, the power unit is the equivalent of the engine and the jet drive is the equivalent of your driveline (the transmission/transaxle). The engine is the assembly that produces the power necessary to move the craft, while the jet drive is the assembly that transmits that power via gears, shafts and a jet impeller.

Your engine is an assembly of long-life components that are protected through proper maintenance. Lubrication, the use of high-quality oils and frequent oil inspection/changes are the most important ways to preserve engine condition. Similarly, proper tune-ups that help maintain proper air/fuel mixture ratios and prevent pinging, knocking or other potentially damaging operating conditions are the next best way to preserve your motor. But, even given the best of conditions, components in a motor begin wearing the first time the motor is started and will continue to do so over the life of the engine.

Eventually, all motors will require some repair. The particular broken or worn component, plus the age and overall condition of the motor may help dictate whether a small repair or major overhaul is warranted. As much as you can generalize about mechanical work, the complexity of the job will vary with 2 major factors:

- The age of the motor (the older OR less well maintained the motor is) the more difficult the repair
- The larger and more complex the motor, the more difficult the repair.

Again, these are generalizations and, working carefully, a skilled do-it-yourself boater can disassemble and repair an engine, as well as a seasoned professional. But both DIYers and professionals must know their limits. These days, many professionals will leave portions of machine work (from cylinder block and piston disassembly, cleaning and inspection to honing and assembly up to a machinist). This is not because they are not capable of the task, but because that's what a machinist does day in and day out. A machinist is naturally going to be more experienced with the procedures.

If a complete overhaul is necessary on your motor, we recommend that you find a local machine shop that has both an excellent reputation and that specializes in marine work. This is just as important and handy a resource to the professional as a DIYer. If possible, consult with the machine shop before disassembly to make sure you follow procedures or mark components, as they would desire. Some machine shops would prefer to perform the disassembly themselves. In these cases, you can usually remove the engine from the vessel and deliver the entire unit to the shop for disassembly, inspection, machining and assembly.

■ **There are a number of major re-manufacturers throughout the U.S. and you may be able to quite cost effectively purchase a replacement unit for installation making it an attractive option. At least it is an option you should consider before undertaking a major overhaul.**

If you decide to perform the entire overhaul yourself, proceed slowly, taking care to follow instructions closely. Consider using a digital camera (if available) to help document component mounting during the removal and disassembly procedures. This can be especially helpful if the overhaul or rebuild is going to take place over an extended amount of time. If this is your first overhaul, don't even *think* about trying to get it done in 1 weekend, YOU WON'T. It is better to proceed slowly, asking for help when necessary from your trusted parts counterman or a tech with experience on these motors.

Keep in mind that anytime pistons, rings and bearings have been replaced, the engine must be broken-in again, as if it were a brand-new motor.

IMPORTANT POINTS FOR ENGINE OVERHAUL

Repair Procedures

Service and repair procedures will vary slightly between individual models, but the basic instructions are quite similar. Special tools may be called out in certain instances. These tools may be purchased from the local marine dealer.

When working on an engine that has been removed, use either a very sturdy workbench or an engine stand to hold the assembly. Although removal and installation is relatively straightforward on most models, the overhaul procedures can be quite involved. Whatever portion you decide to tackle, always, *always*, ALWAYS take good notes and tag as many parts/hoses/connections as you can during the removal process. As they are removed, arrange components along the work surface in the same orientation to each other as they are when installed.

Torque Sequence & Values

A torque sequence means that when installing that component you tighten bolt No. 1 a couple of turns, then No. 2, then 3, and so on. Once the bolts start to snug, you can precede 1 of 2 ways. If you don't have a torque value, you tighten the bolts securely, a 1/4 turn or so at a time still following that sequence until you feel they are securely fastened. However, if you have do have a torque value, you tighten the bolts just to the point when they start to snug, then set a torque wrench to the proper value and continue to tighten each bolt in 1-2 passes of that sequence until the proper amount of torque is placed on each of the bolts.

■ **Sometimes Yamaha will provide stepped torque specs, meaning multiple torque specifications for the same bolts, first a lower, then a middle or final specification. When stepped specs are provided you tighten the bolts until snug, then to the first value, then to the next value etc until you reach the final value, each time following the proper torque sequence.**

ENGINE COMPONENTS

Service procedures for the throttle bodies, fuel pumps, starter, and other engine components are given in their respective sections of this technical guide.

CLEANLINESS

General

Make a determined effort to keep parts and the work area as clean as possible. Parts must be cleaned and thoroughly inspected before they are assembled, installed, or adjusted. Use proper lubricants, or their equivalent, whenever they are recommended.

Assembly Sealants & Lubricants

Over the years, Yamaha seems to vary or change their recommendations for fastener and component installation, not only from model-to-model but from year-to-year on even the same model. Some bolts are installed dry, some are to be coated with engine oil and others with Loctite. Some seals should be lubricated with clean engine oil and others with marine grade grease. And lastly, internal engine components are usually coated with engine oil, but many bearing surfaces may also be coated with a molybdenum disulfide engine assembly lube.

We honestly don't believe that coating a seal with grease instead of oil or vice versa is going to cause any problem. Nor do we believe that Loctite on bolt threads instead of oil is ever an issue. When it comes to engine oil vs. engine assembly lube, the assembly lube may have some advantages and can almost always be substituted for oil on bearing surfaces. Pay attention to the instructions and accompanying illustrations.

Some compounds that Yamaha specifies may be listed under their own part number as opposed to the lubricant, sealant, threadlocker manufacturer's part number. You may need to order them from a Yamaha parts supplier to ensure you're getting the right compound.

Engine

REMOVAL & INSTALLATION

998cc & 1052cc Engines

OEM 3 DIFFICULT

◆ See Figures 1 and 2

1. Remove the seat(s) and cargo container.
2. Disconnect the battery cables.
3. Drain the engine oil as detailed in the Maintenance & Tune-Up section. Remove the oil filter.

4. Remove the 4 bolts and lift off the deck beam if equipped.

5. Tag and disconnect the throttle cable and fuel line as detailed in the Fuel System section.

6. Loosen the clamp attaching the forward exhaust tube to the adaptor at the rear of the #3 exhaust pipe. Do the same with the clamp holding the center exhaust tube to the water lock and then remove the forward tube and water lock as detailed in the Exhaust & Cooling section.

7. Disconnect the electrical box where it attaches to the hull as detailed in the Ignition & Electrical section.

8. On VX models, remove the air filter case as detailed in the Fuel System section. You must also disconnect the ECM and rectifier/regulator bracket and move the assembly out of the way.

9. Tag and disconnect the water inlet and outlet hoses at their connections on the engine. Do the same with each of the water pilot outlet hoses.

10. Disconnect the grease hose at the fitting near the thermostat. Position it out of the way.

11. Tag and disconnect any remaining hoses, lines and electrical leads that may interfere with engine removal.

12. At the back of the engine, remove the coupler cover bolt and lift off the cover. Do not lose the bolt collar.

13. Attach an engine lifting hoist to the lift hooks on the engine - 2 in the rear and 1 at the front. Tighten the hoist lines until they are just taut.

14. Now, loosen each of the 4 engine mount bolts. Increase the tension on the lift so that the engine pressure on the mounts is lessened and to the point that you can reach in and remove the shims from each mount. They are slotted and can be slid out without removing the bolts. It is very important that you remove each mount's shims and then store them for replacement on the *same* mount.

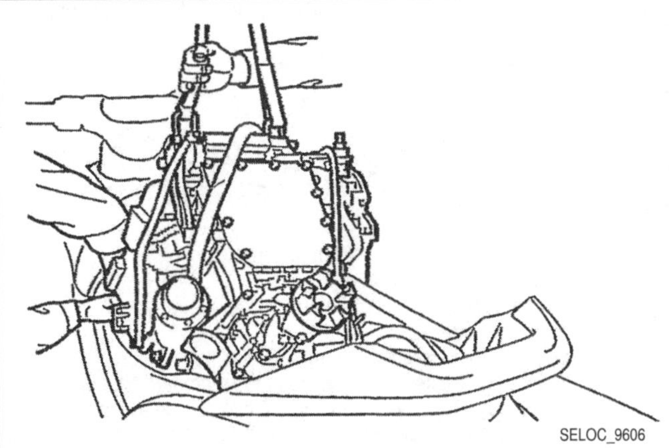

Fig. 1 Always lift the rear of the engine out first - 998cc and 1052cc engines

Fig. 2 Two of the lift hooks are mounted to the oil cooler assembly - 998cc and 1052cc engines

15. Remove the 4 mount bolts and then raise the engine slightly - just enough that you can carefully slide it forward enough to disconnect the engine coupler from the jet pump coupler.

16. Lower the engine until it rest on the mounts again then remove the strap from the front hanger. Slowly begin to lift the engine out of the hull from the rear, rotating it clockwise as you lift. Be very careful to avoid damage to the exhaust pipe or any other components as the engine comes through the opening in the hull. Place the engine on a suitable work surface, preferably an engine stand, before remove the hoist.

To install:

17. Check the rubber engine mounts carefully for cracks, separation or any other damage.

18. Now is always a good time to completely clean out the engine compartment.

19. Attach the hoist to the 2 rear hangers again and slowly, carefully, lower the engine back into the hull. Get the front in and then the rear down and in. Once in, reattach the front strap/chain and then ease the engine back so that the coupler meshes with the jet pump coupler. Once engaged, orient the engine so the mount bolt holes are in alignment and then lower the unit until it is sitting fully on the mounts and the hoist tension has been removed. Allow it to sit on the mounts for a while. . .at least an hour or so.

20. While waiting, you can thread in the 4 mount bolts just enough that the threads have sufficient purchase; not even finger tight though.

21. After allowing everything to settle in, raise the engine just enough to insert the shims back into their ORIGINAL positions on each mount. Lower the engine and remove the hoist. Remove the mount bolts individually, coat the threads with Loctite 572 and then screw them back in, tightening to 12 ft. lbs. (17 Nm).

22. Check the engine coupler alignment as detailed in this section. Once done, position the coupler cover and tighten the bolt to 5.6 ft. lbs./66 inch lbs. (7.8 Nm) except on the 2010-11 VX where it is 5.9 ft. lbs./71 inch lbs. (8 Nm). Don't forget to coat the threads with Loctite 572 or the bolt collar.

23. Reconnect the grease line at the fitting.

24. Reconnect the electrical box.

25. On VX models, install the air filter case and reconnect the ECM and rectifier/regulator back to the engine.

26. Reconnect the water lines you removed during engine removal. Make sure they are all fully on their fittings and that the hose ends are in good condition.

27. Reconnect all other hoses, lines and leads that were disconnected.

28. Reconnect the throttle cable and fuel line.

29. Install the water lock and forward exhaust tube. Make sure that you refer to the appropriate section for correct hose/tube alignment at the exhaust pipe adaptor and lock.

30. Install the deck beam (if removed) and tighten the 4 bolts to 11 ft. lbs. (15 Nm).

31. Install the oil filter and fill the engine with the correct amount of oil. Connect the battery cables.

32. Install the cargo container and seat(s).

FX HO Models & VX Models w/1812cc Engines

OEM 3 DIFFICULT

◆ **See Figures 3, 4 and 5**

1. Remove the seat(s) and cargo container.

2. Disconnect the battery cables.

3. Drain the engine oil as detailed in the Maintenance & Tune-Up section. Remove the oil filter.

4. Remove the 4 bolts and lift off the deck beam if equipped.

5. Remove the engine cover.

6. Remove the air filter case as detailed in the Fuel System section.

7. Tag and disconnect the throttle cable and fuel line as detailed in the Fuel System section. Remove the throttle body.

8. Remove the cover from the electrical box and then remove the assembly as detailed in the Ignition & Electrical section.

9. Remove the exhaust system as detailed in the Exhaust & Cooling section.

10. Tag and disconnect the hoses at the oil breather and then remove the assembly.

11. Tag and disconnect the lines/hoses at the rectifier/regulator and remove it as detailed in the ignition & Electrical section.

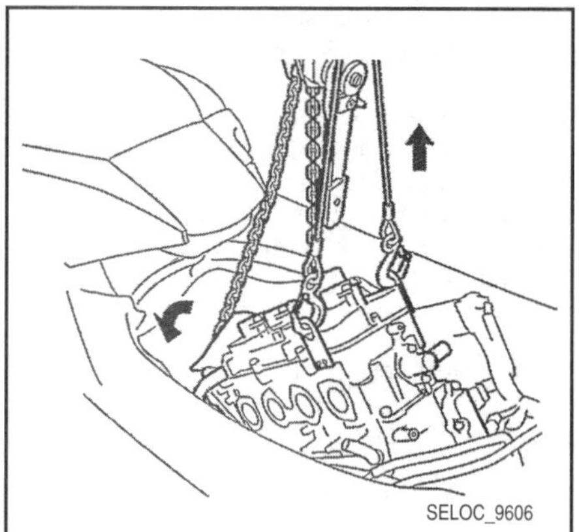

Fig. 3 Always lift the rear of the engine out first - 1812cc engines

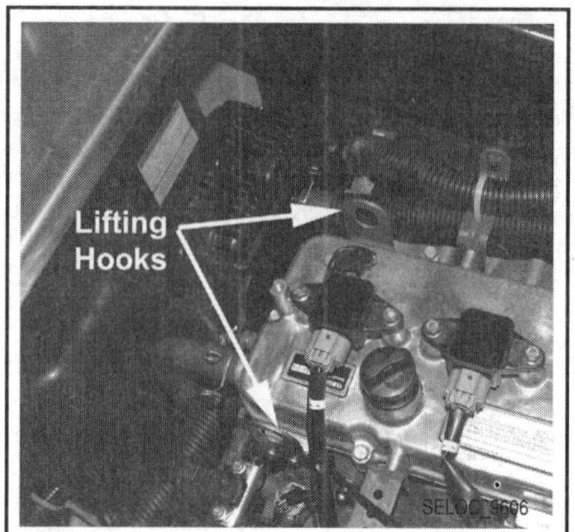

Fig. 4 There are 2 lift hooks at the rear of the engine - 1812cc engines

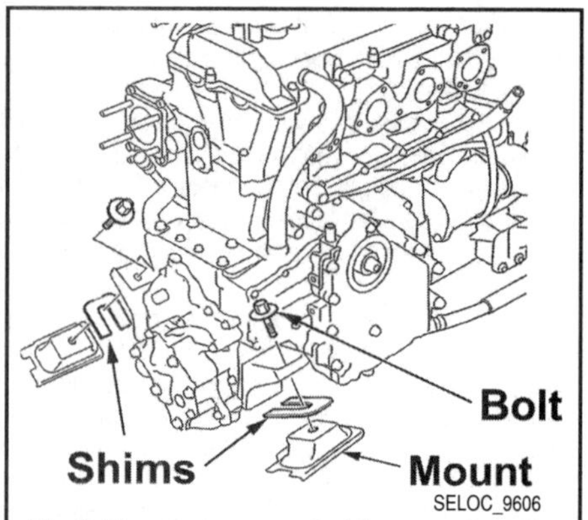

Fig. 5 The engine mount shims are slotted and should slide right out - HO shown, SHO and FZ similar

12. Tag and disconnect all water inlet and outlet hoses at their connections on the engine. Do the same with each of the water pilot outlet hoses.
13. Disconnect the grease hose at the fitting near the thermostat. Position it out of the way.
14. Tag and disconnect any remaining hoses, lines and electrical leads that may interfere with engine removal.
15. At the back of the engine, remove the coupler cover bolt and lift off the cover. Do not lose the bolt collar.
16. Attach an engine lifting hoist to the lift hooks on the engine - 2 in the rear and 1 at the front. Tighten the hoist lines until they are just taut.
17. Now, loosen each of the 4 engine mount bolts. Increase the tension on the lift so that the engine pressure on the mounts is lessened and to the point that you can just reach in (*be careful*) and remove the shims from each mount. They are slotted and can be slid out without removing the bolts. It is very important that you remove each mount's shims and then store them for replacement on the *same* mount.
18. Remove the 4 mount bolts and then raise the engine slightly - just enough that you can carefully slide it forward enough to disconnect the engine coupler from the jet pump coupler.
19. Lower the engine until it rest on the mounts again then add some slack to the strap from the front hanger. Slowly begin to lift the engine out of the hull from the rear. Be very careful to avoid damage to other components as the engine comes through the opening in the hull. Place the engine on a suitable work surface, preferably an engine stand, before remove the hoist.

To install:

20. Check the rubber engine mounts carefully for cracks, separation or any other damage. Do the same with the 2 center dampers.
21. Now is always a good time to completely clean out the engine compartment.
22. Attach the hoist to the front hanger with slack again. Then the 2 rear hangers again and slowly, carefully, lower the engine back into the hull. Get the front in and then the rear down and in. Once in, remove the slack from the front strap/chain and then ease the engine back so that the coupler meshes with the jet pump coupler. Once engaged, orient the engine so the mount bolt holes are in alignment and then lower the unit until it is sitting fully on the mounts and the hoist tension has been removed. Allow it to sit on the mounts for a while. . .at least an hour or so.
23. While waiting, you can thread in the 4 mount bolts just enough that the threads have sufficient purchase; not even finger tight though.
24. After allowing everything to settle in, raise the engine just enough to insert the shims back into their ORIGINAL positions on each mount. Lower the engine and remove the hoist. Remove the mount bolts individually, coat the threads with Loctite 271 and then screw them back in, tightening to 12.5 ft. lbs. (17 Nm).
25. Check the engine coupler alignment as detailed in this section. Once done, position the coupler cover and tighten the bolt to 5.9 ft. lbs./71 inch lbs. (8 Nm). Don't forget to coat the threads with Loctite 572 or the bolt collar.
26. Reconnect the grease line at the fitting.
27. Install the rectifier/regulator and connect the hoses. Do the same with the oil breather on the other side of the engine.
28. Install the exhaust system.
29. Reconnect the electrical box and install the cover.
30. Install the throttle body and reconnect the fuel line and throttle cable.
31. Install the air filter case and reconnect the ECM and rectifier/regulator back to the engine.
32. Install the engine cover.
33. Reconnect the water lines you removed during engine removal. Make sure they are all fully on their fittings and that the hose ends are in good condition.
34. Reconnect all other hoses, lines and leads that were disconnected.
35. Install the deck beam (if removed) and tighten the 4 bolts to 13.3 ft. lbs. (18 Nm).
36. Install the oil filter and fill the engine with the correct amount of oil. Connect the battery cables.
37. Install the cargo container and seat(s).

FX SHO & FZ Models w/1812cc Engines

OEM 3 DIFFICULT

◆ **See Figures 3, 4 and 5**

1. Remove the seat(s) and cargo container.
2. Disconnect the battery cables.
3. Remove the 4 bolts and lift off the deck beam if equipped.
4. Remove the engine cover.
5. Open the hood and remove the front service lid.
6. Locate the large harness running along the starboard side of the engine. It is secured by 8 plastic ties - wrap a piece of tape around the harness at each tie and then cut them all. Tag and disconnect the 14 couplers (APS, multi-function meter, fuel pump, bilge, etc) and fold the harness out of the way on the engine.
7. Remove the air filter case as detailed in the Fuel System section.
8. Tag and disconnect the throttle cable and fuel line as detailed in the Fuel System section. Remove the throttle body.
9. Drain the engine oil as detailed in the Maintenance & Tune-Up section. Remove the oil filter.
10. Tag and disconnect the hoses at the oil breather and then remove the assembly.
11. Tag and disconnect the lines/hoses at the rectifier/regulator and remove it as detailed in the ignition & Electrical section.
12. Remove the exhaust system as detailed in the Exhaust & Cooling section.
13. Tag and disconnect all water inlet and outlet hoses at their connections on the engine. Do the same with each of the water pilot outlet hoses.
14. Tag and disconnect any remaining hoses, lines and electrical leads that may interfere with engine removal.
15. Remove the cover from the electrical box and then remove the assembly as detailed in the Ignition & Electrical section.
16. At the back of the engine, remove the coupler cover bolt and lift off the cover. Do not lose the bolt collar.
17. Attach an engine lifting hoist to the lift hooks on the engine - 2 in the rear and 1 at the front. Tighten the hoist lines until they are just taut.
18. Now, loosen each of the 4 engine mount bolts. Increase the tension on the lift so that the engine pressure on the mounts is lessened and to the

point that you can just reach in (*be careful*) and remove the shims from each mount. They are slotted and can be slid out without removing the bolts. It is very important that you remove each mount's shims and then store them for replacement on the *same* mount.

19. Remove the 4 mount bolts and then raise the engine slightly - just enough that you can carefully slide it forward enough to disconnect the engine coupler from the jet pump coupler.

20. Lower the engine until it rest on the mounts again then add some slack to the strap from the front hanger. Slowly begin to lift the engine out of the hull from the rear. Be very careful to avoid damage to other components as the engine comes through the opening in the hull. Place the engine on a suitable work surface, preferably an engine stand, before remove the hoist.

To install:

21. Check the rubber engine mounts carefully for cracks, separation or any other damage. Do the same with the 2 center dampers.

22. Now is always a good time to completely clean out the engine compartment.

23. Attach the hoist to the front hanger with slack again. Then the 2 rear hangers again and slowly, carefully, lower the engine back into the hull. Get the front in and then the rear down and in. Once in, remove the slack from the front strap/chain and then ease the engine back so that the coupler meshes with the jet pump coupler. Once engaged, orient the engine so the mount bolt holes are in alignment and then lower the unit until it is sitting fully on the mounts and the hoist tension has been removed. Allow it to sit on the mounts for a while. . .at least an hour or so.

24. While waiting, you can thread in the 4 mount bolts just enough that the threads have sufficient purchase; not even finger tight though.

25. After allowing everything to settle in, raise the engine just enough to insert the shims back into their ORIGINAL positions on each mount. Lower the engine and remove the hoist. Remove the mount bolts individually, coat the threads with Loctite 271 and then screw them back in, tightening to 12.5 ft. lbs. (17 Nm).

26. Check the engine coupler alignment as detailed in this section. Once done, position the coupler cover and tighten the bolt to 5.9 ft. lbs./71 inch lbs. (8 Nm). Don't forget to coat the threads with Loctite 572 or the bolt collar.

27. Reconnect the electrical box and install the cover.

28. Install the rectifier/regulator and connect the hoses. Do the same with the oil breather on the other side of the engine.

29. Install the exhaust system.

30. Install the rectifier/regulator and connect the hoses. Do the same with the oil breather on the other side of the engine.

31. Install the throttle body and reconnect the fuel line and throttle cable.

32. Install the air filter case and reconnect the ECM and rectifier/regulator back to the engine.

33. Position the engine harness and pull on new plastic ties in each of the taped locations. Reconnect all of the couplers.

34. Install the service lid and close the hood.

35. Install the engine cover.

36. Reconnect the water lines you removed during engine removal. Make sure they are all fully on their fittings and that the hose ends are in good condition.

37. Reconnect all other hoses, lines and leads that were disconnected.

38. Install the deck beam and tighten the 4 bolts to 13.3 ft. lbs. (18 Nm).

39. Install the oil filter and fill the engine with the correct amount of oil. Connect the battery cables.

40. Install the cargo container and seat(s).

Engine Coupler

ALIGNMENT

OEM ② MODERATE

◆ See Figure 6

■ **Engine coupler removal is included in the Reduction Drive Gear procedures in this section.**

Always check coupler alignment after the engine or jet pump has been removed.

1. Remove the coupler cover bolt and collar. Lift off the cover.

2. Lay a straightedge across the top of the coupler halves (A) and check that the clearance between the two does not exceed 0.020 in. (0.5mm) except on 1812cc engines where is should be 0.039 in. (1mm). If it does, you will have to adjust by adding or removing engine mount shims based on whether the engine coupler needs to be raised or lowered.

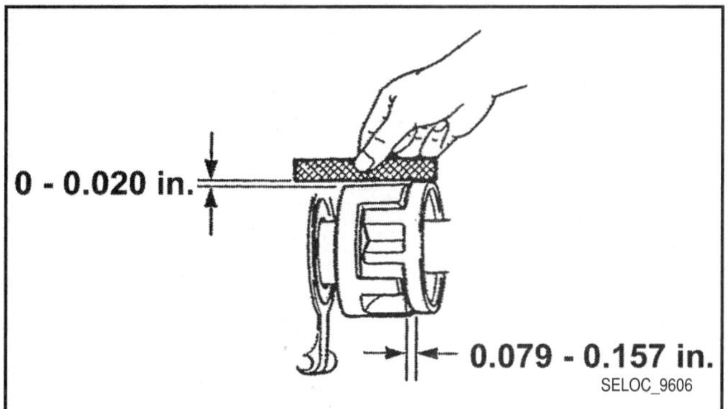

Fig. 6 Engine coupler alignment points

3. Now use a feeler gauge to measure the clearance between the two couplers where they mesh (B). Clearance on this measurement should be 0.079-0.157 in. (2-4mm). If greater than this, loosen the 4 engine mount bolts and move the engine toward the rear until clearance comes within range (yes, you'll have to use the engine hoist again!)

If your clearance is less than indicated, you will have to move the engine forward until the correct clearance is achieved.

■ **Replacement shims are generally available in thicknesses of 0.1mm, 0.3mm, 0.5mm, 1mm and 2mm.**

4. If either of these adjustments fell out of range, you will need to re-coat the engine mount bolts with Loctite 572 again and then tighten them to 12.5 ft. lbs. (17 Nm).

5. Position the coupler cover, coat the bolt threads with Loctite 572 and then tighten to 5.9 ft. lbs./71 inch lbs. (8 Nm). Don't forget the collar!

Engine Mounts

REMOVAL & INSTALLATION

OEM ③ DIFFICULT

◆ See Figures 7 thru 12

■ **As with the mount shims during engine removal, you want to take note of where each mount or damper came from (if you are re-using) so that they can be installed in the same position.**

1. Remove the engine as detailed in this section.

2. Remove the bolt (VX w/1052cc engine) or 2 bolts (all FX models and 2011 VX w/1812cc engines) for each of the center dampers and lift off the damper.

3. FZ and 2010-11 VX (1052cc only) models use a spacer underneath each of the dampers, so remove the bolts and lift them out as well.

4. Remove the 2 mount bolts and lift off the mount; repeat for all 4 mounts of necessary.

5. On 2002-08 FX models with 998cc/1052cc engines, the 2 forward mounts use a liner between them and the hull. All other models use a liner under all 4 mounts. Lift them off if necessary.

6. Check the dampers, spacers and mounts for cracking and other signs of wear.

7. Position the liner on the hull and then the engine mount. Coat the bolt threads with Loctite 572 and tighten each bolt to 12.5 ft. lbs. (17 Nm).

8. On FZ and 2010-11 VX (1052cc only) models, position the damper spacers and tighten the bolts to 3.7 ft. lbs./44 inch lbs. (5 Nm) after applying a little Loctite 572 to the threads.

9. Position the damper onto its seat in the hull or spacer. Coat the bolt threads with Loctite 271 (572 on 1812cc engines) and then tighten each bolt to:

- FX models with 998cc/1052cc engines - 4.6 ft. lbs./55 inch lbs. (6.4 Nm)
- FX and FZ models w/1812cc engines - 4.4 ft. lbs./53 inch lbs. (6 Nm)
- 2005-09 VX models - 4.8 ft. lbs./58 inch lbs. (6.6 Nm)
- 2010-11 VX w/1052cc engines - 5.2 ft. lbs./62 inch lbs. (7 Nm).
- 2011 VX w/1812cc engines - 4 ft. lbs./48 inch lbs. (5.5 Nm).

10. Install the engine.

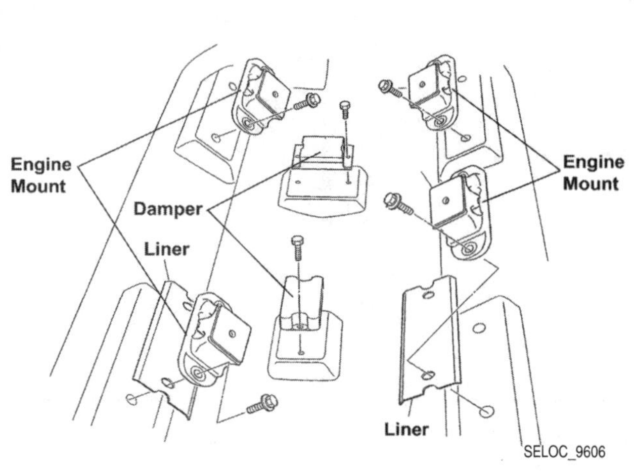

Fig. 7 Removing the engine mounts - 2002-08 FX models with 998cc/1052cc engines

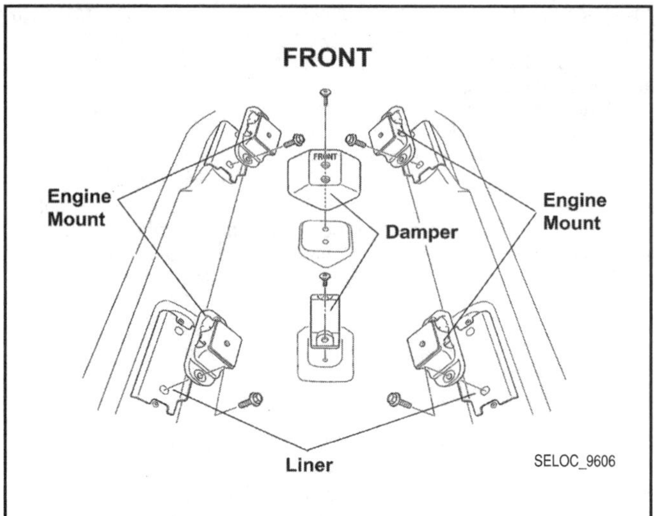

Fig. 8 Removing the engine mounts - FX models w/1812cc engines

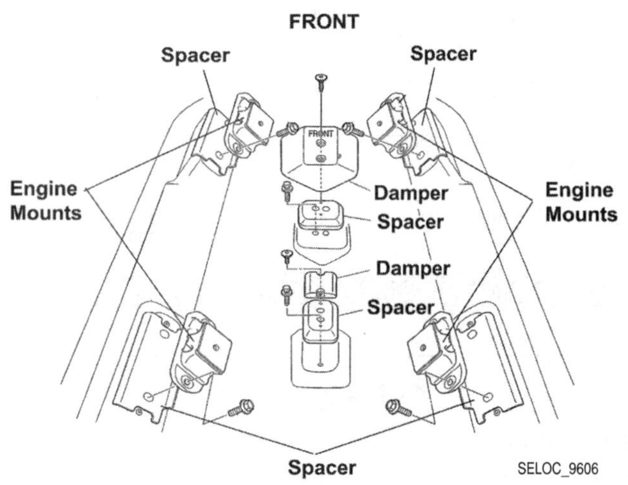

Fig. 9 Removing the engine mounts - FZ models w/1812cc engines

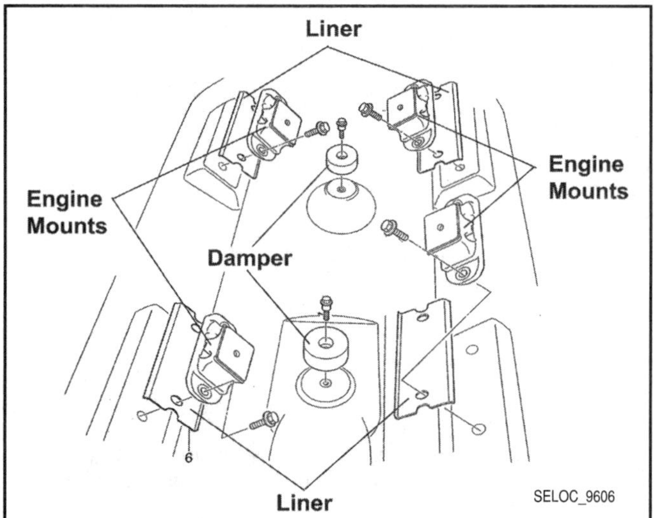

Fig. 10 Removing the engine mounts - 2004-09 VX models

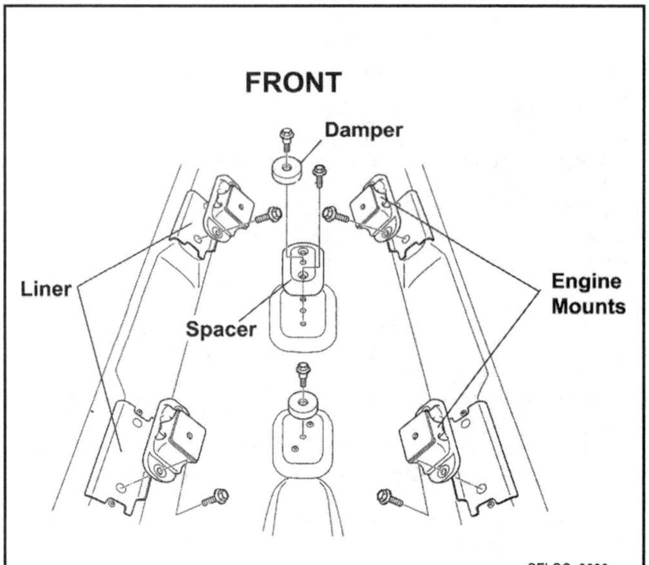

Fig. 11 Removing the engine mounts - 2010-11 VX models w/1052cc engines

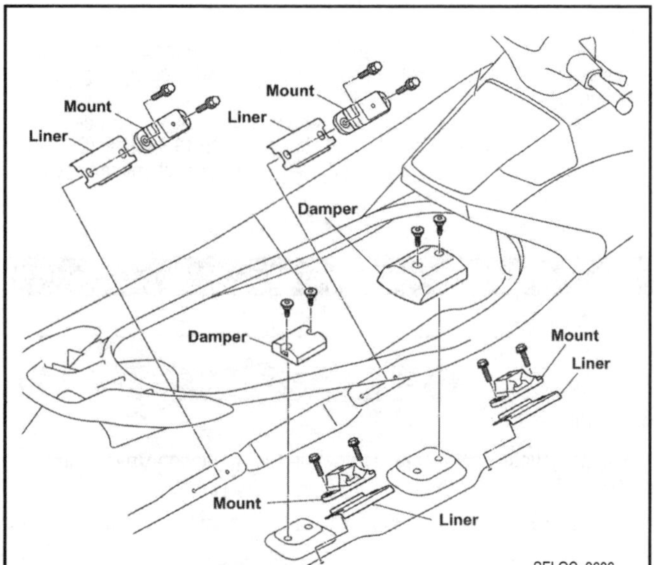

Fig. 12 Removing the engine mounts - 2011 VX models w/1812cc engines

Oil Tank

REMOVAL & INSTALLATION

2002-08 FX Models With 998cc/1052cc Engines

OEM ③ DIFFICULT

◆ See Figures 13 thru 18

The oil pump is located at the rear of the engine.

1. Remove the engine as detailed in this section.
2. Remove the air filter case as detailed in the Fuel System section.
3. On 998cc engines, remove the ignition coil box as detailed in the Ignition & Electrical section.
4. Tag and disconnect any water lines still connected to the thermostat after engine removal. The grease line should already have been removed, but if not, disconnect it. Remove the thermostat housing.
5. Tag and disconnect all other water and breather hoses connected to the oil tank assembly - there are at least 7. If you have a digital camera, take some pictures to help when installing the tank later.
6. Remove the 4 bolts attaching the water jacket to the starboard side of the tank and lift it off. Remove the gasket and throw it away.
7. You should already have disconnected the breather hoses attached to the oil separator, but if not, remove them now being sure too tag them. Remove the 3 mounting bolts (2 on 1052cc engines) and lift off the separator. Pull out the 2 collars and grommets and then clean the separator with a good solvent. Allow to air dry completely before installation, or blow it out with compressed air.
8. Remove the small bolt and collar connecting the band to the port side of the tank. Now remove the 3 mounting bolts and pull off the cover underneath the band along with its gasket. Remove the bolt inside the recess and disconnect the ground lead(s).

■ **Refer to the loosening sequence as shown for these next steps.**

9. On 998cc engines, remove the 4 tank mounting bolts and the 2 nuts and then lift off the assembly.
10. Remove the 2 mounting stay bolts and lift off the stay if necessary.
11. On 1052cc engines, loosen the 2 mounting stay bolts in between the tank and engine.
12. Now loosen the 2 nuts on the back side of the tank attached to the stay and then loosen the 5 remaining oil tank mounting bolts at the bottom of the tank.
13. Remove completely the 2 nuts and lower 5 bolts before lifting the tank off of the mounting stay.
14. Remove the remaining 2 bolts and lift off the stay.
15. With all engines, on the lower mounting bosses, remove the small pin and the 2 connectors with their O-rings. Discard the O-rings.

To install:

16. Clean the tank thoroughly and inspect it for cracks or other damage.
17. Coat *new* O-rings with a quality light marine grease and roll them into their grooves at each end of the 2 connectors. Press the connectors back into the bolt boss bores.
18. Press the 2 dowel pins fully into their recesses on the tank.
19. Position the mounting stay onto the block. Coat the bolt threads with Loctite 572 and then thread them in until finger-tight.
20. Position the tank assembly so it aligns with the mounting stay studs and bolt holes. Press it into place making sure that the dowel pins fit and the mounting stay studs feed through their respective holes on the tank.
21. Coat the stud threads and mounting bolt threads with Loctite 572 and thread the bolts in or the nuts on until finger-tight.
22. Now, following the tightening sequence, tighten the 5 lower bolts in 2 stages - 1st to 11 ft. lbs. (15 Nm) and finally to 28 ft. lbs. (39 Nm) - 29.5 ft. lbs. (40 Nm) on the 2010 models. Tighten the 2 nuts and 2 bolts in 3 stages - 1st to 1.4 ft. lbs./17 inch lbs. (2 Nm), 2nd to 11 ft. lbs. (15 Nm) and finally to 20 ft. lbs. (28 Nm) on 998cc engines, or 28 ft. lbs. (39 Nm) on 1052cc engines.

■ **Tighten all bolts and nuts in the sequence illustrated. This means you'll skip the bottom bolts while tightening the upper bolts/nuts to their 1st pass. Next, in complete order now, you'll tighten them *all* to 11 ft. lbs. (15 Nm) - which is the 1st pass on the 5 lower bolts and the 2nd pass on the upper bolts/nuts. And finally, tighten them all to the ultimate figure.**

23. Reconnect the ground lead and tighten the bolt(s) to 5.5 ft. lbs./66 inch lbs. (7.6 Nm). Install the cover with a *new* gasket. Coat the bolt threads with Loctite 572 and then tighten the 2 bolts to 5.5 ft. lbs./66 inch lbs. (7.6 Nm). Do the same with the bolt for the band, installing it over the cover.
24. Press the 2 collars and grommets into their holes on the oil separator. Move it into position and then thread in the 3 bolts after coating them with Loctite 572. Tighten in 2 passes, 1st to 2.7 ft. lbs./33 inch lbs. (3.7 Nm) and then to 5.5 ft. lbs./66 inch lbs. (7.6 Nm). Install the breather hoses making sure they are fully on the fitting before moving on the pinch clamps. When installing the lower line from the oil pump, make sure that the mark on the hose lines up with the parting line on the separator.
25. Install water jacket with a *new* gasket. Tighten the 4 bolts in 2 passes, 1st to 2.7 ft. lbs./33 inch lbs. (3.7 Nm) and then to 5.5 ft. lbs./66 inch lbs. (7.6 Nm). Make sure you apply Loctite 572 to the threads before installing the bolts.
26. Reconnect all remaining water hoses and breather lines.
27. Install the thermostat housing and grease line.
28. Install the engine.
29. Install the coil box (998cc only) and the air filter housing.

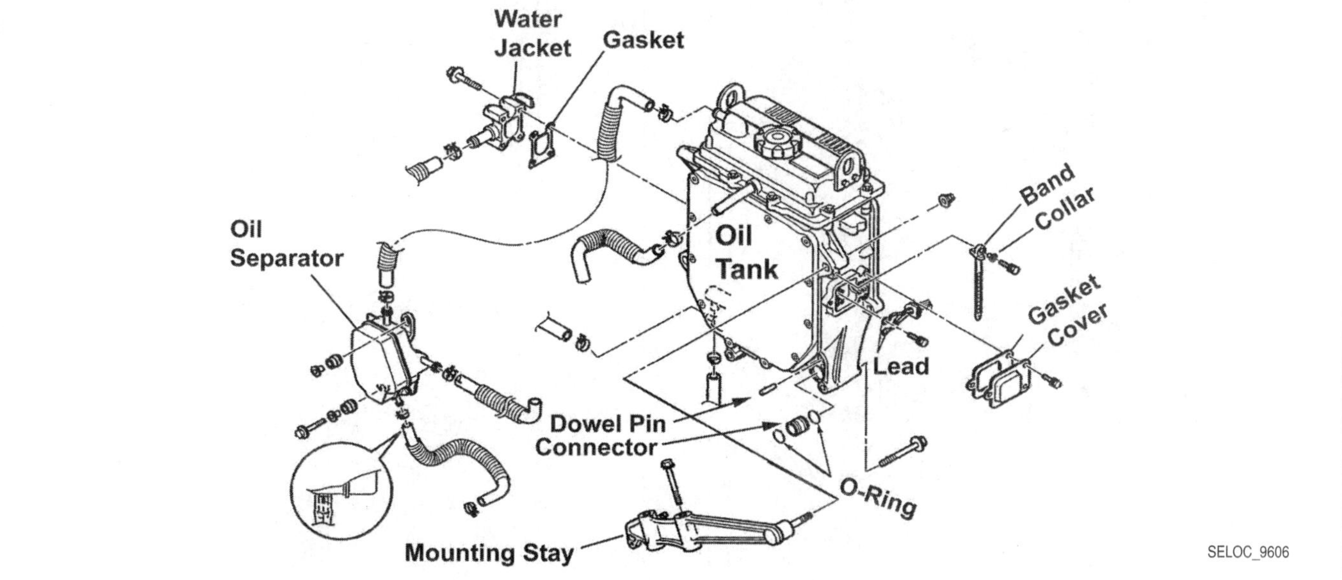

Fig. 13 Removing the oil tank - 998cc engines, 1052cc similar

Fig. 14 A good look at the oil tank with the mounting stay attached (typical)

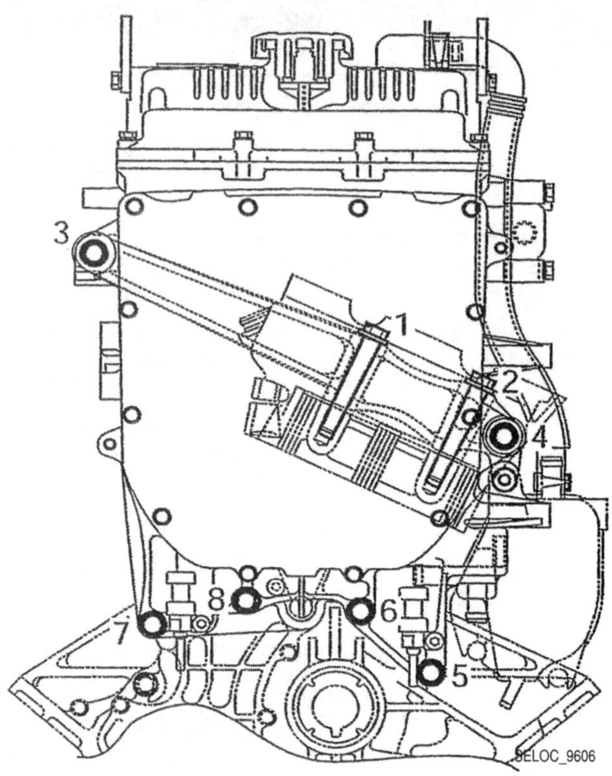

Fig. 15 Oil tank bolt/nut loosening sequence - 998cc engines

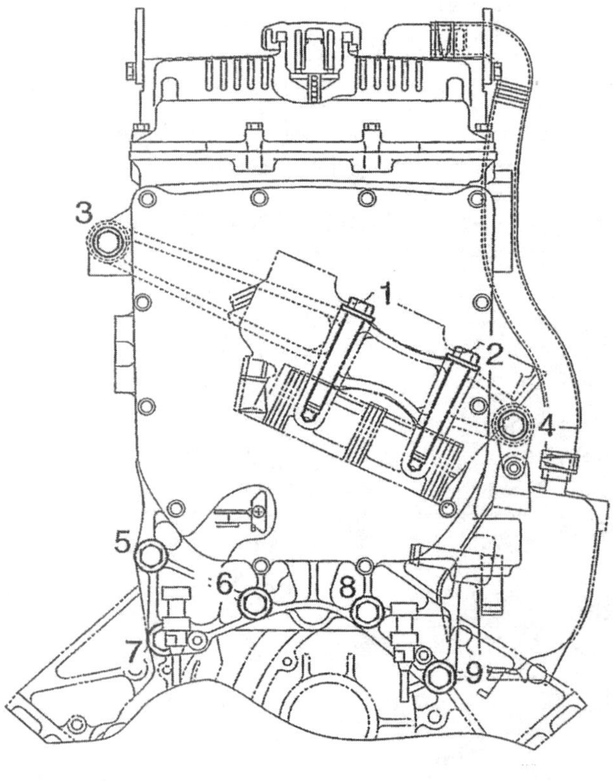

Fig. 16 Oil tank bolt/nut loosening sequence - 1052cc engines

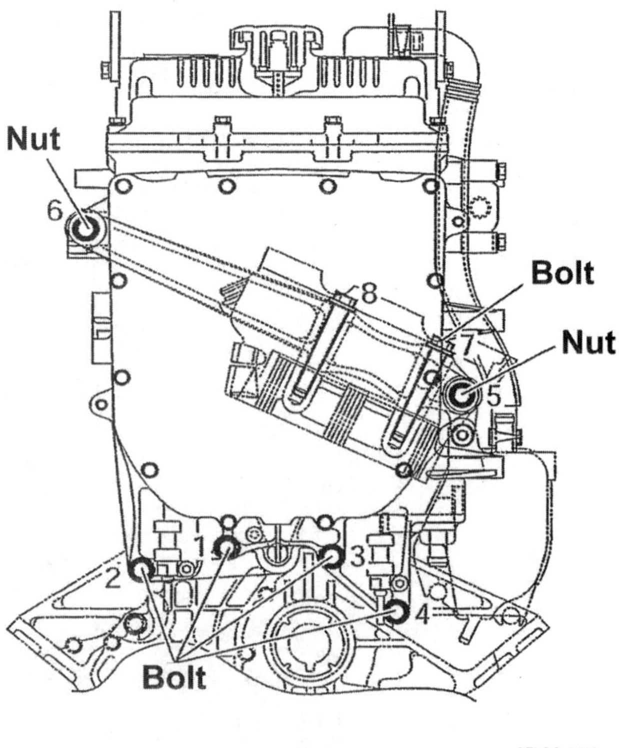

Fig. 17 Oil tank bolt/nut tightening sequence - 998cc engines

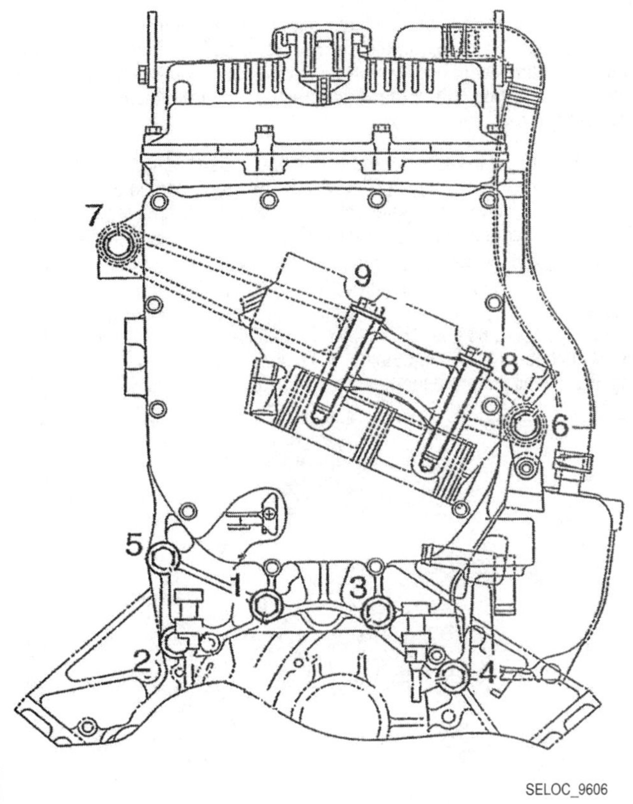

Fig. 18 Oil tank bolt/nut tightening sequence - 1052cc engines

VX Models w/1052cc Engines

OEM ③ DIFFICULT

◆ **See Figures 14 and 19 thru 23**

The oil pump is located at the rear of the engine.

1. Remove the engine as detailed in this section.
2. Tag and disconnect any water lines still connected to the thermostat after engine removal. The grease line should already have been removed, but if not, disconnect it. Remove the thermostat housing.
3. Tag and disconnect all other water and breather hoses connected to the oil tank assembly - there are at least 6. If you have a digital camera, take some pictures to help when installing the tank later.
4. Remove the 4 bolts attaching the water jacket to the starboard side of the tank and lift it off. Remove the gasket and throw it away.
5. You should already have disconnected the breather hoses attached to the oil separator, but if not, remove them now; being sure too tag them. Remove the 2 mounting bolts and lift off the separator. Pull out the 2 collars and grommets and then clean the separator with a good solvent. Allow to air dry completely before installation, or blow it out with compressed air.
6. There are 2 vertical grommets at the bottom of the tank, in back. Remove their mounting bolts and then lift out the small bracket and grommets.
7. Remove the small bolt and collar connecting the band to the port side of the tank. Now remove the 3 mounting bolts and pull off the ground lead box underneath the band along with its O-ring/seal. Remove the bolt inside the recess and disconnect the ground lead(s).

■ **Refer to the loosening sequence as shown for these next steps.**

8. Loosen the 2 mounting stay bolts in between the tank and engine.
9. Now loosen the 2 nuts on the back side of the tank attached to the stay and then loosen the 5 remaining oil tank mounting bolts at the bottom of the tank.
10. Remove completely the 2 nuts and lower 5 bolts before lifting the tank off of the mounting stay.
11. Remove the remaining 2 bolts and lift off the stay.

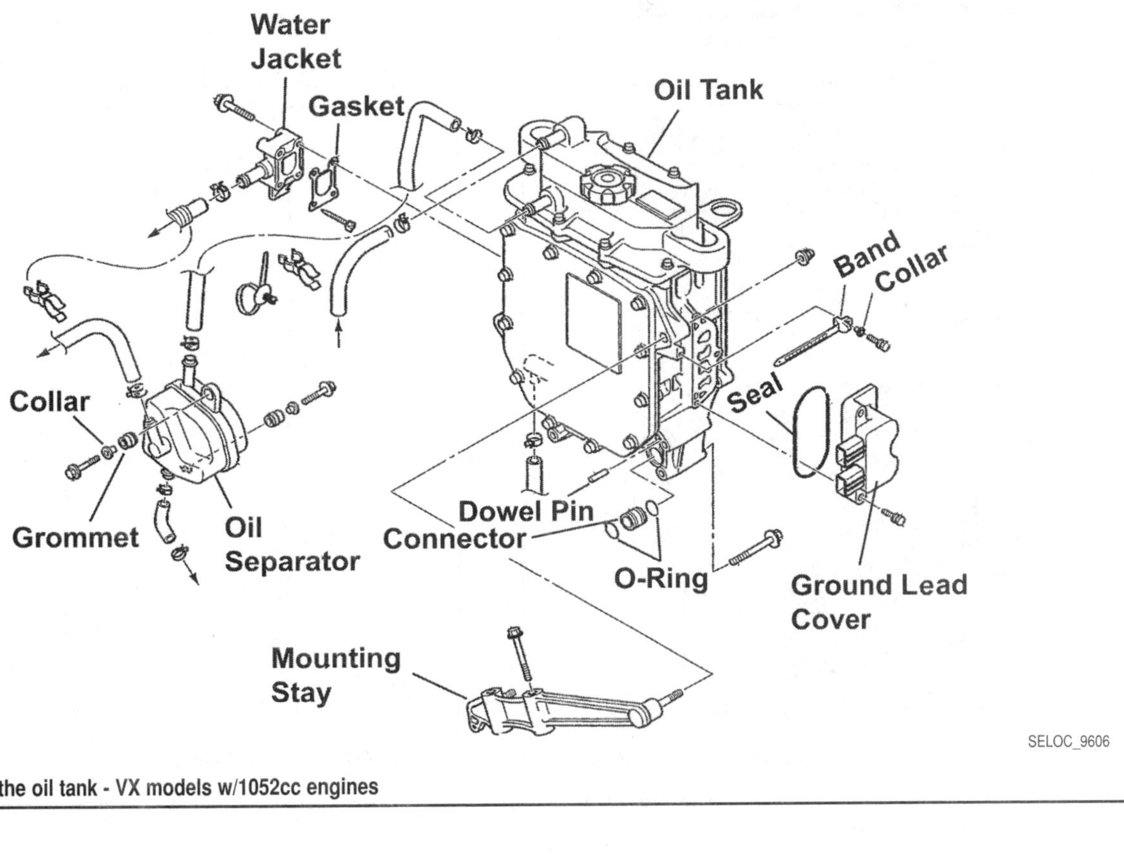

Fig. 19 Removing the oil tank - VX models w/1052cc engines

12. On the lower tank mounting bosses, remove the 2 small pins and the 2 connectors with their O-rings. Discard the O-rings.

To install:

13. Clean the tank thoroughly and inspect it for cracks or other damage.

14. Coat *new* O-rings with a quality light marine grease and roll them into their grooves at each end of the 2 connectors. Press the connectors back into the bolt boss bores.

15. Press the 2 dowel pins fully into their recesses on the tank.

16. Position the mounting stay onto the block. Coat the bolt threads with Loctite 572 and then thread them in until finger-tight.

17. Position the tank assembly so it aligns with the mounting stay studs and bolt holes. Press it into place making sure that the dowel pins fit and the mounting stay studs feed through their respective holes on the tank.

18. Coat the stud threads and mounting bolt threads with Loctite 572 and thread the bolts in or the nuts on until finger-tight.

19. Now, following the tightening sequence, tighten the 5 lower bolts in 2 stages - 1st to 11 ft. lbs. (15 Nm) and finally to 28 ft. lbs. (39 Nm) - 29.5 ft. lbs. (40 Nm) on the 2010 models. Tighten the 2 nuts and 2 bolts in 3 stages - 1st to 1.4 ft. lbs./17 inch lbs. (2 Nm), 2nd to 11 ft. lbs. (15 Nm) and finally to 28 ft. lbs. (39 Nm) - 29.5 ft. lbs. (40 Nm) on the 2010 models.

■ Tighten all bolts and nuts in the sequence illustrated. This means you'll skip the bottom bolts while tightening the upper bolts/nuts to their 1st pass. Next, in complete order now, you'll tighten them *all* to 11 ft. lbs. (15 Nm) - which is the 1st pass on the 5 lower bolts and the 2nd pass on the upper bolts/nuts. And finally, tighten them all to the ultimate figure.

20. Reconnect the ground lead and tighten the bolt(s) to 5.5 ft. lbs./66 inch lbs. (7.6 Nm). Install the cover with a *new* seal. Coat the bolt threads with Loctite 572 and then tighten the 2 bolts to 5.5 ft. lbs./66 inch lbs. (7.6 Nm). Do the same with the bolt for the band, installing it over the cover.

21. Press the 2 collars and grommets into their holes on the oil separator. Move it into position and then thread in the 2 bolts after coating them with Loctite 572. Tighten in 2 passes, 1st to 2.7 ft. lbs./33 inch lbs. (3.7 Nm) and then to 5.5 ft. lbs./66 inch lbs. (7.6 Nm) on 2005-09 engines, or 1st to 3 ft. lbs./36 inch lbs. (4 Nm) and then to 5.9 ft. lbs./71 inch lbs. (8 Nm) on 2010 engines. Install the breather hoses making sure they are fully on the fitting

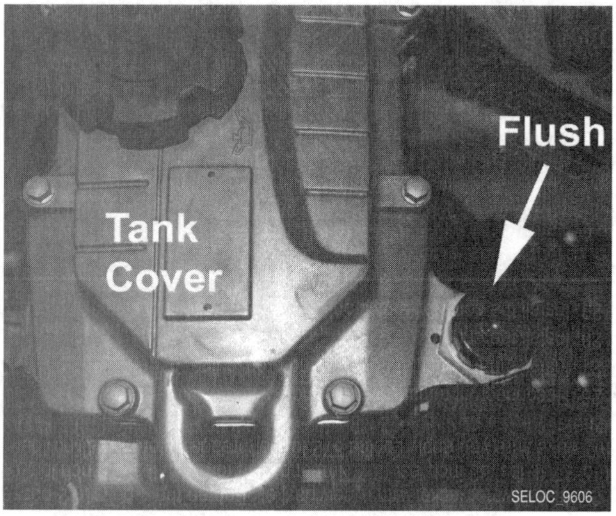

Fig. 20 Remove the flush attachment if not removed already - VX models w/1052cc engines

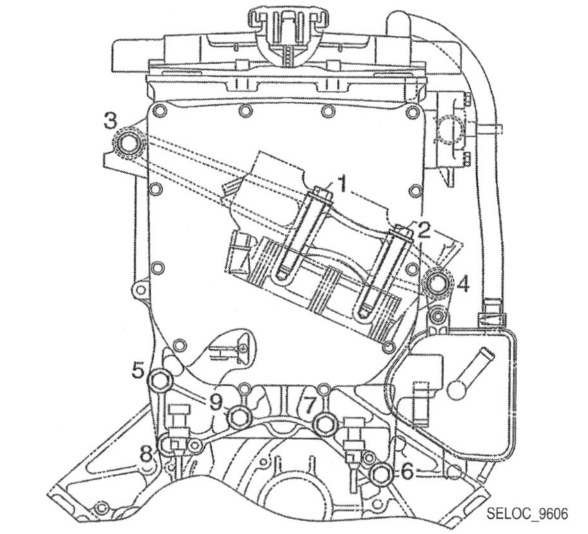

Fig. 21 Oil tank bolt/nut loosening sequence 2005-09 VX models

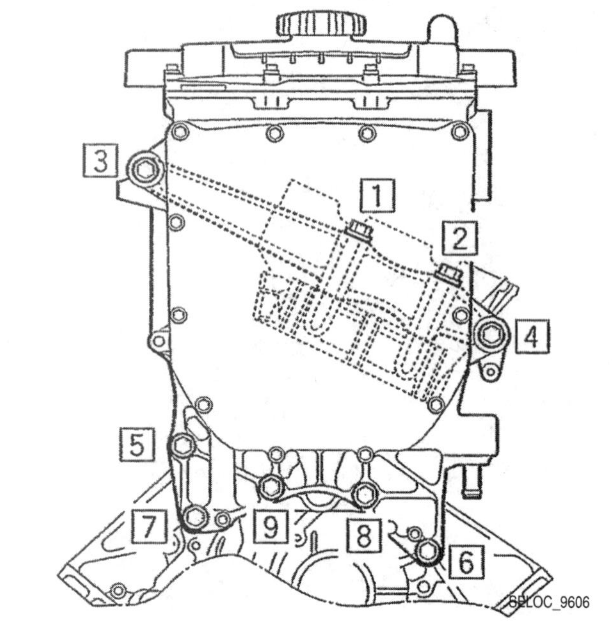

Fig. 22 Oil tank bolt/nut loosening sequence - 2010-11 VX models w/1052cc engines

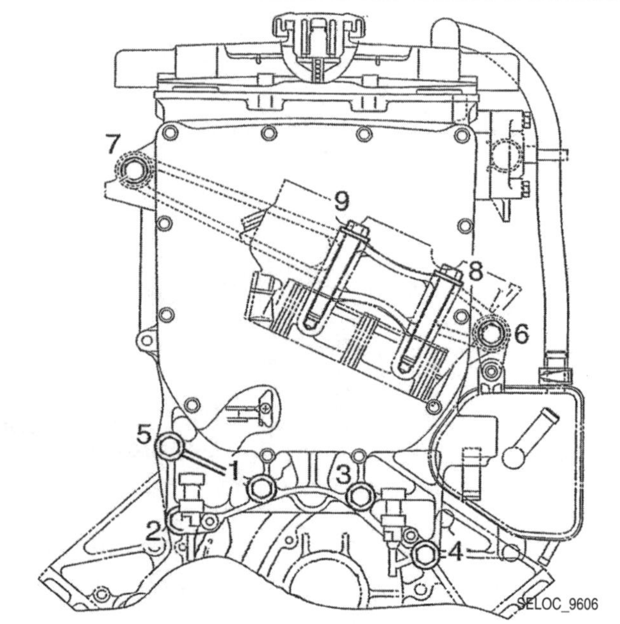

Fig. 23 Oil tank bolt/nut tightening sequence - VX models w/1052cc engines

before moving on the pinch clamps. When installing the lower line from the oil pump, make sure that the mark on the hose lines up with the parting line on the separator.

22. Install water jacket with a *new* gasket. Tighten the 4 bolts in 2 passes, 1st to 2.7 ft. lbs./33 inch lbs. (3.7 Nm) and then to 5.5 ft. lbs./66 inch lbs. (7.6 Nm) on 2005-09 models. On 2010 engines, just tighten them to 5.9 ft. lbs./71 inch lbs. (8 Nm). Make sure you apply Loctite 572 to the threads before installing the bolts and tighten in a criss-cross pattern

23. Reconnect all remaining water hoses and breather lines.

24. Install the thermostat housing and grease line.

25. Install the engine.

DISASSEMBLY & ASSEMBLY

OEM ③ DIFFICULT

◆ **See Figures 24 thru 37**

1. Remove the oil tank as detailed previously.

2. Remove the 2 bolts and washers from the lower end of the tank, and then pull out the bracket and grommet (lower port side).

3. On FX models, remove the 2 mounting bolts and lift the engine hanger from the oil tank cover. Repeat on the other hanger.

4. Remove the oil fill cap. Pull out the gasket/seal, discarding if damaged or worn in any way.

5. Remove the 8 mounting bolts, in the sequence shown, and lift off the tank cover. Remove the gasket and throw it away. Note that the 2010-11 VX uses 1 bolt that is longer than the others (it retains the small holder and uses a collar), so keep it separate.

6. On FX models, turn the cover assembly over and remove the 10 bolts holding the 2 breather plates to the underside of the cover in the sequence shown. Remove the lower and upper plates and throw away the gasket.

7. On VX models, turn the cover assembly over and remove the 9 bolts holding the breather plate to the cover. Remove the plate and discard the gasket.

8. Working inside the tank remove the 3 baffle bolts and lift out the baffle plate.

9. Remove the 2 oil strainer bolts and lift out the strainer. Inspect the screen for tears or blockage.

10. The tank is protected by a front and rear oil cooler cover. Remove the mounting bolts (12/cover, 24 in all), in the sequence shown, and pull off each cover and its gasket; discarding the gaskets.

11. Remove the anode retaining screw on the back side and pull off the sacrificial anode. Replacement of the anode is required if more than 1/3 of it has worn, but why wouldn't you simply spend a couple bucks and replace it after having gone through all this work??

To assemble:

12. Clean all components thoroughly with a good marine grade solvent, making sure everything dries completely prior to reassembly.

13. Inspect all components for cracks, wear or other visible signs of damage, especially the breather/baffle plates and oil screen.

14. Coat the threads of the anode bolt lightly with clean engine oil and then install a *new* anode, tightening the bolt to 2.7 ft. lbs./33 inch lbs. (3.7 Nm) except on the 2010-11 VX where it should be 3 ft. lbs./36 inch lbs. (4 Nm).

15. Position *new* gaskets/seals on the front and rear oil cooler covers. Coat the bolt threads with Loctite 572 and then install all bolts finger-tight. Tighten the bolts for both covers in 2 passes, in the sequence shown, to 2.7 ft. lbs./33 inch lbs. (3.7 Nm) and then again to 5.5 ft. lbs./66 inch lbs. (7.6 Nm). On 2010-11 VX models, tighten 1st to 3 ft. lbs./36 inch lbs. (4 Nm) and then to 5.9 ft. lbs./71 inch lbs. (8 Nm).

16. Coat the threads of the strainer bolts lightly with clean engine oil and then install the strainer, tightening the bolts to 2.7 ft. lbs./33 inch lbs. (3.7 Nm) for a 1st pass and then again to 5.5 ft. lbs./66 inch lbs. (7.6 Nm). On 2010-11 VX models, tighten 1st to 3 ft. lbs./36 inch lbs. (4 Nm) and then to 5.9 ft. lbs./71 inch lbs. (8 Nm).

17. Position the baffle plate into the tank, coat the bolt threads lightly with clean engine oil and then tighten the bolts in 2 passes. First to 1.4 ft. lbs./17 inch lbs. (1.9 Nm), and then to a final pass of 3.2 ft. lbs./15 inch lbs. (4.4 Nm). On 2010-11 VX models, tighten 1st to 1.5 ft. lbs./18 inch lbs. (2 Nm) and then to 3 ft. lbs./36 inch lbs. (4 Nm).

18. Turn the tank cover upside down and position a *new* gasket/seal in it. Install the top and then bottom breather plates (only one on VX models) and then coat the threads of the mounting bolts with clean engine oil. Tighten the 9 (VX) or 10 (FX) bolts, in the sequence shown and in 2 passes; first to 1.4 ft. lbs./17 inch lbs. (1.9 Nm), and then to a final pass of 3.2 ft. lbs./15 inch lbs. (4.4 Nm). On 2010-11 VX models, tighten 1st to 1.5 ft. lbs./18 inch lbs. (2 Nm) and then to 3 ft. lbs./36 inch lbs. (4 Nm).

19. Now position a *new* gasket/seal on the top of the tank and move the cover into position. Coat the bolt threads with Loctite 572 and then tighten the 8 bolts, in the sequence shown and in 2 passes; first to 1.4 ft. lbs./17 inch lbs. (1.9 Nm), and then to a final pass of 3.2 ft. lbs./15 inch lbs. (4.4 Nm). On 2010-11 VX models, tighten 1st to 3 ft. lbs./36 inch lbs. (4 Nm) and

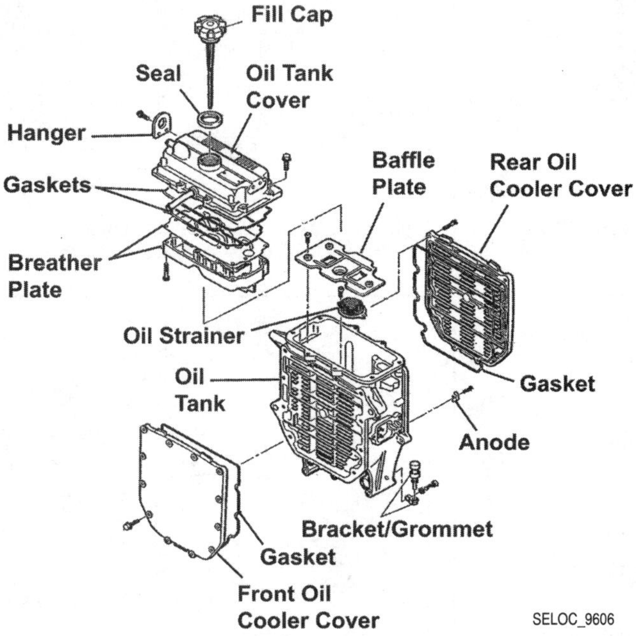

Fig. 24 Exploded view of the oil tank and components - FX models with 998cc/1052cc engines

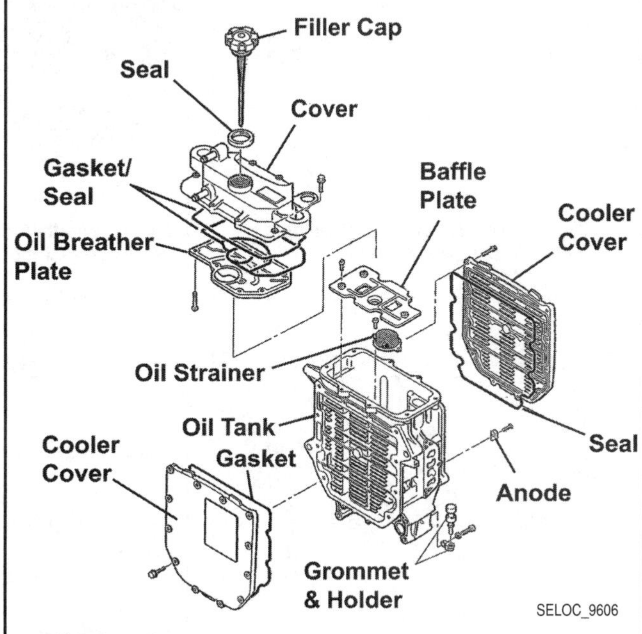

Fig. 25 Exploded view of the oil tank and components - VX models w/1052cc engines

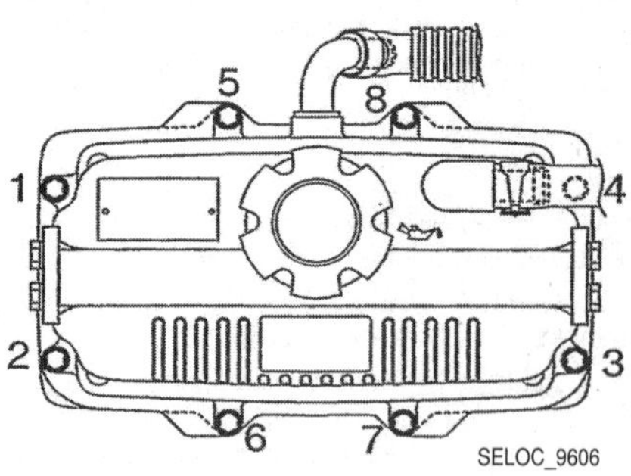

Fig. 26 Oil tank cover bolt loosening sequence - FX models with 998cc/1052cc engines

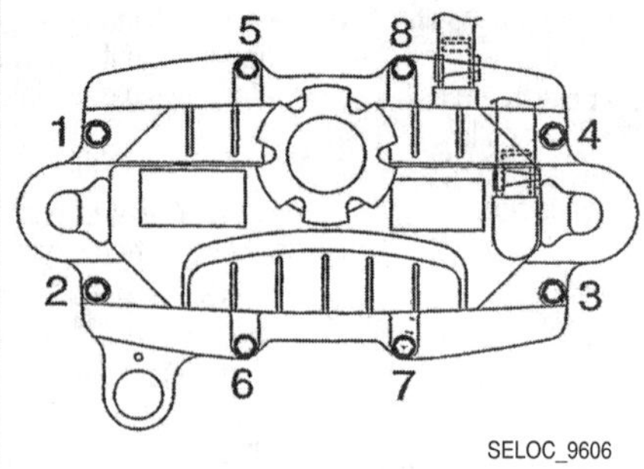

Fig. 27 Oil tank cover bolt loosening sequence - VX models w/1052cc engines

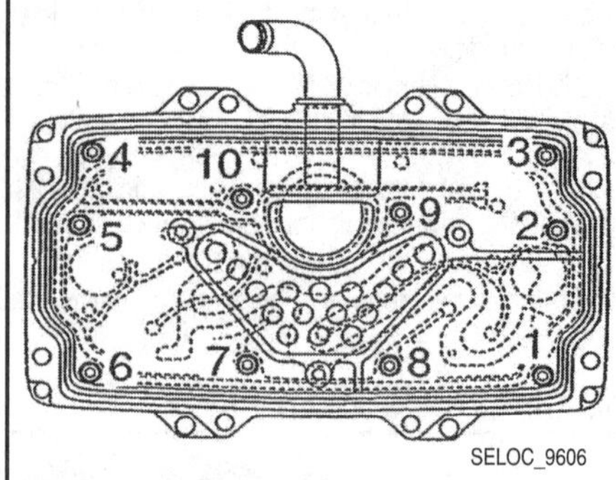

Fig. 28 Breather plate bolt loosening sequence - FX models w/998cc engines

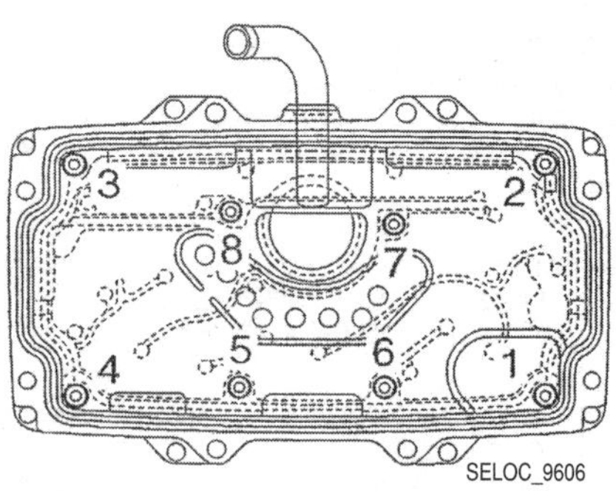

Fig. 29 Breather plate bolt loosening sequence - FX models w/1052cc engines

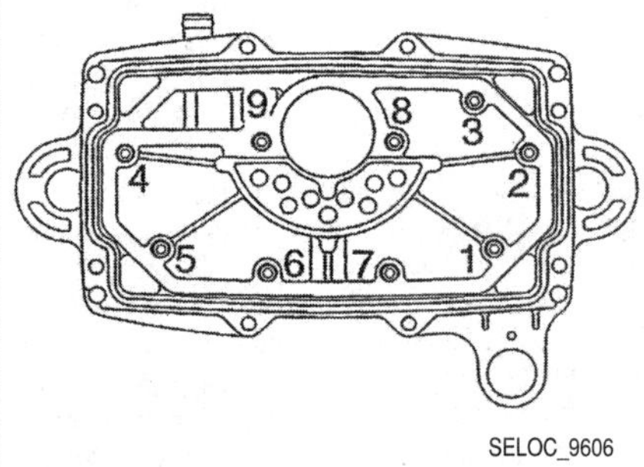

Fig. 30 Breather plate bolt loosening sequence - VX models w/1052cc engines

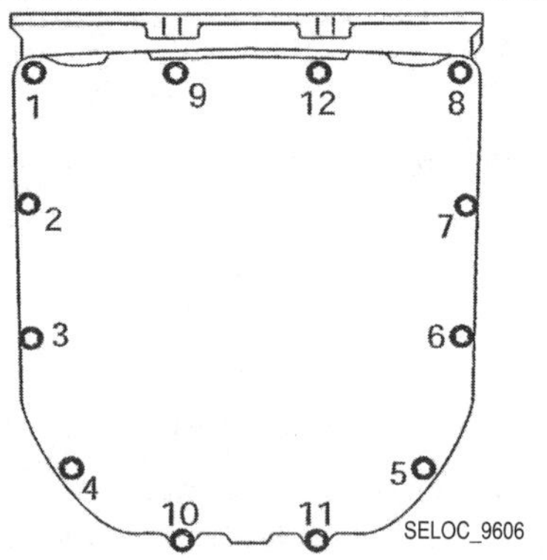

Fig. 31 Oil cooler cover bolt loosening sequence - 998/1052cc engines

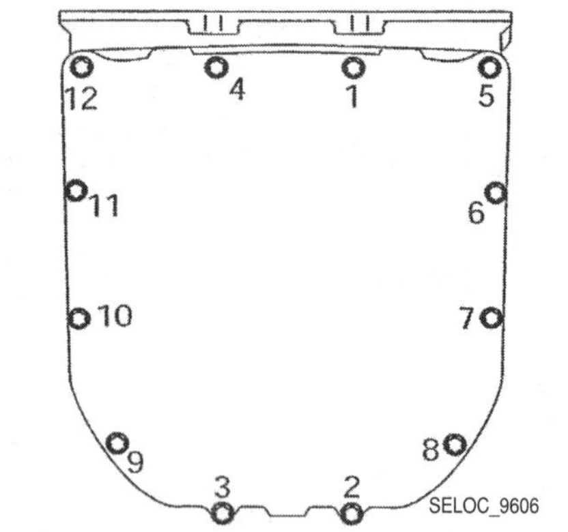

Fig. 32 Oil cooler cover bolt tightening sequence - 998/1052cc engines

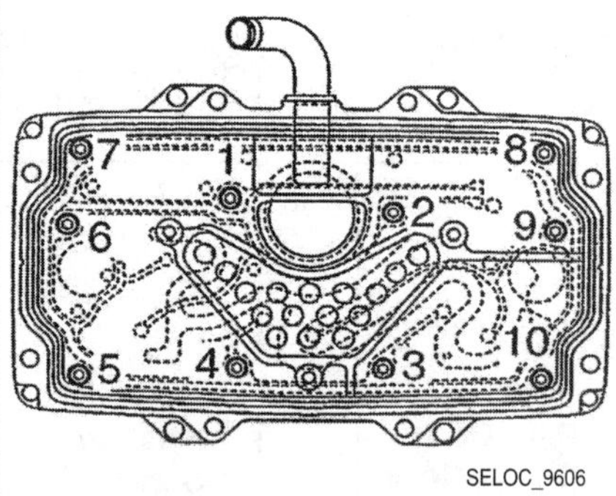

Fig. 33 Breather plate bolt tightening sequence - FX models w/998cc engines

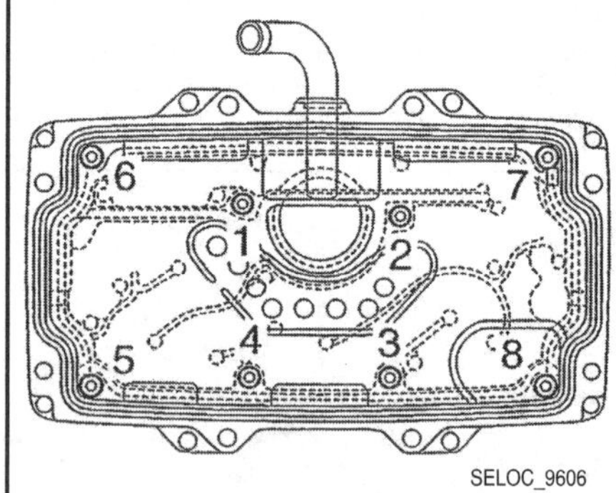

Fig. 34 Breather plate bolt tightening sequence - FX models w/1052cc engines

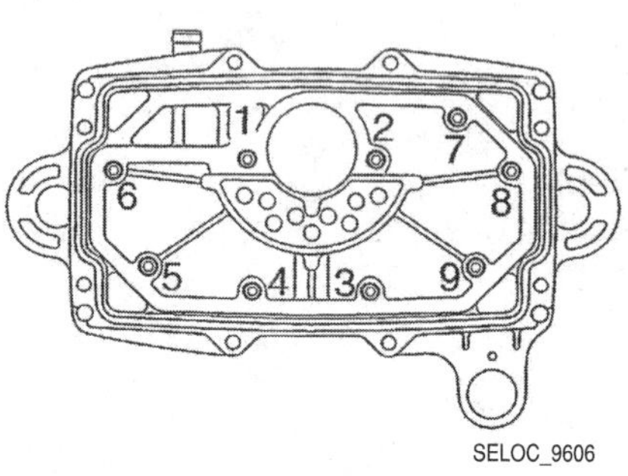

Fig. 35 Breather plate bolt tightening sequence - VX models w/1052cc engines

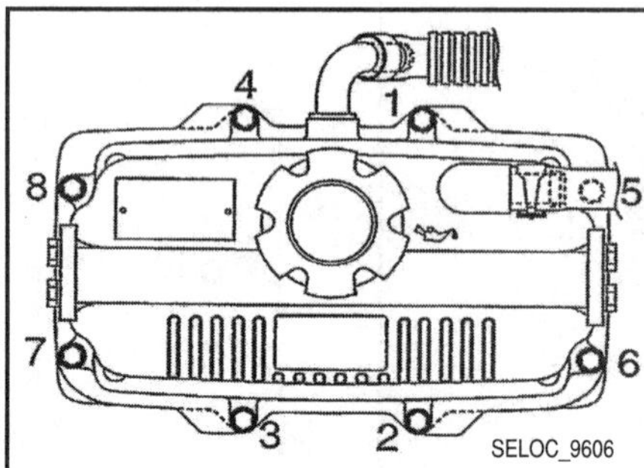

Fig. 36 Oil tank cover bolt tightening sequence - FX models with 998cc/1052cc engines

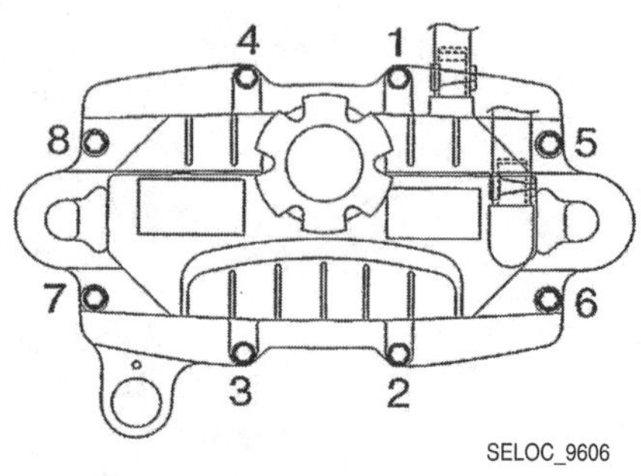

Fig. 37 Oil tank cover bolt tightening sequence - VX models w/1052cc engines

then to 5.9 ft. lbs./71 inch lbs. (8 Nm). . .bet you didn't see that coming did you??

■ **Don't forget the longer bolt and collar/holder on the 2010-11 VX.**

20. Press the dipstick cap gasket into position and then thread in the cap.
21. On FX models, install the 2 hangers on the cover, coating the bolt threads with Loctite 572 and then tighten in 2 passes; first to 2.7 ft. lbs./33 inch lbs. (3.7 Nm) and then finally to 5.5 ft. lbs./66 inch lbs. (7.6 Nm).
22. Install the small bracket and grommet on the tank. Coat the bolt threads with Loctite 572 and then tighten in 2 passes; first to 2.7 ft. lbs./33 inch lbs. (3.7 Nm) and then finally to 5.5 ft. lbs./66 inch lbs. (7.6 Nm).
23. Install the oil tank.

Oil Cooler

REMOVAL & INSTALLATION

All Models w/1812cc Engines

OEM ③ DIFFICULT

◆ **See Figures 38 and 39**

■ **Unlike the other engines, the 1812cc engine uses an oil cooler instead of an oil tank/cooler combination. Further, it is located on the front port side of the engine rather than at the rear.**

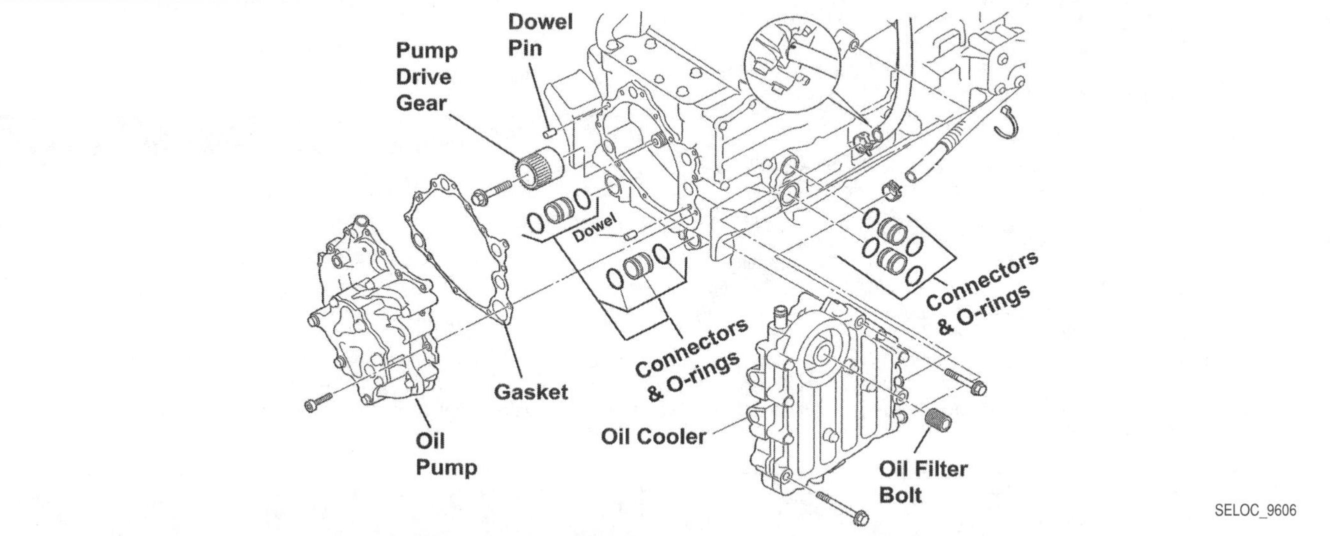

Fig. 38 Removing the oil cooler and oil pump - FX HO models w/1812cc engines (SHO and FZ similar)

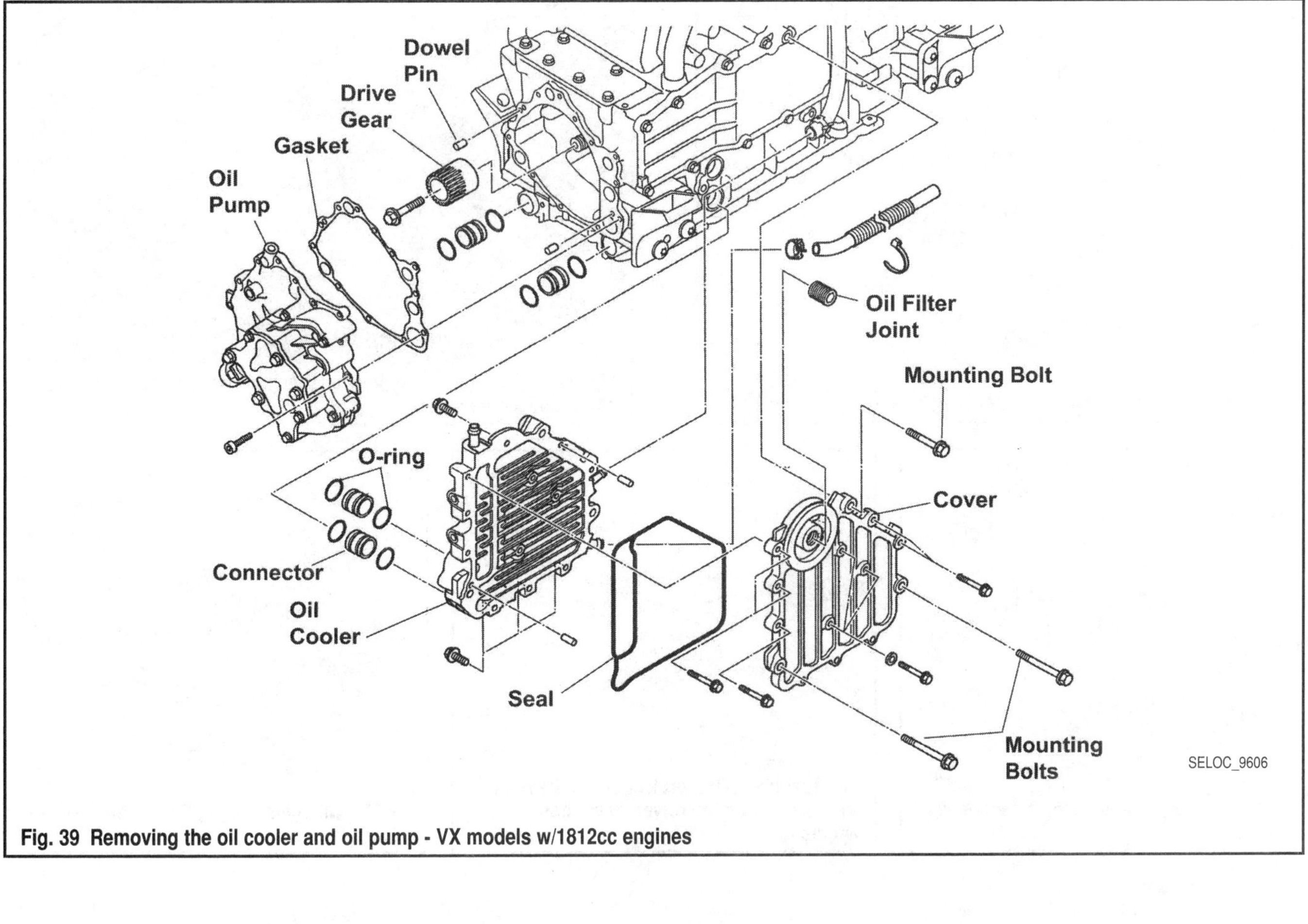

Fig. 39 Removing the oil cooler and oil pump - VX models w/1812cc engines

1. Remove the engine as detailed elsewhere in this section. Removing the exhaust system may allow enough access to get away without removing the entire engine - and since you have to remove the exhaust system anyway when removing the engine, take a look or give it a try.

2. If you decide to go without removing the engine, drain the engine oil and remove the oil filter.

3. Tag and disconnect the lines at the cooler.

4. Remove the 3 mounting bolts (2 long, 1 short) and then carefully pull off the cooler assembly. There are a number of bolts on the cooler, the mounting bolts are the 2 lowest ones on each side (not the bottom) and uppermost one to the right of the oil filter mounting.

5. Remove the 2 passage connectors between the cooler and block. Pull off the O-rings (2 each) with a small pick and throw them away.

6. If necessary, you can remove the cover as well. There are 8 bolts on the front side, 3 with washers, and 4 bolts on the back (engine) side. Remove them all, keeping track of which one went to what hole, and separate the cover. Pull out the large seal and discard it. Make sure that the 2 dowel pins stay with the cooler and not the cover.

To install:

7. If the cooler was taken apart, install a new seal into the cover and then position the cover on the cooler so the dowel pins and bolt holes all line up. Insert all bolts back into the holes that they came out of and thread them in finger-tight after coating with Loctite 572. Tighten the 4 rear bolts and 7 front bolts (M6 x 25mm) to 8.9 ft. lbs./107 inch lbs. (12 Nm). Tighten the remaining front bolt (3rd from the top on the left side, M8 x 25mm) to 14.8 ft. lbs. (20 Nm).

8. Inspect the cooler assembly for cracks, wear or other damage. Do the same with the 2 hose ends.

9. Connect a leak tester (#90890-06840) to the inlet (lower) port on the cooler. Apply 28.5 psi (200 kPa) while holding your fingers over the 2 upper outlet ports and make sure there are no leaks and pressure is maintained. Failure will require replacement of the cooler.

10. Coat the *new* O-rings with a good water resistant marine-grade grease and slide them onto each connector until they seat in the grooves at each end of the connector. Press the connectors into their respective oil cooler bores until they seat.

11. Position the assembly and then push it into place making sure the connectors feed into the bores in the cylinder block without rolling the O-rings out of their grooves.

12. Coat the bolt threads with Loctite 572 and tighten them alternately to 13.3 ft. lbs. (18 Nm) on the FX HO, FZ and VX, or 14 ft. lbs. (19 Nm) on the FX SHO. Remember that the shorter bolt goes in the top right hole as you face the cooler.

13. Push the hose ends over their fittings and slide the pinch clamp into position.

14. If you removed the oil filter bolt for some reason, tighten it to 13.3 ft. lbs. (18 Nm).

15. Install the exhaust system and/or engine.

16. Install the oil filter and refill with oil.

Oil Pump

REMOVAL & INSTALLATION

998cc Engines

OEM 3 DIFFICULT

◆ **See Figures 40 thru 44**

The oil pump is mounted between the oil tank and the reduction drive gear housing, at the rear of the engine.

1. Remove the engine and then the oil tank as described elsewhere in this section.

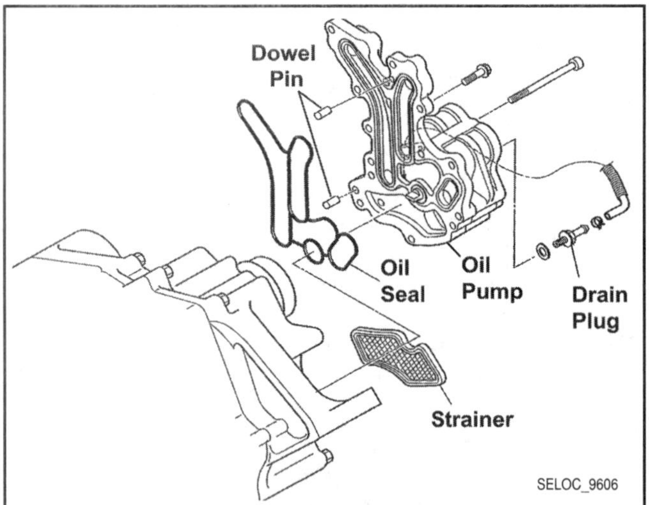

Fig. 40 Removing the oil pump - 998cc engines

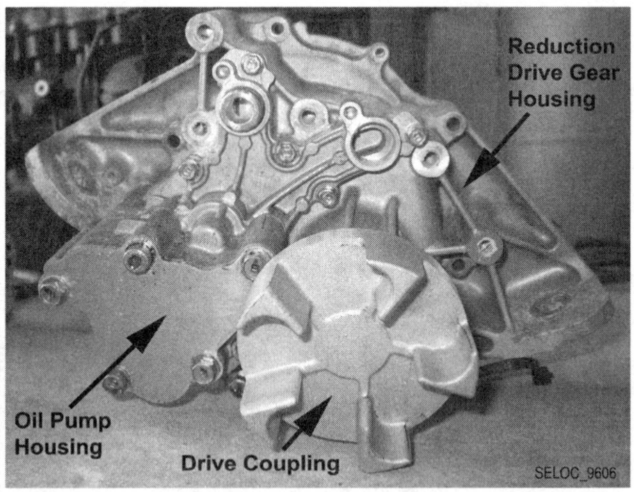

Fig. 41 A nice look at a typical oil pump housing - 1052cc engine shown, 998cc similar

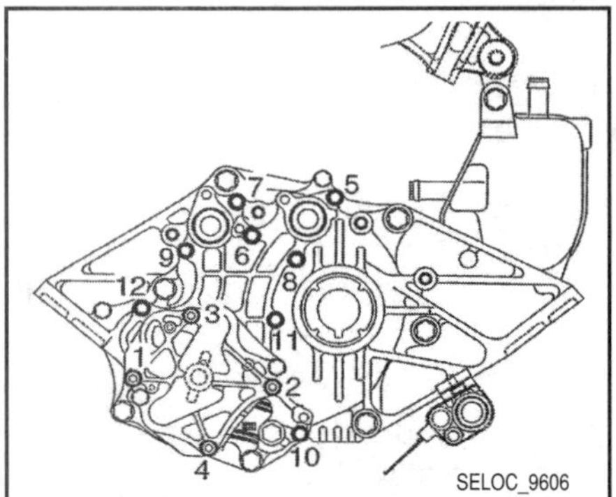

Fig. 42 Oil pump bolt loosening sequence - 998cc engines

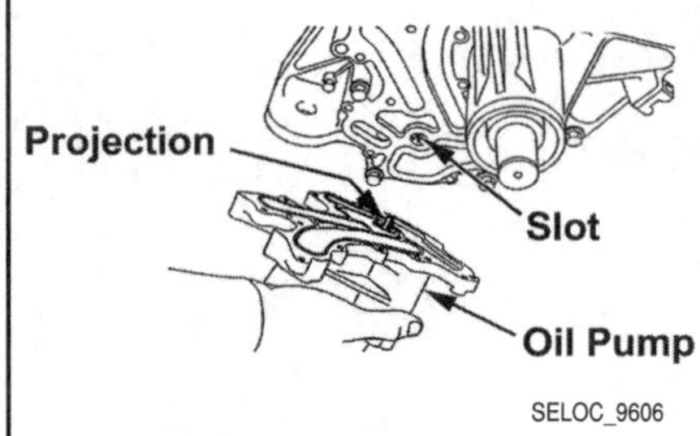

Fig. 43 Line up the projection on the pump with the slot on the driven gear shaft - 998cc engines

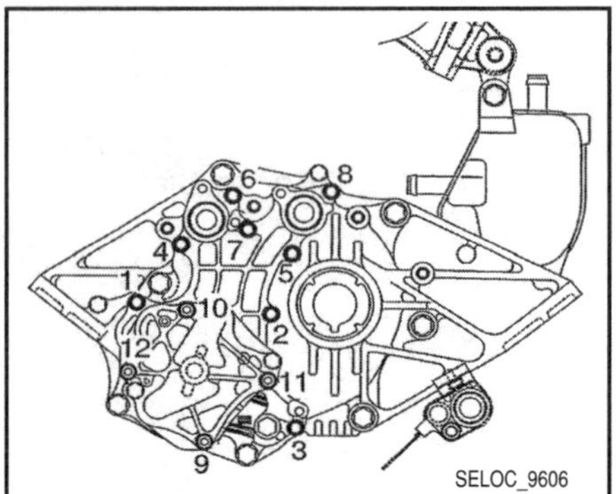

Fig. 44 Oil pump bolt tightening sequence - 998cc engines

2. Position a suitable container under the oil pump and then pinch the clamp and wiggle the breather hose off of its fitting. Unthread the fitting/drain plug and allow all oil to drain completely. Don't lose the washer on the plug.

3. Remove the pump mounting bolts (8 short and 4 long) in the sequence illustrated. Pry off the pump and remove any remaining seal/gasket material from the two mating surfaces. There are 2 dowel pins that should stay in their bores on the inner side of the pump.

4. Lift out the oil strainer and check it for tears or signs of blockage.

To install:

5. Check the oil pump carefully for, wear, cracks or other signs of damage.

6. Insert the strainer screen over the hole in the reduction gear housing.

7. Position the seal/gasket onto the pump mating surface and then install the pump making sure that the seal does not roll and that the pins fit into their respective bores in the gear housing. Make sure that the oil pump shaft feeds into the slot in the driven gear shaft.

8. Coat the bolt threads with Loctite 572 and then thread them in until finger-tight. Make a pass, in the sequence illustrated, to 1 ft. lb./12 inch lbs. (1.4 Nm). Make a final pass, in the same sequence, to 7.2 ft. lbs./87 inch lbs. (10 Nm).

9. Thread in the drain plug/fitting and tighten it to 13 ft. lbs. (18 Nm). Slide on the breather hose and then move the clamp into position. Make sure that there are no cracks along the hose where the clamp is sitting.

10. Install the oil tank and engine.

1052cc Engines

OEM ③ DIFFICULT

◆ See Figures 41, 45, 46 and 47

The oil pump is mounted between the oil tank and the reduction drive gear housing, at the rear of the engine.

1. Remove the engine and then the oil tank as described elsewhere in this section.

2. Position a suitable container under the oil pump and then unthread the drain plug and allow all oil to drain completely. Don't lose the washer on the plug.

3. Remove the 10 pump mounting bolts (6 short, 1 medium, 1 long and 2 very long on the FX; 6 short, 2 medium and 2 long on the VX) alternately and in a criss-cross pattern. Pry off the pump and remove any remaining gasket/seal material from the two mating surfaces. There are 2 dowel pins that should stay in their bores on the inner side of the pump.

4. Remove the 2 small bolts and lift out the oil strainer screen. Check it for tears or signs of blockage.

To install:

5. Check the oil pump carefully for wear, cracks or other signs of damage.

6. Insert the strainer screen over the hole in the reduction gear housing. Thread in the bolts and tighten them to 2.7 ft. lbs./33 inch lbs. (3.7 Nm). Now go back and tighten them again to 5.5 ft. lbs./66 inch lbs. (7.6 Nm). On 2010-11 VX models, tighten 1st to 3 ft. lbs./36 inch lbs. (4 Nm) and then to 5.9 ft. lbs./71 inch lbs. (8 Nm).

7. Position the gasket/seal onto the pump mating surface and then install the pump making sure that the gasket/seal stays in position and that the pins fit into their respective bores in the gear housing. Make sure that the oil pump shaft feeds into the slot in the driven gear shaft.

8. Coat the bolt threads with Loctite 572 and then thread them in until finger-tight. On all but the 2010-11 VX, tighten the short bolts (6), alternately and in a criss-cross pattern, to 7.2 ft. lbs./87 inch lbs. (10 Nm). Now do the same with the remaining bolts except to 11 ft. lbs. (15 Nm). Wait a few minutes and then tighten the 4 long bolts again to 20 ft. lbs. (28 Nm).

On the 2010-11 VX, tighten the short bolts (6), alternately and in a criss-cross pattern, to 7.4 ft. lbs./89 inch lbs. (10 Nm). Now do the same with the remaining 4 bolts, except to 11 ft. lbs. (15 Nm). Wait a few minutes and then tighten again to 21.4 ft. lbs. (29 Nm).

9. Thread in the drain plug (use a *new* sealing washer please) and tighten it to 13 ft. lbs. (18 Nm) on the FX, or 9.4 ft. lbs./113 inch lbs. (13 Nm) on the VX.

10. Fill the scavenge port with 0.1 qt. (0.1 L) of clean SAE 10W-30 engine oil. Use 0.5 qt. (0.5 L) on the feed port.

11. Install the oil tank and engine.

DISASSEMBLY & ASSEMBLY

998cc Engines

OEM ③ DIFFICULT

◆ See Figures 48 and 49

1. Remove the oil pump as described in this section.

2. Remove the 2 mounting bolts and pull off the outer pump housing cover. There is a small clamp on the right bolt (as you face the front of the pump). Pull out the seal/gasket and check for any residual material on the two mating surfaces.

3. Reach into the bore in the pump housing and pull the small pin out of the check valve. There are also 2 small dowel pins, so make sure they are in their respective bores in the cover.

4. Reach in and pull out the two pump scavenge rotors along with the small pin. It wouldn't be a bad idea to make a mark across the top of these two for installation orientation.

5. Pull the inner pump housing cover from the housing. Remove the seal and check for any residual material on the two mating surfaces. Make sure that the 2 dowel pins are in their bores on the cover.

6. Now you can remove the 2 remaining pump feed rotors along with the small pin in the pump shaft. Yes, it is a good idea to mark the mating surfaces here as well.

7. Remove the oil pump shaft and then carefully pry out the check valve seat and the oil seal from the pump housing.

To assemble:

8. Clean all components thoroughly with a suitable marine-grade solvent. Dry thoroughly before assembling, or use compressed air if available.

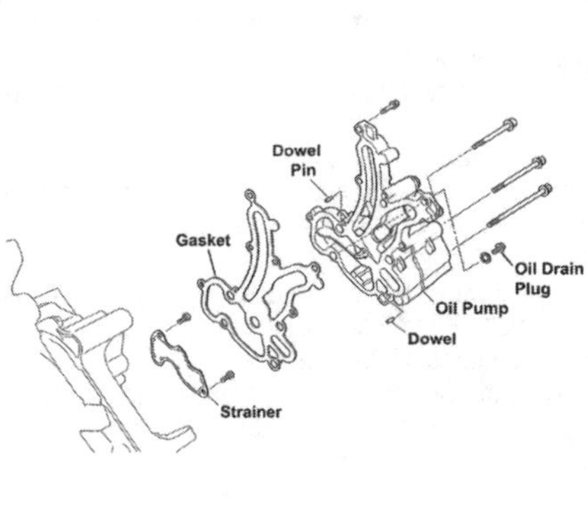

Fig. 45 Removing the oil pump - 1052cc engines

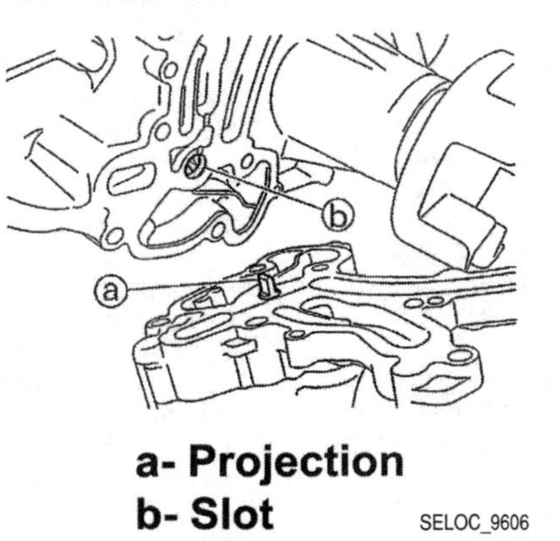

Fig. 46 Line up the projection on the pump with the slot on the driven gear shaft - 1052cc engines

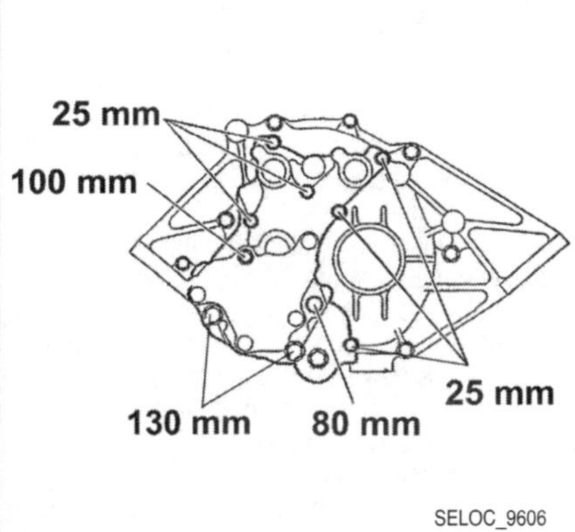

Fig. 47 Use this diagram for proper bolt relocation when installing the pump

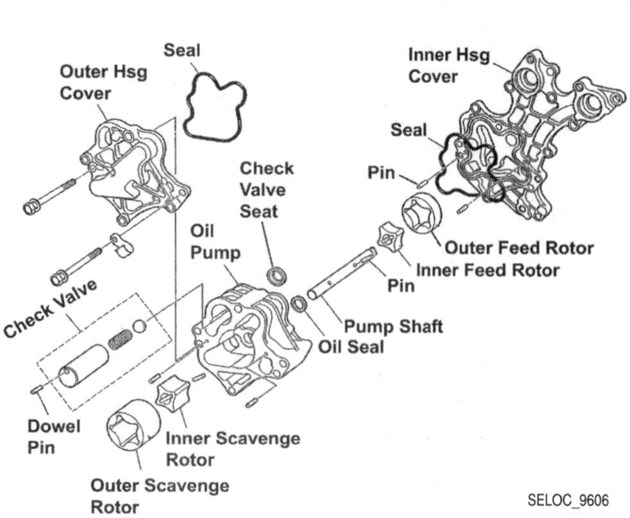

Fig. 48 Exploded view of the oil pump - 998cc engines

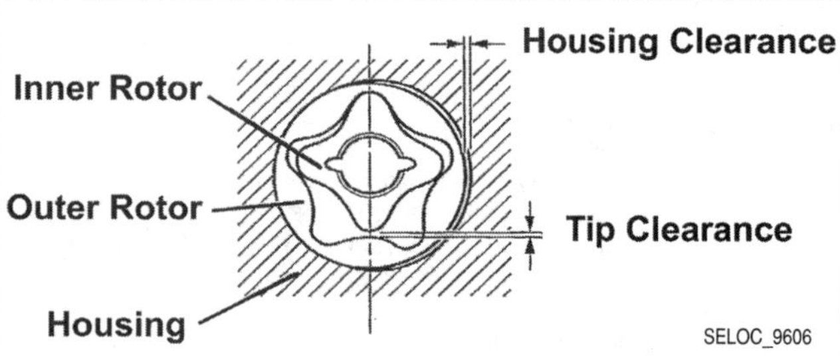

Fig. 49 Checking the oil pump rotor clearances - 998cc engines

9. Check all components for wear, cracks or other signs of damage.
10. Position the scavenge rotors back into the outer side of the pump housing. Use a feeler gauge and measure the inner rotor tip to outer rotor tip clearance by using different size gauges until a slight drag can be felt while inserting and removing it. Correct clearance is 0.004-0.006 in. (0.09-0.15mm) and anything outside this parameter will require oil pump replacement. Repeat this procedure for the feed rotors on the other side of the housing, using the same specification.
11. With the scavenge rotors still inside the outer side of the housing, use a feeler gauge to measure the clearance between the outer rotor and the housing bore. A slight drag should be felt when inserting and removing the feeler gauge and the correct clearance should be 0.0035-0.0075 in. (0.09-0.19mm). Repeat this procedure for the feed rotors on the other side of the housing but use a clearance specification of 0.0035-0.0067 in. (0.09-0.17mm) for this set. Clearance measurements outside the recommended spec will require pump replacement.
12. Remove the rotors from the housing.
13. Press a *new* oil seal into the bore in the housing. Press the check valve seat in as well.
14. Install the pins into the holes on the pump shaft, coat the assembly with clean engine oil and insert it back into the housing so that the end of the shaft with the flat cut-out is facing the inner side of the housing (the side against the engine).
15. Coat the 2 feed rotors and their bore with a little clean engine oil. Insert the outer rotor into the bore until it seats and then slide the inner rotor over the shaft so that the marks made earlier line up and the slots in the bottom of the rotor slide over the pin in the pump shaft.
16. Press a *new* seal into the recess in the inner housing cover and then position it on the housing so that the holes align and the 2 pins fit into their respective bores in the housing body. Be careful that the seal does not come out of the recess in the cover.
17. Now coat the 2 scavenge rotors and their bore with clean engine oil. Insert the outer rotor into the bore until it seats and then slide the inner rotor over the shaft so that the marks made earlier line up.
18. Install the ball, spring and check valve. Once done, insert the small pin into the recess in the check valve.
19. Press a *new* seal into the recess in the outer housing cover and then position it on the housing so that the holes align and the 2 pins fit into their respective bores in the housing body. Be careful that the seal does not come out of the recess in the cover. Slide the clamps over the right bolt and then tread in the 2 pump housing bolts, tightening to 5.7 ft. lbs. (7.9 Nm).
20. Grab the end of the pump shaft carefully with needle-nosed pliers and check to be sure it rotates correctly with no sticking or binding. If so, disassemble everything and try it again. Otherwise, install the oil pump. To the reduction gear housing.

1052cc Engines

◆ See Figures 49 and 50

■ **Yamaha does not provide overhaul procedures for the oil pump after 2005. It appears to be the same pump though, so we suspect these procedures will work for later models as well.**

1. Remove the oil pump as described in this section.
2. Remove the 4 mounting bolts (the lower one on the right will be shorter) and pull off the outer pump housing cover. Pull out the seal/gasket and check for any residual material on the two mating surfaces.
3. Separate the inner (scavenge) pump housing from the outer (feed) pump housing, removing the oil seal from the inner.
4. Reach into the inner housing and pull out the two outer scavenge pump rotors along with the small pin. It wouldn't be a bad idea to make a mark across the top of these two for installation orientation. Now remove the pin and do the same with the 2 inner scavenge rotors.
5. Remove the screw attaching the inner cover to the outer (feed) pump housing. *Carefully* pry of the cover, discarding the oil seal.
6. Remove the spring and relief valve plunger. Remove the check valve assembly
7. Now you can remove the 2 remaining pump feed rotors along with the small pin in the pump shaft. Yes, it is a good idea to mark the mating surfaces here as well.
8. Remove the oil pump shaft and then carefully pry out the check valve seat and the oil seal from the feed pump housing.

To assemble:

9. Clean all components thoroughly with a suitable marine-grade solvent. Dry thoroughly before assembling, or use compressed air if available.
10. Check all components for wear, cracks or other signs of damage.
11. Position the feed rotors back into the outer side of the feed pump housing. Use a feeler gauge and measure the inner rotor tip to outer rotor tip clearance by using different size gauges until a slight drag can be felt while inserting and removing it. Correct clearance is 0.0035-0.0059 in. (0.09-0.15mm) and anything outside this parameter will require oil pump replacement. Repeat this procedure for the scavenge rotors when you install them to the other housing, using the same specification.
12. With the rotors still inside the outer side of the housing, use a feeler gauge to measure the clearance between the outer rotor and the housing bore. A slight drag should be felt when inserting and removing the feeler gauge and the correct clearance should be 0.0035-0.0067 in. (0.09-0.17mm). Repeat this procedure for the scavenge rotors when you install them but use a clearance specification of 0.0035-0.0059 in. (0.09-0.15mm) for this set. Clearance measurements outside the recommended spec will require pump replacement.
13. Install the inner scavenge rotors and perform the clearance measurements just discussed. Now install the outer scavenge rotors and do the same. Once done, remove all 6 rotors from their respective housings.
14. Press a *new* seal into the housing.
15. Coat the pump shaft lightly with clean engine oil and then insert it into the scavenge pump housing, notched end first, until it seats and is protruding from the engine side of the housing. Install the inner pin.
16. Coat the contact surfaces of the inner set of scavenge rotors lightly with clean engine oil and slide them over the pump shaft until the pin fits into the slot in the inner rotor. You may have to rotate the rotors or shaft slightly until the pin lines up with the slot.
17. Repeat the previous step for the outer scavenge rotor set, but leave out the pin at this time.
18. Install the check valve seat in the feed pump housing bore and then drop in the check valve assembly. Do the same with the relief valve plunger and spring, making sure that the end with little holes in it goes in first.

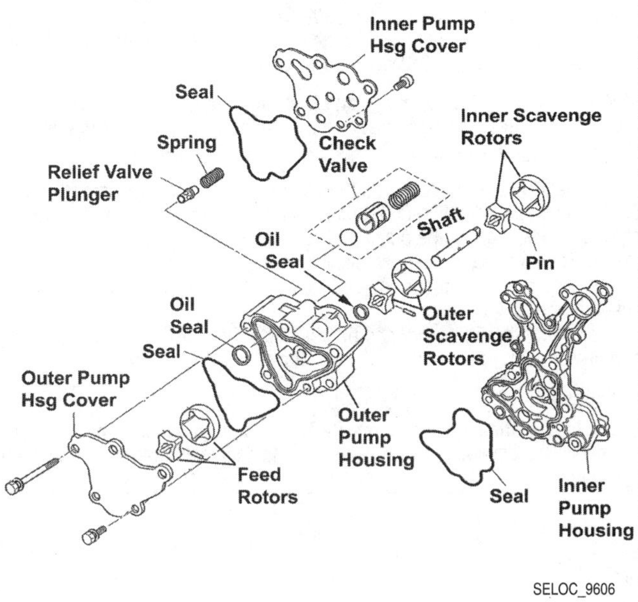

Fig. 50 Exploded view of the oil pump - 1052cc engines

19. Coat a *new* seal lightly with clean engine oil and press it into the groove in the inner feed pump cover. Position the cover on the housing without disturbing the check valve or plunger springs and then tighten the bolt to 1.4 ft. lbs./17 inch lbs. (2 Nm).

20. Coat a *new* seal lightly with clean engine oil and press it into the groove in the scavenge pump housing. Position the housing over the feed pump and then guide the pump shaft through the seal and feed pump until the two pumps come together. Lay the assembly on the scavenge pump housing so the feed pump is up.

21. Insert the pin from the outer scavenge rotor set now and swivel the shaft carefully until it fits into the rotor slot. Make sure that the two pump housings stay together.

22. Coat the contact surfaces of the 2 feed rotors and then slide the outer rotor over the shaft and into position. Now slide the inner rotor over the shaft until it seats over the pin.

23. Coat a *new* seal lightly with clean engine oil and press it into the groove in the feed pump housing. Position the housing cover over the feed pump and then thread in the bolts. Tighten them to 14 ft. lbs. (20 Nm).

24. Grab the end of the pump shaft carefully with needle-nosed pliers and check to be sure it rotates correctly with no sticking or binding. If so, disassemble everything and try it again. Otherwise, install the oil pump. To the reduction gear housing.

Oil Pump & Drive Gear

REMOVAL & INSTALLATION

FX HO & VX Models w/1812cc Engines

OEM ③ DIFFICULT

◆ See Figures 38, 39, 51 and 52

■ **Remember, unlike the other engines, the oil pump is on the *front* of the engine rather than the rear.**

1. Remove the engine and then the oil cooler as described elsewhere in this section.

2. Remove the 12 pump mounting bolts, alternately and in a criss-cross pattern. Pry off the pump (carefully) and remove any remaining gasket/seal material from the two mating surfaces. There are 2 dowel pins that should stay in their bores on the inner side of the pump.

3. Remove the 2 passage connectors between the pump and block. Pull off the O-rings (2 each) with a small pick and throw them away.

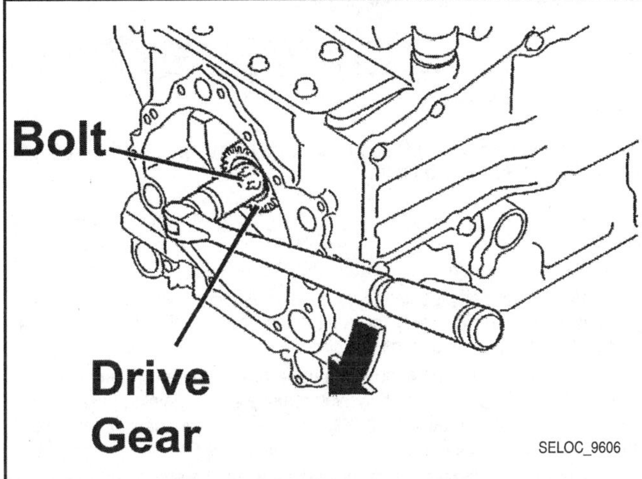

Fig. 51 Remember, the drive gear is a left hand thread and is turned clockwise to loosen it - 1812cc engines

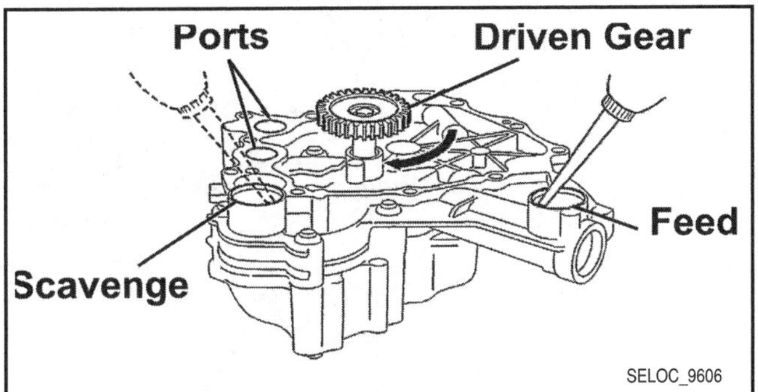

Fig. 52 Make sure you turn the driven gear while filling the oil pump

■ **The following steps should only be preformed if necessary.**

4. Connect a coupler wrench (#90890-06729) to the drive coupling and hold the crankshaft while loosening the drive gear bolt. This bolt is a left-hand thread and must be turned CLOCKWISE to loosen it. Once the bolt is removed, remove the gear.

To install:

5. Inspect the pump assembly for cracks, wear or other damage. Do the same with the drive and driven gears, paying particular attention to the teeth.

6. If you removed the drive gear, install the coupler wrench again and slide the gear onto the end of the crankshaft. Coat the bolt threads lightly with clean engine oil and thread it into the end of the shaft, tightening to 59 ft. lbs. (80 Nm) while holding the crankshaft with the tool. Remember - COUNTERCLOCKWISE to tighten on this bolt!

7. Lay the pump on a flat surface so the inside is facing up. Add 10W-30 engine oil slowly through the scavenge and feed ports while turning the pump driven gear clockwise. Continue filling until oil just begins to seep out of the lower ports.

8. Coat *new* O-rings with a good water resistant marine-grade grease and slide them onto each connector until they seat in the grooves at each end of the connector. Press the connectors into their respective pump bores until they seat.

9. Lay a *new* gasket on the mating surface and then position the assembly over the dowel pins. Push it into place making sure the connectors feed into the bores in the cylinder block without rolling the O-rings out of their grooves.

10. Coat the bolt threads with Loctite 572 and tighten them alternately to 7.4 ft. lbs./88 inch lbs. (10 Nm).

11. If you removed the oil filter bolt for some reason, tighten it to 13.3 ft. lbs. (18 Nm).

12. Install the engine.

13. Install the oil filter and refill with oil.

FZ & FX SHO Models w/1812cc Engines

OEM ③ DIFFICULT

◆ See Figures 52 thru 60

■ **Remember, unlike the other engines, the oil pump is on the *front* of the engine rather than the rear.**

1. Remove the engine and then the oil cooler as described elsewhere in this section.

2. Remove the generator cover on FX SHO models.

3. Remove the oil pipe flange bolts where it connects to the pump and the cylinder block (throw away the 4 washers). Remove the mounting bolt at the center of the pipe and lift off the pipe.

4. Remove the 12 pump mounting bolts, alternately and in a criss-cross pattern. Pry off the pump (carefully) and remove any remaining gasket/seal material from the two mating surfaces. There are 2 dowel pins that should stay in their bores on the inner side of the pump.

5. Remove the 2 passage connectors between the pump and block. Pull off the O-rings (2 each) with a small pick and throw them away.

■ **The following steps should only be performed if necessary.**

6. Connect a holding tool (#90890-06721) to the flywheel on FX SHO models or to the drive coupler (#90890-06729) on FZ models. Hold the crankshaft while loosening the drive gear bolt. This bolt is a left-hand thread and must be turned CLOCKWISE to loosen it.

■ **We're not sure why you couldn't just hold the coupler on SHO models and save having to remove the generator cover, but this is how Yamaha suggests to do it.**

7. Install a bearing puller (#90890-06535), stopper guide plate (#90890-06501) and stand (#90890-06538) through the gear/bearing assembly and onto the block. Tighten the puller and remove the assembly.

a. Remove the fore and aft circlips.

b. Remove the washer and pull the drive gear clutch off of the supercharger drive gear hub.

c. Mount the drive gear and bearing into a universal bearing separator and press the gear out of the bearing. Throw away the washer.

To install:

8. Inspect the pump assembly for cracks, wear or other damage. Do the same with the drive and driven gears, paying particular attention to the teeth.

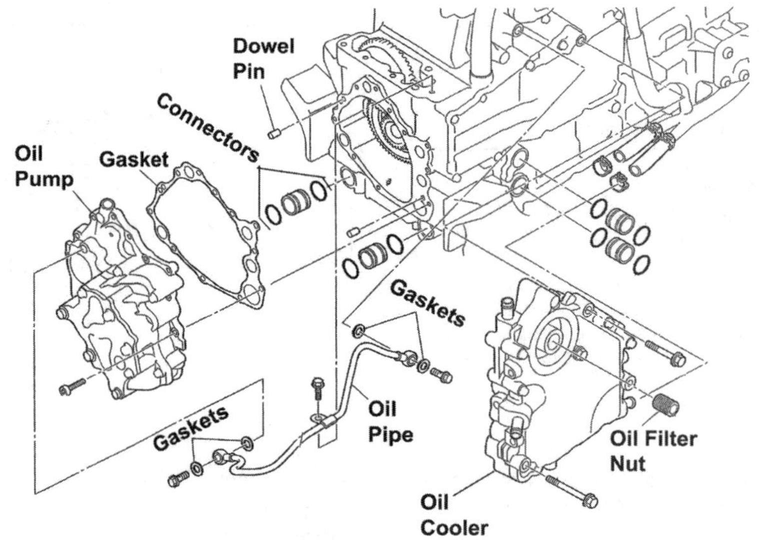

Fig. 53 A good look at the oil pump and cooler - FZ and FX SHO models w/1812cc engines

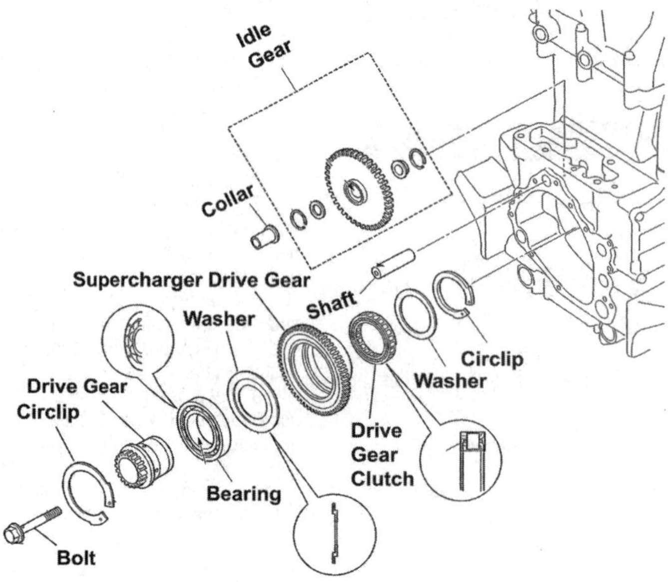

Fig. 54 Exploded view of the drive gear assembly - FZ and FX SHO models w/1812cc engines

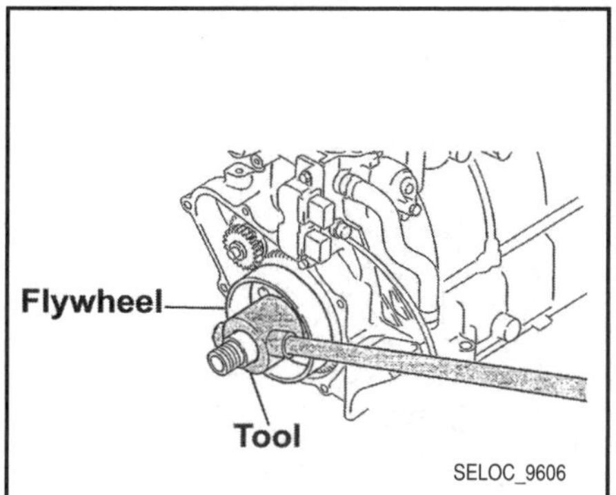

Fig. 55 Hold the flywheel with the special tool on FX SHO models. . .

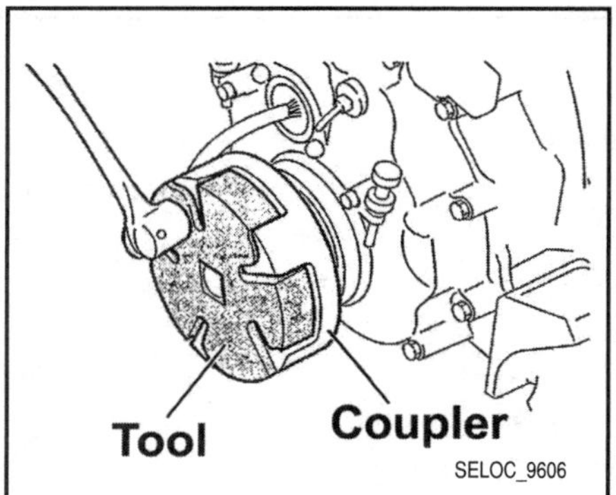

Fig. 56 . . .or the drive coupler on FZ models. . .

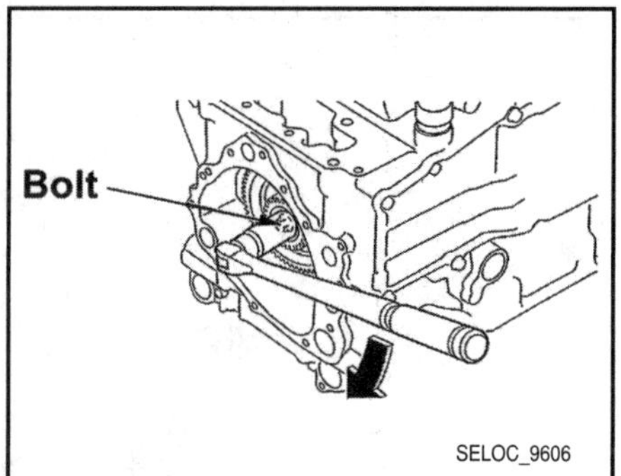

Fig. 57 . . . remove the drive gear bolt (remember, CLOCKWISE to loosen) on all models. . .

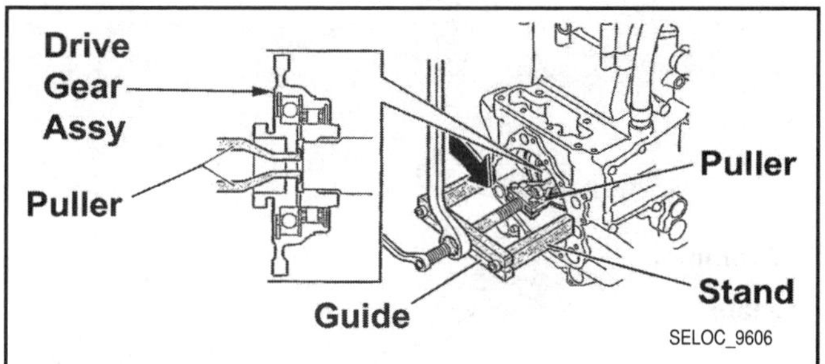

Fig. 58 . . .and then pull out the entire assembly

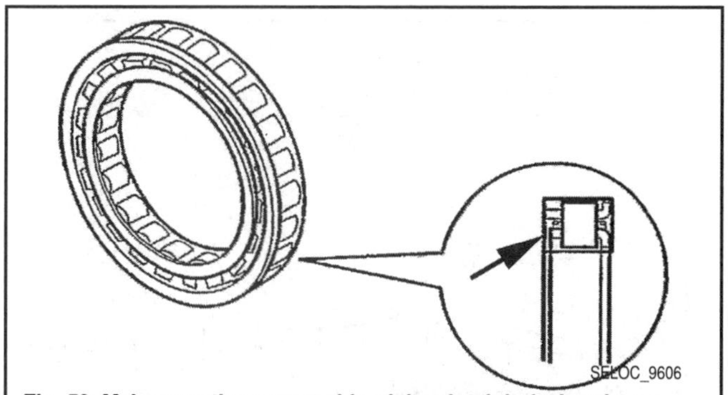

Fig. 59 Make sure the correct side of the clutch is facing the supercharger gear

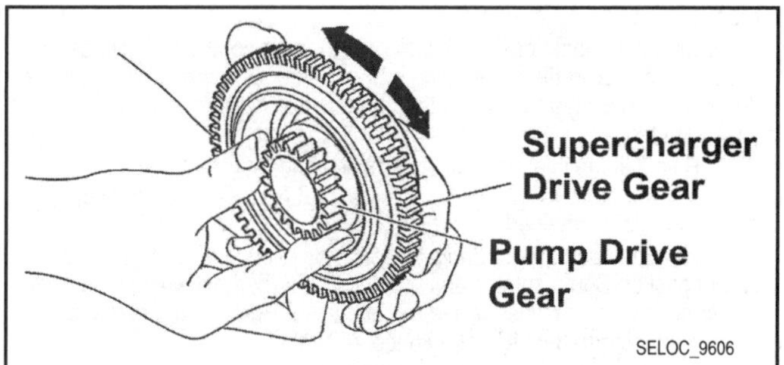

Fig. 60 Checking the drive gear clutch

9. Check the gear teeth for nicks and check the entire drive gear assembly for any other signs of damage.

10. Install the drive gear clutch onto the supercharger hub (as shown). Insert the drive gear (without bearing) into the supercharger gear and clutch. Attempt to rotate the supercharger drive gear counterclockwise while holding the pump drive gear and confirm that it does NOT turn. Now, rotate it clockwise and confirm that it turns smoothly.

11. Use an inner race attachment (#90890-06661) and press the pump gear into the bearing until it seats on the step.

12. Install the drive clutch onto the back of the supercharger drive gear. Install the washer into the bearing recess and then press in the gear/bearing.

13. Install the circlip over the drive gear and then slide the washer over the back side before installing the 2nd circlip.

14. Press the assembly back onto the shaft. Install the holding tool on the flywheel or coupler again. Coat the bolt threads lightly with clean engine oil and thread it into the end of the shaft, tightening to 59 ft. lbs. (80 Nm) while holding the crankshaft with the tool. Remember - COUNTERCLOCKWISE to tighten on this bolt!

15. Lay the pump on a flat surface so the inside is facing up. Add 10W-30 engine oil slowly through the scavenge and feed ports while turning the pump driven gear clockwise. Continue filling until oil just begins to seep out of the lower ports.

16. Coat *new* O-rings with a good water resistant marine-grade grease and slide them onto each connector until they seat in the grooves at each end of the connector. Press the connectors into their respective pump bores until they seat.

17. Lay a *new* gasket on the mating surface and then position the assembly over the dowel pins. Push it into place making sure the connectors feed into the bores in the cylinder block without rolling the O-rings out of their grooves.

18. Coat the bolt threads with Loctite 572 and tighten them alternately to 8.1 ft. lbs./97 inch lbs. (11 Nm) on FX SHO models, or 7.4 ft. lbs./88 inch lbs. (10 Nm) on FZ models.

19. Position the oil pipe and thread in the 2 flange bolts *new* washers. Remember that there is a washer on each side of the flange. Coat the threads of the mount bolt with Loctite 572 and then tighten it to 5.9 ft. lbs./71 inch lbs. (8 Nm). Now go back and tighten the 2 flange bolts to 15.5 ft. lbs. (21 Nm).

20. If you removed the oil filter bolt for some reason, tighten it to 13.3 ft. lbs. (18 Nm).

21. Install the oil cooler.

22. Install the generator cover if removed.

23. Install the engine.

24. Install the oil filter and refill with oil.

Air Cooler & Supercharger

REMOVAL & INSTALLATION

FX SHO & FZ Models w/1812cc Engines

OEM ③ DIFFICULT

◆ See Figures 61 thru 65

1. Remove the engine as detailed in this section. Take a careful look here as you may decide to give this a try without removing the engine but we don't think it's going to work.

2. If you are trying this procedure w/o pulling the engine, remove the air filter housing assembly and it's hoses as detailed in the Fuel System section. If you removed the engine, then this has already been done.

3. Remove the clamp connecting the boost pipe hose to the bottom of the air cooler. Loosen the clamp at the other end of the pipe (on the throttle body). Remove the pipe mounting bolt (don't lose the collar and grommet) and lift off the pipe and hose.

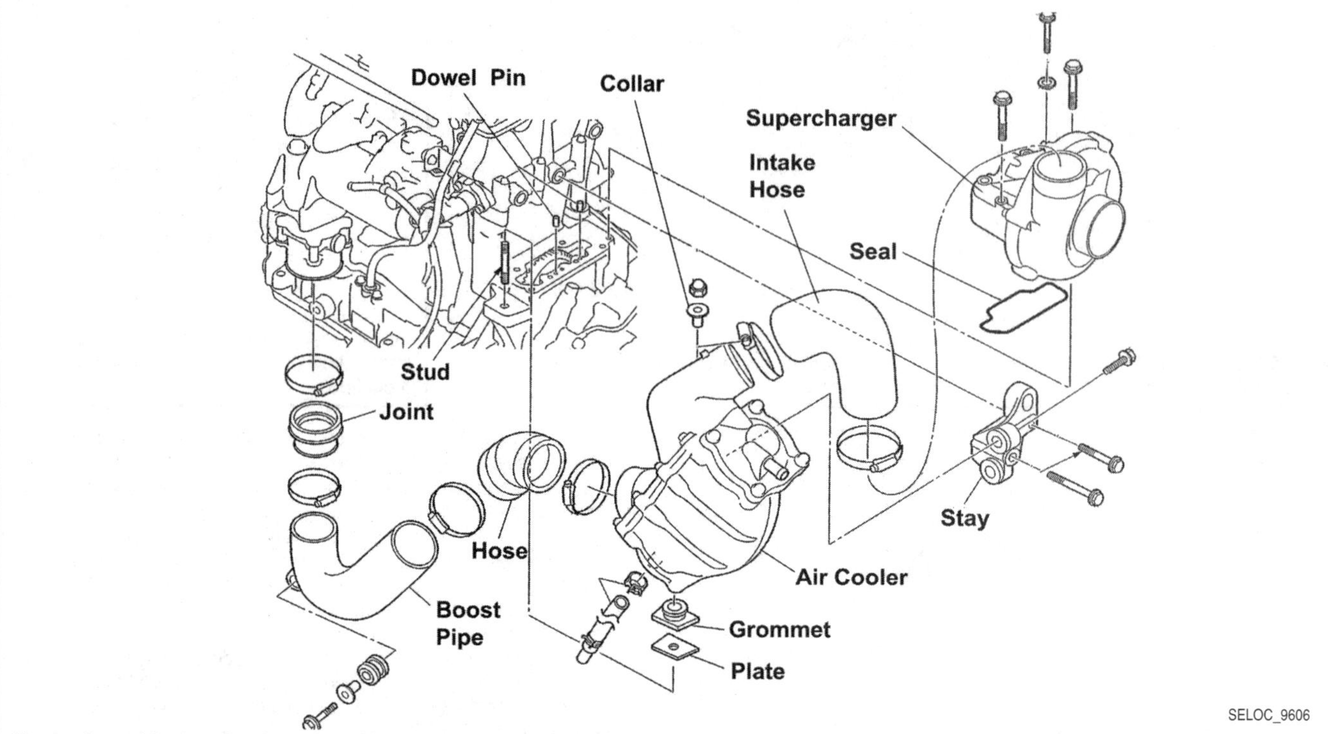

Fig. 61 A good look at the air cooler and supercharger assemblies - FX SHO and FZ models w/1812cc engines

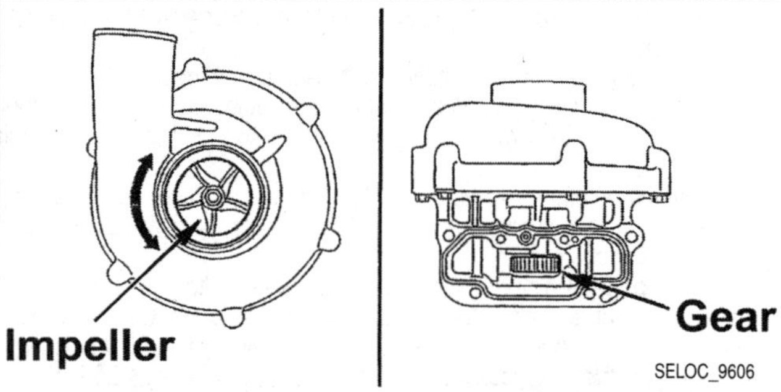

Fig. 62 Checking the impeller and gear on the supercharger

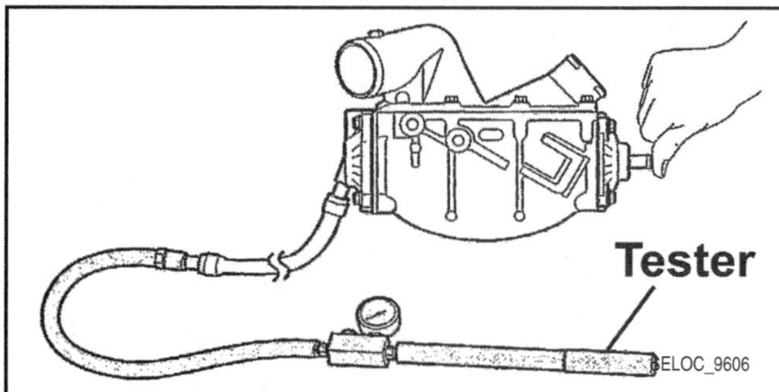

Fig. 63 Checking the water passages. . .

■ **Depending on the extent of the procedures you are performing, you may be able to get away with only loosening the top clamp connecting the hose to the cooler.**

4. Slide the pinch clamp down and wiggle the water hose off of the fitting on the bottom of the cooler. The hose on the top should already have been disconnected during engine removal; if not, remove it as well.
5. Loosen the 2 clamps and remove the intake hose between the cooler and supercharger.
6. Remove the mounting stud nut at the back of the cooler and pull out the collar so it won't get lost in the shuffle.
7. Remove the 2 mounting bolts from the stay and then lift the assembly off of the front of the block. There is a small plate and grommet at the bottom that the stud run through - don't lose them.
8. If necessary, remove the 2 mounting stay bolts and lift off the stay. Be careful, the bolts are two different lengths. . .as you face the front of the engine, the longer one is on the left.
9. Remove the 6 bolts from the top of the supercharger (2 have washers) and then lift the unit from the front of the cylinder block. Pull out the sealing ring and discard it. There are 2 dowel pins and they should stay with their respective bores in the cylinder block mating surface.

To install:

10. Inspect the supercharger and gear for cracks and other obvious signs of damage. Rotate the impeller to make sure it spins smoothly and with no binding.
11. Check the air cooler for cracks and other signs of damage. Check it for leaks:
   a. Connect a leak tester (#90890-06840) to one of the water hose fittings and then plug the other fitting with your thumb. Increase the pressure to 28.5 psi (200 kPa) and confirm that there is no pressure loss after a couple minutes.
   b. Reinstall the inlet and outlet air hoses and then slip on the attachment fittings (#908900-06731). Apply 14.2 psi (100 kPa) and confirm that pressure is maintained.
12. Lay the supercharger on its forward side. Remove the small oil filler bolt and gasket. Squirt some clean engine oil into the hole while turning the impeller so that the shaft is sufficiently lubricated. Reinstall the bolt and gasket, tightening to 3 ft. lbs./36 inch lbs. (4 Nm).
13. Position a new seal into the bottom of the supercharger and then install it onto the block so that it fits over the dowel pins. Coat the bolt threads with Loctite 242 and thread them in until finger-tight. Allow everything to sit for a few minutes and then tighten the bolts with washers (2) to 7.4 ft. lbs./88 inch lbs. and the 4 others to 14.8 ft. lbs. (20 Nm).
14. If removed, install the mounting stay and tighten the bolts to 31 ft. lbs, (42 Nm).
15. Slide the small plate over the air cooler mounting stud so it rests on the block. Press the grommet into the bottom of the cooler and then install the cooler. Slide the top collar over the stud, tighten the nut to 17 ft. lbs. (23 Nm) on the SHO or 16.2 ft. lbs. (22 Nm) on the FZ. Now tighten them again to 31 ft. lbs. (42 Nm).

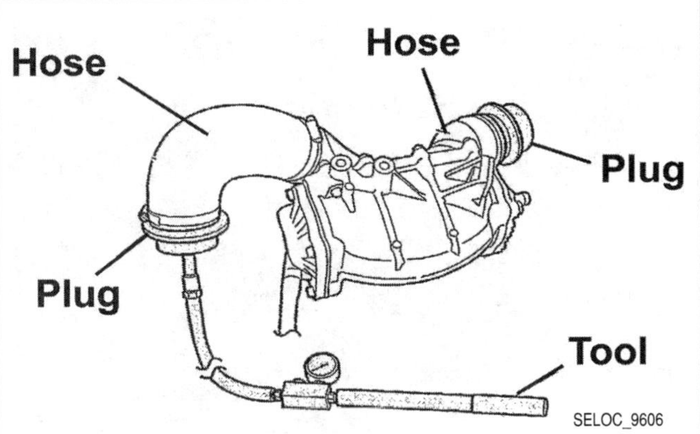

Fig. 64 . . .and the air passages on the air cooler

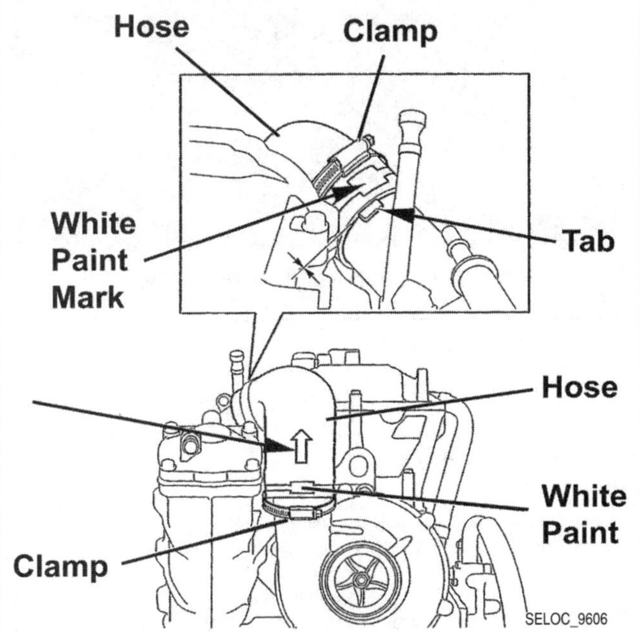

Fig. 65 Installing the intake hose

16. Coat the threads of the 2 stay-to-cooler bolts with Loctite 242 and then tighten them to 17 ft. lbs. (23 Nm) on the SHO or 16.2 ft. lbs. (22 Nm) on the FZ. Now tighten them again to 31 ft. lbs. (42 Nm).

17. Install the intake hose to the cooler and supercharger. Make sure that the arrow is pointing up and away from the supercharger and that the other end of the hose is butted up against the small tab on the air cooler neck. Slide the clamps down into position so that they are riding over the white marks on each end of the hose and that the screw mechanism on the clamp is positioned in the wider portion of the mark. Tighten to 3 ft. lbs./36 inch lbs. (4 Nm).

18. Install the lower water hose to the bottom of the cooler. If you were able to get this far without removing the engine, then do the same with the top hose.

19. Slide the upper hose over the cooler fitting so that the recess in the end of the hose slides over the ridge in the fitting. If you removed the boost pipe and hoses, do the same for the throttle body end and then press the grommet and collar into the bolt hole. Coat the threads with Loctite 242 and tighten the bolt to 14.8 ft. lbs. (20 Nm).

20. Install the engine if removed.

Reduction Drive Gear & Housing

REMOVAL & INSTALLATION

998cc Engines

OEM ③ DIFFICULT

◆ See Figures 41 and 66 thru 69

The reduction drive gear is mounted to the rear of the crankshaft and contained within a housing under the oil tank and pump.

1. Remove the engine, oil tank and oil pump as previously detailed in this section.

2. Remove the generator cover as detailed in the Ignition & Electrical section - don't worry, you'll see why shortly.

3. Working at the front of the engine, connect a sheave holder tool (#YS-01880-A / 90890-01701), or any decent strap wrench, to the flywheel. Now move to the back, install a coupler wrench (#YW-06551 / 90890-06551) so that it meshes with the coupler and remove the drive coupling.

4. Loosen and remove the 9 bolts securing the drive housing to the cylinder block. Carefully remove the housing itself and set it aside - you may need to 'persuade' it, and if so, be very careful. There are 2 dowel pins that might have come out with the cover; make sure that they are secure in their bores in the block.

■ **There are 5 different length bolts securing the housing. It is strongly suggested that you take a digital picture or draw a picture of the housing and then mark each bolt to its respective hole in the housing.**

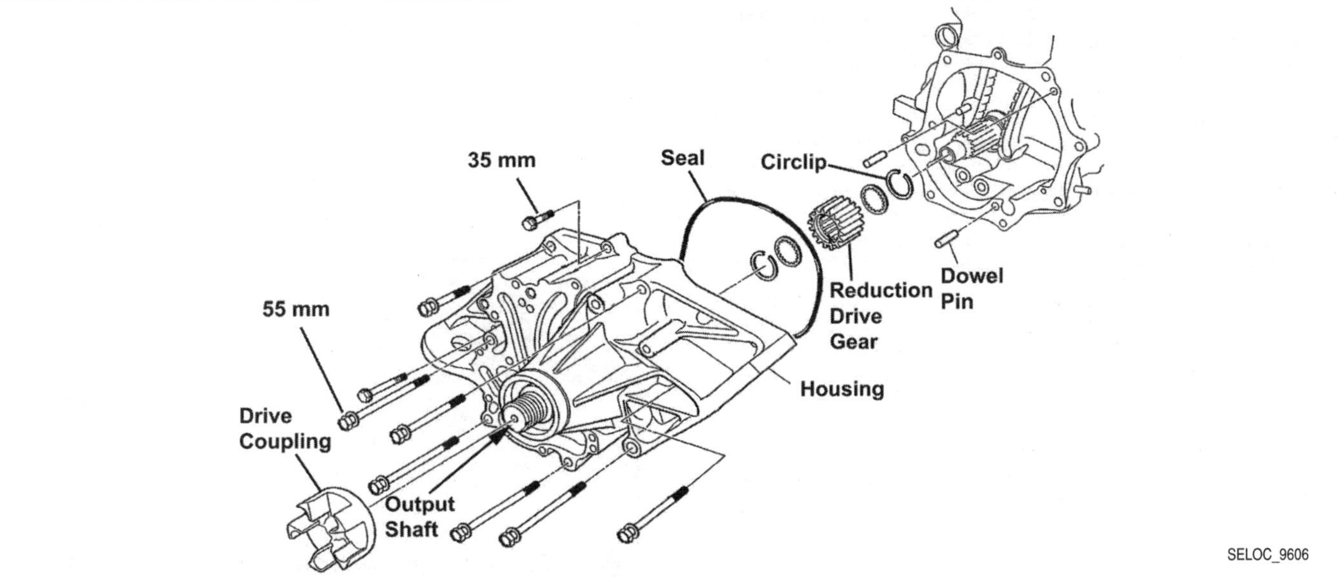

Fig. 66 Removing the reduction drive gear and housing - 998cc engines

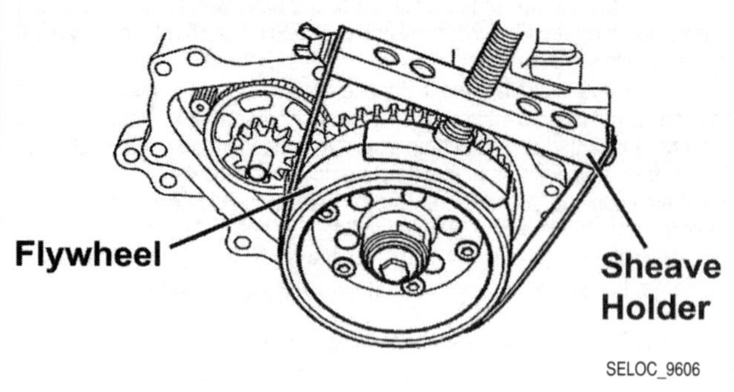

Fig. 67 Hold the flywheel with the special tool. . .

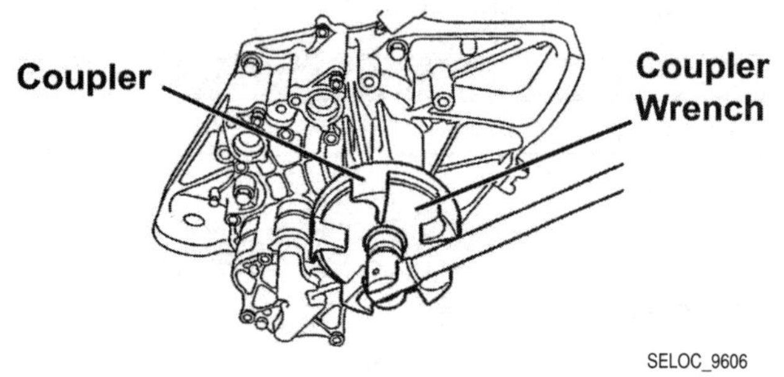

Fig. 68 . . .and then remove the coupler with the coupler wrench - 998cc engines

Fig. 69 The reduction drive gear rides on the splined end of the crankshaft

5. Remove the oil seal from the recess in the inner side of the housing and discard it.

6. Remove the circlip holding the drive gear on the crankshaft. Pull out the washer and then pull off the gear. Reach in and remove the washer and remaining circlip. Throw both of the circlips away.

To install:

7. Check the housing for cracks or other signs of damage. Check the gear for burrs or cracked teeth.

8. Slide a *new* circlip over the crankshaft until it is seated in its groove (inner) on the crank. Position the washer correctly on the splines of the crankshaft and slide it into position against the inner circlip.

9. Lubricate the inner and outer splines of the drive gear lightly with clean engine oil and the slide it onto the crankshaft and up against the inner washer. Slide the outer washer over the crankshaft splines until it is against the outer edge of the gear and then install a *new* outer circlip, making sure that it is fully seated in the groove.

10. Carefully press a *new* seal/gasket into the recess on the inner side of the cover. Move the housing into position and install it onto the block so the bolt holes and dowel pins align correctly. You will probably have to wiggle the output shaft back and forth a bit to get the drive gear to mesh with the driven gear - temporarily installing the coupling could make this a little easier. If you decide not to do this though, be very careful with the threads of the shaft and DO NOT attempt to rotate the shaft using pliers!

11. Once the housing is fully seated on the cylinder block, coat the bolt threads with Loctite 572 and thread them in. Tighten all bolts, alternately and in 2 passes, to 11 ft. lbs. (15 Nm) and then again to 20 ft. lbs. (28 Nm). EXCEPT the 2 bolts marked as 35mm and 55mm in the illustration. These two get tightened to a 1st pass of 2.7 ft. lbs./33 inch lbs. (3.7 Nm), and then to 5.5 ft. lbs./66 inch lbs. (7.6 Nm).

12. Coat the threads of the output shaft with Loctite 572 and then thread on the drive coupling until finger-tight. Install the holding tool on the flywheel and the coupler wrench to the coupling. Tighten the coupling to 20 ft. lbs. (28 Nm).

13. Install the generator cover.

14. Install the oil pump, oil tank and engine.

1052cc Engines

OEM ③ DIFFICULT

◆ See Figures 41 and 69 thru 73

The reduction drive gear is mounted to the rear of the crankshaft and contained within a housing under the oil tank and pump.

1. Remove the engine, oil tank and oil pump as previously detailed in this section.

2. Loosen and remove the 7 bolts securing the drive housing to the cylinder block. Carefully remove the housing itself and set it aside - you may need to 'persuade' it, and if so, be very careful. There are 2 dowel pins that might have come out with the cover; make sure that they are secure in their bores in the block.

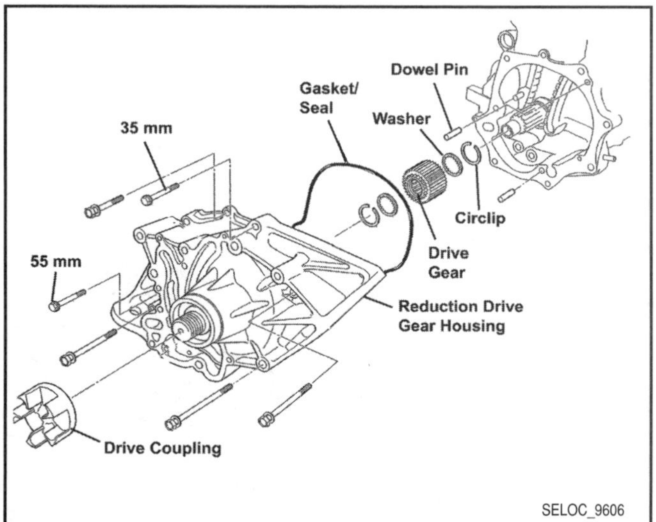

Fig. 70 Removing the reduction drive gear and housing - 1052cc engines

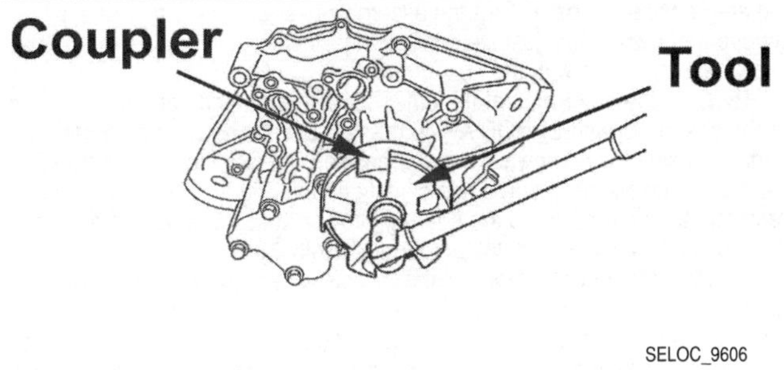

Fig. 71 Hold the drive shaft with the special tool. . .

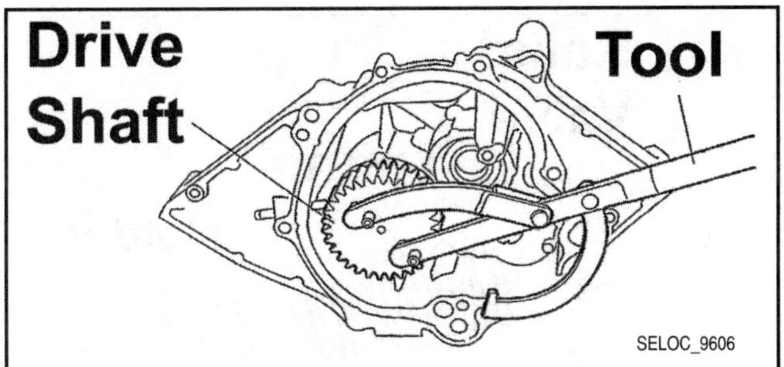

Fig. 72 . . .and then remove the coupler with the coupler wrench - 1052cc engines

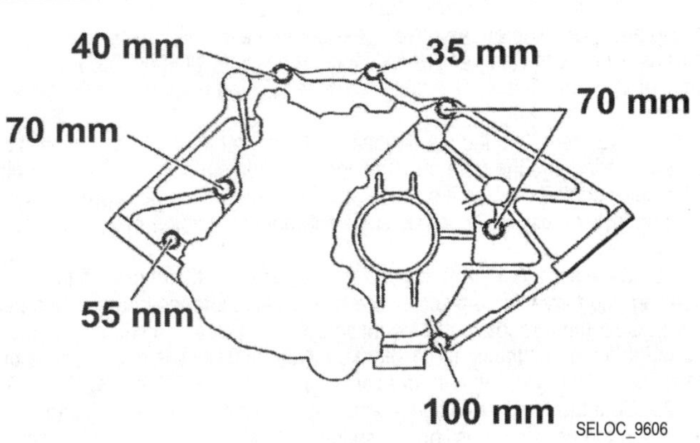

Fig. 73 Use this diagram for proper bolt relocation when installing the pump

■ **There are 5 different length bolts securing the housing. It is strongly suggested that you take a digital picture or draw a picture of the housing and then mark each bolt to its respective hole in the housing.**

3. Remove the oil seal/gasket from the recess in the inner side of the housing and discard it.

4. Remove the circlip holding the drive gear on the crankshaft. Pull out the washer and then pull off the gear. Reach in and remove the washer and remaining circlip. Throw both of the circlips away.

5. Connect a rotor holder tool (all exc. 2008 FX HO and 2010 VX - #YU-01235 / 90890-01235; 2008 FX HO and 2010 VX - #YB-06139 / 90890-06522) to the drive shaft. Now, install a coupler wrench (#YW-06551 / 90890-06551) so that it meshes with the coupler and remove the drive coupling from the end of the shaft.

To install:

6. Check the housing for cracks or other signs of damage. Check the gear for burrs or cracked teeth.

7. Coat the threads of the drive shaft with Loctite 572. Spin the drive coupler onto the shaft and then install it using the same tools from removal, tightening to 20 ft. lbs. (28 Nm). On the 2010 VX, tighten it to 21.4 ft. lbs. (29 Nm).

8. Slide a *new* circlip over the crankshaft until it is seated in its groove (inner) on the crank. Position the washer correctly on the splines of the crankshaft and slide it into position against the inner circlip.

9. Lubricate the inner and outer splines of the drive gear lightly with clean engine oil and the slide it onto the crankshaft and up against the inner washer. Slide the outer washer over the crankshaft splines until it is against the outer edge of the gear and then install a *new* outer circlip, making sure that it is fully seated in the groove.

10. Carefully press a *new* seal into the recess on the inner side of the cover. Move the housing into position and install it onto the block so the bolt holes and dowel pins align correctly. You will probably have to wiggle the output drive shaft (use the drive coupling) back and forth a bit to get the drive gear to mesh with the driven gear.

11. Once the housing is fully seated on the cylinder block, coat the bolt threads with Loctite 572 and thread them in. Tighten all bolts, alternately and in 2 passes, to 11 ft. lbs. (15 Nm) and then again to 20 ft. lbs. (28 Nm) - ok, 21.4 ft. lbs. (29 Nm) n the 2010 VX. EXCEPT the 2 bolts marked as 35mm and 55mm in the illustration. These two get tightened to a 1st pass of 2.7 ft. lbs./33 inch lbs. (3.7 Nm), and then to 5.5 ft. lbs./66 inch lbs. (7.6 Nm). On 2010 VX models, tighten 1st to 3 ft. lbs./36 inch lbs. (4 Nm) and then to 5.9 ft. lbs./71 inch lbs. (8 Nm).

12. Install the oil pump, oil tank and engine.

DISASSEMBLY & ASSEMBLY

998cc Engines

OEM ③ DIFFICULT

◆ **See Figures 41, 69, 74, 75 and 76**

1. Remove the reduction drive gear housing.

2. Grab the oil pump driven gear and pull it out of the recess in the housing.

3. Remove the relief valve mounting bolt and lift out the holder, spring and valve.

4. Loosen the 4 mounting bolts (Allen) and pull off the bearing housing. Keep track of the 2 dowel pins. If necessary, remove the circlip retaining the bearing and then press out the bearing from the other side using a small punch or socket. Once removed the bearing will not be able to be re-used; nor will the circlip.

5. Reach inside the opening of the driven gear and remove the circlip. Slide off the splined washer and then remove the driven gear from the output shaft. Slide off the inner splined washer.

6. Remove the 2 small circlips from their grooves in the output shaft. Do not remove the larger circlip retaining the inner shaft bearing.

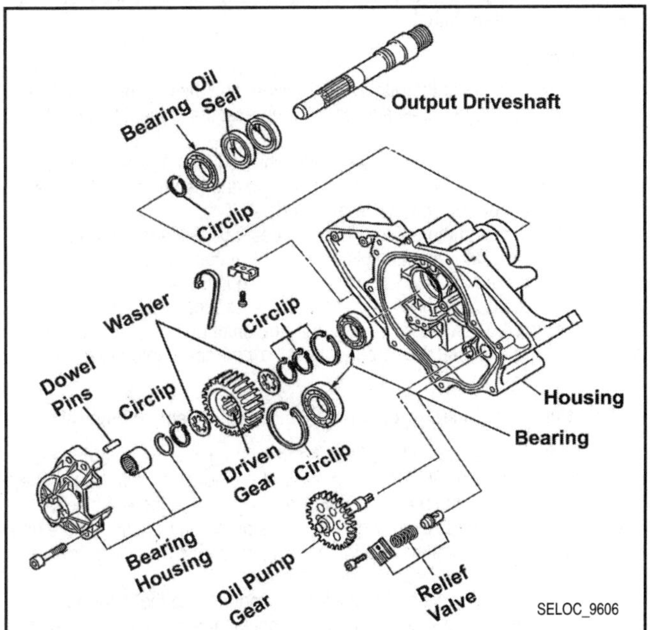

Fig. 74 Exploded view of the reduction drive gear housing components - 998cc engines

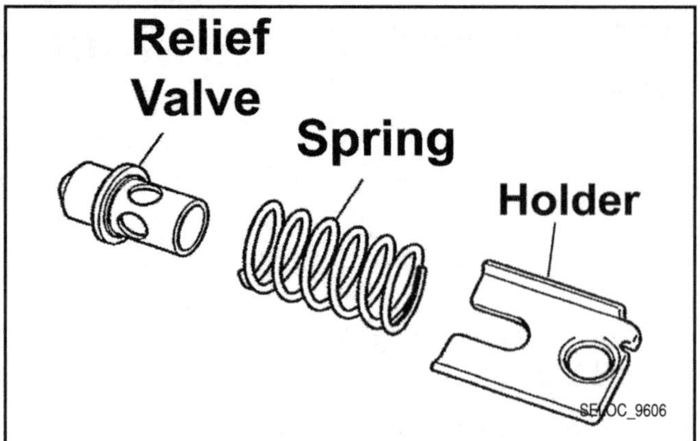

Fig. 75 A good look at the relief valve - 998cc engines

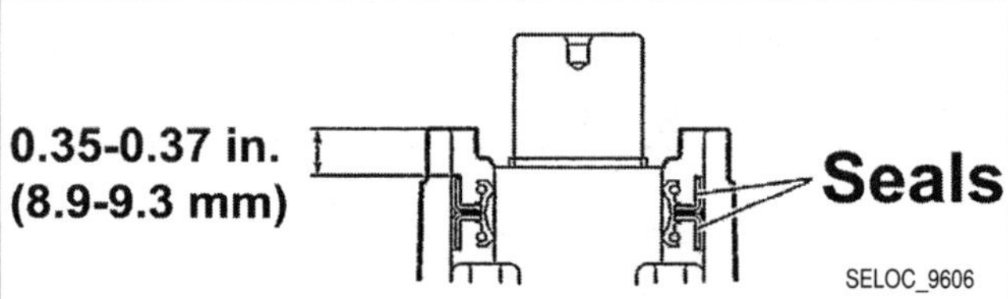

Fig. 76 Installing the 2 bearing oil seals - 998cc engines

■ **There is no reason to perform the remaining procedures unless necessary. Bearing, circlips and oil seals are NOT reusable. If you are not removing the bearings, skip to Step 17 of the installation procedure.**

7. Remove the large circlip retaining the large bearing the end of the crankshaft rides in. Install a suitable slide hammer and pull the bearing out of the housing. We would suggest standing the housing on end and having a friend hold the housing down while removing the bearing.
8. Remove the large circlip securing the inner output shaft bearing. Position the housing in a press (inside facing upward) and press out the output shaft along with the outer bearing and 2 seals. Use a slide hammer and remove the inner shaft bearing as you did with the crank bearing in the previous step.
9. Position the shaft assembly in a press again, and, using a bearing removal tool, press the outer bearing off of the shaft. Remove the 2 seals if they did not come off with the bearing.

To assemble:

10. Clean everything thoroughly with solvent and allow to dry completely prior to assembly. If you have access to compressed air, use it to air in the drying.
11. Inspect everything for cracks, wear, chipped teeth/splines or other signs of damage. Remember that seals, circlips and bearings are NOT re-usable and must be replaced once removed.
12. If you removed the output shaft and bearings, slide the *new* outer bearing onto the shaft from the splined end until it cannot go any further. Make sure that the numbered side of the bearing is facing toward the back of the shaft. Position assembly in a press with the bearing tool in place again and then press the bearing down and onto the shaft until the full width of the circlip groove is exposed. Once done, install a *new* circlip into the groove on the output shaft.

■ **Coat the actual bearing roller balls lightly with clean engine oil.**

13. Coat the outer race of the inner shaft bearing and the inner bore of the shaft hole with EPNOC AP #0 grease. Set the housing on a flat surface, inside facing upward and the position the bearing into the bore. Select a large socket that will ride cleanly on the edges of the outer bearing race and then carefully tap the bearing into the housing bore until the circlip groove is fully exposed. Install a *new* circlip into the groove.
14. Coat the outer race of the outer shaft bearing and the inner bore of the shaft hole with EPNOC AP #0 grease. Position the housing, inner side down, on the press, insert the splined side of the shaft through the bore and the center of the inner bearing until the outer bearing is riding in the bore. Use the press to force the shaft into the cover until the bearing is seated on the step onside the bore. Turn the assembly over and confirm that the groove in the shaft has just cleared the inner bearing - if so, install a *new* circlip. Do the same with the next groove.
15. Coat the outer race of the outer crankshaft bearing and the inner bore of the shaft hole with EPNOC AP #0 grease. Set the housing on a flat surface, inside facing upward and the position the bearing into the bore. Select a large socket that will ride cleanly on the edges of the outer bearing race and then carefully tap the bearing into the housing bore until the circlip groove is fully exposed. Install a *new* circlip into the groove.
16. Coat the inner surfaces of the 2 seals with a good quality marine grease. Position the inner seal in the bore with the lip facing down and press it into partially into position using a seal installation tool or a suitable sized socket. Now position the outer seal so the lip is facing outward (away from the other seal) and use the tool or socket to press them both into the bore until the edge of the outer seal is 0.35-0.37 in. (8.9-9.3mm) *below* the outer edge of the housing.
17. Slide the inner spline washer over the output shaft until it contacts the circlip and then slide on the driven gear after coating the inner and outer splines lightly with clean engine oil. Once in position, slide on the outer washer and then install the circlip so it rides fully in the groove in the output shaft.
18. If you removed the needle bearing from the bearing housing, position a *new* bearing into the bore after coating the outer surface and inner needles with clean engine oil. Use a suitable sized socket and carefully press the bearing into the cap until there is sufficient room to install a *new* circlip into the groove.
19. Make sure that the 2 dowel pins are seated in their bores in the housing and move it into place so the dowels slide into the bores on the gear cover. Coat the threads of the mounting bolts with clean engine oil and then thread them in, tightening them alternately and in 2 passes to a 1st pass of 11 ft. lbs. (15 Nm). Tighten them again to a final pass of 20 ft. lbs. (28 Nm).
20. Coat the relief valve lightly with clean engine oil and slide it into the recess in the housing. Position the spring and then install the holding plate, tightening the bolt to 7.2 ft. lbs./87 inch lbs. (10 Nm).
21. Coat the oil pump gear shaft lightly with clean engine oil and install it in the housing until it seats fully.
22. Install the drive gear housing.

1052cc Engines

OEM ③ DIFFICULT

◆ See Figures 41, 69 and 77 thru 80

1. Remove the reduction drive gear housing.
2. Grab the oil pump drive gear and pull it out of the recess in the housing.
3. Position the housing on a press so that the threaded end of the drive shaft is facing up toward the press. Cover the end of the shaft with a small block of wood and the press the shaft out of the housing.
4. Remove the circlip holding the crankshaft drive gear bearing. On US and Canada models, insert a slide hammer (#YB-06096, or equivalent) and remove the bearing. Other models should use a bearing puller assembly (#90890-06535).
5. Remove the circlip retaining the inner drive shaft bearing. On the other side of the housing, remove the collar and 2 oil seals from the shaft bore.
6. Lay the housing back on the press again, open side down. Select a socket that is just large enough to sit on the bearing race without scratching the walls of the bore and press the outer bearing, sleeve and inner bearing out of the housing. This process will destroy the bearing, so you can discard them.
7. You could use an actual bearing removal tool, but we think the socket works just fine.

To assemble:

8. Clean all components thoroughly and allow to dry completely before installation. Use compressed air if available.
9. Inspect all components for cracks, wear or other damage. Inspect the gears for damaged or worn teeth.

■ **Prior to installing the bearings, be sure to coat the outer race thoroughly with EPNOC grease AP #0.**

10. Move your housing back onto the press so the open side is facing up. Position a *new* drive shaft inner bearing over the bore with the numbers

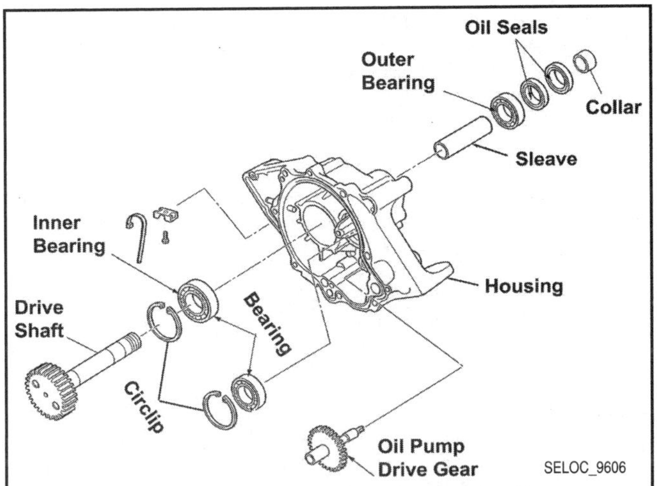

Fig. 77 Exploded view of the reduction drive gear housing components -1052cc engines

Fig. 78 A good look at the reduction drive gear housing and components - 1052cc engines

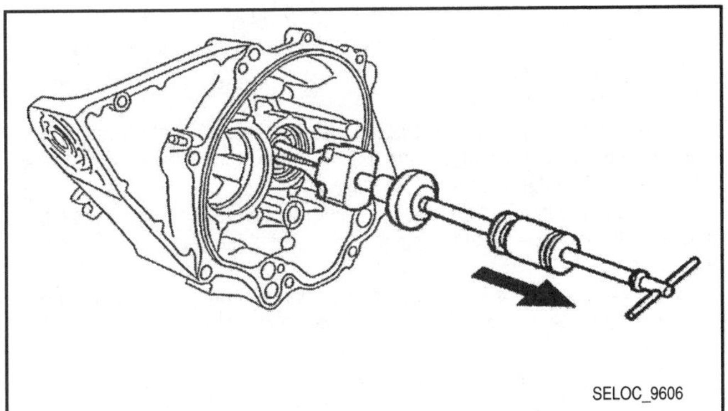

Fig. 79 When removing the drive gear bearing, use a slide hammer in the US and Canada. . .

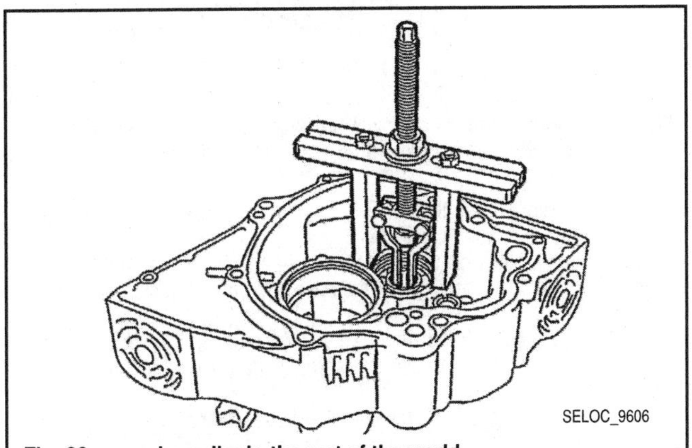

Fig. 80 . . .and a puller in the rest of the world

facing out of the housing (toward you). Use your trusty socket and press the bearing back into the bore until the edge of the bearing is just past the snap ring groove. Install a *new* snap ring.

11. While you've got the housing in this position, you might as well press in a *new* drive gear bearing as well. Install a *new* snap ring.

■ **Before installing the sleeve and outer bearing, it is imperative that you support the inner bearing first. It has been suggested that you find a piece of pipe with an I.D. of 1.18 in. (30mm) and an O.D. of 2.76 in. (70mm), and then cut it to a length of 1.57 in. (40mm) so as to support the inner and outer races of the bearing. We're not completely sure where one might find such a unique piece of pipe, so we suggest that you use one of the old bearings and then add a block of wood or shims underneath it to bridge the gap between it and the press surface; thus supporting the bearing while installing the sleeve and other bearing.**

12. Now flip the housing over and slide the sleeve into the bore until it rests on the inner bearing race. Position a *new* drive shaft outer bearing over the bore with the numbers facing out of the housing (toward you). Use your trusty socket and press the bearing back into the bore until the edge of the bearing's inner race is just touching the upper edge of the sleeve.

13. Use another socket or blunt drift and press a *new* inner seal into the bore until the lip contacts the bearing. Press the outer seal in until it contacts the inner seal. Make sure the seal lip is facing out and that the lip is 0.36-0.38 in. (9.1-9.5mm) BELOW the outer edge of the bore.

14. Flip the housing back over so the open side is facing up and then insert the drive shaft through the bore. Protect the shaft gear with a block of wood or something similar and then press it in until the step on the shaft just under the gear comes in contact with the inner bearing race.

15. Ok, flip it over again and support the gear with a block(s) of wood thick enough the housing is just off of the table. You'll need to find/fabricate a 1.18 in. (30mm) length of pipe with an I.D. of 1.08-1.12 in. (23-32mm) and an O.D. of 1.97 in. (35mm) and then use this to press the collar in until it seats.

16. Coat the oil pimp drive shaft lightly with clean engine oil and then install it into the housing.

17. Install the housing.

Generator/Flywheel Cover

REMOVAL & INSTALLATION

OEM ③ DIFFICULT

■ **Please refer to the Ignition & Electrical section for complete details on the generator/flywheel cover.**

Cylinder Head Cover

REMOVAL & INSTALLATION

998cc & 1052cc Engines

② MODERATE

◆ **See Figure 81**

■ **The cylinder head cover is often called the camshaft cover.**

1. Remove seat and disconnect the battery cables.

2. Remove the air filter and case on FX models with 998cc/1052cc engines as detailed in the Fuel System section.

3. On 998cc engines, tag and disconnect the spark plug boots, moving them out of the way. Do the same on 1052cc engines, but loosen the bolt and remove the ignition coils instead. In most instances, the wires will

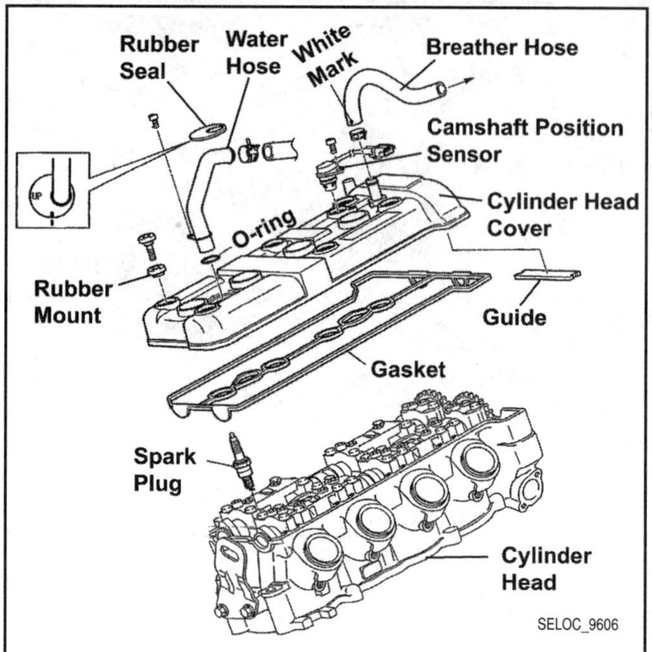

Fig. 81 Exploded view of the cylinder head/camshaft cover and components - 998cc engines, 1052cc engines very similar

already be marked, but tag them if not. Remove the spark plugs as detailed in the Maintenance & Tune-Up section.

4. Loosen the pinch clamp for the breather hose at the rear of the cover and wiggle it off of the fitting. Check the hose end under the clamp for any cracks.

5. Loosen the camshaft position sensor bolt and pop the sensor out of the cover. It isn't necessary to disconnect the electrical lead, just move it out of the way.

6. Loosen the pinch clamp for the water hose toward the front of the cover and wiggle it off of the fitting pipe. Check the hose end under the clamp for any cracks.

7. Remove the bolt and pop the water pipe out of the recess in the cover. Remove the rubber seal and pull out the O-ring.

8. Loosen and remove the 6 cover mounting bolts working alternately from the center bolts outward. Pull out the small rubber grommets in the bolt hole recesses. Lift off the cylinder head cover. Remove the gasket and timing chain guide (at the rear in the cover).

To install:

9. Inspect the cover for cracks, wear or other signs of damage. Make sure that all residual gasket material has been removed from the mating surfaces of the head and cover.

10. Check the rubber grommets carefully for cracks, tears or other signs of deterioration. If ok, pop them back into their respective bores in the cover.

11. Apply a thin bead of Gasket Maker along the front and rear mating surface of the cylinder head, making sure that you coat the 4 recesses thoroughly. Position a *new* gasket on the head and confirm that all bolt holes line up and that the tabs on the gasket are fully seated in each of the recesses on the head.

12. Insert the guide into the rear of the cover and then move the cover onto the head and check that the holes all line up and the gasket has stayed in position. Coat the bolts threads with clean engine oil and then thread them all in. Working in a criss-cross pattern, tighten each bolt to 8.7 ft. lbs./104 inch lbs. (12 Nm). On the 2010-11 VX, tighten to 8.9 ft. lbs./107 inch lbs. (12 Nm) - yes, we know it's the same, but try to reflect factory specs as accurately as possible.

13. Slide a *new* O-ring over the bottom of the water pipe and slide it up until it meets the bottom of the mounting flange. Press the pipe into the cover, coat the bolt threads with Loctite 572 and then tighten it to an initial torque of 2.7 ft. lbs./33 inch lbs. (3.7 Nm). Wait a moment and then tighten it to 5.5 ft. lbs./66 inch lbs. (7.6 Nm). On 2010-11 VX models, 3 ft. lbs./36 inch lbs. (4 Nm). Wait a moment and then tighten it to 5.9 ft. lbs./71 inch lbs. (8 Nm).

14. Slide the rubber seal over the end of the pipe and all the way down against the cover. Swivel the seal so that the small mark on it is lined up with the mark on the cylinder head cover. Needless to say, the **UP** mark should be facing. . .*up*.

15. Press the camshaft position sensor into the recess in the cover and tighten the bolt to 7.2 ft. lbs./86 inch lbs. (10 Nm) after coating the threads with Loctite 572. 7.4 ft. lbs./88 inch lbs. on the 2010-11 VX.

16. Reconnect the water hose to the pipe and slide on the clamp, making sure that the hose is on far enough that the clamp is fully inside the ridge on the end of the pipe. Do the same with the breather hose and make sure that the white mark on the hose is in the position shown in the illustration.

17. Install the spark plugs and then press in the boots (or coils) - make sure that they all go back onto the correct plug. On 1052cc engines, tighten the coil bolt to 5.5 ft. lbs./66 inch lbs. (7.6 Nm); except for the 2010-11 VX, where it should be 5.9 ft. lbs./71 inch lbs. (8 Nm).

18. Install the air filter housing and filter. Connect the battery cables and install the seat.

1812cc Engines

② MODERATE

◆ **See Figures 82 and 83**

■ **The cylinder head cover is often called the camshaft cover.**

1. Remove seat and disconnect the battery cables.

2. Remove the engine cover as detailed in the Fuel System section.

3. If necessary, remove the holder bolt and collar/grommet; there are 2 on the port side.

4. Loosen the bolts and remove the ignition coils. In most instances, the wires will already be marked, but tag them if not. Remove the spark plugs as detailed in the Maintenance & Tune-Up section.

5. Loosen the pinch clamp for the breather hose at the front port side of the cover and wiggle it off of the fitting. Check the hose end under the clamp for any cracks.

6. Loosen the camshaft position sensor bolt and pop the sensor out of the cover. It isn't necessary to disconnect the electrical lead, just move it out of the way.

7. Loosen the pinch clamp for the water hose at the back of the cover and wiggle it off of the fitting pipe. Check the hose end under the clamp for any cracks.

8. Remove the oil fill cap and lift out the rubber seal.

9. Loosen and remove the 8 cover mounting bolts working alternately from the center bolts outward. Lift off the cylinder head cover. Remove the gasket and timing chain guide (at the front in the cover).

To install:

10. Inspect the cover for cracks, wear or other signs of damage. Make sure that all residual gasket material has been removed from the mating surfaces of the head and cover.

11. Apply a thin bead of ThreeBond 1280B along the 2 recesses in the front surface of the cylinder head, making sure that you coat them thoroughly. Position a *new* gasket on the head and confirm that all bolt holes line up and that the tabs on the gasket are fully seated in each of the recesses on the head.

12. Insert the guide into the front of the cover and then move the cover onto the head and check that the holes all line up and the gasket has stayed in position. Coat the bolts threads with Loctite 572 and then thread them all in. Working in a criss-cross pattern, tighten each bolt to 5.9 ft. lbs./71 inch lbs. (8 Nm).

13. Press the camshaft position sensor into the recess in the cover and tighten the bolt to 7.4 ft. lbs./88 inch lbs. (10 Nm) after coating the threads with Loctite 572.

14. Reconnect the water hose to the pipe and slide on the clamp, making sure that the hose is on far enough that the clamp is fully inside the ridge on the end of the pipe. Do the same with the breather hose and make sure that the white mark on the hose is in the position shown in the illustration.

15. Install the spark plugs and then press in the coils - make sure that they all go back onto the correct plug. Coat the threads with Loctite 572 and then tighten the coil bolts to 5.9 ft. lbs./71 inch lbs. (8 Nm).

16. Install a new seal in the oil filler hole and then thread the cap back in.

17. Install the collar/grommets if removed, coat the bolt threads with Loctite 242 and tighten the 2 bolts to 4.4 ft. lbs./53 inch lbs. (6 Nm).

18. Install the engine cover.

19. Connect the battery cables and install the seat.

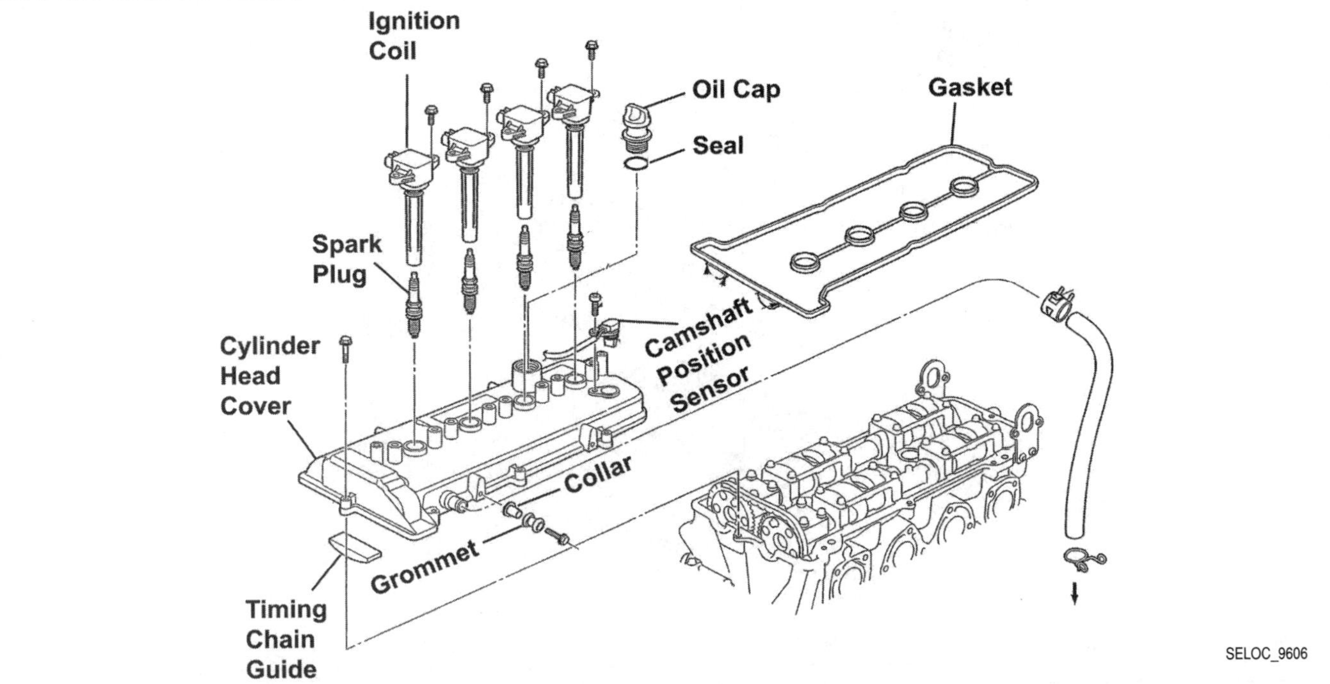

Fig. 82 Exploded view of the cylinder head/camshaft cover and components - 1812cc engines

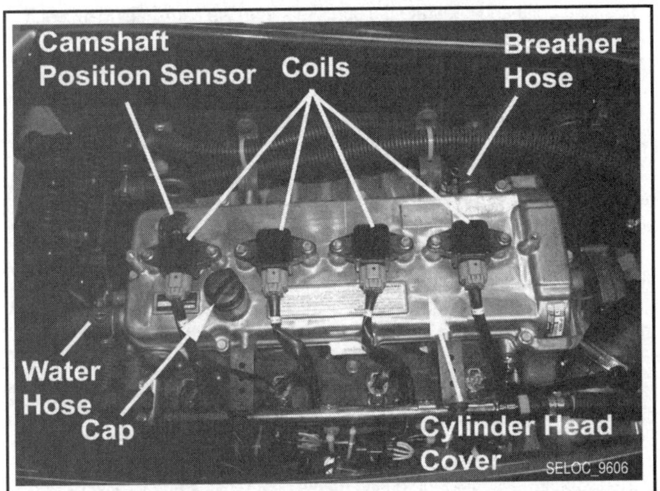

Fig. 83 A good look at the cylinder head cover - 1812cc engines

Camshafts & Sprockets

REMOVAL & INSTALLATION

998cc & 1052cc Engines

OEM ③ DIFFICULT

◆ See Figures 84 thru 95

■ **All Yamaha engines covered here are equipped with dual overhead camshafts (DOHC).**

1. Remove the cylinder head cover as detailed previously in this section.
2. Remove the drive coupling cover.
3. Position the No. 1 piston so it is at TDC of the compression stroke for the No. 1 cylinder (first cylinder at the front of the engine). To do this you could use a remote starter that has been hooked up according to the manufacturer's instructions, or you can simply rotate the engine coupler; either way, the object is to rotate the crankshaft so that the piston in the No.1 cylinder gets to TDC. We think it's easier, and certainly cheaper, to use the coupler.

With your finger over the No.1 cylinder's spark plug hole (the first one), have a friend *slowly* rotate the coupler counterclockwise until you detect pressure in the cylinder, indicating TDC on the compression stroke. You can also check the camshaft lobe positions - they will be pointing away from each other.

4. Thread a dial gauge stand (#90890-06583) into the spark plug hole on the No. 1 cylinder and then install the gauge into the stand. You may also need a gauge needle (#90890-06584, or you can get the set (#90890-06585). Use the gauge to confirm your TDC positioning.
   a. Keeping your eye on the dial gauge indicator, have your friend rotate, *very slowly*, the coupler counterclockwise again until you see the indicator just begin to move in the opposite direction.
   b. Now rotate again in the opposite direction while watching the needle and stop as it just changes direction again.
   c. Do this a few times until you are confident that you have pinpointed that point where the needle changes direction at the first turn of the coupler in either direction.
   d. Make sure that the intake and exhaust cam lobes on the No. 1 cylinder are again pointing away from each other. If not, go back and start over again until you have achieved both confirmations.
   e. Now you can be confident you have found TDC. Use a magic marker and place a small mark on the top of the coupler in case you need to do this again during the procedure.
5. Lock the timing chain tensioner rod and then remove the timing chain tensioner cap bolt and throw away the small O-ring. Remove the 2 mounting bolts and pull the tensioner body out of the cylinder head. Remove the gasket and discard it.
6. Take a Sharpie marker and carefully make a mark across the camshaft sprocket and timing chain; and then across the sprocket and camshaft. Do this for both sprockets. While you've got the marker out, its not a bad idea to mark each camshaft cap as well. Mark each with **IN** or **EX** and then also **F1** (front), **F2** (center) and **F3** (rear) as they all must go back on in the exact same positions as they were removed from. We know you could carefully line them up in order somewhere on the floor, but why take a chance that someone bumps them?

■ **The 4 larger caps will usually have an arrow cast onto them - be careful as this is an alignment arrow and actually points to the rear of the engine when installed, NOT to the front.**

7. Remove each of the 10 exhaust cap mounting bolts in the sequence shown. Lift off each cap and set them aside. There will be 6 dowel pins; make sure they stay in their bores in the cylinder head.

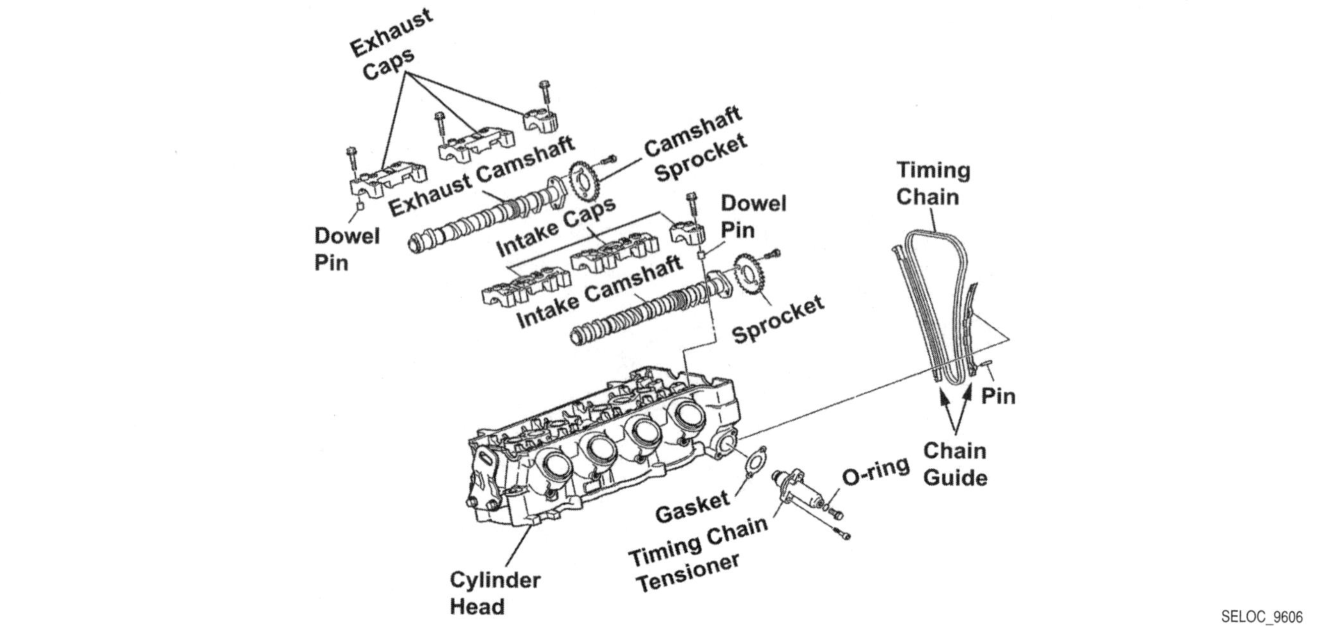

Fig. 84 Exploded view of the camshafts and components - 998/1052cc engines

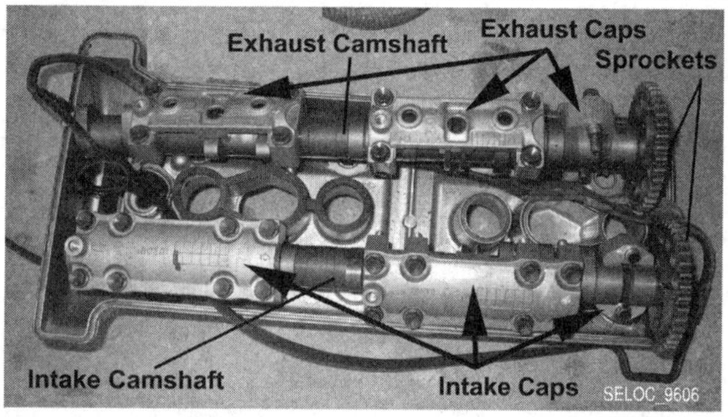

Fig. 85 A closer look at the camshaft components - 998/1052cc engines

Fig. 86 Remove the bolt and lift off the coupling cover - 998/1052cc engines

8. Remove each of the 18 intake cap mounting bolts in the sequence shown. Lift off each cap and set them aside. There will be 6 dowel pins; make sure they stay in their bores in the cylinder head.

9. Tie a piece of string, or wrap a length of wire, around the timing chain between the 2 sprockets and secure this to another part of the engine in order to keep the chain from falling down into the engine.

10. Lift the chain off of the sprocket and remove the intake camshaft. Do the same with the exhaust shaft. Make sure you have sufficient tension on the wire to keep the chain from dropping back into the crankcase.

11. If necessary, install the special holding tool (#YU-01235 / 90890-01235, a flywheel tool; or #90890-06724 on the 2010-11 VX) and loosen the 2 mounting bolts and separate the sprockets from the camshaft. Mark each sprocket as to which camshaft it goes to.

To install:

12. Clean all components thoroughly and inspect them as detailed in the Inspection section.

13. Very carefully check that the #1 piston is still at TDC, although it should be unless the crankshaft has been rotated for some reason. Either way, confirm it to be safe.

14. If you removed either of the camshaft sprockets, coat the bolt threads lightly with clean engine oil and install them while using the holding tool. Make sure that when positioned on the shaft, the punch mark on the sprocket is facing outward and the marks made during removal are lined up before tightening the bolts to 17 ft. lbs. (24 Nm).

15. Lift up on the timing chain and position the intake camshaft sprocket under the chain. Move the cam until the marks made previously line up and then drape the chain over the sprocket teeth. Lower the camshaft into the lower bearing bores and confirm that the chain-to-sprocket marks are still aligned. Confirm also that the punch marks on the camshaft are visible and facing upward.

16. Repeat the previous step for the exhaust camshaft and then position the caps over the camshafts so that they are in the correct locations from which they were removed. Make sure that the arrow on the rear of the forward intake cap and the center exhaust cap is pointing directly at the small punch mark in each camshaft. If the arrow and marks do not line up, check to make sure your chain-to-sprocket marks are still aligned. If they are, then you'll need to go back and reset the engine to TDC again and make another try at installation.

17. Coat all of the cap mounting bolt threads lightly with clean engine oil and then thread them all in until finger-tight. Now tighten them again in a few stages to 7.2 ft. lbs./45 inch lbs. (10 Nm) using the sequence shown in the illustration. There is no spec for the interim passes, just tighten them a little at a time until the final pass is made.

18. Grab the timing chain tensioner and hold the tensioner rod all the way in with your finger. Insert a thin screwdriver and turn the tensioner rod clockwise while keeping pressure on the rod until it has fully set. Hold the screwdriver in the tensioner and have someone else position a *new* gasket on the mounting flange while you insert the plunger and install the unit with the screwdriver still in place. The *UP* mark should be facing. . .up. Tighten the bolts to 7.2 ft. lbs./45 inch lbs. (10 Nm) and then remove the thin screwdriver. Once the screwdriver is removed, the tensioner rod should

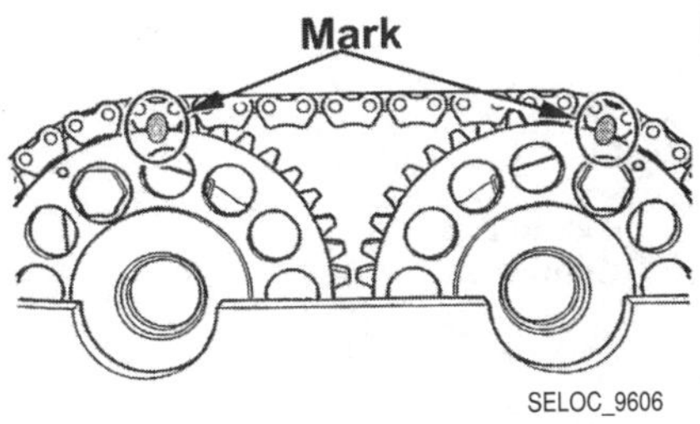

Fig. 87 The cam lobes should both be pointing outward when the #1 cylinder is at TDC on the compression stroke - 998/1052cc engines

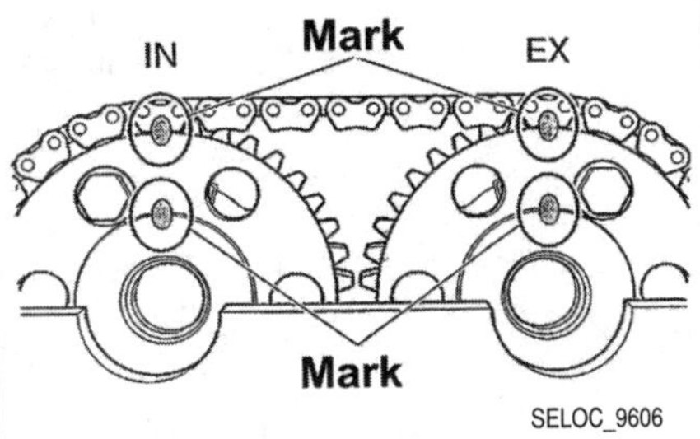

Fig. 88 Make a mark on the chain and sprockets and on the sprockets and camshafts - 998/1052cc engines

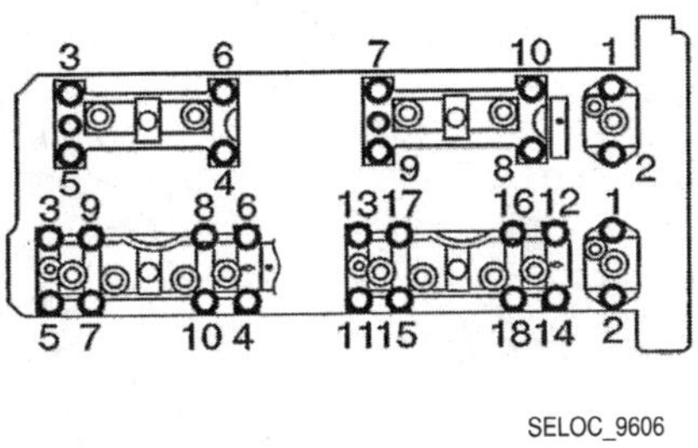

Fig. 89 Camshaft cap bolt loosening sequence - 998/1052cc engines

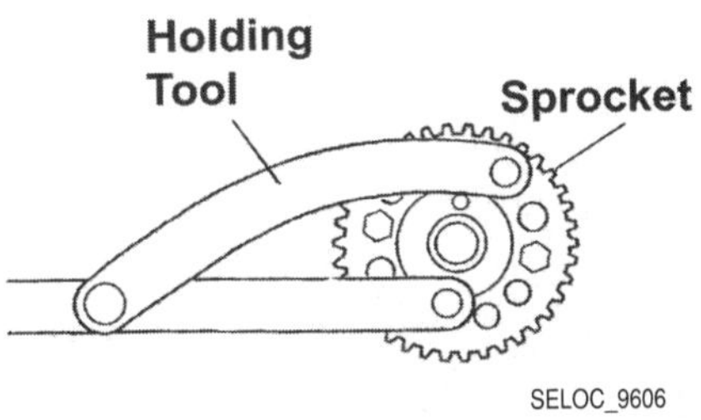

Fig. 90 Use the holding tool to remove or install the camshaft sprockets. . .

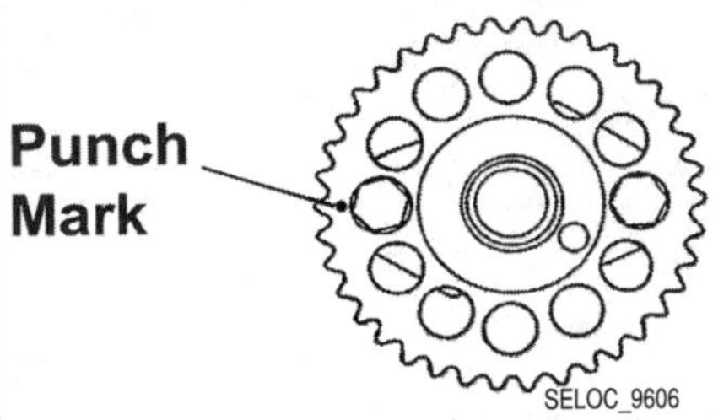

Fig. 91 . . .and make sure that the punch mark is facing outward when installed - 998/1052cc engines

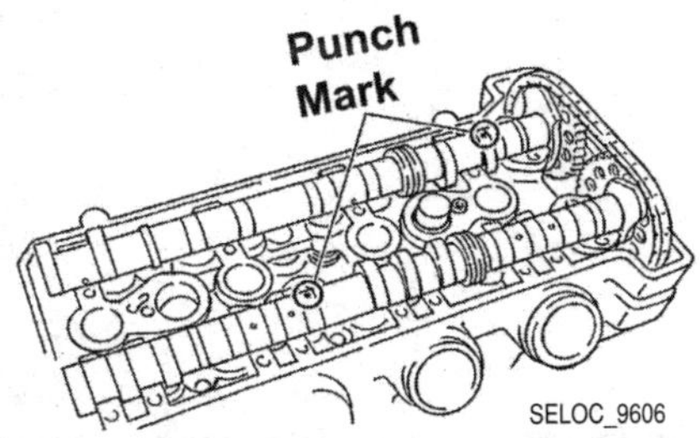

Fig. 92 The punch marks on each camshaft must be visible and facing upward before installing the caps - 998cc/1052cc engines

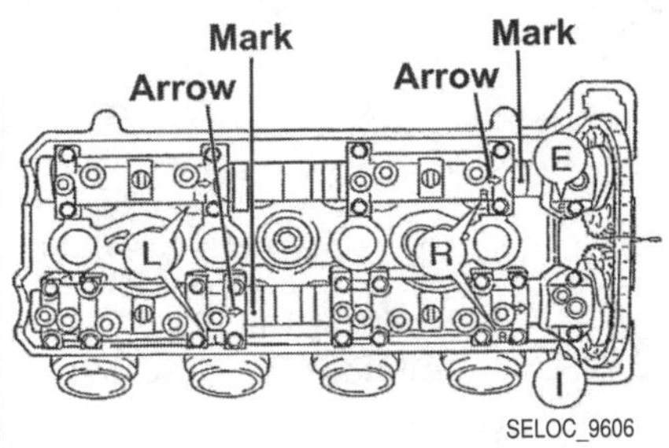

Fig. 93 Ensure that the arrows on the caps point directly to the punch marks and that the L, R, E and I stampings are as shown - 998cc/1052cc engines

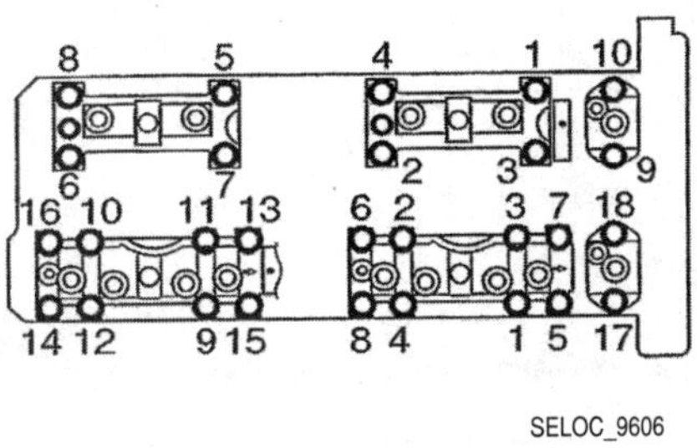

Fig. 94 Camshaft cap bolt tightening sequence - 998cc/1052cc engines

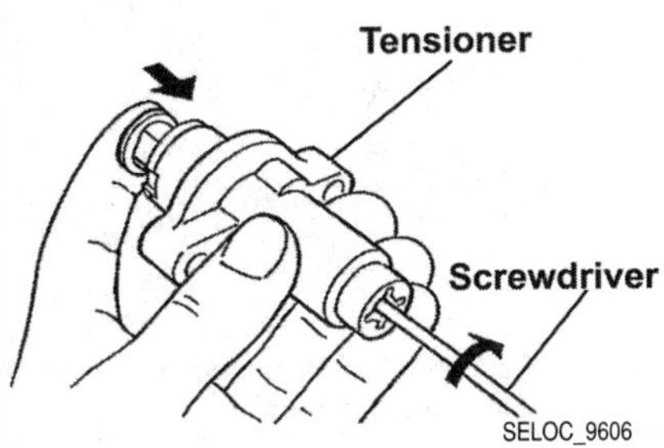

Fig. 95 Setting up the timing chain tensioner - 998cc/1052cc engines

release and press against the tensioner. Insert the small seal/gasket into the opening and then thread in the cap bolt, tightening it to 7.2 ft. lbs./45 inch lbs. (10 Nm).

■ **You can check the tensioner rod before installation if you like. While you have the screwdriver holding and your finger holding the rod in place, remove the screwdriver and allow the tensioner rod to slowly release. If it comes out of the housing smoothly, its fine. If there is rough or sticking movement as it releases, replace the tensioner.**

19. Reach in and turn the coupling a few turns counterclockwise. Reinstall the dial gauge set-up if you removed it and set the engine back to TDC on compression stroke of the # 1 cylinder. Confirm that the arrows on the 2 caps mentioned earlier are still pointing directly at to punch marks on the camshafts. If not, you will need to start all over and set everything up again.
20. Adjust the valve clearance as detailed in the Maintenance & Tune-Up section.
21. Install the coupling cover and tighten the bolt to 5.6 ft. lbs./69 inch lbs. (7.8 Nm).
22. Install the cylinder head cover.
23. Connect the battery cables and install the seat.

1812cc Engines

◆ **See Figures 96 thru 106**

■ **All Yamaha engines covered here are equipped with dual overhead camshafts (DOHC).**

1. Remove the cylinder head cover as detailed previously in this section.
2. Remove the drive coupling cover.
3. Position the No. 1 piston so it is at TDC of the compression stroke for the No. 1 cylinder (first cylinder at the front of the engine). To do this you could use a remote starter that has been hooked up according to the manufacturer's instructions, or you can simply rotate the engine coupler; either way, the object is to rotate the crankshaft so that the piston in the No.1 cylinder gets to TDC. We think it's easier, and certainly cheaper, to use the coupler.

With your finger over the No. 1 cylinder's spark plug hole (the first one), have a friend *slowly* rotate the coupler counterclockwise until you detect pressure in the cylinder, indicating TDC on the compression stroke.

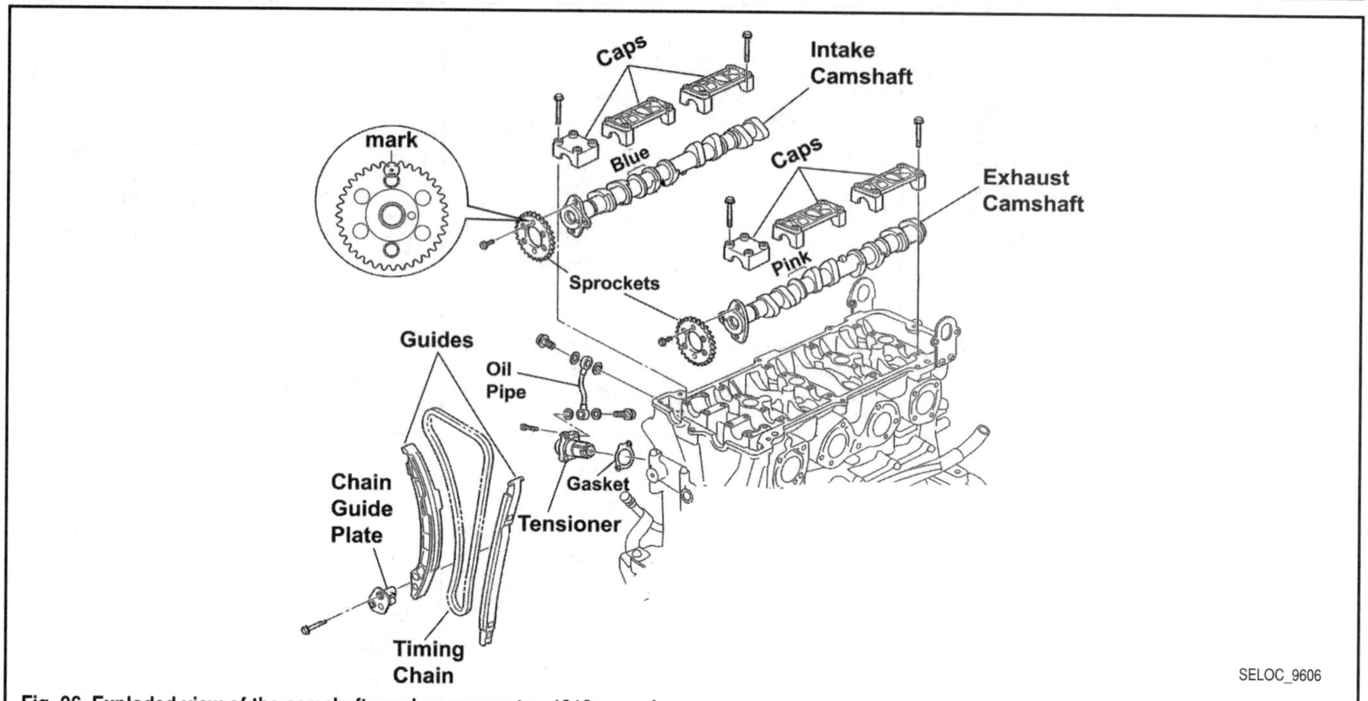

Fig. 96 Exploded view of the camshafts and components - 1812cc engines

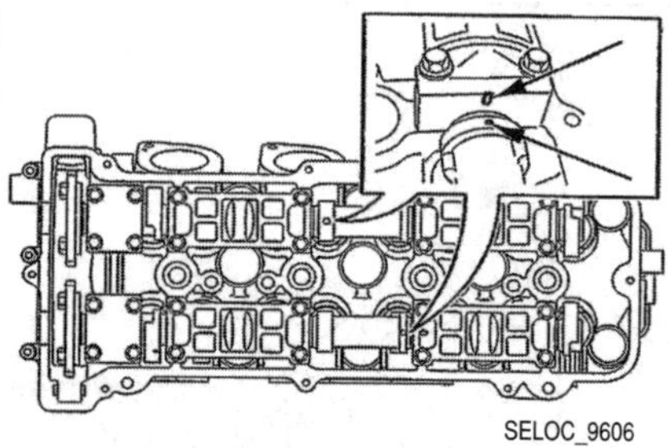

Fig. 97 Make sure that the marks all line up before confirming you are at TDC

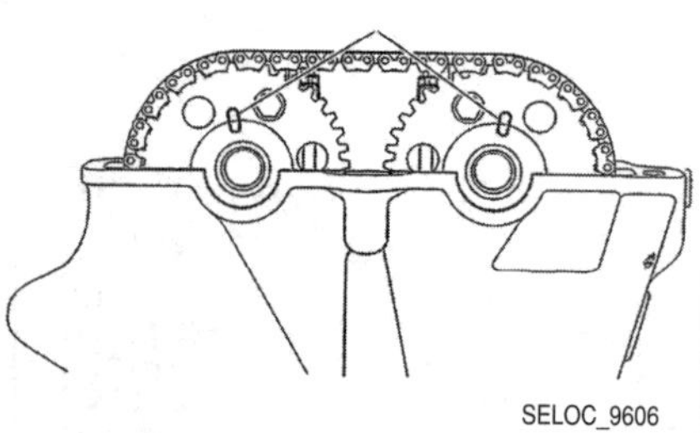

Fig. 98 Make a mark across the shaft and sprocket

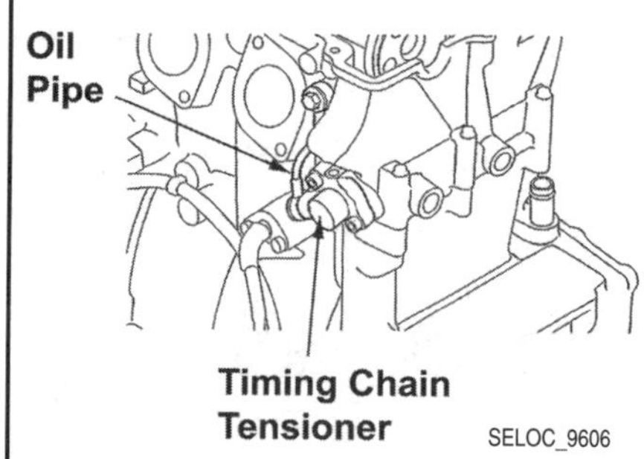

Fig. 99 Remove the chain tensioner

4. Thread a dial gauge stand (#90890-06725) into the spark plug hole on the No. 1 cylinder and then install the gauge into the stand. You may also need a gauge needle (#90890-06584, or you can get the set (#90890-06585). Use the gauge to confirm your TDC positioning.

a. Keeping your eye on the dial gauge indicator, have your friend rotate, *very slowly*, the coupler counterclockwise again until you see the indicator just begin to move in the opposite direction.

b. Now rotate again in the opposite direction while watching the needle and stop as it just changes direction again.

c. Do this a few times until you are confident that you have pinpointed that point where the needle changes direction at the first turn of the coupler in either direction.

d. Make sure that the punch marks on the camshaft are lined up with the marks on the camshaft bearing caps. If not, go back and start over again until you have achieved both confirmations.

e. Now you can be confident you have found TDC. Use a magic marker and place a small mark on the top of the coupler in case you need to do this again during the procedure.

5. Remove the 2 bolts connecting the small oil pipe on the front starboard side and lift off the pipe; this pipe runs between the block and the timing chain tensioner. There will be 4 small gaskets that you can discard.

6. Lock the timing chain tensioner rod and then remove the 2 mounting bolts. Pull the tensioner body out of the cylinder head. Remove the gasket and discard it.

7. Reach in and remove the exhaust side timing chain guide.

8. Take a Sharpie marker and carefully make a mark across the camshaft sprocket and timing chain; and then across the sprocket and camshaft. Do this for both sprockets. While you've got the marker out, it's not a bad idea to mark each camshaft cap as well. Mark each with **IN** or **EX** and then also **F1** (front), **F2** (center) and **F3** (rear) as they all must go back on in the exact same positions as they were removed from. We know you could carefully line them up in order somewhere on the floor, but why take a chance that someone bumps them?

■ **Prior to performing the next 2 steps, take note that there is a separate loosening sequence for each camshaft.**

9. Remove each of the 12 exhaust cap mounting bolts in the sequence shown. Lift off each cap and set them aside.

10. Remove each of the 12 intake cap mounting bolts in the sequence shown. Lift off each cap and set them aside.

11. Tie a piece of string, or wrap a length of wire, around the timing chain between the 2 sprockets and secure this to another part of the engine in order to keep the chain from falling down into the engine.

12. Lift the chain off of the sprocket and remove the intake camshaft. Do the same with the exhaust shaft. Make sure you have sufficient tension on the wire to keep the chain from dropping back into the crankcase.

13. If necessary, install the special holding tool (#90890-06724) and loosen the 2 mounting bolts and separate the sprockets from the camshaft. Mark each sprocket as to which camshaft it goes to.

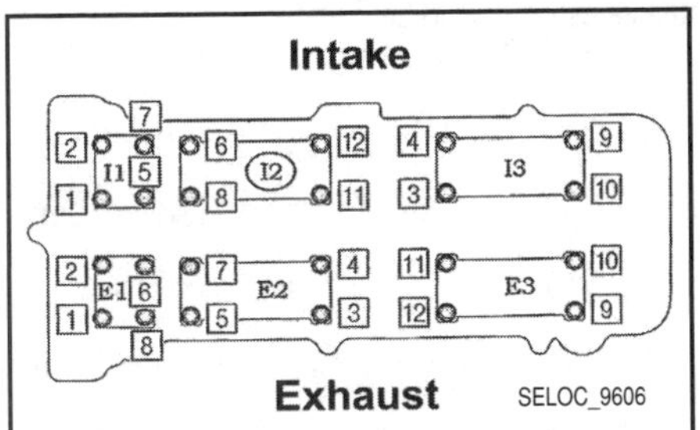

Fig. 100 Camshaft cap bolt loosening sequence - 1812cc engines

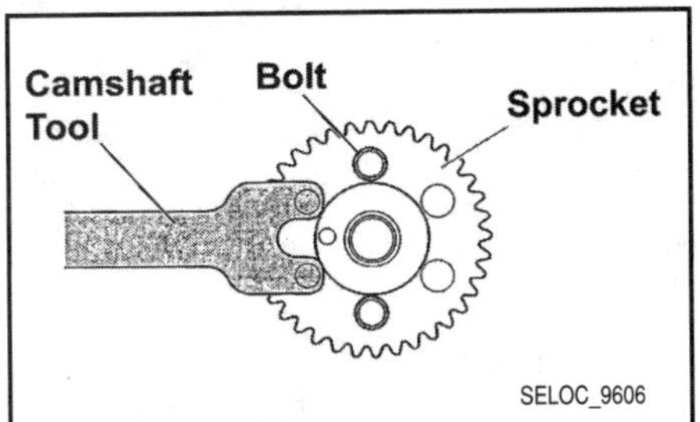

Fig. 101 Use the holding tool to remove or install the camshaft sprockets. . .

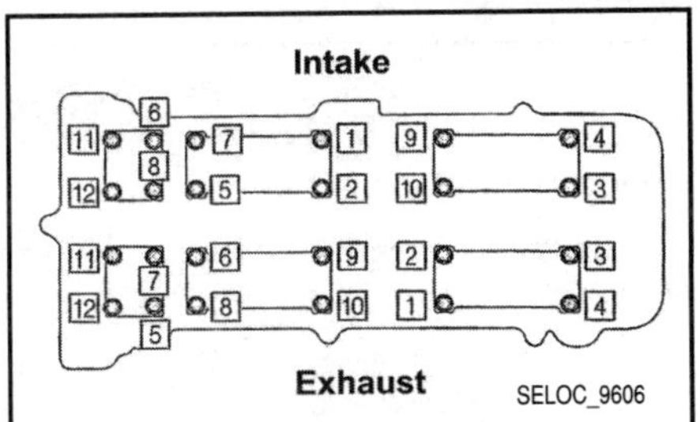

Fig. 102 Camshaft cap bolt tightening sequence - 1st pass, finger-tight. . .

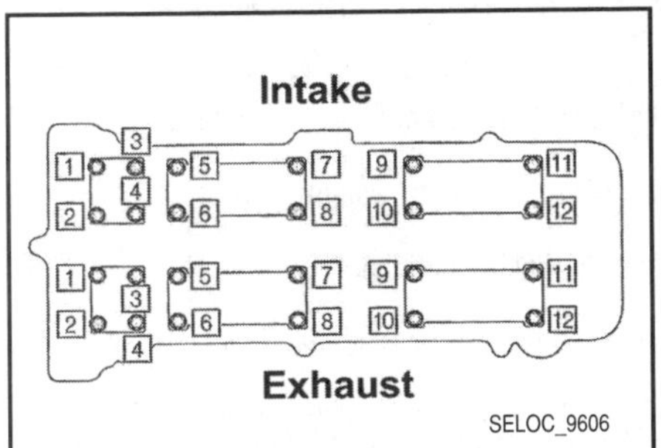

Fig. 103 . . .2nd pass, final torque - 1812cc engines

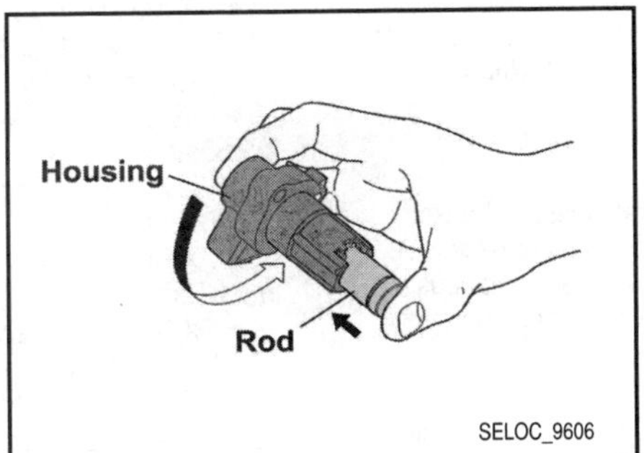

Fig. 104 Press in the tensioner rod. . .

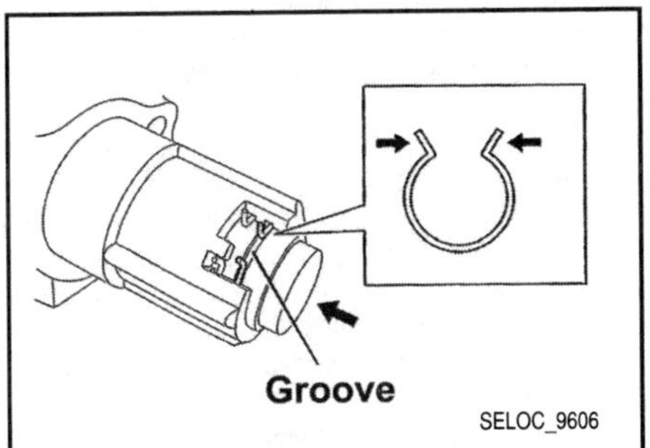

Fig. 105 . . .until the circlip fits into the groove

14. Remove the 2 timing chain guide plate bolts and lift it out. Remove the intake-side timing chain guide.

To install:

15. Clean all components thoroughly and inspect them as detailed in the Inspection section.

16. Very carefully, check that the #1 piston is still at TDC, although it should be unless the crankshaft has been rotated for some reason. Either way, confirm it just to be on the safe side.

17. Insert the intake chain guide back into the tunnel and then install the guide plate. Tighten the 2 bolts to 7.4 ft. lbs./77 inch lbs. (10 Nm).

18. If you removed either of the camshaft sprockets, coat the bolt threads lightly with clean engine oil and install them while using the holding tool. Make sure that when positioned on the shaft, the punch mark on the sprocket is facing outward and the marks made during removal are lined up before tightening the bolts to 13.3 ft. lbs. (18 Nm) on FX HO and VX models, or 17.7 ft. lbs. (24 Nm) on FX SHO and FZ models.

19. Coat the camshaft journals with molybdenum disulfide grease. Lift up on the timing chain and position the intake camshaft sprocket under the chain. Move the cam until the marks made previously line up and then drape the chain over the sprocket teeth. Lower the camshaft into the lower bearing bores and confirm that the chain-to-sprocket marks are still aligned. Confirm also that the punch marks on the camshaft are visible and facing upward.

■ **There should be a yellow paint mark to distinguish the exhaust camshaft.**

20. Repeat the previous step for the exhaust camshaft and then position the caps over the camshafts so that they are in the correct locations from which they were removed. Make sure that the arrows on the caps all point forward and that the mark on each cap is pointing directly at the small punch mark in each camshaft. If the arrow and marks do not line up, check to make sure your chain-to-sprocket marks are still aligned. If they are, then you'll need to go back and reset the engine to TDC again and make another try at installation.

21. Coat all of the cap mounting bolt threads lightly with clean engine oil and then thread them all in until finger-tight and the cap is just contacting the head, using the sequence shown and remembering that there is a separate sequence for each camshaft. Now tighten them again to 11.8 ft. lbs. (16 Nm) using the sequence shown in the illustration.

22. Install the exhaust side timing chain guide and then remove the wire you used to hold up the chain.

23. Grab the timing chain tensioner and press in the tensioner rod while turning the housing clockwise until the groove in the end of the rod is lined up with the snap ring in the housing. Squeeze the snap ring ends until the

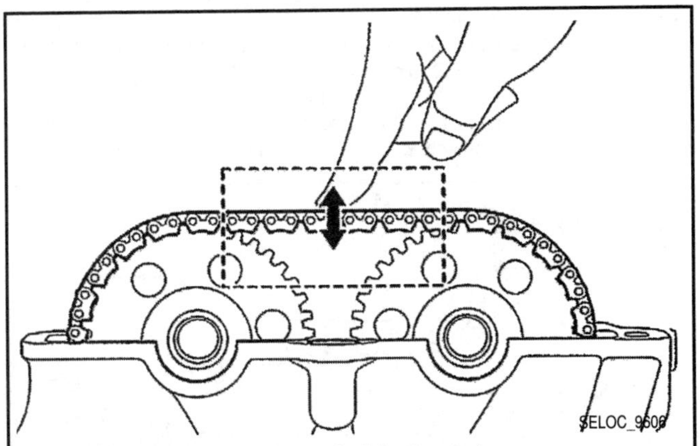

Fig. 106 Make sure there is no slack in the chain

ring fits completely into the groove and then slowly release the rod until it locks in place.

24. Position a *new* gasket on the housing and then install it on the block, tightening the bolts until finger-tight. Do the same with the small oil pipe, using 4 *new* gaskets, one on each side of the fittings.

■ **There should be a yellow paint mark on the end of the pipe connecting to the block.**

25. Tighten the pipe bolts to 14.8 ft. lbs. (20 Nm). Tighten the tensioner bolts to 7.4 ft. lbs./88 inch lbs. (10 Nm).

26. Reach in and turn the coupling a couple (2) complete turns counterclockwise. Reinstall the dial gauge set-up if you removed it and set the engine back to TDC on compression stroke of the # 1 cylinder. Confirm that the arrows on the 2 caps mentioned earlier are still pointing directly at to punch marks on the camshafts. Confirm that the marks you made on the chain are lined up with those on the sprocket and that the chain is taut where it spans the 2 sprockets. If not, you will need to start all over and set everything up again.

27. Adjust the valve clearance as detailed in the Maintenance & Tune-Up section.

28. Install the coupling cover and tighten the bolt to 5.6 ft. lbs./69 inch lbs. (7.8 Nm) on FX HO and VX models, or 5.9 ft. lbs./71 inch lbs. (8 Nm) on FX SHO and FZ models.

29. Install the cylinder head cover.

30. Connect the battery cables and install the seat.

INSPECTION

OEM ② MODERATE

◆ See Figures 107 thru 110

No matter how the camshaft is rotated, it is used to open and close the intake/exhaust valves at the appropriate time allowing the engine to draw air/fuel mixtures into the combustion chamber, seal the chamber when producing power, and open the chamber afterwards to release the unburned remains of the air/fuel charge.

The height of the intake and exhaust lobes is probably the most important specification on a camshaft, as they determine the maximum possible amount the shaft can open the valves. A shaft with worn lobes will not fully open the affected valves, leading to performance problems (by preventing the combustion chamber from breathing properly, preventing it from drawing in sufficient air/fuel charges and/or from completely venting the exhaust gasses).

In addition to camshaft lobe specs, additional journal or run-out specs may be available.

On all engines, the general condition of the shaft itself must be checked. Look for scored or worn areas, signs of discoloration from heat or excessive wear possibly caused by a lack of oiling. Replace the camshaft is obvious defects are found or if any surfaces measure out of specification.

As with all inspection procedures, make sure the surfaces are clean and dry.

1. Check the camshaft sprockets for wear, broken teeth or other signs of damage. A single tooth should not be worn down more than 25% when compared to the other teeth.

2. Check the lobes on each camshaft for blue discoloration, scratches, pitting or other signs of damage.

3. Using a micrometer, measure the height and width of each camshaft lobe. When the lobe (raised portion of the camshaft) is positioned upward, the height is the vertical measurement from base to lobe tip. In the same position, the width is measured across the lobe from one side to the other. Anything outside specifications will necessitate camshaft replacement.

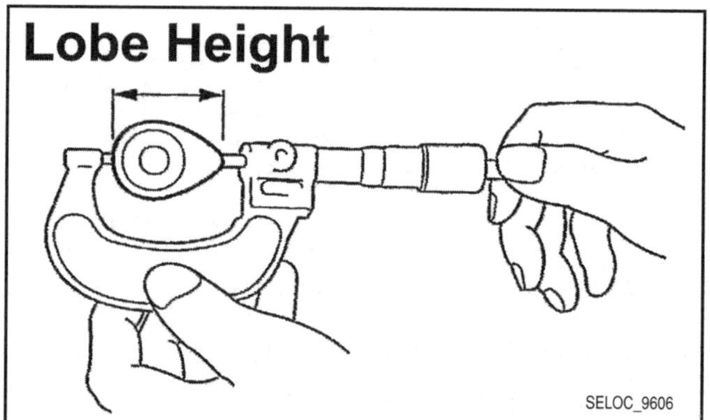

Fig. 107 Measure the camshaft lobe height. . .

Measurements for the intake camshaft should be:

- 998cc engines: Width - 0.984 in. (25mm). Height - 1.270 in. (32.25mm)
- 2004-07 FX w/1052cc engines: Width - 0.984 in. (25mm). Height - 1.281 in. (32.55mm)
- 2008 FX w/1052cc engines: Width - 0.984 in. (25mm). Height - 1.299 in. (33.00mm)
- FX HO w/1812cc engines: Width - 1.260 in. (32mm). Height - 1.630 in. (41.4mm)
- FX SHO w/1812cc engines: Width - 1.260 in. (32mm). Height - 1.610 in. (40.9mm)
- FZ models: Width - 1.260 in. (32mm). Height - 1.610 in. (40.9mm)
- VX models w/1052cc engines: Width - 0.984 in. (25mm). Height - 1.252 in. (31.80mm)
- VX models w/1812cc engines: Width - 1.260 in. (32mm). Height - 1.627-1.631 in. (41.3-41.4mm)

Measurements for the exhaust camshaft should be:

- 998cc engines: Width - 0.984 in. (25mm). Height - 1.285 in. (32.65mm)
- FX w/1052cc engines: Width - 0.984 in. (25mm). Height - 1.299 in. (33.00mm)
- FX HO w/1812cc engines: Width - 1.268 in. (32.2mm). Height - 1.626 in. (41.3mm)
- FX SHO w/1812cc engines: Width - 1.268 in. (32.2mm). Height - 1.614 in. (41mm)
- FZ models: Width - 1.268 in. (32.2mm). Height - 1.614 in. (41mm)
- VX models w/1052cc engines: Width - 1.268 in. (32.2mm). Height - 1.626 in. (41.3mm)
- VX models w/1812cc engines: Width - 1.260 in. (32mm). Height - 1.623-1.627 in. (41.2-41.3mm)

4. Check the camshaft run-out, as follows:

a. Support the camshaft on V-blocks under the top and bottom (end) journals.

b. Position a dial indicator with the tip on one of the middle journals and zero the dial.

c. Rotate the camshaft slowly and record the maximum movement on the dial. Measurements over 0.0012 in. (0.03mm) will require camshaft replacement. On the 2010-11 VX models w/1052cc engines and all models w/1812cc engines, the max is 0.0006 in. (0.015mm).

To check bearing oil clearances you need a way of measuring a tiny gap (thousands of an inch or hundredths of a millimeter) between otherwise inaccessible components. The basic method is to apply a strip of a gauging material such as Plastigage® to the camshaft journal and then to bolt on the camshaft cap. When the components are bolted together the gauging material is compressed (squished) between the bearing and journal. The key is that the material is designed to expand at a uniform rate so that you can then remove the cap and measure the width of the squished gauging material. The width of the material will correspond to a specific clearance measurement (a width-to-clearance conversion scale is provided with the gauging material).

Using Plastigage® is a simple, inexpensive, and fairly reliable method to measure main bearing and connecting rod bearing clearance. However, Plastigage® is soluble in oil, so clean the camshaft thoroughly of all traces of oil. To measure bearing clearances, proceed as follows:

Fig. 108 . . .and then the width

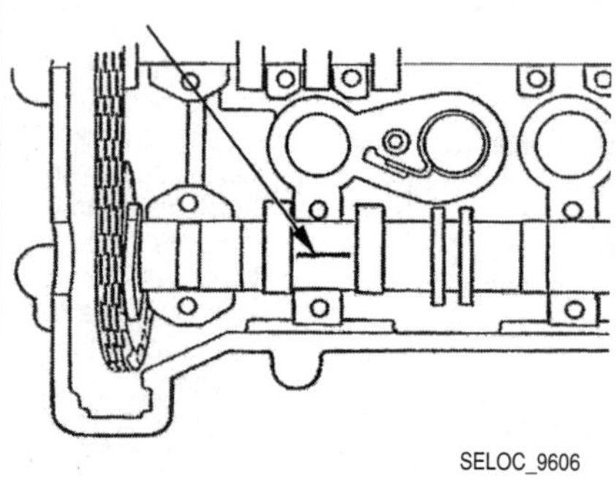

Fig. 109 Apply a thin strip of Plastigage® to the journal. . .

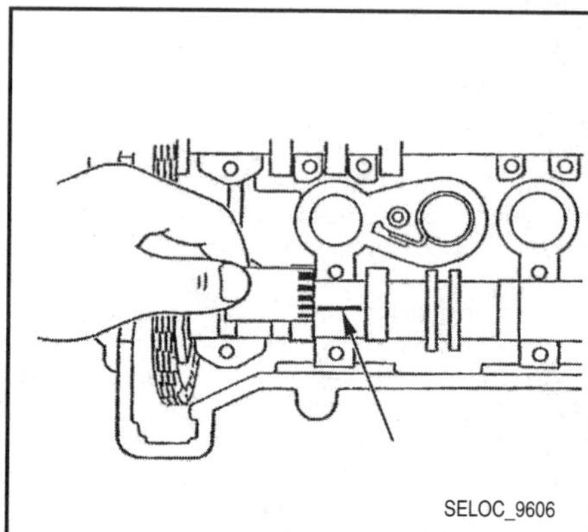

Fig. 110 . . .and then measure the width against the gauge after removing the caps

5. Obtain a package of Plastigage® from the local auto or marine parts outlet.

6. Cut a 1 inch (2.5cm) section of the bead for each journal/bearing combination that is being measured.

7. Place each 1 inch section on each camshaft journal.

■ ***DO NOT*** **rotate the camshaft until the measurements have been completed. Such action will smear the Plastigage® bead and the measurement will be useless.**

8. Install the camshaft bearing caps and tighten to specification. *Take your time* to be sure each is installed back into the same position from which it was removed according to the marks made during disassembling.

9. Remove the camshaft bearing cap bolts again gain access to the Plastigage®.

10. Measure the flattened Plastigage® at its widest point using the scale printed on the Plastigage® envelope. The number within the graduation which most closely corresponds to the width of the Plastigage® indicates the bearing clearance in thousands of an inch (or hundredths of a millimeter, depending upon the scale). A metric scale in mm is normally given on the back of the package. The widest point of the Plastigage® is the minimum clearance and the narrowest point is the maximum clearance. Compare the Plastigage® measurements with the specifications that follow (or in the Engine Specifications chart):

- FZ w/998cc engines: 0.0020-0.0031 in. (0.05-0.08mm)
- FX w/1052cc engines: 0.0012-0.0024 in. (0.03-0.06mm)
- FX w/1812cc engines: 0.0008-0.0024 in. (0.02-0.06mm)
- FZ models: 0.0008-0.0024 in. (0.02-0.06mm)
- 2005-09 VX models: 0.0020-0.0024 in. (0.05-0.06mm)
- 2010-11 VX models w/1052cc engines: 0.0020-0.0033 in. (0.050-0.084mm)
- VX w/1812cc engines: 0.0008-0.0024 in. (0.02-0.06mm)

11. If the clearance is not within specifications, then either the camshaft or the bearing (which means the cylinder head) will require replacement, but you'll need to measure the journal diameter first.

■ **Once you are finished checking clearance, clean all traces of the Plastigage® from the journal/bearing surfaces, and then follow necessary steps again (oiling the bearing surfaces and skipping the Plastigage®) to make a final installation of the camshaft and caps as detailed in the Removal & Installation section.**

■ **If camshaft cap-to-journal clearance was within specifications, there is no need to check journal diameter.**

12. Grab your micrometer again and measure the diameter of each journal on both camshafts. Journal diameter should be:

- FX w/998cc engines: 0.9622-0.9626 in. (24.44-24.45mm)
- FX w/1052cc engines: 0.9630-0.9634 in. (24.46-24.47mm)
- FX w/1812cc engines: 0.9827-0.9835 in. (24.96-24.98mm)
- FZ models: 0.9827-0.9835 in. (24.96-24.98mm)
- 2005-09 VX models: intake - 0.9630-0.9634 in. (24.46-24.47mm), exhaust - 0.9622-0.9626 in. (24.44-24.45mm)
- 2010 VX models w/1052cc engines: 0.9621-0.9626 in. (24.437-24.450mm) for both camshafts
- VX w/1812cc engines: 0.9827-0.9835 in. (24.96-24.98mm)

If journal diameter is within specification and you experienced excessive cap-to journal clearance in the previous inspection, you will need to replace the entire cylinder head and all camshaft caps (they are a matched set). If journal diameter is out of specification, replace the camshaft.

Cylinder Head & Valvetrain

REMOVAL & INSTALLATION

998cc & 1052cc Engines

OEM ③ DIFFICULT

◆ See Figures 111 thru 114

1. Remove the engine as detailed previously in this section.

2. Remove the forward exhaust components as detailed in the Exhaust & Cooling section.

3. Remove the air filter and case on FX models as detailed in the Fuel Injection section.

4. Remove the oil tank as detailed previously in this section.

5. Remove the camshafts as detailed previously in this section.

■ **Ok, Ok, we know this all seems excessive, but this is what Yamaha suggests. You *may* be able to get away without removing the engine if you can get the exhaust manifold bolts out, but this is unlikely. You may also be able to get away with not removing the camshafts, but the following is preceding steps are the recommended means of removing the head. Once done, the remaining steps are simple.**

6. There are 3 bolts on the back of the cylinder head - remove them.

7. There are 2 cap nuts/washers at the front of the head and 8 more Allen head nuts/washers recessed in the head itself, remove all of them using a couple passes and in the sequence illustrated.

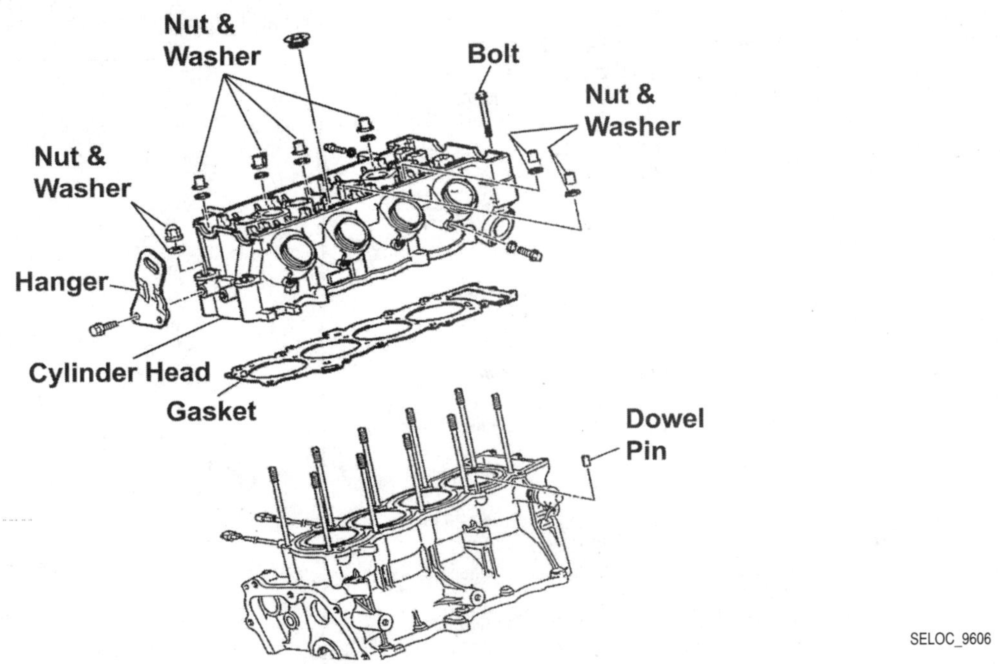

Fig. 111 Removing the cylinder head - 998cc/1052cc engines

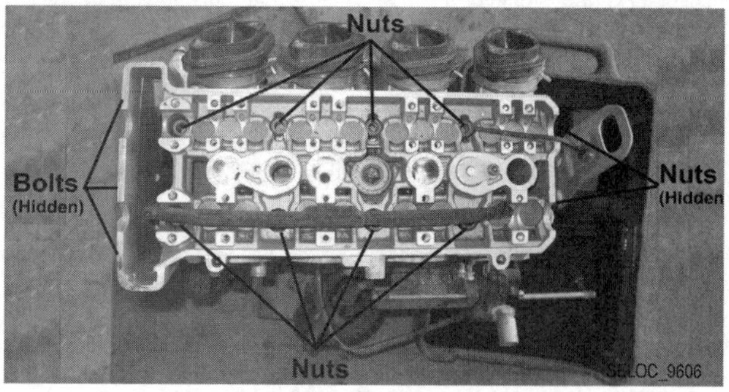

Fig. 112 Many of the cylinder head nuts will be recessed - 998cc/1052cc engines

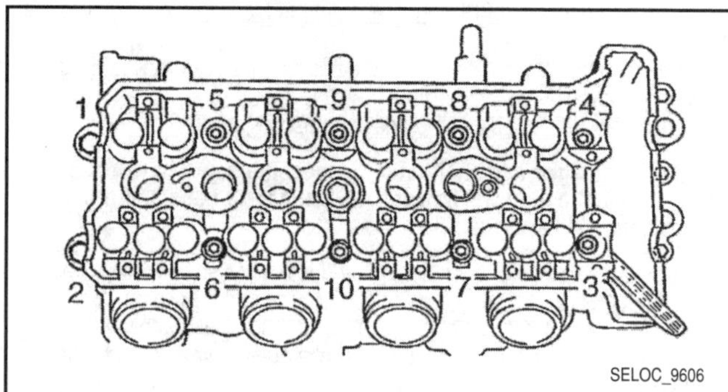

Fig. 113 Cylinder head nut loosening sequence - 998cc/1052cc engines

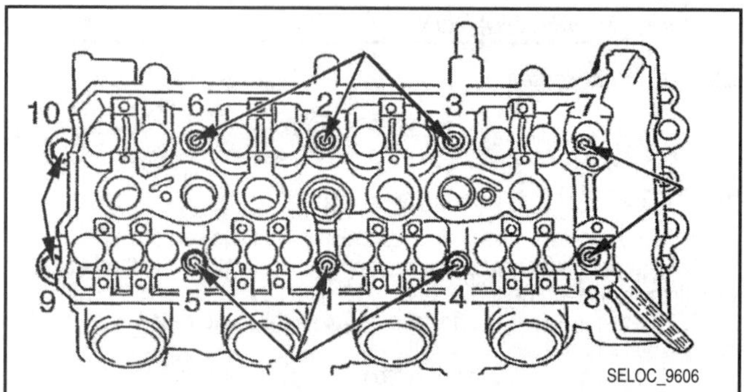

Fig. 114 Cylinder head nut tightening sequence - 998cc/1052cc engines

■ **The Allen head nuts on the exhaust side of the head are usually flanged, while the ones on the intake side are not.**

8. Tap the head in a couple places, *lightly*, with a rubber mallet to break it loose and then lift it up and off of the studs while feeding the timing chain and wire through the chain opening. There are 2 dowel pins that should stay with the cylinder block, so if they come off with the head make sure you reinsert them into their bores in the block.

9. Remove the cylinder head gasket and check to make sure there is no residual gasket material on either of the mating surfaces.

To install:

10. Inspect the cylinder head as detailed in the Inspection section.

11. Position a *new* gasket down over the studs and onto the cylinder block. Make sure that all holes in the gasket and block line up with no restrictions.

12. Do the same with the cylinder head itself while feeding the timing chain through the opening. Make sure the head seats fully over the 2 dowel pins.

13. Coat the cylinder head nut threads lightly with clean engine oil and thread them on until finger-tight. Tighten all 10 nuts to 14 ft. lbs. (20 Nm) in the order shown in the illustration.

a. Next on 998cc engines, tighten the 8 recessed nuts to 35 ft. lbs. (49 Nm) and the 2 nuts at the front of the head to 46 ft. lbs. (64 Nm), in the sequence shown.

b. Next on 1052cc engines, using the torque sequence shown and an angle wrench in place of the torque wrench:

- 2004-07 FX HO w/1052cc engines - tighten the #1 and #2 nuts (in the torque sequence) an additional 121° +/- 5°. Now tighten the remaining recessed nuts (#3 thru #8) an additional 105° +/- 5°. Finally, tighten the front nuts (#9 and #10) an additional 140° +/- 5°.
- 2008 FX HO w/1052cc engines - tighten the #6, #2, #3 and #7 (in the torque sequence) exhaust side nuts an additional 120°. Now tighten the remaining recessed intake side nuts (#5, #1, #4 and #8) an additional 105°. Finally, tighten the front nuts (#9 and #10) an additional 140°.
- VX models - tighten the #1 and #2 nuts (in the torque sequence) an additional 121° +/- 5° (120° on 2010 models). Now tighten the remaining recessed nuts (#3 thru #8) an additional 105° +/- 5°. Finally, tighten the front nuts (#9 and #10) an additional 140° +/- 5°.

■ **Even though there is an allowance of +/- 5°, we suggest simply tightening the bolts to the degree. . .121°, 105°, 140°, etc.**

14. Coat the threads of the 3 remaining bolts with clean engine oil and thread them into the back of the head. Tighten them in one pass to 7.2 ft. lbs./88 inch lbs. (10 Nm).

15. Install the camshafts.

16. Install the oil tank.

17. Install the forward exhaust components.

18. Install the air filter and case.

19. Install the engine.

1812cc Engines

◆ **See Figures 115, 116 and 117**

1. Remove the engine as detailed previously in this section.
2. Remove the engine cover.
3. Remove the intake manifold as detailed in the Fuel System section.
4. On SHO and FZ models, you may want to remove the air cooler and supercharger as detailed elsewhere in this section.
5. Remove the camshafts as detailed previously in this section.
6. Loosen the bolt and remove the front engine hanger. Remove the bolts (2) and lift of the 2 rear hangers.
7. There are 3 bolts on the front edge of the cylinder head - remove them.
8. There are 10 more Allen head nuts/washers recessed in the head itself, remove all of them using a couple passes and in the sequence illustrated. Throw them away because they are not reusable.

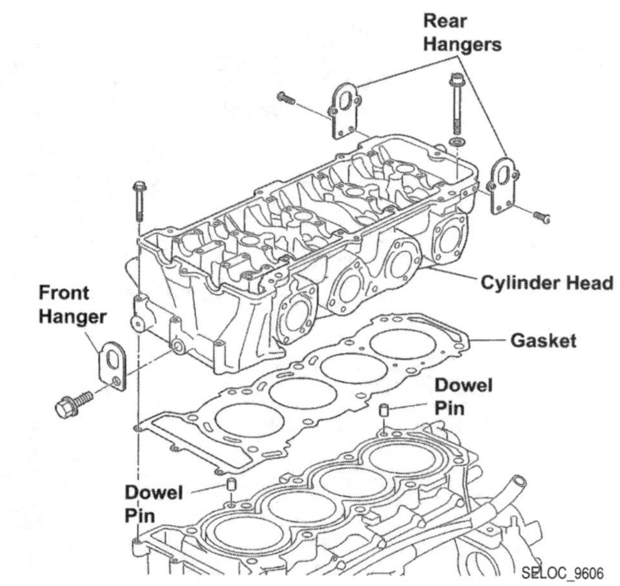

Fig. 115 Removing the cylinder head - 1812cc engines

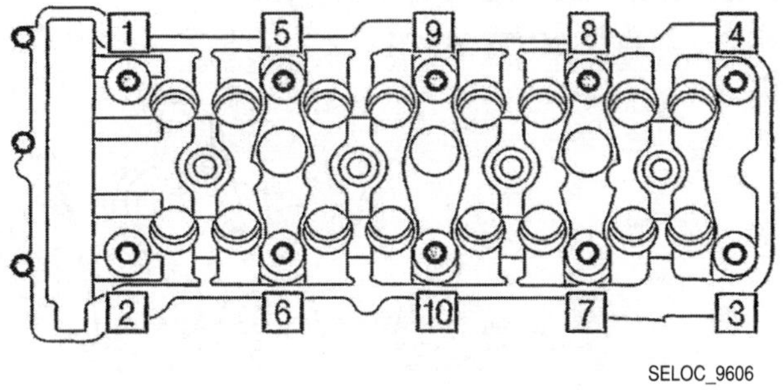

Fig. 116 Cylinder head loosening sequence - 1812cc engines

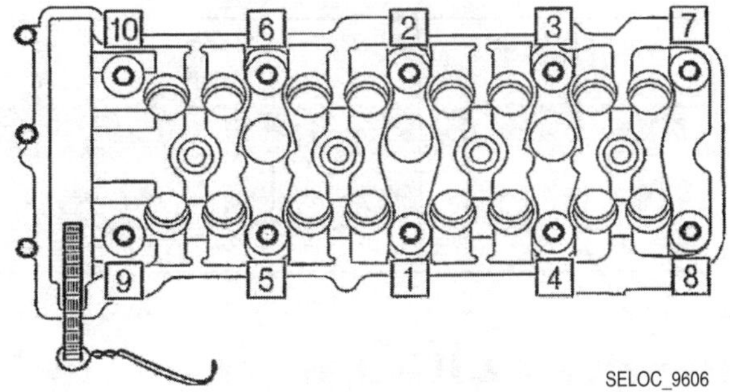

Fig. 117 Cylinder head tightening sequence - 1812cc engines

9. Tap the head in a couple places, *lightly*, with a rubber mallet to break it loose and then lift it up and off of the cylinder block while feeding the timing chain and wire through the chain opening. There are 2 dowel pins that should stay with the cylinder block, so if they come off with the head make sure you reinsert them into their bores in the block.

10. Remove the cylinder head gasket and check to make sure there is no residual gasket material on either of the mating surfaces.

To install:

11. Inspect the cylinder head as detailed in the Inspection section.

12. Position a *new* gasket onto the cylinder block. Make sure that all holes in the gasket and block line up with no restrictions.

13. Do the same with the cylinder head itself while feeding the timing chain through the opening. Make sure the head seats fully over the 2 dowel pins.

14. Coat the 10 *new* cylinder head bolt threads lightly with clean engine oil and thread them on until finger-tight. Tighten all 10 bolts to 35.4 ft. lbs. (48 Nm) in the order shown in the illustration. On VX models, tighten to t40.6 ft. lbs. (55 Nm).

a. Next, tighten them an additional 90°, in order, using an angle wrench.

b. Coat the threads of the 3 remaining bolts with clean engine oil and thread them into the front of the head. Tighten them, in one pass, to 7.4 ft. lbs./89 inch lbs. (10 Nm).

15. Install the 2 rear hangers. Coat the bolt threads with Loctite 242 and tighten them to 8.9 ft. lbs./107 inch lbs. (12 Nm).

16. Coat the front hanger bolt threads with Loctite 242, install the hanger and tighten the bolt to 31 ft. lbs. (42 Nm).

17. Install the camshafts.

18. Install the intake manifold.

19. Install the air cooler and supercharger if removed.

20. Install the engine cover.

21. Install the engine if removed.

DISASSEMBLY & ASSEMBLY

998cc & 1052cc Engines

OEM ③ DIFFICULT

◆ See Figures 118 thru 123

1. Remove the cylinder head as previously detailed and lay it on a clean work surface.

■ **The next step involves removing a large number of pieces from the head and they must be kept in the specific order that they were removed in. We highly recommend making some sort of holder where you can label and store them without getting them mixed up.**

2. Carefully grab the intake valve lifters (12) with pliers and pull them from the head. Make sure you are storing them safely and in the order that they were removed. Go back and pull out each of the lifter pads and drop them into the lifter that they were found under.

3. Install a valve spring compressor tool (#YM-04019 / 90890-04019) over the 1st intake valve using the intake valve attachment (#YM-4114 / 90890-04114). Compress the tool just enough to allow you to remove the 2 keepers.

4. Remove the tool and then lift out the upper spring seat, spring, oil seal and lower spring seat. If reusing these, make sure you keep them together and labeled for their respective valves. Remove the valve.

5. Repeat Steps 2-4 for the 11 remaining intake valves (or any individual ones that require removal).

6. Now move to the exhaust valves and repeat Steps 2-5. When compressing the valve springs use the attachment (#YM-4108 / 90890-04108) for exhaust valves.

7. If replacing the cylinder head, remove the 2 mounting bolts and lift of the forward engine hanger.

To assemble:

8. Inspect the valve train as detailed in the Inspection section.

■ **It is imperative that the valves are installed in their original location and that the correct valve is used. All valves will have a series of characters stamped on the face of a valve. The right and left (or front and back) intake valves (for each cylinder) will show *4XV:* (note the colon after the V), while the center intake valve for each cylinder will show *4XV.* (note the single period after the V). All exhaust valves will show *5LV* (nothing after the V).**

9. Insert the correct valve into the valve guide from the bottom of the cylinder head until it rests in the valve seat. From the top side of the head now, install the lower spring seat over the valve stem making sure that the open side is facing upward. Coat the valve seal lightly with clean engine oil and slide it down the stem, small side facing up, until it seats fully. When looking at the valve spring, you'll notice that the coil spacing is not even - install the spring with the tighter spacing on the bottom and the wider spacing on the top. Slide the upper seat over the valve stem until it contacts the spring (the protruding side should be facing down).

■ **Valve seals on 1052cc engines are not re-usable and must be replaced with *new* ones once removed.**

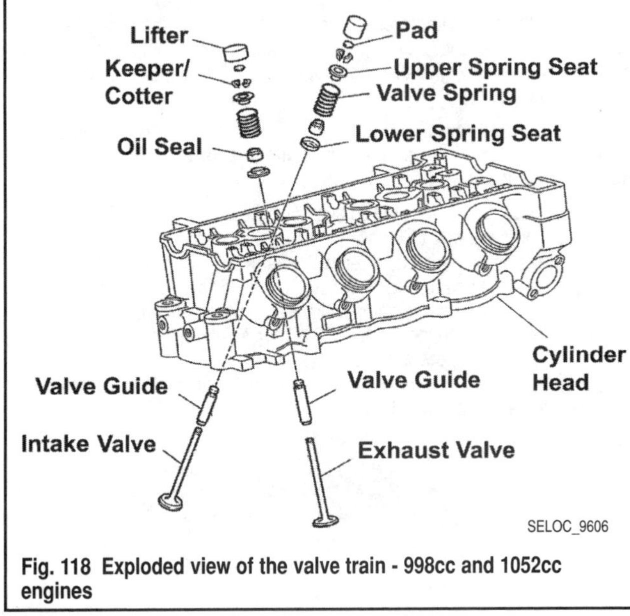

Fig. 118 Exploded view of the valve train - 998cc and 1052cc engines

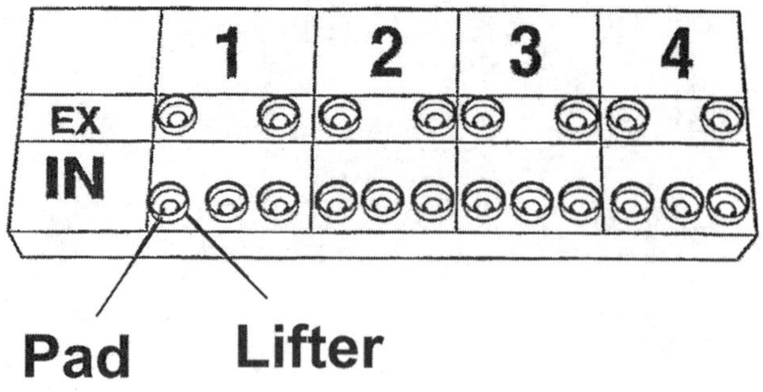

Fig. 119 It is very important to keep the valve lifters and pads in order

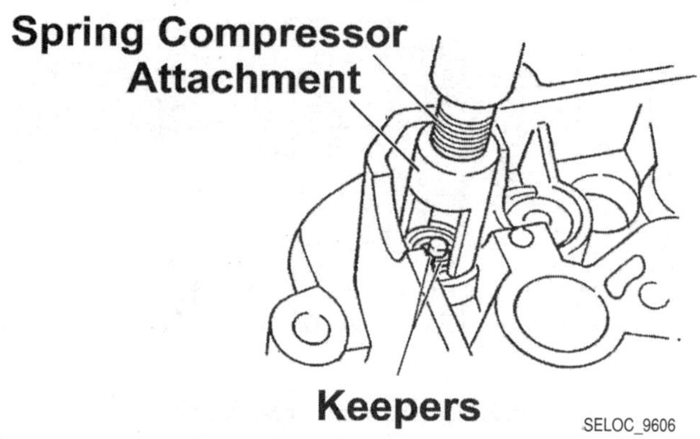

Fig. 120 Use a valve spring compressor to remove the valve keepers

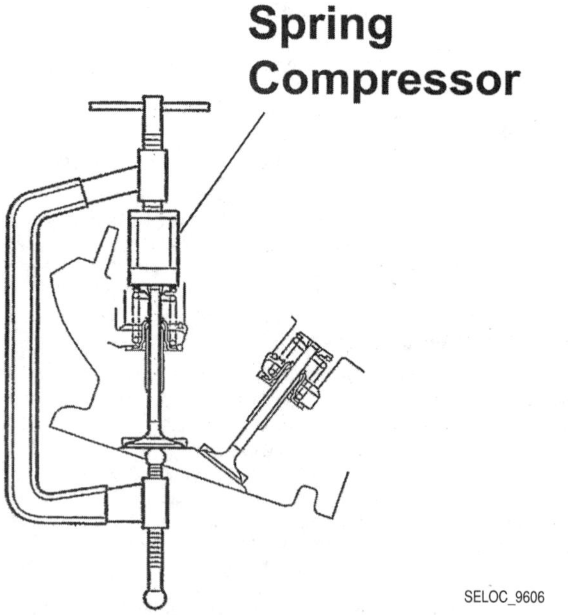

Fig. 121 Another look at the spring compressor

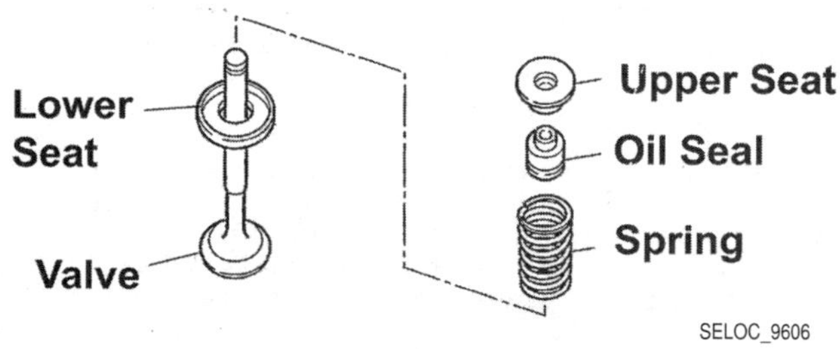

Fig. 122 A good look at the actual valve components

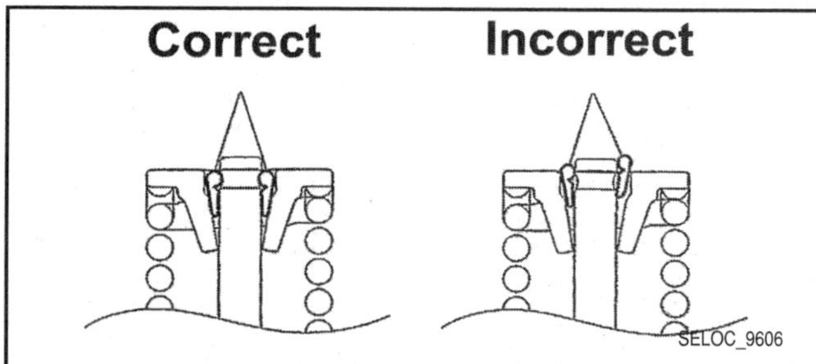

Fig. 123 Make sure that the keepers are properly seated in the groove on the valve stem

10. Install the spring compressor again and compress the spring just enough to fit the 2 keepers through the seat and into the groove in the valve stem. Once in position, give the seat a light tap with a mallet to seat the keepers in the groove.

■ **A nice hint from Yamaha. Put a little grease on the end of a small screwdriver and use the grease to hold the keeper while inserting it with the screwdriver.**

11. Repeat Steps 2-3 for the remaining intake valves.
12. Now you can repeat Steps 2-4 for the exhaust valves.
13. Coat the valve pads and lifters with clean engine oil. Position the correct pad over the keepers for the valve it is associated with and then press in the lifter until it seats on the pad. Repeat this procedure for each valve. It is always a good idea to tap each lifter *lightly* with a plastic mallet to make sure that the keepers are properly seated.
14. Install the hanger if removed and tighten the 2 bolts to 29 ft. lbs. (40 Nm).
15. Install the cylinder head.

1812cc Engines

OEM ③ DIFFICULT

◆ See Figures 119 thru 124

1. Remove the cylinder head as previously detailed and lay it on a clean work surface.

■ **The next step involves removing a large number of pieces from the head and they must be kept in the specific order that they were removed in. We highly recommend making some sort of holder where you can label and store them without getting them mixed up.**

2. Carefully grab the intake valve lifters (8) with pliers and pull them from the head. Make sure you are storing them safely and in the order that they were removed. Go back and pull out each of the lifter pads and drop them into the lifter that they were found under.
3. Install a valve spring compressor tool (#YM-04019 / 90890-04019) over the 1st intake valve using the valve attachment (#YM-4114 / 90890-04114). Compress the tool just enough to allow you to remove the 2 keepers.
4. Remove the tool and then lift out the upper spring seat, spring and lower spring seat. Tilt the head to pull the valve out from the bottom and then remove the valve seal. If reusing these, make sure you keep them together and labeled for their respective valves.
5. Repeat Steps 2-4 for the 7 remaining intake valves (or any individual ones that require removal).
6. Now move to the exhaust valves (8) and repeat Steps 2-5.

To assemble:

7. Inspect the valve train as detailed in the Inspection section.

■ **It is imperative that the valves are installed in their original location and that the correct valve is used.**

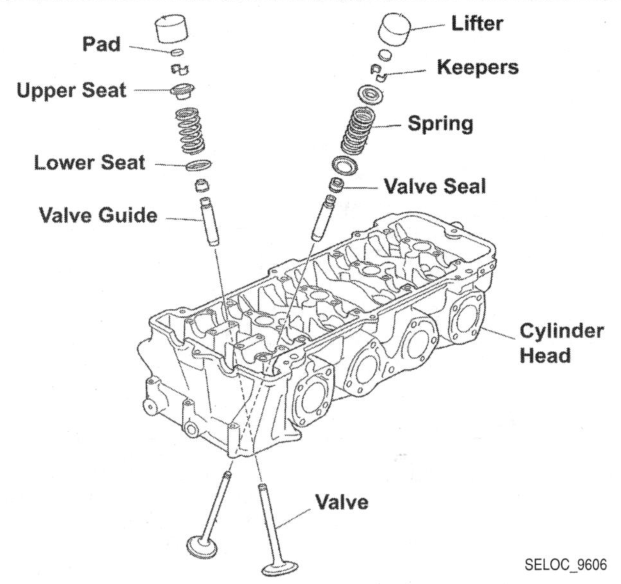

Fig. 124 Exploded view of the valve train - 1812cc engines

8. Coat the valve seal lightly with clean engine oil and insert the correct valve seal into the recess in the head.

■ **Valve seals (and guides) are not re-usable and must be replaced with *new* ones once removed.**

9. Tilt the head back up and insert the correct valve into the valve guide from the bottom of the cylinder head until it rests in the valve seat.
10. From the top side of the head now, install the lower spring seat over the valve stem.
11. When looking at the valve spring, you'll notice that the coil spacing is not even - install the spring with the tighter spacing on the bottom and the wider spacing on the top. Slide the upper seat over the valve stem until it contacts the spring (the protruding side should be facing down).

■ **If installing a new valve, you must also install a new valve seal and guide.**

12. Install the spring compressor again and compress the spring just enough to fit the 2 keepers through the seat and into the groove in the valve stem. Once in position, give the seat a light tap with a mallet to seat the keepers in the groove.

■ **A nice hint from Yamaha. Put a little grease on the end of a small screwdriver and use the grease to hold the keeper while inserting it with the screwdriver.**

13. Repeat Steps 2-5 for the remaining intake valves.
14. Now you can repeat Steps 2-6 for the exhaust valves.
15. Coat the valve pads and lifters with clean engine oil. Position the correct pad over the keepers for the valve it is associated with and then press in the lifter until it seats on the pad. Repeat this procedure for each valve. It is always a good idea to tap each lifter *lightly* with a plastic mallet to make sure that the keepers are properly seated.
16. Install the cylinder head.

INSPECTION

Now that all of the cylinder head components are clean, it's time to inspect them for wear and/or damage. To accurately inspect them, you will need some specialized tools:

- A 0-1 in. micrometer for the valves
- A dial indicator or inside diameter gauge for the valve guides
- A spring pressure test gauge

If you do not have access to the proper tools, you may want to bring the components to a shop that does.

Valves

OEM ③ DIFFICULT

◆ **See Figures 125 thru 128**

The first thing to inspect are the valve heads. Look closely at the head, margin and face for any cracks, excessive wear or burning. The margin is the best place to look for burning. It should have a squared edge with an even width all around the diameter. When a valve burns, the margin will look melted and the edges rounded. Also inspect the valve head for any signs of tulipping. This will show as a lifting of the edges or dishing in the center of the head and will usually not occur to all of the valves. All of the heads should look the same; any that seem dished more than others are probably bad. Next, inspect the valve keeper grooves and valve tips. Check for any burrs around the grooves, especially if you had to file them to remove the valve. Valve tips should appear flat, although slight rounding with high mileage engines is normal. Slightly worn valve tips will need to be machined flat. Last, measure the valve stem diameter with the micrometer. Measure the area that rides within the guide, especially towards the tip where most of the wear occurs. Take several measurements along its length and compare them to each other. Wear should be even along the length with little to no taper. Please refer to the Engine Specifications chart for stem diameters. Any valves that fail these inspections should be replaced.

While the valve is out, you might as well measure the valve margin thickness. Recommended range is:

- FX models with 998cc/1052cc engines - 0.0197-0.0354 in. (0.5-0.9mm)
- FX models w/1812cc engines - 0.031-0.047 in. (0.8-1.2mm)
- FZ models - 0.031-0.047 in. (0.8-1.2mm)
- 2005-09 VX models - 0.0335-0.0453 in. (0.85-1.15mm)
- 2010 VX models w/1052cc engines - 0.0276 in. (0.7mm)
- VX models w/1812cc engines - 0.031-0.047 in. (0.8-1.2mm)

Any measurement outside this range will require valve replacement.

And also check the stem run-out. Position the valve in two V-blocks and use a dial gauge and stand to check while slowly rotating the valve. Maximum run-out should be 0.0004 in. (0.01mm).

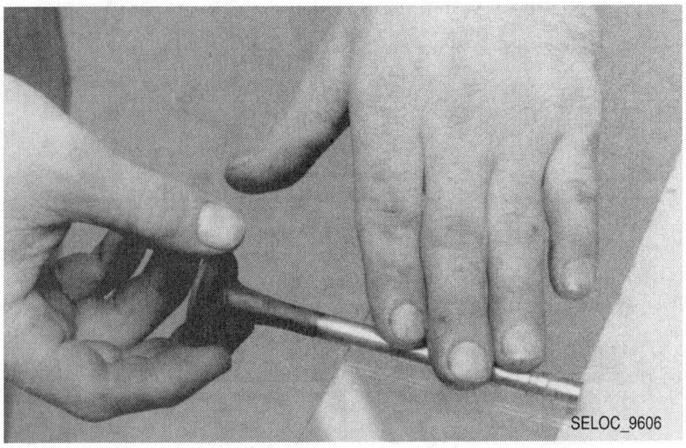

Fig. 125 Valve stems may be rolled on a flat surface to check for bends

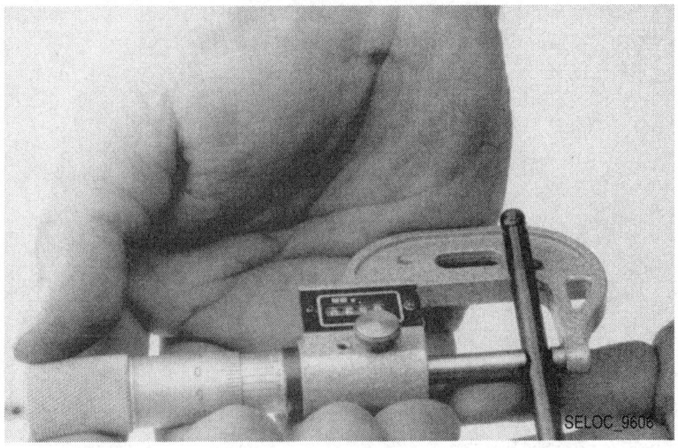

Fig. 126 Use a micrometer to check the valve stem diameter

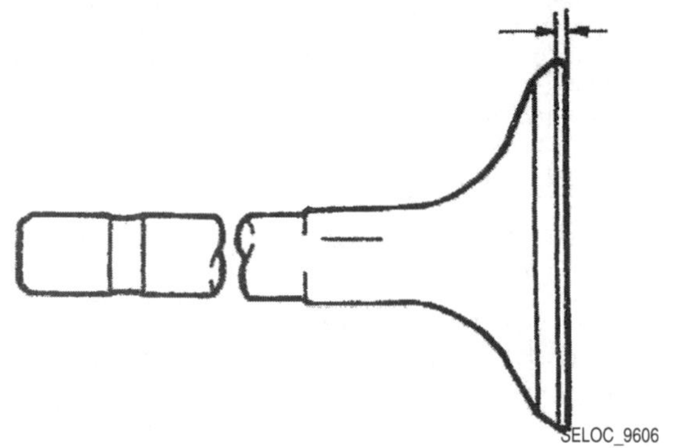

Fig. 127 Measure the valve margin here

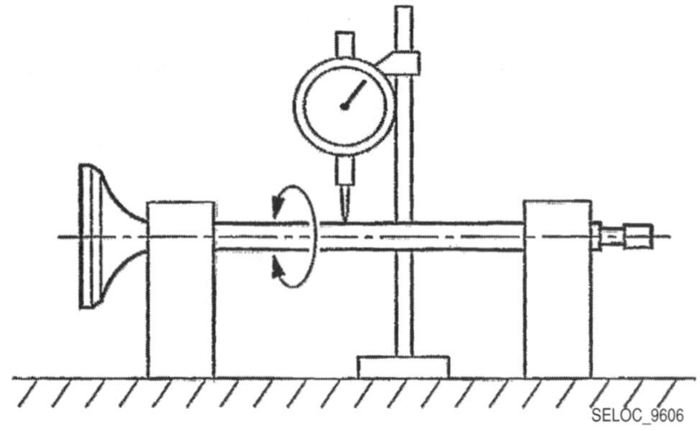

Fig. 128 Checking the valve stem run-out

Springs, Seat & Keepers

OEM ③ DIFFICULT

◆ See Figures 129 and 130

The first thing to check is the most obvious, broken springs. Next check the free length and squareness of each spring. If applicable, insure to distinguish between intake and exhaust springs. Use a ruler and/or carpenter's square to measure the length. A carpenter's square should be used to check the springs for squareness. If a spring pressure test gauge is available, check each springs rating and compare to the specifications chart. Check the readings against the specifications given in the Engine Specifications chart. Any springs that fail these inspections should be replaced.

The spring seats rarely need replacing, however they should still be checked as a precaution. Inspect the spring mating surface and the keeper retention area for any signs of excessive wear. Also check for any signs of cracking. Replace any retainers that are questionable.

Valve keepers should be inspected for excessive wear on the outside contact area as well as on the inner notched surface. Any keepers which appear worn or broken and their respective valve should be replaced.

Valve Guides

OEM ③ DIFFICULT

Now that you know the valves are good, you can use them to check the guides, although a new valve, if available, is preferred. Before you measure anything, look at the guides carefully and inspect them for any cracks, chips or breakage. Also if the guide is a removable style (as in most aluminum heads and those covered here), check them for any looseness or evidence of movement. All of the guides should appear to be at the same height from the spring seat. If any seem lower (or higher) from another, the guide has moved. Mount a dial indicator onto the spring side of the cylinder head. Lightly oil the valve stem and insert it into the cylinder head. Position the dial indicator against the valve stem near the tip and zero the gauge. Grasp the valve stem and wiggle towards and away from the dial indicator and observe the readings. Mount the dial indicator 90 degrees from the initial point and zero the gauge and again take a reading. Compare the two readings for a out of round condition. Check the readings against the specifications given in the Engine Specifications chart. An Inside Diameter (I.D.) gauge designed for valve guides will give you an accurate valve guide bore measurement. If the I.D. gauge is used, compare the readings with the specifications given. Any guides that fail these inspections should be replaced or machined.

Valve Seats

A visual inspection of the valve seats should show a slightly worn and pitted surface where the valve face contacts the seat. Inspect the seat carefully for severe pitting or cracks. Also, a seat that is badly worn will be recessed into the cylinder head. A severely worn or recessed seat may need to be replaced. All cracked seats must be replaced. A seat concentricity gauge, if available, should be used to check the seat run-out. If run-out exceeds specifications the seat must be machined (if no specification is given use 0.002 in. or 0.051mm).

Checking For Valve Leakage

② MODERATE

The cylinder heads on 4-stroke motors should be checked for valve leakage before disassembly for overhaul.

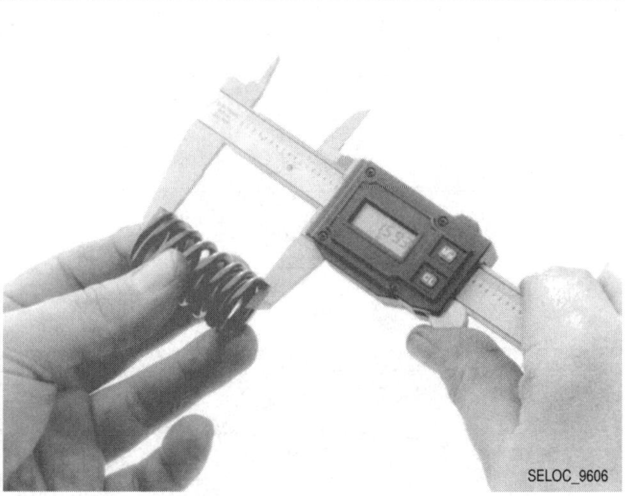

Fig. 129 Use a caliper to check the valve spring free-length

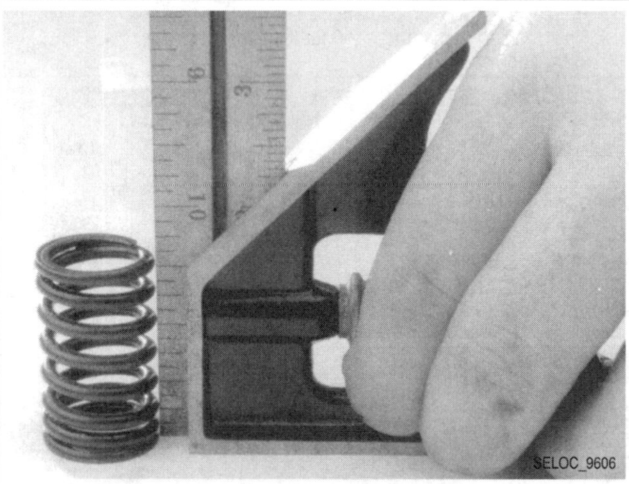

Fig. 130 Check the valve spring for squareness on a flat surface; a carpenter's square can be used

Check the cylinder head for valve leakage. If leakage or an obvious defect is present, disassemble the cylinder head (as detailed in this section) and inspect the valve seating surfaces.

1. Place the cylinder head on a suitable work surface. Make sure the intake manifold ports are located above the valves.
2. Pour solvent or kerosene into the intake and exhaust ports, and watch the valves. There should be little or no leakage from a closed valve.
3. If significant leakage is present, check the valve seats, valve faces and springs for signs of defects, wear or damage.
4. Pour the solvent or kerosene out of the cylinder head before proceeding.

Cylinder Head Surface Flatness

◆ See Figures 131 and 132

After you have cleaned the gasket surface of the cylinder head of any old gasket material, check the head for flatness.

Place a straightedge across the gasket surface. Using feeler gauges, determine the clearance at the center of the straightedge and across the cylinder head at several points. Check along the centerline and diagonally on the head surface. If the warpage exceeds 0.004 in. (0.1mm), the cylinder head must be resurfaced. Always check the specifications chart for exact specifications.

Cracks & Physical Damage

Generally, cracks are limited to the combustion chamber, however, it is not uncommon for the head to crack in a spark plug hole, port, outside of the head or in the valve spring/rocker arm area. The first area to inspect is always the hottest: the exhaust seat/port area.

A visual inspection should be performed, but just because you don't see a crack does not mean it is not there. Some more reliable methods for inspecting for cracks include Magnaflux®, a magnetic process or Zyglo®, a dye penetrant. Magnaflux® is used only on ferrous metal (cast iron) heads. Zyglo® uses a spray on fluorescent mixture along with a black light to reveal the cracks. It is strongly recommended to have your cylinder head checked professionally for cracks, especially if the engine was known to have overheated and/or leaked or consumed coolant. Contact a local shop for availability and pricing of these services.

Physical damage is usually very evident. For example, a broken mounting ear from dropping the head or a bent or broken stud and/or bolt. All of these defects should be fixed or, if unrepairable, the head should be replaced.

REFINISHING & REPAIRING

Many of the procedures given for refinishing and repairing the cylinder head components must be performed by a machine shop. Certain steps, if the inspected part is not worn, can be performed yourself inexpensively. However, you spent a lot of time and effort so far; why risk trying to save a couple bucks if you might have to do it all over again?

Valves

OEM ④ SKILLED

◆ See Figure 133

Any valves that were not replaced should be refaced and the tips ground flat. Unless you have access to a valve grinding machine, this should be done by a machine shop. If the valves are in extremely good condition, as well as the valve seats and guides, they may be lapped in without performing machine work.

It is a recommended practice to lap the valves even after machine work has been performed and/or new valves have been purchased. This insures a positive seal between the valve and seat.

Lapping The Valves

■ **Before lapping the valves to the seats, read the rest of the cylinder head procedure to insure that any related parts are in acceptable enough condition to continue.**

■ **Before any valve seat machining and/or lapping can be performed, the guides must be within factory recommended specifications.**

1. Invert the cylinder head.
2. Lightly lubricate the valve stems and insert them into the cylinder head in their numbered order.
3. Raise the valve from the seat and apply a small amount of fine lapping compound to the seat.
4. Moisten the suction head of a hand-lapping tool and attach it to the head of the valve.
5. Rotate the tool between the palms of both hands, changing the position of the valve on the valve seat and lifting the tool often to prevent grooving.
6. Lap the valve until a smooth, polished circle is evident on the valve and seat.
7. Remove the tool and the valve. Wipe away all traces of the grinding compound and store the valve to maintain its lapped location.

✳✳ WARNING

Do not get the valves out of order after they have been lapped. They must be put back with the same valve seat with which they were lapped.

Springs, Seats & Keepers

There is no repair or refinishing possible with the springs, seats and keepers. If they are found to be worn or defective, they must be replaced with new (or known good) parts.

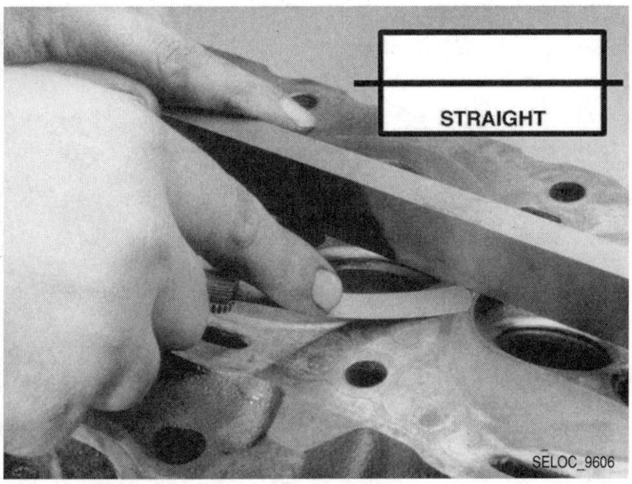

Fig. 131 Check the head for flatness across the center of the head surface using a straightedge and feeler gauge

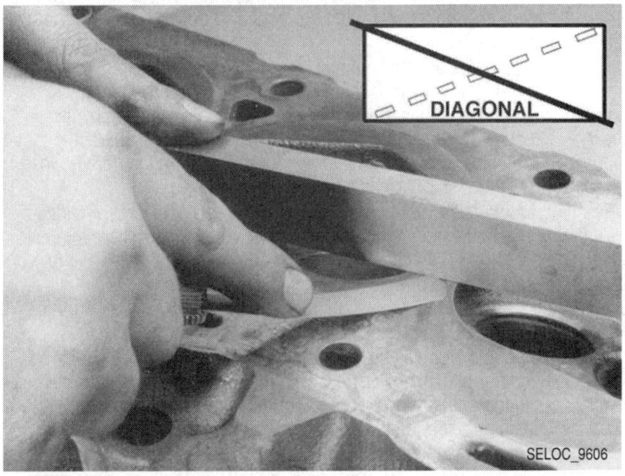

Fig. 132 Checks should also be made along both diagonals of the head surface

Fig. 133 Lapping the valve

Cylinder Head

Most refinishing procedures dealing with the cylinder head must be performed by a machine shop. Read the procedures below and review your inspection data to determine whether or not machining is necessary.

Valve Guide

■ **If any machining or replacements are made to the valve guides, the seats must be machined.**

Unless the valve guides need machining or replacing, the only service to perform is to thoroughly clean them of any dirt or oil residue.

There are only two types of valve guides used on engines: the replaceable-type (all aluminum heads) and the cast-in integral-type (most cast iron heads). There are four recommended methods for repairing worn guides.

- Knurling
- Inserts
- Reaming oversize
- Replacing

Knurling is a process in which metal is displaced and raised, thereby reducing clearance, giving a true center, and providing oil control. It is the least expensive way of repairing the valve guides. However, it is not necessarily the best, and in some cases, a knurled valve guide will not stand up for more than a short time. It requires a special knurlizer and precision reaming tools to obtain proper clearances. It would not be cost effective to purchase these tools, unless you plan on rebuilding several of the same cylinder head.

Installing a guide insert involves machining the guide to accept a bronze insert. One style is the coil-type which is installed into a threaded guide. Another is the thin-walled insert where the guide is reamed oversize to accept a split-sleeve insert. After the insert is installed, a special tool is then run through the guide to expand the insert, locking it to the guide. The insert is then reamed to the standard size for proper valve clearance.

Reaming for oversize valves restores normal clearances and provides a true valve seat. Most cast-in type guides can be reamed to accept an valve with an oversize stem. The cost factor for this can become quite high as you will need to purchase the reamer and new, oversize stem valves for all guides which were reamed. Oversizes are generally 0.003 to 0.030 in. (0.076 to 0.762mm), with 0.015 in. (0.381mm) being the most common.

To replace cast-in type valve guides, they must be drilled out, then reamed to accept replacement guides. This must be done on a fixture which will allow centering and leveling off of the original valve seat or guide, otherwise a serious guide-to-seat misalignment may occur making it impossible to properly machine the seat.

Replaceable-type guides are pressed into the cylinder head. A hammer and a stepped drift or punch (#YM-04111 / 90890-04111 - intake, #YM-04116 / 90890-04116 - exhaust) may be used to install and remove the guides. Before removing the guides, measure the protrusion on the spring side of the head and record it for installation. Use the stepped drift to hammer out the old guide from the combustion chamber side of the head. When installing, determine whether or not the guide also seals a water jacket in the head, and if it does, use the recommended sealing agent. If there is no water jacket, grease the valve guide and its bore. Use the stepped drift, and hammer the new guide into the cylinder head from the spring side of the cylinder head. A stack of washers the same thickness as the measured protrusion may help the installation process.

Valve Seats

■ **Before any valve seat machining can be performed, the guides must be within factory recommended specifications.**

■ **If any machining or replacements were made to the valve guides, the seats must be machined.**

If the seats are in good condition, the valves can be lapped to the seats, and the cylinder head assembled. See the valves procedures for instructions on lapping.

If the valve seats are worn, cracked or damaged, they must be serviced by a machine shop. The valve seat must be perfectly centered to the valve guide, which requires very accurate machining.

Cylinder Head Surface

If the cylinder head is warped, it must be machined flat. If the warpage is extremely severe, the head may need to be replaced. In some instances, it may be possible to straighten a warped head enough to allow machining. In either case, contact a professional machine shop for service.

Cracks & Physical Damage

Certain cracks can be repaired in both cast iron and aluminum heads. For cast iron, a tapered threaded insert is installed along the length of the crack. Aluminum can also use the tapered inserts, however welding is the preferred method. Some physical damage can be repaired through brazing or welding. Contact a machine shop to get expert advice for your particular dilemma.

Oil Separator Tank Cover

REMOVAL & INSTALLATION

◆ **See Figures 134, 135 and 136**

■ **The oil separator tank is only used on the 1812cc engines.**

1. Remove the engine as detailed elsewhere in this section. Removing the exhaust system may allow enough access to get away without removing the entire engine - and since you have to remove the exhaust system anyway when removing the engine, take a look or give it a try.
2. If you decide to go without removing the engine, drain the engine oil, remove the oil filter and remove the oil cooler as detailed elsewhere in this section.

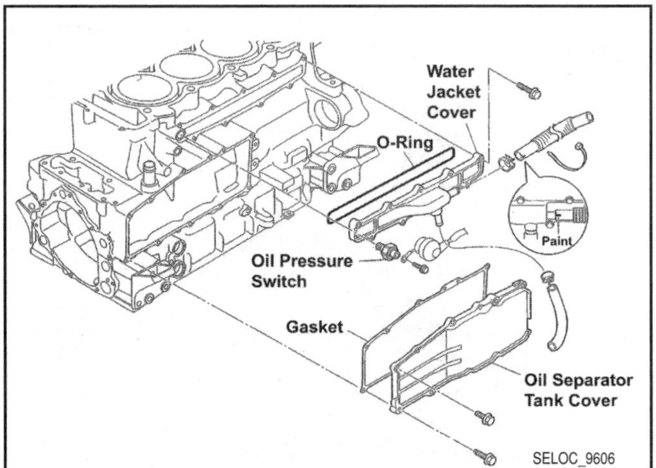

Fig. 134 A good look at the oil separator tank cover and water jacket cover - 1812cc engines

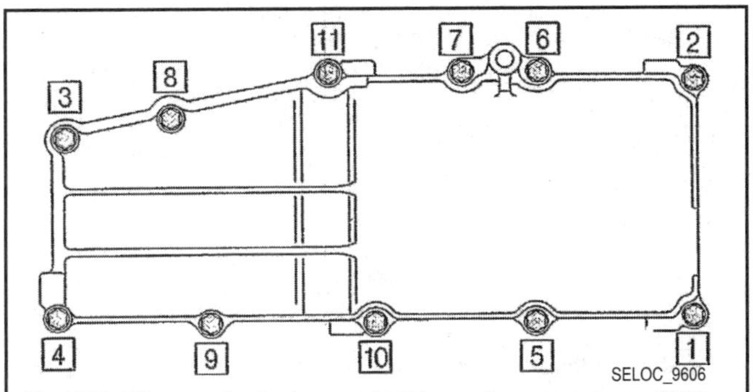

Fig. 135 Oil separator tank cover bolt loosening sequence - 1812cc engines

3. Tag and disconnect any hoses or lines that may interfere with cover removal.
4. Find the tank cover just above where the crankcase and cylinder block mate, toward the front of the engine (port side), and remove the 11 mounting bolts in the sequence shown in the illustration. There are 9 long ones and 2 shorter ones, so pay attention to where each bolt goes the 1st and 2nd bolts running from front to back along the top of the cover are the short ones.
5. Carefully pry the cover off of the block and remove the gasket.
6. If necessary, remove and replace the anode.

To install:

7. Clean the cover-to-block mating surfaces of any residual gasket material and then install a *new* gasket in place on the block.
8. Install the anode if removed.
9. Coat the threads of the mounting bolts with Loctite 242. Position the cover over the gasket and thread in all mounting bolts until finger-tight. Tighten them to 5.9 ft. lbs./71 inch lbs. (8 Nm) in the order shown.
10. Connect any lines or hoses removed to allow access.
11. If you got away without removing the engine, install the oil cooler and exhaust system. Install the oil filter and fill the engine with oil. Otherwise, install the engine.

Water Jacket Cover

REMOVAL & INSTALLATION

◆ See Figure 134

■ **The water jacket is only used on the 1812cc engines.**

1. Remove the engine as detailed elsewhere in this section. Removing the exhaust system may allow enough access to get away without removing the entire engine - and since you have to remove the exhaust system anyway when removing the engine, take a look or give it a try.
2. Tag and disconnect the 2 hoses connected to the cover and move them out of the way.
3. Pull back the cover and remove the electrical lead bolt on the oil pressure switch. Remove the switch.
4. Tag and disconnect any other hoses or lines that may interfere with cover removal.
5. Find the water jacket cover just above starter, at the top of the cylinder block on the port side. Remove the 9 mounting bolts and carefully pry off the cover and O-ring.

To install:

6. Insert a *new* O-ring into the groove in the cover.
7. Thread the oil pressure switch in and tighten it to 5.9 ft. lbs./71 inch lbs. (8 Nm). Insert the small screw through the lead ring and tighten it to 1.5 ft. lbs./18 inch lbs. (2 Nm).
8. Reconnect the 2 water lines at the manifold. Make sure that the white paint mark on the rear hose is facing the port side of the hull.
9. Reconnect any lines or hoses that you may have disconnected.
10. If you got away without removing the engine, install the exhaust system. Install the oil filter and fill the engine with oil. Otherwise, install the engine.

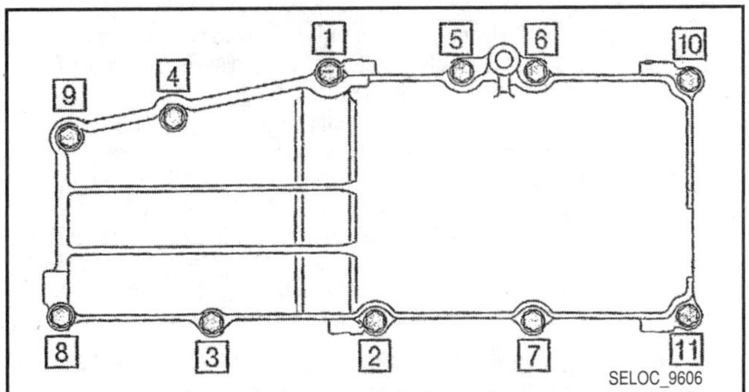

Fig. 136 Oil separator tank cover bolt tightening sequence - 1812cc engines

Cylinder Block, Crankcase & Oil Pan

REMOVAL & INSTALLATION

998cc & 1052cc Engines

◆ **See Figures 137 thru 148**

The upper cylinder block (or upper crankcase), crankcase (or lower crankcase) and oil pan comprise what is commonly referred to as the cylinder block or crankcase. You do not *remove* the crankcase so much as you remove everything else, *separate* the 3 pieces and then take out the components. Removing the engine removes the crankcase.

Separating the 3 pieces, means, basically, removing every other component attached to the unit.

1. Remove the engine as detailed previously in this section and install it to an engine stand.
2. Remove the air filter and case as detailed in the Fuel Injection section.
3. Remove the forward exhaust components as detailed in the Exhaust & Cooling section.
4. Remove the generator cover as detailed in the Ignition & Electrical section.
5. Remove the oil tank, oil pump and reduction drive gear cover as detailed previously in this section.

Phewwww, sure you want to continue?

6. Depending on the extent of what you are doing, you may wish to remove the flywheel and drive coupling. And if you do this, remove the stator and pulser coil.
7. Remove the cylinder head as detailed in this section.
8. Remove the starter and oil filter.
9. Remove the timing chain.
10. Remove the engine temperature sensor and O-ring from the right side of the cylinder block, toward the front.
11. Remove the 2 bolts and pull out the engine thermo-switch (next to the temp sensor).
12. On the other side of the block toward the rear, remove the oil pressure switch.
13. Back on the other side again, near the rear, remove the anode as detailed in the Maintenance & Tune-Up section.
14. Turn the engine over in the stand and loosen the oil pan bolts a little at a time (1/4 turn), in the order shown in the loosening sequence illustration. There are 4 long bolts and 9 shorter ones on 998cc engines, while 1052cc engines use 15 bolts, all the same length. Once all bolts are completely loose, remove them. The bolts on 998cc engines are varying lengths, so make sure you mark them for their respective holes.

✱✱ CAUTION

Be careful, the numbers embossed on the pan (some engines) are for the *tightening* sequence so make sure you loosen the bolts in the *opposite* order. Start at 13 (or 15) and work your way to 1.

15. With the bolts removed, tap the pan lightly with a rubber mallet in a few places until it breaks loose. Remove any residual gasket/sealant material from the pan and crankcase mating surfaces with a plastic scraper; being careful that nothing falls down into the crankcase.

16. The crankcase is attached to the cylinder block using 20 bolts of varying lengths - make sure you identify each bolt before removing it completely. Once again, the numbers that may be embossed on the crankcase are for the *tightening* sequence so make sure you loosen them in the *reverse* order starting at 20 and working down to 1. Loosen the bolts a little at a time (1/4 turn), in the opposite order of the tightening sequence. Once all bolts are completely loose, remove them. Again, the bolts are varying lengths, so make sure you mark them for their respective holes.

■ **The 10 inner bolts (marked 1-10) are not reusable and must be discarded. These bolts also serve as the main bearing bolts on the crankshaft.**

17. With the bolts removed, tap the crankcase lightly with a rubber mallet in a few places until it breaks loose. Remove any residual gasket/sealant material from the crankcase and cylinder block mating surfaces with a plastic scraper; being careful that nothing falls down into the cylinder block.

18. Remove the 3 dowel pins - one will also have an O-ring (rear corner, port side).

19. The 5 lower main crankshaft bearings may stay in the crankcase or remain on the journal; remove them and store in the order that they were removed as they must go back on the same journal; if not replacing the crankcase, insert them back in their positions in the case.

20. If necessary, remove the oil passage cover and O-ring from below the oil filter mounting boss. Do the same with the internal oil filter bolt. If you need to remove the oil pipe, you'll have to remove the crankshaft first.

21. Further disassembly of the cylinder block will be covered in the appropriate component section.

To install:

22. Clean the crankcase and oil pan with solvent and allow to dry completely. Use compressed air if available.

23. Make sure that there is no residual gasket material left on any of the component mating surfaces.

24. Coat the threads of the oil filter bolt with clean engine oil and tighten it to 25 ft. lbs. (35 Nm). Coat the threads of the oil passage cover bolts with Loctite 572 and tighten them to 8.7 ft. lbs./104 inch lbs. (12 Nm). Make sure you use a *new* O-ring!

25. Press the dowel pins firmly into their bores in the cylinder block. Slide a *new* O-ring onto the 3rd dowel and press it into the bore on the left rear corner.

26. Coat the crankshaft journal bearings with clean engine oil and then, if not already done, position them in their original locations in the crankcase making sure that the small tab fits into the recess/notch in the crankcase.

27. Make sure the crankcase-to-cylinder block mating surfaces are clean and dry and then run a bead of Gasket Maker® along the mating surface of

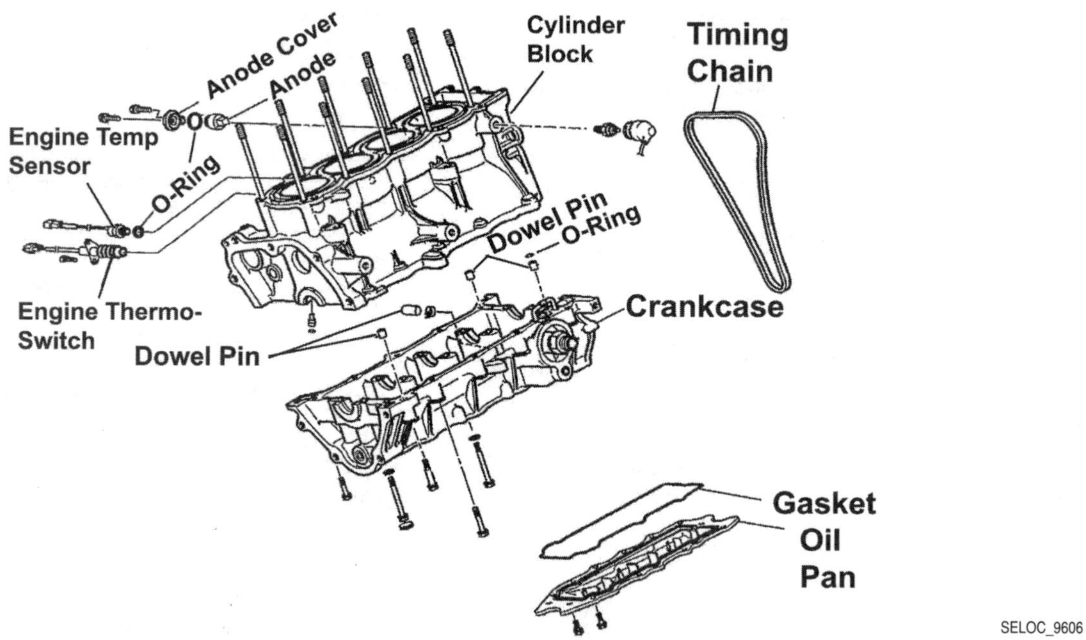

Fig. 137 Cylinder block and crankcase - 998cc engine shown, 1052cc similar

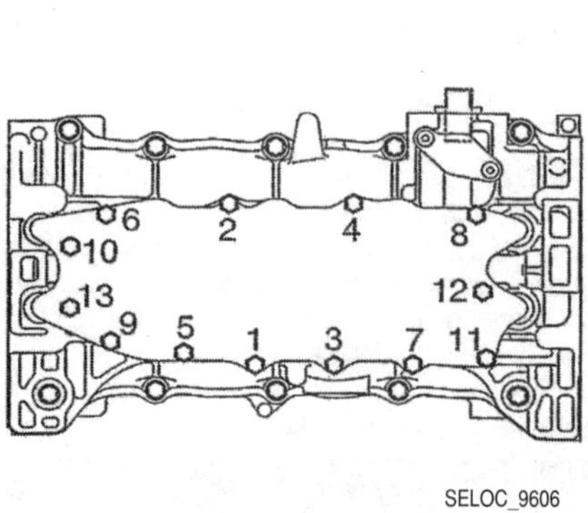

Fig. 138 Oil pan bolt loosening sequence - 998cc engines

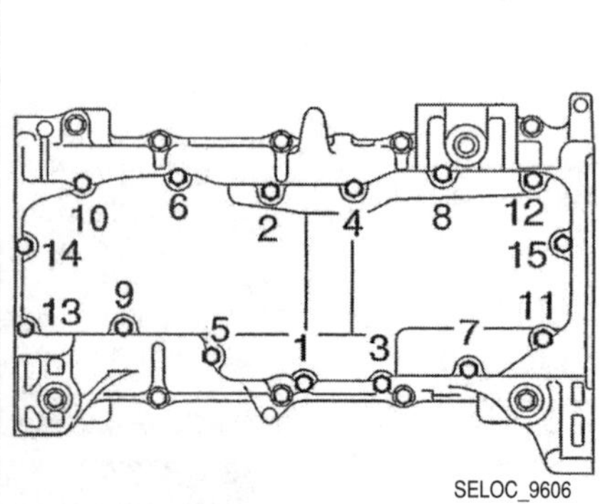

Fig. 139 Oil pan bolt loosening sequence - 2005-09 VX and 2004-07 FX w/1052cc engines

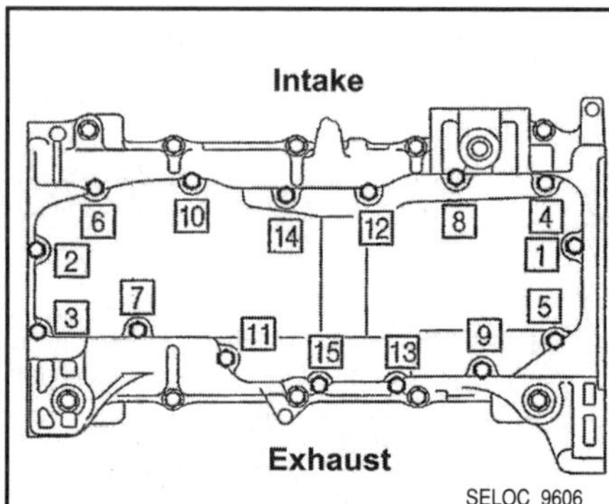

Fig. 140 Oil pan bolt loosening sequence - 2010-11 VX and 2008 FX HO w/1052cc engines

the crankcase. Make sure that you do not get any in the oil galleries or on the crankshaft journal bearings and that the bead completely surrounds each bolt hole. Carefully line up the crankcase over the alignment dowels and place it in position on the block. Coat the threads and washers of the 10 inner bolts (remember these are the new bolts) with clean engine oil and thread them in until just finger-tight. Coat the threads of the remaining bolts with Loctite 572 and do the same.

28. Tighten the 10 inner bolts, in the sequence shown, to 5.6 ft. lbs./67 inch lbs. (7.8 Nm). Loosen them in reverse order and then re-tighten (in order) to 11 ft. lbs. (15 Nm). On 998cc engines, it is not necessary to loosen the bolt, just go right to the 2nd torque figure. On 2010-11 VX engines, the 2nd tightening is to 10.6 ft. lbs. (14.7 Nm).

Now, using an angle torque gauge, tighten each bolt, in order, an additional 49° +/- 5° (just 50° on 2008 FX HO and 2010-11 VX engines). DO NOT over-tighten the bolts or they will require replacement.

■ **There is no tightening sequence illustration for the 2008 HO w/1052cc engine. Use the sequence on the case, or simply reverse the loosening sequence.**

29. Now you can continue on to the remaining bolts, tightening, in order, to 8.7 ft. lbs./106 inch lbs. (12 Nm).

30. Press a *new* gasket into the recess in the oil pan and then position it on the crankcase, making sure all holes line up. Coat the bolt threads with

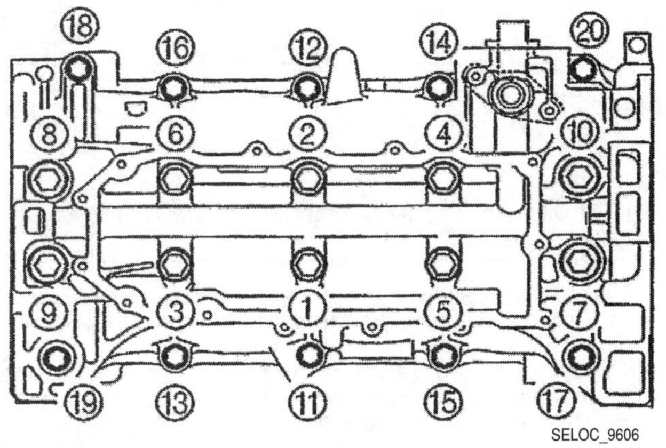

Fig. 141 Crankcase bolt loosening sequence - 2005-09 VX and 2002-07 FX with 998cc/1052cc engines

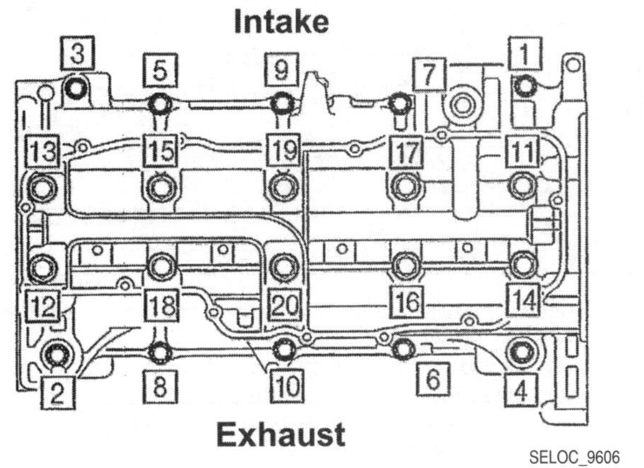

Fig. 142 Crankcase bolt loosening sequence - 2010-11 VX and 2008 FX HO w/1052cc engines

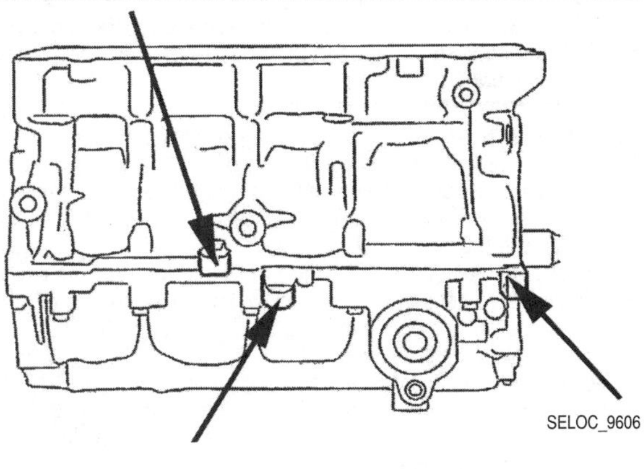

Fig. 143 Use a mallet on the crankcase in reinforced areas only

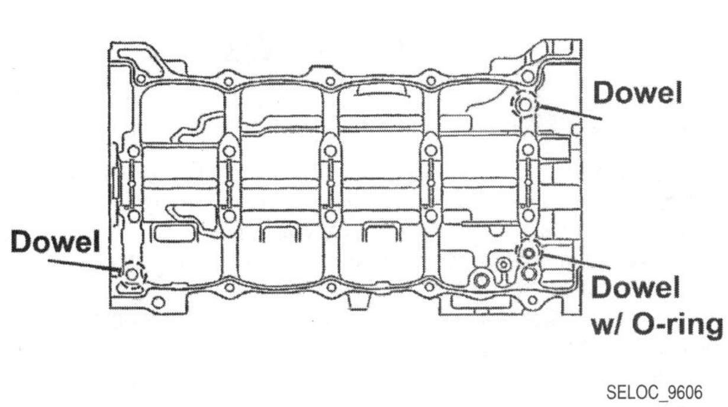

Fig. 144 Proper dowel pin positioning

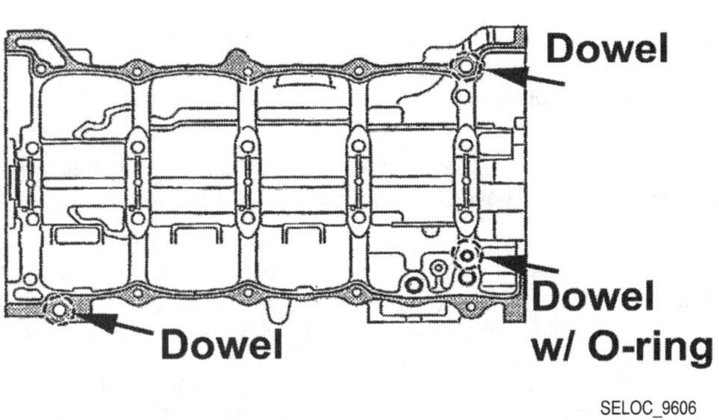

Fig. 145 Run a bead of gasket sealer along the edge of the crankcase mating surface. . .

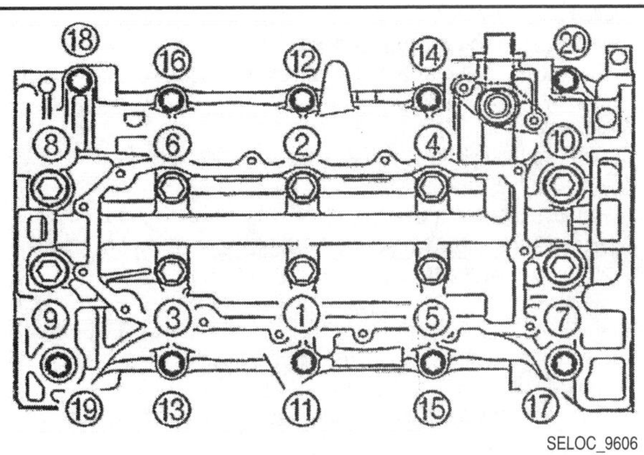

Fig. 146 . . .install the crankcase in position and then tighten the bolts in the order shown

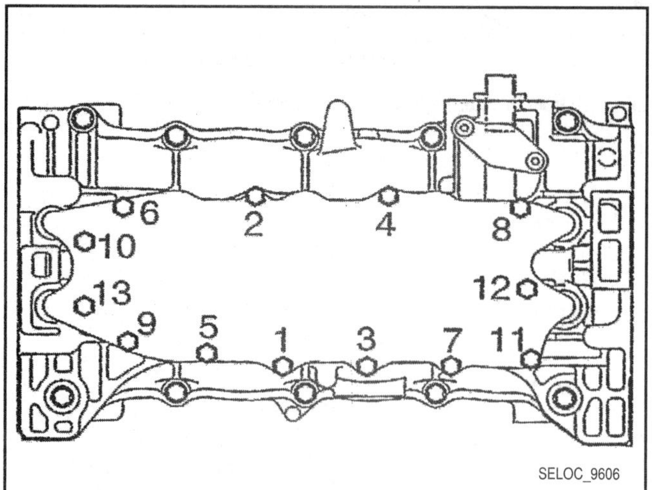

Fig. 147 Oil pan bolt tightening sequence - 998cc engines

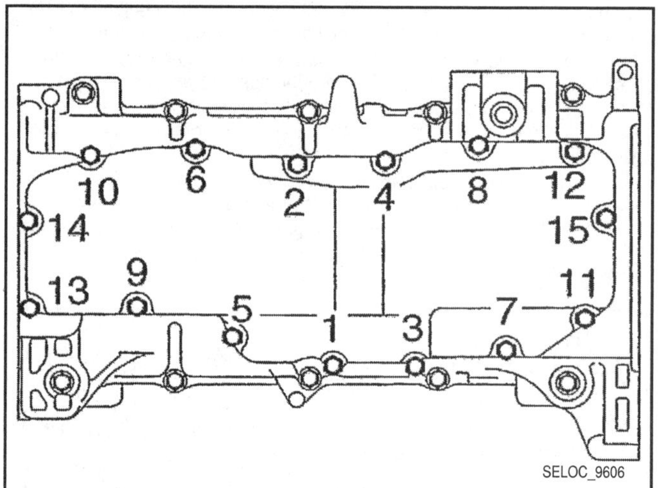

Fig. 148 Oil pan bolt tightening sequence - VX and 2004-08 FX w/1052cc engines

Loctite 572 and thread them in until finger-tight. Tighten all bolts, in the order shown, to 8.7 ft. lbs./106 inch lbs. (12 Nm).

31. Install the anode. Install and tighten the following:

- Oil pressure switch - 6.1 ft. lbs./73 inch lbs. (8.4 Nm)
- Engine thermo-switch - 5.5 ft. lbs./66 inch lbs. (7.6 Nm); 5.9 ft.lbs./71 inch lbs. (8 Nm) on the 2010-11 VX
- Engine temperature switch - 11 ft. lbs. (15 Nm)

32. Install the timing chain back onto the crankshaft sprocket.

33. Install all remaining components as detailed in their respective procedures.

34. Install the engine.

1812cc Engines

OEM 3 DIFFICULT

◆ **See Figures 149 thru 154 and 163**

The upper cylinder block (or upper crankcase), crankcase (or lower crankcase) and oil pan comprise what is commonly referred to as the cylinder block or crankcase. You do not *remove* the crankcase so much as you remove everything else, *separate* the 3 pieces and then take out the components. Removing the engine removes the crankcase.

Separating the 3 pieces, means, basically, removing every other component attached to the unit.

1. Remove the engine as detailed previously in this section and install it to an engine stand.

2. Remove the generator cover as detailed in the Ignition & Electrical section.

3. Remove the oil filter, oil cooler and oil pump as detailed previously in this section.

Phewwww, sure you want to continue?

4. Depending on the extent of what you are doing, you may wish to remove the flywheel and drive coupling. And if you do this, remove the stator and pulser coil.

5. Remove the cylinder head as detailed in this section.

6. Remove the starter.

7. Remove the timing chain.

8. Remove the engine temperature sensor and O-ring from the right side of the cylinder block, toward the rear.

9. Remove the 2 bolts and pull out the engine thermo-switch (right side, toward the front).

10. On supercharged engines, remove the knock sensor from the right side of the cylinder bloc.

11. On the other side of the block toward the rear, remove the oil pressure switch.

12. Back on the other side again, near the rear, remove the anode as detailed in the Maintenance & Tune-Up section.

13. Remove the oil dipstick tube.

14. Loosen the mounting bolts (3 per bracket) and remove all 4 mount brackets. There will be 2 dowel pins per bracket - make sure they stay with the crankcase.

15. Disconnect the oil line at the rear of the oil pan if not done so already.

16. Turn the engine over in the stand and loosen the oil pan bolts a little at a time (1/4 turn), in the *opposite* order of that embossed on the pan (this is the tightening sequence). There are 16 long bolts, 1 medium bolt and 1 short one. Once all bolts are completely loose, remove them. Make sure you mark them for their respective holes and lengths.

✻✻ CAUTION

Be careful, the numbers embossed on the pan are for the *tightening* sequence so make sure you loosen the bolts in the *opposite* order. Start at 18 and work your way to 1.

17. With the bolts removed, tap the pan lightly with a rubber mallet in a few places until it breaks loose. Remove any residual gasket/sealant material from the pan and crankcase mating surfaces with a plastic scraper; being careful that nothing falls down into the crankcase. There are 2 dowel pins, be sure they stay with the pan.

18. If necessary, remove the 10 bolts and lift off the oil pan baffle plate.

19. Remove the 4 bolts connecting the 2 oil pipe flanges to the front of the oil pan. The lower bolt on the starboard side is shorter than the other 3, so take note of this for installation.

20. Remove the 3 pipe mounting bolts and lift out the 2 pipes. Throw the flange gaskets away and pull the grommet off of the port side pipe.

21. On FX HO and VX models, remove the small bolt and lift the inner baffle plate from the crankcase.

22. The crankcase is attached to the cylinder block using 22 bolts of two lengths - make sure you identify each bolt before removing it completely. There are 12 outer bolts and 10 inner bolts (which cannot be re-used). Once again, the numbers that may be embossed on the crankcase are for the *tightening* sequence so make sure you loosen them in the *reverse* order starting at 22 and working down to 1 (or just follow the loosening sequence as shown). Loosen the bolts a little at a time (1/4 turn), in the opposite order of the tightening sequence. Once all bolts are completely loose, remove them.

■ **The 10 inner bolts (marked 1-10 on the case) are not reusable and must be discarded. These bolts also serve as the main bearing bolts on the crankshaft.**

23. With the bolts removed, use a small prybar at the points illustrated until it breaks loose. Remove any residual gasket/sealant material from the crankcase and cylinder block mating surfaces with a plastic scraper; being careful that nothing falls down into the cylinder block.

24. Remove the dowel pin and O-ring (port side, toward the front).

25. The 5 lower main crankshaft bearings may stay in the crankcase or remain on the journal; remove them and store in the order that they were removed as they must go back on the same journal; if not replacing the crankcase, insert them back into their positions in the case.

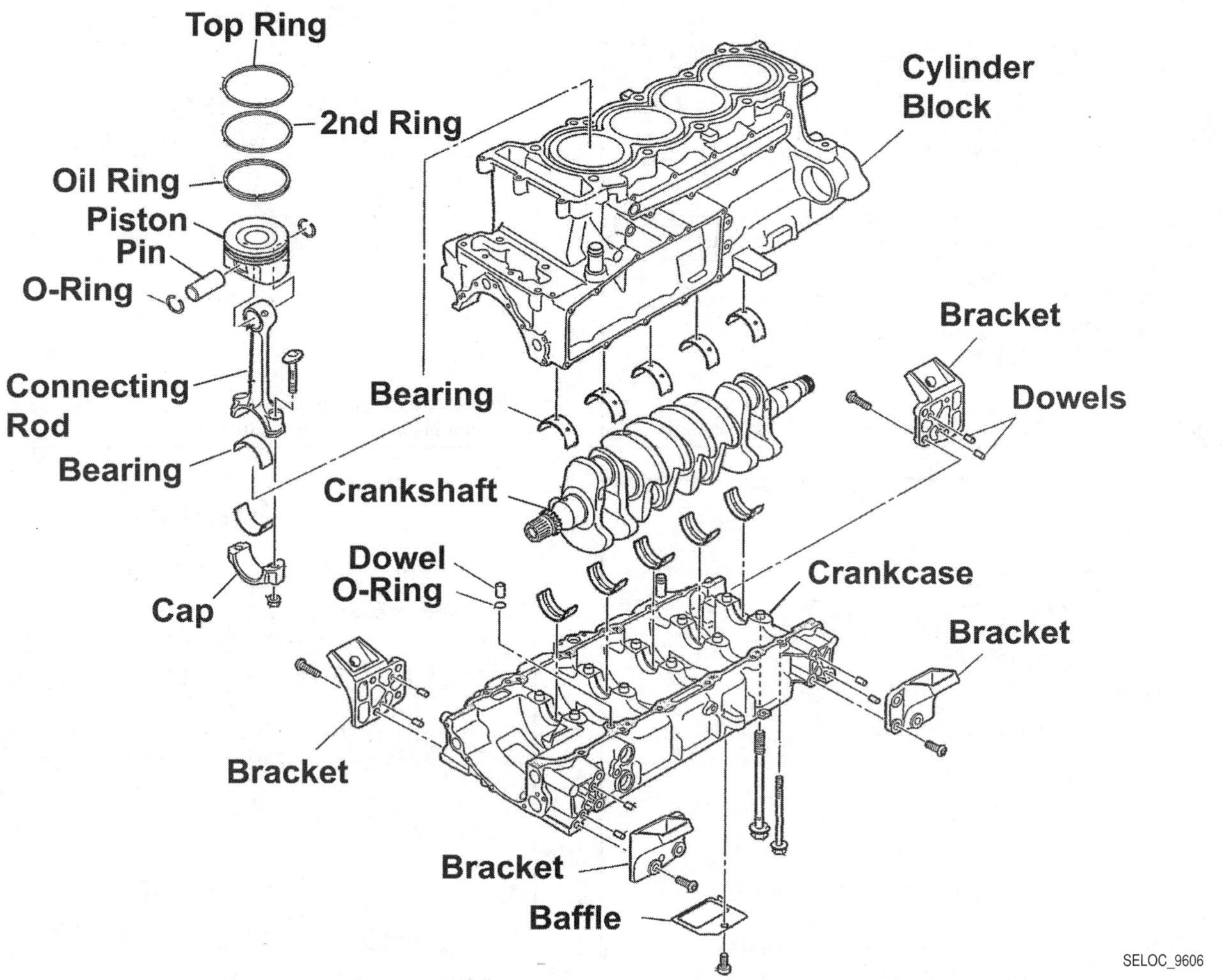

Fig. 149 Exploded view of the cylinder block, crankcase and internal components - 1812cc engines

26. Further disassembly of the cylinder block will be covered in the appropriate component section.

To install:

27. Clean the crankcase and oil pan with solvent and allow to dry completely. Use compressed air if available.

28. Make sure that there is no residual gasket material left on any of the component mating surfaces.

29. Press the dowel pin firmly into its bore in the cylinder block. Slide a *new* O-ring onto the dowel and press it into the bore on the left side.

30. Coat the crankshaft journal bearings with clean engine oil and then, if not already done, position them in their original locations in the crankcase making sure that the small tab fits into the recess/notch in the crankcase.

31. Make sure the crankcase-to-cylinder block mating surfaces on each side are clean and dry, and then run a bead of ThreeBond® 1280B along the mating surfaces of the crankcase. Make sure that you do not get any in the oil galleries or on the crankshaft journal bearings and that the bead completely surrounds each bolt hole. Carefully line up the crankcase over the alignment dowels and place it in position on the block.

32. Coat the threads and washers of the 10 inner bolts (remember these are the *new* bolts) with clean engine oil and thread them in until just finger-tight. Coat the threads of the remaining bolts with oil and do the same.

33. Tighten the 10 inner bolts, in the sequence shown, to 22.1 ft. lbs. (7.8 Nm). Now tighten the 12 outer bolts, in the sequence shown, to 7.4 ft. lbs./88 inch lbs. (10 Nm).

Next, using an angle torque gauge, tighten each of the 10 inner bolts, in order, an additional 90°. DO NOT over-tighten the bolts or they will require replacement.

34. On FX HO and VX models, position the small baffle plate and the tighten the bolt to 7.4 ft. lbs./88 inch lbs. (10 Nm).

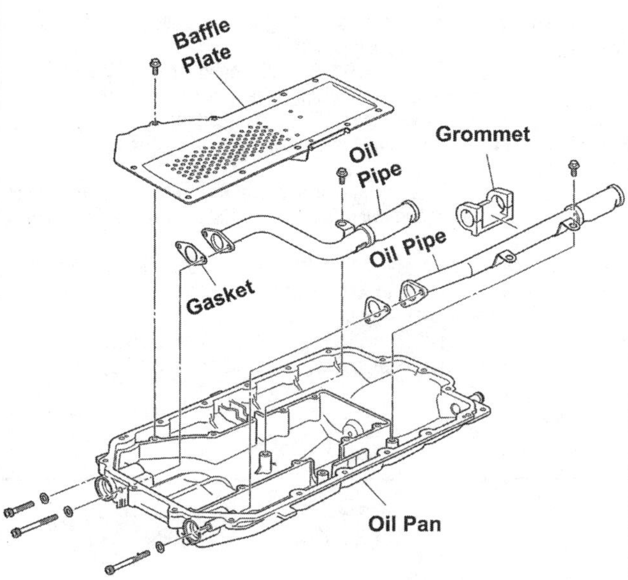

Fig. 150 A good look at the oil pan and components - 1812cc engines

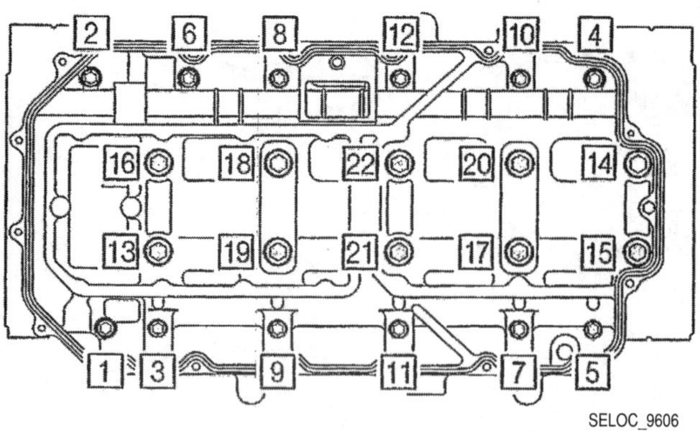

Fig. 151 Crankcase bolt loosening sequence - 1812cc engines

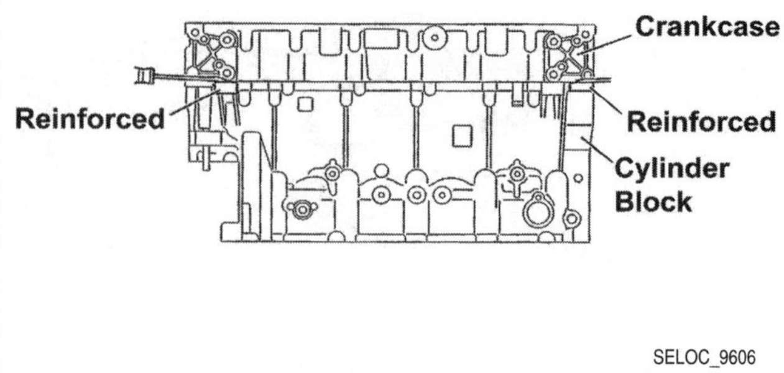

Fig. 152 Use a small prybar to separate the crankcase from the cylinder block on the starboard side. . .

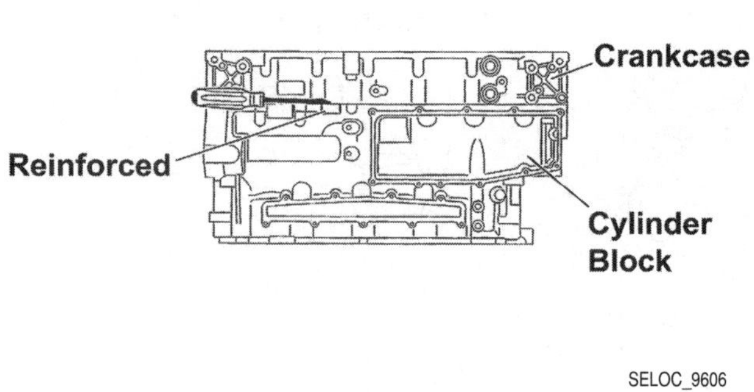

Fig. 153 . . .and the port side - 1812cc engines

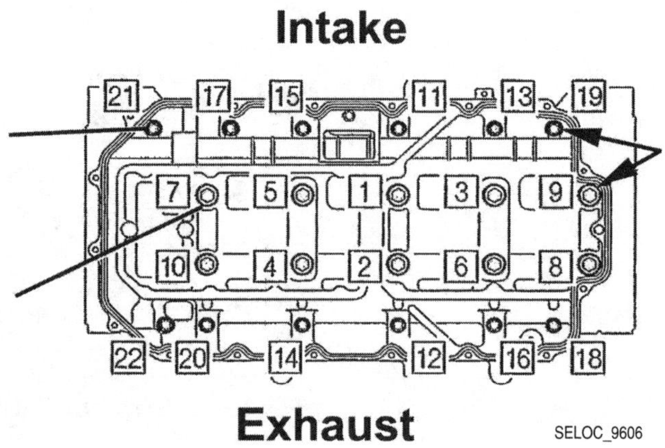

Fig.154 Crankcase bolt tightening sequence - 1812cc engines

35. Pop the grommet back onto the port oil pipe and then position both pipes into the oil pan. Use *new* gaskets on each flange. Tighten the 3 bolts to 8.9 ft. lbs./106 inch lbs. (12 Nm).

36. Coat the threads of the 4 flange bolts with Loctite 271 and then tighten them 6.6 ft. lbs./43 inch lbs. (9 Nm). Remember where the short one goes??

37. Lay the baffle plate in position and tighten the bolts to 8.9 ft. lbs./106 inch lbs. (12 Nm) in an alternating pattern.

38. Run a bead of ThreeBond® 1280B all around the mating surface of the oil pan and then position it on the crankcase, making sure all holes line up. Coat the bolt threads with Loctite 572 and thread them in until finger-tight. Tighten all bolts, in the order shown (on the pan), to 7.4 ft. lbs./88 inch lbs. (10 Nm).

39. Position each of the mount brackets over their respective dowel pins and then coat the bolt threads with Loctite 271. Tighten each bolt to 36.9 ft. lbs. (50 Nm).

40. Install the anode. Install the switches and sensors, tightening each securely (or refer to their respective sections for the actual torque figure).

41. Install the oil dipstick and tube.

42. Install the timing chain back onto the crankshaft sprocket.

43. Install all remaining components as detailed in their respective procedures.

44. Install the engine.

INSPECTION

OEM ④ SKILLED

◆ See Figures 155 thru 158

■ **Images shown are typical and not meant to be representative of your specific engine.**

Inspect the cylinder block and cylinder bores for cracks or other damage. Remove carbon with a fine wire brush on a shaft attached to an electric drill or use a carbon remover solution.

** WARNING

If the cylinder block is to be submerged in a carbon removal solution, all external fuel or electrical components (as well as all anodes) should be removed for their protection. But then, you've probably already done this haven't you?

Use an inside micrometer or telescopic gauge and micrometer to check the cylinders for wear. Check the bore for out-of-round and/or oversize bore. If the bore is tapered, out-of-round, or worn more than the wear limit specified by the manufacturer, the cylinders should be re-bored - provided oversize pistons and rings are available.

Check with a local Yamaha dealer prior to boring. If oversize pistons and matching rings are not available, the block must then be replaced.

■ **Oversize piston weight is approximately the same as a standard size piston. Therefore, it is not usually necessary to re-bore all cylinders in a block just because 1 cylinder requires boring.**

Traditionally, Yamaha cylinder sleeves are an integral part of the die cast cylinder block and cannot be replaced. In other words, the cylinder cannot be sleeved. But it's never a bad idea to double-check with the machine shop.

Four inside cylinder bore measurements must be taken for each cylinder to determine an out-of-round condition, the maximum taper, and the maximum bore diameter.

In the accompanying illustration, measurements **D1** and **D2** are diameters measured at about the top of the bore at right angles to each other. Measurements **D3** and **D4** are diameters measured toward the center of the bore, while measurements **D5** and **D6** are diameters measured toward the bottom the bore, again both at right angles to each other. **A** (20mm), **B**

(40mm) and **C** (60mm) are the depths that the measurements should take place, while **D** and **E** are the directions. Depths on the 1812cc engines are **A** (0.8 in./20mm), **B** (2 in./50mm) and **C** (3.1 in./80mm)

Out-of-Round

Measure the cylinder diameter at **D1** and **D2** and **D5** and **D6**. The manufacturer requires the maximum of the 4 measurements should be less than 0.002 inch (0.05mm) for most models.

Maximum Taper

Measure the cylinder diameter at **D1**, **D2**, **D5**, and **D6**. Take the largest of the **D1** or **D5** measurements and subtract the smallest measurement at **D2** or **D6**. The result is the cylinder taper - it should be less than 0.003 inch (0.08mm).

Bore Wear Limit

The maximum cylinder diameter **D1**, **D2**, **D3**, **D4**, **D5**, and **D6**must not exceed the bore wear limits before the bore is re-bored for the first time:

- 998cc engines - 2.9213 in. (74mm)
- 1052cc engines - 2.9961 in. (76mm)
- 1812cc engines - 3.3858-3.3864 in. (86-86.015mm)

These limits are only imposed on original parts because the sealing ability of the rings would be lost, resulting in power loss, increased engine noise, unnecessary vibration, piston slap, and excessive oil consumption.

On these motors, bore wear limit is determined by piston clearance and the availability of oversize pistons.

Piston Clearance

Piston clearance is the difference between a maximum piston diameter and a minimum cylinder bore diameter. If this clearance is excessive, the engine will develop the same symptoms as for excessive cylinder bore wear - loss of ring sealing ability, loss of power, increased engine noise, unnecessary vibration, and excessive oil consumption.

Maximum piston diameter is described elsewhere in this section. Minimum cylinder bore diameter is usually determined by measurement **D3** or **D4** also described earlier in this section.

If the piston clearance exceeds the limits in the specifications chart, either the piston or the cylinder block must be replaced.

Calculate the piston clearance by subtracting the maximum piston skirt diameter from the maximum cylinder bore measurement and compare the results for the model being serviced.

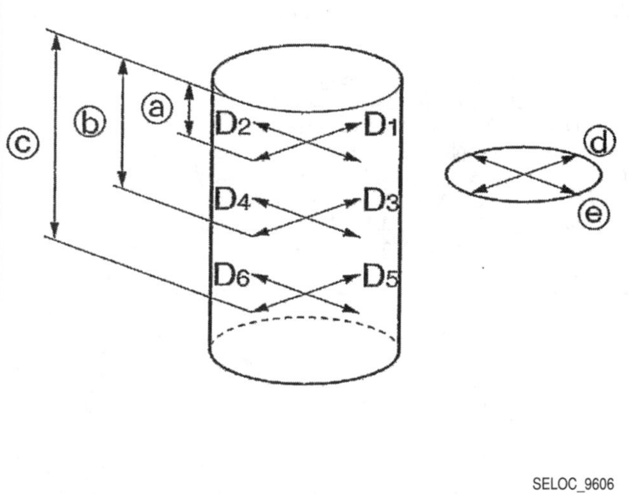

Fig. 155 Top view and cross section of a typical cylinder to indicate where measurements should be made for wear, taper and out of round

Fig. 156 Check the cylinder bore for taper using an inside micrometer. Measure near the top, in the middle and near the bottom

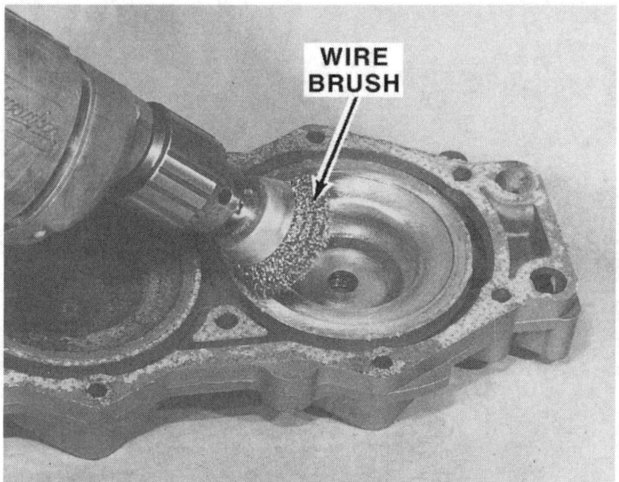

Fig. 157 Using a drill mounted wire brush to clean carbon deposits from the cylinder head

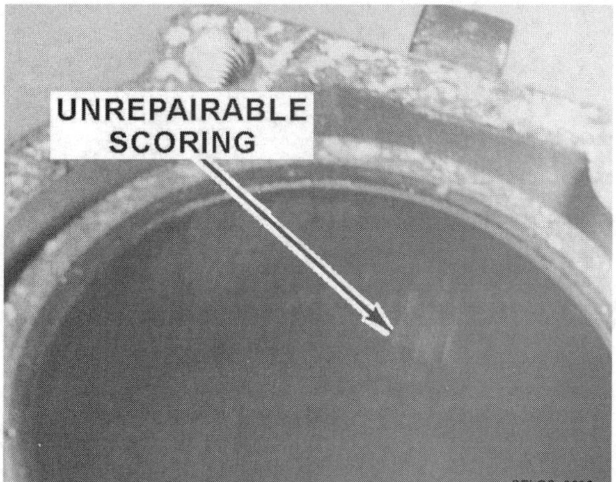

Fig. 158 This cylinder was scored beyond repair when a piston ring broke and worked its way into the combustion chamber

HONING CYLINDER WALLS

OEM ④ SKILLED

◆ See Figures 159 and 160

■ **Images shown are typical and not meant to be representative of your specific engine.**

Hone the cylinder walls lightly to seat the new piston rings, as outlined in this section. If the cylinders have been scored, but are not out-of-round or the bore is rough, clean the surface of the cylinder with a cylinder hone as described in the following procedures.

■ **If overheating has occurred, check and resurface the spark plug end of the cylinder block, if necessary. This can be accomplished with 240-grit sandpaper and a small flat block of wood.**

To ensure satisfactory engine performance and long life following the overhaul work, the honing work should be performed with patience, skill, and in the following sequence:

1. Follow the hone manufacturer's recommendations for use of the hone and for cleaning and lubricating during the honing operation. A Christmas tree hone may also be used.
2. Pump a continuous flow of honing oil into the work area. If pumping is not practical, use an oil can. Apply the oil generously and frequently on both the stones and work surface.
3. Begin the stroking at the smallest diameter. Maintain a firm stone pressure against the cylinder wall to assure fast stock removal and accurate results.
4. Expand the stones as necessary to compensate for stock removal and stone wear. The best cross hatch pattern is obtained using a stroke rate of 30 complete cycles per minute. Again, use the honing oil generously.
5. Hone the cylinder walls only enough to deglaze the walls.
6. After the honing operation has been completed, clean the cylinder bores with hot water and detergent. Scrub the walls with a stiff bristle brush and rinse thoroughly with hot water. The cylinders must be thoroughly cleaned to prevent any abrasive material from remaining in the cylinder bore. Such material will cause rapid wear of new piston rings, the cylinder bore, and the bearings.
7. After cleaning, swab the bores several times with engine oil and a clean cloth, and then wipe them dry with a clean cloth. Never use kerosene or gasoline to clean the cylinders.
8. Clean the remainder of the cylinder block to remove any excess material spread during the honing operation.

■ **If oversize pistons are not available, the local marine shop may have the facilities to knurl the piston, making it larger. If installing oversize or knurled pistons, it should be remembered that a honed cylinder block is just surface finished. It is not parallel and therefore if an oversized or knurled piston is installed in this bore, the piston will soon seize at the lower narrow end of the bore. When installing oversize or knurled pistons, the block must be bored oversize.**

CYLINDER BLOCK WARPAGE

OEM ② MODERATE

◆ See Figures 161 thru 165

■ **Images shown are typical and not meant to be representative of your specific engine.**

First, check to be sure all old gasket material has been removed from the contact surfaces of the block and the cylinder head. Clean both surfaces down to shiny metal, to ensure a true measurement.

Next, place a straightedge across the gasket surface. Check under the straightedge with a suitably sized feeler gauge (see the Engine Specifications chart). The warpage limit is normally 0.004 inch (0.1mm). Move the straightedge to at least eight different locations. If the feeler gauge can pass under the straightedge - anywhere contact with the other is made -

Fig. 159 Refinishing the cylinder wall using a ball hone. Always keep the home moving and constantly flood the cylinder with honing oil

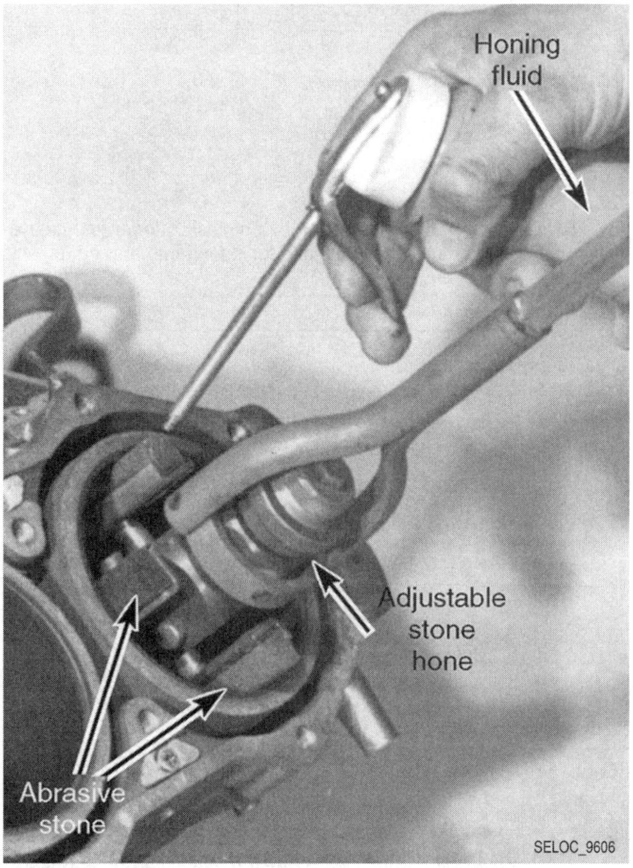

Fig. 160 A machinist can use a hone to cut the bore oversize

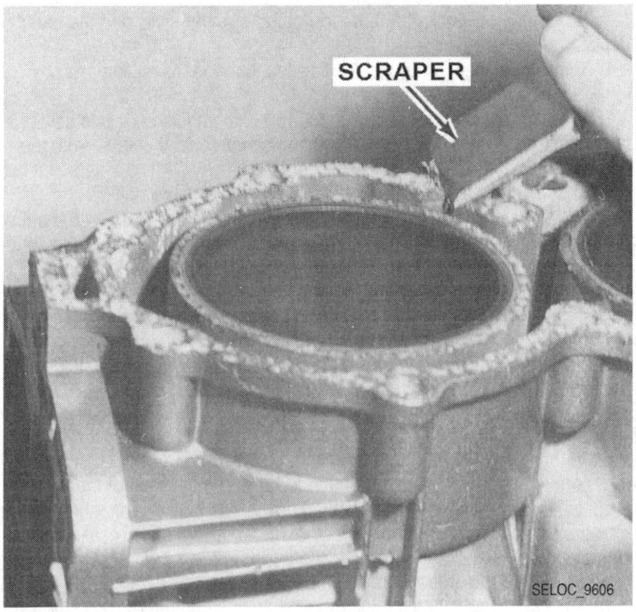

Fig. 161 All traces of gasket material must be removed to ensure accuracy

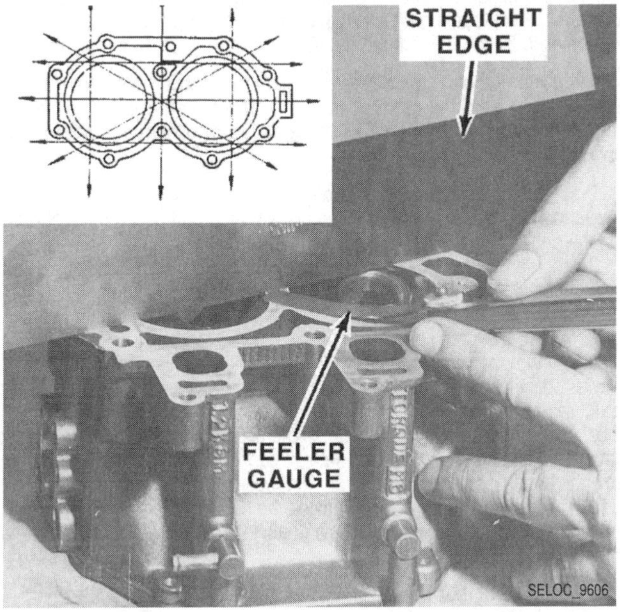

Fig. 162 Use a straightedge and feeler gauge to check the cylinder head-to-powerhead mating surfaces

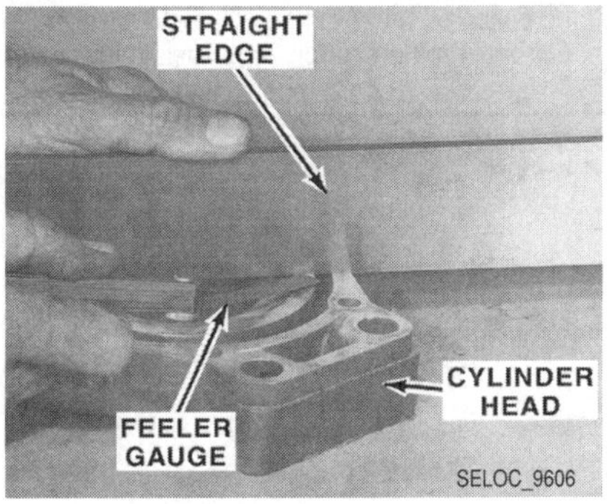

Fig. 163 Use a straightedge and feeler gauge to check for cylinder head warpage

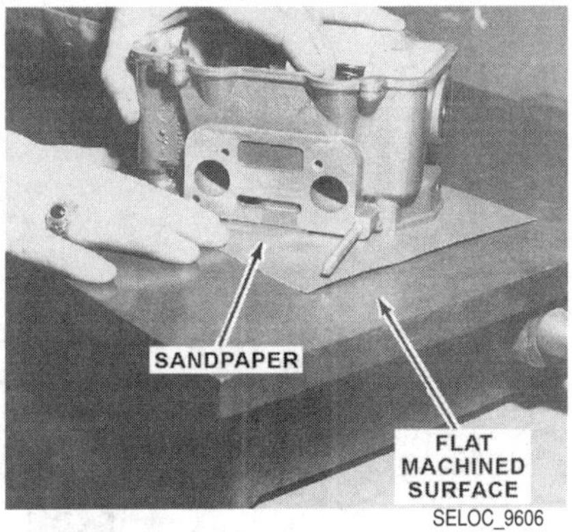

Fig. 164 Use only a flat machined surface and sandpaper to correct warpage on a cylinder head

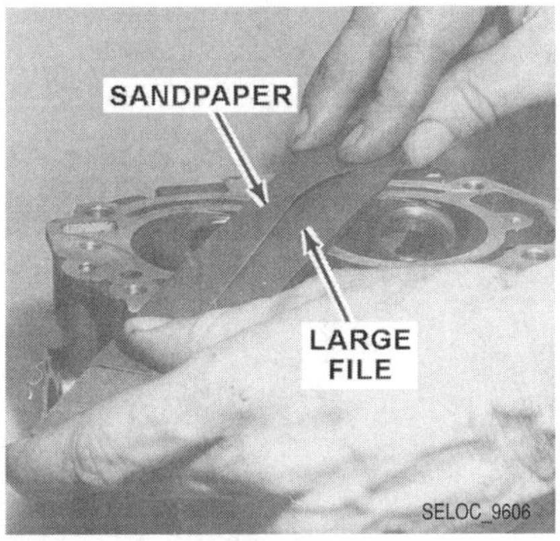

Fig. 165 When a flat machined surface is not available, a large flat file wrapped with sandpaper may be used

the surface will have to be resurfaced.

The block or the cylinder head may be resurfaced by placing the warped surface on 400-600 grit wet sandpaper, with the sandpaper resting on a flat machined surface. If a machined surface is not available a large piece of glass or mirror may be used. Do not attempt to use a workbench or similar surface for this task. A workbench is never perfectly flat and the block or cylinder head will pick up the imperfections of the surface and the warpage will be made worse.

Sand - work the warped surface on the wet sandpaper using large figure 8 motions. Rotate the block, or head, through 180° (turn it end for end) and spend an equal amount of time in each position to avoid removing too much material from one side.

If a suitable flat surface is not available, the next best method is to wrap 400-600 grit wet sandpaper around a large file. Draw the file as evenly as possible in one sweep across the surface. Do not file in one place. Draw the file in many, many directions to get as even a finish as possible.

As the work moves along, check the progress with the straightedge and feeler gauge. Once the 0.004 inch (0.1mm) feeler gauge will no longer slide under the straightedge, consider the work completed.

If the warpage cannot be reduced using one of the described methods, the block or cylinder head should be replaced (or resurfaced by a machine shop, if possible).

Pistons, Connecting Rods & Crankshaft

REMOVAL & INSTALLATION

OEM ③ DIFFICULT

◆ **See Figures 149 and 166 thru 175**

1. Remove the engine and attach it to an engine stand.
2. Remove the cylinder head.
3. Remove the oil pan and separate the crankcase from the cylinder block.
4. Rotate the engine onto its side. Use the crankshaft gear at the rear and rotate the shaft so that you have access to each connecting rod and cap. Use a marker (or scribe) and mark the cylinder number on both the rod and cap. You can also run a small line across the two where they mate. Repeat this for each cylinder.
5. Now you can mark the top of each piston with its respective cylinder number as well.
6. Spin the engine back around so the crankshaft is on top. Rotate the crankshaft until the #1 cylinder connecting rod is at its lowest (which would be highest since you're upside down) and then remove the 2 connecting rod

cap nuts (and bolts). Loosen them alternately and a little at a time. Remove the cap and lower bearing. You may have to tap the cap lightly with a small rubber mallet. Repeat this for the remaining pistons.

■ **As you remove each successive connecting rod cap, you need to have someone hold the crankshaft in position while turning it because with the crankcase removed, the con rods are the only thing holding it in place.**

7. Lift out the crankshaft and upper main bearing inserts, making sure the bearing inserts are identified so as to keep them with the correct journal. Spin the engine around so the block is right side up again.

8. Insert a wooden dowel (or hammer handle) into the cylinder from the top of the block and carefully press the piston assembly out of the block through the bottom. Once out, it's a good idea to reinstall the bearing and cap loosely so they don't get mixed up.

9. Lay the piston and rod assembly on a clean work surface and remove the piston pin snap-rings from the small groove in the pin. Make sure that there are no burrs in the pin groove or piston pin bore. Use a small socket or dowel and press the pin out of the piston and rod. Separate the rod from the piston and lay them aside with all parts staying together - you can throw the snap-rings away as they are not reusable.

10. Although you can probably remove the rings with your fingers, we recommend using a ring expander and removing the 3 rings starting with the top one.

■ **Be very careful that you do not scratch the piston skirt while removing the rings.**

11. Repeat the previous 3 steps on the remaining pistons.

To install:

12. Clean all components in a suitable solvent and allow to dry thoroughly; using compressed air if available.

13. Visually inspect all components for cracks, burrs or other damage. Clean any carbon deposits from the top of the cylinder walls and the top of the pistons.

14. Perform all necessary procedures found in the Inspection and Measurement sections.

15. Use a ring expander and slide the top compression ring into its slot so that the small **1T** near the opening is facing upward. Do the same with the 2nd ring **2T**.

16. Install the lower oil rail, the spacer and then the top rail.

17. On 998cc and 1052cc engines, lay the piston on its side so the arrow is pointing up. Grab the corresponding connecting rod and position in into the piston skirt so that the pin holes line up and the small **YD** marking on the side of the rod is facing to your right as you look into the inside of the piston - remember, when the piston is installed, the marking has to be facing the front of the engine..

On 1812cc engines, lay the piston on its side so the piston pin holes are on each side and the small 'front' mark is facing to one side or the other. Grab the corresponding connecting rod and position it into the piston skirt so that the pin holes line up and the small **Y** marking on the side of the rod is facing to the same side as the 'front' mark on the piston - remember, when the piston is installed, the 2 marks have to be facing the front of the engine..

18. Coat the piston pin with clean engine oil and slowly press through the skirt bore and connecting rod until you can see the snap-ring recess on both sides of the pin. Install *new* snap-rings into the groove on each side of the pin and make sure that they seat fully in the groove.

19. On 998cc and 1052cc engines, hold the piston so the **YD** on the connecting rod is facing to your right. The intake valve depressions in the dome should be away from you and the arrow should be pointing toward you. Holding the piston assembly in this manner, refer to the piston ring positioning illustration and rotate each ring to match the image.

On 1812cc engines, hold the piston so the **Y** on the connecting rod is facing to your left (as does the 'front' mark on the piston). Holding the piston

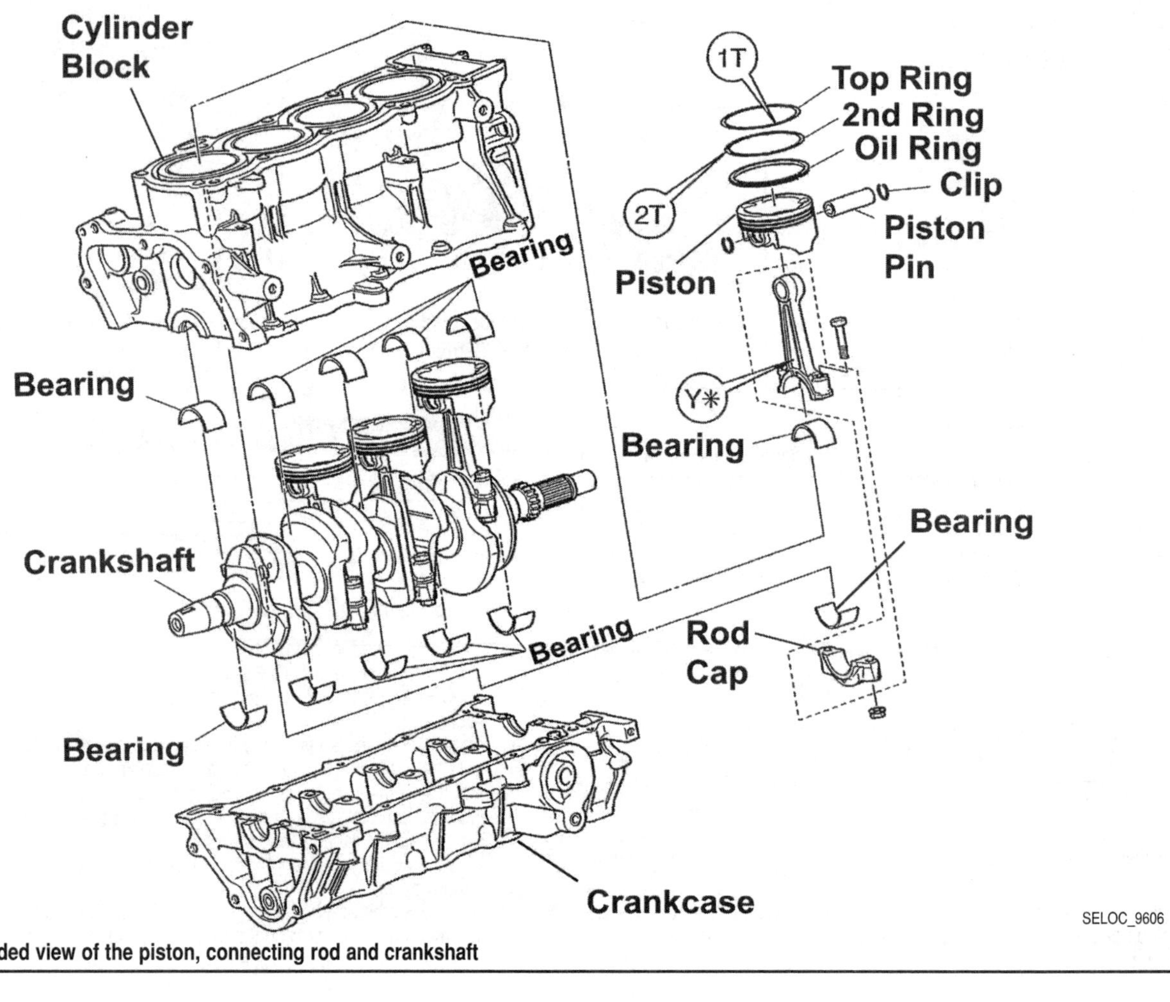

Fig. 166 Exploded view of the piston, connecting rod and crankshaft

assembly in this manner, refer to the piston ring positioning illustration and rotate each ring to match the image.

20. Being very careful not to disturb the ring positions, coat the piston and cylinder bore liberally with clean engine oil.

21. Fit the upper bearing (coat with oil first) into the connecting rod big end so that the small projection in the bearing pops into the notch in the underside of the rod.

22. Install *new* connecting rod cap bolts through the holes in the bottom of the rod.

23. Install a ring compressor (#YM-08037 / 90890-05158) over the piston and rings. The trick is to tighten the compressor just enough to compress the rings into the grooves, but not so tight as to restrict your ability to slide the entire assembly through the tool. Position the piston over its corresponding cylinder so that the mark on the rod is facing forward and then insert it into the cylinder. When the compressor tool seats on the top of the cylinder block, use a wooden dowel to press the piston slowly into the cylinder until it has cleared the top of the bore.

■ **If resistance is felt while pressing the piston into the cylinder, pull everything back out and check the ring positioning before starting again.**

24. Repeat the last few steps for the remaining cylinders. Once all are in, flip the engine back over.

25. Position the upper bearing inserts into their correct spots in the cylinder block so that the small projection fits into the notch. Lay the crankshaft in place. Rotate the crank until the journal for the #1 piston is at its lowest and then press the piston all the way through until the connecting rod big end seats on the crankshaft journal.

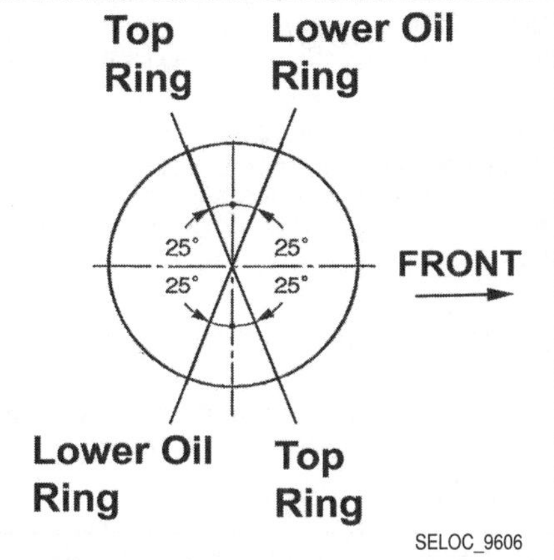

Fig. 167 Piston ring positioning - 998cc engines

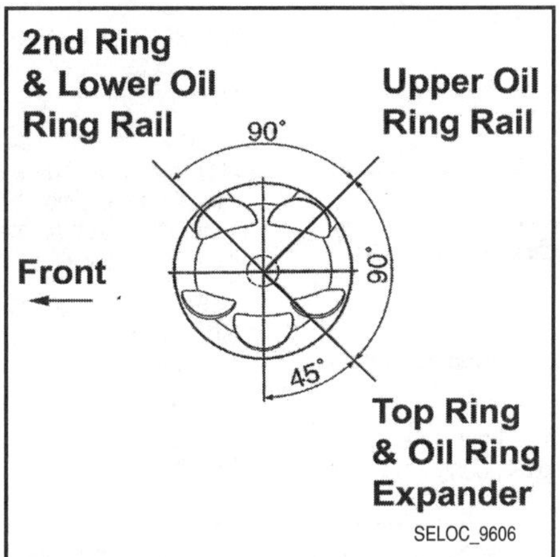

Fig. 168 Piston ring positioning - 1052cc engines

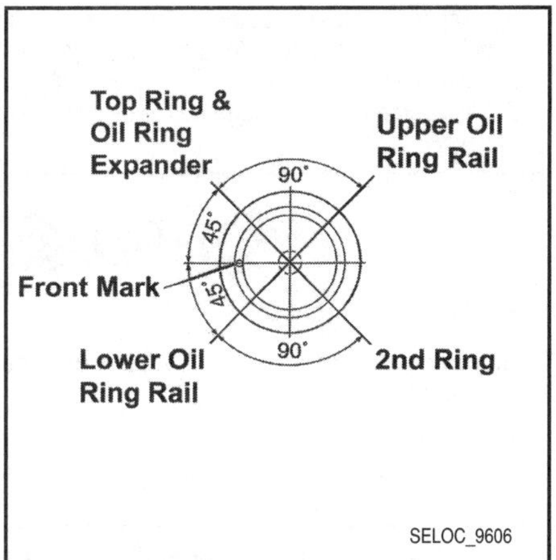

Fig. 169 Piston ring positioning - FZ, FX HO and VX w/1812cc engines

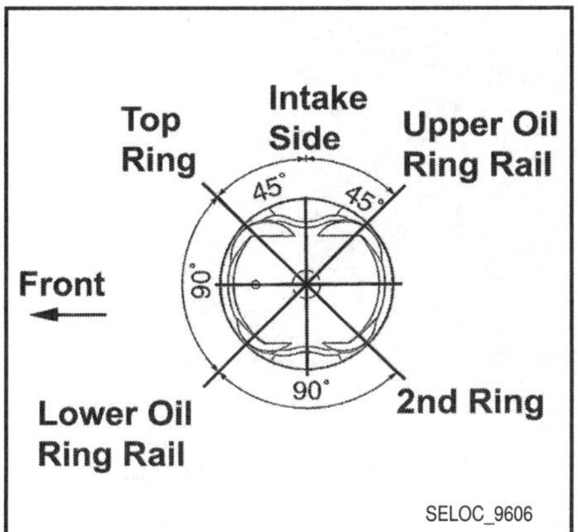

Fig. 170 Piston ring positioning - FX SHO w/1812cc engines

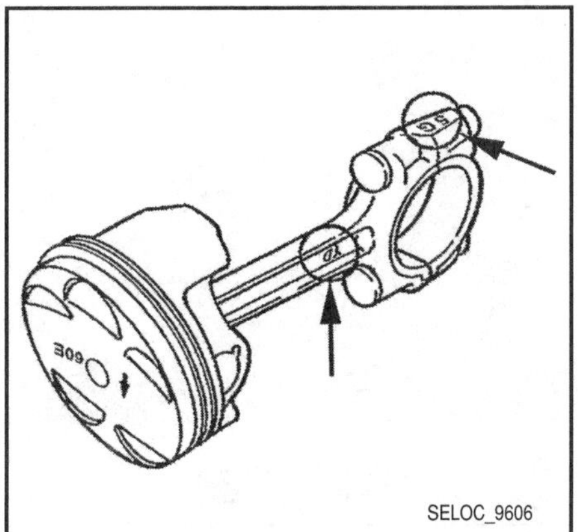

Fig. 171 Setting up the assembly before installation - 998cc and 1052cc engines

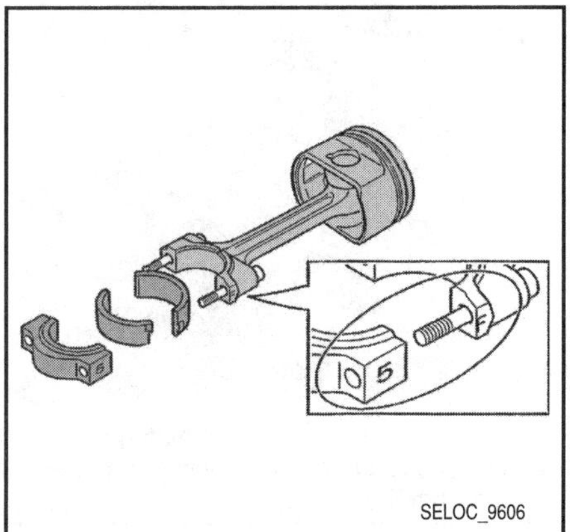

Fig. 172 Setting up the assembly before installation - 1812cc engines

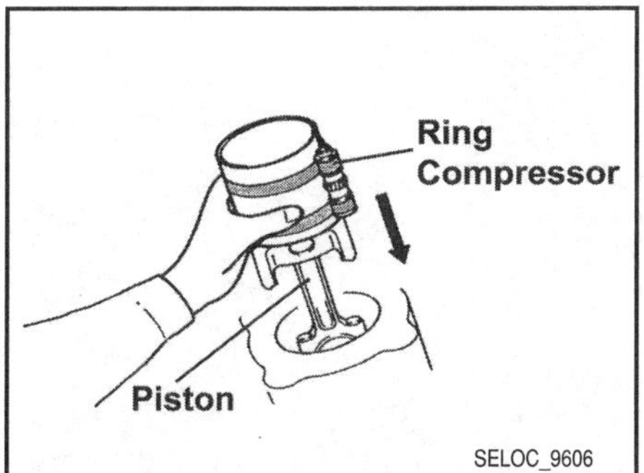

Fig. 173 Use ring compressor when installing the piston assembly

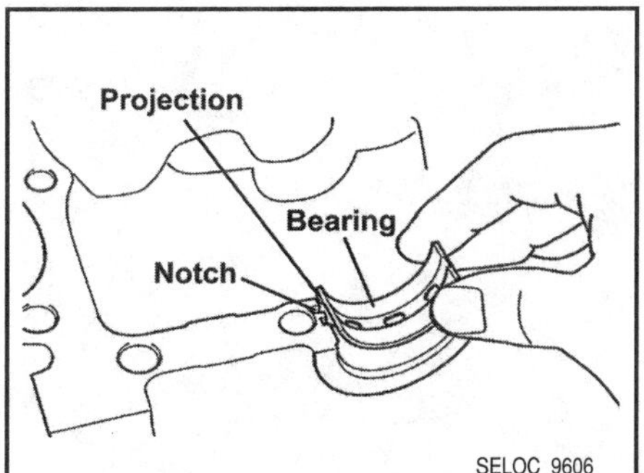

Fig. 174 Installing the crankshaft bearing insert

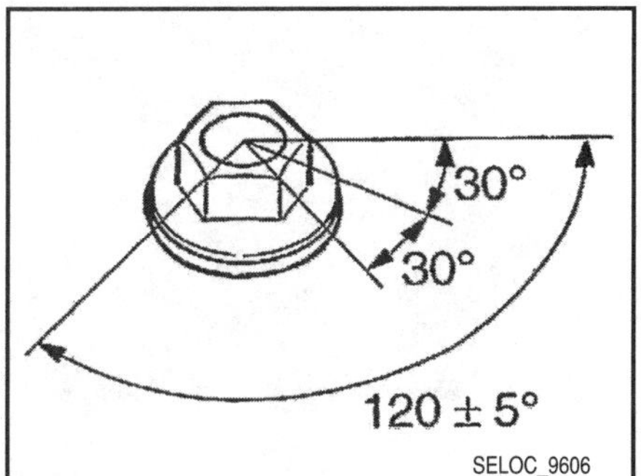

Fig. 175 The angle from one corner of a hexagonal nut to the other is 60°

26. Coat the lower big end bearing lightly with clean engine oil and place it in the rod cap so the projection fits into the notch in the cap. Feed the bolts through the holes in the cap and seat the cap on the crankshaft so that the marks made during removal are aligned. Thread on the *new* nuts and tighten them to:

- Models with 998cc/1052cc engines - 14 ft. lbs. (20 Nm)
- FX HO and VX w/1812 engine - 37 ft. lbs. (50 Nm)
- FX SHO w/1812 engine - 37.6 ft. lbs. (51 Nm)
- FZ models - 37.6 ft. lbs. (51 Nm)

Now, using an angle wrench (*not* a torque wrench!), tighten each nut an additional 120°, except on 1812cc engines where the figure should be 90°.

■ **If you accidentally tighten the nuts past the recommended angle, you cannot just loosen them back to the correct angle. Once over-tightened, you will need to replace the bolts/nuts again.**

27. Repeat the previous step for the remaining pistons, carefully rotating the crankshaft so the journal for that cylinder is at its lowest.
28. Install the crankcase and oil pan.
29. Install the cylinder head and then install the engine.

INSPECTION

OEM ④ SKILLED

◆ **See Figures 176 thru 181**

■ **Images shown are typical and not meant to be representative of your specific engine.**

Inspect each piston for evidence of scoring, cracks, metal damage, cracked piston pin boss, or worn pin boss. If the piston pin is bent, the piston and pin must be replaced as a set for two reasons. First, a bent pin will damage the boss when it is removed. Secondly, a piston pin is not usually sold as a separate item.

Check the piston ring grooves for wear, burns distortion, or loose locating pins. During an overhaul, the rings should be replaced to ensure lasting repair and proper performance after the work is completed. Clean the piston dome, ring grooves, and the piston skirt. Clean carbon deposits from the ring grooves using the recessed end of an old or broken piston ring.

Never use a rectangular ring to clean the groove for a tapered ring, or use a tapered ring to clean the groove for a rectangular ring.

If equipped with locating pins, unless you are VERY cautious, avoid the use an automotive type ring groove cleaner, because such a tool may loosen the piston ring locating pins.

Clean carbon deposits from the top of the piston using a soft wire brush, carbon removal solution or by sand blasting. If a wire brush is used, take care not to burr or round the machined edges. Clean the piston skirt with crocus cloth.

Install the piston pin through the first boss only. Check for vertical free-play. There should be no vertical free-play. The presence of play is an indication the piston boss is worn. The piston is usually manufactured from a softer material than the piston pin. Therefore, the piston boss will normally wear more quickly than the pin.

Excessive piston skirt wear cannot be visually detected. Therefore, good shop practice dictates the piston skirt diameter be measured with a micrometer.

Piston skirt diameter should be measured at right angles to the piston pin axis at a point above the bottom edge of the piston. Refer to the Engine Specifications chart for specifications.

Fig. 176 This piston seized at high rpm (the connecting rod ripped the lower part of the piston from the dome)

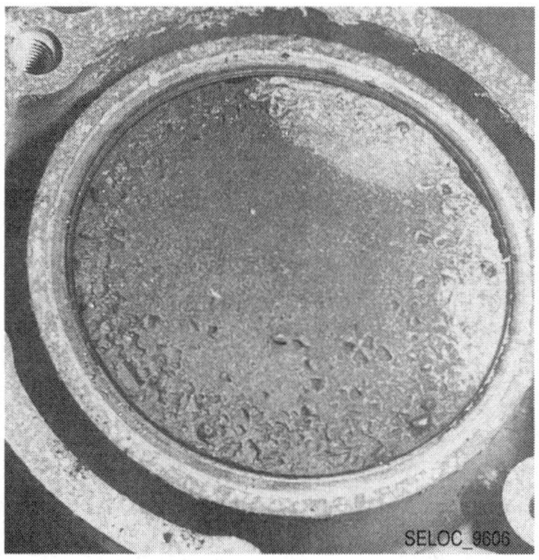

Fig. 177 This pitting is likely a result of foreign matter in the cylinder

Fig. 178 Cleaning the ring groove with a broken ring

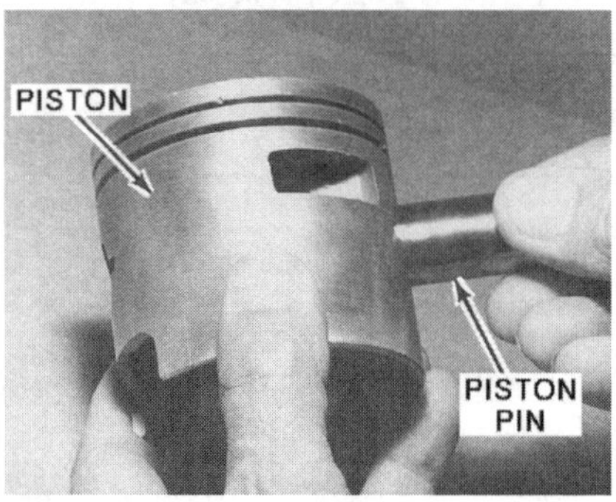

SELOC_9606

Fig. 179 There should be no play on the piston pin

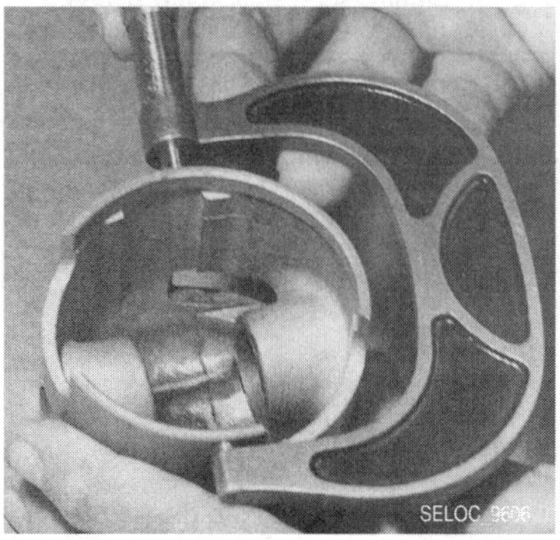

Fig. 180 Use a micrometer to check the piston skirt diameter

Fig. 181 The rings on this piston became stuck due to lack of lubrication

Piston Ring End-Gap Clearance

◆ See Figure 182

■ **Images shown are typical and not meant to be representative of your specific engine.**

Before the piston rings are installed onto the piston, the ring end-gap clearance for each ring must be determined. The purpose of the piston rings is to prevent the blow-by of gases in the combustion chamber. This cannot be achieved unless the correct oil film thickness is left on the cylinder wall.

This thin coating of oil acts as a seal between the cylinder wall and the face of the piston ring. An excessive end-gap will allow blow-by and the cylinder will lose compression. An inadequate end-gap will scrape too much oil from the cylinder wall and limit lubrication. Lack of adequate lubrication will cause excessive heat and wear.

Ideally the ring end-gap measurement should be taken after the cylinder bore has been measured for wear and taper and after any corrective work, such as boring or honing, has been completed.

If the ring end-gap is measured with a taper to the cylinder wall, the diameter at the lower limit of ring travel will be smaller than the diameter at the top of the cylinder.

If the ring is fitted to the upper part of a cylinder with a taper, the ring end-gap will not be great enough at the lower limit of ring travel. Such a condition could result in a broken ring and/or damage to the cylinder wall and/or damage to the piston and/or damage to the cylinder head.

If the cylinder is to be only honed, not bored, or if only cleaned, not honed, the ring end-gap should be measured at the lower limit of ring travel.

Some manufacturers give the precise depth the ring should be inserted into the cylinder and assume the cylinder walls are parallel with no taper. Most motors covered here will require the ring to be inserted 0.2 in. (5mm) below the top of the cylinder.

Rings may be inserted from the top or bottom (depending upon the measurement point). In most cases the piston can be inserted and used to square the ring up in the bore. Once the ring is in the proper position, measure the ring end-gap with a feeler gauge.

If the end-gap is greater than the amount listed in the Engine Specifications chart, replace the entire ring set.

If the end-gap is less than the amount listed, carefully file the ends of the ring - just a little at a time - until the correct end-gap is obtained.

■ **The oil ring expander spacer's end gap cannot be measured. If the ring rail's gap is too large, you will have to replace all 3 rings.**

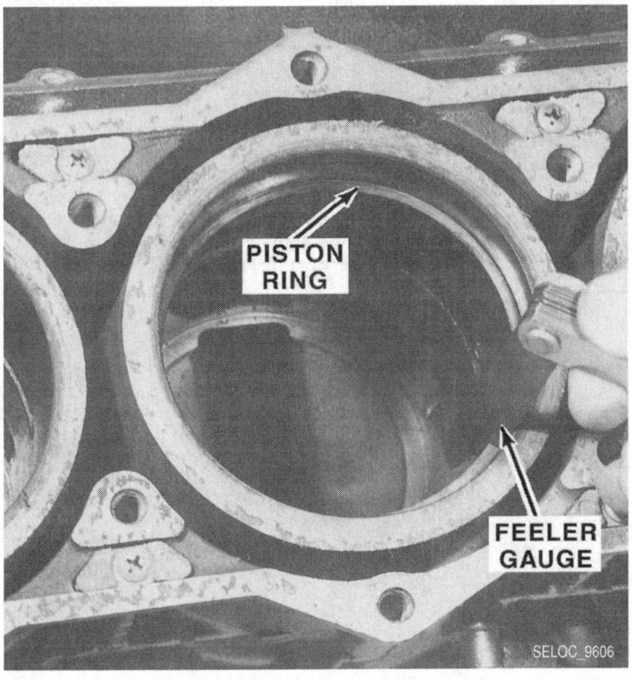

Fig. 182 Using a feeler gauge to check ring end-gap

Piston Ring Side Clearance

◆ See Figure 183

■ **Images shown are typical and not meant to be representative of your specific engine.**

After the rings are installed on the piston, the clearance in the grooves needs to be checked with a feeler gauge. Check the clearance between the ring and the upper land and compare your measurement with the Engine Specifications chart. Ring wear forms a step at the inner portion of the upper land. If the piston grooves have worn to the extent to cause high steps on the upper land, the step will interfere with the operation of new rings and the ring clearance will be too much. Therefore, if steps exist in any of the upper lands, the piston should be replaced.

On a plain ring, this clearance may be measured either above or below the ring. On a keystone ring - one with a taper at the top - the clearance must be measured below the ring.

Piston Pin Clearance

◆ See Figures 184 and 185

■ **Images shown are typical and not meant to be representative of your specific engine.**

1. Use a micrometer to determine the outside diameter of the pin. Check it in several places and confirm that the dimension is:
   - 998cc and 1052cc engines - 0.6689-0.6693 in. (16.991-17.000mm) with a limit of 0.6681 in. (16.971mm)
   - 1812cc engines - 0.8658-0.8661 in. (21.991-22.000mm)

   Failure to meet these specifications will require pin replacement.
2. Now, using an inside micrometer, measure the inside diameter of the piston pin bore in the skirt. Confirm that it is:
   - 998cc and 1052cc engines (exc. 2010-11 VX) - 0.6694-0.6698 in. (17.002-17.013mm)
   - 2010-11 VX - 0.6695-0.6700 in. (17.005-17.018mm).
   - 1812cc engines - 0.8663-0.8667 in. (22.004-22.015mm)

   Failure to meet these specifications will require piston replacement.
3. Subtract the pin diameter from the bore diameter for the clearance, confirming that it is:

Fig. 183 Using a feeler gauge to check piston ring side clearance

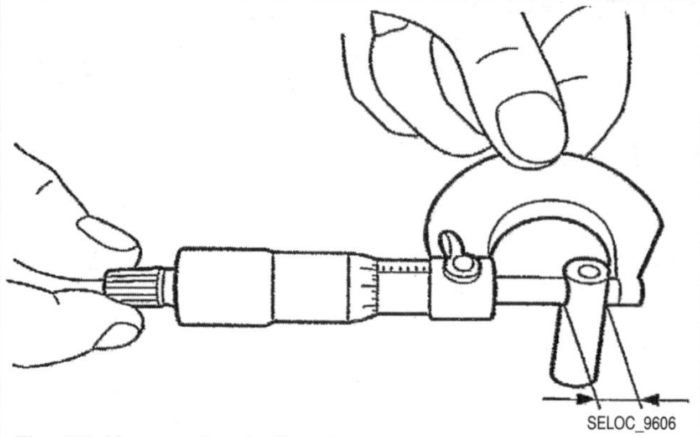

Fig. 184 Measure the pin diameter. . .

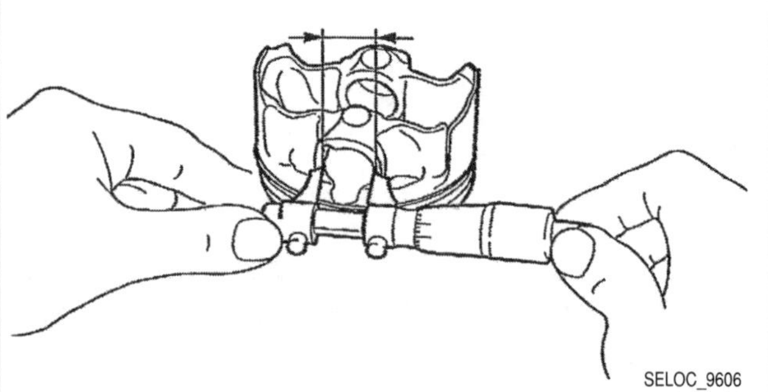

Fig. 185 . . .and then the pin bore diameter

- FX models with 998cc/1052cc engines - 0.0001-0.0017 in. (0.002-0.042mm)
- 2005-09 VX models - 0.0001-0.0009 in. (0.002-0.022mm)
- 2010-11 VX models w/1052cc engines - 0.0006-0.0007 (0.014-0.018mm)
- FX/FZ and VX models w/1812cc engines - 0.0005-0.0006 (0.003-0.015mm)

If not, replace the pin.

Connecting Rods

OEM ③ DIFFICULT

◆ See Figures 186 thru 190

■ Specifications available for the connecting rods vary by model, so check the Engine Specifications charts to determine which steps are applicable to the motor on which you are working.

Each connecting rod should be cleaned and, once cleaned, should be thoroughly inspected to make sure it is not damaged, bent or excessively worn. Obvious defects, such as visible warpage or cracks are signs of un-serviceability. On many motors, the connecting rod small-end (piston end) free-play should be measured using a dial gauge, while connecting rod big-end (crankshaft end) must be checked for Side Clearance using a feeler gauge. On some models measurements may be provided for small end diameter (which will correspond to a measurement slightly larger than piston pin diameter). To thoroughly inspect the connecting rods, proceed as follows:

1. Clean the inside diameter of the piston pin end of the connecting rod with crocus cloth. Clean the connecting rod only enough to remove marks. Do not continue once the marks have disappeared.
2. Assemble the piston end of the connecting rod with the loose needle bearings, piston pin, retainers, and C-lock rings. Insert the piston pin and check for vertical play. The piston pin should have no noticeable vertical play.

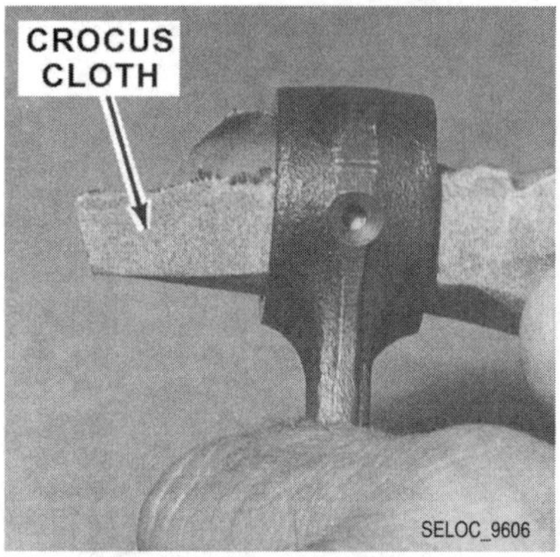

Fig. 186 Cleaning the piston end of a connecting rod with crocus cloth

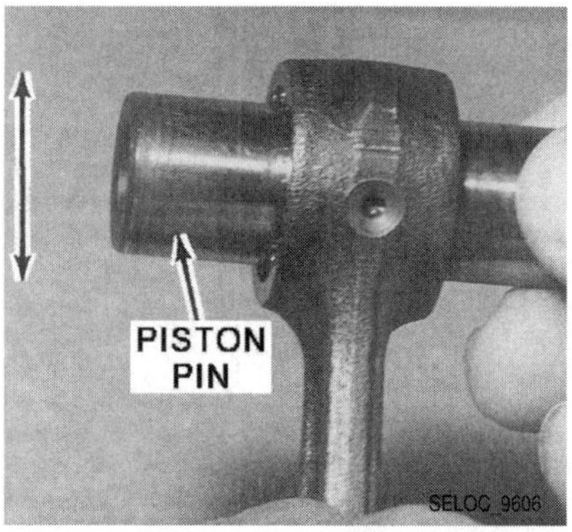

Fig. 187 Checking the piston end of a connecting rod for vertical free-play using a piston pin

Fig. 188 Testing 2 rods for warpage at the piston end using a feeler gauge

Fig. 189 Testing 2 rods for warpage at the crankshaft end using a feeler gauge

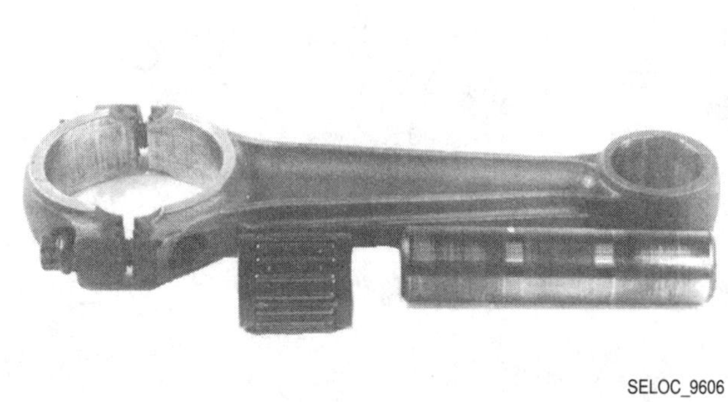

Fig. 190 Visually inspect the connecting rod for signs of damage

3. If the pin is loose or there is vertical play check for and replace the worn parts.

4. Inspect the piston pin and matching rod end for signs of heat discoloration. Overheating is identified as a bluish bearing surface color and is caused by inadequate lubrication or by operating the engine at excessive high rpm.

5. Check the bearing surface of the rod and rod cap for signs of chatter marks. This condition is identified by a rough bearing surface resembling a tiny washboard. The condition is caused by a combination of low-speed low-load operation in cold water. The condition is aggravated by inadequate lubrication and improper fuel.

Under these conditions, the crankshaft journal is hammered by the connecting rod. As ignition occurs in the cylinder, the piston pushes the connecting rod with tremendous force. This force is transferred to the connecting rod journal. Since there is little or no load on the crankshaft, it bounces away from the connecting rod. The crankshaft then remains immobile for a split second, until the piston travel causes the connecting rod to catch up to the waiting crankshaft journal, then hammers it. In some instances, the connecting rod crank pin bore becomes highly polished.

6. Inspect the bearing surface of the rod and rod cap for signs of uneven wear and possible overheating. Overheating is identified as a bluish bearing surface color and is caused by inadequate lubrication or by operating the engine at excessive high rpm.

7. Use a feeler gauge to check big-end side clearance on the crankshaft/connecting rod assembly by inserting a feeler gauge between the connecting rod and the counterweight of the crankshaft.

If the side-clearance measurement is not within specification, measure the crank width (if specifications are provided). Such a measurement may give an indication if the counterweight is walking on the crank pin or if the clearance is due to worn parts. The connecting rod would wear before the edge of the counterweight because the rod is manufactured from a much softer material.

Keep in mind that even if only one connecting rod shows excessive free-play or side-clearance, good shop practice dictates the entire set of rods be replaced. Rods are sold in sets.

8. Most models utilize replaceable automotive style connecting rod caps with plain bearing inserts. Clean the big end bearing halves and the inside of the road and cap thoroughly. Install the bearings into the rod and cap so that the small projection on each bearing fits into the notch on the cap and rod.

a. Lay a small piece of Plastigage® across the crankshaft journal and then install the rod and cap (use the old bolts/nuts) over the crankshaft so that the **Y** or **YD** mark on the rod faces forward and the numbers on the rod and cap are aligned (you should also have made marks yourself during removal).

b. Tighten the nuts to the proper torque figure and then remove them (now you can throw them away) and then measure the compressed Plastigage® strip against the chart that came with it. If clearance is out of specification, you'll need to install replacement bearings.

Crankshaft

◆ **See Figures 191 and 192**

■ **Images shown are typical and not meant to be representative of your specific engine.**

Clean the crankshaft with solvent and wipe the journals dry with a lint free cloth. Inspect the main journals and connecting rod journals for cracks, scratches, grooves, or scores. Inspect the crankshaft oil seal surface for nicks, sharp edges or burrs which might damage the oil seal during installation or might cause premature seal wear. Always handle the crankshaft carefully to avoid damaging the highly finished journal surfaces. Blow out all oil passages with compressed air. The oil passageway leads from the rod to the main bearing journal. Take care not to blow dirt into the main bearing journal bore.

Inspect the internal splines at one end and threads at the other end for signs of abnormal wear. Check the crankshaft run-out by supporting it on 2 V-blocks at the main bearing surfaces.

Install a dial indicator gauge above the main bearing journals. Rotate the crankshaft and measure the run-out (or the out-of-round) and the taper at both ends and in the 2 center journals. Maximum run-out should not exceed 0.0012 in. (0.03mm).

If V-blocks or a dial indicator are not available, a micrometer may be used to measure the diameter of the journal. Make a second measurement at right angles to the first. Check the difference between the first and second measurement for out-of-round condition. If the journals are tapered, ridged, or out-of-round by more than the specification allows, the journals should be reground, or the crankshaft replaced.

Any out-of-round or taper shortens bearing life. Good shop practice dictates new main bearings be installed with a new or reground crankshaft.

Inspect the crankshaft oil seal surfaces to be sure they are not grooved, pitted, or scratched. Replace the crankshaft if it is severely damaged or worn. Check all crankshaft bearing surfaces for rust, water marks, chatter marks, uneven wear, or overheating. Clean the crankshaft surfaces with 320-grit wet/dry sandpaper dampened with solvent. Never spin-dry a crankshaft ball bearing with compressed air.

Clean the crankshaft and, if equipped, crankshaft ball bearing with solvent. Dry the parts, but not the ball bearing, with compressed air. Check the crankshaft surfaces a second time. Replace the crankshaft if the surfaces cannot be cleaned properly for satisfactory service. If the crankshaft is to be installed for service, lubricate the surfaces with light oil. Do not lubricate the crankshaft ball bearing at this time.

On models so equipped, after the crankshaft has been cleaned, grasp the outer race of the crankshaft ball bearing installed on the lower end of the crankshaft, and attempt to work the race back and forth. There should not be excessive play. A very slight amount of side play is acceptable because there is only about 0.001 inch (0.025mm) clearance in the bearing.

Lubricate the ball bearing with light oil. Check the action of the bearing by rotating the outer bearing race. The bearing should have a smooth action and no rust stains. If the ball bearing sounds or feels rough or catches, the bearing should be removed and discarded.

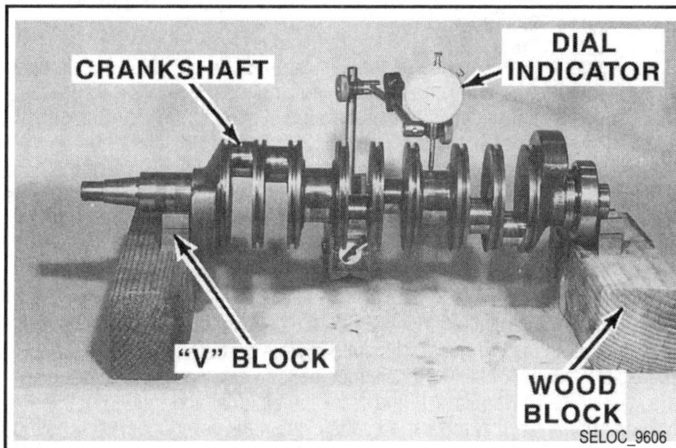

Fig. 191 Crankshaft set up with V-blocks and a dial indicator to measure run-out

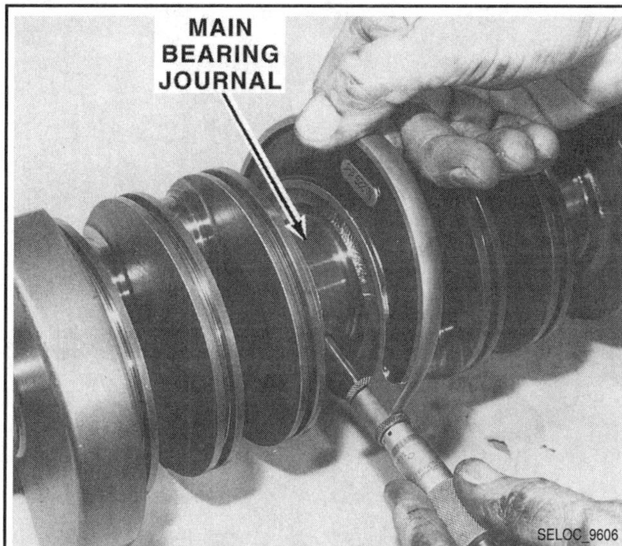

Fig. 192 You can also use a micrometer to check run-out and taper on the main bearing journals (and connecting rod bearing journals depending upon what specs are available)

For models equipped with plain bearings or plain bearing inserts, use a strip of Plastigage® or some equivalent compressible gauging material across the center of the crankshaft journal (lengthwise from flywheel-to-driveshaft end of the journal) and bolt the crankcase in position, then remove again to exam the gauging material. For more details, please refer to the Measuring Bearing Clearance Using Plastigage® section.

Oversize Pistons & Rings

OEM ④ SKILLED

Scored cylinder blocks can be saved for further service by boring and installing oversize pistons and piston rings. However, in most cases, if the scoring is over 0.0075 inch (0.13mm) deep, the block cannot be effectively re-bored for continued use.

Oversize pistons and rings are not available for all engines. Check with the parts department at your local Yamaha dealer for the model being serviced.

If oversize pistons are not available, the local marine shop may have the facilities to knurl the piston, making it larger.

A honed cylinder block is just surface finished. It is not parallel and therefore if an oversized or knurled piston is installed in this bore, the piston will soon seize at the lower narrow end of the bore.

MEASURING BEARING CLEARANCE USING PLASTIGAGE®

OEM ③ DIFFICULT

◆ **See Figures 193, 194 and 195**

Most Yamaha 4-strokes use either plain bearings or automotive style plain bearing inserts for the crankshaft and connecting rods. Besides a visual inspection for obvious defects or flaws, the most important check that can be made to plain bearings is to measure their installed oil clearance (the distance between the bearing and crankshaft journal) to determine if they are still fit for service. On most Yamahas that use bearing inserts there are codes to help find the proper replacement inserts. However, even when selecting replacement bearings using color or number codes provided on the motor components, this check must be made to ensure the assembled motor will be within design tolerances.

To check bearing oil clearances you need a way of measuring a tiny gap (thousands of an inch or hundredths of a millimeter) between otherwise inaccessible components. The basic method is to apply a strip of a gauging material such as Plastigage® to the crankshaft journal and then to bolt the crankcase or connecting rod cap together. When the components are bolted together the gauging material is compressed (squished) between the bearing and journal. The key is that the material is designed to expand at a uniform rate so that you can then remove the bearing and measure the width of the squished gauging material. The width of the material will correspond to a specific clearance measurement (a width-to-clearance conversion scale is provided with the gauging material).

Using Plastigage® is a simple, inexpensive, and fairly reliable method to measure main bearing and connecting rod bearing clearance. However, Plastigage® is soluble in oil, so clean the crankshaft journal thoroughly of all traces of oil, and then turn the crankshaft until the oil hole is down away from the cap to prevent getting any oil on the Plastigage®. To measure bearing clearances, proceed as follows:

1. Obtain a package of Plastigage® from the local auto or marine parts outlet.
2. Cut a 1 in. (2.5cm) section of the bead for each journal/bearing combination that is being measured.
3. Place each 1 inch section on the crankshaft journal surface at the main bearings and on each connecting rod bearing surface.

■ ***Do not* rotate the crankshaft until the measurements have been completed. Such action will smear the Plastigage® bead and the measurement will be useless.**

4. Install the main bearings and tighten the crankcase to specification.
5. Remove the crankcase/bearing retainer bolts to gain access to the Plastigage®.
6. Measure the flattened Plastigage® at its widest point using the scale printed on the Plastigage® envelope. The number within the graduation which most closely corresponds to the width of the Plastigage® indicates the bearing clearance in thousands of an inch (or hundredths of a millimeter, depending upon the scale). A metric scale in mm is normally given on the back of the package. The widest point of the Plastigage® is the minimum clearance and the narrowest point is the maximum clearance. The difference between the readings indicates the taper of the journals. Compare the Plastigage® measurements with the specifications found in the Engine Specifications chart in this section.
7. If the clearance is not within specifications, the bearings or the crankshaft *must* be replaced as required. Obviously, on models that use inserts, you'd start with new bearing inserts to see if one can be obtained to return that particular journal to specified clearances.
8. Once you are finished checking and selecting bearings, clean all traces of the Plastigage® from the bearing surfaces, and then follow necessary steps again (oiling the bearing surfaces and skipping the Plastigage®) to make a final installation of the the crankcase. Check to be *sure* the crankshaft rotates freely. If any binding is discovered, the thrust washer tabs may be misaligned or one or more of the connecting rod caps may be installed backwards.

Fig. 193 To check bearing clearances, apply a strip of gauge material to each journal and bolt the bearings/retainers in place. . .

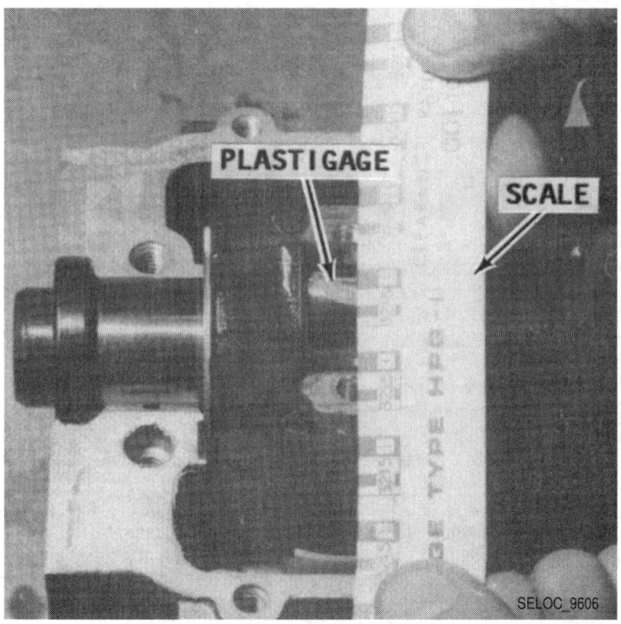

Fig. 194 . . .then remove the crankcase and compare it to the scale

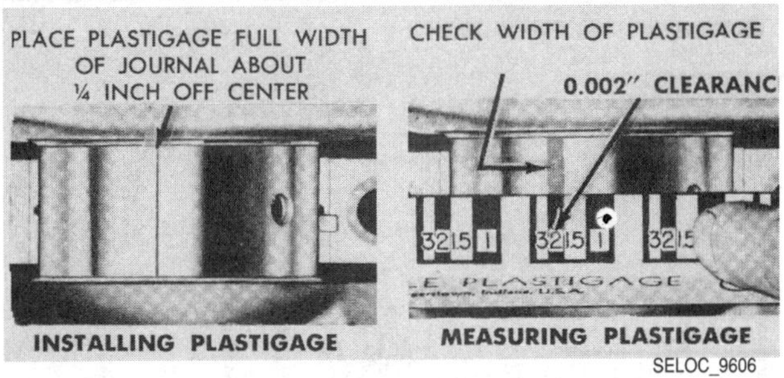

Fig. 195 Once the crankcase and/or rod caps are properly tightened and subsequently removed again compare the now flattened gauging material to the scale that came with the package

Engine Temperature Switch

REMOVAL & INSTALLATION

EASY

◆ **See Figure 196**

Locate the oil pressure switch on the starboard side of the cylinder block, toward the rear.

Disconnect the electrical lead at the coupler and then unscrew the switch.

Tighten the switch to 11.1 ft. lbs. (15 Nm).

TESTING

■ **Please refer to the Exhaust & Cooling section for switch testing procedures.**

Engine Thermo-Switch

REMOVAL & INSTALLATION

EASY

◆ **See Figure 196**

Locate the oil pressure switch on the starboard side of the cylinder block, toward the front.

Disconnect the electrical lead at the coupler and then unscrew the 2 mounting bolts and remove the switch.

Install the switch and tighten the bolts to:

- 998cc engines - 5.5 ft. lbs./66 inch lbs. (7.6 Nm)
- 2004-09 1052cc engines - 5.5 ft. lbs./66 inch lbs. (7.6 Nm)
- 2010-11 1052cc engines - 5.9 ft. lbs./71 inch lbs. (8 Nm).
- 1812cc engines - 5.9 ft. lbs./71 inch lbs. (8 Nm).

Fig. 196 The engine temperature switch is found on the starboard side of the block (1052cc shown)

TESTING

■ **Please refer to the Exhaust & Cooling section for switch testing procedures.**

Knock Sensor

REMOVAL & INSTALLATION

EASY

Knock sensors are used only on supercharged 1812cc engines. Locate the oil sensor on the starboard side of the cylinder block, toward the rear.

Disconnect the electrical lead at the coupler and then unscrew the switch.

Tighten the switch to 11.1 ft. lbs. (15 Nm).

TESTING

MODERATE

◆ **See Figure 197**

Knock sensor resistance must be measured at, or near, 68°F (20°C).

Remove the sensor and connect a DVOM as per the manufacturer's instructions.

Connect the negative meter lead to the end of the sensor and touch the positive lead to the terminal in the sensor coupler. Resistance should be 504-616 kili-ohms or the sensor will require replacement.

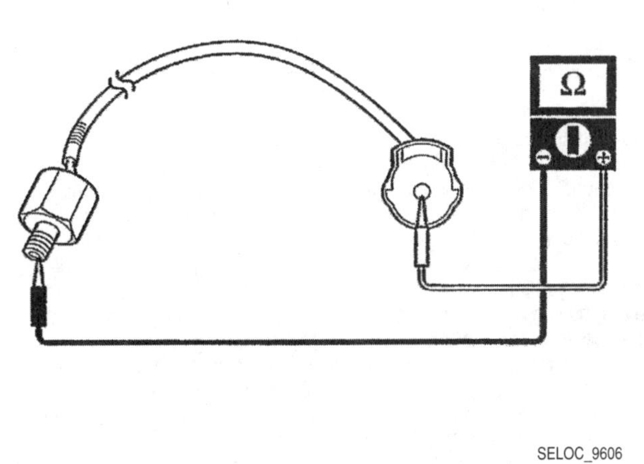

Fig. 197 Testing the knock sensor

Oil Pressure Switch

REMOVAL & INSTALLATION

EASY

◆ **See Figure 198**

Locate the oil pressure switch on the port side of the cylinder block, toward the front on 1812cc engines and toward the rear on all others.

Pull back the rubber cover and remove the electrical lead bolt. Remove the switch.

Coat the switch threads with Loctite 572 and tighten it to 5.9 ft. lbs./71 inch lbs. (8 Nm). Tighten the lead bolt to 1.5 ft. lbs./18 inch lbs. (2 Nm). Install the rubber hood.

Fig. 198 The oil pressure switch is found on the port side of the block (1052cc shown)

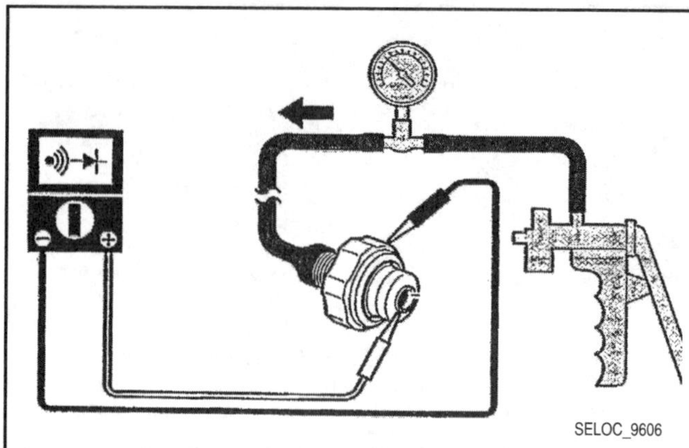

Fig. 199 Testing the oil pressure switch

TESTING

OEM MODERATE

◆ See Figure 199

1. Remove the switch.
2. Connect a DVOM according to the manufacturer's instructions.
3. Connect a vacuum/pressure pump with a gauge.
4. Touch the negative meter lead to the body of the switch and the positive lead to the terminal thread on the end of the switch while applying pressure.
5. The meter should show continuity up to 18.2-23.6 psi (128-166 kPa) and show no continuity above 18.2-23.6 psi (128-166 kPa). Anything outside the range will require switch replacement.

ENGINE REFINISHING & OVERHAUL

■ **Although most refinishing or overhaul procedures have been included with the specific components, the following are general guidelines and suggestion that may be applicable to many of the more specific overhaul procedures detailed elsewhere.**

General Information

BASICS

The success of the overhaul work is largely dependent on how well the cleaning and inspecting tasks are completed. If some parts are not thoroughly cleaned, or if an unsatisfactory unit is allowed to be returned to service through negligent inspection, the time and expense involved in the work will not be justified with peak performance and long operating life. Therefore, the procedures in the following sections should be followed closely and the work performed with patience and attention to detail.

■ **When measuring components, remember that specifications may vary with temperature. Unless otherwise noted, specifications are for components at an ambient room temperature of 68°F (20°C).**

CLEANING

General

All engine components must be clean and free of gasket material, oil and carbon deposits before they are inspected. Take your time when cleaning components. Before using solvent to clean something, make sure the chemical is compatible with the material of which the component is constructed. Also, check each component for matchmarks or ID marks before cleaning (as some solvents may remove even permanent marker). If necessary, re-ID the component as soon it has been cleaned and dried (before moving onto the next component). Whenever you match-marked 2 components in order to ensure exact alignment during installation, if possible, fasten them together during cleaning (in case the marks come off). In this way, they can be re-marked after the cleaning process is through.

When it comes to inspection, most procedures will be included with the individual components. But, keep in mind that not all components will be found on all motors. If in doubt whether a component is used (or should be checked) refer to the Disassembly, Assembly or Overhaul procedures (as applicable). Also, be sure to check the Specifications Chart for your motor. If a component is not listed in the specification chart, it is either not used, or, does not have a specific tolerance for which it must be inspected. Whether or not a tolerance is provided, all components must be clean and free of obvious defects (deep cracks, scoring, excessive carbon deposits that cannot be removed, warpage, etc). If in doubt whether or not a component is serviceable, seek someone with more Yamaha engine rebuilding experience than yourself.

■ **The 4-stroke motors, by design, tend to be much more complicated than a 2-stroke motor, containing a greater number of precision machined parts that require inspection. This is especially true in the cylinder head and related components (camshaft and valve train) most of which are not found on 2-strokes.**

When inspection involves precision measurements, not only must the components be clean, but they must be measured roughly at room temperature, using the appropriate measurement equipment to ensure accurate results.

It is very important to keep in mind that all wear components (pistons, rings, shafts, springs, valves, bearings, etc) *must* be reinstalled in their original locations whenever they are being reused. Wear patterns form on all contact surfaces during use. Mismatching wear patterns will accelerate wear, while matching wear patterns helps ensure a durable and reliable repair.

Cleaning Techniques

MODERATE

◆ See Figures 200 thru 205

**** WARNING**

Avoid removing excessive amounts of metal when removing carbon deposits.

1. Use a blunt-tip scraper or dulled chisel to loosen carbon deposits from various components of the combustion chamber and ports/valves. Work slowly and carefully to prevent damaging or excessively scoring the surfaces. Then use a Scotchbrite® pad and mild solvent to remove most/all of the remaining deposits. Remove deposits from the following components, as applicable:
   - Remove carbon from the combustion chambers in the cylinder head. Use great care on these motors, since the valve seating surface will be exposed after disassembly. DO NOT score or damage the valve seating surfaces. Another option is to protect the valve seating surface by leaving the

valves in place while you are removing the carbon deposits, but the same care must be taken to prevent deep scores on the bottoms of the valves.

- Remove carbon deposits or corrosion from the exhaust cover(s).

■ **Deposits on pistons can be removed while they are still installed in the bores. This is handy when the cylinder head is removed for service without completely disassembling the block and crankcase. If this is to be accomplished, position the piston to be cleaned at TDC and cover the other piston bore(s) using rags and plastic. Thoroughly clean all debris using solvent and compressed air (WHILE WEARING SAFETY GLASSES) before moving to the next piston.**

- Remove carbon deposits from the top of the piston(s). When working on the piston domes, use a light touch to prevent scratching, or worse, gouging the piston.

**** WARNING**

Wire brushes are not really recommended these days for cleaning piston domes since particles of steel could become lodged into the piston surface. If this occurs, they could glow hot when the piston is returned to service, causing pre-ignition or detonation that could damage the piston and combustion chamber.

- Remove carbon from the ring grooves either using a ring groove cleaner, or, better yet, using a broken piece of the piston ring with an angle ground on the end. When using a ring to clean the piston grooves, use the ring actually removed from the groove (if it is being replaced) or one from the same groove on another piston.

**** WARNING**

When cleaning piston ring grooves, use the same caution as with the pistons. Do NOT remove excessive amount of material or the piston will be damaged beyond use. Some ring groove cleaning tools are heavy duty and will easily remove too much material, so use them with care. Believe it or not some manufacturer recommend the filed broken ring method.

2. Inspect all water passages for corrosion deposits, debris or blockage. Remove debris and clean corrosion as needed and accessible. Use low-pressure compressed air to blow out all water passages.
3. Clean and degrease all regularly oiled surfaces (including the crankshaft, pistons and connecting rods) using solvent or degreaser. Use low-pressure compressed air to remove all build-up from shaft and rod oil holes.
4. Remove all traces of gasket or sealant using a spray gasket remover or equivalent gasket removing solvent. Whenever possible, avoid the use of gasket scrapers to help avoid the possibility of scoring and damaging the gasket mating surfaces.

■ **Honing the cylinder walls can wait until after measurements are taken to determine if boring for oversized pistons/rings will be necessary. If honing is pushed off until after inspection, be sure to follow the remaining steps in order to finish cleaning the block, then repeat the appropriate steps again, after honing.**

5. Check the cylinder walls for glazing (a smooth, glassy appearance) and, if found, hone the cylinder walls using a medium grit cylinder hone. Use

Fig. 200 Although a blunt chisel is preferred, a wire brush can be used WITH CARE to clean most carbon deposits

Fig. 201 Pistons CAN be cleaned while still installed, but again use care, and leave NO metallic deposits behind

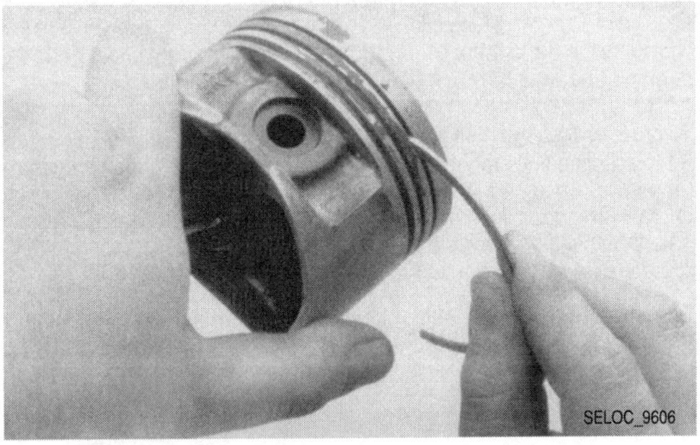

Fig. 202 The preferred method for cleaning piston grooves is to use the filed end of a broken ring. . .

Fig. 203 . . . but a ring groove cleaning tool can be used, with care to prevent damaging the piston

Fig. 204 The cylinder walls must show an obvious cross-hatching (tiny grooves, criss-crossed in a pattern around the bore). . .

Fig. 205 . . . otherwise a cylinder hone must be used to break the smooth glazed surface into a cross-hatch pattern

a slow rpm while raising and lowering the hone through the cylinder in order to cross-hatch the cylinder walls for maximum oil retention.

✻✻ WARNING

Use the cylinder hone slowly, carefully and as little as possible to avoid the possibility of removing too much material from the cylinder walls. If the bores are over-honed, it could cause the pistons to be below specification for the new cylinder measurement, or worse, cause the cylinder bores to be overspec for what pistons are available.

6. Wash the entire cylinder block, crankcase and head using warm, soapy water to remove all traces of contaminants. Use low-pressure compressed air to blow dry all passageways.

7. Apply a light coating of clean engine oil to all machined surfaces that are not about to be measured right away. When you return to the task of measuring components that have been oiled, use a solvent covered rag to wipe away the oil before measurements are taken.

8. Cover all components using a plastic sheet to keep dust, dirt or debris from contaminating the cleaned and especially the oiled surfaces.

ENGINE BREAK-IN

Break-in Procedures

Anytime a new or rebuilt engine is installed (this includes an engine whose wear components such as pistons and rings, main bearings or cylinder head components have been replaced), the motor must undergo proper break-in.

By following break-in procedures largely consisting of specific engine operating limitations during the first 10 hours of operation, you will help you will help ensure a long and trouble-free life. Failure to follow these recommendations may allow components to seat improperly, causing accelerated wear and premature engine failure.

Special attention is required to the engine oil during initial break-in, expect increased oil consumption during that period. Don't be alarmed, but be sure to check the engine oil level frequently to prevent the possibility of engine damage from operation with a low oil level. Oil consumption should lessen as the piston rings seat to the cylinder walls.

Especially during break-in, pay close attention to all pre and post operation checks. This goes double when checking for fuel, oil or water leaks. At each start-up and frequently during operation, check for presence of the cooling indicator stream.

At the completion of break-in, double-check the tightness of all exposed engine fasteners.

Even though Yamaha technically ends the break-in period of a motor after the first 10 hours of operation, it is a good idea to continue to pay closer attention to the motor until the completion of the entire first 20 hours of engine operation (and completion of a 20 hour service). During break-in, one of the most important things you can do is to **vary** the engine speed. This allows parts to wear in under conditions throughout the powerband, not just at idle or mid-throttle.

■ **During break-in, check your hourmeter or a watch frequently and be sure to change the engine speed at least every 15 minutes (that means between every 2-3 tenths on the hourmeter).**

Always be sure to allow the engine to reach operating temperature before setting the throttle anywhere above idle. This means you should always start and run the motor for at least 5 minutes before advancing the throttle.

✻✻ WARNING

NEVER run the engine out of the water, unless a flush fitting is used to provide a source of cooling. Remember that the water pump can be destroyed in less than a minute just from a lack of water. The engine will suffer damage in very little time as well, but even if it is not overheated out of water, reduced cooling from a damaged water pump impeller could destroy it later. Don't risk it.

■ **Proper break-in means running the motor under varied LOADS, not in neutral! Although a flush fitting can be used to run an engine, it CANNOT be used to break-in a motor, as you cannot run an engine much above idle on a flush fitting without risking damage from a runaway engine (the motor running overspeed) and without the motor pushing a prop through water there is no load.**

✻✻ WARNING

Do not operate the engine at full throttle except for very short periods (5 minutes at a time max), until after 10 hours of operation.

To properly break-in a engine, proceed as follows:

1. Each time you start the motor, IMMEDIATELY check to be sure the water pump is operating. If the water pump is operating, a fine stream will be discharged from the pilot outlet hole at the rear of the craft. Also, each time you start the engine, allow it to idle until it reaches normal operating temperature (at least 5 minutes) before increasing speed above idle.

■ **If the water indicator stream is missing, shut the engine down and figure out why before proceeding.**

2. The first time you start the motor, allow it to idle for **10** minutes. Make sure the motor runs at the slowest possible speed (a fast idle in neutral is best).

3. For the remainder of the first hour operate the motor in gear at various speeds, but DO NOT exceed 1/2 throttle or about 2000 rpm.

Remember to vary engine speed AT LEAST every 15 minutes.

■ **DO NOT exceed 2000 rpm for the first hour or 3000 rpm for the second hour.**

4. For the second hour of operation avoid operation of speeds above about 3/4 throttle or 3000 rpm. The one exception is that if the boat to which the motor is rigged planes easily, use sufficient throttle (still avoiding full throttle) to plane the boat, then reduce speed to 3/4 throttle or less. Continue to vary the engine speed and operate the motor under different load conditions.

5. For the remainder of the first 10 hours (for hours 3-10) continue the ritual of warming the motor and varying engine speed. Full throttle operation should NOT exceed 5 minutes allowing sufficient time to slowly cool the pistons and operate under other conditions before the next full throttle run.

■ **Build good habits now. It is NEVER good for the motor to operate at full throttle for any length of time then suddenly drop off to idle. You should always decelerate the motor gradually, allowing the pistons to cool as the speed is reduced.**

6. After the 10th hour of operation, run the engine as normal.

■ **While the engine is operating during the initial period, check the fuel, exhaust, and water systems for leaks. Tilt the motor up out of the water at the end of each day to closely inspect the drive unit and look for signs of damage or leakage (of course, if you're trailering, this should be done to protect the skeg before pulling it out on the trailer anyway, right?).**

SPECIFICATIONS

Engine Specifications - 998cc Models

Component				Standard (in.) ①	Metric (mm) ①
Camshaft	Cap Inside Diameter			0.9646	24.5
	Journal Diameter			0.9622-0.9626	24.44-24.45
	Journal to Cap Clearance			0.0020-0.0031	0.05-0.08
	Run-out			0.0012 Max	0.03 Max
Connecting Rod	Bearing Color Code		#1	Brown	-
			#2	Black	-
			#3	Blue	-
			#4	Green	-
	Big End Oil Clearance			0.0006-0.0016	0.016-0.040
	Small End Inside Diameter			0.6694-0.6699	17.005-17.018
Crankshaft	Bearing Color Code		#1	Brown	-
			#2	Black	-
			#3	Blue	-
			#4	Green	-
			#5	Yellow	-
	Deflection Limit			0.0012 Max	0.03 Max
	Journal Oil Clearance		2002-05	0.0008-0.0022	0.020-0.057
			2006-08	0.0002-0.0011	0.004-0.028
Cylinder	Bore Size			2.9157-2.9163	74.060-74.075
	Out-of-Round Limit			0.002	0.05
	Taper Limit			0.003	0.08
	Wear Limit			2.9213	74.2
Cylinder Head	Compression Pressure			192 psi	1350 kPa
	Warpage Limit			0.004	0.1
Piston & Ring	Piston	Diameter		2.9116-2.9121	73.955-73.970
		Piston-to-Cylinder Clearance		0.0039-0.0043	0.10-0.11
		Wear Limit		0.0067	0.17
	Piston Pin	Boss Inside Diameter		0.6693-0.6698	17.002-17.013
		Outside Diameter		0.6689-0.6693	16.991-17.000
		Wear Limit	2002-05	0.6681	16.971
			2006-08	0.67	16.98
	Top Ring	Groove Clearance		0.0012-0.0026	0.030-0.065
		Gap	2002-05	0.0075-0.0122	0.38-0.55
			2006-08	0.0150-0.0217	0.19-0.31
	2nd Ring	Groove Clearance		0.0008-0.0022	0.020-0.055

SELOC_9606

Engine Specifications - 998cc Models

Component				Standard (in.) ①	Metric (mm) ①
Piston & Ring (cont'd)		Gap	2002-05	0.0193-0.0272	0.49-0.69
			2006-08	0.0118-0.0117	0.30-0.45
	Oil Ring	Groove Clearance		0.0016-0.0063	0.040-0.160
		Gap	2002-05	0.0114-0.0232	0.29-0.59
			2006-08	0.0039-0.0138	0.10-0.35
Timing Chain	Model Number		2002-05		DID SCR-0412SV
			2006-08		92RH2015-130
	Number of Links				130
	Tensioning System				Automatic
Valve System	Spring	Free Length	Exhaust	1.6	40.67
			Intake	1.53	38.9
		Installed Length	Exhaust	1.38	35
			Intake	1.36	34.5
		Spring Limit	Exhaust	0.071 ②	1.8 ②
			Intake	0.067 ②	1.7 ②
	Valve	Clearance (cold)	Exhaust	0.0098-0.0134	0.25-0.34
			Intake	0.0043-0.0079	0.11-0.20
		Guide Inside Diameter	Exhaust	0.1772-0.1776	4.500-4.512
			Intake	0.1575--0.1580	4.000-4.012
		Head Diameter	Exhaust	0.9606-0.9685	24.4-24.6
			Intake	0.9016-0.9094	22.9-23.1
		Seat Width	Exhaust	0.0354-0.0433	0.0-1.1
			Intake	0.0354-0.0433	0.0-1.1
		Stem Diameter	Exhaust	0.1756-0.1762 ③	4.460-4.475 ③
			Intake	0.1565-0.1571	3.975-3.990
		Stem-to-Guide Clearance	Exhaust	0.0010-0.0020 ④	0.025-0.052 ④
			Intake	0.0004-0.0015	0.010-0.037
		Stem Run-Out		0.0004	0.01

① Unless otherwise noted
② 2.5 degrees
③ 2006-08: 0.1758-0.1764 in. (4.465-4.475mm)
④ 2006-08: 0.0008.0.0019 in. (0.020-0.047-4.475mm)

SELOC_9606

Engine Specifications - 1052cc Models

Component				Standard (in.) ①	Metric (mm) ①
Camshaft	Cap Inside Diameter			0.9646	24.5
	Journal Diameter	FX		0.9630-0.9634	24.46-24.47
		VX	Ex.	0.9622-0.9626	24.44-24.45
			In.	0.9630-0.9634	24.46-24.47
	Journal Oill Clearance	VX	VX	0.0020-0.0033	0.050-0.084
	Journal to Cap Clearance	FX		0.0012-0.0024	0.03-0.06
		VX		0.0020-0.0024	0.05-0.06
	Lobe Height	FX	Ex.	1.299	33
			In.	1.281 ③	32.55 ③
		VX	Ex.	1.211	30.75
			In.	1.226 ④	31.15 ④
	Lobe Width		Ex.	0.984	25
			In.	0.984	25
	Run-out	FX		0.0012 Max	0.03 Max
		VX	2005-09	0.0012 Max	0.03 Max
			2010	0.0006 Max	0.015 Max
Connecting Rod	Bearing Color Code	#1		Brown	-
		#2		Black	-
		#3		Blue	-
		#4		Green	-
	Big End Oil Clearance	FX	2004-07	0.0006-0.0016	0.016-0.040
			2008	0.0016-0.0025	0.040-0.064
		VX		0.0006-0.0016	0.016-0.040
	Small End Inside Diameter	FX	2004-07	0.6694-0.6699	17.005-17.018
			2008	0.6695-0.6700	17.005-17.018
		VX		0.6695-0.6700	17.005-17.018
Crankshaft	Bearing Color Code	FX	#1	Brown	-
			#2	Black	-
			#3	Blue	-
			#4	Green	-
			#5	Yellow	-
		VX	#3	Red/Red	-
			#4	Red/Brown	-
			#5	Red/Black	-
			#6	Red/Blue	-

SELOC_9606

Engine Specifications - 1052cc Models

Component					Standard (in.) ①	Metric (mm) ①
Crankshaft (cont'd)				#7	Red/Green	-
	Deflection Limit		FX	2004-07	0.0012 Max	0.03 Max
			VX		0.0012 Max	0.03 Max
	Journal Oil Clearance		FX	2004-07	0.0002-0.0011	0.004-0.028
				2008	0.0010-0.0020	0.026-0.050
			VX		0.0002-0.0011	0.004-0.028
	Journal Diameter		FX	2008	1.3376-1.3386	33.976-34.000
			VX	2010	1.3376-1.3386	33.976-34.000
	Pin Diameter		FX	2008	1.4164-1.4173	35.976-36.000
			VX	2010	1.4164-1.4173	35.976-36.000
	Pin Oil Clearance		FX	2008	0.0016-0.0025	0.040-0.064
			VX	2010	0.0006-0.0016	0.016-0.040
	Run-Out		FX	2008	0.0012 Max	0.03 Max
			VX	2010	0.0012 Max	0.03 Max
Cylinder	Bore Size				2.9921-2.9927	76.000-76.015
	Out-of-Round Limit				0.002	0.05
	Taper Limit				0.003	0.08
	Wear Limit				2.9961	76.1
Cylinder Head	Compression Pressure				164 psi	1150 kPa
	Warpage Limit				0.004	0.1
Piston & Ring	Piston	Diameter			2.9980-2.9886	73.955-73.970
		Piston-to-Cylinder Clearance	FX	2004-07	0.0039-0.0043	0.10-0.11
			VX		0.0039-0.0043	0.10-0.11
		Wear Limit	FX	2004-07	0.0067	0.17
			VX		0.0067	0.17
	Piston Pin	Boss Inside Diameter			0.6693-0.6698	17.002-17.013
		Outside Diameter			0.6689-0.6693	16.991-17.000
		Wear Limit	FX	2004-07	0.67	16.98
			VX		0.67	16.98
	Top Ring	Groove Clearance			0.0012-0.0026	0.030-0.065
		Gap			0.0126-0.0173	0.32-0.44
	2nd Ring	Groove Clearance			0.0008-0.0022	0.020-0.055
		Gap			0.0169-0.0228	0.43-0.58
	Oil Ring	Groove Clearance			0.0016-0.0063	0.040-0.160
		Gap			0.0039-0.0138	0.10-0.35
Timing Chain	Model Number		FX	2004-05	DID SCR-0412SV	-
				2006-08	92RH2015-130	-
			VX	2005-09	92RH2015-130	-
	Number of Links				130	-
	Tensioning System				Automatic	-
Valve System	Spring	Free Length	Exhaust		1.6	40.67
			Intake		1.53	38.9
		Installed Length	Exhaust		1.38	35
			Intake		1.36	34.5
		Spring Limit	Exhaust		0.071 ②	1.8 ②
			Intake		0.067 ②	1.7 ②
	Valve	Clearance (cold)	Exhaust		0.0098-0.0134	0.25-0.34
			Intake		0.0043-0.0079	0.11-0.20
		Guide Inside Diameter	Exhaust		0.1772-0.1776	4.500-4.512
			Intake		0.1575--0.1580	4.000-4.012
		Head Diameter	Exhaust		0.9606-0.9685	24.4-24.6
			Intake		0.9016-0.9094	22.9-23.1
		Seat Width	Exhaust		0.0354-0.0433	0.0-1.1
			Intake		0.0354-0.0433	0.0-1.1
		Stem Diameter	Exhaust		0.1758-0.1764	4.460-4.475
			Intake		0.1565-0.1571	3.975-3.990
		Stem-to-Guide Clearance	Exhaust		0.0008-0.0019	0.020-0.047
			Intake		0.0004-0.0015	0.010-0.037
		Stem Run-Out			0.0004	0.01

SELOC_9606

① Unless otherwise noted
② 2.5 degrees
③ 2006-08: 1.299 in. (33mm)
④ 2010: 1.252 in. (31.8mm)

Engine Specifications - 1812cc Models

Component				Standard (in.) ①	Metric (mm) ①
Camshaft	Cap Inside Diameter			0.9646	24.5
	Journal Diameter			0.9827-0.9835	24.96-24.98
	Journal to Cap Clearance			0.0008-0.0024	0.20-0.061
	Lobe Height	FX HO/FX Cruiser HO	Ex.	1.626	41.3
			In.	1.63	41.4
		FX SHO/FX Cruiser SHO/FZ	Ex.	1.614	41
			In.	1.61	40.9
		VXR/VXS	Ex.	1.623-1.627	41.22-41.32
			In.	1.627-1,631	41.33-41.43
	Lobe Width	FX/FZ	Ex.	1.268	32.2
			In.	1.26	32
		VXR/VXS	Ex.	1.258-1.262	31.96-32.06
			In.	1.258-1.262	31.96-32.06
	Run-out			0.0006 Max	0.015 Max
Connecting Rod	Bearing Color Code	#1		Brown	-
		#2		Black	-
		#3		Blue	-
		#4		Green	-
	Small End Inside Diameter			0.8667-0.8672	22.015-22.028
Crankshaft	Bearing Color Code	#1		Brown	-
		#2		Black	-
		#3		Blue	-
		#4		Green	-
		#5		Yellow	-
	Journal Oil Clearance			0.0009-0.0021	0.024-0.053
	Journal Diameter			1.5739-1.5748	39.976-40
	Pin Diameter			1.6526-1.6535	41.976-42
	Pin Oil Clearance			0.0008-0.0022	0.020-0.056
	Run-Out			0.0012 Max	0.03 Max
Cylinder	Bore Size			3.3858-3.3864	86-86.015
Cylinder Head	Compression Pressure			N/A	N/A
	Warpage Limit			0.004	0.1
Piston & Ring	Piston	Diameter		3.3825-3.3831	85.915-85.930
	Piston Pin	Boss Inside Diameter		0.8663-0.8667	22.004-22.015
		Outside Diameter		0.8658-0.8661	21.991-22
	Top Ring	Groove Clearance		0.0008-0.0024	0.020-0.060
		Gap		0.012-0.018	0.32-0.44
	2nd Ring	Groove Clearance		0.0008-0.0024	0.020-0.06
		Gap		0.018-0.024	0.45-0.60
	Oil Ring	Groove Clearance		0.0016-0.0063	0.040-0.160
		Gap		0.004-0.014	0.10-0.35
Timing Chain	Model Number			97RH2015-152PX ②	
	Number of Links			152 ②	
	Tensioning System			Automatic	
Valve System	Spring	Free Length	Exhaust	1.794	45.58
			Intake	1.794	45.58
		Installed Length	Exhaust	1.339	34
			Intake	1.339	34
	Valve	Clearance (cold)	Exhaust	0.011-0.015	0.28-0.37
			Intake	0.006-0.009	0.14-0.23
		Guide Inside Diameter	Exhaust	0.2167-0.2174	5.504-5.522
			Intake	0.2167-0.2174	5.504-5.522
		Head Diameter	Exhaust	1.138-1.146	28.9-29.1
			Intake	1.335-1.343	33.9-34.1
		Seat Width	Exhaust	0.059-0.067	1.50-1.70
			Intake	0.055-0.063	1.40-1.60
		Stem Diameter	Exhaust	0.2151-0.2157	5.464-5.479
			Intake	0.2156-0.2162	5.477-5.492
		Stem-to-Guide Clearance	Exhaust	0.0010-0.0023	0.025-0.058
			Intake	0.0005-0.0018	0.012-0.045
		Stem Run-Out		0.0004	0.01

① Unless otherwise noted ② FX and FZ models only

SELOC_9606

Torque Specifications - 998cc Engines

Component				Inch Lbs.	ft. lbs.	Nm
Anode Cover	To Anode			104	8.7	12
	To Cylinder Head			-	14	20
Bearing Housing To Reduction Drive Gear Case	1st Pass			-	11	15
	2nd Pass			-	20	28
Camshaft	Camshaft Position Sensor			86	7.2	10
	Exhaust Cam Cap			86	7.2	10
	Exhaust Cam Sprocket			-	17	24
	Intake Cam Cap			86	7.2	10
	Intake Cam Sprocket			-	17	24
Camshaft Position Sensor				86	7.2	10
Coupling Cover				66	5.6	7.8
Crankcase	Connecting Rod Cap	1st Pass		-	14	20
		2nd Pass		-	120° +/- 5°	-
	Cover To Lower			104	8.7	12
	Generator Cover	1st Pass		-	11	15
		2nd Pass		-	36	50
	Oil Pipe To Lower			104	8.7	12
	Upper To Lower	1st Pass		67	5.6	7.8
		2nd Pass		-	11	15
		3rd Pass		-	49° +/- 5°	-
Cylinder Head	Cover			104	8.7	12
	Hanger			-	29	40
	Mounting Bolts			86	7.2	10
	Rear Mounting Nuts (2)	1st Pass		-	14	20
		2nd Pass		-	46	64
	Remaining Mounting Nuts (8)	1st Pass		-	14	20
		2nd Pass		-	35	49
Driveshaft Drive Coupling				-	20	28
Engine Mount				-	12	17
Engine Temperature Sensor				-	11	15
Engine Thermoswitch				66	5.5	7.6
Exhaust System	Exhaust Manifold 1 To Cylinder Head	#1 & 2		-	16	22
		#3		-	25	35
	Exhaust Manifold 2 To Cylinder Head	#1 & 2		-	16	22
		#3		-	25	35
	Exhaust Pipe To Crankcase	#1, 2, 3 & 4		13	1.4	2
		#5, 6, 7 & 8		-	11	15
		#9. 10. 11 & 12		-	28	39
	Joint Clamp, Inner & Outer, Pipe 2-to-Pipe 3			38	3.2	4.4
	Pipe End, Pipe 3	1st Pass		30	2.7	3.7
		2nd Pass		65	5.5	7.6
	Pipe Stay To Crankcase	1st Pass		-	11	15
		2nd Pass		-	30	42
	Pipe Stay, Pipe 1	1st Pass		-	11	15
		2nd Pass		-	30	42
	Pipe 2 To Pipe 1			-	28	39
	Pipe 1 To Manifold 1/2	#1 & 2		-	16	22
		#3		-	25	35
	Thermoswitch, Pipe 3			65	5.5	7.6
Generator Cover To Crankcase	1st Pass			-	11	15
	2nd Pass			-	36	50
Lighting Coil To Generator Cover				-	10	14
Lubrication System	Oil Cooler Cover To Oil Tank	1st Pass		30	2.7	3.7
		2nd Pass		65	5.5	7.6
	Oil Drain Plug			-	13	18
	Oil Filter	Filter Bolt		-	12	17
		Stud Bolt		-	25	35
	Oil Pan			104	8.7	12
	Oil Pressure Switch			73	6.1	8.4
	Oil Pump To Reduction Drive Gear Case	2002	1st Pass	12	1	14
			2nd Pass	-	7.2	10
		2003-08	1st Pass, Short	12	1	14
			2nd Pass, Short	86	7.2	10
			1st Pass, Long	78	6.5	8.8
			2nd Pass, Long	-	13	18
	Oil Strainer To Oil Tank	1st Pass		30	2.7	3.7
		2nd Pass		65	5.5	7.6
	Oil Tank Anode			30	2.7	3.7
	Oil Tank Stay	1st Pass		13	1.4	2
		2nd Pass		-	11	15
		3rd Pass		-	28	39
	Oil Tank To Reduction Drive Gear Case	1st Pass		-	11	15
		2nd Pass		-	20	28
	Water Jacket, Oil Tank	1st Pass		30	2.7	3.7
		2nd Pass		65	5.5	7.6
Oil Drain Plug				-	13	18

SELOC_9606

Torque Specifications - 998cc Engines

Component			Inch Lbs.	ft. lbs.	Nm
Oil Filter	Filter Bolt		-	12	17
	Stud Bolt		-	25	35
Oil Pan			104	8.7	12
Oil Pressure Switch			73	6.1	8.4
Pressure Control Valve		1st Pass	32	2.7	3.7
		2nd Pass	66	5.5	7.6
Pulser Coil To Generator Cover			42	3.5	4.9
Reduction Drive Gear Case To Crankcase	Large Bolt, 1st Pass		-	11	15
	Large Bolt, 2nd Pass		-	20	28
	Small Bolt, 1st Pass		30	2.7	3.7
	Small Bolt, 2nd Pass		65	5.5	7.6
Rotor	To Crankshaft		-	54	75
	To Starter Clutch		-	17	24
Spark Plug			-	9.4	13
Starter Motor	Electrical Lead		42	3.5	4.9
	Mounting Bolts		-	13	18
	Negative Battery Lead		-	13	18
Thermostat Housing Cover	1st Pass		32	2.7	3.7
	2nd Pass		66	5.5	7.6
Timing Chain	Tensioner		86	7.2	10
	Tensioner Cap Bolt		86	7.2	10

SELOC_9606

Torque Specifications - FX Models w/1052cc Engines

Component				Inch lbs.	ft. lbs.	Nm
Anode Cover	To Anode	Short Bolt		104	8.7	12
	To Cylinder Head	Long Bolt		-	14	20
Bearing Housing To Reduction Drive Gear Case	Thru 2007	1st Pass		-	11	15
		2nd Pass		-	20	28
Camshaft	Camshaft Position Sensor			86	7.2	10
	Exhaust Cam Cap			86	7.2	10
	Exhaust Cam Sprocket			-	17	24
	Intake Cam Cap			86	7.2	10
	Intake Cam Sprocket			-	17	24
Camshaft Position Sensor				86	7.2	10
Coupling Cover				66	5.6	7.8
Crankcase	Connecting Rod Cap	1st Pass		-	14	20
		2nd Pass	Thru 2007	-	120° +/- 5°	-
			2008 & Later	-	120°	-
	Cover To Lower			104	8.7	12
	Generator Cover	Nuts	1st Pass	-	11	15
			2nd Pass	-	36	50
		Bolts	1st Pass	-	11	15
			2nd Pass	-	36	50
	Oil Pipe To Lower			104	8.7	12
	Upper To Lower	1st Pass		67	5.6	7.8
		2nd Pass		Loosen Completely		
		3rd Pass		-	11	15
		4th Pass	Thru 2007	-	49° +/- 5°	-
			2008 & Later	-	50°	-
Cylinder Head	Cover			104	8.7	12
	Hanger			-	29	40
	Mounting Nuts (Thru 2007)	Center (2) ①	1st Pass	-	14	20
			2nd Pass	-	121° +/- 5°	-
		All Others (6) ①	1st Pass	-	14	20
			2nd Pass	-	105° +/- 5°	-
		Rear (2) ①	1st Pass	-	14	20
			2nd Pass	-	140° +/- 5°	-
	Mounting Nuts (2008 & Later)	Right Side (3) ①	1st Pass	-	14	20
			2nd Pass	-	105°	-
		All Others (5) ①	1st Pass	-	14	20

SELOC_9606

Torque Specifications - FX Models w/1052cc Engines

Component				Inch lbs.	ft. lbs.	Nm
Cylinder Head (cont'd)			2nd Pass	-	120°	-
		Rear (2) ①	1st Pass	-	14	20
			2nd Pass	-	140°	-
Driveshaft Drive Coupling				-	20	28
Engine Mount	Engine To Mount			-	12	17
	Engine Damper			55	4.6	6.4
	Mount To Vessel			-	12	17
Engine Temperature Sensor				-	11	15
Engine Thermoswitch				66	5.5	7.6
Exhaust System	Exhaust Manifold To Cylinder Head		1st & 2nd Pass	-	16	22
			3rd Pass	-	25	35
	Exhaust Pipe Assembly	2008 & Later	1st & 2nd Pass	-	16	22
			3rd Pass	-	25	35
	Exhaust Pipe To Crankcase	Thru 2007	1st Pass	13	1.4	2
			2nd Pass	-	11	15
			3rd Pass	-	28	39
	Joint Clamp, Inner & Outer, Pipe 2-to-Pipe 3	2 Passes		30	2.5	3.4
	Muffler Assembly	1st Pass		13	1.4	2
		2nd Pass		-	11	15
		3rd Pass		-	28	39
	Pipe End, Pipe 3	1st Pass		30	2.7	0.37
		2nd Pass		65	5.5	7.6
	Pipe Stay To Crankcase	1st Pass		-	11	15
		2nd Pass		-	30	42
	Pipe Stays	1st Pass		-	11	15
		2nd Pass		-	30	42
	Pipe 2 To Pipe 1	Thru 2007		-	28	39
	Pipe 1 To Manifold 1/2		1st & 2nd Pass	-	16	22
			3rd Pass	-	25	35
	Ring Joint			-	28	39
	Temperature Sensor	Calif. Only	2008 & Later	-	28	39
	Thermoswitch			65	5.5	7.6
Flywheel Magneto	2008 & Later	Mounting Bolts		-	54	75
		To Starter Clutch		-	17	24
Generator Cover To Crankcase	Nuts & Bolts	1st Pass		-	11	15
		2nd Pass		-	36	50
Ignition Coil To Cylinder Head Cover				65	5.5	7.6
Lighting Coil To Generator Cover	Thru 2007			-	10	14
Lubrication System	Baffle Plate	1st Pass		17	1.4	1.9
		2nd Pass		38	3.2	4.4
	Hanger	1st Pass		30	2.7	3.7
		2nd Pass		65	5.5	7.6
	Oil Cooler Cover To Oil Tank	1st Pass		30	2.7	3.7
		2nd Pass		65	5.5	7.6
	Oil Drain Plug	Thru 2007		-	14	20
		2008 & Later		-	13	18
	Oil Filter	Filter Bolt		-	12	17
		Stud Bolt		-	25	35
	Oil Pan			104	8.7	12
	Oil Pressure Switch			73	6.1	8.4
	Oil Pump To Reduction Drive Gear Case	Short Bolts		86	7.2	10
		Long Bolts	1st Pass	-	11	15
			2nd Pass	-	20	28
	Oil Separator	2008 & Later	1st Pass	30	2.7	3.7
			2nd Pass	65	5.5	7.6
	Oil Strainer To Oil Tank	1st Pass		30	2.7	3.7
		2nd Pass		65	5.5	7.6
	Oil Tank	2008 & Later	1st Pass, Bolts	-	11	15
			2nd Pass, Bolts	-	28	39
			1st Pass, Nuts	13	1.4	2
			2nd Pass, Nuts		11	15
			3rd Pass, Nuts	-	28	39
	Oil Tank Anode			30	2.7	3.7
	Oil Tank Cover To Oil Tank	1st Pass		30	2.7	3.7
		2nd Pass		65	5.5	7.6
	Oil Tank Stay	1st Pass		13	1.4	2
		2nd Pass		-	11	15
		3rd Pass		-	28	39
	Oil Tank To Reduction Drive Gear Case	Thru 2007	1st Pass	-	11	15
			2nd Pass	-	28	39
	Water Jacket, Oil Tank	1st Pass		30	2.7	3.7
		2nd Pass		65	5.5	7.6
Oil Drain Plug	Thru 2007			-	14	20
	2008 & Later			-	13	18

SELOC_9606

Torque Specifications - FX Models w/1052cc Engines

Component			Inch lbs.	ft. lbs.	Nm
Oil Filter	Filter Bolt		-	12	17
	Stud Bolt		-	25	35
Oil Pan			104	8.7	12
Oil Pressure Switch			73	6.1	8.4
Pressure Control Valve	Thru 2007	1st Pass	32	2.7	3.7
		2nd Pass	66	5.5	7.6
Pick-Up Coil To Generator Cover	Thru 2007		42	3.5	4.9
Reduction Drive Gear Case To Crankcase	Large Bolt, 1st Pass		-	11	15
	Large Bolt, 2nd Pass		-	20	28
	Small Bolt, 1st Pass		30	2.7	3.7
	Small Bolt, 2nd Pass		65	5.5	7.6
Rotor	Thru 2007	To Crankshaft	-	54	75
		To Starter Clutch	-	17	24
Spark Plug			-	9.4	13
Starter Motor	Electrical Lead		42	3.5	4.9
	Mounting Bolts		-	13	18
	Negative Battery Lead		-	13	18
Stator Coil	Allen	2008 & Later	42	3.5	4.9
	Hex	2008 & Later	120	10	14
Thermo Sensor	2008 & Later		-	14	20
Thermostat Housing	Grease Nipple		52	4.3	5.9
	Housing Cover	1st Pass	32	2.7	3.7
		2nd Pass	66	5.5	7.6
	Housing Holder	1st Pass	32	2.7	3.7
		2nd Pass	66	5.5	7.6
Timing Chain	Tensioner		86	7.2	10
	Tensioner Cap Bolt		86	7.2	10

① See procedures in text

SELOC_9606

Torque Specifications - FX/FZ Models w/1812cc Engines

Component				Inch lbs.	ft. lbs.	Nm
Air Filter	Bracket Nut	FZ Models		-	12.5	17
	Mounting	HO/SHO Models		115	9.6	13
		FZ Models		107	8.9	12
Anode	To Anode			71	5.9	8
	To Cylinder Block			-	14.8	20
APS Unit				-	12.5	17
Breather Assembly				71	5.9	8
Camshaft	Camshaft Position Sensor			89	7.4	10
	Cap			-	11.8	16
	Sprocket	HO Models		-	13.3	18
		SHO/FZ Models		-	17.7	24
Camshaft Position Sensor				89	7.4	10
Coupling Cover				71	5.9	8
Crankcase	Connecting Rod Cap	HO Models	1st Pass	-	36.9	50
			2nd Pass	-	90°	-
		SHO/FZ Models	1st Pass	-	37.6	51
			2nd Pass	-	90°	-
	Upper To Lower	Long Bolts (10)	1st Pass	-	22.1	30
			2nd Pass	-	90°	-
		Short Bolts (12)		89	7.4	10
Cylinder Head	Cover			71	5.9	8
	Hanger	HO Models	Front	-	31	42
			Rear	107	8.9	12
		SHO/FZ Models		107	8.9	12
	Mounting Bolts	Long Bolts (10) ①	1st Pass	-	35.4	48
			2nd Pass	-	90°	-
		Short Bolts (3) ①		89	7.4	10
Deck Beam	FZ Models	Bolts/Nuts		-	13.3	18
Drive Coupling	HO Models			-	184.4	250
	SHO/FZ Models	Special Nut		-	184.4	250
		Holder ②		-	59	80
Electrical Box				-	12.5	17
Engine Cover	Cover Bolt			44	3.7	5
	Cover Grommet Bolt			53	4.4	6

SELOC_9606

Torque Specifications - FX/FZ Models w/1812cc Engines

Component					Inch lbs.	ft. lbs.	Nm
Engine Cover (cont'd)	Stay Bolt	HO/SHO Models			89	7.4	10
		FZ Models			72	5.9	8
Engine Mount	Engine To Mount				-	12.5	17
	Engine Damper				53	4.4	6
	Mount To Vessel				-	12.5	17
	Spacer Bolt	FZ Models			44	3.7	5
Engine Temperature Sensor					133	11.1	15
Engine Thermoswitch					71	5.9	8
Exhaust System	Exhaust Manifold	1st Pass			-	14.8	20
		2nd Pass			-	25.8	35
	Exhaust Pipe 1	1st Pass			-	14.8	20
		2nd Pass			-	25.8	35
	Exhaust Pipe 1 Stay	HO/FZ Models	1st Pass		-	16.2	22
			2nd Pass		-	31	42
		SHO Models	1st Pass		-	17	23
			2nd Pass		-	31	42
	Exhaust Pipe 2	1st Pass			133	11.1	15
		2nd Pass			-	21.4	29
	Exhaust Pipe 3	HO/SHO Models			-	44.3	60
		FZ Models			-	46.5	63
	Thermosensor				-	14.8	20
	Thermoswitch				71	5.9	8
Flywheel Magneto					-	17.7	24
Fuse Box Nut					-	12.5	17
Generator Cover					-	19.2	26
Ignition Coil					71	5.9	8
Intake Manifold	IAT Sensor				133	11.1	15
	IAP Sensor	SHO/FZ Models	Bracket Bolt		31	2,6	3.5
			Sensor Bolt		44	3.7	5
	Mounting	Bolts	1st Pass		89	7.4	10
			2nd Pass		-	14.8	20
		Nuts	1st Pass		89	7.4	10
			2nd Pass		-	14.8	20
		Studs	1st Pass		89	7.4	10
			2nd Pass		-	14.8	20
	Bracket Bolt	FZ Models	1st Pass		-	16.2	22
			2nd Pass		-	31	42
		HO Models	1st Pass		133	11.1	15
			2nd Pass		-	22.1	30
		SHO Models	1st Pass		-	17	23
			2nd Pass		-	31	42
Knock Sensor					133	11.1	15
Lubrication System	Baffle Plate				107	8.9	12
	Earth Plate				71	5.9	8
	Oil Cooler	HO/FZ Models			-	13.3	18
		SHO Models			-	14	19
	Oil Filter	Filter Bolt			-	13.3	18
	Oil Pan				89	7.4	10
	Oil Pipe	Pipe Bolt			-	15.5	21
		Pipe Holder Bolt			71	5.9	8
	Oil Pressure Switch				71	5.9	8
	Oil Pump	Mount	HO/FZ Models		89	7.4	10
			SHO Models		97	8.1	11
		Drive Gear ②			-	59	80
	Oil Separator Tank Cover				71	5.9	8
	Water Jacket Cover				71	5.9	8
Oil Filter					-	13.3	18
Oil Pan					89	7.4	10
Oil Pressure Switch					71	5.9	8
Pick-Up Coil					44	3.7	5
Rectifier/Regulator					-	19.2	26
Spark Plug					-	18.4	25
Starter Motor	Electrical Lead				44	3.7	5
	Mounting Bolts				-	13.3	18
Stator Coil	HO/FZ Models				124	10.3	14
	SHO Models				133	11.1	15
Supercharger	Air Cooler	Mount	Bolts	1st Pass	-	17 ③	23
				2nd Pass	-	31	42

SELOC_9606

Torque Specifications - FX/FZ Models w/1812cc Engines

Component					Inch lbs.	ft. lbs.	Nm
Supercharger (cont'd)			Nuts	1st Pass	-	17 ③	23
				2nd Pass	-	31	42
			Stud	1st Pass	89	7.4	10
				2nd Pass	-	14.8	20
		Stay			-	31	42
	Boost Pipe				-	14.8	20
	Oil Filler Bolt				36	3	4
	Supercharger	1st Pass			89	7.4	10
		2nd Pass			-	14.8	20
Thermostat Housing					71	5.9	8
Throttle Body	Joint Bolt				115	9.6	13
	Mount Nut				115	9.6	13
Timing Chain	Guide Plate Bolt				89	7.4	10
	Tensioner				89	7.4	10
	Tensioner Cap Bolt				89	7.4	10
Transfer Shaft	HO/FZ Models	1st Pass			-	36.9	50
		2nd Pass			-	60°	-
		3rd Pass			-	184.4	250
	SHO Models				-	184.4	250

SELOC_9606

① See procedures in text
② Left-hand thread
③ FZ models: 16.2 ft.lbs. (22 Nm)

Torque Specifications - VX Models w/1052cc Engines

Component					Inch lbs.	ft. lbs.	Nm
Anode Cover	To Anode	Short Bolt	2005-09		104	8.7	12
			2010		107	8.9	12
	To Cylinder Head	Long Bolt	2005-09		-	14	20
			2010		-	14.8	20
Camshaft	Camshaft Position Sensor				86	7.2	10
	Cap	2005-09			86	7.2	10
		2010			89	7.4	10
	Sprocket	2005-09			-	17	24
		2010			-	17.7	24
Camshaft Position Sensor					86	7.2	10
Coupling Cover	2005-09				66	5.6	7.8
	2010				71	5.9	8
Crankcase	Connecting Rod Cap	2005-09	1st Pass		-	14	20
			2nd Pass		-	120° +/- 5°	-
		2010	1st Pass		-	14.8	20
			2nd Pass		-	120°	-
	Oil Pipe To Lower	2005-09			104	8.7	12
		2010			107	8.9	12
	Upper To Lower	Short Bolts (10)	2005-09		104	8.7	12
			2010		107	8.9	12
		Long Bolts (10)	1st Pass		67	5.6	7.8
			2nd Pass		Loosen Completely		
			3rd Pass		-	11	15
			4th Pass	2005-09	-	49° +/- 5°	-
				2010	-	50°	-
Cylinder Head	Cover	2005-09			104	8.7	12
		2010			107	8.9	12
	Hanger	2005-09			-	29	40
		2010			-	13.3	18
	Mounting Nuts/Bolt	2005-09	Center (2) ①	1st Pass	-	14	20
				2nd Pass	-	121° +/- 5°	-
			All Others (6) ①	1st Pass	-	14	20
				2nd Pass	-	105° +/- 5°	-
			Rear (2) ①	1st Pass	-	14	20
				2nd Pass	-	140° +/- 5°	-
		2010	Center (2) ①	1st Pass	-	14.8	20

SELOC_9606

Torque Specifications - VX Models w/1052cc Engines

Component					Inch lbs.	ft. lbs.	Nm
Cylinder Head (cont'd)				2nd Pass	-	120°	-
			All Others (6) ①	1st Pass	-	14.8	20
				2nd Pass	-	105°	-
			Rear (2) ①	1st Pass	-	14.8	20
				2nd Pass	-	140°	-
			Front Bolts (3) ①	1st Pass	89	7.4	10
	Plate Bolt	2010	1st Pass		36	3	4
			2nd Pass		72	5.9	8
Driveshaft Drive Coupling	2005-09				-	20	28
	2010				-	21.4	29
Electrical Plate Bolt					-	12.5	17
Engine Mount	Engine To Mount	2005-09			-	12	17
		2010			-	12.5	17
	Engine Damper				55	4.6	6.4
	Mount To Vessel	2005-09			-	12	17
		2010			-	12.5	17
	Spacer Bolt				44	3.7	5
	Stopper Bolt				62	5.2	7
Engine Temperature Sensor					-	11	15
Engine Thermoswitch	2005-09				66	5.5	7.6
	2010				71	5.9	8
Exhaust System	Exhaust Manifold To Cylinder Head	2005-09	1st & 2nd Pass		-	16	22
			3rd Pass		-	25	35
		2010	1st Pass		-	16.2	22
			2nd Pass		-	25.8	35
	Exhaust Pipe 1	2005-09	1st & 2nd Pass		-	16	22
			3rd Pass		-	25	35
		2010	1st Pass		-	16.2	22
			2nd Pass		-	25.8	35
	Exhaust Pipe 2	2005-09	1st Pass		13	1.4	2
			2nd Pass		-	11	15
			3rd Pass		-	28	39
		2010	1st Pass		18	1.5	2
			2nd Pass		133	11.1	15
			3rd Pass		-	29.5	40
	Joint Clamp, Inner & Outer	2005-09	2 Passes		30	2.5	3.4
		2010	Fore		18	1.5	2
			Aft		32	2.7	3.7
	Pipe End	2005-09	1st Pass		30	2.7	3.7
			2nd Pass		65	5.5	7.6
		2010	1st Pass		36	3	4
			2nd Pass		71	5.9	8
	Thermoswitch	2005-09			66	5.5	7.6
		2010			71	5.9	8
Flywheel Magneto	Mounting Bolts	2005-09			-	54	75
		2010			-	55.3	75
	To Starter Clutch	2005-09			-	17	24
		2010			-	17.7	24
Fuse Box Nut					-	12.5	17
Generator Cover	2005-09	1st Pass			-	11	15
		2nd Pass			-	36	50
	2010	1st Pass			133	11.1	15
		2nd Pass			-	36.9	50
Ignition Coil					71	5.9	8
Intake Manifold	Map Sensor				36	3	4
	Mounting Bolts	1st Pass			79	6.6	9
		2nd Pass			-	13.3	18
	Bracket Bolt	1st Pass			133	11.1	13
		2nd Pass			-	29.5	40
Lubrication System	Anode	2010			36	3	4
	Baffle Plate	2005-09	1st Pass		17	1.4	1.9
			2nd Pass		38	3.2	4.4
		2010	1st Pass		18	1.5	2
			2nd Pass		36	3	4
	Earth Plate	2010			71	5.9	8
	Hanger/Holder	1st Pass			36	3	4
		2nd Pass			71	5.9	8
	Oil Breather Plate	2005-09	1st Pass		17	1.4	1.9
			2nd Pass		38	3.2	4.4
		2010	1st Pass		18	1.5	2
			2nd Pass		36	3	4
	Oil Cooler Cover To Oil Tank	2005-09	1st Pass		30	2.7	3.7

SELOC_9606

Torque Specifications - VX Models w/1052cc Engines

Component					Inch lbs.	ft. lbs.	Nm
Lubrication System (cont'd)			2nd Pass		65	5.5	7.6
		2010	1st Pass		36	3	4
			2nd Pass		71	5.9	8
	Oil Drain Plug	2005-09			113	9.4	13
		2010			115	9.6	13
	Oil Filter	Filter Bolt	2005-09		-	12	17
			2010		-	13.3	18
		Stud Bolt	2005-09		-	25	35
			2010		-	25.8	35
	Oil Pan	2005-09			104	8.7	12
		2010			107	8.9	12
	Oil Pressure Switch				73	6.1	8.4
	Oil Pump	2005-09	Short Bolts		86	7.2	10
			Long Bolts	1st Pass	-	11	15
				2nd Pass	-	20	28
		2010	Short Bolts		89	7.4	10
			Long Bolts	1st Pass	133	11.1	15
				2nd Pass	-	21.4	29
	Oil Separator	2005-09	1st Pass		30	2.7	3.7
			2nd Pass		65	5.5	7.6
		2010	1st Pass		36	3	4
			2nd Pass		71	5.9	8
	Oil Strainer To Oil Tank	2005-09	1st Pass		30	2.7	3.7
			2nd Pass		65	5.5	7.6
		2010	1st Pass		36	3	4
			2nd Pass		71	5.9	8
	Oil Tank	2005-09	Bolts	1st Pass	-	11	15
				2nd Pass	-	28	39
			Nuts	1st Pass	13	1.4	2
				2nd Pass		11	15
				3rd Pass	-	28	39
		2010	Bolts	1st Pass	133	11.1	15
				2nd Pass	-	29.5	40
			Nuts	1st Pass	18	1.5	2
				2nd Pass	133	11.1	15
				3rd Pass	-	29.5	40
	Oil Tank Cover To Oil Tank	2005-09	1st Pass		30	2.7	3.7
			2nd Pass		65	5.5	7.6
		2010	1st Pass		36	3	4
			2nd Pass		71	5.9	8
	Oil Tank Stay	2005-09	1st Pass		13	1.4	2
			2nd Pass		-	11	15
			3rd Pass		-	28	39
		2010	1st Pass		18	1.5	2
			2nd Pass		133	11.1	15
			3rd Pass		-	29.5	40
	Water Jacket, Oil Tank	2010			71	5.9	8
Oil Drain Plug	2005-09				113	9.4	13
	2010				115	9.6	13
Oil Filter	Filter Bolt	2005-09			-	12	17
		2010			-	13.3	18
	Stud Bolt	2005-09			-	25	35
		2010			-	25.8	35
Oil Pan	2005-09				104	8.7	12
	2010				107	8.9	12
Oil Pressure Switch					73	6.1	8.4
Pick-Up Coil					44	3.7	5
Reduction Drive Gear Case To Crankcase	2005-09	Large Bolt	1st Pass		-	11	15
			2nd Pass		-	20	28
		Small Bolt	1st Pass		30	2.7	3.7
			2nd Pass		65	5.5	7.6
	2010	Large Bolt	1st Pass		133	11.1	15
			2nd Pass		-	21.4	29
		Small Bolt	1st Pass		36	3	4
			2nd Pass		71	5.9	8
Spark Plug	2005-09				-	9.4	13
	2010				115	9.6	13
Starter Clutch	2010				-	17,7	24
Starter Motor	Electrical Lead	2005-09			42	3.5	4.9
		2010			44	3.7	5
	Mounting Bolts	2005-09			-	13	18
		2010			-	13.3	18
	Negative Battery Lead				-	13	18

SELOC_9606

Torque Specifications - VX Models w/1052cc Engines

Component			Inch lbs.	ft. lbs.	Nm
Stator Coil			133	11.1	15
Timing Chain	Tensioner	2005-09	86	7.2	10
		2010	89	7.4	10
	Tensioner Cap Bolt	2005-09	86	7.2	10
		2010	89	7.4	10

SELOC_9606

① See procedures in text

Torque Specifications - VXR/VXS Models w/1812cc Engines

Component				Inch lbs.	ft. lbs.	Nm
Air Filter	Bracket Nut			-	12.5	17
	Mounting			107	8.9	12
Anode	To Anode			71	5.9	8
	To Cylinder Block			-	14.8	20
Camshaft	Camshaft Position Sensor			89	7.4	10
	Cap			-	11.8	16
	Sprocket			-	13.3	18
Camshaft Position Sensor				89	7.4	10
Coupling Cover				71	5.9	8
Crankcase	Connecting Rod Cap	HO Models	1st Pass	-	36.9	50
			2nd Pass	-	90°	-
	Upper To Lower	Long Bolts (10)	1st Pass	-	22.1	30
			2nd Pass	-	90°	-
		Short Bolts (12)		89	7.4	10
Cylinder Head	Cover			71	5.9	8
	Hanger	Front		-	31	42
		Rear		107	8.9	12
	Mounting Bolts	Long Bolts (10) ①	1st Pass	-	40.6	55
			2nd Pass	-	90°	-
		Short Bolts (3) ①		89	7.4	10
Deck Beam	FZ Models	Bolts/Nuts		-	13.3	18
Drive Coupling	Nut			-	184.4	250
	Holder ②			-	59	80
Electrical Box				-	12.5	17
Engine Cover	Cover Bolt			44	3.7	5
	Cover Grommet Bolt			53	4.4	6
Engine Mount	Engine To Mount			-	12.5	17
	Engine Damper			49	4.1	5.5
	Mount To Vessel			-	12.5	17
Engine Temperature Sensor				133	11.1	15
Engine Thermo-switch				71	5.9	8
Exhaust System	Exhaust Manifold	1st Pass		-	14.8	20
		2nd Pass		-	14.8	20
		3rd Pass		-	25.8	35

SELOC_9606

Torque Specifications - VXR/VXS Models w/1812cc Engines

Component				Inch lbs.	ft. lbs.	Nm
Exhaust System (cont'd)	Exhaust Pipe 1	1st Pass		-	14.8	20
		2nd Pass		-	25.8	35
	Exhaust Pipe 2	1st Pass		133	11.1	15
		2nd Pass		-	21.4	29
	Exhaust Pipe 3			-	147.5	200
	Thermosensor			-	14.8	20
	Thermoswitch			71	5.9	8
Flywheel Magneto	Bolts			-	17.7	24
	Special nut			-	184.4	250
Fuse Box Nut				71	5.9	8
Generator Cover				-	19.2	26
Ignition Coil				71	5.9	8
Intake Manifold	IAT Sensor			133	11.1	15
	IAP Sensor	Bracket Bolt		31	2,6	3.5
		Sensor Bolt		44	3.7	5
	Manifold-to-Engine	Bolts	1st Pass	89	7.4	10
			2nd Pass	-	14.8	20
		Nuts	1st Pass	89	7.4	10
			2nd Pass	-	14.8	20
		Studs	1st Pass	89	7.4	10
			2nd Pass	-	14.8	20
	Bracket Bolts	To Manifold	1st Pass	-	11.1	15
			2nd Pass	-	22.1	30
		To Engine	1st Pass	-	11.1	15
			2nd Pass	-	22.1	30
Lubrication System	Baffle Plate			107	8.9	12
	Oil Cooler Cover	M6		107	8.9	12
		M8		-	14.8	20
	Oil Filter	Filter Bolt		-	13.3	18
	Oil Pan			89	7.4	10
	Oil Pipe	Pipe Bolt		-	14.8	20
		Pipe Holder Bolt		71	5.9	8
	Oil Pressure Switch			71	5.9	8
	Oil Pump	Mount		89	7.4	10
		Drive Gear ②		-	59	80
	Oil Separator Tank Cover			71	5.9	8
	Water Jacket Cover			71	5.9	8
Oil Filter				-	13.3	18
Oil Pan				89	7.4	10
Oil Pressure Switch				71	5.9	8
Pick-Up Coil				44	3.7	5
Rectifier/Regulator	Cover			16	9.6	13
	Mounting				19.2	26
Spark Plug				-	18.4	25
Starter Motor	Electrical Lead			44	3.7	5
	Mounting Bolts			-	13.3	18
Stator Coil				124	10.3	14
Thermostat Housing				71	5.9	8
Throttle Body	Joint Bolt			-	17.7	24
	Mount Nut			115	9.6	13
Timing Chain	Guide Plate Bolt			89	7.4	10
	Tensioner			89	7.4	10
Transfer Shaft	HO/FZ Models	1st Pass		-	36.9	50
		2nd Pass		-	60°	-
		3rd Pass		-	184.4	250

① See procedures in text
② Left-hand thread

SELOC_9606

7

DRIVE SYSTEM

JET DRIVE

General Information

Starting at the rear of the engine, the drive system is connected to the engine crankshaft by a drive coupler. The crankshaft has a coupler, or damper, connected to it which then meshes with a coupler on the forward end of the jet pump intermediate shaft.

On FX models and also VX models w/1812cc engines, the short intermediate shaft runs through an intermediate bearing where it connects to the end of the pump driveshaft, running through the impeller and housing, ending in the jet nozzle at the rear of the pump assembly.

On VX models w/1052cc engines, the long intermediate shaft runs through an intermediate bearing where it connects to the driveshaft at the impeller, which runs through the housing, and ends in the jet nozzle at the rear of the pump assembly.

The jet pump is an axial flow, single stage design. The pump assembly consists of the previously mentioned driveshaft, a forward impeller housing (FX models with 998cc and 1052cc engines), an impeller housed in a main impeller housing, an impeller duct (some people call this the bearing housing), which is connected to the jet nozzle.

A thrust nozzle, or steering nozzle as it is frequently called is connected to the jet nozzle by a nozzle ring (FX models, VX models do not use the ring) and a reverse gate is the final piece of the puzzle.

■ **Interestingly, the VXR and VXS models utilize the FX drive from the Impeller housing/duct forward, and the jet nozzle, thrust nozzle and reverse gate from the smaller engined VX.**

As the impeller rotates (driven off the engine's crankshaft via the driveshaft), water is drawn through the intake grate and duct at the bottom of the craft. The impeller forces water through the housing where it exits through the nozzle assembly. As engine rpm increases, a higher volume of water is forced through the jet pump, thereby increasing the vehicle's speed.

A steering cable runs from the front of the craft to the thrust/steering nozzle, controlling steering movement.

Also on the bottom of the craft is an intake grate/duct, ride plate and speed sensor.

Intake Grate, Intake Duct & Ride Plate

REMOVAL & INSTALLATION

FX Models w/998cc Engines & 2004-07 FX HO Models w/1052cc Engines

2 MODERATE

◆ **See Figures 1 and 2**

1. Remove the craft from the water. It is best to install it in a stand or block, but most people will simply jockey it on the trailer until access is provided. Disconnect the battery cables.
2. Remove the 4 mounting bolts (2 front and 2 back) and lift off the intake grate. Notice that the front and rear bolts are different lengths and make a note of which came from where!
3. The speed sensor is held in position on the ride plate with 4 small bolts. Remove them and disconnect the sensor. On 1052cc models, the sensor also has a holder that can be slid off it. The paddle wheel is connected to the sensor by 2 fasteners that snap over the shaft; the wheel is replaced as a set (wheel, shaft, retainers) and cannot be re-used if removed.

■ **On 998cc models, the sensor is mounted to the rear of the ride plate, on the port side. On 1052cc models, it is still on the port side, but at the front of the plate.**

4. Remove the 4 bolts and drop down the ride plate. You can let the sensor hang there, just be careful not to get anything snagged on it.
5. Remove the 4 bolts and pull out the intake duct. There is a long piece of felt between it and the hull which should come off with the duct - remove it from the duct or the hull and throw it away.

To install:

6. Inspect all components carefully for crack or other signs of damage. Look for burrs on the intake grate teeth.
7. Check the speed sensor paddle wheel for nicks, wear of damage, replacing as necessary. If paddle wheel replacement is necessary, be sure that you install the wheel with cupped surfaces facing you as you look at it from the back of the craft.
8. Coat a *new* length of felt with a marine-grade sealant and then position it on the mating surface of the intake duct. Coat the mounting bolt threads with Loctite 242, position the duct into the hull and then thread in the bolts, tightening them, in a criss-cross pattern, to:
   - 998cc models - 14 ft. lbs. (20 Nm)
   - 1052cc models - 12 ft. lbs. (17 Nm) on 1052cc models
9. Coat the threads of the ride plate bolts with Loctite 242, move the plate into position and tighten the bolts to 12 ft. lbs. (17 Nm) in a criss-cross pattern. Be sure that the speed sensor lead is routed correctly (between the jet pump and bilge hose) and not pinched.
10. Coat the sensor bolts with Loctite 242 and install the sensor into the slot in the ride plate, tightening the bolts, in a criss-cross pattern, to 2.7 ft. lbs./26 inch lbs. (3.7 Nm).
11. Coat the threads of the intake grate bolts with Loctite 242 and tighten them to:
    - 998cc models - 5.6 ft. lbs./68 inch lbs. (7.8 Nm)
    - 1052cc models - 5.5 ft. lbs./66 inch lbs. (7.6 Nm)
12. Connect the battery cables and install the seat.

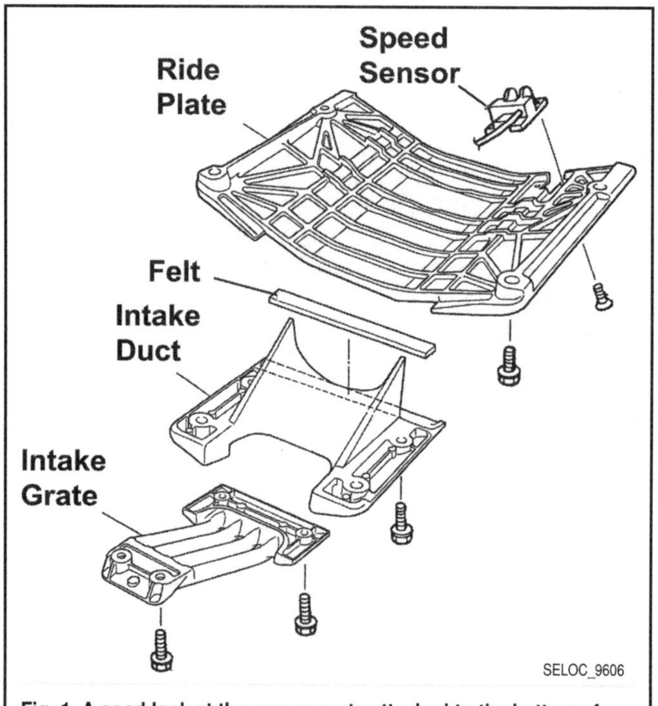

Fig. 1 A good look at the components attached to the bottom of your craft - FX w/998cc shown, FX HO w/1052cc similar

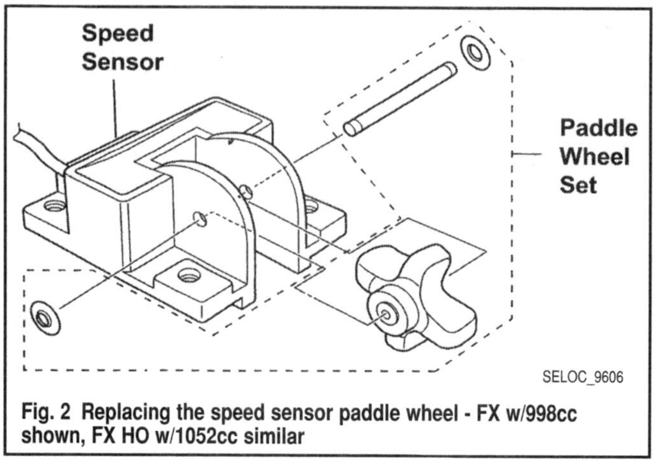

Fig. 2 Replacing the speed sensor paddle wheel - FX w/998cc shown, FX HO w/1052cc similar

VX Models, 2008 FX HO Models & All Models w/1812cc Engines

② MODERATE

◆ **See Figures 2 thru 6**

1. Remove the craft from the water. It is best to install it in a stand or block, but most people will simply jockey it on the trailer until access is provided. Disconnect the battery cables.

2. On FX models, remove the 6 mounting bolts (2 front and 4 back) and lift off the intake grate/duct. Notice that the 2 front bolts are shorter than the others.

3. On FZ and VX models, remove the 4 mounting bolts (2 front and 2 back) and lift off the intake grate/duct. Notice that the 2 front bolts are longer than the others on VX models w/1052cc engines. Do not remove the bracket on VXR/VXS models.

4. The speed sensor (and water temperature sensor on certain models) is held in position on the front port side of the ride plate with 4 small bolts. Remove them and disconnect the sensor. The sensor also has a holder that can be slid off it. The paddle wheel is connected to the sensor by 2 fasteners that snap over the shaft; the wheel is replaced as a set (wheel, shaft, retainers) and cannot be re-used if removed.

5. Remove the 4 bolts and drop down the ride plate. You can let the sensor hang there, just be careful not to get anything snagged on it. Pry off the side plates on VXR/VXS models, but do not remove the brackets unless necessary.

■ **On many later models, you will also need to remove the 2 lower bolts on the rubber plate.**

6. On FX SHO models remove the spacer from the front of the ride plate if necessary. On FZ models there are 2 spacers, one on each side of the plate.

To install:

7. Inspect all components carefully for cracks or other signs of damage. Look for burrs on the intake grate teeth.

8. Check the speed sensor paddle wheel for nicks, wear of damage, replacing as necessary. If paddle wheel replacement is necessary, be sure that you install the wheel with cupped surfaces facing you as you look at it from the back of the craft.

9. Attach the spacer(s) to the front or sides of the ride plate if it was removed. Position the 2 plates over the holes in the ride plate on the VXR/VXS models.

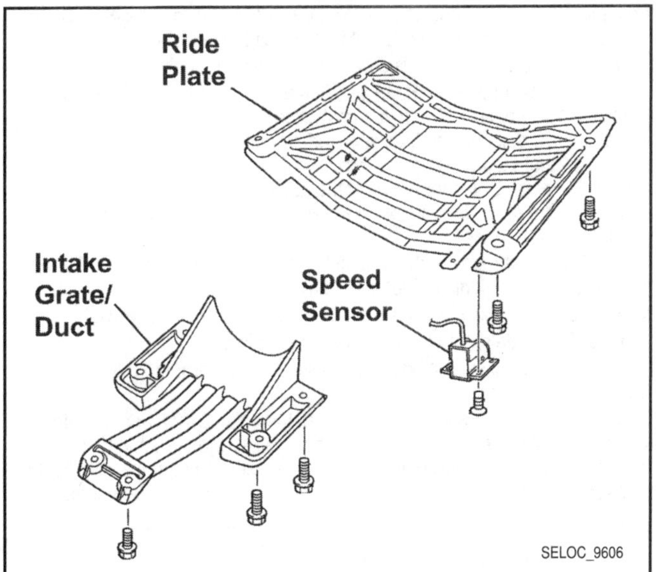

Fig. 3 A good look at the components attached to the bottom of your craft - 2008-10 FX HO and VX models w/1052cc engines

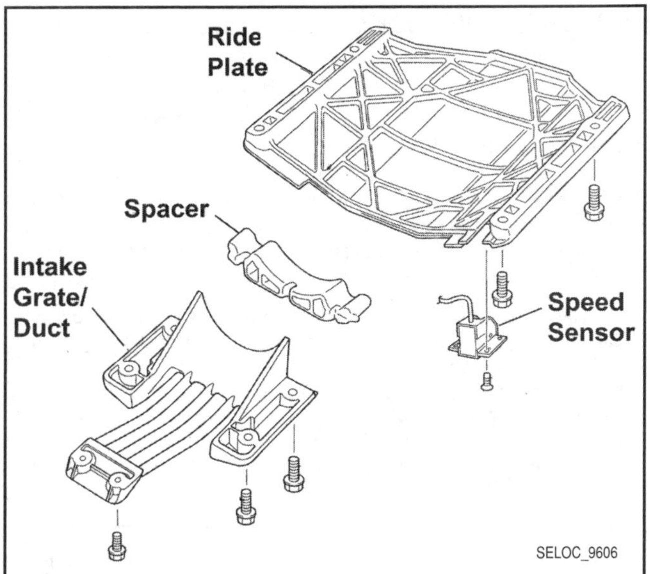

Fig. 4 A good look at the components attached to the bottom of your craft - FX SHO models

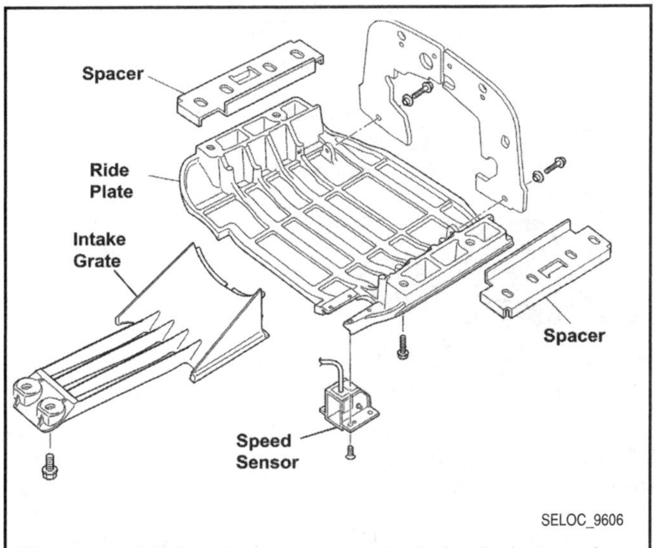

Fig. 5 A good look at the components attached to the bottom of your craft - FZ models

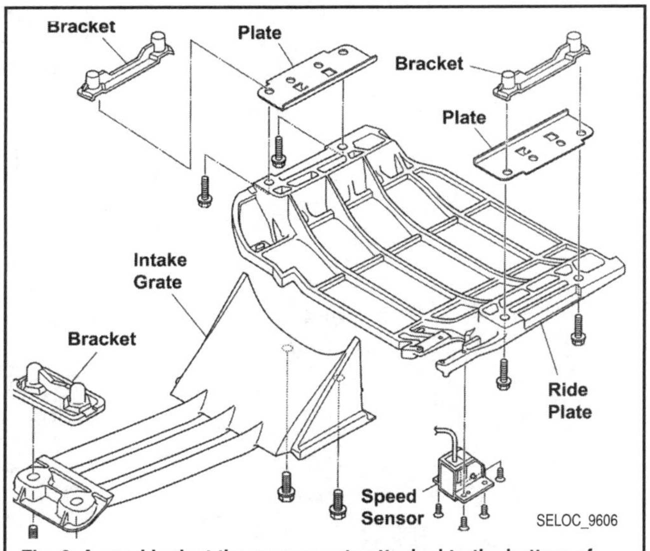

Fig. 6 A good look at the components attached to the bottom of your craft - VX models w/1812cc engines

10. Coat the threads of the ride plate bolts with Loctite 572, move the plate into position and tighten the bolts to 12.5 ft. lbs. (17 Nm) in a criss-cross pattern. Be sure that the speed sensor lead is routed correctly (between the jet pump and bilge hose) and not pinched.

11. Coat the sensor bolts with Loctite 572 and install the sensor into the slot in the ride plate, tightening the bolts, in a criss-cross pattern, to:

- All models except below - 2.7 ft. lbs./26 inch lbs. (3.7 Nm)
- 2010-11 VX models and all models w/1812cc engines - 3 ft. lbs./36 inch lbs. (4 Nm)

12. Coat the threads of the intake grate/duct bolts with Loctite 242 and tighten the 2 front bolts to:

- All models except below - 5.5 ft. lbs./66 inch lbs. (7.6 Nm)
- FX models w/1812cc engines - 5.9 ft. lbs./71 inch lbs. (8 Nm)
- 2010-11 VX models w/1052cc engines - 5.9 ft. lbs./71 inch lbs. (8 Nm)
- FZ and VX models w/1812cc engines - 29 ft. lbs. (40 Nm)

Tighten the remaining bolts, in a criss-cross pattern, to 12 ft. lbs. (17 Nm) on FX models or 29 ft. lbs. (40 Nm) on the FZ and VX models.

13. Connect the battery cables and install the seat.

Jet Pump Assembly

REMOVAL & INSTALLATION

FX Models

③ DIFFICULT

◆ **See Figures 7 thru 13**

■ **Remember that the jet pump assembly is comprised of the impeller duct and housing, the jet nozzle, nozzle ring, thrust nozzle and reverse gate. They all come out as a unit and the individual components are removed or disassembled once the unit is out of the craft.**

1. Remove the craft from the water. It is best to install it in a stand or block, but most people will simply work on it while on the trailer (or ground. Disconnect the battery cables.

2. Disconnect the shift and QSTS cable joints at the rear of the assembly (from the outside). Pull back (toward the engine) on the lock sleeve and then pull them out of the pivot.

3. Slide back the pinch clamp and wiggle the bilge hose from its fitting on the port side of the jet nozzle.

4. Remove the nut and pop the steering cable joint out of the thrust nozzle. Don't lose the 2 washers (upper and lower).

5. Slide the indicator spout hose clamp back and wiggle the hose from the spout fitting.

6. Looking in from the back of the craft, remove the 4 mounting bolts (6 on the 2008 HO and 1812cc models) and their collars from the rubber plates. Remove the 2 plates from the cavity.

7. Remove the 2 bolts and washers and then lift out the plate bracket.

8. Remove the 6mm bolt from the port side of the pump mounting flange, this is the center bolt on the left side.

9. Have someone support the assembly and then remove the 4 remaining 10mm bolts and lift out the pump assembly. The forward housing should come off with the assembly on 998cc and 1052cc engines, but if not, remove it from the transom plate. There should be 2 dowel pins on the inner (forward) impeller housing - if they stayed in the transom plate, pull them out and keep with the assembly. There also 2 dowel pins between the inner housing and the main impeller housing/impeller duct.

To install:

10. On models with a forward housing, press the 2 dowel pins into the main impeller housing bores fully and then run a bead of YamaBond #4 (ThreeBond 1104J on 2008 HO models) around the impeller opening where it contacts the forward housing. Install the forward housing; the dowels and YamaBond should be enough to hold it in place temporarily.

11. Coat the driveshaft splines with molybdenum grease.

12. Run a bead of YamaBond #4 (ThreeBond 1104J on 2008 HO or 1194E on 1812cc models) around the mating surface of the forward housing or main housing, install the 2 dowels and then move the assembly into position so that the driveshaft fits into the intermediate bearing housing and the splines mate. Once done, swivel the assembly until the dowel pins line up with the bores on the transom plate and push the pump all the way into position.

13. Coat the threads of the 5 mounting bolts with Loctite 572 and thread them in until finger-tight. Tighten the smaller bolt to:

- 998cc and 1052cc models - 5.6 ft. lbs./68 inch lbs. (7.8 Nm)
- 1812cc models - 5.9 ft. lbs./71 inch lbs. (8 Nm)

Tighten the remaining 4 bolts, in a criss-cross pattern, to 29 ft. lbs. (40 Nm).

14. Install the plate bracket and tighten the bolts to 10 ft. lbs. (14 Nm). Move the 2 rubber plates into place in the bracket and then thread in the mounting bolts so the collars fit into the plate holes. Tighten the bolts to 4.9 ft. lbs./58 inch lbs. (6.8 Nm). On models w/1812cc engines, the figures are

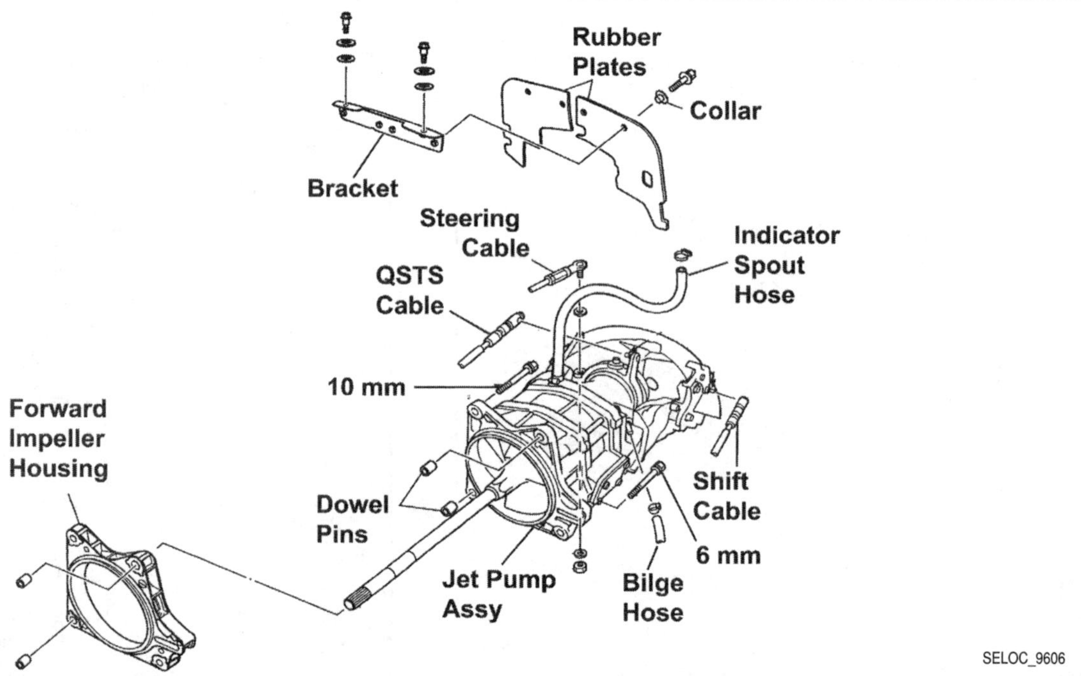

Fig. 7 A good look at the jet pump assembly - FX models with 998cc/1052cc engines, 1812cc engines do not use the forward housing

11 ft. lbs. (15 Nm) for the bracket bolts and 5.2 ft. lbs./63 inch lbs. (7 Nm). Don't forget Loctite 572 on the bolt threads.

15. Check the end of the spout indictor hose for cracks or other signs of deterioration and then wiggle it back on to the fitting. Pinch the clamp and move it back over the hose end.

16. Slide the upper washer over the end of the steering cable and then press the joint through the hole in the nozzle. Move the other washer over the end and tighten the nut to:

- 998cc and 1052cc models - 4.9 ft. lbs./58 inch lbs. (6.8 Nm)
- 1812cc models - 5.2 ft. lbs./63 inch lbs. (7 Nm)

17. Check the end of the bilge hose for cracks or other signs of deterioration and then wiggle it back on to the fitting. Pinch the clamp and move it back over the hose end.

■ **Make sure that the speed sensor lead is between the bilge hose and the pump assembly.**

18. Reconnect the shift and QSTS cables.
19. Adjust all cables as detailed in the Maintenance & Tune-Up section.
20. Connect the batteries.

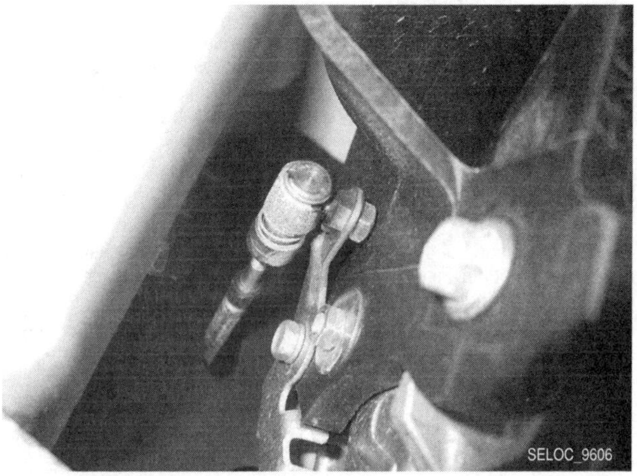

Fig. 8 Pull back (toward engine) on the cable sleeve to disconnect the shift cable joint. . .

Fig. 9 . . .and then do the same with the QSTS cable joint - FX models

Fig. 10 The steering cable is held in place with a nut

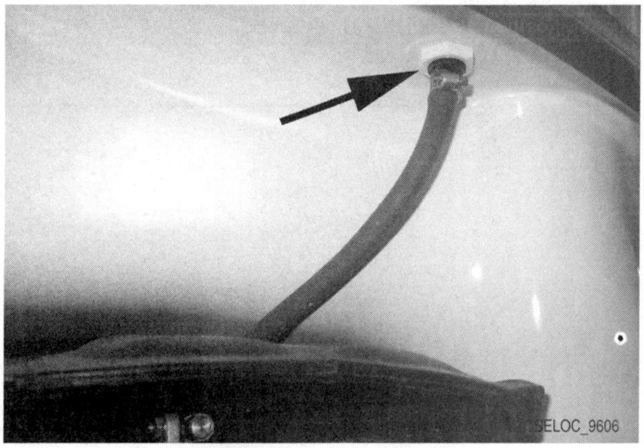

Fig. 11 Disconnect the indicator spout hose

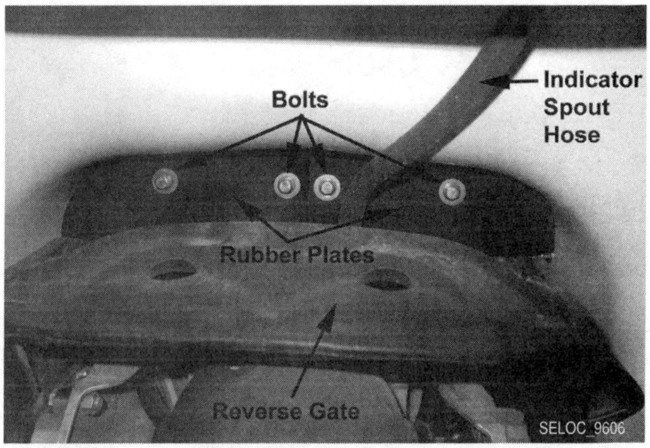

Fig. 12 The 2 rubber plates are held in position with 4 or 6 bolts

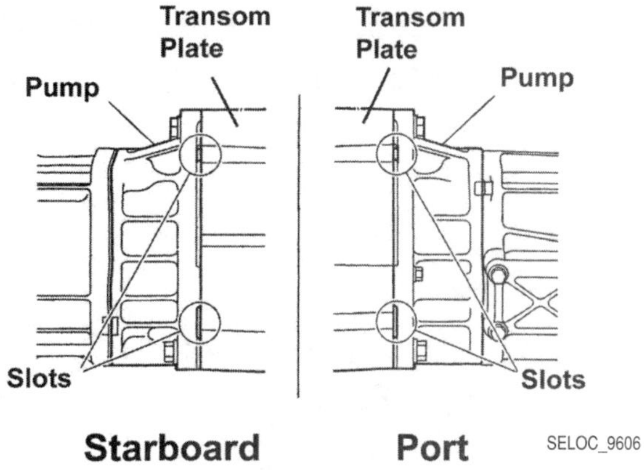

Fig. 13 There are 2 slots between the transom plate and pump to help separate the unit

FZ Models

DIFFICULT

◆ See Figures 8 thru 11 and 14

■ **Remember that the jet pump assembly is comprised of the impeller duct and housing, the jet nozzle, nozzle ring, thrust nozzle and reverse gate. They all come out as a unit and the individual components are removed or disassembled once the unit is out of the craft.**

1. Remove the craft from the water. It is best to install it in a stand or block, but most people will simply work on it while on the trailer (or ground. Disconnect the battery cables.
2. Disconnect the shift cable joint at the rear of the assembly (from the outside). Pull back (toward the engine) on the lock sleeve and then pull them out of the pivot.
3. Slide back the pinch clamp and wiggle the bilge hose from its fitting on top of the jet nozzle.
4. Remove the nut and pop the steering cable joint out of the thrust nozzle. Don't lose the 2 washers (upper and lower).
5. Slide the indicator spout hose clamp back and wiggle the hose from the spout fitting in the hull.
6. Looking in from the back of the craft, remove the 6 (4 upper, 2 lower) mounting bolts and their collars from the rubber plates. Remove the 2 plates from the cavity.
7. Remove the 6mm bolt from the port side of the pump mounting flange, this is the center bolt on the left side.
8. Remove the 2 bolts attaching the nozzle plate to the upper bracket - they should have been removed already.
9. Have someone support the assembly and then remove the 4 remaining 10mm bolts and lift out the pump assembly. There should be 2 dowel pins on the impeller housing/duct - if they stayed in the transom plate, pull them out and keep with the assembly.

■ **If the assembly is difficult to separate from the transom plate, use a small prybar in the slots (2/side) to help.**

10. If necessary, remove the 2 bolts and lift off the upper bracket. Each bolt has a washer and gasket.

To install:

11. If removed, install the upper bracket. Coat the bolt threads with Loctite 572, slide on the washer and then the gasket, and tighten to 10.3 ft. lbs. (14 Nm).
12. Press the 2 dowel pins into the impeller housing bores fully and then run a bead of ThreeBond 1194E around the mating surface of the housing.
13. Coat the driveshaft splines with molybdenum grease.
14. Move the assembly into position so that the driveshaft fits into the end of the intermediate shaft and the splines mate. Once done, swivel the assembly until the dowel pins line up with the bores on the transom plate and push the pump all the way into position.
15. Coat the threads of the 5 mounting bolts with Loctite 572 and thread them in until finger-tight. Install the upper bracket bolts and tighten them to 12.5 ft. lbs. (17 Nm). Now go back and tighten the smaller housing bolt to 5.9 ft. lbs./71 inch lbs. (8 Nm). Tighten the remaining 4 bolts, in a criss-cross pattern, to 29.5 ft. lbs. (40 Nm).
16. Move the 2 rubber plates into place in the bracket and then thread in the mounting bolts so the collars fit into the plate holes. Tighten the bolts to 5.9 ft. lbs./71 inch lbs. (8 Nm).
17. Check the end of the spout indictor hose for cracks or other signs of deterioration and then wiggle it back on to the fitting. Pinch the clamp and move it back over the hose end.
18. Slide the upper washer over the end of the steering cable and then press the joint through the hole in the nozzle. Move the other washer over the end and tighten the nut to 5.2 ft. lbs./63 inch lbs. (7 Nm).
19. Check the end of the bilge hose for cracks or other signs of deterioration and then wiggle it back on to the fitting. Pinch the clamp and move it back over the hose end.

■ **Make sure that the speed sensor lead is between the bilge hose and the pump assembly.**

20. Reconnect the shift cable and adjust as detailed in the Maintenance & Tune-Up section.
21. Connect the batteries.

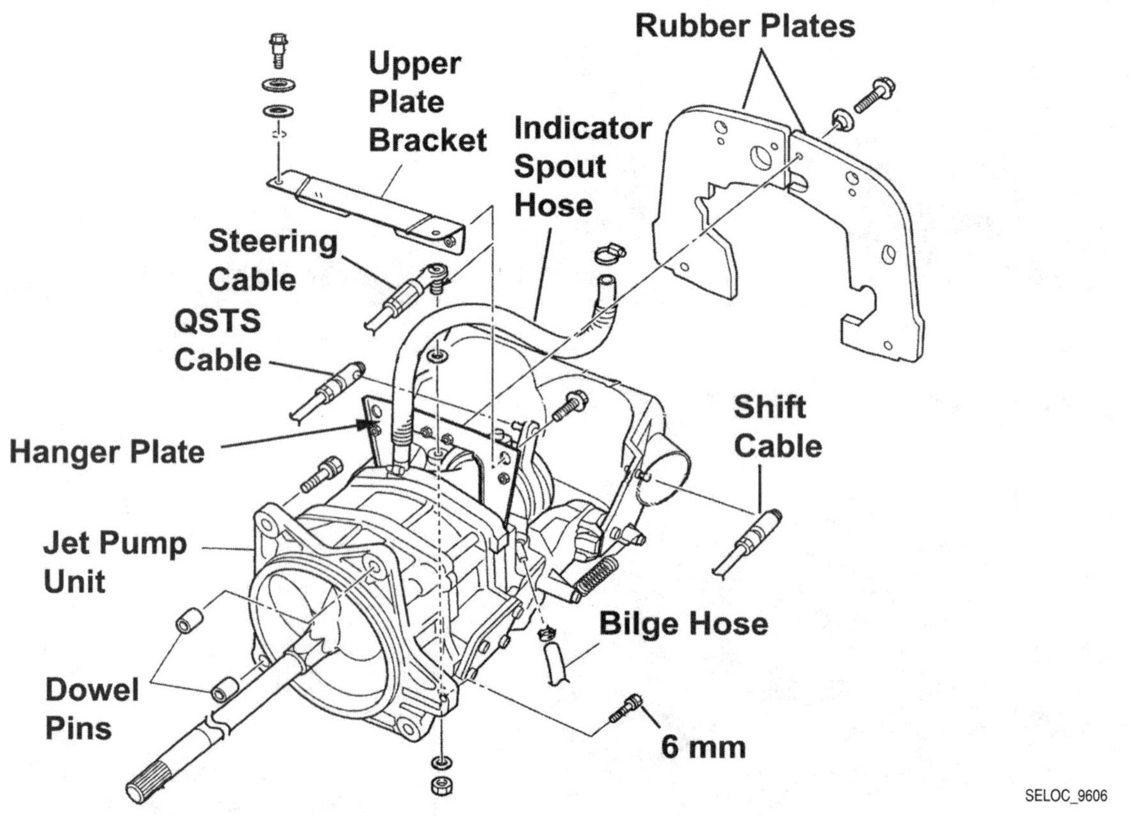

Fig. 14 A good look at the jet pump assembly - FZ models

VX Models

DIFFICULT

◆ See Figures 8, 9, 10, 15 and 16

■ Remember that the jet pump assembly is comprised of the impeller duct and housing, the jet nozzle, thrust nozzle and reverse gate. They all come out as a unit and the individual components are removed or disassembled once the unit is out of the craft.

1. Remove the craft from the water. It is best to install it in a stand or block, but most people will simply work on it while on the trailer (or ground. Disconnect the battery cables.
2. Disconnect the shift cable joint (if equipped) at the rear of the assembly. Pull back (toward the engine) on the lock sleeve and then pull it off of the pivot.
3. Slide back the pinch clamp and wiggle the bilge hose from its fitting on top of the jet nozzle.
4. Remove the nut and pop the steering cable joint out of the thrust nozzle. Don't lose the 2 washers (upper and lower).
5. Slide the indicator spout hose clamp back and wiggle the hose from the spout fitting in the hull.
6. Looking in from the back of the craft, remove the 4 mounting bolts and their collars from the rubber plates. Now remove the nut and collar from the bottom inner side of each plate and pull the 2 bolts out of the boss on the impeller duct. Remove the 2 plates from the cavity.
7. Remove the M6 x 30mm bolt from the port side of the pump mounting flange, this is the center bolt on the left side.
8. Remove the 2 bolts attaching the nozzle plate to the upper bracket - they should have been removed already.
9. Have someone support the assembly and then remove the 4 remaining 45mm bolts and lift out the pump assembly. There should be 2 dowel pins on the impeller housing/duct - if they stayed in the transom plate, pull them out and keep with the assembly.

■ If the assembly is difficult to separate from the transom plate, use a small prybar in the slots (2/side) to help.

10. If necessary, remove the 2 bolts and lift off the upper bracket. Each bolt has a washer and gasket.

To install:

11. If removed, install the upper bracket. Coat the bolt threads with Loctite 572, slide on the washer and then the gasket, and tighten to 10.3 ft. lbs. (14 Nm).
12. Press the 2 dowel pins into the impeller housing bores fully and then run a bead of ThreeBond 1104J around the mating surface of the housing. On 2010-11 models, use 1207B.
13. Coat the intermediate shaft splines with molybdenum grease; driveshaft splines on the 1812cc models.
14. Move the assembly into position so that the intermediate shaft fits into the end of the impeller and the splines mate. Once done, swivel the assembly until the dowel pins line up with the bores on the transom plate and push the pump all the way into position.

■ Remember, on the 1812cc models, the driveshaft splines will fit into the end of the intermediate shaft.

15. Coat the threads of the 5 mounting bolts with Loctite 572 and thread them in until finger-tight. Install the upper bracket bolts and tighten them to 12.5 ft. lbs. (17 Nm). Now go back and tighten the smaller housing bolt to 5.6 ft. lbs./68 inch lbs. (7.8 Nm) on all except the 2010-11 models where the figure is 5.9 ft. lbs./71 inch lbs. (8 Nm). Tighten the remaining 4 bolts, in a criss-cross pattern, to 29.5 ft. lbs. (40 Nm).
16. Move the 2 rubber plates into place in the bracket and then thread in the mounting bolts so the collars fit into the plate holes. Tighten the bolts to 4.9 ft. lbs./58 inch lbs. (6.8 Nm). Reach in and push the 2 lower bolts through the boss and plate, and then tighten the nuts to 4.9 ft. lbs./58 inch lbs. (6.8 Nm). On 2010-11 models, both the nuts and bolts are tightened to 5.2 ft. lbs./63 inch lbs. (7 Nm).
17. Check the end of the spout indictor hose for cracks or other signs of deterioration and then wiggle it back on to the fitting. Pinch the clamp and move it back over the hose end.
18. Slide the upper washer over the end of the steering cable and then press the joint through the hole in the nozzle. Move the other washer over the end and tighten the nut to:
   - 2005-09 - 4.9 ft. lbs./58 inch lbs. (6.8 Nm)

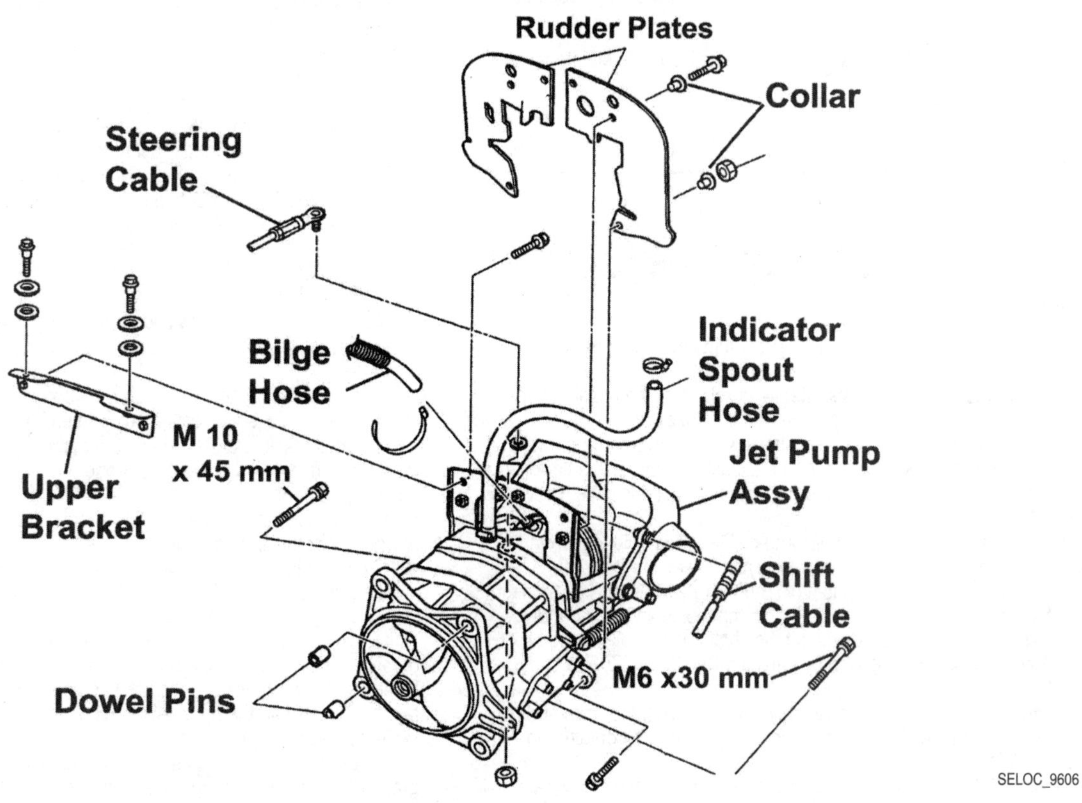

Fig. 15 A good look at the jet pump assembly - VX models w/1052cc engines

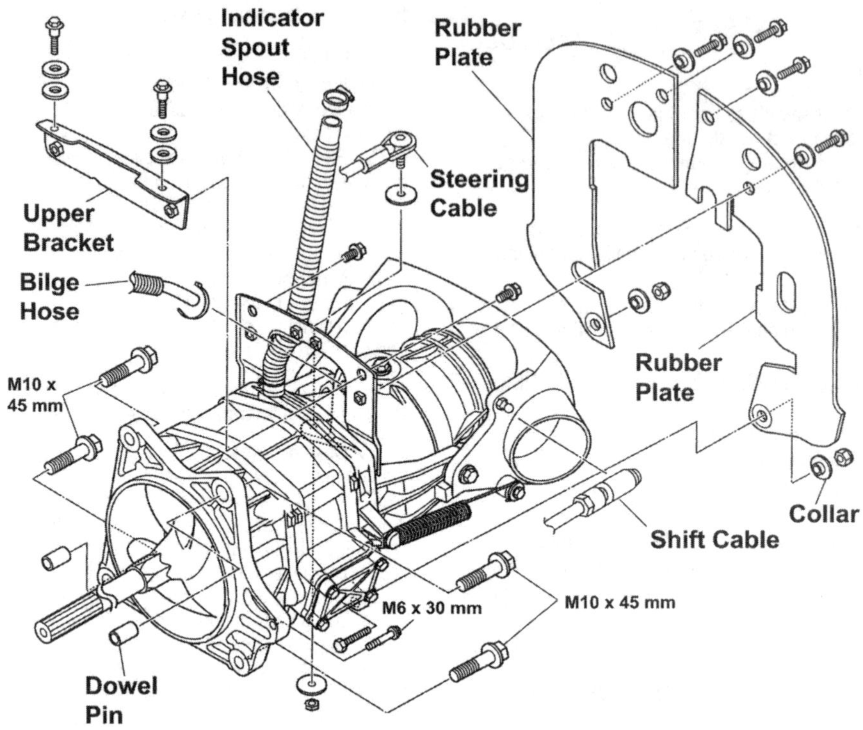

Fig. 16 A good look at the jet pump assembly - VX models w/1812cc engines

• 2010-11 - 5.2 ft. lbs./63 inch lbs. (7 Nm).

19. Check the end of the bilge hose for cracks or other signs of deterioration and then wiggle it back on to the fitting. Pinch the clamp and move it back over the hose end.

■ **Make sure that the speed sensor lead is between the bilge hose and the pump assembly.**

20. Reconnect the shift cable and adjust as detailed in the Maintenance & Tune-Up section.

21. Connect the batteries.

DISASSEMBLY & ASSEMBLY

FX Models With 998cc & 1052cc Engines

OEM ③ DIFFICULT

◆ **See Figures 17 thru 23**

1. Remove the reverse gate, thrust nozzle and nozzle ring as detailed elsewhere in this section.
2. Remove the 4 bolts and separate the nozzle from the impeller duct. There should be 2 dowel pins, keep them with the nozzle.
3. The forward impeller housing extension should have come off with pump removal, but if not, separate it from the main housing at this time.
4. Separate the main impeller housing from the duct and slide it out over the driveshaft.
5. If necessary, remove the 4 bolts and pull the water inlet cover and strainer out of the side of the impeller duct. Throw away the 2 seals.
6. Obtain a driveshaft holding tool (#YB-06151 / 90890-06519) and fit it over the forward end of the driveshaft. Clamp the tool in a vise and then loosen the impeller so you can slide it off the shaft. The impeller uses LEFT-HAND threads and will be loosened by turning it CLOCKWISE. Pull the spacer off of the shaft.
7. Working at the rear of the assembly now, remove the 3 bolts and pull off the cone. Discard the O-ring. Make sure that you have a container to catch the fluid in the cap!
8. Now, with the driveshaft holder still in the vise, remove the nut and washer at the rear of the driveshaft. Remove the assembly from the vise.
9. Position the impeller duct on the table portion of a press so the long part of the driveshaft is facing down through the hole and the back (threaded end) is facing upward. Support the lower side of the driveshaft while using the press to force it through the bearings until you can pull it the remainder of the way out of the duct. Make sure you use a block of wood between the end of the driveshaft and the press.
10. Use a slide hammer (#YB-06096 or equivalent) and remove the rear bearing from the impeller duct on US and Canada models. Use a bearing puller (#90890-06535 or equivalent) for all other countries. Discard the bearing. Pull out the spacer.
11. Lay the duct on a flat surface with the front bearing on the top side. Use a removal tool (#YB-06112 / 90890-06614) and drive rod (#YB-06071 / 90890-06652) to press out the bearing by lightly tapping the tool with a mallet; or you could use a press. Discard the bearing.
12. Now position the duct so that the oil seals are facing down. Assemble a seal driver (#YB-06196 / 90890-06653) and rod (#YB-06071) to press out the 2 seals. You can throw away the seals as well.

To assemble:

13. Clean all components and allow to dry thoroughly.
14. Inspect all components for cracks, wear, nicks or other signs of damage.
15. Check the impeller-to-housing clearance as detailed in the Maintenance & Tune-Up section after assembly.
16. On 2008 HO models, check the main impeller housing bore diameter. Measure horizontally from edge-to-edge and confirm that it is 6.116-6.120 in. (155.35-155.45mm).
17. Lay the duct on a flat surface again so that the seal side is facing upward. Coat the lip of a *new* inner oil seal with Yamaha Grease A (any water resistant marine grease will work) and position it in the bore so the lip is facing up. Use a driver (#YB-06085 / 90890-06634) and rod (#YB-06071 / 90890-06606) to press the seal into the bore until it is slightly below the lip.
18. Now, coat the lip of a *new* outer oil seal with water resistant marine grease and position it in the bore so the lip is facing up and it is sitting on the inner seal lip. Move the driver back into position and using a hammer, lightly tap the 2 seals into the bore until the top lip of the outer seal is 0.19 in. (4.7mm) below the rim of the bore.
19. Coat the races of a *new* front bearing with water resistant marine grease and slide it onto the driveshaft from the threaded (rear) side so that the side with the numbers is facing down (toward the front splined side of the shaft). Turn the driveshaft over so the splined side is on top and position it in

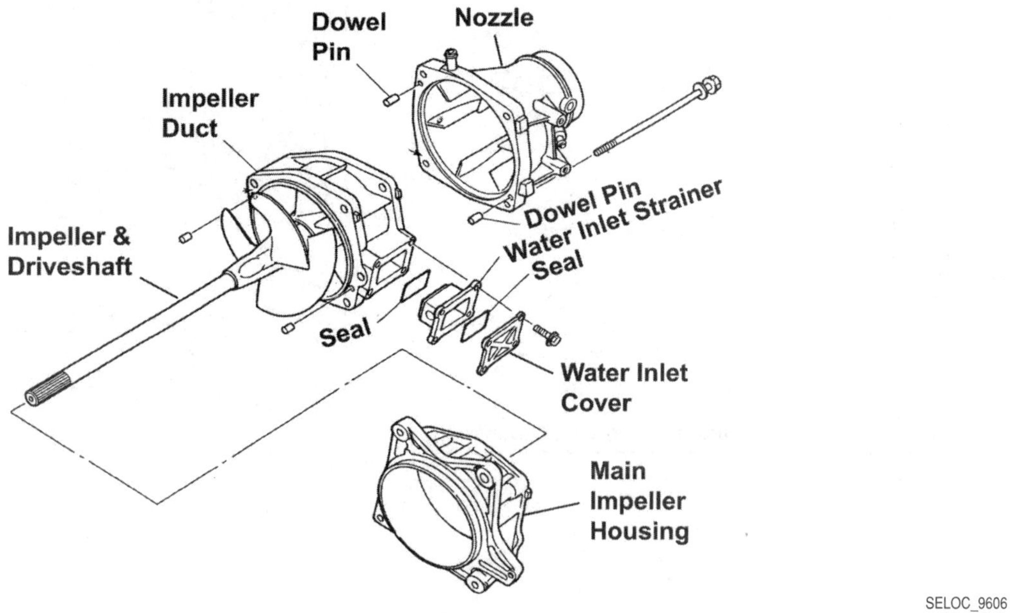

Fig. 17 Exploded view of the main pump assembly - FX models with 998cc/1052cc engines; models w/1812cc engines similar

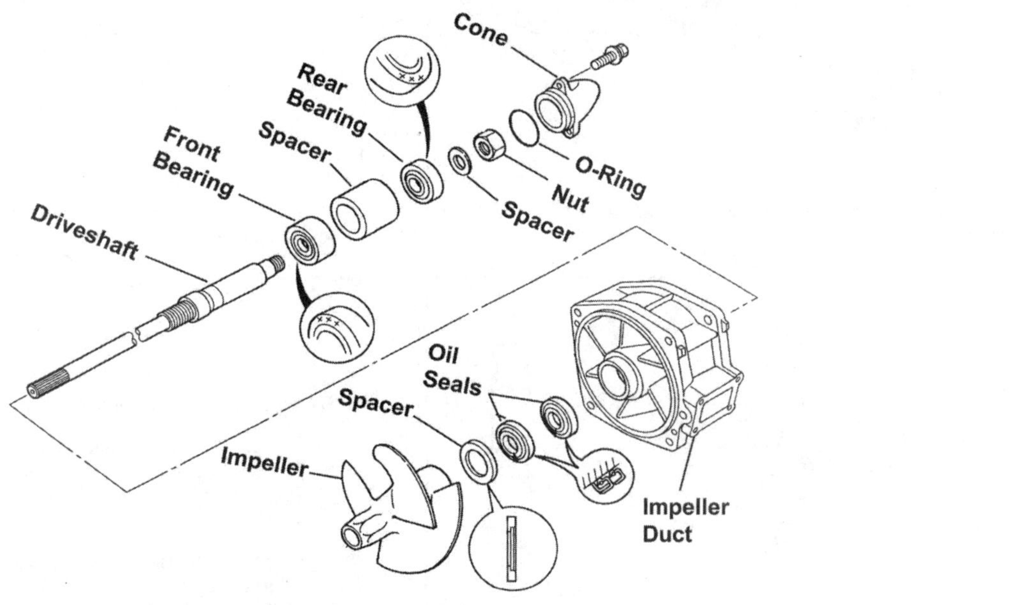

Fig. 18 Exploded view of the driveshaft assembly - FX models with 998cc/1052cc engines; models w/1812cc engines similar

a press so the bearing is resting on the press table. Press the shaft through the bearing until the side of the bearing facing up (the numbered side) is 0.91 in. (23mm) from the edge of the step on the driveshaft.

20. Move the impeller duct onto the press table so the oil seals are facing down and the bore is centered over the hole in the table. Guide the driveshaft through the bore and oil seals, splined end first, until the bearing is in the bore as far as it will go (without forcing it). Slide the spacer over the shaft until it rests on the bearing.

21. Find a pipe with an outer diameter no more than 1.97 in. (50mm) and an inner diameter that is at least 1.02 in. (26mm). Cut it so that it is at least 1.30 in. (33mm) long. Slide the pipe over the end of the shaft until it is resting on the spacer and then press the spacer/bearing in until is rests on the step in the housing just above the oil seals. Remove the tool and fill the cavity between the spacer and driveshaft with EPNOC AP #0 grease until it is about 1/3 full on 998cc models or add 0.7 oz. (20 g) on 1052cc models.

22. Coat the races of a *new* rear bearing with water resistant marine grease and slide it onto the driveshaft from the threaded (rear) side so that the side with the numbers is facing up (toward the rear threaded side of the shaft). Use a driver (#YB-34474) and press the bearing in until it seats on the spacer. If the tool is not available, find a washer or piece of pipe that has an outer diameter 1.81 in. (46mm) and an inner diameter of 0.79 in. (20mm).

23. Position the washer over the end of the shaft and then thread on the nut until finger-tight. Re-install the driveshaft holder to the splined side of the shaft and clamp it in a vise again. Tighten the nut to 50 ft. lbs. (69 Nm) on all models but the 2008 HO, where it should be 54 ft. lbs. (74 Nm).

24. Coat a *new* oil seal with water resistant marine grease and press it into the recess in the mating surface of the cone. Fill the cone with about 1/3 EPNOC AP #0 grease on 998cc models or add 0.7 oz. (20 g) on 1052cc models. Move the assembly so you can install the cone without spilling anything, coat the bolt threads with Loctite 572 and then tighten them securely.

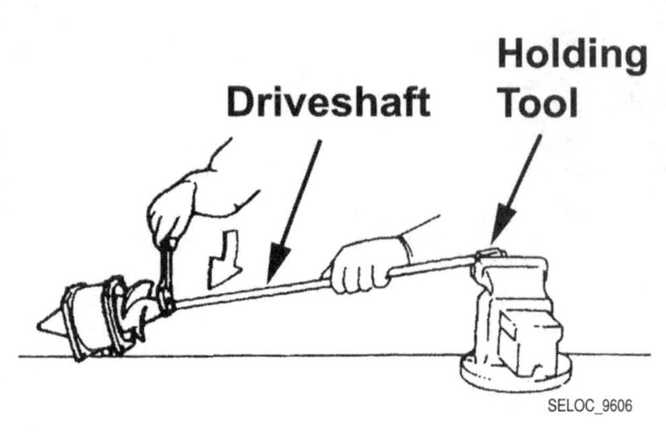

Fig. 19 Remember to loosen the impeller by turning it CLOCKWISE

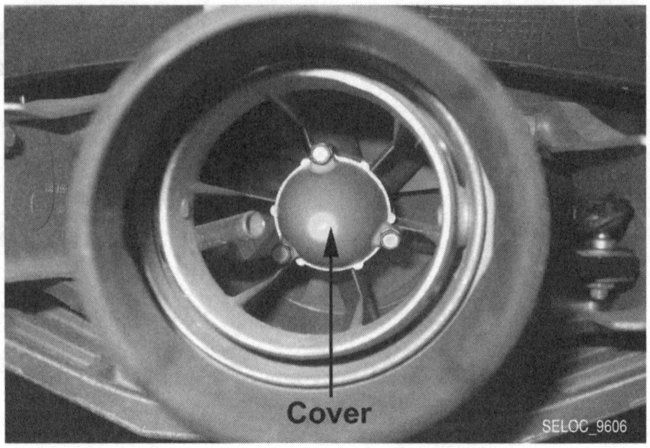

Fig. 20 The cone protects the end of the driveshaft

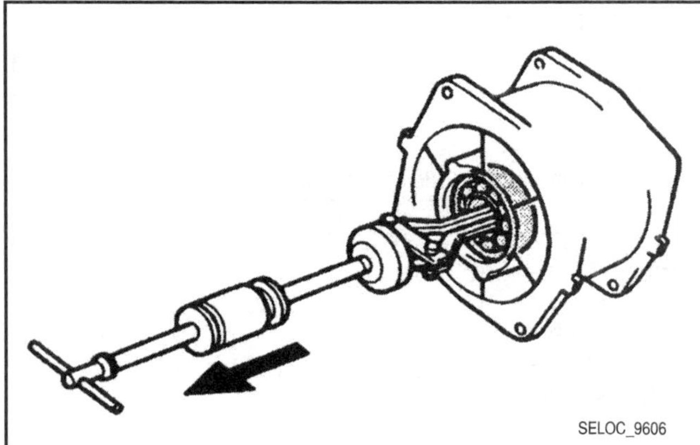

Fig. 21 Use a slide hammer to remove the rear bearing on US and Canada models. . .

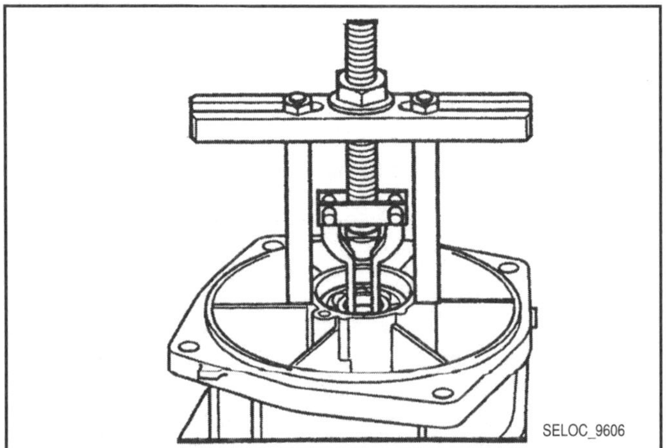

Fig. 22 . . .or a bearing puller for all other countries

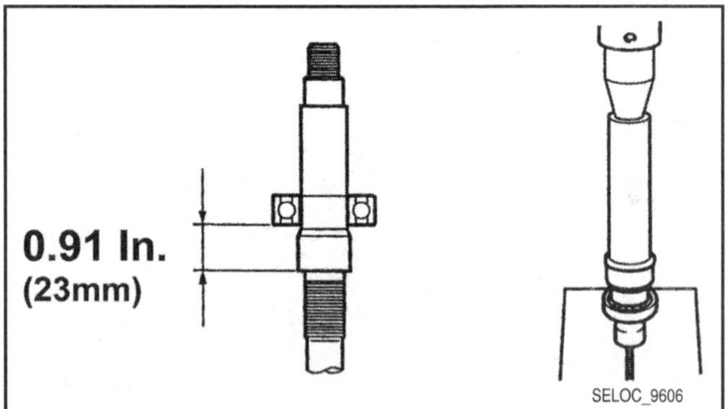

Fig. 23 Position the bearing correctly on the driveshaft

25. Slide the spacer on and then feed the impeller over the splined end of the driveshaft, install the shaft holder again and clamp it in the vise. Tighten the impeller to 19 ft. lbs. (27 Nm) on 998cc models or 54 ft. lbs. (75 Nm) on 1052cc models. Remember, you tighten by turning it COUNTERCLOCKWISE.

26. Install a *new* seal onto the water inlet strainer and insert it into the impeller duct. Lay in the other seal and position the cover so the bolt holes line up. Coat the bolt threads with Loctite 572 and then tighten them, in a criss-cross pattern, to 4.8 ft. lbs./52 inch lbs. (6.6 Nm).

27. Run a bead of YamaBond #4 (ThreeBond 1104J on 2008 HO models) around the mating surface of the impeller duct, install the 2 dowel pins and slide on the main impeller housing. Do the same on the other side and install the nozzle. Coat the bolt threads with Loctite 572 and then tighten them to 29 ft. lbs. (40 Nm).

28. Install the nozzle ring, thrust nozzle and reverse gate.

FX, FZ & VX Models w/1812cc Engines

OEM ③ DIFFICULT

◆ See Figures 17 thru 23 and 24 thru 27

1. Remove the reverse gate, thrust nozzle and nozzle ring as detailed elsewhere in this section.

2. Remove the 4 bolts and separate the nozzle from the impeller duct. There *should* be 2 dowel pins, if so, keep them with the nozzle.

■ **On FZ models, there is a hanger plate secured to the nozzle (identical to the one found on VX models) that must be removed when pulling out the upper bolts.**

3. Separate the main impeller housing from the duct and slide it out over the driveshaft.

4. If necessary, remove the 4 bolts and pull the water inlet cover and strainer out of the side of the impeller duct. Throw away the 2 seals.

5. Obtain a driveshaft holding tool (#YB-06201 / 90890-06520) and fit it over the forward end of the driveshaft. Clamp the tool in a vise and then loosen the impeller so you can slide it off the shaft. The impeller uses LEFT-HAND threads and will be loosened by turning it CLOCKWISE. Pull the spacer off of the shaft.

6. Working at the rear of the assembly now, remove the 3 bolts and pull off the cone. Discard the O-ring. Make sure that you have a container to catch the fluid in the cap!

7. Now, with the driveshaft holder still in the vise, remove the nut and washer at the rear of the driveshaft. Remove the assembly from the vise.

8. Position the impeller duct on the table portion of a press so the long part of the driveshaft is facing down through the hole and the back (threaded end) is facing upward. Support the lower side of the driveshaft while using the press to force it through the bearings until you can pull it the remainder

of the way out of the duct. Make sure you use a block of wood between the end of the driveshaft and the press.

9. Use a slide hammer (#YB-06096 or equivalent) and remove the rear bearing from the impeller duct on US and Canada models. Use a bearing puller (#90890-06535 or equivalent) for all other countries. Discard the bearing. Pull out the spacer.

10. Turn the duct over so the oil seals are facing upward. Use the slide hammer or bearing puller from the previous step and remove the 2 oil seals.

11. Use a removal tool (#YB-06194 / 90890-06609) and drive rod (#YB-06071 / 90890-06652) to press out the forward bearing by lightly tapping the tool with a mallet; or you could use a press. Discard the bearing.

To assemble:

12. Clean all components and allow to dry thoroughly.

13. Inspect all components for cracks, wear, nicks or other signs of damage.

14. Check the impeller-to-housing clearance as detailed in the Maintenance & Tune-Up section after assembly.

15. Check the main impeller housing bore diameter. Measure horizontally from edge-to-edge and confirm that it is 6.116-6.120 in. (155.35-155.45mm).

16. Lay the duct on a flat surface again so that the seal side is facing upward. Coat the lip of a *new* inner oil seal with Yamaha Grease A (any water resistant marine grease will work) and position it in the bore so the lip is facing up. Use an installer tool (#YB-06167 / 90890-06628) and driver (#YB-06071 / 90890-06606) to press the seal into the bore until it is slightly below the lip.

17. Now, coat the lip of a *new* outer oil seal with water resistant marine grease and position it in the bore so the lip is facing up and it is sitting on the inner seal lip. Move the driver back into position and using a hammer, lightly tap the 2 seals into the bore until the top lip of the outer seal is 0.28 in. (7mm) below the rim of the bore. There is a margin of +/- 0.008 in. (+/- 0.2mm).

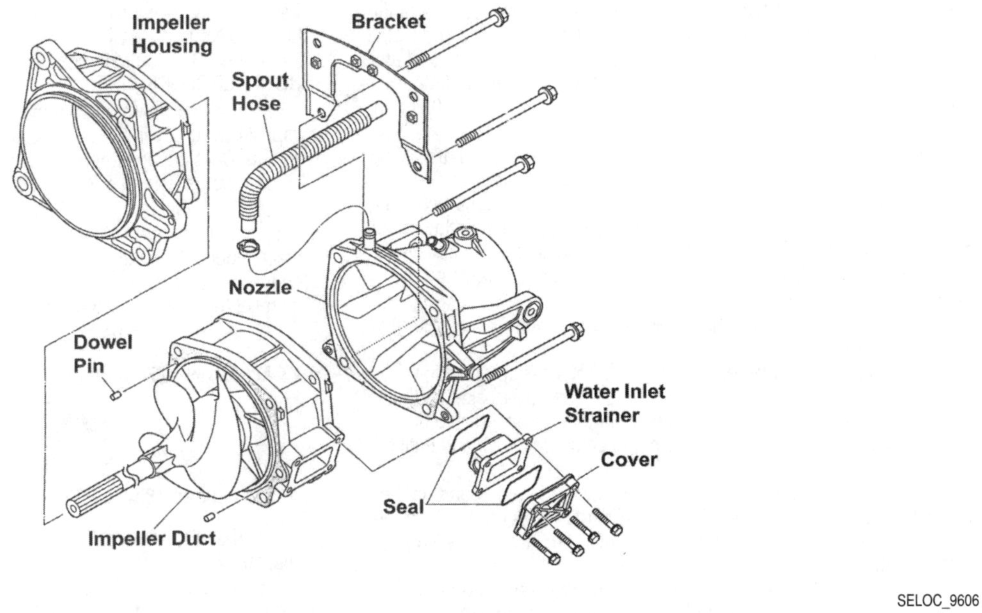

Fig. 24 Exploded view of the main pump assembly - VX models with 1812cc engines; other models w/1812cc engines similar

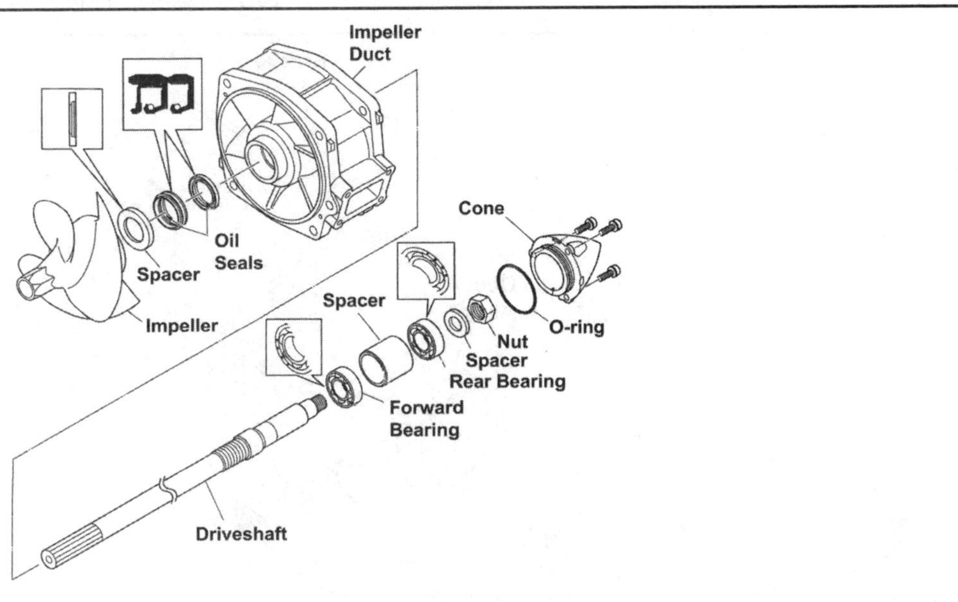

Fig. 25 Exploded view of the driveshaft assembly - VX models with 1812cc engines; other models w/1812cc engines similar

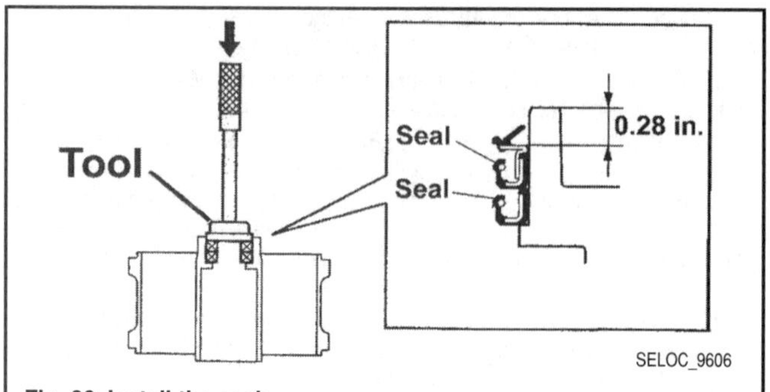

Fig. 26 Install the seals. . .

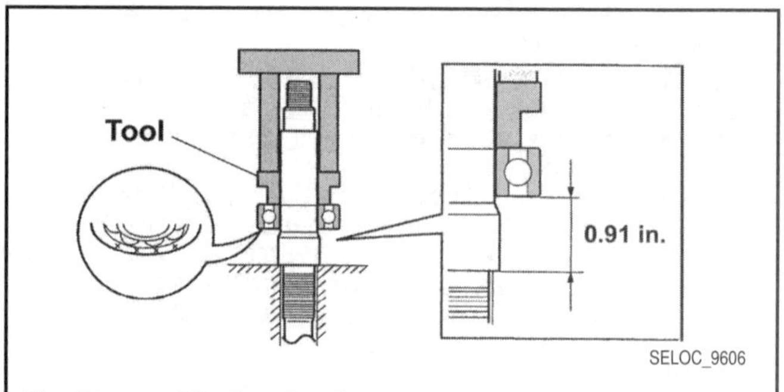

Fig. 27 . . .and the front bearing

18. Coat the races of a *new* front bearing with water resistant marine grease and slide it onto the driveshaft from the threaded (rear) side so that the side with the numbers is facing down (toward the front splined side of the shaft). Position the shaft on a press so the threaded side is on top and install a bearing attachment tool (#90890-06728). Press on the attachment tool until the side of the bearing facing down (the numbered side) is 0.91 in. (23mm) from the edge of the step on the driveshaft. There is a margin of +/- 0.004 in. (+/- 0.1mm).

19. Move the impeller duct onto the press table so the oil seals are facing down and the bore is centered over the hole in the table. Guide the driveshaft through the bore and oil seals, splined end first, until the bearing is in the bore as far as it will go (without forcing it). Use the bearing attachment tool again and press the bearing in until it seats on the step in the housing just above the oil seals. Slide the spacer over the shaft until it rests on the bearing.

20. Fill the cavity between the spacer and driveshaft with 0.7 oz. (20 g) of EPNOC AP #0 grease.

21. Coat the races of a *new* rear bearing with water resistant marine grease and slide it onto the driveshaft from the threaded (rear) side so that the side with the numbers is facing up (toward the rear threaded side of the shaft). Use a bearing attachment tool again and press the bearing in until it seats on the spacer.

22. Position the washer over the end of the shaft and then thread on the nut until finger-tight. Re-install the driveshaft holder to the splined side of the shaft and clamp it in a vise again. Tighten the nut to 50.9 ft. lbs. (69 Nm). Remove from the vise and remove the tool.

23. Coat a *new* oil seal with water resistant marine grease and press it into the recess in the mating surface of the cone. Fill the cone with 0.7 oz. (20 g) of EPNOC AP #0 grease. Move the assembly so you can install the cone without spilling anything, coat the bolt threads with Loctite 572 and then tighten them to 5.9 ft. lbs./71 inch lbs (8 Nm).

24. Slide the spacer onto the splined end of the driveshaft and then feed the impeller over the shaft. Install the shaft holder again and clamp it in the vise. Tighten the impeller to 238 ft. lbs. (330 Nm) on FX HO models, 147.5 ft. lbs. (200 Nm) on FX SHO models, 151 ft. lbs. (205 Nm) on FZ models, or 243 ft. lbs. (330 Nm). Remember, you tighten by turning it COUNTERCLOCKWISE.

25. Install a *new* seal onto the water inlet strainer and insert it into the impeller duct. Lay in the other seal and position the cover so the bolt holes line up. Coat the bolt threads with Loctite 572 and then tighten them, in a criss-cross pattern, to 5.2 ft. lbs./63 inch lbs. (7 Nm).

26. Run a bead of ThreeBond 1194E (use 1207B on the VX) around the mating surface of the impeller duct, install the 2 dowel pins and slide on the main impeller housing. Do the same on the other side (use silicone sealant on the VX) and install the nozzle. Coat the bolt threads with Loctite 572 and then tighten them to 29.5 ft. lbs. (40 Nm).

27. Install the nozzle ring, thrust nozzle and reverse gate.

VX Models w/1052cc Engines

OEM 3 DIFFICULT

◆ **See Figures 19 thru 23 and 28 thru 33**

1. If equipped, remove the reverse gate as detailed elsewhere in this section.

2. Remove the 2 bolts and pull off the thrust nozzle. There should be a collar in the nozzle for each bolt - don't lose them.

3. Loosen, alternately, the 4 mounting bolts on the corners of the nozzle. Remove the 2 top bolts and lift off the bracket. Now remove the 2 bottom bolts and separate the nozzle from the impeller duct.

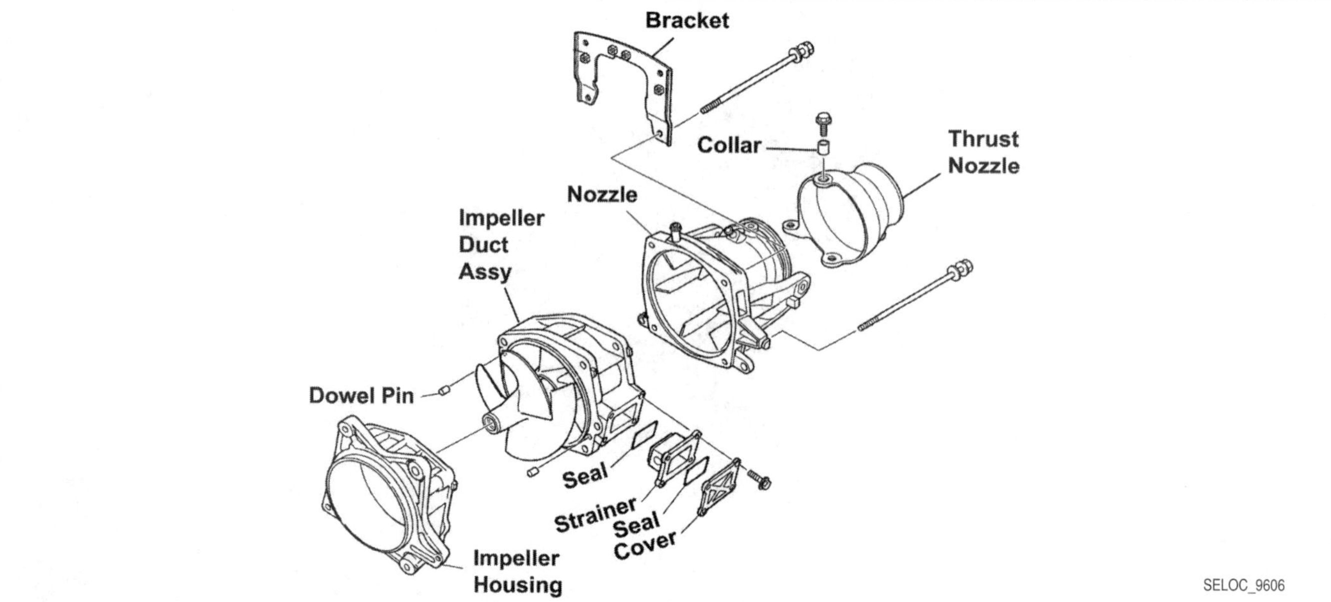

Fig. 28 Exploded view of the main pump assembly - VX models w/1052cc engines

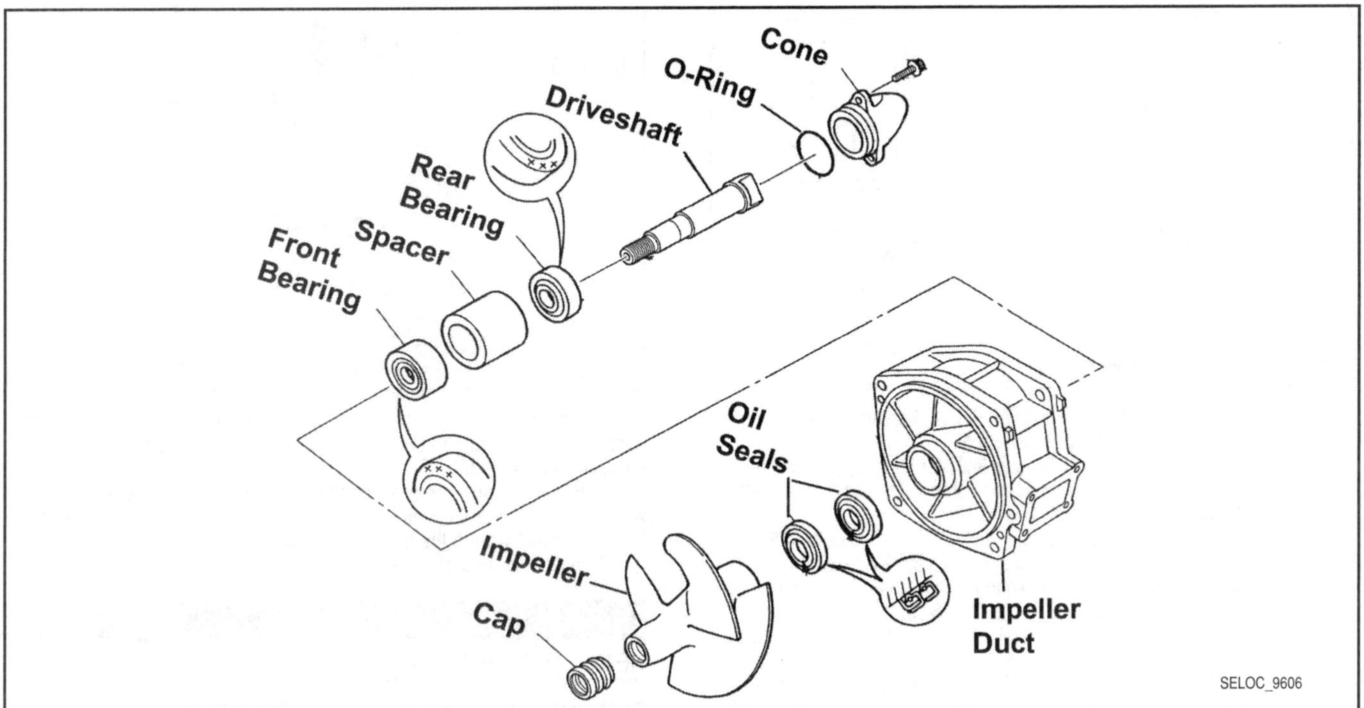

Fig. 29 Exploded view of the driveshaft assembly - VX models w/1052cc engines

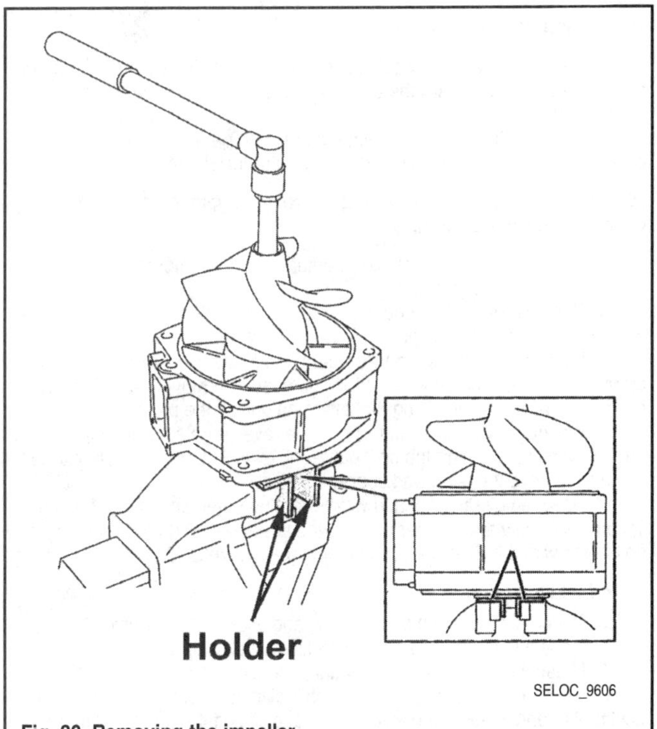

Fig. 30 Removing the impeller

4. Separate the impeller housing from the duct and slide it out over the impeller.

5. If necessary, remove the 4 bolts and pull the water inlet cover and strainer out of the side of the impeller duct. Throw away the 2 seals.

6. Working at the rear of the assembly, remove the 3 bolts and pull off the cone. Discard the O-ring. Make sure that you have a container to catch the fluid in the cap!

7. Pop the cap out of the impeller recess and check it out for cracks or tears.

8. Place the impeller duct into a vise with soft jaws and clamp the vise onto the flat surfaces of the driveshaft end. Obtain a holding tool (#YB-06552

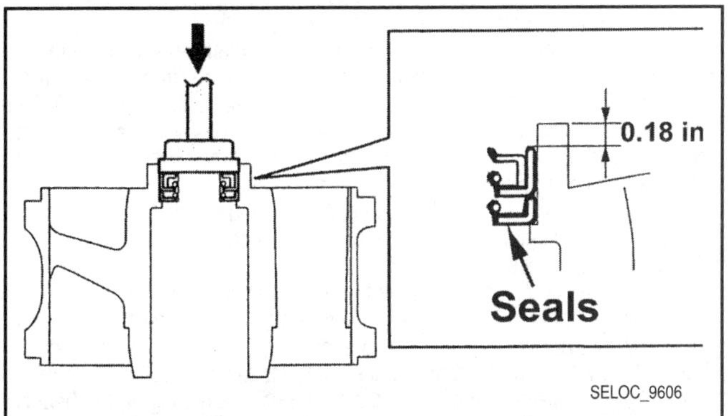

Fig. 31 Installing the oil seals

/ 90890-06552) and insert it through the impeller and into the forward end of the driveshaft. Use the tool to break the impeller loose and then thread it off of the shaft.

9. Position the impeller duct on the table portion of a press so the squared end of the driveshaft is facing down through the hole and the front (threaded) end is facing upward. Support the lower side of the driveshaft while using the press to force it through the bearings until you can pull it the remainder of the way out of the duct. Make sure you use a block of wood between the end of the driveshaft and the press.

10. Flip the duct over and use a slide hammer (#YB-06096 or equivalent) and remove the rear bearing from the impeller duct on US and Canada models. Use a bearing puller (#90890-06535 or equivalent) for all other countries. Discard the bearing. Pull out the spacer.

11. Lay the duct on a flat surface with the oil seals facing up and use a small pick to pull them out of the bore. You can throw away the seals. If the seals are difficult to pull out, you can always use the hammer or puller that you used on the rear bearing in the last step.

12. With the duct still in the same position, use a bearing remover (#YB-06112 / 90890-06614), driver (#YB-06071/ 90890-06652) to press out the bearing by lightly tapping the tool with a mallet; or you could use a press. Discard the bearing.

To assemble:

13. Clean all components and allow to dry thoroughly.

14. Inspect all components for cracks, wear, nicks or other signs of damage.

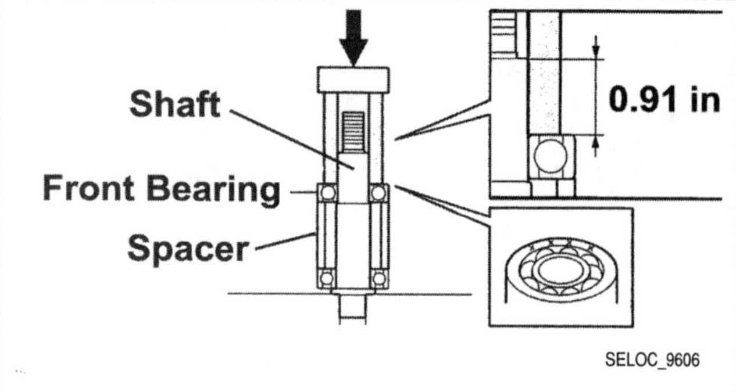

Fig. 32 Installing the front bearing

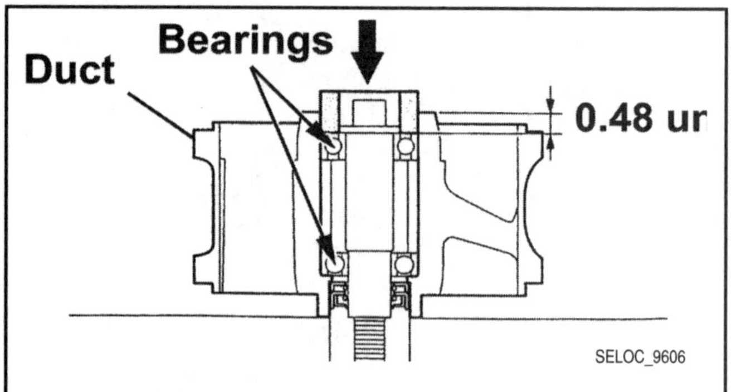

Fig. 33 Pressing in the driveshaft

15. Check the impeller-to-housing clearance as detailed in the Maintenance & Tune-Up section after assembly.

16. On 2010-11 models, check the main impeller housing bore diameter. Measure horizontally from edge-to-edge and confirm that it is 6.116-6.120 in. (155.35-155.45mm).

17. Lay the duct on a flat surface again so that the seal side is facing upward. Coat the lip of a *new* inner oil seal with Yamaha Grease A (any water resistant marine grease will work) and position it in the bore so the lip is facing up. Use a driver (#YB-06071 / 90890-06606) and installer (#YB-06085 / 90890-06634) to press the seal into the bore until it is slightly below the lip. On 2010 models, the installer should be #YB-06167 / 90890-06628.

18. Now, coat the lip of a *new* outer oil seal with water resistant marine grease and position it in the bore so the lip is facing up and it is sitting on the inner seal lip. Move the driver/installer back into position and using a hammer, lightly tap the 2 seals into the bore until the top lip of the outer seal is 0.18 in. (4.6mm) below the rim of the bore.

19. Coat the races of a *new* rear bearing with water resistant marine grease and slide it onto the driveshaft from the threaded (front) side so that the side with the numbers is facing down (toward the back of the shaft). Turn the driveshaft over so the splined side is on top and position it in a press so the bearing is resting on the table. Press the shaft through the bearing until the bearing is seated on the step on the driveshaft.

20. Slide the spacer over the shaft so it is resting on the bearing and then fill the cavity between the spacer and driveshaft with 0.7 oz. (20 g) of EPNOC AP #0 grease.

21. Coat the races of a *new* front bearing with water resistant marine grease and slide it onto the driveshaft from the threaded (front) side so that the side with the numbers is facing up. Find a pipe with an inner diameter that is at least 1.02 in. (26mm). Cut it so that it is at least 2.36 in. (60mm) long. Slide the pipe over the end of the shaft until it is resting on the bearing and then press the spacer/bearing in until is rests on spacer. The distance from the top of the bearing to the step at the end of the shaft (before the threads) should be 0.91 in. (23 Nm).

22. Take the driveshaft assembly and insert it into the rear of the impeller duct, feeding the threaded end through the oil seals as far as it will go. Once again, find a washer or piece of pipe that has an outer diameter 1.97 in. (50mm) and an inner diameter of 1.30 in. (33mm). Position the tool over the rear end of the shaft so it is resting on the rear bearing and then press everything in until the top of the bearing is 0.46 in. (11.7mm) below the lip of the bore. Guess what? On the 2010-11 models, the measurement should be 0.48 in. (12.1mm)!

23. Position the assembly back into the soft-jawed vise and then coat the driveshaft threads with Loctite 572. Spin the impeller back onto the shaft, install the holding tool (same as in removal) and then tighten to 54 ft. lbs. (75 Nm). Yep, you guessed it - 2010-11 models are different, 81 ft. lbs. (110 Nm).

24. Coat a *new* oil seal with water resistant marine grease and press it into the recess in the mating surface of the cone. Fill the cone with 0.7 oz. (20 g) of EPNOC AP #0. Move the assembly so you can install the cone without spilling anything, coat the bolt threads with Loctite 572 and then tighten to 5.6 ft. lbs./68 inch lbs. (7.8 Nm) or 5.9 ft. lbs./71 inch lbs. (8 Nm) on 2010-11 models.

25. Install a *new* seal onto the water inlet strainer and insert it into the impeller duct. Lay in the other seal and position the cover so the bolt holes line up. Coat the bolt threads with Loctite 572 and then tighten them, in a criss-cross pattern, to 4.8 ft. lbs./52 inch lbs. (6.6 Nm) or 5.2 ft. lbs./63 inch lbs. (7 Nm) on the 2010-11 models.

26. Run a bead of ThreeBond 1104J (1207B on the 2010-11 models) around the mating surface of the impeller duct, install the 2 dowel pins and slide on the impeller housing. Do the same on the other side and install the nozzle. Make sure the bracket is in position on the nozzle, coat the bolt threads with Loctite 572 and then tighten them to 29 ft. lbs. (40 Nm).

27. Install the thrust nozzle with the collars in their holes (coat the collars with some marine grade water resistant grease). Coat the bolt threads with Loctite 271 and tighten the bolts to 11 ft. lbs. (15 Nm).

28. Install the reverse gate.

Reverse Gate

REMOVAL & INSTALLATION

FX Models w/998cc Engines & 2004-07 FX HO Models

◆ See Figures 8 and 34 thru 38

② MODERATE

1. Disconnect the shift cable joint at the rear of the assembly (from the outside). Pull back (toward the engine) on the lock sleeve and then pull it off of the pivot.

2. Remove the 2 bolts mounting the reverse gate to the mounting stays on each side - each bolt uses a washer and collar. Remove the gate.

■ **If all you are doing is removing the reverse gate, stop hear and skip to Step 14 for installation.**

3. Remove the rear most bolt and nut on the port mounting stay. Don't lose the collar and 2 washers.

4. Remove the 3 stay mounting bolts and lift it off of the nozzle. FX HO models have a small bracket held in position by the front bolt.

5. Move over to the starboard side, remove the 3 mounting bolts and lift off the other mounting stay as well. If you need to remove the shift cable pivot, remove the inner nut on the lever and pull off the pivot.

6. Working on the inner left side of the reverse gate, remove the inside nut and washer retaining the shift cable link lever to the gate. Pull out the bolt and remove the lever and spacer.

7. Now, remove the nut and washer on the outer left side of the gate. Take note of how the stopper lever spring is situated and then remove the bolt along with the 2 washers, collar spring and lever.

To install:

8. Clean all components thoroughly and allow to dry completely.

9. Inspect everything for wear, cracks or other signs of damage.

10. Install the shift cable pivot and tighten the nut to 5.6 ft. lbs./68 inch lbs. (7.8 Nm). Coat the threads of the shift pivot lever mounting bolts with Loctite 242 and slide on the lever and spacer. Feed the bolt through the hole in the gate and then install the washer and tighten the nut to 5.6 ft. lbs./68 inch lbs. (7.8 Nm).

11. Coat the threads of the stop lever bolt with Loctite 242. Slide on the washer, collar, spring, lever and washer and then insert the bolt through the gate. Slide a washer over the bolt end and thread the nut on until finger-tight. Hopefully you were smart and took a picture of the spring positioning before removing it, but if not, hook the back end of the spring to the lever and the front portion to the gate. Make sure the stop lever tab is resting on the shift lever tab and then tighten the nut to 5.6 ft. lbs./68 inch lbs. (7.8 Nm). Check the lever(s) action to make sure they are not binding or sticking - is they are, try assembling them again.

12. Coat the threads of the 6 mounting stay bolts with Loctite 242 and then install each stay, tightening the bolts to 5.6 ft. lbs./68 inch lbs. (7.8 Nm).

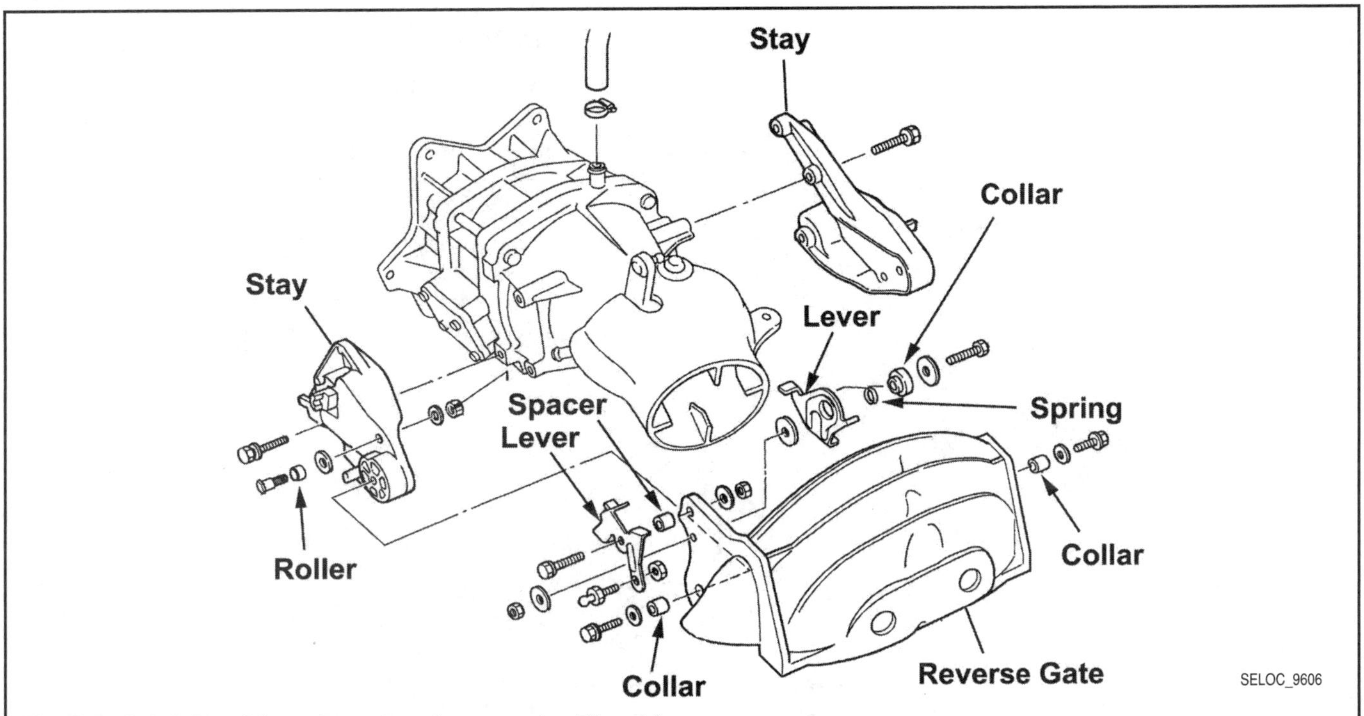

Fig. 34 Exploded view of the reverse gate and components - FX models

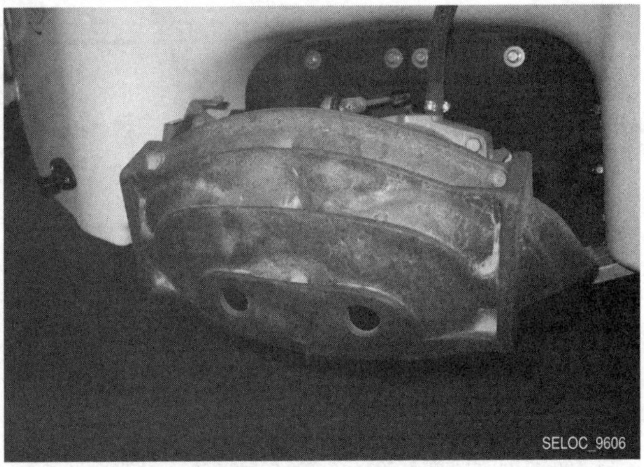

Fig. 35 A good look at the reverse gate. . .

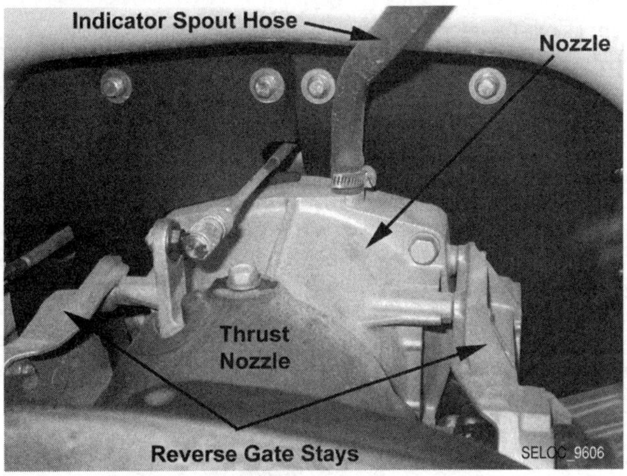

Fig. 36 . . .and it's mounting stays (2008 HO similar)

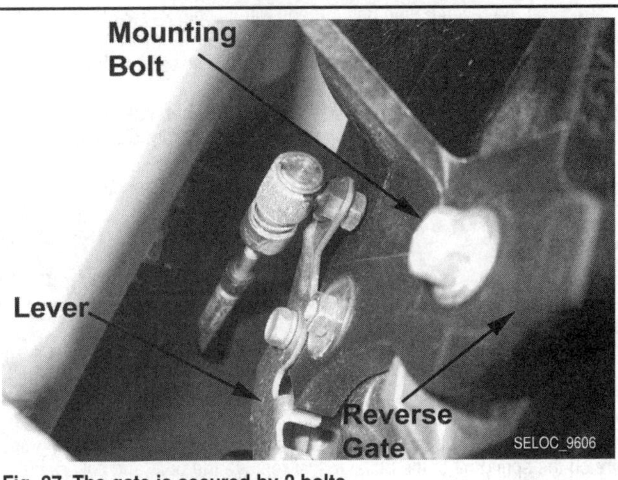

Fig. 37 The gate is secured by 2 bolts

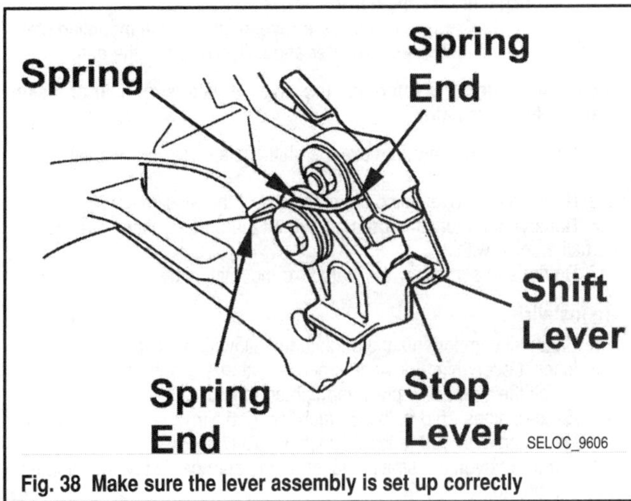

Fig. 38 Make sure the lever assembly is set up correctly

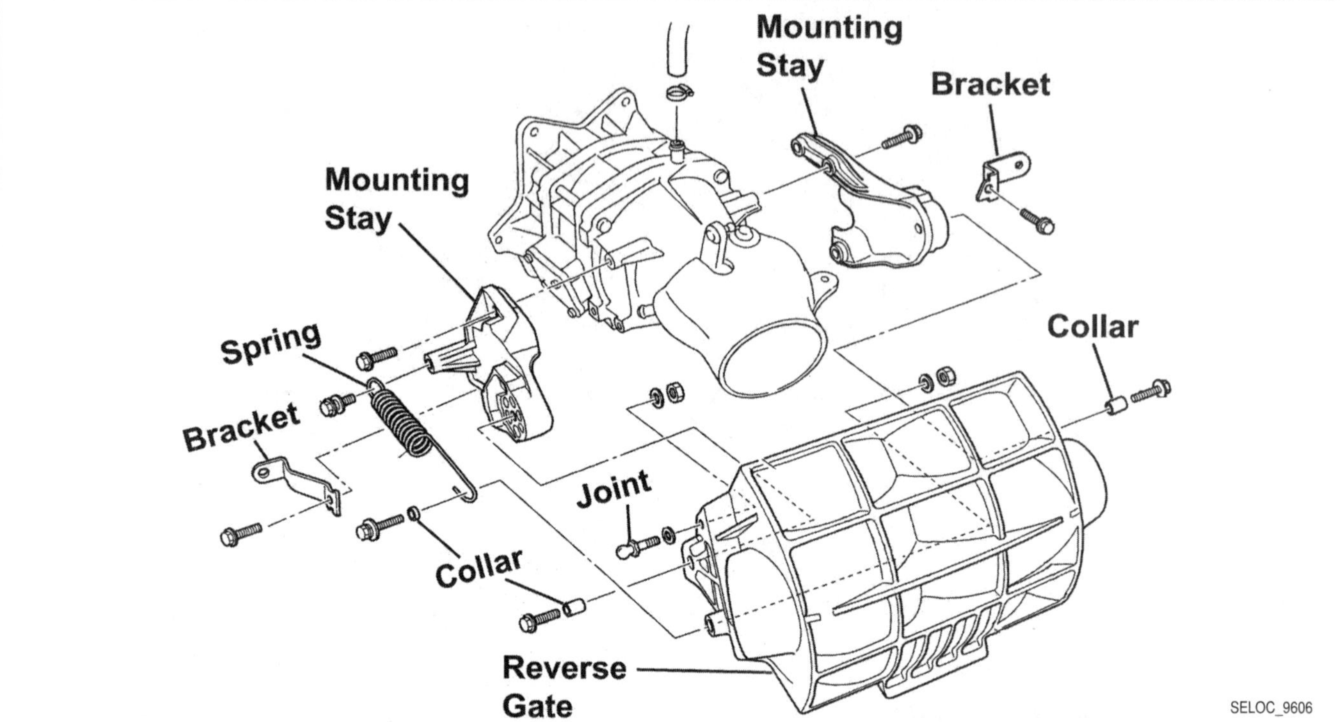

Fig. 39 Exploded view of the reverse gate and components - 2008 FX models and all FX models w/1812cc engines

Don't to forget to reconnect to the small bracket that attaches to the mounting bolt on the port stay of FX HO models.

13. Coat the threads of the remaining bolt on the left stay with Loctite 271, slide on the roller and washer and insert it through the hole in the stay. Slide on the washer and then tighten the nut to 19 ft. lbs. (26 Nm).

14. Coat the reverse gate bolt threads with Loctite 242 and slide on the washers and rollers. Position the gate over the mounting stays so the bolt holes line up and the tighten the bolts to 14 ft. lbs. (20 Nm).

15. Install the shift cable joint to the pivot and adjust the cable as detailed in the Maintenance & Tune-Up section.

2008 FX HO Models & All FX/FZ Models w/1812cc Engines

◆ See Figures 8, 35, 36, 39 and 40

1. Disconnect the shift cable joint at the rear of the assembly (from the outside). Pull back (toward the engine) on the lock sleeve and then pull it off of the pivot.

2. Loosen the bolt on the port side of the gate retaining the spring and stretch the spring off of the bolt. Or, just remove the bolt (and collar), but be careful as spring tension can be high.

3. Remove the 2 bolts mounting the reverse gate to the mounting stays on each side - each bolt uses a washer and collar. Remove the gate.

■ **If all you are doing is removing the reverse gate, stop hear and skip to Step 14 for installation.**

4. If replacing the gate, remove the shift cable pivot from the left side of the gate.

5. Remove the bolt holding the other end of the tension spring.

6. Remove the mounting bolts and lift off the port mounting stay. Keep the small bracket with it.

7. Do the same thing for the starboard mounting stay.

To install:

8. Clean all components thoroughly and allow to dry completely.

9. Inspect everything for wear, cracks or other signs of damage.

10. Install the shift cable pivot and tighten the nut to:

- 1052cc engines - 5.6 ft. lbs./68 inch lbs. (7.8 Nm)
- 1812cc engines - 5.9 ft. lbs./71 inch lbs. (8 Nm)

11. If you removed the small bracket on the starboard stay, install it and tighten the bolt to:

Fig. 40 Spring installation positioning - 2008 FX models and all FX models w/1812cc engines

- 1052cc engines - 5.6 ft. lbs./68 inch lbs. (7.8 Nm)
- 1812cc engines - 5.9 ft. lbs./71 inch lbs. (8 Nm)

Coat the bolt threads with Loctite 242, position the stay on the nozzle and tighten the bolts to as listed above.

12. Now, repeat the previous step for the port mounting stay.

13. Thread the forward spring retaining bolt into the port stay. Loop the spring end around the bolt from the bottom and tighten the bolt to 14 ft. lbs. (20 Nm).

14. Coat the reverse gate bolt threads with Loctite 242 and slide on the washers and rollers. Coat the washers with a little water resistant marine grease. Position the gate over the mounting stays so the bolt holes line up and the tighten the bolts to 14.8 ft. lbs. (20 Nm).

15. Thread the aft spring retaining bolt into the gate, stretch the spring over it from the top and then tighten the bolt/nut securely. Don't forget there is a washer under this nut.

16. Install the shift cable joint to the pivot and adjust the cable as detailed in the Maintenance & Tune-Up section.

VX Models

◆ See Figures 4, 40 and 41

1. Disconnect the shift cable joint at the rear of the assembly (from the outside). Pull back (toward the engine) on the lock sleeve and then pull it off of the pivot.

2. Loosen the bolt on the port side of the gate retaining the spring and stretch the spring off of the bolt. Or, just remove the bolt (and collar), but be careful as spring tension can be high.

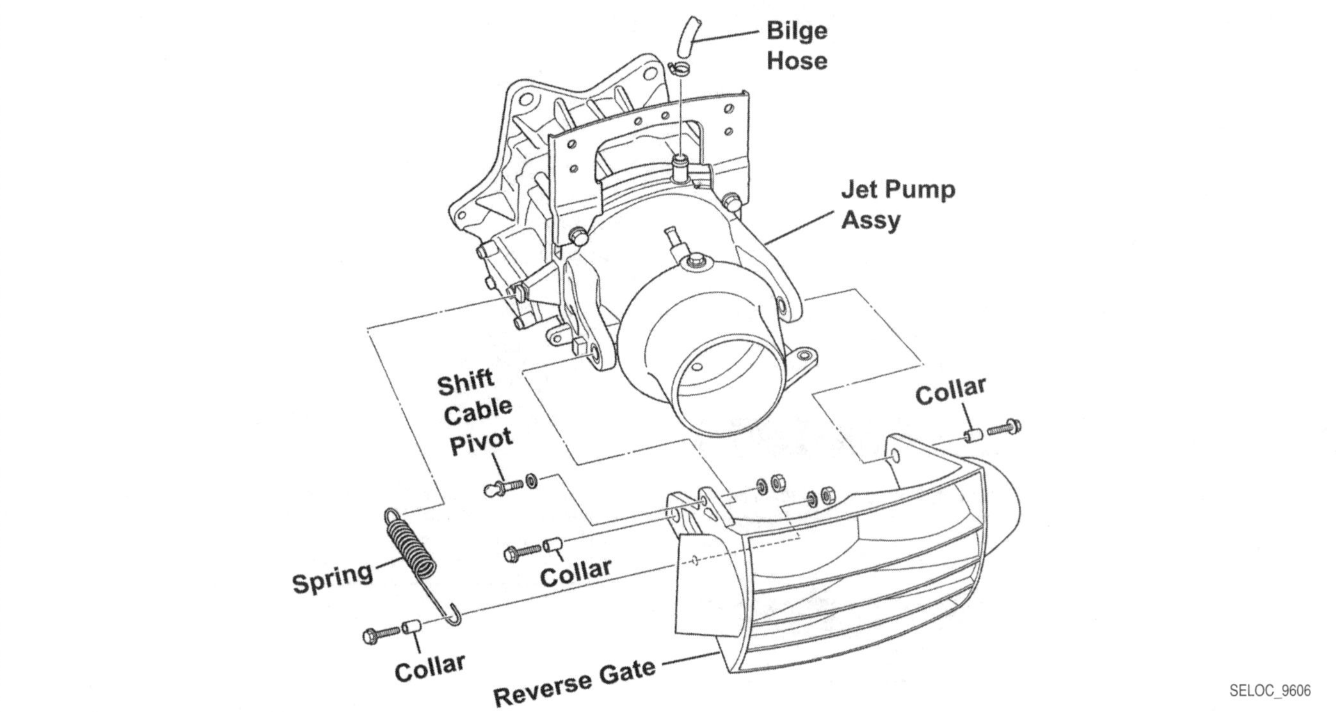

Fig. 41 Exploded view of the reverse gate and components - VX models

3. Remove the 2 bolts mounting the reverse gate to the mounting stays on each side - each bolt uses a washer and collar. Remove the gate.

4. If replacing the gate, remove the shift cable pivot from the left side of the gate.

To install:

5. Clean all components thoroughly and allow to dry completely.

6. Inspect everything for wear, cracks or other signs of damage.

7. Install the shift cable pivot and tighten the nut to 5.6 ft. lbs./68 inch lbs. (7.8 Nm) except on 2010-11 models where it is 5.9 ft. lbs./71 inch lbs. (8 Nm).

8. Loop the spring end around the boss on the nozzle from the top (bottom on 2010-11 models).

9. Coat the reverse gate bolt threads with Loctite 271 and slide on the washers and rollers. Coat the washers with a little water resistant marine grease. Position the gate over the mounting stays so the bolt holes line up and the tighten the bolts to 11 ft. lbs. (15 Nm).

10. Thread in the aft spring retaining bolt, stretch the spring over it from the bottom (top on 2010-11 models) and then tighten the bolt to 5.9 ft. lbs./71 inch lbs. (8 Nm).

11. Install the shift cable joint to the pivot and adjust the cable as detailed in the Maintenance & Tune-Up section.

Thrust Nozzle & Nozzle Ring

REMOVAL & INSTALLATION

FX & FZ Models

② MODERATE

◆ See Figure 42

1. Remove the jet pump assembly as detailed elsewhere in this section.

2. Remove the reverse gate and mounting stays as detailed previously in this section.

3. Remove the upper and lower thrust nozzle mounting bolts and pull the nozzle off of the ring. Remove the bolt collars.

4. Remove the mounting bolt on each side of the nozzle ring and pull the ring off of the pump nozzle. There will be a washer and collar for each bolt between the ring and nozzle.

5. Remove the QSTS pivot if needed.

To install:

6. Clean all components thoroughly and allow to dry completely.

7. Inspect everything for wear, cracks or other signs of damage.

8. If removed, insert the QSTS pivot through the boss and tighten the nut to:

- 998cc and 1052cc engines - 5.6 ft. lbs./68 inch lbs. (7.8 Nm)
- 1812cc engines - 5.9 ft. lbs./71 inch lbs. (8 Nm)

Remember there is a washer on each side of the boss.

9. Position the washer and collar on each side of the nozzle ring bolt holes and slide it onto the pump nozzle trying to keep everything in position so the holes line up - use a little water resistant marine grease on the collars to keep them in place. Coat the bolt threads with Loctite 271 and then tighten them to:

- 998cc and 1052cc engines - 11 ft. lbs. (15 Nm)
- FZ and FX HO w/1812cc engines - 11 ft. lbs. (15 Nm)
- FX SHO - 11.8 ft. lbs. (16 Nm)

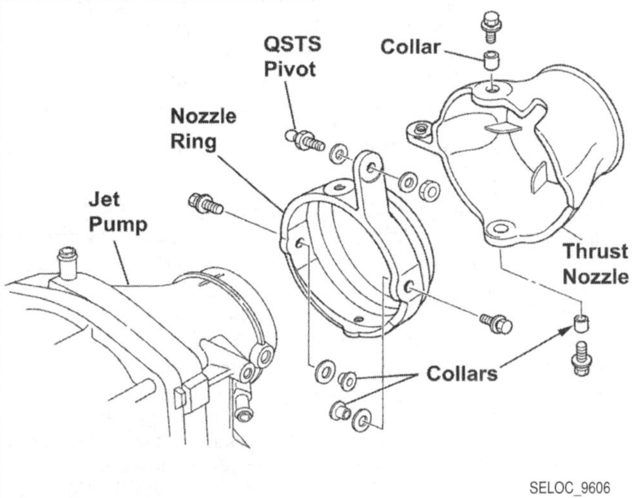

Fig. 42 Exploded view of the thrust nozzle and ring - FX models with 998cc/1052cc engines, 1812cc models similar

10. Coat the thrust nozzle bolt collars with a little water resistant marine grease and insert them into their respective holes; slide the nozzle over the ring. Coat the bolt threads with Loctite 271 and then tighten them to the toque figures listed in the previous step.

11. Install the reverse gate.

12. Install the jet pump assembly.

VX Models

■ VX models do not utilize a nozzle ring and the thrust nozzle removal and installation procedures are included as part of the Disassembly & Assembly procedures in the Jet Pump Assembly section. Be careful here, although the jet pump procedures for the VXR and VXS are included with the other 1812cc models and *not* in the VX section, the thrust nozzle is not covered, so you must read through the jet pump assembly procedures for the VX w/1052cc engines

Impeller & Impeller Housing/Duct

■ Impeller, impeller housing, impeller duct and jet nozzle removal are all covered in the Disassembly & Assembly portion of the Jet Pump Assembly section.

Intermediate Bearing Housing

REMOVAL & INSTALLATION

FX, FZ & VX Models w/1812cc Engines

2 MODERATE

◆ See Figures 43 and 44

1. Remove the engine as detailed previously in this section.
2. You should have already disconnected the grease line during engine removal, but if not, make sure you do it now.
3. Loosen the 3 bearing mounting bolts. Pull it back just enough to slide out the shims, remove the shims and store them in a way that you will be able to install them exactly where they came from.

■ If there are no shims, it's ok. It just means they were not necessary on your particular model.

4. Remove the bolts and pull off the bearing housing. It should slide right off the jet pump driveshaft, but if not, you will probably have to remove the pump as well. There are 2 dowel pins; keep them with the housing.

To install:

5. Position the bearing housing so that you can wiggle it slightly until the driveshaft meshes with the bearing splines. Rotate the housing until the dowel pins align with their holes and then press it all the way on.
6. Coat the bolt threads with Loctite 572 and then thread them in until they just catch. Pull the housing out just enough to slide in the shims and then move it back in position again. Tighten the bolts to 12.5 ft. lbs. (17 Nm).

■ Remember that the bearing housing shims must go back in the same places that they were removed from.

If your shims were bad, or you mixed them up, or you replaced the bearing housing, measure the clearance between the hull and the housing at each bolt hole to determine the thickness of any new shims that you may need. If clearance is:

- 0-0.008 in. (0-0.2mm), you do not need any shims
- 0.012-0.028 in. (0.3-0.7mm), you need a 0.5mm shim
- 0.031-0.047 in. (0.8-1.2mm), you need a 1mm shim
- 0.051-0.079 in. (1.3-2mm), you need a 1.5mm shim

■ The above specs are for the 998cc models. For some reason Yamaha has not provided them for the other models, but we feel pretty comfortable that these will work fine for the 1052cc and 1812cc models as well.

7. Install the engine.

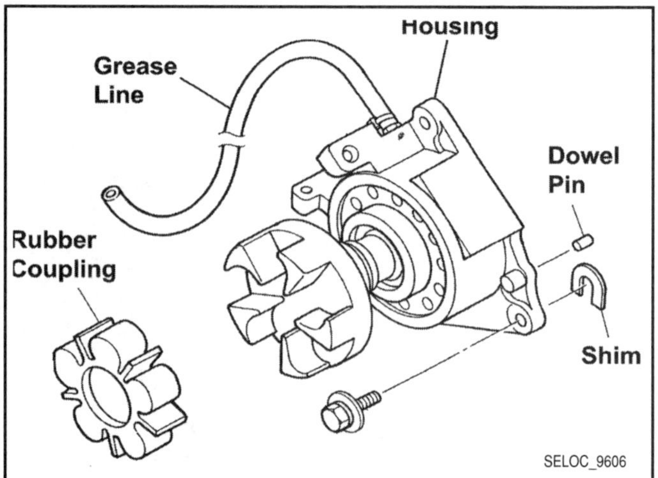

Fig. 43 A good look at the intermediate bearing assembly - FX models with 998cc/1052cc engines, 1812cc engines and 2010-11 VX w/1052cc engines similar

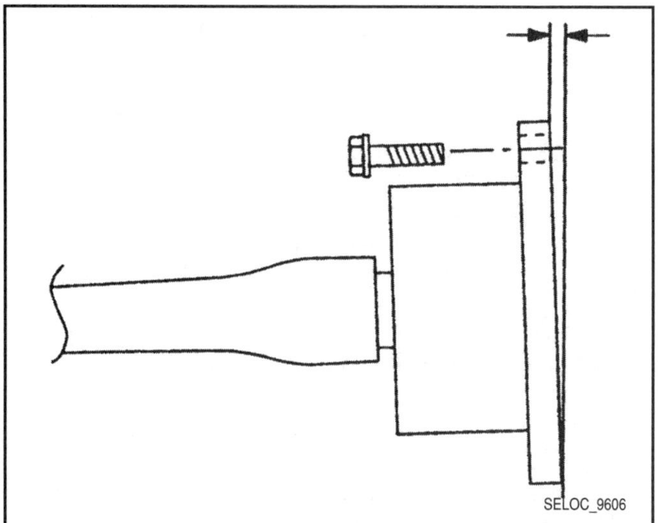

Fig. 44 Checking clearance for shim thickness

2005-09 VX Models

2 MODERATE

◆ See Figures 45 and 46

1. Remove the engine as detailed previously in this section.

■ Although not strictly necessary, may will find it easier to remove the jet pump assembly as well.

2. Remove the 4 housing cover mounting bolts and lift off the cover..
3. Loosen the 2 hose clamp aft of the bearing and then carefully lift/slide the bearing and shaft assembly out of the tunnel.
4. Remove the small rubber tube.

To install:

5. Inspect all components for cracks, wear and other signs of damage.
6. Slide the rubber tube over the neck of the pass-thru tube and slide on both clamps. Move the rear clamp into position and tighten it to 3 ft. lbs./36 inch lbs. (4.2Nm).
7. Coat the intermediate shaft splines marine grade water resistant grease. Guide the shaft into the tube and move it backwards until the bearing is in the housing and the end of the shaft has meshed with the impeller. You may have to wiggle the coupler slightly until the splines mesh with each other. Once seated correctly, the rear of the bearing should be right up against the step in the lower housing.

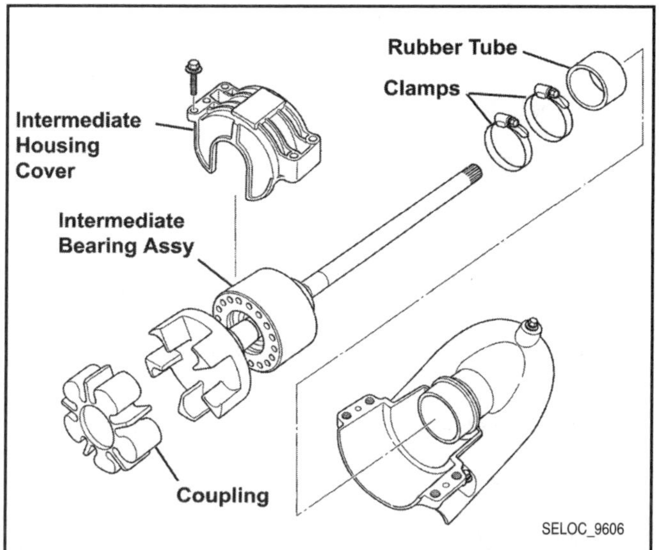

Fig. 45 A good look at the intermediate bearing assembly - 2005-09 VX models

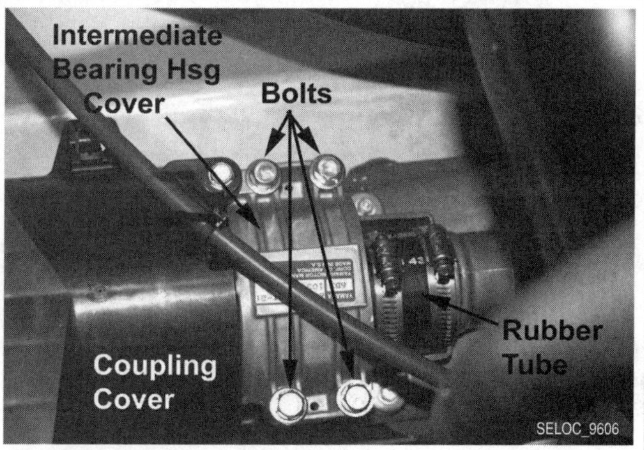

Fig. 46 Remove the coupling cover first

8. Slide the forward clamp into position on the bearing and tighten to 3 ft. lbs./36 inch lbs. (4.2Nm).
9. Position the cover over the bearing. Coat the bolt threads with Loctite 271 and thread them in until they are finger-tight. Now tighten the bolt to 12 ft. lbs. (17 Nm) in a criss-cross pattern.
10. Install the engine.

2010-11 VX Models w/1052cc Engines

◆ **See Figures 43 and 44**

1. Remove the engine as detailed previously in this section.
2. Loosen the 3 bearing mounting bolts. Pull it back just enough to slide out the shims, remove the shims and store them in a way that you will be able to install them exactly where they came from.

■ **If there are no shims, it's ok. It just means they were not necessary on your particular model.**

3. Remove the bolts and pull off the bearing housing. The intermediate shaft should slide right out of the impeller, but if not, you will probably have to remove the pump as well. There are 2 dowel pins; keep them with the housing.

To install:

4. Position the bearing housing so that you can wiggle it until the intermediate shaft meshes with the impeller. Rotate the housing until the dowel pins align with their holes and then press it all the way on.
5. Coat the bolt threads with Loctite 572 and thread them in until they just catch. Pull the housing out just enough to slide in the shims and then press it all the back on again. Tighten the bolts to 12 ft. lbs. (17 Nm).

■ **Remember that the bearing housing shims must go back in the same places that they were removed from.**

6. Install the engine.

DISASSEMBLY & ASSEMBLY

All FX & FZ Models And VX Models w/1812cc Engines

◆ **See Figures 47 thru 52**

■ **Disassembly will ruin the bearing. Do not perform this procedure if it is not necessary.**

1. Slide back the clamp and wiggle the grease line off of the nipple. Remove the nipple.
2. Insert a shaft holding tool (#YB-06552 / 90890-06552 on 998/1052cc engines, and #90890-06730 for 1812cc engines) into the intermediate shaft at the rear of the bearing housing and clamp the tool in a vise. Connect a coupler wrench (#YW-06551 / 90890-06551 on 998/1052cc engines, and #90890-06729 for 1812cc engines) to the driven coupling and break the coupling loose. Do not remove the coupling yet on 2002-07 models. Remove it completely on 2008 and later models.
3. On 2002-07 models, locate a rod or pipe that is no more than 0.63 in. (16mm) thick. Cut it so it is at least 8.27 in. (210mm) long and then insert it into the back of the intermediate shaft until it contacts the coupling. Secure the housing and then slowly tap the end of the rod until you have pressed the shaft/coupling out of the housing. Remove the rod.
4. Now you can remove the coupling from the intermediate shaft. Remove the washer and use a small pick to remove the 2 O-rings at the back of the shaft.
5. On 2008 and later models, move the assembly to a press table with the font side of the housing down and centered over the hole. Remove the thrust washer on 1812cc models. Place a block of wood over the splined side of the intermediate shaft and press the shaft out. If a press table is not available, you can support the housing on block and then use a small mallet on the block of wood, but be careful if you go this route.
6. On all models again, use your pick again and pull out the 2 oil seals in the back of the housing bore. You are going to discard the seals so it's ok if you damage them, but be careful not to scrape the bore itself. Do the same with the front side oil seal.

■ **If you are unable to get the seals out, use a slide hammer or bearing puller.**

7. Remove the circlips on each side of the bearing.
8. Set the housing on the table of a press so the rear side is facing upward and the bottom is either over the hole in your table or supported just enough so the bearing can pop out. Install a driver (#YB-06071 / 90890-06606) and removal tool (#YB-06434 / 90890-06623) into the bore and press out the bearing. We suspect you could also do this with just a mallet as long as you were careful.

To assemble:

9. Clean all components thoroughly and allow to dry completely.
10. Inspect everything for wear, cracks or other signs of damage. Check the shaft splines carefully for cracked or nicked teeth.
11. Install the rear circlip into the recess in the bore. Circlips are not re-usable on 1052cc or 1812cc models, so you'll have to use a *new* one.
12. Position the housing on a flat surface with the front side facing upward. Coat the races of a *new* bearing with water resistant marine grease and position it into the bore. Move the driver tools in so they rest on the edge of the bearing race and then press the bearing in until it seats against the rear circlip.
13. Turn the housing over and install the front circlip (remember, use a new one on 1052cc and 1812cc models).
14. Coat the lip surface of a *new* front oil seal with a good water resistant marine grease and place it into the bore with the lip side facing out. Use your

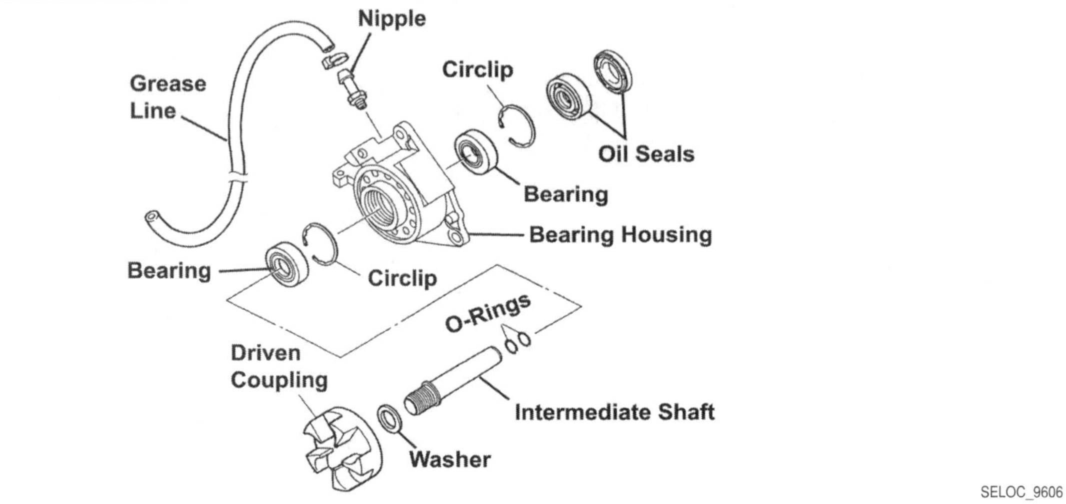

Fig. 47 Exploded view of the intermediate bearing assembly - FX models with 998cc/1052cc engines

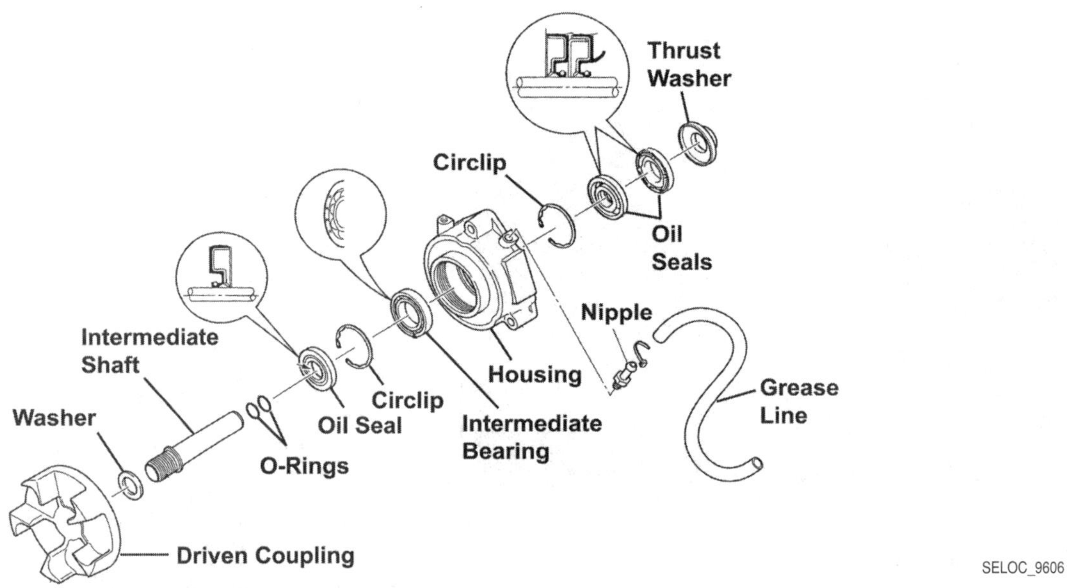

Fig. 48 Exploded view of the intermediate bearing assembly - FX, FZ and VX models w/1812cc engines

bearing drivers and press the seal in until it contacts the circlip. The top of the seal should now be 0.27-0.29 in. (6.8-7.2mm) below the edge of the bore on 998cc and 1052cc models, or 0.25-0.27 in. (6.3-6.7mm) on 1812cc models.

15. Move the assembly over to the press (or other table with a hole in it) so the front side is facing upward. Support the underside of the unit with a couple small blocks and center the bore over the hole.

Insert the intermediate shaft into the front oil seal, splined (rear) end first, and feed it through the bearing. Use a small block of wood on the threaded (front) end of the shaft and use the press or gently tap it through the bearing until the splined end is protruding 1.79-1.183 in. (45.5-46.5mm) *below* the bottom edge of the bearing assembly. On 1812cc models there is a bearing attachment tool (#90890-06728) that you can use, but is not strictly necessary.

16. Turn the assembly over so that it is now resting on its front side with the threaded end of the shaft through the hole in the table (or you can just place blocks under it so there is clearance).

17. Coat the lip surface of a *new* inner rear oil seal with a good water resistant marine grease and place it into the bore with the lip side facing out. Using another piece of pipe that will fit into the bore and around the intermediate shaft, gently press the seal in until it is just a little below the edge of the bore. You can also use a bearing attachment tool (#90890-06728)

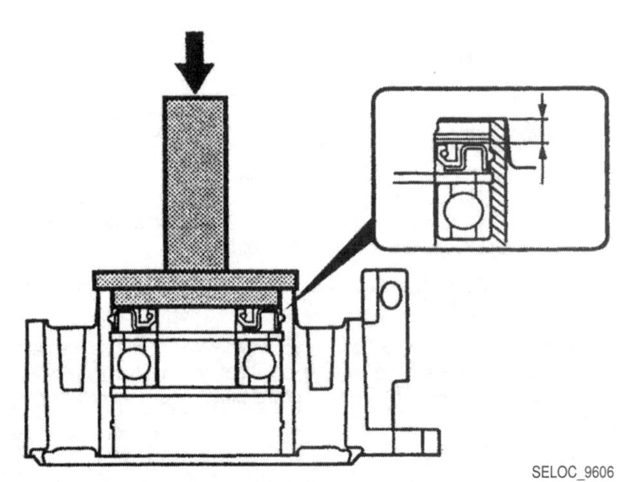

Fig. 49 Press in the front oil seal until it is the correct distance below the top (front) edge of the bore

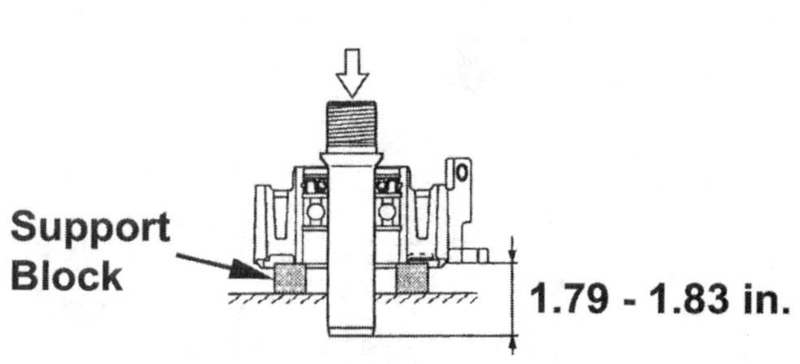

Fig. 50 Measure the distance that the intermediate shaft protrudes from the back of the housing

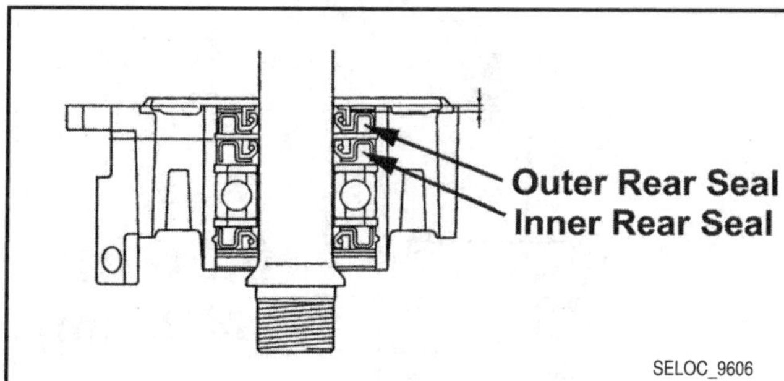

Fig. 51 Press in the rear oil seals until the upper edge of the seal is the correct distance below the top (rear) edge of the bore

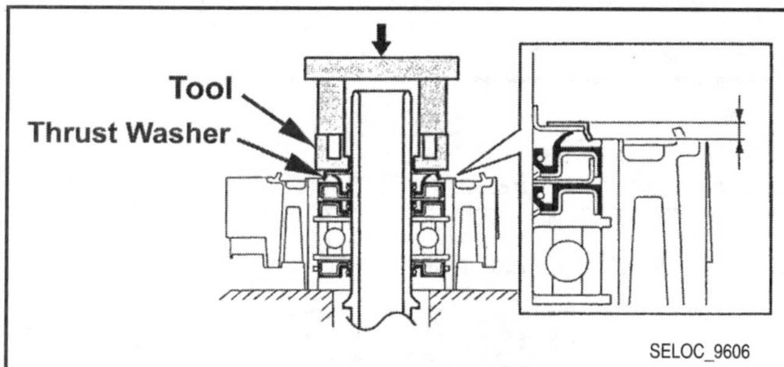

Fig. 52 Installing the thrust washer on 1812cc engines

18. Now, coat the lip surface of a *new* outer rear oil seal with a good water resistant marine grease and place it into the bore with the lip side facing out so it rests on the inner seal. Use your pipe or attachment tool and press both seals in until the lip of the outer seal is 0.02-0.04 in. (0.6-1mm) below the edge of the bore on 998cc and 1052cc models, or 0.05-0.07 in. (1.4-1.8mm) on 1812cc models.

19. Coat 2 *new* O-rings molybdenum grease and roll them into the grooves on the end of the intermediate shaft.

20. Position the thrust washer over the rear of the shaft and use another bearing attachment tool to press it in until the flat surface of the washer is 0.16-0.18 in. (4.1-4.7mm) above the surface of the bearing bore.

21. Coat the threads on the forward end of the intermediate shaft with Loctite 572. Slide on the washer and then spin on the driven coupling until it seats against the washer. Insert the holding tool (same as removal tool) into the splined end of the shaft and clamp it in a vise. Tighten the coupling to 25 ft. lbs. (36 Nm) on 998cc and 1052cc models, 155 ft. lbs. (210 Nm) on FX HO and VX models (1812cc), 147.5 ft. lbs. (200 Nm) on FX SHO models, or 151 ft. lbs. (205 Nm) on FZ models.

22. Coat the threads on the grease line fitting with Loctite 572 and tighten it securely on 998cc and 1052cc models, or 3.7 ft. lbs./44 inch lbs. (5 Nm) on 1812cc models.

23. Install the assembly.

2005-09 VX Models

OEM ② MODERATE

◆ See Figures 53 thru 56

■ **Disassembly will ruin the bearing. Do not perform this procedure if it is not necessary.**

1. Install a shaft holding tool (#YB-06151 / 90890-06519) onto the splined end of the intermediate shaft and clamp the tool in a vise. Connect a coupler wrench (#YW-06551 / 90890-06551) to the driven coupling and break the coupling loose. Remove the coupling

2. Slide the damper off of the bearing and set aside.

3. Position the front side of the assembly onto the table of a press so the threaded end of the intermediate shaft is protruding through the hole.

4. Place a piece of wood on top of the splined end of the shaft and then press it out of the housing. Grab the other side as you are pressing it and guide it through the seals and bearing. Obviously, you could use a small mallet as well, but be careful.

5. Remove the circlip and then turn the housing over so the rear side is now facing up and the front side is positioned over the hole in the table. Install a bearing remover tool (#YB-06432 / 90890-06628) and driver handle (#YB-06071 / 90890-06606) and press out the bearing and 3 oil seals. This will damage them, so you can throw them all away.

To assemble:

6. Clean all components thoroughly and allow to dry completely.

7. Inspect everything for wear, cracks or other signs of damage. Check the shaft splines carefully for cracked or nicked teeth.

8. Position the housing on a flat surface with the front side facing upward; remember, this is the wider side. Coat the lip surface of a *new* outer rear oil seal with a good water resistant marine grease and place it into the bore with the lip side facing out. Use your bearing drivers and press the seal in until it seats on the step at the bottom of the housing.

9. Repeat the previous step with the inner rear seal, pressing in until it seats on the other seal. Be careful not to press to far or you will damage the outer seal.

10. Coat the races of a *new* bearing with water resistant marine grease and position it into the bore so that the numbers on the bearing are facing upward. Move the driver tools in so they rest on the edge of the bearing race and then press the bearing in until it just seats against the lip of the inner rear oil seal. Measure the distance between the top of the bearing and the upper edge of the housing and adjust the bearing so it is 0.55-0.57 in. (12.2-16.2mm).

11. Finally, coat the lip surface of a *new* front oil seal with a good water resistant marine grease and place it into the bore with the lip side facing out. Use your bearing drivers and press the seal in until it just seats on the intermediate bearing and check that the distance between the seal lip and the upper edge of the housing is 0.23-0.25 in. (6-6.4mm).

12. Install a *new* circlip into the groove in the housing.

13. Insert the splined end of the intermediate shaft into the housing. Place a block of wood over the threaded end of the shaft and press it

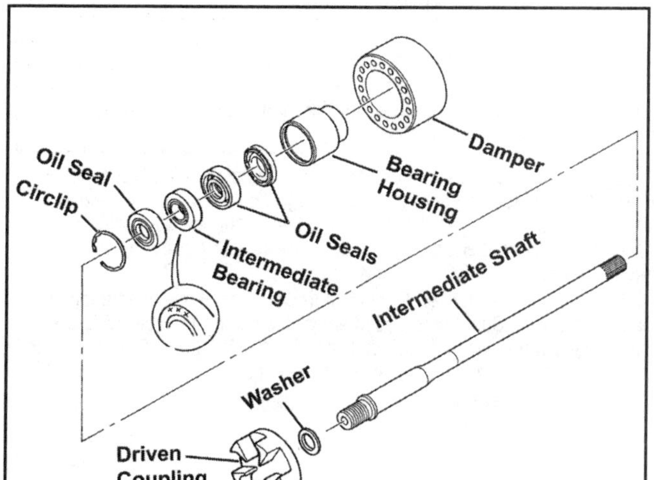

Fig. 53 Exploded view of the intermediate bearing assembly - 2005-09 VX models

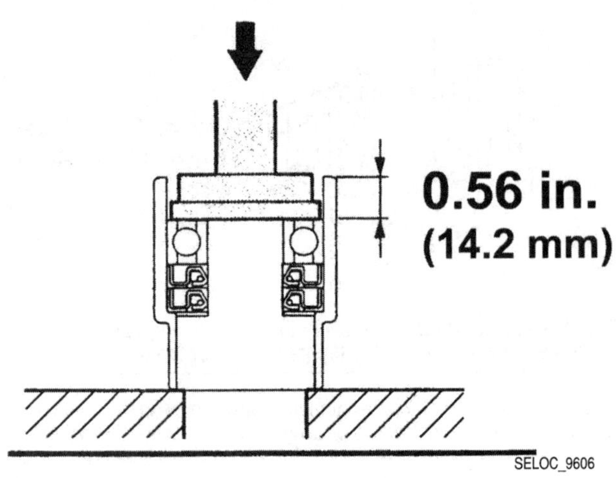

Fig. 54 Press in the rear oil seals and bearing until the bearing is the correct distance below the top edge of the bore. . .

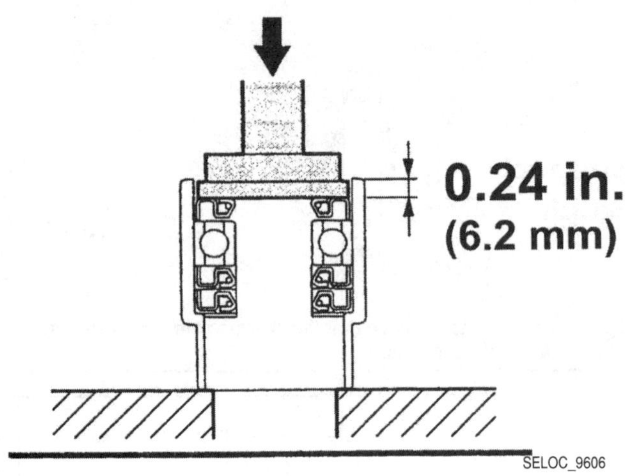

Fig. 55 . . .and then do the same with the front oil seal

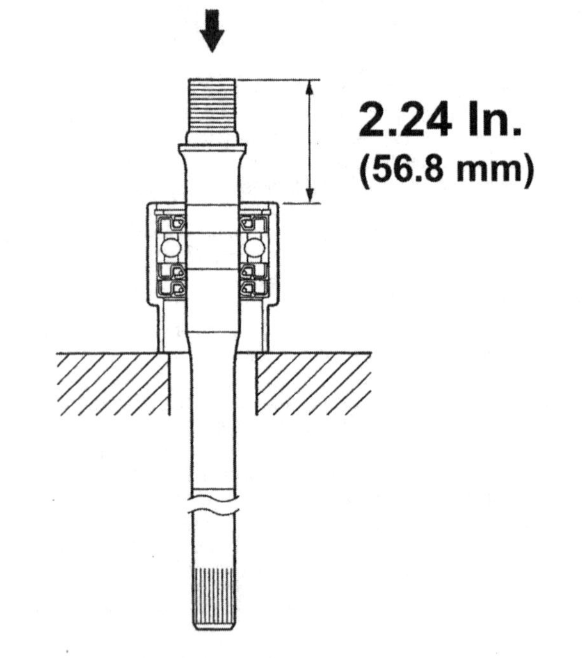

Fig. 56 Measure the distance that the intermediate shaft protrudes from the front of the housing

through the seals and bearing until the end of the shaft is 2.23-2.25 in. (56.6-57mm) from the top edge of the housing. If you go too far, you'll need to disassemble everything (which means another set of new seals and bearing!) and start all over; so press the shaft in slowly, checking you measurement frequently.

14. Reinstall the shaft holding tool to the long end of the shaft again and clamp it in a vise. Coat the threaded end of the shaft with Loctite 572 and then install the washer and coupler, tightening to 26 ft. lbs. (36 Nm) with the same coupler wrench you used during disassembly.
15. Install the assembly.

2010-11 VX Models w/1052cc Engines

OEM ② MODERATE

◆ **See Figures 57 thru 60**

■ **Disassembly will ruin the bearing. Do not perform this procedure if it is not necessary.**

1. Insert a shaft holding tool (#YB-06151 / 90890-06519) onto the rear of the intermediate shaft and clamp the tool in a vise. Connect a coupler wrench (#YW-06551 / 90890-06551) to the driven coupling and break the coupling loose. Do not remove it yet.
2. Position the front side of the assembly onto the table of a press so the threaded end of the intermediate shaft is protruding through the hole.
3. Place a piece of wood on top of the splined end of the shaft and then press it out of the housing. Grab the other side as you are pressing it and guide it through the seals and bearing. Obviously, you could use a small mallet as well, but be careful.
4. Use a slide hammer (#YB-06096 or equivalent) and remove the roil seals from the bearing housing on US and Canada models. Use a bearing puller (#90890-06535 or equivalent) for all other countries. The rear seals can be removed at the same time. Discard the seals as they will be ruined during removal.
5. Remove the 2 circlips and throw them away.
6. Move the housing back to the table and position it over the hole so the front side is facing down. Install a bearing tool (#YB-06085 / 90890-06538) and driver (#YB-06071 / 90890-06606) and then press out the bearing. You can throw it away as well. Obviously, if you do not have a table with a hole available, you can simply prop the sides of the housing up with small blocks of wood - the object is to have enough clearance for the bearing to come out.

To assemble:

7. Clean all components thoroughly and allow to dry completely.
8. Inspect everything for wear, cracks or other signs of damage. Check the shaft splines carefully for cracked or nicked teeth.
9. Install a *new* rear circlip into the recess in the bore.
10. Position the housing on a flat surface with the front side facing upward. Coat the races of a *new* bearing with water resistant marine grease and position it into the bore so that the numbers are facing up. Move the driver tools in so they rest on the edge of the bearing race and then press the bearing in until it seats against the rear circlip.

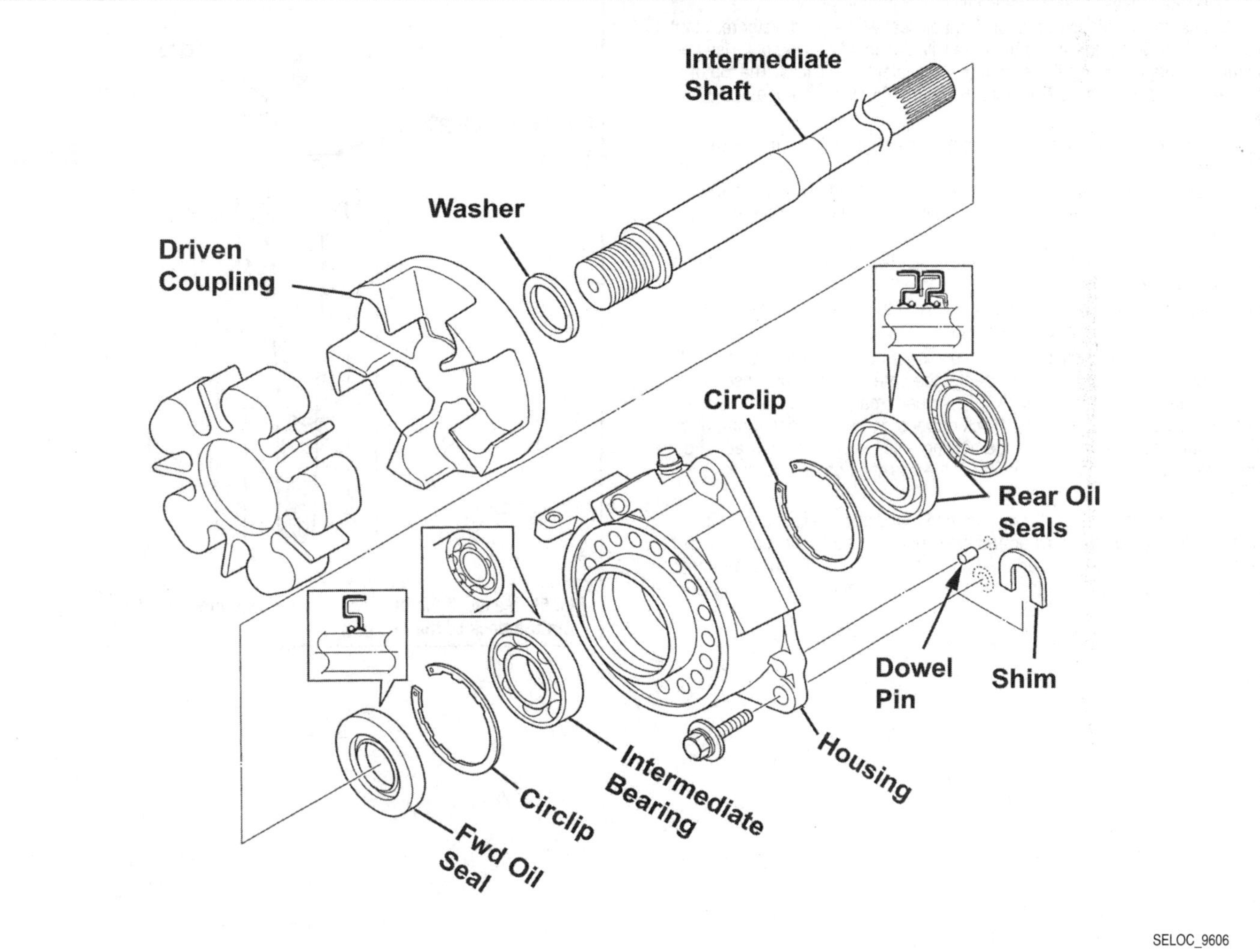

Fig. 57 Exploded view of the intermediate bearing assembly - 2010-11 VX models w/1052cc engines

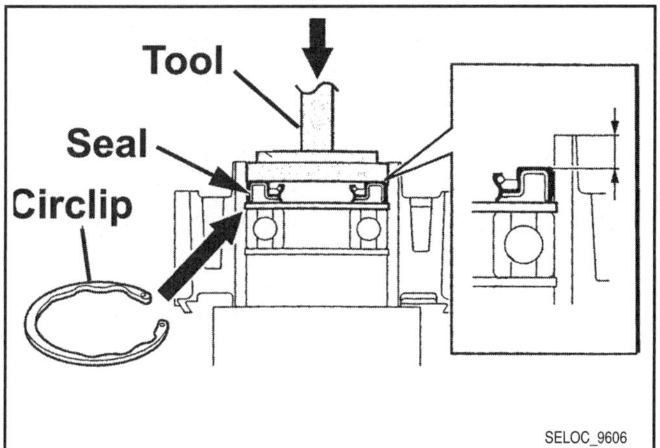

Fig. 58 Press in the front oil seal until it is the correct distance below the top (front) edge of the bore

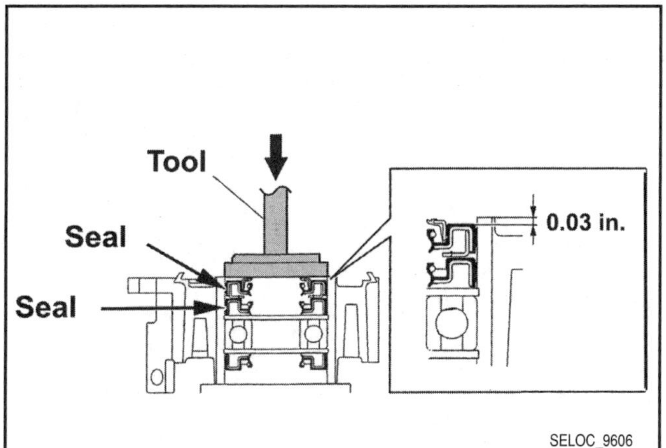

Fig. 59 Press in the rear oil seals until the upper edge of the seal is the correct distance below the top (rear) edge of the bore

11. Turn the housing over and install a *new* front circlip.

12. Coat the lip surface of a *new* front oil seal with a good water resistant marine grease and place it into the bore with the lip side facing out. Use your bearing drivers and press the seal in until it contacts the circlip. The top of the seal should now be 0.27-0.29 in. (6.8-7.2mm) below the edge of the bore.

13. Turn the assembly over so that it is now resting on its front side. Coat the lip surface of a *new* inner rear oil seal with a good water resistant marine grease and place it into the bore with the lip side facing out. Use the remove tools and press the seal in until it is just a little below the edge of the bore.

14. Now, coat the lip surface of a *new* outer rear oil seal with a good water resistant marine grease and place it into the bore with the lip side facing out so it rests on the inner seal. Press both seals in until the lip of the outer seal is 0.02-0.04 in. (0.6-1mm) below the edge of the bore.

15. Move the assembly back to the press (or other table with a hole in it) so the front side is facing upward. Center the bore over the hole. Insert the intermediate shaft into the front oil seal, splined (rear) end first, and feed it through the bearing. Use a small block of wood on the threaded (front) end of the shaft and press it through the bearing until the distance between the threaded end of the shaft and the top of the housing bore is 1.53 in. (38.9mm).

16. Coat the threads on the forward end of the intermediate shaft with Loctite 572. Slide on the washer and then spin on the driven coupling until it seats against the washer. Insert the holding tool (same as removal tool) into the splined end of the shaft and clamp it in a vise. Tighten the coupling to 26 ft. lbs. (36 Nm).

17. Install the assembly.

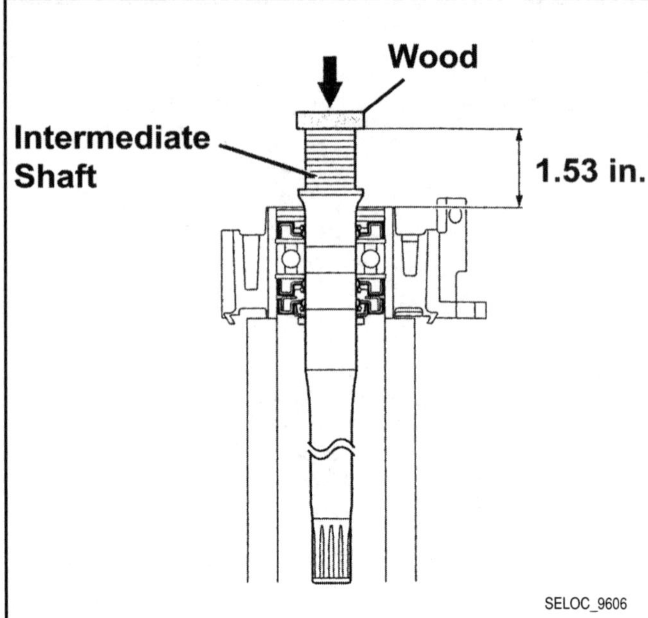

Fig. 60 Measure the distance that the intermediate shaft protrudes from the back of the housing

8

HULL, HELM & STEERING

HULL, HELM & STEERING

Handlebars

REMOVAL & INSTALLATION

All 2002-05 FX Models, 2006-08 FX Models & 2006-07 FX HO Models

MODERATE

◆ See Figures 1 thru 6

1. Remove the seat and cargo container (if equipped). Disconnect the battery cables.
2. Remove the 4 bolts from the underside of the handlebar cover. The cover is one piece, but you can spread apart the forward end where it sits on the lower handle boss cover. Grab the top and bottom side and carefully spread the open end just enough to allow sliding it over the bars.
3. Pull off the steering pad and then remove the 2 bolts attaching the cover bracket to the top of the steering column. Lift out the bracket.
4. Loosen the throttle cable locknut and thread it back until it makes contact with the adjusting nut. Squeeze the throttle a little until you are able to disconnect the cable from the grip control.

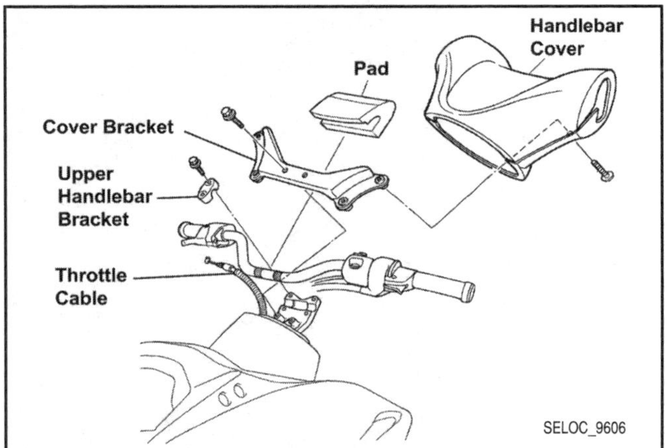

Fig. 1 Disconnecting the handlebars is easy. . .

5. Take a Sharpie or other marker and mark the handlebar at the outer edge of each handlebar bracket where it covers the handle bar. Loosen the 4 bolts (2 per bracket) and remove the 2 upper bar brackets.

■ **If you are replacing the handlebars and not just removing them, you may wish to skip these next few steps and simply remove the QSTS, handlebar and throttles controls as detailed in the Disassembly section. For that matter, you may want to choose that route regardless. The following is Yamaha's preferred procedure - you make the call.**

6. If your unit is equipped with QSTS (quick shift trim selector), find the converter on the inner starboard side of the hull, toward the back. Loosen the locknuts on the 2 lower cables and disconnect them from the converter. Follow the cables forward and take note of any plastic ties or holders that may be securing the 2 cables - you'll need this information shortly.
7. Tag and disconnect the handlebar switch and buzzer couplers.
8. Remove the 4 bolts securing the handle boss cover.
9. Reach in and cut the plastic ties securing the packing to the 2 grommets and pull out the grommets.
10. Grab the handlebars and slowly pull them up and away while having a friend or assistant guide the QSTS cable through the hull and up the steering tunnel. Make note of where you have to cut any plastic ties along the way.

To install:

11. Insert the QSTS cables and electrical leads back through handle boss cover. Do the same with the grommets and packing, sliding the grommets up the cables/leads as you lower the bars into the recess in the steering column. Make sure that the 2 punch marks on the bars are facing upward.
12. Install the 2 bar brackets temporarily to hold everything in place.
13. Reconnect the 3 couplers on the starboard side. Route the 2 QSTS cables back through the hull, installing new plastic ties in correct places. Reconnect the cable ends to the converter, making sure that the cable with the white tape in it is the higher of the two.
14. Install a new plastic ties around the top of each grommet and then pop them back in place so they are secure in the deck.
15. Position the handle boss cover on the helm and then coat the bolt threads with Loctite 242. Tighten the bolts, in a criss-cross pattern, to 0.7 ft. lbs./8.5 inch lbs. (0.9 Nm). . .remember, it is plastic.
16. Now, take a look at the 2 upper handlebar brackets. The small punch mark on each should be facing forward (down); if not, flip them around. Move

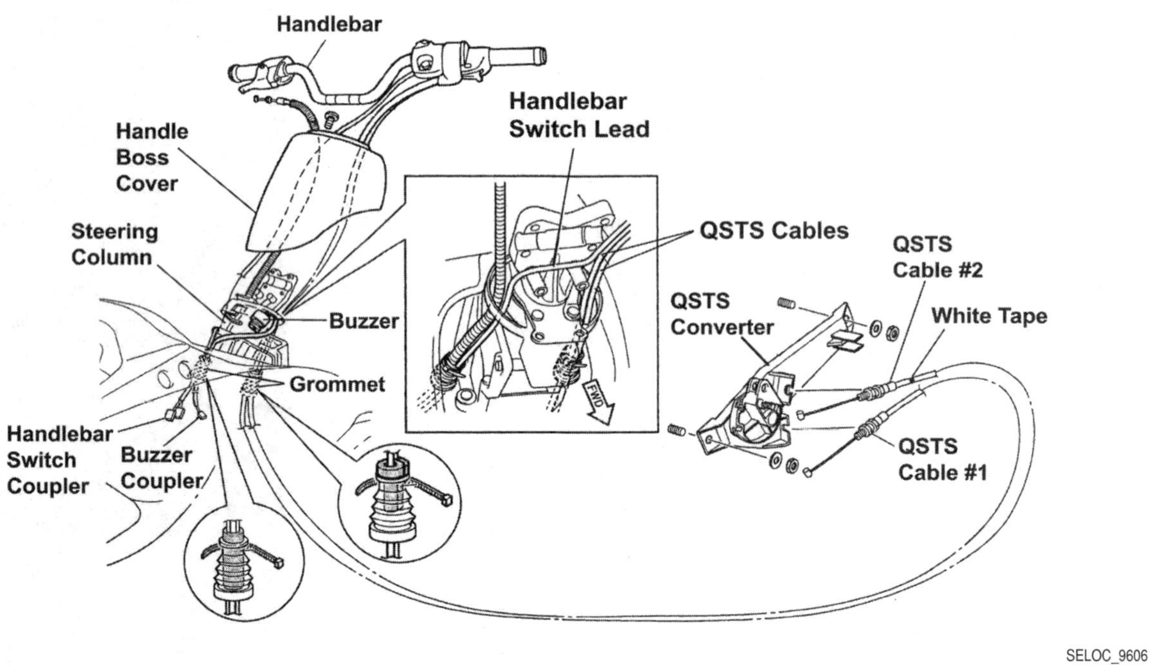

Fig. 2 . . .it's the removal part that slows things up - all 2002-05 FX models, 2006-08 FX models & 2006-07 FX HO models

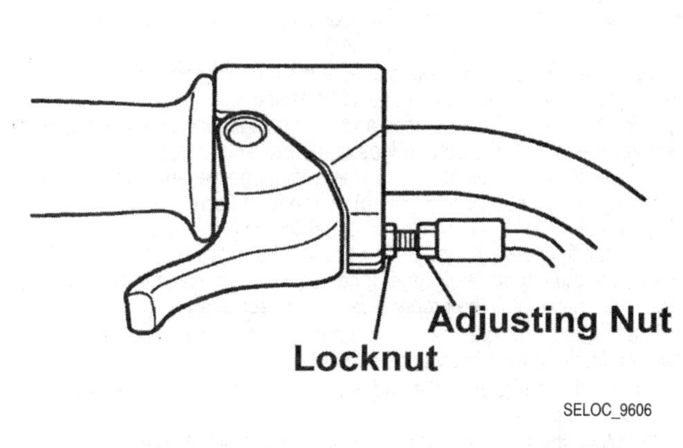

Fig. 3 Loosen the lock nut before disconnecting the throttle cable

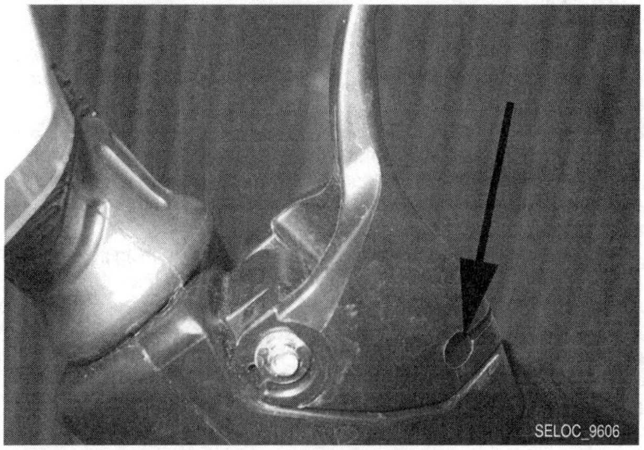

Fig. 4 The cable end will come out of this hole

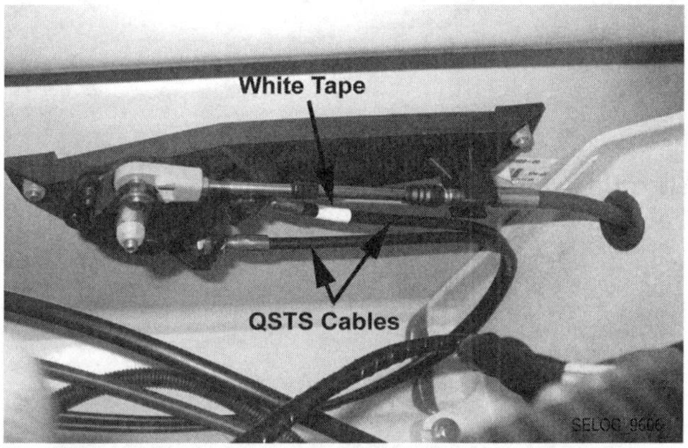

Fig. 5 A good shot at the QSTS converter. The cable with white tape is the rear cable at the handlebar control

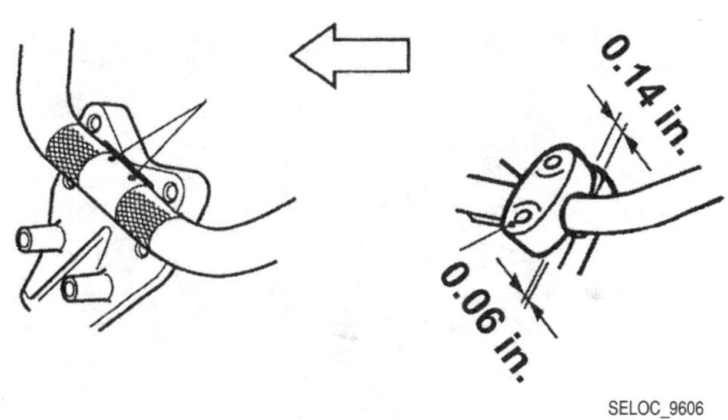

Fig. 6 Aligning the handlebars

the bars back and forth until the marks you made during removal line up with the outer edges of the brackets. Tighten the bolts, alternately, to 14.8 ft. lbs. (20 Nm). Use a feeler gauge and check that you have a clearance of 0.06 in. (1.5mm) on the front (lower) edge and 0.14 in. (3.5mm) at the upper (rear) edge. If not, tighten or loosen the bolts in very small increments until the proper clearance is achieved.

The important thing is that the figure for the lower measurement is less than the figure of the upper measurement.

17. Install the throttle cable and adjust it as detailed in the Maintenance & Tune-Up section.

18. Position the cover bracket, coat the bolt threads with Loctite 271, and then tighten to 0.8 ft. lbs./9.6 inch lbs. (1.1 Nm). Press the pad into position.

19. Spread apart (*carefully*) the lower edges of the handlebar cover and move it into position over the bars. Make sure the bars are riding in the 'arm holes' of the cover correctly and then coat the bolt threads with Loctite 242. Tighten them, in a criss-cross pattern, to 0.8 ft. lbs./9.6 inch lbs. (1.1 Nm).

■ **There should be an even gap running between the handle boss cover and the handlebar cover. If not, recheck your bar alignment gaps.**

20. Adjust the QSTS cables as detailed in the Maintenance & Repair section.

21. Connect the battery cables and install the cargo container (if removed) and seat(s).

2006-08 FX Cruiser & 2006-11 FX Cruiser HO/SHO Models

② MODERATE

◆ **See Figures 3 thru 9**

1. Remove the seat and cargo container (if equipped). Disconnect the battery cables.

2. Remove the 10 bolts from the underside of the handlebar cover. Two of the bolts attach the lower cover to the steering column and not the upper cover. On 1052cc models, these 2 bolts are black, so take note of which holes they came out of. Separate and remove the upper and lower handlebar covers.

3. Remove the 4 steering pad mounting bolts and lift off the pad. If necessary, remove the rubber grommet form the top of the pad bracket.

4. Loosen the throttle cable locknut and thread it back until it makes contact with the adjusting nut. Squeeze the throttle a little until you are able to disconnect the cable from the grip control, making sure that you work small cable seal out as you remove the cable.

5. Take a Sharpie or other marker and mark the handlebar at the outer edge of each handlebar bracket where it covers the handlebar. Loosen the 4 bolts (2 per bracket) and remove the 2 upper bar brackets.

■ **If you are replacing the handlebars on 2008 and later models or even just removing them, skip these next few steps and simply remove the QSTS, handlebar switch assembly and throttle controls as detailed in the Disassembly section. For that matter, even on 2006-07 models you may want to choose that route regardless. The following though, is Yamaha's preferred procedure for the earlier models - you make the call.**

6. Find the QSTS (quick shift trim selector) converter on the inner starboard side of the hull, toward the back. Loosen the locknuts on the 2 lower cables and disconnect them from the converter. Follow the cables forward and take note of any plastic ties or holders that may be securing the 2 cables - you'll need this information shortly.

7. Tag and disconnect the handlebar switch and buzzer couplers.

8. Remove the 4 bolts securing the handle boss cover.

9. Reach in and cut the plastic ties securing the packing to the 2 grommets and pull out the grommets.

10. Grab the handlebars and slowly pull them up and away while having a friend or assistant guide the QSTS cable and switch assembly lead through the hull and up the steering tunnel. Make note of where you have to cut any plastic ties along the way.

To install:

11. On 2008 and later models, install the 2 bar brackets temporarily to hold everything in place.

12. Install the QSTS control grip, handlebar switch and throttle control grip assemblies. Skip to Step 18.

13. On early models, insert the QSTS cables and electrical leads back through handle boss cover. Do the same with the grommets and packing sponges, sliding the grommets/sponges up the cables/leads as you lower the bars into the recess in the steering column. Make sure that the 2 punch marks on the bars are facing upward.

14. Install the 2 bar brackets temporarily to hold everything in place.

15. Reconnect the 2 couplers on the port side. Route the 2 QSTS cables back through the hull, installing new plastic ties in correct places. Reconnect the cable ends to the converter, making sure that the cable with the white tape in it is the higher of the two.

16. Install a new plastic ties around the top of each grommet and then pop them back in place so they are secure in the deck. Make sure that there is sufficient slack in the cables between the handlebar grips and where the grommets.

17. Position the handle boss cover on the helm and then coat the bolt threads with Loctite 242. Tighten the bolts, in a criss-cross pattern, to 0.7 ft. lbs./8.5 inch lbs. (0.9 Nm). . .remember, it is plastic.

18. On all models now, take a look at the 2 upper handlebar brackets. The small punch mark on each should be facing forward (down); if not, flip them around. Move the bars back and forth until the marks you made during removal line up with the outer edges of the brackets. Tighten the bolts, alternately, to 14.8 ft. lbs. (20 Nm). Use a feeler gauge and check that you have a clearance of 0.06 in. (1.5mm) on the front (lower) edge and 0.14 in. (3.5mm) at the upper (rear) edge on 2006-08 models. On 2008 and later models, the figure is 0.02 in. (0.4mm) on the front (lower) edge and 0.09 in. (2.4mm) at the upper (rear) edge. If not, tighten or loosen the bolts in very small increments until the proper clearance is achieved.

The important thing is that the figure for the lower measurement is less than the figure of the upper measurement.

19. Install the throttle cable and adjust it as detailed in the Maintenance & Tune-Up section.

20. Move the steering pad into position, coat the bolt threads with Loctite 271, and then tighten to 0.8 ft. lbs./9.6 inch lbs. (1.1 Nm) on 2006-08 models,

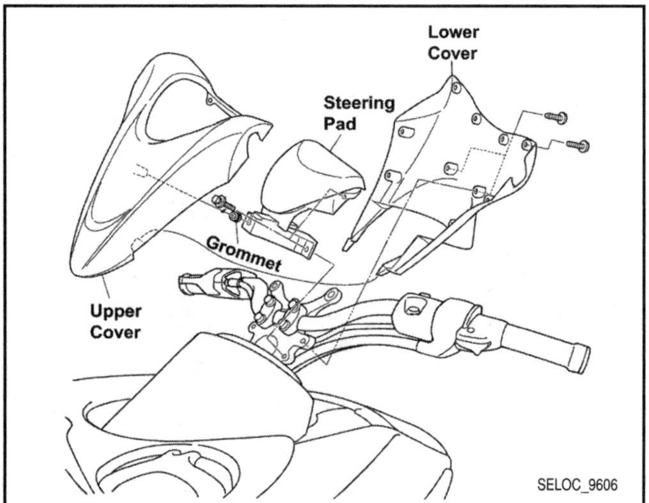

Fig. 7 These models use a 2-piece handlebar cover - 2006-08 FX Cruiser & 2006-11 FX Cruiser HO/SHO Models

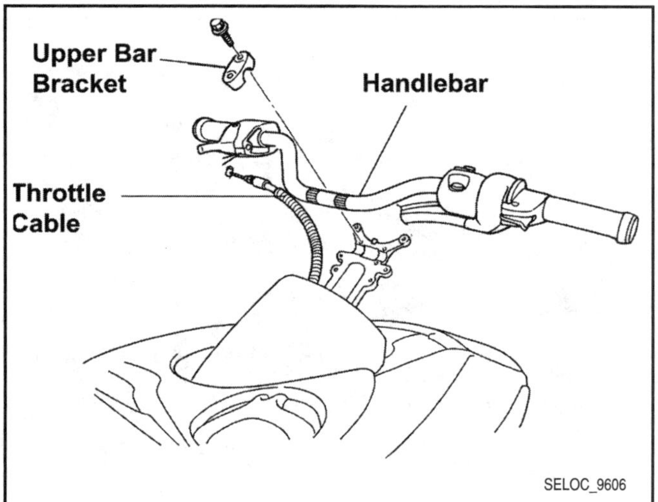

Fig. 8 Disconnecting the handlebars is easy. . .

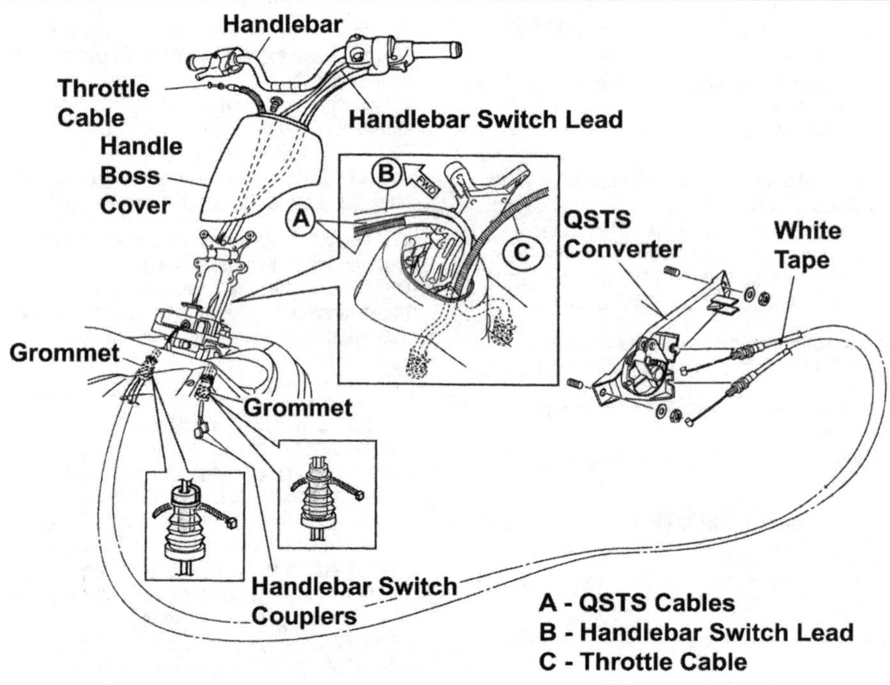

Fig. 9 . . .it's the removal part that slows things up - 2006-07 models shown

or 0.7 ft. lbs./8.4 inch lbs. (1 Nm) on 2009 and later. Press in the grommet if you removed it.

21. Position the lower cover on the steering assembly, coat the bolt (black on 1052cc models) threads with Loctite 242 and tighten to 3.9 ft. lbs./47 inch lbs. (5.4 Nm) on 2006-08 models, or 3.7 ft. lbs./44 inch lbs. (5 Nm) on 2009 and later models.

22. Move the upper cover into place so that the projection on the inner side of the cover lines up with the grommet and press it in until seated in the grommet. Make sure the bars are riding in the 'arm holes' of the covers correctly and then coat the remaining 8 bolt threads with Loctite 242. Tighten them, in a criss-cross pattern, to 0.8 ft. lbs./9.6 inch lbs. (1.1 Nm) on 2006-08 models, or 0.7 ft. lbs./8.4 inch lbs. (1 Nm) on 2009 and later models.

23. On all models again, adjust the QSTS cables as detailed in the Maintenance & Repair section.

24. Connect the battery cables and install the cargo container (if removed) and seat(s).

2008-11 FX HO/SHO & All FZ Models

MODERATE

◆ **See Figures 3 thru 6, 8, 10 and 11**

1. Remove the seat and cargo container (if equipped). Disconnect the battery cables.

2. On FX models, here are 6 bolts on the underside of the lower cover and 2 on the side. On FZ models, there 5 bolts on the upper underside of the lower cover. Remove these and lift off the upper cover and steering pad.

3. Remove the 3 remaining bolts (4 on the FZ) on the lower cover and lift it off.

4. Loosen the throttle cable locknut and thread it back until it makes contact with the adjusting nut. Squeeze the throttle a little until you are able to disconnect the cable from the grip control, making sure that you work small cable seal out as you remove the cable.

5. Remove the QSTS control grip, handlebar switch and throttle control grip assemblies as detailed in the Disassembly & Assembly section.

6. Take a Sharpie or other marker and mark handlebar at the outer edge of each handlebar bracket where it covers the handlebar. Loosen the 4 bolts (2 per bracket) and remove the 2 upper bar brackets. Remove the handlebars.

To install:

7. Install the 2 bar brackets temporarily to hold everything in place.

8. Install the QSTS control grip, handlebar switch and throttle control grip assemblies.

9. Now, take a look at the 2 upper handlebar brackets. The small punch mark on each should be facing forward (down); if not, flip them around. Move the bars back and forth until the marks you made during removal line up with the outer edges of the brackets. Tighten each bolt, a little at a time, to 14.8 ft. lbs. (20 Nm) using a criss-cross pattern across the 4 bolts holding the 2 upper brackets. On SHO models, the figure is 15.5 ft. lbs (21 Nm). Now, check the upper-to-lower bracket clearance at the upper and lower sides using a feeler gauge. You have a clearance of 0.02 in. (0.4mm) on the front (lower) edge and 0.09 in. (2.4mm) at the upper (rear) edge. If not, tighten or loosen the bolts in very small increments until the proper clearance is achieved.

The important thing is that the figure for the lower measurement is less than the figure of the upper measurement. Once the proper clearance has been achieved tighten the bolts

10. Install the throttle cable and adjust it as detailed in the Maintenance & Tune-Up section.

11. Move the lower cover into position, coat the 3 bolts with Loctite 242 and tighten them to 3.9 ft. lbs./47 inch lbs. (5.4 Nm) on models with 1052cc engines or 3.7 ft. lbs./45 inch lbs. (1 Nm) on models with 1812cc engines.

12. Install the steering pad and then move the upper cover into place, making sure the bars are riding in the 'arm holes' of the covers correctly. Coat the remaining bolts with Loctite 242 and tighten them, in a criss-cross pattern, to 0.8 ft. lbs./9.6 inch lbs. (1.1 Nm) on models with 1052cc engines or 0.7 ft. lbs./8.4 inch lbs. (1 Nm) on models with 1812cc engines.

13. Adjust the QSTS cables as detailed in the Maintenance & Repair section.

14. Connect the battery cables and install the cargo container (if removed) and seat(s).

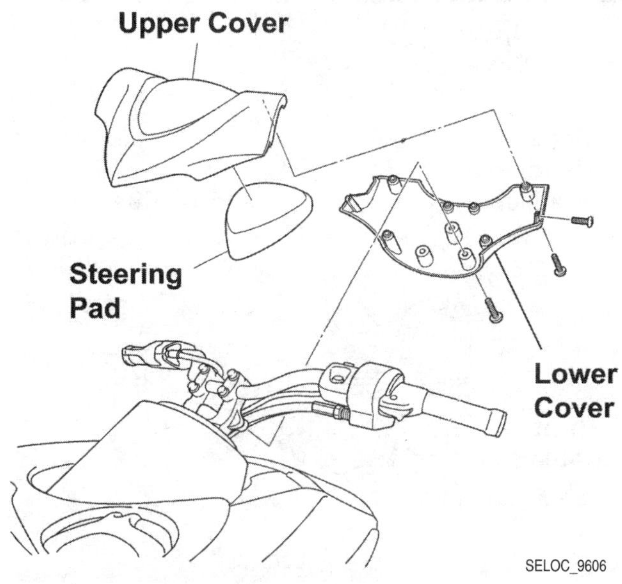

Fig. 10 2008-11 FX HO/SHO models use a different 2-piece cover. . .

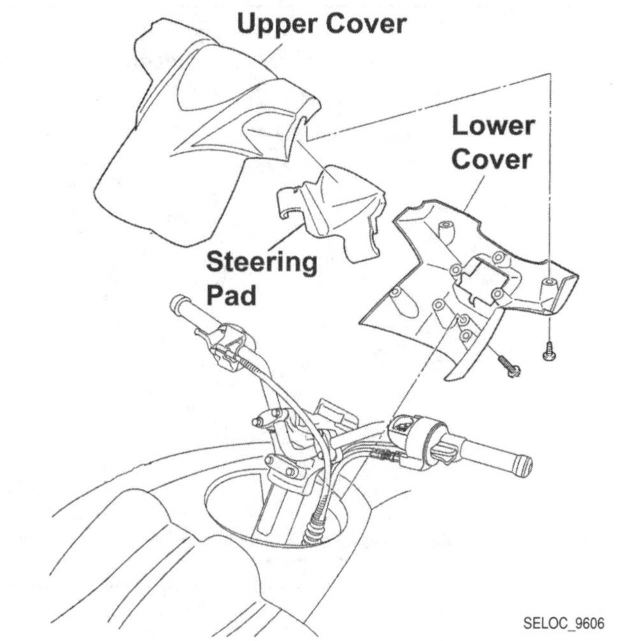

Fig. 11 . . .as do FZ models

All VX Models

MODERATE

◆ **See Figures 3, 4 and 12 thru 16**

1. Remove the seat and cargo container (if equipped). Disconnect the battery cables.

2. On 2005-09 models, remove the 8 bolts from the underside of the handlebar cover. Half of the bolts are slightly longer, so take note of which holes they came out of. On 2010-11 models, remove the 4 bolts.

3. Remove the 2 side bolts and lift off the upper handlebar cover and pull out the pad.

4. Loosen the throttle cable locknut and thread it back until it makes contact with the adjusting nut. Squeeze the throttle a little until you are able to disconnect the cable from the grip control, making sure that you work small cable seal out as you remove the cable.

5. Remove the left handlebar switch assembly as detailed in the Disassembly & Assembly section.

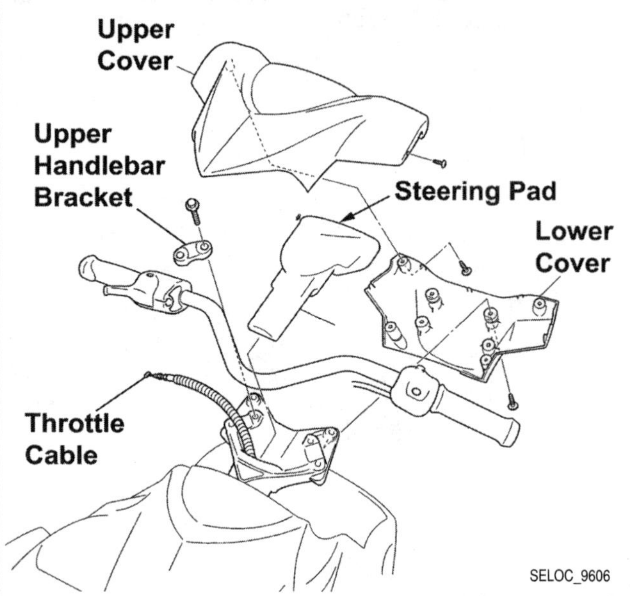

Fig. 12 VX models also use a 2-piece handlebar cover. . .

6. Take a Sharpie or other marker and mark the handlebar at the outer edge of each handlebar bracket where it covers the handlebar. Loosen the 4 bolts (2 per bracket) and remove the 2 upper bar brackets.

7. Grab the handlebars and remove them.

8. Remove the lower cover if necessary.

To install:

9. Move the lower cover into position. On 2010-11 models, coat the bolt threads with Loctite 242 and tighten them to 3 ft. lbs./36 inch lbs. (4 Nm).

10. Position the handlebars in the lower bracket and then install the 2 upper bar brackets. Take a look at the 2 upper handlebar brackets. The small punch mark on each should be facing forward (down); if not, flip them around. Move the bars back and forth until the marks you made during removal line up with the outer edges of the brackets. Tighten the bolts, alternately, to 14.8 ft. lbs. (20 Nm). Use a feeler gauge and check that you have a clearance of 0.06 in. (1.5mm) on the front (lower) edge and 0.14 in. (3.5mm) at the upper (rear) edge on 2005-09 models. On 2010-11 models, clearance should be 0.02 in. 0.4mm) on the bottom and 0.09 in. (1.1mm) on the top. If not, tighten or loosen the bolts in very small increments until the proper clearance is achieved.

The important thing is that the figure for the lower measurement is less than the figure of the upper measurement.

11. Install the handlebar switch assembly.

12. Install the throttle cable and adjust it as detailed in the Maintenance & Tune-Up section.

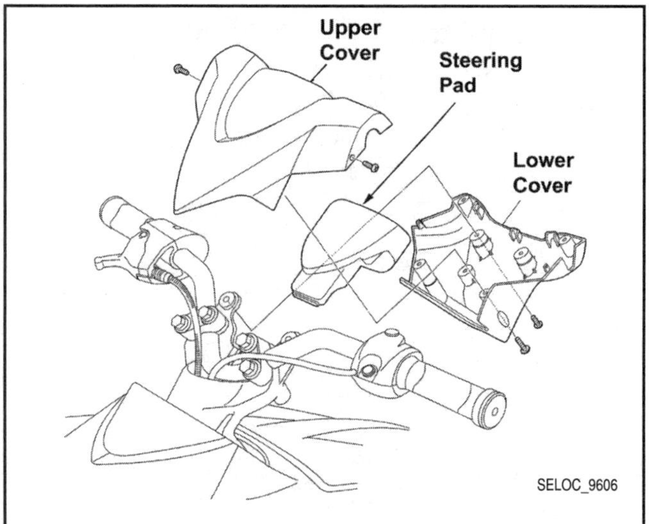

Fig. 13 . . .although the one on 2010-11 models is slightly different

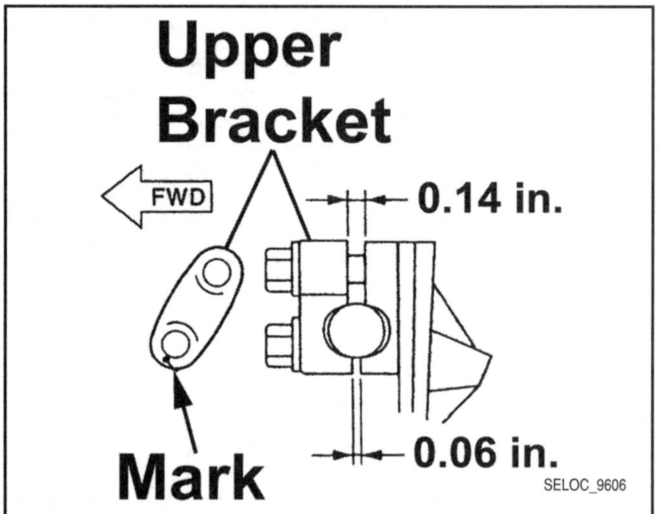

Fig. 14 Measure the upper bracket clearance

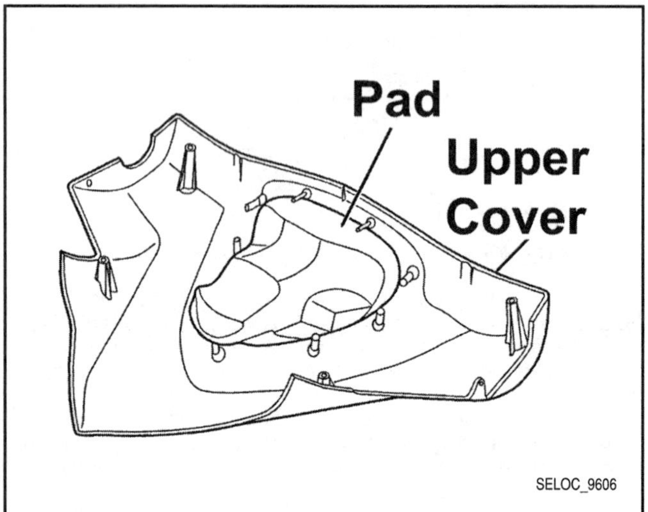

Fig. 15 Move the pad into the cover so it rests between the nibs. . .

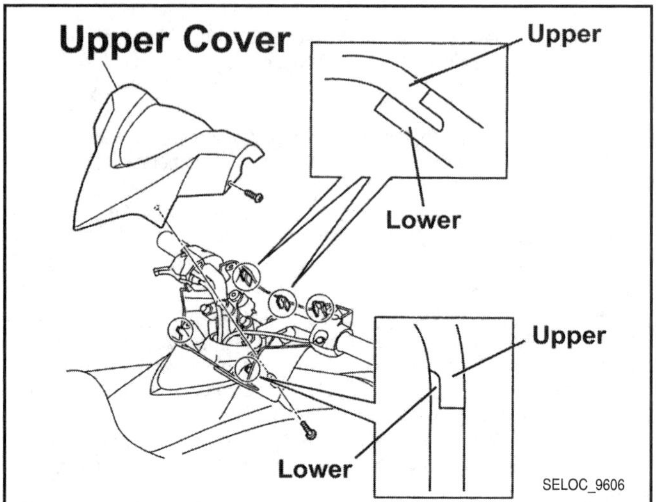

Fig. 16 . . .and then fit the nibs into the connectors - 2010-11 VX models

13. On 2005-09 models, install the pad and then move the upper cover into place. Make sure the bars are riding in the 'arm holes' of the covers correctly and then coat the bolt threads with Loctite 242. Tighten them, in a criss-cross pattern, to 2.7 ft. lbs./33 inch lbs. (3.7 Nm) on the 4 short ones and 0.8 ft. lbs./9.6 inch lbs. (1.1 Nm) on the 4 longer ones. Make sure that they all went back into their respective holes - you did mark where they went, didn't you??

14. On 2010-11 models, position the pad in the upper cover so it is resting between the nibs that mark it place in the cover. Now move both over the lower cover making sure that the nibs feed into the recesses on the lower cover. Tighten all bolts to 0.7 ft. lbs./8.4 inch lbs. (1 Nm).

15. Tighten the 2 side bolts to 0.8 ft. lbs./9.6 inch lbs. (1.1 Nm) on 2005-09 models, or 0.7 ft. lbs./8.4 inch lbs. (1 Nm) on 2010-11 models.

16. Connect the battery cables and install the cargo container (if removed) and seat(s).

DISASSEMBLY & ASSEMBLY

All FX & FZ Models

2 MODERATE

◆ See Figures 17, 18 and 19

1. Remove the bolt in the left handlebar and then use a small pick to pull out the grip end and spacer. There is a good chance you will need to work something thin and dull in between the grip and the bar to break it loose from the adhesive.

2. Remove the bolt from the bottom side of the QSTS control and slide the assembly off of the handlebar end. There is a special nut housed in a slot on the inner edge of the control that will fall out on most models, so cup your hand under the assembly as you slide it off.

3. Scratch a light mark on the handlebar right at the outer edge of the handlebar switch assembly (where it butts against the QSTS control). Loosen the 2 screws on the bottom side of the handlebar switch control (still working on the left bar end) until you can lift of the top half of the assembly while still holding the bottom half in place. Now remove the bottom half and take note of the wire routing in the case.

4. On the right side of the bars now, remove the bolts and pry out the grip end. Remove the grip. As mentioned in Step 2, you may have to 'persuade' the grip loose of the bar.

5. Squeeze the throttle lever to reveal the 2 throttle housing mounting bolts and remove them. Separate the front and rear halves.

To assemble:

6. Check all components for wear and other signs of damage.

7. Position the throttle housing halves around the bar and then thread in the bolts until they just seat.

8. Coat the inner circumference of the right grip with handlebar grip adhesive (commonly available at motorcycle and bike shops) and then slide it onto the bar until the ends are flush.

9. Coat the threads of the grip end bolt with Loctite 242. Slide it through the grip end and tighten the bolt to 0.9 ft. lbs./11 inch lbs. (1.2 Nm) on models with 1052cc engines or 0.7 ft. lbs./8.4 inch lbs. (1 Nm) on models with 1812cc engines.

10. Slide the throttle housing over until it butts against the inner end of the grip and then tighten the bolts to 2.2 ft. lbs./24.5 inch lbs. (3 Nm) on 998cc and 1812cc models or 2.5 ft. lbs./30 inch lbs. (3.4 Nm) on 1052cc models. Tighten the upper bolt first on FZ models.

11. Working on the other side of the bar now, snap the two halves of the handlebar switch assembly together over the bar and thread in the 2 screws until they just seat. Be very careful that the internal wires are routed correctly and not pinched. Move the assembly on the bar until it matches the mark you made on the bar and then tighten the bolts to 2.5 ft. lbs./30 inch lbs. (3.4 Nm) on 1052cc models or 2.2 ft. lbs./24.5 inch lbs. (3 Nm) on 1812cc models.

■ **On FZ models, there is a small positioning hole in the handlebar and a projection in the lower cover. Make sure this projection fits into the hole in the bar before tightening the bolts.**

12. Slide the QSTS assembly onto the bar about half way. Slide the special wedged nut into the slot on the inner side of the assembly and then continue installing the assembly until it butts against the handlebar switch housing. If you did everything correctly, the special nut (if used) did not fall out and the grip end is even with the bar end. Install the bolt and tighten to 2.5 ft. lbs./30 inch lbs. (3.4 Nm) or 2.2 ft. lbs./24.5 inch lbs. (3 Nm) on 1812cc models.

13. Install the spacer and grip end. Coat the bolt threads with Loctite 242 and then tighten to 0.9 ft. lbs./11 inch lbs. (1.2 Nm) on models with 1052cc engines or 0.7 ft. lbs./8.4 inch lbs. (1 Nm) on models with 1812cc engines.

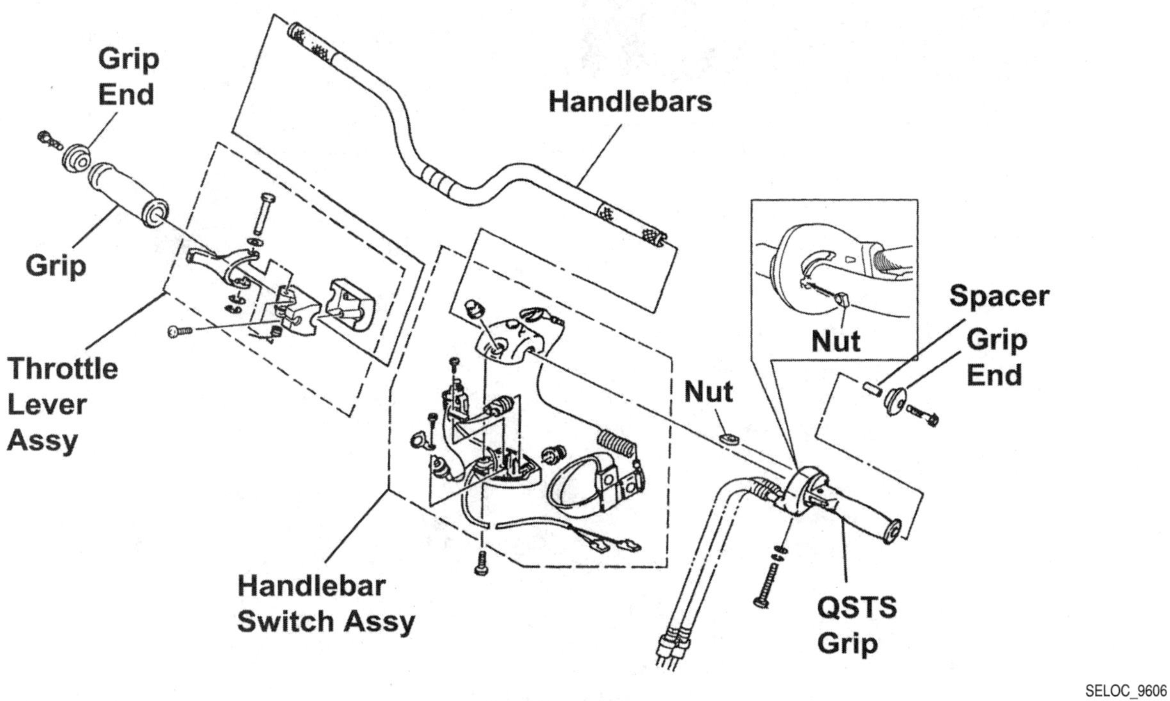

Fig. 17 Exploded view of the handlebar and components, all FX/FZ models similar

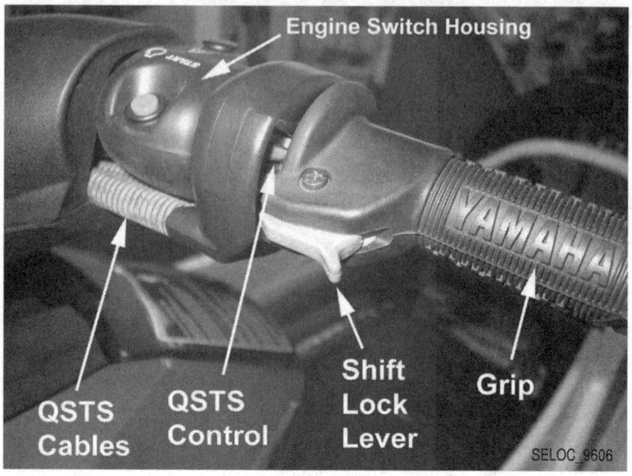

Fig. 18 A good look at the left side of the handlebar. . .

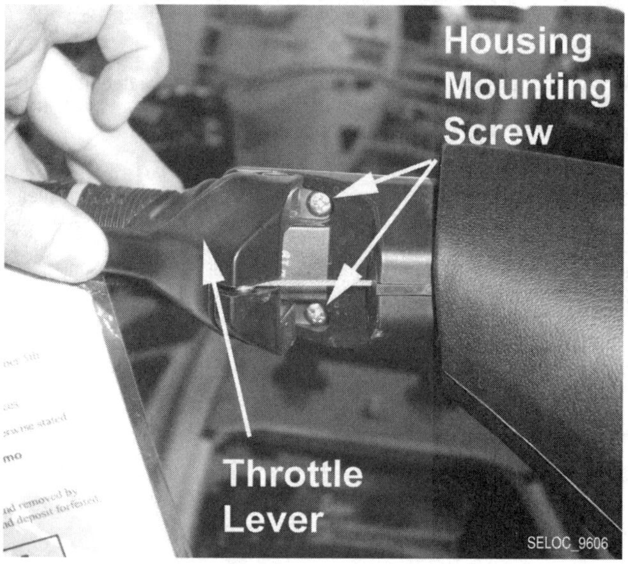

Fig. 19 . . .and the right side

All VX Models

② MODERATE

◆ See Figures 19 and 20

1. Remove the bolt in the left handlebar and then use a small pick to pull out the grip end. There is a good chance you will need to work something thin and dull in between the grip and the bar to break it loose from the adhesive.
2. Remove the end plate on 2005-09 models.
3. Scratch a light mark on the handlebar right at the outer edge of the handlebar switch assembly (where it butts against the end plate). Loosen the 2 screws on the bottom side of the handlebar switch control (still working on the left bar end) until you can lift of the top half of the assembly while still holding the bottom half in place. Now remove the bottom half and take note of the wire routing in the case.
4. On the right side of the bars now, remove the bolts and pry out the grip end. Remove the grip. As mentioned in Step 2, you may have to 'persuade' the grip loose of the bar.
5. Remove the end plate on 2005-09 models.
6. On 2005-09 models, squeeze the throttle lever to reveal the 2 throttle housing mounting bolts and remove them. Separate the front and rear halves. On 2010-11 models, the screws come in from the back side.

To assemble:

7. Check all components for wear and other signs of damage.
8. Position the throttle housing halves around the bar and then thread in the bolts until they just seat.
9. Install the end plate on 2005-09 models.
10. Coat the inner circumference of the right grip with handlebar grip adhesive (commonly available at motorcycle and bike shops) and then slide it onto the bar until the ends are flush.
11. Coat the threads of the grip end bolt with Loctite 242. Slide it through the grip end and tighten the bolt to 0.9 ft. lbs./11 inch lbs. (1.2 Nm) on 2005-09 models or 0.7 ft. lbs./8.4 inch lbs. (1.2 Nm) on 2010-11 models.
12. Slide the throttle housing over until it butts against the inner end of the grip and then tighten the bolts to 2.5 ft. lbs./30 inch lbs. (3.4 Nm) on all models but the 2010-11, where the figure is 2.2 ft. lbs./27 inch lbs. (3 Nm).
13. Working on the other side of the bar now, snap the two halves of the handlebar switch assembly together over the bar and thread in the 2 screws until they just seat. Be very careful that the internal wires are routed correctly and not pinched. Move the assembly on the bar until it matches the mark you made on the bar and then tighten the bolts to 2.5 ft. lbs./30 inch lbs. (3.4 Nm) on all models but the 2010-11, where the figure is 2.2 ft. lbs./27 inch lbs. (3 Nm).
14. Install the end plate on models so equipped.
15. Install the grip and grip end as you did with the right side.

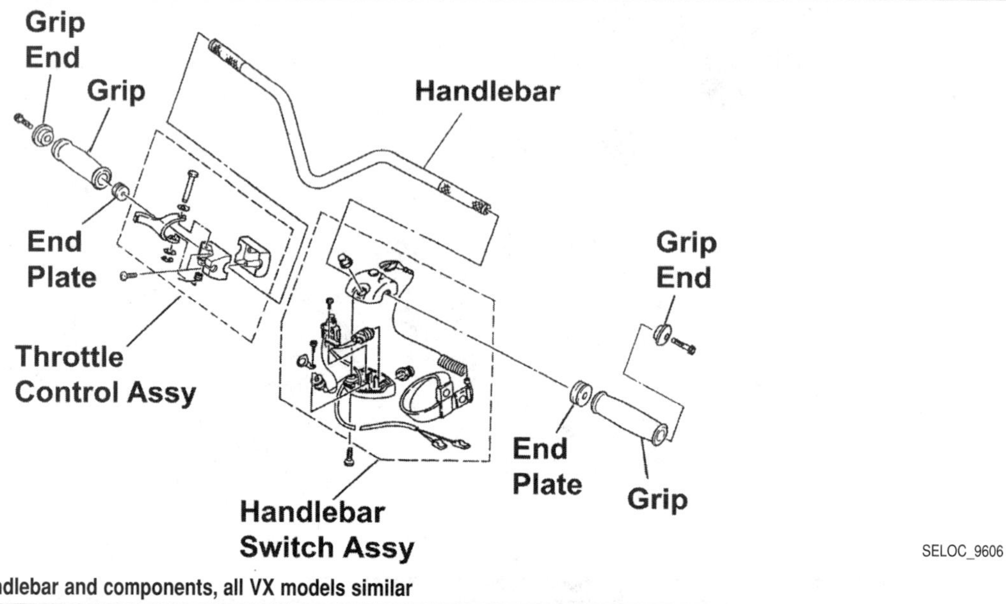

Fig. 20 Exploded view of the handlebar and components, all VX models similar

Steering Assembly

REMOVAL & INSTALLATION

2002-05 FX Models

MODERATE

◆ **See Figures 21 and 22**

1. Remove the seat and disconnect the battery cables.
2. Remove the handlebars as previously detailed.
3. Open the hood and remove the 4 bolts securing the service lid.
4. Reach in and cut the plastic tie securing the harness. Tag and disconnect the multifunction meter coupler.
5. Remove the 4 bolts and lift out the meter back panel.
6. Reach in and cut the plastic tie securing the packing material inside the meter harness grommet.
7. Remove the 4 bolts and lift out the multifunction meter.
8. Remove the 4 steering console cover bolts. Remove the 2 bolts securing the shift handle lever and lift out the handle. Now remove the 2 remaining bolts securing the cover to the glove box and remove the cover.
9. Pull back on the steering cable end and pop it off of the steering lever pivot.
10. Remove the bolt (and washer) and take out the steering lever.

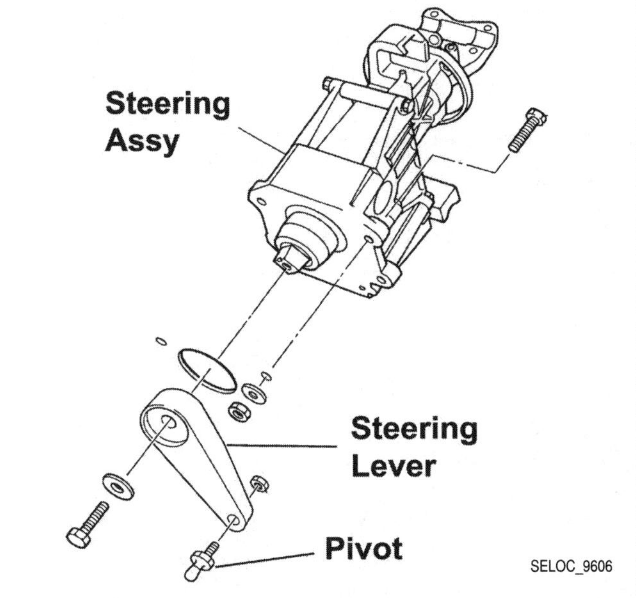

Fig. 21 A look at the steering assembly - 2002-05 998cc models

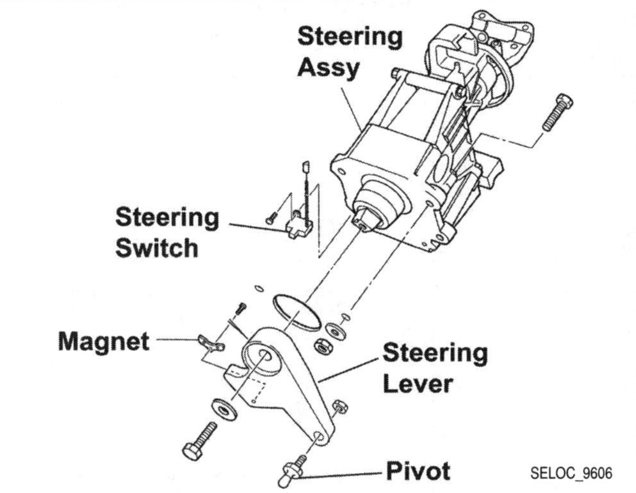

Fig. 22 A look at the steering assembly - 2004-05 FX models w/1052cc engines

11. Now remove the 4 nuts and washers and lift out the steering assembly from the top. You might want to have someone hold the assembly as you remove the nuts.
12. On 1052cc models, and only if necessary, remove the small bolt and lift the magnet off of the steering lever.
13. Again, on 1052cc models, and only if necessary, remove the 2 small bolts and lift off the steering switch.

To install:

14. If you removed the magnet or steering switch on 1052cc models, install then and tighten the bolts to 2.1 ft. lbs./25 inch lbs. (2.8 Nm).
15. Position the steering assembly so the holes line up with those in the hull and the end of the lower shaft is through the hole. Insert the bolts from the top and then tighten the nuts to 14 ft. lbs. (20 Nm) in a criss-cross pattern.
16. Install the steering lever so that the small nib in the lever recess lines up with the notch in the lower steering shaft. Coat the bolt threads with Loctite 271 and then tighten the bolt to 13 ft. lbs. (18 Nm) while having someone hold steering column from turning.
17. Connect the steering cable to the lever and remember too check the adjustment after you complete the installation.
18. Move the console cover into position, coat the threads of the 2 rear bolts with Loctite 242 and then tighten to 3.9 ft. lbs./47 inch lbs. (5.4 Nm).
19. Install the shift lever handle. Coat the bolt threads with Loctite 271 and then tighten the 2 bolts to 3.9 ft. lbs./47 inch lbs. (5.4 Nm).
20. Install the remaining 4 bolts and tighten the nuts to 3.9 ft. lbs./47 inch lbs. (5.4 Nm). Don't forget the washers.
21. Feed the meter harness through the grommet until the meter is in position. Coat the bolt threads with Loctite 242 and then tighten them to 2.8 ft. lbs./33 inch lbs. (3.9 Nm). Pull a new plastic tie around the harness packing, just below the bottom of the grommet.
22. Coat the meter panel bolts with Loctite 242, install them and tighten to 2.8 ft. lbs./33 inch lbs. (3.9 Nm) using a criss-cross pattern.
23. Reconnect the meter lead and then install a new plastic tie. Remember that the other couplers will not be installed until the handlebars are back on.
24. Install the service lid. Coat the bolts with Loctite 271 and then tighten them, in a criss-cross pattern, to 3.9 ft. lbs./47 inch lbs. (5.4 Nm).
25. Close the hood and install the handlebars.

2006-11 FX Models

MODERATE

◆ **See Figures 23 thru 29**

1. Remove the seat and disconnect the battery cables.
2. Remove the handlebars as previously detailed.
3. Remove the hood as detailed previously in this section.

■ **The next 4 steps are for 1052cc models only.**

4. Reach into the hull and remove the forward bolt on the starboard side mirror assembly.
5. The center bolt comes from the top, remove the nut and washer from inside the hull.
6. Remove the rear bolt from the top and lift off the mirror assembly. There is no nut on this one.
7. Repeat Steps 4-6 for the port mirror.

■ **You may wish to remove the steering console cover for easier access.**

8. Remove the 4 rivets (5 on 1812cc models) securing the service lid. To remove, push in the rivet pin in until just below the surface of the rivet and it clicks.
9. Reach in and cut the plastic tie securing the harness. Tag and disconnect the multifunction meter coupler (1052cc engines).
10. Remove the 4 bolts and lift out the meter back panel (1052cc engines).
11. Pull back on the steering cable end and pop it off of the steering lever pivot.
12. Remove the bolt (and washer) and take out the steering lever. Remove the pivot if necessary. Instead of a washer, 2008-11 models utilize a small retaining plate in its place.
13. Tag and disconnect the couplers at the steering sensor and warning buzzer.
14. Loosen the pinch bolt and pull the cable stopper off of the shaft.
15. Now remove the 4 nuts and washers and lift out the steering assembly from the top. You might want to have someone hold the assembly as you remove the nuts.

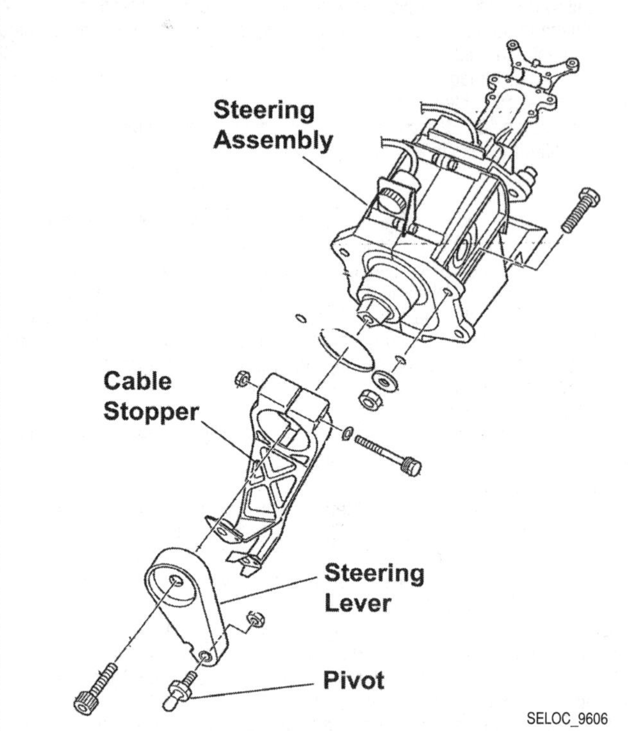

Fig. 23 A look at the steering assembly - 2006-11 FX models (2008 and later FX HO/SHO models utilize a retaining plate on the steering lever bolt)

To install:

16. Position the steering assembly so the holes line up with those in the hull and the end of the lower shaft is through the hole. Insert the bolts from the top and then tighten the nuts to 14.8 ft. lbs. (20 Nm) in a criss-cross pattern. On SHO models the figure is 15.5 ft. lbs. (21 Nm).

17. There is a small rib on the back of the steering shaft; install the cable stopper so that the groove in the bore fits over this rib and then tighten the pinch bolt to 5.2 ft. lbs./63 ft. lbs. (7 Nm).

18. Install the steering lever as shown in the illustration (rounded corner of the square shaft faces bow). Coat the bolt threads with Loctite 271 and then tighten the bolt to 13 ft. lbs. (18 Nm) while having someone hold steering column from turning. On 2008 and later FX HO/SHO models only, position the retainer plate so the large tabs are resting against the edge of the cut-out in the lever and then tighten the bolt to:

- 2008 FX HO w/1052cc engines -11 ft. lbs. (15 Nm)
- 2008-11 FX HO/SHO w/1812cc engines - 11.8 ft. lbs. (16 Nm)

Once tight, tweak the bolt slightly so you can bend one or more of the small tabs up and against the flat(s) on the bolt head.

19. If you removed the pivot from the steering lever, tighten the nut to 5.2 ft. lbs./63 ft. lbs. (7 Nm).

20. Reconnect the couplers for the sensor and buzzer.

21. Connect the steering cable to the lever and remember to check the adjustment after you complete the installation.

22. Coat the meter panel bolts with Loctite 242, install them and tighten to 2.8 ft. lbs./33 inch lbs. (3.9 Nm) using a criss-cross pattern (1052cc engines).

23. Reconnect the meter lead and then install a new plastic tie. Remember that the other couplers will not be installed until the handlebars are back on (1052cc engines).

24. Install the service lid. Hold the flange of the rivet with your fingers and then press the end of the pin against a hard surface until the pins comes out the top of the flange. Now you can insert the rivet through the holes and press the pin in until it clicks and is flush with the flange.

25. Install the hood and install the handlebars.

26. Connect the battery cables and install the seat.

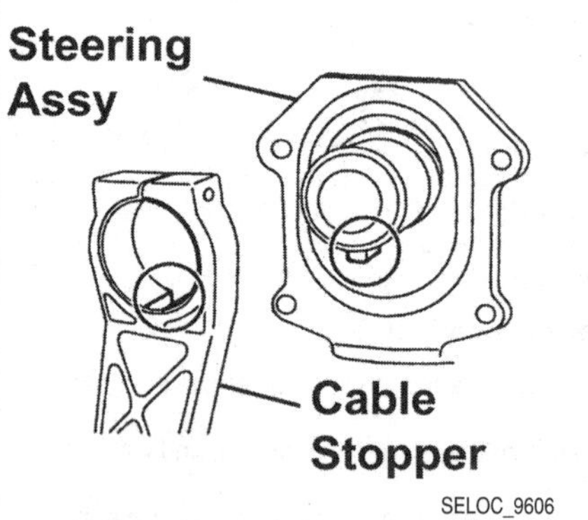

Fig. 24 Make sure that the rib on the shaft fits into the groove in the cable stopper

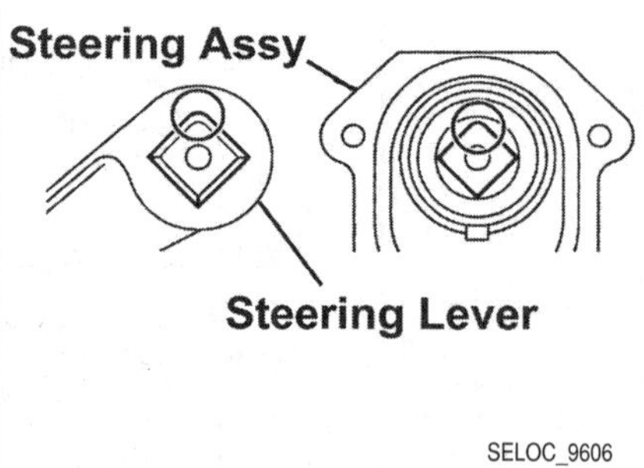

Fig. 25 Installing the steering lever

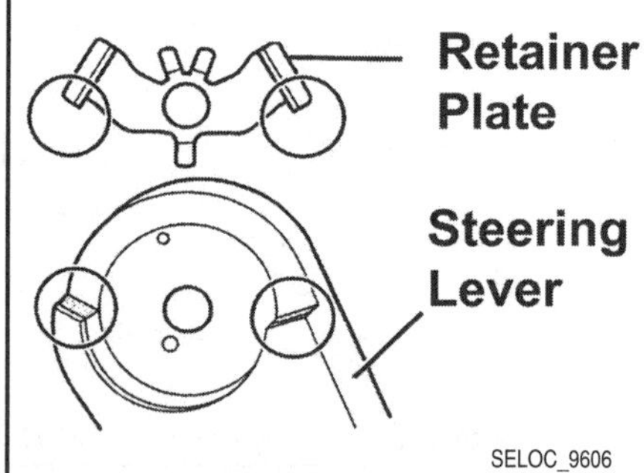

Fig. 26 On 2008-11 FX HO/SHO models, bend one or more of the *small* tabs over a flat on the bolt head

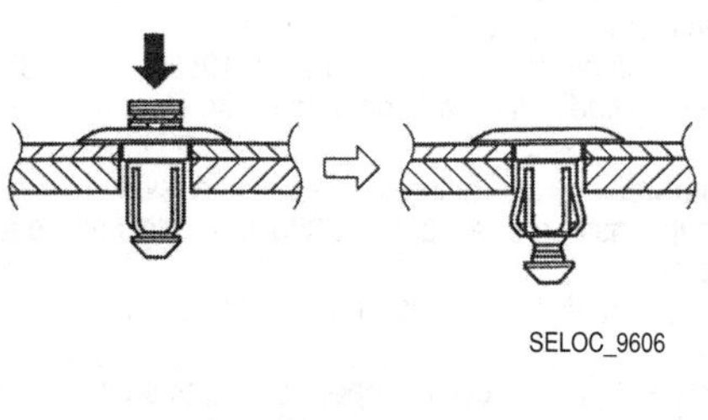

Fig. 27 Insert the rivet and then press in the pin until it clicks

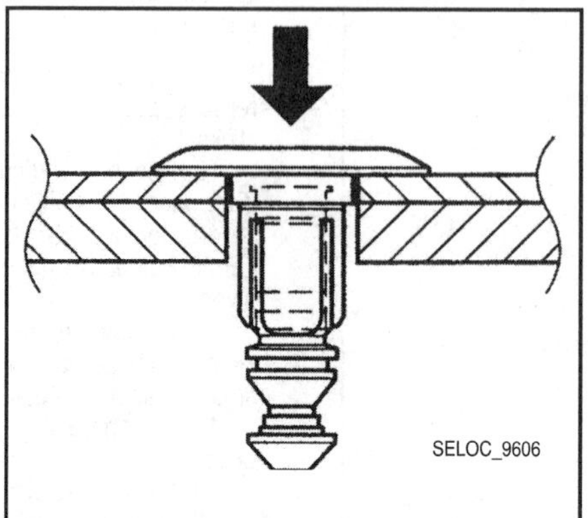

Fig. 28 When removing, press the pin in below the flange until it clicks again

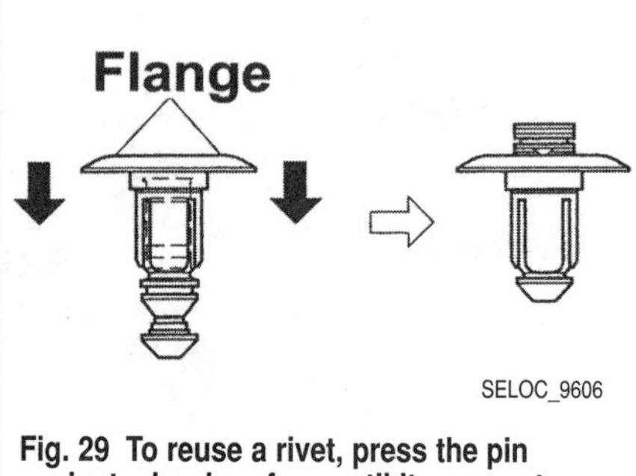

Fig. 29 To reuse a rivet, press the pin against a hard surface until it pops out above the flange

FZ Models

MODERATE

◆ See Figures 27 thru 35

1. Remove the seat and disconnect the battery cables.
2. Remove the handlebars as previously detailed.
3. Remove the hood as detailed previously in this section.
4. Remove the 5 rivets securing the rear service lid. To remove, push in the rivet pin in until just below the surface of the rivet and it clicks.
5. Reach in and cut the plastic tie securing the harness. Tag and disconnect the multifunction meter couplers.
6. Remove the 2 rivets and lift out the meter back panel (just above the service lid). Disconnect the coupler.
7. Remove the 2 screws and pull off the meter cover.
8. Remove the 4 bolts (they come from inside the hull) and take out the meter.

■ We think you ought to be able to remove the assembly without removing the side and center covers, but have included the steps here just in case. Take a look and decide for yourself. If you decide to give it a try, skip to Step 13.

9. Remove the 2 bolts and pull out the right side shift grip/handle. Remove the drink holder.
10. Remove the 9 bolts (one is inside the grip recess) and take off the side covers.
11. Remove the 2 induction boxes.
12. Remove the 6 nuts and 5 bolts and then lift off the center cover.
13. Pull back on the steering cable end and pop it off of the steering lever pivot.
14. Tag and disconnect the couplers at the steering sensor and warning buzzer.
15. Now remove the 4 bolts and lift out the steering assembly from the top. You might want to have someone hold the assembly as you remove the nuts.
16. Remove the large grommet.

To install:

17. Position the steering assembly so the holes line up with those in the hull. Insert the bolts from the top and then tighten to 14.8 ft. lbs. (20 Nm) in a criss-cross pattern.
18. Reconnect the couplers for the sensor and buzzer.
19. Connect the steering cable to the lever and remember to check the adjustment after you complete the installation.

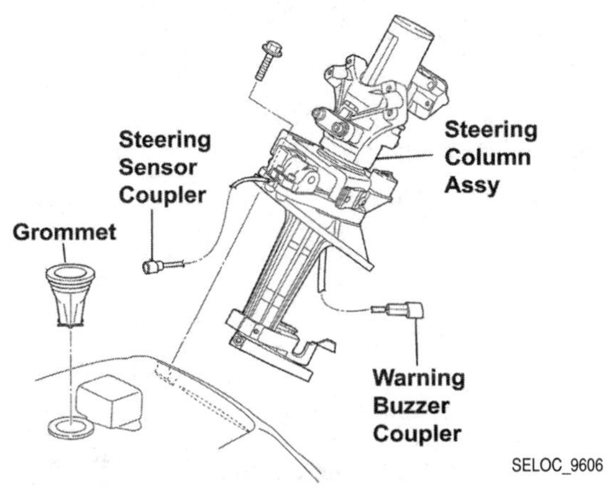

Fig. 30 A look at the steering assembly - FZ models

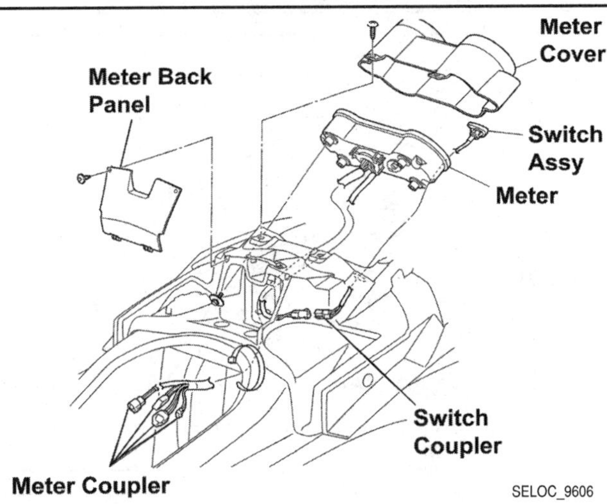

Fig. 31 Removing the meter

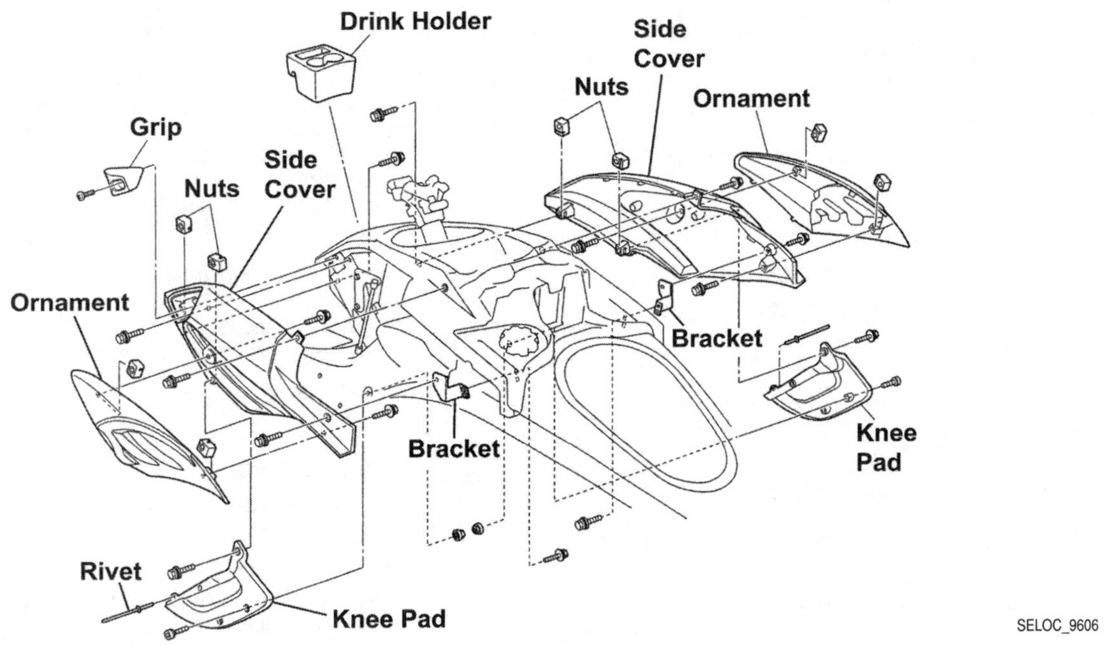

Fig. 32 Removing the side covers

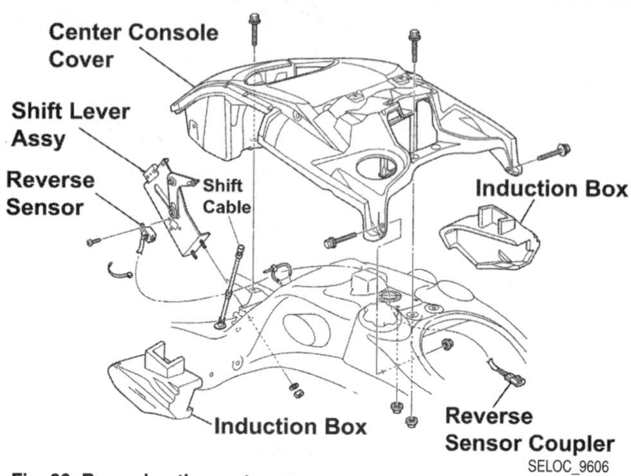

Fig. 33 Removing the center cover

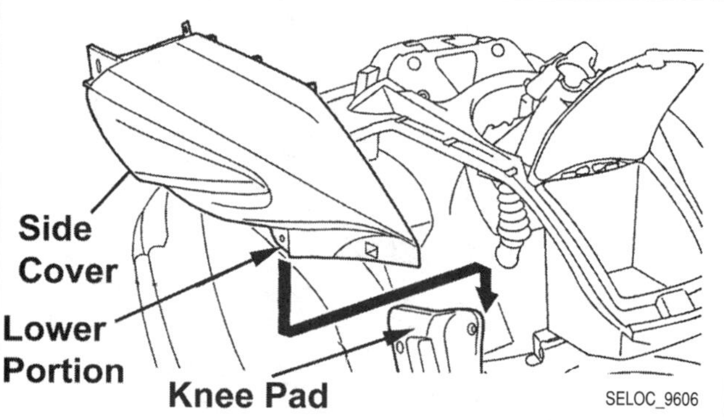

Fig. 34 When installing the side cover, fit the lower portion between the hull and knee pad. . .

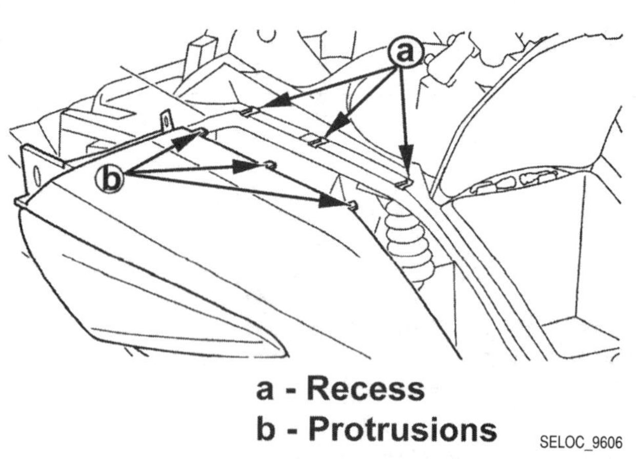

Fig. 35 . . .and the tabs on the cover edge into the recesses on the center cover

20. Install the center cover and tighten all bolts and nuts to 3.7 ft. lbs./44 inch lbs. (5 Nm).
21. Install the side covers and tighten all bolts to 3.7 ft. lbs./44 inch lbs. (5 Nm).
22. Install the grip and tighten the screws to 3.7 ft. lbs./44 inch lbs. (5 Nm).
23. Install the meter and tighten the bolts to 3 ft. lbs./36 inch lbs. (4 Nm). Install the switch and then connect the coupler.
24. Slide on the meter cover and tighten the screws to 1.5 ft. lbs./18 inch lbs. (2 Nm).
25. Reconnect the couplers and install a new plastic tie.
26. Install meter panel. Hold the flange of the rivet with your fingers and then press the end of the pin against a hard surface until the pins comes out the top of the flange. Now you can insert the rivet through the holes and press the pin in until it clicks and is flush with the flange.
27. Install the service lid. Hold the flange of the rivet with your fingers and then press the end of the pin against a hard surface until the pins comes out the top of the flange. Now you can insert the rivet through the holes and press the pin in until it clicks and is flush with the flange.
28. Install the hood and handlebars.
29. Connect the battery cables and install the seat.

2005-09 VX Models

(2) MODERATE

◆ **See Figures 36, 37 and 38**

1. Remove the seat and disconnect the battery cables.
2. Remove the handlebars as previously detailed.
3. Open the hood.

■ **Removal of the rear service lid may make provide easier access to some bolts in the following steps.**

4. Remove the plastic tie and pull out the grommet.
5. Tag and disconnect the steering sensor and warning buzzer at their couplers.
6. Remove the 4 forward engine hatch mounting bolts with their nuts and washers.
7. Remove the 4 rear mounting bolts - 2 are on the outside and 2 are underneath. Deluxe models will also have an additional bolt in the back.
8. Also on Deluxe models, pop off the damper stopper at the shift handle. Remove the 2 bolts and lift out the shift lever.
9. Lift off the cover.
10. Pull back on the steering cable lock casing and remove the cable from the steering lever pivot.
11. Remove the bolt and take out the steering lever. If necessary, remove the pivot (the nut is welded on the lever).
12. Remove the 4 mounting bolts and lift out the steering column assembly.

To install:

13. Position the steering column assembly so the holes line up with those in the plate. Coat the bolt threads with Loctite 572, insert the bolts from the top and then tighten to 12 ft. lbs. (17 Nm) in a criss-cross pattern.
14. If removed, install the steering lever pivot and tighten to 5 ft. lbs./60 inch lbs. (6.9 Nm).
15. Install the steering lever so the hole tang in the lever fits into the recess in the shaft and the holes line up. Coat the bolt threads with Loctite 242 and tighten to 14 ft. lbs. (20 Nm).
16. Reconnect the sensor and buzzer couplers.
17. Install the grommet and secure with a plastic tie.
18. Reconnect the steering cable and remember to check the adjustment (as detailed in Maintenance & Tune-Up) after everything is back together.
19. On Deluxe models, coat the shift lever bolt threads with Loctite 572 and then tighten to 2.7 ft. lbs./33 inch lbs. (3.7 Nm). Press the damper stopper into the recess in the handle from the bottom and then slide it up and into position.
20. Install the engine hatch cover and tighten all bolts and nuts to 5 ft. lbs./60 inch lbs. (6.9 Nm). Don't forget the rear bolt on Deluxe models.
21. Install the handlebars.
22. Connect the battery cables and install the seat.

2010-11 VX Models

(2) MODERATE

◆ **See Figures 39 and 40**

1. Remove the seat and disconnect the battery cables.
2. Remove the handlebars as previously detailed.
3. Open the hood.

■ **Removal of the rear service lid may make provide easier access to some bolts in the following steps.**

4. Pull out the cup holder on Cruiser and Deluxe models. Lift out the glove box.

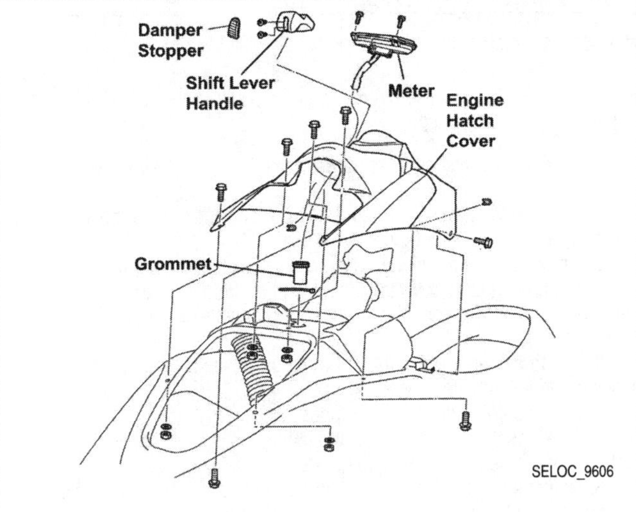

Fig. 36 Removing the engine hatch cover - 2005-09 VX models

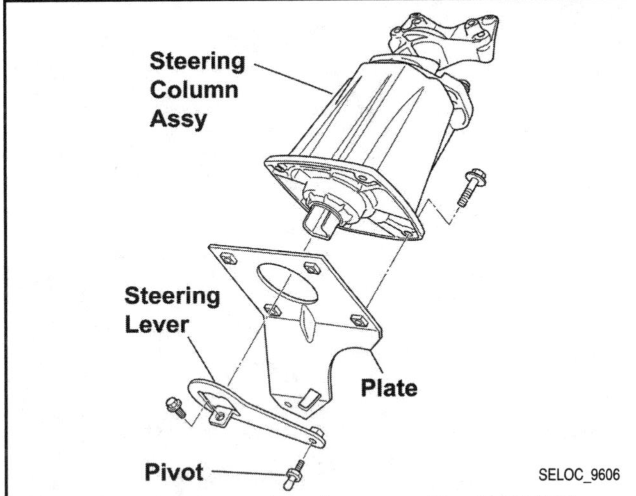

Fig. 37 Removing the steering column assembly - 2005-09 VX models

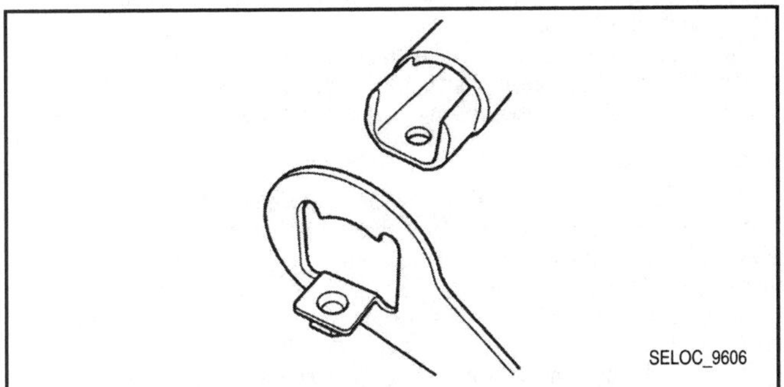

Fig. 38 Installing the steering lever is a little different on these models

5. Pop out the 2 hinge pins and lift off the center console box cover.
6. Remove the 2 damper bolts, the damper and then the shift lever grip.
7. Remove the plastic tie and pull out the grommet.
8. Tag and disconnect the steering sensor and warning buzzer at their couplers.
9. Remove the 14 bolts and 6 nuts and then lift off the engine hatch cover.
10. Pull back on the steering cable lock casing and remove the cable from the shift lever pivot. Remove the bolt and take out the cable stopper.
11. Remove the 4 mounting bolts and lift out the steering column assembly.

■ **If cable stopper bracket removal is necessary, it is held in place by 2 rivets.**

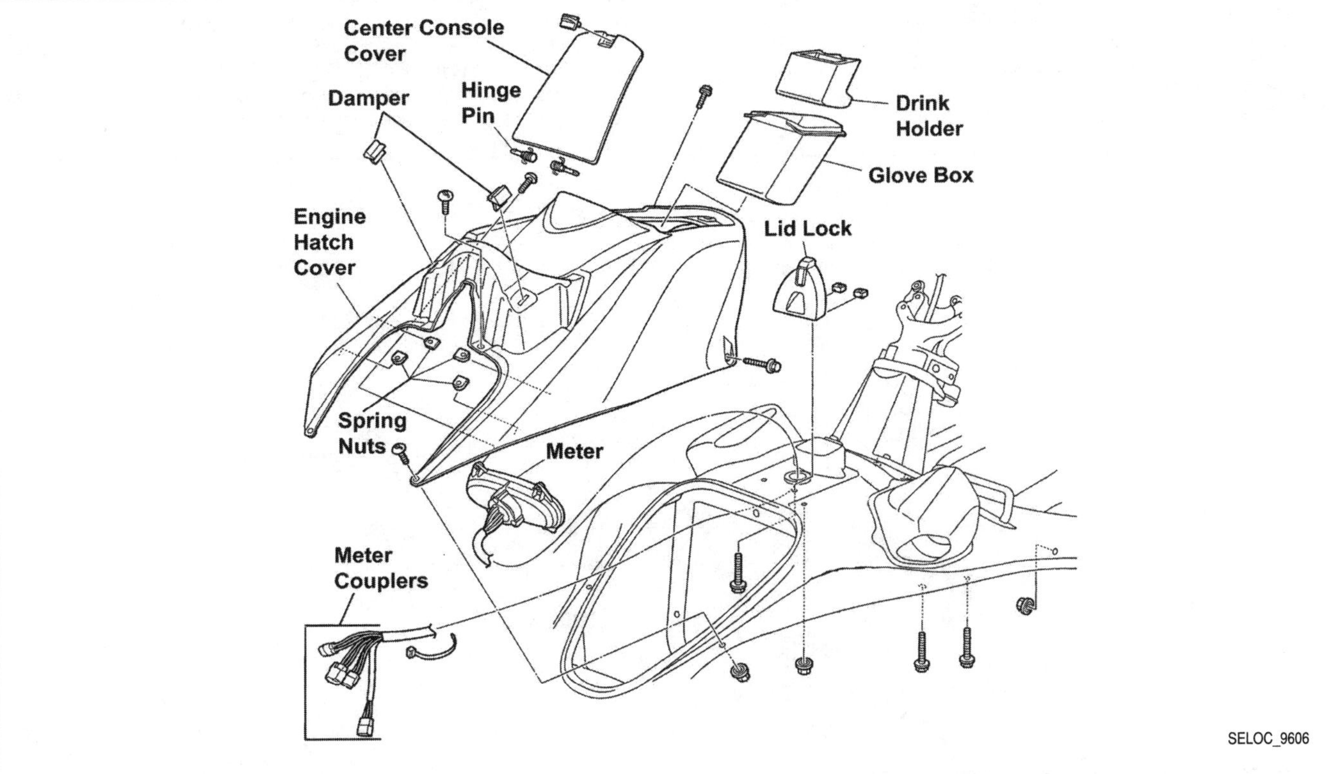

Fig. 39 Removing the engine hatch cover - 2010-11 VX models

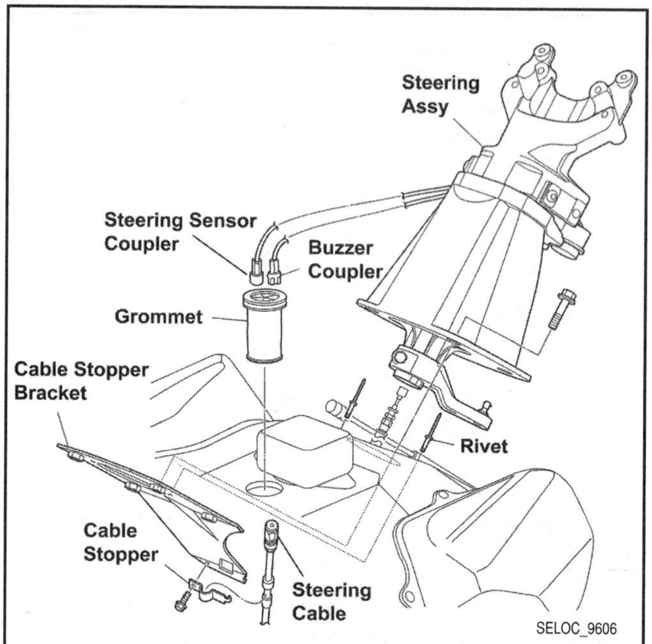

Fig. 40 Removing the steering column assembly - 2010-11VX models

To install:

12. Position the steering column assembly so the holes line up with those in the plate. Coat the bolt threads with Loctite 572, insert the bolts from the top and then tighten to 12.5 ft. lbs. (17 Nm) in a criss-cross pattern.
13. Reconnect the sensor and buzzer couplers.
14. Install the grommet and secure with a plastic tie.
15. Reconnect the steering cable and remember to check the adjustment (as detailed in Maintenance & Tune-Up) after everything is back together.
16. Install the engine hatch cover and tighten all bolts and nuts to 3.7 ft. lbs./45 inch lbs. (5 Nm).
17. Install the grip and damper, tightening the bolts to 3 ft. lbs./36 inch lbs. (4 Nm) after coating the threads with Loctite 572.
18. Install the console cover.
19. Install the glove box and cup holder on models so equipped.
20. Install the handlebars.
21. Connect the battery cables and install the seat.

DISASSEMBLY & ASSEMBLY

2002-05 FX Models

3 DIFFICULT

◆ **See Figures 41 and 42**

1. Remove the assembly as detailed previously and lay it on a clean, flat surface.
2. Remove the 2 bolts holding the upper housing stay at the front of the assembly and then pull out the stay.
3. Remove the 6 mounting bolts/washers and lift off the upper housing.
4. Reach in and remove the smaller spring from its seat in the back of the lower housing. Remove the larger spring from its seat in the bottom of the lower housing.
5. Pull the steering shaft and tube out of the bushings in the lower housing and set it aside.
6. Remove the tilt lever and stopper.
7. If necessary, pull the upper and lower bushings out of the bore in the bottom of the lower housing.
8. Remove the 4 shaft-to-tube retainer bolts and then pull out the nuts and holders. Pull out the 2 retainers.
9. Pull the upper shaft out of the steering tube. Reach in and pull the cross piece out of the top of the lower shaft - if tight, wait until you remove the cap in the next step.
10. Pull the lower shaft out of the steering tube while pushing from the bottom. Remove the 2 lower shaft cap bolts and separate the lower shaft from the cap. Each bolt uses a conventional washer and a spring washer.

To assemble:

11. Inspect all components for wear or other signs of damage.
12. Coat the lower shaft cap bolt threads with Loctite 271 and slide on the 2 washers. Mount the cap on the shaft, thread in the bolts and tighten them to 4.9 ft. lbs./59 inch lbs. (6.8 Nm). Coat the cross piece with a good marine grade water resistant grease and slide it into position between the cap and shaft.
13. Coat the bottom (flat portion of shaft) with grease and then insert the shaft through the hole in the bottom of the steering tube until it seats on the bottom and the square portion protrudes.
14. Position the upper shaft so the handlebar retainer recess is facing forward and insert into the top of the lower shaft and over the cross piece.
15. Insert the retainer nuts into the holders and then position them in their slots. Install the retainer halves around the shaft and onto the steering tube flange. Install the bolts and tighten them to 4.9 ft. lbs./59 inch lbs. (6.8 Nm).
16. If you removed the lower housing bushings, coat their lips with grease and then press them into place from the top and bottom.

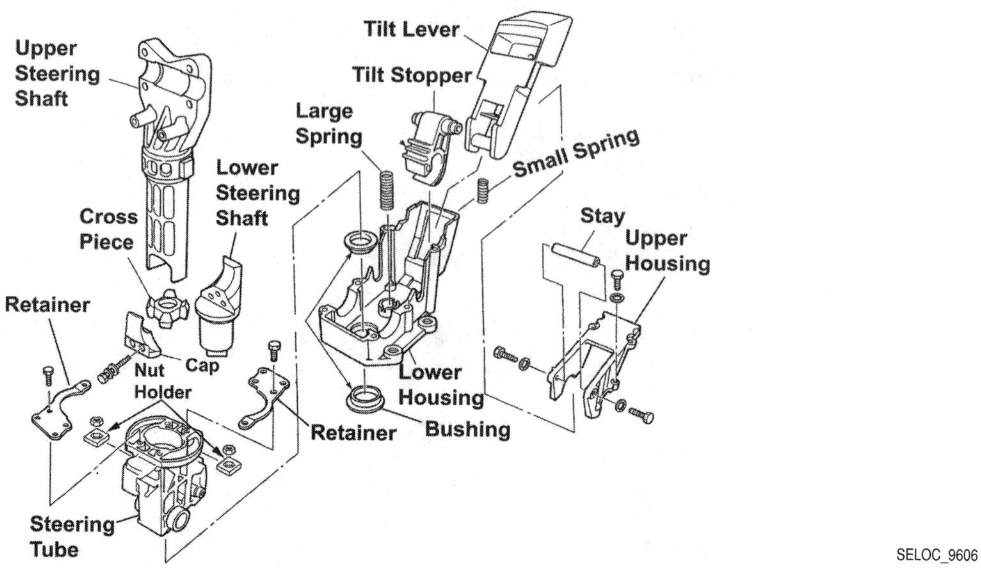

Fig. 41 Exploded view of the steering assembly - 2002-05 FX models

Fig. 42 The tilt lever is part of the steering assembly

17. Position the large spring back into the recess in the lower housing after coating it with grease and then install the tube and shaft assembly into the housing making sure that it seats fully and correctly in the cut-outs on each side.

18. Coat the teeth of the tilt stop with grease and install it into the case so it is sitting in the cut-outs on each side of the housing.

19. Insert the tilt lever into the housing so that it is resting in the cut-out and then install the smaller spring.

20. Install the upper housing over the lower and around the steering shafts. Tighten the bolts to 4.9 ft. lbs./59 inch lbs. (6.8 Nm).

21. Move the front stay into position between the sides of the upper housing and then tighten the bolts to 4.9 ft. lbs./59 inch lbs. (6.8 Nm).

22. Install the assembly to the hull.

2006-08 FX/Cruiser & 2006-07 FX HO/Cruiser HO Models

◆ **See Figures 42 thru 45**

1. Remove the assembly as detailed previously and lay it on a clean, flat surface.

2. Grab the steering shaft and pull it out of the steering tube. The upper bushing may come off with it, but if not, remove it now. Do the same with the lower bushing.

3. Unscrew the threaded cap and lift the warning buzzer out of its holder. Remove the bolt and lift of the holder if necessary, otherwise you can do this when separating the housings.

4. Remove the 4 steering sensor mounting bolts. Remove the bolt securing the starboard bracket and lift the bracket and sensor off of the housing/steering tube.

5. Remove the bolt and take off the port bracket.

6. Remove the 5 housing nuts. Lay the assembly on its port side and then lift off the port side housing half.

7. Reach in and remove the spring and its holder. Remove the other spring. They may already have fallen out when you separated the housing halves.

8. Lift out the tilt lever and stopper.

9. Remove the steering tube. Pull off the bushing on each side and then remove the large rubber seal from the bottom (it will probably be damaged, but you have to replace it anyway).

To assemble:

10. Clean all components and allow to dry thoroughly prior to assembly. Obviously, make sure you don't use soap or cleaner on the sensor or buzzer.

11. Inspect everything for cracks, tears or other signs of damage. Pay particular attention to rubber seals and bushings.

12. Coat the inner and outer circumferences of the steering tube bushings with a good marine grade water resistant grease and then press them over the lip on each side of the tube. You can also insert them into each of the housing halves if you prefer, either way will work.

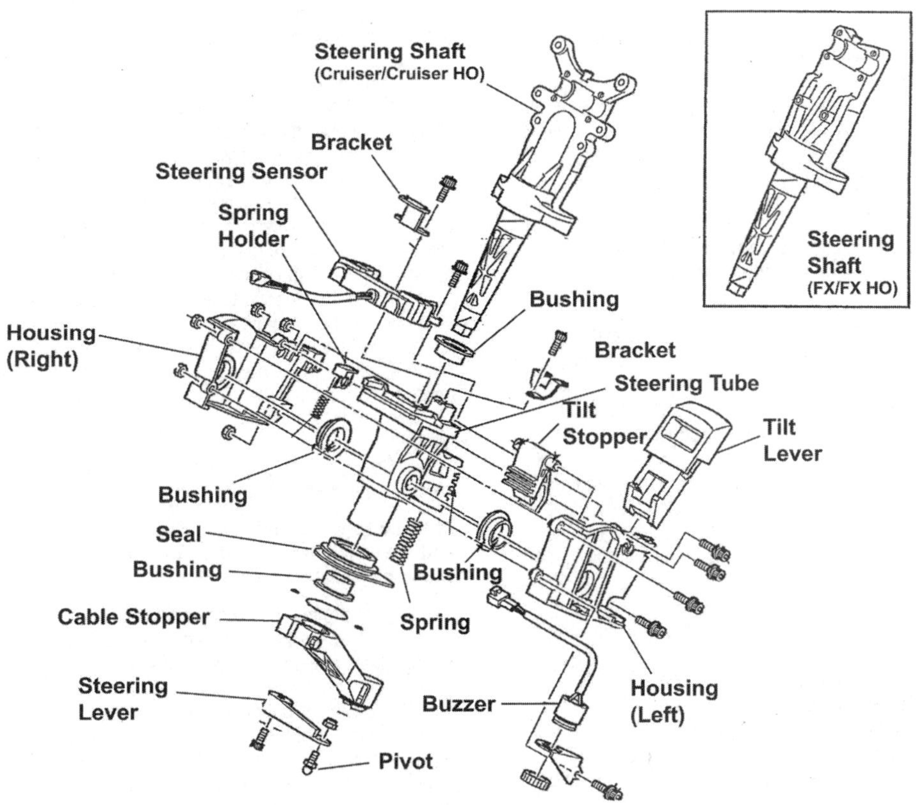

Fig. 43 Exploded view of the steering assembly - 2006-08 FX/Cruiser & 2006-07 FX HO/Cruiser HO Models

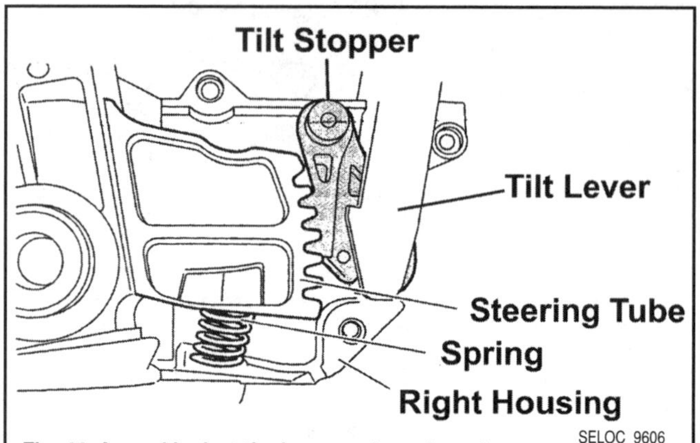

Fig. 44 A good look at the lower spring orientation. . .

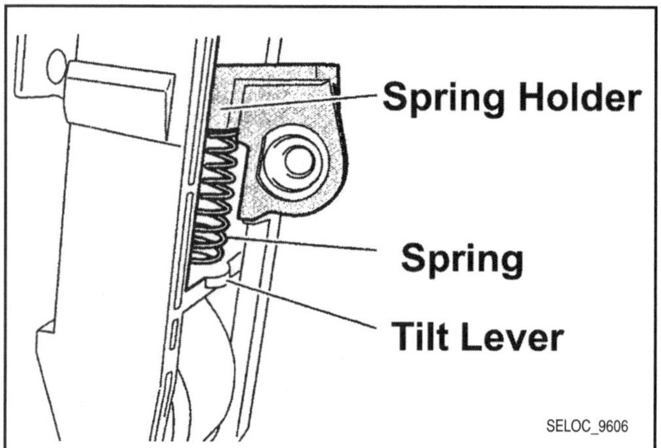

Fig. 45 . . .and the upper spring

13. Place the steering tube into the starboard housing so that the pivot boss with the bushing fits into the hole in the housing half (or the pivot boss fits into the hole with the bushing if that is how you installed the bushings).

14. Coat the pivot bosses on the tilt stopper with the grease and lay it into the housing so the boss is riding in the recess in the housing and the splines are engaged with those on the back of the steering tube.

15. Coat the mating surfaces of the tilt lever with your grease and then install it so it hooks into the stopper.

■ **Personally, we think it is easier to connect the stopper and lever *first* and the install them together. Take a look and see which you think would be easier.**

16. Coat the end of the small spring with a little grease and insert it into the spring bracket. Position them over the bolt hole in the cover so it is seated on the step of the tilt lever.

17. Now coat the end of the other spring with a little grease and insert it into the hole in the bottom of the steering tube spline protrusion.

18. Move the other housing half into position and slide the bolts through from the port side. Make sure the hole in the housing fits over the steering tube bushing and the stopper pivot is in the other recess. Don't forget that the small spring bracket is held in place by the upper bolt and the warning buzzer bracket is held by the lower front bolt. Thread on the nuts and tighten securely using a criss-cross pattern.

19. Coat the inner and outer circumferences of the upper and lower bushing with a little grease and then press them into their respective openings in the housing.

20. Install the port bracket on the back of the housing, coat the bolt threads with Loctite 572 and tighten it securely.

21. Lay the steering sensor over the housings. Coat the bolt threads with Loctite 572 and tighten them securely.

22. Install the starboard bracket on the back of the housing, coat the bolt threads with Loctite 572 and tighten it securely.

23. Coat the bottom of the steering shaft with grease. Do the same with the top of the shaft just below the flat spot and then insert it through the upper bushing, the steering tube and then out the lower bushing until it seats against the top of the tube/housing. Check the tilt lever is working properly - if not, pull everything apart and start over.

24. Insert the buzzer into the bracket and thread on the cap.

25. Coat the upper surface of a *new* rubber seal (where it rests on the assembly) with instant adhesive. Slide it over the shaft so the longer side of the lip is toward the rear and then hold it in position for a moment.

26. Install the assembly.

2008-11 FX HO/SHO & FX Cruiser HO/SHO Models

◆ **See Figures 44, 45 and 46** — ③ DIFFICULT

1. Remove the assembly as detailed previously and lay it on a clean, flat surface.

2. Remove the bolt and lift off the starboard side steering sensor bracket.

3. Grab the steering shaft and pull it out of the steering tube. The upper bushing may come off with it, but if not, remove it now. Do the same with the lower bushing. Remove the pipe seal.

**** CAUTION**

Do not remove the 2 board springs from the shaft arms.

4. Unscrew the threaded cap and lift the warning buzzer out of its holder. Remove the bolt and lift of the holder if necessary, otherwise you can do this when separating the housings.

5. Remove the 4 steering sensor mounting bolts with their washers and lift of the sensor. Remove the bolt securing the port bracket and lift off the bracket.

6. Remove the 5 housing nuts. Lay the assembly on its port side and then lift off the port side housing half. Remember that the buzzer bracket is secured with the lower front bolt so it will come off as well.

7. Reach in and remove the spring and its holder. Remove the other spring. They may already have fallen out when you separated the housing halves.

8. Lift out the tilt lever and stopper. Remove the 4 small bushings

9. Remove the steering tube. Pull off the bushing on each side and then remove the large rubber seal from the bottom (it will probably be damaged, but you have to replace it anyway).

To assemble:

10. Clean all components and allow to dry thoroughly prior to assembly. Obviously, make sure you don't use soap or cleaner on the sensor or buzzer.

11. Inspect everything for cracks, tears or other signs of damage. Pay particular attention to rubber seals and bushings.

12. Coat the inner and outer circumferences of the steering tube bushings with a good marine grade water resistant grease and then press them over the lip on each side of the tube. You can also insert them into each of the housing halves if you prefer, either way will work.

13. Place the steering tube into the starboard housing so that the pivot boss with the bushing fits into the hole in the housing half (or the pivot boss fits into the hole with the bushing if that is how you installed the bushings).

14. Coat the small bushings and pivot bosses on the tilt stopper with the grease and lay it into the housing so the boss is riding in the recess in the housing and the splines are engaged with those on the back of the steering tube.

15. Coat the mating surfaces of the tilt lever with your grease and then install it so it hooks into the stopper.

■ **Personally, we think it is easier to connect the stopper and lever *first* and the install them together. Take a look and see which you think would be easier.**

16. Coat the end of the small spring with a little grease and insert it into the spring bracket. Position them over the bolt hole in the cover and so that the bottom is resting on the step of the steering lever.

17. Now coat the end of the other spring with a little grease and insert it into the hole in the bottom of the steering tube spline protrusion.

18. Move the other housing half into position and slide the bolts through from the port side. Make sure the hole in the housing fits over the steering tube bushing and the stopper pivot is in the other recess. Don't forget that the small spring bracket is held in place by the upper bolt and the warning buzzer bracket is held by the lower front bolt. Thread on the nuts and tighten to 5.2 ft. lbs./50 inch lbs. (7 Nm) using a criss-cross pattern.

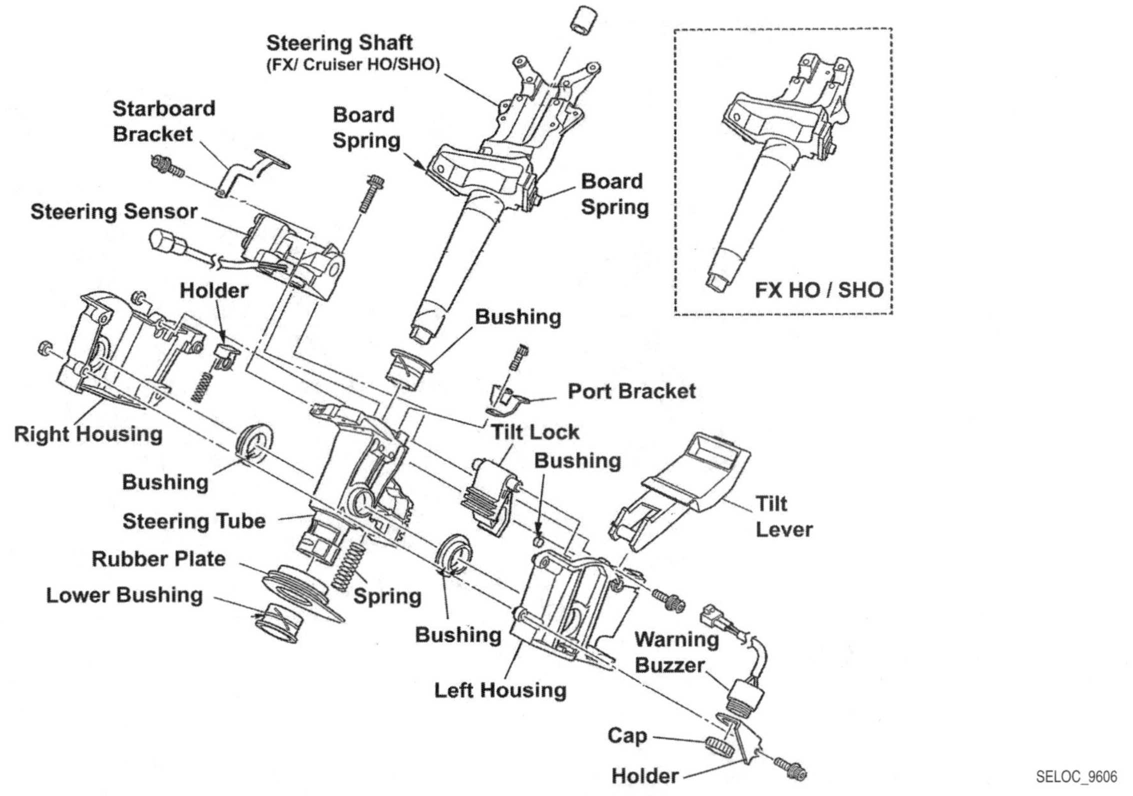

Fig. 46 Exploded view of the steering assembly - FX models w/1812cc engines and 2008 FX HO/Cruiser HO w/1052cc engines

19. Coat the inner and outer circumferences of the upper and lower bushing with a little grease and then press them into their respective openings in the housing.

20. Install the port bracket on the back of the housing, coat the bolt threads with Loctite 572 and tighten it to 3 ft. lbs./36 inch lbs (4 Nm).

21. Lay the steering sensor over the housings. Coat the bolt threads with Loctite 572 and tighten them to 5.2 ft. lbs./50 inch lbs. (7 Nm), using a criss-cross pattern.

22. Coat the bottom of the steering shaft with grease. Do the same with the top of the shaft just below the flat spot and then insert it through the upper bushing, the steering tube and then out the lower bushing until it seats against the top of the tube/housing. Check the tilt lever is working properly - if not, pull everything apart and start over.

23. Install the starboard bracket on the back of the housing, coat the bolt threads with Loctite 572 and tighten it to 3 ft. lbs./36 inch lbs (4 Nm).

24. Insert the buzzer into the bracket and thread on the cap.

25. Coat the upper surface of a *new* rubber seal (where it rests on the assembly) with instant adhesive. Slide it over the shaft so the longer side of the lip is toward the rear and then hold it in position for a moment.

26. Install the assembly.

FZ Models

3 DIFFICULT

◆ See Figures 44, 45, 47 and 48

1. Remove the assembly as detailed previously and lay it on a clean, flat surface.

2. Bend back the bolt retainer tab on the steering lever pinch bolt, loosen the bolt and work the lever off of the shaft. If necessary, remove the nut and pull off the steering cable pivot.

3. Grab the top of the steering shaft assembly and carefully pull it out of the lower housing. If the upper bushing comes off with it, slide it off of the shaft. If not, pry it out of the bore.

4. If necessary, remove the 3 bolts and lift off the lock assembly. Do the same with the grease nipple and upper shaft cap if you need to remove them.

■ **Do not remove the handlebar bracket (a) or either of the board springs (b).**

5. Loosen the cable stopper pinch bolt and remove it. Now you can remove the lower bushing.

6. Remove the 4 bolts and lift the steering sensor off of the base plate. Pop the warning buzzer out of its holder on the plate as well.

7. Remove the rubber seal from the bottom of the lower housing flange.

8. Remove the 4 mounting bolts and lift the sensor base and seal from the housing.

To assemble:

9. Clean all components and allow to dry thoroughly prior to assembly. Obviously, make sure you don't use soap or cleaner on the sensor or buzzer.

10. Inspect everything for cracks, tears or other signs of damage. Pay particular attention to rubber seals and bushings.

11. Position a *new* seal on the top of the lower housing and move the sensor base into place. Coat the bolt threads with Loctite 242 and tighten them to 12.5 ft. lbs. (17 Nm).

12. Coat the sensor bolt threads with Loctite 242 and tighten them to 5.2 ft. lbs./63 inch lbs. (7 Nm). Pop the warning buzzer into the holder on the base.

13. Slide a *new* seal into position under the flange on the lower housing.

14. Coat the inner and outer circumferences of the bushings with a good water resistant marine grade grease and press them into their respective bores.

15. Slide the cable stopper into place and tighten the pinch bolt to 5.2 ft. lbs./63 inch lbs. (7 Nm).

16. If you removed the cap from the top of the shaft, press it back into place. Thread in the grease nipple and tighten it to 1.5 ft. lbs./18 inch lbs. (2 Nm).

17. Coat the lock assembly bolt threads with Loctite 242 and attach it to the shaft, tightening to 5.2 ft. lbs./63 inch lbs. (7 Nm).

18. Insert the shaft back into the housing.

19. Install the cable pivot to the steering lever and tighten it to 5.2 ft. lbs./63 inch lbs. (7 Nm).

20. Slide the steering lever onto the bottom of the shaft. Position the bolt retainer plate over the hole in the lever so the long flat tab is butting up

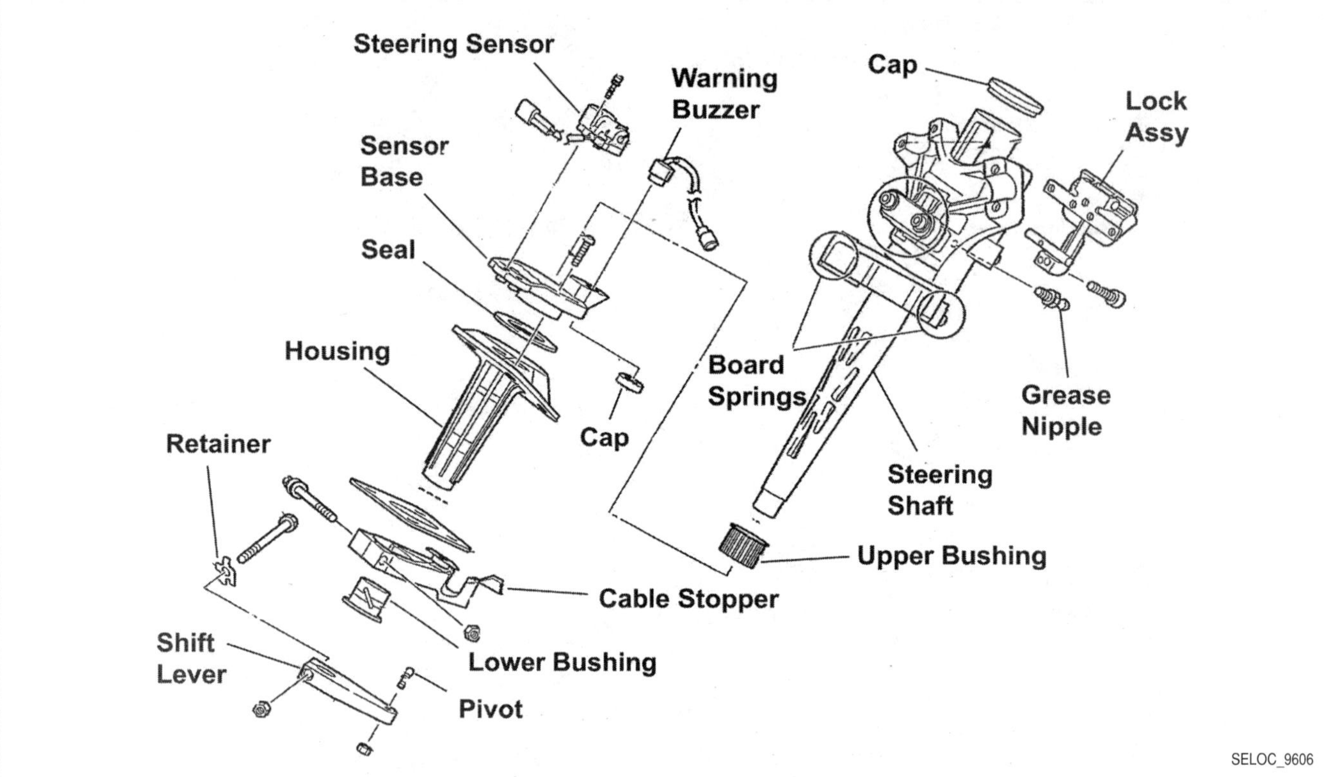

Fig. 47 Exploded view of the steering assembly - FZ models

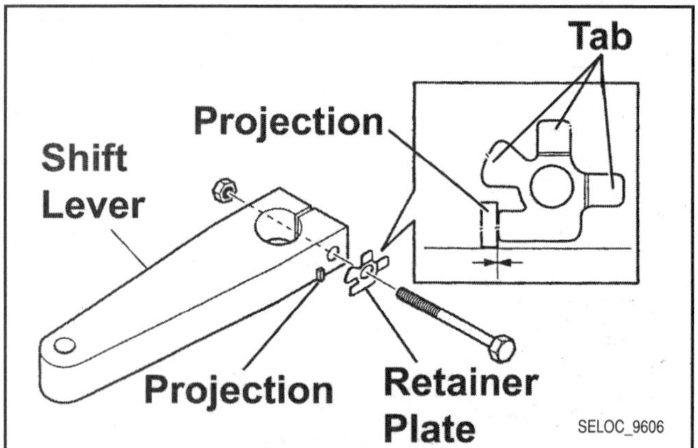

Fig. 48 Positioning the shift lever bolt retainer

against the small ridge on the lever. Slide in the bolt and tighten to 12.5 ft. lbs. (17 Nm). Move the bolt slightly in either direction until you are able to bend one or more of the 3 retainer tabs up and against a flat on the bolt head.

21. Install the assembly.

22. Once installed, but before installing the handlebar covers, shoot some Yamaha Grease A into the nipple until it just begins to come out the bottom of the handlebar bracket (meaning the entire bracket assembly, not the 2 brackets that actually hold the bar in position).

2005-09 VX Models

③ DIFFICULT

◆ **See Figure 49**

1. Remove the assembly as detailed previously and lay it on a clean, flat surface.

2. Grab the top of the steering column and carefully pull it out of the housing.

3. Remove the rubber seal(s) from the underside of the housing.

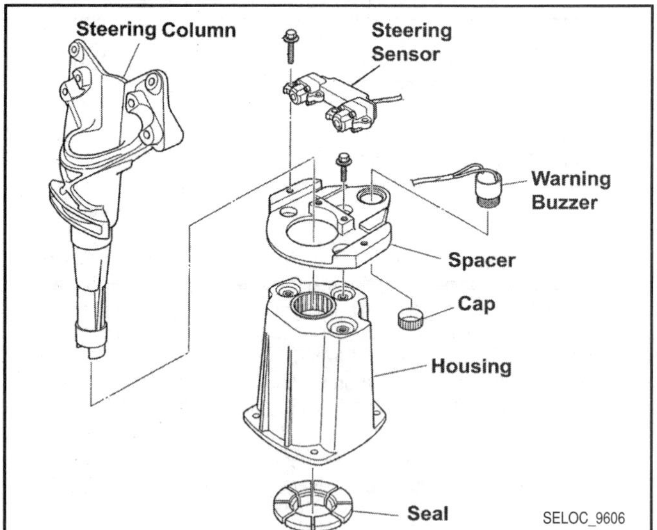

Fig. 49 Exploded view of the steering assembly - 2005-09 VX models

4. Remove the 4 mounting bolts and lift off the steering sensor.

5. Remove the 3 mounting bolts and lift off the spacer. Unscrew the warning buzzer if necessary.

To assemble:

6. Thread the buzzer back into the spacer if removed.

7. Position the spacer on the housing, coat the bolt threads with Loctite 572 and tighten to 12 ft. lbs. (16 Nm).

8. Coat the sensor bolt threads with Loctite 572 and tighten them to 12 ft. lbs. (16 Nm).

9. Insert the bottom of the steering column into the housing and press it in until it seats fully.

10. Install the assembly.

All 2010-11 VX Models

◆ **See Figures 50, 51 and 52**

1. Remove the assembly as detailed previously and lay it on a clean, flat surface.
2. Bend back the bolt retainer tab on the steering lever pinch bolt, loosen the bolt and work the lever off of the shaft. If necessary, remove the nut and pull off the steering cable pivot.
3. Grab the top of the steering shaft assembly and carefully pull it out of the lower housing. If the upper bushing comes off with it, slide it off of the shaft. If not, pry it out of the bore.
4. Cut the 2 plastic ties holding the warning buzzer and steering sensor leads together. Spin off the buzzer cap and remove it from the bracket. If necessary, remove the 2 bolts and lift off the bracket.
5. Remove the 4 bolts and lift off the steering sensor.
6. Remove the lower gasket.

To assemble:

7. Clean all components and allow to dry thoroughly prior to assembly. Obviously, make sure you don't use soap or cleaner on the sensor or buzzer.
8. Inspect everything for cracks, tears or other signs of damage. Pay particular attention to rubber seals and bushings.
9. Coat the sensor bolt threads with Loctite 242 and tighten them to 5.2 ft. lbs./63 inch lbs. (7 Nm).

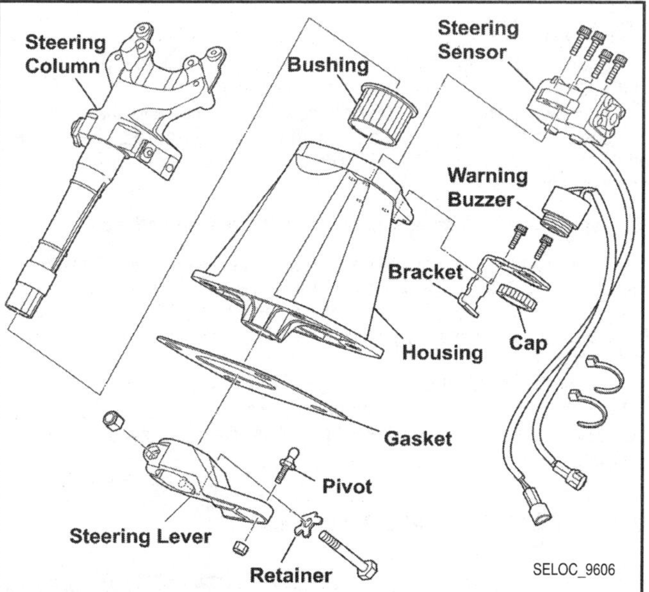

Fig. 50 Exploded view of the steering assembly - 2010-11 VX models

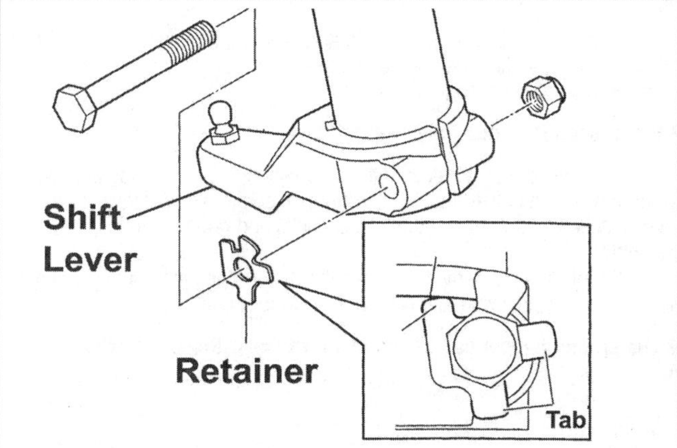

Fig. 51 Positioning the shift lever bolt retainer

10. Install the buzzer bracket, coat the threads with Loctite 242 and tighten to 3 ft. lbs./36 inch lbs. (4 Nm). Pop the warning buzzer into the holder on the base and thread on the cap.
11. Slide a *new* seal/gasket into position under the flange on the lower housing.
12. Coat the inner and outer circumferences of the bushing with a good water resistant marine grade grease and press it into the bore.
13. Insert the shaft back into the housing.
14. Install the cable pivot to the shift lever and tighten it to 5.2 ft. lbs./63 inch lbs. (7 Nm).
15. Slide the steering lever onto the bottom of the shaft. Position the bolt retainer plate over the hole in the lever so the long flat tab is butting up against the small ridge on the lever. Slide in the bolt and tighten to 12.5 ft. lbs. (17 Nm). Move the bolt slightly in either direction until you are able to bend one or more of the 3 retainer tabs up and against a flat on the bolt head.
16. Install the assembly.
17. Once installed, but before installing the handlebar covers, shoot some Yamaha Grease A into the nipple until it just begins to come out the bottom of the handlebar bracket (meaning the entire bracket assembly, not the 2 brackets that actually hold the bar in position).

Steering Sensor

TESTING

◆ **See Figures 53, 54 and 55**

A steering sensor is used on most 2006 and later FX/FZ models and all VX models. Removal and installation procedures are included in the Steering Assembly section under Disassembly & Assembly. Testing procedures require Yamaha's YDIS software, but we have included it here on the off-chance that you decided to spring for the software.

1. Connect your laptop to the YDIS system.
2. Loosely wrap a plastic tie around each handlebar grip so it is in the middle of the grip. Turn the bars all the way to the left.
3. Connect the hook of a spring tension gauge to the plastic tie on the right grip.
4. Push the **Unlock** button and then use the gauge to put 22 lbs. (10 kg) on the bars while pulling it at a right angle to the bars - this will move the button on the steering sensor in the direction shown.
5. Confirm that the Engine Monitor window on the software is showing:
   Not Pulled - Off
   Pulled - On
6. Turn the bars all the way to the right and then repeat the above steps.
7. Disconnect the steering sensor at the coupler. Push the **Unlock** button and then use a multi-meter to measure the input voltage at the harness side of the coupler (the side *not* connected to the sensor). Voltage between the Black/Orange and Orange/Red wires should be 4.75-5.25 V. If it is, replace the sensor. If not, check the harness for a problem.

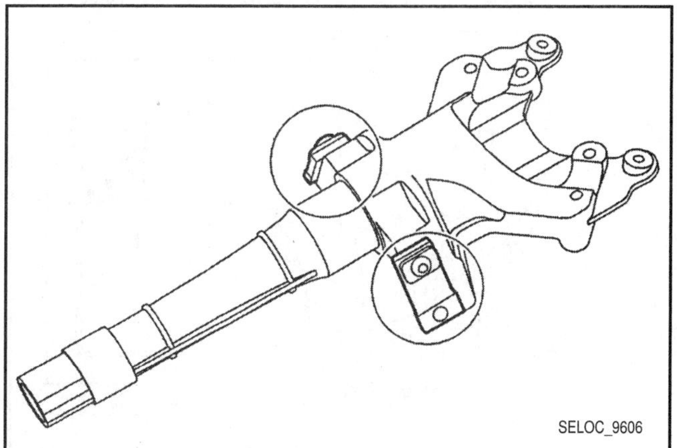

Fig. 52 Do not remove the board spring assemblies on the shaft

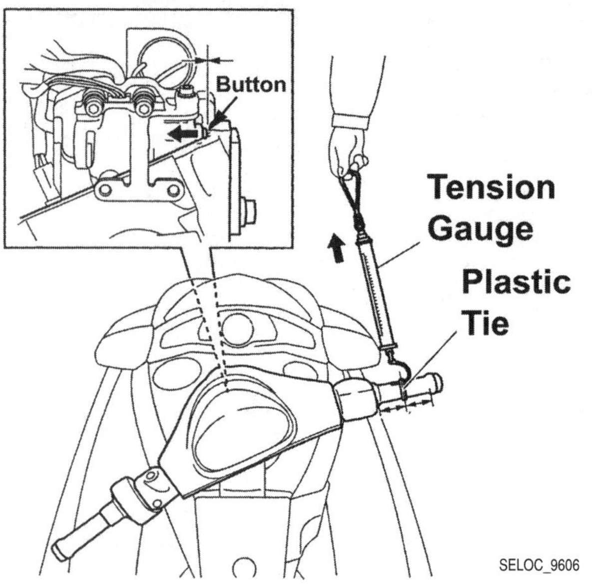

Fig. 53 Testing the steering sensor - FX models

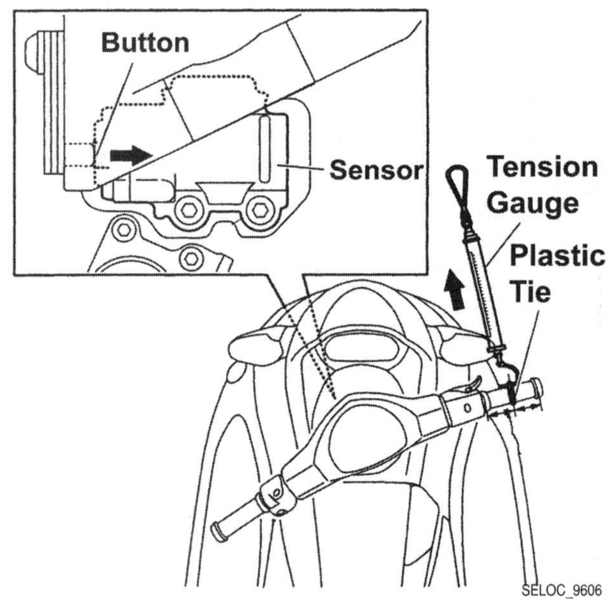

Fig. 54 Testing the steering sensor - VX models

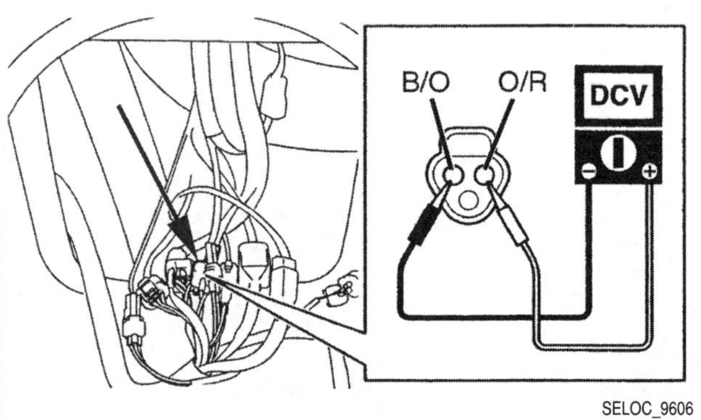

Fig. 55 Checking input voltage on the steering sensor - FX shown, other model locations may be different

Hood

REMOVAL & INSTALLATION

2002-08 FX Models With 998cc/1052cc Engines

② MODERATE

◆ **See Figures 56, 57 and 58**

1. Open the hood.
2. Reach into the hull and remove the 4 hinge retaining nuts and their washers. Lift off the hood. Make sure you have someone holding the hood while removing the nuts.

■ **If removing the hood only, skip to Step 8.**

3. If it is necessary to remove the hinge/spring assembly, lay the hood upside down on a soft surface so it will not be scratched. Remove the 2 upper bracket bolts. Remove the 4 hinge bolts and remove the hinge assembly.
4. If replacing the damper, remove the circlips and press or pull out each pin. The lower damper pin is longer than the upper and there is a small spacer in each grommet that may or may not come out. Cut the plastic tie and slide of the rubber boot.

■ **The spring damper can be removed with or without removing the hood.**

To install:

5. Slide the damper over the bottom of the spring with the lipped side up until it is over the casing and install a new plastic tie.
6. Press the small collars through the grommet and move the damper into position between the upper bracket and hinge bracket so that the arrow on the damper is pointing upward. Slide the pin through from the port side and install the circlip.

■ **If replacing the grommets, press them in from the port side.**

7. Position the hinge on the hood so the holes line up and then tighten the 4 bolts to 3.9 ft. lbs./47 inch lbs. (5.4 Nm). Do the same with the upper bracket and tighten these bolts to 3.9 ft. lbs./47 inch lbs. (5.4 Nm) - you may have to extend the damper slightly to get the bracket to line up with the holes in the hood.
8. Position the hood over the hull so the hinge studs fit through the holes in the hull and then thread on the nuts, tightening to 3.9 ft. lbs./47 inch lbs. (5.4 Nm), except on the 2008 FX HO where the figure is 4.6 ft. lbs./55 inch lbs. (6.4 Nm).
9. Close the hood.

2008-11 FZ & FX Models w/1812cc Engines

② MODERATE

◆ **See Figure 59**

1. Open the hood.
2. Reach into the hull and remove the 4 hinge retaining nuts and their washers.
3. Remove the lover hinge pin nut and slide out the pin. Lift off the hood with the mirrors. Make sure you have someone holding the hood while removing the nuts and hinge pin.

■ **If removing the hood only, skip to Step 8.**

4. If it is necessary to remove the hinge/spring assembly, lay the hood upside down on a soft surface so it will not be scratched. Remove the 2 upper bracket bolts. Remove the 4 hinge bolts and remove the hinge assembly.
5. If replacing the damper, remove the upper hinge pin nut and pull out the pin. Cut the plastic tie and remove the damper boot.

■ **The spring damper can be removed with or without removing the hood.**

To install:

6. Slide the damper over the bottom of the spring with the lipped side up until it is over the casing and install a new plastic tie.

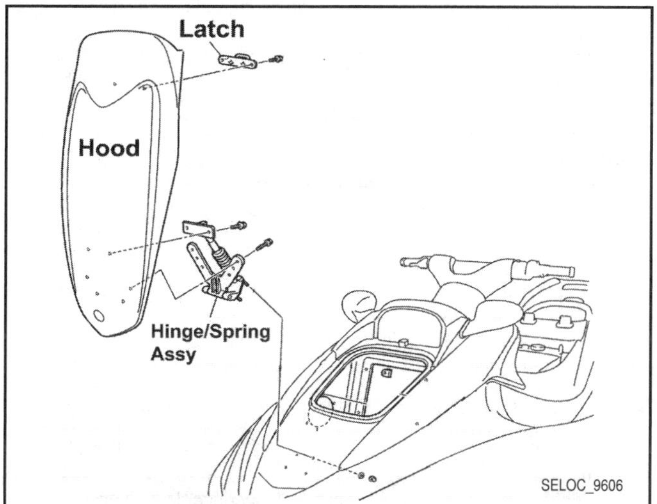

Fig. 56 Removing the hood is a fairly simple process - 2002-08 FX models with 998cc/1052cc engines

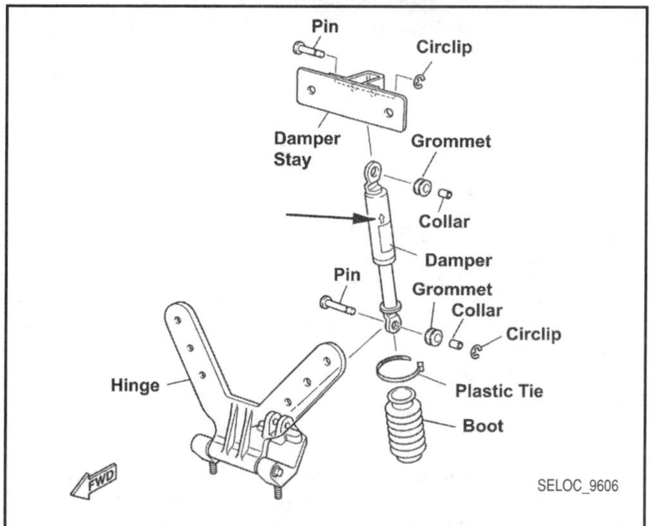

Fig. 57 A closer look at the hood hinge components

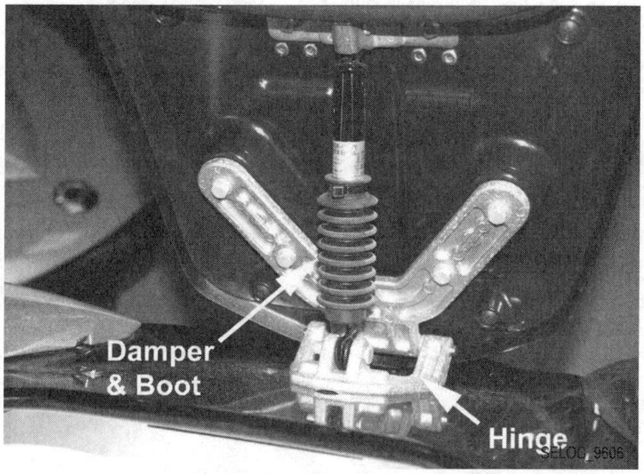

Fig. 58 It is easier to remove the hood with the hinge attached than disconnecting the hood from the hinge

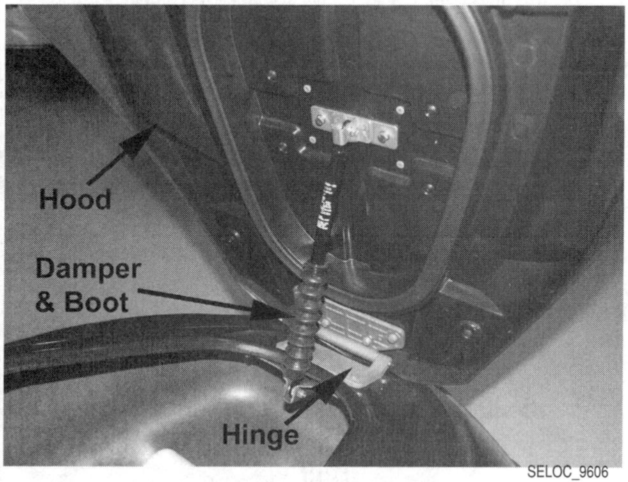

SELOC_9606

Fig. 59 A good shot of the hood hinge assembly - FX models w/1812cc engines

7. Move the damper into position between the upper bracket and slide the pin through from the starboard side. Tighten the nut to 3.7 ft. lbs./44 inch lbs. (5 Nm).

8. Position the hinge on the hood so the holes line up and then tighten the 4 bolts to 3.7 ft. lbs./44 inch lbs. (5 Nm). Do the same with the upper bracket and tighten these bolts to 3.7 ft. lbs./44 inch lbs. (5 Nm). Use Loctite 572 on the bolt threads.

9. Position the hood over the hull so the hinge studs fit through the holes in the hull and then thread on the nuts, tightening to 4.4 ft. lbs./53 inch lbs. (6 Nm).

10. Stretch the damper down to the lower bracket and install the hinge pin. Tighten the nut to 3.7 ft. lbs./44 inch lbs. (5 Nm).

11. Close the hood.

VX Models

② MODERATE

◆ **See Figure 60**

1. Open the hood.

2. Have an assistant hold the hood while you remove the 2 upper hinge bolts with their nuts and washers.

3. Lift off the hood.

4. On installation, tighten the bolts to 5 ft. lbs./60 inch lbs. (6.9 Nm) on 2005-09 models or 5.2 ft. lbs./63 inch lbs. (7 Nm) on 2010-11 models.

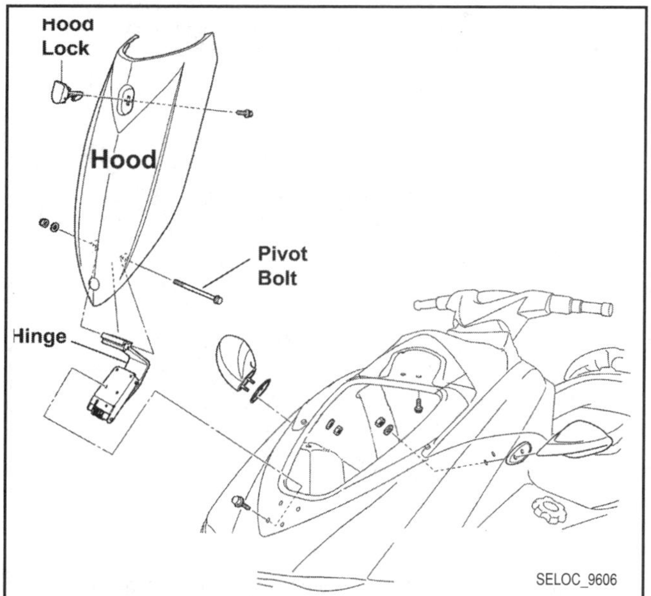

Fig. 60 shown, 2010-11 models similar

Shift Lever Assembly

REMOVAL & INSTALLATION

2002-05 FX Models

② MODERATE

◆ **See Figures 61 thru 64**

1. Remove the steering console cover as detailed in the Steering Assembly section. This is not absolutely necessary, but it will make everything much easier.

2. Reach into the hull and remove the starboard side cover bolts and lift off the cover. There are 4 bolts for each cover and they thread into clips on the underside of the cover.

3. Pull the shift cable sleeve back on the cable until you can pop it off of the shift pivot.

4. Remove the 2 bolts and take the shift handle off of the lever (if necessary).

5. Remove the center bolt and washer at the base of the shift lever. Remove the lever - it may have to persuaded a little by using 2 small pry bars between the lever and the hull plate, but be very *careful* if you try this.

6. Take a look at the hull plate. There are 6 bolts in it - 3 secure the shifter base and the other 3 mount the plate to the hull (using nuts and washers on the inside of the hull). Reach inside the hull and hold the base while removing the 3 retaining bolts.

7. If necessary, remove the 3 remaining bolts (and their nuts/washers) to remove the plate from the hull.

8. If shifter base assembly is required, remove the 2 bolts securing the inner retainer plate to the base and lift if off. We suggest laying the shift base assembly hull-side down on your workbench.

9. Remove the spring and small actuator and then pull the roller out of the slot in the shift shaft.

10. Lift out the shift shaft and inner shift lever.

11. Remove the shift cable pivot retaining nut.

To install:

12. Clean all components thoroughly and allow to dry completely prior to reassembly or installation.

13. Inspect all components for cracks, wear or other signs of damage. Check the teeth/splines on the shaft and inner lever carefully.

■ **If you did not disassemble the shifter base, please skip to Step 20.**

14. Install the cable pivot and tighten the nut to 4.9 ft. lbs./59 inch lbs. (6.8 Nm).

15. Coat the splines/teeth of the shift shaft and inner lever lightly with a good marine grade water resistant grease. Do the same with the ends of each of these where it rides in the recesses in the base or retainer plate.

16. Position the inner lever into the base so the boss on the base-side of the lever is seated in the recess in the base. Move the lever so it is resting against the upper or lower stop on the base.

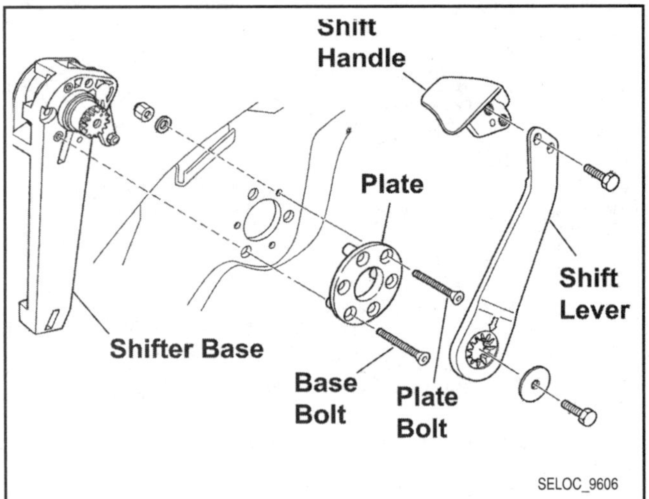

Fig. 61 The shift lever is on the starboard side of the craft

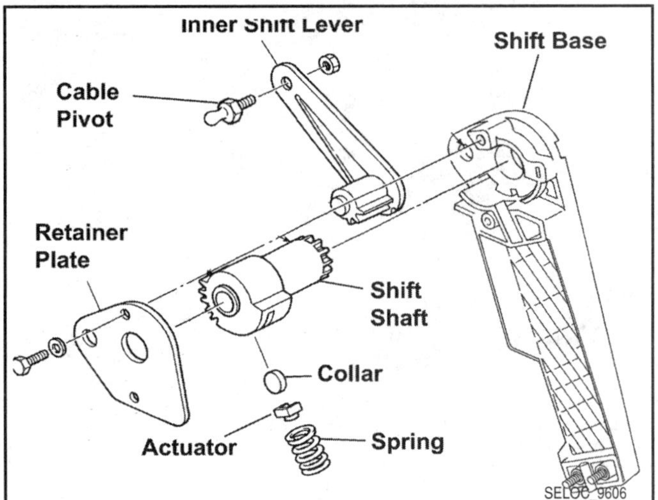

Fig. 62 Exploded view of the shifter base and its components - 2002-05 FX models

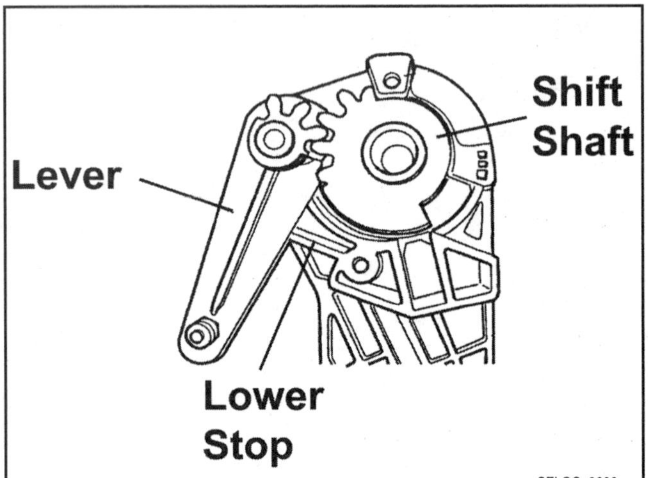

Fig. 63 If the inner shift lever is installed against the lower stop, the splines on the shift shaft should align like this

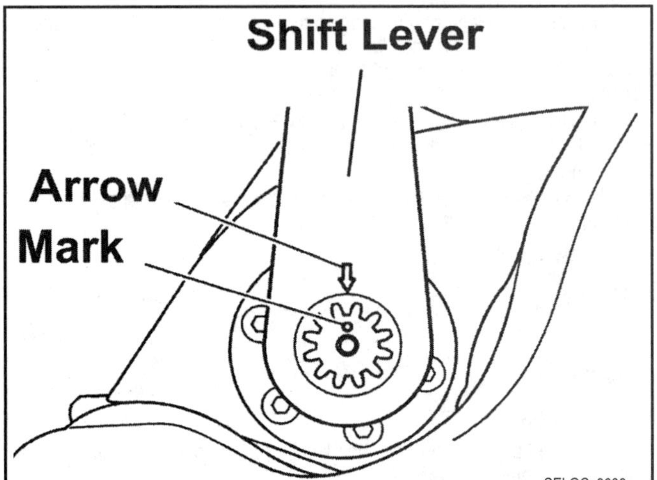

Fig. 64 Make sure the arrow on the lever lines up with the mark on the shift shaft

17. Position the shift shaft so the teeth/splines are facing forward (toward the splines on the lever). If the lever is against the upper stop, insert the shaft so the *upper* tooth meshes with the *upper* tooth on the lever. If the lever was positioned against the lower stop, obviously you'll want to engage the lower splines.

18. Insert the roller into the slot on the side of the shift shaft. Hold the actuator over it so the open side is facing the base and squeeze the spring into position.

19. Position the retainer plate over the assembly so the openings fit over the bosses on the end of the lever and shaft, coat the bolts threads with Loctite 271 and then tighten the bolts securely in a criss-cross pattern.

20. If you removed the hull plate, move it into place so the base retaining bolt sleeves fit through the larger holes (in the hull) and then tighten the nuts to 3.9 ft. lbs./47 inch lbs. (5.4 Nm).

21. Move the base into position so the bolt sleeves in the plate lone up with the holes in the base. Coat the bolt threads with Loctite 242 and then tighten them, in a criss-cross pattern, to 3.9 ft. lbs./47 inch lbs. (5.4 Nm).

22. Line the arrow on the shift lever up with the small punch mark on the face of the shift shaft splines and then press the lever into position. Coat the bolt threads with Loctite 271, slide on the washer and then tighten the bolt to 3.9 ft. lbs./47 inch lbs. (5.4 Nm).

23. If you removed the shift handle, coat the bolt threads with Loctite 271 and then tighten to 3.9 ft. lbs./47 inch lbs. (5.4 Nm).

24. Install the shift cable and check adjustment as detailed in the Maintenance & Tune-Up section.

25. Install the side cover. Coat the bolt threads with Loctite 271 and then tighten to 3.9 ft. lbs./47 inch lbs. (5.4 Nm).

26. Install the console cover if removed.

2006-08 FX Models With 998cc/1052cc Engines

② MODERATE

◆ See Figure 65

1. Remove the seat and disconnect the battery cables.

2. Reach into the hull and remove the forward bolt on the starboard side mirror assembly.

3. The center bolt comes from the top, remove the nut and washer from inside the hull.

4. Remove the rear bolt from the top and lift off the mirror assembly. There is no nut on this one.

5. On the starboard side cover, remove the forward bolt with its nut and washer. Remove the 4 remaining bolts, 3 nuts and 2 screws. Lift off the side cover.

6. Pull back (down) on the shift cable coupler until you can pop it off of the pivot on the lever assembly.

7. Reach inside the hull and remove the 3 nuts/washers securing the assembly to the hull. Lift off the shift lever assembly.

To install:

8. Clean the assembly thoroughly and check for any signs of damage.

9. Position the assembly on the hull so the bolt holes line up. Insert the bolts, thread on the nuts with their washers and tighten to 3.9 ft. lbs./47 inch lbs. (5.4 Nm) in an alternating pattern.

10. Connect the shift cable at the pivot and check adjustment as detailed in the Maintenance & Tune-Up section.

11. Install the side cover. Tighten the bolts and nuts to 3.9 ft. lbs./47 inch lbs. (5.4 Nm) and the screws to 1.6 ft. lbs./19 inch lbs. (2.2 Nm).

12. Install the mirror assembly and tighten the nuts/bolts to 3.9 ft. lbs./47 inch lbs. (5.4 Nm).

13. Install the hood.

2008-11 FX Models w/1812cc Engines

② MODERATE

◆ See Figures 27, 28, 29 and 66

1. Remove the seat and disconnect the battery cables.

2. Open the hood and remove the 5 rivets securing the rear service lid. To remove, push in the rivet pin in until just below the surface of the rivet and it clicks.

3. Remove the 2 rivets attaching the upper panel cover just above the service lid.

4. Reach in and remove the 3 starboard side cover bolts. Remove the single bolt at the top of the side panel and then the lower one (which runs through the ornament panel. Lift off the panel.

5. Remove the 3 bolts (Allen head) and rivet so you can pull off the starboard ornament panel.

6. Pull back (down) on the shift cable coupler until you can pop it off of the pivot on the lever assembly.

7. Disconnect the reverse sensor coupler.

8. Reach inside the hull and remove the 3 nuts/washers securing the assembly to the hull. Lift off the shift lever assembly.

To install:

9. Clean the assembly thoroughly and check for any signs of damage.

10. Position the assembly on the hull so the bolt holes line up. Insert the bolts, thread on the nuts with their washers and tighten to 3.7 ft. lbs./45 inch lbs. (5 Nm) in an alternating pattern.

11. Connect the reverse sensor coupler.

12. Connect the shift cable at the pivot and check adjustment as detailed in the Maintenance & Tune-Up section.

13. Install ornament panel. Tighten to 3.7 ft. lbs./45 inch lbs. (5 Nm) in an alternating pattern. Remember the upper rear bolt will be installed with the side cover. Press the riviet into the bottom hole.

14. Install the side cover. Tighten the bolts and nuts to 3.7 ft. lbs./45 inch lbs. (5 Nm).

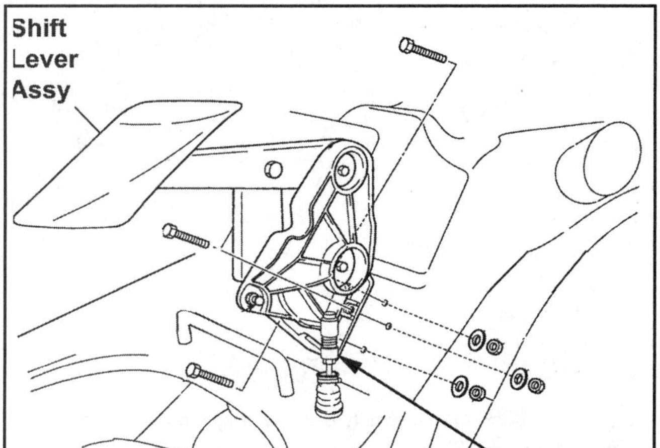

Fig. 65 A good look at the shift lever assembly - 2006-08 FX models with 998cc/1052cc engines

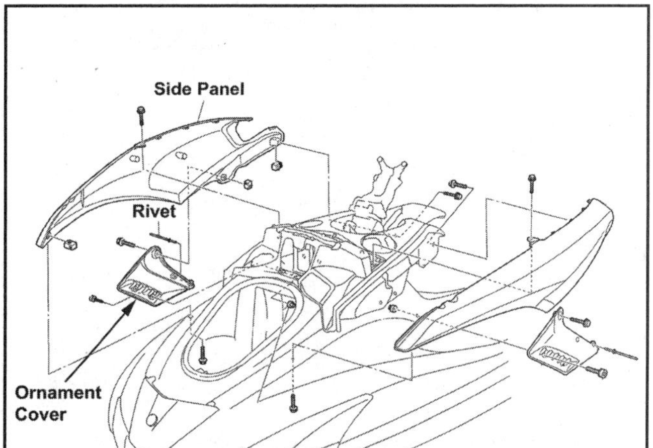

Fig. 66 Removing the side cover and ornament panel - FX models w/1812cc models

15. Install the upper panel cover and service lid. Hold the flange of the rivet with your fingers and then press the end of the pin against a hard surface until the pins comes out the top of the flange. Now you can insert the rivet through the holes and press the pin in until it clicks and is flush with the flange.

16. Close the hood. Connect the battery cables and install the seat.

FZ Models

② MODERATE

◆ **See Figures 27, 28, 29, 31 thru 35 and 67**

1. Remove the seat and disconnect the battery cables.
2. Remove the handlebars as previously detailed.
3. Remove the hood as detailed previously in this section.
4. Remove the 5 rivets securing the rear service lid. To remove, push in the rivet pin in until just below the surface of the rivet and it clicks.
5. Reach in and cut the plastic tie securing the harness. Tag and disconnect the multifunction meter couplers.
6. Remove the 2 rivets and lift out the meter back panel (just above the service lid). Disconnect the coupler.
7. Remove the 2 screws and pull off the meter cover.
8. Remove the 4 bolts (they come from inside the hull) and take out the meter.

■ **We think you ought to be able to remove the assembly without removing the side and center covers, but have included the steps here just in case. Take a look and decide for yourself. If you decide to give it a try, skip to Step 13.**

9. Remove the 2 bolts and pull out the right side shift grip/handle. Remove the drink holder.
10. Remove the 9 bolts (one is inside the grip recess) and take off the side covers.
11. Remove the 2 induction boxes.
12. Remove the 6 nuts and 5 bolts and then lift off the center cover.

To install:

13. Install the center cover and tighten all bolts and nuts to 3.7 ft. lbs./44 inch lbs. (5 Nm).
14. Install the side covers and tighten all bolts to 3.7 ft. lbs./44 inch lbs. (5 Nm).
15. Install the grip and tighten the screws to 3.7 ft. lbs./44 inch lbs. (5 Nm).
16. Install the meter and tighten the bolts to 3 ft. lbs./36 inch lbs. (4 Nm). Install the switch and then connect the coupler.
17. Slide on the meter cover and tighten the screws to 1.5 ft. lbs./18 inch lbs. (2 Nm).
18. Reconnect the couplers and install a new plastic tie.
19. Install meter panel. Hold the flange of the rivet with your fingers and then press the end of the pin against a hard surface until the pins comes out the top of the flange. Now you can insert the rivet through the holes and press the pin in until it clicks and is flush with the flange.
20. Install the service lid. Hold the flange of the rivet with your fingers and then press the end of the pin against a hard surface until the pins comes out the top of the flange. Now you can insert the rivet through the holes and press the pin in until it clicks and is flush with the flange.
21. Install the hood and handlebars.
22. Connect the battery cables and install the seat.

VX Models

② MODERATE

◆ **See Figures 36 and 68**

1. Remove the seat and disconnect the battery cables.
2. Remove the handlebars as previously detailed.
3. Remove the engine hatch cover as previously detailed.
4. On 2005-09 models, remove the 2 bolts and pull off the detent plate. Grab the shift lever and twist it to disconnect it from the end of the shift cable.
5. If necessary, remove the 2 nuts and lift off the bracket.
6. On 2010-11 models, simply disconnect the cable and remove the 2 nuts and washers and lift off the assembly - there is no disassembly.

To install:

7. On 2010-11 models, install the assembly and tighten the nuts to 3.7 ft. lbs./44 inch lbs. (5 Nm). Connect the shift cable.
8. On all others, if removed, position the shift lever bracket over the studs and tighten the nuts to 11 ft. lbs. (15 Nm).
9. Feed the end of the shift cable through the hole in the shift lever and then move it to the bracket.
10. Position the detent plate over the shift lever and bracket so all holes line up. Coat the bolt threads with Loctite 572 and tighten them to 5.8 ft. lbs./70 inch lbs.
11. On all models again, install the engine hatch cover and handlebars.
12. Connect the battery cables and install the seat.

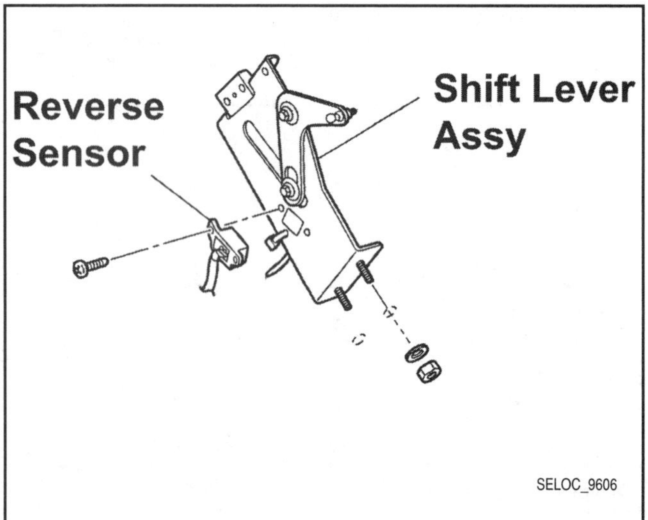

Fig. 67 A closer look at the shift lever assembly - FZ models

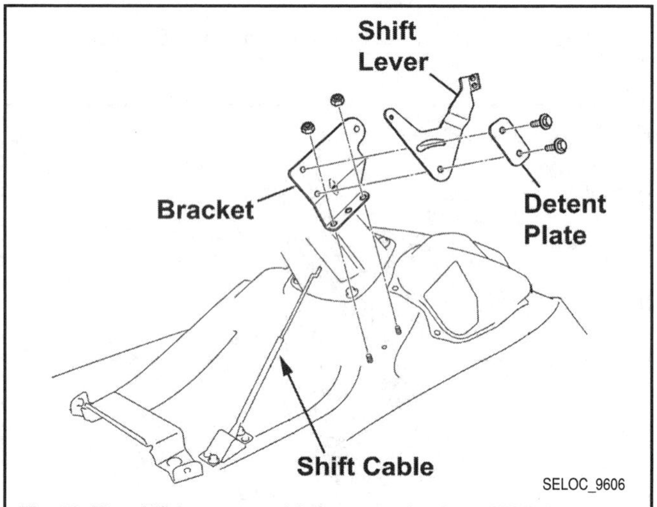

Fig. 68 The shift lever assembly is pretty simple on VX Deluxe models - 2005-09 shown